PEARSON

ALWAYS LEARNING

Raymond A. Barnett • Michael R. Ziegler • Karl E. Byleen

Calculus for Business, Economics, Life Sciences, and Social Sciences

Fourth Special Edition for DePaul University

Taken from:
*Calculus for Business, Economics, Life Sciences,
and Social Sciences,* Twelfth Edition
by Raymond A. Barnett, Michael R. Ziegler, and Karl E. Byleen

*Additional Calculus Topics: To Accompany Calculus,
12e and College Mathematics, 12e,* Twelfth Edition
by Raymond A. Barnett, Michael R. Ziegler, and Karl E. Byleen

Cover Art: Chicago 53, Chicago 74, Chicago 50-9 by Patrick Linehan.

Taken from:

Calculus for Business, Economics, Life Sciences, and Social Sciences, Twelfth Edition
by Raymond A. Barnett, Michael R. Ziegler, and Karl E. Byleen
Copyright © 2011, 2008, 2005 by Pearson Education, Inc.
Published by Prentice Hall
Boston, Massachusetts 02116

Additional Calculus Topics: To Accompany Calculus, 12e and College Mathematics, 12e, Twelfth Edition
by Raymond A. Barnett, Michael R. Ziegler, and Karl E. Byleen
Copyright © 2011, 2008, 2006 by Pearson Education, Inc.
Published by Prentice Hall
Boston, Massachusetts 02116

Pearson Learning Solutions, 501 Boylston Street, Suite 900, Boston, MA 02116
A Pearson Education Company
www.pearsoned.com

Printed in the United States of America

3 4 5 6 7 8 9 10 V092 17 16 15 14 13

000200010270757790

MC

ISBN 10: 1-256-80501-7
ISBN 13: 978-1-256-80501-4

You are enrolled in MAT 135/136–Business Calculus.

This course covers differential calculus of one or more variables with business applications (MAT 135) and integral calculus, matrix algebra, and probability theory with business applications (MAT 136).

The Department of Mathematical Sciences has worked with Pearson Learning Solutions to produce this book exclusively for students enrolled in MATH 137. It is composed of chapters from Calculus for Business, Economics, Life Sciences, and Social Sciences, 12/e by Barnett, Ziegler, and Byleen. In doing so, we have removed parts of the book that are not used in the course. Thus providing you with just the book you need at a reduced price.

CONTENTS

Chapters 1–8, Appendix A, B, C, Tables, I, II, and Answer for ch 1–8 taken from *Calculus for Business, Economics, Life Sciences, and Social Sciences,* Twelfth Edition by Raymond A. Barnett, Michael R. Ziegler, and Karl E. Byleen

Chapters 9, Table III, and Answers for ch 9 taken from *Additional Calculus Topics: To Accompany Caluculus, 12e and College Mathematics, 12e,* Twelfth Edition by Raymond A. Barnett, Michael R. Ziegler, and Karl E. Byleen

Accuracy Check

Because of the careful checking and proofing by a number of mathematics instructors (acting independently), the authors and publisher believe this book to be substantially error free. If an error should be found, the authors would be grateful if notification were sent to Karl E. Byleen, 9322 W. Garden Court, Hales Corners, WI 53130; or by e-mail, to kbyleen@wi.rr.com.

STUDENT SUPPLEMENTS

Student's Solutions Manual

By Garret J. Etgen, University of Houston

This manual contains detailed, carefully worked-out solutions to all odd-numbered section exercises and all Chapter Review exercises. Each section begins with Things to Remember, a list of key material for review.

ISBN 13: 978-0-321-65498-4; ISBN 10: 0-321-65498-6

Additional Calculus Topics to Accompany Calculus, 12e and College Mathematics, 12e

This separate book contains three unique chapters: Differential Equations, Taylor Polynomials and Infinite Series, and Probability and Calculus.

ISBN 13: 978-0-321-65509-7; ISBN 10: 0-321-65509-5

Worksheets for Classroom or Lab Practice

These Worksheets provide students with a structured place to take notes, define key concepts and terms, and work through unique examples to reinforce what is taught in the lecture.

ISBN 13: 978-0-321-65398-7; ISBN 10: 0-321-65398-X

Videos on DVD-ROM with Optional Captioning

The video lectures with optional captioning for this text make it easy and convenient for students to watch videos from a computer at home or on campus. The complete digitized set, affordable and portable for students, is ideal for distance learning or supplemental instruction. There is a video for every text example.

ISBN 13: 0-978-0-321-70869-4; ISBN 10: 0-321-70869-5

INSTRUCTOR SUPPLEMENTS

Instructor's Edition

This book contains answers to all exercises in the text.

ISBN 13: 978-0-321-64543-2; ISBN 10: 0-321-64543-X

Online Instructor's Solutions Manual (downloadable)

By Jason Aubrey, University of Missouri—Columbia

This manual contains detailed solutions to all even-numbered section problems.

Available in MyMathLab or through http://www.pearson highered.com.

Mini Lectures (downloadable)

Mini Lectures are provided for the teaching assistant, adjunct, part-time, or even full-time instructor for lecture preparation by providing learning objectives, examples (and answers) not found in the text, and teaching notes.

Available in MyMathLab or through http://www.pearson highered.com.

TestGen®

TestGen® (www.pearsoned.com/testgen) enables instructors to build, edit, print, and administer tests using a computerized bank of questions developed to cover all the objectives of the text. TestGen is algorithmically based, allowing instructors to create multiple but equivalent versions of the same question or test with the click of a button. Instructors can also modify test bank questions or add new questions. The software and testbank are available for download from Pearson Education's online catalog.

PowerPoint® Lecture Slides

These slides present key concepts and definitions from the text. They are available in MyMathLab or at http://www.pearsonhighered.com/educator.

MyMathLab® Online Course (access code required)

MyMathLab is a text-specific, easily customizable online course that integrates interactive multimedia instruction with textbook content. MyMathLab gives you the tools you need to deliver all or a portion of your course online, whether your students are in a lab setting or working from home.

- **Interactive homework exercises**, correlated to your textbook at the objective level, are algorithmically generated for unlimited practice and mastery. Most exercises are free-response and provide guided solutions, sample problems, and tutorial learning aids for extra help.

- **Personalized Study Plan**, generated when students complete a test or quiz, indicates which topics have been mastered and links to tutorial exercises for topics students have not mastered. You can customize the Study Plan so that the topics available match your course contents or so that students' homework results also determine mastery.

- **Multimedia learning aids, such as videos for every example in the text**, provide help for students when they need it. Other student-help features include Help Me Solve This and Additional Examples. You can assign these multimedia learning aids as homework to help your students grasp the concepts.

- **Homework and Test Manager** lets you assign homework, quizzes, and tests that are automatically graded. Select just the right mix of questions from the MyMathLab exercise bank, instructor-created custom exercises, and/or TestGen® test items.

- **Gradebook,** designed specifically for mathematics and statistics, automatically tracks students' results, lets you stay on top of student performance, and gives you control over how to calculate final grades. You can also add offline (paper-and-pencil) grades to the gradebook.

- **MathXL Exercise Builder** allows you to create static and algorithmic exercises for your online assignments. You can use the library of sample exercises as an easy starting point, or you can edit any course-related exercise.

- **Pearson Tutor Center** (www.pearsontutorservices.com) access is automatically included with MyMathLab. The Tutor Center is staffed by qualified math instructors who provide textbook-specific tutoring for students via toll-free phone, fax, email, and interactive Web sessions.

Students do the assignments in the new Flash®-based MathXL Player, which is compatible with almost any browser (Firefox®, Safari™, or Internet Explorer®) on almost any platform (Macintosh® or Windows®). MyMathLab is powered by CourseCompass™, Pearson Education's online teaching and learning environment, and by MathXL®, our online homework, tutorial, and assessment system. MyMathLab is available to qualified adopters. For more information, visit www.mymathlab.com or contact your Pearson representative.

MathXL® Online Course (access code required)

MathXL® is an online homework, tutorial, and assessment system that accompanies Pearson's textbooks in mathematics or statistics.

- **Interactive homework exercises**, correlated to your textbook at the objective level, are algorithmically generated for unlimited practice and mastery. Most exercises are free-response and provide guided solutions, sample problems, and learning aids for extra help.

- **Personalized Study Plan,** generated when students complete a test or quiz, indicates which topics have been mastered and links to tutorial exercises for topics students have not mastered. Instructors can customize the available topics in the study plan to match their course concepts.

- **Multimedia learning aids, such as videos for every example in the text**, provide help for students when they need it. Other student-help features include Help Me Solve This and Additional Examples. These are assignable as homework, to further encourage their use.

- **Gradebook,** designed specifically for mathematics and statistics, automatically tracks students' results, lets you stay on top of student performance, and gives you control over how to calculate final grades.

- **MathXL Exercise Builder** allows you to create static and algorithmic exercises for your online assignments. You can use the library of sample exercises as an easy starting point or use the Exercise Builder to edit any of the course-related exercises.

- **Homework and Test Manager** lets you create online homework, quizzes, and tests that are automatically graded. Select just the right mix of questions from the MathXL exercise bank, instructor-created custom exercises, and/or TestGen test items.

The new Flash®-based MathXL Player is compatible with almost any browser (Firefox®, Safari™, or Internet

Explorer®) on almost any platform (Macintosh® or Windows®). MathXL is available to qualified adopters. For more information, visit our website at www.mathxl. com, or contact your Pearson sales representative.

InterAct Math Tutorial Website:
www.interactmath.com

Get practice and tutorial help online! This interactive tutorial website provides algorithmically generated practice exercises that correlate directly to the exercises in the textbook. Students can retry an exercise as many times as they like with new values each time for unlimited practice and mastery. Every exercise is accompanied by an interactive guided solution that provides helpful feedback for incorrect answers, and students can also view a worked-out sample problem that steps them through an exercise similar to the one they're working on.

Acknowledgments

In addition to the authors many others are involved in the successful publication of a book. We wish to thank the following reviewers of the 11th and 12th editions:

Christine Cosgrove, *Fitchburg State College*
Darryl Egley, *North Harris College*
Lauren Fern, *University of Montana*
Gregory Goeckel, *Presbyterian College*
Garland Guyton, *Montgomery College*
Virginia Hanning, *San Jacinto College*
Bruce Hedman, *University of Connecticut*
Yvette Hester, *Texas A&M University*
Fritz Keinert, *Iowa State University*
Steven Klassen, *Missouri Western State University*
Wesley W. Maiers, *Valparaiso University*
James Martin, *Christopher Newport University*
Gary R. Penner, *Richland College*
Jon Prewett, *University of Wyoming*
Cynthia Schultz, *Illinois Valley Community College*
Maria Terrell, *Cornell University*
Fred M. Wright, *Iowa State University*
Amy Ann Yielding, *Washington State University*

We also wish to thank our colleagues who have provided input on previous editions:

Chris Boldt, Bob Bradshaw, Bruce Chaffee, Robert Chaney, Dianne Clark, Charles E. Cleaver, Barbara Cohen, Richard L. Conlon, Catherine Cron, Lou D'Alotto, Madhu Deshpande, Kenneth A. Dodaro, Michael W. Ecker, Jerry R. Ehman, Lucina Gallagher, Martha M. Harvey, Sue Henderson, Lloyd R. Hicks, Louis F. Hoelzle, Paul Hutchins, K. Wayne James, Jeffrey Lynn Johnson, Robert H. Johnston, Robert Krystock, Inessa Levi, James T. Loats, Frank Lopez, Roy H. Luke, Wayne Miller, Mel Mitchell, Linda M. Neal, Ronald Persky, Kenneth A. Peters, Jr., Dix Petty, Tom Plavchak, Bob Prielipp, Thomas Riedel, Stephen Rodi, Arthur Rosenthal, Sheldon Rothman, Elaine Russell, John Ryan, Daniel E. Scanlon, George R. Schriro, Arnold L. Schroeder, Hari Shanker, Joan Smith, J. Sriskandarajah, Steven Terry, Beverly Vredevelt, Delores A. Williams, Caroline Woods, Charles W. Zimmerman, Pat Zrolka, and Cathleen A. Zucco-Tevelot.

We also express our thanks to:

Caroline Woods, Anthony Gagliardi, Damon Demas, John Samons, Theresa Schille, Blaise DeSesa, and Debra McGivney for providing a careful and thorough accuracy check of the text, problems and answers.

Garret Etgen, Jason Aubrey, Dale R. Buske, and Karla Neal for developing the supplemental materials so important to the success of a text.

All the people at Pearson Education who contributed their efforts to the production of this book.

PART 1

A LIBRARY
OF ELEMENTARY
FUNCTIONS

Linear Equations and Graphs

Introduction

We begin by discussing some algebraic methods for solving equations and inequalities. Next, we introduce coordinate systems that allow us to explore the relationship between algebra and geometry. Finally, we use this algebraic–geometric relationship to find equations that can be used to describe real-world data sets. For example, in Section 1-3 you will learn how to find the equation of a line that fits data on winning times in an Olympic swimming event (see Problems 27 and 28 on page 38). We also consider many applied problems that can be solved using the concepts discussed in this chapter.

1-1   Linear Equations and Inequalities

- Linear Equations
- Linear Inequalities
- Applications

The equation

$$3 - 2(x + 3) = \frac{x}{3} - 5$$

and the inequality

$$\frac{x}{2} + 2(3x - 1) \geq 5$$

are both first degree in one variable. In general, a **first-degree**, or **linear**, **equation** in one variable is any equation that can be written in the form

$$\text{Standard form:} \quad ax + b = 0 \quad a \neq 0 \tag{1}$$

If the equality symbol, =, in (1) is replaced by $<$, $>$, $\leq$, or $\geq$, the resulting expression is called a **first-degree**, or **linear**, **inequality**.

A **solution** of an equation (or inequality) involving a single variable is a number that when substituted for the variable makes the equation (or inequality) true. The set of all solutions is called the **solution set**. When we say that we **solve an equation** (or inequality), we mean that we find its solution set.

Knowing what is meant by the solution set is one thing; finding it is another. We start by recalling the idea of equivalent equations and equivalent inequalities. If we perform an operation on an equation (or inequality) that produces another equation (or inequality) with the same solution set, then the two equations (or inequalities) are said to be **equivalent**. The basic idea in solving equations or inequalities is to perform operations that produce simpler equivalent equations or inequalities and to continue the process until we obtain an equation or inequality with an obvious solution.

Linear Equations

Linear equations are generally solved using the following equality properties.

THEOREM 1   Equality Properties

An equivalent equation will result if

1. The same quantity is added to or subtracted from each side of a given equation.

2. Each side of a given equation is multiplied by or divided by the same nonzero quantity.

EXAMPLE 1   **Solving a Linear Equation** Solve and check:

$$8x - 3(x - 4) = 3(x - 4) + 6$$

SOLUTION

$8x - 3(x - 4) = 3(x - 4) + 6$	Use the distributive property.
$8x - 3x + 12 = 3x - 12 + 6$	Combine like terms.
$5x + 12 = 3x - 6$	Subtract 3x from both sides.
$2x + 12 = -6$	Subtract 12 from both sides.
$2x = -18$	Divide both sides by 2.
$x = -9$	

CHECK

$$8x - 3(x - 4) = 3(x - 4) + 6$$

$$8(-9) - 3[(-9) - 4] \stackrel{?}{=} 3[(-9) - 4] + 6$$

$$-72 - 3(-13) \stackrel{?}{=} 3(-13) + 6$$

$$-33 \stackrel{\checkmark}{=} -33$$

Matched Problem 1   Solve and check: $3x - 2(2x - 5) = 2(x + 3) - 8$

EXPLORE & DISCUSS 1

According to equality property 2, multiplying both sides of an equation by a nonzero number always produces an equivalent equation. What is the smallest positive number that you could use to multiply both sides of the following equation to produce an equivalent equation without fractions?

$$\frac{x + 1}{3} - \frac{x}{4} = \frac{1}{2}$$

EXAMPLE 2   **Solving a Linear Equation**   Solve and check: $\dfrac{x + 2}{2} - \dfrac{x}{3} = 5$

SOLUTION   What operations can we perform on

$$\frac{x + 2}{2} - \frac{x}{3} = 5$$

to eliminate the denominators? If we can find a number that is exactly divisible by each denominator, we can use the multiplication property of equality to clear the denominators. The LCD (least common denominator) of the fractions, 6, is exactly what we are looking for! Actually, any common denominator will do, but the LCD results in a simpler equivalent equation. So, we multiply both sides of the equation by 6:

$$6\left(\frac{x + 2}{2} - \frac{x}{3}\right) = 6 \cdot 5 \quad *$$

$$\overset{3}{6} \cdot \frac{(x + 2)}{2} - \overset{2}{6} \cdot \frac{x}{3} = 30$$
$$\underset{1}{} \qquad \underset{1}{}$$

$$3(x + 2) - 2x = 30 \qquad \text{Use the distributive property.}$$
$$3x + 6 - 2x = 30 \qquad \text{Combine like terms.}$$
$$x + 6 = 30 \qquad \text{Subtract 6 from both sides.}$$
$$x = 24$$

CHECK

$$\frac{x + 2}{2} - \frac{x}{3} = 5$$

$$\frac{24 + 2}{2} - \frac{24}{3} \overset{?}{=} 5$$

$$13 - 8 \overset{?}{=} 5$$

$$5 \overset{\checkmark}{=} 5$$

Matched Problem 2   Solve and check: $\dfrac{x + 1}{3} - \dfrac{x}{4} = \dfrac{1}{2}$

In many applications of algebra, formulas or equations must be changed to alternative equivalent forms. The following example is typical.

EXAMPLE 3   **Solving a Formula for a Particular Variable**   If you deposit a principal P in an account that earns simple interest at an annual rate r, then the amount A in the account after t years is given by $A = P + Prt$. Solve for

(A) r in terms of $A, P,$ and t

(B) P in terms of $A, r,$ and t

---

*Dashed boxes are used throughout the book to denote steps that are usually performed mentally.

SOLUTION   (A)
$$A = P + Prt \qquad \text{Reverse equation.}$$
$$P + Prt = A \qquad \text{Subtract } P \text{ from both sides.}$$
$$Prt = A - P \qquad \text{Divide both members by } Pt.$$
$$r = \frac{A - P}{Pt}$$

(B)
$$A = P + Prt \qquad \text{Reverse equation.}$$
$$P + Prt = A \qquad \text{Factor out } P \text{ (note the use of the distributive property).}$$
$$P(1 + rt) = A \qquad \text{Divide by } (1 + rt).$$
$$P = \frac{A}{1 + rt}$$

Matched Problem 3   If a cardboard box has length L, width W, and height H, then its surface area is given by the formula $S = 2LW + 2LH + 2WH$. Solve the formula for

(A) L in terms of S, W, and H

(B) H in terms of S, L, and W

Linear Inequalities

Before we start solving linear inequalities, let us recall what we mean by $<$ (less than) and $>$ (greater than). If a and b are real numbers, we write

$$a < b \qquad a \text{ is less than } b$$

if there exists a positive number p such that $a + p = b$. Certainly, we would expect that if a positive number was added to any real number, the sum would be larger than the original. That is essentially what the definition states. If $a < b$, we may also write

$$b > a \qquad b \text{ is greater than } a.$$

EXAMPLE 4   **Inequalities**

(A) $\qquad\qquad 3 < 5 \qquad$ Since $3 + 2 = 5$

(B) $\qquad\qquad -6 < -2 \qquad$ Since $-6 + 4 = -2$

(C) $\qquad\qquad 0 > -10 \qquad$ Since $-10 < 0$ (because $-10 + 10 = 0$)

Matched Problem 4   Replace each question mark with either $<$ or $>$.

(A) $2\,?\,8$ \qquad (B) $-20\,?\,0$ \qquad (C) $-3\,?\,-30$

Figure 1 $a < b, c > d$

The inequality symbols have a very clear geometric interpretation on the real number line. If $a < b$, then a is to the left of b on the number line; if $c > d$, then c is to the right of d on the number line (Fig. 1). Check this geometric property with the inequalities in Example 4.

EXPLORE & DISCUSS 2   Replace $?$ with $<$ or $>$ in each of the following:

(A) $-1\,?\,3$ \qquad and \qquad $2(-1)\,?\,2(3)$

(B) $-1\,?\,3$ \qquad and \qquad $-2(-1)\,?\,-2(3)$

(C) $12\,?\,-8$ \qquad and \qquad $\dfrac{12}{4}\,?\,\dfrac{-8}{4}$

(D) $12\,?\,-8$ \qquad and \qquad $\dfrac{12}{-4}\,?\,\dfrac{-8}{-4}$

Based on these examples, describe verbally the effect of multiplying both sides of an inequality by a number.

The procedures used to solve linear inequalities in one variable are almost the same as those used to solve linear equations in one variable, but with one important exception, as noted in item 3 of Theorem 2.

THEOREM 2   Inequality Properties

An equivalent inequality will result, and the **sense or direction will remain the same** if each side of the original inequality

1. has the same real number added to or subtracted from it.
2. is multiplied or divided by the same *positive* number.

An equivalent inequality will result, and the **sense or direction will reverse** if each side of the original inequality

3. is multiplied or divided by the same *negative* number.

NOTE: Multiplication by 0 and division by 0 are not permitted.

Therefore, we can perform essentially the same operations on inequalities that we perform on equations, with the exception that **the sense of the inequality reverses if we multiply or divide both sides by a negative number**. Otherwise, the sense of the inequality does not change. For example, if we start with the true statement

$$-3 > -7$$

and multiply both sides by 2, we obtain

$$-6 > -14$$

and the sense of the inequality stays the same. But if we multiply both sides of $-3 > -7$ by -2, the left side becomes 6 and the right side becomes 14, so we must write

$$6 < 14$$

to have a true statement. The sense of the inequality reverses.

If $a < b$, the **double inequality** $a < x < b$ means that $a < x$ **and** $x < b$; that is, x is between a and b. **Interval notation** is also used to describe sets defined by inequalities, as shown in Table 1.

The numbers a and b in Table 1 are called the **endpoints** of the interval. An interval is **closed** if it contains all its endpoints and **open** if it does not contain any of its endpoints. The intervals $[a, b]$, $(-\infty, a]$, and $[b, \infty)$ are closed, and the intervals (a, b), $(-\infty, a)$, and (b, ∞) are open. Note that the symbol ∞ (read infinity) is not a number. When we write $[b, \infty)$, we are simply referring to the interval that starts at b and continues indefinitely to the right. We never refer to ∞ as an endpoint, and we never write $[b, \infty]$. The interval $(-\infty, \infty)$ is the entire real number line.

Note that an endpoint of a line graph in Table 1 has a square bracket through it if the endpoint is included in the interval; a parenthesis through an endpoint indicates that it is not included.

CONCEPTUAL INSIGHT

The notation $(2, 7)$ has two common mathematical interpretations: the ordered pair with first coordinate 2 and second coordinate 7, and the open interval consisting of all real numbers between 2 and 7. The choice of interpretation is usually determined by the context in which the notation is used. The notation $(2, -7)$ could be interpreted as an ordered pair but not as an interval. In interval notation, the left endpoint is always written first. So, $(-7, 2)$ is correct interval notation, but $(2, -7)$ is not.

Table 1 Interval Notation

Interval Notation	Inequality Notation	Line Graph
$[a, b]$	$a \leq x \leq b$	
$[a, b)$	$a \leq x < b$	
$(a, b]$	$a < x \leq b$	
(a, b)	$a < x < b$	
$(-\infty, a]$	$x \leq a$	
$(-\infty, a)$	$x < a$	
$[b, \infty)$	$x \geq b$	
(b, ∞)	$x > b$	

EXAMPLE 5 **Interval and Inequality Notation, and Line Graphs**

(A) Write $[-2, 3)$ as a double inequality and graph.

(B) Write $x \geq -5$ in interval notation and graph.

SOLUTION (A) $[-2, 3)$ is equivalent to $-2 \leq x < 3$.

(B) $x \geq -5$ is equivalent to $[-5, \infty)$.

Matched Problem 5 (A) Write $(-7, 4]$ as a double inequality and graph.

(B) Write $x < 3$ in interval notation and graph.

EXPLORE & DISCUSS 3 The solution to Example 5B shows the graph of the inequality $x \geq -5$. What is the graph of $x < -5$? What is the corresponding interval? Describe the relationship between these sets.

EXAMPLE 6 **Solving a Linear Inequality** Solve and graph:

$$2(2x + 3) < 6(x - 2) + 10$$

SOLUTION

$2(2x + 3) < 6(x - 2) + 10$	Remove parentheses.
$4x + 6 < 6x - 12 + 10$	Combine like terms.
$4x + 6 < 6x - 2$	Subtract $6x$ from both sides.
$-2x + 6 < -2$	Subtract 6 from both sides.
$-2x < -8$	Divide both sides by -2 and reverse the sense of the inequality.

$$x > 4 \quad \text{or} \quad (4, \infty)$$

Notice that in the graph of $x > 4$, we use a parenthesis through 4, since the point 4 is not included in the graph.

Matched Problem 6 Solve and graph: $3(x - 1) \leq 5(x + 2) - 5$

EXAMPLE 7 **Solving a Double Inequality** Solve and graph: $-3 < 2x + 3 \leq 9$

SOLUTION We are looking for all numbers x such that $2x + 3$ is between -3 and 9, including 9 but not -3. We proceed as before except that we try to isolate x in the middle:

$$-3 < 2x + 3 \leq 9$$
$$-3 \; - \; 3 < 2x + 3 \; - \; 3 \leq 9 \; - \; 3$$
$$-6 < 2x \leq 6$$
$$\frac{-6}{2} < \frac{2x}{2} \leq \frac{6}{2}$$
$$-3 < x \leq 3 \quad \text{or} \quad (-3, 3]$$

Matched Problem 7 Solve and graph: $-8 \leq 3x - 5 < 7$

Note that a linear equation usually has exactly one solution, while a linear inequality usually has infinitely many solutions.

Applications

To realize the full potential of algebra, we must be able to translate real-world problems into mathematics. In short, we must be able to do word problems.

Here are some suggestions that will help you get started:

Procedure for Solving Word Problems

1. Read the problem carefully and introduce a variable to represent an unknown quantity in the problem. Often the question asked in a problem will indicate the best way to introduce this variable.

2. Identify other quantities in the problem (known or unknown), and whenever possible, express unknown quantities in terms of the variable you introduced in Step 1.

3. Write a verbal statement using the conditions stated in the problem and then write an equivalent mathematical statement (equation or inequality).

4. Solve the equation or inequality and answer the questions posed in the problem.

5. Check the solution(s) in the original problem.

EXAMPLE 8 **Purchase Price** John purchases a computer from an online store for $851.26, including a $57 shipping charge and 5.2% state sales tax. What is the purchase price of the computer?

SOLUTION **Step 1** **Introduce a variable for the unknown quantity.** After reading the problem, we decide to let x represent the purchase price of the computer.

Step 2 **Identify quantities in the problem.**

Shipping charges: $57

Sales tax: $0.052x$

Total cost: $851.26

Step 3   **Write a verbal statement and an equation.**

$$\text{Price} + \text{Shipping Charges} + \text{Sales Tax} = \text{Total Order Cost}$$
$$x \quad + \qquad\quad 57 \qquad + \quad 0.052x \;\; = 851.26$$

Step 4   **Solve the equation and answer the question.**

$$x + 57 + 0.052x = 851.26 \qquad \textit{Combine like terms.}$$
$$1.052x + 57 = 851.26 \qquad \textit{Subtract 57 from both sides.}$$
$$1.052x = 794.26 \qquad \textit{Divide both sides by 1.052.}$$
$$x = 755$$

The price of the computer is $755.

Step 5   **Check the answer in the original problem.**

$$\text{Price} = \$755.00$$
$$\text{Shipping charges} = \;\;\$57.00$$
$$\underline{\text{Tax } 0.052 \cdot 755 = \;\;\$39.26}$$
$$\text{Total} = \$851.26$$

Matched Problem 8   Mary paid 8.5% sales tax and a $190 title and license fee when she bought a new car for a total of $28,400. What is the purchase price of the car?

The next example involves the important concept of **break-even analysis**, which is encountered in several places in this text. Any manufacturing company has **costs**, C, and **revenues**, R. The company will have a **loss** if $R < C$, will **break even** if $R = C$, and will have a **profit** if $R > C$. Costs involve **fixed costs**, such as plant overhead, product design, setup, and promotion, and **variable costs**, which are dependent on the number of items produced at a certain cost per item.

EXAMPLE 9   **Break-Even Analysis** A multimedia company produces DVDs. One-time fixed costs for a particular DVD are $48,000, which include costs such as filming, editing, and promotion. Variable costs amount to $12.40 per DVD and include manufacturing, packaging, and distribution costs for each DVD actually sold to a retailer. The DVD is sold to retail outlets at $17.40 each. How many DVDs must be manufactured and sold in order for the company to break even?

SOLUTION   **Step 1**   Let $x =$ number of DVDs manufactured and sold.

Step 2
$$C = \text{cost of producing } x \text{ DVDs}$$
$$R = \text{revenue (return) on sales of } x \text{ DVDs}$$
$$\text{Fixed costs} = \$48,000$$
$$\text{Variable costs} = \$12.40x$$
$$C = \text{Fixed costs} + \text{variable costs}$$
$$= \$48,000 + \$12.40x$$
$$R = \$17.40x$$

Step 3   The company breaks even if $R = C$; that is, if

$$\$17.40x = \$48,000 + \$12.40x$$

Step 4 $17.4x = 48,000 + 12.4x$ *Subtract 12.4x from both sides.*
$$5x = 48,000 \qquad\qquad \textit{Divide both sides by 5.}$$
$$x = 9,600$$

The company must make and sell 9,600 DVDs to break even.

Step 5 Check:

Costs	Revenue
$48,000 + 12.4(9,600)$	$17.4(9,600)$
$= \$167,040$	$= \$167,040$

Matched Problem 9 How many DVSs would a multimedia company have to make and sell to break even if the fixed costs are $36,000, variable costs are $10.40 per DVD, and the DVDs are sold to retailers for $15.20 each?

EXAMPLE 10 **Consumer Price Index** The Consumer Price Index (CPI) is a measure of the average change in prices over time from a designated reference period, which equals 100. The index is based on prices of basic consumer goods and services. Table 2 lists the CPI for several years from 1960 to 2005. What net annual salary in 2005 would have the same purchasing power as a net annual salary of $13,000 in 1960? Compute the answer to the nearest dollar. (*Source*: U.S. Bureau of Labor Statistics)

Table 2 CPI (1982–1984 = 100)

Year	Index
1960	29.6
1975	53.8
1990	130.7
2005	195.3

SOLUTION

Step 1 Let $x = $ the purchasing power of an annual salary in 2005.

Step 2 Annual salary in 1960 = $13,000

$$\text{CPI in 1960} = 29.6$$
$$\text{CPI in 2005} = 195.3$$

Step 3 The ratio of a salary in 2005 to a salary in 1960 is the same as the ratio of the CPI in 2005 to the CPI in 1960.

$$\frac{x}{13,000} = \frac{195.3}{29.6} \quad \text{Multiply both sides by 13,000.}$$

Step 4
$$x = 13,000 \cdot \frac{195.3}{29.6}$$
$$= \$85,774 \text{ per year}$$

Step 5

Salary Ratio	**CPI Ratio**
$\dfrac{85,774}{13,000} = 6.598$	$\dfrac{195.3}{29.6} = 6.59797$

Note: The slight difference in these ratios is due to rounding the 2005 salary to the nearest dollar.

Matched Problem 10 What net annual salary in 1975 would have had the same purchasing power as a net annual salary of $100,000 in 2005? Compute the answer to the nearest dollar.

Exercises 1-1

A

Solve Problems 1–6.

1. $2m + 9 = 5m - 6$ **2.** $3y - 4 = 6y - 19$

3. $2x + 3 < -4$ **4.** $5x + 2 > 1$

5. $-3x \geq -12$ **6.** $-4x \leq 8$

Solve Problems 7–10 and graph.

7. $-4x - 7 > 5$ **8.** $-2x + 8 < 4$

9. $2 \leq x + 3 \leq 5$ **10.** $-4 < 2y - 3 < 9$

Solve Problems 11–24.

11. $\dfrac{x}{4} + \dfrac{1}{2} = \dfrac{1}{8}$ **12.** $\dfrac{m}{3} - 4 = \dfrac{2}{3}$

13. $\dfrac{y}{-5} > \dfrac{3}{2}$

14. $\dfrac{x}{-4} < \dfrac{5}{6}$

15. $2u + 4 = 5u + 1 - 7u$

16. $-3y + 9 + y = 13 - 8y$

B

17. $10x + 25(x - 3) = 275$

18. $-3(4 - x) = 5 - (x + 1)$

19. $3 - y \le 4(y - 3)$

20. $x - 2 \ge 2(x - 5)$

21. $\dfrac{x}{5} - \dfrac{x}{6} = \dfrac{6}{5}$

22. $\dfrac{y}{4} - \dfrac{y}{3} = \dfrac{1}{2}$

23. $\dfrac{m}{5} - 3 < \dfrac{3}{5} - \dfrac{m}{2}$

24. $\dfrac{u}{2} - \dfrac{2}{3} < \dfrac{u}{3} + 2$

Solve Problems 25–28 and graph.

25. $2 \le 3x - 7 < 14$

26. $-4 \le 5x + 6 < 21$

27. $-4 \le \frac{9}{5}C + 32 \le 68$

28. $-1 \le \frac{2}{3}t + 5 \le 11$

C

Solve Problems 29–34 for the indicated variable.

29. $3x - 4y = 12$; for y

30. $y = -\frac{2}{3}x + 8$; for x

31. $Ax + By = C$; for y $(B \ne 0)$

32. $y = mx + b$; for m

33. $F = \frac{9}{5}C + 32$; for C

34. $C = \frac{5}{9}(F - 32)$; for F

Solve Problems 35 and 36 and graph.

35. $-3 \le 4 - 7x < 18$

36. $-10 \le 8 - 3u \le -6$

37. What can be said about the signs of the numbers a and b in each case?

(A) $ab > 0$

(B) $ab < 0$

(C) $\dfrac{a}{b} > 0$

(D) $\dfrac{a}{b} < 0$

38. What can be said about the signs of the numbers a, b, and c in each case?

(A) $abc > 0$

(B) $\dfrac{ab}{c} < 0$

(C) $\dfrac{a}{bc} > 0$

(D) $\dfrac{a^2}{bc} < 0$

39. If both a and b are positive numbers and b/a is greater than 1, then is $a - b$ positive or negative?

40. If both a and b are negative numbers and b/a is greater than 1, then is $a - b$ positive or negative?

In Problems 41–46, discuss the validity of each statement. If the statement is true, explain why. If not, give a counterexample.

41. If the intersection of two open intervals is nonempty, then their intersection is an open interval.

42. If the intersection of two closed intervals is nonempty, then their intersection is a closed interval.

43. The union of any two open intervals is an open interval.

44. The union of any two closed intervals is a closed interval.

45. If the intersection of two open intervals is nonempty, then their union is an open interval.

46. If the intersection of two closed intervals is nonempty, then their union is a closed interval.

Applications

47. **Ticket sales.** A rock concert brought in $432,500 on the sale of 9,500 tickets. If the tickets sold for $35 and $55 each, how many of each type of ticket were sold?

48. **Parking meter coins.** An all-day parking meter takes only dimes and quarters. If it contains 100 coins with a total value of $14.50, how many of each type of coin are in the meter?

49. **IRA.** You have $500,000 in an IRA (Individual Retirement Account) at the time you retire. You have the option of investing this money in two funds: Fund A pays 5.2% annually and Fund B pays 7.7% annually. How should you divide your money between Fund A and Fund B to produce an annual interest income of $34,000?

50. **IRA.** Refer to Problem 49. How should you divide your money between Fund A and Fund B to produce an annual interest income of $30,000?

51. **Car prices.** If the price change of cars parallels the change in the CPI (see Table 2 in Example 10), what would a car sell for (to the nearest dollar) in 2005 if a comparable model sold for $10,000 in 1990?

52. **Home values.** If the price change in houses parallels the CPI (see Table 2 in Example 10), what would a house valued at $200,000 in 2005 be valued at (to the nearest dollar) in 1960?

53. **Retail and wholesale prices.** Retail prices in a department store are obtained by marking up the wholesale price by 40%. That is, retail price is obtained by adding 40% of the wholesale price to the wholesale price.

(A) What is the retail price of a suit if the wholesale price is $300?

(B) What is the wholesale price of a pair of jeans if the retail price is $77?

54. **Retail and sale prices.** Sale prices in a department store are obtained by marking down the retail price by 15%. That is, sale price is obtained by subtracting 15% of the retail price from the retail price.

(A) What is the sale price of a hat that has a retail price of $60?

(B) What is the retail price of a dress that has a sale price of $136?

55. **Equipment rental.** A golf course charges $52 for a round of golf using a set of their clubs, and $44 if you have your own clubs. If you buy a set of clubs for $270, how many rounds must you play to recover the cost of the clubs?

56. **Equipment rental.** The local supermarket rents carpet cleaners for $20 a day. These cleaners use shampoo in a special cartridge that sells for $16 and is available only from the supermarket. A home carpet cleaner can be purchased for $300. Shampoo for the home cleaner is readily available for $9 a bottle. Past experience has shown that it takes two shampoo cartridges to clean the 10-foot-by-12- foot carpet in your living room with the rented cleaner. Cleaning the same area with the home cleaner will consume three bottles of shampoo. If you buy the home cleaner, how many times must you clean the living-room carpet to make buying cheaper than renting?

57. Sales commissions. One employee of a computer store is paid a base salary of $2,000 a month plus an 8% commission on all sales over $7,000 during the month. How much must the employee sell in one month to earn a total of $4,000 for the month?

58. Sales commissions. A second employee of the computer store in Problem 57 is paid a base salary of $3,000 a month plus a 5% commission on all sales during the month.

  (A) How much must this employee sell in one month to earn a total of $4,000 for the month?

  (B) Determine the sales level at which both employees receive the same monthly income.

  (C) If employees can select either of these payment methods, how would you advise an employee to make this selection?

59. Break-even analysis. A publisher for a promising new novel figures fixed costs (overhead, advances, promotion, copy editing, typesetting) at $55,000, and variable costs (printing, paper, binding, shipping) at $1.60 for each book produced. If the book is sold to distributors for $11 each, how many must be produced and sold for the publisher to break even?

60. Break-even analysis. The publisher of a new book figures fixed costs at $92,000 and variable costs at $2.10 for each book produced. If the book is sold to distributors for $15 each, how many must be sold for the publisher to break even?

61. Break-even analysis. The publisher in Problem 59 finds that rising prices for paper increase the variable costs to $2.10 per book.

  (A) Discuss possible strategies the company might use to deal with this increase in costs.

  (B) If the company continues to sell the books for $11, how many books must they sell now to make a profit?

  (C) If the company wants to start making a profit at the same production level as before the cost increase, how much should they sell the book for now?

62. Break-even analysis. The publisher in Problem 60 finds that rising prices for paper increase the variable costs to $2.70 per book.

  (A) Discuss possible strategies the company might use to deal with this increase in costs.

  (B) If the company continues to sell the books for $15, how many books must they sell now to make a profit?

  (C) If the company wants to start making a profit at the same production level as before the cost increase, how much should they sell the book for now?

63. Wildlife management. A naturalist estimated the total number of rainbow trout in a certain lake using the capture–mark–recapture technique. He netted, marked, and released 200 rainbow trout. A week later, allowing for thorough mixing, he again netted 200 trout, and found 8 marked ones among them. Assuming that the proportion of marked fish in the second sample was the same as the proportion of all marked fish in the total population, estimate the number of rainbow trout in the lake.

64. Temperature conversion. If the temperature for a 24-hour period at an Antarctic station ranged between $-49°F$ and $14°F$ (that is, $-49 \le F \le 14$), what was the range in degrees Celsius? [*Note:* $F = \frac{9}{5}C + 32$.]

65. Psychology. The IQ (intelligence quotient) is found by dividing the mental age (MA), as indicated on standard tests, by the chronological age (CA) and multiplying by 100. For example, if a child has a mental age of 12 and a chronological age of 8, the calculated IQ is 150. If a 9-year-old girl has an IQ of 140, compute her mental age.

66. Psychology. Refer to Problem 65. If the IQ of a group of 12-year-old children varies between 80 and 140, what is the range of their mental ages?

67. Anthropology. In their study of genetic groupings, anthropologists use a ratio called the **cephalic index**. This is the ratio of the breadth B of the head to its length L (looking down from above) expressed as a percentage. A study of the Gurung community of Nepal published in the *Kathmandu University Medical Journal* in 2005 found that the average head length of males was 18 cm, and their head breadths varied between 12 and 18 cm. Find the range of the cephalic index for males. Round endpoints to one decimal place.

Figure
for 67–68

68. Anthropology. Refer to Problem 67. The same study found that the average head length of females was 17.4 cm, and their head breadths varied between 15 and 20 cm. Find the range of the cephalic index for females. Round endpoints to one decimal place.

Answers to Matched Problems

1. $x = 4$ **2.** $x = 2$

3. (A) $L = \dfrac{S - 2WH}{2W + 2H}$ (B) $H = \dfrac{S - 2LW}{2L + 2W}$

4. (A) $<$ (B) $<$ (C) $>$

5. (A) $-7 < x \le 4$;

  (B) $(-\infty, 3)$

6. $x \ge -4$ or $[-4, \infty)$

7. $-1 \le x < 4$ or $[-1, 4)$

8. $26,000   **9.** 7,500 DVDs   **10.** $27,547

1-2   Graphs and Lines

In this section, we will consider one of the most basic geometric figures—a line. When we use the term *line* in this book, we mean *straight line*. We will learn how to recognize and graph a line, and how to use information concerning a line to find its equation. Examining the graph of any equation often results in additional insight into the nature of the equation's solutions.

Cartesian Coordinate System

Recall that to form a **Cartesian** or **rectangular coordinate system**, we select two real number lines—one horizontal and one vertical—and let them cross through their origins as indicated in Figure 1. Up and to the right are the usual choices for the positive directions. These two number lines are called the **horizontal axis** and the **vertical axis**, or, together, the **coordinate axes**. The horizontal axis is usually referred to as the ***x* axis** and the vertical axis as the ***y* axis**, and each is labeled accordingly. The coordinate axes divide the plane into four parts called **quadrants**, which are numbered counterclockwise from I to IV (see Fig. 1).

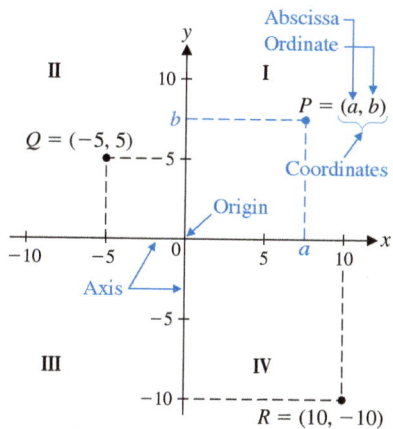

Figure 1 The Cartesian (rectangular) coordinate system

Now we want to assign *coordinates* to each point in the plane. Given an arbitrary point P in the plane, pass horizontal and vertical lines through the point (Fig. 1). The vertical line will intersect the horizontal axis at a point with coordinate a, and the horizontal line will intersect the vertical axis at a point with coordinate b. These two numbers, written as the **ordered pair** (a, b)* form the **coordinates** of the point P. The first coordinate, a, is called the **abscissa** of P; the second coordinate, b, is called the **ordinate** of P. The abscissa of Q in Figure 1 is -5, and the ordinate of Q is 5. The coordinates of a point can also be referenced in terms of the axis labels. The ***x* coordinate** of R in Figure 1 is 10, and the ***y* coordinate** of R is -10. The point with coordinates $(0, 0)$ is called the **origin**.

The procedure we have just described assigns to each point P in the plane a unique pair of real numbers (a, b). Conversely, if we are given an ordered pair of real numbers (a, b), then, reversing this procedure, we can determine a unique point P in the plane. Thus,

There is a one-to-one correspondence between the points in a plane and the elements in the set of all ordered pairs of real numbers.

This is often referred to as the **fundamental theorem of analytic geometry**.

---

*Here we use (a, b) as the coordinates of a point in a plane. In Section 1-1, we used (a, b) to represent an interval on a real number line. These concepts are not the same. You must always interpret the symbol (a, b) in terms of the context in which it is used.

Graphs of $Ax + By = C$

In Section 1-1, we called an equation of the form $ax + b = 0$ $(a \neq 0)$ a linear equation in one variable. Now we want to consider linear equations in two variables:

DEFINITION   Linear Equations in Two Variables

A **linear equation in two variables** is an equation that can be written in the standard form

$$Ax + By = C$$

where $A, B,$ and C are constants (A and B not both 0), and x and y are variables.

A **solution** of an equation in two variables is an ordered pair of real numbers that satisfies the equation. For example, $(4, 3)$ is a solution of $3x - 2y = 6$. The **solution set** of an equation in two variables is the set of all solutions of the equation. The **graph** of an equation is the graph of its solution set.

EXPLORE & DISCUSS 1

(A) As noted earlier, $(4, 3)$ is a solution of the equation

$$3x - 2y = 6$$

Find three more solutions of this equation. Plot these solutions in a Cartesian coordinate system. What familiar geometric shape could be used to describe the solution set of this equation?

(B) Repeat part (A) for the equation $x = 2$.

(C) Repeat part (A) for the equation $y = -3$.

In Explore & Discuss 1, you may have recognized that the graph of each equation is a (straight) line. Theorem 1 confirms this fact.

THEOREM 1   Graph of a Linear Equation in Two Variables

The graph of any equation of the form

$$Ax + By = C \quad (A \text{ and } B \text{ not both } 0) \tag{1}$$

is a line, and any line in a Cartesian coordinate system is the graph of an equation of this form.

If $A \neq 0$ and $B \neq 0$, then equation (1) can be written as

$$y = -\frac{A}{B}x + \frac{C}{B} = mx + b, m \neq 0$$

If $A = 0$ and $B \neq 0$, then equation (1) can be written as

$$y = \frac{C}{B}$$

and its graph is a **horizontal line**. If $A \neq 0$ and $B = 0$, then equation (1) can be written as

$$x = \frac{C}{A}$$

and its graph is a **vertical line**. To graph equation (1), or any of its special cases, plot any two points in the solution set and use a straightedge to draw the line through these two points. The points where the line crosses the axes are often the easiest to find. The *y* intercept* is the *y* coordinate of the point where the graph crosses the *y*

---

*If the *x* intercept is *a* and the *y* intercept is *b*, then the graph of the line passes through the points $(a, 0)$ and $(0, b)$. It is common practice to refer to both the numbers *a* and *b* and the points $(a, 0)$ and $(0, b)$ as the *x* and *y* intercepts of the line.

axis, and the x intercept is the x coordinate of the point where the graph crosses the x axis. To find the y intercept, let $x = 0$ and solve for y. To find the x intercept, let $y = 0$ and solve for x. It is a good idea to find a third point as a check point.

EXAMPLE 1   **Using Intercepts to Graph a Line**   Graph: $3x - 4y = 12$

SOLUTION

x	y	
0	–3	y intercept
4	0	x intercept
8	3	Check point

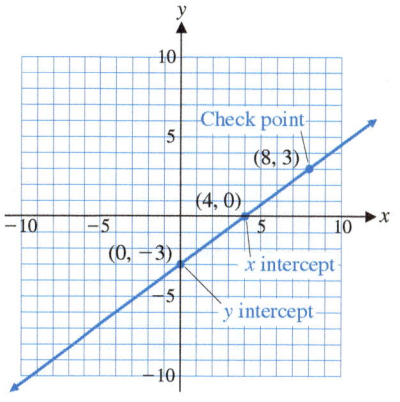

Matched Problem 1   Graph: $4x - 3y = 12$

   The icon in the margin is used throughout this book to identify optional graphing calculator activities that are intended to give you additional insight into the concepts under discussion. You may have to consult the manual for your calculator* for the details necessary to carry out these activities.

 EXAMPLE 2   **Using a Graphing Calculator**   Graph $3x - 4y = 12$ on a graphing calculator and find the intercepts.

SOLUTION   First, we solve $3x - 4y = 12$ for y.

$$3x - 4y = 12 \qquad \text{Add } -3x \text{ to both sides.}$$
$$-4y = -3x + 12 \qquad \text{Divide both sides by } -4.$$
$$y = \frac{-3x + 12}{-4} \qquad \text{Simplify.}$$
$$y = \frac{3}{4}x - 3 \tag{2}$$

Now we enter the right side of equation (2) in a calculator (Fig. 2A), enter values for the window variables (Fig. 2B), and graph the line (Fig. 2C). (The numerals to the left and right of the screen in Figure 2C are Xmin and Xmax, respectively. Similarly, the numerals below and above the screen are Ymin and Ymax.)

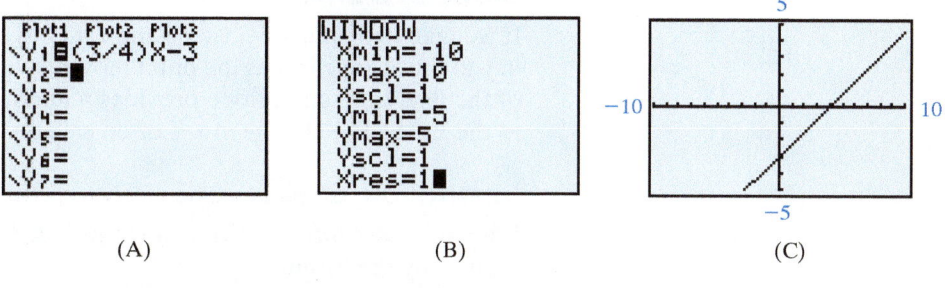

| (A) | (B) | (C) |

Figure 2   Graphing a line on a graphing calculator

---

*We used a Texas Instruments graphing calculator from the TI-83/84 family to produce the graphing calculator screens in the book. Manuals for most graphing calculators are readily available on the Internet.

Figure 3 Using TRACE and zero on a graphing calculator

Next we use two calculator commands to find the intercepts: TRACE (Fig. 3A) and zero (Fig. 3B).

Matched Problem 2 Graph $4x - 3y = 12$ on a graphing calculator and find the intercepts.

EXAMPLE 3 **Horizontal and Vertical Lines**

(A) Graph $x = -4$ and $y = 6$ simultaneously in the same rectangular coordinate system.

(B) Write the equations of the vertical and horizontal lines that pass through the point $(7, -5)$.

SOLUTION (A)

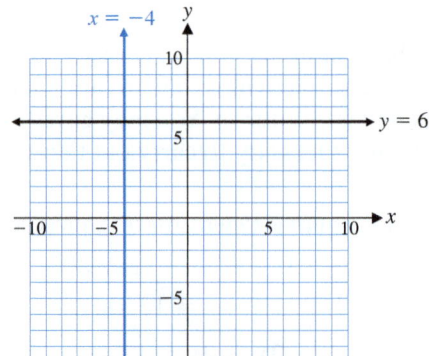

(B) Horizontal line through $(7, -5)$: $y = -5$
Vertical line through $(7, -5)$: $x = 7$

Matched Problem 3 (A) Graph $x = 5$ and $y = -3$ simultaneously in the same rectangular coordinate system.

(B) Write the equations of the vertical and horizontal lines that pass through the point $(-8, 2)$.

Slope of a Line

If we take two points, $P_1(x_1, y_1)$ and $P_2(x_2, y_2)$, on a line, then the ratio of the change in y to the change in x as the point moves from point P_1 to point P_2 is called the **slope** of the line. In a sense, slope provides a measure of the "steepness" of a line relative to the x axis. The change in x is often called the **run**, and the change in y is the **rise**.

DEFINITION Slope of a Line

If a line passes through two distinct points, $P_1(x_1, y_1)$ and $P_2(x_2, y_2)$, then its slope is given by the formula

$$m = \frac{y_2 - y_1}{x_2 - x_1} \qquad x_1 \neq x_2$$

$$= \frac{\text{vertical change (rise)}}{\text{horizontal change (run)}}$$

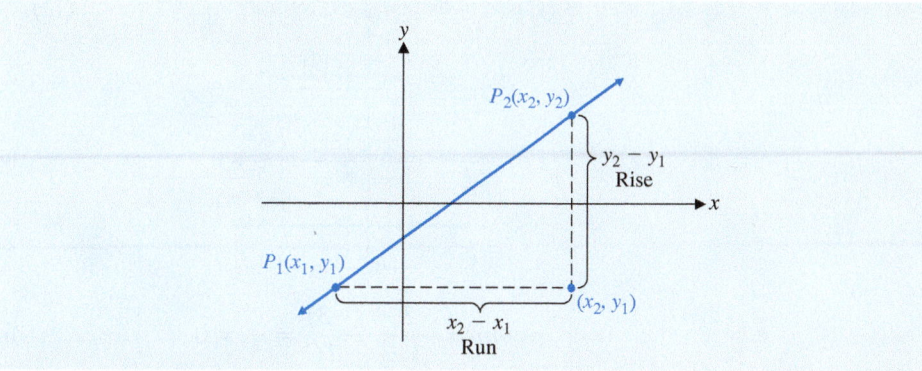

For a horizontal line, y does not change; its slope is 0. For a vertical line, x does not change; $x_1 = x_2$ so its slope is not defined. In general, the slope of a line may be positive, negative, 0, or not defined. Each case is illustrated geometrically in Table 1.

Table 1 Geometric Interpretation of Slope

Line	Rising as x moves from left to right	Falling as x moves from left to right	Horizontal	Vertical
Slope	Positive	Negative	0	Not defined
Example				

 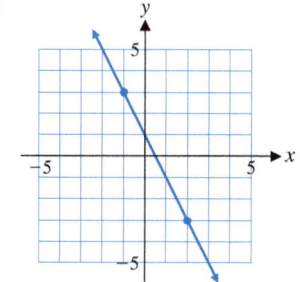

CONCEPTUAL INSIGHT

One property of real numbers discussed in Appendix A, Section A-1, is

$$\frac{-a}{-b} = -\frac{-a}{b} = -\frac{a}{-b} = \frac{a}{b}, \quad b \neq 0$$

This property implies that it does not matter which point we label as P_1 and which we label as P_2 in the slope formula. For example, if $A = (4, 3)$ and $B = (1, 2)$, then

$$B = P_2 = (1, 2) \qquad A = P_2 = (4, 3)$$
$$A = P_1 = (4, 3) \qquad B = P_1 = (1, 2)$$
$$m = \frac{2-3}{1-4} = \frac{-1}{-3} = \frac{1}{3} = \frac{3-2}{4-1}$$

A property of similar triangles (see Table I in Appendix C) ensures that the slope of a line is the same for any pair of distinct points on the line (Fig. 4).

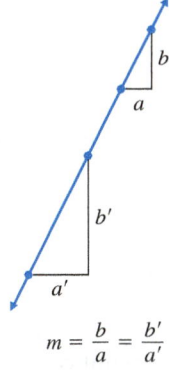

$$m = \frac{b}{a} = \frac{b'}{a'}$$

Figure 4

EXAMPLE 4 **Finding Slopes** Sketch a line through each pair of points, and find the slope of each line.

(A) $(-3, -2), (3, 4)$ (B) $(-1, 3), (2, -3)$

(C) $(-2, -3), (3, -3)$ (D) $(-2, 4), (-2, -2)$

SOLUTION (A)   (B)

$$m = \frac{4 - (-2)}{3 - (-3)} = \frac{6}{6} = 1 \qquad\qquad m = \frac{-3 - 3}{2 - (-1)} = \frac{-6}{3} = -2$$

(C)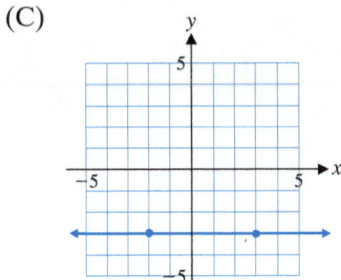

$$m = \frac{-3 - (-3)}{3 - (-2)} = \frac{0}{5} = 0$$

(D)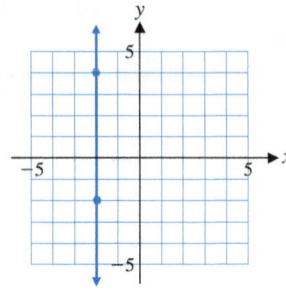

$$m = \frac{-2 - 4}{-2 - (-2)} = \frac{-6}{0}$$

Slope is not defined.

Matched Problem 4   Find the slope of the line through each pair of points.

(A) $(-2, 4), (3, 4)$ (B) $(-2, 4), (0, -4)$

(C) $(-1, 5), (-1, -2)$ (D) $(-1, -2), (2, 1)$

Equations of Lines: Special Forms

Let us start by investigating why $y = mx + b$ is called the *slope-intercept form* for a line.

EXPLORE & DISCUSS 2

(A) Graph $y = x + b$ for $b = -5, -3, 0, 3,$ and 5 simultaneously in the same coordinate system. Verbally describe the geometric significance of b.

(B) Graph $y = mx - 1$ for $m = -2, -1, 0, 1,$ and 2 simultaneously in the same coordinate system. Verbally describe the geometric significance of m.

(C) Using a graphing calculator, explore the graph of $y = mx + b$ for different values of m and b.

As you may have deduced from Explore & Discuss 2, constants m and b in $y = mx + b$ have the following geometric interpretations.

If we let $x = 0$, then $y = b$. So the graph of $y = mx + b$ crosses the y axis at $(0, b)$. The constant b is the y *intercept*. For example, the y intercept of the graph of $y = -4x - 1$ is -1.

To determine the geometric significance of m, we proceed as follows: If $y = mx + b$, then by setting $x = 0$ and $x = 1$, we conclude that $(0, b)$ and $(1, m + b)$ lie on its graph (Fig. 5). The slope of this line is given by:

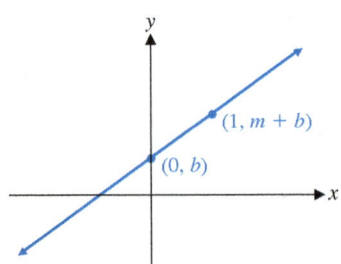

Figure 5

$$\text{Slope} = \frac{y_2 - y_1}{x_2 - x_1} = \frac{(m + b) - b}{1 - 0} = m$$

So m is the slope of the line given by $y = mx + b$.

DEFINITION Slope-Intercept Form

The equation

$$y = mx + b \quad m = \text{slope}, b = y \text{ intercept} \tag{3}$$

is called the **slope-intercept form** of an equation of a line.

EXAMPLE 5 **Using the Slope-Intercept Form**

(A) Find the slope and y intercept, and graph $y = -\frac{2}{3}x - 3$.

(B) Write the equation of the line with slope $\frac{2}{3}$ and y intercept -2.

SOLUTION (A) Slope $= m = -\frac{2}{3}$ (B) $m = \frac{2}{3}$ and $b = -2$;

y intercept $= b = -3$ so, $y = \frac{2}{3}x - 2$

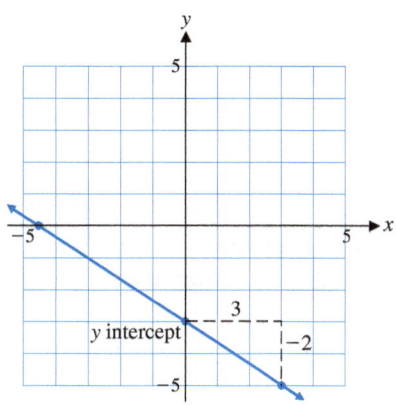

Matched Problem 5 Write the equation of the line with slope $\frac{1}{2}$ and y intercept -1. Graph.

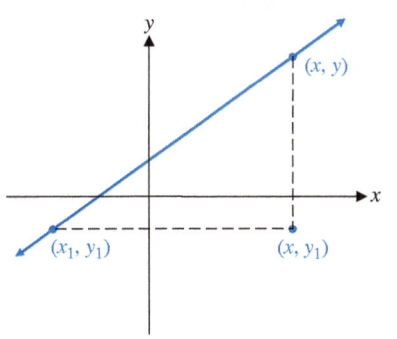

Figure 6

Suppose that a line has slope m and passes through a fixed point (x_1, y_1). If the point (x, y) is any other point on the line (Fig. 6), then

$$\frac{y - y_1}{x - x_1} = m$$

That is,

$$y - y_1 = m(x - x_1) \qquad (4)$$

We now observe that (x_1, y_1) also satisfies equation (4) and conclude that equation (4) is an equation of a line with slope m that passes through (x_1, y_1).

DEFINITION Point-Slope Form

An equation of a line with slope m that passes through (x_1, y_1) is

$$y - y_1 = m(x - x_1) \qquad (4)$$

which is called the **point-slope form** of an equation of a line.

The point-slope form is extremely useful, since it enables us to find an equation for a line if we know its slope and the coordinates of a point on the line or if we know the coordinates of two points on the line.

EXAMPLE 6 **Using the Point-Slope Form**

(A) Find an equation for the line that has slope $\frac{1}{2}$ and passes through $(-4, 3)$. Write the final answer in the form $Ax + By = C$.

(B) Find an equation for the line that passes through the points $(-3, 2)$ and $(-4, 5)$. Write the resulting equation in the form $y = mx + b$.

SOLUTION (A) Use $y - y_1 = m(x - x_1)$. Let $m = \frac{1}{2}$ and $(x_1, y_1) = (-4, 3)$. Then

$$y - 3 = \tfrac{1}{2}[x - (-4)]$$
$$y - 3 = \tfrac{1}{2}(x + 4) \qquad \text{Multiply both sides by 2.}$$
$$2y - 6 = x + 4$$
$$-x + 2y = 10 \quad \text{or} \quad x - 2y = -10$$

(B) First, find the slope of the line by using the slope formula:

$$m = \frac{y_2 - y_1}{x_2 - x_1} = \frac{5 - 2}{-4 - (-3)} = \frac{3}{-1} = -3$$

Now use $y - y_1 = m(x - x_1)$ with $m = -3$ and $(x_1, y_1) = (-3, 2)$:

$$y - 2 = -3[x - (-3)]$$
$$y - 2 = -3(x + 3)$$
$$y - 2 = -3x - 9$$
$$y = -3x - 7$$

Matched Problem 6 (A) Find an equation for the line that has slope $\frac{2}{3}$ and passes through $(6, -2)$. Write the resulting equation in the form $Ax + By = C, A > 0$.

(B) Find an equation for the line that passes through $(2, -3)$ and $(4, 3)$. Write the resulting equation in the form $y = mx + b$.

The various forms of the equation of a line that we have discussed are summarized in Table 2 for quick reference.

Table 2   Equations of a Line

Standard form	$Ax + By = C$	A and B not both 0
Slope-intercept form	$y = mx + b$	Slope: m; y intercept: b
Point-slope form	$y - y_1 = m(x - x_1)$	Slope: m; point: (x_1, y_1)
Horizontal line	$y = b$	Slope: 0
Vertical line	$x = a$	Slope: undefined

Applications

We will now see how equations of lines occur in certain applications.

EXAMPLE 7   **Cost Equation** The management of a company that manufactures skateboards has fixed costs (costs at 0 output) of $300 per day and total costs of $4,300 per day at an output of 100 skateboards per day. Assume that cost C is linearly related to output x.

(A) Find the slope of the line joining the points associated with outputs of 0 and 100; that is, the line passing through $(0, 300)$ and $(100, 4,300)$.

(B) Find an equation of the line relating output to cost. Write the final answer in the form $C = mx + b$.

(C) Graph the cost equation from part (B) for $0 \le x \le 200$.

SOLUTION   (A) $m = \dfrac{y_2 - y_1}{x_2 - x_1}$

$$= \frac{4,300 - 300}{100 - 0}$$

$$= \frac{4,000}{100} = 40$$

(B) We must find an equation of the line that passes through $(0, 300)$ with slope 40. We use the slope-intercept form:

$$C = mx + b$$
$$C = 40x + 300$$

(C)

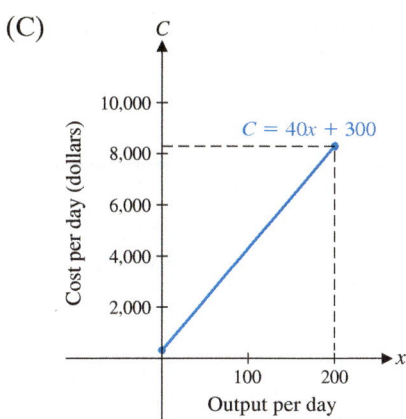

In Example 7, the *fixed cost* of $300 per day covers plant cost, insurance, and so on. This cost is incurred whether or not there is any production. The *variable cost* is $40x$, which depends on the day's output. Since increasing production from x to $x + 1$ will increase the cost by $40 (from $40x + 300$ to $40x + 340$), the slope 40 can be interpreted as the **rate of change** of the cost function with respect to production x.

Matched Problem 7   Answer parts (A) and (B) in Example 7 for fixed costs of $250 per day and total costs of $3,450 per day at an output of 80 skateboards per day.

In a free competitive market, the price of a product is determined by the relationship between supply and demand. If there is a surplus—that is, the supply is greater than the demand—the price tends to come down. If there is a shortage—that is, the demand is greater than the supply—the price tends to go up. The price tends to move toward an equilibrium price at which the supply and demand are equal. Example 8 introduces the basic concepts.

EXAMPLE 8   **Supply and Demand**   At a price of $9.00 per box of oranges, the supply is 320,000 boxes and the demand is 200,000 boxes. At a price of $8.50 per box, the supply is 270,000 boxes and the demand is 300,000 boxes.

(A) Find a price–supply equation of the form $p = mx + b$, where p is the price in dollars and x is the corresponding supply in thousands of boxes.

(B) Find a price–demand equation of the form $p = mx + b$, where p is the price in dollars and x is the corresponding demand in thousands of boxes.

(C) Graph the price–supply and price–demand equations in the same coordinate system and find their point of intersection.

SOLUTION   (A) To find a price–supply equation of the form $p = mx + b$, we must find two points of the form (x, p) that are on the supply line. From the given supply data, $(320, 9)$ and $(270, 8.5)$ are two such points. First, find the slope of the line:

$$m = \frac{9 - 8.5}{320 - 270} = \frac{0.5}{50} = 0.01$$

Now use the point-slope form to find the equation of the line:

$$p - p_1 = m(x - x_1)$$
$$p - 9 = 0.01(x - 320) \qquad (x_1, p_1) = (320, 9)$$
$$p - 9 = 0.01x - 3.2$$
$$p = 0.01x + 5.8 \qquad \text{Price–supply equation}$$

(B) From the given demand data, $(200, 9)$ and $(300, 8.5)$ are two points on the demand line.

$$m = \frac{8.5 - 9}{300 - 200} = \frac{-0.5}{100} = -0.005$$

$$p - p_1 = m(x - x_1)$$
$$p - 9 = -0.005(x - 200) \qquad (x_1, p_1) = (200, 9)$$
$$p - 9 = -0.005x + 1$$
$$p = -0.005x + 10 \qquad \textit{Price–demand equation}$$

(C) From part (A), we plot the points $(320, 9)$ and $(270, 8.5)$ and then draw the line through them. We do the same with the points $(200, 9)$ and $(300, 8.5)$ from part (B) (Fig. 7). (Note that we restricted the axes to intervals that contain these data points.) To find the intersection point of the two lines, we equate the right-hand sides of the price–supply and price–demand equations and solve for x:

$$\begin{array}{cc} \textbf{Price–supply} & \textbf{Price–demand} \end{array}$$
$$0.01x + 5.8 = -0.005x + 10$$
$$0.015x = 4.2$$
$$x = 280$$

Figure 7 Graphs of price–supply and price–demand equations

Now use the price–supply equation to find p when $x = 280$:

$$p = 0.01x + 5.8$$
$$p = 0.01(280) + 5.8 = 8.6$$

As a check, we use the price–demand equation to find p when $x = 280$:

$$p = -0.005x + 10$$
$$p = -0.005(280) + 10 = 8.6$$

The lines intersect at $(280, 8.6)$. The intersection point of the price–supply and price–demand equations is called the **equilibrium point**, and its coordinates are the **equilibrium quantity** (280) and the **equilibrium price** ($8.60). These terms are illustrated in Figure 7. The intersection point can also be found by using the INTERSECT command on a graphing calculator (Fig. 8). To summarize, the price of a box of oranges tends toward the equilibrium price of $8.60, at which the supply and demand are both equal to 280,000 boxes.

Figure 8 Finding an intersection point

Matched Problem 8 At a price of \$12.59 per box of grapefruit, the supply is 595,000 boxes and the demand is 650,000 boxes. At a price of \$13.19 per box, the supply is 695,000 boxes and the demand is 590,000 boxes. Assume that the relationship between price and supply is linear and that the relationship between price and demand is linear.

(A) Find a price–supply equation of the form $p = mx + b$.

(B) Find a price–demand equation of the form $p = mx + b$.

(C) Find the equilibrium point.

Exercises 1-2

A

Problems 1–4 refer to graphs (A)–(D).

(A) (B)

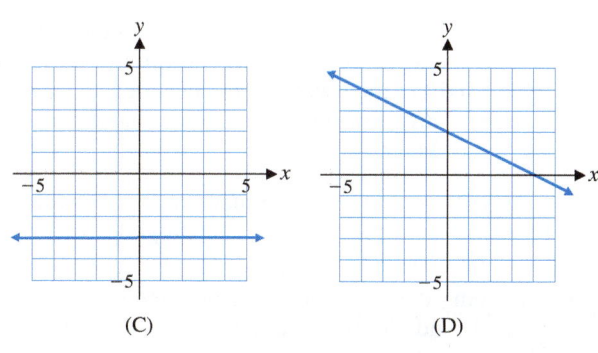

(C) (D)

1. Identify the graph(s) of lines with a negative slope.

2. Identify the graph(s) of lines with a positive slope.

3. Identify the graph(s) of any lines with slope zero.

4. Identify the graph(s) of any lines with undefined slope.

In Problems 5–8, sketch a graph of each equation in a rectangular coordinate system.

5. $y = 2x - 3$

6. $y = \dfrac{x}{2} + 1$

7. $2x + 3y = 12$

8. $8x - 3y = 24$

In Problems 9–14, find the slope and y intercept of the graph of each equation.

9. $y = 5x - 7$

10. $y = 3x + 2$

11. $y = -\dfrac{5}{2}x - 9$

12. $y = -\dfrac{10}{3}x + 4$

13. $y = \dfrac{x}{4} + \dfrac{2}{3}$

14. $y = \dfrac{x}{5} - \dfrac{1}{2}$

In Problems 15–18, write an equation of the line with the indicated slope and y intercept.

15. Slope = 2
 y intercept = 1

16. Slope = 1
 y intercept = 5

17. Slope = $-\dfrac{1}{3}$
 y intercept = 6

18. Slope = $\dfrac{6}{7}$
 y intercept = $-\dfrac{9}{2}$

In Problems 19–22, use the graph of each line to find the x intercept, y intercept, and slope. Write the slope-intercept form of the equation of the line.

19.

20.

21.

22.

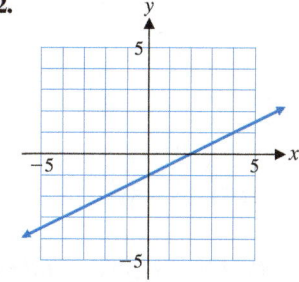

Sketch a graph of each equation or pair of equations in Problems 23–28 in a rectangular coordinate system.

23. $y = -\dfrac{2}{3}x - 2$

24. $y = -\dfrac{3}{2}x + 1$

25. $3x - 2y = 10$

26. $5x - 6y = 15$

27. $x = 3; y = -2$

28. $x = -3; y = 2$

In Problems 29–34, find the slope of the graph of each equation.

29. $4x + y = 3$

30. $5x - y = -2$

31. $3x + 5y = 15$

32. $2x - 3y = 18$

33. $-4x + 2y = 9$

34. $-x + 8y = 4$

35. Given $Ax + By = 12$, graph each of the following three cases in the same coordinate system.

(A) $A = 2$ and $B = 0$

(B) $A = 0$ and $B = 3$

(C) $A = 3$ and $B = 4$

36. Given $Ax + By = 24$, graph each of the following three cases in the same coordinate system.

(A) $A = 6$ and $B = 0$

(B) $A = 0$ and $B = 8$

(C) $A = 2$ and $B = 3$

37. Graph $y = 25x + 200, x \geq 0$.

38. Graph $y = 40x + 160, x \geq 0$.

39. (A) Graph $y = 1.2x - 4.2$ in a rectangular coordinate system.

(B) Find the x and y intercepts algebraically to one decimal place.

 (C) Graph $y = 1.2x - 4.2$ in a graphing calculator.

 (D) Find the x and y intercepts to one decimal place using TRACE and the zero command.

40. (A) Graph $y = -0.8x + 5.2$ in a rectangular coordinate system.

(B) Find the x and y intercepts algebraically to one decimal place.

 (C) Graph $y = -0.8x + 5.2$ in a graphing calculator.

 (D) Find the x and y intercepts to one decimal place using TRACE and the zero command.

(E) Using the results of parts (A) and (B), or (C) and (D), find the solution set for the linear inequality

$$-0.8x + 5.2 < 0$$

In Problems 41–44, write the equations of the vertical and horizontal lines through each point.

41. $(4, -3)$

42. $(-5, 6)$

43. $(-1.5, -3.5)$

44. $(2.6, 3.8)$

In Problems 45–50, write the equation of the line through each indicated point with the indicated slope. Write the final answer in the form $y = mx + b$.

45. $m = -4; (2, -3)$

46. $m = -6; (-4, 1)$

47. $m = \frac{3}{2}; (-4, -5)$

48. $m = \frac{4}{3}; (-6, 2)$

49. $m = 0; (-1.5, 4.6)$

50. $m = 0; (3.1, -2.7)$

In Problems 51–58,

(A) *Find the slope of the line that passes through the given points.*

(B) *Find the standard form of the equation of the line.*

(C) *Find the slope-intercept form of the equation of the line.*

51. $(2, 5)$ and $(5, 7)$

52. $(1, 2)$ and $(3, 5)$

53. $(-2, -1)$ and $(2, -6)$

54. $(2, 3)$ and $(-3, 7)$

55. $(5, 3)$ and $(5, -3)$

56. $(1, 4)$ and $(0, 4)$

57. $(-2, 5)$ and $(3, 5)$

58. $(2, 0)$ and $(2, -3)$

 59. Discuss the relationship among the graphs of the lines with equation $y = mx + 2$, where m is any real number.

60. Discuss the relationship among the graphs of the lines with equation $y = -0.5x + b$, where b is any real number.

Applications

61. **Cost analysis.** A donut shop has a fixed cost of $124 per day and a variable cost of $0.12 per donut. Find the total daily cost of producing x donuts. How many donuts can be produced for a total daily cost of $250?

62. **Cost analysis.** A small company manufactures picnic tables. The weekly fixed cost is $1,200 and the variable cost is $45 per table. Find the total weekly cost of producing x picnic tables. How many picnic tables can be produced for a total weekly cost of $4,800?

63. **Cost analysis.** A plant can manufacture 80 golf clubs per day for a total daily cost of $7,647 and 100 golf clubs per day for a total daily cost of $9,147.

(A) Assuming that daily cost and production are linearly related, find the total daily cost of producing x golf clubs.

(B) Graph the total daily cost for $0 \leq x \leq 200$.

(C) Interpret the slope and y intercept of this cost equation.

64. **Cost analysis.** A plant can manufacture 50 tennis rackets per day for a total daily cost of $3,855 and 60 tennis rackets per day for a total daily cost of $4,245.

(A) Assuming that daily cost and production are linearly related, find the total daily cost of producing x tennis rackets.

(B) Graph the total daily cost for $0 \leq x \leq 100$.

(C) Interpret the slope and y intercept of this cost equation.

65. **Business—Markup policy.** A drugstore sells a drug costing $85 for $112 and a drug costing $175 for $238.

(A) If the markup policy of the drugstore is assumed to be linear, write an equation that expresses retail price R in terms of cost C (wholesale price).

(B) What does a store pay (to the nearest dollar) for a drug that retails for $185?

66. Business—Markup policy. A clothing store sells a shirt costing $20 for $33 and a jacket costing $60 for $93.

(A) If the markup policy of the store is assumed to be linear, write an equation that expresses retail price R in terms of cost C (wholesale price).

(B) What does a store pay for a suit that retails for $240?

67. Business—Depreciation. A farmer buys a new tractor for $157,000 and assumes that it will have a trade-in value of $82,000 after 10 years. The farmer uses a constant rate of depreciation (commonly called **straight-line depreciation**—one of several methods permitted by the IRS) to determine the annual value of the tractor.

(A) Find a linear model for the depreciated value V of the tractor t years after it was purchased.

(B) What is the depreciated value of the tractor after 6 years?

(C) When will the depreciated value fall below $70,000?

(D) Graph V for $0 \le t \le 20$ and illustrate the answers from parts (A) and (B) on the graph.

68. Business—Depreciation. A charter fishing company buys a new boat for $224,000 and assumes that it will have a trade-in value of $115,200 after 16 years.

(A) Find a linear model for the depreciated value V of the boat t years after it was purchased.

(B) What is the depreciated value of the boat after 10 years?

(C) When will the depreciated value fall below $100,000?

(D) Graph V for $0 \le t \le 30$ and illustrate the answers from (A) and (B) on the graph.

69. Boiling point. The temperature at which water starts to boil is called its **boiling point** and is linearly related to the altitude. Water boils at 212°F at sea level and at 193.6°F at an altitude of 10,000 feet. (*Source*: biggreenegg.com)

(A) Find a relationship of the form $T = mx + b$ where T is degrees Fahrenheit and x is altitude in thousands of feet.

(B) Find the boiling point at an altitude of 3,500 feet.

(C) Find the altitude if the boiling point is 200°F.

(D) Graph T and illustrate the answers to (B) and (C) on the graph.

70. Boiling point. The temperature at which water starts to boil is also linearly related to barometric pressure. Water boils at 212°F at a pressure of 29.9 inHg (inches of mercury) and at 191°F at a pressure of 28.4 inHg. (*Source*: biggreenegg.com)

(A) Find a relationship of the form $T = mx + b$, where T is degrees Fahrenheit and x is pressure in inches of mercury.

(B) Find the boiling point at a pressure of 31 inHg.

(C) Find the pressure if the boiling point is 199°F.

(D) Graph T and illustrate the answers to (B) and (C) on the graph.

71. Flight conditions. In stable air, the air temperature drops about 3.6°F for each 1,000-foot rise in altitude. (*Source*: Federal Aviation Administration)

(A) If the temperature at sea level is 70°F, write a linear equation that expresses temperature T in terms of altitude A in thousands of feet.

(B) At what altitude is the temperature 34°F?

72. Flight navigation. The airspeed indicator on some aircraft is affected by the changes in atmospheric pressure at different altitudes. A pilot can estimate the true airspeed by observing the indicated airspeed and adding to it about 1.6% for every 1,000 feet of altitude. (*Source*: Megginson Technologies Ltd.)

(A) A pilot maintains a constant reading of 200 miles per hour on the airspeed indicator as the aircraft climbs from sea level to an altitude of 10,000 feet. Write a linear equation that expresses true airspeed T (in miles per hour) in terms of altitude A (in thousands of feet).

(B) What would be the true airspeed of the aircraft at 6,500 feet?

73. Demographics. The average number of persons per household in the United States has been shrinking steadily for as long as statistics have been kept and is approximately linear with respect to time. In 1980 there were about 2.76 persons per household, and in 2008 about 2.55. (*Source*: U.S. Census Bureau)

(A) If N represents the average number of persons per household and t represents the number of years since 1980, write a linear equation that expresses N in terms of t.

(B) Use this equation to estimate household size in the year 2020.

74. Demographics. The **median** household income divides the households into two groups: the half whose income is less than or equal to the median, and the half whose income is greater than the median. The median household income in the United States grew from about $30,000 in 1990 to about $48,000 in 2006. (*Source*: U.S. Census Bureau)

(A) If I represents the median household income and t represents the number of years since 1990, write a linear equation that expresses I in terms of t.

(B) Use this equation to estimate median household income in the year 2030.

75. Cigarette smoking. The percentage of female cigarette smokers in the United States declined from 21% in 2000 to 18% in 2006. (*Source*: Centers for Disease Control)

(A) Find a linear equation relating percentage of female smokers (f) to years since 2000 (t).

(B) Find the year in which the percentage of female smokers falls below 10%.

76. Cigarette smoking. The percentage of male cigarette smokers in the United States declined from 25.2% in 2001 to 23.9% in 2006. (*Source*: Centers for Disease Control)

(A) Find a linear equation relating percentage of male smokers (m) to years since 2000 (t).

(B) Find the year in which the percentage of male smokers falls below 20%.

77. Supply and demand. At a price of $2.28 per bushel, the supply of barley is 7,500 million bushels and the demand is 7,900 million bushels. At a price of $2.37 per bushel, the supply is 7,900 million bushels and the demand is 7,800 million bushels.

(A) Find a price–supply equation of the form $p = mx + b$.

(B) Find a price–demand equation of the form $p = mx + b$.

(C) Find the equilibrium point.

(D) Graph the price–supply equation, price–demand equation, and equilibrium point in the same coordinate system.

78. Supply and demand. At a price of $1.94 per bushel, the supply of corn is 9,800 million bushels and the demand is 9,300 million bushels. At a price of $1.82 per bushel, the supply is 9,400 million bushels and the demand is 9,500 million bushels.

(A) Find a price–supply equation of the form $p = mx + b$.

(B) Find a price–demand equation of the form $p = mx + b$.

(C) Find the equilibrium point.

(D) Graph the price–supply equation, price–demand equation, and equilibrium point in the same coordinate system.

79. Physics. Hooke's law states that the relationship between the stretch s of a spring and the weight w causing the stretch is linear. For a particular spring, a 5-pound weight causes a stretch of 2 inches, while with no weight, the stretch of the spring is 0.

(A) Find a linear equation that expresses s in terms of w.

(B) What is the stretch for a weight of 20 pounds?

(C) What weight will cause a stretch of 3.6 inches?

80. Physics. The distance d between a fixed spring and the floor is a linear function of the weight w attached to the bottom of the spring. The bottom of the spring is 18 inches from the floor when the weight is 3 pounds, and 10 inches from the floor when the weight is 5 pounds.

(A) Find a linear equation that expresses d in terms of w.

(B) Find the distance from the bottom of the spring to the floor if no weight is attached.

(C) Find the smallest weight that will make the bottom of the spring touch the floor. (Ignore the height of the weight.)

81. Energy consumption. Table 3 lists U.S. oil imports as a percentage* of total energy consumption for selected years. Let x represent years since 1960 and y represent the corresponding percentage of oil imports.

(A) Find the equation of the line through $(0, 9)$ and $(40, 29)$, the first and last data points in Table 3.

(B) Find the equation of the line through $(0, 9)$ and $(10, 12)$, the first and second data points in Table 3.

(C) Graph the lines from parts (A) and (B) and the data points (x, y) from Table 3 in the same coordinate system.

*We use percentage because energy is measured in quadrillion BTUs (British Thermal Units) and one quadrillion is 1,000,000,000,000,000. Percentages are easier to comprehend.

Table 3

Year	Oil Imports (%)
1960	9
1970	12
1980	21
1990	24
2000	29

Source: Energy Information Administration

(D) Use each equation to predict oil imports as a percentage of total consumption in 2020.

(E) Which of the two lines seems to better represent the data in Table 3? Discuss.

82. Energy production. Table 4 lists U.S. crude oil production as a percentage of total U.S. energy production for selected years. Let x represent years since 1960 and y represent the corresponding percentage of oil production.

(A) Find the equation of the line through $(0, 35)$ and $(40, 17)$, the first and last data points in Table 4.

(B) Find the equation of the line through $(0, 35)$ and $(10, 32)$, the first and second data points in Table 4.

(C) Graph the lines from parts (A) and (B) and the data points (x, y) from Table 4 in the same coordinate system.

(D) Use each equation to predict oil imports as a percentage of total consumption in 2020.

(E) Which of the two lines better represents the data in Table 4? Discuss.

Table 4

Year	Oil Production (%)
1960	35
1970	32
1980	27
1990	22
2000	17

Source: Energy Information Administration

Answers to Matched Problems

1.

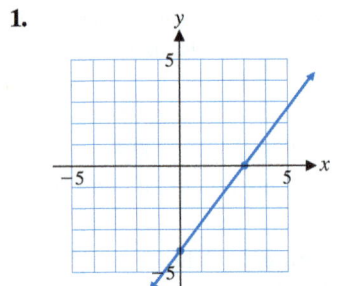

2. y intercept $= -4$, x intercept $= 3$

3. (A)

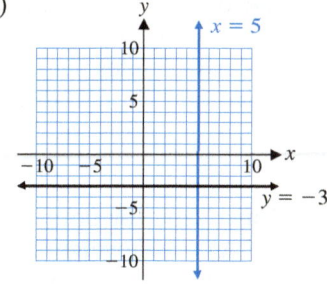

(B) Horizontal line: $y = 2$;
 vertical line: $= -8$

4. (A) 0 (B) -4
 (C) Not defined (D) 1

5. $y = \frac{1}{2}x - 1$

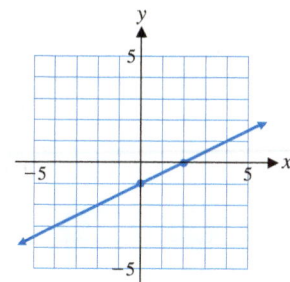

6. (A) $2x - 3y = 18$ (B) $y = 3x - 9$

7. (A) $m = 40$ (B) $C = 40x + 250$

8. (A) $p = 0.006x + 9.02$ (B) $p = -0.01x + 19.09$
 (C) $(629, 12.80)$

1-3 Linear Regression

- Slope as a Rate of Change
- Linear Regression

Mathematical modeling is the process of using mathematics to solve real-world problems. This process can be broken down into three steps (Fig. 1):

Step 1 *Construct* the **mathematical model** (that is, a mathematics problem that, when solved, will provide information about the real-world problem).

Step 2 *Solve* the mathematical model.

Step 3 *Interpret* the solution to the mathematical model in terms of the original real-world problem.

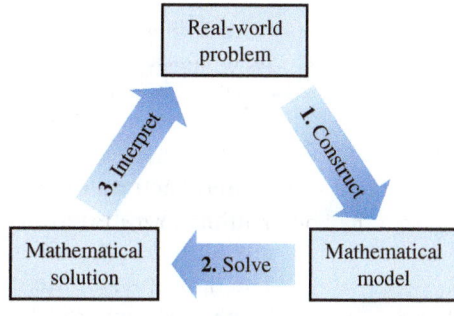

Figure 1

In more complex problems, this cycle may have to be repeated several times to obtain the required information about the real-world problem. In this section, we will discuss one of the simplest mathematical models, a linear equation. With the aid of a graphing calculator or computer, we also will learn how to analyze a linear model based on real-world data.

Slope as a Rate of Change

If x and y are related by the equation $y = mx + b$, where m and b are constants with $m \neq 0$, then x and y are **linearly related**. If (x_1, y_1) and (x_2, y_2) are two distinct points on this line, then the slope of the line is

$$m = \frac{y_2 - y_1}{x_2 - x_1} = \frac{\text{Change in } y}{\text{Change in } x} \tag{1}$$

In applications, ratio (1) is called the **rate of change** of y with respect to x. Since the slope of a line is unique, **the rate of change of two linearly related variables is constant**. Here are some examples of familiar rates of change: miles per hour, revolutions per minute, price per pound, passengers per plane, and so on. If the relationship between x and y is not linear, ratio (1) is called the **average rate of change** of y with respect to x.

EXAMPLE 1   **Estimating Body Surface Area**   Appropriate doses of medicine for both animals and humans are often based on body surface area (BSA). Since weight is much easier to determine than BSA, veterinarians use the weight of an animal to estimate BSA. The following linear equation expresses BSA for canines in terms of weight:

$$a = 16.12w + 375.6$$

where a is BSA in square inches and w is weight in pounds. (*Source*: Veterinary Oncology Consultants, PTY LTD)

(A) Interpret the slope of the BSA equation.

(B) What is the effect of a one-pound increase in weight?

SOLUTION   (A) The rate-of-change BSA with respect to weight is 16.12 square inches per pound.

(B) Since slope is the ratio of rise to run, increasing w by 1 pound (run) increases a by 16.12 square inches (rise).

Matched Problem 1   The equation $a = 28.55w + 118.7$ expresses BSA for felines in terms of weight, where a is BSA in square inches and w is weight in pounds.

(A) Interpret the slope of the BSA equation.

(B) What is the effect of a one-pound increase in weight?

EXPLORE & DISCUSS 1   As illustrated in Example 1A, the slope m of a line with equation $y = mx + b$ has two interpretations:

1. m is the rate of change of y with respect to x.
2. Increasing x by one unit will change y by m units.

How are these two interpretations related?

Parachutes are used to deliver cargo to areas that cannot be reached by other means. The **rate of descent** of the cargo is the rate of change of altitude with respect to time. The absolute value of the rate of descent is called the **speed** of the cargo. At low altitudes, the altitude of the cargo and the time in the air are linearly related. The appropriate rate of descent varies widely with the item. Bulk food (rice, flour, beans, etc.) and clothing can tolerate nearly any rate of descent under 40 ft/sec. Machinery and electronics (pumps, generators, radios, etc.) should generally be dropped at 15 ft/sec or less. Butler Tactical Parachute Systems in Roanoke, Virginia, manufactures a variety of canopies for dropping cargo. The following example uses information taken from the company's brochures.

EXAMPLE 2   **Finding the Rate of Descent**   A 100-pound cargo of delicate electronic equipment is dropped from an altitude of 2,880 feet and lands 200 seconds later. (*Source*: Butler Tactical Parachute Systems)

(A) Find a linear model relating altitude a (in feet) and time in the air t (in seconds).

(B) How fast is the cargo moving when it lands?

SOLUTION   (A) If $a = mt + b$ is the linear equation relating altitude a and time in air t, then the graph of this equation must pass through the following points:

$$(t_1, a_1) = (0, 2,880) \qquad \text{Cargo is dropped from plane.}$$
$$(t_2, a_2) = (200, 0) \qquad \text{Cargo lands.}$$

The slope of this line is

$$m = \frac{a_2 - a_1}{t_2 - t_1} = \frac{0 - 2,880}{200 - 0} = -14.4$$

and the equation of this line is

$$a - 0 = -14.4(t - 200)$$
$$a = -14.4t + 2,880$$

(B) The rate of descent is the slope $m = -14.4$, so the speed of the cargo at landing is $|-14.4| = 14.4$ ft/sec.

Matched Problem 2 A 400-pound load of grain is dropped from an altitude of 2,880 feet and lands 80 seconds later.

(A) Find a linear model relating altitude a (in feet) and time in the air t (in seconds).

(B) How fast is the cargo moving when it lands?

Linear Regression

In real-world applications, we often encounter numerical data in the form of a table. **Regression analysis** is a process for finding a function that provides a useful model for a set of data points. Graphs of equations are often called **curves**, and regression analysis is also referred to as **curve fitting**. In the next example, we use a linear model obtained by using **linear regression** on a graphing calculator.

EXAMPLE 3 **Diamond Prices** Prices for round-shaped diamonds taken from an online trader are given in Table 1.

(A) A linear model for the data in Table 1 is given by

$$p = 6,140c - 480 \qquad (2)$$

where p is the price of a diamond weighing c carats. (We will discuss the source of models like this later in this section.) Plot the points in Table 1 on a Cartesian coordinate system, producing a *scatter plot*, and graph the model on the same axes.

(B) Interpret the slope of the model in (2).

(C) Use the model to estimate the cost of a 0.85-carat diamond and the cost of a 1.2-carat diamond. Round answers to the nearest dollar.

(D) Use the model to estimate the weight of a diamond (to two decimal places) that sells for $4,000.

Table 1 Round-Shaped Diamond Prices

Weight (carats)	Price
0.5	$2,790
0.6	$3,191
0.7	$3,694
0.8	$4,154
0.9	$5,018
1.0	$5,898

Source: www.tradeshop.com

SOLUTION (A) A **scatter plot** is simply a graph of the points in Table 1 (Fig. 2A). To add the graph of the model to the scatter plot, we find any two points that satisfy equation (2) [we choose (0.4, 1,976) and (1.1, 6,274)]. Plotting these points and drawing a line through them gives us Figure 2B.

(A) Scatter plot

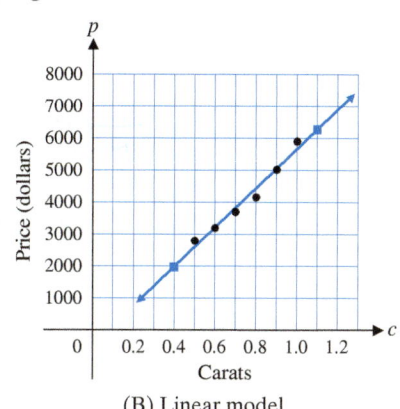

(B) Linear model

Figure 2

(B) The rate of change of the price of a diamond with respect to its weight is 6,140. Increasing the weight by one carat will increase the price by about $6,140.

(C) The graph of the model (Fig. 2B) does not pass through any of the points in the scatter plot, but it comes close to all of them. [Verify this by evaluating equation (2) at $c = 0.5, 0.6, \ldots, 1.$] So we can use equation (2) to approximate points not in Table 1.

$$c = 0.85 \qquad\qquad c = 1.2$$
$$p \approx 6,140(0.85) - 480 \qquad p \approx 6,140(1.2) - 480$$
$$= \$4,739 \qquad\qquad = \$6,888$$

A 0.85-carat diamond will cost about $4,739, and a 1.2-carat diamond will cost about $6,888.

(D) To find the weight of a $4,000 diamond, we solve the following equation for c:

$$6,140c - 480 = 4,000 \qquad \textit{Add 480 to both sides.}$$
$$6,140c = 4,480 \qquad \textit{Divide both sides by 6,140.}$$
$$c = \frac{4,480}{6,140} \approx 0.73 \qquad \textit{Rounded to two decimal places.}$$

A $4,000 diamond will weigh about 0.73 carats.

Matched Problem 3

Prices for emerald-shaped diamonds from an online trader are given in Table 2. Repeat Example 3 for this data with the linear model

$$p = 5,600c - 1,100$$

where p is the price of an emerald-shaped diamond weighing c carats.

Table 2 Emerald-Shaped Diamond Prices

Weight (carats)	Price
0.5	$1,677
0.6	$2,353
0.7	$2,718
0.8	$3,218
0.9	$3,982
1.0	$4,510

Source: www.tradeshop.com

The model we used in Example 3 was obtained using a technique called **linear regression**, and the model is called the **regression line**. This technique produces a line that is the **best fit** for a given data set. Although you can find a linear regression line by hand, we prefer to leave the calculations to a graphing calculator or a computer. Don't be concerned if you don't have either of these electronic devices. We will supply the regression model in most of the applications we discuss, as we did in Example 3.

EXPLORE & DISCUSS 2

As stated above, we used linear regression to produce the model in Example 3. If you have a graphing calculator that supports linear regression, then you can find this model. The linear regression process varies greatly from one calculator to another. Consult the user's manual* for the details of linear regression. The screens in Figure 3 are related to the construction of the model in Example 3 on a Texas Instruments TI-84 Plus.

(A) Produce similar screens on your graphing calculator.

(B) Do the same for Matched Problem 3.

(A) Entering the data

(B) Finding the model

(C) Graphing the data and the model

Figure 3 Linear regression on a graphing calculator

---

*Manuals for most graphing calculators are readily available on the Internet.

In Example 3, we used the regression model to approximate points that were not given in Table 1 but would fit between points in the table. This process is called **interpolation**. In the next example, we use a regression model to approximate points outside the given data set. This process is called **extrapolation**, and the approximations are often referred to as **predictions**.

EXAMPLE 4 **Carbon Monoxide Emissions** Table 3 contains information about U.S. carbon monoxide emissions. The linear regression model for carbon monoxide emissions from all sources (after rounding) is

$$C = 188 - 3.6t$$

where C is the carbon monoxide emissions (in millions of short tons*) and t is time in years with $t = 0$ corresponding to 1980.

(A) Interpret the slope of the regression line as a rate of change.

(B) Use the regression model to predict carbon monoxide emissions from all sources in 2018.

Table 3 U.S. Carbon Monoxide Emissions (millions of short tons)

	1980	1985	1990	1995	2000	2005
All sources	185	177	154	129	114	103
Highway vehicles	144	134	110	84	68	56

Source: U.S. Bureau of Transportation Statistics

SOLUTION (A) The slope $m = -3.6$ is the rate of change of emissions with respect to time. Since the slope is negative and the emissions are given in millions of tons, the emissions are decreasing at a rate of $3.6(1,000,000) = 3,600,000$ tons per year.

(B) If $t = 38$, then

$$C = 188 - 3.6(38) = 51.2 \qquad \text{or} \qquad 51,200,000 \text{ tons}$$

So approximately 51,200,000 tons of carbon monoxide will be emitted in 2018.

Matched Problem 4 Repeat Example 4 using the linear regression model

$$C = -3.8t + 147$$

where C is the carbon monoxide emissions from highway vehicles (in millions of short tons) and t is time in years with $t = 0$ corresponding to 1980.

Forest managers estimate growth, volume, yield, and forest potential. One common measure is the diameter of a tree at breast height (Dbh), which is defined as the diameter of the tree at a point 4.5 feet above the ground on the uphill side of the tree. Example 5 uses Dbh to estimate the height of balsam fir trees.

EXAMPLE 5 **Forestry** A linear regression model for the height of balsam fir trees is

$$h = 3.8d + 18.73$$

where d is Dbh in inches and h is the height in feet.

(A) Interpret the slope of this model.

(B) What is the effect of a 1-inch increase in Dbh?

(C) Estimate the height of a balsam fir with a Dbh of 8 inches. Round your answer to the nearest foot.

---

*A short ton (or a U.S. ton) is 2,000 pounds, a long ton (or a U.K. ton) is 2,240 pounds, and a metric tonne is 1,000 kilograms, or 2204.6 pounds.

(D) Estimate the Dbh of a balsam fir that is 30 feet tall. Round your answer to the nearest inch.

SOLUTION

(A) The rate of change of height with respect to breast height diameter is 3.8 feet per inch.

(B) Height increases by 3.8 feet.

(C) We must find h when $d = 8$:

$$h = 3.8d + 18.73 \qquad \text{Substitute } d = 8.$$
$$h = 3.8(8) + 18.73 \qquad \text{Evaluate.}$$
$$h = 49.13 \approx 49 \text{ ft}$$

(D) We must find d when $h = 30$:

$$h = 3.8d + 18.73 \qquad \text{Substitute } h = 30.$$
$$30 = 3.8d + 18.73 \qquad \text{Subtract 18.73 from both sides.}$$
$$11.27 = 3.8d \qquad \text{Divide both sides by 3.8.}$$
$$d = \frac{11.27}{3.8} \approx 3 \text{ in}$$

The data used to produce the regression model in Example 5 are from the Jack Haggerty Forest at Lakehead University in Canada (Table 4). We used the popular spreadsheet Excel to produce a scatter plot of the data in Table 4 and to find the regression model (Fig. 4).

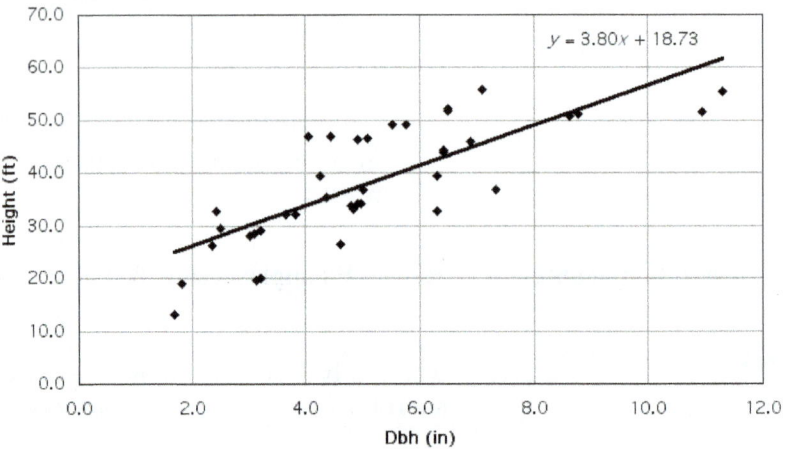

Figure 4 Linear regression with a spreadsheet

Table 4 Height and Diameter of the Balsam Fir

Dbh (in)	Height (ft)	Dbh (in)	Height (ft)	Dbh (in)	Height (ft)	Dbh (in)	Height (ft)
6.5	51.8	6.4	44.0	3.1	19.7	4.6	26.6
8.6	50.9	4.4	46.9	7.1	55.8	4.8	33.1
5.7	49.2	6.5	52.2	6.3	32.8	3.1	28.5
4.9	46.3	4.1	46.9	2.4	26.2	3.2	29.2
6.4	44.3	8.8	51.2	2.5	29.5	5.0	34.1
4.1	46.9	5.0	36.7	6.9	45.9	3.0	28.2
1.7	13.1	4.9	34.1	2.4	32.8	4.8	33.8
1.8	19.0	3.8	32.2	4.3	39.4	4.4	35.4
3.2	20.0	5.5	49.2	7.3	36.7	11.3	55.4
5.1	46.6	6.3	39.4	10.9	51.5	3.7	32.2

(*Source*: Jack Haggerty Forest, Lakehead University, Canada)

Figure 5 Linear regression for white spruce trees

  Matched Problem 5 Figure 5 shows the scatter plot for white spruce trees in the Jack Haggerty Forest at Lakehead University in Canada. A regression model produced by a spreadsheet (Fig. 5), after rounding, is

$$h = 1.8d + 34$$

where d is Dbh in inches and h is the height in feet.

(A) Interpret the slope of this model.

(B) What is the effect of a 1-inch increase in Dbh?

(C) Estimate the height of a white spruce with a Dbh of 10 inches. Round your answer to the nearest foot.

(D) Estimate the Dbh of a white spruce that is 65 feet tall. Round your answer to the nearest inch.

Exercises 1-3

Applications

1. **Ideal weight.** In 1983, Dr. J. D. Robinson published the following estimate of the ideal body weight of a woman:

   49 kg + 1.7 kg for each inch over 5 ft

   (A) Find a linear model for Robinson's estimate of the ideal weight of a woman using w for ideal body weight (in kilograms) and h for height over 5 ft (in inches).

   (B) Interpret the slope of the model.

   (C) If a woman is 5′4″ tall, what does the model predict her weight to be?

   (D) If a woman weighs 60 kg, what does the model predict her height to be?

2. **Ideal weight.** Dr. J. D. Robinson also published the following estimate of the ideal body weight of a man:

   52 kg + 1.9 kg for each inch over 5 ft

   (A) Find a linear model for Robinson's estimate of the ideal weight of a man using w for ideal body weight (in kilograms) and h for height over 5 ft (in inches).

   (B) Interpret the slope of the model.

   (C) If a man is 5′8″ tall, what does the model predict his weight to be?

   (D) If a man weighs 70 kg, what does the model predict his height to be?

3. **Underwater pressure.** At sea level, the weight of the atmosphere exerts a pressure of 14.7 pounds per square inch, commonly referred to as 1 **atmosphere of pressure**. As an object descends in water, pressure P and depth d are linearly related. In salt water, the pressure at a depth of 33 ft is 2 atms, or 29.4 pounds per square inch.

   (A) Find a linear model that relates pressure P (in pounds per square inch) to depth d (in feet).

   (B) Interpret the slope of the model.

   (C) Find the pressure at a depth of 50 ft.

   (D) Find the depth at which the pressure is 4 atms.

4. **Underwater pressure.** Refer to Problem 3. In fresh water, the pressure at a depth of 34 ft is 2 atms, or 29.4 pounds per square inch.

   (A) Find a linear model that relates pressure P (in pounds per square inch) to depth d (in feet).

(B) Interpret the slope of the model.

(C) Find the pressure at a depth of 50 ft.

(D) Find the depth at which the pressure is 4 atms.

5. **Rate of descent—Parachutes.** At low altitudes, the altitude of a parachutist and time in the air are linearly related. A jump at 2,880 ft using the U.S. Army's T-10 parachute system lasts 120 secs.

(A) Find a linear model relating altitude a (in feet) and time in the air t (in seconds).

(B) Find the rate of descent for a T-10 system.

(C) Find the speed of the parachutist at landing.

6. **Rate of descent—Parachutes.** The U.S Army is considering a new parachute, the Advanced Tactical Parachute System (ATPS). A jump at 2,880 ft using the ATPS system lasts 180 secs.

(A) Find a linear model relating altitude a (in feet) and time in the air t (in seconds).

(B) Find the rate of descent for an ATPS system parachute.

(C) Find the speed of the parachutist at landing.

7. **Speed of sound.** The speed of sound through air is linearly related to the temperature of the air. If sound travels at 331 m/sec at 0°C and at 343 m/sec at 20°C, construct a linear model relating the speed of sound (s) and the air temperature (t). Interpret the slope of this model. (*Source:* Engineering Toolbox)

8. **Speed of sound.** The speed of sound through sea water is linearly related to the temperature of the water. If sound travels at 1,403 m/sec at 0°C and at 1,481 m/sec at 20°C, construct a linear model relating the speed of sound (s) and the air temperature (t). Interpret the slope of this model. (*Source:* Engineering Toolbox)

9. **Energy production.** Table 5 lists U.S. fossil fuel production as a percentage of total energy production for selected years. A linear regression model for this data is

$$y = -0.3x + 84.6$$

where x represents years since 1985 and y represents the corresponding percentage of total energy production.

Table 5 U.S. Fossil Fuel Production

Year	Production (%)
1985	85
1990	83
1995	81
2000	80
2005	79

Source: Energy Information Administration

(A) Draw a scatter plot of the data and a graph of the model on the same axes.

(B) Interpret the slope of the model.

(C) Use the model to predict fossil fuel production in 2020.

(D) Use the model to estimate the first year for which fossil fuel production is less than 70% of total energy production.

10. **Energy consumption.** Table 6 lists U.S. fossil fuel consumption as a percentage of total energy consumption for selected years. A linear regression model for this data is

$$y = -0.06x + 85.6$$

where x represents years since 1985 and y represents the corresponding percentage of fossil fuel consumption.

Table 6 U.S. Fossil Fuel Consumption

Year	Consumption (%)
1985	86
1990	85
1995	85
2000	84
2005	85

Source: Energy Information Administration

(A) Draw a scatter plot of the data and a graph of the model on the same axes.

(B) Interpret the slope of the model.

(C) Use the model to predict fossil fuel consumption in 2020.

(D) Use the model to estimate the first year for which fossil fuel consumption is less than 80% of total energy consumption.

11. **Cigarette smoking.** The data in Table 7 shows that the percentage of female cigarette smokers in the U.S. declined from 22.1% in 1997 to 18.0% in 2006.

Table 7 Percentage of Smoking Prevalence among U.S. Adults

Year	Males (%)	Females (%)
1997	27.6	22.1
2000	25.7	21.0
2003	24.1	19.2
2006	23.9	18.0

Source: Centers for Disease Control

(A) Applying linear regression to the data for females in Table 7 produces the model

$$f = -0.47t + 22.19$$

where f is percentage of female smokers and t is time in years since 1997. Draw a scatter plot of the female smoker data and a graph of the regression model on the same axes for $0 \le t \le 30$.

(B) Estimate the first year in which the percentage of female smokers is less than 10%.

12. **Cigarette smoking.** The data in Table 7 shows that the percentage of male cigarette smokers in the U.S. declined from 27.6% in 1997 to 23.9% in 2006.

(A) Applying linear regression to the data for males in Table 7 produces the model

$$m = -0.42t + 27.23$$

where m is percentage of male smokers and t is time in years since 1997. Draw a scatter plot of the male smoker data and a graph of the regression model for $0 \le t \le 15$.

(B) Estimate the first year in which the percentage of male smokers is less than 15%.

13. Undergraduate enrollment. Table 8 lists enrollment in U.S. degree-granting institutions for both undergraduate and graduate students. A linear regression model for undergraduate enrollment is

$$y = 0.17x + 10.1$$

where x represents years since 1980 and y is undergraduate enrollment in millions of students.

Table 8 Fall Undergraduate and Graduate Enrollment (millions of students)

Year	Undergraduate	Graduate
1980	10.48	1.34
1985	10.60	1.38
1990	11.96	1.59
1995	12.23	1.73
2000	13.16	1.85
2005	14.96	2.19

Source: National Center for Education Statistics

(A) Draw a scatter plot of the undergraduate enrollment data and a graph of the model on the same axes.

(B) Predict the undergraduate student enrollment in 2016 (to the nearest 100,000).

(C) Interpret the slope of the model.

14. Graduate student enrollment. A linear regression model for the graduate student enrollment in Table 8 is

$$y = 0.033x + 1.27$$

where x represents years since 1980 and y is graduate enrollment in millions of students.

(A) Draw a scatter plot of the graduate enrollment data and a graph of the model on the same axes.

(B) Predict the graduate student enrollment in 2016 (to the nearest 100,000).

(C) Interpret the slope of the model.

15. Licensed drivers. Table 9 contains the state population and the number of licensed drivers in the state (both in millions) for the states with population under 1 million in 2006. The regression model for this data is

$$y = 0.72x + 0.03$$

where x is the state population and y is the number of licensed drivers in the state.

Table 9 Licensed Drivers in 2006

State	Population	Licensed Drivers
Alaska	0.67	0.49
Delaware	0.85	0.62
Montana	0.94	0.72
North Dakota	0.64	0.47
South Dakota	0.78	0.58
Vermont	0.62	0.53
Wyoming	0.52	0.39

Source: Bureau of Transportation Statistics

(A) Draw a scatter plot of the data and a graph of the model on the same axes.

(B) If the population of Idaho in 2006 was about 1.5 million, use the model to estimate the number of licensed drivers in Idaho in 2006 to the nearest thousand.

(C) If the number of licensed drivers in Rhode Island in 2006 was about 0.74 million, use the model to estimate the population of Rhode Island in 2006 to the nearest thousand.

16. Licensed drivers. Table 10 contains the state population and the number of licensed drivers in the state (both in millions) for the states with population over 10 million in 2006. The regression model for this data is

$$y = 0.60x + 1.15$$

where x is the state population and y is the number of licensed drivers in the state.

Table 10 Licensed Drivers in 2006

State	Population	Licensed Drivers
California	36	23
Florida	18	14
Illinois	13	8
Michigan	10	7
New York	19	11
Ohio	11	8
Pennsylvania	12	9
Texas	24	15

Source: Bureau of Transportation Statistics

(A) Draw a scatter plot of the data and a graph of the model on the same axes.

(B) If the population of Minnesota in 2006 was about 5 million, use the model to estimate the number of licensed drivers in Minnesota in 2006 to the nearest thousand.

(C) If the number of licensed drivers in Wisconsin in 2006 was about 4 million, use the model to estimate the population of Wisconsin in 2006 to the nearest thousand.

17. Net sales. A linear regression model for the net sales data in Table 11 is

$$S = 30.7t + 128$$

where S is net sales and t is time since 2000 in years.

Table 11 Wal-Mart Stores, Inc.

Billions of U.S. Dollars	2004	2005	2006	2007	2008
Net sales	253	282	304	345	375
Operating income	15.3	17.3	18.7	20.5	22.0

Source: Wal-Mart Stores, Inc.

(A) Draw a scatter plot of the data and a graph of the model on the same axes.

(B) Predict Wal-Mart's net sales for 2017.

18. Operating income. A linear regression model for the operating income data in Table 11 is

$$I = 1.66t + 8.80$$

where I is operating income and t is time since 2000 in years.

(A) Draw a scatter plot of the data and a graph of the model on the same axes.

(B) Predict Wal-Mart's annual operating income for 2018.

19. Freezing temperature. Ethylene glycol and propylene glycol are liquids used in antifreeze and deicing solutions. Ethylene glycol is listed as a hazardous chemical by the Environmental Protection Agency, while propylene glycol is generally regarded as safe. Table 12 lists the freezing temperature for various concentrations (as a percentage of total weight) of each chemical in a solution used to deice airplanes. A linear regression model for the ethylene glycol data in Table 12 is

$$E = -0.55T + 31$$

where E is the percentage of ethylene glycol in the deicing solution and T is the temperature at which the solution freezes.

Table 12 Freezing Temperatures

Freezing Temperature (°F)	Ethylene Glycol (% Wt.)	Propylene Glycol (% Wt.)
−50	56	58
−40	53	55
−30	49	52
−20	45	48
−10	40	43
0	33	36
10	25	29
20	16	19

Source: T. Labuza, University of Minnesota

(A) Draw a scatter plot of the data and a graph of the model on the same axes.

(B) Use the model to estimate the freezing temperature to the nearest degree of a solution that is 30% ethylene glycol.

(C) Use the model to estimate the percentage of ethylene glycol in a solution that freezes at 15°F.

20. Freezing temperature. A linear regression model for the propylene glycol data in Table 12 is

$$P = -0.54T + 34$$

where P is the percentage of propylene glycol in the deicing solution and T is the temperature at which the solution freezes.

(A) Draw a scatter plot of the data and a graph of the model on the same axes.

(B) Use the model to estimate the freezing temperature to the nearest degree of a solution that is 30% propylene glycol.

(C) Use the model to estimate the percentage of propylene glycol in a solution that freezes at 15°F.

21. Forestry. The figure contains a scatter plot of 100 data points for black spruce trees and the linear regression model for this data.

(A) Interpret the slope of the model.

(B) What is the effect of a 1-in. increase in Dbh?

(C) Estimate the height of a black spruce with a Dbh of 15 in. Round your answer to the nearest foot.

(D) Estimate the Dbh of a black spruce that is 25 ft tall. Round your answer to the nearest inch.

Figure for 21; black spruce
Source: Lakehead University

22. Forestry. The figure contains a scatter plot of 100 data points for black walnut trees and the linear regression model for this data.

(A) Interpret the slope of the model.

(B) What is the effect of a 1-in. increase in Dbh?

(C) Estimate the height of a black walnut with a Dbh of 12 in. Round your answer to the nearest foot.

(D) Estimate the Dbh of a black walnut that is 25 ft tall. Round your answer to the nearest inch.

Figure for 22; black walnut
Source: Kagen Research

23. Cable television. Table 13 shows the increase in both price and revenue for cable television in the United States. The figure shows a scatter plot and a linear regression model for the average monthly price data in Table 13.

(A) Interpret the slope of the model.

(B) Use the model to predict the average monthly price (to the nearest dollar) in 2020.

Table 13 Cable Television Price and Revenue

Year	Average Monthly Price (dollars)	Annual Total Revenue (billions of dollars)
2000	30.08	40.86
2001	31.58	43.52
2002	34.52	49.43
2003	36.59	51.30
2004	38.23	57.60
2005	39.96	62.27
2006	41.17	68.57
2007	42.76	75.21

24. Cable television. The figure shows a scatter plot and a linear regression model for the annual revenue data in Table 13.

(A) Interpret the slope of the model.

(B) Use the model to predict the annual revenue (to the nearest billion dollars) in 2020.

25. College enrollment. Table 14 lists the fall enrollment in degree-granting institutions by gender, and the figure contains a scatter plot and a regression line for each data set.

Table 14 Fall Enrollment (millions of students)

Year	Male	Female
1970	5.04	3.54
1975	6.15	5.04
1980	5.87	6.22
1985	5.82	6.43
1990	6.28	7.53
1995	6.34	7.92
2000	6.72	8.59
2005	7.46	10.03

Source: National Center for Education Statistics

(A) Interpret the slope of each model.

(B) Predict both the male enrollment and the female enrollment in 2018.

(C) Estimate the first year for which female enrollment will exceed male enrollment by at least 5 million.

Figure for 25; fall enrollment by gender

26. Telephone expenditures. Table 15 lists average annual telephone expenditures (in dollars) per consumer unit on residential phone service and cellular phone service, and the figure contains a scatter plot and regression line for each data set.

Table 15 Telephone Expenditures

Year	Residential Service ($)	Cellular Service ($)
2001	686	210
2002	641	294
2003	620	316
2004	592	378
2005	570	455
2006	542	524
2007	482	608

Source: Bureau of Labor Statistics

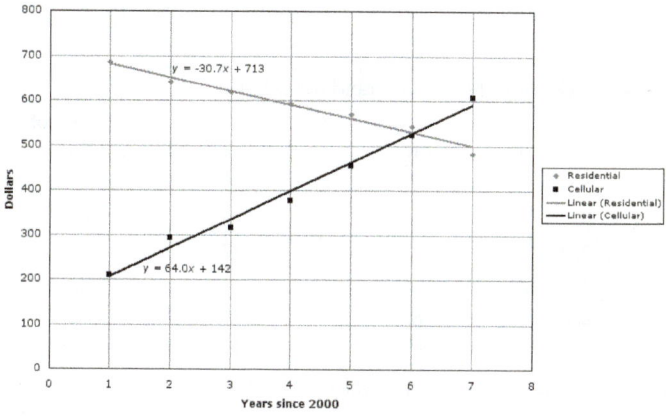

Figure for 26; telephone expenditures

(A) Interpret the slope of each model.

(B) Predict (to the nearest dollar) the average annual residential and cellular expenditures in 2015.

(C) Would the linear regression models give reasonable predictions for the year 2025? Explain.

Problems 27–30 require a graphing calculator or a computer that can calculate the linear regression line for a given data set.

27. Olympic Games. Find a linear regression model for the men's 100-meter freestyle data given in Table 16, where x is years since 1980 and y is winning time (in seconds). Do the same for the women's 100-meter freestyle data. (Round regression coefficients to three decimal places.) Do these models indicate that the women will eventually catch up with the men?

Table 16 Winning Times in Olympic Swimming Events

	100-Meter Freestyle		200-Meter Backstroke	
	Men	**Women**	**Men**	**Women**
1980	50.40	54.79	2:01.93	2:11.77
1984	49.80	55.92	2:00.23	2:12.38
1988	48.63	54.93	1:59.37	2:09.29
1992	49.02	54.65	1:58.47	2:07.06
1996	48.74	54.50	1.58.54	2:07.83
2000	48.30	53.83	1:56.76	2:08.16
2004	48.17	53.84	1:54.76	2:09.16
2008	47.21	53.12	1:53.94	2:05.24

Source: www.infoplease.com

28. Olympic Games. Find a linear regression model for the men's 200-meter backstroke data given in Table 16, where x is years since 1980 and y is winning time (in seconds). Do the same for the women's 200-meter backstroke data. (Round regression coefficients to four decimal places.) Do these models indicate that the women will eventually catch up with the men?

29. Supply and demand. Table 17 contains price–supply data and price–demand data for corn. Find a linear regression model for the price–supply data where x is supply (in billions of bushels) and y is price (in dollars). Do the same for the price–demand data. (Round regression coefficients to two decimal places.) Find the equilibrium price for corn.

Table 17 Supply and Demand for U.S. Corn

Price ($/bu)	Supply (billion bu)	Price ($/bu)	Demand (billion bu)
2.15	6.29	2.07	9.78
2.29	7.27	2.15	9.35
2.36	7.53	2.22	8.47
2.48	7.93	2.34	8.12
2.47	8.12	2.39	7.76
2.55	8.24	2.47	6.98

Source: www.usda.gov/nass/pubs/histdata.htm

30. Supply and demand. Table 18 contains price–supply data and price–demand data for soybeans. Find a linear regression model for the price–supply data where x is supply (in billions of bushels) and y is price (in dollars). Do the same for the price–demand data. (Round regression coefficients to two decimal places.) Find the equilibrium price for soybeans.

Table 18 Supply and Demand for U.S. Soybeans

Price ($/bu)	Supply (billion bu)	Price ($/bu)	Demand (billion bu)
5.15	1.55	4.93	2.60
5.79	1.86	5.48	2.40
5.88	1.94	5.71	2.18
6.07	2.08	6.07	2.05
6.15	2.15	6.40	1.95
6.25	2.27	6.66	1.85

Source: www.usda.gov/nass/pubs/histdata.htm

Answers to Matched Problems

1. (A) The rate of change BSA with respect to weight is 28.55 square inches per pound.

(B) Increasing w by 1 pound increases a by 28.55 square inches.

2. (A) $a = -36t + 2,880$

(B) 36 ft/sec

3. (A)

(B) The rate of change of the price of a diamond with respect to its weight is $5,600. Increasing the weight by one carat will increase the price by about $5,600.

(C) $3,660; $5,620

(D) 0.91 carats

4. (A) The slope $m = -3.8$ is the rate of change of emissions with respect to time. Since the slope is negative and the emissions are given in millions of tons, the emissions are decreasing at a rate of $3.8(1,000,000) = 3,800,000$ tons per year.

(B) Approximately 2,600,000 tons of carbon monoxide will be emitted in 2018.

5. (A) The slope is 1.8, so the rate of change of height with respect to breast height diameter is 1.8 feet per inch.

(B) Height increases by 1.8 feet.

(C) 52 ft

(D) 17 in.

Chapter 1   Review

Important Terms, Symbols, and Concepts

1-1 Linear Equations and Inequalities

- A **first-degree**, or **linear**, **equation** in one variable is any equation that can be written in the form

    Standard form: $ax + b = 0$ $a \neq 0$

    If the equality sign in the standard form is replaced by $<$, $>$, $\leq$, or $\geq$, the resulting expression is called a **first-degree**, or **linear**, **inequality**.

- A **solution** of an equation (or inequality) involving a single variable is a number that when substituted for the variable makes the equation (inequality) true. The set of all solutions is called the **solution set**.

- If we perform an operation on an equation (or inequality) that produces another equation (or inequality) with the same solution set, then the two equations (or inequalities) are **equivalent**. Equations are solved by adding or subtracting the same quantity to both sides, or by multiplying both sides by the same *nonzero* quantity until an equation with an obvious solution is obtained.   Ex. 1, p. 3
    Ex. 2, p. 4

- The **interval notation** $[a, b)$, for example, represents the solution of the **double inequality** $a \leq x < b$.   Ex. 5, p. 7

- Inequalities are solved in the same manner as equations with one important exception. If both sides of an inequality are multiplied by the same *negative* number or divided by the same *negative* number, then the direction or sense of the inequality will reverse ($<$ becomes $>$, $\geq$ becomes $\leq$, and so on).   Ex. 6, p. 7
    Ex. 7, p. 8

- A suggested strategy (p. 8) can be used to solve many word problems.   Ex. 8, p. 8

- A company breaks even if revenues R = costs C, makes a profit if $R > C$, and incurs a loss if $R < C$.   Ex. 9, p. 9

1-2 Graphs and Lines

- A **Cartesian or rectangular coordinate system** is formed by the intersection of a horizontal real number line, usually called the **x axis**, and a vertical real number line, usually called the **y axis**, at their origins. The axes determine a plane and divide this plane into four **quadrants**. Each point in the plane corresponds to its **coordinates**—an ordered pair (a, b) determined by passing horizontal and vertical lines through the point. The **abscissa** or **x coordinate** a is the coordinate of the intersection of the vertical line and the x axis, and the **ordinate** or **y coordinate** b is the coordinate of the intersection of the horizontal line and the y axis. The point with coordinates $(0, 0)$ is called the **origin**.   Fig 1, p. 13

- The **standard form** for a linear equation in two variables is $Ax + By = C$, with A and B not both zero. The graph of this equation is a line, and every line in a Cartesian coordinate system is the graph of a linear equation.   Ex. 1, p. 15
    Ex. 2, p. 15

- The graph of the equation $x = a$ is a **vertical line** and the graph of $y = b$ is a **horizontal line**.   Ex. 3, p. 16

- If (x_1, y_1) and (x_2, y_2) are two distinct points on a line, then $m = (y_2 - y_1)/(x_2 - x_1)$ is the **slope** of the line.   Ex. 4, p. 17

- The equation $y = mx + b$ is the **slope-intercept form** of the equation of the line with slope m and y intercept b.   Ex. 5, p. 19

- The **point-slope form** of the equation of the line with slope m that passes through (x_1, y_1) is $y - y_1 = m(x - x_1)$.   Ex. 6, p. 19

- In a competitive market, the intersection of the supply equation and the demand equation is called the **equilibrium point**, the corresponding price is called the **equilibrium price**, and the common value of supply and demand is called the **equilibrium quantity**.   Ex. 8, p. 21

1-3 Linear Regression

- A **mathematical model** is a mathematics problem that, when solved, will provide information about a real-world problem.

- If the variables x and y are related by the equation $y = mx + b$, then x and y are **linearly related** and the slope m is the **rate of change** of y with respect to x.   Ex. 1, p. 28
    Ex. 2, p. 28

- A graph of the points in a data set is called a **scatter plot**. **Linear regression** is used to find the line that is the **best fit** for a data set. A regression model can be used to **interpolate** between points in a data set or to **extrapolate** or predict points outside the data set.   Ex. 3, p. 29
    Ex. 4, p. 31
    Ex. 5, p. 31

Review Exercises

Work through all the problems in this chapter review and check answers in the back of the book. Following each answer you will find a number in italics indicating the section where that type of problem is discussed. Where weaknesses show up, review appropriate sections in the text.

A

1. Solve $2x + 3 = 7x - 11$.   **2.** Solve $\dfrac{x}{12} - \dfrac{x - 3}{3} = \dfrac{1}{2}$.

3. Solve $2x + 5y = 9$ for y.   **4.** Solve $3x - 4y = 7$ for x.

Solve Problems 5–7 and graph on a real number line.

5. $4y - 3 < 10$ **6.** $-1 < -2x + 5 \le 3$

7. $1 - \dfrac{x - 3}{3} \le \dfrac{1}{2}$

8. Sketch a graph of $3x + 2y = 9$.

9. Write an equation of a line with x intercept 6 and y intercept 4. Write the final answer in the form $Ax + By = C$.

10. Sketch a graph of $2x - 3y = 18$. What are the intercepts and slope of the line?

11. Write an equation in the form $y = mx + b$ for a line with slope $-\dfrac{2}{3}$ and y intercept 6.

12. Write the equations of the vertical line and the horizontal line that pass through $(-6, 5)$.

13. Write the equation of a line through each indicated point with the indicated slope. Write the final answer in the form $y = mx + b$.

(A) $m = -\dfrac{2}{3};\,(-3, 2)$ (B) $m = 0;\,(3, 3)$

14. Write the equation of the line through the two indicated points. Write the final answer in the form $Ax + By = C$.

(A) $(-3, 5),\,(1, -1)$ (B) $(-1, 5),\,(4, 5)$

(C) $(-2, 7),\,(-2, -2)$

B

Solve Problems 15–19.

15. $3x + 25 = 5x$ **16.** $\dfrac{u}{5} = \dfrac{u}{6} + \dfrac{6}{5}$

17. $\dfrac{5x}{3} - \dfrac{4 + x}{2} = \dfrac{x - 2}{4} + 1$

18. $0.05x + 0.25(30 - x) = 3.3$

19. $0.2(x - 3) + 0.05x = 0.4$

Solve Problems 20–24 and graph on a real number line.

20. $2(x + 4) > 5x - 4$ **21.** $3(2 - x) - 2 \le 2x - 1$

22. $\dfrac{x + 3}{8} - \dfrac{4 + x}{2} > 5 - \dfrac{2 - x}{3}$

23. $-5 \le 3 - 2x < 1$ **24.** $-1.5 \le 2 - 4x \le 0.5$

25. Given $Ax + By = 30$, graph each of the following cases on the same coordinate axes.

(A) $A = 5$ and $B = 0$ (B) $A = 0$ and $B = 6$

(C) $A = 6$ and $B = 5$

26. Describe the graphs of $x = -3$ and $y = 2$. Graph both simultaneously in the same coordinate system.

27. Describe the lines defined by the following equations:

(A) $3x + 4y = 0$ (B) $3x + 4 = 0$

(C) $4y = 0$ (D) $3x + 4y - 36 = 0$

C

Solve Problems 28 and 29 for the indicated variable.

28. $A = \dfrac{1}{2}(a + b)h$; for $a\,(h \ne 0)$

29. $S = \dfrac{P}{1 - dt}$; for $d\,(dt \ne 1)$

30. For what values of a and b is the inequality $a + b < b - a$ true?

31. If a and b are negative numbers and $a > b$, then is a/b greater than 1 or less than 1?

32. Graph $y = mx + b$ and $y = -\dfrac{1}{m}x + b$ simultaneously in the same coordinate system for b fixed and several different values of m, $m \ne 0$. Describe the apparent relationship between the graphs of the two equations.

Applications

33. Investing. An investor has $300,000 to invest. If part is invested at 5% and the rest at 9%, how much should be invested at 5% to yield 8% on the total amount?

34. Break-even analysis. A producer of educational DVDs is producing an instructional DVD. She estimates that it will cost $90,000 to record the DVD and $5.10 per unit to copy and distribute the DVD. If the wholesale price of the DVD is $14.70, how many DVDs must be sold for the producer to break even?

35. Sports medicine. A simple rule of thumb for determining your maximum safe heart rate (in beats per minute) is to subtract your age from 220. While exercising, you should maintain a heart rate between 60% and 85% of your maximum safe rate.

(A) Find a linear model for the minimum heart rate m that a person of age x years should maintain while exercising.

(B) Find a linear model for the maximum heart rate M that a person of age x years should maintain while exercising.

(C) What range of heartbeats should you maintain while exercising if you are 20 years old?

(D) What range of heartbeats should you maintain while exercising if you are 50 years old?

36. Linear depreciation. A bulldozer was purchased by a construction company for $224,000 and has a depreciated value of $100,000 after 8 years. If the value is depreciated linearly from $224,000 to $100,000,

(A) Find the linear equation that relates value V (in dollars) to time t (in years).

(B) What would be the depreciated value after 12 years?

37. Business — Pricing. A sporting goods store sells tennis rackets that cost $130 for $208 and court shoes that cost $50 for $80.

(A) If the markup policy of the store for items that cost over $10 is linear and is reflected in the pricing of these two items, write an equation that expresses retail price R in terms of cost C.

(B) What would be the retail price of a pair of in-line skates that cost $120?

(C) What would be the cost of a pair of cross-country skis that had a retail price of $176?

(D) What is the slope of the graph of the equation found in part (A)? Interpret the slope relative to the problem.

38. Income. A salesperson receives a base salary of $400 per week and a commission of 10% on all sales over $6,000 during the week. Find the weekly earnings for weekly sales of $4,000 and for weekly sales of $10,000.

39. Price-demand. The weekly demand for mouthwash in a chain of drug stores is 1,160 bottles at a price of $3.79 each. If the price is lowered to $3.59, the weekly demand increases to 1,320 bottles. Assuming that the relationship between the weekly demand x and price per bottle p is linear, express p in terms of x. How many bottles would the stores sell each week if the price were lowered to $3.29?

40. Freezing temperature. Methanol, also known as wood alcohol, can be used as a fuel for suitably equipped vehicles. Table 1 lists the freezing temperature for various concentrations (as a percentage of total weight) of methanol in water. A linear regression model for the data in Table 1 is

$$T = 40 - 2M$$

where M is the percentage of methanol in the solution and T is the temperature at which the solution freezes.

(A) Draw a scatter plot of the data and a graph of the model on the same axes.

Table 1

Methanol (%Wt)	Freezing temperature (°F)
0	32
10	20
20	0
30	−15
40	−40
50	−65
60	−95

Source: Ashland Inc.

(B) Use the model to estimate the freezing temperature to the nearest degree of a solution that is 35% methanol.

(C) Use the model to estimate the percentage of methanol in a solution that freezes at −50°F.

41. High school dropout rates. Table 2 gives U.S. high school dropout rates as percentages for selected years since 1980. A linear regression model for the data is

$$r = -0.164t + 13.9$$

where t represents years since 1980 and r is the dropout rate.

Table 2 High School Dropout Rates (%)

1980	1985	1990	1995	2000	2005
14.1	12.6	12.1	12.0	10.9	9.4

(A) Interpret the slope of the model.

(B) Draw a scatter plot of the data and the model in the same coordinate system.

(C) Use the model to predict the first year for which the dropout rate is less than 8%.

42. Consumer Price Index. The U.S. Consumer Price Index (CPI) in recent years is given in Table 3. A scatter plot of the data and linear regression line are shown in the figure, where x represents years since 2000.

Table 3 Consumer Price Index (1982–1984 = 100)

Year	CPI
2000	172.2
2001	177.1
2002	179.9
2003	184.0
2004	188.9
2005	195.3
2006	198.3
2007	202.4
2008	211.1

Source: U.S. Bureau of Labor Statistics

Figure for 42

(A) Interpret the slope of the model.

(B) Predict the CPI in 2020.

43. Forestry. The figure contains a scatter plot of 20 data points for white pine trees and the linear regression model for this data.

(A) Interpret the slope of the model.

(B) What is the effect of a 1-in. increase in Dbh?

(C) Estimate the height of a white pine tree with a Dbh of 25 in. Round your answer to the nearest foot.

(D) Estimate the Dbh of a white pine tree that is 15 ft tall. Round your answer to the nearest inch.

Figure for 43

Functions and Graphs

2

Introduction

The function concept is one of the most important ideas in mathematics. The study of mathematics beyond the elementary level requires a firm understanding of a basic list of elementary functions, their properties, and their graphs. See the inside back cover of this book for a list of the functions that form our library of elementary functions. Most functions in the list will be introduced to you by the end of Chapter 2. For example, in Section 2-3 you will learn how to apply quadratic functions to model the effect of tire pressure on mileage (see Problems 57 and 59 on page 82).

2-1 Functions

We introduce the general notion of a *function* as a correspondence between two sets. Then we restrict attention to functions for which the two sets are both sets of real numbers. The most useful are those functions that are specified by equations in two variables. We discuss the terminology and notation associated with functions, graphs of functions, and applications.

Equations in Two Variables

In Chapter 1, we found that the graph of an equation of the form $Ax + By = C$, where A and B are not both zero, is a line. Because a line is determined by any two of its points, such an equation is easy to graph: Just plot *any* two points in its solution set and sketch the unique line through them.

More complicated equations in two variables, such as $y = 9 - x^2$ or $x^2 = y^4$, are more difficult to graph. To **sketch the graph** of an equation, we plot enough points from its solution set in a rectangular coordinate system so that the total graph is apparent, and then we connect these points with a smooth curve. This process is called **point-by-point plotting**.

EXAMPLE 1 **Point-by-Point Plotting** Sketch the graph of each equation.

(A) $y = 9 - x^2$ (B) $x^2 = y^4$

SOLUTIONS (A) Make up a table of solutions—that is, ordered pairs of real numbers that satisfy the given equation. For easy mental calculation, choose integer values for x.

x	-4	-3	-2	-1	0	1	2	3	4
y	-7	0	5	8	9	8	5	0	-7

After plotting these solutions, if there are any portions of the graph that are unclear, plot additional points until the shape of the graph is apparent. Then join all the plotted points with a smooth curve (Fig. 1). Arrowheads are used to indicate that the graph continues beyond the portion shown here with no significant changes in shape.

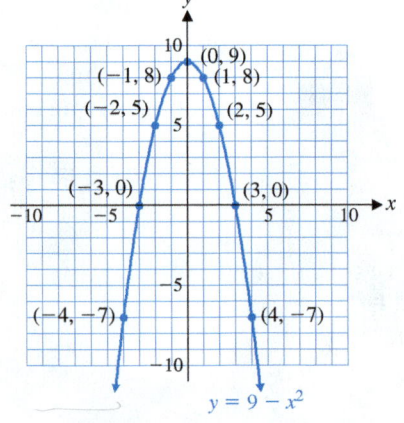

Figure 1 $y = 9 - x^2$

(B) Again we make a table of solutions—here it may be easier to choose integer values for y and calculate values for x. Note, for example, that if $y = 2$, then $x = \pm 4$; that is, the ordered pairs $(4, 2)$ and $(-4, 2)$ are both in the solution set.

x	± 9	± 4	± 1	0	± 1	± 4	± 9
y	-3	-2	-1	0	1	2	3

We plot these points and join them with a smooth curve (Fig. 2).

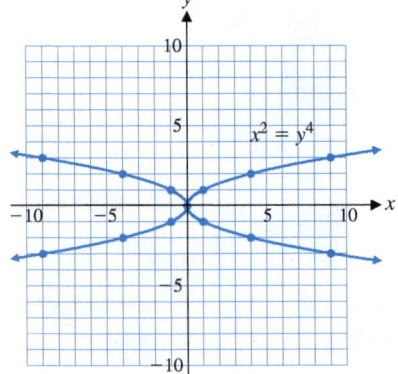

Figure 2 $x^2 = y^4$

Matched Problem 1   Sketch the graph of each equation.

(A) $y = x^2 - 4$ (B) $y^2 = \dfrac{100}{x^2 + 1}$

EXPLORE & DISCUSS 1

To graph the equation $y = -x^3 + 3x$, we use point-by-point plotting to obtain

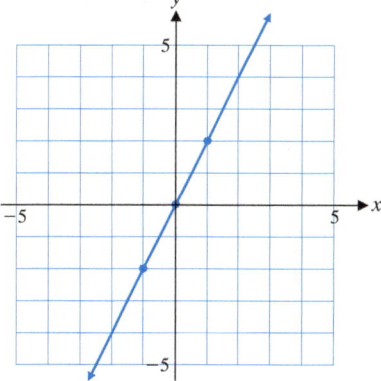

x	y
-1	-2
0	0
1	2

(A) Do you think this is the correct graph of the equation? Why or why not?

(B) Add points on the graph for $x = -2, -1.5, -0.5, 0.5, 1.5,$ and 2.

(C) Now, what do you think the graph looks like? Sketch your version of the graph, adding more points as necessary.

(D) Graph this equation on a graphing calculator and compare it with your graph from part (C).

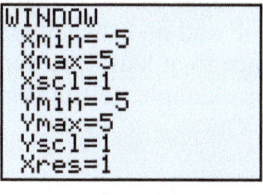

(A)

```
WINDOW
 Xmin=-5
 Xmax=5
 Xscl=1
 Ymin=-5
 Ymax=5
 Yscl=1
 Xres=1
```

(B)

Figure 3

The icon in the margin is used throughout this book to identify optional graphing calculator activities that are intended to give you additional insight into the concepts under discussion. You may have to consult the manual for your graphing calculator for the details necessary to carry out these activities. For example, to graph the equation in Explore & Discuss 1 on most graphing calculators, you must enter the equation (Fig. 3A) and the window variables (Fig. 3B).

As Explore & Discuss 1 illustrates, the shape of a graph may not be apparent from your first choice of points. Using point-by-point plotting, it may be difficult to find points in the solution set of the equation, and it may be difficult to determine when you have found enough points to understand the shape of the graph. We will supplement the technique of point-by-point plotting with a detailed analysis of several basic equations, giving you the ability to sketch graphs with accuracy and confidence.

Definition of a Function

Central to the concept of function is correspondence. You are familiar with correspondences in daily life. For example,

To each person, there corresponds an annual income.

To each item in a supermarket, there corresponds a price.

To each student, there corresponds a grade-point average.

To each day, there corresponds a maximum temperature.

For the manufacture of x items, there corresponds a cost.

For the sale of x items, there corresponds a revenue.

To each square, there corresponds an area.

To each number, there corresponds its cube.

One of the most important aspects of any science is the establishment of correspondences among various types of phenomena. Once a correspondence is known, predictions can be made. A cost analyst would like to predict costs for various levels of output in a manufacturing process; a medical researcher would like to know the correspondence between heart disease and obesity; a psychologist would like to predict the level of performance after a subject has repeated a task a given number of times; and so on.

What do all of these examples have in common? Each describes the matching of elements from one set with the elements in a second set.

Consider Tables 1–3. Tables 1 and 2 specify functions, but Table 3 does not. Why not? The definition of the term *function* will explain.

TABLE 1		**TABLE 2**		**TABLE 3**	
Domain	Range	Domain	Range	Domain	Range
Number	*Cube*	*Number*	*Square*	*Number*	*Square root*

Domain	Range
$-2 \longrightarrow -8$
$-1 \longrightarrow -1$
$0 \longrightarrow 0$
$1 \longrightarrow 1$
$2 \longrightarrow 8$

TABLE 2: Domain $-2, -1, 0, 1, 2$; Range $4, 1, 0$

TABLE 3: Domain $0, 1, 4, 9$; Range $0, 1, 1, 2, 2, 3, 3$

DEFINITION Function

A **function** is a correspondence between two sets of elements such that to each element in the first set, there corresponds one and only one element in the second set.

The first set is called the **domain**, and the set of corresponding elements in the second set is called the **range**.

Tables 1 and 2 specify functions since to each domain value, there corresponds exactly one range value (for example, the cube of -2 is -8 and no other number). On the other hand, Table 3 does not specify a function since to at least one domain value, there corresponds more than one range value (for example, to the domain value 9, there corresponds -3 and 3, both square roots of 9).

EXPLORE & DISCUSS 2

Consider the set of students enrolled in a college and the set of faculty members at that college. Suppose we define a correspondence between the two sets by saying that a student corresponds to a faculty member if the student is currently enrolled in a course taught by that faculty member. Is this correspondence a function? Discuss.

Functions Specified by Equations

Most of the functions in this book will have domains and ranges that are (infinite) sets of real numbers. The **graph** of such a function is the set of all points (x, y) in the Cartesian plane such that x is an element of the domain and y is the corresponding element in the range. The correspondence between domain and range elements is often specified by an equation in two variables. Consider, for example, the equation for the area of a rectangle with width 1 inch less than its length (Fig. 4). If x is the length, then the area y is given by

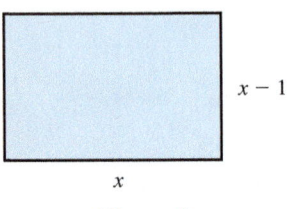

Figure 4

$$y = x(x - 1) \qquad x \geq 1$$

For each **input** x (length), we obtain an **output** y (area). For example,

$$\text{If} \quad x = 5, \quad \text{then} \quad y = 5\,(5 - 1) = 5 \cdot 4 = 20.$$
$$\text{If} \quad x = 1, \quad \text{then} \quad y = 1\,(1 - 1) = 1 \cdot 0 = 0.$$
$$\text{If} \quad x = \sqrt{5}, \quad \text{then} \quad y = \sqrt{5}\,(\sqrt{5} - 1) = 5 - \sqrt{5}$$
$$\approx 2.7639.$$

The input values are domain values, and the output values are range values. The equation assigns each domain value x a range value y. The variable x is called an *independent variable* (since values can be "independently" assigned to x from the domain), and y is called a *dependent variable* (since the value of y "depends" on the value assigned to x). In general, any variable used as a placeholder for domain values is called an **independent variable**; any variable that is used as a placeholder for range values is called a **dependent variable**.

When does an equation specify a function?

DEFINITION Functions Specified by Equations

If in an equation in two variables, we get exactly one output (value for the dependent variable) for each input (value for the independent variable), then the equation specifies a function. The graph of such a function is just the graph of the specifying equation.

If we get more than one output for a given input, the equation does not specify a function.

EXAMPLE 2   **Functions and Equations** Determine which of the following equations specify functions with independent variable x.

(A) $4y - 3x = 8$, x a real number   (B) $y^2 - x^2 = 9$, x a real number

SOLUTION   (A) Solving for the dependent variable y, we have

$$4y - 3x = 8$$
$$4y = 8 + 3x \qquad (1)$$
$$y = 2 + \frac{3}{4}x$$

Since each input value x corresponds to exactly one output value $(y = 2 + \frac{3}{4}x)$, we see that equation (1) specifies a function.

(B) Solving for the dependent variable y, we have

$$y^2 - x^2 = 9$$
$$y^2 = 9 + x^2 \qquad (2)$$
$$y = \pm\sqrt{9 + x^2}$$

Since $9 + x^2$ is always a positive real number for any real number x, and since each positive real number has two square roots,* then to each input value x there corresponds two output values $(y = -\sqrt{9 + x^2}$ and $y = \sqrt{9 + x^2})$. For example, if $x = 4$, then equation (2) is satisfied for $y = 5$ and for $y = -5$. So equation (2) does not specify a function.

Matched Problem 2   Determine which of the following equations specify functions with independent variable x.

(A) $y^2 - x^4 = 9$, x a real number   (B) $3y - 2x = 3$, x a real number

Since the graph of an equation is the graph of all the ordered pairs that satisfy the equation, it is very easy to determine whether an equation specifies a function by examining its graph. The graphs of the two equations we considered in Example 2 are shown in Figure 5.

In Figure 5A, notice that any vertical line will intersect the graph of the equation $4y - 3x = 8$ in exactly one point. This shows that to each x value, there corresponds

---

*Recall that each positive real number N has two square roots: $\sqrt{N}$, the principal square root; and $-\sqrt{N}$, the negative of the principal square root (see Appendix A, Section A-6).

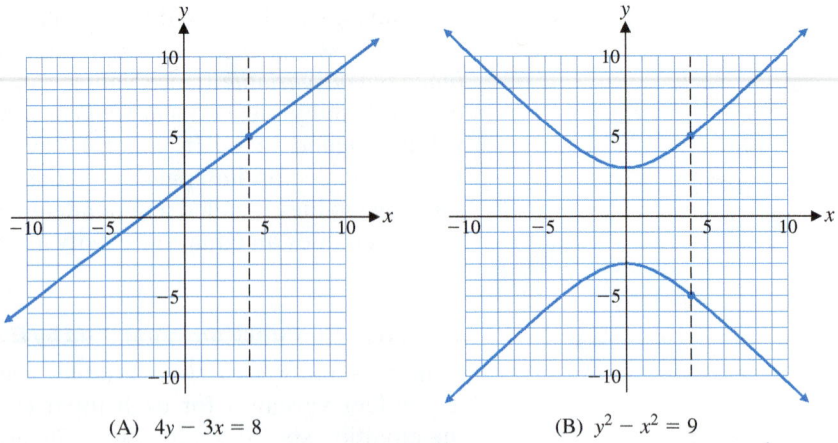

(A) $4y - 3x = 8$ (B) $y^2 - x^2 = 9$

Figure 5

exactly one y value, confirming our conclusion that this equation specifies a function. On the other hand, Figure 5B shows that there exist vertical lines that intersect the graph of $y^2 - x^2 = 9$ in two points. This indicates that there exist x values to which there correspond two different y values and verifies our conclusion that this equation does not specify a function. These observations are generalized in Theorem 1.

THEOREM 1   Vertical-Line Test for a Function

An equation specifies a function if each vertical line in the coordinate system passes through, at most, one point on the graph of the equation.

If any vertical line passes through two or more points on the graph of an equation, then the equation does not specify a function.

The function graphed in Figure 5A is an example of a *linear function*. The vertical-line test implies that equations of the form $y = mx + b$, where $m \neq 0$, specify functions; they are called **linear functions**. Similarly, equations of the form $y = b$ specify functions; they are called **constant functions**, and their graphs are horizontal lines. The vertical-line test implies that equations of the form $x = a$ do not specify functions; note that the graph of $x = a$ is a vertical line.

In Example 2, the domains were explicitly stated along with the given equations. In many cases, this will not be done. Unless stated to the contrary, we shall adhere to the following convention regarding domains and ranges for functions specified by equations:

> **If a function is specified by an equation and the domain is not indicated, then we assume that the domain is the set of all real-number replacements of the independent variable (inputs) that produce real values for the dependent variable (outputs). The range is the set of all outputs corresponding to input values.**

EXAMPLE 3   **Finding a Domain** Find the domain of the function specified by the equation $y = \sqrt{4 - x}$, assuming that x is the independent variable.

SOLUTION   For y to be real, $4 - x$ must be greater than or equal to 0; that is,

$$4 - x \geq 0$$
$$-x \geq -4$$
$$x \leq 4 \qquad \text{Sense of inequality reverses when both sides are divided by } -1.$$

Domain: $x \leq 4$ (inequality notation) or $(-\infty, 4]$ (interval notation)

Matched Problem 3   Find the domain of the function specified by the equation $y = \sqrt{x - 2}$, assuming x is the independent variable.

Function Notation

We have seen that a function involves two sets, a domain and a range, and a correspondence that assigns to each element in the domain exactly one element in the range. Just as we use letters as names for numbers, now we will use letters as names for functions. For example, f and g may be used to name the functions specified by the equations $y = 2x + 1$ and $y = x^2 + 2x - 3$:

$$f: \quad y = 2x + 1$$
$$g: \quad y = x^2 + 2x - 3 \tag{3}$$

If x represents an element in the domain of a function f, then we frequently use the symbol

$$f(x)$$

in place of y to designate the number in the range of the function f to which x is paired (Fig. 6). This symbol does *not* represent the product of f and x. The symbol $f(x)$ is read as "f of x," "f at x," or "the value of f at x." Whenever we write $y = f(x)$, we assume that the variable x is an independent variable and that both y and $f(x)$ are dependent variables.

DOMAIN   RANGE

Figure 6

Using function notation, we can now write functions f and g in equation (3) as

$$f(x) = 2x + 1 \qquad \text{and} \qquad g(x) = x^2 + 2x - 3$$

Let us find $f(3)$ and $g(-5)$. To find $f(3)$, we replace x with 3 wherever x occurs in $f(x) = 2x + 1$ and evaluate the right side:

$$f(x) = 2x + 1$$
$$f(\mathbf{3}) = 2 \cdot \mathbf{3} + 1$$
$$= 6 + 1 = 7 \qquad \text{\textit{For input 3, the output is 7.}}$$

Therefore,

$$f(3) = 7 \qquad \text{\textit{The function f assigns the range value 7 to the domain value 3.}}$$

To find $g(-5)$, we replace each x by -5 in $g(x) = x^2 + 2x - 3$ and evaluate the right side:

$$g(x) = x^2 + 2x - 3$$
$$g(\mathbf{-5}) = (\mathbf{-5})^2 + 2(\mathbf{-5}) - 3$$
$$= 25 - 10 - 3 = 12 \qquad \text{\textit{For input} -5\textit{, the output is 12.}}$$

Therefore,

$$g(-5) = 12 \qquad \text{\textit{The function g assigns the range value 12 to the domain value} -5.}$$

It is very important to understand and remember the definition of $f(x)$:

For any element x in the domain of the function f, the symbol $f(x)$ represents the element in the range of f corresponding to x in the domain of f. If x is an input value, then $f(x)$ is the corresponding output value. If x is an element that is not in the domain of f, then f is not defined at x and $f(x)$ does not exist.

EXAMPLE 4 **Function Evaluation** For $f(x) = 12/(x - 2)$, $g(x) = 1 - x^2$, and $h(x) = \sqrt{x - 1}$, evaluate:

(A) $f(6)$ (B) $g(-2)$ (C) $h(-2)$ (D) $f(0) + g(1) - h(10)$

SOLUTION (A) $f(\mathbf{6}) = \dfrac{12}{\mathbf{6} - 2}^{*} = \dfrac{12}{4} = 3$

(B) $g(\mathbf{-2}) = 1 - (\mathbf{-2})^2 = 1 - 4 = -3$

(C) $h(\mathbf{-2}) = \sqrt{\mathbf{-2} - 1} = \sqrt{-3}$

---

*Dashed boxes are used throughout the book to represent steps that are usually performed mentally.

But $\sqrt{-3}$ is not a real number. Since we have agreed to restrict the domain of a function to values of x that produce real values for the function, -2 is not in the domain of h, and $h(-2)$ does not exist.

(D) $f(0) + g(1) - h(10) = \dfrac{12}{0 - 2} + (1 - 1^2) - \sqrt{10 - 1}$

$$= \dfrac{12}{-2} + 0 - \sqrt{9}$$

$$= -6 - 3 = -9$$

Matched Problem 4 Use the functions in Example 4 to find

(A) $f(-2)$ (B) $g(-1)$ (C) $h(-8)$ (D) $\dfrac{f(3)}{h(5)}$

EXAMPLE 5 **Finding Domains** Find the domains of functions f, g, and h:

$$f(x) = \dfrac{12}{x - 2} \qquad g(x) = 1 - x^2 \qquad h(x) = \sqrt{x - 1}$$

SOLUTION *Domain of f:* $12/(x - 2)$ represents a real number for all replacements of x by real numbers except for $x = 2$ (division by 0 is not defined). Thus, $f(2)$ does not exist, and the domain of f is the set of all real numbers except 2. We often indicate this by writing

$$f(x) = \dfrac{12}{x - 2} \qquad x \neq 2$$

Domain of g: The domain is R, the set of all real numbers, since $1 - x^2$ represents a real number for all replacements of x by real numbers.

Domain of h: The domain is the set of all real numbers x such that $\sqrt{x - 1}$ is a real number, so

$$x - 1 \geq 0$$

$$x \geq 1 \quad \text{or, in interval notation,} \quad [1, \infty)$$

Matched Problem 5 Find the domains of functions F, G, and H:

$$F(x) = x^2 - 3x + 1 \qquad G(x) = \dfrac{5}{x + 3} \qquad H(x) = \sqrt{2 - x}$$

In addition to evaluating functions at specific numbers, it is important to be able to evaluate functions at expressions that involve one or more variables. For example, the **difference quotient**

$$\dfrac{f(x + h) - f(x)}{h} \qquad x \text{ and } x + h \text{ in the domain of } f, h \neq 0$$

is studied extensively in calculus.

CONCEPTUAL INSIGHT

In algebra, you learned to use parentheses for grouping variables. For example,

$$2(x + h) = 2x + 2h$$

Now we are using parentheses in the function symbol $f(x)$. For example, if $f(x) = x^2$, then

$$f(x + h) = (x + h)^2 = x^2 + 2xh + h^2$$

Note that $f(x) + f(h) = x^2 + h^2 \neq f(x + h)$. That is, the function name f does not distribute across the grouped variables $(x + h)$, as the "2" does in $2(x + h)$ (see Appendix A, Section A-2).

EXAMPLE 6 **Using Function Notation** For $f(x) = x^2 - 2x + 7$, find

(A) $f(a)$ (B) $f(a + h)$ (C) $f(a + h) - f(a)$ (D) $\dfrac{f(a + h) - f(a)}{h}$, $h \neq 0$

SOLUTION (A) $f(a) = a^2 - 2a + 7$

(B) $f(a + h) = (a + h)^2 - 2(a + h) + 7 = a^2 + 2ah + h^2 - 2a - 2h + 7$

(C) $f(a + h) - f(a) = (a^2 + 2ah + h^2 - 2a - 2h + 7) - (a^2 - 2a + 7)$

$$= 2ah + h^2 - 2h$$

(D) $\dfrac{f(a + h) - f(a)}{h} = \dfrac{2ah + h^2 - 2h}{h} = \dfrac{h(2a + h - 2)}{h}$ *Because $h \neq 0$, $\dfrac{h}{h} = 1$.*

$$= 2a + h - 2$$

Matched Problem 6 Repeat Example 6 for $f(x) = x^2 - 4x + 9$.

Applications

We now turn to the important concepts of **break-even** and **profit–loss** analysis, which we will return to a number of times in this book. Any manufacturing company has **costs**, C, and **revenues**, R. The company will have a **loss** if $R < C$, will **break even** if $R = C$, and will have a **profit** if $R > C$. Costs include **fixed costs** such as plant overhead, product design, setup, and promotion; and **variable costs**, which are dependent on the number of items produced at a certain cost per item. In addition, **price–demand** functions, usually established by financial departments using historical data or sampling techniques, play an important part in profit–loss analysis. We will let x, the number of units manufactured and sold, represent the independent variable. Cost functions, revenue functions, profit functions, and price–demand functions are often stated in the following forms, where $a, b, m,$ and n are constants determined from the context of a particular problem:

Cost Function

$$C = \text{(fixed costs)} + \text{(variable costs)}$$
$$= a + bx$$

Price–Demand Function

$$p = m - nx \qquad \text{x is the number of items that can be sold at \$p per item.}$$

Revenue Function

$$R = \text{(number of items sold)} \times \text{(price per item)}$$
$$= xp = x(m - nx)$$

Profit Function

$$P = R - C$$
$$= x(m - nx) - (a + bx)$$

Example 7 and Matched Problem 7 explore the relationships among the algebraic definition of a function, the numerical values of the function, and the graphical representation of the function. The interplay among algebraic, numeric, and graphic viewpoints is an important aspect of our treatment of functions and their use. In Example 7, we will see how a function can be used to describe data from the real world, a process that is often referred to as *mathematical modeling*. Note that the domain of such a function is determined by practical considerations within the problem.

EXAMPLE 7 **Price–Demand and Revenue Modeling** A manufacturer of a popular digital camera wholesales the camera to retail outlets throughout the United States. Using statistical methods, the financial department in the company produced the price–demand data in Table 4, where p is the wholesale price per camera at which x million cameras are sold. Notice that as the price goes down, the number sold goes up.

Table 4 Price–Demand

x (Millions)	p ($)
2	87
5	68
8	53
12	37

Using special analytical techniques (regression analysis), an analyst obtained the following price–demand function to model the Table 4 data:

$$p(x) = 94.8 - 5x \qquad 1 \le x \le 15 \qquad (4)$$

(A) Plot the data in Table 4. Then sketch a graph of the price–demand function in the same coordinate system.

(B) What is the company's revenue function for this camera, and what is its domain?

(C) Complete Table 5, computing revenues to the nearest million dollars.

(D) Plot the data in Table 5. Then sketch a graph of the revenue function using these points.

(E) Plot the revenue function on a graphing calculator.

Table 5 Revenue

x (Millions)	R(x) (Million $)
1	90
3	
6	
9	
12	
15	

SOLUTION (A)

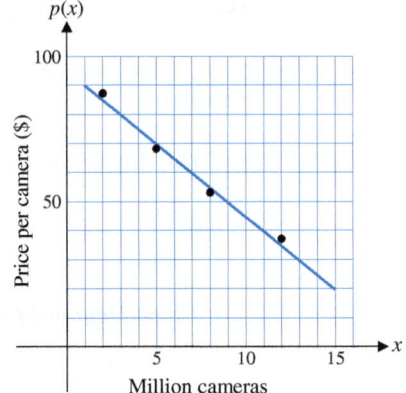

Figure 7 Price–demand

In Figure 7, notice that the model approximates the actual data in Table 4, and it is assumed that it gives realistic and useful results for all other values of x between 1 million and 15 million.

(B) $R(x) = xp(x) = x(94.8 - 5x)$ million dollars

Domain: $1 \le x \le 15$

[Same domain as the price–demand function, equation (4).]

(C)

Table 5 Revenue

x (Millions)	R(x) (Million $)
1	90
3	239
6	389
9	448
12	418
15	297

(D)

(E)

Matched Problem 7 The financial department in Example 7, using statistical techniques, produced the data in Table 6, where $C(x)$ is the cost in millions of dollars for manufacturing and selling x million cameras.

Table 6 Cost Data

x (Millions)	$C(x)$ (Million \$)
1	175
5	260
8	305
12	395

Using special analytical techniques (regression analysis), an analyst produced the following cost function to model the Table 6 data:

$$C(x) = 156 + 19.7x \qquad 1 \le x \le 15 \qquad (5)$$

(A) Plot the data in Table 6. Then sketch a graph of equation (5) in the same coordinate system.

(B) What is the company's profit function for this camera, and what is its domain?

(C) Complete Table 7, computing profits to the nearest million dollars.

Table 7 Profit

x (Millions)	$P(x)$ (Million \$)
1	-86
3	
6	
9	
12	
15	

(D) Plot the data in Table 7. Then sketch a graph of the profit function using these points.

(E) Plot the profit function on a graphing calculator.

Exercises 2-1

A

In Problems 1–8, use point-by-point plotting to sketch the graph of each equation.

1. $y = x + 1$

2. $x = y + 1$

3. $x = y^2$

4. $y = x^2$

5. $y = x^3$

6. $x = y^3$

7. $xy = -6$

8. $xy = 12$

Indicate whether each table in Problems 9–14 specifies a function.

9.

10.

11.

12.

13.
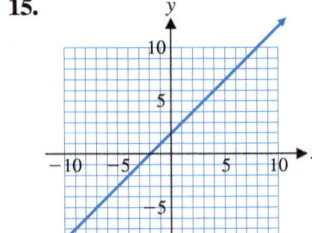

14.

Indicate whether each graph in Problems 15–20 specifies a function.

15.

16.

17.

18.

19.
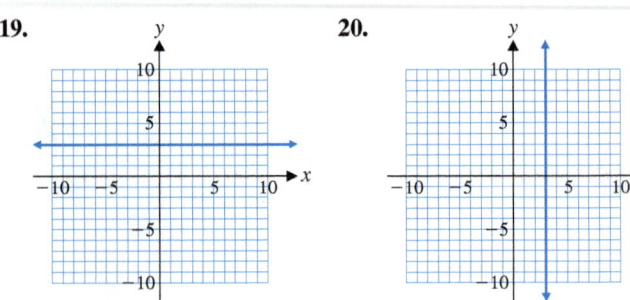

20.

In Problems 21–28, each equation specifies a function with independent variable x. Determine whether the function is linear, constant, or neither.

21. $y - 2x = 7$

22. $y = 10 - 3x$

23. $xy - 4 = 0$

24. $x^2 - y = 8$

25. $y = 5x + \frac{1}{2}(7 - 10x)$

26. $y = \frac{2 + x}{3} + \frac{2 - x}{3}$

27. $3x + 4y = 5$

28. $9x - 2y + 6 = 0$

In Problems 29–36, use point-by-point plotting to sketch the graph of each function.

29. $f(x) = 1 - x$

30. $f(x) = \frac{x}{2} - 3$

31. $f(x) = x^2 - 1$

32. $f(x) = 3 - x^2$

33. $f(x) = 4 - x^3$

34. $f(x) = x^3 - 2$

35. $f(x) = \frac{8}{x}$

36. $f(x) = \frac{-6}{x}$

In Problems 37 and 38, the three points in the table are on the graph of the indicated function f. Do these three points provide sufficient information for you to sketch the graph of $y = f(x)$? Add more points to the table until you are satisfied that your sketch is a good representation of the graph of $y = f(x)$ on the interval $[-5, 5]$.

37.

x	-1	0	1
$f(x)$	-1	0	1

$f(x) = \frac{2x}{x^2 + 1}$

38.

x	0	1	2
$f(x)$	0	1	2

$f(x) = \frac{3x^2}{x^2 + 2}$

In Problems 39–46, use the following graph of a function f to determine x or y to the nearest integer, as indicated. Some problems may have more than one answer.

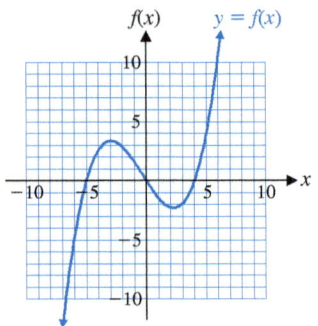

39. $y = f(-5)$ **40.** $y = f(4)$

41. $y = f(5)$ **42.** $y = f(-2)$

43. $0 = f(x)$ **44.** $3 = f(x), x < 0$

45. $-4 = f(x)$ **46.** $4 = f(x)$

In Problems 47–54, find each expression if $f(x) = 2x - 3$ and $g(x) = x^2 + 2x$.

47. $f(5)$ **48.** $f(-6)$

49. $g(-8)$ **50.** $f(9)$

51. $f(10 + 4)$ **52.** $g(3 + 7)$

53. $f(10) + f(4)$ **54.** $g(3) + g(7)$

B

In Problems 55–60, find the domain of each function.

55. $F(x) = 2x^3 - x^2 + 3$ **56.** $H(x) = 7 - 2x^2 - x^4$

57. $f(x) = \dfrac{x - 2}{x + 4}$ **58.** $g(x) = \dfrac{x + 1}{x - 2}$

59. $g(x) = \sqrt{7 - x}$ **60.** $F(x) = \dfrac{1}{\sqrt{5 + x}}$

In Problems 61–68, determine which of the equations specify functions with independent variable x. For those that do, find the domain. For those that do not, find a value of x to which there corresponds more than one value of y.

61. $x^2 - y = 1$ **62.** $x - y^2 = 1$

63. $x + y^2 = 10$ **64.** $x^2 + y = 10$

65. $xy - 4y = 1$ **66.** $xy + y - x = 5$

67. $x^2 + y^2 = 25$ **68.** $x^2 - y^2 = 16$

In Problems 69–76, find and simplify each expression if $f(x) = x^2 - 1$.

69. $f(2x)$ **70.** $f(-3x)$

71. $f(x - 3)$ **72.** $f(x + 2)$

73. $f(4 + h)$ **74.** $f(5 + h)$

75. $f(4 + h) - f(4)$ **76.** $f(5 + h) - f(5)$

C

In Problems 77–82, find and simplify each of the following, assuming $h \neq 0$ in (C).

(A) $f(x + h)$

(B) $f(x + h) - f(x)$

(C) $\dfrac{f(x + h) - f(x)}{h}$

77. $f(x) = 4x - 3$ **78.** $f(x) = -3x + 9$

79. $f(x) = 4x^2 - 7x + 6$ **80.** $f(x) = 3x^2 + 5x - 8$

81. $f(x) = x(20 - x)$ **82.** $f(x) = x(x + 40)$

Problems 83–86 refer to the area A and perimeter P of a rectangle with length l and width w (see the figure).

$A = lw$

$P = 2l + 2w$

w

l

83. The area of a rectangle is 25 sq/in. Express the perimeter $P(w)$ as a function of the width w, and state the domain of this function.

84. The area of a rectangle is 81 sq/in. Express the perimeter $P(l)$ as a function of the length l, and state the domain of this function.

85. The perimeter of a rectangle is 100 m. Express the area $A(l)$ as a function of the length l, and state the domain of this function.

86. The perimeter of a rectangle is 160 m. Express the area $A(w)$ as a function of the width w, and state the domain of this function.

Applications

87. Price–demand. A company manufactures memory chips for microcomputers. Its marketing research department, using statistical techniques, collected the data shown in Table 8, where p is the wholesale price per chip at which x million chips can be sold. Using special analytical techniques (regression analysis), an analyst produced the following price–demand function to model the data:

$$p(x) = 75 - 3x \qquad 1 \le x \le 20$$

Table 8 Price–Demand

x (Millions)	p ($)
1	72
4	63
9	48
14	33
20	15

(A) Plot the data points in Table 8, and sketch a graph of the price–demand function in the same coordinate system.

(B) What would be the estimated price per chip for a demand of 7 million chips? For a demand of 11 million chips?

88. Price–demand. A company manufactures notebook computers. Its marketing research department, using statistical techniques, collected the data shown in Table 9, where p is the wholesale price per computer at which x thousand computers can be sold. Using special analytical techniques (regression analysis), an analyst produced the following price–demand function to model the data:

$$p(x) = 2,000 - 60x \quad 1 \le x \le 25$$

Table 9 Price–Demand

x (Thousands)	p ($)
1	1,940
8	1,520
16	1,040
21	740
25	500

(A) Plot the data points in Table 9, and sketch a graph of the price–demand function in the same coordinate system.

(B) What would be the estimated price per computer for a demand of 11,000 computers? For a demand of 18,000 computers?

89. Revenue.

(A) Using the price–demand function

$$p(x) = 75 - 3x \quad 1 \le x \le 20$$

from Problem 87, write the company's revenue function and indicate its domain.

(B) Complete Table 10, computing revenues to the nearest million dollars.

Table 10 Revenue

x (Millions)	R(x) (Million $)
1	72
4	
8	
12	
16	
20	

(C) Plot the points from part (B) and sketch a graph of the revenue function using these points. Choose millions for the units on the horizontal and vertical axes.

90. Revenue.

(A) Using the price–demand function

$$p(x) = 2,000 - 60x \quad 1 \le x \le 25$$

from Problem 88, write the company's revenue function and indicate its domain.

(B) Complete Table 11, computing revenues to the nearest thousand dollars.

Table 11 Revenue

x (Thousands)	R(x) (Thousand $)
1	1,940
5	
10	
15	
20	
25	

(C) Plot the points from part (B) and sketch a graph of the revenue function using these points. Choose thousands for the units on the horizontal and vertical axes.

91. Profit. The financial department for the company in Problems 87 and 89 established the following cost function for producing and selling x million memory chips:

$$C(x) = 125 + 16x \text{ million dollars}$$

(A) Write a profit function for producing and selling x million memory chips and indicate its domain.

(B) Complete Table 12, computing profits to the nearest million dollars.

Table 12 Profit

x (Millions)	P(x) (Million $)
1	−69
4	
8	
12	
16	
20	

(C) Plot the points in part (B) and sketch a graph of the profit function using these points.

92. Profit. The financial department for the company in Problems 88 and 90 established the following cost function for producing and selling x thousand notebook computers:

$$C(x) = 4,000 + 500x \text{ thousand dollars}$$

(A) Write a profit function for producing and selling x thousand notebook computers and indicate its domain.

(B) Complete Table 13, computing profits to the nearest thousand dollars.

Table 13 Profit

x (Thousands)	P(x) (Thousand $)
1	−2,560
5	
10	
15	
20	
25	

(C) Plot the points in part (B) and sketch a graph of the profit function using these points.

93. Packaging. A candy box will be made out of a piece of cardboard that measures 8 by 12 in. Equal-sized squares x inches on a side will be cut out of each corner, and then the ends and sides will be folded up to form a rectangular box.

(A) Express the volume of the box $V(x)$ in terms of x.

(B) What is the domain of the function V (determined by the physical restrictions)?

(C) Complete Table 14.

Table 14	Volume
x	$V(x)$
1	
2	
3	

(D) Plot the points in part (C) and sketch a graph of the volume function using these points.

94. Packaging. Refer to Problem 93.

(A) Table 15 shows the volume of the box for some values of x between 1 and 2. Use these values to estimate to one decimal place the value of x between 1 and 2 that would produce a box with a volume of 65 cu/in.

Table 15	Volume
x	$V(x)$
1.1	62.524
1.2	64.512
1.3	65.988
1.4	66.976
1.5	67.5
1.6	67.584
1.7	67.252

(B) Describe how you could refine this table to estimate x to two decimal places.

(C) Carry out the refinement you described in part (B) and approximate x to two decimal places.

95. Packaging. Refer to Problems 93 and 94.

(A) Examine the graph of $V(x)$ from Problem 93D and discuss the possible locations of other values of x that would produce a box with a volume of 65 cu/in.

(B) Construct a table like Table 15 to estimate any such value to one decimal place.

(C) Refine the table you constructed in part (B) to provide an approximation to two decimal places.

96. Packaging. A parcel delivery service will only deliver packages with length plus girth (distance around) not exceeding 108 in. A rectangular shipping box with square ends x inches on a side is to be used.

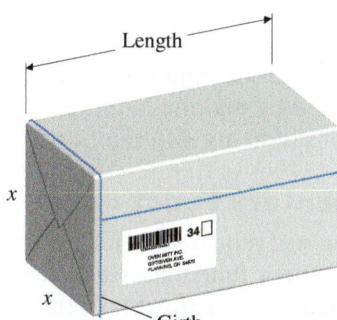

(A) If the full 108 in. is to be used, express the volume of the box $V(x)$ in terms of x.

(B) What is the domain of the function V (determined by the physical restrictions)?

(C) Complete Table 16.

Table 16	Volume
x	$V(x)$
5	
10	
15	
20	
25	

(D) Plot the points in part (C) and sketch a graph of the volume function using these points.

97. Muscle contraction. In a study of the speed of muscle contraction in frogs under various loads, British biophysicist A. W. Hill determined that the weight w (in grams) placed on the muscle and the speed of contraction v (in centimeters per second) are approximately related by an equation of the form

$$(w + a)(v + b) = c$$

where a, b, and c are constants. Suppose that for a certain muscle, $a = 15$, $b = 1$, and $c = 90$. Express v as a function of w. Find the speed of contraction if a weight of 16 g is placed on the muscle.

98. Politics. The percentage s of seats in the House of Representatives won by Democrats and the percentage v of votes cast for Democrats (when expressed as decimal fractions) are related by the equation

$$5v - 2s = 1.4 \qquad 0 < s < 1, \quad 0.28 < v < 0.68$$

(A) Express v as a function of s and find the percentage of votes required for the Democrats to win 51% of the seats.

(B) Express s as a function of v and find the percentage of seats won if Democrats receive 51% of the votes.

Answers to Matched Problems

1. (A)

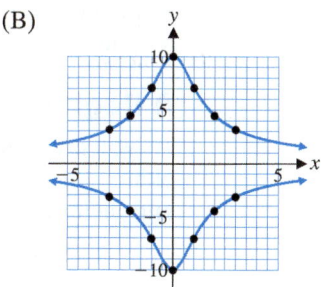

(B)

2. (A) Does not specify a function

(B) Specifies a function

3. $x \geq 2$ (inequality notation) or $[2, \infty)$ (interval notation)

4. (A) -3 (B) 0 (C) Does not exist   (D) 6

5. Domain of F: R; domain of G: all real numbers except -3; domain of H: $x \leq 2$ (inequality notation) or $(-\infty, 2]$ (interval notation)

6. (A) $a^2 - 4a + 9$ (B) $a^2 + 2ah + h^2 - 4a - 4h + 9$
(C) $2ah + h^2 - 4h$ (D) $2a + h - 4$

7. (A) $C(x)$

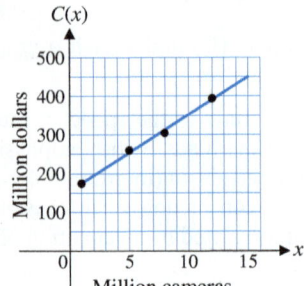

Million dollars / Million cameras

(B) $P(x) = R(x) - C(x) = x(94.8 - 5x) - (156 + 19.7x)$; domain: $1 \leq x \leq 15$

(C)

Table 7 Profit

x (Millions)	$P(x)$ (Million \$)
1	-86
3	24
6	115
9	115
12	25
15	-155

(D) $P(x)$

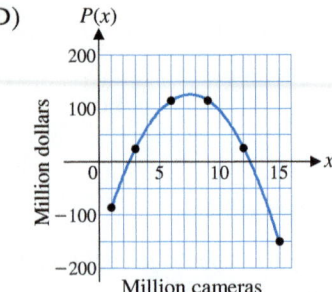

Million dollars / Million cameras

(E)

2-2 Elementary Functions: Graphs and Transformations

- A Beginning Library of Elementary Functions
- Vertical and Horizontal Shifts
- Reflections, Stretches, and Shrinks
- Piecewise-Defined Functions

Each of the functions

$$g(x) = x^2 - 4 \qquad h(x) = (x - 4)^2 \qquad k(x) = -4x^2$$

can be expressed in terms of the function $f(x) = x^2$:

$$g(x) = f(x) - 4 \qquad h(x) = f(x - 4) \qquad k(x) = -4f(x)$$

In this section, we will see that the graphs of functions $g, h,$ and k are closely related to the graph of function f. Insight gained by understanding these relationships will help us analyze and interpret the graphs of many different functions.

A Beginning Library of Elementary Functions

As you progress through this book, you will repeatedly encounter a small number of elementary functions. We will identify these functions, study their basic properties, and include them in a library of elementary functions (see the inside front cover).

This library will become an important addition to your mathematical toolbox and can be used in any course or activity where mathematics is applied.

We begin by placing six basic functions in our library.

DEFINITION Basic Elementary Functions

$$f(x) = x \qquad \text{Identity function}$$
$$h(x) = x^2 \qquad \text{Square function}$$
$$m(x) = x^3 \qquad \text{Cube function}$$
$$n(x) = \sqrt{x} \qquad \text{Square root function}$$
$$p(x) = \sqrt[3]{x} \qquad \text{Cube root function}$$
$$g(x) = |x| \qquad \text{Absolute value function}$$

These elementary functions can be evaluated by hand for certain values of x and with a calculator for all values of x for which they are defined.

EXAMPLE 1   **Evaluating Basic Elementary Functions** Evaluate each basic elementary function at

(A) $x = 64$ (B) $x = -12.75$

Round any approximate values to four decimal places.

SOLUTION   (A) $f(64) = 64$
$\quad h(64) = 64^2 = 4{,}096$ Use a calculator.
$\quad m(64) = 64^3 = 262{,}144$ Use a calculator.
$\quad n(64) = \sqrt{64} = 8$
$\quad p(64) = \sqrt[3]{64} \; = 4$
$\quad g(64) = |64| = 64$

(B) $f(-12.75) = -12.75$
$\quad h(-12.75) = (-12.75)^2 = 162.5625$ Use a calculator.
$\quad m(-12.75) = (-12.75)^3 \approx -2{,}072.6719$ Use a calculator.
$\quad n(-12.75) = \sqrt{-12.75}$ Not a real number.
$\quad p(-12.75) = \sqrt[3]{-12.75} \approx -2.3362$ Use a calculator.
$\quad g(-12.75) = |-12.75| = 12.75$

Matched Problem 1   Evaluate each basic elementary function at

(A) $x = 729$ (B) $x = -5.25$

Round any approximate values to four decimal places.

Remark—Most computers and graphing calculators use ABS(x) to represent the absolute value function. The following representation can also be useful:

$$|x| = \sqrt{x^2}$$

Figure 1 shows the graph, range, and domain of each of the basic elementary functions.

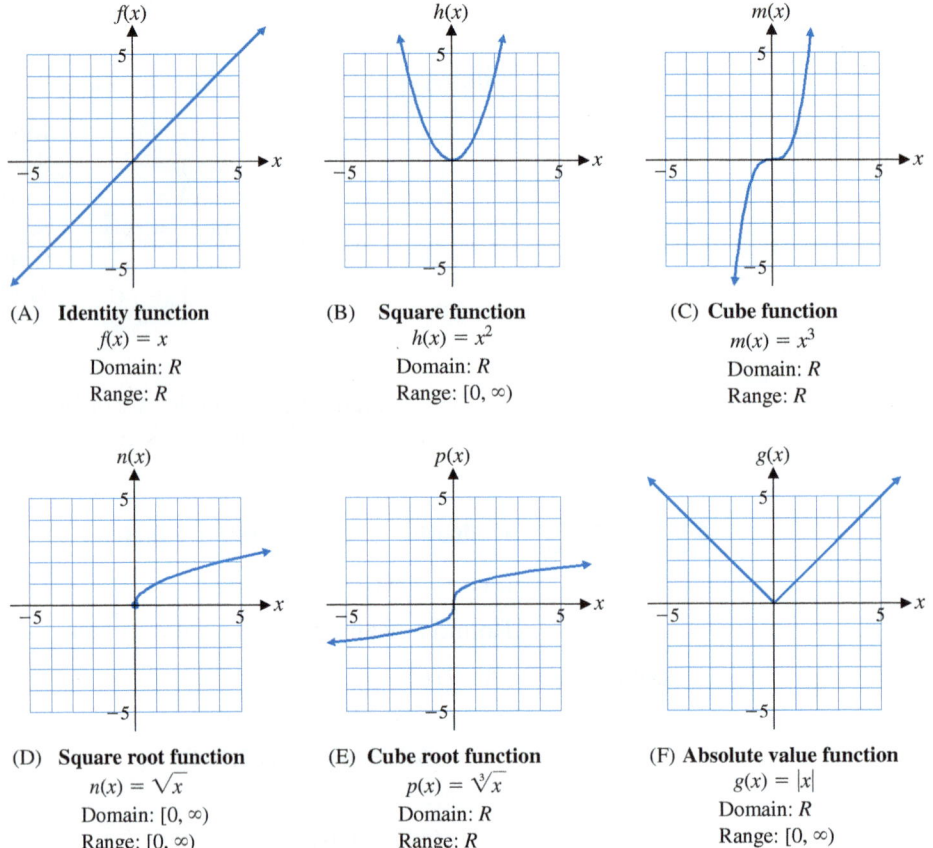

(A) **Identity function**
$f(x) = x$
Domain: R
Range: R

(B) **Square function**
$h(x) = x^2$
Domain: R
Range: $[0, \infty)$

(C) **Cube function**
$m(x) = x^3$
Domain: R
Range: R

(D) **Square root function**
$n(x) = \sqrt{x}$
Domain: $[0, \infty)$
Range: $[0, \infty)$

(E) **Cube root function**
$p(x) = \sqrt[3]{x}$
Domain: R
Range: R

(F) **Absolute value function**
$g(x) = |x|$
Domain: R
Range: $[0, \infty)$

Figure 1 Some basic functions and their graphs*

CONCEPTUAL INSIGHT

Absolute Value In beginning algebra, absolute value is often interpreted as distance from the origin on a real number line (see Appendix A, Section A-1).

$$\text{distance} = 6 = -(-6) \qquad \text{distance} = 5$$

If $x < 0$, then $-x$ is the *positive* distance from the origin to x, and if $x > 0$, then x is the positive distance from the origin to x. Thus,

$$|x| = \begin{cases} -x & \text{if } x < 0 \\ x & \text{if } x \geq 0 \end{cases}$$

Vertical and Horizontal Shifts

If a new function is formed by performing an operation on a given function, then the graph of the new function is called a **transformation** of the graph of the original function. For example, graphs of $y = f(x) + k$ and $y = f(x + h)$ are transformations of the graph of $y = f(x)$.

---

*Note: Letters used to designate these functions may vary from context to context; R is the set of all real numbers.

EXPLORE & DISCUSS 1

Let $f(x) = x^2$.

(A) Graph $y = f(x) + k$ for $k = -4, 0$, and 2 simultaneously in the same coordinate system. Describe the relationship between the graph of $y = f(x)$ and the graph of $y = f(x) + k$ for any real number k.

(B) Graph $y = f(x + h)$ for $h = -4, 0$, and 2 simultaneously in the same coordinate system. Describe the relationship between the graph of $y = f(x)$ and the graph of $y = f(x + h)$ for any real number h.

EXAMPLE 2   **Vertical and Horizontal Shifts**

(A) How are the graphs of $y = |x| + 4$ and $y = |x| - 5$ related to the graph of $y = |x|$? Confirm your answer by graphing all three functions simultaneously in the same coordinate system.

(B) How are the graphs of $y = |x + 4|$ and $y = |x - 5|$ related to the graph of $y = |x|$? Confirm your answer by graphing all three functions simultaneously in the same coordinate system.

SOLUTION   (A) The graph of $y = |x| + 4$ is the same as the graph of $y = |x|$ shifted upward 4 units, and the graph of $y = |x| - 5$ is the same as the graph of $y = |x|$ shifted downward 5 units. Figure 2 confirms these conclusions. [It appears that the graph of $y = f(x) + k$ is the graph of $y = f(x)$ shifted up if k is positive and down if k is negative.]

(B) The graph of $y = |x + 4|$ is the same as the graph of $y = |x|$ shifted to the left 4 units, and the graph of $y = |x - 5|$ is the same as the graph of $y = |x|$ shifted to the right 5 units. Figure 3 confirms these conclusions. [It appears that the graph of $y = f(x + h)$ is the graph of $y = f(x)$ shifted right if h is negative and left if h is positive—the opposite of what you might expect.]

Figure 2 Vertical shifts

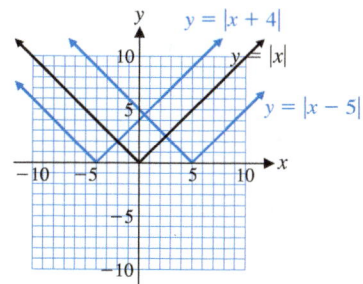

Figure 3 Horizontal shifts

Matched Problem 2

(A) How are the graphs of $y = \sqrt{x} + 5$ and $y = \sqrt{x} - 4$ related to the graph of $y = \sqrt{x}$? Confirm your answer by graphing all three functions simultaneously in the same coordinate system.

(B) How are the graphs of $y = \sqrt{x + 5}$ and $y = \sqrt{x - 4}$ related to the graph of $y = \sqrt{x}$? Confirm your answer by graphing all three functions simultaneously in the same coordinate system.

Comparing the graphs of $y = f(x) + k$ with the graph of $y = f(x)$, we see that the graph of $y = f(x) + k$ can be obtained from the graph of $y = f(x)$ by **vertically translating** (shifting) the graph of the latter upward k units if k is positive and downward $|k|$ units if k is negative. Comparing the graphs of $y = f(x + h)$ with the graph of $y = f(x)$, we see that the graph of $y = f(x + h)$ can be obtained from the graph of $y = f(x)$ by **horizontally translating** (shifting) the graph of the latter h units to the left if h is positive and $|h|$ units to the right if h is negative.

EXAMPLE 3 **Vertical and Horizontal Translations (Shifts)** The graphs in Figure 4 are either horizontal or vertical shifts of the graph of $f(x) = x^2$. Write appropriate equations for functions H, G, M, and N in terms of f.

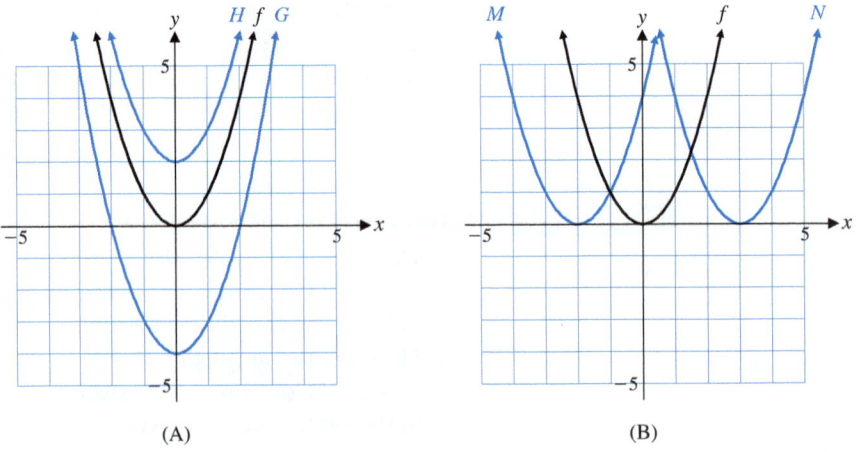

(A)   (B)

Figure 4 Vertical and horizontal shifts

SOLUTION Functions H and G are vertical shifts given by

$$H(x) = x^2 + 2 \qquad G(x) = x^2 - 4$$

Functions M and N are horizontal shifts given by

$$M(x) = (x + 2)^2 \qquad N(x) = (x - 3)^2$$

Matched Problem 3 The graphs in Figure 5 are either horizontal or vertical shifts of the graph of $f(x) = \sqrt[3]{x}$. Write appropriate equations for functions H, G, M, and N in terms of f.

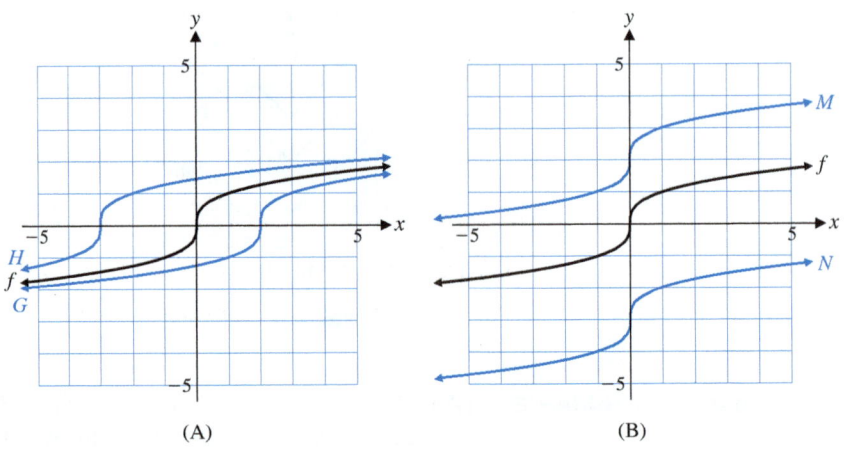

(A)   (B)

Figure 5 Vertical and horizontal shifts

Reflections, Stretches, and Shrinks

We now investigate how the graph of $y = Af(x)$ is related to the graph of $y = f(x)$ for different real numbers A.

EXPLORE & DISCUSS 2

(A) Graph $y = Ax^2$ for $A = 1, 4$, and $\frac{1}{4}$ simultaneously in the same coordinate system.

(B) Graph $y = Ax^2$ for $A = -1, -4$, and $-\frac{1}{4}$ simultaneously in the same coordinate system.

(C) Describe the relationship between the graph of $h(x) = x^2$ and the graph of $G(x) = Ax^2$ for any real number A.

Comparing $y = Af(x)$ to $y = f(x)$, we see that the graph of $y = Af(x)$ can be obtained from the graph of $y = f(x)$ by multiplying each ordinate value of the latter by A. The result is a **vertical stretch** of the graph of $y = f(x)$ if $A > 1$, a **vertical shrink** of the graph of $y = f(x)$ if $0 < A < 1$, and a **reflection in the x axis** if $A = -1$. If A is a negative number other than -1, then the result is a combination of a reflection in the x axis and either a vertical stretch or a vertical shrink.

EXAMPLE 4   **Reflections, Stretches, and Shrinks**

(A) How are the graphs of $y = 2|x|$ and $y = 0.5|x|$ related to the graph of $y = |x|$? Confirm your answer by graphing all three functions simultaneously in the same coordinate system.

(B) How is the graph of $y = -2|x|$ related to the graph of $y = |x|$? Confirm your answer by graphing both functions simultaneously in the same coordinate system.

SOLUTION   (A) The graph of $y = 2|x|$ is a vertical stretch of the graph of $y = |x|$ by a factor of 2, and the graph of $y = 0.5|x|$ is a vertical shrink of the graph of $y = |x|$ by a factor of 0.5. Figure 6 confirms this conclusion.

(B) The graph of $y = -2|x|$ is a reflection in the x axis and a vertical stretch of the graph of $y = |x|$. Figure 7 confirms this conclusion.

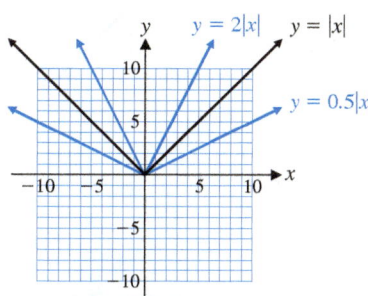

Figure 6 Vertical stretch and shrink

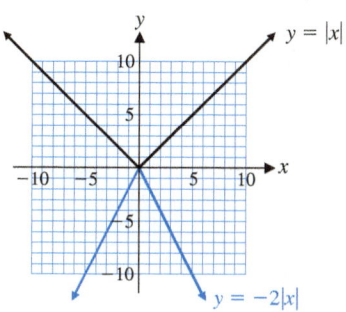

Figure 7 Reflection and vertical stretch

Matched Problem 4   (A) How are the graphs of $y = 2x$ and $y = 0.5x$ related to the graph of $y = x$? Confirm your answer by graphing all three functions simultaneously in the same coordinate system.

(B) How is the graph of $y = -0.5x$ related to the graph of $y = x$? Confirm your answer by graphing both functions in the same coordinate system.

The various transformations considered above are summarized in the following box for easy reference:

SUMMARY   Graph Transformations

Vertical Translation:

$$y = f(x) + k \quad \begin{cases} k > 0 & \text{Shift graph of } y = f(x) \text{ up } k \text{ units.} \\ k < 0 & \text{Shift graph of } y = f(x) \text{ down } |k| \text{ units.} \end{cases}$$

Horizontal Translation:

$$y = f(x + h) \quad \begin{cases} h > 0 & \text{Shift graph of } y = f(x) \text{ left } h \text{ units.} \\ h < 0 & \text{Shift graph of } y = f(x) \text{ right } |h| \text{ units.} \end{cases}$$

Reflection:

$$y = -f(x) \quad \text{Reflect the graph of } y = f(x) \text{ in the } x \text{ axis.}$$

Vertical Stretch and Shrink:

$$y = Af(x) \begin{cases} A > 1 & \text{Stretch graph of } y = f(x) \text{ vertically} \\ & \text{by multiplying each ordinate value by } A. \\ 0 < A < 1 & \text{Shrink graph of } y = f(x) \text{ vertically} \\ & \text{by multiplying each ordinate value by } A. \end{cases}$$

EXPLORE & DISCUSS 3

Use a graphing calculator to explore the graph of $y = A(x + h)^2 + k$ for various values of the constants A, h, and k. Discuss how the graph of $y = A(x + h)^2 + k$ is related to the graph of $y = x^2$.

EXAMPLE 5 **Combining Graph Transformations** Discuss the relationship between the graphs of $y = -|x - 3| + 1$ and $y = |x|$. Confirm your answer by graphing both functions simultaneously in the same coordinate system.

SOLUTION The graph of $y = -|x - 3| + 1$ is a reflection in the x axis, a horizontal translation of 3 units to the right, and a vertical translation of 1 unit upward of the graph of $y = |x|$. Figure 8 confirms this description.

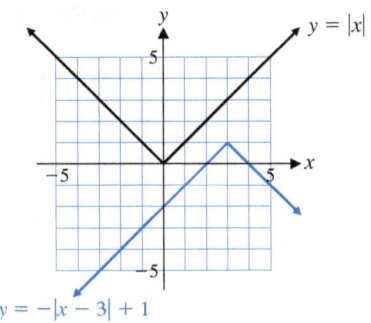

Figure 8 Combined transformations

Matched Problem 5 The graph of $y = G(x)$ in Figure 9 involves a reflection and a translation of the graph of $y = x^3$. Describe how the graph of function G is related to the graph of $y = x^3$ and find an equation of the function G.

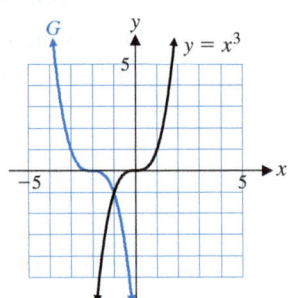

Figure 9 Combined transformations

Piecewise-Defined Functions

Earlier we noted that the absolute value of a real number x can be defined as

$$|x| = \begin{cases} -x & \text{if } x < 0 \\ x & \text{if } x \geq 0 \end{cases}$$

Notice that this function is defined by different rules for different parts of its domain. Functions whose definitions involve more than one rule are called **piecewise-defined functions**. Graphing one of these functions involves graphing each rule over the appropriate portion of the domain (Fig. 10). In Figure 10C, notice that an open dot is used to show that the point $(0, -2)$ is not part of the graph and a solid dot is used to show that $(0, 2)$ is part of the graph.

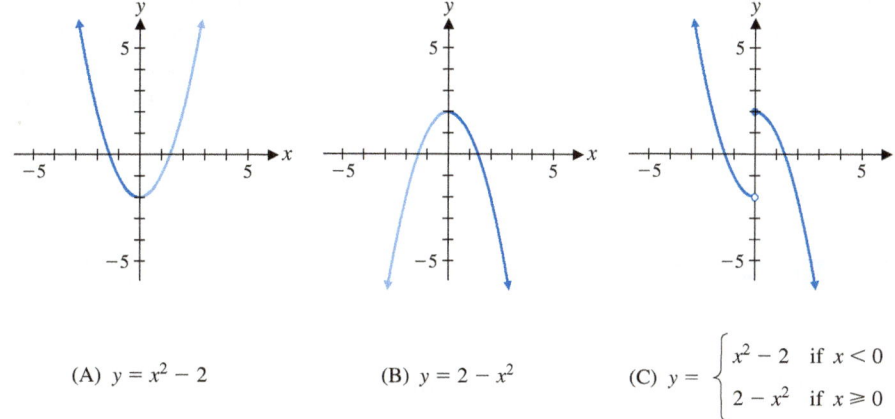

(A) $y = x^2 - 2$ (B) $y = 2 - x^2$ (C) $y = \begin{cases} x^2 - 2 & \text{if } x < 0 \\ 2 - x^2 & \text{if } x \geqslant 0 \end{cases}$

Figure 10 Graphing a piecewise-defined function

As the next example illustrates, piecewise-defined functions occur naturally in many applications.

EXAMPLE 6   **Natural Gas Rates** Easton Utilities uses the rates shown in Table 1 to compute the monthly cost of natural gas for each customer. Write a piecewise definition for the cost of consuming x CCF (cubic hundred feet) of natural gas and graph the function.

Table 1   Charges per Month
$0.7866 per CCF for the first 5 CCF
$0.4601 per CCF for the next 35 CCF
$0.2508 per CCF for all over 40 CCF

SOLUTION   If $C(x)$ is the cost, in dollars, of using x CCF of natural gas in one month, then the first line of Table 1 implies that

$$C(x) = 0.7866x \quad \text{if } 0 \leq x \leq 5$$

Note that $C(5) = 3.933$ is the cost of 5 CCF. If $5 < x \leq 40$, then $x - 5$ represents the amount of gas that cost $0.4601 per CCF, $0.4601(x - 5)$ represents the cost of this gas, and the total cost is

$$C(x) = 3.933 + 0.4601(x - 5)$$

If $x > 40$, then

$$C(x) = 20.0365 + 0.2508(x - 40)$$

where $20.0365 = C(40)$, the cost of the first 40 CCF. Combining all these equations, we have the following piecewise definition for $C(x)$:

$$C(x) = \begin{cases} 0.7866x & \text{if } 0 \leq x \leq 5 \\ 3.933 + 0.4601(x - 5) & \text{if } 5 < x \leq 40 \\ 20.0365 + 0.2508(x - 40) & \text{if } x > 40 \end{cases}$$

To graph C, first note that each rule in the definition of C represents a transformation of the identity function $f(x) = x$. Graphing each transformation over the indicated interval produces the graph of C shown in Figure 11.

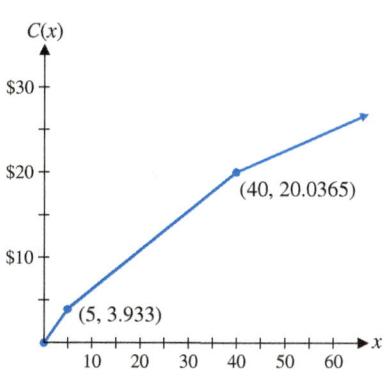

Figure 11 Cost of purchasing x CCF of natural gas

Matched Problem 6 Trussville Utilities uses the rates shown in Table 2 to compute the monthly cost of natural gas for residential customers. Write a piecewise definition for the cost of consuming x CCF of natural gas and graph the function.

Table 2 Charges per Month

$0.7675 per CCF for the first 50 CCF
$0.6400 per CCF for the next 150 CCF
$0.6130 per CCF for all over 200 CCF

Exercise 2-2

A

In Problems 1–8, give the domain and range of each function.

1. $f(x) = x^2 + 3$ **2.** $g(x) = x^2 - 4$

3. $h(x) = -5|x|$ **4.** $k(x) = -2|x|$

5. $m(x) = \sqrt{x} - 10$ **6.** $n(x) = -6\sqrt{x}$

7. $r(x) = -\sqrt[3]{8x}$ **8.** $s(x) = 1 + \sqrt[3]{x}$

In Problems 9–20, graph each of the functions using the graphs of functions f and g below.

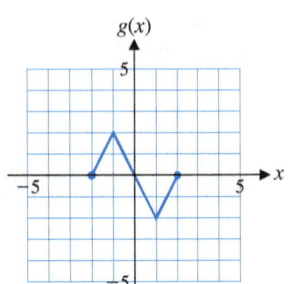

9. $y = f(x) + 2$ **10.** $y = g(x) - 1$

11. $y = f(x + 2)$ **12.** $y = g(x - 1)$

13. $y = g(x - 3)$ **14.** $y = f(x + 3)$

15. $y = g(x) - 3$ **16.** $y = f(x) + 3$

17. $y = -f(x)$ **18.** $y = -g(x)$

19. $y = 0.5g(x)$ **20.** $y = 2f(x)$

B

In Problems 21–28, indicate verbally how the graph of each function is related to the graph of one of the six basic functions in Figure 1 on page 60. Sketch a graph of each function.

21. $g(x) = -|x + 3|$ **22.** $h(x) = -|x - 5|$

23. $f(x) = (x - 4)^2 - 3$ **24.** $m(x) = (x + 3)^2 + 4$

25. $f(x) = 7 - \sqrt{x}$ **26.** $g(x) = -6 + \sqrt[3]{x}$

27. $h(x) = -3|x|$ **28.** $m(x) = -0.4x^2$

Each graph in Problems 29–36 is the result of applying a sequence of transformations to the graph of one of the six basic

functions in Figure 1 on page 60. Identify the basic function and describe the transformation verbally. Write an equation for the given graph.

29.

30.

31.

32.

33.

34.

35.

36.
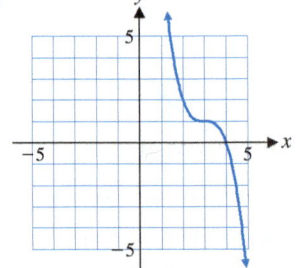

In Problems 37–42, the graph of the function g is formed by applying the indicated sequence of transformations to the given function f. Find an equation for the function g and graph g using −5 ≤ x ≤ 5 and −5 ≤ y ≤ 5.

37. The graph of $f(x) = \sqrt{x}$ is shifted 2 units to the right and 3 units down.

38. The graph of $f(x) = \sqrt[3]{x}$ is shifted 3 units to the left and 2 units up.

39. The graph of $f(x) = |x|$ is reflected in the *x* axis and shifted to the left 3 units.

40. The graph of $f(x) = |x|$ is reflected in the *x* axis and shifted to the right 1 unit.

41. The graph of $f(x) = x^3$ is reflected in the *x* axis and shifted 2 units to the right and down 1 unit.

42. The graph of $f(x) = x^2$ is reflected in the *x* axis and shifted to the left 2 units and up 4 units.

Graph each function in Problems 43–48.

43. $f(x) = \begin{cases} 2 - 2x & \text{if } x < 2 \\ x - 2 & \text{if } x \geq 2 \end{cases}$

44. $g(x) = \begin{cases} x + 1 & \text{if } x < -1 \\ 2 + 2x & \text{if } x \geq -1 \end{cases}$

45. $h(x) = \begin{cases} 5 + 0.5x & \text{if } 0 \leq x \leq 10 \\ -10 + 2x & \text{if } x > 10 \end{cases}$

46. $h(x) = \begin{cases} 10 + 2x & \text{if } 0 \leq x \leq 20 \\ 40 + 0.5x & \text{if } x > 20 \end{cases}$

47. $h(x) = \begin{cases} 2x & \text{if } 0 \leq x \leq 20 \\ x + 20 & \text{if } 20 < x \leq 40 \\ 0.5x + 40 & \text{if } x > 40 \end{cases}$

48. $h(x) = \begin{cases} 4x + 20 & \text{if } 0 \leq x \leq 20 \\ 2x + 60 & \text{if } 20 < x \leq 100 \\ -x + 360 & \text{if } x > 100 \end{cases}$

C

 Each of the graphs in Problems 49–54 involves a reflection in the x axis and/or a vertical stretch or shrink of one of the basic functions in Figure 1 on page 60. Identify the basic function, and describe the transformation verbally. Write an equation for the given graph.

49.

50.

51.

52.

53.

54.

 Changing the order in a sequence of transformations may change the final result. Investigate each pair of transformations in Problems 55–60 to determine if reversing their order can produce a different result. Support your conclusions with specific examples and/or mathematical arguments.

55. Vertical shift; horizontal shift

56. Vertical shift; reflection in *y* axis

57. Vertical shift; reflection in *x* axis

58. Vertical shift; vertical stretch

59. Horizontal shift; reflection in *y* axis

60. Horizontal shift; vertical shrink

Applications

61. Price–demand. A retail chain sells DVD players. The retail price $p(x)$ (in dollars) and the weekly demand *x* for a particular model are related by

$$p(x) = 115 - 4\sqrt{x} \qquad 9 \leq x \leq 289$$

(A) Describe how the graph of function *p* can be obtained from the graph of one of the basic functions in Figure 1 on page 60.

(B) Sketch a graph of function *p* using part (A) as an aid.

62. Price–supply. The manufacturers of the DVD players in Problem 61 are willing to supply *x* players at a price of $p(x)$ as given by the equation

$$p(x) = 4\sqrt{x} \qquad 9 \leq x \leq 289$$

(A) Describe how the graph of function *p* can be obtained from the graph of one of the basic functions in Figure 1 on page 60.

(B) Sketch a graph of function *p* using part (A) as an aid.

63. Hospital costs. Using statistical methods, the financial department of a hospital arrived at the cost equation

$$C(x) = 0.00048(x - 500)^3 + 60,000 \qquad 100 \le x \le 1,000$$

where $C(x)$ is the cost in dollars for handling x cases per month.

(A) Describe how the graph of function C can be obtained from the graph of one of the basic functions in Figure 1 on page 60.

(B) Sketch a graph of function C using part (A) and a graphing calculator as aids.

64. Price–demand. A company manufactures and sells in-line skates. Its financial department has established the price–demand function

$$p(x) = 190 - 0.013(x - 10)^2 \qquad 10 \le x \le 100$$

where $p(x)$ is the price at which x thousand pairs of in-line skates can be sold.

(A) Describe how the graph of function p can be obtained from the graph of one of the basic functions in Figure 1 on page 60.

(B) Sketch a graph of function p using part (A) and a graphing calculator as aids.

65. Electricity rates. Table 3 shows the electricity rates charged by Monroe Utilities in the summer months. The base is a fixed monthly charge, independent of the kWh (kilowatt-hours) used during the month.

(A) Write a piecewise definition of the monthly charge $S(x)$ for a customer who uses x kWh in a summer month.

(B) Graph $S(x)$.

Table 3   Summer (July–October)
Base charge, $8.50
First 700 kWh or less at 0.0650/kWh
Over 700 kWh at 0.0900/kWh

66. Electricity rates. Table 4 shows the electricity rates charged by Monroe Utilities in the winter months.

(A) Write a piecewise definition of the monthly charge $W(x)$ for a customer who uses x kWh in a winter month.

Table 4   Winter (November–June)
Base charge, $8.50
First 700 kWh or less at 0.0650/kWh
Over 700 kWh at 0.0530/kWh

(B) Graph $W(x)$.

67. State income tax. Table 5 shows a recent state income tax schedule for married couples filing a joint return in Kansas.

(A) Write a piecewise definition for the tax due $T(x)$ on an income of x dollars.

(B) Graph $T(x)$.

(C) Find the tax due on a taxable income of $40,000. Of $70,000.

Table 5 Kansas State Income Tax		
SCHEDULE I—MARRIED FILING JOINT		
If taxable income is		
Over	But Not Over	Tax Due Is
$0	$30,000	3.50% of taxable income
$30,000	$60,000	$1,050 plus 6.25% of excess over $30,000
$60,000		$2,925 plus 6.45% of excess over $60,000

68. State income tax. Table 6 shows a recent state income tax schedule for individuals filing a return in Kansas.

Table 6 Kansas State Income Tax		
SCHEDULE II—SINGLE, HEAD OF HOUSEHOLD, OR MARRIED FILING SEPARATE		
If taxable income is		
Over	But Not Over	Tax Due Is
$0	$15,000	3.50% of taxable income
$15,000	$30,000	$525 plus 6.25% of excess over $15,000
$30,000		$1,462.50 plus 6.45% of excess over $30,000

(A) Write a piecewise definition for the tax due $T(x)$ on an income of x dollars.

(B) Graph $T(x)$.

(C) Find the tax due on a taxable income of $20,000. Of $35,000.

(D) Would it be better for a married couple in Kansas with two equal incomes to file jointly or separately? Discuss.

69. Human weight. A good approximation of the normal weight of a person 60 inches or taller but not taller than 80 inches is given by $w(x) = 5.5x - 220$, where x is height in inches and $w(x)$ is weight in pounds.

(A) Describe how the graph of function w can be obtained from the graph of one of the basic functions in Figure 1, page 60.

(B) Sketch a graph of function w using part (A) as an aid.

70. Herpetology. The average weight of a particular species of snake is given by $w(x) = 463x^3$, $0.2 \le x \le 0.8$, where x is length in meters and $w(x)$ is weight in grams.

✏️ (A) Describe how the graph of function w can be obtained from the graph of one of the basic functions in Figure 1, page 60.

(B) Sketch a graph of function w using part (A) as an aid.

71. Safety research. Under ideal conditions, if a person driving a vehicle slams on the brakes and skids to a stop, the speed of the vehicle $v(x)$ (in miles per hour) is given approximately by $v(x) = C\sqrt{x}$, where x is the length of skid marks (in feet) and C is a constant that depends on the road conditions and the weight of the vehicle. For a particular vehicle, $v(x) = 7.08\sqrt{x}$ and $4 \le x \le 144$.

✏️ (A) Describe how the graph of function v can be obtained from the graph of one of the basic functions in Figure 1, page 60.

(B) Sketch a graph of function v using part (A) as an aid.

72. Learning. A production analyst has found that on average it takes a new person $T(x)$ minutes to perform a particular assembly operation after x performances of the operation, where $T(x) = 10 - \sqrt[3]{x}$, $0 \le x \le 125$.

✏️ (A) Describe how the graph of function T can be obtained from the graph of one of the basic functions in Figure 1, page 60.

(B) Sketch a graph of function T using part (A) as an aid.

Answers to Matched Problems

1. (A) $f(729) = 729$, $h(729) = 531{,}441$,
   $m(729) = 387{,}420{,}489$, $n(729) = 27$, $p(729) = 9$,
   $g(729) = 729$

   (B) $f(-5.25) = -5.25$, $h(-5.25) = 27.5625$,
   $m(-5.25) = -144.7031$, $n(-5.25)$ is not a real number,
   $p(-5.25) = -1.7380$, $g(-5.25) = 5.25$

2. (A) The graph of $y = \sqrt{x} + 5$ is the same as the graph of $y = \sqrt{x}$ shifted upward 5 units, and the graph of $y = \sqrt{x} - 4$ is the same as the graph of $y = \sqrt{x}$ shifted downward 4 units. The figure confirms these conclusions.

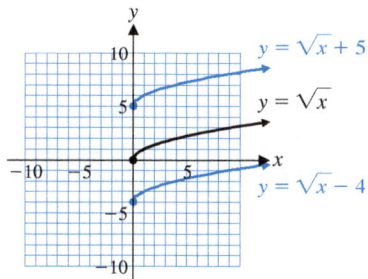

(B) The graph of $y = \sqrt{x + 5}$ is the same as the graph of $y = \sqrt{x}$ shifted to the left 5 units, and the graph of $y = \sqrt{x - 4}$ is the same as the graph of $y = \sqrt{x}$ shifted to the right 4 units. The figure confirms these conclusions.

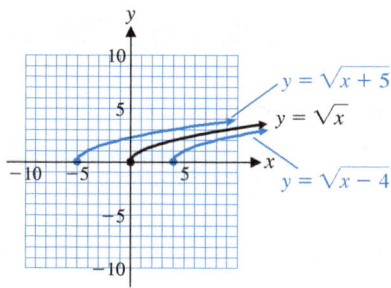

3. $H(x) = \sqrt[3]{x + 3}$, $G(x) = \sqrt[3]{x - 2}$, $M(x) = \sqrt[3]{x} + 2$,
   $N(x) = \sqrt[3]{x} - 3$

4. (A) The graph of $y = 2x$ is a vertical stretch of the graph of $y = x$, and the graph of $y = 0.5x$ is a vertical shrink of the graph of $y = x$. The figure confirms these conclusions.

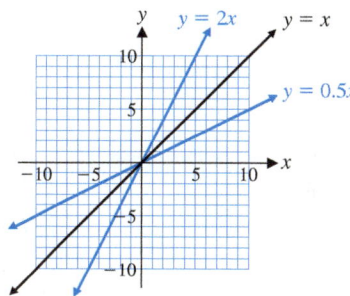

(B) The graph of $y = -0.5x$ is a vertical shrink and a reflection in the x axis of the graph of $y = x$. The figure confirms this conclusion.

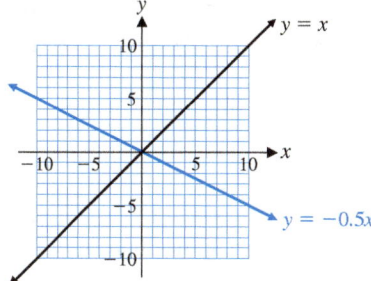

5. The graph of function G is a reflection in the x axis and a horizontal translation of 2 units to the left of the graph of $y = x^3$. An equation for G is $G(x) = -(x + 2)^3$.

6. $C(x) = \begin{cases} 0.7675x & \text{if } 0 \le x \le 50 \\ 38.375 + 0.64(x - 50) & \text{if } 50 < x \le 200 \\ 134.375 + 0.613(x - 200) & \text{if } 200 < x \end{cases}$

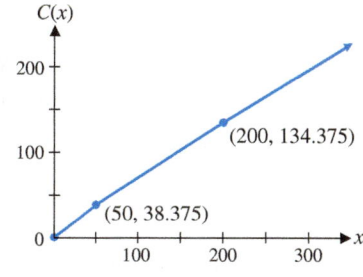

2-3   Quadratic Functions

- Quadratic Functions, Equations, and Inequalities
- Properties of Quadratic Functions and Their Graphs
- Applications

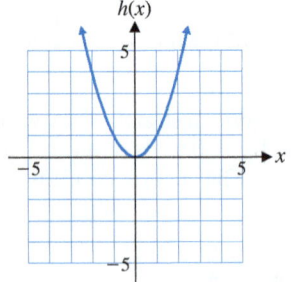

Figure 1 Square function $h(x) = x^2$

If the degree of a linear function is increased by one, we obtain a *second-degree function*, usually called a *quadratic function*, another basic function that we will need in our library of elementary functions. We will investigate relationships between quadratic functions and the solutions to quadratic equations and inequalities. Other important properties of quadratic functions will also be investigated, including maxima and minima. We will then be in a position to solve important practical problems such as finding production levels that will generate maximum revenue or maximum profit.

Quadratic Functions, Equations, and Inequalities

The graph of the square function $h(x) = x^2$ is shown in Figure 1. Notice that the graph is symmetric with respect to the y axis and that $(0, 0)$ is the lowest point on the graph. Let's explore the effect of applying a sequence of basic transformations to the graph of h.

EXPLORE & DISCUSS 1

Indicate how the graph of each function is related to the graph of the function $h(x) = x^2$. Find the highest or lowest point, whichever exists, on each graph.

(A) $f(x) = (x - 3)^2 - 7 = x^2 - 6x + 2$

(B) $g(x) = 0.5(x + 2)^2 + 3 = 0.5x^2 + 2x + 5$

(C) $m(x) = -(x - 4)^2 + 8 = -x^2 + 8x - 8$

(D) $n(x) = -3(x + 1)^2 - 1 = -3x^2 - 6x - 4$

Graphing the functions in Explore & Discuss 1 produces figures similar in shape to the graph of the square function in Figure 1. These figures are called *parabolas*. The functions that produce these parabolas are examples of the important class of *quadratic functions*.

DEFINITION   Quadratic Functions

If a, b, and c are real numbers with $a \neq 0$, then the function

$$f(x) = ax^2 + bx + c \qquad \text{Standard form}$$

is a **quadratic function** and its graph is a **parabola**.

CONCEPTUAL INSIGHT

If x is any real number, then $ax^2 + bx + c$ is also a real number. According to the agreement on domain and range in Section 2-1, the domain of a quadratic function is R, the set of real numbers.

We will discuss methods for determining the range of a quadratic function later in this section. Typical graphs of quadratic functions are illustrated in Figure 2.

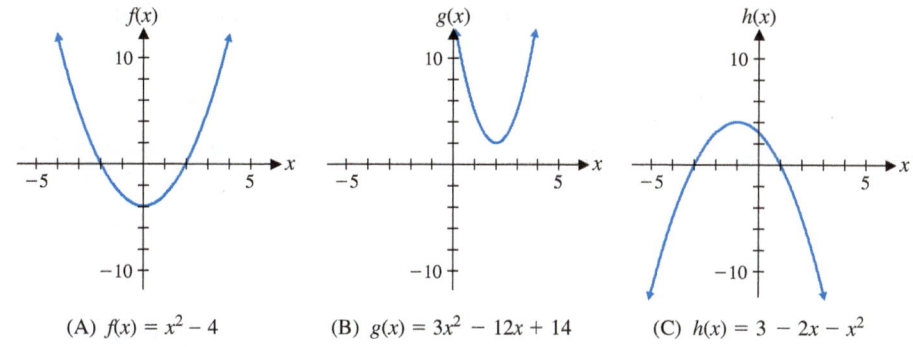

(A) $f(x) = x^2 - 4$ (B) $g(x) = 3x^2 - 12x + 14$ (C) $h(x) = 3 - 2x - x^2$

Figure 2 Graphs of quadratic functions

CONCEPTUAL INSIGHT

An x intercept of a function is also called a **zero** of the function. The x intercepts of a linear function can be found by solving the linear equation $y = mx + b = 0$ for x, $m \neq 0$ (see Section 1-2). Similarly, the x intercepts of a quadratic function can be found by solving the quadratic equation $y = ax^2 + bx + c = 0$ for x, $a \neq 0$. Several methods for solving quadratic equations are discussed in Appendix A, Section A-7. The most popular of these is the **quadratic formula**.

If $ax^2 + bx + c = 0$, $a \neq 0$, then

$$x = \frac{-b \pm \sqrt{b^2 - 4ac}}{2a}, \text{ provided } b^2 - 4ac \geq 0$$

EXAMPLE 1   **Intercepts, Equations, and Inequalities**

(A) Sketch a graph of $f(x) = -x^2 + 5x + 3$ in a rectangular coordinate system.

(B) Find x and y intercepts algebraically to four decimal places.

(C) Graph $f(x) = -x^2 + 5x + 3$ in a standard viewing window.

(D) Find the x and y intercepts to four decimal places using TRACE and ZERO on your graphing calculator.

(E) Solve the quadratic inequality $-x^2 + 5x + 3 \geq 0$ graphically to four decimal places using the results of parts (A) and (B) or (C) and (D).

(F) Solve the equation $-x^2 + 5x + 3 = 4$ graphically to four decimal places using INTERSECT on your graphing calculator.

SOLUTION   (A) Hand-sketching a graph of f:

x	y
-1	-3
0	3
1	7
2	9
3	9
4	7
5	3
6	-3

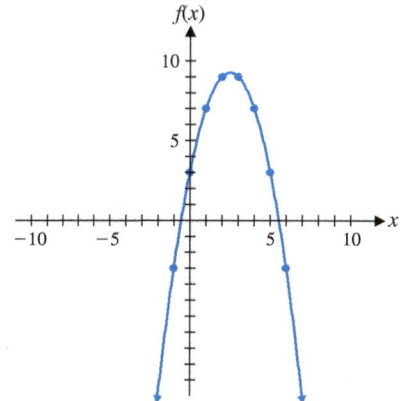

(B) Finding intercepts algebraically:

$$y \text{ intercept: } f(0) = -(0)^2 + 5(0) + 3 = 3$$

x intercepts: $f(x) = 0$

$$-x^2 + 5x + 3 = 0 \qquad \textcolor{blue}{\text{Quadratic equation}}$$

$$x = \frac{-b \pm \sqrt{b^2 - 4ac}}{2a} \qquad \textcolor{blue}{\text{Quadratic formula (see Appendix A-7)}}$$

$$x = \frac{-(5) \pm \sqrt{5^2 - 4(-1)(3)}}{2(-1)}$$

$$= \frac{-5 \pm \sqrt{37}}{-2} = -0.5414 \quad \text{or} \quad 5.5414$$

(C) Graphing in a graphing calculator:

(D) Finding intercepts graphically using a graphing calculator:

x intercept: -0.5414 $\qquad$ x intercept: 5.5414 $\qquad$ y intercept: 3

(E) Solving $-x^2 + 5x + 3 \geq 0$ graphically: The quadratic inequality

$$-x^2 + 5x + 3 \geq 0$$

holds for those values of x for which the graph of $f(x) = -x^2 + 5x + 3$ in the figures in parts (A) and (C) is at or above the x axis. This happens for x between the two x intercepts [found in part (B) or (D)], including the two x intercepts. The solution set for the quadratic inequality is $-0.5414 \leq x \leq 5.5414$ or $[-0.5414, 5.5414]$.

(F) Solving the equation $-x^2 + 5x + 3 = 4$ using a graphing calculator:

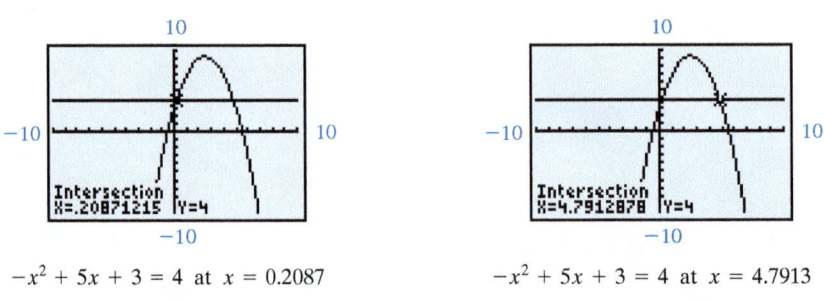

$-x^2 + 5x + 3 = 4$ at $x = 0.2087$ $\qquad$ $-x^2 + 5x + 3 = 4$ at $x = 4.7913$

Matched Problem 1

(A) Sketch a graph of $g(x) = 2x^2 - 5x - 5$ in a rectangular coordinate system.

(B) Find x and y intercepts algebraically to four decimal places.

(C) Graph $g(x) = 2x^2 - 5x - 5$ in a standard viewing window.

(D) Find the x and y intercepts to four decimal places using TRACE and the ZERO command on your graphing calculator.

(E) Solve $2x^2 - 5x - 5 \geq 0$ graphically to four decimal places using the results of parts (A) and (B) or (C) and (D).

(F) Solve the equation $2x^2 - 5x - 5 = -3$ graphically to four decimal places using INTERSECT on your graphing calculator.

EXPLORE & DISCUSS 2

How many x intercepts can the graph of a quadratic function have? How many y intercepts? Explain your reasoning.

Properties of Quadratic Functions and Their Graphs

Many useful properties of the quadratic function can be uncovered by transforming

$$f(x) = ax^2 + bx + c \qquad a \neq 0$$

into the **vertex form**

$$f(x) = a(x - h)^2 + k$$

The process of *completing the square* (see Appendix A-7) is central to the transformation. We illustrate the process through a specific example and then generalize the results.

Consider the quadratic function given by

$$f(x) = -2x^2 + 16x - 24 \tag{1}$$

We use completing the square to transform this function into vertex form:

$$
\begin{aligned}
f(x) &= -2x^2 + 16x - 24 && \text{Factor the coefficient of } x^2 \\
&= -2(x^2 - 8x) - 24 && \text{out of the first two terms.} \\
&= -2(x^2 - 8x + \textbf{?}) - 24 \\
\\
&= -2(x^2 - 8x + \textbf{16}) - 24 + \textbf{32} && \text{Add 16 to complete the} \\
&&& \text{square inside the parentheses.} \\
&&& \text{Because of the } -2 \text{ outside} \\
&&& \text{the parentheses, we have} \\
&&& \text{actually added } -32, \text{ so we} \\
&&& \text{must add 32 to the outside.} \\
\\
&= -2(x - 4)^2 + 8 && \text{The transformation is} \\
&&& \text{complete and can be checked} \\
&&& \text{by multiplying out.}
\end{aligned}
$$

Therefore,

$$f(x) = -2(x - 4)^2 + 8 \tag{2}$$

If $x = 4$, then $-2(x - 4)^2 = 0$ and $f(4) = 8$. For any other value of x, the negative number $-2(x - 4)^2$ is added to 8, making it smaller. Therefore,

$$f(4) = 8$$

is the *maximum value* of $f(x)$ for all x. Furthermore, if we choose any two x values that are the same distance from 4, we will obtain the same function value. For example, $x = 3$ and $x = 5$ are each one unit from $x = 4$ and their function values are

$$f(3) = -2(3 - 4)^2 + 8 = 6$$
$$f(5) = -2(5 - 4)^2 + 8 = 6$$

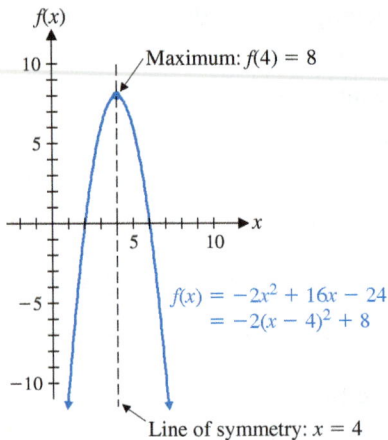

Figure 3 Graph of a quadratic function

Therefore, the vertical line $x = 4$ is a line of symmetry. That is, if the graph of equation (1) is drawn on a piece of paper and the paper is folded along the line $x = 4$, then the two sides of the parabola will match exactly. All these results are illustrated by graphing equations (1) and (2) and the line $x = 4$ simultaneously in the same coordinate system (Fig. 3).

From the preceding discussion, we see that as x moves from left to right, $f(x)$ is increasing on $(-\infty, 4]$, and decreasing on $[4, \infty)$, and that $f(x)$ can assume no value greater than 8. Thus,

$$\text{Range of } f: \quad y \le 8 \quad \text{or} \quad (-\infty, 8]$$

In general, the graph of a quadratic function is a parabola with line of symmetry parallel to the vertical axis. The lowest or highest point on the parabola, whichever exists, is called the **vertex**. The maximum or minimum value of a quadratic function always occurs at the vertex of the parabola. The line of symmetry through the vertex is called the **axis** of the parabola. In the example above, $x = 4$ is the axis of the parabola and $(4, 8)$ is its vertex.

CONCEPTUAL INSIGHT

Applying the graph transformation properties discussed in Section 2-2 to the transformed equation,

$$f(x) = -2x^2 + 16x - 24$$
$$= -2(x - 4)^2 + 8$$

we see that the graph of $f(x) = -2x^2 + 16x - 24$ is the graph of $g(x) = x^2$ vertically stretched by a factor of 2, reflected in the x axis, and shifted to the right 4 units and up 8 units, as shown in Figure 4.

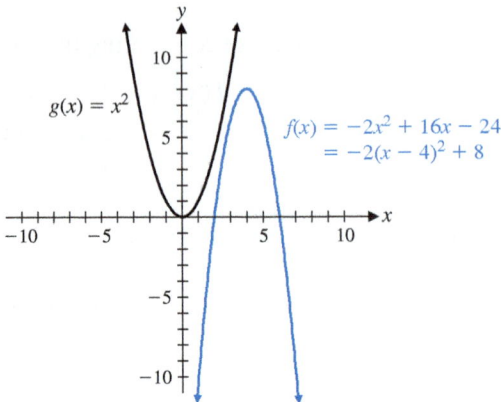

Figure 4 Graph of f is the graph of g transformed

Note the important results we have obtained from the vertex form of the quadratic function f:

- The vertex of the parabola
- The axis of the parabola
- The maximum value of $f(x)$
- The range of the function f
- The relationship between the graph of $g(x) = x^2$ and the graph of $f(x) = -2x^2 + 16x - 24$

The preceding discussion is generalized to all quadratic functions in the following summary:

SUMMARY **Properties of a Quadratic Function and Its Graph**

Given a quadratic function and the vertex form obtained by completing the square

$$f(x) = ax^2 + bx + c \qquad a \neq 0 \qquad \textit{Standard form}$$
$$= a(x - h)^2 + k \qquad \textit{Vertex form}$$

we summarize general properties as follows:

1. The graph of f is a parabola:

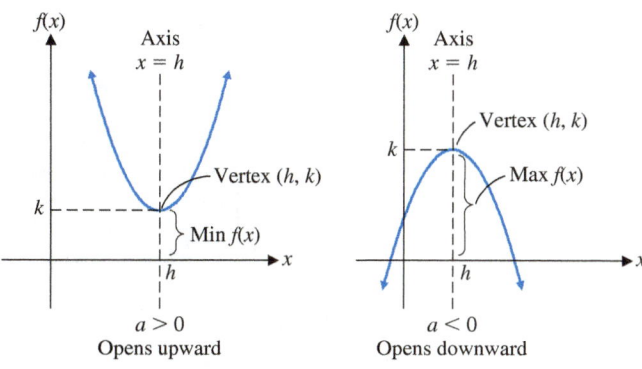

$$a > 0$$
Opens upward

$$a < 0$$
Opens downward

2. Vertex: (h, k) (parabola increases on one side of the vertex and decreases on the other)
3. Axis (of symmetry): $x = h$ (parallel to y axis)
4. $f(h) = k$ is the minimum if $a > 0$ and the maximum if $a < 0$
5. Domain: All real numbers.   Range: $(-\infty, k]$ if $a < 0$ or $[k, \infty)$ if $a > 0$
6. The graph of f is the graph of $g(x) = ax^2$ translated horizontally h units and vertically k units.

EXAMPLE 2   **Analyzing a Quadratic Function**   Given the quadratic function

$$f(x) = 0.5x^2 - 6x + 21$$

(A) Find the vertex form for f.

(B) Find the vertex and the maximum or minimum. State the range of f.

(C) Describe how the graph of function f can be obtained from the graph of $g(x) = x^2$ using transformations.

(D) Sketch a graph of function f in a rectangular coordinate system.

(E) Graph function f using a suitable viewing window.

(F) Find the vertex and the maximum or minimum using the appropriate graphing calculator command.

SOLUTION   (A) Complete the square to find the vertex form:

$$f(x) = 0.5x^2 - 6x + 21$$
$$= 0.5(x^2 - 12x + \,?\,) + 21$$
$$= 0.5(x^2 - 12x + 36) + 21 - 18$$
$$= 0.5(x - 6)^2 + 3$$

(B) From the vertex form, we see that $h = 6$ and $k = 3$. Thus, vertex: $(6, 3)$; minimum: $f(6) = 3$; range: $y \geq 3$ or $[3, \infty)$.

(C) The graph of $f(x) = 0.5(x - 6)^2 + 3$ is the same as the graph of $g(x) = x^2$ vertically shrunk by a factor of 0.5, and shifted to the right 6 units and up 3 units.

(D) Graph in a rectangular coordinate system:

 (E) Graph in a graphing calculator:

(F) Find the vertex and minimum using the minimum command:

Vertex: $(6, 3)$; minimum: $f(6) = 3$

Matched Problem 2 Given the quadratic function $f(x) = -0.25x^2 - 2x + 2$

(A) Find the vertex form for f.

(B) Find the vertex and the maximum or minimum. State the range of f.

(C) Describe how the graph of function f can be obtained from the graph of $g(x) = x^2$ using transformations.

(D) Sketch a graph of function f in a rectangular coordinate system.

(E) Graph function f using a suitable viewing window.

(F) Find the vertex and the maximum or minimum using the appropriate graphing calculator command.

Applications

EXAMPLE 3 **Maximum Revenue** This is a continuation of Example 7 in Section 2-1. Recall that the financial department in the company that produces a digital camera arrived at the following price–demand function and the corresponding revenue function:

$$p(x) = 94.8 - 5x \qquad \text{Price–demand function}$$
$$R(x) = xp(x) = x(94.8 - 5x) \qquad \text{Revenue function}$$

where $p(x)$ is the wholesale price per camera at which x million cameras can be sold and $R(x)$ is the corresponding revenue (in millions of dollars). Both functions have domain $1 \le x \le 15$.

(A) Find the value of x to the nearest thousand cameras that will generate the maximum revenue. What is the maximum revenue to the nearest thousand dollars? Solve the problem algebraically by completing the square.

(B) What is the wholesale price per camera (to the nearest dollar) that generates the maximum revenue?

 (C) Graph the revenue function using an appropriate viewing window.

(D) Find the value of x to the nearest thousand cameras that will generate the maximum revenue. What is the maximum revenue to the nearest thousand dollars? Solve the problem graphically using the maximum command.

SOLUTION (A) Algebraic solution:

$$\begin{aligned} R(x) &= x(94.8 - 5x) \\ &= -5x^2 + 94.8x \\ &= -5(x^2 - 18.96x + \,?) \\ &= -5(x^2 - 18.96x + 89.8704) + 449.352 \\ &= -5(x - 9.48)^2 + 449.352 \end{aligned}$$

The maximum revenue of 449.352 million dollars ($449,352,000) occurs when $x = 9.480$ million cameras (9,480,000 cameras).

(B) Finding the wholesale price per camera: Use the price–demand function for an output of 9.480 million cameras:

$$\begin{aligned} p(x) &= 94.8 - 5x \\ p(9.480) &= 94.8 - 5(9.480) \\ &= \$47 \text{ per camera} \end{aligned}$$

 (C) Graph on a graphing calculator:

 (D) Graphical solution using a graphing calculator:

The manufacture and sale of 9.480 million cameras (9,480,000 cameras) will generate a maximum revenue of 449.352 million dollars ($449,352,000).

Matched Problem 3 The financial department in Example 3, using statistical and analytical techniques (see Matched Problem 7 in Section 2-1), arrived at the cost function

$$C(x) = 156 + 19.7x \qquad \textit{Cost function}$$

where $C(x)$ is the cost (in millions of dollars) for manufacturing and selling x million cameras.

(A) Using the revenue function from Example 3 and the preceding cost function, write an equation for the profit function.

(B) Find the value of x to the nearest thousand cameras that will generate the maximum profit. What is the maximum profit to the nearest thousand dollars? Solve the problem algebraically by completing the square.

(C) What is the wholesale price per camera (to the nearest dollar) that generates the maximum profit?

 (D) Graph the profit function using an appropriate viewing window.

(E) Find the output to the nearest thousand cameras that will generate the maximum profit. What is the maximum profit to the nearest thousand dollars? Solve the problem graphically using the maximum command.

EXAMPLE 4   **Break-Even Analysis** Use the revenue function from Example 3 and the cost function from Matched Problem 3:

$$R(x) = x(94.8 - 5x) \qquad \text{Revenue function}$$
$$C(x) = 156 + 19.7x \qquad \text{Cost function}$$

Both have domain $1 \le x \le 15$.

(A) Sketch the graphs of both functions in the same coordinate system.

(B) **Break-even points** are the production levels at which $R(x) = C(x)$. Find the break-even points algebraically to the nearest thousand cameras.

(C) Plot both functions simultaneously in the same viewing window.

(D) Use INTERSECT to find the break-even points graphically to the nearest thousand cameras.

(E) Recall that a loss occurs if $R(x) < C(x)$ and a profit occurs if $R(x) > C(x)$. For what values of x (to the nearest thousand cameras) will a loss occur? A profit?

SOLUTION   (A) Sketch of functions:

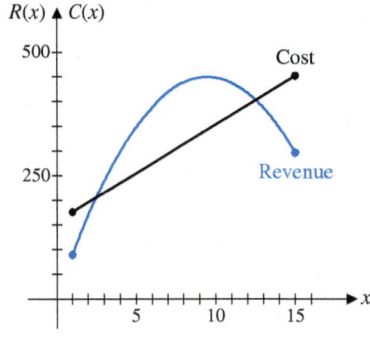

(B) Algebraic solution:
Find x such that $R(x) = C(x)$:

$$x(94.8 - 5x) = 156 + 19.7x$$
$$-5x^2 + 75.1x - 156 = 0$$
$$x = \frac{-75.1 \pm \sqrt{75.1^2 - 4(-5)(-156)}}{2(-5)}$$
$$= \frac{-75.1 \pm \sqrt{2,520.01}}{-10}$$
$$x = 2.490 \quad \text{and} \quad 12.530$$

The company breaks even at $x = 2.490$ million cameras (2,490,000 cameras) and at $x = 12.530$ million cameras (12,530,000 cameras).

(C) Graph on a graphing calculator:

(D) Graphical solution:

The company breaks even at $x = 2.490$ million cameras (2,490,000 cameras) and at $x = 12.530$ million cameras (12,530,000 cameras).

(E) Use the results from parts (A) and (B) or (C) and (D):

$$\text{Loss:}\quad 1 \leq x < 2.490 \quad \text{or} \quad 12.530 < x \leq 15$$
$$\text{Profit:}\quad 2.490 < x < 12.530$$

Matched Problem 4   Use the profit equation from Matched Problem 3:

$$P(x) = R(x) - C(x)$$
$$= -5x^2 + 75.1x - 156 \qquad \textit{Profit function}$$

Domain: $1 \leq x \leq 15$

(A) Sketch a graph of the profit function in a rectangular coordinate system.

(B) Break-even points occur when $P(x) = 0$. Find the break-even points algebraically to the nearest thousand cameras.

 (C) Plot the profit function in an appropriate viewing window.

 (D) Find the break-even points graphically to the nearest thousand cameras.

(E) A loss occurs if $P(x) < 0$, and a profit occurs if $P(x) > 0$. For what values of x (to the nearest thousand cameras) will a loss occur? A profit?

A visual inspection of the plot of a data set might indicate that a parabola would be a better model of the data than a straight line. In that case, rather than using linear regression to fit a linear model to the data, we would use **quadratic regression** on a graphing calculator to find the function of the form $y = ax^2 + bx + c$ that best fits the data.

 EXAMPLE 5   **Outboard Motors** Table 1 gives performance data for a boat powered by an Evinrude outboard motor. Use quadratic regression to find the best model of the form $y = ax^2 + bx + c$ for fuel consumption y (in miles per gallon) as a function of speed x (in miles per hour). Estimate the fuel consumption (to one decimal place) at a speed of 12 miles per hour.

Table 1

rpm	mph	mpg
2,500	10.3	4.1
3,000	18.3	5.6
3,500	24.6	6.6
4,000	29.1	6.4
4,500	33.0	6.1
5,000	36.0	5.4
5,400	38.9	4.9

SOLUTION Enter the data in a graphing calculator (Fig. 5A) and find the quadratic regression equation (Fig. 5B). The data set and the regression equation are graphed in Figure 5C. Using TRACE, we see that the estimated fuel consumption at a speed of 12 mph is 4.5 mpg.

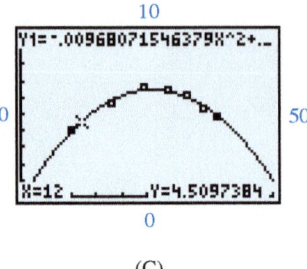

(A)   (B)   (C)

Figure 5

Matched Problem 5 Refer to Table 1. Use quadratic regression to find the best model of the form $y = ax^2 + bx + c$ for boat speed y (in miles per hour) as a function of engine speed x (in revolutions per minute). Estimate the boat speed (in miles per hour, to one decimal place) at an engine speed of 3,400 rpm.

Exercises 2-3

A

In Problems 1–4, complete the square and find the vertex form of each quadratic function.

1. $f(x) = x^2 - 4x + 3$

2. $g(x) = x^2 - 2x - 5$

3. $m(x) = -x^2 + 6x - 4$

4. $n(x) = -x^2 + 8x - 9$

In Problems 5–8, write a brief verbal description of the relationship between the graph of the indicated function (from Problems 1–4) and the graph of $y = x^2$.

5. $f(x) = x^2 - 4x + 3$

6. $g(x) = x^2 - 2x - 5$

7. $m(x) = -x^2 + 6x - 4$

8. $n(x) = -x^2 + 8x - 9$

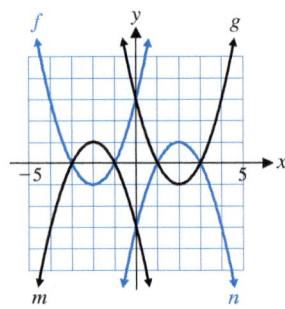

Figure for 9

9. Match each equation with a graph of one of the functions $f, g, m,$ or n in the figure.

(A) $y = -(x + 2)^2 + 1$ (B) $y = (x - 2)^2 - 1$

(C) $y = (x + 2)^2 - 1$ (D) $y = -(x - 2)^2 + 1$

10. Match each equation with a graph of one of the functions $f, g, m,$ or n in the figure.

(A) $y = (x - 3)^2 - 4$ (B) $y = -(x + 3)^2 + 4$

(C) $y = -(x - 3)^2 + 4$ (D) $y = (x + 3)^2 - 4$

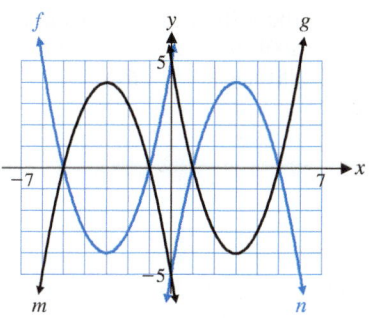

Figure for 10

For the functions indicated in Problems 11–14, find each of the following to the nearest integer by referring to the graphs for Problems 9 and 10.

(A) *Intercepts* (B) *Vertex*

(C) *Maximum or minimum* (D) *Range*

11. Function n in the figure for Problem 9

12. Function m in the figure for Problem 10

13. Function f in the figure for Problem 9

14. Function g in the figure for Problem 10

In Problems 15–18, find each of the following:

(A) *Intercepts* (B) *Vertex*

(C) *Maximum or minimum* (D) *Range*

15. $f(x) = -(x - 3)^2 + 2$ **16.** $g(x) = -(x + 2)^2 + 3$

17. $m(x) = (x + 1)^2 - 2$ **18.** $n(x) = (x - 4)^2 - 3$

B

In Problems 19–22, write an equation for each graph in the form $y = a(x - h)^2 + k$, where a is either 1 or −1 and h and k are integers.

19.

20.

21.

22.

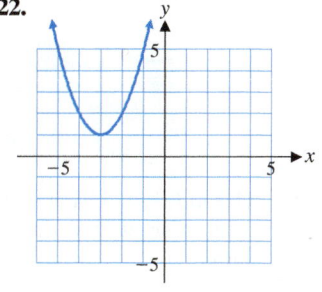

In Problems 23–28, find the vertex form for each quadratic function. Then find each of the following:

(A) *Intercepts* (B) *Vertex*

(C) *Maximum or minimum* (D) *Range*

23. $f(x) = x^2 - 8x + 12$ **24.** $g(x) = x^2 - 6x + 5$

25. $r(x) = -4x^2 + 16x - 15$ **26.** $s(x) = -4x^2 - 8x - 3$

27. $u(x) = 0.5x^2 - 2x + 5$ **28.** $v(x) = 0.5x^2 + 4x + 10$

 29. Let $f(x) = 0.3x^2 - x - 8$. Solve each equation graphically to two decimal places.

(A) $f(x) = 4$ (B) $f(x) = -1$ (C) $f(x) = -9$

 30. Let $g(x) = -0.6x^2 + 3x + 4$. Solve each equation graphically to two decimal places.

(A) $g(x) = -2$ (B) $g(x) = 5$ (C) $g(x) = 8$

31. Let $f(x) = 125x - 6x^2$. Find the maximum value of f to four decimal places graphically.

32. Let $f(x) = 100x - 7x^2 - 10$. Find the maximum value of f to four decimal places graphically.

C

In Problems 33–36, first write each function in vertex form; then find each of the following (to two decimal places):

(A) *Intercepts* (B) *Vertex*

(C) *Maximum or minimum* (D) *Range*

33. $g(x) = 0.25x^2 - 1.5x - 7$

34. $m(x) = 0.20x^2 - 1.6x - 1$

35. $f(x) = -0.12x^2 + 0.96x + 1.2$

36. $n(x) = -0.15x^2 - 0.90x + 3.3$

Solve Problems 37–42 graphically to two decimal places using a graphing calculator.

37. $2 - 5x - x^2 = 0$ **38.** $7 + 3x - 2x^2 = 0$

39. $1.9x^2 - 1.5x - 5.6 < 0$ **40.** $3.4 + 2.9x - 1.1x^2 \geq 0$

41. $2.8 + 3.1x - 0.9x^2 \leq 0$ **42.** $1.8x^2 - 3.1x - 4.9 > 0$

43. Given that f is a quadratic function with minimum $f(x) = f(2) = 4$, find the axis, vertex, range, and x intercepts.

44. Given that f is a quadratic function with maximum $f(x) = f(-3) = -5$, find the axis, vertex, range, and x intercepts.

In Problems 45–48,

(A) *Graph f and g in the same coordinate system.*

(B) *Solve $f(x) = g(x)$ algebraically to two decimal places.*

(C) *Solve $f(x) > g(x)$ using parts (A) and (B).*

(D) *Solve $f(x) < g(x)$ using parts (A) and (B).*

45. $f(x) = -0.4x(x - 10)$

$g(x) = 0.3x + 5$

$0 \le x \le 10$

46. $f(x) = -0.7x(x - 7)$

$g(x) = 0.5x + 3.5$

$0 \le x \le 7$

47. $f(x) = -0.9x^2 + 7.2x$

$g(x) = 1.2x + 5.5$

$0 \le x \le 8$

48. $f(x) = -0.7x^2 + 6.3x$

$g(x) = 1.1x + 4.8$

$0 \le x \le 9$

49. How can you tell from the graph of a quadratic function whether it has exactly one real zero?

50. How can you tell from the graph of a quadratic function whether it has no real zeros?

51. How can you tell from the standard form $y = ax^2 + bx + c$ whether a quadratic function has two real zeros?

52. How can you tell from the standard form $y = ax^2 + bx + c$ whether a quadratic function has exactly one real zero?

53. How can you tell from the vertex form $y = a(x - h)^2 + k$ whether a quadratic function has no real zeros?

54. How can you tell from the vertex form $y = a(x - h)^2 + k$ whether a quadratic function has two real zeros?

In Problems 55 and 56, assume that a, b, c, h, and k are constants with a ≠ 0 such that

$$ax^2 + bx + c = a(x - h)^2 + k$$

for all real numbers x.

55. Show that $h = -\dfrac{b}{2a}$.

56. Show that $k = \dfrac{4ac - b^2}{4a}$.

Applications

57. **Tire mileage.** An automobile tire manufacturer collected the data in the table relating tire pressure x (in pounds per square inch) and mileage (in thousands of miles):

x	Mileage
28	45
30	52
32	55
34	51
36	47

A mathematical model for the data is given by

$$f(x) = -0.518x^2 + 33.3x - 481$$

(A) Complete the following table. Round values of $f(x)$ to one decimal place.

x	Mileage	f(x)
28	45	
30	52	
32	55	
34	51	
36	47	

(B) Sketch the graph of f and the mileage data in the same coordinate system.

(C) Use values of the modeling function rounded to two decimal places to estimate the mileage for a tire pressure of 31 lbs/sq in. and for 35 lbs/sq in.

(D) Write a brief description of the relationship between tire pressure and mileage.

58. **Automobile production.** The table shows the retail market share of passenger cars from Ford Motor Company as a percentage of the U.S. market.

Year	Market Share
1980	17.2%
1985	18.8%
1990	20.0%
1995	20.7%
2000	20.2%
2005	17.4%

A mathematical model for this data is given by

$$f(x) = -0.0206x^2 + 0.548x + 16.9$$

where $x = 0$ corresponds to 1980.

(A) Complete the following table. Round values of $f(x)$ to one decimal place.

x	Market Share	f(x)
0	17.2	
5	18.8	
10	20.0	
15	20.7	
20	20.2	
25	17.4	

(B) Sketch the graph of f and the market share data in the same coordinate system.

(C) Use values of the modeling function f to estimate Ford's market share in 2015 and in 2020.

(D) Write a brief verbal description of Ford's market share from 1980 to 2005.

59. **Tire mileage.** Using quadratic regression on a graphing calculator, show that the quadratic function that best fits the data on tire mileage in Problem 57 is

$$f(x) = -0.518x^2 + 33.3x - 481$$

60. Automobile production. Using quadratic regression on a graphing calculator, show that the quadratic function that best fits the data on market share in Problem 58 is

$$f(x) = -0.0206x^2 + 0.548x + 16.9$$

61. Revenue. The marketing research department for a company that manufactures and sells memory chips for microcomputers established the following price–demand and revenue functions:

$p(x) = 75 - 3x$ Price–demand function

$R(x) = xp(x) = x(75 - 3x)$ Revenue function

where $p(x)$ is the wholesale price in dollars at which x million chips can be sold, and $R(x)$ is in millions of dollars. Both functions have domain $1 \le x \le 20$.

(A) Sketch a graph of the revenue function in a rectangular coordinate system.

(B) Find the value of x that will produce the maximum revenue. What is the maximum revenue?

(C) What is the wholesale price per chip that produces the maximum revenue?

62. Revenue. The marketing research department for a company that manufactures and sells notebook computers established the following price–demand and revenue functions:

$p(x) = 2,000 - 60x$ Price–demand function

$R(x) = xp(x)$ Revenue function

$\quad\quad = x(2,000 - 60x)$

where $p(x)$ is the wholesale price in dollars at which x thousand computers can be sold, and $R(x)$ is in thousands of dollars. Both functions have domain $1 \le x \le 25$.

(A) Sketch a graph of the revenue function in a rectangular coordinate system.

(B) Find the value of x that will produce the maximum revenue. What is the maximum revenue to the nearest thousand dollars?

(C) What is the wholesale price per computer (to the nearest dollar) that produces the maximum revenue?

63. Break-even analysis. Use the revenue function from Problem 61 in this exercise and the given cost function:

$R(x) = x(75 - 3x)$ Revenue function

$C(x) = 125 + 16x$ Cost function

where x is in millions of chips, and $R(x)$ and $C(x)$ are in millions of dollars. Both functions have domain $1 \le x \le 20$.

(A) Sketch a graph of both functions in the same rectangular coordinate system.

(B) Find the break-even points to the nearest thousand chips.

(C) For what values of x will a loss occur? A profit?

64. Break-even analysis. Use the revenue function from Problem 62, in this exercise and the given cost function:

$R(x) = x(2,000 - 60x)$ Revenue function

$C(x) = 4,000 + 500x$ Cost function

where x is thousands of computers, and $C(x)$ and $R(x)$ are in thousands of dollars. Both functions have domain $1 \le x \le 25$.

(A) Sketch a graph of both functions in the same rectangular coordinate system.

(B) Find the break-even points.

(C) For what values of x will a loss occur? A profit?

65. Profit–loss analysis. Use the revenue and cost functions from Problem 63 in this exercise:

$R(x) = x(75 - 3x)$ Revenue function

$C(x) = 125 + 16x$ Cost function

where x is in millions of chips, and $R(x)$ and $C(x)$ are in millions of dollars. Both functions have domain $1 \le x \le 20$.

(A) Form a profit function P, and graph R, C, and P in the same rectangular coordinate system.

(B) Discuss the relationship between the intersection points of the graphs of R and C and the x intercepts of P.

(C) Find the x intercepts of P and the break-even points to the nearest thousand chips.

(D) Refer to the graph drawn in part (A). Does the maximum profit appear to occur at the same value of x as the maximum revenue? Are the maximum profit and the maximum revenue equal? Explain.

(E) Verify your conclusion in part (D) by finding the value of x (to the nearest thousand chips) that produces the maximum profit. Find the maximum profit (to the nearest thousand dollars), and compare with Problem 61B.

66. Profit–loss analysis. Use the revenue and cost functions from Problem 64 in this exercise:

$R(x) = x(2,000 - 60x)$ Revenue function

$C(x) = 4,000 + 500x$ Cost function

where x is thousands of computers, and $R(x)$ and $C(x)$ are in thousands of dollars. Both functions have domain $1 \le x \le 25$.

(A) Form a profit function P, and graph R, C, and P in the same rectangular coordinate system.

(B) Discuss the relationship between the intersection points of the graphs of R and C and the x intercepts of P.

(C) Find the x intercepts of P and the break-even points.

(D) Refer to the graph drawn in part (A). Does the maximum profit appear to occur at the same value of x as the maximum revenue? Are the maximum profit and the maximum revenue equal? Explain.

(E) Verify your conclusion in part (D) by finding the value of x that produces the maximum profit. Find the maximum profit and compare with Problem 62B.

67. Medicine. The French physician Poiseuille was the first to discover that blood flows faster near the center of an artery than near the edge. Experimental evidence has shown that the rate of flow v (in centimeters per second) at a point x centimeters from the center of an artery (see the figure) is given by

$$v = f(x) = 1,000(0.04 - x^2) \qquad 0 \leq x \leq 0.2$$

Find the distance from the center that the rate of flow is 20 centimeters per second. Round answer to two decimal places.

Figure for 67 and 68

68. Medicine. Refer to Problem 67. Find the distance from the center that the rate of flow is 30 centimeters per second. Round answer to two decimal places.

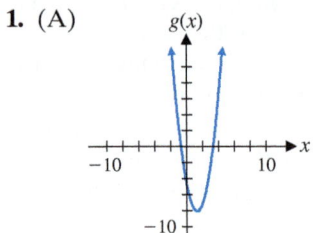**69. Outboard motors.** The table gives performance data for a boat powered by an Evinrude outboard motor. Find a quadratic regression model ($y = ax^2 + bx + c$) for boat speed y (in miles per hour) as a function of engine speed (in revolutions per minute). Estimate the boat speed at an engine speed of 3,100 revolutions per minute.

Table for 69 and 70

rpm	mph	mpg
1,500	4.5	8.2
2,000	5.7	6.9
2,500	7.8	4.8
3,000	9.6	4.1
3,500	13.4	3.7

70. Outboard motors. The table gives performance data for a boat powered by an Evinrude outboard motor. Find a quadratic regression model ($y = ax^2 + bx + c$) for fuel consumption y (in miles per gallon) as a function of engine speed (in revolutions per minute). Estimate the fuel consumption at an engine speed of 2,300 revolutions per minute.

Answers to Matched Problems

1. (A)

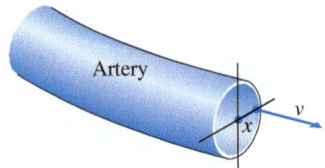

(B) x intercepts: -0.7656, 3.2656; y intercept: -5

(C)

(D) x intercepts: -0.7656, 3.2656; y intercept: -5

(E) $x \leq -0.7656$ or $x \geq 3.2656$; or $(-\infty, -0.7656]$ or $[3.2656, \infty)$

(F) $x = -0.3508, 2.8508$

2. (A) $f(x) = -0.25(x + 4)^2 + 6$.

(B) Vertex: $(-4, 6)$; maximum: $f(-4) = 6$; range: $y \leq 6$ or $(-\infty, 6]$

(C) The graph of $f(x) = -0.25(x + 4)^2 + 6$ is the same as the graph of $g(x) = x^2$ vertically shrunk by a factor of 0.25, reflected in the x axis, and shifted 4 units to the left and 6 units up.

(D)

(E)

(F) Vertex: $(-4, 6)$; maximum: $f(-4) = 6$

3. (A) $P(x) = R(x) - C(x) = -5x^2 + 75.1x - 156$

(B) $P(x) = R(x) - C(x) = -5(x - 7.51)^2 + 126.0005$; the manufacture and sale of 7,510,000 million cameras will produce a maximum profit of $126,001,000.

(C) $p(7.510) = \$57$

(D)

(E)

The manufacture and sale of 7,510,000 million cameras will produce a maximum profit of \$126,001,000. (Notice that maximum profit does not occur at the same value of x where maximum revenue occurs.)

4. (A)

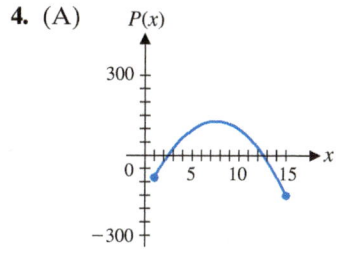

(B) $x = 2.490$ million cameras (2,490,000 cameras) and $x = 12.530$ million cameras. (12,530,000 cameras)

(C)

(D) $x = 2.490$ million cameras (2,490,000 cameras) and $x = 12.530$ million cameras (12,530,000 cameras)

(E) Loss: $1 \le x < 2.490$ or $12.530 < x \le 15$; profit: $2.490 < x < 12.530$

5. 22.9 mph

2-4 Polynomial and Rational Functions

- Polynomial Functions
- Regression Polynomials
- Rational Functions
- Applications

Linear and quadratic functions are special cases of the more general class of *polynomial functions*. Polynomial functions are a special case of an even larger class of functions, the *rational functions*. We will describe the basic features of the graphs of polynomial and rational functions. We will use these functions to solve real-world problems where linear or quadratic models are inadequate; for example, to determine the relationship between length and weight of a species of fish, or to model the training of new employees.

Polynomial Functions

A linear function has the form $f(x) = mx + b$ (where $m \ne 0$) and is a polynomial function of degree 1. A quadratic function has the form $f(x) = ax^2 + bx + c$ (where $a \ne 0$) and is a polynomial function of degree 2. Here is the general definition of a polynomial function.

> **DEFINITION Polynomial Function**
>
> A **polynomial function** is a function that can be written in the form
>
> $$f(x) = a_n x^n + a_{n-1} x^{n-1} + \cdots + a_1 x + a_0$$
>
> for n a nonnegative integer, called the **degree** of the polynomial. The coefficients $a_0, a_1, \ldots, a_n$ are real numbers with $a_n \ne 0$. The **domain** of a polynomial function is the set of all real numbers.

Figure 1 shows graphs of representative polynomial functions of degrees 1 through 6. The figure suggests some general properties of graphs of polynomial functions.

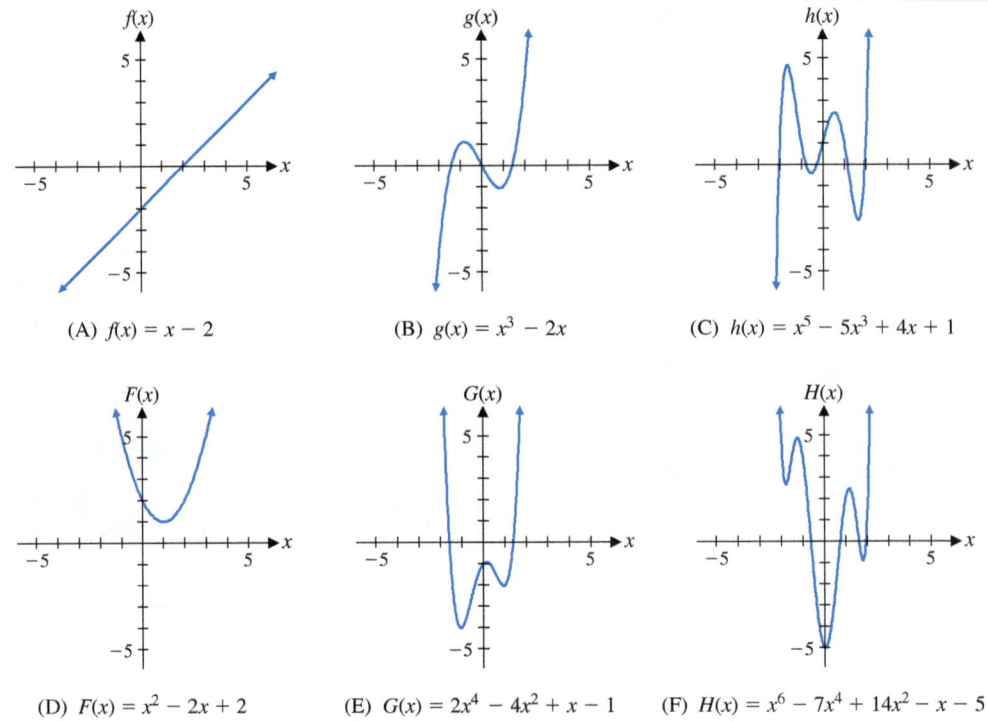

(A) $f(x) = x - 2$ (B) $g(x) = x^3 - 2x$ (C) $h(x) = x^5 - 5x^3 + 4x + 1$

(D) $F(x) = x^2 - 2x + 2$ (E) $G(x) = 2x^4 - 4x^2 + x - 1$ (F) $H(x) = x^6 - 7x^4 + 14x^2 - x - 5$

Figure 1 Graphs of polynomial functions

Notice that the odd-degree polynomial graphs start negative, end positive, and cross the x axis at least once. The even-degree polynomial graphs start positive, end positive, and may not cross the x axis at all. In all cases in Figure 1, the **leading coefficient**—that is, the coefficient of the highest-degree term—was chosen positive. If any leading coefficient had been chosen negative, then we would have a similar graph but reflected in the x axis.

A polynomial of degree n can have, at most, n linear factors. Therefore, the graph of a polynomial function of positive degree n can intersect the x axis at most n times. Note from Figure 1 that a polynomial of degree n may intersect the x axis fewer than n times. An x intercept of a function is also called a **zero*** or **root** of the function.

The graph of a polynomial function is **continuous**, with no holes or breaks. That is, the graph can be drawn without removing a pen from the paper. Also, the graph of a polynomial has no sharp corners. Figure 2 shows the graphs of two functions—one that is not continuous, and the other that is continuous but with a sharp corner. Neither function is a polynomial.

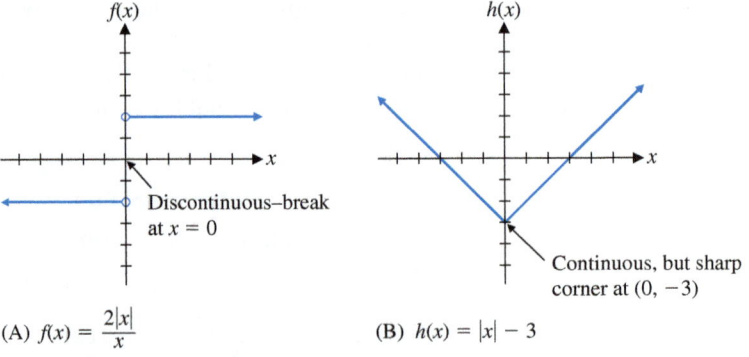

(A) $f(x) = \dfrac{2|x|}{x}$ (B) $h(x) = |x| - 3$

Figure 2 Discontinuous and sharp-corner functions

*Only real numbers can be x intercepts. Functions may have complex zeros that are not real numbers, but such zeros, which are not x intercepts, will not be discussed in this book.

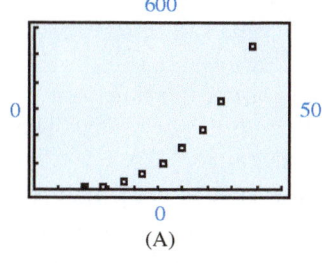 Regression Polynomials

In Chapter 1, we saw that regression techniques can be used to fit a straight line to a set of data. Linear functions are not the only ones that can be applied in this manner. Most graphing calculators have the ability to fit a variety of curves to a given set of data. We will discuss polynomial regression models in this section and other types of regression models in later sections.

EXAMPLE 1

Estimating the Weight of a Fish   Using the length of a fish to estimate its weight is of interest to both scientists and sport anglers. The data in Table 1 give the average weights of lake trout for certain lengths. Use the data and regression techniques to find a polynomial model that can be used to estimate the weight of a lake trout for any length. Estimate (to the nearest ounce) the weights of lake trout of lengths 39, 40, 41, 42, and 43 inches, respectively.

Table 1   Lake Trout

Length (in.)	Weight (oz)	Length (in.)	Weight (oz)
x	y	x	y
10	5	30	152
14	12	34	226
18	26	38	326
22	56	44	536
26	96		

SOLUTION   The graph of the data in Table 1 (Fig. 3A) indicates that a linear regression model would not be appropriate in this case. And, in fact, we would not expect a linear relationship between length and weight. Instead, it is more likely that the weight would be related to the cube of the length. We use a cubic regression polynomial to model the data (Fig. 3B). (Consult your manual for the details of calculating regression polynomials on your graphing utility.) Figure 3C adds the graph of the polynomial model to the graph of the data. The graph in Figure 3C shows that this cubic polynomial does provide a good fit for the data. (We will have more to say about the choice of functions and the accuracy of the fit provided by regression analysis later in the book.) Figure 3D shows the estimated weights for the lengths requested.

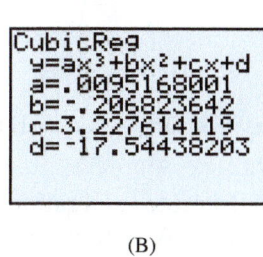

(A)   (B)   (C)   (D)

Figure 3

Matched Problem 1   The data in Table 2 give the average weights of pike for certain lengths. Use a cubic regression polynomial to model the data. Estimate (to the nearest ounce) the weights of pike of lengths 39, 40, 41, 42, and 43 inches, respectively.

Table 2   Pike

Length (in.)	Weight (oz)	Length (in.)	Weight (oz)
x	y	x	y
10	5	30	108
14	12	34	154
18	26	38	210
22	44	44	326
26	72	52	522

Rational Functions

Just as rational numbers are defined in terms of quotients of integers, *rational functions* are defined in terms of quotients of polynomials. The following equations specify rational functions:

$$f(x) = \frac{1}{x} \quad g(x) = \frac{x - 2}{x^2 - x - 6} \quad h(x) = \frac{x^3 - 8}{x}$$

$$p(x) = 3x^2 - 5x \quad q(x) = 7 \quad r(x) = 0$$

DEFINITION Rational Function

A **rational function** is any function that can be written in the form

$$f(x) = \frac{n(x)}{d(x)} \quad d(x) \neq 0$$

where $n(x)$ and $d(x)$ are polynomials. The **domain** is the set of all real numbers such that $d(x) \neq 0$.

Figure 4 shows the graphs of representative rational functions. Note, for example, that in Figure 4A the line $x = 2$ is a *vertical asymptote* for the function. The graph of f gets closer to this line as x gets closer to 2. The line $y = 1$ in Figure 4A is a *horizontal asymptote* for the function. The graph of f gets closer to this line as x increases or decreases without bound.

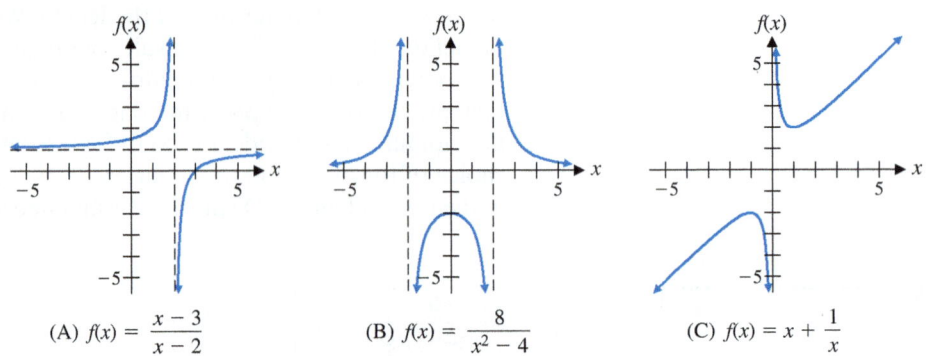

$$\text{(A) } f(x) = \frac{x - 3}{x - 2} \qquad \text{(B) } f(x) = \frac{8}{x^2 - 4} \qquad \text{(C) } f(x) = x + \frac{1}{x}$$

Figure 4 Graphs of rational functions

The number of vertical asymptotes of a rational function $f(x) = n(x)/d(x)$ is at most equal to the degree of $d(x)$. A rational function has at most one horizontal asymptote (note that the graph in Fig. 4C does not have a horizontal asymptote). Moreover, the graph of a rational function approaches the horizontal asymptote (when one exists) both as x increases and decreases without bound.

EXAMPLE 2   **Graphing Rational Functions** Given the rational function:

$$f(x) = \frac{3x}{x^2 - 4}$$

(A) Find the domain.

(B) Find the x and y intercepts.

(C) Find the equations of all vertical asymptotes.

(D) If there is a horizontal asymptote, find its equation.

(E) Using the information from (A)–(D) and additional points as necessary, sketch a graph of f for $-7 \leq x \leq 7$.

SOLUTION   (A) $x^2 - 4 = (x - 2)(x + 2)$, so the denominator is 0 if $x = -2$ or $x = 2$. Therefore the domain is the set of all real numbers except -2 and 2.

(B) *x intercepts:* $f(x) = 0$ only if $3x = 0$, or $x = 0$. So the only x intercept is 0.

y intercept:

$$f(0) = \frac{3 \cdot 0}{0^2 - 4} = \frac{0}{-4} = 0$$

So the *y* intercept is 0.

(C) Consider individually the values of x for which the denominator is 0, namely, 2 and -2, found in part (A).

(i) If $x = 2$, the numerator is 6, and the denominator is 0, so $f(2)$ is undefined. But for numbers just to the right of 2 (like 2.1, 2.01, 2.001), the numerator is close to 6, and the denominator is a positive number close to 0, so the fraction $f(x)$ is large and positive. For numbers just to the left of 2 (like 1.9, 1.99, 1.999), the numerator is close to 6, and the denominator is a negative number close to 0, so the fraction $f(x)$ is large (in absolute value) and negative. Therefore, the line $x = 2$ is a vertical asymptote, and $f(x)$ is positive to the right of the asymptote, and negative to the left.

(ii) If $x = -2$, the numerator is -6, and the denominator is 0, so $f(2)$ is undefined. But for numbers just to the right of -2 (like $-1.9, -1.99, -1.999$), the numerator is close to -6, and the denominator is a negative number close to 0, so the fraction $f(x)$ is large and positive. For numbers just to the left of 2 (like $-2.1, -2.01, -2.001$), the numerator is close to -6, and the denominator is a positive number close to 0, so the fraction $f(x)$ is large (in absolute value) and negative. Therefore, the line $x = -2$ is a vertical asymptote, and $f(x)$ is positive to the right of the asymptote and negative to the left.

(D) Rewrite $f(x)$ by dividing each term in the numerator and denominator by the highest power of x in $f(x)$.

$$f(x) = \frac{3x}{x^2 - 4} = \frac{\dfrac{3x}{x^2}}{\dfrac{x^2}{x^2} - \dfrac{4}{x^2}} = \frac{\dfrac{3}{x}}{1 - \dfrac{4}{x^2}}$$

As x increases or decreases without bound, the numerator tends to 0 and the denominator tends to 1; so, $f(x)$ tends to 0. The line $y = 0$ is a horizontal asymptote.

(E) Use the information from parts (A)–(D) and plot additional points as necessary to complete the graph, as shown in Figure 5.

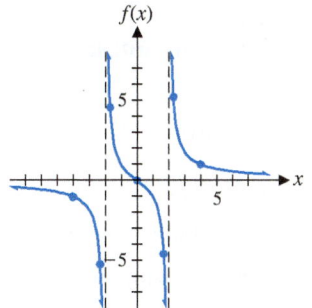

x	$f(x)$
-4	-1
-2.3	-5.3
-1.7	4.6
0	0
1.7	-4.6
2.3	5.3
4	1

Figure 5

Matched Problem 2   Given the rational function $g(x) = \dfrac{3x + 3}{x^2 - 9}$,

(A) Find the domain.

(B) Find the x and y intercepts.

(C) Find the equations of all vertical asymptotes.

(D) If there is a horizontal asymptote, find its equation.

(E) Using the information from parts (A)–(D) and additional points as necessary, sketch a graph of g for $-10 \le x \le 10$.

CONCEPTUAL INSIGHT

Consider the rational function

$$g(x) = \frac{3x^2 - 12x}{x^3 - 4x^2 - 4x + 16} = \frac{3x(x - 4)}{(x^2 - 4)(x - 4)}$$

The numerator and denominator of g have a common zero, $x = 4$. If $x \ne 4$, then we can cancel the factor $x - 4$ from the numerator and denominator, leaving the function $f(x)$ of Example 2. So the graph of g (Fig. 6) is identical to the graph of f (Fig. 5), except that the graph of g has an open dot at $(4, 1)$, indicating that 4 is not in the domain of g. In particular, f and g have the same asymptotes. Note that the line $x = 4$ is *not* a vertical asymptote of g, even though 4 is a zero of its denominator.

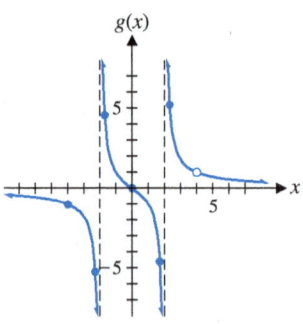

Figure 6

Graphing rational functions is aided by locating vertical and horizontal asymptotes first, if they exist. The following general procedure is suggested by Example 2 and the Conceptual Insight above.

PROCEDURE   Vertical and Horizontal Asymptotes of Rational Functions

Consider the rational function

$$f(x) = \frac{n(x)}{d(x)}$$

where $n(x)$ and $d(x)$ are polynomials.

Vertical asymptotes:

Case 1. Suppose $n(x)$ and $d(x)$ have no real zero in common. If c is a real number such that $d(c) = 0$, then the line $x = c$ is a vertical asymptote of the graph of f.

Case 2. If $n(x)$ and $d(x)$ have one or more real zeros in common, cancel common linear factors, as in Example 2, and apply Case 1 to the reduced function. (The reduced function has the same asymptotes as f.)

Horizontal asymptote:

Case 1. If degree $n(x)$ < degree $d(x)$, then $y = 0$ is the horizontal asymptote.

Case 2. If degree $n(x)$ = degree $d(x)$, then $y = a/b$ is the horizontal asymptote, where a is the leading coefficient of $n(x)$, and b is the leading coefficient of $d(x)$.

Case 3. If degree $n(x)$ > degree $d(x)$, there is no horizontal asymptote.

EXAMPLE 3   **Finding Asymptotes** Find the vertical and horizontal asymptotes of the rational function

$$f(x) = \frac{3x^2 + 3x - 6}{2x^2 - 2}$$

SOLUTION   **Vertical asymptotes** We factor the numerator $n(x)$ and the denominator $d(x)$:

$$n(x) = 3(x^2 + x - 2) = 3(x - 1)(x + 2)$$
$$d(x) = 2(x^2 - 1) = 2(x - 1)(x + 1)$$

The reduced function is

$$\frac{3(x + 2)}{2(x + 1)}$$

which, by the procedure, has the vertical asymptote $x = -1$. Therefore, $x = -1$ is the only vertical asymptote of f.

Horizontal asymptote Both $n(x)$ and $d(x)$ have degree 2 (Case 2 of the procedure for horizontal asymptotes). The leading coefficient of the numerator $n(x)$ is 3, and the leading coefficient of the denominator $d(x)$ is 2. So $y = 3/2$ is the horizontal asymptote.

Matched Problem 3   Find the vertical and horizontal asymptotes of the rational function

$$f(x) = \frac{x^3 - 4x}{x^2 + 5x}$$

EXPLORE & DISCUSS 1   A function f is **bounded** if the entire graph of f lies between two horizontal lines. The only polynomials that are bounded are the constant functions, but there are many rational functions that are bounded. Give an example of a bounded rational function, with domain the set of all real numbers, that is not a constant function.

Applications

Rational functions occur naturally in many types of applications.

EXAMPLE 4   **Employee Training** A company that manufactures computers has established that, on the average, a new employee can assemble $N(t)$ components per day after t days of on-the-job training, as given by

$$N(t) = \frac{50t}{t + 4} \qquad t \geq 0$$

Sketch a graph of N, $0 \leq t \leq 100$, including any vertical or horizontal asymptotes. What does $N(t)$ approach as t increases without bound?

SOLUTION   **Vertical asymptotes** None for $t \geq 0$

Horizontal asymptote

$$N(t) = \frac{50t}{t + 4} = \frac{50}{1 + \dfrac{4}{t}}$$

$N(t)$ approaches 50 (the leading coefficient of $50t$ divided by the leading coefficient of $t + 4$) as t increases without bound. So $y = 50$ is a horizontal asymptote.

Sketch of graph The graph is shown in the margin.

$N(t)$ approaches 50 as t increases without bound. It appears that 50 components per day would be the upper limit that an employee would be expected to assemble.

Matched Problem 4   Repeat Example 4 for $N(t) = \dfrac{25t + 5}{t + 5} \qquad t \geq 0$.

Exercises 2-4

A

In Problems 1–10, for each polynomial function find the following:

(A) Degree of the polynomial

(B) All x intercepts

(C) The y intercept

1. $f(x) = 5x + 3$

2. $f(x) = 4 - 3x$

3. $f(x) = x^2 - 9$

4. $f(x) = 5x - x^2 - 6$

5. $f(x) = (x - 2)(x + 3)(x - 5)$

6. $f(x) = (x + 7)(x - 4)$

7. $f(x) = (2x - 9)(3x + 4)$

8. $f(x) = (2 - 5x)(x - 6)(x + 1)$

9. $f(x) = x^6 + 1$

10. $f(x) = (x^8 + 5)(x^{12} + 7)$

Each graph in Problems 11–18 is the graph of a polynomial function. Answer the following questions for each graph:

(A) What is the minimum degree of a polynomial function that could have the graph?

(B) Is the leading coefficient of the polynomial negative or positive?

11.

12.

13.

14.

15.

16.

17.   **18.**

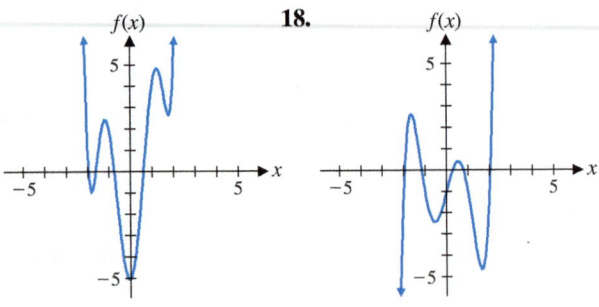

19. What is the maximum number of x intercepts that a polynomial of degree 10 can have?

20. What is the maximum number of x intercepts that a polynomial of degree 7 can have?

21. What is the minimum number of x intercepts that a polynomial of degree 9 can have? Explain.

22. What is the minimum number of x intercepts that a polynomial of degree 6 can have? Explain.

B

For each rational function in Problems 23–28,

(A) Find the intercepts for the graph.

(B) Determine the domain.

(C) Find any vertical or horizontal asymptotes for the graph.

(D) Sketch any asymptotes as dashed lines. Then sketch a graph of $y = f(x)$ for $-10 \le x \le 10$ and $-10 \le y \le 10$.

(E) Graph $y = f(x)$ in a standard viewing window using a graphing calculator.

23. $f(x) = \dfrac{x + 2}{x - 2}$

24. $f(x) = \dfrac{x - 3}{x + 3}$

25. $f(x) = \dfrac{3x}{x + 2}$

26. $f(x) = \dfrac{2x}{x - 3}$

27. $f(x) = \dfrac{4 - 2x}{x - 4}$

28. $f(x) = \dfrac{3 - 3x}{x - 2}$

29. Compare the graph of $y = 2x^4$ to the graph of $y = 2x^4 - 5x^2 + x + 2$ in the following two viewing windows:

(A) $-5 \le x \le 5, -5 \le y \le 5$

(B) $-5 \le x \le 5, -500 \le y \le 500$

30. Compare the graph of $y = x^3$ to the graph of $y = x^3 - 2x + 2$ in the following two viewing windows:

(A) $-5 \le x \le 5, -5 \le y \le 5$

(B) $-5 \le x \le 5, -500 \le y \le 500$

31. Compare the graph of $y = -x^5$ to the graph of $y = -x^5 + 4x^3 - 4x + 1$ in the following two viewing windows:

(A) $-5 \le x \le 5, -5 \le y \le 5$

(B) $-5 \le x \le 5, -500 \le y \le 500$

32. Compare the graph of $y = -x^5$ to the graph of $y = -x^5 + 5x^3 - 5x + 2$ in the following two viewing windows:

(A) $-5 \leq x \leq 5, -5 \leq y \leq 5$

(B) $-5 \leq x \leq 5, -500 \leq y \leq 500$

In Problems 33–40, find the equation of any horizontal asymptote.

33. $f(x) = \dfrac{5x^3 + 2x - 3}{6x^3 - 7x + 1}$ **34.** $f(x) = \dfrac{6x^4 - x^3 + 2}{4x^4 + 10x + 5}$

35. $f(x) = \dfrac{1 - 5x + x^2}{2 + 3x + 4x^2}$ **36.** $f(x) = \dfrac{8 - x^3}{1 + 2x^3}$

37. $f(x) = \dfrac{x^4 + 2x^2 + 1}{1 - x^5}$ **38.** $f(x) = \dfrac{3 + 5x}{x^2 + x + 3}$

39. $f(x) = \dfrac{x^2 + 6x + 1}{x - 5}$ **40.** $f(x) = \dfrac{x^2 + x^4 + 1}{x^3 + 2x - 4}$

In Problems 41–46, find the equations of any vertical asymptotes.

41. $f(x) = \dfrac{x^2 + 1}{(x^2 - 1)(x^2 - 9)}$ **42.** $f(x) = \dfrac{2x + 5}{(x^2 - 4)(x^2 - 16)}$

43. $f(x) = \dfrac{x^2 - x - 6}{x^2 - 3x - 10}$ **44.** $f(x) = \dfrac{x^2 - 8x + 7}{x^2 + 7x - 8}$

45. $f(x) = \dfrac{x^2 + 3x}{x^3 - 36x}$ **46.** $f(x) = \dfrac{x^2 + x - 2}{x^3 - 3x^2 + 2x}$

C

For each rational function in Problems 47–52,

(A) *Find any intercepts for the graph.*

(B) *Find any vertical and horizontal asymptotes for the graph.*

(C) *Sketch any asymptotes as dashed lines. Then sketch a graph of f for $-10 \leq x \leq 10$ and $-10 \leq y \leq 10$.*

 (D) *Graph the function in a standard viewing window using a graphing calculator.*

47. $f(x) = \dfrac{2x^2}{x^2 - x - 6}$ **48.** $f(x) = \dfrac{3x^2}{x^2 + x - 6}$

49. $f(x) = \dfrac{6 - 2x^2}{x^2 - 9}$ **50.** $f(x) = \dfrac{3 - 3x^2}{x^2 - 4}$

51. $f(x) = \dfrac{-4x}{x^2 + x - 6}$ **52.** $f(x) = \dfrac{5x}{x^2 + x - 12}$

53. Write an equation for the lowest-degree polynomial function with the graph and intercepts shown in the figure.

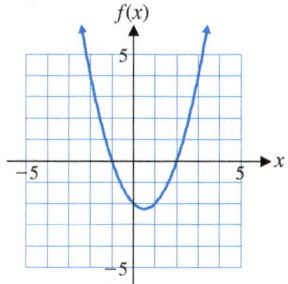

Figure for 53

54. Write an equation for the lowest-degree polynomial function with the graph and intercepts shown in the figure.

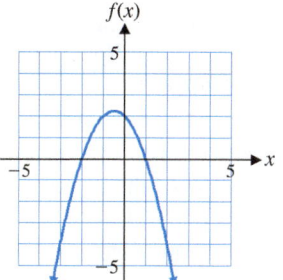

Figure for 54

55. Write an equation for the lowest-degree polynomial function with the graph and intercepts shown in the figure.

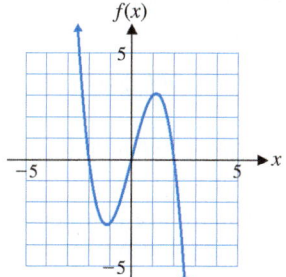

Figure for 55

56. Write an equation for the lowest-degree polynomial function with the graph and intercepts shown in the figure.

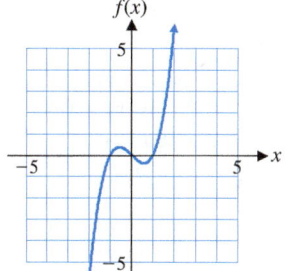

Figure for 56

Applications

57. Average cost. A company manufacturing snowboards has fixed costs of $200 per day and total costs of $3,800 per day at a daily output of 20 boards.

(A) Assuming that the total cost per day, $C(x)$, is linearly related to the total output per day, x, write an equation for the cost function.

(B) The average cost per board for an output of x boards is given by $\overline{C}(x) = C(x)/x$. Find the average cost function.

(C) Sketch a graph of the average cost function, including any asymptotes, for $1 \leq x \leq 30$.

(D) What does the average cost per board tend to as production increases?

58. Average cost. A company manufacturing surfboards has fixed costs of $300 per day and total costs of $5,100 per day at a daily output of 20 boards.

(A) Assuming that the total cost per day, $C(x)$, is linearly related to the total output per day, x, write an equation for the cost function.

(B) The average cost per board for an output of x boards is given by $\overline{C}(x) = C(x)/x$. Find the average cost function.

(C) Sketch a graph of the average cost function, including any asymptotes, for $1 \le x \le 30$.

(D) What does the average cost per board tend to as production increases?

59. Replacement time. An office copier has an initial price of $2,500. A service contract costs $200 for the first year and increases $50 per year thereafter. It can be shown that the total cost of the copier after n years is given by

$$C(n) = 2{,}500 + 175n + 25n^2$$

The average cost per year for n years is given by $\overline{C}(n) = C(n)/n$.

(A) Find the rational function $\overline{C}$.

(B) Sketch a graph of $\overline{C}$ for $2 \le n \le 20$.

(C) When is the average cost per year at a minimum, and what is the minimum average annual cost? [*Hint:* Refer to the sketch in part (B) and evaluate $\overline{C}(n)$ at appropriate integer values until a minimum value is found.] The time when the average cost is minimum is frequently referred to as the **replacement time** for the piece of equipment.

(D) Graph the average cost function $\overline{C}$ on a graphing calculator and use an appropriate command to find when the average annual cost is at a minimum.

60. Minimum average cost. Financial analysts in a company that manufactures DVD players arrived at the following daily cost equation for manufacturing x DVD players per day:

$$C(x) = x^2 + 2x + 2{,}000$$

The average cost per unit at a production level of x players per day is $\overline{C}(x) = C(x)/x$.

(A) Find the rational function $\overline{C}$.

(B) Sketch a graph of $\overline{C}$ for $5 \le x \le 150$.

(C) For what daily production level (to the nearest integer) is the average cost per unit at a minimum, and what is the minimum average cost per player (to the nearest cent)? [*Hint:* Refer to the sketch in part (B) and evaluate $\overline{C}(x)$ at appropriate integer values until a minimum value is found.]

(D) Graph the average cost function $\overline{C}$ on a graphing calculator and use an appropriate command to find the daily production level (to the nearest integer) at which the average cost per player is at a minimum. What is the minimum average cost to the nearest cent?

61. Minimum average cost. A consulting firm, using statistical methods, provided a veterinary clinic with the cost equation

$$C(x) = 0.00048(x - 500)^3 + 60{,}000$$

$$100 \le x \le 1{,}000$$

where $C(x)$ is the cost in dollars for handling x cases per month. The average cost per case is given by $\overline{C}(x) = C(x)/x$.

(A) Write the equation for the average cost function $\overline{C}$.

(B) Graph $\overline{C}$ on a graphing calculator.

(C) Use an appropriate command to find the monthly caseload for the minimum average cost per case. What is the minimum average cost per case?

62. Minimum average cost. The financial department of a hospital, using statistical methods, arrived at the cost equation

$$C(x) = 20x^3 - 360x^2 + 2{,}300x - 1{,}000$$

$$1 \le x \le 12$$

where $C(x)$ is the cost in thousands of dollars for handling x thousand cases per month. The average cost per case is given by $\overline{C}(x) = C(x)/x$.

(A) Write the equation for the average cost function $\overline{C}$.

(B) Graph $\overline{C}$ on a graphing calculator.

(C) Use an appropriate command to find the monthly caseload for the minimum average cost per case. What is the minimum average cost per case to the nearest dollar?

63. Diet. Table 3 shows the per capita consumption of ice cream and eggs in the United States for selected years since 1980.

(A) Let x represent the number of years since 1980 and find a cubic regression polynomial for the per capita consumption of ice cream.

(B) Use the polynomial model from part (A) to estimate (to the nearest tenth of a pound) the per capita consumption of ice cream in 2020.

64. Diet. Refer to Table 3.

(A) Let x represent the number of years since 1980 and find a cubic regression polynomial for the per capita consumption of eggs.

(B) Use the polynomial model from part (A) to estimate (to the nearest integer) the per capita consumption of eggs in 2020.

Table 3 Per Capita Consumption of Ice Cream and Eggs

Year	Ice Cream (pounds)	Eggs (number)
1980	17.5	266
1985	18.1	251
1990	15.8	231
1995	15.5	229
2000	16.5	247
2005	14.4	255

65. Physiology. In a study on the speed of muscle contraction in frogs under various loads, researchers W. O. Fems and J. Marsh found that the speed of contraction decreases with increasing loads. In particular, they found that the relationship between speed of contraction v (in centimeters per second) and load x (in grams) is given approximately by

$$v(x) = \frac{26 + 0.06x}{x} \qquad x \ge 5$$

(A) What does $v(x)$ approach as x increases?

(B) Sketch a graph of function v.

66. Learning theory. In 1917, L. L. Thurstone, a pioneer in quantitative learning theory, proposed the rational function

$$f(x) = \frac{a(x + c)}{(x + c) + b}$$

to model the number of successful acts per unit time that a person could accomplish after x practice sessions. Suppose that for a particular person enrolled in a typing class,

$$f(x) = \frac{55(x + 1)}{(x + 8)} \quad x \geq 0$$

where $f(x)$ is the number of words per minute the person is able to type after x weeks of lessons.

(A) What does $f(x)$ approach as x increases?

(B) Sketch a graph of function f, including any vertical or horizontal asymptotes.

67. Marriage. Table 4 shows the marriage and divorce rates per 1,000 population for selected years since 1960.

(A) Let x represent the number of years since 1960 and find a cubic regression polynomial for the marriage rate.

(B) Use the polynomial model from part (A) to estimate the marriage rate (to one decimal place) for 2020.

Table 4 Marriages and Divorces (per 1,000 Population)

Date	Marriages	Divorces
1960	8.5	2.2
1970	10.6	3.5
1980	10.6	5.2
1990	9.8	4.7
2000	8.5	4.1

68. Divorce. Refer to Table 4.

(A) Let x represent the number of years since 1950 and find a cubic regression polynomial for the divorce rate.

(B) Use the polynomial model from part (A) to estimate the divorce rate (to one decimal place) for 2020.

Answers to Matched Problems

1.

2. (A) Domain: all real numbers except -3 and 3

(B) x intercept: -1; y intercept: $-\dfrac{1}{3}$

(C) Vertical asymptotes: $x = -3$ and $x = 3$;

(D) Horizontal asymptote: $y = 0$

(E)

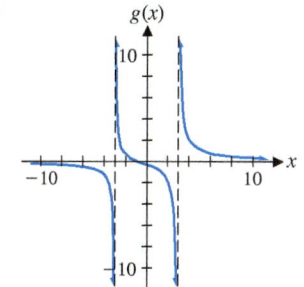

3. Vertical asymptote: $x = -5$

Horizontal asymptote: none

4. No vertical asymptotes for $t \leq 0$; $y = 25$ is a horizontal asymptote. $N(t)$ approaches 25 as t increases without bound. It appears that 25 components per day would be the upper limit that an employee would be expected to assemble.

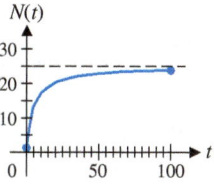

2-5 Exponential Functions

- Exponential Functions
- Base e Exponential Functions
- Growth and Decay Applications
- Compound Interest

This section introduces an important class of functions called *exponential functions*. These functions are used extensively in modeling and solving a wide variety of real-world problems, including growth of money at compound interest, growth of populations, radioactive decay and learning associated with the mastery of such devices as a new computer or an assembly process in a manufacturing plant.

Exponential Functions

We start by noting that

$$f(x) = 2^x \quad \text{and} \quad g(x) = x^2$$

are not the same function. Whether a variable appears as an exponent with a constant base or as a base with a constant exponent makes a big difference. The function g is a quadratic function, which we have already discussed. The function f is a new type of function called an *exponential function*. In general,

> **DEFINITION Exponential Function**
>
> The equation
>
> $$f(x) = b^x \quad b > 0, b \neq 1$$
>
> defines an **exponential function** for each different constant b, called the **base**. The **domain** of f is the set of all real numbers, and the **range** of f is the set of all positive real numbers.

We require the base b to be positive to avoid imaginary numbers such as $(-2)^{1/2} = \sqrt{-2} = i\sqrt{2}$. We exclude $b = 1$ as a base, since $f(x) = 1^x = 1$ is a constant function, which we have already considered.

When asked to hand-sketch graphs of equations such as $y = 2^x$ or $y = 2^{-x}$, many students do not hesitate. [*Note:* $2^{-x} = 1/2^x = (1/2)^x$.] They make tables by assigning integers to x, plot the resulting points, and then join these points with a smooth curve as in Figure 1. The only catch is that we have not defined 2^x for all real numbers. From Appendix A, Section A-7, we know what $2^5, 2^{-3}, 2^{2/3}, 2^{-3/5}, 2^{1.4}$, and $2^{-3.14}$ mean (that is, 2^p, where p is a rational number), but what does

$$2^{\sqrt{2}}$$

mean? The question is not easy to answer at this time. In fact, a precise definition of $2^{\sqrt{2}}$ must wait for more advanced courses, where it is shown that

$$2^x$$

names a positive real number for x any real number, and that the graph of $y = 2^x$ is as indicated in Figure 1.

It is useful to compare the graphs of $y = 2^x$ and $y = 2^{-x}$ by plotting both on the same set of coordinate axes, as shown in Figure 2A. The graph of

$$f(x) = b^x \quad b > 1 \text{ (Fig. 2B)}$$

looks very much like the graph of $y = 2^x$, and the graph of

$$f(x) = b^x \quad 0 < b < 1 \text{ (Fig. 2B)}$$

looks very much like the graph of $y = 2^{-x}$. Note that in both cases the x axis is a horizontal asymptote for the graphs.

The graphs in Figure 2 suggest the following general properties of exponential functions, which we state without proof:

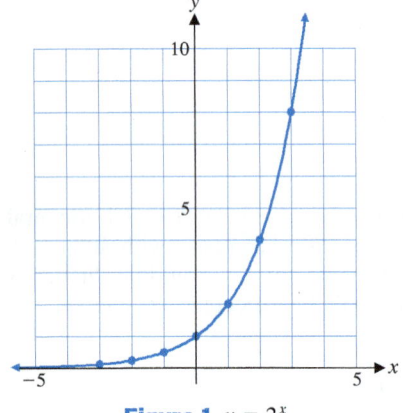

Figure 1 $y = 2^x$

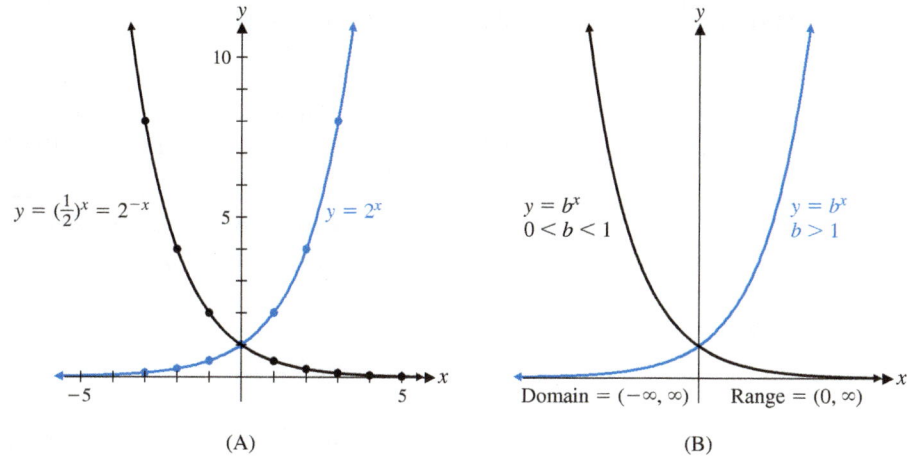

Figure 2 Exponential functions

THEOREM 1   Basic Properties of the Graph of $f(x) = b^x, b > 0, b \neq 1$

1. All graphs will pass through the point $(0, 1)$. $b^0 = 1$ for any permissible base b.

2. All graphs are continuous curves, with no holes or jumps.

3. The x axis is a horizontal asymptote.

4. If $b > 1$, then b^x increases as x increases.

5. If $0 < b < 1$, then b^x decreases as x increases.

CONCEPTUAL INSIGHT

Recall that the graph of a rational function has at most one horizontal asymptote and that it approaches the horizontal asymptote (if one exists) both as $x \to \infty$ *and* as $x \to -\infty$ (see Section 2-4). The graph of an exponential function, on the other hand, approaches its horizontal asymptote as $x \to \infty$ *or* as $x \to -\infty$, but not both. In particular, there is no rational function that has the same graph as an exponential function.

The use of a calculator with the key $\boxed{y^x}$, or its equivalent, makes the graphing of exponential functions almost routine. Example 1 illustrates the process.

EXAMPLE 1   **Graphing Exponential Functions** Sketch a graph of $y = \left(\frac{1}{2}\right)4^x, -2 \leq x \leq 2$.

SOLUTION   Use a calculator to create the table of values shown. Plot these points, and then join them with a smooth curve as in Figure 3.

x	y
-2	0.031
-1	0.125
0	0.50
1	2.00
2	8.00

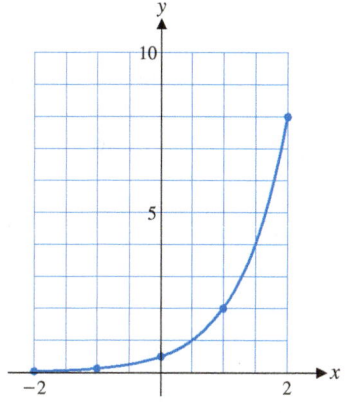

Figure 3 Graph of $y = \left(\frac{1}{2}\right)4^x$

Matched Problem 1   Sketch a graph of $y = \left(\frac{1}{2}\right)4^{-x}, -2 \leq x \leq 2$.

Exponential functions, whose domains include irrational numbers, obey the familiar laws of exponents discussed in Appendix A, Section A-6 for rational exponents. We summarize these exponent laws here and add two other important and useful properties.

THEOREM 2   **Properties of Exponential Functions**

For a and b positive, $a \neq 1, b \neq 1$, and x and y real,

1. Exponent laws:

$$a^x a^y = a^{x+y} \quad \frac{a^x}{a^y} = a^{x-y} \qquad \qquad \frac{4^{2y}}{4^{5y}} = 4^{2y-5y} = 4^{-3y}$$

$$(a^x)^y = a^{xy} \quad (ab)^x = a^x b^x \quad \left(\frac{a}{b}\right)^x = \frac{a^x}{b^x}$$

2. $a^x = a^y$ if and only if $x = y$

If $7^{5t+1} = 7^{3t-3}$, then
$5t + 1 = 3t - 3$, and $t = -2$.

3. For $x \neq 0$,

$a^x = b^x$ if and only if $a = b$

If $a^5 = 2^5$, then $a = 2$.

Base e Exponential Functions

Of all the possible bases b we can use for the exponential function $y = b^x$, which ones are the most useful? If you look at the keys on a calculator, you will probably see $\boxed{10^x}$ and $\boxed{e^x}$. It is clear why base 10 would be important, because our number system is a base 10 system. But what is e, and why is it included as a base? It turns out that base e is used more frequently than all other bases combined. The reason for this is that certain formulas and the results of certain processes found in calculus and more advanced mathematics take on their simplest form if this base is used. This is why you will see e used extensively in expressions and formulas that model real-world phenomena. In fact, its use is so prevalent that you will often hear people refer to $y = e^x$ as *the* exponential function.

The base e is an irrational number, and like π, it cannot be represented exactly by any finite decimal or fraction. However, e can be approximated as closely as we like by evaluating the expression

$$\left(1 + \frac{1}{x}\right)^x \tag{1}$$

for sufficiently large values of x. What happens to the value of expression (1) as x increases without bound? Think about this for a moment before proceeding. Maybe you guessed that the value approaches 1, because

$$1 + \frac{1}{x}$$

approaches 1, and 1 raised to any power is 1. Let us see if this reasoning is correct by actually calculating the value of the expression for larger and larger values of x. Table 1 summarizes the results.

Table 1

x	$\left(1 + \dfrac{1}{x}\right)^x$
1	2
10	2.593 74 . . .
100	2.704 81 . . .
1,000	2.716 92 . . .
10,000	2.718 14 . . .
100,000	2.718 27 . . .
1,000,000	2.718 28 . . .

Interestingly, the value of expression (1) is never close to 1 but seems to be approaching a number close to 2.7183. In fact, as x increases without bound, the value of expression (1) approaches an irrational number that we call e. The irrational number e to 12 decimal places is

$$e = \textbf{2.718 281 828 459}$$

Compare this value of e with the value of e^1 from a calculator.

> **DEFINITION Exponential Function with Base e**
>
> Exponential functions with base e and base $1/e$, respectively, are defined by
>
> $$y = e^x \quad \text{and} \quad y = e^{-x}$$
>
> Domain: $(-\infty, \infty)$
>
> Range: $(0, \infty)$
>
>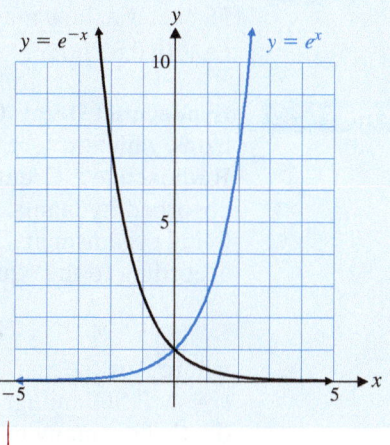

EXPLORE & DISCUSS 1

Graph the functions $f(x) = e^x, g(x) = 2^x$, and $h(x) = 3^x$ on the same set of coordinate axes. At which values of x do the graphs intersect? For positive values of x, which of the three graphs lies above the other two? Below the other two? How does your answer change for negative values of x?

Growth and Decay Applications

Functions of the form $y = ce^{kt}$, where c and k are constants and the independent variable t represents time, are often used to model population growth and radioactive decay. Note that if $t = 0$, then $y = c$. So the constant c represents the initial population (or initial amount). The constant k is called the **relative growth rate** and has the following interpretation: Suppose that $y = ce^{kt}$ models the population growth of a country, where y is the number of persons and t is time in years. If the relative growth rate is $k = 0.02$, then at any time t, the population is growing at a rate of $0.02y$ persons (that is, 2% of the population) per year.

We say that **population is growing continuously at relative growth rate k** to mean that the population y is given by the model $y = ce^{kt}$.

EXAMPLE 2 **Exponential Growth** Cholera, an intestinal disease, is caused by a cholera bacterium that multiplies exponentially. The number of bacteria grows continuously at relative growth rate 1.386, that is,

$$N = N_0 e^{1.386t}$$

where N is the number of bacteria present after t hours and N_0 is the number of bacteria present at the start ($t = 0$). If we start with 25 bacteria, how many bacteria (to the nearest unit) will be present:

(A) In 0.6 hour?   (B) In 3.5 hours?

SOLUTION   Substituting $N_0 = 25$ into the preceding equation, we obtain

$$N = 25e^{1.386t}$$ The graph is shown in Figure 4.

(A) Solve for N when $t = 0.6$:

$$N = 25e^{1.386(0.6)}$$ Use a calculator.

$$= 57 \text{ bacteria}$$

(B) Solve for N when $t = 3.5$:

$$N = 25e^{1.386(3.5)}$$ Use a calculator.

$$= 3{,}197 \text{ bacteria}$$

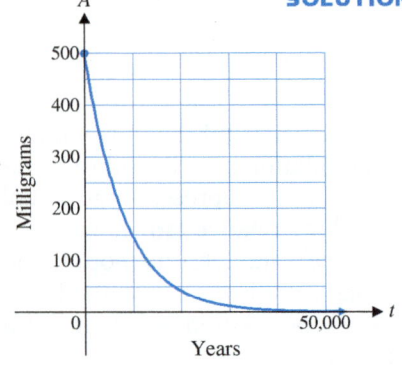

Figure 4

Matched Problem 2   Refer to the exponential growth model for cholera in Example 2. If we start with 55 bacteria, how many bacteria (to the nearest unit) will be present
(A) In 0.85 hour?   (B) In 7.25 hours?

EXAMPLE 3   **Exponential Decay** Cosmic-ray bombardment of the atmosphere produces neutrons, which in turn react with nitrogen to produce radioactive carbon-14 (^{14}C). Radioactive ^{14}C enters all living tissues through carbon dioxide, which is first absorbed by plants. As long as a plant or animal is alive, ^{14}C is maintained in the living organism at a constant level. Once the organism dies, however, ^{14}C decays according to the equation

$$A = A_0 e^{-0.000124t}$$

where A is the amount present after t years and A_0 is the amount present at time $t = 0$. If 500 milligrams of ^{14}C is present in a sample from a skull at the time of death, how many milligrams will be present in the sample in
(A) 15,000 years?   (B) 45,000 years?
Compute answers to two decimal places.

SOLUTION   Substituting $A_0 = 500$ in the decay equation, we have

$$A = 500e^{-0.000124t}$$ See the graph in Figure 5.

(A) Solve for A when $t = 15{,}000$:

$$A = 500e^{-0.000124(15{,}000)}$$ Use a calculator.

$$= 77.84 \text{ milligrams}$$

(B) Solve for A when $t = 45{,}000$:

$$A = 500e^{-0.000124(45{,}000)}$$ Use a calculator.

$$= 1.89 \text{ milligrams}$$

Figure 5

Matched Problem 3   Refer to the exponential decay model in Example 3. How many milligrams of ^{14}C would have to be present at the beginning in order to have 25 milligrams present after 18,000 years? Compute the answer to the nearest milligram.

If you buy a new car, it is likely to depreciate in value by several thousand dollars during the first year you own it. You would expect the value of the car to decrease in each subsequent year, but not by as much as in the previous year. If you drive the car long enough, its resale value will get close to zero. An exponential decay function will often be a good model of depreciation; a linear or quadratic function would not be suitable (why?). We can use **exponential regression** on a graphing calculator to find the function of the form $y = ab^x$ that best fits a data set.

EXAMPLE 4 **Depreciation** Table 2 gives the market value of a hybrid sedan (in dollars) x years after its purchase. Find an exponential regression model of the form $y = ab^x$ for this data set. Estimate the purchase price of the hybrid. Estimate the value of the hybrid 10 years after its purchase. Round answers to the nearest dollar.

Table 2

x	Value ($)
1	12,575
2	9,455
3	8,115
4	6,845
5	5,225
6	4,485

SOLUTION Enter the data into a graphing calculator (Fig. 6A) and find the exponential regression equation (Fig. 6B). The estimated purchase price is $y_1(0) = \$14,910$. The data set and the regression equation are graphed in Figure 6C. Using TRACE, we see that the estimated value after 10 years is $1,959.

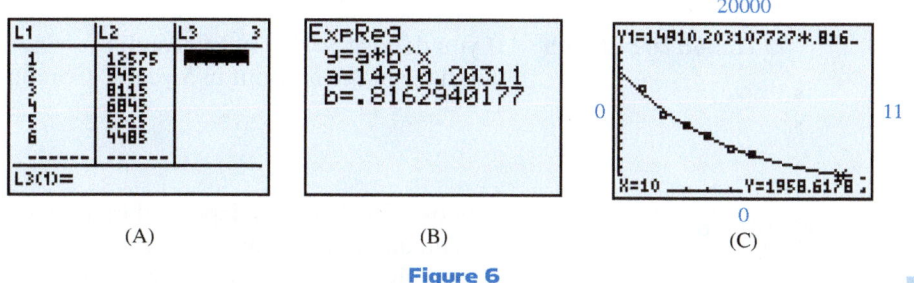

(A)   (B)   (C)

Figure 6

Matched Problem 4 Table 3 gives the market value of a midsize sedan (in dollars) x years after its purchase. Find an exponential regression model of the form $y = ab^x$ for this data set. Estimate the purchase price of the sedan. Estimate the value of the sedan 10 years after its purchase. Round answers to the nearest dollar.

Table 3

x	Value ($)
1	23,125
2	19,050
3	15,625
4	11,875
5	9,450
6	7,125

Compound Interest

The fee paid to use another's money is called **interest**. It is usually computed as a percent (called **interest rate**) of the principal over a given period of time. If, at the end of a payment period, the interest due is reinvested at the same rate, then the interest earned as well as the principal will earn interest during the next payment period. Interest paid on interest reinvested is called **compound interest** and may be calculated using the following compound interest formula:

If a **principal** P **(present value)** is invested at an annual **rate** r (expressed as a decimal) compounded m times a year, then the **amount** A **(future value)** in the account at the end of t years is given by

$$A = P\left(1 + \frac{r}{m}\right)^{mt} \qquad \textit{Compound interest formula}$$

For given r and m, the amount A is equal to the principal P multiplied by the exponential function b^t, where $b = (1 + r/m)^m$.

EXAMPLE 5   **Compound Growth** If $1,000 is invested in an account paying 10% compounded monthly, how much will be in the account at the end of 10 years? Compute the answer to the nearest cent.

SOLUTION   We use the compound interest formula as follows:

$$A = P\left(1 + \frac{r}{m}\right)^{mt}$$

$$= 1,000\left(1 + \frac{0.10}{12}\right)^{(12)(10)} \quad \textit{Use a calculator.}$$

$$= \$2,707.04$$

The graph of

$$A = 1,000\left(1 + \frac{0.10}{12}\right)^{12t}$$

for $0 \le t \le 20$ is shown in Figure 7.

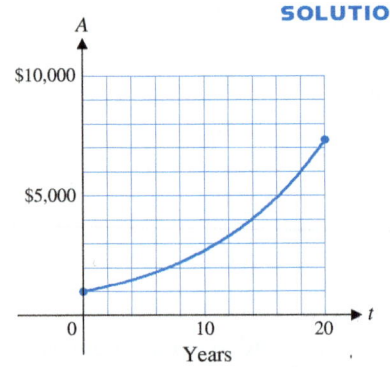

Figure 7

Matched Problem 5   If you deposit $5,000 in an account paying 9% compounded daily, how much will you have in the account in 5 years? Compute the answer to the nearest cent.

EXPLORE & DISCUSS 2   Suppose that $1,000 is deposited in a savings account at an annual rate of 5%. Guess the amount in the account at the end of 1 year if interest is compounded (1) quarterly, (2) monthly, (3) daily, (4) hourly. Use the compound interest formula to compute the amounts at the end of 1 year to the nearest cent. Discuss the accuracy of your initial guesses.

Explore & Discuss 2 suggests that if $1,000 were deposited in a savings account at an annual interest rate of 5%, then the amount at the end of 1 year would be less than $1,051.28, even if interest were compounded every minute or every second. The limiting value, approximately $1,051.271096, is said to be the amount in the account if interest were compounded continuously.

If a principal, P, is invested at an annual rate, r, and compounded continuously, then the amount in the account at the end of t years is given by

$$A = Pe^{rt} \quad \textit{Continuous compound interest formula}$$

where the constant $e \approx 2.71828$ is the base of the exponential function.

EXAMPLE 6   **Continuous Compound Interest** If $1,000 is invested in an account paying 10% compounded continuously, how much will be in the account at the end of 10 years? Compute the answer to the nearest cent.

SOLUTION   We use the continuous compound interest formula:

$$A = Pe^{rt} = 1000e^{0.10(10)} = 1000e = \$2,718.28$$

Compare with the answer to Example 5.

Matched Problem 6   If you deposit $5,000 in an account paying 9% compounded continuously, how much will you have in the account in 5 years? Compute the answer to the nearest cent.

The formulas for compound interest and continuous compound interest are summarized below for convenient reference.

SUMMARY

Compound Interest: $A = P\left(1 + \dfrac{r}{m}\right)^{mt}$

Continuous Compound Interest: $A = Pe^{rt}$

where A = amount (future value) at the end of t years

P = principal (present value)

r = annual rate (expressed as a decimal)

m = number of compounding periods per year

t = time in years

Exercises 2-5

A

1. Match each equation with the graph of f, g, h, or k in the figure.

(A) $y = 2^x$ (B) $y = (0.2)^x$

(C) $y = 4^x$ (D) $y = \left(\frac{1}{3}\right)^x$

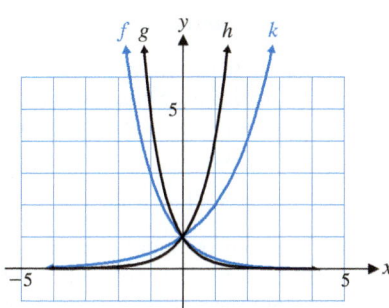

2. Match each equation with the graph of f, g, h, or k in the figure.

(A) $y = \left(\frac{1}{4}\right)^x$ (B) $y = (0.5)^x$

(C) $y = 5^x$ (D) $y = 3^x$

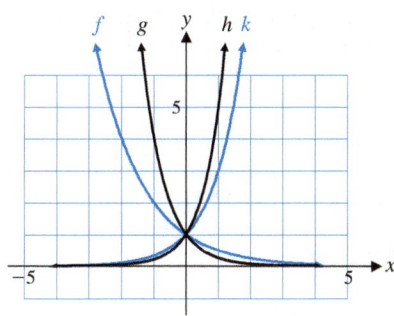

Graph each function in Problems 3–10 over the indicated interval.

3. $y = 5^x; [-2, 2]$ **4.** $y = 3^x; [-3, 3]$

5. $y = \left(\frac{1}{5}\right)^x = 5^{-x}; [-2, 2]$ **6.** $y = \left(\frac{1}{3}\right)^x = 3^{-x}; [-3, 3]$

7. $f(x) = -5^x; [-2, 2]$ **8.** $g(x) = -3^{-x}; [-3, 3]$

9. $y = -e^{-x}; [-3, 3]$ **10.** $y = -e^x; [-3, 3]$

B

In Problems 11–18, describe verbally the transformations that can be used to obtain the graph of g from the graph of f (see Section 2-2).

11. $g(x) = -2^x; f(x) = 2^x$ **12.** $g(x) = 2^{x-2}; f(x) = 2^x$

13. $g(x) = 3^{x+1}; f(x) = 3^x$ **14.** $g(x) = -3^x; f(x) = 3^x$

15. $g(x) = e^x + 1; f(x) = e^x$ **16.** $g(x) = e^x - 2; f(x) = e^x$

17. $g(x) = 2e^{-(x+2)}; f(x) = e^{-x}$

18. $g(x) = 0.5e^{-(x-1)}; f(x) = e^{-x}$

19. Use the graph of f shown in the figure to sketch the graph of each of the following.

(A) $y = f(x) - 1$ (B) $y = f(x + 2)$

(C) $y = 3f(x) - 2$ (D) $y = 2 - f(x - 3)$

20. Use the graph of f shown in the figure to sketch the graph of each of the following.

(A) $y = f(x) + 2$ (B) $y = f(x - 3)$

(C) $y = 2f(x) - 4$ (D) $y = 4 - f(x + 2)$

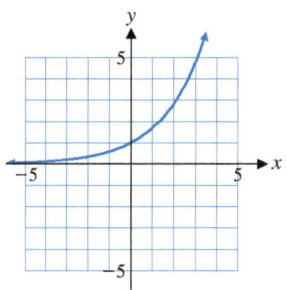

Figure for 19 and 20

In Problems 21–26, graph each function over the indicated interval.

21. $f(t) = 2^{t/10}; [-30, 30]$ **22.** $G(t) = 3^{t/100}; [-200, 200]$

23. $y = -3 + e^{1+x}; [-4, 2]$ **24.** $y = 2 + e^{x-2}; [-1, 5]$

25. $y = e^{|x|}; [-3, 3]$ **26.** $y = e^{-|x|}; [-3, 3]$

27. Find all real numbers a such that $a^2 = a^{-2}$. Explain why this does not violate the second exponential function property in Theorem 2 on page 98.

28. Find real numbers a and b such that $a \neq b$ but $a^4 = b^4$. Explain why this does not violate the third exponential function property in Theorem 2 on page 98.

Solve each equation in Problems 29–34 for x.

29. $10^{2-3x} = 10^{5x-6}$ **30.** $5^{3x} = 5^{4x-2}$

31. $4^{5x-x^2} = 4^{-6}$ **32.** $7^{x^2} = 7^{2x+3}$

33. $5^3 = (x + 2)^3$ **34.** $(1 - x)^5 = (2x - 1)^5$

C

Solve each equation in Problems 35–38 for x. (Remember: $e^x \neq 0$ and $e^{-x} \neq 0$.)

35. $(x - 3)e^x = 0$ **36.** $2xe^{-x} = 0$

37. $3xe^{-x} + x^2e^{-x} = 0$ **38.** $x^2e^x - 5xe^x = 0$

Graph each function in Problems 39–42 over the indicated interval.

39. $h(x) = x(2^x)$; $[-5, 0]$ **40.** $m(x) = x(3^{-x})$; $[0, 3]$

41. $N = \dfrac{100}{1 + e^{-t}}$; $[0, 5]$ **42.** $N = \dfrac{200}{1 + 3e^{-t}}$; $[0, 5]$

Applications

In all problems involving days, a 365-day year is assumed.

43. Continuous compound interest. Find the value of an investment of $10,000 in 12 years if it earns an annual rate of 3.95% compounded continuously.

44. Continuous compound interest. Find the value of an investment of $24,000 in 7 years if it earns an annual rate of 4.35% compounded continuously.

45. Compound growth. Suppose that $2,500 is invested at 7% compounded quarterly. How much money will be in the account in

(A) $\frac{3}{4}$ year? (B) 15 years?

Compute answers to the nearest cent.

46. Compound growth. Suppose that $4,000 is invested at 6% compounded weekly. How much money will be in the account in

(A) $\frac{1}{2}$ year? (B) 10 years?

Compute answers to the nearest cent.

47. Finance. A person wishes to have $15,000 cash for a new car 5 years from now. How much should be placed in an account now, if the account pays 6.75% compounded weekly? Compute the answer to the nearest dollar.

48. Finance. A couple just had a baby. How much should they invest now at 5.5% compounded daily in order to have $40,000 for the child's education 17 years from now? Compute the answer to the nearest dollar.

49. Money growth. BanxQuote operates a network of Web sites providing real-time market data from leading financial providers. The following rates for 12-month certificates of deposit were taken from the Web sites:

(A) Stonebridge Bank, 5.40% compounded monthly

(B) DeepGreen Bank, 4.95% compounded daily

(C) Provident Bank, 5.15% compounded quarterly

Compute the value of $10,000 invested in each account at the end of 1 year.

50. Money growth. Refer to Problem 49. The following rates for 60-month certificates of deposit were also taken from BanxQuote Web sites:

(A) Oriental Bank & Trust, 5.50% compounded quarterly

(B) BMW Bank of North America, 5.12% compounded monthly

(C) BankFirst Corporation, 4.86% compounded daily

Compute the value of $10,000 invested in each account at the end of 5 years.

51. Advertising. A company is trying to introduce a new product to as many people as possible through television advertising in a large metropolitan area with 2 million possible viewers. A model for the number of people N (in millions) who are aware of the product after t days of advertising was found to be

$$N = 2(1 - e^{-0.037t})$$

Graph this function for $0 \leq t \leq 50$. What value does N approach as t increases without bound?

52. Learning curve. People assigned to assemble circuit boards for a computer manufacturing company undergo on-the-job training. From past experience, the learning curve for the average employee is given by

$$N = 40(1 - e^{-0.12t})$$

where N is the number of boards assembled per day after t days of training. Graph this function for $0 \leq t \leq 30$. What is the maximum number of boards an average employee can be expected to produce in 1 day?

53. Sports salaries. Table 4 shows the average salaries for players in Major League Baseball (MLB) and the National Basketball Association (NBA) in selected years since 1990.

(A) Let x represent the number of years since 1990 and find an exponential regression model ($y = ab^x$) for the average salary in MLB. Use the model to estimate the average salary (to the nearest thousand dollars) in 2020.

 (B) The average salary in MLB in 2000 was 1.984 million. How does this compare with the value given by the model of part (A)? How would the inclusion of the year 2000 data affect the estimated average salary in 2020? Explain.

Table 4	Average Salary (thousand $)	
Year	**MLB**	**NBA**
1990	589	750
1993	1,062	1,300
1996	1,101	2,000
1999	1,724	2,400
2002	2,300	4,500
2005	2,633	5,000
2008	3,155	5,585

54. Sports salaries. Refer to Table 4.

(A) Let x represent the number of years since 1990 and find an exponential regression model ($y = ab^x$) for the average salary in the NBA. Use the model to estimate the average salary (to the nearest thousand dollars) in 2020.

(B) The average salary in the NBA in 1997 was $2.2 million. How does this compare with the value given by the model of part (A)? How would the inclusion of the year 1997 data affect the estimated average salary in 2020? Explain.

55. Marine biology. Marine life depends on the microscopic plant life that exists in the photic zone, a zone that goes to a depth where only 1% of surface light remains. In some waters with a great deal of sediment, the photic zone may go down only 15 to 20 feet. In some murky harbors, the intensity of light d feet below the surface is given approximately by

$$I = I_0 e^{-0.23d}$$

What percentage of the surface light will reach a depth of

(A) 10 feet?   (B) 20 feet?

56. Marine biology. Refer to Problem 55. Light intensity I relative to depth d (in feet) for one of the clearest bodies of water in the world, the Sargasso Sea, can be approximated by

$$I = I_0 e^{-0.00942d}$$

where I_0 is the intensity of light at the surface. What percentage of the surface light will reach a depth of

(A) 50 feet?   (B) 100 feet?

57. World population growth. From the dawn of humanity to 1830, world population grew to one billion people. In 100 more years (by 1930) it grew to two billion, and 3 billion more were added in only 60 years (by 1990). In 2009, the estimated world population was 6.8 billion with a relative growth rate of 1.14%.

(A) Write an equation that models the world population growth, letting 2009 be year 0.

(B) Based on the model, what is the expected world population (to the nearest hundred million) in 2020? In 2030?

58. Population growth in Ethiopia. In 2008, the estimated population in Ethiopia was 83 million people with a relative growth rate of 3.2%.

(A) Write an equation that models the population growth in Ethiopia, letting 2008 be year 0.

(B) Based on the model, what is the expected population in Ethiopia (to the nearest million) in 2020? In 2030?

59. Internet growth. The number of Internet hosts grew very rapidly from 1994 to 2009 (Table 5).

(A) Let x represent the number of years since 1994. Find an exponential regression model ($y = ab^x$) for this data set and estimate the number of hosts in 2018 (to the nearest million).

(B) Discuss the implications of this model if the number of Internet hosts continues to grow at this rate.

Table 5	Internet Hosts (Millions)
Year	**Hosts**
1994	2.4
1997	16.1
2000	72.4
2003	171.6
2006	394.0
2009	625.2

Source: Internet Software Consortium

60. Life expectancy. Table 6 shows the life expectancy (in years) at birth for residents of the United States from 1970 to 2005. Let x represent years since 1970. Find an exponential regression model for this data and use it to estimate the life expectancy for a person born in 2015.

Table 6	
Year of Birth	**Life Expectancy**
1970	70.8
1975	72.6
1980	73.7
1985	74.7
1990	75.4
1995	75.9
2000	76.9
2005	77.7

Answers to Matched Problems

1.

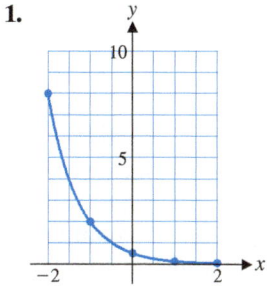

2. (A) 179 bacteria   (B) 1,271,659 bacteria

3. 233 mg

4. Purchase price: $30,363; value after 10 yr: $2,864

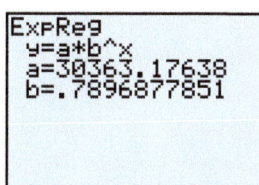

5. $7,841.13   **6.** $7,841.56

2-6   Logarithmic Functions

Find the exponential function keys $\boxed{10^x}$ and $\boxed{e^x}$ on your calculator. Close to these keys you will find the $\boxed{\text{LOG}}$ and $\boxed{\text{LN}}$ keys. The latter two keys represent *logarithmic functions,* and each is closely related to its nearby exponential function. In fact, the exponential function and the corresponding logarithmic function are said to be *inverses* of each other. In this section we will develop the concept of inverse functions and use it to define a logarithmic function as the inverse of an exponential function. We will then investigate basic properties of logarithmic functions, use a calculator to evaluate them for particular values of x, and apply them to real-world problems.

Logarithmic functions are used in modeling and solving many types of problems. For example, the decibel scale is a logarithmic scale used to measure sound intensity, and the Richter scale is a logarithmic scale used to measure the strength of the force of an earthquake. An important business application has to do with finding the time it takes money to double if it is invested at a certain rate compounded a given number of times a year or compounded continuously. This requires the solution of an exponential equation, and logarithms play a central role in the process.

Inverse Functions

Look at the graphs of $f(x) = \dfrac{x}{2}$ and $g(x) = \dfrac{|x|}{2}$ in Figure 1:

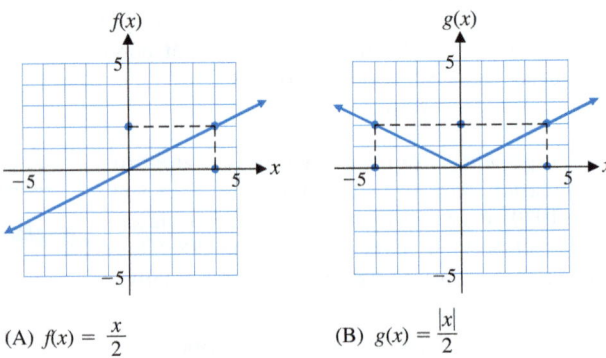

(A) $f(x) = \dfrac{x}{2}$ (B) $g(x) = \dfrac{|x|}{2}$

Figure 1

Because both f and g are functions, each domain value corresponds to exactly one range value. For which function does each range value correspond to exactly one domain value? This is the case only for function f. Note that for function f, the range value 2 corresponds to the domain value 4. For function g the range value 2 corresponds to both -4 and 4. Function f is said to be *one-to-one.*

> **DEFINITION One-to-One Functions**
>
> A function f is said to be **one-to-one** if each range value corresponds to exactly one domain value.

It can be shown that any continuous function that is either increasing* or decreasing for all domain values is one-to-one. If a continuous function increases for some domain values and decreases for others, then it cannot be one-to-one. Figure 1 shows an example of each case.

---

*Formally, we say that the function f is **increasing** on an interval (a, b) if $f(x_2) > f(x_1)$ whenever $a < x_1 < x_2 < b$; and f is **decreasing** on (a, b) if $f(x_2) < f(x_1)$ whenever $a < x_1 < x_2 < b$.

EXPLORE & DISCUSS 1

Graph $f(x) = 2^x$ and $g(x) = x^2$. For a range value of 4, what are the corresponding domain values for each function? Which of the two functions is one-to-one? Explain why.

Starting with a one-to-one function f, we can obtain a new function called the *inverse* of f.

DEFINITION   Inverse of a Function

If f is a one-to-one function, then the **inverse** of f is the function formed by interchanging the independent and dependent variables for f. Thus, if (a, b) is a point on the graph of f, then (b, a) is a point on the graph of the inverse of f.

Note: If f is not one-to-one, then f **does not have an inverse**.

In this course, we are interested in the inverses of exponential functions, called *logarithmic functions*.

Logarithmic Functions

If we start with the exponential function f defined by

$$y = 2^x \tag{1}$$

and interchange the variables, we obtain the inverse of f:

$$x = 2^y \tag{2}$$

We call the inverse the **logarithmic function with base 2**, and write

$$y = \log_2 x \quad \text{if and only if} \quad x = 2^y$$

We can graph $y = \log_2 x$ by graphing $x = 2^y$ since they are equivalent. Any ordered pair of numbers on the graph of the exponential function will be on the graph of the logarithmic function if we interchange the order of the components. For example, $(3, 8)$ satisfies equation (1) and $(8, 3)$ satisfies equation (2). The graphs of $y = 2^x$ and $y = \log_2 x$ are shown in Figure 2. Note that if we fold the paper along the dashed line $y = x$ in Figure 2, the two graphs match exactly. The line $y = x$ is a line of symmetry for the two graphs.

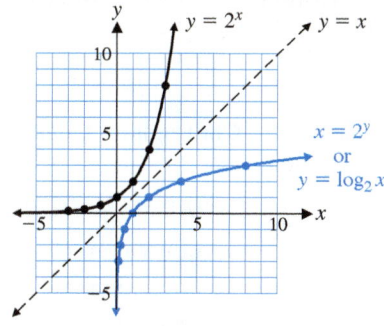

Exponential Function		Logarithmic Function	
x	$y = 2^x$	$x = 2^y$	y
-3	$\frac{1}{8}$	$\frac{1}{8}$	-3
-2	$\frac{1}{4}$	$\frac{1}{4}$	-2
-1	$\frac{1}{2}$	$\frac{1}{2}$	-1
0	1	1	0
1	2	2	1
2	4	4	2
3	8	8	3

Figure 2

$$\begin{bmatrix} \text{Ordered} \\ \text{pairs} \\ \text{reversed} \end{bmatrix}$$

In general, since the graphs of all exponential functions of the form $f(x) = b^x, b \neq 1, b > 0$, are either increasing or decreasing, exponential functions have inverses.

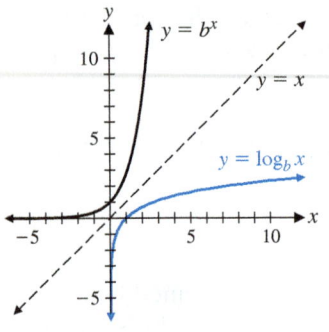

Figure 3

DEFINITION **Logarithmic Functions**

The inverse of an exponential function is called a **logarithmic function**. For $b > 0$ and $b \neq 1$,

Logarithmic form		Exponential form
$y = \log_b x$	is equivalent to	$x = b^y$

The **log to the base b of x** is the exponent to which b must be raised to obtain x. [*Remember:* A logarithm is an exponent.] The **domain** of the logarithmic function is the set of all positive real numbers, which is also the range of the corresponding exponential function; and the **range** of the logarithmic function is the set of all real numbers, which is also the domain of the corresponding exponential function. Typical graphs of an exponential function and its inverse, a logarithmic function, are shown in Figure 3.

CONCEPTUAL INSIGHT

Because the domain of a logarithmic function consists of the positive real numbers, the entire graph of a logarithmic function lies to the right of the y axis. In contrast, the graphs of polynomial and exponential functions intersect every vertical line, and the graphs of rational functions intersect all but a finite number of vertical lines.

The following examples involve converting logarithmic forms to equivalent exponential forms, and vice versa.

EXAMPLE 1   **Logarithmic–Exponential Conversions**   Change each logarithmic form to an equivalent exponential form:

(A) $\log_5 25 = 2$ (B) $\log_9 3 = \frac{1}{2}$ (C) $\log_2\left(\frac{1}{4}\right) = -2$

SOLUTION   (A) $\log_5 25 = 2$ is equivalent to $25 = 5^2$
(B) $\log_9 3 = \frac{1}{2}$ is equivalent to $3 = 9^{1/2}$
(C) $\log_2\left(\frac{1}{4}\right) = -2$ is equivalent to $\frac{1}{4} = 2^{-2}$

Matched Problem 1   Change each logarithmic form to an equivalent exponential form:

(A) $\log_3 9 = 2$ (B) $\log_4 2 = \frac{1}{2}$ (C) $\log_3\left(\frac{1}{9}\right) = -2$

EXAMPLE 2   **Exponential–Logarithmic Conversions**   Change each exponential form to an equivalent logarithmic form:

(A) $64 = 4^3$ (B) $6 = \sqrt{36}$ (C) $\frac{1}{8} = 2^{-3}$

SOLUTION   (A) $64 = 4^3$ is equivalent to $\log_4 64 = 3$
(B) $6 = \sqrt{36}$ is equivalent to $\log_{36} 6 = \frac{1}{2}$
(C) $\frac{1}{8} = 2^{-3}$ is equivalent to $\log_2\left(\frac{1}{8}\right) = -3$

Matched Problem 2   Change each exponential form to an equivalent logarithmic form:

(A) $49 = 7^2$ (B) $3 = \sqrt{9}$ (C) $\frac{1}{3} = 3^{-1}$

To gain a deeper understanding of logarithmic functions and their relationship to exponential functions, we consider a few problems where we want to find x, b, or y in $y = \log_b x$, given the other two values. All values are chosen so that the problems can be solved exactly without a calculator.

EXAMPLE 3 Solutions of the Equation $y = \log_b x$ Find $y, b,$ or x, as indicated.
(A) Find y: $y = \log_4 16$ (B) Find x: $\log_2 x = -3$
(C) Find b: $\log_b 100 = 2$

SOLUTION (A) $y = \log_4 16$ is equivalent to $16 = 4^y$. So,
$$y = 2$$

(B) $\log_2 x = -3$ is equivalent to $x = 2^{-3}$. So,
$$x = \frac{1}{2^3} = \frac{1}{8}$$

(C) $\log_b 100 = 2$ is equivalent to $100 = b^2$. So,
$$b = 10 \quad \textit{Recall that b cannot be negative.}$$

Matched Problem 3 Find $y, b,$ or x, as indicated.
(A) Find y: $y = \log_9 27$ (B) Find x: $\log_3 x = -1$
(C) Find b: $\log_b 1{,}000 = 3$

Properties of Logarithmic Functions

Logarithmic functions have many powerful and useful properties. We list eight basic properties in Theorem 1.

THEOREM 1 Properties of Logarithmic Functions
If $b, M,$ and N are positive real numbers, $b \neq 1$, and p and x are real numbers, then

1. $\log_b 1 = 0$
2. $\log_b b = 1$
3. $\log_b b^x = x$
4. $b^{\log_b x} = x, \quad x > 0$
5. $\log_b MN = \log_b M + \log_b N$
6. $\log_b \frac{M}{N} = \log_b M - \log_b N$
7. $\log_b M^p = p \log_b M$
8. $\log_b M = \log_b N$ if and only if $M = N$

EXAMPLE 4 Using Logarithmic Properties

(A) $\log_b \dfrac{wx}{yz}$ $= \log_b wx - \log_b yz$
$= \log_b w + \log_b x - (\log_b y + \log_b z)$
$= \log_b w + \log_b x - \log_b y - \log_b z$

(B) $\log_b(wx)^{3/5}$ $= \frac{3}{5}\log_b wx = \frac{3}{5}(\log_b w + \log_b x)$

(C) $e^{x \log_e b} = e^{\log_e b^x} = b^x$

(D) $\dfrac{\log_e x}{\log_e b} = \dfrac{\log_e(b^{\log_b x})}{\log_e b} = \dfrac{(\log_b x)(\log_e b)}{\log_e b} = \log_b x$

Matched Problem 4 Write in simpler forms, as in Example 4.

(A) $\log_b \dfrac{R}{ST}$ (B) $\log_b\left(\dfrac{R}{S}\right)^{2/3}$ (C) $2^{u \log_2 b}$ (D) $\dfrac{\log_2 x}{\log_2 b}$

The following examples and problems will give you additional practice in using basic logarithmic properties.

EXAMPLE 5   **Solving Logarithmic Equations**   Find x so that

$$\tfrac{3}{2}\log_b 4 - \tfrac{2}{3}\log_b 8 + \log_b 2 = \log_b x$$

SOLUTION

$$\tfrac{3}{2}\log_b 4 - \tfrac{2}{3}\log_b 8 + \log_b 2 = \log_b x$$

$$\log_b 4^{3/2} - \log_b 8^{2/3} + \log_b 2 = \log_b x \qquad \textcolor{blue}{\text{Property 7}}$$

$$\log_b 8 - \log_b 4 + \log_b 2 = \log_b x$$

$$\log_b \frac{8 \cdot 2}{4} = \log_b x \qquad \textcolor{blue}{\text{Properties 5 and 6}}$$

$$\log_b 4 = \log_b x$$

$$x = 4 \qquad \textcolor{blue}{\text{Property 8}}$$

Matched Problem 5   Find x so that $3 \log_b 2 + \tfrac{1}{2}\log_b 25 - \log_b 20 = \log_b x$.

EXAMPLE 6   **Solving Logarithmic Equations**   Solve: $\log_{10} x + \log_{10}(x + 1) = \log_{10} 6$.

SOLUTION

$$\log_{10} x + \log_{10}(x + 1) = \log_{10} 6$$

$$\log_{10}[x(x + 1)] = \log_{10} 6 \qquad \textcolor{blue}{\text{Property 5}}$$

$$x(x + 1) = 6 \qquad \textcolor{blue}{\text{Property 8}}$$

$$x^2 + x - 6 = 0 \qquad \textcolor{blue}{\text{Solve by factoring.}}$$

$$(x + 3)(x - 2) = 0$$

$$x = -3, 2$$

We must exclude $x = -3$, since the domain of the function $\log_{10}(x + 1)$ is $x > -1$ or $(-1, \infty)$; so $x = 2$ is the only solution.

Matched Problem 6   Solve: $\log_3 x + \log_3(x - 3) = \log_3 10$.

Calculator Evaluation of Logarithms

Of all possible logarithmic bases, e and 10 are used almost exclusively. Before we can use logarithms in certain practical problems, we need to be able to approximate the logarithm of any positive number either to base 10 or to base e. And conversely, if we are given the logarithm of a number to base 10 or base e, we need to be able to approximate the number. Historically, tables were used for this purpose, but now calculators make computations faster and far more accurate.

Common logarithms are logarithms with base 10. **Natural logarithms** are logarithms with base e. Most calculators have a key labeled "log" (or "LOG") and a key labeled "ln" (or "LN"). The former represents a common (base 10) logarithm and the latter a natural (base e) logarithm. In fact, "log" and "ln" are both used extensively in mathematical literature, and whenever you see either used in this book without a base indicated, they will be interpreted as follows:

Common logarithm: $\log x$ means $\log_{10} x$

Natural logarithm: $\ln x$ means $\log_e x$

Finding the common or natural logarithm using a calculator is very easy. On some calculators, you simply enter a number from the domain of the function and press $\boxed{\text{LOG}}$ or $\boxed{\text{LN}}$. On other calculators, you press either $\boxed{\text{LOG}}$ or $\boxed{\text{LN}}$, enter a number from the domain, and then press $\boxed{\text{ENTER}}$. Check the user's manual for your calculator.

EXAMPLE 7   **Calculator Evaluation of Logarithms**   Use a calculator to evaluate each to six decimal places:

(A) $\log 3{,}184$ (B) $\ln 0.000\,349$ (C) $\log(-3.24)$

SOLUTION   (A) $\log 3{,}184 = 3.502\,973$

(B) $\ln 0.000\,349 = -7.960\,439$

(C) $\log(-3.24) = $ Error*   *−3.24 is not in the domain of the log function.*

Matched Problem 7   Use a calculator to evaluate each to six decimal places:

(A) $\log 0.013\,529$

(B) $\ln 28.693\,28$

(C) $\ln(-0.438)$

Given the logarithm of a number, how do you find the number? We make direct use of the logarithmic–exponential relationships, which follow from the definition of logarithmic function given at the beginning of this section.

$$\log x = y \quad \textbf{is equivalent to} \quad x = 10^{y}$$
$$\ln x = y \quad \textbf{is equivalent to} \quad x = e^{y}$$

EXAMPLE 8   **Solving $\log_b x = y$ for x**   Find x to four decimal places, given the indicated logarithm:

(A) $\log x = -2.315$ (B) $\ln x = 2.386$

SOLUTION   (A) $\log x = -2.315$ *Change to equivalent exponential form.*

$\qquad\quad x = 10^{-2.315}$ *Evaluate with a calculator.*

$\qquad\quad\ = 0.0048$

(B) $\ \ln x = 2.386$ *Change to equivalent exponential form.*

$\qquad\quad x = e^{2.386}$ *Evaluate with a calculator.*

$\qquad\quad\ = 10.8699$

Matched Problem 8   Find x to four decimal places, given the indicated logarithm:

(A) $\ln x = -5.062$ (B) $\log x = 2.0821$

We can use logarithms to solve exponential equations.

EXAMPLE 9   **Solving Exponential Equations**   Solve for x to four decimal places:

(A) $10^{x} = 2$ (B) $e^{x} = 3$ (C) $3^{x} = 4$

SOLUTION   (A) $\qquad 10^{x} = 2$ *Take common logarithms of both sides.*

$\qquad \log 10^{x} = \log 2$ *Property 3*

$\qquad\qquad\ x = \log 2$ *Use a calculator.*

$\qquad\qquad\quad\ = 0.3010$ *To four decimal places*

---

*Some calculators use a more advanced definition of logarithms involving complex numbers and will display an ordered pair of real numbers as the value of $\log(-3.24)$. You should interpret such a result as an indication that the number entered is not in the domain of the logarithm function as we have defined it.

(B) $e^x = 3$ Take natural logarithms of both sides.

$\ln e^x = \ln 3$ Property 3

$x = \ln 3$ Use a calculator.

$= 1.0986$ To four decimal places

(C) $3^x = 4$ Take either natural or common logarithms of both sides. (We choose common logarithms.)

$\log 3^x = \log 4$ Property 7

$x \log 3 = \log 4$ Solve for x.

$x = \dfrac{\log 4}{\log 3}$ Use a calculator.

$= 1.2619$ To four decimal places

Matched Problem 9   Solve for x to four decimal places:

(A) $10^x = 7$ (B) $e^x = 6$ (C) $4^x = 5$

 Exponential equations can also be solved graphically by graphing both sides of an equation and finding the points of intersection. Figure 4 illustrates this approach for the equations in Example 9.

(A) $y_1 = 10^x$ (B) $y_1 = e^x$ (C) $y_1 = 3^x$
    $y_2 = 2$ $y_2 = 3$ $y_2 = 4$

Figure 4 Graphical solution of exponential equations

Explore & Discuss 2   Discuss how you could find $y = \log_5 38.25$ using either natural or common logarithms on a calculator. [*Hint:* Start by rewriting the equation in exponential form.]

Remark—In the usual notation for natural logarithms, the simplifications of Example 4, parts (C) and (D) on page 109, become

$$e^{x \ln b} = b^x \qquad \text{and} \qquad \frac{\ln x}{\ln b} = \log_b x$$

With these formulas, we can change an exponential function with base b, or a logarithmic function with base b, to expressions involving exponential or logarithmic functions, respectively, to the base e. Such **change-of-base formulas** are useful in calculus.

Applications

A convenient and easily understood way of comparing different investments is to use their **doubling times**—the length of time it takes the value of an investment to double. Logarithm properties, as you will see in Example 10, provide us with just the right tool for solving some doubling-time problems.

EXAMPLE 10   **Doubling Time for an Investment** How long (to the next whole year) will it take money to double if it is invested at 20% compounded annually?

SOLUTION   We use the compound interest formula discussed in Section 2-5:

$$A = P\left(1 + \frac{r}{m}\right)^{mt} \qquad \textit{Compound interest}$$

The problem is to find t, given $r = 0.20$, $m = 1$, and $A = 2p$; that is,

$$2P = P(1 + 0.2)^t$$
$$2 = 1.2^t \qquad \textit{Solve for t by taking the natural or}$$
$$1.2^t = 2 \qquad \textit{common logarithm of both sides (we choose}$$
$$\ln 1.2^t = \ln 2 \qquad \textit{the natural logarithm).}$$
$$t \ln 1.2 = \ln 2 \qquad \textit{Property 7}$$
$$t = \frac{\ln 2}{\ln 1.2} \qquad \textit{Use a calculator.}$$
$$\approx 3.8 \text{ years} \qquad \textit{[Note: (ln 2)/(ln 1.2)} \neq \textit{ln 2} - \textit{ln 1.2]}$$

$$\approx 4 \text{ years} \qquad \textit{To the next whole year}$$

When interest is paid at the end of 3 years, the money will not be doubled; when paid at the end of 4 years, the money will be slightly more than doubled.

Example 10 can also be solved graphically by graphing both sides of the equation $2 = 1.2^t$, and finding the intersection point (Fig. 5).

Figure 5 $y_1 = 1.2^x$, $y_2 = 2$

Matched Problem 10   How long (to the next whole year) will it take money to triple if it is invested at 13% compounded annually?

It is interesting and instructive to graph the doubling times for various rates compounded annually. We proceed as follows:

$$A = P(1 + r)^t$$
$$2P = P(1 + r)^t$$
$$2 = (1 + r)^t$$
$$(1 + r)^t = 2$$
$$\ln(1 + r)^t = \ln 2$$
$$t \ln(1 + r) = \ln 2$$
$$t = \frac{\ln 2}{\ln(1 + r)}$$

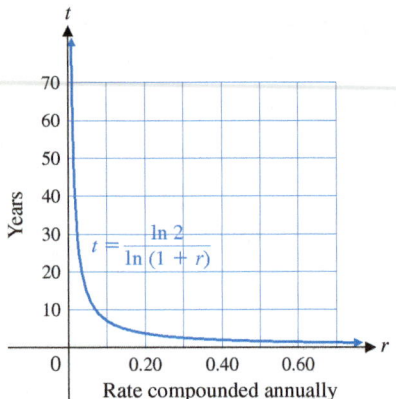

Figure 6

Figure 6 shows the graph of this equation (doubling time in years) for interest rates compounded annually from 1 to 70% (expressed as decimals). Note the dramatic change in doubling time as rates change from 1 to 20% (from 0.01 to 0.20).

Among increasing functions, the logarithmic functions (with bases $b > 1$) increase much more slowly for large values of x than either exponential or polynomial functions. When a visual inspection of the plot of a data set indicates a slowly increasing function, a logarithmic function often provides a good model. We use **logarithmic regression** on a graphing calculator to find the function of the form $y = a + b \ln x$ that best fits the data.

EXAMPLE 11   **Home Ownership Rates**   The U.S. Census Bureau published the data in Table 1 on home ownership rates. Let x represent time in years with $x = 0$ representing 1900. Use logarithmic regression to find the best model of the form $y = a + b \ln x$ for the home ownership rate y as a function of time x. Use the model to predict the home ownership rate in the United States in 2015 (to the nearest tenth of a percent).

Table 1	Home Ownership Rates
Year	**Rate (%)**
1950	55.0
1960	61.9
1970	62.9
1980	64.4
1990	64.2
2000	67.4

SOLUTION   Enter the data in a graphing calculator (Fig. 7A) and find the logarithmic regression equation (Fig. 7B). The data set and the regression equation are graphed in Figure 7C. Using TRACE, we predict that the home ownership rate in 2015 would be 69.4%.

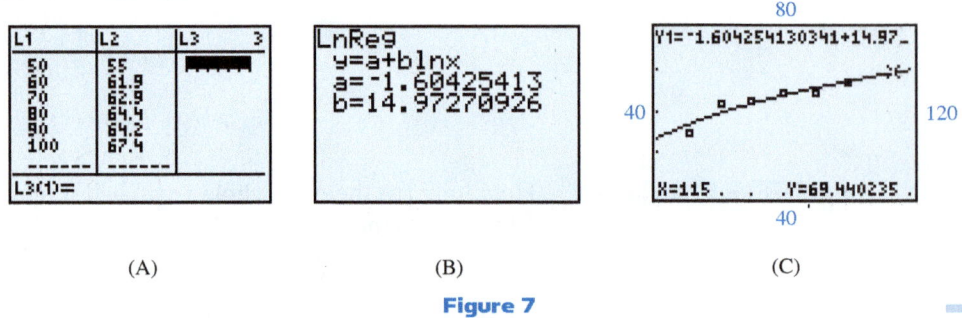

(A)   (B)   (C)

Figure 7

Matched Problem 11   The home-ownership rate in 2005 was 69.1%. Add this point to the data of Table 1, and, with $x = 0$ representing 1900, use logarithmic regression to find the best model of the form $y = a + b \ln x$ for the home-ownership rate y as a function of time x. Use the model to predict the home ownership rate in the United States in 2015 (to the nearest tenth of a percent).

⚠ **CAUTION** Note that in Example 11 we let $x = 0$ represent 1900. If we let $x = 0$ represent 1940, for example, we would obtain a different logarithmic regression equation, but the prediction for 2015 would be the same. We would *not* let $x = 0$ represent 1950 (the first year in Table 1) or any later year, because logarithmic functions are undefined at 0.

Exercises 2-6

A

For Problems 1–6, rewrite in equivalent exponential form.

1. $\log_3 27 = 3$ **2.** $\log_2 32 = 5$

3. $\log_{10} 1 = 0$ **4.** $\log_e 1 = 0$

5. $\log_4 8 = \frac{3}{2}$ **6.** $\log_9 27 = \frac{3}{2}$

For Problems 7–12, rewrite in equivalent logarithmic form.

7. $49 = 7^2$ **8.** $36 = 6^2$

9. $8 = 4^{3/2}$ **10.** $9 = 27^{2/3}$

11. $A = b^u$ **12.** $M = b^x$

For Problems 13–20, evaluate without a calculator.

13. $\log_{10} 1$ **14.** $\log_e 1$

15. $\log_e e$ **16.** $\log_{10} 10$

17. $\log_2 2^{-3}$ **18.** $\log_3 3^5$

19. $\log_{10} 1,000$ **20.** $\log_6 36$

For Problems 21–26, write in terms of simpler forms, as in Example 4.

21. $\log_b \dfrac{P}{Q}$ **22.** $\log_b FG$

23. $\log_b L^5$ **24.** $\log_b w^{15}$

25. $3^{p \log_3 q}$ **26.** $\dfrac{\log_3 P}{\log_3 R}$

B

For Problems 27–34, find x, y, or b without a calculator.

27. $\log_3 x = 2$ **28.** $\log_2 x = 2$

29. $\log_7 49 = y$ **30.** $\log_3 27 = y$

31. $\log_b 10^{-4} = -4$ **32.** $\log_b e^{-2} = -2$

33. $\log_4 x = \frac{1}{2}$ **34.** $\log_{25} x = \frac{1}{2}$

In Problems 35–42, discuss the validity of each statement. If the statement is always true, explain why. If not, give a counterexample.

35. Every polynomial function is one-to-one.

36. Every polynomial function of odd degree is one-to-one.

37. If g is the inverse of a function f, then g is one-to-one.

38. The graph of a one-to-one function intersects each vertical line exactly once.

39. The inverse of $f(x) = 2x$ is $g(x) = x/2$.

40. The inverse of $f(x) = x^2$ is $g(x) = \sqrt{x}$.

41. If f is one-to-one, then the domain of f is equal to the range of f.

42. If g is the inverse of a function f, then f is the inverse of g.

Find x in Problems 43–50.

43. $\log_b x = \frac{2}{3}\log_b 8 + \frac{1}{2}\log_b 9 - \log_b 6$

44. $\log_b x = \frac{2}{3}\log_b 27 + 2\log_b 2 - \log_b 3$

45. $\log_b x = \frac{3}{2}\log_b 4 - \frac{2}{3}\log_b 8 + 2\log_b 2$

46. $\log_b x = 3\log_b 2 + \frac{1}{2}\log_b 25 - \log_b 20$

47. $\log_b x + \log_b(x - 4) = \log_b 21$

48. $\log_b(x + 2) + \log_b x = \log_b 24$

49. $\log_{10}(x - 1) - \log_{10}(x + 1) = 1$

50. $\log_{10}(x + 6) - \log_{10}(x - 3) = 1$

Graph Problems 51 and 52 by converting to exponential form first.

51. $y = \log_2(x - 2)$

52. $y = \log_3(x + 2)$

53. Explain how the graph of the equation in Problem 51 can be obtained from the graph of $y = \log_2 x$ using a simple transformation (see Section 2-2).

54. Explain how the graph of the equation in Problem 52 can be obtained from the graph of $y = \log_3 x$ using a simple transformation (see Section 2-2).

55. What are the domain and range of the function defined by $y = 1 + \ln(x + 1)$?

56. What are the domain and range of the function defined by $y = \log(x - 1) - 1$?

For Problems 57 and 58, evaluate to five decimal places using a calculator.

57. (A) $\log 3{,}527.2$ (B) $\log 0.006\,913\,2$
    (C) $\ln 277.63$ (D) $\ln 0.040\,883$

58. (A) $\log 72.604$ (B) $\log 0.033\,041$
    (C) $\ln 40{,}257$ (D) $\ln 0.005\,926\,3$

For Problems 59 and 60, find x to four decimal places.

59. (A) $\log x = 1.1285$ (B) $\log x = -2.0497$
    (C) $\ln x = 2.7763$ (D) $\ln x = -1.8879$

60. (A) $\log x = 2.0832$ (B) $\log x = -1.1577$
    (C) $\ln x = 3.1336$ (D) $\ln x = -4.3281$

For Problems 61–66, solve each equation to four decimal places.

61. $10^x = 12$ **62.** $10^x = 153$

63. $e^x = 4.304$ **64.** $e^x = 0.3059$

65. $1.005^{12t} = 3$ **66.** $1.02^{4t} = 2$

Graph Problems 67–74 using a calculator and point-by-point plotting. Indicate increasing and decreasing intervals.

67. $y = \ln x$ **68.** $y = -\ln x$

69. $y = |\ln x|$

70. $y = \ln|x|$

71. $y = 2 \ln(x + 2)$

72. $y = 2 \ln x + 2$

73. $y = 4 \ln x - 3$

74. $y = 4 \ln(x - 3)$

75. Explain why the logarithm of 1 for any permissible base is 0.

76. Explain why 1 is not a suitable logarithmic base.

77. Let $p(x) = \ln x$, $q(x) = \sqrt{x}$, and $r(x) = x$. Use a graphing calculator to draw graphs of all three functions in the same viewing window for $1 \le x \le 16$. Discuss what it means for one function to be larger than another on an interval, and then order the three functions from largest to smallest for $1 < x \le 16$.

78. Let $p(x) = \log x$, $q(x) = \sqrt[3]{x}$, and $r(x) = x$. Use a graphing calculator to draw graphs of all three functions in the same viewing window for $1 \le x \le 16$. Discuss what it means for one function to be smaller than another on an interval, and then order the three functions from smallest to largest for $1 < x \le 16$.

Applications

79. Doubling time. In its first 10 years the Gabelli Growth Fund produced an average annual return of 21.36%. Assume that money invested in this fund continues to earn 21.36% compounded annually. How long will it take money invested in this fund to double?

80. Doubling time. In its first 10 years the Janus Flexible Income Fund produced an average annual return of 9.58%. Assume that money invested in this fund continues to earn 9.58% compounded annually. How long will it take money invested in this fund to double?

81. Investing. How many years (to two decimal places) will it take $1,000 to grow to $1,800 if it is invested at 6% compounded quarterly? Compounded daily?

82. Investing. How many years (to two decimal places) will it take $5,000 to grow to $7,500 if it is invested at 8% compounded semiannually? Compounded monthly?

83. Continuous compound interest. How many years (to two decimal places) will it take an investment of $35,000 to grow to $50,000 if it is invested at 4.75% compounded continuously?

84. Continuous compound interest. How many years (to two decimal places) will it take an investment of $17,000 to grow to $41,000 if it is invested at 2.95% compounded continuously?

85. Supply and demand. A cordless screwdriver is sold through a national chain of discount stores. A marketing company established price–demand and price–supply tables (Tables 1 and 2), where x is the number of screwdrivers people are willing to buy and the store is willing to sell each month at a price of p dollars per screwdriver.

(A) Find a logarithmic regression model ($y = a + b \ln x$) for the data in Table 1. Estimate the demand (to the nearest unit) at a price level of $50.

Table 1	Price–Demand
x	$p = D(x)$ ($)
1,000	91
2,000	73
3,000	64
4,000	56
5,000	53

(B) Find a logarithmic regression model ($y = a + b \ln x$) for the data in Table 2. Estimate the supply (to the nearest unit) at a price level of $50.

Table 2	Price–Supply
x	$p = S(x)$ ($)
1,000	9
2,000	26
3,000	34
4,000	38
5,000	41

(C) Does a price level of $50 represent a stable condition, or is the price likely to increase or decrease? Explain.

86. Equilibrium point. Use the models constructed in Problem 85 to find the equilibrium point. Write the equilibrium price to the nearest cent and the equilibrium quantity to the nearest unit.

87. Sound intensity: decibels. Because of the extraordinary range of sensitivity of the human ear (a range of over 1,000 million millions to 1), it is helpful to use a logarithmic scale, rather than an absolute scale, to measure sound intensity over this range. The unit of measure is called the *decibel,* after the inventor of the telephone, Alexander Graham Bell. If we let N be the number of decibels, I the power of the sound in question (in watts per square centimeter), and I_0 the power of sound just below the threshold of hearing (approximately 10^{-16} watt per square centimeter), then

$$I = I_0 10^{N/10}$$

Show that this formula can be written in the form

$$N = 10 \log \frac{I}{I_0}$$

88. Sound intensity: decibels. Use the formula in Problem 87 (with $I_0 = 10^{-16}$ W/cm^2) to find the decibel ratings of the following sounds:

(A) Whisper: 10^{-13} W/cm^2

(B) Normal conversation: 3.16×10^{-10} W/cm^2

(C) Heavy traffic: 10^{-8} W/cm^2

(D) Jet plane with afterburner: 10^{-1} W/cm^2

89. Agriculture. Table 3 shows the yield (in bushels per acre) and the total production (in millions of bushels) for corn in the United States for selected years since 1950. Let x represent years since 1900. Find a logarithmic regression model ($y = a + b \ln x$) for the yield. Estimate (to one decimal place) the yield in 2020.

Table 3   United States Corn Production

Year	x	Yield (bushels per acre)	Total Production (million bushels)
1950	50	37.6	2,782
1960	60	55.6	3,479
1970	70	81.4	4,802
1980	80	97.7	6,867
1990	90	115.6	7,802
2000	100	139.6	10,192

 90. Agriculture. Refer to Table 3. Find a logarithmic regression model ($y = a + b \ln x$) for the total production. Estimate (to the nearest million) the production in 2020.

91. World population. If the world population is now 6.8 billion people and if it continues to grow at an annual rate of 1.14% compounded continuously, how long (to the nearest year) will it take before there is only 1 square yard of land per person? (The Earth contains approximately 1.68×10^{14} square yards of land.)

92. Archaeology: carbon-14 dating. The radioactive carbon-14 (^{14}C) in an organism at the time of its death decays according to the equation

$$A = A_0 e^{-0.000124t}$$

where t is time in years and A_0 is the amount of ^{14}C present at time $t = 0$. (See Example 3 in Section 2-5.) Estimate the age of a skull uncovered in an archaeological site if 10% of the original amount of ^{14}C is still present. [*Hint:* Find t such that $A = 0.1A_0$.]

Answers to Matched Problems

1. (A) $9 = 3^2$ (B) $2 = 4^{1/2}$ (C) $\frac{1}{9} = 3^{-2}$
2. (A) $\log_7 49 = 2$
   (B) $\log_9 3 = \frac{1}{2}$
   (C) $\log_3(\frac{1}{3}) = -1$
3. (A) $y = \frac{3}{2}$ (B) $x = \frac{1}{3}$ (C) $b = 10$
4. (A) $\log_b R - \log_b S - \log_b T$
   (B) $\frac{2}{3}(\log_b R - \log_b S)$
   (C) b^u
   (D) $\log_b x$
5. $x = 2$ 6. $x = 5$
7. (A) $-1.868\ 734$
   (B) $3.356\ 663$
   (C) Not defined
8. (A) 0.0063 (B) 120.8092
9. (A) 0.8451 (B) 1.7918 (C) 1.1610
10. 9 yr
11. 69.9%

```
LnReg
y=a+blnx
a=-4.545195557
b=15.68361102
```

Chapter 2   Review

Important Terms, Symbols, and Concepts

2-1 Functions

EXAMPLES

- **Point-by-point plotting** may be used to **sketch the graph** of an equation in two variables: Plot enough points from its solution set in a rectangular cordinate system so that the total graph is apparent and then connect these points with a smooth curve.

  Ex. 1, p. 44

- A **function** is a correspondence between two sets of elements such that to each element in the first set there corresponds one and only one element in the second set. The first set is called the **domain** and the set of corresponding elements in the second set is called the **range**.

- If x is a placeholder for the elements in the domain of a function, then x is called the **independent variable** or the **input**. If y is a placeholder for the elements in the range, then y is called the **dependent variable** or the **output**.

- If in an equation in two variables we get exactly one output for each input, then the equation specifies a function. The graph of such a function is just the graph of the specifying equation. If we get more than one output for a given input, then the equation does not specify a function.

  Ex. 2, p. 47

- The **vertical-line test** can be used to determine whether or not an equation in two variables specifies a function (Theorem 1, p. 48).

- The functions specified by equations of the form $y = mx + b$, where $m \neq 0$, are called **linear functions**. Functions specified by equations of the form $y = b$ are called **constant functions**.

- If a function is specified by an equation and the domain is not indicated, we agree to assume that the domain is the set of all inputs that produce outputs that are real numbers.

Ex. 3, p. 48
Ex. 5, p. 50

- The symbol $f(x)$ represents the element in the range of f that corresponds to the element x of the domain.

Ex. 4, p. 49
Ex. 6, p. 51

- **Break-even** and **profit–loss** analysis use a cost function C and a revenue function R to determine when a company will have a loss ($R < C$), will break even ($R = C$), or will have a profit ($R > C$). Typical **cost**, **revenue**, **profit**, and **price–demand functions** are given on page 51.

Ex. 7, p. 52

2-2 Elementary Functions: Graphs and Transformations

- The graphs of **six basic elementary functions** (the identity function, the square and cube functions, the square root and cube root functions, and the absolute value function) are shown on page 60.

Ex. 1, p. 59

- Performing an operation on a function produces a **transformation** of the graph of the function. The basic graph transformations, **vertical and horizontal translations** (shifts), **reflection in the x axis**, and **vertical stretches and shrinks**, are summarized on pages 63 and 64.

Ex. 2, p. 61
Ex. 3, p. 62
Ex. 4, p. 63
Ex. 5, p. 64

- A **piecewise-defined function** is a function whose definition involves more than one formula.

Ex. 6, p. 65

2-3 Quadratic Functions

- If $a, b,$ and c are real numbers with $a \neq 0$, then the function

$$f(x) = ax^2 + bx + c \qquad \textit{Standard form}$$

is a **quadratic function** in **standard form** and its graph is a **parabola**.

- The quadratic formula

Ex. 1, p. 71

$$x = \frac{-b \pm \sqrt{b^2 - 4ac}}{2a} \qquad b^2 - 4ac \geq 0$$

can be used to find the x intercepts of a quadratic function.

- Completing the square in the standard form of a quadratic function produces the **vertex form**

$$f(x) = a(x - h)^2 + k \qquad \textit{Vertex form}$$

- From the vertex form of a quadratic function, we can read off the vertex, axis of symmetry, maximum or minimum, and range, and can easily sketch the graph (page 75).

Ex. 2, p. 75
Ex. 3, p. 76

- If a revenue function $R(x)$ and a cost function $C(x)$ intersect at a point (x_0, y_0), then both this point and its x coordinate x_0 are referred to as **break-even points**.

Ex. 4, p. 78

- **Quadratic regression** on a graphing calculator produces the function of the form $y = ax^2 + bx + c$ that best fits a data set.

Ex. 5, p. 79

2-4 Polynomial and Rational Functions

- A **polynomial function** is a function that can be written in the form

$$f(x) = a_n x^n + a_{n-1} x^{n-1} + \cdots + a_1 x + a_0$$

for n a nonnegative integer called the **degree** of the polynomial. The coefficients $a_0, a_1, \ldots, a_n$ are real numbers with **leading coefficient** $a_n \neq 0$. The **domain** of a polynomial function is the set of all real numbers. Graphs of representative polynomial functions are shown on page 86 and inside the front cover.

- The graph of a polynomial function of degree n can intersect the x axis at most n times. An x intercept is also called a **zero** or **root**.

- The graph of a polynomial function has no sharp corners and is **continuous**, that is, it has no holes or breaks.

- **Polynomial regression** produces a polynomial of specified degree that best fits a data set.

Ex. 1, p. 87

- A **rational function** is any function that can be written in the form

$$f(x) = \frac{n(x)}{d(x)} \qquad d(x) \neq 0$$

where $n(x)$ and $d(x)$ are polynomials. The **domain** is the set of all real numbers such that $d(x) \neq 0$. Graphs of representative rational functions are shown on page 88 and inside the front cover.

- Unlike polynomial functions, a rational function can have vertical asymptotes [but not more than the degree of the denominator $d(x)$] and at most one horizontal asymptote.   Ex. 2, p. 88

- A procedure for finding the vertical and horizontal asymptotes of a rational function is given on page 90.   Ex. 3, p. 91

2-5 Exponential Functions

- An **exponential function** is a function of the form

$$f(x) = b^x$$

where $b \neq 1$ is a positive constant called the **base**. The **domain** of f is the set of all real numbers, and the **range** is the set of positive real numbers.

- The graph of an exponential function is continuous, passes through $(0, 1)$, and has the x axis as a horizontal asymptote. If $b > 1$, then b^x increases as x increases; if $0 < b < 1$, then b^x decreases as x increases (Theorem 1, p. 97).   Ex. 1, p. 97

- Exponential functions obey the familiar laws of exponents and satisfy additional properties (Theorem 2, p. 98).

- The base that is used most frequently in mathematics is the irrational number $e \approx 2.7183$.   Ex. 2, p. 99

- Exponential functions can be used to model population growth and radioactive decay.   Ex. 3, p. 100

- **Exponential regression** on a graphing calculator produces the function of the form $y = ab^x$ that best fits a data set.   Ex. 4, p. 101

- Exponential functions are used in computations of **compound interest** and **continuous compound interest**:

$$A = P\left(1 + \frac{r}{m}\right)^{mt} \qquad \text{Compound interest}$$

$$A = Pe^{rt} \qquad \text{Continuous compound interest}$$

(see summary on page 103).

2-6 Logarithmic Functions

- A function is said to be **one-to-one** if each range value corresponds to exactly one domain value.

- The **inverse** of a one-to-one function f is the function formed by interchanging the independent and dependent variables of f. That is, (a, b) is a point on the graph of f if and only if (b, a) is a point on the graph of the inverse of f. A function that is not one-to-one does not have an inverse.

- The inverse of the exponential function with base b is called the **logarithmic function with base b**, denoted $y = \log_b x$. The **domain** of $\log_b x$ is the set of all positive real numbers (which is the range of b^x), and the range of $\log_b x$ is the set of all real numbers (which is the domain of b^x).

- Because $\log_b x$ is the inverse of the function b^x,   Ex. 1, p. 108

   Logarithmic form   Exponential form   Ex. 2, p. 108

   $y = \log_b x$ is equivalent to $x = b^y$ Ex. 3, p. 109

- Properties of logarithmic functions can be obtained from corresponding properties of exponential functions (Theorem 1, p. 109).   Ex. 4, p. 109
Ex. 5, p. 110

- Logarithms to the base 10 are called **common logarithms**, often denoted simply by $\log x$. Logarithms to the base e are called **natural logarithms**, often denoted by $\ln x$.   Ex. 6, p. 110
Ex. 7, p. 111
Ex. 8, p. 111
Ex. 9, p. 111

- Logarithms can be used to find an investment's **doubling time**—the length of time it takes for the value of an investment to double.   Ex. 10, p. 113

- **Logarithmic regression** on a graphing calculator produces the function of the form $y = a + b \ln x$ that best fits a data set.   Ex. 11, p. 114

Review Exercises

Work through all the problems in this chapter review and check your answers in the back of the book. Answers to all review problems are there along with section numbers in italics to indicate where each type of problem is discussed. Where weaknesses show up, review appropriate sections in the text.

A

In Problems 1–3, use point-by-point plotting to sketch the graph of each equation.

1. $y = 5 - x^2$

2. $x^2 = y^2$

3. $y^2 = 4x^2$

4. Indicate whether each graph specifies a function:

(A) (B)

(C) (D)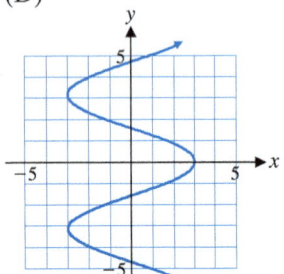

5. For $f(x) = 2x - 1$ and $g(x) = x^2 - 2x$, find:

(A) $f(-2) + g(-1)$ (B) $f(0) \cdot g(4)$

(C) $\dfrac{g(2)}{f(3)}$ (D) $\dfrac{f(3)}{g(2)}$

6. Write in logarithmic form using base e: $u = e^v$.

7. Write in logarithmic form using base 10: $x = 10^y$.

8. Write in exponential form using base e: $\ln M = N$.

9. Write in exponential form using base 10: $\log u = v$.

Solve Problems 10–12 for x exactly without using a calculator.

10. $\log_3 x = 2$ 11. $\log_x 36 = 2$

12. $\log_2 16 = x$

Solve Problems 13–16 for x to three decimal places.

13. $10^x = 143.7$ 14. $e^x = 503{,}000$

15. $\log x = 3.105$ 16. $\ln x = -1.147$

17. Use the graph of function f in the figure to determine (to the nearest integer) x or y as indicated.

(A) $y = f(0)$ (B) $4 = f(x)$

(C) $y = f(3)$ (D) $3 = f(x)$

(E) $y = f(-6)$ (F) $-1 = f(x)$

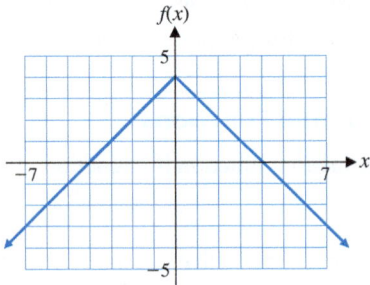

Figure for 17

18. Sketch a graph of each of the functions in parts (A)–(D) using the graph of function f in the figure below.

(A) $y = -f(x)$ (B) $y = f(x) + 4$

(C) $y = f(x - 2)$ (D) $y = -f(x + 3) - 3$

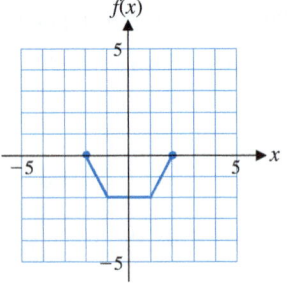

Figure for 18

19. Complete the square and find the standard form for the quadratic function

$$f(x) = -x^2 + 4x$$

Then write a brief verbal description of the relationship between the graph of f and the graph of $y = x^2$.

20. Match each equation with a graph of one of the functions f, g, m, or n in the figure.

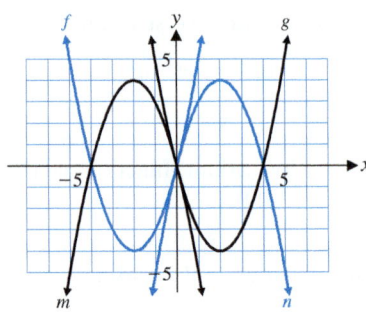

Figure for 20

(A) $y = (x - 2)^2 - 4$ (B) $y = -(x + 2)^2 + 4$

(C) $y = -(x - 2)^2 + 4$ (D) $y = (x + 2)^2 - 4$

21. Referring to the graph of function f in the figure for Problem 20 and using known properties of quadratic functions, find each of the following to the nearest integer:

(A) Intercepts (B) Vertex

(C) Maximum or minimum (D) Range

In Problems 22–25, each equation specifies a function. Determine whether the function is linear, quadratic, constant, or none of these.

22. $y = 4 - x + 3x^2$

23. $y = \dfrac{1 + 5x}{6}$

24. $y = \dfrac{7 - 4x}{2x}$

25. $y = 8x + 2(10 - 4x)$

Solve Problems 26–33 for x exactly without using a calculator.

26. $\log(x + 5) = \log(2x - 3)$ **27.** $2\ln(x - 1) = \ln(x^2 - 5)$

28. $9^{x-1} = 3^{1+x}$ **29.** $e^{2x} = e^{x^2 - 3}$

30. $2x^2 e^x = 3xe^x$ **31.** $\log_{1/3} 9 = x$

32. $\log_x 8 = -3$ **33.** $\log_9 x = \frac{3}{2}$

Solve Problems 34–41 for x to four decimal places.

34. $x = 3(e^{1.49})$ **35.** $x = 230(10^{-0.161})$

36. $\log x = -2.0144$ **37.** $\ln x = 0.3618$

38. $35 = 7(3^x)$ **39.** $0.01 = e^{-0.05x}$

40. $8,000 = 4,000(1.08^x)$ **41.** $5^{2x-3} = 7.08$

42. Find the domain of each function:

(A) $f(x) = \dfrac{2x - 5}{x^2 - x - 6}$ (B) $g(x) = \dfrac{3x}{\sqrt{5 - x}}$

43. Find the vertex form for $f(x) = 4x^2 + 4x - 3$ and then find the intercepts, the vertex, the maximum or minimum, and the range.

 44. Let $f(x) = e^x - 1$ and $g(x) = \ln(x + 2)$. Find all points of intersection for the graphs of f and g. Round answers to two decimal places.

In Problems 45 and 46, use point-by-point plotting to sketch the graph of each function.

45. $f(x) = \dfrac{50}{x^2 + 1}$ **46.** $f(x) = \dfrac{-66}{2 + x^2}$

If $f(x) = 5x + 1$, find and simplify each of the following in Problems 47–50.

47. $f(f(0))$ **48.** $f(f(-1))$

49. $f(2x - 1)$ **50.** $f(4 - x)$

51. Let $f(x) = 3 - 2x$. Find

(A) $f(2)$ (B) $f(2 + h)$

(C) $f(2 + h) - f(2)$ (D) $\dfrac{f(2 + h) - f(2)}{h}, h \neq 0$

52. Let $f(x) = x^2 - 3x + 1$. Find

(A) $f(a)$ (B) $f(a + h)$

(C) $f(a + h) - f(a)$ (D) $\dfrac{f(a + h) - f(a)}{h}, h \neq 0$

53. Explain how the graph of $m(x) = -|x - 4|$ is related to the graph of $y = |x|$.

54. Explain how the graph of $g(x) = 0.3x^3 + 3$ is related to the graph of $y = x^3$.

55. The following graph is the result of applying a sequence of transformations to the graph of $y = x^2$. Describe the transformations verbally and write an equation for the graph.

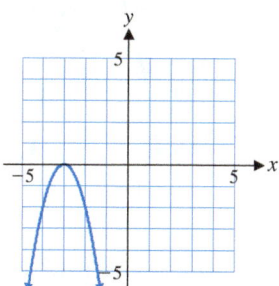

Figure for 55

56. The graph of a function f is formed by vertically stretching the graph of $y = \sqrt{x}$ by a factor of 2, and shifting it to the left 3 units and down 1 unit. Find an equation for function f and graph it for $-5 \leq x \leq 5$ and $-5 \leq y \leq 5$.

In Problems 57–59, find the equation of any horizontal asymptote.

57. $f(x) = \dfrac{5x + 4}{x^2 - 3x + 1}$ **58.** $f(x) = \dfrac{3x^2 + 2x - 1}{4x^2 - 5x + 3}$

59. $f(x) = \dfrac{x^2 + 4}{100x + 1}$

In Problems 60 and 61, find the equations of any vertical asymptotes.

60. $f(x) = \dfrac{x^2 + 100}{x^2 - 100}$ **61.** $f(x) = \dfrac{x^2 + 3x}{x^2 + 2x}$

In Problems 62–67, discuss the validity of each statement. If the statement is always true, explain why. If not, give a counter example.

62. Every polynomial function is a rational function.

63. Every rational function is a polynomial function.

64. The graph of every rational function has at least one vertical asymptote.

65. The graph of every exponential function has a horizontal asymptote.

66. The graph of every logarithmic function has a vertical asymptote.

67. There exists a rational function that has both a vertical and horizontal asymptote.

68. Sketch the graph of f for $x \geq 0$.

$$f(x) = \begin{cases} 9 + 0.3x & \text{if } 0 \leq x \leq 20 \\ 5 + 0.2x & \text{if } x > 20 \end{cases}$$

69. Sketch the graph of g for $x \geq 0$.

$$g(x) = \begin{cases} 0.5x + 5 & \text{if } 0 \leq x \leq 10 \\ 1.2x - 2 & \text{if } 10 < x \leq 30 \\ 2x - 26 & \text{if } x > 30 \end{cases}$$

70. Write an equation for the graph shown in the form $y = a(x - h)^2 + k$, where a is either -1 or $+1$ and h and k are integers.

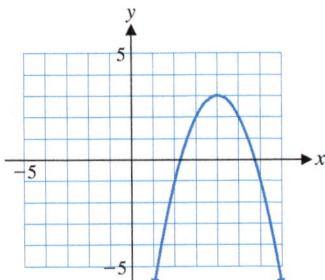

Figure for 77

71. Given $f(x) = -0.4x^2 + 3.2x + 1.2$, find the following algebraically (to three decimal places) without referring to a graph:

(A) Intercepts

(B) Vertex

(C) Maximum or minimum

(D) Range

 72. Graph $f(x) = -0.4x^2 + 3.2x + 1.2$ in a graphing calculator and find the following (to three decimal places) using TRACE and appropriate commands:

(A) Intercepts

(B) Vertex

(C) Maximum or minimum

(D) Range

C

 73. Noting that $\pi = 3.141\,592\,654\ldots$ and $\sqrt{2} = 1.414\,213\,562\ldots$ explain why the calculator results shown here are obvious. Discuss similar connections between the natural logarithmic function and the exponential function with base e.

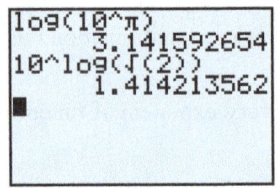

Solve Problems 74–77 exactly without using a calculator.

74. $\log x - \log 3 = \log 4 - \log(x + 4)$

75. $\ln(2x - 2) - \ln(x - 1) = \ln x$

76. $\ln(x + 3) - \ln x = 2 \ln 2$

77. $\log 3x^2 = 2 + \log 9x$

78. Write $\ln y = -5t + \ln c$ in an exponential form free of logarithms. Then solve for y in terms of the remaining variables.

79. Explain why 1 cannot be used as a logarithmic base.

80. The following graph is the result of applying a sequence of transformations to the graph of $y = \sqrt[3]{x}$. Describe the transformations verbally, and write an equation for the graph.

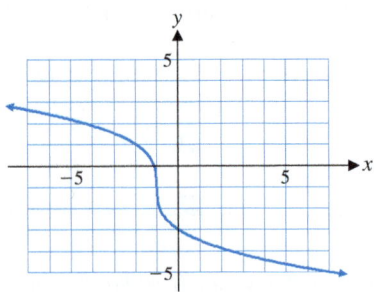

Figure for 80

81. Given $G(x) = 0.3x^2 + 1.2x - 6.9$, find the following algebraically (to three decimal places) without the use of a graph:

(A) Intercepts   (B) Vertex

(C) Maximum or minimum   (D) Range

82. Graph $G(x) = 0.3x^2 + 1.2x - 6.9$ in a standard viewing window. Then find each of the following (to three decimal places) using appropriate commands.

(A) Intercepts   (B) Vertex

(C) Maximum or minimum   (D) Range

Applications

In all problems involving days, a 365-day year is assumed.

83. Electricity rates. The table shows the electricity rates charged by Easton Utilities in the summer months.

(A) Write a piecewise definition of the monthly charge $S(x)$ (in dollars) for a customer who uses x kWh in a summer month.

(B) Graph $S(x)$.

Energy Charge (June–September)
\$3.00 for the first 20 kWh or less
5.70¢ per kWh for the next 180 kWh
3.46¢ per kWh for the next 800 kWh
2.17¢ per kWh for all over 1,000 kWh

84. Money growth. Provident Bank of Cincinnati, Ohio recently offered a certificate of deposit that paid 5.35% compounded quarterly. If a \$5,000 CD earns this rate for 5 years, how much will it be worth?

85. Money growth. Capital One Bank of Glen Allen, Virginia recently offered a certificate of deposit that paid 4.82% compounded daily. If a \$5,000 CD earns this rate for 5 years, how much will it be worth?

86. Money growth. How long will it take for money invested at 6.59% compounded monthly to triple?

87. Money growth. How long will it take for money invested at 7.39% compounded continuously to double?

88. Break-even analysis. The research department in a company that manufactures AM/FM clock radios established the following price–demand, cost, and revenue functions:

$$p(x) = 50 - 1.25x \qquad \textit{Price–demand function}$$
$$C(x) = 160 + 10x \qquad \textit{Cost function}$$
$$R(x) = xp(x)$$
$$\quad = x(50 - 1.25x) \qquad \textit{Revenue function}$$

where x is in thousands of units, and $C(x)$ and $R(x)$ are in thousands of dollars. All three functions have domain $1 \le x \le 40$.

(A) Graph the cost function and the revenue function simultaneously in the same coordinate system.

(B) Determine algebraically when $R = C$. Then, with the aid of part (A), determine when $R < C$ and $R > C$ to the nearest unit.

(C) Determine algebraically the maximum revenue (to the nearest thousand dollars) and the output (to the nearest unit) that produces the maximum revenue. What is the wholesale price of the radio (to the nearest dollar) at this output?

 89. Profit–loss analysis. Use the cost and revenue functions from Problem 88.

(A) Write a profit function and graph it in a graphing calculator.

(B) Determine graphically when $P = 0, P < 0$, and $P > 0$ to the nearest unit.

(C) Determine graphically the maximum profit (to the nearest thousand dollars) and the output (to the nearest unit) that produces the maximum profit. What is the wholesale price of the radio (to the nearest dollar) at this output? [Compare with Problem 88C.]

90. Construction. A construction company has 840 feet of chain-link fence that is used to enclose storage areas for equipment and materials at construction sites. The supervisor wants to set up two identical rectangular storage areas sharing a common fence (see the figure).

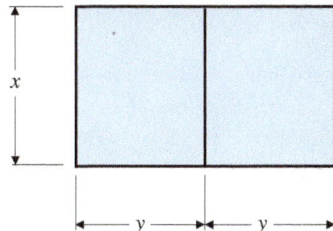

Assuming that all fencing is used,

(A) Express the total area $A(x)$ enclosed by both pens as a function of x.

(B) From physical considerations, what is the domain of the function A?

(C) Graph function A in a rectangular coordinate system.

(D) Use the graph to discuss the number and approximate locations of values of x that would produce storage areas with a combined area of 25,000 square feet.

(E) Approximate graphically (to the nearest foot) the values of x that would produce storage areas with a combined area of 25,000 square feet.

(F) Determine algebraically the dimensions of the storage areas that have the maximum total combined area. What is the maximum area?

91. Equilibrium point. A company is planning to introduce a 10-piece set of nonstick cookware. A marketing company established price–demand and price–supply tables for selected prices (Tables 1 and 2), where x is the number of cookware sets people are willing to buy and the company is willing to sell each month at a price of p dollars per set.

Table 1   Price–Demand

x	$p = D(x)$ ($)
985	330
2,145	225
2,950	170
4,225	105
5,100	50

Table 2   Price–Supply

x	$p = S(x)$ ($)
985	30
2,145	75
2,950	110
4,225	155
5,100	190

(A) Find a quadratic regression model for the data in Table 1. Estimate the demand at a price level of $180.

(B) Find a linear regression model for the data in Table 2. Estimate the supply at a price level of $180.

(C) Does a price level of $180 represent a stable condition, or is the price likely to increase or decrease? Explain.

(D) Use the models in parts (A) and (B) to find the equilibrium point. Write the equilibrium price to the nearest cent and the equilibrium quantity to the nearest unit.

92. Crime statistics. According to data published by the FBI, the crime index in the United States has shown a downward trend since the early 1990s (Table 3).

Table 3   Crime Index

Year	Crimes per 100,000 Inhabitants
1987	5,550
1992	5,660
1997	4,930
2002	4,119
2007	3,016

(A) Find a cubic regression model for the crime index if $x = 0$ represents 1987.

(B) Use the cubic regression model to predict the crime index in 2017.

93. Medicine. One leukemic cell injected into a healthy mouse will divide into 2 cells in about $\frac{1}{2}$ day. At the end of the day these 2 cells will divide into 4. This doubling continues until 1 billion cells are formed; then the animal dies with leukemic cells in every part of the body.

(A) Write an equation that will give the number N of leukemic cells at the end of t days.

(B) When, to the nearest day, will the mouse die?

94. Marine biology. The intensity of light entering water is reduced according to the exponential equation

$$I = I_0 e^{-kd}$$

where I is the intensity d feet below the surface, I_0 is the intensity at the surface, and k is the coefficient of extinction. Measurements in the Sargasso Sea have indicated that half of the surface light reaches a depth of 73.6 feet. Find k (to five decimal places), and find the depth (to the nearest foot) at which 1% of the surface light remains.

95. Agriculture. The number of dairy cows on farms in the United States is shown in Table 4 for selected years since 1950. Let 1940 be year 0.

Table 4 Dairy Cows on Farms in the United States

Year	Dairy Cows (thousands)
1950	23,853
1960	19,527
1970	12,091
1980	10,758
1990	10,015
2000	9,190

(A) Find a logarithmic regression model ($y = a + b \ln x$) for the data. Estimate (to the nearest thousand) the number of dairy cows in 2020.

(B) Explain why it is not a good idea to let 1950 be year 0.

96. Population growth. The population of some countries has a relative growth rate of 3% (or more) per year. At this rate, how many years (to the nearest tenth of a year) will it take a population to double?

97. Medicare. The annual expenditures for Medicare (in billions of dollars) by the U.S. government for selected years since 1980 are shown in Table 5. Let x represent years since 1980.

Table 5 Medicare Expenditures

Year	Billion $
1980	37
1985	72
1990	111
1995	181
2000	197
2005	330

(A) Find an exponential regression model ($y = ab^x$) for the data. Estimate (to the nearest billion) the annual expenditures in 2017.

(B) When will the annual expenditures reach one trillion dollars?

CALCULUS

Limits and the Derivative

3

Introduction

How do algebra and calculus differ? The two words *static* and *dynamic* probably come as close as any to expressing the difference between the two disciplines. In algebra, we solve equations for a particular value of a variable—a static notion. In calculus, we are interested in how a change in one variable affects another variable—a dynamic notion.

Isaac Newton (1642–1727) of England and Gottfried Wilhelm von Leibniz (1646–1716) of Germany developed calculus independently to solve problems concerning motion. Today calculus is used not just in the physical sciences, but also in business, economics, life sciences, and social sciences—any discipline that seeks to understand dynamic phenomena.

In Chapter 3 we introduce the *derivative*, one of the two key concepts of calculus. The second, the *integral*, is the subject of Chapter 6. Both key concepts depend on the notion of *limit*, which is explained in Sections 3–1 and 3–2. We consider many applications of limits and derivatives. See, for example, Problems 69 and 70 in Section 3–3 on the concentration of a drug in the bloodstream.

3-1 Introduction to Limits

Basic to the study of calculus is the concept of a *limit*. This concept helps us to describe, in a precise way, the behavior of $f(x)$ when x is close, but not equal, to a particular value c. In this section, we develop an intuitive and informal approach to evaluating limits.

Functions and Graphs: Brief Review

The graph of the function $y = f(x) = x + 2$ is the graph of the set of all ordered pairs $(x, f(x))$. For example, if $x = 2$, then $f(2) = 4$ and $(2, f(2)) = (2, 4)$ is a point on the graph of f. Figure 1 shows $(-1, f(-1))$, $(1, f(1))$, and $(2, f(2))$ plotted on the graph of f. Notice that the domain values -1, 1, and 2 are associated with the x axis and the range values $f(-1) = 1$, $f(1) = 3$, and $f(2) = 4$ are associated with the y axis.

Given x, it is sometimes useful to read $f(x)$ directly from the graph of f. Example 1 reviews this process.

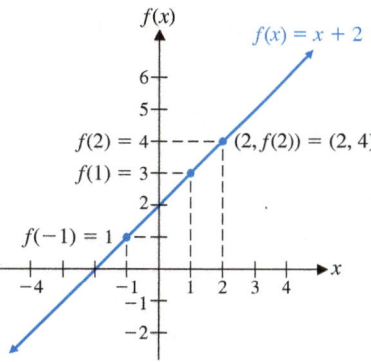

Figure 1

EXAMPLE 1 **Finding Values of a Function from Its Graph** Complete the following table, using the given graph of the function g.

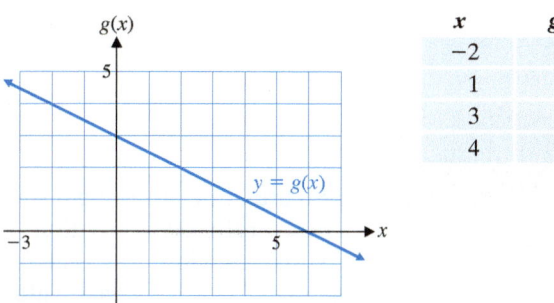

x	$g(x)$
-2	
1	
3	
4	

SOLUTION To determine $g(x)$, proceed vertically from the x value on the x axis to the graph of g and then horizontally to the corresponding y value $g(x)$ on the y axis (as indicated by the dashed lines).

x	$g(x)$
-2	4.0
1	2.5
3	1.5
4	1.0

Matched Problem 1   Complete the following table, using the given graph of the function h.

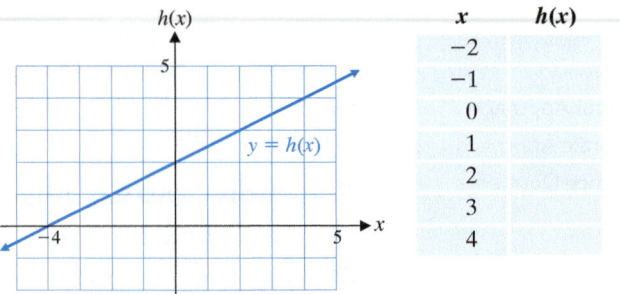

x	$h(x)$
-2	
-1	
0	
1	
2	
3	
4	

Limits: A Graphical Approach

We introduce the important concept of a *limit* through an example, which leads to an intuitive definition of the concept.

EXAMPLE 2   **Analyzing a Limit** Let $f(x) = x + 2$. Discuss the behavior of the values of $f(x)$ when x is close to 2.

SOLUTION   We begin by drawing a graph of f that includes the domain value $x = 2$ (Fig. 2).

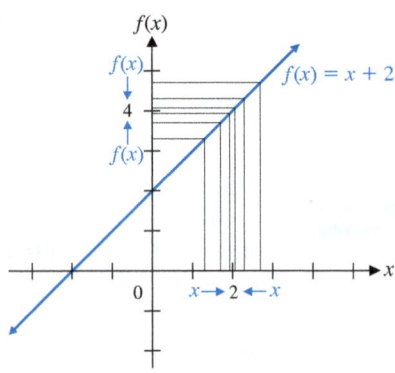

Figure 2

In Figure 2, we are using a static drawing to describe a dynamic process. This requires careful interpretation. The thin vertical lines in Figure 2 represent values of x that are close to 2. The corresponding horizontal lines identify the value of $f(x)$ associated with each value of x. [Example 1 dealt with the relationship between x and $f(x)$ on a graph.] The graph in Figure 2 indicates that as the values of x get closer and closer to 2 on either side of 2, the corresponding values of $f(x)$ get closer and closer to 4. Symbolically, we write

$$\lim_{x \to 2} f(x) = 4$$

This equation is read as "The limit of $f(x)$ as x approaches 2 is 4." Note that $f(2) = 4$. That is, the value of the function at 2 and the limit of the function as x approaches 2 are the same. This relationship can be expressed as

$$\lim_{x \to 2} f(x) = f(2) = 4$$

Graphically, this means that there is no hole or break in the graph of f at $x = 2$.

Matched Problem 2   Let $f(x) = x + 1$. Discuss the behavior of the values of $f(x)$ when x is close to 1.

We now present an informal definition of the important concept of a limit. A precise definition is not needed for our discussion, but one is given in a footnote.*

DEFINITION Limit

We write

$$\lim_{x \to c} f(x) = L \quad \text{or} \quad f(x) \to L \text{ as } x \to c$$

if the functional value $f(x)$ is close to the single real number L whenever x is close, but not equal, to c (on either side of c).

Note: The existence of a limit at c has nothing to do with the value of the function at c. In fact, c may not even be in the domain of f. However, the function must be defined on both sides of c.

The next example involves the **absolute value function**:

$$f(x) = |x| = \begin{cases} -x & \text{if } x < 0 \\ x & \text{if } x \geq 0 \end{cases}$$

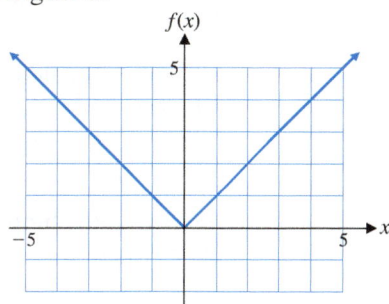

$$f(-2) = |-2| = -(-2) = 2$$
$$f(3) = |3| = 3$$

The graph of f is shown in Figure 3.

$$f(x)$$

Figure 3 $f(x) = |x|$

EXAMPLE 3 **Analyzing a Limit** Let $h(x) = |x|/x$. Explore the behavior of $h(x)$ for x near, but not equal, to 0. Find $\lim_{x \to 0} h(x)$ if it exists.

SOLUTION The function h is defined for all real numbers except 0. For example,

$$h(-2) = \frac{|-2|}{-2} = \frac{2}{-2} = -1$$

$$h(0) = \frac{|0|}{0} = \frac{0}{0} \qquad \text{Not defined}$$

$$h(2) = \frac{|2|}{2} = \frac{2}{2} = 1$$

In general, $h(x)$ is -1 for all negative x and 1 for all positive x. Figure 4 illustrates the behavior of $h(x)$ for x near 0. Note that the absence of a solid dot on the vertical axis indicates that h is not defined when $x = 0$.

When x is near 0 (on either side of 0), is $h(x)$ near one specific number? The answer is "No," because $h(x)$ is -1 for $x < 0$ and 1 for $x > 0$. Consequently, we say that

$$\lim_{x \to 0} \frac{|x|}{x} \text{ does not exist}$$

Neither $h(x)$ nor the limit of $h(x)$ exists at $x = 0$. However, the limit from the left and the limit from the right both exist at 0, but they are not equal.

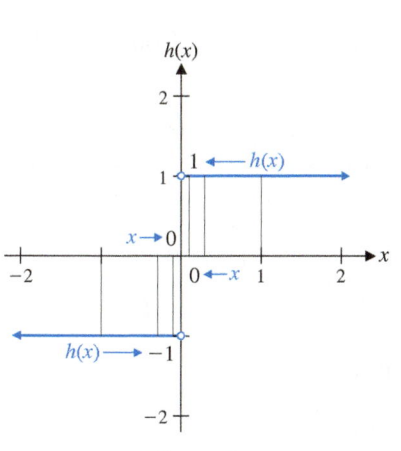

Figure 4

---

*To make the informal definition of *limit* precise, we must make the word *close* more precise. This is done as follows: We write $\lim_{x \to c} f(x) = L$ if, for each $e > 0$, there exists a $d > 0$ such that $|f(x) - L| < e$ whenever $0 < |x - c| < d$. This definition is used to establish particular limits and to prove many useful properties of limits that will be helpful in finding particular limits.

Matched Problem 3   Graph

$$h(x) = \frac{x - 2}{|x - 2|}$$

and find $\lim_{x \to 2} h(x)$ if it exists.

In Example 3, we see that the values of the function $h(x)$ approach two different numbers, depending on the direction of approach, and it is natural to refer to these values as "the limit from the left" and "the limit from the right." These experiences suggest that the notion of **one-sided limits** will be very useful in discussing basic limit concepts.

DEFINITION One-Sided Limits

We write

$$\lim_{x \to c^-} f(x) = K \quad x \to c^- \text{ is read "x approaches } c \text{ from the left" and means } x \to c \text{ and } x < c.$$

and call K the **limit from the left** or the **left-hand limit** if $f(x)$ is close to K whenever x is close to, but to the left of, c on the real number line. We write

$$\lim_{x \to c^+} f(x) = L \quad x \to c^+ \text{ is read "x approaches } c \text{ from the right" and means } x \to c \text{ and } x > c.$$

and call L the **limit from the right** or the **right-hand limit** if $f(x)$ is close to L whenever x is close to, but to the right of, c on the real number line.

If no direction is specified in a limit statement, we will always assume that the limit is **two-sided** or **unrestricted**. Theorem 1 states an important relationship between one-sided limits and unrestricted limits.

THEOREM 1   On the Existence of a Limit

For a (two-sided) limit to exist, the limit from the left and the limit from the right must exist and be equal. That is,

$$\lim_{x \to c} f(x) = L \text{ if and only if } \lim_{x \to c^-} f(x) = \lim_{x \to c^+} f(x) = L$$

In Example 3,

$$\lim_{x \to 0^-} \frac{|x|}{x} = -1 \quad \text{and} \quad \lim_{x \to 0^+} \frac{|x|}{x} = 1$$

Since the left- and right-hand limits are *not* the same,

$$\lim_{x \to 0} \frac{|x|}{x} \text{ does not exist}$$

EXAMPLE 4   **Analyzing Limits Graphically** Given the graph of the function f in Figure 5, discuss the behavior of $f(x)$ for x near (A) -1, (B) 1, and (C) 2.

SOLUTION   (A) Since we have only a graph to work with, we use vertical and horizontal lines to relate the values of x and the corresponding values of $f(x)$. For any x near -1 on either side of -1, we see that the corresponding value of $f(x)$, determined by a horizontal line, is close to 1.

Figure 5

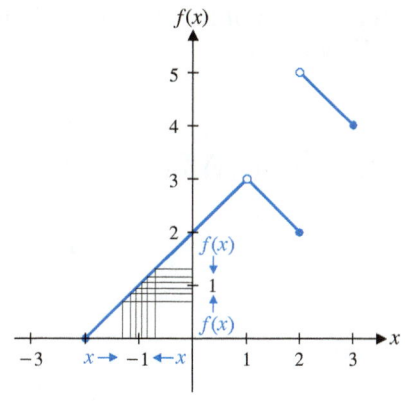

$$\lim_{x \to -1^-} f(x) = 1$$

$$\lim_{x \to -1^+} f(x) = 1$$

$$\lim_{x \to -1} f(x) = 1$$

$$f(-1) = 1$$

(B) Again, for any x near, but not equal to, 1, the vertical and horizontal lines indicate that the corresponding value of $f(x)$ is close to 3. The open dot at $(1, 3)$, together with the absence of a solid dot anywhere on the vertical line through $x = 1$, indicates that $f(1)$ is not defined.

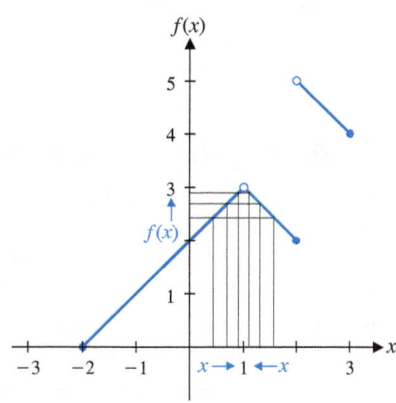

$$\lim_{x \to 1^-} f(x) = 3$$

$$\lim_{x \to 1^+} f(x) = 3$$

$$\lim_{x \to 1} f(x) = 3$$

$$f(1) \text{ not defined}$$

(C) The abrupt break in the graph at $x = 2$ indicates that the behavior of the graph near $x = 2$ is more complicated than in the two preceding cases. If x is close to 2 on the left side of 2, the corresponding horizontal line intersects the y axis at a point close to 2. If x is close to 2 on the right side of 2, the corresponding horizontal line intersects the y axis at a point close to 5. This is a case where the one-sided limits are different.

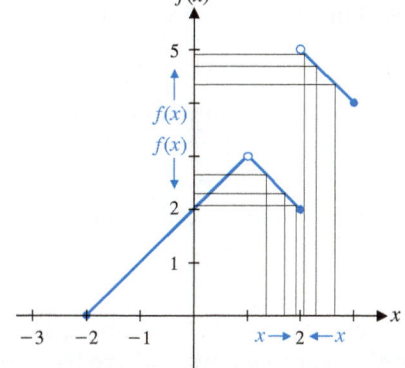

$$\lim_{x \to 2^-} f(x) = 2$$

$$\lim_{x \to 2^+} f(x) = 5$$

$$\lim_{x \to 2} f(x) \text{ does not exist}$$

$$f(2) = 2$$

Matched Problem 4   Given the graph of the function f shown in Figure 6, discuss the following, as we did in Example 4:

(A) Behavior of $f(x)$ for x near 0

(B) Behavior of $f(x)$ for x near 1

(C) Behavior of $f(x)$ for x near 3

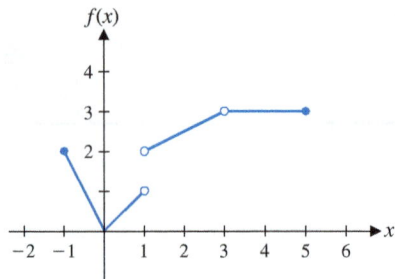

Figure 6

CONCEPTUAL INSIGHT

In Example 4B, note that $\lim\limits_{x \to 1} f(x)$ exists even though f is not defined at $x = 1$ and the graph has a hole at $x = 1$. In general, the value of a function at $x = c$ has no effect on the limit of the function as x approaches c.

Limits: An Algebraic Approach

Graphs are very useful tools for investigating limits, especially if something unusual happens at the point in question. However, many of the limits encountered in calculus are routine and can be evaluated quickly with a little algebraic simplification, some intuition, and basic properties of limits. The following list of properties of limits forms the basis for this approach:

THEOREM 2   Properties of Limits

Let f and g be two functions, and assume that

$$\lim_{x \to c} f(x) = L \qquad \lim_{x \to c} g(x) = M$$

where L and M are real numbers (both limits exist). Then

1. $\lim\limits_{x \to c} k = k$ for any constant k

2. $\lim\limits_{x \to c} x = c$

3. $\lim\limits_{x \to c} [f(x) + g(x)] = \lim\limits_{x \to c} f(x) + \lim\limits_{x \to c} g(x) = L + M$

4. $\lim\limits_{x \to c} [f(x) - g(x)] = \lim\limits_{x \to c} f(x) - \lim\limits_{x \to c} g(x) = L - M$

5. $\lim\limits_{x \to c} kf(x) = k \lim\limits_{x \to c} f(x) = kL$ for any constant k

6. $\lim\limits_{x \to c} [f(x) \cdot g(x)] = [\lim\limits_{x \to c} f(x)][\lim\limits_{x \to c} g(x)] = LM$

7. $\lim\limits_{x \to c} \dfrac{f(x)}{g(x)} = \dfrac{\lim\limits_{x \to c} f(x)}{\lim\limits_{x \to c} g(x)} = \dfrac{L}{M}$ if $M \neq 0$

8. $\lim\limits_{x \to c} \sqrt[n]{f(x)} = \sqrt[n]{\lim\limits_{x \to c} f(x)} = \sqrt[n]{L}$ $L > 0$ for n even

---

Each property in Theorem 2 is also valid if $x \to c$ is replaced everywhere by $x \to c^-$ or replaced everywhere by $x \to c^+$.

EXPLORE & DISCUSS 1 The properties listed in Theorem 2 can be paraphrased in brief verbal statements. For example, property 3 simply states that *the limit of a sum is equal to the sum of the limits.* Write brief verbal statements for the remaining properties in Theorem 2.

EXAMPLE 5   **Using Limit Properties** Find $\lim_{x \to 3}(x^2 - 4x)$.

SOLUTION

$$\lim_{x \to 3}(x^2 - 4x) = \lim_{x \to 3} x^2 - \lim_{x \to 3} 4x \qquad \text{Property 4}$$

$$= \left(\lim_{x \to 3} x\right) \cdot \left(\lim_{x \to 3} x\right) - 4\lim_{x \to 3} x \qquad \text{Properties 5 and 6}$$

$$= 3 \cdot 3 - 4 \cdot 3 = -3 \qquad \text{Property 2}$$

With a little practice, you will soon be able to omit the steps in the dashed boxes and simply write

$$\lim_{x \to 3}(x^2 - 4x) = 3 \cdot 3 - 4 \cdot 3 = -3$$

Matched Problem 5   Find $\lim_{x \to -2}(x^2 + 5x)$.

What happens if we try to evaluate a limit like the one in Example 5, but with x approaching an unspecified number, such as c? Proceeding as we did in Example 5, we have

$$\lim_{x \to c}(x^2 - 4x) = c \cdot c - 4 \cdot c = c^2 - 4c$$

If we let $f(x) = x^2 - 4x$, we have

$$\lim_{x \to c} f(x) = \lim_{x \to c}(x^2 - 4x) = c^2 - 4c = f(c)$$

That is, this limit can be evaluated simply by evaluating the function f at c. It would certainly simplify the process of evaluating limits if we could identify the functions for which

$$\lim_{x \to c} f(x) = f(c) \tag{1}$$

since we could use this fact to evaluate the limit. It turns out that there are many functions that satisfy equation (1). We postpone a detailed discussion of these functions until the next section. For now, we note that if

$$f(x) = a_n x^n + a_{n-1} x^{n-1} + \cdots + a_0$$

is a polynomial function, then, by the properties in Theorem 1,

$$\lim_{x \to c} f(x) = \lim_{x \to c}(a_n x^n + a_{n-1} x^{n-1} + \cdots + a_0)$$

$$= a_n c^n + a_{n-1} c^{n-1} + \cdots + a_0 = f(c)$$

and if

$$r(x) = \frac{n(x)}{d(x)}$$

is a rational function, where $n(x)$ and $d(x)$ are polynomials with $d(c) \neq 0$, then by property 7 and the fact that polynomials $n(x)$ and $d(x)$ satisfy equation (1),

$$\lim_{x \to c} r(x) = \lim_{x \to c} \frac{n(x)}{d(x)} = \frac{\lim_{x \to c} n(x)}{\lim_{x \to c} d(x)} = \frac{n(c)}{d(c)} = r(c)$$

These results are summarized in Theorem 3.

> **THEOREM 3   Limits of Polynomial and Rational Functions**
>
> 1. $\lim\limits_{x \to c} f(x) = f(c)$ for f any polynomial function.
> 2. $\lim\limits_{x \to c} r(x) = r(c)$ for r any rational function with a nonzero denominator at $x = c$.

EXAMPLE 6   **Evaluating Limits** Find each limit.

(A) $\lim\limits_{x \to 2} (x^3 - 5x - 1)$ (B) $\lim\limits_{x \to -1} \sqrt{2x^2 + 3}$ (C) $\lim\limits_{x \to 4} \dfrac{2x}{3x + 1}$

SOLUTION   (A) $\lim\limits_{x \to 2} (x^3 - 5x - 1) = 2^3 - 5 \cdot 2 - 1 = -3$ *Theorem 3*

(B) $\lim\limits_{x \to -1} \sqrt{2x^2 + 3} = \sqrt{\lim\limits_{x \to -1} (2x^2 + 3)}$ *Property 8*

$$= \sqrt{2(-1)^2 + 3} \qquad \text{Theorem 3}$$

$$= \sqrt{5}$$

(C) $\lim\limits_{x \to 4} \dfrac{2x}{3x + 1} = \dfrac{2 \cdot 4}{3 \cdot 4 + 1}$ *Theorem 3*

$$= \frac{8}{13}$$

Matched Problem 6   Find each limit.

(A) $\lim\limits_{x \to -1} (x^4 - 2x + 3)$ (B) $\lim\limits_{x \to 2} \sqrt{3x^2 - 6}$ (C) $\lim\limits_{x \to -2} \dfrac{x^2}{x^2 + 1}$

EXAMPLE 7   **Evaluating Limits** Let

$$f(x) = \begin{cases} x^2 + 1 & \text{if } x < 2 \\ x - 1 & \text{if } x > 2 \end{cases}$$

Find each limit.

(A) $\lim\limits_{x \to 2^-} f(x)$ (B) $\lim\limits_{x \to 2^+} f(x)$ (C) $\lim\limits_{x \to 2} f(x)$ (D) $f(2)$

SOLUTION   (A) $\lim\limits_{x \to 2^-} f(x) = \lim\limits_{x \to 2^-} (x^2 + 1)$ If $x < 2$, $f(x) = x^2 + 1$.

$$= 2^2 + 1 = 5$$

(B) $\lim\limits_{x \to 2^+} f(x) = \lim\limits_{x \to 2^+} (x - 1)$ If $x > 2$, $f(x) = x - 1$.

$$= 2 - 1 = 1$$

(C) Since the one-sided limits are not equal, $\lim\limits_{x \to 2} f(x)$ does not exist.

(D) Because the definition of f does not assign a value to f for $x = 2$, only for $x < 2$ and $x > 2$, $f(2)$ does not exist.

Matched Problem 7   Let

$$f(x) = \begin{cases} 2x + 3 & \text{if } x < 5 \\ -x + 12 & \text{if } x > 5 \end{cases}$$

Find each limit.

(A) $\lim\limits_{x \to 5^-} f(x)$ (B) $\lim\limits_{x \to 5^+} f(x)$ (C) $\lim\limits_{x \to 5} f(x)$ (D) $f(5)$

It is important to note that there are restrictions on some of the limit properties. In particular, if

$$\lim_{x \to c} f(x) = 0 \quad \text{and} \quad \lim_{x \to c} g(x) = 0, \quad \text{then finding} \lim_{x \to c} \frac{f(x)}{g(x)}$$

may present some difficulties, since limit property 7 (the limit of a quotient) does not apply when $\lim_{x \to c} g(x) = 0$. The next example illustrates some techniques that can be useful in this situation.

EXAMPLE 8   **Evaluating Limits** Find each limit.

(A) $\lim_{x \to 2} \dfrac{x^2 - 4}{x - 2}$

(B) $\lim_{x \to -1} \dfrac{x|x + 1|}{x + 1}$

SOLUTION   (A) Algebraic simplification is often useful when the numerator and denominator are both approaching 0.

$$\lim_{x \to 2} \frac{x^2 - 4}{x - 2} = \lim_{x \to 2} \frac{(x - 2)(x + 2)}{x - 2} = \lim_{x \to 2} (x + 2) = 4$$

(B) One-sided limits are helpful for limits involving the absolute value function.

$$\lim_{x \to -1^+} \frac{x|x + 1|}{x + 1} = \lim_{x \to -1^+} (x) = -1 \quad \text{If } x > -1, \text{ then } \frac{|x + 1|}{x + 1} = 1.$$

$$\lim_{x \to -1^-} \frac{x|x + 1|}{x + 1} = \lim_{x \to -1^-} (-x) = 1 \quad \text{If } x < -1, \text{ then } \frac{|x + 1|}{x + 1} = -1.$$

Since the limit from the left and the limit from the right are not the same, we conclude that

$$\lim_{x \to -1} \frac{x|x + 1|}{x + 1} \quad \text{does not exist}$$

Matched Problem 8   Find each limit.

(A) $\lim_{x \to -3} \dfrac{x^2 + 4x + 3}{x + 3}$

(B) $\lim_{x \to 4} \dfrac{x^2 - 16}{|x - 4|}$

CONCEPTUAL INSIGHT

In the solution to Example 8A we used the following algebraic identity:

$$\frac{x^2 - 4}{x - 2} = \frac{(x - 2)(x + 2)}{x - 2} = x + 2, \qquad x \neq 2$$

The restriction $x \neq 2$ is necessary here because the first two expressions are not defined at $x = 2$. Why didn't we include this restriction in the solution? When x approaches 2 in a limit problem, it is assumed that x is close, but not equal, to 2. It is important that you understand that both of the following statements are valid:

$$\lim_{x \to 2} \frac{x^2 - 4}{x - 2} = \lim_{x \to 2} (x + 2) \quad \text{and} \quad \frac{x^2 - 4}{x - 2} = x + 2, \; x \neq 2$$

Limits like those in Example 8 occur so frequently in calculus that they are given a special name.

DEFINITION Indeterminate Form

If $\lim\limits_{x \to c} f(x) = 0$ and $\lim\limits_{x \to c} g(x) = 0$, then $\lim\limits_{x \to c} \dfrac{f(x)}{g(x)}$ is said to be **indeterminate**, or, more specifically, a **0/0 indeterminate form**.

The term *indeterminate* is used because the limit of an indeterminate form may or may not exist (see Example 8A and 8B).

! **CAUTION** The expression 0/0 does not represent a real number and should never be used as the value of a limit. If a limit is a 0/0 indeterminate form, further investigation is always required to determine whether the limit exists and to find its value if it does exist.

If the denominator of a quotient approaches 0 and the numerator approaches a nonzero number, then the limit of the quotient is not an indeterminate form. In fact, a limit of this form never exists.

THEOREM 4 Limit of a Quotient

If $\lim\limits_{x \to c} f(x) = L$, $L \neq 0$, and $\lim\limits_{x \to c} g(x) = 0$,

then

$$\lim\limits_{x \to c} \frac{f(x)}{g(x)} \qquad \text{does not exist}$$

EXPLORE & DISCUSS 2

Use algebraic and/or graphical techniques to analyze each of the following indeterminate forms.

(A) $\lim\limits_{x \to 1} \dfrac{x - 1}{x^2 - 1}$

(B) $\lim\limits_{x \to 1} \dfrac{(x - 1)^2}{x^2 - 1}$

(C) $\lim\limits_{x \to 1} \dfrac{x^2 - 1}{(x - 1)^2}$

Limits of Difference Quotients

Let the function f be defined in an open interval containing the number a. One of the most important limits in calculus is the limit of the **difference quotient**,

$$\lim\limits_{h \to 0} \frac{f(a + h) - f(a)}{h} \tag{2}$$

If

$$\lim\limits_{h \to 0} [f(a + h) - f(a)] = 0$$

as it often does, then limit (2) is an indeterminate form.

EXAMPLE 9 **Limit of a Difference Quotient** Find the following limit for $f(x) = 4x - 5$:

$$\lim\limits_{h \to 0} \frac{f(3 + h) - f(3)}{h}$$

SOLUTION

$$\lim\limits_{h \to 0} \frac{f(3 + h) - f(3)}{h} = \lim\limits_{h \to 0} \frac{[4(3 + h) - 5] - [4(3) - 5]}{h}$$

$$= \lim\limits_{h \to 0} \frac{12 + 4h - 5 - 12 + 5}{h}$$

$$= \lim\limits_{h \to 0} \frac{4h}{h} = \lim\limits_{h \to 0} 4 = 4$$

Since this is a 0/0 indeterminate form and property 7 in Theorem 2 does not apply, we proceed with algebraic simplification.

Matched Problem 9 Find the following limit for $f(x) = 7 - 2x$: $\lim\limits_{h \to 0} \dfrac{f(4 + h) - f(4)}{h}$.

EXAMPLE 10 **Limit of a Difference Quotient** Find the following limit for $f(x) = |x + 5|$:

$$\lim_{h \to 0} \frac{f(-5 + h) - f(-5)}{h}$$

SOLUTION

$$\lim_{h \to 0} \frac{f(-5 + h) - f(-5)}{h} = \lim_{h \to 0} \frac{|(-5 + h) + 5| - |-5 + 5|}{h}$$

$$= \lim_{h \to 0} \frac{|h|}{h} \text{ does not exist}$$

Since this is a 0/0 indeterminate form and property 7 in Theorem 2 does not apply, we proceed with algebraic simplification.

Matched Problem 10 Find the following limit for $f(x) = |x - 1|$: $\lim\limits_{h \to 0} \dfrac{f(1 + h) - f(1)}{h}$.

EXAMPLE 11 **Limit of a Difference Quotient** Find the following limit for $f(x) = \sqrt{x}$:

$$\lim_{h \to 0} \frac{f(2 + h) - f(2)}{h}$$

SOLUTION

$$\lim_{h \to 0} \frac{f(2 + h) - f(2)}{h} = \lim_{h \to 0} \frac{\sqrt{2 + h} - \sqrt{2}}{h}$$

$$= \lim_{h \to 0} \frac{\sqrt{2 + h} - \sqrt{2}}{h} \cdot \frac{\sqrt{2 + h} + \sqrt{2}}{\sqrt{2 + h} + \sqrt{2}}$$

$$= \lim_{h \to 0} \frac{2 + h - 2}{h(\sqrt{2 + h} + \sqrt{2})}$$

$$= \lim_{h \to 0} \frac{1}{\sqrt{2 + h} + \sqrt{2}}$$

$$= \frac{1}{\sqrt{2} + \sqrt{2}} = \frac{1}{2\sqrt{2}}$$

This is a 0/0 indeterminate form, so property 7 in Theorem 2 does not apply. Rationalizing the numerator will help.

$(A - B)(A + B) = A^2 - B^2$

Matched Problem 11 Find the following limit for $f(x) = \sqrt{x}$: $\lim\limits_{h \to 0} \dfrac{f(3 + h) - f(3)}{h}$.

Exercises 3-1

A

In Problems 1–8, use the graph of the function f shown to estimate the indicated limits and function values.

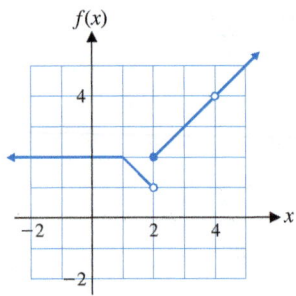

Figure for 1–8

1. $f(-0.5)$ 2. $f(-1.5)$

3. $f(1.75)$ 4. $f(1.25)$

5. (A) $\lim\limits_{x\to 0^-} f(x)$ (B) $\lim\limits_{x\to 0^+} f(x)$
   (C) $\lim\limits_{x\to 0} f(x)$ (D) $f(0)$

6. (A) $\lim\limits_{x\to 1^-} f(x)$ (B) $\lim\limits_{x\to 1^+} f(x)$
   (C) $\lim\limits_{x\to 1} f(x)$ (D) $f(1)$

7. (A) $\lim\limits_{x\to 2^-} f(x)$ (B) $\lim\limits_{x\to 2^+} f(x)$
   (C) $\lim\limits_{x\to 2} f(x)$ (D) $f(2)$
    (E) Is it possible to redefine $f(2)$ so that $\lim\limits_{x\to 2} f(x) = f(2)$? Explain.

8. (A) $\lim\limits_{x\to 4^-} f(x)$ (B) $\lim\limits_{x\to 4^+} f(x)$
   (C) $\lim\limits_{x\to 4} f(x)$ (D) $f(4)$
    (E) Is it possible to define $f(4)$ so that $\lim\limits_{x\to 4} f(x) = f(4)$? Explain.

In Problems 9–16, use the graph of the function g shown to estimate the indicated limits and function values.

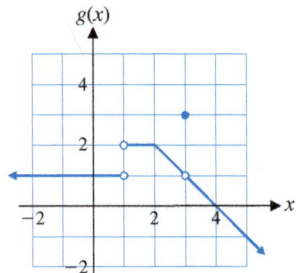

Figure for 9–16

9. $g(1.9)$ 10. $g(0.1)$

11. $g(3.5)$ 12. $g(2.5)$

13. (A) $\lim\limits_{x\to 1^-} g(x)$ (B) $\lim\limits_{x\to 1^+} g(x)$
    (C) $\lim\limits_{x\to 1} g(x)$ (D) $g(1)$
     (E) Is it possible to define $g(1)$ so that $\lim\limits_{x\to 1} g(x) = g(1)$? Explain.

14. (A) $\lim\limits_{x\to 2^-} g(x)$ (B) $\lim\limits_{x\to 2^+} g(x)$
    (C) $\lim\limits_{x\to 2} g(x)$ (D) $g(2)$

15. (A) $\lim\limits_{x\to 3^-} g(x)$ (B) $\lim\limits_{x\to 3^+} g(x)$
    (C) $\lim\limits_{x\to 3} g(x)$ (D) $g(3)$
     (E) Is it possible to redefine $g(3)$ so that $\lim\limits_{x\to 3} g(x) = g(3)$? Explain.

16. (A) $\lim\limits_{x\to 4^-} g(x)$ (B) $\lim\limits_{x\to 4^+} g(x)$
    (C) $\lim\limits_{x\to 4} g(x)$ (D) $g(4)$

In Problems 17–20, use the graph of the function f shown to estimate the indicated limits and function values.

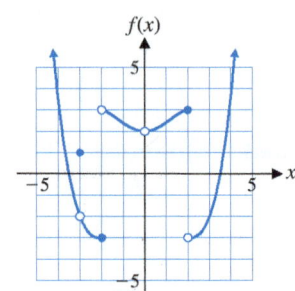

Figure for 17–20

17. (A) $\lim\limits_{x\to -3^+} f(x)$ (B) $\lim\limits_{x\to -3^-} f(x)$
    (C) $\lim\limits_{x\to -3} f(x)$ (D) $f(-3)$
     (E) Is it possible to redefine $f(-3)$ so that $\lim\limits_{x\to -3} f(x) = f(-3)$? Explain.

18. (A) $\lim\limits_{x\to -2^+} f(x)$ (B) $\lim\limits_{x\to -2^-} f(x)$
    (C) $\lim\limits_{x\to -2} f(x)$ (D) $f(-2)$
     (E) Is it possible to define $f(-2)$ so that $\lim\limits_{x\to -2} f(x) = f(-2)$? Explain.

19. (A) $\lim\limits_{x\to 0^+} f(x)$ (B) $\lim\limits_{x\to 0^-} f(x)$
    (C) $\lim\limits_{x\to 0} f(x)$ (D) $f(0)$
     (E) Is it possible to redefine $f(0)$ so that $\lim\limits_{x\to 0} f(x) = f(0)$? Explain.

20. (A) $\lim\limits_{x\to 2^+} f(x)$ (B) $\lim\limits_{x\to 2^-} f(x)$
    (C) $\lim\limits_{x\to 2} f(x)$ (D) $f(2)$
     (E) Is it possible to redefine $f(2)$ so that $\lim\limits_{x\to 2} f(x) = f(2)$? Explain.

In Problems 21–30, find each limit if it exists.

21. $\lim\limits_{x\to 3} 4x$ 22. $\lim\limits_{x\to -2} 3x$

23. $\lim\limits_{x\to -4} (x + 5)$ 24. $\lim\limits_{x\to 5}(x - 3)$

25. $\lim\limits_{x\to 2} x(x - 4)$ 26. $\lim\limits_{x\to -1} x(x + 3)$

27. $\lim\limits_{x\to -3} \dfrac{x}{x + 5}$ 28. $\lim\limits_{x\to 4} \dfrac{x - 2}{x}$

29. $\lim\limits_{x\to 1} \sqrt{5x + 4}$ 30. $\lim\limits_{x\to 0} \sqrt{16 - 7x}$

Given that $\lim_{x \to 1} f(x) = -5$ and $\lim_{x \to 1} g(x) = 4$, find the indicated limits in Problems 31–38.

31. $\lim_{x \to 1}(-3)f(x)$

32. $\lim_{x \to 1} 2g(x)$

33. $\lim_{x \to 1}[2f(x) + g(x)]$

34. $\lim_{x \to 1}[g(x) - 3f(x)]$

35. $\lim_{x \to 1} \dfrac{2 - f(x)}{x + g(x)}$

36. $\lim_{x \to 1} \dfrac{3 - f(x)}{1 - 4g(x)}$

37. $\lim_{x \to 1} \sqrt{g(x) - f(x)}$

38. $\lim_{x \to 1} \sqrt[3]{2x + 2f(x)}$

In Problems 39–42, sketch a possible graph of a function that satisfies the given conditions.

39. $f(0) = 1$; $\lim_{x \to 0^-} f(x) = 3$; $\lim_{x \to 0^+} f(x) = 1$

40. $f(1) = -2$; $\lim_{x \to 1^-} f(x) = 2$; $\lim_{x \to 1^+} f(x) = -2$

41. $f(-2) = 2$; $\lim_{x \to -2^-} f(x) = 1$; $\lim_{x \to -2^+} f(x) = 1$

42. $f(0) = -1$; $\lim_{x \to 0^-} f(x) = 2$; $\lim_{x \to 0^+} f(x) = 2$

B

In Problems 43–58, find each indicated quantity if it exists.

43. Let $f(x) = \begin{cases} 1 - x^2 & \text{if } x \le 0 \\ 1 + x^2 & \text{if } x > 0 \end{cases}$. Find

(A) $\lim_{x \to 0^+} f(x)$ (B) $\lim_{x \to 0^-} f(x)$

(C) $\lim_{x \to 0} f(x)$ (D) $f(0)$

44. Let $f(x) = \begin{cases} 2 + x & \text{if } x \le 0 \\ 2 - x & \text{if } x > 0 \end{cases}$. Find

(A) $\lim_{x \to 0^+} f(x)$ (B) $\lim_{x \to 0^-} f(x)$

(C) $\lim_{x \to 0} f(x)$ (D) $f(0)$

45. Let $f(x) = \begin{cases} x^2 & \text{if } x < 1 \\ 2x & \text{if } x > 1 \end{cases}$. Find

(A) $\lim_{x \to 1^+} f(x)$ (B) $\lim_{x \to 1^-} f(x)$

(C) $\lim_{x \to 1} f(x)$ (D) $f(1)$

46. Let $f(x) = \begin{cases} x + 3 & \text{if } x < -2 \\ \sqrt{x + 2} & \text{if } x > -2 \end{cases}$. Find

(A) $\lim_{x \to -2^+} f(x)$ (B) $\lim_{x \to -2^-} f(x)$

(C) $\lim_{x \to -2} f(x)$ (D) $f(-2)$

47. Let $f(x) = \begin{cases} \dfrac{x^2 - 9}{x + 3} & \text{if } x < 0 \\ \dfrac{x^2 - 9}{x - 3} & \text{if } x > 0 \end{cases}$. Find

(A) $\lim_{x \to -3} f(x)$ (B) $\lim_{x \to 0} f(x)$

(C) $\lim_{x \to 3} f(x)$

48. Let $f(x) = \begin{cases} \dfrac{x}{x + 3} & \text{if } x < 0 \\ \dfrac{x}{x - 3} & \text{if } x > 0 \end{cases}$. Find

(A) $\lim_{x \to -3} f(x)$ (B) $\lim_{x \to 0} f(x)$

(C) $\lim_{x \to 3} f(x)$

49. Let $f(x) = \dfrac{|x - 1|}{x - 1}$. Find

(A) $\lim_{x \to 1^+} f(x)$ (B) $\lim_{x \to 1^-} f(x)$

(C) $\lim_{x \to 1} f(x)$ (D) $f(1)$

50. Let $f(x) = \dfrac{x - 3}{|x - 3|}$. Find

(A) $\lim_{x \to 3^+} f(x)$ (B) $\lim_{x \to 3^-} f(x)$

(C) $\lim_{x \to 3} f(x)$ (D) $f(3)$

51. Let $f(x) = \dfrac{x - 2}{x^2 - 2x}$. Find

(A) $\lim_{x \to 0} f(x)$ (B) $\lim_{x \to 2} f(x)$

(C) $\lim_{x \to 4} f(x)$

52. Let $f(x) = \dfrac{x + 3}{x^2 + 3x}$. Find

(A) $\lim_{x \to -3} f(x)$ (B) $\lim_{x \to 0} f(x)$

(C) $\lim_{x \to 3} f(x)$

53. Let $f(x) = \dfrac{x^2 - x - 6}{x + 2}$. Find

(A) $\lim_{x \to -2} f(x)$ (B) $\lim_{x \to 0} f(x)$

(C) $\lim_{x \to 3} f(x)$

54. Let $f(x) = \dfrac{x^2 + x - 6}{x + 3}$. Find

(A) $\lim_{x \to -3} f(x)$ (B) $\lim_{x \to 0} f(x)$

(C) $\lim_{x \to 2} f(x)$

55. Let $f(x) = \dfrac{(x + 2)^2}{x^2 - 4}$. Find

(A) $\lim_{x \to -2} f(x)$ (B) $\lim_{x \to 0} f(x)$

(C) $\lim_{x \to 2} f(x)$

56. Let $f(x) = \dfrac{x^2 - 1}{(x + 1)^2}$. Find

(A) $\lim_{x \to -1} f(x)$ (B) $\lim_{x \to 0} f(x)$

(C) $\lim_{x \to 1} f(x)$

57. Let $f(x) = \dfrac{2x^2 - 3x - 2}{x^2 + x - 6}$. Find

(A) $\lim_{x \to 2} f(x)$ (B) $\lim_{x \to 0} f(x)$

(C) $\lim_{x \to 1} f(x)$

58. Let $f(x) = \dfrac{3x^2 + 2x - 1}{x^2 + 3x + 2}$. Find

(A) $\lim_{x \to -3} f(x)$ (B) $\lim_{x \to -1} f(x)$

(C) $\lim_{x \to 2} f(x)$

In Problems 59–64, discuss the validity of each statement. If the statement is always true, explain why. If not, give a counterexample.

59. If $\lim_{x \to 1} f(x) = 0$ and $\lim_{x \to 1} g(x) = 0$, then $\lim_{x \to 1} \dfrac{f(x)}{g(x)} = 0$.

60. If $\lim_{x \to 1} f(x) = 1$ and $\lim_{x \to 1} g(x) = 1$, then $\lim_{x \to 1} \dfrac{f(x)}{g(x)} = 1$.

61. If f is a polynomial, then, as x approaches 0, the right-hand limit exists and is equal to the left-hand limit.

62. If f is a rational function, then, as x approaches 0, the right-hand limit exists and is equal to the left-hand limit.

63. If f is a function such that $\lim_{x \to 0} f(x)$ exists, then $f(0)$ exists.

64. If f is a function such that $f(0)$ exists, then $\lim_{x \to 0} f(x)$ exists.

Compute the following limit for each function in Problems 65–68.

$$\lim_{h\to 0} \frac{f(2 + h) - f(2)}{h}$$

65. $f(x) = 3x + 1$ **66.** $f(x) = 5x - 1$

67. $f(x) = x^2 + 1$ **68.** $f(x) = x^2 - 2$

C

69. Let f be defined by

$$f(x) = \begin{cases} 1 + mx & \text{if } x \le 1 \\ 4 - mx & \text{if } x > 1 \end{cases}$$

where m is a constant.

(A) Graph f for $m = 1$, and find

$$\lim_{x\to 1^-} f(x) \quad \text{and} \quad \lim_{x\to 1^+} f(x)$$

(B) Graph f for $m = 2$, and find

$$\lim_{x\to 1^-} f(x) \quad \text{and} \quad \lim_{x\to 1^+} f(x)$$

(C) Find m so that

$$\lim_{x\to 1^-} f(x) = \lim_{x\to 1^+} f(x)$$

and graph f for this value of m.

✎ (D) Write a brief verbal description of each graph. How does the graph in part (C) differ from the graphs in parts (A) and (B)?

70. Let f be defined by

$$f(x) = \begin{cases} -3m + 0.5x & \text{if } x \le 2 \\ 3m - x & \text{if } x > 2 \end{cases}$$

where m is a constant.

(A) Graph f for $m = 0$, and find

$$\lim_{x\to 2^-} f(x) \quad \text{and} \quad \lim_{x\to 2^+} f(x)$$

(B) Graph f for $m = 1$, and find

$$\lim_{x\to 2^-} f(x) \quad \text{and} \quad \lim_{x\to 2^+} f(x)$$

(C) Find m so that

$$\lim_{x\to 2^-} f(x) = \lim_{x\to 2^+} f(x)$$

and graph f for this value of m.

✎ (D) Write a brief verbal description of each graph. How does the graph in part (C) differ from the graphs in parts (A) and (B)?

Find each limit in Problems 71–74, where a is a real constant.

71. $\lim_{h\to 0} \dfrac{(a + h)^2 - a^2}{h}$

72. $\lim_{h\to 0} \dfrac{[3(a + h) - 2] - (3a - 2)}{h}$

73. $\lim_{h\to 0} \dfrac{\sqrt{a + h} - \sqrt{a}}{h}, \quad a > 0$

74. $\lim_{h\to 0} \dfrac{\dfrac{1}{a + h} - \dfrac{1}{a}}{h}, \quad a \ne 0$

Applications

75. Telephone rates. A long-distance telephone service charges \$0.99 for the first 20 minutes or less of a call and \$0.07 per minute for each additional minute or fraction thereof.

(A) Write a piecewise definition of the charge $F(x)$ for a long-distance call lasting x minutes.

(B) Graph $F(x)$ for $0 < x \le 40$.

(C) Find $\lim_{x\to 20^-} F(x)$, $\lim_{x\to 20^+} F(x)$, and $\lim_{x\to 20} F(x)$, whichever exist.

76. Telephone rates. A second long-distance telephone service charges \$0.09 per minute or fraction thereof for calls lasting 10 minutes or more and \$0.18 per minute or fraction thereof for calls lasting less than 10 minutes.

(A) Write a piecewise definition of the charge $G(x)$ for a long-distance call lasting x minutes.

(B) Graph $G(x)$ for $0 < x \le 40$.

(C) Find $\lim_{x\to 10^-} G(x)$, $\lim_{x\to 10^+} G(x)$, and $\lim_{x\to 10} G(x)$, whichever exist.

✎ **77. Telephone rates.** Refer to Problems 75 and 76. Write a brief verbal comparison of the two services described for calls lasting 20 minutes or less.

✎ **78. Telephone rates.** Refer to Problems 75 and 76. Write a brief verbal comparison of the two services described for calls lasting more than 20 minutes.

A company sells custom embroidered apparel and promotional products. Table 1 shows the volume discounts offered by the company, where x is the volume of a purchase in dollars. Problems 79 and 80 deal with two different interpretations of this discount method.

Table 1

Volume Discount (Excluding Tax)

Volume (\$x)	Discount Amount
\$300 $\le x <$ \$1,000	3%
\$1,000 $\le x <$ \$3,000	5%
\$3,000 $\le x <$ \$5,000	7%
\$5,000 $\le x$	10%

79. Volume discount. Assume that the volume discounts in Table 1 apply to the entire purchase. That is, if the volume x satisfies \$300 $\le x <$ \$1,000, then the entire purchase is discounted 3%. If the volume x satisfies \$1,000 $\le x <$ \$3,000, the entire purchase is discounted 5%, and so on.

(A) If x is the volume of a purchase before the discount is applied, then write a piecewise definition for the discounted price $D(x)$ of this purchase.

(B) Use one-sided limits to investigate the limit of $D(x)$ as x approaches \$1,000. As x approaches \$3,000.

80. Volume discount. Assume that the volume discounts in Table 1 apply only to that portion of the volume in each interval. That is, the discounted price for a \$4,000 purchase would be computed as follows:

$$300 + 0.97(700) + 0.95(2,000) + 0.93(1,000) = 3,809$$

(A) If x is the volume of a purchase before the discount is applied, then write a piecewise definition for the discounted price $P(x)$ of this purchase.

(B) Use one-sided limits to investigate the limit of $P(x)$ as x approaches \$1,000. As x approaches \$3,000.

 (C) Compare this discount method with the one in Problem 79. Does one always produce a lower price than the other? Discuss.

81. Pollution. A state charges polluters an annual fee of \$20 per ton for each ton of pollutant emitted into the atmosphere, up to a maximum of 4,000 tons. No fees are charged for emissions beyond the 4,000-ton limit. Write a piecewise definition of the fees $F(x)$ charged for the emission of x tons of pollutant in a year. What is the limit of $F(x)$ as x approaches 4,000 tons? As x approaches 8,000 tons?

82. Pollution. Refer to Problem 81. The fee per ton of pollution is given by $A(x) = F(x)/x$. Write a piecewise definition of $A(x)$. What is the limit of $A(x)$ as x approaches 4,000 tons? As x approaches 8,000 tons?

83. Voter turnout. Statisticians often use piecewise-defined functions to predict outcomes of elections. For the following functions f and g, find the limit of each function as x approaches 5 and as x approaches 10.

$$f(x) = \begin{cases} 0 & \text{if } x \leq 5 \\ 0.8 - 0.08x & \text{if } 5 < x < 10 \\ 0 & \text{if } 10 \leq x \end{cases}$$

$$g(x) = \begin{cases} 0 & \text{if } x \leq 5 \\ 0.8x - 0.04x^2 - 3 & \text{if } 5 < x < 10 \\ 1 & \text{if } 10 \leq x \end{cases}$$

Answers to Matched Problems

1.

x	-2	-1	0	1	2	3	4
$h(x)$	1.0	1.5	2.0	2.5	3.0	3.5	4.0

2. $\lim\limits_{x \to 1} f(x) = 2$

3. 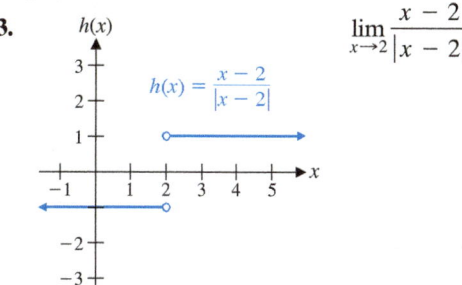 $\lim\limits_{x \to 2} \dfrac{x-2}{|x-2|}$ does not exist

4. (A) $\lim\limits_{x \to 0^-} f(x) = 0$ (B) $\lim\limits_{x \to 1^-} f(x) = 1$

   $\lim\limits_{x \to 0^+} f(x) = 0$ $\lim\limits_{x \to 1^+} f(x) = 2$

   $\lim\limits_{x \to 0} f(x) = 0$ $\lim\limits_{x \to 1} f(x)$ does not exist

   $f(0) = 0$ $f(1)$ not defined

   (C) $\lim\limits_{x \to 3^-} f(x) = 3$ 5. -6

   $\lim\limits_{x \to 3^+} f(x) = 3$ 6. (A) 6

   $\lim\limits_{x \to 3} f(x) = 3$ (B) $\sqrt{6}$

   $f(3)$ not defined   (C) $\frac{4}{5}$

7. (A) 13   8. (A) -2

   (B) 7   (B) Does not exist

   (C) Does not exist

   (D) Not defined

9. -2 10. Does not exist   11. $1/(2\sqrt{3})$

3-2 Infinite Limits and Limits at Infinity

- Infinite Limits
- Locating Vertical Asymptotes
- Limits at Infinity
- Finding Horizontal Asymptotes

In this section, we consider two new types of limits: infinite limits and limits at infinity. Infinite limits and vertical asymptotes are used to describe the behavior of functions that are unbounded near $x = a$. Limits at infinity and horizontal asymptotes are used to describe the behavior of functions as x assumes arbitrarily large positive values or arbitrarily large negative values. Although we will include graphs to illustrate basic concepts, we postpone a discussion of graphing techniques until Chapter 5.

Infinite Limits

The graph of $f(x) = \dfrac{1}{x-1}$ (Fig. 1) indicates that

$$\lim_{x \to 1^+} \frac{1}{x-1}$$

does not exist. There does not exist a real number L that the values of $f(x)$ approach as x approaches 1 from the right. Instead, as x approaches 1 from the right, the values of $f(x)$ are positive and become larger and larger; that is, $f(x)$ increases without bound (Table 1). We express this behavior symbolically as

$$\lim_{x \to 1^+} \frac{1}{x-1} = \infty \quad \text{or} \quad f(x) = \frac{1}{x-1} \to \infty \quad \text{as} \quad x \to 1^+ \tag{1}$$

Since ∞ is a not a real number, *the limit in (1) does not exist.* We are using the symbol ∞ to describe the manner in which the limit fails to exist, and we call this situation an **infinite limit**. If x approaches 1 from the left, the values of $f(x)$ are negative and become larger and larger in absolute value; that is, $f(x)$ decreases through negative values without bound (Table 2). We express this behavior symbolically as

$$\lim_{x \to 1^-} \frac{1}{x - 1} = -\infty \quad \text{or} \quad f(x) = \frac{1}{x - 1} \to -\infty \quad \text{as} \quad x \to 1^- \tag{2}$$

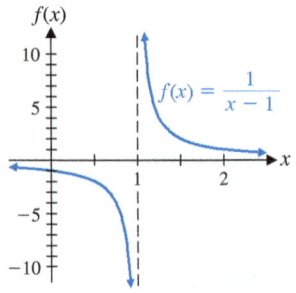

$f(x)$

$$f(x) = \frac{1}{x - 1}$$

Figure 1

Table 1

x	$f(x) = \dfrac{1}{x - 1}$
1.1	10
1.01	100
1.001	1,000
1.0001	10,000
1.00001	100,000
1.000001	1,000,000

Table 2

x	$f(x) = \dfrac{1}{x - 1}$
0.9	-10
0.99	-100
0.999	$-1,000$
0.9999	$-10,000$
0.99999	$-100,000$
0.999999	$-1,000,000$

The one-sided limits in (1) and (2) describe the behavior of the graph as $x \to 1$ (Fig. 1). Does the two-sided limit of $f(x)$ as $x = 1$ exist? No, because neither of the one-sided limits exists. Also, there is no reasonable way to use the symbol ∞ to describe the behavior of $f(x)$ as $x \to 1$ on both sides of 1. We say that

$$\lim_{x \to 1} \frac{1}{x - 1} \text{ does not exist}$$

EXPLORE & DISCUSS 1

Let $g(x) = \dfrac{1}{(x - 1)^2}$

Construct tables for $g(x)$ as $x \to 1^+$ and as $x \to 1^-$. Use these tables and infinite limits to discuss the behavior of $g(x)$ near $x = 1$.

We used the dashed vertical line $x = 1$ in Figure 1 to illustrate the infinite limits as x approaches 1 from the right and from the left. We call this line a *vertical asymptote.*

> **DEFINITION Infinite Limits and Vertical Asymptotes**
>
> The vertical line $x = a$ is a **vertical asymptote** for the graph of $y = f(x)$ if
>
> $$f(x) \to \infty \quad \text{or} \quad f(x) \to -\infty \quad \text{as} \quad x \to a^+ \quad \text{or} \quad x \to a^-$$
>
> [That is, if $f(x)$ either increases or decreases without bound as x approaches a from the right or from the left].

Locating Vertical Asymptotes

How do we locate vertical asymptotes? If f is a polynomial function, then $\lim_{x \to a} f(x)$ is equal to the real number $f(a)$ [Theorem 3, Section 3-1]. So *a polynomial function has no vertical asymptotes.* Similarly (again by Theorem 3, Section 3-1), *a vertical asymptote of a rational function can occur only at a zero of its denominator.* Theorem 1 provides a simple procedure for locating the vertical asymptotes of a rational function.

> **THEOREM 1   Locating Vertical Asymptotes of Rational Functions**
> If $f(x) = n(x)/d(x)$ is a rational function, $d(c) = 0$ and $n(c) \neq 0$, then the line $x = c$ is a vertical asymptote of the graph of f.

If $f(x) = n(x)/d(x)$ and both $n(c) = 0$ and $d(c) = 0$, then the limit of $f(x)$ as x approaches c involves an indeterminate form and Theorem 1 does not apply:

$$\lim_{x \to c} f(x) = \lim_{x \to c} \frac{n(x)}{d(x)} \quad \frac{0}{0} \text{ indeterminate form}$$

Algebraic simplification is often useful in this situation.

EXAMPLE 1   **Locating Vertical Asymptotes** Let $f(x) = \dfrac{x^2 + x - 2}{x^2 - 1}$

Describe the behavior of f at each zero of the denominator. Use ∞ and $-\infty$ when appropriate. Identify all vertical asymptotes.

SOLUTION   Let $n(x) = x^2 + x - 2$ and $d(x) = x^2 - 1$. Factoring the denominator, we see that

$$d(x) = x^2 - 1 = (x - 1)(x + 1)$$

has two zeros: $x = -1$ and $x = 1$.

First, we consider $x = -1$. Since $d(-1) = 0$ and $n(-1) = -2 \neq 0$, Theorem 1 tells us that the line $x = -1$ is a vertical asymptote. So at least one of the one-sided limits at $x = -1$ must be either ∞ or $-\infty$. Examining tables of values of f for x near -1 or a graph on a graphing calculator will show which is the case. From Tables 3 and 4, we see that

$$\lim_{x \to -1^-} \frac{x^2 + x - 2}{x^2 - 1} = -\infty \quad \text{and} \quad \lim_{x \to -1^+} \frac{x^2 + x - 2}{x^2 - 1} = \infty$$

Table 3

x	$f(x) = \dfrac{x^2 + x - 2}{x^2 - 1}$
−1.1	−9
−1.01	−99
−1.001	−999
−1.0001	−9,999
−1.00001	−99,999

Table 4

x	$f(x) = \dfrac{x^2 + x - 2}{x^2 - 1}$
−0.9	11
−0.99	101
−0.999	1,001
−0.9999	10,001
−0.99999	100,001

Now we consider the other zero of $d(x)$, $x = 1$. This time $n(1) = 0$ and Theorem 1 does not apply. We use algebraic simplification to investigate the behavior of the function at $x = 1$:

$$\lim_{x \to 1} f(x) = \lim_{x \to 1} \frac{x^2 + x - 2}{x^2 - 1} \qquad \frac{0}{0} \text{ indeterminate form}$$

$$= \lim_{x \to 1} \frac{(x - 1)(x + 2)}{(x - 1)(x + 1)}$$

$$= \lim_{x \to 1} \frac{x + 2}{x + 1} \qquad \text{Reduced to lowest terms}$$
$$\qquad\qquad\qquad\qquad \text{(see Appendix A-4)}$$

$$= \frac{3}{2}$$

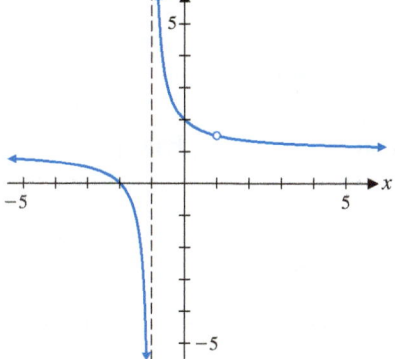

Figure 2 $f(x) = \dfrac{x^2 + x - 2}{x^2 - 1}$

Since the limit exists as x approaches 1, f does not have a vertical asymptote at $x = 1$. The graph of f (Fig. 2) shows the behavior at the vertical asymptote $x = -1$ and also at $x = 1$.

Matched Problem 1   Let $f(x) = \dfrac{x - 3}{x^2 - 4x + 3}$.

Describe the behavior of f at each zero of the denominator. Use ∞ and $-\infty$ when appropriate. Identify all vertical asymptotes.

EXAMPLE 2   **Locating Vertical Asymptotes** Let $f(x) = \dfrac{x^2 + 20}{5(x - 2)^2}$.

Describe the behavior of f at each zero of the denominator. Use ∞ and $-\infty$ when appropriate. Identify all vertical asymptotes.

SOLUTION   Let $n(x) = x^2 + 20$ and $d(x) = 5(x - 2)^2$. The only zero of $d(x)$ is $x = 2$. Since $n(2) = 24 \neq 0$, f has a vertical asymptote at $x = 2$ (Theorem 1). Tables 5 and 6 show that $f(x) \to \infty$ as $x \to 2$ from either side, and we have

$$\lim_{x \to 2^+} \frac{x^2 + 20}{5(x - 2)^2} = \infty \quad \text{and} \quad \lim_{x \to 2^-} \frac{x^2 + 20}{5(x - 2)^2} = \infty$$

Table 5

x	$f(x) = \dfrac{x^2 + 20}{5(x - 2)^2}$
2.1	488.2
2.01	48,080.02
2.001	4,800,800.2

Table 6

x	$f(x) = \dfrac{x^2 + 20}{5(x - 2)^2}$
1.9	472.2
1.99	47,920.02
1.999	4,799,200.2

$f(x) = \dfrac{x^2 + 20}{5(x - 2)^2}$

Figure 3

The denominator d has no other zeros, so f does not have any other vertical asymptotes. The graph of f (Fig. 3) shows the behavior at the vertical asymptote $x = 2$. Because the left- and right-hand limits are both infinite, we write

$$\lim_{x \to 2} \frac{x^2 + 20}{5(x - 2)^2} = \infty$$

Matched Problem 2   Let $f(x) = \dfrac{x - 1}{(x + 3)^2}$.

Describe the behavior of f at each zero of the denominator. Use ∞ and $-\infty$ when appropriate. Identify all vertical asymptotes.

CONCEPTUAL INSIGHT

When is it correct to say that a limit does not exist, and when is it correct to use $\pm\infty$? It depends on the situation. Table 7 lists the infinite limits that we discussed in Examples 1 and 2.

Table 7

Right-Hand Limit	Left-Hand Limit	Two-Sided Limit
$\lim\limits_{x \to -1^+} \dfrac{x^2 + x - 2}{x^2 - 1} = \infty$	$\lim\limits_{x \to -1^-} \dfrac{x^2 + x - 2}{x^2 - 1} = -\infty$	$\lim\limits_{x \to -1} \dfrac{x^2 + x - 2}{x^2 - 1}$ does not exist
$\lim\limits_{x \to -2^+} \dfrac{x^2 + 20}{5(x - 2)^2} = \infty$	$\lim\limits_{x \to 2^-} \dfrac{x^2 + 20}{5(x - 2)^2} = \infty$	$\lim\limits_{x \to 2} \dfrac{x^2 + 20}{5(x - 2)^2} = \infty$

The instructions in Examples 1 and 2 said that we should use infinite limits to describe the behavior at vertical asymptotes. If we had been asked to *evaluate* the limits, with no mention of ∞ or asymptotes, then the correct answer would be that **all of these limits do not exist**. Remember, ∞ is a symbol used to describe the behavior of functions at vertical asymptotes.

Limits at Infinity

The symbol ∞ can also be used to indicate that an independent variable is increasing or decreasing without bound. We write $x \to \infty$ to indicate that x is increasing without bound through positive values and $x \to -\infty$ to indicate that x is decreasing without bound through negative values. We begin by considering power functions of the form x^p and $1/x^p$ where p is a positive real number.

If p is a positive real number, then x^p increases as x increases. There is no upper bound on the values of x^p. We indicate this behavior by writing

$$\lim_{x \to \infty} x^p = \infty \quad \text{or} \quad x^p \to \infty \quad \text{as} \quad x \to \infty$$

Since the reciprocals of very large numbers are very small numbers, it follows that $1/x^p$ approaches 0 as x increases without bound. We indicate this behavior by writing

$$\lim_{x \to \infty} \frac{1}{x^p} = 0 \quad \text{or} \quad \frac{1}{x^p} \to 0 \quad \text{as} \quad x \to \infty$$

Figure 4 illustrates the preceding behavior for $f(x) = x^2$ and $g(x) = 1/x^2$, and we write

$$\lim_{x \to \infty} f(x) = \infty \quad \text{and} \quad \lim_{x \to \infty} g(x) = 0$$

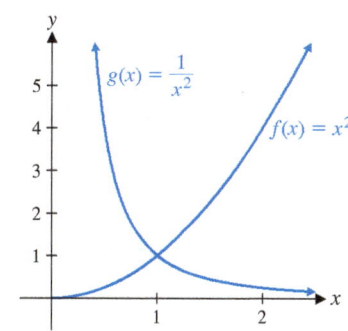

Figure 4

Limits of power forms as x decreases without bound behave in a similar manner, with two important differences. First, if x is negative, then x^p is not defined for all values of p. For example, $x^{1/2} = \sqrt{x}$ is not defined for negative values of x. Second, if x^p is defined, then it may approach ∞ or $-\infty$, depending on the value of p. For example,

$$\lim_{x \to -\infty} x^2 = \infty \quad \text{but} \quad \lim_{x \to -\infty} x^3 = -\infty$$

For the function g in Figure 4, the line $y = 0$ (the x axis) is called a *horizontal asymptote*. In general, a line $y = b$ is a **horizontal asymptote** of the graph of $y = f(x)$ if $f(x)$ approaches b as either x increases without bound or x decreases without bound. Symbolically, $y = b$ is a horizontal asymptote if either

$$\lim_{x \to -\infty} f(x) = b \quad \text{or} \quad \lim_{x \to \infty} f(x) = b$$

In the first case, the graph of f will be close to the horizontal line $y = b$ for large (in absolute value) negative x. In the second case, the graph will be close to the horizontal line $y = b$ for large positive x. Figure 5 shows the graph of a function with two horizontal asymptotes: $y = 1$ and $y = -1$.

Theorem 2 summarizes the various possibilities for limits of power functions as x increases or decreases without bound.

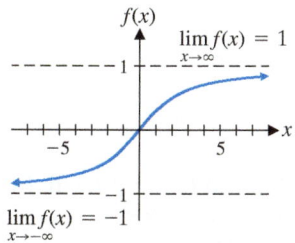

Figure 5

> **THEOREM 2   Limits of Power Functions at Infinity**
>
> If p is a positive real number and k is any real number except 0, then
>
> **1.** $\displaystyle\lim_{x \to -\infty} \frac{k}{x^p} = 0$ **2.** $\displaystyle\lim_{x \to \infty} \frac{k}{x^p} = 0$
>
> **3.** $\displaystyle\lim_{x \to -\infty} kx^p = \pm\infty$ **4.** $\displaystyle\lim_{x \to \infty} kx^p = \pm\infty$
>
> provided that x^p is a real number for negative values of x. The limits in 3 and 4 will be either $-\infty$ or ∞, depending on k and p.

How can we use Theorem 2 to evaluate limits at infinity? It turns out that the limit properties listed in Theorem 2, Section 3-1, are also valid if we replace the statement $x \to c$ with $x \to \infty$ or $x \to -\infty$.

EXAMPLE 3   **Limit of a Polynomial Function at Infinity** Let $p(x) = 2x^3 - x^2 - 7x + 3$. Find the limit of $p(x)$ as x approaches ∞ and as x approaches $-\infty$.

SOLUTION   Since limits of power functions of the form $1/x^p$ approach 0 as x approaches ∞ or $-\infty$, it is convenient to work with these reciprocal forms whenever possible. If we factor out the term involving the highest power of x, then we can write $p(x)$ as

$$p(x) = 2x^3\left(1 - \frac{1}{2x} - \frac{7}{2x^2} + \frac{3}{2x^3}\right)$$

Using Theorem 2 in Section 3-3 and Theorem 2 in Section 3-1, we write

$$\lim_{x \to \infty}\left(1 - \frac{1}{2x} - \frac{7}{2x^2} + \frac{3}{2x^3}\right) = 1 - 0 - 0 + 0 = 1$$

For large values of x,

$$\left(1 - \frac{1}{2x} - \frac{7}{2x^2} + \frac{3}{2x^3}\right) \approx 1$$

and

$$p(x) = 2x^3\left(1 - \frac{1}{2x} - \frac{7}{2x^2} + \frac{3}{2x^3}\right) \approx 2x^3$$

Since $2x^3 \to \infty$ as $x \to \infty$, it follows that

$$\lim_{x \to \infty} p(x) = \lim_{x \to \infty} 2x^3 = \infty$$

Similarly, $2x^3 \to -\infty$ as $x \to -\infty$ implies that

$$\lim_{x \to -\infty} p(x) = \lim_{x \to -\infty} 2x^3 = -\infty$$

So the behavior of $p(x)$ for large values is the same as the behavior of the highest-degree term, $2x^3$.

Matched Problem 3   Let $p(x) = -4x^4 + 2x^3 + 3x$. Find the limit of $p(x)$ as x approaches ∞ and as x approaches $-\infty$.

The term with highest degree in a polynomial is called the **leading term**. In the solution to Example 3, the limits at infinity of $p(x) = 2x^3 - x^2 - 7x + 3$ were the same as the limits of the leading term $2x^3$. Theorem 3 states that this is true for any polynomial of degree greater than or equal to 1.

> **THEOREM 3   Limits of Polynomial Functions at Infinity**
>
> If
>
> $$p(x) = a_n x^n + a_{n-1} x^{n-1} + \cdots + a_1 x + a_0, \, a_n \neq 0, \, n \geq 1$$
>
> then
>
> $$\lim_{x \to \infty} p(x) = \lim_{x \to \infty} a_n x^n = \pm\infty$$
>
> and
>
> $$\lim_{x \to -\infty} p(x) = \lim_{x \to -\infty} a_n x^n = \pm\infty$$
>
> Each limit will be either $-\infty$ or ∞, depending on a_n and n.

A polynomial of degree 0 is a constant function $p(x) = a_0$, and its limit as x approaches ∞ or $-\infty$ is the number a_0. For any polynomial of degree 1 or greater, Theorem 3 states that the limit as x approaches ∞ or $-\infty$ cannot be equal to a number. This means that **polynomials of degree 1 or greater never have horizontal asymptotes**.

A description of the behavior of the limits of any function f at infinity is called the **end behavior** of f. $\text{Lim}_{x \to \infty} f(x)$ determines the right end behavior, and $\lim_{x \to -\infty} f(x)$ determines the left end behavior. According to Theorem 3, the right and left end behavior of a nonconstant polynomial $p(x)$ is always an infinite limit.

EXAMPLE 4   **End Behavior of a Polynomial** Describe the end behavior of each polynomial.

(A) $p(x) = 3x^3 - 500x^2$ (B) $p(x) = 3x^3 - 500x^4$

SOLUTION   (A) According to Theorem 3, as x increases without bound to the right, the right end behavior is

$$\lim_{x \to \infty} (3x^3 - 500x^2) = \lim_{x \to \infty} 3x^3 = \infty$$

and as x increases without bound to the left, the end behavior is

$$\lim_{x \to -\infty} (3x^3 - 500x^2) = \lim_{x \to -\infty} 3x^3 = -\infty$$

(B) As x increases without bound to the right, the end behavior is

$$\lim_{x \to \infty} (3x^3 - 500x^4) = \lim_{x \to \infty} (-500x^4) = -\infty$$

and as x increases without bound to the left, the end behavior is

$$\lim_{x \to -\infty} (3x^3 - 500x^4) = \lim_{x \to -\infty} (-500x^4) = -\infty$$

Matched Problem 4   Describe the end behavior of each polynomial.

(A) $p(x) = 300x^2 - 4x^5$ (B) $p(x) = 300x^6 - 4x^5$

Finding Horizontal Asymptotes

Since a rational function is the ratio of two polynomials, it is not surprising that reciprocals of powers of x can be used to analyze limits of rational functions at infinity. For example, consider the rational function

$$f(x) = \frac{3x^2 - 5x + 9}{2x^2 + 7}$$

Factoring the highest-degree term out of the numerator and the denominator, we write

$$f(x) = \frac{3x^2}{2x^2} \cdot \frac{1 - \dfrac{5}{3x} + \dfrac{3}{x^2}}{1 + \dfrac{7}{2x^2}}$$

$$\lim_{x \to \infty} f(x) = \lim_{x \to \infty} \frac{3x^2}{2x^2} \cdot \lim_{x \to \infty} \frac{1 - \dfrac{5}{3x} + \dfrac{3}{x^2}}{1 + \dfrac{7}{2x^2}} \;\boxed{= \frac{3}{2} \cdot \frac{1 - 0 + 0}{1 + 0}} = \frac{3}{2}$$

The behavior of this rational function as x approaches infinity is determined by the ratio of the highest-degree term in the numerator ($3x^2$) to the highest-degree term in the denominator ($2x^2$). Theorem 4 generalizes this result to any rational function and lists the three possible outcomes.

THEOREM 4   Limits of Rational Functions at Infinity and Horizontal Asymptotes of Rational Functions

(A) If $f(x) = \dfrac{a_m x^m + a_{m-1} x^{m-1} + \cdots + a_1 x + a_0}{b_n x^n + b_{n-1} x^{n-1} + \cdots + b_1 x + b_0}$, $a_m \neq 0, b_n \neq 0$

then $\displaystyle\lim_{x \to \infty} f(x) = \lim_{x \to \infty} \frac{a_m x^m}{b_n x^n}$ and $\displaystyle\lim_{x \to -\infty} f(x) = \lim_{x \to -\infty} \frac{a_m x^m}{b_n x^n}$

(B) There are three possible cases for these limits:

1. If $m < n$, then $\displaystyle\lim_{x \to \infty} f(x) = \lim_{x \to -\infty} f(x) = 0$, and the line $y = 0$ (the x axis) is a horizontal asymptote of $f(x)$.

2. If $m = n$, then $\displaystyle\lim_{x \to \infty} f(x) = \lim_{x \to -\infty} f(x) = \frac{a_m}{b_n}$, and the line $y = \dfrac{a_m}{b_n}$ is a horizontal asymptote of $f(x)$.

3. If $m > n$, then each limit will be ∞ or $-\infty$, depending on m, n, a_m, and b_n, and $f(x)$ does not have a horizontal asymptote.

Notice that in cases 1 and 2 of Theorem 4, the limit is the same if x approaches ∞ or $-\infty$. So, **a rational function can have at most one horizontal asymptote** (see Fig. 6).

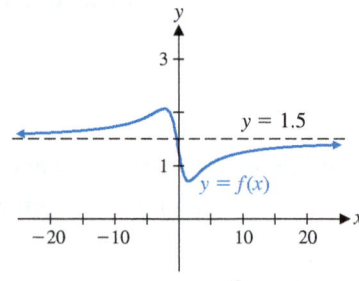

Figure 6 $f(x) = \dfrac{3x^2 - 5x + 9}{2x^2 + 7}$

CONCEPTUAL INSIGHT

The graph of f in Figure 6 dispels the misconception that the graph of a function cannot cross a horizontal asymptote. Horizontal asymptotes give us information about the graph of a function only as $x \to \infty$ and $x \to -\infty$, not at any specific value of x.

EXAMPLE 5 **Finding Horizontal Asymptotes** Find all horizontal asymptotes, if any, of each function.

(A) $f(x) = \dfrac{5x^3 - 2x^2 + 1}{4x^3 + 2x - 7}$ (B) $f(x) = \dfrac{3x^4 - x^2 + 1}{8x^6 - 10}$

(C) $f(x) = \dfrac{2x^5 - x^3 - 1}{6x^3 + 2x^2 - 7}$

SOLUTION We will make use of part A of Theorem 4.

(A) $\displaystyle\lim_{x \to \infty} f(x) = \lim_{x \to \infty} \frac{5x^3 - 2x^2 + 1}{4x^3 + 2x - 7} = \lim_{x \to \infty} \frac{5x^3}{4x^3} = \frac{5}{4}$

The line $y = 5/4$ is a horizontal asymptote of $f(x)$.

(B) $\displaystyle\lim_{x \to \infty} f(x) = \lim_{x \to \infty} \frac{3x^4 - x^2 + 1}{8x^6 - 10} = \lim_{x \to \infty} \frac{3x^4}{8x^6} = \lim_{x \to \infty} \frac{3}{8x^2} = 0$

The line $y = 0$ (the x axis) is a horizontal asymptote of $f(x)$.

(C) $\displaystyle\lim_{x \to \infty} f(x) = \lim_{x \to \infty} \frac{2x^5 - x^3 - 1}{6x^3 + 2x^2 - 7} = \lim_{x \to \infty} \frac{2x^5}{6x^3} = \lim_{x \to \infty} \frac{x^2}{3} = \infty$

The function $f(x)$ has no horizontal asymptotes.

Matched Problem 5 Find all horizontal asymptotes, if any, of each function.

(A) $f(x) = \dfrac{4x^3 - 5x + 8}{2x^4 - 7}$ (B) $f(x) = \dfrac{5x^6 + 3x}{2x^5 - x - 5}$

(C) $f(x) = \dfrac{2x^3 - x + 7}{4x^3 + 3x^2 - 100}$

An accurate sketch of the graph of a rational function requires knowledge of both vertical and horizontal asymptotes. As we mentioned earlier, we are postponing a detailed discussion of graphing techniques until Section 5-4.

EXAMPLE 6 **Finding the Asymptotes of a Rational Function** Find all the asymptotes of the function

$$f(x) = \frac{2x^2 - 5}{x^2 + 4x + 4}$$

SOLUTION Let $n(x) = 2x^2 - 5$ and $d(x) = x^2 + 4x + 4 = (x + 2)^2$. Since $d(x) = 0$ only at $x = -2$ and $n(-2) = 3$, the only vertical asymptote of f is the line $x = -2$ (Theorem 1). Since

$$\lim_{x \to \infty} f(x) = \lim_{x \to \infty} \frac{2x^2 - 5}{x^2 + 4x + 4} = \lim_{x \to \infty} \frac{2x^2}{x^2} = 2$$

the horizontal asymptote is the line $y = 2$ (Theorem 3).

Matched Problem 6 Find all the asymptotes of the function $f(x) = \dfrac{x^2 - 9}{x^2 - 4}$.

Exercises 3-2

A

Problems 1–8 refer to the following graph of $y = f(x)$.

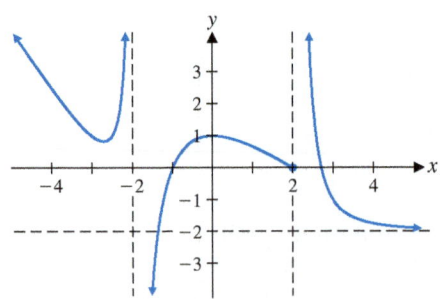

Figure for 1–8

1. $\lim\limits_{x \to \infty} f(x) = ?$
2. $\lim\limits_{x \to -\infty} f(x) = ?$
3. $\lim\limits_{x \to -2^+} f(x) = ?$
4. $\lim\limits_{x \to -2^-} f(x) = ?$
5. $\lim\limits_{x \to -2} f(x) = ?$
6. $\lim\limits_{x \to 2^+} f(x) = ?$
7. $\lim\limits_{x \to 2^-} f(x) = ?$
8. $\lim\limits_{x \to 2} f(x) = ?$

In Problems 9–16, find each limit. Use $-\infty$ and ∞ when appropriate.

9. $f(x) = \dfrac{x}{x - 5}$

   (A) $\lim\limits_{x \to 5^-} f(x)$ (B) $\lim\limits_{x \to 5^+} f(x)$ (C) $\lim\limits_{x \to 5} f(x)$

10. $f(x) = \dfrac{x^2}{x + 3}$

    (A) $\lim\limits_{x \to -3^-} f(x)$ (B) $\lim\limits_{x \to -3^+} f(x)$ (C) $\lim\limits_{x \to -3} f(x)$

11. $f(x) = \dfrac{2x - 4}{(x - 4)^2}$

    (A) $\lim\limits_{x \to 4^-} f(x)$ (B) $\lim\limits_{x \to 4^+} f(x)$ (C) $\lim\limits_{x \to 4} f(x)$

12. $f(x) = \dfrac{2x + 2}{(x + 2)^2}$

    (A) $\lim\limits_{x \to -2^-} f(x)$ (B) $\lim\limits_{x \to -2^+} f(x)$ (C) $\lim\limits_{x \to -2} f(x)$

13. $f(x) = \dfrac{x^2 + x - 2}{x - 1}$

    (A) $\lim\limits_{x \to 1^-} f(x)$ (B) $\lim\limits_{x \to 1^+} f(x)$ (C) $\lim\limits_{x \to 1} f(x)$

14. $f(x) = \dfrac{x^2 + x + 2}{x - 1}$

    (A) $\lim\limits_{x \to 1^-} f(x)$ (B) $\lim\limits_{x \to 1^+} f(x)$ (C) $\lim\limits_{x \to 1} f(x)$

15. $f(x) = \dfrac{x^2 - 3x + 2}{x + 2}$

    (A) $\lim\limits_{x \to -2^-} f(x)$ (B) $\lim\limits_{x \to -2^+} f(x)$ (C) $\lim\limits_{x \to -2} f(x)$

16. $f(x) = \dfrac{x^2 + x - 2}{x + 2}$

    (A) $\lim\limits_{x \to -2^-} f(x)$ (B) $\lim\limits_{x \to -2^+} f(x)$ (C) $\lim\limits_{x \to -2} f(x)$

In Problems 17–20, find the limit of each polynomial p(x) (A) as x approaches ∞ (B) as x approaches $-\infty$

17. $p(x) = 4x^5 - 3x^4 + 1$
18. $p(x) = 4x^3 - 3x^4 + x^2$
19. $p(x) = 2x^5 - 2x^6 - 11$
20. $p(x) = 2x^4 - 2x^3 + 9x$

B

In Problems 21–30, use $-\infty$ or ∞ where appropriate to describe the behavior at each zero of the denominator and identify all vertical asymptotes.

21. $f(x) = \dfrac{1}{x + 3}$
22. $g(x) = \dfrac{x}{4 - x}$

23. $h(x) = \dfrac{x^2 + 4}{x^2 - 4}$
24. $k(x) = \dfrac{x^2 - 9}{x^2 + 9}$

25. $F(x) = \dfrac{x^2 - 4}{x^2 + 4}$
26. $G(x) = \dfrac{x^2 + 9}{9 - x^2}$

27. $H(x) = \dfrac{x^2 - 2x - 3}{x^2 - 4x + 3}$
28. $K(x) = \dfrac{x^2 + 2x - 3}{x^2 - 4x + 3}$

29. $T(x) = \dfrac{8x - 16}{x^4 - 8x^3 + 16x^2}$
30. $S(x) = \dfrac{6x + 9}{x^4 + 6x^3 + 9x^2}$

In Problems 31–38, find each function value and limit. Use $-\infty$ or ∞ where appropriate.

31. $f(x) = \dfrac{4x + 7}{5x - 9}$

    (A) $f(10)$ (B) $f(100)$ (C) $\lim\limits_{x \to \infty} f(x)$

32. $f(x) = \dfrac{2 - 3x^3}{7 + 4x^3}$

    (A) $f(5)$ (B) $f(10)$ (C) $\lim\limits_{x \to \infty} f(x)$

33. $f(x) = \dfrac{5x^2 + 11}{7x - 2}$

    (A) $f(20)$ (B) $f(50)$ (C) $\lim\limits_{x \to \infty} f(x)$

34. $f(x) = \dfrac{5x + 11}{7x^3 - 2}$

    (A) $f(8)$ (B) $f(16)$ (C) $\lim\limits_{x \to -\infty} f(x)$

35. $f(x) = \dfrac{7x^4 - 14x^2}{6x^5 + 3}$

    (A) $f(-6)$ (B) $f(-12)$ (C) $\lim\limits_{x \to -\infty} f(x)$

36. $f(x) = \dfrac{4x^7 - 8x}{6x^4 + 9x^2}$

    (A) $f(-3)$ (B) $f(-6)$ (C) $\lim\limits_{x \to -\infty} f(x)$

37. $f(x) = \dfrac{10 - 7x^3}{4 + x^3}$

    (A) $f(-10)$ (B) $f(-20)$ (C) $\lim\limits_{x \to -\infty} f(x)$

38. $f(x) = \dfrac{3 + x}{5 + 4x}$

    (A) $f(-50)$ (B) $f(-100)$ (C) $\lim\limits_{x \to -\infty} f(x)$

In Problems 39–52, find all horizontal and vertical asymptotes.

39. $f(x) = \dfrac{2x}{x + 2}$

40. $f(x) = \dfrac{3x + 2}{x - 4}$

41. $f(x) = \dfrac{x^2 + 1}{x^2 - 1}$

42. $f(x) = \dfrac{x^2 - 1}{x^2 + 2}$

43. $f(x) = \dfrac{x^3}{x^2 + 6}$

44. $f(x) = \dfrac{x}{x^2 - 4}$

45. $f(x) = \dfrac{x}{x^2 + 4}$

46. $f(x) = \dfrac{x^2 + 9}{x}$

47. $f(x) = \dfrac{x^2}{x - 3}$

48. $f(x) = \dfrac{x + 5}{x^2}$

49. $f(x) = \dfrac{2x^2 + 3x - 2}{x^2 - x - 2}$

50. $f(x) = \dfrac{2x^2 + 7x + 12}{2x^2 + 5x - 12}$

51. $f(x) = \dfrac{2x^2 - 5x + 2}{x^2 - x - 2}$

52. $f(x) = \dfrac{x^2 - x - 12}{2x^2 + 5x - 12}$

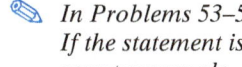

In Problems 53–58, discuss the validity of each statement. If the statement is always true, explain why. If not, give a counterexample.

53. A rational function has at least one vertical asymptote.

54. A rational function has at most one vertical asymptote.

55. A rational function has at least one horizontal asymptote.

56. A rational function has at most one horizontal asymptote.

57. A polynomial function of degree ≥ 1 has neither horizontal nor vertical asymptotes.

58. The graph of a rational function cannot cross a horizontal asymptote.

C

59. Theorem 3 states that

$$\lim_{x \to \infty} (a_n x^n + a_{n-1} x^{n-1} + \cdots + a_0) = \pm\infty.$$

What conditions must n and a_n satisfy for the limit to be ∞? For the limit to be $-\infty$?

60. Theorem 3 also states that

$$\lim_{x \to -\infty} (a_n x^n + a_{n-1} x^{n-1} + \cdots + a_0) = \pm\infty.$$

What conditions must n and a_n satisfy for the limit to be ∞? For the limit to be $-\infty$?

Describe the end behavior of each function in Problems 61–68.

61. $f(x) = 2x^4 - 5x + 11$

62. $f(x) = 3x - 7x^5 - 2$

63. $f(x) = 7x^2 + 9x^3 + 5x$

64. $f(x) = 4x^3 - 5x^2 - 6x^6$

65. $f(x) = \dfrac{x^2 - 5x - 7}{x + 11}$

66. $f(x) = \dfrac{5x^4 + 7x^7 - 10}{x^2 + 6x^4 + 3}$

67. $f(x) = \dfrac{5x^5 + 7x^4 - 10}{-x^3 + 6x^2 + 3}$

68. $f(x) = \dfrac{-3x^6 + 4x^4 + 2}{2x^2 - 2x - 3}$

Applications

69. Average cost. A company manufacturing snowboards has fixed costs of $200 per day and total costs of $3,800 per day for a daily output of 20 boards.

  (A) Assuming that the total cost per day $C(x)$ is linearly related to the total output per day x, write an equation for the cost function.

  (B) The average cost per board for an output of x boards is given by $\overline{C}(x) = C(x)/x$. Find the average cost function.

  (C) Sketch a graph of the average cost function, including any asymptotes, for $1 \leq x \leq 30$.

  (D) What does the average cost per board tend to as production increases?

70. Average cost. A company manufacturing surfboards has fixed costs of $300 per day and total costs of $5,100 per day for a daily output of 20 boards.

  (A) Assuming that the total cost per day $C(x)$ is linearly related to the total output per day x, write an equation for the cost function.

  (B) The average cost per board for an output of x boards is given by $\overline{C}(x) = C(x)/x$. Find the average cost function.

  (C) Sketch a graph of the average cost function, including any asymptotes, for $1 \leq x \leq 30$.

  (D) What does the average cost per board tend to as production increases?

71. Energy costs. Most appliance manufacturers produce conventional and energy-efficient models. The energy-efficient models are more expensive to make but cheaper to operate. The costs of purchasing and operating a 23-cubic-foot refrigerator of each type are given in Table 8. These costs do not include maintenance charges or changes in electricity prices.

Table 8   23-ft³ Refrigerators

	Energy-Efficient Model	Conventional Model
Initial cost	$950	$900
Total volume	23 ft³	23 ft³
Annual cost of electricity	$56	$66

  (A) Express the total cost $C_e(x)$ and the average cost $\overline{C}_e(x) = C_e(x)/x$ of purchasing and operating an energy-efficient model for x years.

  (B) Express the total cost $C_c(x)$ and the average cost $\overline{C}_c(x) = C_c(x)/x$ of purchasing and operating a conventional model for x years.

  (C) Are the total costs for an energy-efficient model and for a conventional model ever the same? If so, when?

  (D) Are the average costs for an energy-efficient model and for a conventional model ever the same? If so, when?

  (E) Find the limit of each average cost function as $x \to \infty$ and discuss the implications of the results.

72. Energy costs. Most appliance manufacturers produce conventional and energy-efficient models. The energy-efficient models are more expensive to make but cheaper to operate. The costs of purchasing and operating a 36,000-Btu central air conditioner of each type are given in Table 9. These costs do not include maintenance charges or changes in electricity prices.

Table 9   36,000 Btu Central Air Conditioner

	Energy-Efficient Model	Conventional Model
Initial cost	$4,000	$2,700
Total capacity	36,000 Btu	36,000 Btu
Annual cost of electricity	$932	$1,332

(A) Express the total cost $C_e(x)$ and the average cost $\overline{C}_e(x) = C_e(x)/x$ of purchasing and operating an energy-efficient model for x years.

(B) Express the total cost $C_c(x)$ and the average cost $\overline{C}_c(x) = C_c(x)/x$ of purchasing and operating a conventional model for x years.

(C) Are the total costs for an energy-efficient model and for a conventional model ever the same? If so, when?

(D) Are the average costs for an energy-efficient model and for a conventional model ever the same? If so, when?

(E) Find the limit of each average cost function as $x \to \infty$ and discuss the implications of the results.

73. Drug concentration. A drug is administered to a patient through an injection. The drug concentration (in milligrams/milliliter) in the bloodstream t hours after the injection is given by $C(t) = \dfrac{5t^2(t + 50)}{t^3 + 100}$. Find and interpret $\lim\limits_{t \to \infty} C(t)$.

74. Drug concentration. A drug is administered to a patient through an IV drip. The drug concentration (in milligrams/milliliter) in the bloodstream t hours after the drip was started is given by $C(t) = \dfrac{5t(t + 50)}{t^3 + 100}$. Find and interpret $\lim\limits_{t \to \infty} C(t)$.

75. Pollution. In Silicon Valley, a number of computer-related manufacturing firms were contaminating underground water supplies with toxic chemicals stored in leaking underground containers. A water quality control agency ordered the companies to take immediate corrective action and contribute to a monetary pool for the testing and cleanup of the underground contamination. Suppose that the monetary pool (in millions of dollars) for the testing and cleanup is given by

$$P(x) = \frac{2x}{1 - x} \qquad 0 \le x < 1$$

where x is the percentage (expressed as a decimal) of the total contaminant removed.

(A) How much must be in the pool to remove 90% of the contaminant?

(B) How much must be in the pool to remove 95% of the contaminant?

(C) Find $\lim\limits_{x \to 1^-} P(x)$ and discuss the implications of this limit.

76. Employee training. A company producing computer components has established that, on average, a new employee can assemble $N(t)$ components per day after t days of on-the-job training, as given by

$$N(t) = \frac{100t}{t + 9} \qquad t \ge 0$$

(A) How many components per day can a new employee assemble after 6 days of on-the-job training?

(B) How many days of on-the-job training will a new employee need to reach the level of 70 components per day?

(C) Find $\lim\limits_{t \to \infty} N(t)$ and discuss the implications of this limit.

77. Biochemistry. In 1913, biochemists Leonor Michaelis and Maude Menten proposed the rational function model (see figure)

$$v(s) = \frac{V_{\max} \, s}{K_M + s}$$

for the velocity of the enzymatic reaction v, where s is the substrate concentration. The constants $V_{\max}$ and K_M are determined from experimental data.

Figure for 77

(A) Show that $\lim\limits_{s \to \infty} v(s) = V_{\max}$.

(B) Show that $v(K_M) = \dfrac{V_{\max}}{2}$.

(C) Table 10* lists data for an enzyme treated with the substrate saccharose.

Plot the preceding points on graph paper and estimate $V_{\max}$ to the nearest integer. To estimate K_M, add the

Table 10

s	v
5.2	0.866
10.4	1.466
20.8	2.114
41.6	2.666
83.3	3.236
167	3.636
333	3.636

*Michaelis and Menten (1913) *Biochem. Z.* 49, 333–369.

horizontal line $v = \dfrac{V_{max}}{2}$ to your graph, connect successive points on the graph with straight-line segments, and estimate the value of s (to the nearest multiple of 10) that satisfies $v(s) = \dfrac{V_{max}}{2}$.

(D) Use the constants V_{max} and K_M from part C to form a Michaelis–Menten function for the data in Table 10.

(E) Use the function from part D to estimate the velocity of the enzyme reaction when the saccharose is 15 and to estimate the saccharose when the velocity is 3.

78. Biochemistry. Table 11* lists data for the enzyme invertase treated with the substrate sucrose. We want to model these data with a Michaelis–Menten function.

Table 11

s	v
2.92	18.2
5.84	26.5
8.76	31.1
11.7	33
14.6	34.9
17.5	37.2
23.4	37.1

(A) Plot the points in Table 11 on graph paper and estimate V_{max} to the nearest integer. To estimate K_M, add the horizontal line $v = \dfrac{V_{max}}{2}$ to your graph, connect successive points on the graph with straight-line segments, and estimate the value of s (to the nearest integer) that satisfies $v(s) = \dfrac{V_{max}}{2}$.

(B) Use the constants V_{max} and K_M from part (A) to form a Michaelis–Menten function for the data in Table 11.

(C) Use the function from part (B) to estimate the velocity of the enzyme reaction when the sucrose is 9 and to estimate the sucrose when the velocity is 32.

79. Physics. The coefficient of thermal expansion (CTE) is a measure of the expansion of an object subjected to extreme temperatures. To model this coefficient, we use a Michaelis–Menten function of the form

$$C(T) = \frac{C_{max}\,T}{M + T} \qquad \text{(Problem 77)}$$

where C = CTE, T is temperature in K (degrees Kelvin), and C_{max} and M are constants. Table 12† lists the coeffi-

cients of thermal expansion for nickel and for copper at various temperatures.

Table 12 Coefficients of Thermal Expansion

T (K)	Nickel	Copper
100	6.6	10.3
200	11.3	15.2
293	13.4	16.5
500	15.3	18.3
800	16.8	20.3
1,100	17.8	23.7

(A) Plot the points in columns 1 and 2 of Table 12 on graph paper and estimate C_{max} to the nearest integer. To estimate M, add the horizontal line $\text{CTE} = \dfrac{C_{max}}{2}$ to your graph, connect successive points on the graph with straight-line segments, and estimate the value of T (to the nearest multiple of fifty) that satisfies $C(T) = \dfrac{C_{max}}{2}$.

(B) Use the constants $\dfrac{C_{max}}{2}$ and M from part (A) to form a Michaelis–Menten function for the CTE of nickel.

(C) Use the function from part (B) to estimate the CTE of nickel at 600 K and to estimate the temperature when the CTE of nickel is 12.

80. Physics. Repeat Problem 79 for the CTE of copper (column 3 of Table 12).

Answers to Matched Problems

1. Vertical asymptote: $x = 1$; $\lim\limits_{x \to 1^+} f(x) = \infty$, $\lim\limits_{x \to 1^-} f(x) = -\infty$
   $\lim\limits_{x \to 3} f(x) = 1/2$ so f does not have a vertical asymptote at $x = 3$

2. Vertical asymptote: $x = -3$; $\lim\limits_{x \to -3^+} f(x) = \lim\limits_{x \to -3^-} f(x) = -\infty$

3. $\lim\limits_{x \to \infty} p(x) = \lim\limits_{x \to -\infty} p(x) = -\infty$

4. (A) $\lim\limits_{x \to \infty} p(x) = -\infty$, $\lim\limits_{x \to -\infty} p(x) = \infty$
   (B) $\lim\limits_{x \to \infty} p(x) = \infty$, $\lim\limits_{x \to -\infty} p(x) = \infty$

5. (A) $y = 0$ (B) No horizontal asymptotes
   (C) $y = 1/2$

6. Vertical asymptotes: $x = -2$, $x = 2$; horizontal asymptote: $y = 1$

---

*Institute of Chemistry, Macedonia.

†National Physical Laboratory

3-3   Continuity

Theorem 3 in Section 3-1 states that if f is a polynomial function or a rational function with a nonzero denominator at $x = c$, then

$$\lim_{x \to c} f(x) = f(c) \qquad (1)$$

Functions that satisfy equation (1) are said to be *continuous* at $x = c$. A firm understanding of continuous functions is essential for sketching and analyzing graphs. We will also see that continuity properties provide a simple and efficient method for solving inequalities—a tool that we will use extensively in later sections.

Continuity

Compare the graphs shown in Figure 1. Notice that two of the graphs are broken; that is, they cannot be drawn without lifting a pen off the paper. Informally, a function is *continuous over an interval* if its graph over the interval can be drawn without removing a pen from the paper. A function whose graph is broken (disconnected) at $x = c$ is said to be *discontinuous* at $x = c$. Function f (Fig. 1A) is continuous for all x. Function g (Fig. 1B) is discontinuous at $x = 2$ but is continuous over any interval that does not include 2. Function h (Fig. 1C) is discontinuous at $x = 0$ but is continuous over any interval that does not include 0.

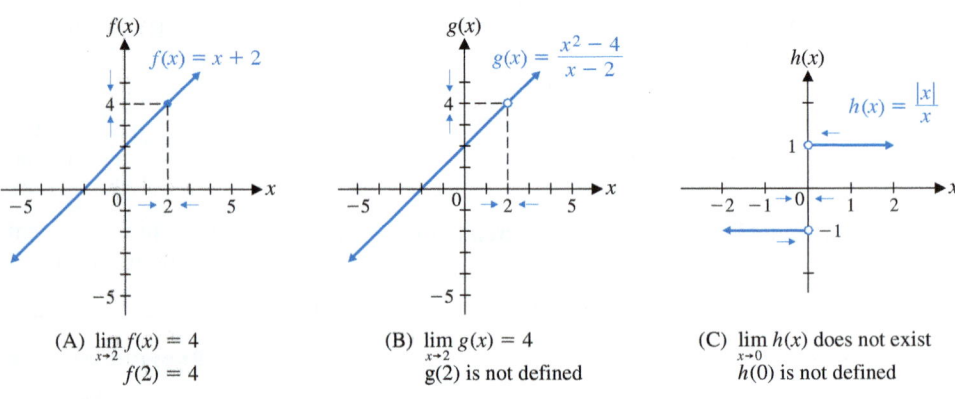

(A) $\lim_{x \to 2} f(x) = 4$
$f(2) = 4$

(B) $\lim_{x \to 2} g(x) = 4$
$g(2)$ is not defined

(C) $\lim_{x \to 0} h(x)$ does not exist
$h(0)$ is not defined

Figure 1

Most graphs of natural phenomena are continuous, whereas many graphs in business and economics applications have discontinuities. Figure 2A illustrates temperature variation over a 24-hour period—a continuous phenomenon. Figure 2B illustrates warehouse inventory over a 1-week period—a discontinuous phenomenon.

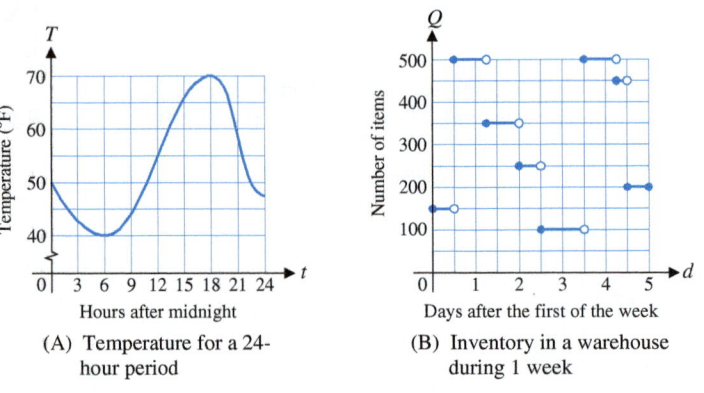

(A) Temperature for a 24-hour period

(B) Inventory in a warehouse during 1 week

Figure 2

(A) Write a brief verbal description of the temperature variation illustrated in Figure 2A, including estimates of the high and low temperatures during the period shown and the times at which they occurred.

(B) Write a brief verbal description of the changes in inventory illustrated in Figure 2B, including estimates of the changes in inventory and the times at which those changes occurred.

The preceding discussion leads to the following formal definition of continuity:

> **DEFINITION Continuity**
>
> A function f is **continuous at the point $x = c$ if**
>
>   **1.** $\lim\limits_{x \to c} f(x)$ exists   **2.** $f(c)$ exists   **3.** $\lim\limits_{x \to c} f(x) = f(c)$
>
> A function is **continuous on the open interval* (a, b)** if it is continuous at each point on the interval.
>
> _____
> *See Section 1-1 for a review of interval notation.

If one or more of the three conditions in the definition fails, then the function is **discontinuous** at $x = c$.

EXAMPLE 1   **Continuity of a Function Defined by a Graph**   Use the definition of continuity to discuss the continuity of the function whose graph is shown in Figure 3.

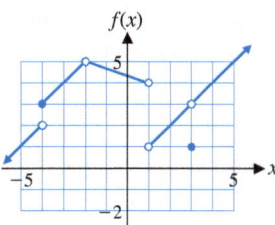

Figure 3

SOLUTION   We begin by identifying the points of discontinuity. Examining the graph, we see breaks or holes at $x = -4, -2, 1,$ and 3. Now we must determine which conditions in the definition of continuity are not satisfied at each of these points. In each case, we find the value of the function and the limit of the function at the point in question.

Discontinuity at $x = -4$:

$$\lim_{x \to -4^-} f(x) = 2$$ *Since the one-sided limits are different, the limit does not exist (Section 3-1).*

$$\lim_{x \to -4^+} f(x) = 3$$

$$\lim_{x \to -4} f(x) \text{ does not exist}$$

$$f(-4) = 3$$

So, f is not continuous at $x = -4$ because condition 1 is not satisfied.

Discontinuity at $x = -2$:

$$\lim_{x \to -2^-} f(x) = 5$$ *The hole at $(-2, 5)$ indicates that 5 is not the value of f at -2. Since there is no solid dot elsewhere on the vertical line $x = -2$, $f(-2)$ is not defined.*

$$\lim_{x \to -2^+} f(x) = 5$$

$$\lim_{x \to -2} f(x) = 5$$

$$f(-2) \text{ does not exist}$$

So f is not continuous at $x = -2$ because condition 2 is not satisfied.

Discontinuity at $x = 1$:

$$\lim_{x \to 1^-} f(x) = 4$$

$$\lim_{x \to 1^+} f(x) = 1$$

$$\lim_{x \to 1} f(x) \text{ does not exist}$$

$$f(1) \text{ does not exist}$$

This time, f is not continuous at $x = 1$ because neither of conditions 1 and 2 is satisfied.

Discontinuity at $x = 3$:

$$\lim_{x \to 3^-} f(x) = 3 \qquad \text{The solid dot at } (3, 1) \text{ indicates that } f(3) = 1.$$

$$\lim_{x \to 3^+} f(x) = 3$$

$$\lim_{x \to 3} f(x) = 3$$

$$f(3) = 1$$

Conditions 1 and 2 are satisfied, but f is not continuous at $x = 3$ because condition 3 is not satisfied.

Having identified and discussed all points of discontinuity, we can now conclude that f is continuous except at $x = -4, -2, 1,$ and 3.

CONCEPTUAL INSIGHT

Rather than list the points where a function is discontinuous, sometimes it is useful to state the intervals on which the function is continuous. Using the set operation **union,** denoted by $\cup$, we can express the set of points where the function in Example 1 is continuous as follows:

$$(-\infty, -4) \cup (-4, -2) \cup (-2, 1) \cup (1, 3) \cup (3, \infty)$$

Matched Problem 1   Use the definition of continuity to discuss the continuity of the function whose graph is shown in Figure 4.

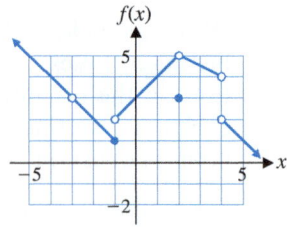

Figure 4

For functions defined by equations, it is important to be able to locate points of discontinuity by examining the equation.

EXAMPLE 2   **Continuity of Functions Defined by Equations**   Using the definition of continuity, discuss the continuity of each function at the indicated point(s).

(A) $f(x) = x + 2$ at $x = 2$ 　　　　(B) $g(x) = \dfrac{x^2 - 4}{x - 2}$ at $x = 2$

(C) $h(x) = \dfrac{|x|}{x}$ at $x = 0$ and   at $x = 1$

SOLUTION   (A) f is continuous at $x = 2$, since

$$\lim_{x \to 2} f(x) = 4 = f(2) \quad \textit{See Figure 1A.}$$

(B) g is not continuous at $x = 2$, since $g(2) = 0/0$ is not defined (see Fig. 1B).

(C) h is not continuous at $x = 0$, since $h(0) = |0|/0$ is not defined; also, $\lim_{x \to 0} h(x)$ does not exist.

h is continuous at $x = 1$, since

$$\lim_{x \to 1} \frac{|x|}{x} = 1 = h(1) \quad \textit{See Figure 1C.}$$

Matched Problem 2   Using the definition of continuity, discuss the continuity of each function at the indicated point(s).

(A) $f(x) = x + 1$ at $x = 1$ (B) $g(x) = \dfrac{x^2 - 1}{x - 1}$ at $x = 1$

(C) $h(x) = \dfrac{x - 2}{|x - 2|}$ at $x = 2$ and at $x = 0$

We can also talk about one-sided continuity, just as we talked about one-sided limits. For example, a function is said to be **continuous on the right** at $x = c$ if $\lim_{x \to c^+} f(x) = f(c)$ and **continuous on the left** at $x = c$ if $\lim_{x \to c^-} f(x) = f(c)$. A function is **continuous on the closed interval [a, b]** if it is continuous on the open interval (a, b) and is continuous both on the right at a and on the left at b.

Figure 5A illustrates a function that is continuous on the closed interval $[-1, 1]$. Figure 5B illustrates a function that is continuous on the half-closed interval $[0, \infty)$.

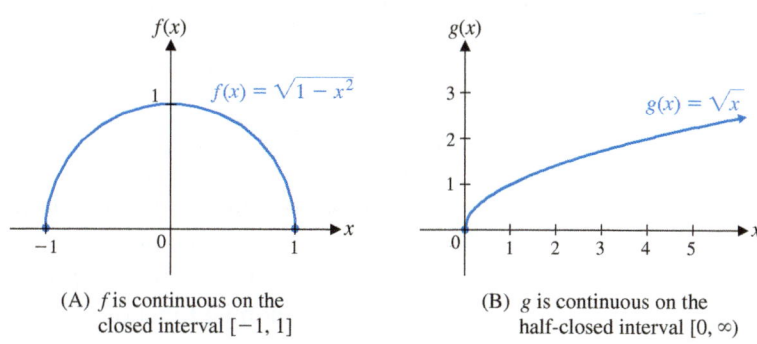

(A) f is continuous on the
closed interval $[-1, 1]$

(B) g is continuous on the
half-closed interval $[0, \infty)$

Figure 5 Continuity on closed and half-closed intervals

Continuity Properties

Functions have some useful **general continuity properties**:

> **If two functions are continuous on the same interval, then their sum, difference, product, and quotient are continuous on the same interval except for values of x that make a denominator 0.**

These properties, along with Theorem 1, enable us to determine intervals of continuity for some important classes of functions without having to look at their graphs or use the three conditions in the definition.

THEOREM 1 **Continuity Properties of Some Specific Functions**

(A) A constant function $f(x) = k$, where k is a constant, is continuous for all x.

$f(x) = 7$ is continuous for all x.

(B) For n a positive integer, $f(x) = x^n$ is continuous for all x.

$f(x) = x^5$ is continuous for all x.

(C) A polynomial function is continuous for all x.

$2x^3 - 3x^2 + x - 5$ is continuous for all x.

(D) A rational function is continuous for all x except those values that make a denominator 0.

$\dfrac{x^2 + 1}{x - 1}$ is continuous for all x except $x = 1$, a value that makes the denominator 0.

(E) For n an odd positive integer greater than 1, $\sqrt[n]{f(x)}$ is continuous wherever $f(x)$ is continuous.

$\sqrt[3]{x^2}$ is continuous for all x.

(F) For n an even positive integer, $\sqrt[n]{f(x)}$ is continuous wherever $f(x)$ is continuous and nonnegative.

$\sqrt[4]{x}$ is continuous on the interval $[0, \infty)$.

Parts C and D of Theorem 1 are the same as Theorem 3 in Section 3-1. They are repeated here to emphasize their importance.

EXAMPLE 3 **Using Continuity Properties** Using Theorem 1 and the general properties of continuity, determine where each function is continuous.

(A) $f(x) = x^2 - 2x + 1$

(B) $f(x) = \dfrac{x}{(x + 2)(x - 3)}$

(C) $f(x) = \sqrt[3]{x^2 - 4}$

(D) $f(x) = \sqrt{x - 2}$

SOLUTION

(A) Since f is a polynomial function, f is continuous for all x.

(B) Since f is a rational function, f is continuous for all x except -2 and 3 (values that make the denominator 0).

(C) The polynomial function $x^2 - 4$ is continuous for all x. Since $n = 3$ is odd, f is continuous for all x.

(D) The polynomial function $x - 2$ is continuous for all x and nonnegative for $x \geq 2$. Since $n = 2$ is even, f is continuous for $x \geq 2$, or on the interval $[2, \infty)$.

Matched Problem 3 Using Theorem 1 and the general properties of continuity, determine where each function is continuous.

(A) $f(x) = x^4 + 2x^2 + 1$

(B) $f(x) = \dfrac{x^2}{(x + 1)(x - 4)}$

(C) $f(x) = \sqrt{x - 4}$

(D) $f(x) = \sqrt[3]{x^3 + 1}$

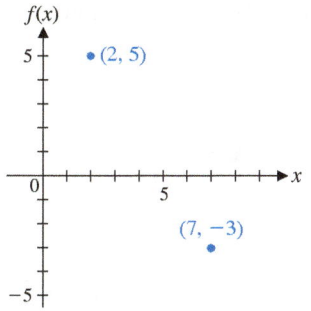

Figure 6

Solving Inequalities Using Continuity Properties

One of the basic tools for analyzing graphs in calculus is a special line graph called a *sign chart*. We will make extensive use of this type of chart in later sections. In the discussion that follows, we use continuity properties to develop a simple and efficient procedure for constructing sign charts.

Suppose that a function f is continuous over the interval $(1, 8)$ and $f(x) \neq 0$ for any x in $(1, 8)$. Suppose also that $f(2) = 5$, a positive number. Is it possible for $f(x)$ to be negative for any x in the interval $(1, 8)$? The answer is "no." If $f(7)$ were -3, for example, as shown in Figure 6, then how would it be possible to join the points $(2, 5)$ and $(7, -3)$ with the graph of a continuous function without crossing the x axis between 1 and 8 at least once? [Crossing the x axis would violate our assumption that $f(x) \neq 0$ for any x in $(1, 8)$.] We conclude that $f(x)$ must be positive for all x in $(1, 8)$. If $f(2)$ were negative, then, using the same type of reasoning, $f(x)$ would have to be negative over the entire interval $(1, 8)$.

In general, **if f is continuous and $f(x) \neq 0$ on the interval (a, b), then $f(x)$ cannot change sign on (a, b).** This is the essence of Theorem 2.

> **THEOREM 2 Sign Properties on an Interval (a, b)**
> If f is continuous on (a, b) and $f(x) \neq 0$ for all x in (a, b), then either $f(x) > 0$ for all x in (a, b) or $f(x) < 0$ for all x in (a, b).

Theorem 2 provides the basis for an effective method of solving many types of inequalities. Example 4 illustrates the process.

EXAMPLE 4 **Solving an Inequality** Solve $\dfrac{x + 1}{x - 2} > 0$.

SOLUTION We start by using the left side of the inequality to form the function f.

$$f(x) = \frac{x + 1}{x - 2}$$

Figure 7

The rational function f is discontinuous at $x = 2$, and $f(x) = 0$ for $x = -1$ (a fraction is 0 when the numerator is 0 and the denominator is not 0). We plot $x = 2$ and $x = -1$, which we call *partition numbers,* on a real-number line (Fig. 7). (Note that the dot at 2 is open because the function is not defined at $x = 2$.) The partition numbers 2 and -1 determine three open intervals: $(-\infty, -1), (-1, 2)$, and $(2, \infty)$. The function f is continuous and nonzero on each of these intervals. From Theorem 2, we know that $f(x)$ does not change sign on any of these intervals. We can find the sign of $f(x)$ on each of the intervals by selecting a **test number** in each interval and evaluating $f(x)$ at that number. Since any number in each subinterval will do, we choose test numbers that are easy to evaluate: $-2, 0$, and 3. The table in the margin shows the results.

The sign of $f(x)$ at each test number is the same as the sign of $f(x)$ over the interval containing that test number. Using this information, we construct a **sign chart** for $f(x)$ as shown in Figure 8.

Test Numbers

x	$f(x)$
-2	$\frac{1}{4}\,(+)$
0	$-\frac{1}{2}\,(-)$
3	$4\,(+)$

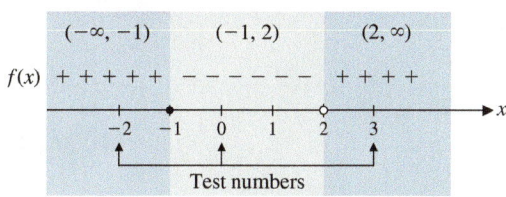

Figure 8

From the sign chart, we can easily write the solution of the given nonlinear inequality:

$$f(x) > 0 \quad \text{for} \quad \begin{array}{ll} x < -1 \quad \text{or} \quad x > 2 & \text{Inequality notation} \\ (-\infty, -1) \cup (2, \infty) & \text{Interval notation} \end{array}$$

Matched Problem 4   Solve $\dfrac{x^2 - 1}{x - 3} < 0.$

Most of the inequalities we encounter will involve strict inequalities ($>$ or $<$). If it is necessary to solve inequalities of the form $\geq$ or $\leq$, we simply include the endpoint x of any interval if f is defined at x and $f(x)$ satisfies the given inequality. For example, from the sign chart in Figure 8, the solution of the inequality

$$\dfrac{x + 1}{x - 2} \geq 0 \quad \text{is} \quad \begin{array}{ll} x \leq -1 \quad \text{or} \quad x > 2 & \text{Inequality notation} \\ (-\infty, -1] \cup (2, \infty) & \text{Interval notation} \end{array}$$

In general, given a function f, a **partition number** is a value of x such that f is discontinuous at x or $f(x) = 0$. **Partition numbers determine open intervals in which $f(x)$ does not change sign**. By using a test number from each interval, we can construct a sign chart for $f(x)$ on the real-number line. It is then easy to solve the inequality $f(x) < 0$ or $f(x) > 0$.

We summarize the procedure for constructing sign charts in the following box:

PROCEDURE Constructing Sign Charts

Given a function f,

Step 1   Find all partition numbers:

(A) Find all numbers such that f is discontinuous. (Rational functions are discontinuous for values of x that make a denominator 0.)

(B) Find all numbers such that $f(x) = 0$. (For a rational function, this occurs where the numerator is 0 and the denominator is not 0.)

Step 2   Plot the numbers found in step 1 on a real-number line, dividing the number line into intervals.

Step 3   Select a test number in each open interval determined in step 2 and evaluate $f(x)$ at each test number to determine whether $f(x)$ is positive ($+$) or negative ($-$) in each interval.

Step 4   Construct a sign chart, using the real-number line in step 2. This will show the sign of $f(x)$ on each open interval.

Exercises 3-3

A

In Problems 1–6, sketch a possible graph of a function that satisfies the given conditions at $x = 1$ and discuss the continuity of f at $x = 1$.

1. $f(1) = 2$ and $\lim\limits_{x \to 1} f(x) = 2$

2. $f(1) = -2$ and $\lim\limits_{x \to 1} f(x) = 2$

3. $f(1) = 2$ and $\lim\limits_{x \to 1} f(x) = -2$

4. $f(1) = -2$ and $\lim\limits_{x \to 1} f(x) = -2$

5. $f(1) = -2$, $\lim\limits_{x \to 1^-} f(x) = 2$, and $\lim\limits_{x \to 1^+} f(x) = -2$

6. $f(1) = 2$, $\lim\limits_{x \to 1^-} f(x) = 2$, and $\lim\limits_{x \to 1^+} f(x) = -2$

Problems 7–14 refer to the function f shown in the figure. Use the graph to estimate the indicated function values and limits.

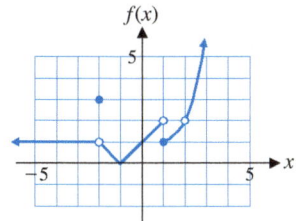

$f(x)$

Figure for 7–14

7. $f(0.9)$ **8.** $f(0.1)$

9. $f(-1.9)$ **10.** $f(-0.9)$

11. (A) $\lim\limits_{x \to 1^-} f(x)$ (B) $\lim\limits_{x \to 1^+} f(x)$
    (C) $\lim\limits_{x \to 1} f(x)$ (D) $f(1)$
    (E) Is f continuous at $x = 1$? Explain.

12. (A) $\lim\limits_{x \to 2^-} f(x)$ (B) $\lim\limits_{x \to 2^+} f(x)$
    (C) $\lim\limits_{x \to 2} f(x)$ (D) $f(2)$
    (E) Is f continuous at $x = 2$? Explain.

13. (A) $\lim\limits_{x \to -2^-} f(x)$ (B) $\lim\limits_{x \to -2^+} f(x)$
    (C) $\lim\limits_{x \to -2} f(x)$ (D) $f(-2)$
    (E) Is f continuous at $x = -2$? Explain.

14. (A) $\lim\limits_{x \to -1^-} f(x)$ (B) $\lim\limits_{x \to -1^+} f(x)$
    (C) $\lim\limits_{x \to -1} f(x)$ (D) $f(-1)$
    (E) Is f continuous at $x = -1$? Explain.

Problems 15–22 refer to the function g shown in the figure. Use the graph to estimate the indicated function values and limits.

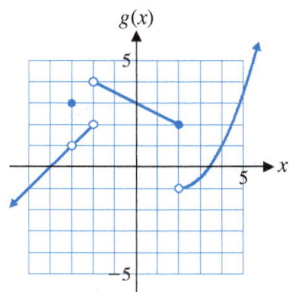

$g(x)$

Figure for 15–22

15. $g(-3.1)$ **16.** $g(-2.1)$

17. $g(1.9)$ **18.** $g(-1.9)$

19. (A) $\lim\limits_{x \to -3^-} g(x)$ (B) $\lim\limits_{x \to -3^+} g(x)$
    (C) $\lim\limits_{x \to -3} g(x)$ (D) $g(-3)$
    (E) Is g continuous at $x = -3$? Explain.

20. (A) $\lim\limits_{x \to -2^-} g(x)$ (B) $\lim\limits_{x \to -2^+} g(x)$
    (C) $\lim\limits_{x \to -2} g(x)$ (D) $g(-2)$
    (E) Is g continuous at $x = -2$? Explain.

21. (A) $\lim\limits_{x \to 2^-} g(x)$ (B) $\lim\limits_{x \to 2^+} g(x)$
    (C) $\lim\limits_{x \to 2} g(x)$ (D) $g(2)$
    (E) Is g continuous at $x = 2$? Explain.

22. (A) $\lim\limits_{x \to 4^-} g(x)$ (B) $\lim\limits_{x \to 4^+} g(x)$
    (C) $\lim\limits_{x \to 4} g(x)$ (D) $g(4)$
    (E) Is g continuous at $x = 4$? Explain.

Use Theorem 1 to determine where each function in Problems 23–32 is continuous.

23. $f(x) = 3x - 4$ **24.** $h(x) = 4 - 2x$

25. $g(x) = \dfrac{3x}{x + 2}$ **26.** $k(x) = \dfrac{2x}{x - 4}$

27. $m(x) = \dfrac{x + 1}{(x - 1)(x + 4)}$ **28.** $n(x) = \dfrac{x - 2}{(x - 3)(x + 1)}$

29. $F(x) = \dfrac{2x}{x^2 + 9}$ **30.** $G(x) = \dfrac{1 - x^2}{x^2 + 1}$

31. $M(x) = \dfrac{x - 1}{4x^2 - 9}$ **32.** $N(x) = \dfrac{x^2 + 4}{4 - 25x^2}$

B

33. Given the function

$$f(x) = \begin{cases} 2 & \text{if } x \text{ is an integer} \\ 1 & \text{if } x \text{ is not an integer} \end{cases}$$

(A) Graph f.

(B) $\lim\limits_{x \to 2} f(x) = ?$

(C) $f(2) = ?$

(D) Is f continuous at $x = 2$?

(E) Where is f discontinuous?

34. Given the function

$$g(x) = \begin{cases} -1 & \text{if } x \text{ is an even integer} \\ 1 & \text{if } x \text{ is not an even integer} \end{cases}$$

(A) Graph g.

(B) $\lim\limits_{x \to 1} g(x) = ?$

(C) $g(1) = ?$

(D) Is g continuous at $x = 1$?

(E) Where is g discontinuous?

In Problems 35–42, use a sign chart to solve each inequality. Express answers in inequality and interval notation.

35. $x^2 - x - 12 < 0$

36. $x^2 - 2x - 8 < 0$

37. $x^2 + 21 > 10x$

38. $x^2 + 7x > -10$

39. $x^3 < 4x$

40. $x^4 - 9x^2 > 0$

41. $\dfrac{x^2 + 5x}{x - 3} > 0$

42. $\dfrac{x - 4}{x^2 + 2x} < 0$

43. Use the graph of f to determine where

(A) $f(x) > 0$ (B) $f(x) < 0$

Express answers in interval notation.

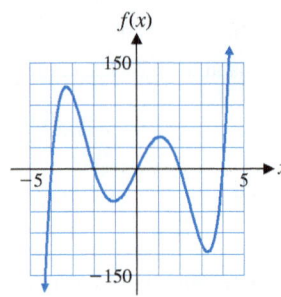

Figure for 43

44. Use the graph of g to determine where

(A) $g(x) > 0$ (B) $g(x) < 0$

Express answers in interval notation.

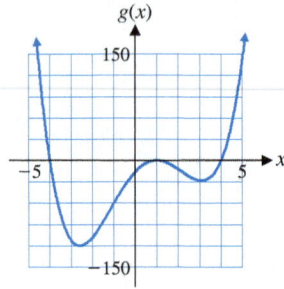

Figure for 44

In Problems 45–48, use a graphing calculator to approximate the partition numbers of each function $f(x)$ to four decimal places. Then solve the following inequalities:

(A) $f(x) > 0$ (B) $f(x) < 0$

Express answers in interval notation.

45. $f(x) = x^4 - 6x^2 + 3x + 5$

46. $f(x) = x^4 - 4x^2 - 2x + 2$

47. $f(x) = \dfrac{3 + 6x - x^3}{x^2 - 1}$ **48.** $f(x) = \dfrac{x^3 - 5x + 1}{x^2 - 1}$

Use Theorem 1 to determine where each function in Problems 49–56 is continuous. Express the answer in interval notation.

49. $\sqrt{x - 6}$ **50.** $\sqrt{7 - x}$

51. $\sqrt[3]{5 - x}$ **52.** $\sqrt[3]{x - 8}$

53. $\sqrt{x^2 - 9}$ **54.** $\sqrt{4 - x^2}$

55. $\sqrt{x^2 + 1}$ **56.** $\sqrt[3]{x^2 + 2}$

In Problems 57–62, graph f, locate all points of discontinuity, and discuss the behavior of f at these points.

57. $f(x) = \begin{cases} 1 + x & \text{if } x < 1 \\ 5 - x & \text{if } x \ge 1 \end{cases}$

58. $f(x) = \begin{cases} x^2 & \text{if } x \le 1 \\ 2x & \text{if } x > 1 \end{cases}$

59. $f(x) = \begin{cases} 1 + x & \text{if } x \le 2 \\ 5 - x & \text{if } x > 2 \end{cases}$

60. $f(x) = \begin{cases} x^2 & \text{if } x \le 2 \\ 2x & \text{if } x > 2 \end{cases}$

61. $f(x) = \begin{cases} -x & \text{if } x < 0 \\ 1 & \text{if } x = 0 \\ x & \text{if } x > 0 \end{cases}$

62. $f(x) = \begin{cases} 1 & \text{if } x < 0 \\ 0 & \text{if } x = 0 \\ 1 + x & \text{if } x > 0 \end{cases}$

C

Problems 63 and 64 refer to the **greatest integer function**, which is denoted by $[\![x]\!]$ and is defined as

$$[\![x]\!] = \text{greatest integer} \le x$$

For example,

$$[\![-3.6]\!] = \text{greatest integer} \le -3.6 = -4$$

$$[\![2]\!] = \text{greatest integer} \le 2 = 2$$

$$[\![2.5]\!] = \text{greatest integer} \le 2.5 = 2$$

The graph of $f(x) = [\![x]\!]$ is shown. There, we can see that

$$[\![x]\!] = -2 \quad \text{for} \quad -2 \le x < -1$$

$$[\![x]\!] = -1 \quad \text{for} \quad -1 \le x < 0$$

$$[\![x]\!] = 0 \quad \text{for} \quad 0 \le x < 1$$

$$[\![x]\!] = 1 \quad \text{for} \quad 1 \le x < 2$$

$$[\![x]\!] = 2 \quad \text{for} \quad 2 \le x < 3$$

and so on.

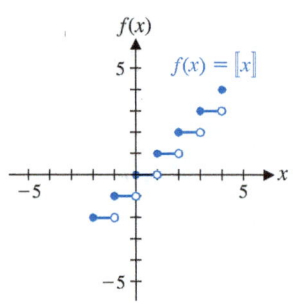

Figure for 63 and 64

63. (A) Is f continuous from the right at $x = 0$?

(B) Is f continuous from the left at $x = 0$?

(C) Is f continuous on the open interval $(0, 1)$?

(D) Is f continuous on the closed interval $[0, 1]$?

(E) Is f continuous on the half-closed interval $[0, 1)$?

64. (A) Is f continuous from the right at $x = 2$?

(B) Is f continuous from the left at $x = 2$?

(C) Is f continuous on the open interval $(1, 2)$?

(D) Is f continuous on the closed interval $[1, 2]$?

(E) Is f continuous on the half-closed interval $[1, 2)$?

In Problems 65–70, discuss the validity of each statement. If the statement is always true, explain why. If not, give a counterexample.

65. A polynomial function is continuous for all real numbers.

66. A rational function is continuous for all but finitely many real numbers.

67. If f is a function that is continuous at $x = 0$ and $x = 2$, then f is continuous at $x = 1$.

68. If f is a function that is continuous on the open interval $(0, 2)$, then f is continuous at $x = 1$.

69. If f is a function that has no partition numbers in the interval (a, b), then f is continuous on (a, b).

70. The greatest integer function (see Problem 61) is a rational function.

In Problems 71–74, sketch a possible graph of a function f that is continuous for all real numbers and satisfies the given conditions. Find the x intercepts of f.

71. $f(x) < 0$ on $(-\infty, -5)$ and $(2, \infty)$; $f(x) > 0$ on $(-5, 2)$

72. $f(x) > 0$ on $(-\infty, -4)$ and $(3, \infty)$; $f(x) < 0$ on $(-4, 3)$

73. $f(x) < 0$ on $(-\infty, -6)$ and $(-1, 4)$; $f(x) > 0$ on $(-6, -1)$ and $(4, \infty)$

74. $f(x) > 0$ on $(-\infty, -3)$ and $(2, 7)$; $f(x) < 0$ on $(-3, 2)$ and $(7, \infty)$

75. The function $f(x) = 2/(1 - x)$ satisfies $f(0) = 2$ and $f(2) = -2$. Is f equal to 0 anywhere on the interval $(-1, 3)$? Does this contradict Theorem 2? Explain.

76. The function $f(x) = 6/(x - 4)$ satisfies $f(2) = -3$ and $f(7) = 2$. Is f equal to 0 anywhere on the interval $(0, 9)$? Does this contradict Theorem 2? Explain.

Applications

77. Postal rates. First-class postage in 2009 was $0.44 for the first ounce (or any fraction thereof) and $0.17 for each additional ounce (or fraction thereof) up to a maximum weight of 3.5 ounces.

(A) Write a piecewise definition of the first-class postage $P(x)$ for a letter weighing x ounces.

(B) Graph $P(x)$ for $0 < x \le 3.5$.

(C) Is $P(x)$ continuous at $x = 2.5$? At $x = 3$? Explain.

78. Telephone rates. A long-distance telephone service charges $0.07 for the first minute (or any fraction thereof) and $0.05 for each additional minute (or fraction thereof).

(A) Write a piecewise definition of the charge $R(x)$ for a long-distance call lasting x minutes.

(B) Graph $R(x)$ for $0 < x \le 6$.

(C) Is $R(x)$ continuous at $x = 3.5$? At $x = 3$? Explain.

79. Postal rates. Discuss the differences between the function $Q(x) = 0.44 + 0.17[x]$ and the function $P(x)$ defined in Problem 77.

80. Telephone rates. Discuss the differences between the function $S(x) = 0.07 + 0.05[x]$ and the function $R(x)$ defined in Problem 78.

81. Natural-gas rates. Table 1 shows the rates for natural gas charged by the Middle Tennessee Natural Gas Utility District during summer months. The customer charge is a fixed monthly charge, independent of the amount of gas used per month.

Table 1 Summer (May–September)	
Base charge	$5.00
First 50 therms	0.63 per therm
Over 50 therms	0.45 per therm

(A) Write a piecewise definition of the monthly charge $S(x)$ for a customer who uses x therms* in a summer month.

(B) Graph $S(x)$.

(C) Is $S(x)$ continuous at $x = 50$? Explain.

82. Natural-gas rates. Table 2 shows the rates for natural gas charged by the Middle Tennessee Natural Gas Utility District during winter months. The customer charge is a fixed monthly charge, independent of the amount of gas used per month.

Table 2 Winter (October–April)	
Base charge	$5.00
First 5 therms	0.69 per therm
Next 45 therms	0.65 per therm
Over 50 therms	0.63 per therm

*A British thermal unit (Btu) is the amount of heat required to raise the temperature of 1 pound of water 1 degree Fahrenheit, and a therm is 100,000 Btu.

(A) Write a piecewise definition of the monthly charge $S(x)$ for a customer who uses x therms in a winter month.

(B) Graph $S(x)$.

(C) Is $S(x)$ continuous at $x = 5$? At $x = 50$? Explain.

83. Income. A personal-computer salesperson receives a base salary of $1,000 per month and a commission of 5% of all sales over $10,000 during the month. If the monthly sales are $20,000 or more, then the salesperson is given an additional $500 bonus. Let $E(s)$ represent the person's earnings per month as a function of the monthly sales s.

(A) Graph $E(s)$ for $0 \le s \le 30,000$.

(B) Find $\lim_{s \to 10,000} E(s)$ and $E(10,000)$.

(C) Find $\lim_{s \to 20,000} E(s)$ and $E(20,000)$.

(D) Is E continuous at $s = 10,000$? At $s = 20,000$?

84. Equipment rental. An office equipment rental and leasing company rents copiers for $10 per day (and any fraction thereof) or for $50 per 7-day week. Let $C(x)$ be the cost of renting a copier for x days.

(A) Graph $C(x)$ for $0 \le x \le 10$.

(B) Find $\lim_{x \to 4.5} C(x)$ and $C(4.5)$.

(C) Find $\lim_{x \to 8} C(x)$ and $C(8)$.

(D) Is C continuous at $x = 4.5$? At $x = 8$?

85. Animal supply. A medical laboratory raises its own rabbits. The number of rabbits $N(t)$ available at any time t depends on the number of births and deaths. When a birth or death occurs, the function N generally has a discontinuity, as shown in the figure.

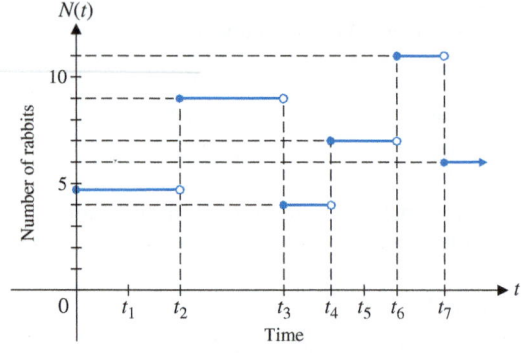

Figure for 85

(A) Where is the function N discontinuous?

(B) $\lim_{t \to t_5} N(t) = ?$; $N(t_5) = ?$

(C) $\lim_{t \to t_3} N(t) = ?$; $N(t_3) = ?$

86. Learning. The graph shown represents the history of a person learning the material on limits and continuity in this book. At time t_2, the student's mind goes blank during a quiz. At time t_4, the instructor explains a concept particularly well, then suddenly a big jump in understanding takes place.

(A) Where is the function p discontinuous?

(B) $\lim_{t \to t_1} p(t) = ?$; $p(t_1) = ?$

(C) $\lim_{t \to t_2} p(t) = ?$; $p(t_2) = ?$

(D) $\lim_{t \to t_4} p(t) = ?$; $p(t_4) = ?$

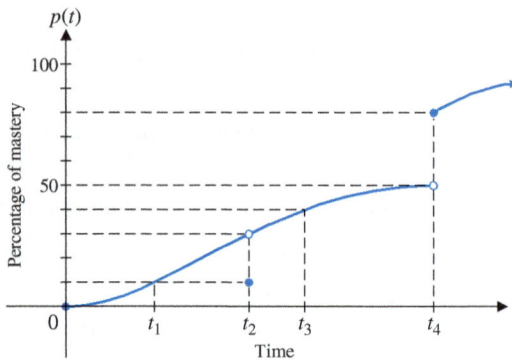

Figure for 86

Answers to Matched Problems

1. f is not continuous at $x = -3, -1, 2,$ and 4.

$x = -3$: $\lim_{x \to -3} f(x) = 3$, but $f(-3)$ does not exist

$x = -1$: $f(-1) = 1$, but $\lim_{x \to -1} f(x)$ does not exist

$x = 2$: $\lim_{x \to 2} f(x) = 5$, but $f(2) = 3$

$x = 4$: $\lim_{x \to 4} f(x)$ does not exist, and $f(4)$ does not exist

2. (A) f is continuous at $x = 1$, since $\lim_{x \to 1} f(x) = 2 = f(1)$.

(B) g is not continuous at $x = 1$, since $g(1)$ is not defined.

(C) h is not continuous at $x = 2$ for two reasons: $h(2)$ does not exist and $\lim_{x \to 2} h(x)$ does not exist.
h is continuous at $x = 0$, since $\lim_{x \to 0} h(x) = -1 = h(0)$.

3. (A) Since f is a polynomial function, f is continuous for all x.

(B) Since f is a rational function, f is continuous for all x except -1 and 4 (values that make the denominator 0).

(C) The polynomial function $x - 4$ is continuous for all x and nonnegative for $x \ge 4$. Since $n = 2$ is even, f is continuous for $x \ge 4$, or on the interval $[4, \infty)$.

(D) The polynomial function $x^3 + 1$ is continuous for all x. Since $n = 3$ is odd, f is continuous for all x.

4. $-\infty < x < -1$ or $1 < x < 3$; $(-\infty, -1) \cup (1, 3)$

3-4   The Derivative

- Rate of Change
- Slope of the Tangent Line
- The Derivative
- Nonexistence of the Derivative

We will now make use of the limit concepts developed in Sections 3-1, 3-2, and 3-3 to solve the two important problems illustrated in Figure 1. The solution of each of these apparently unrelated problems involves a common concept called the *derivative*.

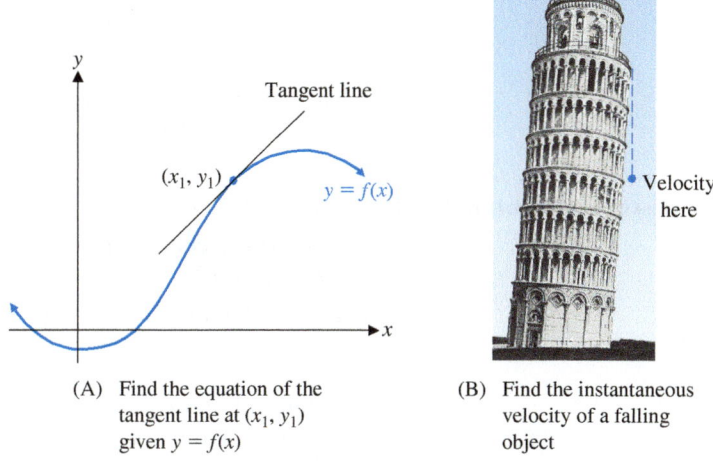

(A) Find the equation of the tangent line at (x_1, y_1) given $y = f(x)$

(B) Find the instantaneous velocity of a falling object

Figure 1 Two basic problems of calculus

Rate of Change

Let us start by considering a simple example.

EXAMPLE 1   **Revenue Analysis** The revenue (in dollars) from the sale of x plastic planter boxes is given by

$$R(x) = 20x - 0.02x^2 \qquad 0 \le x \le 1,000$$

and is graphed in Figure 2.

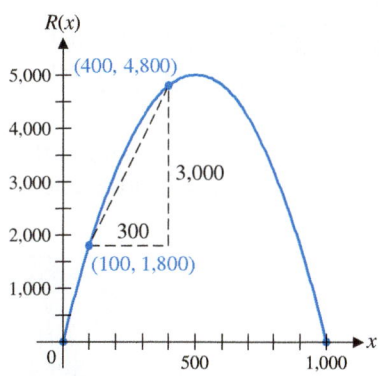

Figure 2 $R(x) = 20x - 0.02x^2$

(A) What is the change in revenue if production is changed from 100 planters to 400 planters?

(B) What is the average change in revenue for this change in production?

SOLUTION (A) The change in revenue is given by

$$R(400) - R(100) = 20(400) - 0.02(400)^2 - [20(100) - 0.02(100)^2]$$
$$= 4,800 - 1,800 = \$3,000$$

Increasing production from 100 planters to 400 planters will increase revenue by $3,000.

(B) To find the average change in revenue, we divide the change in revenue by the change in production:

$$\frac{R(400) - R(100)}{400 - 100} = \frac{3,000}{300} = \$10$$

The average change in revenue is $10 per planter when production is increased from 100 to 400 planters.

Matched Problem 1 Refer to the revenue function in Example 1.

(A) What is the change in revenue if production is changed from 600 planters to 800 planters?

(B) What is the average change in revenue for this change in production?

In general, if we are given a function $y = f(x)$ and if x is changed from a to $a + h$, then y will change from $f(a)$ to $f(a + h)$. The *average rate of change is the ratio of the change in y to the change in x.*

DEFINITION Average Rate of Change

For $y = f(x)$, the **average rate of change from $x = a$ to $x = a + h$** is

$$\frac{f(a + h) - f(a)}{(a + h) - a} = \frac{f(a + h) - f(a)}{h} \qquad h \neq 0 \qquad (1)$$

As we noted in Section 3-1, equation (1) is called the **difference quotient**. The preceding discussion shows that the difference quotient can be interpreted as an average rate of change. The next example illustrates another interpretation of this quotient: the velocity of a moving object.

EXAMPLE 2 **Velocity** A small steel ball dropped from a tower will fall a distance of y feet in x seconds, as given approximately by the formula

$$y = f(x) = 16x^2$$

Figure 3 shows the position of the ball on a coordinate line (positive direction down) at the end of 0, 1, 2, and 3 seconds.

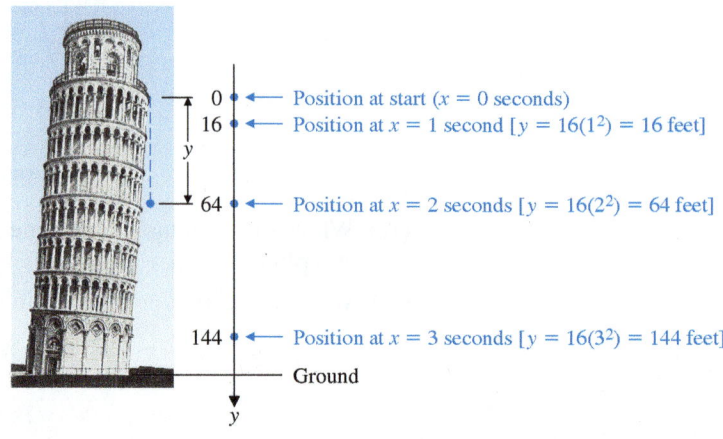

0 ← Position at start ($x = 0$ seconds)
16 ← Position at $x = 1$ second [$y = 16(1^2) = 16$ feet]
64 ← Position at $x = 2$ seconds [$y = 16(2^2) = 64$ feet]
144 ← Position at $x = 3$ seconds [$y = 16(3^2) = 144$ feet]
Ground

Figure 3 Note: Positive y direction is down.

(A) Find the average velocity from $x = 2$ seconds to $x = 3$ seconds.

(B) Find and simplify the average velocity from $x = 2$ seconds to $x = 2 + h$ seconds, $h \neq 0$.

(C) Find the limit of the expression from part B as $h \rightarrow 0$ if that limit exists.

(D) Discuss possible interpretations of the limit from part C.

SOLUTION   (A) Recall the formula $d = rt$, which can be written in the form

$$r = \frac{d}{t} = \frac{\text{Distance covered}}{\text{Elapsed time}} = \text{Average velocity}$$

For example, if a person drives from San Francisco to Los Angeles (a distance of about 420 miles) in 7 hours, then the average velocity is

$$r = \frac{d}{t} = \frac{420}{7} = 60 \text{ miles per hour}$$

Sometimes the person will be traveling faster and sometimes slower, but the average velocity is 60 miles per hour. In our present problem, the average velocity of the steel ball from $x = 2$ seconds to $x = 3$ seconds is

$$\begin{aligned}\text{Average velocity} &= \frac{\text{Distance covered}}{\text{Elapsed time}} \\ &= \frac{f(3) - f(2)}{3 - 2} \\ &= \frac{16(3)^2 - 16(2)^2}{1} = 80 \text{ feet per second}\end{aligned}$$

We see that if $y = f(x)$ is the position of the falling ball, then the average velocity is simply the average rate of change of $f(x)$ with respect to time x, and we have another interpretation of the difference quotient (1).

(B) Proceeding as in part A, we have

$$\begin{aligned}\text{Average velocity} &= \frac{\text{Distance covered}}{\text{Elapsed time}} \\ &= \frac{f(2 + h) - f(2)}{h} &&\text{\textit{Difference quotient}} \\ &= \frac{16(2 + h)^2 - 16(2)^2}{h} &&\text{\textit{Simplify this 0/0}} \\ &&&\text{\textit{indeterminate form.}} \\ &= \frac{64 + 64h + 16h^2 - 64}{h} \\ &= \frac{h(64 + 16h)}{h} = 64 + 16h \qquad h \neq 0\end{aligned}$$

Notice that if $h = 1$, the average velocity is 80 feet per second, which is the result in part A.

(C) The limit of the average velocity expression from part B as $h \rightarrow 0$ is

$$\lim_{h \to 0} \frac{f(2 + h) - f(2)}{h} = \lim_{h \to 0} (64 + 16h)$$

$$= 64 \text{ feet per second}$$

(D) The average velocity over smaller and smaller time intervals approaches 64 feet per second. This limit can be interpreted as the velocity of the ball at the *instant* that the ball has been falling for exactly 2 seconds. Therefore, 64 feet per second is referred to as the **instantaneous velocity** at $x = 2$ seconds, and we have solved one of the basic problems of calculus (see Fig. 1B).

Matched Problem 2   For the falling steel ball in Example 2, find

(A) The average velocity from $x = 1$ second to $x = 2$ seconds

(B) The average velocity (in simplified form) from $x = 1$ second to $x = 1 + h$ seconds, $h \neq 0$

(C) The instantaneous velocity at $x = 1$ second

The ideas introduced in Example 2 are not confined to average velocity, but can be applied to the average rate of change of any function.

> **DEFINITION   Instantaneous Rate of Change**
>
> For $y = f(x)$, the **instantaneous rate of change at $x = a$** is
>
> $$\lim_{h \to 0} \frac{f(a + h) - f(a)}{h} \qquad (2)$$
>
> if the limit exists.

The adjective *instantaneous* is often omitted with the understanding that the phrase **rate of change** always refers to the instantaneous rate of change and not the average rate of change. Similarly, **velocity** always refers to the instantaneous rate of change of distance with respect to time.

Slope of the Tangent Line

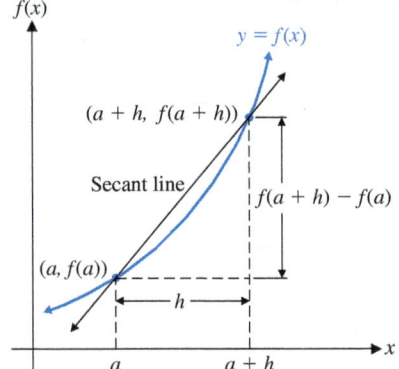

Figure 4 Secant line

So far, our interpretations of the difference quotient have been numerical in nature. Now we want to consider a geometric interpretation. A line through two points on the graph of a function is called a **secant line**. If $(a, f(a))$ and $(a + h, f(a + h))$ are two points on the graph of $y = f(x)$, then we can use the slope formula from Section 1-2 to find the slope of the secant line through these points (Fig. 4).

$$\text{Slope of secant line} = \frac{f(a + h) - f(a)}{(a + h) - a}$$

$$= \frac{f(a + h) - f(a)}{h} \qquad \textit{Difference quotient}$$

The difference quotient can be interpreted as both the average rate of change and the slope of the secant line.

> **CONCEPTUAL INSIGHT**
>
> If (x_1, y_1) and (x_2, y_2) are two points in the plane with $x_1 \neq x_2$ and L is the line passing through these two points (see Section 1-2), then
>
> $$\text{Slope of L} \qquad \text{Point–slope form for L}$$
>
> $$m = \frac{y_2 - y_1}{x_2 - x_1} \qquad y - y_1 = m(x - x_1)$$
>
> These formulas will be used extensively in the remainder of this chapter.

EXAMPLE 3 **Slope of a Secant Line** Given $f(x) = x^2$,

(A) Find the slope of the secant line for $a = 1$ and $h = 2$ and 1, respectively. Graph $y = f(x)$ and the two secant lines.

(B) Find and simplify the slope of the secant line for $a = 1$ and h any nonzero number.

(C) Find the limit of the expression in part B.

(D) Discuss possible interpretations of the limit in part C.

SOLUTION (A) For $a = 1$ and $h = 2$, the secant line goes through $(1, f(1)) = (1, 1)$ and $(3, f(3)) = (3, 9)$, and its slope is

$$\frac{f(1 + 2) - f(1)}{2} = \frac{3^2 - 1^2}{2} = 4$$

For $a = 1$ and $h = 1$, the secant line goes through $(1, f(1)) = (1, 1)$ and $(2, f(2)) = (2, 4)$, and its slope is

$$\frac{f(1 + 1) - f(1)}{1} = \frac{2^2 - 1^2}{1} = 3$$

The graphs of $y = f(x)$ and the two secant lines are shown in Figure 5.

(B) For $a = 1$ and h any nonzero number, the secant line goes through $(1, f(1)) = (1, 1)$ and $(1 + h, f(1 + h)) = (1 + h, (1 + h)^2)$, and its slope is

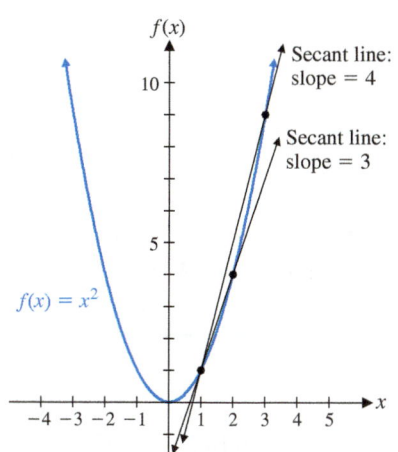

Figure 5 Secant lines

$$\frac{f(1 + h) - f(1)}{h} = \frac{(1 + h)^2 - 1^2}{h}$$ *Square the binomial.*

$$= \frac{1 + 2h + h^2 - 1}{h}$$ *Combine like terms and factor the numerator.*

$$= \frac{h(2 + h)}{h}$$ *Cancel.*

$$= 2 + h \qquad h \neq 0$$

(C) The limit of the secant line slope from part B is

$$\lim_{h \to 0} \frac{f(1 + h) - f(1)}{h} = \lim_{h \to 0} (2 + h)$$
$$= 2$$

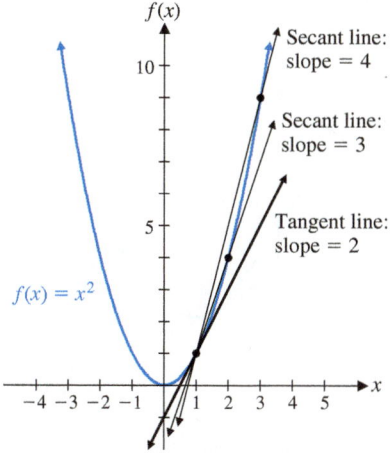

Figure 6 Tangent line

(D) In part C, we saw that the limit of the slopes of the secant lines through the point $(1, f(1))$ is 2. If we graph the line through $(1, f(1))$ with slope 2 (Fig. 6), then this line is the limit of the secant lines. The slope obtained from the limit of slopes of secant lines is called the *slope of the graph* at $x = 1$. The line through the point $(1, f(1))$ with this slope is called the *tangent line*. We have solved another basic problem of calculus (see Fig. 1A on page 165).

Matched Problem 3 Given $f(x) = x^2$,

(A) Find the slope of the secant line for $a = 2$ and $h = 2$ and 1, respectively.

(B) Find and simplify the slope of the secant line for $a = 2$ and h any nonzero number.

(C) Find the limit of the expression in part B.

(D) Find the slope of the graph and the slope of the tangent line at $a = 2$.

The ideas introduced in the preceding example are summarized next:

DEFINITION   Slope of a Graph

Given $y = f(x)$, the **slope of the graph** at the point $(a, f(a))$ is given by

$$\lim_{h \to 0} \frac{f(a + h) - f(a)}{h} \qquad (3)$$

provided that the limit exists. The slope of the graph is also the **slope of the tangent line** at the point $(a, f(a))$.

CONCEPTUAL INSIGHT

If the function f is continuous at a, then

$$\lim_{h \to 0} f(a + h) = f(a)$$

and limit (3) will be a 0/0 indeterminate form. As we saw in Examples 2 and 3, evaluating this type of limit typically involves algebraic simplification.

From plane geometry, we know that a line tangent to a circle is a line that passes through one and only one point of the circle (Fig. 7A). Although this definition cannot be extended to graphs of functions in general, the visual relationship between graphs of functions and their tangent lines (Fig. 7B) is similar to the circle case. Limit (3) provides both a mathematically sound definition of a tangent line and a method for approximating the slope of the tangent line.

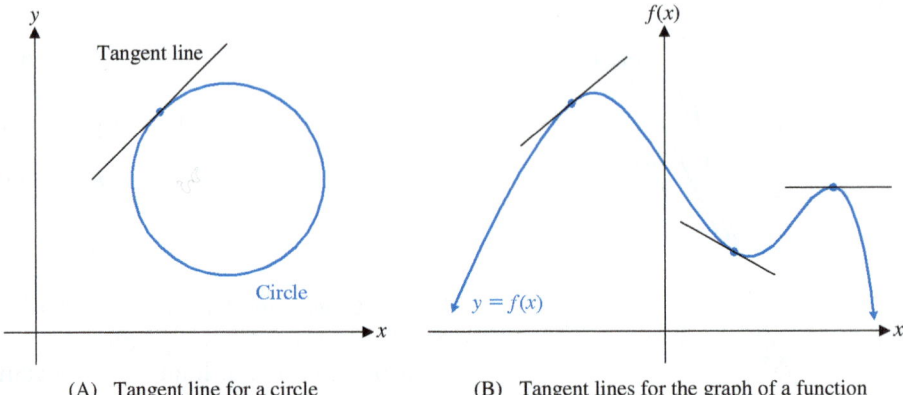

(A)   Tangent line for a circle   (B)   Tangent lines for the graph of a function

Figure 7

The Derivative

We have seen that the limit of a difference quotient can be interpreted as a rate of change, as a velocity, or as the slope of a tangent line. In addition, this limit provides solutions to two of the three basic problems stated at the beginning of the chapter. We are now ready to introduce some terms that refer to that limit. To follow customary practice, we use x in place of a and think of the difference quotient

$$\frac{f(x + h) - f(x)}{h}$$

as a function of h, with x held fixed as h tends to 0.

DEFINITION The Derivative

For $y = f(x)$, we define the **derivative of f at x**, denoted by $f'(x)$, to be

$$f'(x) = \lim_{h \to 0} \frac{f(x + h) - f(x)}{h} \text{ if the limit exists}$$

If $f'(x)$ exists for each x in the open interval (a, b), then f is said to be **differentiable** over (a, b).

(Differentiability from the left or from the right is defined by using $h \to 0^-$ or $h \to 0^+$, respectively, in place of $h \to 0$ in the preceding definition.)

The process of finding the derivative of a function is called **differentiation**. The derivative of a function is obtained by **differentiating** the function.

SUMMARY Interpretations of the Derivative

The derivative of a function f is a new function f'. The domain of f' is a subset of the domain of f. The derivative has various applications and interpretations, including the following:

1. *Slope of the tangent line.* For each x in the domain of f', $f'(x)$ is the slope of the line tangent to the graph of f at the point $(x, f(x))$.
2. *Instantaneous rate of change.* For each x in the domain of f', $f'(x)$ is the instantaneous rate of change of $y = f(x)$ with respect to x.
3. *Velocity.* If $f(x)$ is the position of a moving object at time x, then $v = f'(x)$ is the velocity of the object at that time.

Example 4 illustrates the **four-step process** that we use to find derivatives in this section. In subsequent sections, we develop rules for finding derivatives that do not involve limits. However, it is important that you master the limit process in order to fully comprehend and appreciate the various applications we will consider.

EXAMPLE 4   **Finding a Derivative** Find $f'(x)$, the derivative of f at x, for $f(x) = 4x - x^2$.

SOLUTION   To find $f'(x)$, we use a four-step process.

Step 1 Find $f(x + h)$.
$$f(x + h) = 4(x + h) - (x + h)^2$$
$$= 4x + 4h - x^2 - 2xh - h^2$$

Step 2 Find $f(x + h) - f(x)$.
$$f(x + h) - f(x) = 4x + 4h - x^2 - 2xh - h^2 - (4x - x^2)$$
$$= 4h - 2xh - h^2$$

Step 3 Find $\dfrac{f(x + h) - f(x)}{h}$.
$$\frac{f(x + h) - f(x)}{h} = \frac{4h - 2xh - h^2}{h} = \frac{h(4 - 2x - h)}{h}$$
$$= 4 - 2x - h, \quad h \neq 0$$

Step 4 Find $f'(x) = \lim\limits_{h \to 0} \dfrac{f(x + h) - f(x)}{h}$.
$$f'(x) = \lim_{h \to 0} \frac{f(x + h) - f(x)}{h} = \lim_{h \to 0} (4 - 2x - h) = 4 - 2x$$

So if $f(x) = 4x - x^2$, then $f'(x) = 4 - 2x$. The function f' is a new function derived from the function f.

Matched Problem 4   Find $f'(x)$, the derivative of f at x, for $f(x) = 8x - 2x^2$.

The four-step process used in Example 4 is summarized as follows for easy reference:

PROCEDURE   The four-step process for finding the derivative of a function f:

Step 1   Find $f(x + h)$.

Step 2   Find $f(x + h) - f(x)$.

Step 3   Find $\dfrac{f(x + h) - f(x)}{h}$.

Step 4   Find $\lim\limits_{h \to 0} \dfrac{f(x + h) - f(x)}{h}$.

EXAMPLE 5   **Finding Tangent Line Slopes** In Example 4, we started with the function $f(x) = 4x - x^2$ and found the derivative of f at x to be $f'(x) = 4 - 2x$. So the slope of a line tangent to the graph of f at any point $(x, f(x))$ on the graph is

$$m = f'(x) = 4 - 2x$$

(A) Find the slope of the graph of f at $x = 0$, $x = 2$, and $x = 3$.

(B) Graph $y = f(x) = 4x - x^2$ and use the slopes found in part (A) to make a rough sketch of the lines tangent to the graph at $x = 0$, $x = 2$, and $x = 3$.

SOLUTION   (A) Using $f'(x) = 4 - 2x$, we have

$$f'(0) = 4 - 2(0) = 4 \qquad \textit{Slope at } x = 0$$
$$f'(2) = 4 - 2(2) = 0 \qquad \textit{Slope at } x = 2$$
$$f'(3) = 4 - 2(3) = -2 \quad \textit{Slope at } x = 3$$

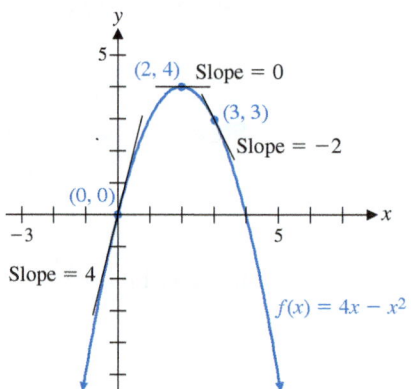

Matched Problem 5   In Matched Problem 4, we started with the function $f(x) = 8x - 2x^2$. Using the derivative found there,

(A) Find the slope of the graph of f at $x = 1$, $x = 2$, and $x = 4$.

(B) Graph $y = f(x) = 8x - 2x^2$, and use the slopes from part (A) to make a rough sketch of the lines tangent to the graph at $x = 1$, $x = 2$, and $x = 4$.

EXPLORE & DISCUSS 1

In Example 4, we found that the derivative of $f(x) = 4x - x^2$ is $f'(x) = 4 - 2x$. In Example 5, we graphed $f(x)$ and several tangent lines.

(A) Graph f and f' on the same set of axes.

(B) The graph of f' is a straight line. Is it a tangent line for the graph of f? Explain.

(C) Find the x intercept for the graph of f'. What is the slope of the line tangent to the graph of f for this value of x? Write a verbal description of the relationship between the slopes of the tangent lines of a function and the x intercepts of the derivative of the function.

EXAMPLE 6

Finding a Derivative Find $f'(x)$, the derivative of f at x, for $f(x) = \sqrt{x} + 2$.

SOLUTION

We use the four-step process to find $f'(x)$.

Step 1 Find $f(x + h)$.

$$f(x + h) = \sqrt{x + h} + 2$$

Step 2 Find $f(x + h) - f(x)$.

$$f(x + h) - f(x) = \sqrt{x + h} + 2 - (\sqrt{x} + 2) \quad \textit{Combine like terms.}$$
$$= \sqrt{x + h} - \sqrt{x}$$

Step 3 Find $\dfrac{f(x + h) - f(x)}{h}$.

$$\frac{f(x + h) - f(x)}{h} = \frac{\sqrt{x + h} - \sqrt{x}}{h}$$

$$= \frac{\sqrt{x + h} - \sqrt{x}}{h} \cdot \frac{\sqrt{x + h} + \sqrt{x}}{\sqrt{x + h} + \sqrt{x}}$$

$$= \frac{x + h - x}{h(\sqrt{x + h} + \sqrt{x})}$$

$$= \frac{h}{h(\sqrt{x + h} + \sqrt{x})}$$

$$= \frac{1}{\sqrt{x + h} + \sqrt{x}} \quad h \neq 0$$

We rationalize the numerator (Appendix A, Section A-6) to change the form of this fraction.
Combine like terms.
Cancel.

Step 4 Find $f'(x) = \lim\limits_{h \to 0} \dfrac{f(x + h) - f(x)}{h}$.

$$\lim_{h \to 0} \frac{f(x + h) - f(x)}{h} = \lim_{h \to 0} \frac{1}{\sqrt{x + h} + \sqrt{x}}$$

$$= \frac{1}{\sqrt{x} + \sqrt{x}} = \frac{1}{2\sqrt{x}} \quad x > 0$$

So the derivative of $f(x) = \sqrt{x} + 2$ is $f'(x) = 1/(2\sqrt{x})$, a new function. The domain of f is $[0, \infty)$. Since $f'(0)$ is not defined, the domain of f' is $(0, \infty)$, a subset of the domain of f.

Matched Problem 6

Find $f'(x)$ for $f(x) = \sqrt{x + 4}$.

EXAMPLE 7

Sales Analysis A company's total sales (in millions of dollars) t months from now are given by $S(t) = \sqrt{t} + 2$. Find and interpret $S(25)$ and $S'(25)$. Use these results to estimate the total sales after 26 months and after 27 months.

SOLUTION   The total sales function S has the same form as the function f in Example 6. Only the letters used to represent the function and the independent variable have been changed. It follows that S' and f' also have the same form:

$$S(t) = \sqrt{t} + 2 \qquad f(x) = \sqrt{x} + 2$$

$$S'(t) = \frac{1}{2\sqrt{t}} \qquad f'(x) = \frac{1}{2\sqrt{x}}$$

Evaluating S and S' at $t = 25$, we have

$$S(25) = \sqrt{25} + 2 = 7 \qquad S'(25) = \frac{1}{2\sqrt{25}} = 0.1$$

So 25 months from now, the total sales will be $7 million and will be increasing at the rate of $0.1 million ($100,000) per month. If this instantaneous rate of change of sales remained constant, the sales would grow to $7.1 million after 26 months, $7.2 million after 27 months, and so on. Even though $S'(t)$ is not a constant function in this case, these values provide useful estimates of the total sales.

Matched Problem 7   A company's total sales (in millions of dollars) t months from now are given by $S(t) = \sqrt{t} + 4$. Find and interpret $S(12)$ and $S'(12)$. Use these results to estimate the total sales after 13 months and after 14 months. (Use the derivative found in Matched Problem 6.)

In Example 7, we can compare the estimates of total sales by using the derivative with the corresponding exact values of $S(t)$:

| Exact values | Estimated values |

$$S(26) = \sqrt{26} + 2 = 7.099\ldots \approx 7.1$$
$$S(27) = \sqrt{27} + 2 = 7.196\ldots \approx 7.2$$

For this function, the estimated values provide very good approximations to the exact values of $S(t)$. For other functions, the approximation might not be as accurate.

Using the instantaneous rate of change of a function at a point to estimate values of the function at nearby points is an important application of the derivative.

Nonexistence of the Derivative

The existence of a derivative at $x = a$ depends on the existence of a limit at $x = a$, that is, on the existence of

$$f'(a) = \lim_{h \to 0} \frac{f(a + h) - f(a)}{h} \qquad (4)$$

If the limit does not exist at $x = a$, we say that the function f is **nondifferentiable at** $x = a$, or $f'(a)$ **does not exist.**

EXPLORE & DISCUSS 2   Let $f(x) = |x - 1|$.
(A) Graph f.
(B) Complete the following table:

h	−0.1	−0.01	−0.001	→0←	0.001	0.01	0.1
$\dfrac{f(1 + h) - f(1)}{h}$	?	?	?	→?←	?	?	?

(C) Find the following limit if it exists:

$$\lim_{h \to 0} \frac{f(1 + h) - f(1)}{h}$$

(D) Use the results of parts (A)–(C) to discuss the existence of $f'(1)$.

(E) Repeat parts (A)–(D) for $\sqrt[3]{x - 1}$.

How can we recognize the points on the graph of f where $f'(a)$ does not exist? It is impossible to describe all the ways that the limit of a difference quotient can fail to exist. However, we can illustrate some common situations where $f'(a)$ fails to exist (see Fig. 8):

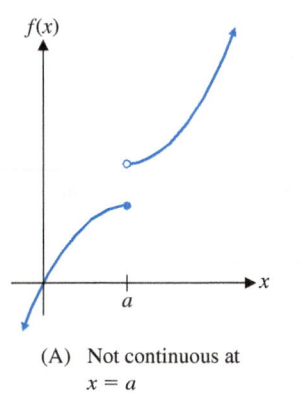

(A) Not continuous at $x = a$

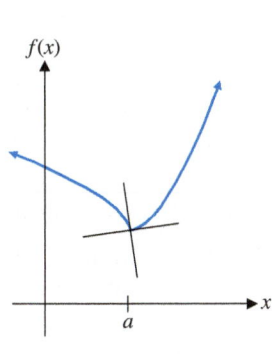

(B) Graph has sharp corner at $x = a$

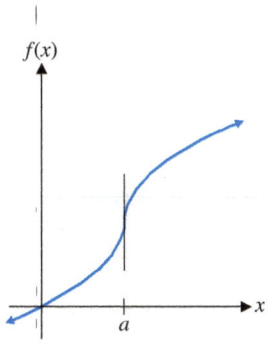

(C) Vertical tangent at $x = a$

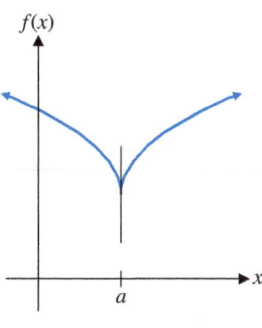

(D) Vertical tangent at $x = a$

Figure 8 The function f is nondifferentiable at $x = a$.

1. If the graph of f has a hole or a break at $x = a$, then $f'(a)$ does not exist (Fig. 8A).

2. If the graph of f has a sharp corner at $x = a$, then $f'(a)$ does not exist, and the graph has no tangent line at $x = a$ (Fig. 8B). (In Fig. 8B, the left- and right-hand derivatives exist but are not equal.)

3. If the graph of f has a vertical tangent line at $x = a$, then $f'(a)$ does not exist (Fig. 8C and D).

Exercises 3-4

A

In Problems 1 and 2, find the indicated quantity for $y = f(x) = 5 - x^2$ and interpret that quantity in terms of the following graph.

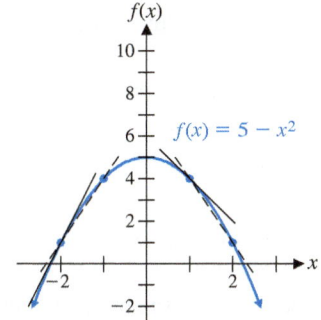

1. (A) $\dfrac{f(2) - f(1)}{2 - 1}$ (B) $\dfrac{f(1 + h) - f(1)}{h}$

(C) $\displaystyle\lim_{h \to 0} \dfrac{f(1 + h) - f(1)}{h}$

2. (A) $\dfrac{f(-1) - f(-2)}{-1 - (-2)}$

(B) $\dfrac{f(-2 + h) - f(-2)}{h}$

(C) $\displaystyle\lim_{h \to 0} \dfrac{f(-2 + h) - f(-2)}{h}$

3. Find the indicated quantities for $f(x) = 3x^2$.

(A) The average rate of change of $f(x)$ if x changes from 1 to 4.

(B) The slope of the secant line through the points $(1, f(1))$ and $(4, f(4))$ on the graph of $y = f(x)$.

(C) The slope of the secant line through the points $(1, f(1))$ and $(1 + h, f(1 + h))$, $h \neq 0$. Simplify your answer.

(D) The slope of the graph at $(1, f(1))$.

(E) The instantaneous rate of change of $y = f(x)$ with respect to x at $x = 1$.

(F) The slope of the tangent line at $(1, f(1))$.

(G) The equation of the tangent line at $(1, f(1))$.

4. Find the indicated quantities for $f(x) = 3x^2$.

(A) The average rate of change of $f(x)$ if x changes from 2 to 5.

(B) The slope of the secant line through the points $(2, f(2))$ and $(5, f(5))$ on the graph of $y = f(x)$.

(C) The slope of the secant line through the points $(2, f(2))$ and $(2 + h, f(2 + h))$, $h \neq 0$. Simplify your answer.

(D) The slope of the graph at $(2, f(2))$.

(E) The instantaneous rate of change of $y = f(x)$ with respect to x at $x = 2$.

(F) The slope of the tangent line at $(2, f(2))$.

(G) The equation of the tangent line at $(2, f(2))$.

In Problems 5–26, use the four-step process to find $f'(x)$ and then find $f'(1)$, $f'(2)$, and $f'(3)$.

5. $f(x) = -5$

6. $f(x) = 9$

7. $f(x) = 3x - 7$

8. $f(x) = 4 - 6x$

9. $f(x) = 2 - 3x^2$

10. $f(x) = 2x^2 + 8$

11. $f(x) = x^2 + 6x - 10$

12. $f(x) = x^2 + 4x + 7$

13. $f(x) = 2x^2 - 7x + 3$

14. $f(x) = 2x^2 + 5x + 1$

15. $f(x) = -x^2 + 4x - 9$

16. $f(x) = -x^2 + 9x - 2$

17. $f(x) = 2x^3 + 1$

18. $f(x) = -2x^3 + 5$

19. $f(x) = 4 + \dfrac{4}{x}$

20. $f(x) = \dfrac{6}{x} - 2$

21. $f(x) = 5 + 3\sqrt{x}$

22. $f(x) = 3 - 7\sqrt{x}$

23. $f(x) = 10\sqrt{x} + 5$

24. $f(x) = 16\sqrt{x} + 9$

25. $f(x) = \dfrac{3x}{x + 2}$

26. $f(x) = \dfrac{5x}{3 + x}$

B

Problems 27 and 28 refer to the graph of $y = f(x) = x^2 + x$ shown.

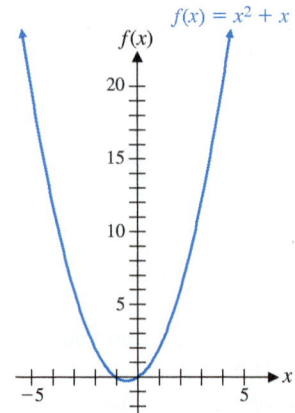

$f(x) = x^2 + x$

27. (A) Find the slope of the secant line joining $(1, f(1))$ and $(3, f(3))$.

(B) Find the slope of the secant line joining $(1, f(1))$ and $(1 + h, f(1 + h))$.

(C) Find the slope of the tangent line at $(1, f(1))$.

(D) Find the equation of the tangent line at $(1, f(1))$.

28. (A) Find the slope of the secant line joining $(2, f(2))$ and $(4, f(4))$.

(B) Find the slope of the secant line joining $(2, f(2))$ and $(2 + h, f(2 + h))$.

(C) Find the slope of the tangent line at $(2, f(2))$.

(D) Find the equation of the tangent line at $(2, f(2))$.

In Problems 29 and 30, suppose an object moves along the y axis so that its location is $y = f(x) = x^2 + x$ at time x (y is in meters and x is in seconds). Find

29. (A) The average velocity (the average rate of change of y with respect to x) for x changing from 1 to 3 seconds

(B) The average velocity for x changing from 1 to $1 + h$ seconds

(C) The instantaneous velocity at $x = 1$ second

30. (A) The average velocity (the average rate of change of y with respect to x) for x changing from 2 to 4 seconds

(B) The average velocity for x changing from 2 to $2 + h$ seconds

(C) The instantaneous velocity at $x = 2$ seconds

Problems 31–38 refer to the function F in the graph shown. Use the graph to determine whether $F'(x)$ exists at each indicated value of x.

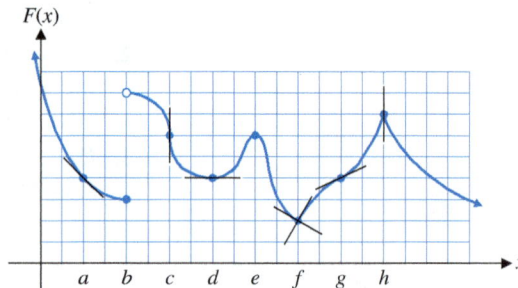

$F(x)$

31. $x = a$

32. $x = b$

33. $x = c$

34. $x = d$

35. $x = e$

36. $x = f$

37. $x = g$

38. $x = h$

39. Given $f(x) = x^2 - 4x$,

(A) Find $f'(x)$.

(B) Find the slopes of the lines tangent to the graph of f at $x = 0, 2$, and 4.

(C) Graph f and sketch in the tangent lines at $x = 0, 2$, and 4.

40. Given $f(x) = x^2 + 2x$,

(A) Find $f'(x)$.

(B) Find the slopes of the lines tangent to the graph of f at $x = -2, -1$, and 1.

(C) Graph f and sketch in the tangent lines at $x = -2, -1$, and 1.

41. If an object moves along a line so that it is at $y = f(x) = 4x^2 - 2x$ at time x (in seconds), find the instantaneous velocity function $v = f'(x)$ and find the velocity at times $x = 1, 3$, and 5 seconds (y is measured in feet).

42. Repeat Problem 41 with $f(x) = 8x^2 - 4x$.

43. Let $f(x) = x^2$, $g(x) = x^2 - 1$, and $h(x) = x^2 + 2$.

(A) How are the graphs of these functions related? How would you expect the derivatives of these functions to be related?

(B) Use the four-step process to find the derivative of $m(x) = x^2 + C$, where C is any real constant.

44. Let $f(x) = -x^2$, $g(x) = -x^2 - 1$, and $h(x) = -x^2 + 2$.

(A) How are the graphs of these functions related? How would you expect the derivatives of these functions to be related?

(B) Use the four-step process to find the derivative of $m(x) = -x^2 + C$, where C is any real constant.

In Problems 45–50, discuss the validity of each statement. If the statement is always true, explain why. If not, give a counterexample.

45. If $f(x) = C$ is a constant function, then $f'(x) = 0$.

46. If $f(x) = mx + b$ is a linear function, then $f'(x) = m$.

47. If a function f is continuous on the interval (a, b), then f is differentiable on (a, b).

48. If a function f is differentiable on the interval (a, b), then f is continuous on (a, b).

49. The average rate of change of a function f from $x = a$ to $x = a + h$ is less than the instantaneous rate of change at $x = a + \dfrac{h}{2}$.

50. If the graph of f has a sharp corner at $x = a$, then f is not continuous at $x = a$.

C

In Problems 51–54, sketch the graph of f and determine where f is nondifferentiable.

51. $f(x) = \begin{cases} 2x & \text{if } x < 1 \\ 2 & \text{if } x \ge 1 \end{cases}$

52. $f(x) = \begin{cases} 2x & \text{if } x < 2 \\ 6 - x & \text{if } x \ge 2 \end{cases}$

53. $f(x) = \begin{cases} x^2 + 1 & \text{if } x < 0 \\ 1 & \text{if } x \ge 0 \end{cases}$

54. $f(x) = \begin{cases} 2 - x^2 & \text{if } x \le 0 \\ 2 & \text{if } x > 0 \end{cases}$

In Problems 55–60, determine whether f is differentiable at $x = 0$ by considering

$$\lim_{h \to 0} \frac{f(0 + h) - f(0)}{h}$$

55. $f(x) = |x|$

56. $f(x) = 1 - |x|$

57. $f(x) = x^{1/3}$

58. $f(x) = x^{2/3}$

59. $f(x) = \sqrt{1 - x^2}$

60. $f(x) = \sqrt{1 + x^2}$

61. A ball dropped from a balloon falls $y = 16x^2$ feet in x seconds. If the balloon is 576 feet above the ground when the ball is dropped, when does the ball hit the ground? What is the velocity of the ball at the instant it hits the ground?

62. Repeat Problem 61 if the balloon is 1,024 feet above the ground when the ball is dropped.

Applications

63. Revenue. The revenue (in dollars) from the sale of x infant car seats is given by

$$R(x) = 60x - 0.025x^2 \qquad 0 \le x \le 2{,}400$$

(A) Find the average change in revenue if production is changed from 1,000 car seats to 1,050 car seats.

(B) Use the four-step process to find $R'(x)$.

(C) Find the revenue and the instantaneous rate of change of revenue at a production level of 1,000 car seats, and write a brief verbal interpretation of these results.

64. Profit. The profit (in dollars) from the sale of x infant car seats is given by

$$P(x) = 45x - 0.025x^2 - 5{,}000 \qquad 0 \le x \le 2{,}400$$

(A) Find the average change in profit if production is changed from 800 car seats to 850 car seats.

(B) Use the four-step process to find $P'(x)$.

(C) Find the profit and the instantaneous rate of change of profit at a production level of 800 car seats, and write a brief verbal interpretation of these results.

65. Sales analysis. A company's total sales (in millions of dollars) t months from now are given by

$$S(t) = 2\sqrt{t} + 10$$

(A) Use the four-step process to find $S'(t)$.

(B) Find $S(15)$ and $S'(15)$. Write a brief verbal interpretation of these results.

(C) Use the results in part (B) to estimate the total sales after 16 months and after 17 months.

66. Sales analysis. A company's total sales (in millions of dollars) t months from now are given by

$$S(t) = 2\sqrt{t} + 6$$

(A) Use the four-step process to find $S'(t)$.

(B) Find $S(10)$ and $S'(10)$. Write a brief verbal interpretation of these results.

(C) Use the results in part (B) to estimate the total sales after 11 months and after 12 months.

67. Mineral consumption. The U.S. consumption of tungsten (in metric tons) is given approximately by

$$p(t) = 164t^2 + 161t + 12{,}326$$

where t is time in years and $t = 0$ corresponds to 2005.

(A) Use the four-step process to find $p'(t)$.

(B) Find the annual production in 2015 and the instantaneous rate of change of production in 2015, and write a brief verbal interpretation of these results.

68. Mineral consumption. The U.S. consumption of copper (in thousands of metric tons) is given approximately by

$$p(t) = 29t^2 - 258t + 4{,}658$$

where t is time in years and $t = 0$ corresponds to 2005.

(A) Use the four-step process to find $p'(t)$.

(B) Find the annual production in 2017 and the instantaneous rate of change of production in 2017, and write a brief verbal interpretation of these results.

69. Electricity consumption. Table 1 gives the retail sales of electricity (in billions of kilowatt-hours) for the residential and commercial sectors in the United States. (*Source:* Energy Information Administration)

Table 1   Electricity Sales

Year	Residential	Commercial
2000	1,192	1,055
2002	1,265	1,104
2004	1,292	1,230
2006	1,352	1,300
2008	1,379	1,352

(A) Let x represent time (in years) with $x = 0$ corresponding to 2000, and let y represent the corresponding residential sales. Enter the appropriate data set in a graphing calculator and find a quadratic regression equation for the data.

(B) If $y = R(x)$ denotes the regression equation found in part (A), find $R(20)$ and $R'(20)$, and write a brief verbal interpretation of these results. Round answers to the nearest tenth of a billion.

 70. Electricity consumption. Refer to the data in Table 1.

(A) Let x represent time (in years) with $x = 0$ corresponding to 2000, and let y represent the corresponding commercial sales. Enter the appropriate data set in a graphing calculator and find a quadratic regression equation for the data.

(B) If $y = C(x)$ denotes the regression equation found in part (A), find $C(20)$ and $C'(20)$, and write a brief verbal interpretation of these results. Round answers to the nearest tenth of a billion.

71. Air pollution. The ozone level (in parts per billion) on a summer day in a metropolitan area is given by

$$P(t) = 80 + 12t - t^2$$

where t is time in hours and $t = 0$ corresponds to 9 A.M.

(A) Use the four-step process to find $P'(t)$.

(B) Find $P(3)$ and $P'(3)$. Write a brief verbal interpretation of these results.

72. Medicine. The body temperature (in degrees Fahrenheit) of a patient t hours after taking a fever-reducing drug is given by

$$F(t) = 98 + \frac{4}{t + 1}$$

(A) Use the four-step process to find $F'(t)$.

(B) Find $F(3)$ and $F'(3)$. Write a brief verbal interpretation of these results.

Answers to Matched Problems

1. (A) $-\$1,600$ (B) $-\$8$ per planter

2. (A) 48 ft/s
   (B) $32 + 16h$
   (C) 32 ft/s

3. (A) 6, 5   (B) $4 + h$
   (C) 4   (D) Both are 4

4. $f'(x) = 8 - 4x$

5. (A) $f'(1) = 4, f'(2) = 0, f'(4) = -8$
   (B)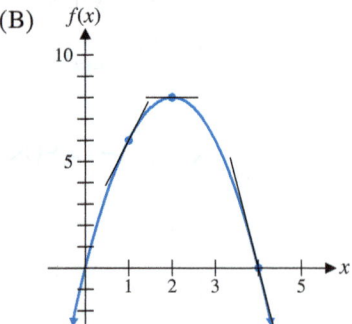

6. $f'(x) = 1/(2\sqrt{x + 4})$

7. $S(12) = 4, S'(12) = 0.125$; 12 months from now, the total sales will be \$4 million and will be increasing at the rate of \$0.125 million (\$125,000) per month. The estimated total sales are \$4.125 million after 13 months and \$4.25 million after 14 months.

3-5   Basic Differentiation Properties

- Constant Function Rule
- Power Rule
- Constant Multiple Property
- Sum and Difference Properties
- Applications

In Section 3-4, we defined the derivative of f at x as

$$f'(x) = \lim_{h \to 0} \frac{f(x + h) - f(x)}{h}$$

if the limit exists, and we used this definition and a four-step process to find the derivatives of several functions. Now we want to develop some rules of differentiation. These rules will enable us to find the derivative of many functions without using the four-step process.

Before exploring these rules, we list some symbols that are often used to represent derivatives.

> **NOTATION   The Derivative**
> If $y = f(x)$, then
> $$f'(x) \qquad y' \qquad \frac{dy}{dx}$$
> all represent the derivative of f at x.

Each of these derivative symbols has its particular advantage in certain situations. All of them will become familiar to you after a little experience.

Constant Function Rule

If $f(x) = C$ is a constant function, then the four-step process can be used to show that $f'(x) = 0$. Therefore,

The derivative of any constant function is 0.

> **THEOREM 1   Constant Function Rule**
> If $y = f(x) = C$, then
> $$f'(x) = 0$$
> Also, $y' = 0$ and $dy/dx = 0$.
>
> **Note:** When we write $C' = 0$ or $\dfrac{d}{dx}C = 0$, we mean that $y' = \dfrac{dy}{dx} = 0$ when $y = C$.

CONCEPTUAL INSIGHT

The graph of $f(x) = C$ is a horizontal line with slope 0 (Fig. 1), so we would expect that $f'(x) = 0$.

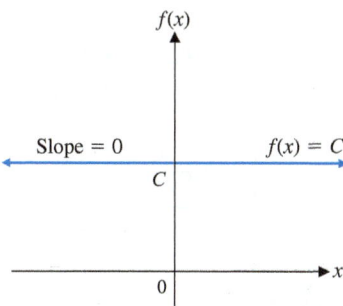

Figure 1

EXAMPLE 1   **Differentiating Constant Functions**

(A) If $f(x) = 3$,   then $f'(x) = 0$.

(B) If $y = -1.4$,   then $y' = 0$.

(C) If $y = \pi$,   then $\dfrac{dy}{dx} = 0$.

(D) $\dfrac{d}{dx}23 = 0$

Matched Problem 1   Find

(A) $f'(x)$ for $f(x) = -24$

(B) y' for $y = 12$

(C) $\dfrac{dy}{dx}$ for $y = -\sqrt{7}$

(D) $\dfrac{d}{dx}(-\pi)$

Power Rule

A function of the form $f(x) = x^k$, where k is a real number, is called a **power function**. The following elementary functions are examples of power functions:

$$f(x) = x \qquad h(x) = x^2 \qquad m(x) = x^3$$
$$n(x) = \sqrt{x} \qquad p(x) = \sqrt[3]{x} \tag{1}$$

EXPLORE & DISCUSS 1

(A) It is clear that the functions f, h, and m in (1) are power functions. Explain why the functions n and p are also power functions.

(B) The domain of a power function depends on the power. Discuss the domain of each of the following power functions:

$$r(x) = x^4 \qquad s(x) = x^{-4} \qquad t(x) = x^{1/4}$$
$$u(x) = x^{-1/4} \qquad v(x) = x^{1/5} \qquad w(x) = x^{-1/5}$$

The definition of the derivative and the four-step process introduced in Section 3-4 can be used to find the derivatives of many power functions. For example, it can be shown that

$$\text{If} \quad f(x) = x^2, \quad \text{then} \quad f'(x) = 2x.$$
$$\text{If} \quad f(x) = x^3, \quad \text{then} \quad f'(x) = 3x^2.$$
$$\text{If} \quad f(x) = x^4, \quad \text{then} \quad f'(x) = 4x^3.$$
$$\text{If} \quad f(x) = x^5, \quad \text{then} \quad f'(x) = 5x^4.$$

Notice the pattern in these derivatives. In each case, the power in f becomes the coefficient in f' and the power in f' is 1 less than the power in f. In general, for any positive integer n,

$$\text{If} \quad f(x) = x^n, \quad \text{then} \quad f'(x) = nx^{n-1}. \tag{2}$$

In fact, more advanced techniques can be used to show that (2) holds for *any* real number n. We will assume this general result for the remainder of the book.

THEOREM 2   Power Rule

If $y = f(x) = x^n$, where n is a real number, then

$$f'(x) = nx^{n-1}$$

Also, $y' = nx^{n-1}$ and $dy/dx = nx^{n-1}$.

EXAMPLE 2 **Differentiating Power Functions**

(A) If $f(x) = x^5$, then $f'(x) = 5x^{5-1} = 5x^4$.

(B) If $y = x^{25}$, then $y' = 25x^{25-1} = 25x^{24}$.

(C) If $y = t^{-3}$, then $\dfrac{dy}{dt} = -3t^{-3-1} = -3t^{-4} = -\dfrac{3}{t^4}$.

(D) $\dfrac{d}{dx}x^{5/3} = \dfrac{5}{3}x^{(5/3)-1} = \dfrac{5}{3}x^{2/3}$.

Matched Problem 2 Find

(A) $f'(x)$ for $f(x) = x^6$

(B) y' for $y = x^{30}$

(C) $\dfrac{dy}{dt}$ for $y = t^{-2}$

(D) $\dfrac{d}{dx}x^{3/2}$

In some cases, properties of exponents must be used to rewrite an expression before the power rule is applied.

EXAMPLE 3 **Differentiating Power Functions**

(A) If $f(x) = 1/x^4$, we can write $f(x) = x^{-4}$ and

$$f'(x) = -4x^{-4-1} = -4x^{-5}, \quad \text{or} \quad \frac{-4}{x^5}$$

(B) If $y = \sqrt{u}$, we can write $y = u^{1/2}$ and

$$y' = \frac{1}{2}u^{(1/2)-1} = \frac{1}{2}u^{-1/2}, \quad \text{or} \quad \frac{1}{2\sqrt{u}}$$

(C) $\dfrac{d}{dx}\dfrac{1}{\sqrt[3]{x}} = \dfrac{d}{dx}x^{-1/3} = -\dfrac{1}{3}x^{(-1/3)-1} = -\dfrac{1}{3}x^{-4/3}, \text{ or } \dfrac{-1}{3\sqrt[3]{x^4}}$

Matched Problem 3 Find

(A) $f'(x)$ for $f(x) = \dfrac{1}{x}$ (B) y' for $y = \sqrt[3]{u^2}$ (C) $\dfrac{d}{dx}\dfrac{1}{\sqrt{x}}$

Constant Multiple Property

Let $f(x) = ku(x)$, where k is a constant and u is differentiable at x. Using the four-step process, we have the following:

Step 1 $f(x + h) = ku(x + h)$

Step 2 $f(x + h) - f(x) = ku(x + h) - ku(x) = k[u(x + h) - u(x)]$

Step 3 $\dfrac{f(x + h) - f(x)}{h} = \dfrac{k[u(x + h) - u(x)]}{h} = k\left[\dfrac{u(x + h) - u(x)}{h}\right]$

Step 4 $f'(x) = \lim\limits_{h \to 0}\dfrac{f(x + h) - f(x)}{h}$

$$= \lim_{h \to 0} k\left[\frac{u(x + h) - u(x)}{h}\right] \qquad \lim_{x \to c} kg(x) = k\lim_{x \to c} g(x)$$

$$= k\lim_{h \to 0}\left[\frac{u(x + h) - u(x)}{h}\right] \qquad \text{Definition of } u'(x)$$

$$= ku'(x)$$

Therefore,

The derivative of a constant times a differentiable function is the constant times the derivative of the function.

THEOREM 3   Constant Multiple Property

If $y = f(x) = ku(x)$, then

$$f'(x) = ku'(x)$$

Also,

$$y' = ku' \qquad \frac{dy}{dx} = k\frac{du}{dx}$$

EXAMPLE 4   **Differentiating a Constant Times a Function**

(A) If $f(x) = 3x^2$, then $f'(x) \boxed{= 3 \cdot 2x^{2-1}} = 6x$.

(B) If $y = \dfrac{t^3}{6} = \dfrac{1}{6}t^3$, then $\dfrac{dy}{dt} \boxed{= \dfrac{1}{6} \cdot 3t^{3-1}} = \dfrac{1}{2}t^2$.

(C) If $y = \dfrac{1}{2x^4} = \dfrac{1}{2}x^{-4}$, then $y' \boxed{= \dfrac{1}{2}(-4x^{-4-1})} = -2x^{-5}$, or $\dfrac{-2}{x^5}$.

(D) $\dfrac{d}{dx}\dfrac{0.4}{\sqrt{x^3}} = \dfrac{d}{dx}\dfrac{0.4}{x^{3/2}} = \dfrac{d}{dx}0.4x^{-3/2} \boxed{= 0.4\left[-\dfrac{3}{2}x^{(-3/2)-1}\right]}$

$= -0.6x^{-5/2}$, or $-\dfrac{0.6}{\sqrt{x^5}}$

Matched Problem 4   Find

(A) $f'(x)$ for $f(x) = 4x^5$

(B) $\dfrac{dy}{dt}$ for $y = \dfrac{t^4}{12}$

(C) y' for $y = \dfrac{1}{3x^3}$

(D) $\dfrac{d}{dx}\dfrac{0.9}{\sqrt[3]{x}}$

Sum and Difference Properties

Let $f(x) = u(x) + v(x)$, where $u'(x)$ and $v'(x)$ exist. Using the four-step process, we have the following:

Step 1 $f(x + h) = u(x + h) + v(x + h)$

Step 2 $f(x + h) - f(x) = u(x + h) + v(x + h) - [u(x) + v(x)]$

$= u(x + h) - u(x) + v(x + h) - v(x)$

Step 3 $\dfrac{f(x + h) - f(x)}{h} = \dfrac{u(x + h) - u(x) + v(x + h) - v(x)}{h}$

$= \dfrac{u(x + h) - u(x)}{h} + \dfrac{v(x + h) - v(x)}{h}$

Step 4 $f'(x) = \lim\limits_{h \to 0} \dfrac{f(x + h) - f(x)}{h}$

$= \lim\limits_{h \to 0}\left[\dfrac{u(x + h) - u(x)}{h} + \dfrac{v(x + h) - v(x)}{h}\right]$

$\lim\limits_{x \to c}[g(x) + h(x)] = \lim\limits_{x \to c} g(x) + \lim\limits_{x \to c} h(x)$

$= \lim\limits_{h \to 0} \dfrac{u(x + h) - u(x)}{h} + \lim\limits_{h \to 0} \dfrac{v(x + h) - v(x)}{h}$

$= u'(x) + v'(x)$

Therefore,

The derivative of the sum of two differentiable functions is the sum of the derivatives of the functions.

Similarly, we can show that

The derivative of the difference of two differentiable functions is the difference of the derivatives of the functions.

Together, we have the **sum and difference property** for differentiation:

> **THEOREM 4   Sum and Difference Property**
> If $y = f(x) = u(x) \pm v(x)$, then
> $$f'(x) = u'(x) \pm v'(x)$$
> Also,
> $$y' = u' \pm v' \qquad \frac{dy}{dx} = \frac{du}{dx} \pm \frac{dv}{dx}$$
>
> **Note:** This rule generalizes to the sum and difference of any given number of functions.

With Theorems 1 through 4, we can compute the derivatives of all polynomials and a variety of other functions.

EXAMPLE 5   **Differentiating Sums and Differences**

(A) If $f(x) = 3x^2 + 2x$,   then
$$f'(x) \;\boxed{= (3x^2)' + (2x)' = 3(2x) + 2(1)} = 6x + 2$$

(B) If $y = 4 + 2x^3 - 3x^{-1}$,   then
$$y' \;\boxed{= (4)' + (2x^3)' - (3x^{-1})' = 0 + 2(3x^2) - 3(-1)x^{-2}} = 6x^2 + 3x^{-2}$$

(C) If $y = \sqrt[3]{w} - 3w$,   then
$$\frac{dy}{dw} = \frac{d}{dw}w^{1/3} - \frac{d}{dw}3w = \frac{1}{3}w^{-2/3} - 3 = \frac{1}{3w^{2/3}} - 3$$

(D) $\dfrac{d}{dx}\left(\dfrac{5}{3x^2} - \dfrac{2}{x^4} + \dfrac{x^3}{9}\right) \;\boxed{= \dfrac{d}{dx}\dfrac{5}{3}x^{-2} - \dfrac{d}{dx}2x^{-4} + \dfrac{d}{dx}\dfrac{1}{9}x^3}$

$$= \frac{5}{3}(-2)x^{-3} - 2(-4)x^{-5} + \frac{1}{9} \cdot 3x^2$$

$$= -\frac{10}{3x^3} + \frac{8}{x^5} + \frac{1}{3}x^2$$

Matched Problem 5   Find

(A) $f'(x)$ for $f(x) = 3x^4 - 2x^3 + x^2 - 5x + 7$

(B) y' for $y = 3 - 7x^{-2}$

(C) $\dfrac{dy}{dv}$ for $y = 5v^3 - \sqrt[4]{v}$

(D) $\dfrac{d}{dx}\left(-\dfrac{3}{4x} + \dfrac{4}{x^3} - \dfrac{x^4}{8}\right)$

Applications

EXAMPLE 6   **Instantaneous Velocity** An object moves along the y axis (marked in feet) so that its position at time x (in seconds) is
$$f(x) = x^3 - 6x^2 + 9x$$

(A) Find the instantaneous velocity function v.

(B) Find the velocity at $x = 2$ and $x = 5$ seconds.

(C) Find the time(s) when the velocity is 0.

SOLUTION

(A) $v = f'(x) = (x^3)' - (6x^2)' + (9x)' = 3x^2 - 12x + 9$

(B) $f'(2) = 3(2)^2 - 12(2) + 9 = -3$ feet per second

$f'(5) = 3(5)^2 - 12(5) + 9 = 24$ feet per second

(C) $v = f'(x) = 3x^2 - 12x + 9 = 0$ *Factor 3 out of each term.*

$3(x^2 - 4x + 3) = 0$ *Factor the quadratic term.*

$3(x - 1)(x - 3) = 0$ *Use the zero property.*

$x = 1, 3$

So, $v = 0$ at $x = 1$ and $x = 3$ seconds.

Matched Problem 6   Repeat Example 6 for $f(x) = x^3 - 15x^2 + 72x$.

EXAMPLE 7   **Tangents** Let $f(x) = x^4 - 6x^2 + 10$.

(A) Find $f'(x)$.

(B) Find the equation of the tangent line at $x = 1$.

(C) Find the values of x where the tangent line is horizontal.

SOLUTION

(A) $f'(x) = (x^4)' - (6x^2)' + (10)'$

$= 4x^3 - 12x$

(B) $y - y_1 = m(x - x_1)$ $y_1 = f(x_1) = f(1) = (1)^4 - 6(1)^2 + 10 = 5$

$y - 5 = -8(x - 1)$ $m = f'(x_1) = f'(1) = 4(1)^3 - 12(1) = -8$

$y = -8x + 13$ *Tangent line at* $x = 1$

(C) Since a horizontal line has 0 slope, we must solve $f'(x) = 0$ for x:

$f'(x) = 4x^3 - 12x = 0$ *Factor 4x out of each term.*

$4x(x^2 - 3) = 0$ *Factor the difference of two squares.*

$4x(x + \sqrt{3})(x - \sqrt{3}) = 0$ *Use the zero property.*

$x = 0, -\sqrt{3}, \sqrt{3}$

Matched Problem 7   Repeat Example 7 for $f(x) = x^4 - 8x^3 + 7$.

Exercises 3-5

A

Find the indicated derivatives in Problems 1–18.

1. $f'(x)$ for $f(x) = 7$

2. $\dfrac{d}{dx} 3$

3. $\dfrac{dy}{dx}$ for $y = x^9$

4. y' for $y = x^6$

5. $\dfrac{d}{dx} x^3$

6. $g'(x)$ for $g(x) = x^5$

7. y' for $y = x^{-4}$

8. $\dfrac{dy}{dx}$ for $y = x^{-8}$

9. $g'(x)$ for $g(x) = x^{8/3}$

10. $f'(x)$ for $f(x) = x^{9/2}$

11. $\dfrac{dy}{dx}$ for $y = \dfrac{1}{x^{10}}$

12. y' for $y = \dfrac{1}{x^{12}}$

13. $f'(x)$ for $f(x) = 5x^2$

14. $\dfrac{d}{dx}(-2x^3)$

15. y' for $y = 0.4x^7$

16. $f'(x)$ for $f(x) = 0.8x^4$

17. $\dfrac{d}{dx}\left(\dfrac{x^3}{18}\right)$

18. $\dfrac{dy}{dx}$ for $y = \dfrac{x^5}{25}$

Problems 19–24 refer to functions f and g that satisfy $f'(2) = 3$ and $g'(2) = -1$. In each problem, find $h'(2)$ for the indicated function h.

19. $h(x) = 4f(x)$

20. $h(x) = 5g(x)$

21. $h(x) = f(x) + g(x)$

22. $h(x) = g(x) - f(x)$

23. $h(x) = 2f(x) - 3g(x) + 7$

24. $h(x) = -4f(x) + 5g(x) - 9$

B

Find the indicated derivatives in Problems 25–48.

25. $\dfrac{d}{dx}(2x - 5)$

26. $\dfrac{d}{dx}(-4x + 9)$

27. $f'(t)$ if $f(t) = 2t^2 - 3t + 1$

28. $\dfrac{dy}{dt}$ if $y = 2 + 5t - 8t^3$

29. y' for $y = 5x^{-2} + 9x^{-1}$

30. $g'(x)$ if $g(x) = 5x^{-7} - 2x^{-4}$

31. $\dfrac{d}{du}(5u^{0.3} - 4u^{2.2})$

32. $\dfrac{d}{du}(2u^{4.5} - 3.1u + 13.2)$

33. $h'(t)$ if $h(t) = 2.1 + 0.5t - 1.1t^3$

34. $F'(t)$ if $F(t) = 0.2t^3 - 3.1t + 13.2$

35. y' if $y = \dfrac{2}{5x^4}$

36. w' if $w = \dfrac{7}{5u^2}$

37. $\dfrac{d}{dx}\left(\dfrac{3x^2}{2} - \dfrac{7}{5x^2}\right)$

38. $\dfrac{d}{dx}\left(\dfrac{5x^3}{4} - \dfrac{2}{5x^3}\right)$

39. $G'(w)$ if $G(w) = \dfrac{5}{9w^4} + 5\sqrt[3]{w}$

40. $H'(w)$ if $H(w) = \dfrac{5}{w^6} - 2\sqrt{w}$

41. $\dfrac{d}{du}(3u^{2/3} - 5u^{1/3})$

42. $\dfrac{d}{du}(8u^{3/4} + 4u^{-1/4})$

43. $h'(t)$ if $h(t) = \dfrac{3}{t^{3/5}} - \dfrac{6}{t^{1/2}}$

44. $F'(t)$ if $F(t) = \dfrac{5}{t^{1/5}} - \dfrac{8}{t^{3/2}}$

45. y' if $y = \dfrac{1}{\sqrt[3]{x}}$

46. w' if $w = \dfrac{10}{\sqrt[5]{u}}$

47. $\dfrac{d}{dx}\left(\dfrac{1.2}{\sqrt{x}} - 3.2x^{-2} + x\right)$

48. $\dfrac{d}{dx}\left(2.8x^{-3} - \dfrac{0.6}{\sqrt[3]{x^2}} + 7\right)$

For Problems 49–52, find

(A) $f'(x)$

(B) The slope of the graph of f at $x = 2$ and $x = 4$

(C) The equations of the tangent lines at $x = 2$ and $x = 4$

(D) The value(s) of x where the tangent line is horizontal

49. $f(x) = 6x - x^2$

50. $f(x) = 2x^2 + 8x$

51. $f(x) = 3x^4 - 6x^2 - 7$

52. $f(x) = x^4 - 32x^2 + 10$

If an object moves along the y axis (marked in feet) so that its position at time x (in seconds) is given by the indicated functions in Problems 53–56, find

(A) The instantaneous velocity function $v = f'(x)$

(B) The velocity when $x = 0$ and $x = 3$ seconds

(C) The time(s) when $v = 0$

53. $f(x) = 176x - 16x^2$

54. $f(x) = 80x - 10x^2$

55. $f(x) = x^3 - 9x^2 + 15x$

56. $f(x) = x^3 - 9x^2 + 24x$

Problems 57–64 require the use of a graphing calculator. For each problem, find $f'(x)$ and approximate (to four decimal places) the value(s) of x where the graph of f has a horizontal tangent line.

57. $f(x) = x^2 - 3x - 4\sqrt{x}$

58. $f(x) = x^2 + x - 10\sqrt{x}$

59. $f(x) = 3\sqrt[3]{x^4} - 1.5x^2 - 3x$

60. $f(x) = 3\sqrt[3]{x^4} - 2x^2 + 4x$

61. $f(x) = 0.05x^4 + 0.1x^3 - 1.5x^2 - 1.6x + 3$

62. $f(x) = 0.02x^4 - 0.06x^3 - 0.78x^2 + 0.94x + 2.2$

63. $f(x) = 0.2x^4 - 3.12x^3 + 16.25x^2 - 28.25x + 7.5$

64. $f(x) = 0.25x^4 - 2.6x^3 + 8.1x^2 - 10x + 9$

65. Let $f(x) = ax^2 + bx + c, a \neq 0$. Recall that the graph of $y = f(x)$ is a parabola. Use the derivative $f'(x)$ to derive a formula for the x coordinate of the vertex of this parabola.

66. Now that you know how to find derivatives, explain why it is no longer necessary for you to memorize the formula for the x coordinate of the vertex of a parabola.

67. Give an example of a cubic polynomial function that has

(A) No horizontal tangents

(B) One horizontal tangent

(C) Two horizontal tangents

 68. Can a cubic polynomial function have more than two horizontal tangents? Explain.

C

Find the indicated derivatives in Problems 69–76.

69. $f'(x)$ if $f(x) = (2x - 1)^2$

70. y' if $y = (2x - 5)^2$

71. $\dfrac{d}{dx} \dfrac{10x + 20}{x}$

72. $\dfrac{dy}{dx}$ if $y = \dfrac{x^2 + 25}{x^2}$

73. $\dfrac{dy}{dx}$ if $y = \dfrac{3x - 4}{12x^2}$

74. $f'(x)$ if $f(x) = \dfrac{2x^5 - 4x^3 + 2x}{x^3}$

 In Problems 75-80, discuss the validity of each statement. If the statement is always true, explain why. If not, give a counterexample.

75. The derivative of a sum is the sum of the derivatives.

76. The derivative of a difference is the difference of the derivatives.

77. The derivative of a product is the product of the derivatives.

78. The derivative of a quotient is the quotient of the derivatives.

79. The derivative of a constant is 0.

80. The derivative of a constant times a function is 0.

Applications

 81. Sales analysis. A company's total sales (in millions of dollars) t months from now are given by

$$S(t) = 0.03t^3 + 0.5t^2 + 2t + 3$$

(A) Find $S'(t)$.

(B) Find $S(5)$ and $S'(5)$ (to two decimal places). Write a brief verbal interpretation of these results.

(C) Find $S(10)$ and $S'(10)$ (to two decimal places). Write a brief verbal interpretation of these results.

 82. Sales analysis. A company's total sales (in millions of dollars) t months from now are given by

$$S(t) = 0.015t^4 + 0.4t^3 + 3.4t^2 + 10t - 3$$

(A) Find $S'(t)$.

(B) Find $S(4)$ and $S'(4)$ (to two decimal places). Write a brief verbal interpretation of these results.

(C) Find $S(8)$ and $S'(8)$ (to two decimal places). Write a brief verbal interpretation of these results.

83. Advertising. A marine manufacturer will sell $N(x)$ power boats after spending $\$x$ thousand on advertising, as given by

$$N(x) = 1,000 - \frac{3,780}{x} \qquad 5 \le x \le 30$$

(see figure).

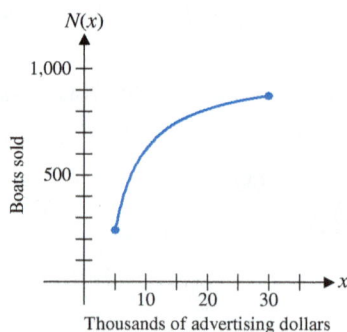

Figure for 83

(A) Find $N'(x)$.

 (B) Find $N'(10)$ and $N'(20)$. Write a brief verbal interpretation of these results.

84. Price–demand equation. Suppose that, in a given gourmet food store, people are willing to buy x pounds of chocolate candy per day at $\$p$ per quarter pound, as given by the price–demand equation

$$x = 10 + \frac{180}{p} \qquad 2 \le p \le 10$$

This function is graphed in the figure. Find the demand and the instantaneous rate of change of demand with respect to price when the price is $\$5$. Write a brief verbal interpretation of these results.

Figure for 84

85. College enrollment. The percentages of male high-school graduates who enrolled in college are given in the second column of Table 1.

Table 1 College enrollment percentages

Year	Male	Female
1970	55.2	48.5
1980	46.7	51.8
1990	58.0	62.2
2000	59.9	66.2
2006	65.8	66.1

(A) Let x represent time (in years) since 1970, and let y represent the corresponding percentage of male high-school graduates who enrolled in college. Enter the data in a graphing calculator and find a cubic regression equation for the data.

 (B) If $y = M(x)$ denotes the regression equation found in part A, find $M(46)$ and $M'(46)$ (to the nearest tenth), and write a brief verbal interpretation of these results.

 86. College enrollment. The percentages of female high-school graduates who enrolled in college are given in the third column of Table 1.

(A) Let x represent time (in years) since 1970, and let y represent the corresponding percentage of female high-school graduates who enrolled in college. Enter the data in a graphing calculator and find a cubic regression equation for the data.

 (B) If $y = F(x)$ denotes the regression equation found in part A, find $F(46)$ and $F'(46)$ (to the nearest tenth), and write a brief verbal interpretation of these results.

87. Medicine. A person x inches tall has a pulse rate of y beats per minute, as given approximately by

$$y = 590x^{-1/2} \qquad 30 \le x \le 75$$

What is the instantaneous rate of change of pulse rate at the

(A) 36-inch level?

(B) 64-inch level?

88. Ecology. A coal-burning electrical generating plant emits sulfur dioxide into the surrounding air. The concentration $C(x)$, in parts per million, is given approximately by

$$C(x) = \frac{0.1}{x^2}$$

where x is the distance from the plant in miles. Find the instantaneous rate of change of concentration at

(A) $x = 1$ mile

(B) $x = 2$ miles

89. Learning. Suppose that a person learns y items in x hours, as given by

$$y = 50\sqrt{x} \qquad 0 \le x \le 9$$

(see figure). Find the rate of learning at the end of

(A) 1 hour   (B) 9 hours

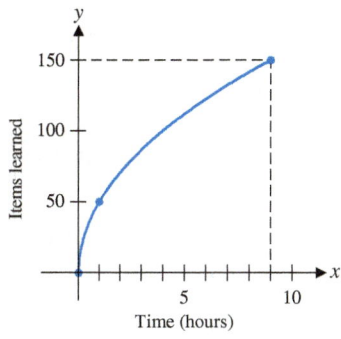

Figure for 89

90. Learning. If a person learns y items in x hours, as given by

$$y = 21\sqrt[3]{x^2} \qquad 0 \le x \le 8$$

find the rate of learning at the end of

(A) 1 hour   (B) 8 hours

Answers to Matched Problems

1. All are 0.

2. (A) $6x^5$ (B) $30x^{29}$
   (C) $-2t^{-3} = -2/t^3$ (D) $\frac{3}{2}x^{1/2}$

3. (A) $-x^{-2}$, or $-1/x^2$ (B) $\frac{2}{3}u^{-1/3}$, or $2/(3\sqrt[3]{u})$
   (C) $-\frac{1}{2}x^{-3/2}$, or $-1/(2\sqrt{x^3})$

4. (A) $20x^4$ (B) $t^3/3$
   (C) $-x^{-4}$, or $-1/x^4$ (D) $-0.3x^{-4/3}$, or $-0.3/\sqrt[3]{x^4}$

5. (A) $12x^3 - 6x^2 + 2x - 5$
   (B) $14x^{-3}$, or $14/x^3$
   (C) $15v^2 - \frac{1}{4}v^{-3/4}$, or $15v^2 - 1/(4v^{3/4})$
   (D) $3/(4x^2) - (12/x^4) - (x^3/2)$

6. (A) $v = 3x^2 - 30x + 72$
   (B) $f'(2) = 24$ ft/s; $f'(5) = -3$ ft/s
   (C) $x = 4$ and $x = 6$ seconds

7. (A) $f'(x) = 4x^3 - 24x^2$
   (B) $y = -20x + 20$
   (C) $x = 0$ and $x = 6$

3-6 Differentials

- Increments
- Differentials
- Approximations Using Differentials

In this section, we introduce increments and differentials. Increments are useful and they provide an alternative notation for defining the derivative. Differentials are often easier to compute than increments and can be used to approximate increments.

Increments

In Section 3-4, we defined the derivative of f at x as the limit of the difference quotient

$$f'(x) = \lim_{h \to 0} \frac{f(x+h) - f(x)}{h}$$

We considered various interpretations of this limit, including slope, velocity, and instantaneous rate of change. Increment notation enables us to interpret the numerator and denominator of the difference quotient separately.

Given $y = f(x) = x^3$, if x changes from 2 to 2.1, then y will change from $y = f(2) = 2^3 = 8$ to $y = f(2.1) = 2.1^3 = 9.261$. The change in x is called the *increment in x* and is denoted by Δx (read as "delta x").* Similarly, the change in y is called the *increment in y* and is denoted by Δy. In terms of the given example, we write

$$\Delta x = 2.1 - 2 = 0.1 \qquad \text{\textcolor{blue}{Change in x}}$$
$$\Delta y = f(2.1) - f(2) \qquad \text{\textcolor{blue}{$f(x) = x^3$}}$$
$$= 2.1^3 - 2^3 \qquad \text{\textcolor{blue}{Use a calculator.}}$$
$$= 9.261 - 8$$
$$= 1.261 \qquad \text{\textcolor{blue}{Corresponding change in y}}$$

CONCEPTUAL INSIGHT

The symbol Δx does not represent the product of Δ and x but is the symbol for a single quantity: the *change in x*. Likewise, the symbol Δy represents a single quantity: the *change in y*.

DEFINITION Increments

For $y = f(x)$, $\quad \Delta x = x_2 - x_1$, $\quad$ so $\quad x_2 = x_1 + \Delta x$, $\quad$ and

$$\Delta y = y_2 - y_1$$
$$= f(x_2) - f(x_1)$$
$$= f(x_1 + \Delta x) - f(x_1)$$

Δy represents the change in y corresponding to a change Δx in x. Δx can be either positive or negative.

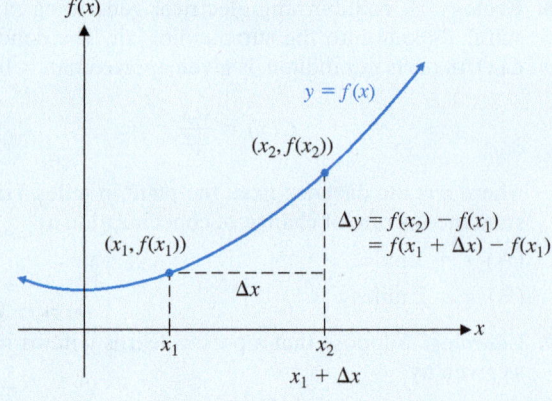

[**Note:** Δy depends on the function f, the input x_1, and the increment Δx.]

EXAMPLE 1   **Increments** Given the function $y = f(x) = \dfrac{x^2}{2}$,

(A) Find Δx, Δy, and $\Delta y / \Delta x$ for $x_1 = 1$ and $x_2 = 2$.

(B) Find $\dfrac{f(x_1 + \Delta x) - f(x_1)}{\Delta x}$ for $x_1 = 1$ and $\Delta x = 2$.

SOLUTION   (A) $\Delta x = x_2 - x_1 = 2 - 1 = 1$

$$\Delta y = f(x_2) - f(x_1)$$
$$= f(2) - f(1) = \frac{4}{2} - \frac{1}{2} = \frac{3}{2}$$
$$\frac{\Delta y}{\Delta x} = \frac{f(x_2) - f(x_1)}{x_2 - x_1} = \frac{\frac{3}{2}}{1} = \frac{3}{2}$$

*Δ is the symbol for the Greek letter delta.

(B) $\dfrac{f(x_1 + \Delta x) - f(x_1)}{\Delta x} = \dfrac{f(1 + 2) - f(1)}{2}$

$\qquad\qquad\qquad\quad = \dfrac{f(3) - f(1)}{2} = \dfrac{\frac{9}{2} - \frac{1}{2}}{2} = \dfrac{4}{2} = 2$

Matched Problem 1   Given the function $y = f(x) = x^2 + 1$,

(A) Find Δx, Δy, and $\Delta y / \Delta x$ for $x_1 = 2$ and $x_2 = 3$.

(B) Find $\dfrac{f(x_1 + \Delta x) - f(x_1)}{\Delta x}$ for $x_1 = 1$ and $\Delta x = 2$.

In Example 1, we observe another notation for the difference quotient

$$\frac{f(x + h) - f(x)}{h} \tag{1}$$

It is common to refer to h, the change in x, as Δx. Then the difference quotient (1) takes on the form

$$\frac{f(x + \Delta x) - f(x)}{\Delta x} \qquad \text{or} \qquad \frac{\Delta y}{\Delta x} \qquad \color{blue}{\Delta y = f(x + \Delta x) - f(x)}$$

and the derivative is defined by

$$f'(x) = \lim_{\Delta x \to 0} \frac{f(x + \Delta x) - f(x)}{\Delta x}$$

or

$$f'(x) = \lim_{\Delta x \to 0} \frac{\Delta y}{\Delta x} \tag{2}$$

if the limit exists.

EXPLORE & DISCUSS 1   Suppose that $y = f(x)$ defines a function whose domain is the set of all real numbers. If every increment Δy is equal to 0, then what is the range of f?

Differentials

Assume that the limit in equation (2) exists. Then, for small Δx, the difference quotient $\Delta y / \Delta x$ provides a good approximation for $f'(x)$. Also, $f'(x)$ provides a good approximation for $\Delta y / \Delta x$. We write

$$\frac{\Delta y}{\Delta x} \approx f'(x) \qquad \color{blue}{\Delta x \text{ is small, but} \neq 0} \tag{3}$$

Multiplying both sides of (3) by Δx gives us

$$\Delta y \approx f'(x)\,\Delta x \qquad \color{blue}{\Delta x \text{ is small, but} \neq 0} \tag{4}$$

From equation (4), we see that $f'(x)\Delta x$ provides a good approximation for Δy when Δx is small.

Because of the practical and theoretical importance of $f'(x)\,\Delta x$, we give it the special name **differential** and represent it with the special symbol dy or df:

$$\boldsymbol{dy = f'(x)\Delta x} \qquad \textbf{or} \qquad \boldsymbol{df = f'(x)\Delta x}$$

For example,

$$d(2x^3) = (2x^3)'\,\Delta x = 6x^2\,\Delta x$$

$$d(x) = (x)'\,\Delta x = 1\,\Delta x = \Delta x$$

In the second example, we usually drop the parentheses in $d(x)$ and simply write

$$\boldsymbol{dx = \Delta x}$$

In summary, we have the following:

DEFINITION Differentials

If $y = f(x)$ defines a differentiable function, then the **differential dy, or df**, is defined as the product of $f'(x)$ and dx, where $dx = \Delta x$. Symbolically,

$$dy = f'(x)\,dx, \qquad \text{or} \qquad df = f'(x)\,dx$$

where

$$dx = \Delta x$$

Note: The differential dy (or df) is actually a function involving two independent variables, x and dx. A change in either one or both will affect dy (or df).

EXAMPLE 2   **Differentials** Find dy for $f(x) = x^2 + 3x$. Evaluate dy for

(A) $x = 2$ and $dx = 0.1$

(B) $x = 3$ and $dx = 0.1$

(C) $x = 1$ and $dx = 0.02$

SOLUTION $dy = f'(x)\,dx$

$\qquad = (2x + 3)\,dx$

(A) When $x = 2$ and $dx = 0.1$,   (B) When $x = 3$ and $dx = 0.1$,

$\qquad dy = [2(2) + 3]0.1 = 0.7$ $dy = [2(3) + 3]0.1 = 0.9$

(C) When $x = 1$ and $dx = 0.02$,

$\qquad dy = [2(1) + 3]0.02 = 0.1$

Matched Problem 2   Find dy for $f(x) = \sqrt{x} + 3$. Evaluate dy for

(A) $x = 4$ and $dx = 0.1$

(B) $x = 9$ and $dx = 0.12$

(C) $x = 1$ and $dx = 0.01$

We now have two interpretations of the symbol dy/dx. Referring to the function $y = f(x) = x^2 + 3x$ in Example 2 with $x = 2$ and $dx = 0.1$, we have

$$\frac{dy}{dx} = f'(2) = 7 \qquad \textit{Derivative}$$

and

$$\frac{dy}{dx} = \frac{0.7}{0.1} = 7 \qquad \textit{Ratio of differentials}$$

Approximations Using Differentials

Earlier, we noted that for small Δx,

$$\frac{\Delta y}{\Delta x} \approx f'(x) \qquad \text{and} \qquad \Delta y \approx f'(x)\Delta x$$

Also, since

$$dy = f'(x)\,dx$$

it follows that

$$\Delta y \approx dy$$

and dy can be used to approximate Δy.

To interpret this result geometrically, we need to recall a basic property of the slope. The vertical change in a line is equal to the product of the slope and the horizontal change, as shown in Figure 1.

Now consider the line tangent to the graph of $y = f(x)$, as shown in Figure 2. Since $f'(x)$ is the slope of the tangent line and dx is the horizontal change in the tangent line, it follows that the vertical change in the tangent line is given by $dy = f'(x)\,dx$, as indicated in Figure 2.

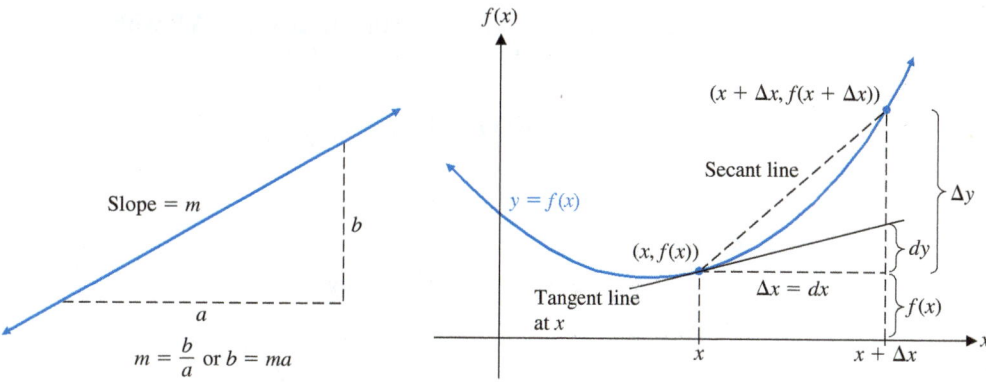

Figure 1

Figure 2

| **EXAMPLE 3** | **Comparing Increments and Differentials** Let $y = f(x) = 6x - x^2$. |

(A) Find Δy and dy when $x = 2$.

(B) Graph Δy and dy from part A for $-1 \le \Delta x \le 1$.

(C) Compare Δy and dy from part A for $\Delta x = 0.1, 0.2,$ and 0.3.

SOLUTION (A) $\Delta y = f(2 + \Delta x) - f(2)$

$\qquad = 6(2 + \Delta x) - (2 + \Delta x)^2 - (6 \cdot 2 - 2^2)$ *Remove parentheses.*

$\qquad = 12 + 6\Delta x - 4 - 4\Delta x - \Delta x^2 - 12 + 4$ *Collect like terms.*

$\qquad = 2\Delta x - \Delta x^2$

Since $f'(x) = 6 - 2x$, $f'(2) = 2$, and $dx = \Delta x$, $dy = f'(2)\,dx = 2\Delta x$

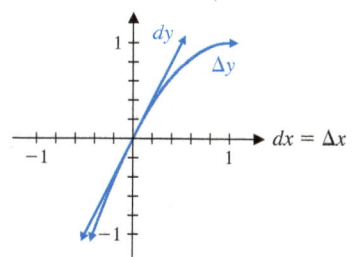

Figure 3 $\Delta y = 2\Delta x - \Delta x^2, dy = 2\Delta x$

(B) The graphs of $\Delta y = 2\Delta x - \Delta x^2$ and $dy = 2\Delta x$ are shown in Figure 3. Examining the graphs, we conclude that the differential provides a linear approximation of the increments and that the approximation is better for values of $\Delta x = dx$ close to 0.

(C) Table 1 compares the values of Δx and dy for the indicated values of Δx.

Table 1

Δx	Δy	dy
0.1	0.19	0.2
0.2	0.36	0.4
0.3	0.51	0.6

Matched Problem 3   Repeat Example 3 for $x = 4$ and $\Delta x = dx = -0.1, -0.2,$ and -0.3.

CONCEPTUAL INSIGHT

The error in the approximation $\Delta y \approx dy$ is usually small when $\Delta x = dx$ is small (see Example 3C), but it can be quite substantial in some cases.

EXAMPLE 4 **Cost–Revenue** A company manufactures and sells x transistor radios per week. If the weekly cost and revenue equations are

$$C(x) = 5,000 + 2x \qquad R(x) = 10x - \frac{x^2}{1,000} \qquad 0 \le x \le 8,000$$

then find the approximate changes in revenue and profit if production is increased from 2,000 to 2,010 units per week.

SOLUTION We will approximate ΔR and ΔP with dR and dP, respectively, using $x = 2,000$ and $dx = 2,010 - 2,000 = 10$.

$$R(x) = 10x - \frac{x^2}{1,000} \qquad\qquad P(x) = R(x) - C(x) = 10x - \frac{x^2}{1,000} - 5,000 - 2x$$

$$dR = R'(x)\,dx \qquad\qquad\qquad\qquad = 8x - \frac{x^2}{1,000} - 5,000$$

$$= \left(10 - \frac{x}{500}\right)dx \qquad\qquad dP = P'(x)\,dx$$

$$= \left(10 - \frac{2,000}{500}\right)10 \qquad\qquad = \left(8 - \frac{x}{500}\right)dx$$

$$= \$60 \text{ per week} \qquad\qquad\qquad = \left(8 - \frac{2,000}{500}\right)10$$

$$\qquad\qquad\qquad\qquad\qquad\qquad = \$40 \text{ per week}$$

Matched Problem 4 Repeat Example 4 with production increasing from 6,000 to 6,010.

Comparing the results in Example 4 and Matched Problem 4, we see that an increase in production results in a revenue and profit increase at the 2,000 production level but a revenue and profit loss at the 6,000 production level.

Exercises 3-6

A

In Problems 1–6, find the indicated quantities for $y = f(x) = 3x^2$.

1. Δx, Δy, and $\Delta y / \Delta x$; given $x_1 = 1$ and $x_2 = 4$

2. Δx, Δy, and $\Delta y / \Delta x$; given $x_1 = 2$ and $x_2 = 5$

3. $\dfrac{f(x_1 + \Delta x) - f(x_1)}{\Delta x}$; given $x_1 = 1$ and $\Delta x = 2$

4. $\dfrac{f(x_1 + \Delta x) - f(x_1)}{\Delta x}$; given $x_1 = 2$ and $\Delta x = 1$

5. $\Delta y / \Delta x$; given $x_1 = 1$ and $x_2 = 3$

6. $\Delta y / \Delta x$; given $x_1 = 2$ and $x_2 = 3$

In Problems 7–12, find dy for each function.

7. $y = 30 + 12x^2 - x^3$

8. $y = 200x - \dfrac{x^2}{30}$

9. $y = x^2\left(1 - \dfrac{x}{9}\right)$

10. $y = x^3(60 - x)$

11. $y = \dfrac{590}{\sqrt{x}}$

12. $y = 52\sqrt{x}$

B

In Problems 13 and 14, find the indicated quantities for $y = f(x) = 3x^2$.

13. (A) $\dfrac{f(2 + \Delta x) - f(2)}{\Delta x}$ (simplify)

(B) What does the quantity in part (A) approach as Δx approaches 0?

14. (A) $\dfrac{f(3 + \Delta x) - f(3)}{\Delta x}$ (simplify)

(B) What does the quantity in part (A) approach as Δx approaches 0?

In Problems 15–18, find dy for each function.

15. $y = (2x + 1)^2$

16. $y = (3x + 5)^2$

17. $y = \dfrac{x^2 + 9}{x}$

18. $y = \dfrac{(x - 1)^2}{x^2}$

In Problems 19–22, evaluate dy and Δy for each function for the indicated values.

19. $y = f(x) = x^2 - 3x + 2; x = 5, dx = \Delta x = 0.2$

20. $y = f(x) = 30 + 12x^2 - x^3; x = 2, dx = \Delta x = 0.1$

21. $y = f(x) = 75\left(1 - \dfrac{2}{x}\right); x = 5, dx = \Delta x = -0.5$

22. $y = f(x) = 100\left(x - \dfrac{4}{x^2}\right); x = 2, dx = \Delta x = -0.1$

23. A cube with 10-inch sides is covered with a coat of fiberglass 0.2 inch thick. Use differentials to estimate the volume of the fiberglass shell.

24. A sphere with a radius of 5 centimeters is coated with ice 0.1 centimeter thick. Use differentials to estimate the volume of the ice. [Recall that $V = \frac{4}{3}\pi r^3$.]

C

In Problems 25–28,

(A) Find Δy and dy for the function f at the indicated value of x.

(B) Graph Δy and dy from part A.

(C) Compare the values of Δy and dy from part A at the indicated values of Δx.

25. $f(x) = x^2 + 2x + 3; x = -0.5, \Delta x = dx = 0.1, 0.2, 0.3$

26. $f(x) = x^2 + 2x + 3; x = -2, \Delta x = dx = -0.1, -0.2, -0.3$

27. $f(x) = x^3 - 2x^2; x = 1, \Delta x = dx = 0.05, 0.10, 0.15$

28. $f(x) = x^3 - 2x^2; x = 2, \Delta x = dx = -0.05, -0.10, -0.15$

In Problems 29–32, discuss the validity of each statement. If the statement is always true, explain why. If not, give a counterexample.

29. If the graph of the function $y = f(x)$ is a line, then the functions Δy and dy (of the independent variable $\Delta x = dx$) for $f(x)$ at $x = 3$ are identical.

30. If the graph of the function $y = f(x)$ is a parabola, then the functions Δy and dy (of the independent variable $\Delta x = dx$) for $f(x)$ at $x = 0$ are identical.

31. Suppose that $y = f(x)$ defines a differentiable function whose domain is the set of all real numbers. If every differential dy at $x = 2$ is equal to 0, then $f(x)$ is a constant function.

32. Suppose that $y = f(x)$ defines a function whose domain is the set of all real numbers. If every increment at $x = 2$ is equal to 0, then $f(x)$ is a constant function.

33. Find dy if $y = (1 - 2x)\sqrt[3]{x^2}$.

34. Find dy if $y = (2x^2 - 4)\sqrt{x}$.

35. Find dy and Δy for $y = 52\sqrt{x}, x = 4$, and $\Delta x = dx = 0.3$.

36. Find dy and Δy for $y = 590/\sqrt{x}, x = 64$, and $\Delta x = dx = 1$.

Applications

Use differential approximations in the following problems.

37. Advertising. A company will sell N units of a product after spending x thousand in advertising, as given by

$$N = 60x - x^2 \qquad 5 \le x \le 30$$

Approximately what increase in sales will result by increasing the advertising budget from $10,000 to $11,000? From $20,000 to $21,000?

38. Price–demand. Suppose that the daily demand (in pounds) for chocolate candy at $x per pound is given by

$$D = 1,000 - 40x^2 \qquad 1 \le x \le 5$$

If the price is increased from $3.00 per pound to $3.20 per pound, what is the approximate change in demand?

39. Average cost. For a company that manufactures tennis rackets, the average cost per racket $\overline{C}$ is

$$\overline{C} = \frac{400}{x} + 5 + \frac{1}{2}x \qquad x \ge 1$$

where x is the number of rackets produced per hour. What will the approximate change in average cost per racket be if production is increased from 20 per hour to 25 per hour? From 40 per hour to 45 per hour?

40. Revenue and profit. A company manufactures and sells x televisions per month. If the cost and revenue equations are

$$C(x) = 72,000 + 60x$$

$$R(x) = 200x - \frac{x^2}{30} \qquad 0 \le x \le 6,000$$

what will the approximate changes in revenue and profit be if production is increased from 1,500 to 1,510? From 4,500 to 4,510?

41. Pulse rate. The average pulse rate y (in beats per minute) of a healthy person x inches tall is given approximately by

$$y = \frac{590}{\sqrt{x}} \qquad 30 \le x \le 75$$

Approximately how will the pulse rate change for a change in height from 36 to 37 inches? From 64 to 65 inches?

42. Measurement. An egg of a particular bird is nearly spherical. If the radius to the inside of the shell is 5 millimeters and the radius to the outside of the shell is 5.3 millimeters, approximately what is the volume of the shell? [Remember that $V = \frac{4}{3}\pi r^3$.]

43. Medicine. A drug is given to a patient to dilate her arteries. If the radius of an artery is increased from 2 to 2.1 millimeters, approximately how much is the cross-sectional area increased? [Assume that the cross section of the artery is circular; that is, $A = \pi r^2$.]

44. Drug sensitivity. One hour after x milligrams of a particular drug are given to a person, the change in body temperature T (in degrees Fahrenheit) is given by

$$T = x^2\left(1 - \frac{x}{9}\right) \qquad 0 \le x \le 6$$

Approximate the changes in body temperature produced by the following changes in drug dosages:

(A) From 2 to 2.1 milligrams

(B) From 3 to 3.1 milligrams

(C) From 4 to 4.1 milligrams

45. Learning. A particular person learning to type has an achievement record given approximately by

$$N = 75\left(1 - \frac{2}{t}\right) \qquad 3 \le t \le 20$$

where N is the number of words per minute typed after t weeks of practice. What is the approximate improvement from 5 to 5.5 weeks of practice?

46. Learning. If a person learns y items in x hours, as given approximately by

$$y = 52\sqrt{x} \qquad 0 \le x \le 9$$

what is the approximate increase in the number of items learned when x changes from 1 to 1.1 hours? From 4 to 4.1 hours?

47. Politics. In a new city, the voting population (in thousands) is given by

$$N(t) = 30 + 12t^2 - t^3 \qquad 0 \le t \le 8$$

where t is time in years. Find the approximate change in votes for the following changes in time:

(A) From 1 to 1.1 years

(B) From 4 to 4.1 years

(C) From 7 to 7.1 years

Answers to Matched Problems

1. (A) $\Delta x = 1$, $\Delta y = 5$, $\Delta y / \Delta x = 5$

  (B) 4

2. $dy = \dfrac{1}{2\sqrt{x}}\, dx$

  (A) 0.025

  (B) 0.02

  (C) 0.005

3. (A) $\Delta y = -2\Delta x - \Delta x^2$; $dy = -2\Delta x$

  (B)

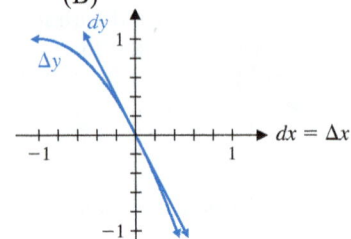

  (C)

Δx	Δy	dy
−0.1	0.19	0.2
−0.2	0.36	0.4
−0.3	0.51	0.6

4. $dR = -\$20/\text{wk}$; $dP = -\$40/\text{wk}$

3-7 Marginal Analysis in Business and Economics

- Marginal Cost, Revenue, and Profit
- Application
- Marginal Average Cost, Revenue, and Profit

Marginal Cost, Revenue, and Profit

One important application of calculus to business and economics involves *marginal analysis*. In economics, the word *marginal* refers to a rate of change—that is, to a derivative. Thus, if $C(x)$ is the total cost of producing x items, then $C'(x)$ is called the *marginal cost* and represents the instantaneous rate of change of total cost with respect to the number of items produced. Similarly, the *marginal revenue* is the derivative of the total revenue function, and the *marginal profit* is the derivative of the total profit function.

> **DEFINITION** Marginal Cost, Revenue, and Profit
>
> If x is the number of units of a product produced in some time interval, then
>
> $$\text{total cost} = C(x)$$
>
> $$\text{marginal cost} = C'(x)$$
>
> $$\text{total revenue} = R(x)$$
>
> $$\text{marginal revenue} = R'(x)$$
>
> $$\text{total profit} = P(x) = R(x) - C(x)$$
>
> $$\text{marginal profit} = P'(x) = R'(x) - C'(x)$$
>
> $$= (\text{marginal revenue}) - (\text{marginal cost})$$
>
> Marginal cost (or revenue or profit) is the instantaneous rate of change of cost (or revenue or profit) relative to production at a given production level.

To begin our discussion, we consider a cost function $C(x)$. It is important to remember that $C(x)$ represents the *total* cost of producing x items, not the cost of producing a *single* item. To find the cost of producing a single item, we use the difference of two successive values of $C(x)$:

$$\text{Total cost of producing } x + 1 \text{ items} = C(x + 1)$$
$$\text{Total cost of producing } x \text{ items} = C(x)$$
$$\text{Exact cost of producing the } (x + 1)\text{st item} = C(x + 1) - C(x)$$

EXAMPLE 1   **Cost Analysis** A company manufactures fuel tanks for cars. The total weekly cost (in dollars) of producing x tanks is given by

$$C(x) = 10{,}000 + 90x - 0.05x^2$$

(A) Find the marginal cost function.

(B) Find the marginal cost at a production level of 500 tanks per week.

(C) Interpret the results of part B.

(D) Find the exact cost of producing the 501st item.

SOLUTION   (A) $C'(x) = 90 - 0.1x$

(B) $C'(500) = 90 - 0.1(500) = \$40$ *Marginal cost*

(C) At a production level of 500 tanks per week, the total production costs are increasing at the rate of $40 per tank.

(D)
$$C(501) = 10{,}000 + 90(501) - 0.05(501)^2$$
$$= \$42{,}539.95 \quad \textit{Total cost of producing 501 tanks per week}$$
$$C(500) = 10{,}000 + 90(500) - 0.05(500)^2$$
$$= \$42{,}500.00 \quad \textit{Total cost of producing 500 tanks per week}$$
$$C(501) - C(500) = 42{,}539.95 - 42{,}500.00$$
$$= \$39.95 \quad \textit{Exact cost of producing the 501st tank}$$

Matched Problem 1   A company manufactures automatic transmissions for cars. The total weekly cost (in dollars) of producing x transmissions is given by

$$C(x) = 50{,}000 + 600x - 0.75x^2$$

(A) Find the marginal cost function.

(B) Find the marginal cost at a production level of 200 transmissions per week.

(C) Interpret the results of part B.

(D) Find the exact cost of producing the 201st transmission.

In Example 1, we found that the cost of the 501st tank and the marginal cost at a production level of 500 tanks differ by only a nickel. Increments and differentials will help us understand the relationship between marginal cost and the cost of a single item. If $C(x)$ is any total cost function, then

$$C'(x) \approx \frac{C(x + \Delta x) - C(x)}{\Delta x} \quad \textit{See Section 3-6}$$

$$C'(x) \approx C(x + 1) - C(x) \quad \Delta x = 1$$

We see that the marginal cost $C'(x)$ approximates $C(x + 1) - C(x)$, the exact cost of producing the $(x + 1)$st item. These observations are summarized next and are illustrated in Figure 1.

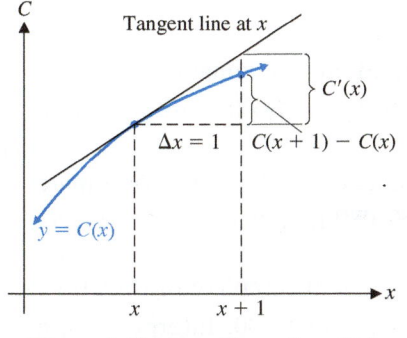

Figure 1 $C'(x) \approx C(x + 1) - C(x)$

> **THEOREM 1   Marginal Cost and Exact Cost**
> If $C(x)$ is the total cost of producing x items, then the marginal cost function approximates the exact cost of producing the $(x + 1)$st item:
>
> Marginal cost   Exact cost
> $$C'(x) \approx C(x + 1) - C(x)$$
>
> Similar statements can be made for total revenue functions and total profit functions.

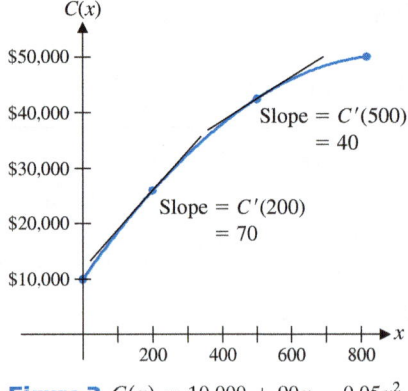

Figure 2 $C(x) = 10,000 + 90x - 0.05x^2$

CONCEPTUAL INSIGHT

Theorem 1 states that the marginal cost at a given production level x approximates the cost of producing the $(x + 1)$st, or *next*, item. In practice, the marginal cost is used more frequently than the exact cost. One reason for this is that the marginal cost is easily visualized when one is examining the graph of the total cost function. Figure 2 shows the graph of the cost function discussed in Example 1, with tangent lines added at $x = 200$ and $x = 500$. The graph clearly shows that as production increases, the slope of the tangent line decreases. Thus, the cost of producing the next tank also decreases, a desirable characteristic of a total cost function. We will have much more to say about graphical analysis in Chapter 5.

Application

Now we discuss how price, demand, revenue, cost, and profit are tied together in typical applications. Although either price or demand can be used as the independent variable in a price–demand equation, it is common to use demand as the independent variable when marginal revenue, cost, and profit are also involved.

EXAMPLE 2   **Production Strategy** A company's market research department recommends the manufacture and marketing of a new headphone set for MP3 players. After suitable test marketing, the research department presents the following **price–demand equation**:

$$x = 10,000 - 1,000p \qquad \text{x is demand at price p.} \qquad (1)$$

Solving (1) for p gives

$$p = 10 - 0.001x \qquad (2)$$

where x is the number of headphones that retailers are likely to buy at $\$p$ per set. The financial department provides the **cost function**

$$C(x) = 7,000 + 2x \qquad (3)$$

where $\$7,000$ is the estimate of fixed costs (tooling and overhead) and $\$2$ is the estimate of variable costs per headphone set (materials, labor, marketing, transportation, storage, etc.).

(A) Find the domain of the function defined by the price–demand equation (2).

(B) Find and interpret the marginal cost function $C'(x)$.

(C) Find the revenue function as a function of x and find its domain.

(D) Find the marginal revenue at $x = 2,000$, $5,000$, and $7,000$. Interpret these results.

(E) Graph the cost function and the revenue function in the same coordinate system. Find the intersection points of these two graphs and interpret the results.

(F) Find the profit function and its domain and sketch the graph of the function.

(G) Find the marginal profit at $x = 1,000$, $4,000$, and $6,000$. Interpret these results.

SOLUTION (A) Since price p and demand x must be non-negative, we have $x \geq 0$ and

$$p = 10 - 0.001x \geq 0$$
$$10 \geq 0.001x$$
$$10{,}000 \geq x$$

Thus, the permissible values of x are $0 \leq x \leq 10{,}000$.

(B) The marginal cost is $C'(x) = 2$. Since this is a constant, it costs an additional $2 to produce one more headphone set at any production level.

(C) The **revenue** is the amount of money R received by the company for manufacturing and selling x headphone sets at $$p$ per set and is given by

$$R = (\text{number of headphone sets sold})(\text{price per headphone set}) = xp$$

In general, the revenue R can be expressed as a function of p using equation (1) or as a function of x using equation (2). As we mentioned earlier, when using marginal functions, we will always use the number of items x as the independent variable. Thus, the **revenue function** is

$$R(x) = xp = x(10 - 0.001x) \quad \text{Using equation (2)} \quad (4)$$
$$= 10x - 0.001x^2$$

Since equation (2) is defined only for $0 \leq x \leq 10{,}000$, it follows that the domain of the revenue function is $0 \leq x \leq 10{,}000$.

(D) The **marginal revenue** is

$$R'(x) = 10 - 0.002x$$

For production levels of $x = 2{,}000$, $5{,}000$, and $7{,}000$, we have

$$R'(2{,}000) = 6 \qquad R'(5{,}000) = 0 \qquad R'(7{,}000) = -4$$

This means that at production levels of $2{,}000$, $5{,}000$, and $7{,}000$, the respective approximate changes in revenue per unit change in production are $6, $0, and $-$4. That is, at the $2{,}000$ output level, revenue increases as production increases; at the $5{,}000$ output level, revenue does not change with a "small" change in production; and at the $7{,}000$ output level, revenue decreases with an increase in production.

(E) Graphing $R(x)$ and $C(x)$ in the same coordinate system results in Figure 3. The intersection points are called the **break-even points**, because revenue equals cost at these production levels. The company neither makes nor loses money, but just breaks even. The break-even points are obtained as follows:

$$C(x) = R(x)$$
$$7{,}000 + 2x = 10x - 0.001x^2$$
$$0.001x^2 - 8x + 7{,}000 = 0 \quad \text{Solve by the quadratic formula}$$
$$x^2 - 8{,}000x + 7{,}000{,}000 = 0 \quad \text{(see Appendix A-7).}$$

$$x = \frac{8{,}000 \pm \sqrt{8{,}000^2 - 4(7{,}000{,}000)}}{2}$$

$$= \frac{8{,}000 \pm \sqrt{36{,}000{,}000}}{2}$$

$$= \frac{8{,}000 \pm 6{,}000}{2}$$

$$= 1{,}000, \quad 7{,}000$$

$$R(1{,}000) = 10(1{,}000) - 0.001(1{,}000)^2 = 9{,}000$$

$$C(1{,}000) = 7{,}000 + 2(1{,}000) = 9{,}000$$

$$R(7{,}000) = 10(7{,}000) - 0.001(7{,}000)^2 = 21{,}000$$

$$C(7{,}000) = 7{,}000 + 2(7{,}000) = 21{,}000$$

The break-even points are (1,000, 9,000) and (7,000, 21,000), as shown in Figure 3. Further examination of the figure shows that cost is greater than revenue for production levels between 0 and 1,000 and also between 7,000 and 10,000. Consequently, the company incurs a loss at these levels. By contrast, for production levels between 1,000 and 7,000, revenue is greater than cost, and the company makes a profit.

(F) The **profit function** is

$$P(x) = R(x) - C(x)$$
$$= (10x - 0.001x^2) - (7,000 + 2x)$$
$$= -0.001x^2 + 8x - 7,000$$

The domain of the cost function is $x \geq 0$, and the domain of the revenue function is $0 \leq x \leq 10,000$. The domain of the profit function is the set of x values for which both functions are defined—that is, $0 \leq x \leq 10,000$. The graph of the profit function is shown in Figure 4. Notice that the x coordinates of the break-even points in Figure 3 are the x intercepts of the profit function. Furthermore, the intervals on which cost is greater than revenue and on which revenue is greater than cost correspond, respectively, to the intervals on which profit is negative and on which profit is positive.

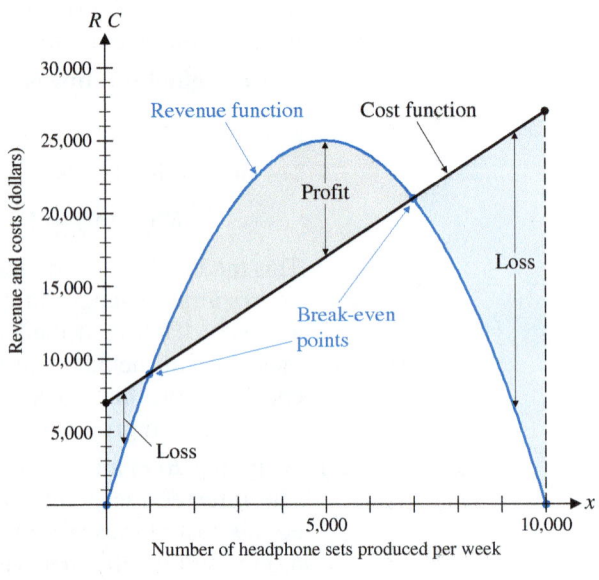

Figure 3

(G) The **marginal profit** is

$$P'(x) = -0.002x + 8$$

For production levels of 1,000, 4,000, and 6,000, we have

$$P'(1,000) = 6 \qquad P'(4,000) = 0 \qquad P'(6,000) = -4$$

This means that at production levels of 1,000, 4,000, and 6,000, the respective approximate changes in profit per unit change in production are $6, $0, and −$4. That is, at the 1,000 output level, profit will be increased if production is increased; at the 4,000 output level, profit does not change for "small" changes in production; and at the 6,000 output level, profits will decrease if production is increased. It seems that the best production level to produce a maximum profit is 4,000.

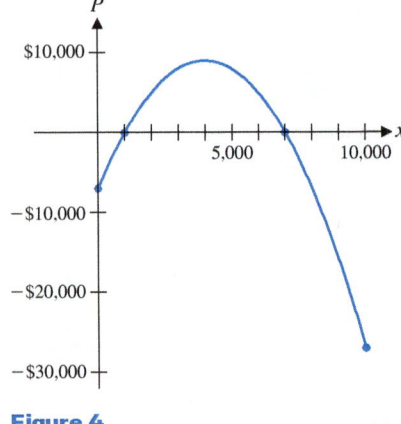

Figure 4

Example 2 requires careful study since a number of important ideas in economics and calculus are involved. In the next chapter, we will develop a systematic procedure for finding the production level (and, using the demand equation, the selling price) that will maximize profit.

Matched Problem 2   Refer to the revenue and profit functions in Example 2.

(A) Find $R'(3,000)$ and $R'(6,000)$. Interpret the results.

(B) Find $P'(2,000)$ and $P'(7,000)$. Interpret the results.

Marginal Average Cost, Revenue, and Profit

Sometimes it is desirable to carry out marginal analysis relative to **average cost (cost per unit), average revenue (revenue per unit),** and **average profit (profit per unit).**

> **DEFINITION   Marginal Average Cost, Revenue, and Profit**
>
> If x is the number of units of a product produced in some time interval, then
>
> **Cost per unit:**   average cost $= \overline{C}(x) = \dfrac{C(x)}{x}$
>
> marginal average cost $= \overline{C}'(x) = \dfrac{d}{dx}\overline{C}(x)$
>
> **Revenue per unit:**   average revenue $= \overline{R}(x) = \dfrac{R(x)}{x}$
>
> marginal average revenue $= \overline{R}'(x) = \dfrac{d}{dx}\overline{R}(x)$
>
> **Profit per unit:**   average profit $= \overline{P}(x) = \dfrac{P(x)}{x}$
>
> marginal average profit $= \overline{P}'(x) = \dfrac{d}{dx}\overline{P}(x)$

EXAMPLE 3   **Cost Analysis**   A small machine shop manufactures drill bits used in the petroleum industry. The manager estimates that the total daily cost (in dollars) of producing x bits is

$$C(x) = 1,000 + 25x - 0.1x^2$$

(A) Find $\overline{C}(x)$ and $\overline{C}'(x)$.

(B) Find $\overline{C}(10)$ and $\overline{C}'(10)$. Interpret these quantities.

(C) Use the results in part (B) to estimate the average cost per bit at a production level of 11 bits per day.

SOLUTION   (A) $\overline{C}(x) = \dfrac{C(x)}{x} = \dfrac{1,000 + 25x - 0.1x^2}{x}$

$$= \dfrac{1,000}{x} + 25 - 0.1x \qquad \textcolor{blue}{\textit{Average cost function}}$$

$$\overline{C}'(x) = \dfrac{d}{dx}\overline{C}(x) = -\dfrac{1,000}{x^2} - 0.1 \qquad \textcolor{blue}{\textit{Marginal average cost function}}$$

(B) $\overline{C}(10) = \dfrac{1,000}{10} + 25 - 0.1(10) = \$124$

$\overline{C}'(10) = -\dfrac{1,000}{10^2} - 0.1 = -\$10.10$

At a production level of 10 bits per day, the average cost of producing a bit is $124. This cost is decreasing at the rate of $10.10 per bit.

(C) If production is increased by 1 bit, then the average cost per bit will decrease by approximately $10.10. So, the average cost per bit at a production level of 11 bits per day is approximately $124 − $10.10 = $113.90.

Matched Problem 3   Consider the cost function for the production of headphone sets from Example 2:

$$C(x) = 7,000 + 2x$$

(A) Find $\overline{C}(x)$ and $\overline{C}'(x)$.

(B) Find $\overline{C}(100)$ and $\overline{C}'(100)$. Interpret these quantities.

(C) Use the results in part (B) to estimate the average cost per headphone set at a production level of 101 headphone sets.

EXPLORE & DISCUSS 1

A student produced the following solution to Matched Problem 3:

$$C(x) = 7,000 + 2x \quad \textit{Cost}$$
$$C'(x) = 2 \quad \textit{Marginal cost}$$
$$\frac{C'(x)}{x} = \frac{2}{x} \quad \textit{"Average" of the marginal cost}$$

Explain why the last function is not the same as the marginal average cost function.

⚠ **CAUTION**

1. The marginal average cost function is computed by first finding the average cost function and then finding its derivative. As Explore & Discuss 1 illustrates, reversing the order of these two steps produces a different function that does not have any useful economic interpretations.

2. Recall that the marginal cost function has two interpretations: the usual interpretation of any derivative as an instantaneous rate of change and the special interpretation as an approximation to the exact cost of the $(x + 1)$st item. This special interpretation does not apply to the marginal average cost function. Referring to Example 3, we would be incorrect to interpret $\overline{C}'(10) = -\$10.10$ to mean that the average cost of the next bit is approximately $-\$10.10$. In fact, the phrase "average cost of the next bit" does not even make sense. Averaging is a concept applied to a collection of items, not to a single item.

These remarks also apply to revenue and profit functions.

Exercises 3-7

A

In Problems 1–4, find the marginal cost function.

1. $C(x) = 175 + 0.8x$

2. $C(x) = 4,500 + 9.5x$

3. $C(x) = 210 + 4.6x - 0.01x^2$

4. $C(x) = 790 + 13x - 0.2x^2$

In Problems 5–8, find the marginal revenue function.

5. $R(x) = 4x - 0.01x^2$

6. $R(x) = 36x - 0.03x^2$

7. $R(x) = x(12 - 0.04x)$

8. $R(x) = x(25 - 0.05x)$

In Problems 9-12, find the marginal profit function if the cost and revenue, respectively, are those in the indicated problems.

9. Problem 1 and Problem 5

10. Problem 2 and Problem 6

11. Problem 3 and Problem 7

12. Problem 4 and Problem 8

B

In Problems 13–20, find the indicated function if cost and revenue are given by $C(x) = 145 + 1.1x$ and $R(x) = 5x - 0.02x^2$, respectively.

13. Average cost function

14. Average revenue function

15. Marginal average cost function

16. Marginal average revenue function

17. Profit function

18. Marginal profit function

19. Average profit function

20. Marginal average profit function

C

In Problems 21–24, discuss the validity of each statement. If the statement is always true, explain why. If not, give a counterexample.

21. If a cost function is linear, then the marginal cost is a constant.

22. If a price–demand equation is linear, then the marginal revenue function is linear.

23. Marginal profit is equal to marginal cost minus marginal revenue.

24. Marginal average cost is equal to average marginal cost.

Applications

25. **Cost analysis.** The total cost (in dollars) of producing x food processors is
$$C(x) = 2,000 + 50x - 0.5x^2$$
   (A) Find the exact cost of producing the 21st food processor.
   (B) Use marginal cost to approximate the cost of producing the 21st food processor.

26. **Cost analysis.** The total cost (in dollars) of producing x electric guitars is
$$C(x) = 1,000 + 100x - 0.25x^2$$
   (A) Find the exact cost of producing the 51st guitar.
   (B) Use marginal cost to approximate the cost of producing the 51st guitar.

27. **Cost analysis.** The total cost (in dollars) of manufacturing x auto body frames is
$$C(x) = 60,000 + 300x$$
   (A) Find the average cost per unit if 500 frames are produced.
   (B) Find the marginal average cost at a production level of 500 units and interpret the results.
   (C) Use the results from parts (A) and (B) to estimate the average cost per frame if 501 frames are produced.

28. **Cost analysis.** The total cost (in dollars) of printing x dictionaries is
$$C(x) = 20,000 + 10x$$
   (A) Find the average cost per unit if 1,000 dictionaries are produced.
   (B) Find the marginal average cost at a production level of 1,000 units and interpret the results.
   (C) Use the results from parts (A) and (B) to estimate the average cost per dictionary if 1,001 dictionaries are produced.

29. **Profit analysis.** The total profit (in dollars) from the sale of x skateboards is
$$P(x) = 30x - 0.3x^2 - 250 \qquad 0 \le x \le 100$$
   (A) Find the exact profit from the sale of the 26th skateboard.
   (B) Use marginal profit to approximate the profit from the sale of the 26th skateboard.

30. **Profit analysis.** The total profit (in dollars) from the sale of x calendars is
$$P(x) = 22x - 0.2x^2 - 400 \qquad 0 \le x \le 100$$
   (A) Find the exact profit from the sale of the 41st calendar.
   (B) Use the marginal profit to approximate the profit from the sale of the 41st calendar.

31. **Profit analysis.** The total profit (in dollars) from the sale of x DVDs is
$$P(x) = 5x - 0.005x^2 - 450 \qquad 0 \le x \le 1,000$$
   Evaluate the marginal profit at the given values of x, and interpret the results.
   (A) $x = 450$ (B) $x = 750$

32. **Profit analysis.** The total profit (in dollars) from the sale of x cameras is
$$P(x) = 12x - 0.02x^2 - 1,000 \qquad 0 \le x \le 600$$
   Evaluate the marginal profit at the given values of x, and interpret the results.
   (A) $x = 200$ (B) $x = 350$

33. **Profit analysis.** The total profit (in dollars) from the sale of x lawn mowers is
$$P(x) = 30x - 0.03x^2 - 750 \qquad 0 \le x \le 1,000$$
   (A) Find the average profit per mower if 50 mowers are produced.
   (B) Find the marginal average profit at a production level of 50 mowers and interpret the results.
   (C) Use the results from parts (A) and (B) to estimate the average profit per mower if 51 mowers are produced.

34. **Profit analysis.** The total profit (in dollars) from the sale of x gas grills is
$$P(x) = 20x - 0.02x^2 - 320 \qquad 0 \le x \le 1,000$$
   (A) Find the average profit per grill if 40 grills are produced.
   (B) Find the marginal average profit at a production level of 40 grills and interpret the results.
   (C) Use the results from parts (A) and (B) to estimate the average profit per grill if 41 grills are produced.

35. **Revenue analysis.** The price p (in dollars) and the demand x for a brand of running shoes are related by the equation
$$x = 4,000 - 40p$$
   (A) Express the price p in terms of the demand x, and find the domain of this function.
   (B) Find the revenue $R(x)$ from the sale of x pairs of running shoes. What is the domain of R?
   (C) Find the marginal revenue at a production level of 1,600 pairs and interpret the results.
   (D) Find the marginal revenue at a production level of 2,500 pairs, and interpret the results.

36. Revenue analysis. The price p (in dollars) and the demand x for a particular steam iron are related by the equation

$$x = 1{,}000 - 20p$$

(A) Express the price p in terms of the demand x, and find the domain of this function.

(B) Find the revenue $R(x)$ from the sale of x steam irons. What is the domain of R?

(C) Find the marginal revenue at a production level of 400 steam irons and interpret the results.

(D) Find the marginal revenue at a production level of 650 steam irons and interpret the results.

37. Revenue, cost, and profit. The price–demand equation and the cost function for the production of table saws are given, respectively, by

$$x = 6{,}000 - 30p \quad \text{and} \quad C(x) = 72{,}000 + 60x$$

where x is the number of saws that can be sold at a price of $\$p$ per saw and $C(x)$ is the total cost (in dollars) of producing x saws.

(A) Express the price p as a function of the demand x, and find the domain of this function.

(B) Find the marginal cost.

(C) Find the revenue function and state its domain.

(D) Find the marginal revenue.

(E) Find $R'(1{,}500)$ and $R'(4{,}500)$ and interpret these quantities.

(F) Graph the cost function and the revenue function on the same coordinate system for $0 \le x \le 6{,}000$. Find the break-even points, and indicate regions of loss and profit.

(G) Find the profit function in terms of x.

(H) Find the marginal profit.

(I) Find $P'(1{,}500)$ and $P'(3{,}000)$ and interpret these quantities.

38. Revenue, cost, and profit. The price–demand equation and the cost function for the production of HDTVs are given, respectively, by

$$x = 9{,}000 - 30p \quad \text{and} \quad C(x) = 150{,}000 + 30x$$

where x is the number of HDTVs that can be sold at a price of $\$p$ per TV and $C(x)$ is the total cost (in dollars) of producing x TVs.

(A) Express the price p as a function of the demand x, and find the domain of this function.

(B) Find the marginal cost.

(C) Find the revenue function and state its domain.

(D) Find the marginal revenue.

(E) Find $R'(3{,}000)$ and $R'(6{,}000)$ and interpret these quantities.

(F) Graph the cost function and the revenue function on the same coordinate system for $0 \le x \le 9{,}000$. Find the break-even points and indicate regions of loss and profit.

(G) Find the profit function in terms of x.

(H) Find the marginal profit.

(I) Find $P'(1{,}500)$ and $P'(4{,}500)$ and interpret these quantities.

39. Revenue, cost, and profit. A company is planning to manufacture and market a new two-slice electric toaster. After conducting extensive market surveys, the research department provides the following estimates: a weekly demand of 200 toasters at a price of $16 per toaster and a weekly demand of 300 toasters at a price of $14 per toaster. The financial department estimates that weekly fixed costs will be $1,400 and variable costs (cost per unit) will be $4.

(A) Assume that the relationship between price p and demand x is linear. Use the research department's estimates to express p as a function of x and find the domain of this function.

(B) Find the revenue function in terms of x and state its domain.

(C) Assume that the cost function is linear. Use the financial department's estimates to express the cost function in terms of x.

(D) Graph the cost function and revenue function on the same coordinate system for $0 \le x \le 1{,}000$. Find the break-even points and indicate regions of loss and profit.

(E) Find the profit function in terms of x.

(F) Evaluate the marginal profit at $x = 250$ and $x = 475$ and interpret the results.

40. Revenue, cost, and profit. The company in Problem 39 is also planning to manufacture and market a four-slice toaster. For this toaster, the research department's estimates are a weekly demand of 300 toasters at a price of $25 per toaster and a weekly demand of 400 toasters at a price of $20. The financial department's estimates are fixed weekly costs of $5,000 and variable costs of $5 per toaster.

(A) Assume that the relationship between price p and demand x is linear. Use the research department's estimates to express p as a function of x, and find the domain of this function.

(B) Find the revenue function in terms of x and state its domain.

(C) Assume that the cost function is linear. Use the financial department's estimates to express the cost function in terms of x.

(D) Graph the cost function and revenue function on the same coordinate system for $0 \le x \le 800$. Find the break-even points and indicate regions of loss and profit.

(E) Find the profit function in terms of x.

(F) Evaluate the marginal profit at $x = 325$ and $x = 425$ and interpret the results.

41. Revenue, cost, and profit. The total cost and the total revenue (in dollars) for the production and sale of x ski jackets are given, respectively, by

$$C(x) = 24x + 21{,}900 \quad \text{and} \quad R(x) = 200x - 0.2x^2$$
$$0 \le x \le 1{,}000$$

(A) Find the value of x where the graph of $R(x)$ has a horizontal tangent line.

(B) Find the profit function $P(x)$.

(C) Find the value of x where the graph of $P(x)$ has a horizontal tangent line.

(D) Graph $C(x)$, $R(x)$, and $P(x)$ on the same coordinate system for $0 \le x \le 1{,}000$. Find the break-even points. Find the x intercepts of the graph of $P(x)$.

42. Revenue, cost, and profit. The total cost and the total revenue (in dollars) for the production and sale of x hair dryers are given, respectively, by

$$C(x) = 5x + 2,340 \quad \text{and} \quad R(x) = 40x - 0.1x^2$$
$$0 \le x \le 400$$

(A) Find the value of x where the graph of $R(x)$ has a horizontal tangent line.

(B) Find the profit function $P(x)$.

(C) Find the value of x where the graph of $P(x)$ has a horizontal tangent line.

(D) Graph $C(x)$, $R(x)$, and $P(x)$ on the same coordinate system for $0 \le x \le 400$. Find the break-even points. Find the x intercepts of the graph of $P(x)$.

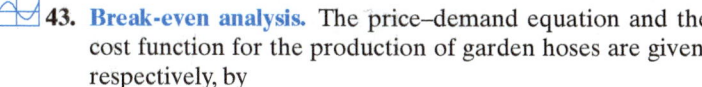 **43. Break-even analysis.** The price–demand equation and the cost function for the production of garden hoses are given, respectively, by

$$p = 20 - \sqrt{x} \quad \text{and} \quad C(x) = 500 + 2x$$

where x is the number of garden hoses that can be sold at a price of \$$p$ per unit and $C(x)$ is the total cost (in dollars) of producing x garden hoses.

(A) Express the revenue function in terms of x.

(B) Graph the cost function and revenue function in the same viewing window for $0 \le x \le 400$. Use approximation techniques to find the break-even points correct to the nearest unit.

44. Break-even analysis. The price–demand equation and the cost function for the production of handwoven silk scarves are given, respectively, by

$$p = 60 - 2\sqrt{x} \quad \text{and} \quad C(x) = 3,000 + 5x$$

where x is the number of scarves that can be sold at a price of \$$p$ per unit and $C(x)$ is the total cost (in dollars) of producing x scarves.

(A) Express the revenue function in terms of x.

(B) Graph the cost function and the revenue function in the same viewing window for $0 \le x \le 900$. Use approximation techniques to find the break-even points correct to the nearest unit.

45. Break-even analysis. Table 2 contains price–demand and total cost data for the production of projectors, where p is the wholesale price (in dollars) of a projector for an annual demand of x projectors and C is the total cost (in dollars) of producing x projectors.

Table 2		
x	$p(\$)$	$C(\$)$
3,190	581	1,130,000
4,570	405	1,241,000
5,740	181	1,410,000
7,330	85	1,620,000

(A) Find a quadratic regression equation for the price–demand data, using x as the independent variable.

(B) Find a linear regression equation for the cost data, using x as the independent variable. Use this equation

to estimate the fixed costs and variable costs per projector. Round answers to the nearest dollar.

(C) Find the break-even points. Round answers to the nearest integer.

(D) Find the price range for which the company will make a profit. Round answers to the nearest dollar.

46. Break-even analysis. Table 3 contains price–demand and total cost data for the production of treadmills, where p is the wholesale price (in dollars) of a treadmill for an annual demand of x treadmills and C is the total cost (in dollars) of producing x treadmills.

Table 3		
x	$p(\$)$	$C(\$)$
2,910	1,435	3,650,000
3,415	1,280	3,870,000
4,645	1,125	4,190,000
5,330	910	4,380,000

(A) Find a linear regression equation for the price–demand data, using x as the independent variable.

(B) Find a linear regression equation for the cost data, using x as the independent variable. Use this equation to estimate the fixed costs and variable costs per treadmill. Round answers to the nearest dollar.

(C) Find the break-even points. Round answers to the nearest integer.

(D) Find the price range for which the company will make a profit. Round answers to the nearest dollar.

Answers to Matched Problems

1. (A) $C'(x) = 600 - 1.5x$

   (B) $C'(200) = 300$. At a production level of 200 transmissions, total costs are increasing at the rate of \$300 per transmission.

   (C) $C(201) - C(200) = \$299.25$

2. (A) $R'(3,000) = 4$. At a production level of 3,000, a unit increase in production will increase revenue by approximately \$4.
   $R'(6,000) = -2$. At a production level of 6,000, a unit increase in production will decrease revenue by approximately \$2.

   (B) $P'(2,000) = 4$. At a production level of 2,000, a unit increase in production will increase profit by approximately \$4.
   $P'(7,000) = -6$. At a production level of 7,000, a unit increase in production will decrease profit by approximately \$6.

3. (A) $\overline{C}(x) = \dfrac{7,000}{x} + 2; \overline{C}'(x) = -\dfrac{7,000}{x^2}$

   (B) $\overline{C}(100) = \$72; \overline{C}'(100) = -\$0.70$. At a production level of 100 headphone sets, the average cost per headphone set is \$72. This average cost is decreasing at a rate of \$0.70 per headphone set.

   (C) Approx. \$71.30.

Chapter 3 Review

Important Terms, Symbols, and Concepts

3-1 Introduction to Limits

- The graph of the function $y = f(x)$ is the graph of the set of all ordered pairs $(x, f(x))$.

- The limit of the function $y = f(x)$ as x approaches c is L, written as $\lim_{x \to c} f(x) = L$, if the functional value $f(x)$ is close to the single real number L whenever x is close, but not equal, to c (on either side of c).

- The limit of the function $y = f(x)$ as x approaches c from the left is K, written as $\lim_{x \to c^-} f(x) = K$, if $f(x)$ is close to K whenever x is close to, but to the left of, c on the real-number line.

- The limit of the function $y = f(x)$ as x approaches c from the right is L, written as $\lim_{x \to c^+} f(x) = L$, if $f(x)$ is close to L whenever x is close to, but to the right of, c on the real-number line.

- The limit of the difference quotient $[f(a + h) - f(a)]/h$ always results in a 0/0 indeterminate form. Algebraic simplification is often required to evaluate this type of limit.

3-2 Infinite Limits and Limits at Infinity

- If $f(x)$ increases or decreases without bound as x approaches a from either side of a, then the line $x = a$ is a **vertical asymptote** of the graph of $y = f(x)$.

- If $f(x)$ gets close to L as x increases without bound or decreases without bound, then L is called the limit of f at ∞ or $-\infty$.

- The end behavior of a polynomial is described in terms of limits at infinity.

- If $f(x)$ approaches L as $x \to \infty$ or as $x \to -\infty$, then the line $y = L$ is a **horizontal asymptote** of the graph of $y = f(x)$. Polynomial functions never have horizontal asymptotes. A rational function can have at most one.

3-3 Continuity

- Intuitively, the graph of a continuous function can be drawn without lifting a pen off the paper. Algebraically, a function f is **continuous at c** if

  1. $\lim_{x \to c} f(x)$ exists, 2. $f(c)$ exists, and 3. $\lim_{x \to c} f(x) = f(c)$

- Continuity properties are useful for determining where a function is continuous and where it is discontinuous.

- Continuity properties are also useful for solving inequalities.

3-4 The Derivative

- Given a function $y = f(x)$, the **average rate of change** is the ratio of the change in y to the change in x.

- The **instantaneous rate of change** is the limit of the average rate of change as the change in x approaches 0.

- The slope of the secant line through two points on the graph of a function $y = f(x)$ is the ratio of the change in y to the change in x. The slope of the tangent line at the point $(a, f(a))$ is the limit of the slope of the secant line through the points $(a, f(a))$ and $(a + h, f(a + h))$ as h approaches 0.

- The **derivative of $y = f(x)$ at x**, denoted as $f'(x)$, is the limit of the difference quotient $[f(x + h) - f(x)]/h$ as $h \to 0$ (if the limit exists).

- The four-step method is used to find derivatives.

- If the limit of the difference quotient does not exist at $x = a$, then f is nondifferentiable at a and $f'(a)$ does not exist.

3-5 Basic Differentiation Properties

- The derivative of a constant function is 0.

- For any real number n, the derivative of $f(x) = x^n$ is nx^{n-1}.

- If f is a differential function, then the derivative of $kf(x)$ is $kf'(x)$.

- The derivative of the sum or difference of two differential functions is the sum or difference of the derivatives of the functions.

3-6 Differentials

- Given the function $y = f(x)$, the change in x is also called the **increment of x** and is denoted as Δx. The corresponding change in y is called the **increment of y** and is given by $\Delta y = f(x + \Delta x) - f(x)$.

Ex. 1, p. 188
Ex. 2, p. 190
Ex. 3, p. 191

- If $y = f(x)$ is differentiable at x, then the **differential of x** is $dx = \Delta x$ and the **differential of $y = f(x)$** is $dy = f'(x)dx$, or $df = f'(x)dx$. In this context, x and dx are both independent variables.

3-7 Marginal Analysis in Business and Economics

- If $y = C(x)$ is the total cost of producing x items, then $y = C'(x)$ is the **marginal cost** and $C(x + 1) - C(x)$ is the exact cost of producing item $x + 1$. Furthermore, $C'(x) \approx C(x + 1) - C(x)$. Similar statements can be made regarding total revenue and total profit functions.

Ex. 1, p. 195
Ex. 2, p. 196

- If $y = C(x)$ is the total cost of producing x items, then the **average cost**, or cost per unit, is $\overline{C}(x) = \dfrac{C(x)}{x}$ and the **marginal average cost** is $\overline{C}'(x) = \dfrac{d}{dx}\overline{C}(x)$. Similar statements can be made regarding total revenue and total profit functions.

Ex. 3, p. 199

Review Exercises

Work through all the problems in this chapter review, and check your answers in the back of the book. Answers to all review problems are there, along with section numbers in italics to indicate where each type of problem is discussed. Where weaknesses show up, review appropriate sections of the text.

  Many of the problems in this exercise set ask you to find a derivative. Most of the answers to these problems contain both an unsimplified form and a simplified form of the derivative. When checking your work, first check that you applied the rules correctly, and then check that you performed the algebraic simplification correctly.

A

1. Find the indicated quantities for $y = f(x) = 2x^2 + 5$:

(A) The change in y if x changes from 1 to 3

(B) The average rate of change of y with respect to x if x changes from 1 to 3

(C) The slope of the secant line through the points $(1, f(1))$ and $(3, f(3))$ on the graph of $y = f(x)$

(D) The instantaneous rate of change of y with respect to x at $x = 1$

(E) The slope of the line tangent to the graph of $y = f(x)$ at $x = 1$

(F) $f'(1)$

2. Use the four-step process to find $f'(x)$ for $f(x) = -3x + 2$.

3. If $\lim\limits_{x \to 1} f(x) = 2$ and $\lim\limits_{x \to 1} g(x) = 4$, find

(A) $\lim\limits_{x \to 1}(5f(x) + 3g(x))$

(B) $\lim\limits_{x \to 1}[f(x)g(x)]$

(C) $\lim\limits_{x \to 1} \dfrac{g(x)}{f(x)}$

(D) $\lim\limits_{x \to 1}[5 + 2x - 3g(x)]$

In Problems 4–10, use the graph of f to estimate the indicated limits and function values.

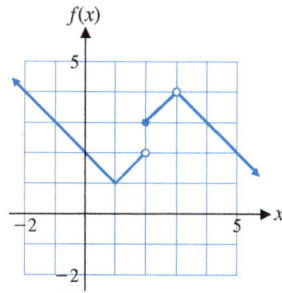

Figure for 4–10

4. $f(1.5)$

5. $f(2.5)$

6. $f(2.75)$

7. $f(3.25)$

8. (A) $\lim\limits_{x \to 1^-} f(x)$ (B) $\lim\limits_{x \to 1^+} f(x)$

    (C) $\lim\limits_{x \to 1} f(x)$ (D) $f(1)$

9. (A) $\lim\limits_{x \to 2^-} f(x)$ (B) $\lim\limits_{x \to 2^+} f(x)$

    (C) $\lim\limits_{x \to 2} f(x)$ (D) $f(2)$

10. (A) $\lim\limits_{x \to 3^-} f(x)$ (B) $\lim\limits_{x \to 3^+} f(x)$

    (C) $\lim\limits_{x \to 3} f(x)$ (D) $f(3)$

In Problems 11–13, use the graph of the function f shown in the figure to answer each question.

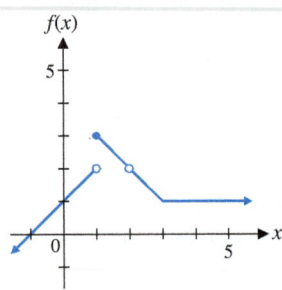

Figure for 11–13

11. (A) $\lim_{x \to 1} f(x) = ?$ (B) $f(1) = ?$
(C) Is f continuous at $x = 1$?

12. (A) $\lim_{x \to 2} f(x) = ?$ (B) $f(2) = ?$
(C) Is f continuous at $x = 2$?

13. (A) $\lim_{x \to 3} f(x) = ?$ (B) $f(3) = ?$
(C) Is f continuous at $x = 3$?

In Problems 14–23, refer to the following graph of $y = f(x)$:

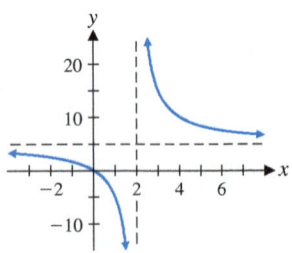

Figure for 14–23

14. $\lim_{x \to \infty} f(x) = ?$ **15.** $\lim_{x \to -\infty} f(x) = ?$

16. $\lim_{x \to 2^+} f(x) = ?$ **17.** $\lim_{x \to 2} f(x) = ?$

18. $\lim_{x \to 0^-} f(x) = ?$ **19.** $\lim_{x \to 0^+} f(x) = ?$

20. $\lim_{x \to 0} f(x) = ?$

21. Identify any vertical asymptotes.

22. Identify any horizontal asymptotes.

23. Where is $y = f(x)$ discontinuous?

24. Use the four-step process to find $f'(x)$ for $f(x) = 5x^2$.

25. If $f(5) = 4, f'(5) = -1, g(5) = 2$, and $g'(5) = -3$, then find $h'(5)$ for each of the following functions:
(A) $h(x) = 3f(x)$
(B) $h(x) = -2g(x)$
(C) $h(x) = 2f(x) + 5$
(D) $h(x) = -g(x) - 1$
(E) $h(x) = 2f(x) + 3g(x)$

In Problems 26–31, find $f'(x)$ and simplify.

26. $f(x) = \frac{1}{3}x^3 - 5x^2 + 1$ **27.** $f(x) = 2x^{1/2} - 3x$

28. $f(x) = 5$ **29.** $f(x) = \frac{3}{2x} + \frac{5x^3}{4}$

30. $f(x) = \frac{0.5}{x^4} + 0.25x^4$

31. $f(x) = (3x^3 - 2)(x + 1)$ (*Hint:* Multiply and then differentiate.)

In Problems 32–35, find the indicated quantities for $y = f(x) = x^2 + x$.

32. $\Delta x, \Delta y$, and $\Delta y/\Delta x$ for $x_1 = 1$ and $x_2 = 3$.

33. $[f(x_1 + \Delta x) - f(x_1)]/\Delta x$ for $x_1 = 1$ and $\Delta x = 2$.

34. dy for $x_1 = 1$ and $x_2 = 3$.

35. Δy and dy for $x = 1, \Delta x = dx = 0.2$.

B

Problems 36–38 refer to the function.

$$f(x) = \begin{cases} x^2 & \text{if } 0 \le x < 2 \\ 8 - x & \text{if } x \ge 2 \end{cases}$$

which is graphed in the figure.

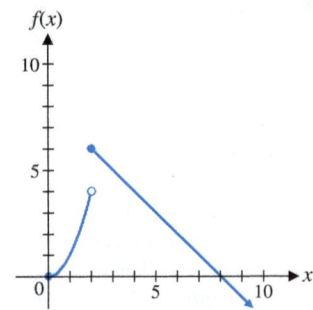

Figure for 36–38

36. (A) $\lim_{x \to 2^-} f(x) = ?$ (B) $\lim_{x \to 2^+} f(x) = ?$
(C) $\lim_{x \to 2} f(x) = ?$ (D) $f(2) = ?$
(E) Is f continuous at $x = 2$?

37. (A) $\lim_{x \to 5^-} f(x) = ?$ (B) $\lim_{x \to 5^+} f(x) = ?$
(C) $\lim_{x \to 5} f(x) = ?$ (D) $f(5) = ?$
(E) Is f continuous at $x = 5$?

38. Solve each inequality. Express answers in interval notation.
(A) $f(x) < 0$ (B) $f(x) \ge 0$

In Problems 39–41, solve each inequality. Express the answer in interval notation. Use a graphing calculator in Problem 41 to approximate partition numbers to four decimal places.

39. $x^2 - x < 12$ **40.** $\frac{x - 5}{x^2 + 3x} > 0$

41. $x^3 + x^2 - 4x - 2 > 0$

42. Let $f(x) = 0.5x^2 - 5$.
(A) Find the slope of the secant line through $(2, f(2))$ and $(4, f(4))$.
(B) Find the slope of the secant line through $(2, f(2))$ and $(2 + h, f(2 + h)), h \ne 0$.
(C) Find the slope of the tangent line at $x = 2$.

In Problems 43–46, find the indicated derivative and simplify.

43. $\dfrac{dy}{dx}$ for $y = \dfrac{1}{3}x^{-3} - 5x^{-2} + 1$

44. y' for $y = \dfrac{3\sqrt{x}}{2} + \dfrac{5}{3\sqrt{x}}$

45. $g'(x)$ for $g(x) = 1.8\sqrt[3]{x} + \dfrac{0.9}{\sqrt[3]{x}}$

46. $\dfrac{dy}{dx}$ for $y = \dfrac{2x^3 - 3}{5x^3}$

47. For $y = f(x) = x^2 + 4$, find
  (A) The slope of the graph at $x = 1$
  (B) The equation of the tangent line at $x = 1$ in the form $y = mx + b$

In Problems 48 and 49, find the value(s) of x where the tangent line is horizontal.

48. $f(x) = 10x - x^2$

49. $f(x) = x^3 + 3x^2 - 45x - 135$

 In Problems 50 and 51, approximate (to four decimal places) the value(s) of x where the graph of f has a horizontal tangent line.

50. $f(x) = x^4 - 2x^3 - 5x^2 + 7x$

51. $f(x) = x^5 - 10x^3 - 5x + 10$

52. If an object moves along the y axis (scale in feet) so that it is at $y = f(x) = 8x^2 - 4x + 1$ at time x (in seconds), find
  (A) The instantaneous velocity function
  (B) The velocity at time $x = 3$ seconds

53. An object moves along the y axis (scale in feet) so that at time x (in seconds) it is at $y = f(x) = -5x^2 + 16x + 3$. Find
  (A) The instantaneous velocity function
  (B) The time(s) when the velocity is 0

 54. Let $f(x) = x^3$, $g(x) = (x - 4)^3$, and $h(x) = (x + 3)^3$.
  (A) How are the graphs of f, g, and h related? Illustrate your conclusion by graphing f, g, and h on the same coordinate axes.
  (B) How would you expect the graphs of the derivatives of these functions to be related? Illustrate your conclusion by graphing f', g', and h' on the same coordinate axes.

In Problems 55–59, determine where f is continuous. Express the answer in interval notation.

55. $f(x) = x^2 - 4$ **56.** $f(x) = \dfrac{x + 1}{x - 2}$

57. $f(x) = \dfrac{x + 4}{x^2 + 3x - 4}$ **58.** $f(x) = \sqrt[3]{4 - x^2}$

59. $f(x) = \sqrt{4 - x^2}$

In Problems 60–69, evaluate the indicated limits if they exist.

60. Let $f(x) = \dfrac{2x}{x^2 - 3x}$. Find
  (A) $\lim\limits_{x \to 1} f(x)$ (B) $\lim\limits_{x \to 3} f(x)$ (C) $\lim\limits_{x \to 0} f(x)$

61. Let $f(x) = \dfrac{x + 1}{(3 - x)^2}$. Find
  (A) $\lim\limits_{x \to 1} f(x)$ (B) $\lim\limits_{x \to -1} f(x)$ (C) $\lim\limits_{x \to 3} f(x)$

62. Let $f(x) = \dfrac{|x - 4|}{x - 4}$. Find
  (A) $\lim\limits_{x \to 4^-} f(x)$ (B) $\lim\limits_{x \to 4^+} f(x)$ (C) $\lim\limits_{x \to 4} f(x)$

63. Let $f(x) = \dfrac{x - 3}{9 - x^2}$. Find
  (A) $\lim\limits_{x \to 3} f(x)$ (B) $\lim\limits_{x \to -3} f(x)$ (C) $\lim\limits_{x \to 0} f(x)$

64. Let $f(x) = \dfrac{x^2 - x - 2}{x^2 - 7x + 10}$. Find
  (A) $\lim\limits_{x \to -1} f(x)$ (B) $\lim\limits_{x \to 2} f(x)$ (C) $\lim\limits_{x \to 5} f(x)$

65. Let $f(x) = \dfrac{2x}{3x - 6}$. Find
  (A) $\lim\limits_{x \to \infty} f(x)$ (B) $\lim\limits_{x \to -\infty} f(x)$ (C) $\lim\limits_{x \to 2} f(x)$

66. Let $f(x) = \dfrac{2x^3}{3(x - 2)^2}$. Find
  (A) $\lim\limits_{x \to \infty} f(x)$ (B) $\lim\limits_{x \to -\infty} f(x)$ (C) $\lim\limits_{x \to 2} f(x)$

67. Let $f(x) = \dfrac{2x}{3(x - 2)^3}$. Find
  (A) $\lim\limits_{x \to \infty} f(x)$ (B) $\lim\limits_{x \to -\infty} f(x)$ (C) $\lim\limits_{x \to 2} f(x)$

68. $\lim\limits_{h \to 0} \dfrac{f(2 + h) - f(2)}{h}$ for $f(x) = x^2 + 4$

69. $\lim\limits_{h \to 0} \dfrac{f(x + h) - f(x)}{h}$ for $f(x) = \dfrac{1}{x + 2}$

In Problems 70 and 71, use the definition of the derivative and the four-step process to find $f'(x)$.

70. $f(x) = x^2 - x$ **71.** $f(x) = \sqrt{x} - 3$

Problems 72–77 refer to the function f in the figure. Determine whether f is differentiable at the indicated value of x.

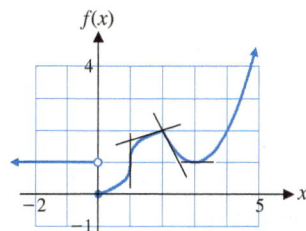

72. $x = -1$ **73.** $x = 0$

74. $x = 1$ **75.** $x = 2$

76. $x = 3$ **77.** $x = 4$

In Problems 78–82, find all horizontal and vertical asymptotes.

78. $f(x) = \dfrac{5x}{x - 7}$ **79.** $f(x) = \dfrac{-2x + 5}{(x - 4)^2}$

80. $f(x) = \dfrac{x^2 + 9}{x - 3}$ **81.** $f(x) = \dfrac{x^2 - 9}{x^2 + x - 2}$

82. $f(x) = \dfrac{x^3 - 1}{x^3 - x^2 - x + 1}$

83. The domain of the power function $f(x) = x^{1/5}$ is the set of all real numbers. Find the domain of the derivative $f'(x)$. Discuss the nature of the graph of $y = f(x)$ for any x values excluded from the domain of $f'(x)$.

84. Let f be defined by

$$f(x) = \begin{cases} x^2 - m & \text{if } x \le 1 \\ -x^2 + m & \text{if } x > 1 \end{cases}$$

where m is a constant.

(A) Graph f for $m = 0$, and find

$$\lim_{x \to 1^-} f(x) \quad \text{and} \quad \lim_{x \to 1^+} f(x)$$

(B) Graph f for $m = 2$, and find

$$\lim_{x \to 1^-} f(x) \quad \text{and} \quad \lim_{x \to 1^+} f(x)$$

(C) Find m so that

$$\lim_{x \to 1^-} f(x) = \lim_{x \to 1^+} f(x)$$

and graph f for this value of m.

(D) Write a brief verbal description of each graph. How does the graph in part (C) differ from the graphs in parts (A) and (B)?

85. Let $f(x) = 1 - |x - 1|, 0 \le x \le 2$ (see the figure).

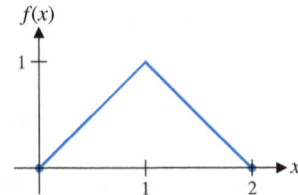

Figure for 85

(A) $\displaystyle \lim_{h \to 0^-} \frac{f(1 + h) - f(1)}{h} = ?$

(B) $\displaystyle \lim_{h \to 0^+} \frac{f(1 + h) - f(1)}{h} = ?$

(C) $\displaystyle \lim_{h \to 0} \frac{f(1 + h) - f(1)}{h} = ?$

(D) Does $f'(1)$ exist?

Applications

86. Natural-gas rates. Table 1 shows the winter rates for natural gas charged by the Bay State Gas Company. The customer charge is a fixed monthly charge, independent of the amount of gas used per month.

Table 1 Natural Gas Rates

Monthly customer charge	$7.47
First 90 therms	$0.4000 per therm
All usage over 90 therms	$0.2076 per therm

(A) Write a piecewise definition of the monthly charge $S(x)$ for a customer who uses x therms in a winter month.

(B) Graph $S(x)$.

(C) Is $S(x)$ continuous at $x = 90$? Explain.

87. Cost analysis. The total cost (in dollars) of producing x HDTVs is

$$C(x) = 10,000 + 200x - 0.1x^2$$

(A) Find the exact cost of producing the 101st TV.

(B) Use the marginal cost to approximate the cost of producing the 101st TV.

88. Cost analysis. The total cost (in dollars) of producing x bicycles is

$$C(x) = 5,000 + 40x + 0.05x^2$$

(A) Find the total cost and the marginal cost at a production level of 100 bicycles and interpret the results.

(B) Find the average cost and the marginal average cost at a production level of 100 bicycles and interpret the results.

89. Cost analysis. The total cost (in dollars) of producing x laser printers per week is shown in the figure. Which is greater, the approximate cost of producing the 201st printer or the approximate cost of producing the 601st printer? Does this graph represent a manufacturing process that is becoming more efficient or less efficient as production levels increase? Explain.

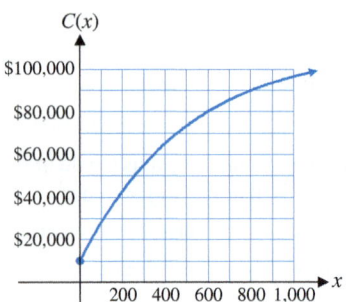

Figure for 89

90. Cost analysis. Let

$$p = 25 - 0.01x \quad \text{and} \quad C(x) = 2x + 9,000$$
$$0 \le x \le 2,500$$

be the price–demand equation and cost function, respectively, for the manufacture of umbrellas.

(A) Find the marginal cost, average cost, and marginal average cost functions.

(B) Express the revenue in terms of x, and find the marginal revenue, average revenue, and marginal average revenue functions.

(C) Find the profit, marginal profit, average profit, and marginal average profit functions.

(D) Find the break-even point(s).

(E) Evaluate the marginal profit at $x = 1,000, 1,150$, and $1,400$, and interpret the results.

(F) Graph $R = R(x)$ and $C = C(x)$ on the same coordinate system, and locate regions of profit and loss.

91. Employee training. A company producing computer components has established that, on average, a new employee can assemble $N(t)$ components per day after t days of on-the-job training, as given by

$$N(t) = \frac{40t - 80}{t}, t \geq 2$$

(A) Find the average rate of change of $N(t)$ from 2 days to 5 days.

(B) Find the instantaneous rate of change of $N(t)$ at 2 days.

92. Sales analysis. The total number of swimming pools, N (in thousands), sold during a year is given by

$$N(t) = 2t + \frac{1}{3}t^{3/2}$$

where t is the number of months since the beginning of the year. Find $N(9)$ and $N'(9)$, and interpret these quantities.

93. Natural-gas consumption. The data in Table 2 give the U.S. consumption of natural gas in trillions of cubic feet.

Table 2

Year	Natural-Gas Consumption
1960	12.0
1970	21.1
1980	19.9
1990	18.7
2000	21.9

(A) Let x represent time (in years), with $x = 0$ corresponding to 1960, and let y represent the corresponding U.S. consumption of natural gas. Enter the data set in a graphing calculator and find a cubic regression equation for the data.

(B) If $y = N(x)$ denotes the regression equation found in part (A), find $N(50)$ and $N'(50)$, and write a brief verbal interpretation of these results.

94. Break-even analysis. Table 3 contains price–demand and total cost data from a bakery for the production of kringles (a Danish pastry), where p is the price (in dollars) of a kringle for a daily demand of x kringles and C is the total cost (in dollars) of producing x kringles.

Table 3

x	$p(\$)$	$C(\$)$
125	9	740
140	8	785
170	7	850
200	6	900

(A) Find a linear regression equation for the price–demand data, using x as the independent variable.

(B) Find a linear regression equation for the cost data, using x as the independent variable. Use this equation to estimate the fixed costs and variable costs per kringle.

(C) Find the break-even points.

(D) Find the price range for which the bakery will make a profit.

95. Pollution. A sewage treatment plant uses a pipeline that extends 1 mile toward the center of a large lake. The concentration of effluent $C(x)$ in parts per million, x meters from the end of the pipe is given approximately by

$$C(x) = \frac{500}{x^2}, x \geq 1$$

What is the instantaneous rate of change of concentration at 10 meters? At 100 meters?

96. Medicine. The body temperature (in degrees Fahrenheit) of a patient t hours after taking a fever-reducing drug is given by

$$F(t) = 0.16t^2 - 1.6t + 102$$

Find $F(4)$ and $F'(4)$. Write a brief verbal interpretation of these quantities.

97. Learning. If a person learns N items in t hours, as given by

$$N(t) = 20\sqrt{t}$$

find the rate of learning after

(A) 1 hour   (B) 4 hours

98. Physics: The coefficient of thermal expansion (CTE) is a measure of the expansion of an object subjected to extreme temperatures. We want to use a Michaelis–Menten function of the form

$$C(T) = \frac{C_{max}T}{M + T}$$

where C = CTE, T is temperature in K (degrees Kelvin), and C_{max} and M are constants. Table 4 lists the coefficients of thermal expansion for titanium at various temperatures.

Table 4   Coefficients of Thermal Expansion

T (K)	Titanium
100	4.5
200	7.4
293	8.6
500	9.9
800	11.1
1100	11.7

(A) Plot the points in columns 1 and 2 of Table 4 on graph paper and estimate C_{max} to the nearest integer. To estimate M, add the horizontal line CTE = $\frac{C_{max}}{2}$ to your graph, connect successive points on the graph with straight-line segments, and estimate the value of T (to the nearest multiple of fifty) that satisfies

$$C(T) = \frac{C_{max}}{2}.$$

(B) Use the constants $\frac{C_{max}}{2}$ and M from part (A) to form a Michaelis–Menten function for the CTE of titanium.

(C) Use the function from part (B) to estimate the CTE of titanium at 600 K and to estimate the temperature when the CTE of titanium is 10.

Additional Derivative Topics

Introduction

In this chapter, we develop techniques for finding derivatives of a wide variety of functions, including exponential and logarithmic functions. There are straightforward procedures—the product rule, quotient rule, and chain rule—for writing down the derivative of any function that is the product, quotient, or composite of functions whose derivatives are known. With the ability to calculate derivatives easily, we consider a wealth of applications involving rates of change. For example, we apply the derivative to study population growth, radioactive decay, elasticity of demand, and environmental crises (see Problem 31 in Section 4-6 or Problem 71 in Section 4-7). Before starting this chapter, you may find it helpful to review the basic properties of exponential and logarithmic functions in Sections 2-5 and 2-6.

4-1 The Constant e and Continuous Compound Interest

- The Constant e
- Continuous Compound Interest

In Chapter 2, both the exponential function with base e and continuous compound interest were introduced informally. Now, with an understanding of limit concepts, we can give precise definitions of e and continuous compound interest.

The Constant e

The irrational number e is a particularly suitable base for both exponential and logarithmic functions. The reasons for choosing this number as a base will become clear as we develop differentiation formulas for the exponential function e^x and the natural logarithmic function $\ln x$.

In precalculus treatments (Chapter 2), the number e is defined informally as the irrational number that can be approximated by the expression $[1 + (1/n)]^n$ for n sufficiently large. Now we will use the limit concept to formally define e as either of the following two limits. [*Note:* If $s = 1/n$, then as $n \to \infty$, $s \to 0$.]

> **DEFINITION The Number e**
>
> $$e = \lim_{n \to \infty} \left(1 + \frac{1}{n} \right)^n \qquad \text{or, alternatively,} \qquad e = \lim_{s \to 0} (1 + s)^{1/s}$$

Both limits are equal to $e = 2.718\ 281\ 828\ 459 \ldots$

Proof that the indicated limits exist and represent an irrational number between 2 and 3 is not easy and is omitted.

> **CONCEPTUAL INSIGHT**
>
> The two limits used to define e are unlike any we have encountered so far. Some people reason (incorrectly) that both limits are 1, since $1 + s \to 1$ as $s \to 0$ and 1 to any power is 1. An ordinary scientific calculator with a y^x key can convince you otherwise. Consider the following table of values for s and $f(s) = (1 + s)^{1/s}$ and Figure 1 for s close to 0. Compute the table values with a calculator yourself, and try several values of s even closer to 0. Note that the function is discontinuous at $s = 0$.
>
>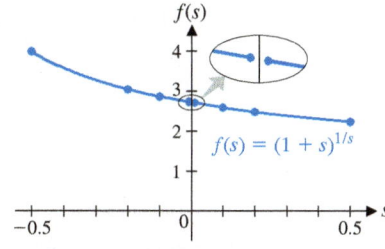
>
> **Figure 1**
>
> s approaches 0 from the left $\to 0 \leftarrow$ s approaches 0 from the right
>
> | s | -0.5 | -0.2 | -0.1 | $-0.01 \to 0 \leftarrow 0.01$ | 0.1 | 0.2 | 0.5 |
> |---|---|---|---|---|---|---|---|
> | $(1 + s)^{1/s}$ | 4.0000 | 3.0518 | 2.8680 | $2.7320 \to e \leftarrow 2.7048$ | 2.5937 | 2.4883 | 2.2500 |

Continuous Compound Interest

Now we can see how e appears quite naturally in the important application of compound interest. Let us start with simple interest, move on to compound interest, and then proceed on to continuous compound interest.

On one hand, if a principal P is borrowed at an annual rate r,* then after t years at simple interest, the borrower will owe the lender an amount A given by

$$A = P + Prt = P(1 + rt) \quad \text{Simple interest} \qquad (1)$$

On the other hand, if interest is compounded n times a year, then the borrower will owe the lender an amount A given by

$$A = P\left(1 + \frac{r}{n}\right)^{nt} \quad \text{Compound interest} \qquad (2)$$

where r/n is the interest rate per compounding period and nt is the number of compounding periods. Suppose that P, r, and t in equation (2) are held fixed and n is increased. Will the amount A increase without bound, or will it tend to approach some limiting value?

Let us perform a calculator experiment before we attack the general limit problem. If $P = \$100$, $r = 0.06$, and $t = 2$ years, then

$$A = 100\left(1 + \frac{0.06}{n}\right)^{2n}$$

We compute A for several values of n in Table 1. The biggest gain appears in the first step, then the gains slow down as n increases. The amount A appears to approach \$112.75 as n gets larger and larger.

Table 1

Compounding Frequency	n	$A = 100\left(1 + \dfrac{0.06}{n}\right)^{2n}$
Annually	1	\$112.3600
Semiannually	2	112.5509
Quarterly	4	112.6493
Monthly	12	112.7160
Weekly	52	112.7419
Daily	365	112.7486
Hourly	8,760	112.7496

Keeping P, r, and t fixed in equation (2), we compute the following limit and observe an interesting and useful result:

$$\lim_{n \to \infty} P\left(1 + \frac{r}{n}\right)^{nt} = P \lim_{n \to \infty} \left(1 + \frac{r}{n}\right)^{(n/r)rt} \quad \begin{array}{l}\text{Insert } r/r \text{ in the exponent and} \\ \text{let } s = r/n. \text{ Note that} \\ n \to \infty \text{ implies } s \to 0.\end{array}$$

$$= P \lim_{s \to 0} [(1 + s)^{1/s}]^{rt} \quad \text{Use a limit property.}^{\dagger}$$

$$= P[\lim_{s \to 0} (1 + s)^{1/s}]^{rt} \quad \lim_{s \to 0} (1 + s)^{1/s} = e$$

$$= Pe^{rt}$$

The resulting formula is called the **continuous compound interest formula**, a widely used formula in business and economics.

> **THEOREM 1 Continuous Compound Interest Formula**
>
> If a principal P is invested at an annual rate r (expressed as a decimal) compounded continuously, then the amount A in the account at the end of t years is given by
>
> $$A = Pe^{rt}$$

---

*If r is the interest rate written as a decimal, then $100r\%$ is the rate in percent. For example, if $r = 0.12$, then $100r\% = 100(0.12)\% = 12\%$. The expressions 0.12 and 12% are equivalent. Unless stated otherwise, all formulas in this book use r in decimal form.

†The following new limit property is used: If $\lim_{x \to c} f(x)$ exists, then $\lim_{x \to c} [f(x)]^p = [\lim_{x \to c} f(x)]^p$, provided that the last expression names a real number.

EXAMPLE 1 **Computing Continuously Compounded Interest** If $100 is invested at 6% compounded continuously,* what amount will be in the account after 2 years? How much interest will be earned?

SOLUTION
$$A = Pe^{rt}$$
$$= 100e^{(0.06)(2)} \qquad \textit{6\% is equivalent to r = 0.06.}$$
$$\approx \$112.7497$$

Compare this result with the values calculated in Table 1. The interest earned is $112.7497 − $100 = $12.7497.

Matched Problem 1 What amount (to the nearest cent) will be in an account after 5 years if $100 is invested at an annual nominal rate of 8% compounded annually? Semiannually? Continuously?

EXAMPLE 2 **Graphing the Growth of an Investment** Union Savings Bank offers a 5-year certificate of deposit (CD) that earns 5.75% compounded continuously. If $1,000 is invested in one of these CDs, graph the amount in the account as a function of time for a period of 5 years.

SOLUTION We want to graph
$$A = 1,000e^{0.0575t} \qquad 0 \le t \le 5$$

Using a calculator, we construct a table of values (Table 2). Then we graph the points from the table and join the points with a smooth curve (Fig. 2).

Table 2

t	$A(\$)$
0	1,000
1	1,059
2	1,122
3	1,188
4	1,259
5	1,333

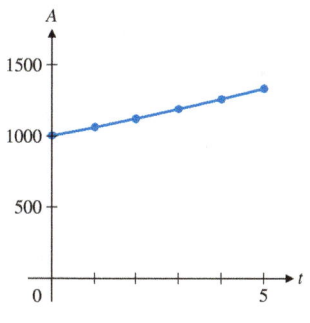

Figure 2

> **CONCEPTUAL INSIGHT**
>
> Depending on the domain, the graph of an exponential function can appear to be linear. Table 2 shows that the graph in Figure 2 is *not* linear. The slope determined by the first two points (for $t = 0$ and $t = 1$) is 59 but the slope determined by the first and third points (for $t = 0$ and $t = 2$) is 61. For a linear graph, the slope determined by any two points is constant.

Matched Problem 2 If $5,000 is invested in a Union Savings Bank 4-year CD that earns 5.61% compounded continuously, graph the amount in the account as a function of time for a period of 4 years.

---

*Following common usage, we will often write "at 6% compounded continuously," understanding that this means "at an annual nominal rate of 6% compounded continuously."

EXAMPLE 3   **Computing Growth Time**   How long will it take an investment of $5,000 to grow to $8,000 if it is invested at 5% compounded continuously?

SOLUTION   Starting with the continous compound interest formula $A = Pe^{rt}$, we must solve for t:

$$A = Pe^{rt}$$
$$8,000 = 5,000e^{0.05t} \qquad \text{Divide both sides by 5,000 and}$$
$$e^{0.05t} = 1.6 \qquad \text{reverse the equation.}$$
$$\ln e^{0.05t} = \ln 1.6 \qquad \text{Take the natural logarithm of both}$$
$$0.05t = \ln 1.6 \qquad \text{sides—recall that } \log_b b^x = x.$$
$$t = \frac{\ln 1.6}{0.05}$$
$$t \approx 9.4 \text{ years}$$

Figure 3
$y_1 = 5,000e^{0.05x}$
$y_2 = 8,000$

Figure 3 shows an alternative method for solving Example 3 on a graphing calculator.

Matched Problem 3   How long will it take an investment of $10,000 to grow to $15,000 if it is invested at 9% compounded continuously?

EXAMPLE 4   **Computing Doubling Time**   How long will it take money to double if it is invested at 6.5% compounded continuously?

SOLUTION   Starting with the continuous compound interest formula $A = Pe^{rt}$, we solve for t, given $A = 2P$ and $R = 0.065$:

$$2P = Pe^{0.065t} \qquad \text{Divide both sides by } P \text{ and reverse the equation.}$$
$$e^{0.065t} = 2 \qquad \text{Take the natural logarithm of both sides.}$$
$$\ln e^{0.065t} = \ln 2$$
$$0.065t = \ln 2$$
$$t = \frac{\ln 2}{0.065}$$
$$t \approx 10.66 \text{ years}$$

Matched Problem 4   How long will it take money to triple if it is invested at 5.5% compounded continuously?

EXPLORE & DISCUSS 1   You are considering three options for investing $10,000: at 7% compounded annually, at 6% compounded monthly, and at 5% compounded continuously.

(A) Which option would be the best for investing $10,000 for 8 years?

(B) How long would you need to invest your money for the third option to be the best?

Exercises 4-1

A

Use a calculator to evaluate A to the nearest cent in Problems 1 and 2.

1. $A = \$1,000e^{0.1t}$ for $t = 2, 5,$ and 8

2. $A = \$5,000e^{0.08t}$ for $t = 1, 4,$ and 10

3. If \$6,000 is invested at 10% compounded continuously, graph the amount in the account as a function of time for a period of 8 years.

4. If \$4,000 is invested at 8% compounded continuously, graph the amount in the account as a function of time for a period of 6 years.

B

In Problems 5–10, solve for t or r to two decimal places.

5. $2 = e^{0.06t}$

6. $2 = e^{0.03t}$

7. $3 = e^{0.1r}$

8. $3 = e^{0.25t}$

9. $2 = e^{5r}$

10. $3 = e^{10r}$

C

In Problems 11 and 12, use a calculator to complete each table to five decimal places.

11.

n	$[1 + (1/n)]^n$
10	2.593 74
100	
1,000	
10,000	
100,000	
1,000,000	
10,000,000	
↓	↓
∞	$e = 2.718\ 281\ 828\ 459\ldots$

12.

s	$(1 + s)^{1/s}$
0.01	2.704 81
−0.01	
0.001	
−0.001	
0.000 1	
−0.000 1	
0.000 01	
−0.000 01	
↓	↓
0	$e = 2.718\ 281\ 828\ 459\ldots$

13. Use a calculator and a table of values to investigate

$$\lim_{n \to \infty} (1 + n)^{1/n}$$

Do you think this limit exists? If so, what do you think it is?

14. Use a calculator and a table of values to investigate

$$\lim_{s \to 0^+} \left(1 + \frac{1}{s}\right)^s$$

Do you think this limit exists? If so, what do you think it is?

15. It can be shown that the number *e* satisfies the inequality

$$\left(1 + \frac{1}{n}\right)^n < e < \left(1 + \frac{1}{n}\right)^{n+1} \qquad n \geq 1$$

Illustrate this condition by graphing

$$y_1 = (1 + 1/n)^n$$
$$y_2 = 2.718\ 281\ 828 \approx e$$
$$y_3 = (1 + 1/n)^{n+1}$$

in the same viewing window, for $1 \leq n \leq 20$.

16. It can be shown that

$$e^s = \lim_{n \to \infty} \left(1 + \frac{s}{n}\right)^n$$

for any real number *s*. Illustrate this equation graphically for $s = 2$ by graphing

$$y_1 = (1 + 2/n)^n$$
$$y_2 = 7.389\ 056\ 099 \approx e^2$$

in the same viewing window, for $1 \leq n \leq 50$.

Applications

17. **Continuous compound interest.** Provident Bank offers a 10-year CD that earns 4.15% compounded continuously.

   (A) If \$10,000 is invested in this CD, how much will it be worth in 10 years?

   (B) How long will it take for the account to be worth \$18,000?

18. **Continuous compound interest.** Provident Bank also offers a 3-year CD that earns 3.64% compounded continuously.

   (A) If \$10,000 is invested in this CD, how much will it be worth in 3 years?

   (B) How long will it take for the account to be worth \$11,000?

19. **Present value.** A note will pay \$20,000 at maturity 10 years from now. How much should you be willing to pay for the note now if money is worth 5.2% compounded continuously?

20. **Present value.** A note will pay \$50,000 at maturity 5 years from now. How much should you be willing to pay for the note now if money is worth 6.4% compounded continuously?

21. **Continuous compound interest.** An investor bought stock for \$20,000. Five years later, the stock was sold for \$30,000. If interest is compounded continuously, what annual nominal rate of interest did the original \$20,000 investment earn?

22. Continuous compound interest. A family paid $99,000 cash for a house. Fifteen years later, the house was sold for $195,000. If interest is compounded continuously, what annual nominal rate of interest did the original $99,000 investment earn?

23. Present value. Solving $A = Pe^{rt}$ for P, we obtain

$$P = Ae^{-rt}$$

which is the present value of the amount A due in t years if money earns interest at an annual nominal rate r compounded continuously.

(A) Graph $P = 10,000e^{-0.08t}$, $0 \le t \le 50$.

(B) $\lim_{t \to \infty} 10,000e^{-0.08t} = $? [Guess, using part (A).]

[*Conclusion:* The longer the time until the amount A is due, the smaller is its present value, as we would expect.]

24. Present value. Referring to Problem 23, in how many years will the $10,000 be due in order for its present value to be $5,000?

25. Doubling time. How long will it take money to double if it is invested at 4% compounded continuously?

26. Doubling time. How long will it take money to double if it is invested at 5% compounded continuously?

27. Doubling rate. At what nominal rate compounded continuously must money be invested to double in 8 years?

28. Doubling rate. At what nominal rate compounded continuously must money be invested to double in 10 years?

29. Growth time. A man with $20,000 to invest decides to diversify his investments by placing $10,000 in an account that earns 7.2% compounded continuously and $10,000 in an account that earns 8.4% compounded annually. Use graphical approximation methods to determine how long it will take for his total investment in the two accounts to grow to $35,000.

30. Growth time. A woman invests $5,000 in an account that earns 8.8% compounded continuously and $7,000 in an account that earns 9.6% compounded annually. Use graphical approximation methods to determine how long it will take for her total investment in the two accounts to grow to $20,000.

31. Doubling times

(A) Show that the doubling time t (in years) at an annual rate r compounded continuously is given by

$$t = \frac{\ln 2}{r}$$

(B) Graph the doubling-time equation from part (A) for $0.02 \le r \le 0.30$. Is this restriction on r reasonable? Explain.

(C) Determine the doubling times (in years, to two decimal places) for $r = 5\%, 10\%, 15\%, 20\%, 25\%$, and 30%.

32. Doubling rates

(A) Show that the rate r that doubles an investment at continuously compounded interest in t years is given by

$$r = \frac{\ln 2}{t}$$

(B) Graph the doubling-rate equation from part (A) for $1 \le t \le 20$. Is this restriction on t reasonable? Explain.

(C) Determine the doubling rates for $t = 2, 4, 6, 8, 10$, and 12 years.

33. Radioactive decay. A mathematical model for the decay of radioactive substances is given by

$$Q = Q_0 e^{rt}$$

where

$Q_0 = $ amount of the substance at time $t = 0$
$r = $ continuous compound rate of decay
$t = $ time in years
$Q = $ amount of the substance at time t

If the continuous compound rate of decay of radium per year is $r = -0.000\ 433\ 2$, how long will it take a certain amount of radium to decay to half the original amount? (This period is the *half-life* of the substance.)

34. Radioactive decay. The continuous compound rate of decay of carbon-14 per year is $r = -0.000\ 123\ 8$. How long will it take a certain amount of carbon-14 to decay to half the original amount? (Use the radioactive decay model in Problem 33.)

35. Radioactive decay. A cesium isotope has a half-life of 30 years. What is the continuous compound rate of decay? (Use the radioactive decay model in Problem 33.)

36. Radioactive decay. A strontium isotope has a half-life of 90 years. What is the continuous compound rate of decay? (Use the radioactive decay model in Problem 33.)

37. World population. A mathematical model for world population growth over short intervals is given by

$$P = P_0 e^{rt}$$

where

$P_0 = $ population at time $t = 0$
$r = $ continuous compound rate of growth
$t = $ time in years
$P = $ population at time t

How long will it take world population to double if it continues to grow at its current continuous compound rate of 1.3% per year?

38. U.S. population. How long will it take for the U.S. population to double if it continues to grow at a rate of 0.975% per year?

39. Population growth. Some underdeveloped nations have population doubling times of 50 years. At what continuous compound rate is the population growing? (Use the population growth model in Problem 37.)

40. Population growth. Some developed nations have population doubling times of 200 years. At what continuous compound rate is the population growing? (Use the population growth model in Problem 37.)

3. 4.51 yr

4. 19.97 yr

1. $146.93; $148.02; $149.18

2. $A = 5,000e^{0.0561t}$

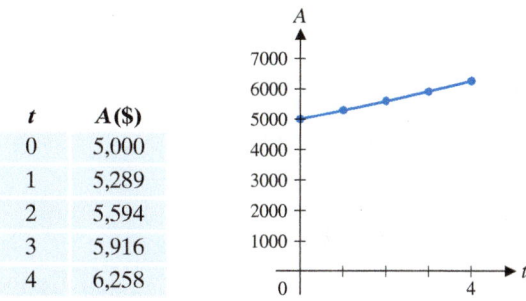

t	$A(\$)$
0	5,000
1	5,289
2	5,594
3	5,916
4	6,258

4-2 Derivatives of Exponential and Logarithmic Functions

- The Derivative of e^x
- The Derivative of $\ln x$
- Other Logarithmic and Exponential Functions
- Exponential and Logarithmic Models

In this section, we find formulas for the derivatives of logarithmic and exponential functions. A review of Sections 2–5 and 2–6 may prove helpful. In particular, recall that $f(x) = e^x$ is the exponential function with base $e \approx 2.718$, and the inverse of the function e^x is the natural logarithm function $\ln x$. More generally, if b is a positive real number, $b \neq 1$, then the exponential function b^x with base b, and the logarithmic function $\log_b x$ with base b, are inverses of each other.

The Derivative of e^x

In the process of finding the derivative of e^x, we use (without proof) the fact that

$$\lim_{h \to 0} \frac{e^h - 1}{h} = 1 \qquad (1)$$

EXPLORE & DISCUSS 1

Complete Table 1.

Table 1

h	−0.1	−0.01	−0.001	→ 0 ←	0.001	0.01	0.1
$\dfrac{e^h - 1}{h}$							

Do your calculations make it reasonable to conclude that

$$\lim_{h \to 0} \frac{e^h - 1}{h} = 1?$$

Discuss.

We now apply the four-step process (Section 3-4) to the exponential function $f(x) = e^x$.

Step 1 Find $f(x + h)$.

$$f(x + h) = e^{x+h} = e^x e^h \qquad \textit{See Section 2-4.}$$

Step 2 Find $f(x + h) - f(x)$.

$$f(x + h) - f(x) = e^x e^h - e^x \qquad \textit{Factor out } e^x.$$

$$= e^x(e^h - 1)$$

Step 3 Find $\dfrac{f(x+h)-f(x)}{h}$.

$$\frac{f(x+h)-f(x)}{h} = \frac{e^x(e^h-1)}{h} = e^x\left(\frac{e^h-1}{h}\right)$$

Step 4 Find $f'(x) = \lim\limits_{h\to 0}\dfrac{f(x+h)-f(x)}{h}$.

$$
\begin{aligned}
f'(x) &= \lim_{h\to 0}\frac{f(x+h)-f(x)}{h}\\
&= \lim_{h\to 0}e^x\left(\frac{e^h-1}{h}\right)\\
&= e^x\lim_{h\to 0}\left(\frac{e^h-1}{h}\right) \qquad \text{Use the limit in (1).}\\
&= e^x\cdot 1 = e^x
\end{aligned}
$$

Therefore,

$$\frac{d}{dx}e^x = e^x \qquad \text{The derivative of the exponential function is the exponential function.}$$

EXAMPLE 1 **Finding Derivatives** Find $f'(x)$ for

(A) $f(x) = 5e^x - 3x^4 + 9x + 16$ (B) $f(x) = -7x^e + 2e^x + e^2$

SOLUTIONS (A) $f'(x) = 5e^x - 12x^3 + 9$ (B) $f'(x) = -7ex^{e-1} + 2e^x$

Remember that e is a real number, so the power rule (Section 3–5) is used to find the derivative of x^e. The derivative of the exponential function e^x, however, is e^x. Note that $e^2 \approx 7.389$ is a constant, so its derivative is 0.

Matched Problem 1 Find $f'(x)$ for

(A) $f(x) = 4e^x + 8x^2 + 7x - 14$ (B) $f(x) = x^7 - x^5 + e^3 - x + e^x$

 CAUTION

$$\frac{d}{dx}e^x \neq xe^{x-1} \qquad \frac{d}{dx}e^x = e^x$$

The power rule cannot be used to differentiate the exponential function. The power rule applies to exponential forms x^n, where the exponent is a constant and the base is a variable. In the exponential form e^x, the base is a constant and the exponent is a variable.

The Derivative of ln x

We summarize some important facts about logarithmic functions from Section 2–6:

SUMMARY

Recall that the inverse of an exponential function is called a **logarithmic function**. For $b > 0$ and $b \neq 1$,

Logarithmic form		Exponential form
$y = \log_b x$	is equivalent to	$x = b^y$
Domain: $(0, \infty)$		Domain: $(-\infty, \infty)$
Range: $(-\infty, \infty)$		Range: $(0, \infty)$

The graphs of $y = \log_b x$ and $y = b^x$ are symmetric with respect to the line $y = x$. (See Figure 1.)

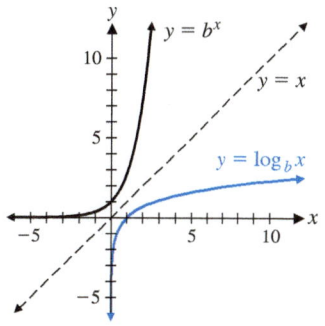

Figure 1

Of all the possible bases for logarithmic functions, the two most widely used are

$$\log x = \log_{10} x \quad \textit{Common logarithm (base 10)}$$

$$\ln x = \log_e x \quad \textit{Natural logarithm (base e)}$$

We are now ready to use the definition of the derivative and the four-step process discussed in Section 3-4 to find a formula for the derivative of $\ln x$. Later we will extend this formula to include $\log_b x$ for any base b.

Let $f(x) = \ln x$, $x > 0$.

Step 1 Find $f(x + h)$.

$$f(x + h) = \ln(x + h) \quad \textit{In(x + h) cannot be simplified.}$$

Step 2 Find $f(x + h) - f(x)$.

$$f(x + h) - f(x) = \ln(x + h) - \ln x \qquad \textit{Use ln A - ln B = ln}\frac{A}{B}.$$

$$= \ln\frac{x + h}{x}$$

Step 3 Find $\dfrac{f(x + h) - f(x)}{h}$.

$$\frac{f(x + h) - f(x)}{h} = \frac{\ln(x + h) - \ln x}{h}$$

$$= \frac{1}{h}\ln\frac{x + h}{x}$$

$$= \frac{x}{x} \cdot \frac{1}{h}\ln\frac{x + h}{x} \qquad \textit{Multiply by 1 = x/x to change form.}$$

$$= \frac{1}{x}\left[\frac{x}{h}\ln\left(1 + \frac{h}{x}\right)\right] \qquad \text{Use } p \ln A = \ln A^p.$$

$$= \frac{1}{x}\ln\left(1 + \frac{h}{x}\right)^{x/h}$$

Step 4 Find $f'(x) = \lim_{h\to 0}\dfrac{f(x + h) - f(x)}{h}$.

$$
\begin{aligned}
f'(x) &= \lim_{h\to 0}\frac{f(x + h) - f(x)}{h} \\
&= \lim_{h\to 0}\left[\frac{1}{x}\ln\left(1 + \frac{h}{x}\right)^{x/h}\right] && \text{Let } s = h/x. \text{ Note that } h\to 0 \\
&&& \text{implies } s\to 0. \\
&= \frac{1}{x}\lim_{s\to 0}\left[\ln(1 + s)^{1/s}\right] && \text{Use a new limit property.}^* \\
&= \frac{1}{x}\ln\left[\lim_{s\to 0}(1 + s)^{1/s}\right] && \text{Use the definition of } e. \\
&= \frac{1}{x}\ln e && \ln e = \log_e e = 1 \\
&= \frac{1}{x}
\end{aligned}
$$

Therefore,

$$\frac{d}{dx}\ln x = \frac{1}{x}$$

CONCEPTUAL INSIGHT

In finding the derivative of $\ln x$, we used the following properties of logarithms:

$$\ln\frac{A}{B} = \ln A - \ln B \qquad \ln A^p = p \ln A$$

We also noted that there is no property that simplifies $\ln(A + B)$. (See Theorem 1 in Section 2-6 for a list of properties of logarithms.)

EXAMPLE 2 **Finding Derivatives** Find y' for

(A) $y = 3e^x + 5 \ln x$ (B) $y = x^4 - \ln x^4$

SOLUTIONS (A) $y' = 3e^x + \dfrac{5}{x}$

(B) Before taking the derivative, we use a property of logarithms (see Theorem 1, Section 2-6) to rewrite y.

$$
\begin{aligned}
y &= x^4 - \ln x^4 && \text{Use } \ln M^p = p \ln M. \\
y &= x^4 - 4 \ln x && \text{Now take the derivative of both sides.} \\
y' &= 4x^3 - \frac{4}{x}
\end{aligned}
$$

Matched Problem 2 Find y' for

(A) $y = 10x^3 - 100 \ln x$ (B) $y = \ln x^5 + e^x - \ln e^2$

---

*The following new limit property is used: If $\lim_{x\to c} f(x)$ exists and is positive, then $\lim_{x\to c}[\ln f(x)] = \ln[\lim_{x\to c} f(x)]$.

Other Logarithmic and Exponential Functions

In most applications involving logarithmic or exponential functions, the number e is the preferred base. However, in some situations it is convenient to use a base other than e. Derivatives of $y = \log_b x$ and $y = b^x$ can be obtained by expressing these functions in terms of the natural logarithmic and exponential functions.

We begin by finding a relationship between $\log_b x$ and $\ln x$ for any base b such that $b > 0$ and $b \neq 1$.

$$y = \log_b x \qquad \textit{Change to exponential form.}$$
$$b^y = x \qquad \textit{Take the natural logarithm of both sides.}$$
$$\ln b^y = \ln x \qquad \textit{Recall that } \ln b^y = y \ln b.$$
$$y \ln b = \ln x \qquad \textit{Solve for y.}$$
$$y = \frac{1}{\ln b} \ln x$$

Therefore,

$$\log_b x = \frac{1}{\ln b} \ln x \qquad \textit{Change-of-base formula for logarithms}^* \qquad (2)$$

Similarly, we can find a relationship between b^x and e^x for any base b such that $b > 0, b \neq 1$.

$$y = b^x \qquad \textit{Take the natural logarithm of both sides.}$$
$$\ln y = \ln b^x \qquad \textit{Recall that } \ln b^x = x \ln b.$$
$$\ln y = x \ln b \qquad \textit{Take the exponential function of both sides.}$$
$$y = e^{x \ln b}$$

Therefore,

$$b^x = e^{x \ln b} \qquad \textit{Change-of-base formula for exponential functions} \qquad (3)$$

Differentiating both sides of equation (2) gives

$$\frac{d}{dx} \log_b x = \frac{1}{\ln b} \frac{d}{dx} \ln x = \frac{1}{\ln b} \left(\frac{1}{x} \right)$$

It can be shown that the derivative of the function e^{cx}, where c is a constant, is the function ce^{cx} (see Problems 49–50 in Exercise 4-2 or the more general results of Section 4–4). Therefore, differentiating both sides of equation (3), we have

$$\frac{d}{dx} b^x = e^{x \ln b} \ln b = b^x \ln b$$

For convenience, we list the derivative formulas for exponential and logarithmic functions:

Derivatives of Exponential and Logarithmic Functions

For $b > 0, b \neq 1$,

$$\frac{d}{dx} e^x = e^x \qquad\qquad \frac{d}{dx} b^x = b^x \ln b$$

$$\frac{d}{dx} \ln x = \frac{1}{x} \qquad \frac{d}{dx} \log_b x = \frac{1}{\ln b} \left(\frac{1}{x} \right)$$

---

*Equation (2) is a special case of the **general change-of-base formula** for logarithms (which can be derived in the same way): $\log_b x = (\log_a x)/(\log_a b)$.

EXAMPLE 3 **Finding Derivatives** Find $g'(x)$ for

(A) $g(x) = 2^x - 3^x$ (B) $g(x) = \log_4 x^5$

SOLUTIONS (A) $g'(x) = 2^x \ln 2 - 3^x \ln 3$

(B) First, use a property of logarithms to rewrite $g(x)$.

$$g(x) = \log_4 x^5 \qquad \text{Use } \log_b M^p = p \log_b M.$$

$$g(x) = 5 \log_4 x \qquad \text{Take the derivative of both sides.}$$

$$g'(x) = \frac{5}{\ln 4}\left(\frac{1}{x}\right)$$

Matched Problem 3 Find $g'(x)$ for

(A) $g(x) = x^{10} + 10^x$ (B) $g(x) = \log_2 x - 6 \log_5 x$

EXPLORE & DISCUSS 2

(A) The graphs of $f(x) = \log_2 x$ and $g(x) = \log_4 x$ are shown in Figure 2. Which graph belongs to which function?

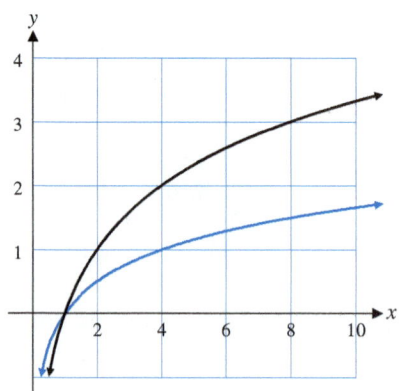

Figure 2

(B) Sketch graphs of $f'(x)$ and $g'(x)$.

(C) The function $f(x)$ is related to $g(x)$ in the same way that $f'(x)$ is related to $g'(x)$. What is that relationship?

Exponential and Logarithmic Models

EXAMPLE 4 **Price–Demand Model** An Internet store sells Australian wool blankets. If the store sells x blankets at a price of $\$p$ per blanket, then the price–demand equation is $p = 350(0.999)^x$. Find the rate of change of price with respect to demand when the demand is 800 blankets and interpret the result.

SOLUTION $\dfrac{dp}{dx} = 350(0.999)^x \ln 0.999$

If $x = 800$, then

$$\frac{dp}{dx} = 350(0.999)^{800} \ln 0.999 \approx -0.157, \text{ or } -\$0.16$$

When the demand is 800 blankets, the price is decreasing by $\$0.16$ per blanket.

Matched Problem 4   The store in Example 4 also sells a reversible fleece blanket. If the price–demand equation for reversible fleece blankets is $p = 200(0.998)^x$, find the rate of change of price with respect to demand when the demand is 400 blankets and interpret the result.

EXAMPLE 5   **Cable TV Subscribers**   A statistician used data from the U.S. Census Bureau to construct the model

$$S(t) = 21 \ln t + 2$$

where $S(t)$ is the number of cable TV subscribers (in millions) in year t ($t = 0$ corresponds to 1980). Use this model to estimate the number of cable TV subscribers in 2015 and the rate of change of the number of subscribers in 2015. Round both to the nearest tenth of a million. Interpret these results.

SOLUTION   Since 2015 corresponds to $t = 35$, we must find $S(35)$ and $S'(35)$.

$$S(35) = 21 \ln 35 + 2 = 76.7 \text{ million}$$

$$S'(t) = 21\frac{1}{t} = \frac{21}{t}$$

$$S'(35) = \frac{21}{35} = 0.6 \text{ million}$$

In 2015 there will be approximately 76.7 million subscribers, and this number is growing at the rate of 0.6 million subscribers per year.

Matched Problem 5   A model for newspaper circulation is

$$C(t) = 83 - 9 \ln t$$

where $C(t)$ is newspaper circulation (in millions) in year t ($t = 0$ corresponds to 1980). Use this model to estimate the circulation and the rate of change of circulation in 2015. Round both to the nearest tenth of a million. Interpret these results.

CONCEPTUAL INSIGHT

On most graphing calculators, exponential regression produces a function of the form $y = a \cdot b^x$. Formula (3) on page 221 allows you to change the base b (chosen by the graphing calculator) to the more familiar base e:

$$y = a \cdot b^x = a \cdot e^{x \ln b}$$

On most graphing calculators, logarithmic regression produces a function of the form $y = a + b \ln x$. Formula (2) on page 221 allows you to write the function in terms of logarithms to any base d that you may prefer:

$$y = a + b \ln x = a + b(\ln d) \log_d x$$

Exercises 4-2

A

In Problems 1–14, find $f'(x)$.

1. $f(x) = 5e^x + 3x + 1$

2. $f(x) = -7e^x - 2x + 5$

3. $f(x) = -2 \ln x + x^2 - 4$

4. $f(x) = 6 \ln x - x^3 + 2$

5. $f(x) = x^3 - 6e^x$

6. $f(x) = 9e^x + 2x^2$

7. $f(x) = e^x + x - \ln x$

8. $f(x) = \ln x + 2e^x - 3x^2$

9. $f(x) = \ln x^3$

10. $f(x) = \ln x^8$

11. $f(x) = 5x - \ln x^5$

12. $f(x) = 4 + \ln x^9$

13. $f(x) = \ln x^2 + 4e^x$

14. $f(x) = \ln x^{10} + 2 \ln x$

B

In Problems 15–22, find the equation of the line tangent to the graph of f at the indicated value of x.

15. $f(x) = 3 + \ln x; x = 1$

16. $f(x) = 2 \ln x; x = 1$

17. $f(x) = 3e^x; x = 0$

18. $f(x) = e^x + 1; x = 0$

19. $f(x) = \ln x^3; x = e$

20. $f(x) = 1 + \ln x^4; x = e$

21. $f(x) = 2 + e^x; x = 1$

22. $f(x) = 5e^x; x = 1$

23. A student claims that the line tangent to the graph of $f(x) = e^x$ at $x = 3$ passes through the point $(2, 0)$ (see the figure). Is she correct? Will the line tangent at $x = 4$ pass through $(3, 0)$? Explain.

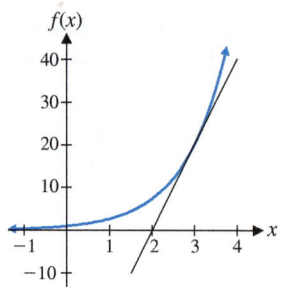

Figure for 23

24. Refer to Problem 23. Does the line tangent to the graph of $f(x) = e^x$ at $x = 1$ pass through the origin? Are there any other lines tangent to the graph of f that pass through the origin? Explain.

25. A student claims that the line tangent to the graph of $g(x) = \ln x$ at $x = 3$ passes through the origin (see the figure). Is he correct? Will the line tangent at $x = 4$ pass through the origin? Explain.

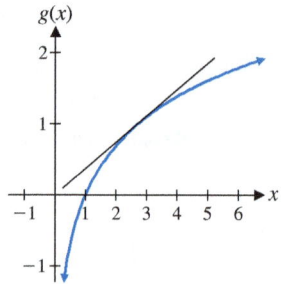

Figure for 25

26. Refer to Problem 25. Does the line tangent to the graph of $f(x) = \ln x$ at $x = e$ pass through the origin? Are there any other lines tangent to the graph of f that pass through the origin? Explain.

In Problems 27–30, first use appropriate properties of logarithms to rewrite f(x), and then find $f'(x)$.

27. $f(x) = 10x + \ln 10x$

28. $f(x) = 2 + 3 \ln \dfrac{1}{x}$

29. $f(x) = \ln \dfrac{4}{x^3}$

30. $f(x) = x + 5 \ln 6x$

C

In Problems 31–42, find $\dfrac{dy}{dx}$ for the indicated function y.

31. $y = \log_2 x$

32. $y = 3 \log_5 x$

33. $y = 3^x$

34. $y = 4^x$

35. $y = 2x - \log x$

36. $y = \log x + 4x^2 + 1$

37. $y = 10 + x + 10^x$

38. $y = x^5 - 5^x$

39. $y = 3 \ln x + 2 \log_3 x$

40. $y = - \log_2 x + 10 \ln x$

41. $y = 2^x + e^2$

42. $y = e^3 - 3^x$

In Problems 43–48, use graphical approximation methods to find the points of intersection of f(x) and g(x) (to two decimal places).

43. $f(x) = e^x; g(x) = x^4$

[Note that there are three points of intersection and that e^x is greater than x^4 for large values of x.]

44. $f(x) = e^x; g(x) = x^5$

[Note that there are two points of intersection and that e^x is greater than x^5 for large values of x.]

45. $f(x) = (\ln x)^2$; $g(x) = x$

46. $f(x) = (\ln x)^3$; $g(x) = \sqrt{x}$

47. $f(x) = \ln x$; $g(x) = x^{1/5}$

48. $f(x) = \ln x$; $g(x) = x^{1/4}$

49. Explain why $\lim\limits_{h \to 0} \dfrac{e^{ch} - 1}{h} = c$.

50. Use the result of Problem 49 and the four-step process to show that if $f(x) = e^{cx}$, then $f'(x) = ce^{cx}$.

Applications

51. Salvage value. The estimated salvage value S (in dollars) of a company airplane after t years is given by

$$S(t) = 300,000(0.9)^t$$

What is the rate of depreciation (in dollars per year) after 1 year? 5 years? 10 years?

52. Resale value. The estimated resale value R (in dollars) of a company car after t years is given by

$$R(t) = 20,000(0.86)^t$$

What is the rate of depreciation (in dollars per year) after 1 year? 2 years? 3 years?

53. Bacterial growth. A single cholera bacterium divides every 0.5 hour to produce two complete cholera bacteria. If we start with a colony of 5,000 bacteria, then after t hours, there will be

$$A(t) = 5,000 \cdot 2^{2t} = 5,000 \cdot 4^t$$

bacteria. Find $A'(t)$, $A'(1)$, and $A'(5)$, and interpret the results.

54. Bacterial growth. Repeat Problem 53 for a starting colony of 1,000 bacteria such that a single bacterium divides every 0.25 hour.

55. Blood pressure. An experiment was set up to find a relationship between weight and systolic blood pressure in children. Using hospital records for 5,000 children, the experimenters found that the systolic blood pressure was given approximately by

$$P(x) = 17.5(1 + \ln x) \qquad 10 \le x \le 100$$

where $P(x)$ is measured in millimeters of mercury and x is measured in pounds. What is the rate of change of blood pressure with respect to weight at the 40-pound weight level? At the 90-pound weight level?

56. Blood pressure. Refer to Problem 55. Find the weight (to the nearest pound) at which the rate of change of blood pressure with respect to weight is 0.3 millimeter of mercury per pound.

57. Psychology: stimulus/response. In psychology, the Weber–Fechner law for the response to a stimulus is

$$R = k \ln \frac{S}{S_0}$$

where R is the response, S is the stimulus, and S_0 is the lowest level of stimulus that can be detected. Find dR/dS.

58. Psychology: learning. A mathematical model for the average of a group of people learning to type is given by

$$N(t) = 10 + 6 \ln t \qquad t \ge 1$$

where $N(t)$ is the number of words per minute typed after t hours of instruction and practice (2 hours per day, 5 days per week). What is the rate of learning after 10 hours of instruction and practice? After 100 hours?

Answers to Matched Problems

1. (A) $4e^x + 16x + 7$
   (B) $7x^6 - 5x^4 - 1 + e^x$
2. (A) $30x^2 - \dfrac{100}{x}$
   (B) $\dfrac{5}{x} + e^x$
3. (A) $10x^9 + 10^x \ln 10$
   (B) $\left(\dfrac{1}{\ln 2} - \dfrac{6}{\ln 5} \right) \dfrac{1}{x}$
4. The price is decreasing at the rate of $0.18 per blanket.
5. The circulation in 2015 is approximately 51.0 million and is decreasing at the rate of 0.3 million per year.

4-3   Derivatives of Products and Quotients

- Derivatives of Products
- Derivatives of Quotients

The derivative properties discussed in Section 3-5 add substantially to our ability to compute and apply derivatives to many practical problems. In this and the next two sections, we add a few more properties that will increase this ability even further.

Derivatives of Products

In Section 3-5, we found that the derivative of a sum is the sum of the derivatives. Is the derivative of a product the product of the derivatives?

Let $F(x) = x^2$, $S(x) = x^3$, and $f(x) = F(x)S(x) = x^5$. Which of the following is $f'(x)$?

(A) $F'(x)S'(x)$ (B) $F(x)S'(x)$

(C) $F'(x)S(x)$ (D) $F(x)S'(x) + F'(x)S(x)$

Comparing the various expressions computed in Explore & Discuss 1, we see that the derivative of a product is not the product of the derivatives.

Using the definition of the derivative and the four-step process, we can show that

The derivative of the product of two functions is the first function times the derivative of the second function, plus the second function times the derivative of the first function.

THEOREM 1   Product Rule

If
$$y = f(x) = F(x)S(x)$$
and if $F'(x)$ and $S'(x)$ exist, then
$$f'(x) = F(x)S'(x) + S(x)F'(x)$$
Using simplified notation,
$$y' = FS' + SF' \qquad \text{or} \qquad \frac{dy}{dx} = F\frac{dS}{dx} + S\frac{dF}{dx}$$

EXAMPLE 1   **Differentiating a Product** Use two different methods to find $f'(x)$ for
$$f(x) = 2x^2(3x^4 - 2).$$

SOLUTION   **Method 1.** Use the product rule:

$$f'(x) = 2x^2(3x^4 - 2)' + (3x^4 - 2)(2x^2)' \qquad \text{\textit{First times derivative of}}$$
$$= 2x^2(12x^3) + (3x^4 - 2)(4x) \qquad \text{\textit{second, plus second times}}$$
$$= 24x^5 + 12x^5 - 8x \qquad \text{\textit{derivative of first}}$$
$$= 36x^5 - 8x$$

Method 2. Multiply first; then take derivatives:
$$f(x) = 2x^2(3x^4 - 2) = 6x^6 - 4x^2$$
$$f'(x) = 36x^5 - 8x$$

Matched Problem 1   Use two different methods to find $f'(x)$ for $f(x) = 3x^3(2x^2 - 3x + 1)$.

Some products we encounter can be differentiated by either method illustrated in Example 1. In other situations, the product rule *must* be used. Unless instructed otherwise, you should use the product rule to differentiate all products in this section in order to gain experience with this important differentiation rule.

EXAMPLE 2   **Tangent Lines** Let $f(x) = (2x - 9)(x^2 + 6)$.

(A) Find the equation of the line tangent to the graph of $f(x)$ at $x = 3$.

(B) Find the value(s) of x where the tangent line is horizontal.

SOLUTION   (A) First, find $f'(x)$:

$$f'(x) = (2x - 9)(x^2 + 6)' + (x^2 + 6)(2x - 9)'$$
$$= (2x - 9)(2x) + (x^2 + 6)(2)$$

Then, find $f(3)$ and $f'(3)$:

$$f(3) = [2(3) - 9](3^2 + 6) = (-3)(15) = -45$$
$$f'(3) = [2(3) - 9]2(3) + (3^2 + 6)(2) = -18 + 30 = 12$$

Now, find the equation of the tangent line at $x = 3$:

$$y - y_1 = m(x - x_1) \qquad y_1 = f(x_1) = f(3) = -45$$
$$y - (-45) = 12(x - 3) \qquad m = f'(x_1) = f'(3) = 12$$
$$y = 12x - 81 \qquad \text{Tangent line at } x = 3$$

(B) The tangent line is horizontal at any value of x such that $f'(x) = 0$, so

$$f'(x) = (2x - 9)2x + (x^2 + 6)2 = 0$$
$$6x^2 - 18x + 12 = 0$$
$$x^2 - 3x + 2 = 0$$
$$(x - 1)(x - 2) = 0$$
$$x = 1, 2$$

The tangent line is horizontal at $x = 1$ and at $x = 2$.

Matched Problem 2   Repeat Example 2 for $f(x) = (2x + 9)(x^2 - 12)$.

CONCEPTUAL INSIGHT

As Example 2 illustrates, the way we write $f'(x)$ depends on what we want to do. If we are interested only in evaluating $f'(x)$ at specified values of x, then the form in part (A) is sufficient. However, if we want to solve $f'(x) = 0$, we must multiply and collect like terms, as we did in part (B).

EXAMPLE 3   **Finding Derivatives** Find $f'(x)$ for

(A) $f(x) = 2x^3 e^x$ (B) $f(x) = 6x^4 \ln x$

SOLUTIONS   (A) $f'(x) = 2x^3(e^x)' + e^x(2x^3)'$ (B) $f'(x) = 6x^4 (\ln x)' + (\ln x) (6x^4)'$

$$= 2x^3 e^x + e^x(6x^2) \qquad\qquad = 6x^4 \frac{1}{x} + (\ln x)(24x^3)$$
$$= 2x^2 e^x(x + 3) \qquad\qquad = 6x^3 + 24x^3 \ln x$$
$$\qquad\qquad\qquad\qquad\qquad\qquad = 6x^3 (1 + 4 \ln x)$$

Matched Problem 3   Find $f'(x)$ for

(A) $f(x) = 5x^8 e^x$ (B) $f(x) = x^7 \ln x$

Derivatives of Quotients

The derivative of a quotient of two functions is not the quotient of the derivatives of the two functions.

Let $T(x) = x^5$, $B(x) = x^2$, and

$$f(x) = \frac{T(x)}{B(x)} = \frac{x^5}{x^2} = x^3$$

Which of the following is $f'(x)$?

(A) $\dfrac{T'(x)}{B'(x)}$ (B) $\dfrac{T'(x)B(x)}{[B(x)]^2}$ (C) $\dfrac{T(x)B'(x)}{[B(x)]^2}$

(D) $\dfrac{T'(x)B(x)}{[B(x)]^2} - \dfrac{T(x)B'(x)}{[B(x)]^2} = \dfrac{T'(x)B(x) - T(x)B'(x)}{[B(x)]^2}$

The expressions in Explore & Discuss 2 suggest that the derivative of a quotient leads to a more complicated quotient than expected.

If $T(x)$ and $B(x)$ are any two differentiable functions and

$$f(x) = \frac{T(x)}{B(x)}$$

then

$$f'(x) = \frac{B(x)T'(x) - T(x)B'(x)}{[B(x)]^2}$$

Therefore,

The derivative of the quotient of two functions is the denominator function times the derivative of the numerator function, minus the numerator function times the derivative of the denominator function, divided by the denominator function squared.

THEOREM 2 Quotient Rule

If

$$y = f(x) = \frac{T(x)}{B(x)}$$

and if $T'(x)$ and $B'(x)$ exist, then

$$f'(x) = \frac{B(x)T'(x) - T(x)B'(x)}{[B(x)]^2}$$

Using simplified notation,

$$y' = \frac{BT' - TB'}{B^2} \quad \text{or} \quad \frac{dy}{dx} = \frac{B\dfrac{dT}{dx} - T\dfrac{dB}{dx}}{B^2}$$

EXAMPLE 4 **Differentiating Quotients**

(A) If $f(x) = \dfrac{x^2}{2x - 1}$, find $f'(x)$.   (B) If $y = \dfrac{t^2 - t}{t^3 + 1}$, find y'.

(C) Find $\dfrac{d}{dx} \dfrac{x^2 - 3}{x^2}$ by using the quotient rule and also by splitting the fraction into two fractions.

SOLUTION (A) $f'(x) = \dfrac{(2x - 1)(x^2)' - x^2(2x - 1)'}{(2x - 1)^2}$

$= \dfrac{(2x - 1)(2x) - x^2(2)}{(2x - 1)^2}$

$= \dfrac{4x^2 - 2x - 2x^2}{(2x - 1)^2}$

$= \dfrac{2x^2 - 2x}{(2x - 1)^2}$

The denominator times the derivative of the numerator, minus the numerator times the derivative of the denominator, divided by the square of the denominator

(B) $y' = \dfrac{(t^3 + 1)(t^2 - t)' - (t^2 - t)(t^3 + 1)'}{(t^3 + 1)^2}$

$= \dfrac{(t^3 + 1)(2t - 1) - (t^2 - t)(3t^2)}{(t^3 + 1)^2}$

$= \dfrac{2t^4 - t^3 + 2t - 1 - 3t^4 + 3t^3}{(t^3 + 1)^2}$

$= \dfrac{-t^4 + 2t^3 + 2t - 1}{(t^3 + 1)^2}$

(C) **Method 1.** Use the quotient rule:

$$\dfrac{d}{dx} \dfrac{x^2 - 3}{x^2} = \dfrac{x^2 \dfrac{d}{dx}(x^2 - 3) - (x^2 - 3)\dfrac{d}{dx}x^2}{(x^2)^2}$$

$$= \dfrac{x^2(2x) - (x^2 - 3)2x}{x^4}$$

$$= \dfrac{2x^3 - 2x^3 + 6x}{x^4} = \dfrac{6x}{x^4} = \dfrac{6}{x^3}$$

Method 2. Split into two fractions:

$$\dfrac{x^2 - 3}{x^2} = \dfrac{x^2}{x^2} - \dfrac{3}{x^2} = 1 - 3x^{-2}$$

$$\dfrac{d}{dx}(1 - 3x^{-2}) = 0 - 3(-2)x^{-3} = \dfrac{6}{x^3}$$

Comparing methods 1 and 2, we see that it often pays to change an expression algebraically before choosing a differentiation formula.

Matched Problem 4 Find

(A) $f'(x)$ for $f(x) = \dfrac{2x}{x^2 + 3}$ (B) y' for $y = \dfrac{t^3 - 3t}{t^2 - 4}$

(C) $\dfrac{d}{dx} \dfrac{2 + x^3}{x^3}$ in two ways

EXAMPLE 5 **Finding Derivatives** Find $f'(x)$ for

(A) $f(x) = \dfrac{3e^x}{1 + e^x}$

(B) $f(x) = \dfrac{\ln x}{2x + 5}$

SOLUTIONS (A) $f'(x) = \dfrac{(1 + e^x)(3e^x)' - 3e^x(1 + e^x)'}{(1 + e^x)^2}$

$= \dfrac{(1 + e^x)3e^x - 3e^xe^x}{(1 + e^x)^2}$

$= \dfrac{3e^x}{(1 + e^x)^2}$

(B) $f'(x) = \dfrac{(2x + 5)(\ln x)' - (\ln x)(2x + 5)'}{(2x + 5)^2}$

$= \dfrac{(2x + 5) \cdot \dfrac{1}{x} - (\ln x)(2)}{(2x + 5)^2}$ Multiply by $\dfrac{x}{x}$

$= \dfrac{2x + 5 - 2x \ln x}{x(2x + 5)^2}$

Matched Problem 5 Find $f'(x)$ for

(A) $f(x) = \dfrac{x^3}{e^x + 2}$

(B) $f(x) = \dfrac{4x}{1 + \ln x}$

EXAMPLE 6 **Sales Analysis** The total sales S (in thousands of games) of a video game t months after the game is introduced are given by

$$S(t) = \frac{125t^2}{t^2 + 100}$$

(A) Find $S'(t)$.

(B) Find $S(10)$ and $S'(10)$. Write a brief interpretation of these results.

(C) Use the results from part (B) to estimate the total sales after 11 months.

SOLUTION (A) $S'(t) = \dfrac{(t^2 + 100)(125t^2)' - 125t^2(t^2 + 100)'}{(t^2 + 100)^2}$

$= \dfrac{(t^2 + 100)(250t) - 125t^2(2t)}{(t^2 + 100)^2}$

$= \dfrac{250t^3 + 25,000t - 250t^3}{(t^2 + 100)^2}$

$= \dfrac{25,000t}{(t^2 + 100)^2}$

(B) $S(10) = \dfrac{125(10)^2}{10^2 + 100} = 62.5$ and $S'(10) = \dfrac{25,000(10)}{(10^2 + 100)^2} = 6.25.$

Total sales after 10 months are 62,500 games, and sales are increasing at the rate of 6,250 games per month.

(C) Total sales will increase by approximately 6,250 games during the next month, so the estimated total sales after 11 months are $62,500 + 6,250 = 68,750$ games.

Matched Problem 6   Refer to Example 6. Suppose that the total sales S (in thousands of games) t months after the game is introduced are given by

$$S(t) = \frac{150t}{t + 3}$$

(A) Find $S'(t)$.

(B) Find $S(12)$ and $S'(12)$. Write a brief interpretation of these results.

(C) Use the results from part (B) to estimate the total sales after 13 months.

Exercises 4-3

Answers to most of the problems in this exercise set contain both an unsimplified form and a simplified form of the derivative. When checking your work, first check that you applied the rules correctly and then check that you performed the algebraic simplification correctly. Unless instructed otherwise, when differentiating a product, use the product rule rather than performing the multiplication first.

A

In Problems 1–26, find $f'(x)$ and simplify.

1. $f(x) = 2x^3(x^2 - 2)$ **2.** $f(x) = 5x^2(x^3 + 2)$

3. $f(x) = (x - 3)(2x - 1)$

4. $f(x) = (3x + 2)(4x - 5)$

5. $f(x) = \dfrac{x}{x - 3}$

6. $f(x) = \dfrac{3x}{2x + 1}$ **7.** $f(x) = \dfrac{2x + 3}{x - 2}$

8. $f(x) = \dfrac{3x - 4}{2x + 3}$ **9.** $f(x) = 3xe^x$

10. $f(x) = x^2 e^x$ **11.** $f(x) = x^3 \ln x$

12. $f(x) = 5x \ln x$

13. $f(x) = (x^2 + 1)(2x - 3)$

14. $f(x) = (3x + 5)(x^2 - 3)$

15. $f(x) = (0.4x + 2)(0.5x - 5)$

16. $f(x) = (0.5x - 4)(0.2x + 1)$

17. $f(x) = \dfrac{x^2 + 1}{2x - 3}$ **18.** $f(x) = \dfrac{3x + 5}{x^2 - 3}$

19. $f(x) = (x^2 + 2)(x^2 - 3)$

20. $f(x) = (x^2 - 4)(x^2 + 5)$

21. $f(x) = \dfrac{x^2 + 2}{x^2 - 3}$ **22.** $f(x) = \dfrac{x^2 - 4}{x^2 + 5}$

23. $f(x) = \dfrac{e^x}{x^2 + 1}$ **24.** $f(x) = \dfrac{1 - e^x}{1 + e^x}$

25. $f(x) = \dfrac{\ln x}{1 + x}$ **26.** $f(x) = \dfrac{2x}{1 + \ln x}$

In Problems 27–38, find $h'(x)$, where $f(x)$ is an unspecified differentiable function.

27. $h(x) = xf(x)$ **28.** $h(x) = x^2 f(x)$

29. $h(x) = x^3 f(x)$ **30.** $h(x) = \dfrac{f(x)}{x}$

31. $h(x) = \dfrac{f(x)}{x^2}$ **32.** $h(x) = \dfrac{f(x)}{x^3}$

33. $h(x) = \dfrac{x}{f(x)}$ **34.** $h(x) = \dfrac{x^2}{f(x)}$

35. $h(x) = e^x f(x)$ **36.** $h(x) = \dfrac{e^x}{f(x)}$

37. $h(x) = \dfrac{\ln x}{f(x)}$ **38.** $h(x) = \dfrac{f(x)}{\ln x}$

B

In Problems 39–48, find the indicated derivatives and simplify.

39. $f'(x)$ for $f(x) = (2x + 1)(x^2 - 3x)$

40. y' for $y = (x^3 + 2x^2)(3x - 1)$

41. $\dfrac{dy}{dt}$ for $y = (2.5t - t^2)(4t + 1.4)$

42. $\dfrac{d}{dt}[(3 - 0.4t^3)(0.5t^2 - 2t)]$

43. y' for $y = \dfrac{5x - 3}{x^2 + 2x}$

44. $f'(x)$ for $f(x) = \dfrac{3x^2}{2x - 1}$

45. $\dfrac{d}{dw}\dfrac{w^2 - 3w + 1}{w^2 - 1}$

46. $\dfrac{dy}{dw}$ for $y = \dfrac{w^4 - w^3}{3w - 1}$

47. y' for $y = (1 + x - x^2)e^x$

48. $\dfrac{dy}{dt}$ for $y = (1 + e^t)\ln t$

In Problems 49–54, find $f'(x)$ and find the equation of the line tangent to the graph of f at $x = 2$.

49. $f(x) = (1 + 3x)(5 - 2x)$

50. $f(x) = (7 - 3x)(1 + 2x)$

51. $f(x) = \dfrac{x - 8}{3x - 4}$

52. $f(x) = \dfrac{2x - 5}{2x - 3}$

53. $f(x) = \dfrac{x}{2^x}$

54. $f(x) = (x - 2)\ln x$

In Problems 55–58, find $f'(x)$ and find the value(s) of x where $f'(x) = 0$.

55. $f(x) = (2x - 15)(x^2 + 18)$

56. $f(x) = (2x - 3)(x^2 - 6)$

57. $f(x) = \dfrac{x}{x^2 + 1}$

58. $f(x) = \dfrac{x}{x^2 + 9}$

In Problems 59–62, find $f'(x)$ in two ways: (1) using the product or quotient rule and (2) simplifying first.

59. $f(x) = x^3(x^4 - 1)$

60. $f(x) = x^4(x^3 - 1)$

61. $f(x) = \dfrac{x^3 + 9}{x^3}$

62. $f(x) = \dfrac{x^4 + 4}{x^4}$

C

In Problems 63–82, find each indicated derivative and simplify.

63. $f(w) = (w + 1)2^w$

64. $g(w) = (w - 5)\log_3 w$

65. $\dfrac{d}{dx}\dfrac{3x^2 - 2x + 3}{4x^2 + 5x - 1}$

66. y' for $y = \dfrac{x^3 - 3x + 4}{2x^2 + 3x - 2}$

67. $\dfrac{dy}{dx}$ for $y = 9x^{1/3}(x^3 + 5)$

68. $\dfrac{d}{dx}[(4x^{1/2} - 1)(3x^{1/3} + 2)]$

69. y' for $y = \dfrac{\log_2 x}{1 + x^2}$

70. $\dfrac{dy}{dx}$ for $y = \dfrac{10^x}{1 + x^4}$

71. $f'(x)$ for $f(x) = \dfrac{6\sqrt[3]{x}}{x^2 - 3}$

72. y' for $y = \dfrac{2\sqrt{x}}{x^2 - 3x + 1}$

73. $g'(t)$ if $g(t) = \dfrac{0.2t}{3t^2 - 1}$

74. $h'(t)$ if $h(t) = \dfrac{-0.05t^2}{2t + 1}$

75. $\dfrac{d}{dx}[4x\log x^5]$

76. $\dfrac{d}{dt}[10^t \log t]$

77. $\dfrac{d}{dx}\dfrac{x^3 - 2x^2}{\sqrt[3]{x^2}}$

78. $\dfrac{dy}{dx}$ for $y = \dfrac{x^2 - 3x + 1}{\sqrt[4]{x}}$

79. $f'(x)$ for $f(x) = \dfrac{(2x^2 - 1)(x^2 + 3)}{x^2 + 1}$

80. y' for $y = \dfrac{2x - 1}{(x^3 + 2)(x^2 - 3)}$

81. $\dfrac{dy}{dt}$ for $y = \dfrac{t\ln t}{e^t}$

82. $\dfrac{dy}{du}$ for $y = \dfrac{u^2 e^u}{1 + \ln u}$

Applications

83. Sales analysis. The total sales S (in thousands of DVDs) of a DVD are given by

$$S(t) = \dfrac{90t^2}{t^2 + 50}$$

where t is the number of months since the release of the DVD.

(A) Find $S'(t)$.

(B) Find $S(10)$ and $S'(10)$. Write a brief interpretation of these results.

(C) Use the results from part (B) to estimate the total sales after 11 months.

84. Sales analysis. A communications company has installed a new cable television system in a city. The total number N (in thousands) of subscribers t months after the installation of the system is given by

$$N(t) = \dfrac{180t}{t + 4}$$

(A) Find $N'(t)$.

(B) Find $N(16)$ and $N'(16)$. Write a brief interpretation of these results.

(C) Use the results from part (B) to estimate the total number of subscribers after 17 months.

85. Price–demand equation. According to economic theory, the demand x for a quantity in a free market decreases as the price p increases (see the figure). Suppose that the number x of DVD players people are willing to buy per week from a retail chain at a price of \$$p$ is given by

$$x = \dfrac{4,000}{0.1p + 1} \qquad 10 \le p \le 70$$

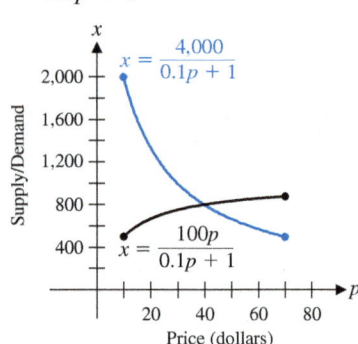

Figure for 85 and 86

(A) Find dx/dp.

(B) Find the demand and the instantaneous rate of change of demand with respect to price when the price is \$40. Write a brief interpretation of these results.

(C) Use the results from part (B) to estimate the demand if the price is increased to \$41.

86. Price–supply equation. According to economic theory, the supply x of a quantity in a free market increases as the price p increases (see the figure). Suppose that the number x of DVD players a retail chain is willing to sell per week at a price of $\$p$ is given by

$$x = \frac{100p}{0.1p + 1} \qquad 10 \leq p \leq 70$$

(A) Find dx/dp.

(B) Find the supply and the instantaneous rate of change of supply with respect to price when the price is $40. Write a brief verbal interpretation of these results.

(C) Use the results from part (B) to estimate the supply if the price is increased to $41.

87. Medicine. A drug is injected into a patient's bloodstream through her right arm. The drug concentration (in milligrams per cubic centimeter) in the bloodstream of the left arm t hours after the injection is given by

$$C(t) = \frac{0.14t}{t^2 + 1}$$

(A) Find $C'(t)$.

(B) Find $C'(0.5)$ and $C'(3)$, and interpret the results.

88. Drug sensitivity. One hour after a dose of x milligrams of a particular drug is administered to a person, the change in body temperature $T(x)$, in degrees Fahrenheit, is given approximately by

$$T(x) = x^2 \left(1 - \frac{x}{9}\right) \qquad 0 \leq x \leq 7$$

The rate $T'(x)$ at which T changes with respect to the size of the dosage x is called the *sensitivity* of the body to the dosage.

(A) Use the product rule to find $T'(x)$.

(B) Find $T'(1)$, $T'(3)$, and $T'(6)$.

89. Learning. In the early twentieth century, L. L. Thurstone found that a given person successfully accomplished $N(x)$ acts after x practice acts, as given by

$$N(x) = \frac{100x + 200}{x + 32}$$

(A) Find the instantaneous rate of change of learning, $N'(x)$, with respect to the number of practice acts, x.

(B) Find $N'(4)$ and $N'(68)$.

Answers to Matched Problems

1. $30x^4 - 36x^3 + 9x^2$
2. (A) $y = 84x - 297$
   (B) $x = -4, x = 1$
3. (A) $5x^8 e^x + e^x(40x^7) = 5x^7(x + 8)e^x$
   (B) $x^7 \cdot \dfrac{1}{x} + \ln x\,(7x^6) = x^6\,(1 + 7\ln x)$
4. (A) $\dfrac{(x^2 + 3)2 - (2x)(2x)}{(x^2 + 3)^2} = \dfrac{6 - 2x^2}{(x^2 + 3)^2}$
   (B) $\dfrac{(t^2 - 4)(3t^2 - 3) - (t^3 - 3t)(2t)}{(t^2 - 4)^2} = \dfrac{t^4 - 9t^2 + 12}{(t^2 - 4)^2}$
   (C) $-\dfrac{6}{x^4}$
5. (A) $\dfrac{(e^x + 2)\,3x^2 - x^3 e^x}{(e^x + 2)^2}$
   (B) $\dfrac{(1 + \ln x)\,4 - 4x\,\dfrac{1}{x}}{(1 + \ln x)^2} = \dfrac{4\ln x}{(1 + \ln x)^2}$
6. (A) $S'(t) = \dfrac{450}{(t + 3)^2}$
   (B) $S(12) = 120; S'(12) = 2$. After 12 months, the total sales are 120,000 games, and sales are increasing at the rate of 2,000 games per month.
   (C) 122,000 games

4-4 The Chain Rule

- Composite Functions
- General Power Rule
- The Chain Rule

The word *chain* in the name "chain rule" comes from the fact that a function formed by composition involves a chain of functions—that is, a function of a function. The *chain rule* enables us to compute the derivative of a composite function in terms of the derivatives of the functions making up the composition. In this section, we review composite functions, introduce the chain rule by means of a special case known as the *general power rule,* and then discuss the chain rule itself.

Composite Functions

The function $m(x) = (x^2 + 4)^3$ is a combination of a quadratic function and a cubic function. To see this more clearly, let

$$y = f(u) = u^3 \qquad \text{and} \qquad u = g(x) = x^2 + 4$$

We can express y as a function of x:

$$y = f(u) = f[g(x)] = [x^2 + 4]^3 = m(x)$$

The function m is the *composite* of the two functions f and g.

DEFINITION **Composite Functions**

A function m is a **composite** of functions f and g if

$$m(x) = f[g(x)]$$

The domain of m is the set of all numbers x such that x is in the domain of g, and $g(x)$ is in the domain of f.

EXAMPLE 1 **Composite Functions** Let $f(u) = e^u$ and $g(x) = -3x$. Find $f[g(x)]$ and $g[f(u)]$.

SOLUTION
$$f[g(x)] = f(-3x) = e^{-3x}$$
$$g[f(u)] = g(e^u) = -3e^u$$

Matched Problem 1 Let $f(u) = 2u$ and $g(x) = e^x$. Find $f[g(x)]$ and $g[f(u)]$.

EXAMPLE 2 **Composite Functions** Write each function as a composition of two simpler functions.

(A) $y = 100e^{0.04x}$ (B) $y = \sqrt{4 - x^2}$

SOLUTION (A) Let
$$y = f(u) = 100e^u$$
$$u = g(x) = 0.04x$$

Check: $y = f[g(x)] = 100e^{g(x)} = 100e^{0.04x}$

(B) Let
$$y = f(u) = \sqrt{u}$$
$$u = g(x) = 4 - x^2$$

Check: $y = f[g(x)] = \sqrt{g(x)} = \sqrt{4 - x^2}$

Matched Problem 2 Write each function as a composition of two simpler functions.

(A) $y = 50e^{-2x}$ (B) $y = \sqrt[3]{1 + x^3}$

CONCEPTUAL INSIGHT

There can be more than one way to express a function as a composition of simpler functions. Choosing $y = f(u) = 100u$ and $u = g(x) = e^{0.04x}$ in Example 2A produces the same result:

$$y = f[g(x)] = 100g(x) = 100e^{0.04x}$$

Since we will be using composition as a means to an end (finding a derivative), usually it will not matter what functions you choose for the composition.

General Power Rule

We have already made extensive use of the power rule,

$$\frac{d}{dx}x^n = nx^{n-1} \tag{1}$$

Now we can generalize this rule in order to differentiate composite functions of the form $[u(x)]^n$, where $u(x)$ is a differentiable function. Is rule (1) still valid if we replace x with a function $u(x)$?

EXPLORE & DISCUSS 1

Let $u(x) = 2x^2$ and $f(x) = [u(x)]^3 = 8x^6$. Which of the following is $f'(x)$?

(A) $3[u(x)]^2$ (B) $3[u'(x)]^2$ (C) $3[u(x)]^2u'(x)$

The calculations in Explore & Discuss 1 show that we cannot generalize the power rule simply by replacing x with $u(x)$ in equation (1).

How can we find a formula for the derivative of $[u(x)]^n$, where $u(x)$ is an arbitrary differentiable function? Let's begin by considering the derivatives of $[u(x)]^2$ and $[u(x)]^3$ to see if a general pattern emerges. Since $[u(x)]^2 = u(x)u(x)$, we use the product rule to write

$$\frac{d}{dx}[u(x)]^2 = \frac{d}{dx}[u(x)u(x)]$$

$$= u(x)u'(x) + u(x)u'(x)$$

$$= 2u(x)u'(x) \tag{2}$$

Because $[u(x)]^3 = [u(x)]^2u(x)$, we use the product rule and the result in equation (2) to write

$$\frac{d}{dx}[u(x)]^3 = \frac{d}{dx}\{[u(x)]^2u(x)\}$$ 　　　　　*Use equation (2) to substitute for*

$$= [u(x)]^2\frac{d}{dx}u(x) + u(x)\frac{d}{dx}[u(x)]^2$$ 　*$\frac{d}{dx}[u(x)]^2$.*

$$= [u(x)]^2u'(x) + u(x)[2u(x)u'(x)]$$

$$= 3[u(x)]^2u'(x)$$

Continuing in this fashion, we can show that

$$\frac{d}{dx}[u(x)]^n = n[u(x)]^{n-1}u'(x) \qquad n \text{ a positive integer} \tag{3}$$

Using more advanced techniques, we can establish formula (3) for all real numbers n, obtaining the **general power rule**.

THEOREM 1 General Power Rule

If $u(x)$ is a differentiable function, n is any real number, and

$$y = f(x) = [u(x)]^n$$

then

$$f'(x) = n[u(x)]^{n-1}u'(x)$$

Using simplified notation,

$$y' = nu^{n-1}u' \qquad \text{or} \qquad \frac{d}{dx}u^n = nu^{n-1}\frac{du}{dx} \qquad \text{where } u = u(x)$$

EXAMPLE 3 **Using the General Power Rule** Find the indicated derivatives:

(A) $f'(x)$ if $f(x) = (3x + 1)^4$ (B) y' if $y = (x^3 + 4)^7$

(C) $\dfrac{d}{dt} \dfrac{1}{(t^2 + t + 4)^3}$ (D) $\dfrac{dh}{dw}$ if $h(w) = \sqrt{3 - w}$

SOLUTION (A) $f(x) = (3x + 1)^4$ Let $u = 3x + 1, n = 4$.

$f'(x) = 4(3x + 1)^3(3x + 1)'$ $nu^{n-1}\dfrac{du}{dx}$

$\qquad = 4(3x + 1)^3\, 3$ $\dfrac{du}{dx} = 3$

$\qquad = 12(3x + 1)^3$

(B) $y = (x^3 + 4)^7$ Let $u = (x^3 + 4), n = 7$.

$y' = 7(x^3 + 4)^6(x^3 + 4)'$ $nu^{n-1}\dfrac{du}{dx}$

$\qquad = 7(x^3 + 4)^6\, 3x^2$ $\dfrac{du}{dx} = 3x^2$

$\qquad = 21x^2(x^3 + 4)^6$

(C) $\dfrac{d}{dt} \dfrac{1}{(t^2 + t + 4)^3}$

$\qquad = \dfrac{d}{dt}(t^2 + t + 4)^{-3}$ Let $u = t^2 + t + 4, n = -3$.

$\qquad = -3(t^2 + t + 4)^{-4}(t^2 + t + 4)'$ $nu^{n-1}\dfrac{du}{dt}$

$\qquad = -3(t^2 + t + 4)^{-4}(2t + 1)$ $\dfrac{du}{dt} = 2t + 1$

$\qquad = \dfrac{-3(2t + 1)}{(t^2 + t + 4)^4}$

(D) $h(w) = \sqrt{3 - w} = (3 - w)^{1/2}$ Let $u = 3 - w, n = \dfrac{1}{2}$.

$\dfrac{dh}{dw} = \dfrac{1}{2}(3 - w)^{-1/2}(3 - w)'$ $nu^{n-1}\dfrac{du}{dw}$

$\qquad = \dfrac{1}{2}(3 - w)^{-1/2}(-1)$ $\dfrac{du}{dw} = -1$

$\qquad = -\dfrac{1}{2(3 - w)^{1/2}}$ or $-\dfrac{1}{2\sqrt{3 - w}}$

Matched Problem 3 Find the indicated derivatives:

(A) $h'(x)$ if $h(x) = (5x + 2)^3$ (B) y' if $y = (x^4 - 5)^5$

(C) $\dfrac{d}{dt} \dfrac{1}{(t^2 + 4)^2}$ (D) $\dfrac{dg}{dw}$ if $g(w) = \sqrt{4 - w}$

Notice that we used two steps to differentiate each function in Example 3. First, we applied the general power rule, and then we found du/dx. As you gain experience with the general power rule, you may want to combine these two steps. If you do this, be certain to multiply by du/dx. For example,

$$\dfrac{d}{dx}(x^5 + 1)^4 = 4(x^5 + 1)^3 5x^4 \qquad \text{Correct}$$

$$\dfrac{d}{dx}(x^5 + 1)^4 \neq 4(x^5 + 1)^3 \qquad du/dx = 5x^4 \text{ is missing}$$

CONCEPTUAL INSIGHT

If we let $u(x) = x$, then $du/dx = 1$, and the general power rule reduces to the (ordinary) power rule discussed in Section 3-5. Compare the following:

$$\frac{d}{dx}x^n = nx^{n-1} \qquad \text{Yes—power rule}$$

$$\frac{d}{dx}u^n = nu^{n-1}\frac{du}{dx} \qquad \text{Yes—general power rule}$$

$$\frac{d}{dx}u^n \neq nu^{n-1} \qquad \text{Unless } u(x) = x + k, \text{ so that } du/dx = 1$$

The Chain Rule

We have used the general power rule to find derivatives of composite functions of the form $f(g(x))$, where $f(u) = u^n$ is a power function. But what if f is not a power function? Then a more general rule, the *chain rule,* enables us to compute the derivatives of many composite functions of the form $f(g(x))$.

Suppose that

$$y = m(x) = f[g(x)]$$

is a composite of f and g, where

$$y = f(u) \qquad \text{and} \qquad u = g(x)$$

To express the derivative dy/dx in terms of the derivatives of f and g, we use the definition of a derivative (see Section 3-4).

$$\frac{dy}{dx} = \lim_{h\to 0}\frac{m(x+h) - m(x)}{h} \qquad \begin{array}{l}\text{Substitute } m(x + h) = f[g(x + h)]\\ \text{and } m(x) = f[g(x)].\end{array}$$

$$= \lim_{h\to 0}\frac{f[g(x+h)] - f[g(x)]}{h} \qquad \text{Multiply by } 1 = \frac{g(x+h) - g(x)}{g(x+h) - g(x)}.$$

$$= \lim_{h\to 0}\left[\frac{f[g(x+h)] - f[g(x)]}{h} \cdot \frac{g(x+h) - g(x)}{g(x+h) - g(x)}\right]$$

$$= \lim_{h\to 0}\left[\frac{f[g(x+h)] - f[g(x)]}{g(x+h) - g(x)} \cdot \frac{g(x+h) - g(x)}{h}\right] \qquad (4)$$

We recognize the second factor in equation (4) as the difference quotient for $g(x)$. To interpret the first factor as the difference quotient for $f(u)$, we let $k = g(x + h) - g(x)$. Since $u = g(x)$, we write

$$u + k = g(x) + g(x + h) - g(x) = g(x + h)$$

Substituting in equation (1), we have

$$\frac{dy}{dx} = \lim_{h\to 0}\left[\frac{f(u + k) - f(u)}{k} \cdot \frac{g(x + h) - g(x)}{h}\right] \qquad (5)$$

If we assume that $k = [g(x + h) - g(x)] \to 0$ as $h \to 0$, we can find the limit of each difference quotient in equation (5):

$$\frac{dy}{dx} = \left[\lim_{k\to 0}\frac{f(u + k) - f(u)}{k}\right]\left[\lim_{h\to 0}\frac{g(x + h) - g(x)}{h}\right]$$

$$= f'(u)g'(x)$$

$$= \frac{dy}{du}\frac{du}{dx}$$

This result is correct under general conditions and is called the *chain rule,* but our "derivation" is superficial because it ignores some hidden problems. Since a formal proof of the chain rule is beyond the scope of this book, we simply state it as follows:

THEOREM 2 Chain Rule

If $y = f(u)$ and $u = g(x)$ define the composite function

$$y = m(x) = f[g(x)]$$

then

$$\frac{dy}{dx} = \frac{dy}{du}\frac{du}{dx} \qquad \text{provided that } \frac{dy}{du} \text{ and } \frac{du}{dx} \text{ exist}$$

or, equivalently,

$$m'(x) = f'[g(x)]g'(x) \qquad \text{provided that } f'[g(x)] \text{ and } g'(x) \text{ exist}$$

EXAMPLE 4 **Using the Chain Rule** Find dy/du, du/dx, and dy/dx (express dy/dx as a function of x) for

(A) $y = u^{3/2}$ and $u = 3x^2 + 1$
(B) $y = e^u$ and $u = 2x^3 + 5$
(C) $y = \ln u$ and $u = x^2 - 4x + 2$

SOLUTION (A) $\dfrac{dy}{du} = \dfrac{3}{2}u^{1/2}$ and $\dfrac{du}{dx} = 6x$ *Basic derivative rules*

$\dfrac{dy}{dx} = \dfrac{dy}{du}\dfrac{du}{dx}$ *Chain rule*

$= \dfrac{3}{2}u^{1/2}(6x) = 9x(3x^2 + 1)^{1/2}$ *Since $u = 3x^2 + 1$*

(B) $\dfrac{dy}{du} = e^u$ and $\dfrac{du}{dx} = 6x^2$ *Basic derivative rules*

$\dfrac{dy}{dx} = \dfrac{dy}{du}\dfrac{du}{dx}$ *Chain rule*

$= e^u(6x^2) = 6x^2 e^{2x^3+5}$ *Since $u = 2x^3 + 5$*

(C) $\dfrac{dy}{du} = \dfrac{1}{u}$ and $\dfrac{du}{dx} = 2x - 4$ *Basic derivative rules*

$\dfrac{dy}{dx} = \dfrac{dy}{du}\dfrac{du}{dx}$ *Chain rule*

$= \dfrac{1}{u}(2x - 4) = \dfrac{2x - 4}{x^2 - 4x + 2}$ *Since $u = x^2 - 4x + 2$*

Matched Problem 4 Find dy/du, du/dx, and dy/dx (express dy/dx as a function of x) for

(A) $y = u^{-5}$ and $u = 2x^3 + 4$
(B) $y = e^u$ and $u = 3x^4 + 6$
(C) $y = \ln u$ and $u = x^2 + 9x + 4$

EXPLORE & DISCUSS 2

Let $m(x) = f[g(x)]$. Use the chain rule and Figures 1 and 2 to find

(A) $f(4)$ (B) $g(6)$ (C) $m(6)$

(D) $f'(4)$ (E) $g'(6)$ (F) $m'(6)$

Figure 1

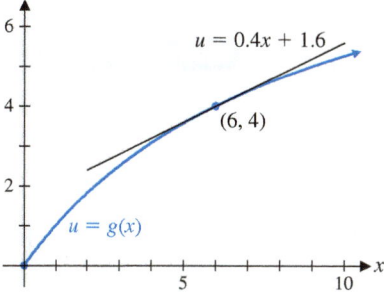

Figure 2

The chain rule can be extended to compositions of three or more functions. For example, if $y = f(w)$, $w = g(u)$, and $u = h(x)$, then

$$\frac{dy}{dx} = \frac{dy}{dw}\frac{dw}{du}\frac{du}{dx}$$

EXAMPLE 5   **Using the Chain Rule**   For $y = h(x) = e^{1+(\ln x)^2}$, find dy/dx.

SOLUTION   Note that h is of the form $y = e^w$, where $w = 1 + u^2$ and $u = \ln x$.

$$\frac{dy}{dx} = \frac{dy}{dw}\frac{dw}{du}\frac{du}{dx}$$

$$= e^w(2u)\left(\frac{1}{x}\right)$$

$$= e^{1+u^2}(2u)\left(\frac{1}{x}\right) \qquad \text{Since } w = 1 + u^2$$

$$= e^{1+(\ln x)^2}(2\ln x)\left(\frac{1}{x}\right) \qquad \text{Since } u = \ln x$$

$$= \frac{2}{x}(\ln x)e^{1+(\ln x)^2}$$

Matched Problem 5   For $y = h(x) = [\ln(1 + e^x)]^3$, find dy/dx.

The chain rule generalizes basic derivative rules. We list three general derivative rules here for convenient reference [the first, equation (6), is the general power rule of Theorem 1].

General Derivative Rules

$$\frac{d}{dx}[f(x)]^n = n[f(x)]^{n-1}f'(x) \tag{6}$$

$$\frac{d}{dx}\ln[f(x)] = \frac{1}{f(x)}f'(x) \tag{7}$$

$$\frac{d}{dx}e^{f(x)} = e^{f(x)}f'(x) \tag{8}$$

Unless directed otherwise, you now have a choice between the chain rule and the general derivative rules. However, practicing with the chain rule will help prepare you for concepts that appear later in the text. Examples 4 and 5 illustrate the chain rule method, and the next example illustrates the general derivative rules method.

EXAMPLE 6 **Using General Derivative Rules**

(A) $\dfrac{d}{dx}e^{2x} = e^{2x}\dfrac{d}{dx}2x$ *Using equation (5)*

$\qquad\qquad = e^{2x}(2) = 2e^{2x}$

(B) $\dfrac{d}{dx}\ln(x^2 + 9) = \dfrac{1}{x^2 + 9}\dfrac{d}{dx}(x^2 + 9)$ *Using equation (4)*

$\qquad\qquad = \dfrac{1}{x^2 + 9}2x = \dfrac{2x}{x^2 + 9}$

(C) $\dfrac{d}{dx}(1 + e^{x^2})^3 = 3(1 + e^{x^2})^2\dfrac{d}{dx}(1 + e^{x^2})$ *Using equation (3)*

$\qquad\qquad = 3(1 + e^{x^2})^2 e^{x^2}\dfrac{d}{dx}x^2$ *Using equation (5)*

$\qquad\qquad = 3(1 + e^{x^2})^2 e^{x^2}(2x)$

$\qquad\qquad = 6xe^{x^2}(1 + e^{x^2})^2$

Matched Problem 6 Find

(A) $\dfrac{d}{dx}\ln(x^3 + 2x)$ (B) $\dfrac{d}{dx}e^{3x^2+2}$ (C) $\dfrac{d}{dx}(2 + e^{-x^2})^4$

Exercises 4-4

For many of the problems in this exercise set, the answers in the back of the book include both an unsimplified form and a simplified form. When checking your work, first check that you applied the rules correctly, and then check that you performed the algebraic simplification correctly.

A

In Problems 1–4, find f[g(x)].

1. $f(u) = u^3;$ $g(x) = 3x^2 + 2$

2. $f(u) = u^4;$ $g(x) = 1 - 4x^3$

3. $f(u) = e^u;$ $g(x) = -x^2$

4. $g(u) = e^u;$ $g(x) = 3x^3$

In Problems 5–8, write each composite function in the form $y = f(u)$ and $u = g(x)$.

5. $y = (3x^2 - x + 5)^4$

6. $y = (2x^3 + x + 3)^5$

7. $y = e^{1+x+x^2}$

8. $y = e^{x^4+2x^2+5}$

In Problems 9–16, replace ? with an expression that will make the indicated equation valid.

9. $\dfrac{d}{dx}(3x + 4)^4 = 4(3x + 4)^3\underline{\quad?\quad}$

10. $\dfrac{d}{dx}(5 - 2x)^6 = 6(5 - 2x)^5\underline{\quad?\quad}$

11. $\dfrac{d}{dx}(4 - 2x^2)^3 = 3(4 - 2x^2)^2\underline{\quad?\quad}$

12. $\dfrac{d}{dx}(3x^2 + 7)^5 = 5(3x^2 + 7)^4\underline{\quad?\quad}$

13. $\dfrac{d}{dx}e^{x^2+1} = e^{x^2+1}\underline{\quad?\quad}$

14. $\dfrac{d}{dx}e^{4x-2} = e^{4x-2}\underline{\quad?\quad}$

15. $\dfrac{d}{dx}\ln(x^4 + 1) = \dfrac{1}{x^4 + 1}\underline{\quad?\quad}$

16. $\dfrac{d}{dx}\ln(x - x^3) = \dfrac{1}{x - x^3}\underline{\quad?\quad}$

In Problems 17–42, find $f'(x)$ and simplify.

17. $f(x) = (x + 3)^2$

18. $f(x) = (x - 6)^3$

19. $f(x) = (2x + 5)^3$

20. $f(x) = (3x - 7)^5$

21. $f(x) = (5 - 2x)^4$

22. $f(x) = (9 - 5x)^2$

23. $f(x) = (4 + 0.2x)^5$

24. $f(x) = (6 - 0.5x)^4$

25. $f(x) = (3x^2 + 5)^5$

26. $f(x) = (5x^2 - 3)^6$

27. $f(x) = 5e^x$

28. $f(x) = 10 - 4e^x$

29. $f(x) = e^{5x}$

30. $f(x) = 6e^{-2x}$

31. $f(x) = 3e^{-6x}$

32. $f(x) = e^{x^2 + 3x + 1}$

33. $f(x) = (2x - 5)^{1/2}$

34. $f(x) = (4x + 3)^{1/2}$

35. $f(x) = (x^4 + 1)^{-2}$

36. $f(x) = (x^5 + 2)^{-3}$

37. $f(x) = 4 - 2 \ln x$

38. $f(x) = 8 \ln x$

39. $f(x) = 3 \ln(1 + x^2)$

40. $f(x) = 2 \ln(x^2 - 3x + 4)$

41. $f(x) = (1 + \ln x)^3$

42. $f(x) = (x - 2 \ln x)^4$

In Problems 43–48, find $f'(x)$ and the equation of the line tangent to the graph of f at the indicated value of x. Find the value(s) of x where the tangent line is horizontal.

43. $f(x) = (2x - 1)^3$; $x = 1$

44. $f(x) = (3x - 1)^4$; $x = 1$

45. $f(x) = (4x - 3)^{1/2}$; $x = 3$

46. $f(x) = (2x + 8)^{1/2}$; $x = 4$

47. $f(x) = 5e^{x^2 - 4x + 1}$; $x = 0$

48. $f(x) = \ln(1 - x^2 + 2x^4)$; $x = 1$

B

In Problems 49–64, find the indicated derivative and simplify.

49. y' if $y = 3(x^2 - 2)^4$

50. y' if $y = 2(x^3 + 6)^5$

51. $\dfrac{d}{dt} 2(t^2 + 3t)^{-3}$

52. $\dfrac{d}{dt} 3(t^3 + t^2)^{-2}$

53. $\dfrac{dh}{dw}$ if $h(w) = \sqrt{w^2 + 8}$

54. $\dfrac{dg}{dw}$ if $g(w) = \sqrt[3]{3w - 7}$

55. $g'(x)$ if $g(x) = 4xe^{3x}$

56. $h'(x)$ if $h(x) = \dfrac{e^{2x}}{x^2 + 9}$

57. $\dfrac{d}{dx} \dfrac{\ln(1 + x)}{x^3}$

58. $\dfrac{d}{dx} [x^4 \ln(1 + x^4)]$

59. $F'(t)$ if $F(t) = (e^{t^2 + 1})^3$

60. $G'(t)$ if $G(t) = (1 - e^{2t})^2$

61. y' if $y = \ln(x^2 + 3)^{3/2}$

62. y' if $y = [\ln(x^2 + 3)]^{3/2}$

63. $\dfrac{d}{dw} \dfrac{1}{(w^3 + 4)^5}$

64. $\dfrac{d}{dw} \dfrac{1}{(w^2 - 2)^6}$

In Problems 65–70, find $f'(x)$ and find the equation of the line tangent to the graph of f at the indicated value of x.

65. $f(x) = x(4 - x)^3$; $x = 2$

66. $f(x) = x^2(1 - x)^4$; $x = 2$

67. $f(x) = \dfrac{x}{(2x - 5)^3}$; $x = 3$

68. $f(x) = \dfrac{x^4}{(3x - 8)^2}$; $x = 4$

69. $f(x) = \sqrt{\ln x}$; $x = e$

70. $f(x) = e\sqrt{x}$; $x = 1$

In Problems 71–76, find $f'(x)$ and find the value(s) of x where the tangent line is horizontal.

71. $f(x) = x^2(x - 5)^3$

72. $f(x) = x^3(x - 7)^4$

73. $f(x) = \dfrac{x}{(2x + 5)^2}$

74. $f(x) = \dfrac{x - 1}{(x - 3)^3}$

75. $f(x) = \sqrt{x^2 - 8x + 20}$

76. $f(x) = \sqrt{x^2 + 4x + 5}$

 77. A student reasons that the functions $f(x) = \ln[5(x^2 + 3)^4]$ and $g(x) = 4 \ln(x^2 + 3)$ must have the same derivative since he has entered $f(x)$, $g(x)$, $f'(x)$, and $g'(x)$ into a graphing calculator, but only three graphs appear (see the figure). Is his reasoning correct? Are $f'(x)$ and $g'(x)$ the same function? Explain.

(A)   (B)

Figure for 77

 78. A student reasons that the functions $f(x) = (x + 1) \ln(x + 1) - x$ and $g(x) = (x + 1)^{1/3}$ must have the same derivative since she has entered $f(x)$, $g(x)$, $f'(x)$, and $g'(x)$ into a graphing calculator, but only three graphs appear (see the figure). Is her reasoning correct? Are $f'(x)$ and $g'(x)$ the same function? Explain.

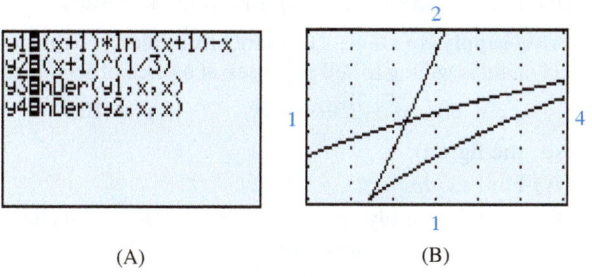

(A)   (B)

Figure for 78

C

In Problems 79–90, find each derivative and simplify.

79. $\dfrac{d}{dx}[3x(x^2 + 1)^3]$

80. $\dfrac{d}{dx}[2x^2(x^3 - 3)^4]$

81. $\dfrac{d}{dx}\dfrac{(x^3 - 7)^4}{2x^3}$

82. $\dfrac{d}{dx}\dfrac{3x^2}{(x^2 + 5)^3}$

83. $\dfrac{d}{dx}\log_2(3x^2 - 1)$

84. $\dfrac{d}{dx}\log(x^3 - 1)$

85. $\dfrac{d}{dx}10^{x^2+x}$

86. $\dfrac{d}{dx}8^{1-2x^2}$

87. $\dfrac{d}{dx}\log_3(4x^3 + 5x + 7)$

88. $\dfrac{d}{dx}\log_5(5^{x^2-1})$

89. $\dfrac{d}{dx}2^{x^3-x^2+4x+1}$

90. $\dfrac{d}{dx}10^{\ln x}$

Applications

91. Cost function. The total cost (in hundreds of dollars) of producing x cell phones per day is

$$C(x) = 10 + \sqrt{2x + 16} \qquad 0 \le x \le 50$$

(see the figure).

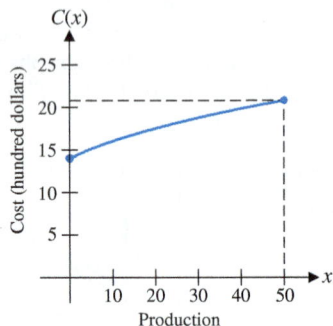

Figure for 91

(A) Find $C'(x)$.

(B) Find $C'(24)$ and $C'(42)$. Interpret the results.

92. Cost function. The total cost (in hundreds of dollars) of producing x cameras per week is

$$C(x) = 6 + \sqrt{4x + 4} \qquad 0 \le x \le 30$$

(A) Find $C'(x)$.

(B) Find $C'(15)$ and $C'(24)$. Interpret the results.

93. Price–supply equation. The number x of bicycle helmets a retail chain is willing to sell per week at a price of $\$p$ is given by

$$x = 80\sqrt{p + 25} - 400 \qquad 20 \le p \le 100$$

(see the figure).

(A) Find dx/dp.

(B) Find the supply and the instantaneous rate of change of supply with respect to price when the price is $\$75$. Write a brief interpretation of these results.

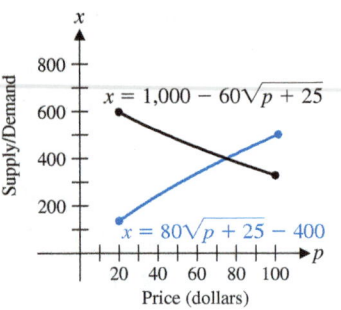

Figure for 93 and 94

94. Price–demand equation. The number x of bicycle helmets people are willing to buy per week from a retail chain at a price of $\$p$ is given by

$$x = 1,000 - 60\sqrt{p + 25} \qquad 20 \le p \le 100$$

(see the figure).

(A) Find dx/dp.

(B) Find the demand and the instantaneous rate of change of demand with respect to price when the price is $\$75$. Write a brief interpretation of these results.

95. Drug concentration. The drug concentration in the bloodstream t hours after injection is given approximately by

$$C(t) = 4.35e^{-t} \qquad 0 \le t \le 5$$

where $C(t)$ is concentration in milligrams per milliliter.

(A) What is the rate of change of concentration after 1 hour? After 4 hours?

(B) Graph C.

96. Water pollution. The use of iodine crystals is a popular way of making small quantities of water safe to drink. Crystals placed in a 1-ounce bottle of water will dissolve until the solution is saturated. After saturation, half of the solution is poured into a quart container of water, and after about an hour, the water is usually safe to drink. The half-empty 1-ounce bottle is then refilled, to be used again in the same way. Suppose that the concentration of iodine in the 1-ounce bottle t minutes after the crystals are introduced can be approximated by

$$C(t) = 250(1 - e^{-t}) \qquad t \ge 0$$

where $C(t)$ is the concentration of iodine in micrograms per milliliter.

(A) What is the rate of change of the concentration after 1 minute? After 4 minutes?

(B) Graph C for $0 \le t \le 5$.

97. Blood pressure and age. A research group using hospital records developed the following mathematical model relating systolic blood pressure and age:

$$P(x) = 40 + 25 \ln(x + 1) \qquad 0 \le x \le 65$$

$P(x)$ is pressure, measured in millimeters of mercury, and x is age in years. What is the rate of change of pressure at the end of 10 years? At the end of 30 years? At the end of 60 years?

98. Biology. A yeast culture at room temperature (68°F) is placed in a refrigerator set at a constant temperature of 38°F. After t hours, the temperature T of the culture is given approximately by

$$T = 30e^{-0.58t} + 38 \qquad t \geq 0$$

What is the rate of change of temperature of the culture at the end of 1 hour? At the end of 4 hours?

99. Learning. In 1930, L. L. Thurstone developed the following formula to indicate how learning time T depends on the length of a list n:

$$T = f(n) = \frac{c}{k}n\sqrt{n - a}$$

Here, a, c, and k are empirical constants. Suppose that, for a particular person, the time T (in minutes) required to learn a list of length n is

$$T = f(n) = 2n\sqrt{n - 2}$$

(A) Find dT/dn.

✎ (B) Find $f'(11)$ and $f'(27)$. Interpret the results.

Answers to Matched Problems

1. $f[g(x)] = 2e^x$, $g[f(u)] = e^{2u}$
2. (A) $f(u) = 50e^u$, $u = -2x$
   (B) $f(u) = \sqrt[3]{u}$, $u = 1 + x^3$

   [*Note:* There are other correct answers.]
3. (A) $15(5x + 2)^2$
   (B) $20x^3(x^4 - 5)^4$
   (C) $-4t/(t^2 + 4)^3$
   (D) $-1/(2\sqrt{4 - w})$
4. (A) $\dfrac{dy}{dx} = -5u^{-4}, \dfrac{du}{dx} = 6x^2, \dfrac{dy}{dx} = -30x^2(2x^3 + 4)^{-6}$
   (B) $\dfrac{dy}{du} = e^u, \dfrac{du}{dx} = 12x^3, \dfrac{dy}{dx} = 12x^3e^{3x^4+6}$
   (C) $\dfrac{dy}{du} = \dfrac{1}{u}, \dfrac{du}{dx} = 2x + 9, \dfrac{dy}{dx} = \dfrac{2x + 9}{x^2 + 9x + 4}$
5. $\dfrac{3e^x[\ln(1 + e^x)]^2}{1 + e^x}$
6. (A) $\dfrac{3x^2 + 2}{x^3 + 2x}$ (B) $6xe^{3x^2+2}$ (C) $-8xe^{-x^2}(2 + e^{-x^2})^3$

4-5   Implicit Differentiation

- Special Function Notation
- Implicit Differentiation

Special Function Notation

The equation

$$y = 2 - 3x^2 \tag{1}$$

defines a function f with y as a dependent variable and x as an independent variable. Using function notation, we would write

$$y = f(x) \qquad \text{or} \qquad f(x) = 2 - 3x^2$$

In order to minimize the number of symbols, we will often write equation (1) in the form

$$y = 2 - 3x^2 = y(x)$$

where y is *both* a dependent variable and a function symbol. This is a convenient notation, and no harm is done as long as one is aware of the double role of y. Other examples are

$$x = 2t^2 - 3t + 1 = x(t)$$
$$z = \sqrt{u^2 - 3u} = z(u)$$
$$r = \frac{1}{(s^2 - 3s)^{2/3}} = r(s)$$

Until now, we have considered functions involving only one independent variable. There is no reason to stop there: The concept can be generalized to functions involving two or more independent variables, and this will be done in detail in

Chapter 8. For now, we will "borrow" the notation for a function involving two independent variables. For example,

$$F(x, y) = x^2 - 2xy + 3y^2 - 5$$

specifies a function F involving two independent variables.

Implicit Differentiation

Consider the equation

$$3x^2 + y - 2 = 0 \tag{2}$$

and the equation obtained by solving equation (2) for y in terms of x,

$$y = 2 - 3x^2 \tag{3}$$

Both equations define the same function with x as the independent variable and y as the dependent variable. For equation (3), we write

$$y = f(x)$$

where

$$f(x) = 2 - 3x^2 \tag{4}$$

and we have an **explicit** (directly stated) rule that enables us to determine y for each value of x. On the other hand, the y in equation (2) is the same y as in equation (3), and equation (2) **implicitly** gives (implies, though does not directly express) y as a function of x. We say that equations (3) and (4) define the function f explicitly and equation (2) defines f implicitly.

The direct use of an equation that defines a function implicitly to find the derivative of the dependent variable with respect to the independent variable is called **implicit differentiation**. Let's differentiate equation (2) implicitly and equation (3) directly, and compare results.

Starting with

$$3x^2 + y - 2 = 0$$

we think of y as a function of x and write

$$3x^2 + y(x) - 2 = 0$$

Then we differentiate both sides with respect to x:

$$\frac{d}{dx}[(3x^2 + y(x) - 2)] = \frac{d}{dx}0$$

$$\frac{d}{dx}3x^2 + \frac{d}{dx}y(x) - \frac{d}{dx}2 = 0$$

$$6x + y' - 0 = 0$$

Since y is a function of x, but is not explicitly given, we simply write $\frac{d}{dx}y(x) = y'$ to indicate its derivative.

Now we solve for y':

$$y' = -6x$$

Note that we get the same result if we start with equation (3) and differentiate directly:

$$y = 2 - 3x^2$$
$$y' = -6x$$

Why are we interested in implicit differentiation? Why not solve for y in terms of x and differentiate directly? The answer is that there are many equations of the form

$$F(x, y) = 0 \tag{5}$$

that are either difficult or impossible to solve for y explicitly in terms of x (try it for $x^2y^5 - 3xy + 5 = 0$ or for $e^y - y = 3x$, for example). But it can be shown that, under fairly general conditions on F, equation (5) will define one or more functions in which y is a dependent variable and x is an independent variable. To find y' under these conditions, we differentiate equation (5) implicitly.

<table>
<tr>
<td>EXPLORE & DISCUSS 1</td>
<td>

(A) How many tangent lines are there to the graph in Figure 1 when $x = 0$? When $x = 1$? When $x = 2$? When $x = 4$? When $x = 6$?

</td>
</tr>
</table>

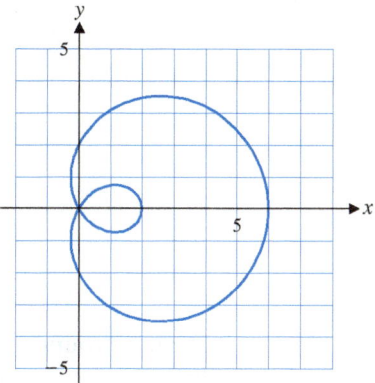

Figure 1

(B) Sketch the tangent lines referred to in part (A), and estimate each of their slopes.

(C) Explain why the graph in Figure 1 is not the graph of a function.

EXAMPLE 1   **Differentiating Implicitly** Given

$$F(x, y) = x^2 + y^2 - 25 = 0 \tag{6}$$

find y' and the slope of the graph at $x = 3$.

SOLUTION   We start with the graph of $x^2 + y^2 - 25 = 0$ (a circle, as shown in Fig. 2) so that we can interpret our results geometrically. From the graph, it is clear that equation (6) does not define a function. But with a suitable restriction on the variables, equation (6) can define two or more functions. For example, the upper half and the lower half of the circle each define a function. On each half-circle, a point that corresponds to $x = 3$ is found by substituting $x = 3$ into equation (6) and solving for y:

$$x^2 + y^2 - 25 = 0$$
$$(3)^2 + y^2 = 25$$
$$y^2 = 16$$
$$y = \pm 4$$

The point $(3, 4)$ is on the upper half-circle, and the point $(3, -4)$ is on the lower half-circle. We will use these results in a moment. We now differentiate equation (6) implicitly, treating y as a function of x [i.e., $y = y(x)$]:

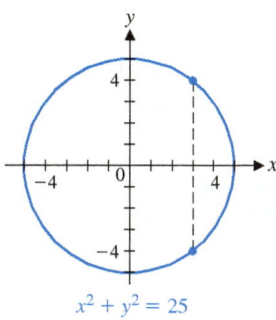

$$x^2 + y^2 = 25$$

Figure 2

$$x^2 + y^2 - 25 = 0$$

$$x^2 + [y(x)]^2 - 25 = 0$$

$$\frac{d}{dx}\{x^2 + [y(x)]^2 - 25\} = \frac{d}{dx}0$$

$$\frac{d}{dx}x^2 + \frac{d}{dx}[y(x)]^2 - \frac{d}{dx}25 = 0 \qquad \text{Use the chain rule.}$$

$$2x + 2[y(x)]^{2-1}y'(x) - 0 = 0$$

$$2x + 2yy' = 0 \qquad \text{Solve for } y' \text{ in terms of } x \text{ and } y.$$

$$y' = -\frac{2x}{2y}$$

$$y' = -\frac{x}{y} \qquad \text{Leave the answer in terms of } x \text{ and } y.$$

We have found y' without first solving $x^2 + y^2 - 25 = 0$ for y in terms of x. And by leaving y' in terms of x and y, we can use $y' = -x/y$ to find y' for *any* point on the graph of $x^2 + y^2 - 25 = 0$ (except where $y = 0$). In particular, for $x = 3$, we found that $(3, 4)$ and $(3, -4)$ are on the graph. The slope of the graph at $(3, 4)$ is

$$y'|_{(3, 4)} = -\tfrac{3}{4} \qquad \text{The slope of the graph at } (3, 4)$$

and the slope at $(3, -4)$ is

$$y'|_{(3, -4)} = -\tfrac{3}{-4} = \tfrac{3}{4} \qquad \text{The slope of the graph at } (3, -4)$$

The symbol

$$y'|_{(a, b)}$$

is used to indicate that we are evaluating y' at $x = a$ and $y = b$.

The results are interpreted geometrically in Figure 3 on the original graph.

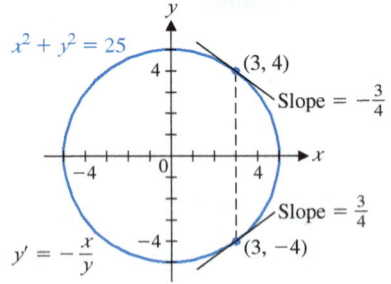

$x^2 + y^2 = 25$

$(3, 4)$

Slope $= -\frac{3}{4}$

Slope $= \frac{3}{4}$

$(3, -4)$

$y' = -\frac{x}{y}$

Figure 3

Matched Problem 1 Graph $x^2 + y^2 - 169 = 0$, find y' by implicit differentiation, and find the slope of the graph when $x = 5$.

CONCEPTUAL INSIGHT

When differentiating implicitly, the derivative of y^2 is $2yy'$, not just $2y$. This is because y represents a function of x, so the chain rule applies. Suppose, for example, that y represents the function $y = 5x + 4$. Then

$$(y^2)' = [(5x + 4)^2]' = 2(5x + 4) \cdot 5 = 2yy'$$

So, when differenting implicitly, the derivative of y is y', the derivative of y^2 is $2yy'$, the derivative of y^3 is $3y^2y'$, and so on.

EXAMPLE 2 **Differentiating Implicitly** Find the equation(s) of the tangent line(s) to the graph of

$$y - xy^2 + x^2 + 1 = 0 \qquad (7)$$

at the point(s) where $x = 1$.

SOLUTION We first find y when $x = 1$:

$$y - xy^2 + x^2 + 1 = 0$$
$$y - (1)y^2 + (1)^2 + 1 = 0$$
$$y - y^2 + 2 = 0$$
$$y^2 - y - 2 = 0$$
$$(y - 2)(y + 1) = 0$$
$$y = -1, 2$$

So there are two points on the graph of (7) where $x = 1$, namely, $(1, -1)$ and $(1, 2)$. We next find the slope of the graph at these two points by differentiating equation (7) implicitly:

$$y - xy^2 + x^2 + 1 = 0$$

Use the product rule and the

$$\frac{d}{dx}y - \frac{d}{dx}xy^2 + \frac{d}{dx}x^2 + \frac{d}{dx}1 = \frac{d}{dx}0$$

chain rule for $\frac{d}{dx}xy^2$.

$$y' - (x \cdot 2yy' + y^2) + 2x = 0$$
$$y' - 2xyy' - y^2 + 2x = 0$$ *Solve for y' by getting all*
$$y' - 2xyy' = y^2 - 2x$$ *terms involving y' on one side.*
$$(1 - 2xy)y' = y^2 - 2x$$
$$y' = \frac{y^2 - 2x}{1 - 2xy}$$

Now find the slope at each point:

$$y'|_{(1,-1)} = \frac{(-1)^2 - 2(1)}{1 - 2(1)(-1)} = \frac{1-2}{1+2} = \frac{-1}{3} = -\frac{1}{3}$$

$$y'|_{(1,2)} = \frac{(2)^2 - 2(1)}{1 - 2(1)(2)} = \frac{4-2}{1-4} = \frac{2}{-3} = -\frac{2}{3}$$

Equation of tangent line at $(1, -1)$:   Equation of tangent line at $(1, 2)$:

$$y - y_1 = m(x - x_1) \qquad\qquad y - y_1 = m(x - x_1)$$
$$y + 1 = -\tfrac{1}{3}(x - 1) \qquad\qquad y - 2 = -\tfrac{2}{3}(x - 1)$$
$$y + 1 = -\tfrac{1}{3}x + \tfrac{1}{3} \qquad\qquad y - 2 = -\tfrac{2}{3}x + \tfrac{2}{3}$$
$$y = -\tfrac{1}{3}x - \tfrac{2}{3} \qquad\qquad y = -\tfrac{2}{3}x + \tfrac{8}{3}$$

Matched Problem 2 Repeat Example 2 for $x^2 + y^2 - xy - 7 = 0$ at $x = 1$.

EXAMPLE 3 **Differentiating Implicitly** Find x' for $x = x(t)$ defined implicitly by

$$t \ln x = xe^t - 1$$

and evaluate x' at $(t, x) = (0, 1)$.

SOLUTION It is important to remember that x is the dependent variable and t is the independent variable. Therefore, we differentiate both sides of the equation with respect to t (using product and chain rules where appropriate) and then solve for x':

$$t \ln x = xe^t - 1 \qquad \text{Differentiate implicitly with respect to } t.$$

$$\frac{d}{dt}(t \ln x) = \frac{d}{dt}(xe^t) - \frac{d}{dt}1$$

$$t\frac{x'}{x} + \ln x = xe^t + x'e^t \qquad \text{Clear fractions.}$$

$$x \cdot t\frac{x'}{x} + x \cdot \ln x = x \cdot xe^t + x \cdot e^tx' \qquad x \ne 0$$

$$tx' + x \ln x = x^2e^t + xe^tx' \qquad \text{Solve for } x'.$$

$$tx' - xe^tx' = x^2e^t - x \ln x \qquad \text{Factor out } x'.$$

$$(t - xe^t)x' = x^2e^t - x \ln x$$

$$x' = \frac{x^2e^t - x \ln x}{t - xe^t}$$

Now we evaluate x' at $(t, x) = (0, 1)$, as requested:

$$x'|_{(0,1)} = \frac{(1)^2e^0 - 1 \ln 1}{0 - 1e^0}$$

$$= \frac{1}{-1} = -1$$

Matched Problem 3   Find x' for $x = x(t)$ defined implicitly by

$$1 + x \ln t = te^x$$

and evaluate x' at $(t, x) = (1, 0)$.

Exercises 4-5

A

In Problems 1–4, find y' in two ways:

(A) *Differentiate the given equation implicitly and then solve for y'.*

(B) *Solve the given equation for y and then differentiate directly.*

1. $3x + 5y + 9 = 0$

2. $-2x + 6y - 4 = 0$

3. $3x^2 - 4y - 18 = 0$

4. $2x^3 + 5y - 2 = 0$

In Problems 5–22, use implicit differentiation to find y' and evaluate y' at the indicated point.

5. $y - 5x^2 + 3 = 0$; $(1, 2)$

6. $5x^3 - y - 1 = 0$; $(1, 4)$

7. $x^2 - y^3 - 3 = 0$; $(2, 1)$

8. $y^2 + x^3 + 4 = 0$; $(-2, 2)$

9. $y^2 + 2y + 3x = 0$; $(-1, 1)$

10. $y^2 - y - 4x = 0$; $(0, 1)$

B

11. $xy - 6 = 0$; $(2, 3)$

12. $3xy - 2x - 2 = 0$; $(2, 1)$

13. $2xy + y + 2 = 0$; $(-1, 2)$

14. $2y + xy - 1 = 0$; $(-1, 1)$

15. $x^2y - 3x^2 - 4 = 0$; $(2, 4)$

16. $2x^3y - x^3 + 5 = 0$; $(-1, 3)$

17. $e^y = x^2 + y^2$; $(1, 0)$

18. $x^2 - y = 4e^y$; $(2, 0)$

19. $x^3 - y = \ln y$; $(1, 1)$

20. $\ln y = 2y^2 - x$; $(2, 1)$

21. $x \ln y + 2y = 2x^3$; $(1, 1)$

22. $xe^y - y = x^2 - 2$; $(2, 0)$

In Problems 23 and 24, find x' for $x = x(t)$ defined implicitly by the given equation. Evaluate x' at the indicated point.

23. $x^2 - t^2x + t^3 + 11 = 0$; $(-2, 1)$

24. $x^3 - tx^2 - 4 = 0$; $(-3, -2)$

Problems 25 and 26 refer to the equation and graph shown in the figure.

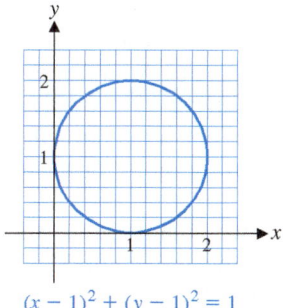

$$(x - 1)^2 + (y - 1)^2 = 1$$

Figure for 25 and 26

25. Use implicit differentiation to find the slopes of the tangent lines at the points on the graph where $x = 1.6$. Check your answers by visually estimating the slopes on the graph in the figure.

26. Find the slopes of the tangent lines at the points on the graph where $x = 0.2$. Check your answers by visually estimating the slopes on the graph in the figure.

In Problems 27–30, find the equation(s) of the tangent line(s) to the graphs of the indicated equations at the point(s) with the given value of x.

27. $xy - x - 4 = 0; x = 2$

28. $3x + xy + 1 = 0; x = -1$

29. $y^2 - xy - 6 = 0; x = 1$

30. $xy^2 - y - 2 = 0; x = 1$

31. If $xe^y = 1$, find y' in two ways, first by differentiating implicitly and then by solving for y explicitly in terms of x. Which method do you prefer? Explain.

32. Explain the difficulty that arises in solving $x^3 + y + xe^y = 1$ for y as an explicit function of x. Find the slope of the tangent line to the graph of the equation at the point $(0, 1)$.

C

In Problems 33–40, find y' and the slope of the tangent line to the graph of each equation at the indicated point.

33. $(1 + y)^3 + y = x + 7; (2, 1)$

34. $(y - 3)^4 - x = y; (-3, 4)$

35. $(x - 2y)^3 = 2y^2 - 3; (1, 1)$

36. $(2x - y)^4 - y^3 = 8; (-1, -2)$

37. $\sqrt{7 + y^2} - x^3 + 4 = 0; (2, 3)$

38. $6\sqrt{y^3 + 1} - 2x^{3/2} - 2 = 0; (4, 2)$

39. $\ln(xy) = y^2 - 1; (1, 1)$

40. $e^{xy} - 2x = y + 1; (0, 0)$

41. Find the equation(s) of the tangent line(s) at the point(s) on the graph of the equation

$$y^3 - xy - x^3 = 2$$

where $x = 1$. Round all approximate values to two decimal places.

42. Refer to the equation in Problem 41. Find the equation(s) of the tangent line(s) at the point(s) on the graph where $y = -1$. Round all approximate values to two decimal places.

Applications

For the demand equations in Problems 43–46, find the rate of change of p with respect to x by differentiating implicitly (x is the number of items that can be sold at a price of $p).

43. $x = p^2 - 2p + 1,000$

44. $x = p^3 - 3p^2 + 200$

45. $x = \sqrt{10,000 - p^2}$

46. $x = \sqrt[3]{1,500 - p^3}$

47. Biophysics. In biophysics, the equation

$$(L + m)(V + n) = k$$

is called the *fundamental equation of muscle contraction,* where m, n, and k are constants and V is the velocity of the shortening of muscle fibers for a muscle subjected to a load L. Find dL/dV by implicit differentiation.

48. Biophysics. In Problem 47, find dV/dL by implicit differentiation.

49. Speed of sound. The speed of sound in air is given by the formula

$$v = k\sqrt{T}$$

where v is the velocity of sound, T is the temperature of the air, and k is a constant. Use implicit differentiation to find $\dfrac{dT}{dv}$.

50. Gravity. The equation

$$F = G\frac{m_1 m_2}{r^2}$$

is Newton's law of universal gravitation. G is a constant and F is the gravitational force between two objects having masses m_1 and m_2 that are a distance r from each other. Use implicit differentiation to find $\dfrac{dr}{dF}$. Assume that m_1 and m_2 are constant.

51. Speed of sound. Refer to Problem 49. Find $\dfrac{dv}{dT}$ and discuss the connection between $\dfrac{dv}{dT}$ and $\dfrac{dT}{dv}$.

52. Gravity. Refer to Problem 50. Find $\dfrac{dF}{dr}$ and discuss the connection between $\dfrac{dF}{dr}$ and $\dfrac{dr}{dF}$.

Answers to Matched Problems

1. $y' = -x/y$. When $x = 5, y = \pm 12$; thus, $y'|_{(5,12)} = -\frac{5}{12}$ and $y'|_{(5,-12)} = \frac{5}{12}$

2. $y' = \dfrac{y - 2x}{2y - x}$; $y = \frac{4}{5}x - \frac{14}{5}, y = \frac{1}{5}x + \frac{14}{5}$

3. $x' = \dfrac{te^x - x}{t \ln t - t^2 e^x}$; $x'|_{(1,0)} = -1$

4-6   Related Rates

Union workers are concerned that the rate at which wages are increasing is lagging behind the rate of increase in the company's profits. An automobile dealer wants to predict how badly an anticipated increase in interest rates will decrease his rate of sales. An investor is studying the connection between the rate of increase in the Dow Jones average and the rate of increase in the gross domestic product over the past 50 years.

In each of these situations, there are two quantities—wages and profits, for example—that are changing with respect to time. We would like to discover the precise relationship between the rates of increase (or decrease) of the two quantities. We begin our discussion of such *related rates* by considering familiar situations in which the two quantities are distances and the two rates are velocities.

EXAMPLE 1   **Related Rates and Motion**   A 26-foot ladder is placed against a wall (Fig. 1). If the top of the ladder is sliding down the wall at 2 feet per second, at what rate is the bottom of the ladder moving away from the wall when the bottom of the ladder is 10 feet away from the wall?

SOLUTION   Many people think that since the ladder is a constant length, the bottom of the ladder will move away from the wall at the rate that the top of the ladder is moving down the wall. This is not the case, however.

At any moment in time, let x be the distance of the bottom of the ladder from the wall and let y be the distance of the top of the ladder from the ground (see Fig. 1). Both x and y are changing with respect to time and can be thought of as functions of time; that is, $x = x(t)$ and $y = y(t)$. Furthermore, x and y are related by the Pythagorean relationship:

$$x^2 + y^2 = 26^2 \tag{1}$$

Differentiating equation (1) implicitly with respect to time t and using the chain rule where appropriate, we obtain

$$2x\frac{dx}{dt} + 2y\frac{dy}{dt} = 0 \tag{2}$$

Figure 1

The rates dx/dt and dy/dt are related by equation (2). This is a **related-rates problem**.

Our problem is to find dx/dt when $x = 10$ feet, given that $dy/dt = -2$ (y is decreasing at a constant rate of 2 feet per second). We have all the quantities we need in equation (2) to solve for dx/dt, except y. When $x = 10$, y can be found from equation (1):

$$10^2 + y^2 = 26^2$$

$$y = \sqrt{26^2 - 10^2} = 24 \text{ feet}$$

Substitute $dy/dt = -2$, $x = 10$, and $y = 24$ into (2). Then solve for dx/dt:

$$2(10)\frac{dx}{dt} + 2(24)(-2) = 0$$

$$\frac{dx}{dt} = \frac{-2(24)(-2)}{2(10)} = 4.8 \text{ feet per second}$$

The bottom of the ladder is moving away from the wall at a rate of 4.8 feet per second.

CONCEPTUAL INSIGHT

In the solution to Example 1, we used equation (1) in two ways: first, to find an equation relating dy/dt and dx/dt, and second, to find the value of y when $x = 10$. These steps must be done in this order. Substituting $x = 10$ and then differentiating does not produce any useful results:

$$x^2 + y^2 = 26^2 \quad \text{Substituting 10 for x has the effect of}$$
$$100 + y^2 = 26^2 \quad \text{stopping the ladder.}$$
$$0 + 2yy' = 0 \quad \text{The rate of change of a stationary}$$
$$y' = 0 \quad \begin{array}{l}\text{object is always 0, but that is not the}\\ \text{rate of change of the moving ladder.}\end{array}$$

Matched Problem 1   Again, a 26-foot ladder is placed against a wall (Fig. 1). If the bottom of the ladder is moving away from the wall at 3 feet per second, at what rate is the top moving down when the top of the ladder is 24 feet above ground?

EXPLORE & DISCUSS 1

(A) For which values of x and y in Example 1 is dx/dt equal to 2 (i.e., the same rate that the ladder is sliding down the wall)?

(B) When is dx/dt greater than 2? Less than 2?

DEFINITION   Suggestions for Solving Related-Rates Problems

Step 1   Sketch a figure if helpful.

Step 2   Identify all relevant variables, including those whose rates are given and those whose rates are to be found.

Step 3   Express all given rates and rates to be found as derivatives.

Step 4   Find an equation connecting the variables identified in step 2.

Step 5   Implicitly differentiate the equation found in step 4, using the chain rule where appropriate, and substitute in all given values.

Step 6   Solve for the derivative that will give the unknown rate.

EXAMPLE 2   **Related Rates and Motion**   Suppose that two motorboats leave from the same point at the same time. If one travels north at 15 miles per hour and the other travels east at 20 miles per hour, how fast will the distance between them be changing after 2 hours?

SOLUTION   First, draw a picture, as shown in Figure 2.

All variables, x, y, and z, are changing with time. They can be considered as functions of time: $x = x(t)$, $y = y(t)$, and $z = z(t)$, given implicitly. It now makes sense to take derivatives of each variable with respect to time. From the Pythagorean theorem,

$$z^2 = x^2 + y^2 \tag{3}$$

We also know that

$$\frac{dx}{dt} = 20 \text{ miles per hour} \qquad \text{and} \qquad \frac{dy}{dt} = 15 \text{ miles per hour}$$

Figure 2

We want to find dz/dt at the end of 2 hours—that is, when $x = 40$ miles and $y = 30$ miles. To do this, we differentiate both sides of equation (3) with respect to t and solve for dz/dt:

$$2z\frac{dz}{dt} = 2x\frac{dx}{dt} + 2y\frac{dy}{dt} \tag{4}$$

We have everything we need except z. From equation (3), when $x = 40$ and $y = 30$, we find z to be 50. Substituting the known quantities into equation (4), we obtain

$$2(50)\frac{dz}{dt} = 2(40)(20) + 2(30)(15)$$

$$\frac{dz}{dt} = 25 \text{ miles per hour}$$

The boats will be separating at a rate of 25 miles per hour.

Matched Problem 2   Repeat Example 2 for the same situation at the end of 3 hours.

EXAMPLE 3   **Related Rates and Motion** Suppose that a point is moving along the graph of $x^2 + y^2 = 25$ (Fig. 3). When the point is at $(-3, 4)$, its x coordinate is increasing at the rate of 0.4 unit per second. How fast is the y coordinate changing at that moment?

SOLUTION   Since both x and y are changing with respect to time, we can consider each as a function of time, namely,

$$x = x(t) \qquad \text{and} \qquad y = y(t)$$

but restricted so that

$$x^2 + y^2 = 25 \tag{5}$$

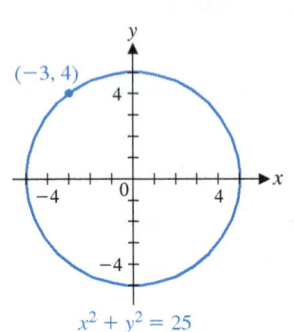

$x^2 + y^2 = 25$

Figure 3

We want to find dy/dt, given $x = -3$, $y = 4$, and $dx/dt = 0.4$. Implicitly differentiating both sides of equation (5) with respect to t, we have

$$x^2 + y^2 = 25$$

$$2x\frac{dx}{dt} + 2y\frac{dy}{dt} = 0 \qquad \text{Divide both sides by 2.}$$

$$x\frac{dx}{dt} + y\frac{dy}{dt} = 0 \qquad \begin{array}{l}\text{Substitute } x = -3, y = 4, \text{ and} \\ dx/dt = 0.4, \text{ and solve for } dy/dt.\end{array}$$

$$(-3)(0.4) + 4\frac{dy}{dt} = 0$$

$$\frac{dy}{dt} = 0.3 \text{ unit per second}$$

Matched Problem 3   A point is moving on the graph of $y^3 = x^2$. When the point is at $(-8, 4)$, its y coordinate is decreasing by 2 units per second. How fast is the x coordinate changing at that moment?

EXAMPLE 4   **Related Rates and Business** Suppose that for a company manufacturing flash drives, the cost, revenue, and profit equations are given by

$$C = 5,000 + 2x \qquad \text{Cost equation}$$
$$R = 10x - 0.001x^2 \qquad \text{Revenue equation}$$
$$P = R - C \qquad \text{Profit equation}$$

where the production output in 1 week is x flash drives. If production is increasing at the rate of 500 flash drives per week when production is 2,000 flash drives, find the rate of increase in

(A) Cost   (B) Revenue   (C) Profit

SOLUTION   If production x is a function of time (it must be, since it is changing with respect to time), then C, R, and P must also be functions of time. These functions are given implicitly (rather than explicitly). Letting t represent time in weeks, we differentiate both sides of each of the preceding three equations with respect to t and then substitute $x = 2,000$ and $dx/dt = 500$ to find the desired rates.

(A) $C = 5,000 + 2x$ Think: $C = C(t)$ and $x = x(t)$.

$$\frac{dC}{dt} = \frac{d}{dt}(5,000) + \frac{d}{dt}(2x) \quad \text{Differentiate both sides with respect to } t.$$

$$\frac{dC}{dt} = 0 + 2\frac{dx}{dt} = 2\frac{dx}{dt}$$

Since $dx/dt = 500$ when $x = 2,000$,

$$\frac{dC}{dt} = 2(500) = \$1,000 \text{ per week}$$

Cost is increasing at a rate of $1,000 per week.

(B) $R = 10x - 0.001x^2$

$$\frac{dR}{dt} = \frac{d}{dt}(10x) - \frac{d}{dt}0.001x^2$$

$$\frac{dR}{dt} = 10\frac{dx}{dt} - 0.002x\frac{dx}{dt}$$

$$\frac{dR}{dt} = (10 - 0.002x)\frac{dx}{dt}$$

Since $dx/dt = 500$ when $x = 2,000$,

$$\frac{dR}{dt} = [10 - 0.002(2,000)](500) = \$3,000 \text{ per week}$$

Revenue is increasing at a rate of $3,000 per week.

(C) $P = R - C$

$$\frac{dP}{dt} = \frac{dR}{dt} - \frac{dC}{dt} \qquad \text{Results from parts (A) and (B)}$$
$$= \$3,000 - \$1,000$$
$$= \$2,000 \text{ per week}$$

Profit is increasing at a rate of $2,000 per week.

Matched Problem 4   Repeat Example 4 for a production level of 6,000 flash drives per week.

Exercises 4-6

In Problems 1–6, assume that $x = x(t)$ and $y = y(t)$. Find the indicated rate, given the other information.

1. $y = x^2 + 2$; $dx/dt = 3$ when $x = 5$; find dy/dt

2. $y = x^3 - 3$; $dx/dt = -2$ when $x = 2$; find dy/dt

3. $x^2 + y^2 = 1$; $dy/dt = -4$ when $x = -0.6$ and $y = 0.8$; find dx/dt

4. $x^2 + y^2 = 4$; $dy/dt = 5$ when $x = 1.2$ and $y = -1.6$; find dx/dt

5. $x^2 + 3xy + y^2 = 11$; $dx/dt = 2$ when $x = 1$ and $y = 2$; find dy/dt

6. $x^2 - 2xy - y^2 = 7$; $dy/dt = -1$ when $x = 2$ and $y = -1$; find dx/dt

7. A point is moving on the graph of $xy = 36$. When the point is at $(4, 9)$, its x coordinate is increasing by 4 units per second. How fast is the y coordinate changing at that moment?

8. A point is moving on the graph of $4x^2 + 9y^2 = 36$. When the point is at $(3, 0)$, its y coordinate is decreasing by 2 units per second. How fast is its x coordinate changing at that moment?

9. A boat is being pulled toward a dock as shown in the figure. If the rope is being pulled in at 3 feet per second, how fast is the distance between the dock and the boat decreasing when it is 30 feet from the dock?

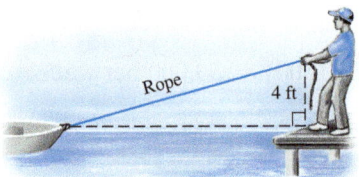

Rope

4 ft

Figure for 9 and 10

10. Refer to Problem 9. Suppose that the distance between the boat and the dock is decreasing by 3.05 feet per second. How fast is the rope being pulled in when the boat is 10 feet from the dock?

11. A rock thrown into a still pond causes a circular ripple. If the radius of the ripple is increasing by 2 feet per second, how fast is the area changing when the radius is 10 feet? [Use $A = \pi R^2$, $\pi \approx 3.14$.]

12. Refer to Problem 11. How fast is the circumference of a circular ripple changing when the radius is 10 feet? [Use $C = 2\pi R$, $\pi \approx 3.14$.]

13. The radius of a spherical balloon is increasing at the rate of 3 centimeters per minute. How fast is the volume changing when the radius is 10 centimeters? [Use $V = \frac{4}{3}\pi R^3$, $\pi \approx 3.14$.]

14. Refer to Problem 13. How fast is the surface area of the sphere increasing when the radius is 10 centimeters? [Use $S = 4\pi R^2$, $\pi \approx 3.14$.]

15. Boyle's law for enclosed gases states that if the volume is kept constant, the pressure P and temperature T are related by the equation

$$\frac{P}{T} = k$$

where k is a constant. If the temperature is increasing at 3 kelvins per hour, what is the rate of change of pressure when the temperature is 250 kelvins and the pressure is 500 pounds per square inch?

16. Boyle's law for enclosed gases states that if the temperature is kept constant, the pressure P and volume V of a gas are related by the equation

$$VP = k$$

where k is a constant. If the volume is decreasing by 5 cubic inches per second, what is the rate of change of pressure when the volume is 1,000 cubic inches and the pressure is 40 pounds per square inch?

17. A 10-foot ladder is placed against a vertical wall. Suppose that the bottom of the ladder slides away from the wall at a constant rate of 3 feet per second. How fast is the top of the ladder sliding down the wall (negative rate) when the bottom is 6 feet from the wall? [*Hint:* Use the Pythagorean theorem, $a^2 + b^2 = c^2$, where c is the length of the hypotenuse of a right triangle and a and b are the lengths of the two shorter sides.]

18. A weather balloon is rising vertically at the rate of 5 meters per second. An observer is standing on the ground 300 meters from where the balloon was released. At what rate is the distance between the observer and the balloon changing when the balloon is 400 meters high?

19. A streetlight is on top of a 20-foot pole. A person who is 5 feet tall walks away from the pole at the rate of 5 feet per second. At what rate is the tip of the person's shadow moving away from the pole when he is 20 feet from the pole?

20. Refer to Problem 19. At what rate is the person's shadow growing when he is 20 feet from the pole?

21. Helium is pumped into a spherical balloon at a constant rate of 4 cubic feet per second. How fast is the radius increasing after 1 minute? After 2 minutes? Is there any time at which the radius is increasing at a rate of 100 feet per second? Explain.

22. A point is moving along the x axis at a constant rate of 5 units per second. At which point is its distance from $(0, 1)$ increasing at a rate of 2 units per second? At 4 units per second? At 5 units per second? At 10 units per second? Explain.

23. A point is moving on the graph of $y = e^x + x + 1$ in such a way that its x coordinate is always increasing at a rate of 3 units per second. How fast is the y coordinate changing when the point crosses the x axis?

24. A point is moving on the graph of $x^3 + y^2 = 1$ in such a way that its y coordinate is always increasing at a rate of 2 units per second. At which point(s) is the x coordinate increasing at a rate of 1 unit per second?

Applications

25. Cost, revenue, and profit rates. Suppose that for a company manufacturing calculators, the cost, revenue, and profit equations are given by

$$C = 90{,}000 + 30x \qquad R = 300x - \frac{x^2}{30}$$

$$P = R - C$$

where the production output in 1 week is x calculators. If production is increasing at a rate of 500 calculators per week when production output is 6,000 calculators, find the rate of increase (decrease) in

(A) Cost   (B) Revenue   (C) Profit

26. Cost, revenue, and profit rates. Repeat Problem 25 for

$$C = 72{,}000 + 60x \qquad R = 200x - \frac{x^2}{30}$$

$$P = R - C$$

where production is increasing at a rate of 500 calculators per week at a production level of 1,500 calculators.

27. Advertising. A retail store estimates that weekly sales s and weekly advertising costs x (both in dollars) are related by

$$s = 60{,}000 - 40{,}000e^{-0.0005x}$$

The current weekly advertising costs are $2,000, and these costs are increasing at the rate of $300 per week. Find the current rate of change of sales.

28. Advertising. Repeat Problem 27 for

$$s = 50{,}000 - 20{,}000e^{-0.0004x}$$

29. Price–demand. The price p (in dollars) and demand x for a product are related by

$$2x^2 + 5xp + 50p^2 = 80{,}000$$

(A) If the price is increasing at a rate of $2 per month when the price is $30, find the rate of change of the demand.

(B) If the demand is decreasing at a rate of 6 units per month when the demand is 150 units, find the rate of change of the price.

30. Price–demand. Repeat Problem 29 for

$$x^2 + 2xp + 25p^2 = 74{,}500$$

31. Pollution. An oil tanker aground on a reef is forming a circular oil slick about 0.1 foot thick (see the figure). To estimate the rate dV/dt (in cubic feet per minute) at which the oil is leaking from the tanker, it was found that the radius of the slick was increasing at 0.32 foot per minute ($dR/dt = 0.32$) when the radius R was 500 feet. Find dV/dt, using $\pi \approx 3.14$.

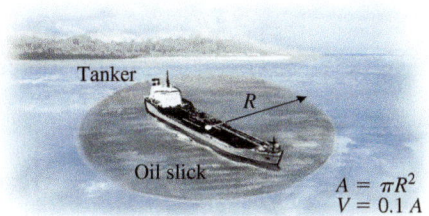

Figure for 31

32. Learning. A person who is new on an assembly line performs an operation in T minutes after x performances of the operation, as given by

$$T = 6\left(1 + \frac{1}{\sqrt{x}}\right)$$

If $dx/dt = 6$ operations per hours, where t is time in hours, find dT/dt after 36 performances of the operation.

Answers to Matched Problems

1. $dy/dt = -1.25$ ft/sec

2. $dz/dt = 25$ mi/hr

3. $dx/dt = 6$ units/sec

4. (A) $dC/dt = \$1{,}000/\text{wk}$

   (B) $dR/dt = -\$1{,}000/\text{wk}$

   (C) $dP/dt = -\$2{,}000/\text{wk}$

4-7 Elasticity of Demand

- Relative Rate of Change
- Elasticity of Demand

When will a price increase lead to an increase in revenue? To answer this question and study relationships among price, demand, and revenue, economists use the notion of *elasticity of demand*. In this section, we define the concepts of *relative rate of change*, *percentage rate of change*, and *elasticity of demand*.

Relative Rate of Change

EXPLORE & DISCUSS 1

A broker is trying to sell you two stocks: Biotech and Comstat. The broker estimates that Biotech's earnings will increase $2 per year over the next several years, while Comstat's earnings will increase only $1 per year. Is this sufficient information for you to choose between the two stocks? What other information might you request from the broker to help you decide?

Interpreting rates of change is a fundamental application of calculus. In Explore & Discuss 1, Biotech's earnings are increasing at twice the rate of Comstat's, but that does not automatically make Biotech the better buy. The obvious information that is missing is the cost of each stock. If Biotech costs $100 a share and Comstat costs $25 share, then which stock is the better buy? To answer this question, we introduce two new concepts: *relative rate of change* and *percentage rate of change*.

DEFINITION Relative and Percentage Rates of Change

The **relative rate of change** of a function $f(x)$ is $\dfrac{f'(x)}{f(x)}$.

The **percentage rate of change** is $100 \times \dfrac{f'(x)}{f(x)}$.

Because

$$\frac{d}{dx}\ln f(x) = \frac{f'(x)}{f(x)}$$

the relative rate of change of $f(x)$ is the derivative of the logarithm of $f(x)$. This is also referred to as the **logarithmic derivative** of $f(x)$. Returning to Explore & Discuss 1, we can now write

	Relative rate of change		Percentage rate of change
Biotech	$\dfrac{2}{100} = 0.02$	or	2%
Comstat	$\dfrac{1}{25} = 0.04$	or	4%

EXAMPLE 1   **Percentage Rate of Change** Table 1 lists the GDP (gross domestic product expressed in billions of 2005 dollars) and U.S. population from 2000 to 2008. A model for the GDP is

$$f(t) = 280t + 11{,}147$$

where t is years since 2000. Find and graph the percentage rate of change of $f(t)$ for $0 \le t \le 8$.

Table 1

Year	Real GDP (billions of 2005 dollars)	Population (in millions)
2000	$11,226	282.2
2002	$11,553	287.7
2004	$12,264	292.9
2006	$12,976	298.4
2008	$13,312	304.1

SOLUTION   If $p(t)$ is the percentage rate of change of $f(t)$, then

$$p(t) = 100 \times \frac{d}{dx}\ln(280t + 11{,}147)$$

$$= \frac{28{,}000}{280t + 11{,}147}$$

The graph of $p(t)$ is shown in Figure 1 (graphing details omitted). Notice that $p(t)$ is decreasing, even though the GDP is increasing.

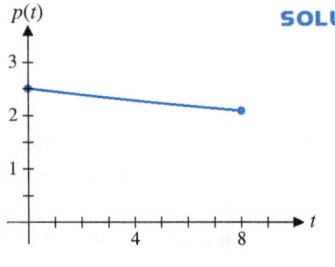

Figure 1

Matched Problem 1   A model for the population data in Table 1 is

$$f(t) = 2.7t + 282$$

where t is years since 2000. Find and graph $p(t)$, the percentage rate of change of $f(t)$ for $0 \leq t \leq 8$.

Elasticity of Demand

Economists use logarithmic derivatives and relative rates of change to study the relationship among price changes, demand, and revenue. Suppose the price $\$p$ and the demand x for a certain product are related by the price–demand equation

$$x + 500p = 10{,}000 \tag{1}$$

In problems involving revenue, cost, and profit, it is customary to use the demand equation to express price as a function of demand. Since we want to know the effects that changes in price have on demand, it is more convenient to express demand as a function of price. Solving (1) for x, we have

$$x = 10{,}000 - 500p$$

$$= 500(20 - p) \quad \textit{Demand as a function of price}$$

or

$$x = f(p) = 500(20 - p) \qquad 0 \leq p \leq 20 \tag{2}$$

Since x and p both represent nonnegative quantities, we must restrict p so that $0 \leq p \leq 20$. For most products, demand is assumed to be a decreasing function of price. That is, price increases result in lower demand, and price decreases result in higher demand (see Figure 2).

Figure 2

Economists use the *elasticity of demand* to study the relationship between changes in price and changes in demand. The **elasticity of demand** is the negative of the ratio of the relative rate of change of demand to the relative rate of change of price. If price and demand are related by a price–demand equation of the form $x = f(p)$, then the elasticity of demand can be expressed as

$$-\frac{\text{relative rate of change of demand}}{\text{relative rate of change of price}} = -\frac{\dfrac{d}{dp}\ln f(p)}{\dfrac{d}{dp}\ln p}$$

$$= -\frac{\dfrac{f'(p)}{f(p)}}{\dfrac{1}{p}}$$

$$= -\frac{pf'(p)}{f(p)}$$

> **THEOREM 1   Elasticity of Demand**
> If price and demand are related by $x = f(p)$, then the elasticity of demand is given by
>
> $$E(p) = -\frac{pf'(p)}{f(p)}$$

CONCEPTUAL INSIGHT

Since p and $f(p)$ are nonnegative and $f'(p)$ is negative (remember, demand is usually a decreasing function of price), $E(p)$ is nonnegative. This is why elasticity of demand is defined as the negative of a ratio.

The next example illustrates interpretations of the elasticity of demand.

EXAMPLE 2   **Elasticity of Demand**   Find $E(p)$ for the price–demand equation

$$x = f(p) = 500(20 - p)$$

Find and interpret each of the following:
(A) $E(4)$ (B) $E(16)$ (C) $E(10)$

SOLUTION $E(p) = -\dfrac{pf'(p)}{f(p)} = -\dfrac{p(-500)}{500(20 - p)} = \dfrac{p}{20 - p}$

In order to interpret values of $E(p)$, we must recall the definition of elasticity:

$$E(p) = -\frac{\text{relative rate of change of demand}}{\text{relative rate of change of price}}$$

or

$$-\left(\begin{array}{c}\text{relative rate of}\\\text{change of demand}\end{array}\right) \approx E(p)\left(\begin{array}{c}\text{relative rate of}\\\text{change of price}\end{array}\right)$$

(A) $E(4) = \frac{4}{16} = 0.25 < 1$. If the \$4 price changes by 10%, then the demand will change by approximately $0.25(10\%) = 2.5\%$.

(B) $E(16) = \frac{16}{4} = 4 > 1$. If the \$16 price changes by 10%, then the demand will change by approximately $4(10\%) = 40\%$.

(C) $E(10) = \frac{10}{10} = 1$. If the \$10 price changes by 10%, then the demand will also change by approximately 10%.

CONCEPTUAL INSIGHT

Do not be concerned with the omission of any negative signs in these interpretations. We already know that if price increases, then demand decreases, and vice versa (Fig. 2 on page 257). $E(p)$ is a measure of *how much* the demand changes for a given change in price.

Matched Problem 2   Find $E(p)$ for the price–demand equation

$$x = f(p) = 1{,}000(40 - p)$$

Find and interpret each of the following:
(A) $E(8)$ (B) $E(30)$ (C) $E(20)$

The three cases illustrated in the solution to Example 2 are referred to as **inelastic demand, elastic demand**, and **unit elasticity**, as indicated in Table 2.

Table 2

E(p)	Demand	Interpretation
$0 < E(p) < 1$	Inelastic	Demand is not sensitive to changes in price. A change in price produces a smaller change in demand.
$E(p) > 1$	Elastic	Demand is sensitive to changes in price. A change in price produces a larger change in demand.
$E(p) = 1$	Unit	A change in price produces the same change in demand.

Now we want to see how revenue and elasticity are related. We have used the following model for revenue many times before:

$$\text{Revenue} = (\text{demand}) \times (\text{price})$$

Since we are looking for a connection between $E(p)$ and revenue, we will use a price–demand equation written in the form $x = f(p)$, where x is demand and p is price.

$$R(p) = xp = pf(p) \qquad \textit{Revenue as a function of price}$$
$$R'(p) = pf'(p) + f(p)$$
$$= f(p)\left[\frac{pf'(p)}{f(p)} + 1\right] \qquad E(p) = -\frac{pf'(p)}{f(p)}$$
$$= f(p)[1 - E(p)]$$

Since $x = f(p) > 0$, it follows that $R'(p)$ and $[1 - E(p)]$ always have the same sign (see Table 3).

Table 3 Revenue and Elasticity of Demand

All Are True or All Are False	All Are True or All Are False
$R'(p) > 0$	$R'(p) < 0$
$E(p) < 1$	$E(p) > 1$
Demand is inelastic	Demand is elastic

These facts are interpreted in the following summary and in Figure 3:

SUMMARY Revenue and Elasticity of Demand

When demand is inelastic,

   A price increase will increase revenue.
   A price decrease will decrease revenue.

When demand is elastic,

   A price increase will decrease revenue.
   A price decrease will increase revenue.

CONCEPTUAL INSIGHT

We know that a price increase will decrease demand at all price levels (Fig. 2). As Figure 3 illustrates, the effects of a price increase on revenue depends on the price level. If demand is elastic, then a price increase decreases revenue. If demand is inelastic, then a price increase increases revenue.

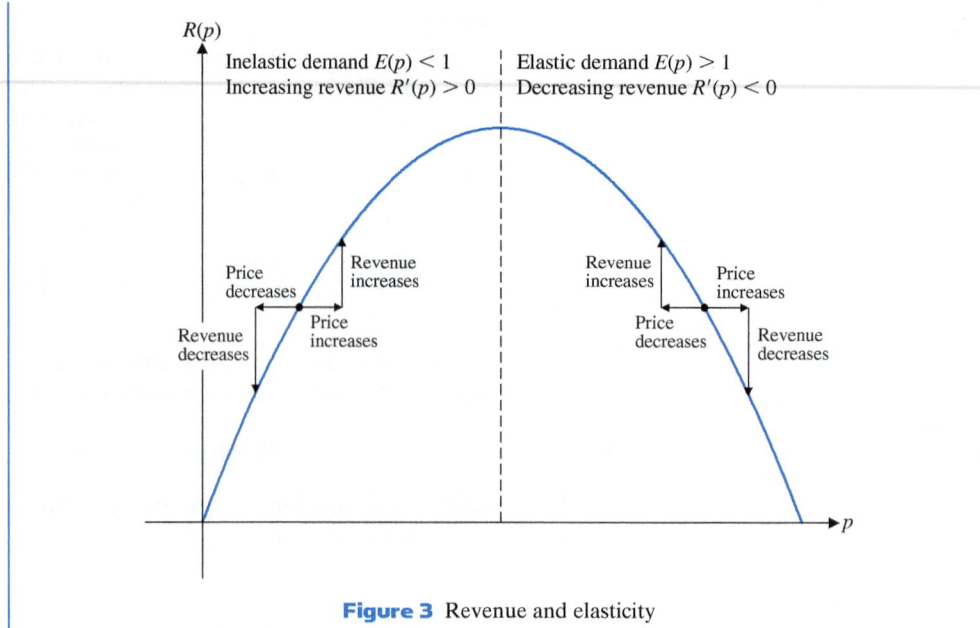

Figure 3 Revenue and elasticity

EXAMPLE 3 **Elasticity and Revenue** A manufacturer of sunglasses currently sells one type for $15 a pair. The price p and the demand x for these glasses are related by

$$x = f(p) = 9{,}500 - 250p$$

If the current price is increased, will revenue increase or decrease?

SOLUTION

$$E(p) = -\frac{pf'(p)}{f(p)}$$

$$= -\frac{p(-250)}{9{,}500 - 250p}$$

$$= \frac{p}{38 - p}$$

$$E(15) = \frac{15}{23} \approx 0.65$$

At the $15 price level, demand is inelastic and a price increase will increase revenue.

Matched Problem 3 Repeat Example 3 if the current price for sunglasses is $21 a pair.

Exercises 4-7

A

In Problems 1–6, find the relative rate of change of f(x).

1. $f(x) = 35x - 0.4x^2$

2. $f(x) = 60x - 1.2x^2$

3. $f(x) = 7 + 4e^{-x}$

4. $f(x) = 15 - 3e^{-0.5x}$

5. $f(x) = 12 + 5 \ln x$

6. $f(x) = 25 - 2 \ln x$

In Problems 7–16, find the relative rate of change of f(x) at the indicated value of x. Round to three decimal places.

7. $f(x) = 45; x = 100$

8. $f(x) = 580; x = 300$

9. $f(x) = 420 - 5x; x = 25$

10. $f(x) = 500 - 6x; x = 40$

11. $f(x) = 420 - 5x; x = 55$

12. $f(x) = 500 - 6x; x = 75$

13. $f(x) = 4x^2 - \ln x; x = 2$

14. $f(x) = 9x - 5 \ln x; x = 3$

15. $f(x) = 4x^2 - \ln x; x = 5$

16. $f(x) = 9x - 5 \ln x; x = 7$

In Problems 17–24, find the percentage rate of change of f(x) at the indicated value of x. Round to the nearest tenth of a percent.

17. $f(x) = 225 + 65x; x = 5$

18. $f(x) = 75 + 110x; x = 4$

19. $f(x) = 225 + 65x; x = 15$

20. $f(x) = 75 + 110x; x = 16$

21. $f(x) = 5,100 - 3x^2; x = 35$

22. $f(x) = 3,000 - 8x^2; x = 12$

23. $f(x) = 5,100 - 3x^2; x = 41$

24. $f(x) = 3,000 - 8x^2; x = 18$

In Problems 25–30, use the price–demand equation to find E(p), the elasticity of demand.

25. $x = f(p) = 25,000 - 450p$

26. $x = f(p) = 10,000 - 190p$

27. $x = f(p) = 4,800 - 4p^2$

28. $x = f(p) = 8,400 - 7p^2$

29. $x = f(p) = 98 - 0.6e^p$

30. $x = f(p) = 160 - 35 \ln p$

B

In Problems 31–34, use the price–demand equation to determine whether demand is elastic, is inelastic, or has unit elasticity at the indicated values of p.

31. $x = f(p) = 12,000 - 10p^2$

  (A) $p = 10$ (B) $p = 20$

  (C) $p = 30$

32. $x = f(p) = 1,875 - p^2$

  (A) $p = 15$ (B) $p = 25$

  (C) $p = 40$

33. $x = f(p) = 950 - 2p - 0.1p^2$

  (A) $p = 30$ (B) $p = 50$

  (C) $p = 70$

34. $x = f(p) = 875 - p - 0.05p^2$

  (A) $p = 50$ (B) $p = 70$

  (C) $p = 100$

35. Given the price–demand equation

$$p + 0.005x = 30$$

(A) Express the demand x as a function of the price p.

(B) Find the elasticity of demand, $E(p)$.

(C) What is the elasticity of demand when $p = \$10$? If this price is increased by 10%, what is the approximate change in demand?

(D) What is the elasticity of demand when $p = \$25$? If this price is increased by 10%, what is the approximate change in demand?

(E) What is the elasticity of demand when $p = \$15$? If this price is increased by 10%, what is the approximate change in demand?

36. Given the price–demand equation

$$p + 0.01x = 50$$

(A) Express the demand x as a function of the price p.

(B) Find the elasticity of demand, $E(p)$.

(C) What is the elasticity of demand when $p = \$10$? If this price is decreased by 5%, what is the approximate change in demand?

(D) What is the elasticity of demand when $p = \$45$? If this price is decreased by 5%, what is the approximate change in demand?

(E) What is the elasticity of demand when $p = \$25$? If this price is decreased by 5%, what is the approximate change in demand?

37. Given the price–demand equation

$$0.02x + p = 60$$

(A) Express the demand x as a function of the price p.

(B) Express the revenue R as a function of the price p.

(C) Find the elasticity of demand, $E(p)$.

(D) For which values of p is demand elastic? Inelastic?

(E) For which values of p is revenue increasing? Decreasing?

(F) If $p = \$10$ and the price is decreased, will revenue increase or decrease?

(G) If $p = \$40$ and the price is decreased, will revenue increase or decrease?

38. Repeat Problem 37 for the price–demand equation

$$0.025x + p = 50$$

In Problems 39–46, use the price–demand equation to find the values of p for which demand is elastic and the values for which demand is inelastic. Assume that price and demand are both positive.

39. $x = f(p) = 210 - 30p$

40. $x = f(p) = 480 - 8p$

41. $x = f(p) = 3,125 - 5p^2$

42. $x = f(p) = 2,400 - 6p^2$

43. $x = f(p) = \sqrt{144 - 2p}$

44. $x = f(p) = \sqrt{324 - 2p}$

45. $x = f(p) = \sqrt{2,500 - 2p^2}$

46. $x = f(p) = \sqrt{3,600 - 2p^2}$

In Problems 47–52, use the demand equation to find the revenue function. Sketch the graph of the revenue function, and indicate the regions of inelastic and elastic demand on the graph.

47. $x = f(p) = 20(10 - p)$

48. $x = f(p) = 10(16 - p)$

49. $x = f(p) = 40(p - 15)^2$

50. $x = f(p) = 10(p - 9)^2$

51. $x = f(p) = 30 - 10\sqrt{p}$

52. $x = f(p) = 30 - 5\sqrt{p}$

C

If a price–demand equation is solved for p, then price is expressed as $p = g(x)$ and x becomes the independent variable. In this case, it can be shown that the elasticity of demand is given by

$$E(x) = -\frac{g(x)}{xg'(x)}$$

In Problems 53–56, use the price–demand equation to find $E(x)$ at the indicated value of x.

53. $p = g(x) = 50 - 0.1x, x = 200$

54. $p = g(x) = 30 - 0.05x, x = 400$

55. $p = g(x) = 50 - 2\sqrt{x}, x = 400$

56. $p = g(x) = 20 - \sqrt{x}, x = 100$

In Problems 57–60, use the price–demand equation to find the values of x for which demand is elastic and for which demand is inelastic.

57. $p = g(x) = 180 - 0.3x$

58. $p = g(x) = 640 - 0.4x$

59. $p = g(x) = 90 - 0.1x^2$

60. $p = g(x) = 540 - 0.2x^2$

61. Find $E(p)$ for $x = f(p) = Ap^{-k}$, where A and k are positive constants.

62. Find $E(p)$ for $x = f(p) = Ae^{-kp}$, where A and k are positive constants.

Applications

63. Rate of change of cost. A fast-food restaurant can produce a hamburger for $2.50. If the restaurant's daily sales are increasing at the rate of 30 hamburgers per day, how fast is its daily cost for hamburgers increasing?

64. Rate of change of cost. The fast-food restaurant in Problem 63 can produce an order of fries for $0.80. If the restaurant's daily sales are increasing at the rate of 45 orders of fries per day, how fast is its daily cost for fries increasing?

65. Revenue and elasticity. The price–demand equation for hamburgers at a fast-food restaurant is

$$x + 400p = 3,000$$

Currently, the price of a hamburger is $3.00. If the price is increased by 10%, will revenue increase or decrease?

66. Revenue and elasticity. Refer to Problem 65. If the current price of a hamburger is $4.00, will a 10% price increase cause revenue to increase or decrease?

67. Revenue and elasticity. The price–demand equation for an order of fries at a fast-food restaurant is

$$x + 1,000p = 2,500$$

Currently, the price of an order of fries is $0.99. If the price is decreased by 10%, will revenue increase or decrease?

68. Revenue and elasticity. Refer to Problem 67. If the current price of an order of fries is $1.29, will a 10% price decrease cause revenue to increase or decrease?

69. Maximum revenue. Refer to Problem 65. What price will maximize the revenue from selling hamburgers?

70. Maximum revenue. Refer to Problem 67. What price will maximize the revenue from selling fries?

71. Population growth. A model for Canada's population growth (Table 4) is

$$f(t) = 0.31t + 18.5$$

where t is years since 1960. Find and graph the percentage rate of change of $f(t)$ for $0 \le t \le 50$.

72. Population growth. A model for Mexico's population growth (Table 4) is

Table 4	Population Growth	
Year	Canada (millions)	Mexico (millions)
1960	18	39
1970	22	53
1980	25	68
1990	28	85
2000	31	100
2010	34	112

$$f(t) = 1.49t + 38.8$$

where t is years since 1960. Find and graph the percentage rate of change of $f(t)$ for $0 \le t \le 50$.

73. Crime. A model for the number of robberies in the United States (Table 5) is

$$r(t) = 9.7 - 2.7 \ln t$$

where t is years since 1990. Find the relative rate of change for robberies in 2008.

Table 5	Number of Victimizations per 1,000 Population Age 12 and Over	
	Robbery	Aggravated Assault
1995	5.4	9.5
1997	4.3	8.6
1999	3.6	6.7
2001	2.8	5.3
2003	2.5	4.6
2005	2.6	4.3

74. Crime. A model for the number of assaults in the United States (Table 5) is

$$a(t) = 18.2 - 5.2 \ln t$$

where t is years since 1990. Find the relative rate of change for assaults in 2008.

Answers to Matched Problems

1. $p(t) = \dfrac{270}{2.7t + 282}$

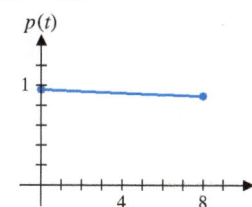

2. $E(p) = \dfrac{p}{40 - p}$

(A) $E(8) = 0.25$; demand is inelastic.

(B) $E(30) = 3$; demand is elastic.

(C) $E(20) = 1$; demand has unit elasticity.

3. $E(21) = \dfrac{21}{17} \approx 1.2$; demand is elastic. Increasing price will decrease revenue.

Chapter 4   Review

Important Terms, Symbols, and Concepts

4-1 The Constant e and Continuous Compound Interest

EXAMPLES

• The number e is defined as

$$\lim_{x \to \infty} \left(1 + \frac{1}{n}\right)^n = \lim_{x \to 0}(1 + s)^{1/s} = 2.718\ 281\ 828\ 459\ldots$$

• If a principal P is invested at an annual rate r (expressed as a decimal) compounded continuously, then the amount A in the account at the end of t years is given by the **compound interest formula**

$$A = Pe^{rt}$$

Ex. 1, p. 213
Ex. 2, p. 213
Ex. 3, p. 214
Ex. 4, p. 214

4-2 Derivatives of Exponential and Logarithmic Functions

• For $b > 0, b \neq 1,$

$$\frac{d}{dx}e^x = e^x \qquad \frac{d}{dx}b^x = b^x \ln b$$

$$\frac{d}{dx}\ln x = \frac{1}{x} \qquad \frac{d}{dx}\log_b x = \frac{1}{\ln b}\frac{1}{x}$$

• The **change-of-base formulas** allow conversion from base e to any base $b, b > 0, b \neq 1$:

$$b^x = e^{x \ln b} \qquad \log_b x = \frac{\ln x}{\ln b}$$

Ex. 1, p. 218
Ex. 2, p. 220
Ex. 3, p. 222
Ex. 4, p. 222
Ex. 5, p. 223

4-3 Derivatives of Products and Quotients

• Product rule. If $y = f(x) = F(x)\,S(x)$, then $f'(x) = F(x)S'(x) + S(x)F'(x)$, provided that both $F'(x)$ and $S'(x)$ exist.

• Quotient rule. If $y = f(x) = \dfrac{T(x)}{B(x)}$, then $f'(x) = \dfrac{B(x)\,T'(x) - T(x)\,B'(x)}{[B(x)]^2}$ provided that both $T'(x)$ and $B'(x)$ exist.

Ex. 1, p. 226
Ex. 2, p. 227
Ex. 3, p. 227
Ex. 4, p. 229
Ex. 5, p. 230
Ex. 6, p. 230

4-4 The Chain Rule

• A function m is a **composite** of functions f and g if $m(x) = f[g(x)]$.

• The **chain rule** gives a formula for the derivative of the composite function $m(x) = f[g(x)]$:

$$m'(x) = f'[g(x)]g'(x)$$

• A special case of the chain rule is called the **general power rule:**

$$\frac{d}{dx}[f(x)]^n = n[f(x)]^{n-1}f'(x)$$

• Other special cases of the chain rule are the following **general derivative rules:**

$$\frac{d}{dx}\ln [f(x)] = \frac{1}{f(x)}f'(x)$$

$$\frac{d}{dx}e^{f(x)} = e^{f(x)}f'(x)$$

Ex. 1, p. 234
Ex. 2, p. 234
Ex. 4, p. 238
Ex. 5, p. 239
Ex. 3, p. 236

Ex. 6, p. 240

4-5 Implicit Differentiation

- If $y = y(x)$ is a function defined by the equation $F(x, y) = 0$, then we use **implicit differentiation** to find an equation in x, y, and y'.

Ex. 1, p. 245
Ex. 2, p. 246
Ex. 3, p. 247

4-6 Related Rates

- If x and y represent quantities that are changing with respect to time and are related by the equation $F(x, y) = 0$, then implicit differentiation produces an equation that relates x, y, dy/dt, and dx/dt. Problems of this type are called **related-rates problems**.
- Suggestions for solving related-rates problems are given on page 251.

Ex. 1, p. 250
Ex. 2, p. 251
Ex. 3, p. 252

4-7 Elasticity of Demand

- The **relative rate of change**, or the **logarithmic derivative**, of a function $f(x)$ is $f'(x)/f(x)$, and the **percentage rate of change** is $100 \times [f'(x)/f(x)]$.
- If price and demand are related by $x = f(p)$, then the **elasticity of demand** is given by

$$E(p) = -\frac{pf'(p)}{f(p)} = -\frac{\text{relative rate of change of demand}}{\text{relative rate of change of price}}$$

- **Demand is inelastic** if $0 < E(p) < 1$. (Demand is not sensitive to changes in price; a change in price produces a smaller change in demand.) **Demand is elastic** if $E(p) > 1$. (Demand is sensitive to changes in price; a change in price produces a larger change in demand.) **Demand has unit elasticity** if $E(p) = 1$. (A change in price produces the same change in demand.)
- If $R(p) = pf(p)$ is the revenue function, then $R'(p)$ and $[1 - E(p)]$ always have the same sign.

Ex. 1, p. 256

Ex. 2, p. 258

Ex. 3, p. 260

Review Exercises

Work through all the problems in this chapter review, and check your answers in the back of the book. Answers to all review problems are there, along with section numbers in italics to indicate where each type of problem is discussed. Where weaknesses show up, review appropriate sections of the text.

A

1. Use a calculator to evaluate $A = 2{,}000e^{0.09t}$ to the nearest cent for $t = 5, 10,$ and 20.

Find the indicated derivatives in Problems 2–4.

2. $\dfrac{d}{dx}(2 \ln x + 3e^x)$ **3.** $\dfrac{d}{dx}e^{2x-3}$

4. y' for $y = \ln(2x + 7)$

5. Let $y = \ln u$ and $u = 3 + e^x$.
  (A) Express y in terms of x.
  (B) Use the chain rule to find dy/dx, and then express dy/dx in terms of x.

6. Find y' for $y = y(x)$ defined implicity by the equation $2y^2 - 3x^3 - 5 = 0$, and evaluate at $(x, y) = (1, 2)$.

7. For $y = 3x^2 - 5$, where $x = x(t)$ and $y = y(t)$, find dy/dt if $dx/dt = 3$ when $x = 12$.

8. Given the demand equation $25p + x = 1{,}000$,
  (A) Express the demand x as a function of the price p.
  (B) Find the elasticity of demand, $E(p)$.
  (C) Find $E(15)$ and interpret.
  (D) Express the revenue function as a function of price p.
  (E) If $p = \$25$, what is the effect of a price cut on revenue?

B

9. Find the slope of the line tangent to $y = 100e^{-0.1x}$ when $x = 0$.

10. Use a calculator and a table of values to investigate

$$\lim_{n \to \infty} \left(1 + \frac{2}{n}\right)^n$$

Do you think the limit exists? If so, what do you think it is?

Find the indicated derivatives in Problems 11–16.

11. $\dfrac{d}{dz}[(\ln z)^7 + \ln z^7]$ **12.** $\dfrac{d}{dx}(x^6 \ln x)$

13. $\dfrac{d}{dx}\dfrac{e^x}{x^6}$ **14.** y' for $y = \ln(2x^3 - 3x)$

15. $f'(x)$ for $f(x) = e^{x^3 - x^2}$ **16.** dy/dx for $y = e^{-2x} \ln 5x$

17. Find the equation of the line tangent to the graph of $y = f(x) = 1 + e^{-x}$ at $x = 0$. At $x = -1$.

18. Find y' for $y = y(x)$ defined implicitly by the equation $x^2 - 3xy + 4y^2 = 23$, and find the slope of the graph at $(-1, 2)$.

19. Find x' for $x = x(t)$ defined implicitly by $x^3 - 2t^2x + 8 = 0$, and evaluate at $(t, x) = (-2, 2)$.

20. Find y' for $y = y(x)$ defined implicitly by $x - y^2 = e^y$, and evaluate at $(1, 0)$.

21. Find y' for $y = y(x)$ defined implicitly by $\ln y = x^2 - y^2$, and evaluate at $(1, 1)$.

22. A point is moving on the graph of $y^2 - 4x^2 = 12$ so that its x coordinate is decreasing by 2 units per second when $(x, y) = (1, 4)$. Find the rate of change of the y coordinate.

23. A 17-foot ladder is placed against a wall. If the foot of the ladder is pushed toward the wall at 0.5 foot per second, how fast is the top of the ladder rising when the foot is 8 feet from the wall?

24. Water is leaking onto a floor. The resulting circular pool has an area that is increasing at the rate of 24 square inches per minute. How fast is the radius R of the pool increasing when the radius is 12 inches? $[A = \pi R^2]$

C

25. Find the values of p for which demand is elastic and the values for which demand is inelastic if the price–demand equation is
$$x = f(p) = 20(p - 15)^2 \qquad 0 \le p \le 15$$

26. Graph the revenue function and indicate the regions of inelastic and elastic demand of the graph if the price–demand equation is
$$x = f(p) = 5(20 - p) \qquad 0 \le p \le 20$$

27. Let $y = w^3$, $w = \ln u$, and $u = 4 - e^x$.

(A) Express y in terms of x.

(B) Use the chain rule to find dy/dx, and then express dy/dx in terms of x.

Find the indicated derivatives in Problems 28–30.

28. y' for $y = 5^{x^2 - 1}$

29. $\dfrac{d}{dx} \log_5(x^2 - x)$ **30.** $\dfrac{d}{dx} \sqrt{\ln(x^2 + x)}$

31. Find y' for $y = y(x)$ defined implicitly by the equation $e^{xy} = x^2 + y + 1$, and evaluate at $(0, 0)$.

32. A rock thrown into a still pond causes a circular ripple. The radius is increasing at a constant rate of 3 feet per second. Show that the area does not increase at a constant rate. When is the rate of increase of the area the smallest? The largest? Explain.

33. A point moves along the graph of $y = x^3$ in such a way that its y coordinate is increasing at a constant rate of 5 units per second. Does the x coordinate ever increase at a faster rate than the y coordinate? Explain.

Applications

34. Doubling time. How long will it take money to double if it is invested at 5% interest compounded

(A) Annually?   (B) Continuously?

35. Continuous compound interest. If $100 is invested at 10% interest compounded continuously, then the amount (in dollars) at the end of t years is given by
$$A = 100e^{0.1t}$$
Find $A'(t)$, $A'(1)$, and $A'(10)$.

36. Marginal analysis. The price–demand equation for 14-cubic-foot refrigerators at an appliance store is
$$p(x) = 1,000e^{-0.02x}$$

where x is the monthly demand and p is the price in dollars. Find the marginal revenue equation.

37. Demand equation. Given the demand equation
$$x = \sqrt{5,000 - 2p^3}$$
find the rate of change of p with respect to x by implicit differentiation (x is the number of items that can be sold at a price of p per item).

38. Rate of change of revenue. A company is manufacturing kayaks and can sell all that it manufactures. The revenue (in dollars) is given by
$$R = 750x - \frac{x^2}{30}$$
where the production output in 1 day is x kayaks. If production is increasing at 3 kayaks per day when production is 40 kayaks per day, find the rate of increase in revenue.

39. Revenue and elasticity. The price–demand equation for home-delivered large pizzas is
$$p = 38.2 - 0.002x$$
where x is the number of pizzas delivered weekly. The current price of one pizza is $21. In order to generate additional revenue from the sale of large pizzas, would you recommend a price increase or a price decrease? Explain.

40. Average income. A model for the average income per household before taxes are paid is
$$f(t) = 1,700t + 20,500$$
where t is years since 1980. Find the relative rate of change of household income in 2015.

41. Drug concentration. The drug concentration in the bloodstream t hours after injection is given approximately by
$$C(t) = 5e^{-0.3t}$$
where $C(t)$ is concentration in milligrams per milliliter. What is the rate of change of concentration after 1 hour? After 5 hours?

42. Wound healing. A circular wound on an arm is healing at the rate of 45 square millimeters per day (the area of the wound is decreasing at this rate). How fast is the radius R of the wound decreasing when $R = 15$ millimeters? $[A = \pi R^2]$

43. Psychology: learning. In a computer assembly plant, a new employee, on the average, is able to assemble
$$N(t) = 10(1 - e^{-0.4t})$$
units after t days of on-the-job training.

(A) What is the rate of learning after 1 day? After 5 days?

(B) Find the number of days (to the nearest day) after which the rate of learning is less than 0.25 unit per day.

44. Learning. A new worker on the production line performs an operation in T minutes after x performances of the operation, as given by
$$T = 2\left(1 + \frac{1}{x^{3/2}}\right)$$
If, after performing the operation 9 times, the rate of improvement is $dx/dt = 3$ operations per hour, find the rate of improvement in time dT/dt in performing each operation.

Graphing and Optimization

Introduction

Since the derivative is associated with the slope of the graph of a function at a point, we might expect that it is also related to other properties of a graph. As we will see in this chapter, the derivative can tell us a great deal about the shape of the graph of a function. In particular, we will study methods for finding absolute maximum and minimum values. Manufacturing companies can use these methods to find production levels that will minimize cost or maximize profit, pharmacologists can use them to find levels of drug dosages that will produce maximum sensitivity, and advertisers can use them to determine the number of ads that will maximize the rate of change of sales (see, for example, Problem 93 in Section 5-2).

5-1   First Derivative and Graphs

Increasing and Decreasing Functions

Sign charts will be used throughout this chapter. You may find it helpful to review the terminology and techniques for constructing sign charts in Section 3-2.

EXPLORE & DISCUSS 1

Figure 1 shows the graph of $y = f(x)$ and a sign chart for $f'(x)$, where

$$f(x) = x^3 - 3x$$

and

$$f'(x) = 3x^2 - 3 = 3(x + 1)(x - 1)$$

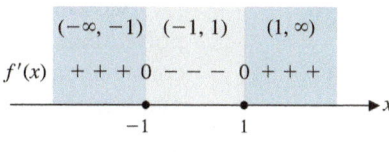

Figure 1

Discuss the relationship between the graph of f and the sign of $f'(x)$ over each interval on which $f'(x)$ has a constant sign. Also, describe the behavior of the graph of f at each partition number for f'.

As they are scanned from left to right, graphs of functions generally have rising and falling sections. If you scan the graph of $f(x) = x^3 - 3x$ in Figure 1 from left to right, you will observe the following:

- On the interval $(-\infty, -1)$, the graph of f is rising, $f(x)$ is increasing,* and the slope of the graph is positive $[f'(x) > 0]$.
- On the interval $(-1, 1)$, the graph of f is falling, $f(x)$ is decreasing, and the slope of the graph is negative $[f'(x) < 0]$.
- On the interval $(1, \infty)$, the graph of f is rising, $f(x)$ is increasing, and the slope of the graph is positive $[f'(x) > 0]$.
- At $x = -1$ and $x = 1$, the slope of the graph is 0 $[f'(x) = 0]$.

If $f'(x) > 0$ (is positive) on the interval (a, b) (Fig. 2), then $f(x)$ increases ($\nearrow$) and the graph of f rises as we move from left to right over the interval. If $f'(x) < 0$ (is negative) on an interval (a, b), then $f(x)$ decreases ($\searrow$) and the graph of f falls as we move from left to right over the interval. We summarize these important results in Theorem 1.

Figure 2

*Formally, we say that the function f is **increasing** on an interval (a, b) if $f(x_2) > f(x_1)$ whenever $a < x_1 < x_2 < b$, and f is **decreasing** on (a, b) if $f(x_2) < f(x_1)$ whenever $a < x_1 < x_2 < b$.

> **THEOREM 1** **Increasing and Decreasing Functions**
> For the interval (a, b),
>
> | $f'(x)$ | $f(x)$ | Graph of f | Examples |
> |---------|--------|--------------|----------|
> | $+$ | Increases ↗ | Rises ↗ | |
> | $-$ | Decreases ↘ | Falls ↘ | |

EXAMPLE 1 **Finding Intervals on Which a Function Is Increasing or Decreasing** Given the function $f(x) = 8x - x^2$,

(A) Which values of x correspond to horizontal tangent lines?

(B) For which values of x is $f(x)$ increasing? Decreasing?

(C) Sketch a graph of f. Add any horizontal tangent lines.

SOLUTION (A) $f'(x) = 8 - 2x = 0$

$$x = 4$$

So, a horizontal tangent line exists at $x = 4$ only.

(B) We will construct a sign chart for $f'(x)$ to determine which values of x make $f'(x) > 0$ and which values make $f'(x) < 0$. Recall from Section 3-2 that the partition numbers for a function are the points where the function is 0 or discontinuous. When constructing a sign chart for $f'(x)$, we must locate all points where $f'(x) = 0$ or $f'(x)$ is discontinuous. From part (A), we know that $f'(x) = 8 - 2x = 0$ at $x = 4$. Since $f'(x) = 8 - 2x$ is a polynomial, it is continuous for all x. So, 4 is the only partition number. We construct a sign chart for the intervals $(-\infty, 4)$ and $(4, \infty)$, using test numbers 3 and 5:

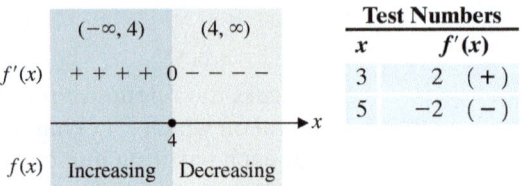

	Test Numbers
x	$f'(x)$
3	2 $(+)$
5	-2 $(-)$

Therefore, $f(x)$ is increasing on $(-\infty, 4)$ and decreasing on $(4, \infty)$.

x	$f(x)$
0	0
2	12
4	16
6	12
8	0

Matched Problem 1 Repeat Example 1 for $f(x) = x^2 - 6x + 10$.

As Example 1 illustrates, the construction of a sign chart will play an important role in using the derivative to analyze and sketch the graph of a function f. The partition numbers for f' are central to the construction of these sign charts and also to the analysis of the graph of $y = f(x)$. We already know that if $f'(c) = 0$, then the graph of $y = f(x)$ will have a horizontal tangent line at $x = c$. But the partition numbers for f' also include the numbers c such that $f'(c)$ does not exist.* There are two possibilities at this type of number: (1) $f(c)$ does not exist; or (2) $f(c)$ exists but the slope of the tangent line at $x = c$ is undefined.

> **DEFINITION   Critical Values**
>
> The values of x in the domain of f where $f'(x) = 0$ or where $f'(x)$ does not exist are called the **critical values** of f.

> **CONCEPTUAL INSIGHT**
>
> The critical values of f are always in the domain of f and are also partition numbers for f', but f' may have partition numbers that are not critical values.
>
> If f is a polynomial, then both the partition numbers for f' and the critical values of f are the solutions of $f'(x) = 0$.

EXAMPLE 2   **Partition Numbers and Critical Values** Find the critical values of f, the intervals on which f is increasing, and those on which f is decreasing, for $f(x) = 1 + x^3$.

SOLUTION   Begin by finding the partition number for $f'(x)$:

$$f'(x) = 3x^2 = 0, \quad \text{only at } x = 0$$

The partition number 0 is in the domain of f, so 0 is the only critical value of f. The sign chart for $f'(x) = 3x^2$ (partition number is 0) is

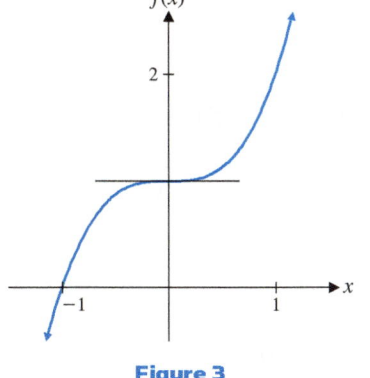

	$(-\infty, 0)$	$(0, \infty)$
$f'(x)$	+ + + + + 0 + + + + +	
$f(x)$	Increasing	Increasing

Test Numbers	
x	$f'(x)$
-1	3 (+)
1	3 (+)

Figure 3

The sign chart indicates that $f(x)$ is increasing on $(-\infty, 0)$ and $(0, \infty)$. Since f is continuous at $x = 0$, it follows that $f(x)$ is increasing for all x. The graph of f is shown in Figure 3.

Matched Problem 2   Find the critical values of f, the intervals on which f is increasing, and those on which f is decreasing, for $f(x) = 1 - x^3$.

EXAMPLE 3   **Partition Numbers and Critical Values** Find the critical values of f, the intervals on which f is increasing, and those on which f is decreasing, for $f(x) = (1 - x)^{1/3}$.

SOLUTION

$$f'(x) = -\frac{1}{3}(1 - x)^{-2/3} = \frac{-1}{3(1 - x)^{2/3}}$$

---

*We are assuming that $f'(c)$ does not exist at any point of discontinuity of f'. There do exist functions f such that f' is discontinuous at $x = c$, yet $f'(c)$ exists. However, we do not consider such functions in this book.

To find partition numbers for f', we note that f' is continuous for all x, except for values of x for which the denominator is 0; that is, $f'(1)$ does not exist and f' is discontinuous at $x = 1$. Since the numerator is the constant -1, $f'(x) \neq 0$ for any value of x. Thus, $x = 1$ is the only partition number for f'. Since 1 is in the domain of f, $x = 1$ is also the only critical value of f. When constructing the sign chart for f' we use the abbreviation ND to note the fact that $f'(x)$ is *not defined* at $x = 1$.

The sign chart for $f'(x) = -1/[3(1 - x)^{2/3}]$ (partition number is 1) is as follows:

$f(x)$

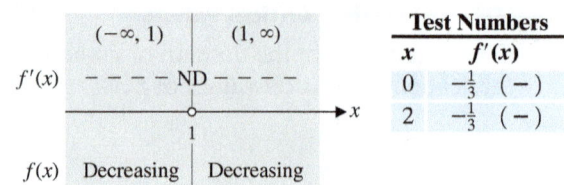

The sign chart indicates that f is decreasing on $(-\infty, 1)$ and $(1, \infty)$. Since f is continuous at $x = 1$, it follows that $f(x)$ is decreasing for all x. **A continuous function can be decreasing (or increasing) on an interval containing values of x where $f'(x)$ does not exist.** The graph of f is shown in Figure 4. Notice that the undefined derivative at $x = 1$ results in a vertical tangent line at $x = 1$. **A vertical tangent will occur at $x = c$ if f is continuous at $x = c$ and if $|f'(x)|$ becomes larger and larger as x approaches c.**

Figure 4

Matched Problem 3 Find the critical values of f, the intervals on which f is increasing, and those on which f is decreasing, for $f(x) = (1 + x)^{1/3}$.

EXAMPLE 4 **Partition Numbers and Critical Values** Find the critical values of f, the intervals on which f is increasing, and those on which f is decreasing, for $f(x) = \dfrac{1}{x - 2}$.

SOLUTION

$$f(x) = \frac{1}{x - 2} = (x - 2)^{-1}$$

$$f'(x) = -(x - 2)^{-2} = \frac{-1}{(x - 2)^2}$$

To find the partition numbers for f', note that $f'(x) \neq 0$ for any x and f' is not defined at $x = 2$. Thus, $x = 2$ is the only partition number for f'. However, $x = 2$ is *not* in the domain of f. Consequently, $x = 2$ is not a critical value of f. This function has no critical values.

The sign chart for $f'(x) = -1/(x - 2)^2$ (partition number is 2) is as follows:

$f(x)$

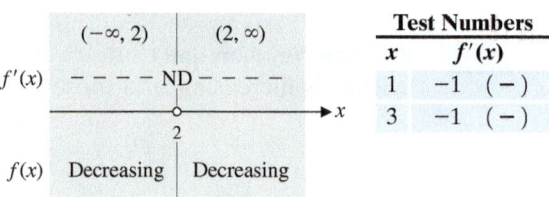

Therefore, f is decreasing on $(-\infty, 2)$ and $(2, \infty)$. The graph of f is shown in Figure 5.

Figure 5

Matched Problem 4   Find the critical values for f, the intervals on which f is increasing, and those on which f is decreasing, for $f(x) = \dfrac{1}{x}$.

EXAMPLE 5   **Partition Numbers and Critical Values**   Find the critical values of f, the intervals on which f is increasing, and those on which f is decreasing, for $f(x) = 8 \ln x - x^2$.

SOLUTION   The natural logarithm function $\ln x$ is defined on $(0, \infty)$, or $x > 0$, so $f(x)$ is defined only for $x > 0$.

$$f(x) = 8 \ln x - x^2, x > 0$$

$$f'(x) = \frac{8}{x} - 2x \qquad \text{\textit{Find a common denominator.}}$$

$$= \frac{8}{x} - \frac{2x^2}{x} \qquad \text{\textit{Subtract numerators.}}$$

$$= \frac{8 - 2x^2}{x} \qquad \text{\textit{Factor numerator.}}$$

$$= \frac{2(2 - x)(2 + x)}{x}, \quad x > 0$$

Note that $f'(x) = 0$ at -2 and at 2, and $f'(x)$ is discontinuous at 0. These are the partition numbers for $f'(x)$. Since the domain of f is $(0, \infty)$, 0 and -2 are not critical values. The remaining partition number, 2, is the only critical value for $f(x)$.

The sign chart for $f'(x) = \dfrac{2(2 - x)(2 + x)}{x}, x > 0$ (partition number is 2), is as follows:

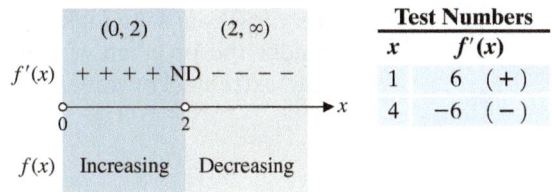

Therefore, f is increasing on $(0, 2)$ and decreasing on $(2, \infty)$. The graph of f is shown in Figure 6.

Figure 6

Matched Problem 5   Find the critical values of f, the intervals on which f is increasing, and those on which f is decreasing, for $f(x) = 5 \ln x - x$.

CONCEPTUAL INSIGHT

Examples 4 and 5 illustrate two important ideas:

1. Do not assume that all partition numbers for the derivative f' are critical values of the function f. To be a critical value, a partition number must also be in the domain of f.

2. The values for which a function is increasing or decreasing must always be expressed in terms of open intervals that are subsets of the domain of the function.

Local Extrema

When the graph of a continuous function changes from rising to falling, a high point, or *local maximum,* occurs. When the graph changes from falling to rising, a low point, or *local minimum,* occurs. In Figure 7, high points occur at c_3 and c_6, and low points occur at c_2 and c_4. In general, we call $f(c)$ a **local maximum** if there exists an interval (m, n) containing c such that

$$f(x) \leq f(c) \qquad \text{for all } x \text{ in } (m, n)$$

Note that this inequality need hold only for values of x near c, which is why we use the term *local.*

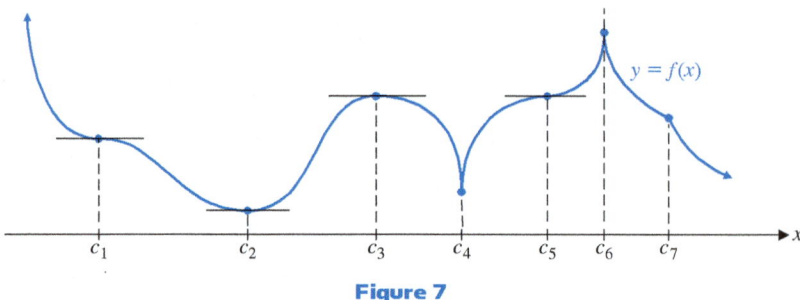

Figure 7

The quantity $f(c)$ is called a **local minimum** if there exists an interval (m, n) containing c such that

$$f(x) \geq f(c) \qquad \text{for all } x \text{ in } (m, n)$$

The quantity $f(c)$ is called a **local extremum** if it is either a local maximum or a local minimum. A point on a graph where a local extremum occurs is also called a **turning point.** In Figure 7 we see that local maxima occur at c_3 and c_6, local minima occur at c_2 and c_4, and all four values produce local extrema. Also, the local maximum $f(c_3)$ is not the highest point on the graph in Figure 7. Later in this chapter, we consider the problem of finding the highest and lowest points on a graph, or absolute extrema. For now, we are concerned only with locating *local* extrema.

EXAMPLE 6 **Analyzing a Graph** Use the graph of f in Figure 8 to find the intervals on which f is increasing, those on which f is decreasing, any local maxima, and any local minima.

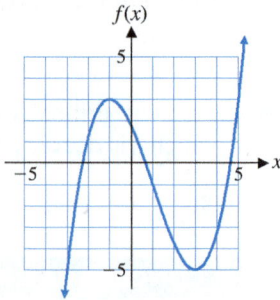

Figure 8

SOLUTION The function f is increasing (the graph is rising) on $(-\infty, -1)$ and on $(3, \infty)$ and is decreasing (the graph is falling) on $(-1, 3)$. Because the graph changes from rising to falling at $x = -1$, $f(-1) = 3$ is a local maximum. Because the graph changes from falling to rising at $x = 3$, $f(3) = -5$ is a local minimum.

Matched Problem 6 Use the graph of g in Figure 9 to find the intervals on which g is increasing, those on which g is decreasing, any local maxima, and any local minima.

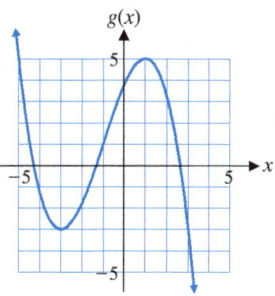

Figure 9

How can we locate local maxima and minima if we are given the equation of a function and not its graph? The key is to examine the critical values of the function. The local extrema of the function f in Figure 7 occur either at points where the derivative is 0 (c_2 and c_3) or at points where the derivative does not exist (c_4 and c_6). In other words, local extrema occur only at critical values of f. Theorem 2 shows that this is true in general.

THEOREM 2 Existence of Local Extrema

If f is continuous on the interval (a, b), c is a number in (a, b), and $f(c)$ is a local extremum, then either $f'(c) = 0$ or $f'(c)$ does not exist (is not defined).

Theorem 2 states that a local extremum can occur only at a critical value, but it does not imply that every critical value produces a local extremum. In Figure 7, c_1 and c_5 are critical values (the slope is 0), but the function does not have a local maximum or local minimum at either of these values.

Our strategy for finding local extrema is now clear: We find all critical values of f and test each one to see if it produces a local maximum, a local minimum, or neither.

First-Derivative Test

If $f'(x)$ exists on both sides of a critical value c, the sign of $f'(x)$ can be used to determine whether the point $(c, f(c))$ is a local maximum, a local minimum, or neither. The various possibilities are summarized in the following box and are illustrated in Figure 10:

PROCEDURE First-Derivative Test for Local Extrema

Let c be a critical value of f [$f(c)$ is defined and either $f'(c) = 0$ or $f'(c)$ is not defined]. Construct a sign chart for $f'(x)$ close to and on either side of c.

Sign Chart

$f(c)$

$f(c)$ is a local minimum.

If $f'(x)$ changes from negative to positive at c, then $f(c)$ is a local minimum.

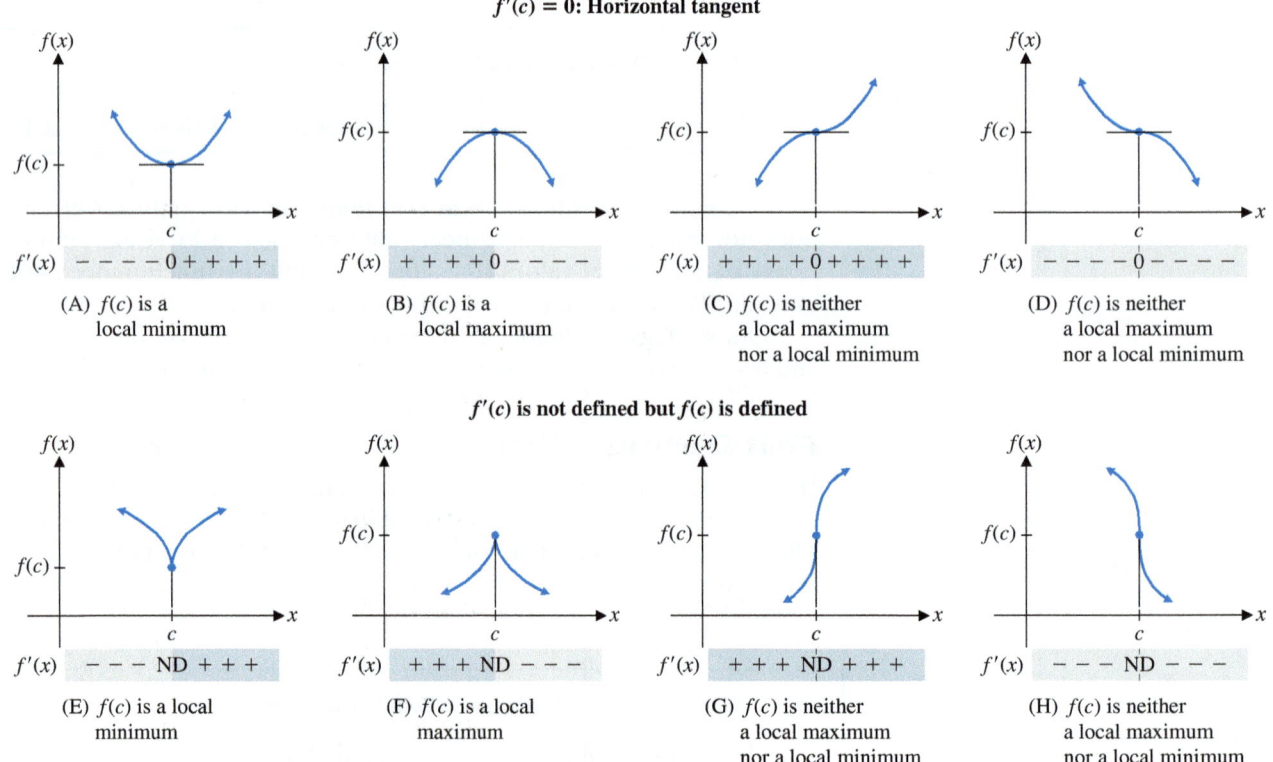

Figure 10 Local extrema

EXAMPLE 7 **Locating Local Extrema** Given $f(x) = x^3 - 6x^2 + 9x + 1$,

(A) Find the critical values of f.

(B) Find the local maxima and minima.

(C) Sketch the graph of f.

SOLUTION   (A) Find all numbers x in the domain of f where $f'(x) = 0$ or $f'(x)$ does not exist.

$$f'(x) = 3x^2 - 12x + 9 = 0$$
$$3(x^2 - 4x + 3) = 0$$
$$3(x - 1)(x - 3) = 0$$
$$x = 1 \quad \text{or} \quad x = 3$$

$f'(x)$ exists for all x; the critical values are $x = 1$ and $x = 3$.

(B) The easiest way to apply the first-derivative test for local maxima and minima is to construct a sign chart for $f'(x)$ for all x. Partition numbers for $f'(x)$ are $x = 1$ and $x = 3$ (which also happen to be critical values of f).

Sign chart for $f'(x) = 3(x - 1)(x - 3)$:

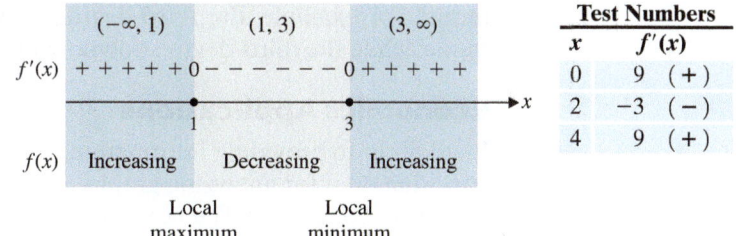

Test Numbers	
x	$f'(x)$
0	9 (+)
2	−3 (−)
4	9 (+)

The sign chart indicates that f increases on $(-\infty, 1)$, has a local maximum at $x = 1$, decreases on $(1, 3)$, has a local minimum at $x = 3$, and increases on $(3, \infty)$. These facts are summarized in the following table:

x	$f'(x)$	$f(x)$	Graph of f
$(-\infty, 1)$	+	Increasing	Rising
$x = 1$	0	Local maximum	Horizontal tangent
$(1, 3)$	−	Decreasing	Falling
$x = 3$	0	Local minimum	Horizontal tangent
$(3, \infty)$	+	Increasing	Rising

(C) We sketch a graph of f, using the information from part (B) and point-by-point plotting.

x	$f(x)$
0	1
1	5
2	3
3	1
4	5

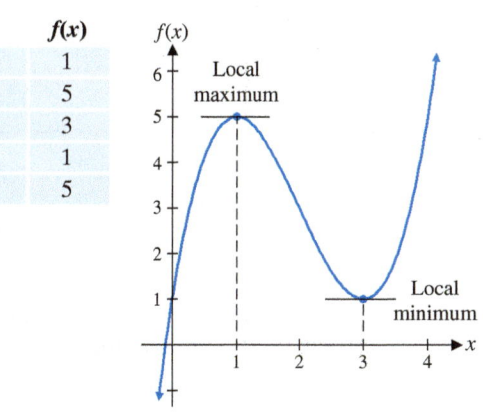

Matched Problem 7   Given $f(x) = x^3 - 9x^2 + 24x - 10$,

(A) Find the critical values of f.

(B) Find the local maxima and minima.

(C) Sketch a graph of f.

How can you tell if you have found all the local extrema of a function? In general, this can be a difficult question to answer. However, in the case of a polynomial function, there is an easily determined upper limit on the number of local extrema. Since the local extrema are the x intercepts of the derivative, this limit is a consequence of the number of x intercepts of a polynomial. The relevant information is summarized in the following theorem, which is stated without proof:

THEOREM 3 Intercepts and Local Extrema of Polynomial Functions

If $f(x) = a_n x^n + a_{n-1} x^{n-1} + \cdots + a_1 x + a_0$, $a_n \neq 0$, is an nth-degree polynomial, then f has at most n x intercepts and at most $n - 1$ local extrema.

Theorem 3 does not guarantee that every nth-degree polynomial has exactly $n - 1$ local extrema; it says only that there can never be more than $n - 1$ local extrema. For example, the third-degree polynomial in Example 7 has two local extrema, while the third-degree polynomial in Example 2 does not have any.

Economics Applications

In addition to providing information for hand-sketching graphs, the derivative is an important tool for analyzing graphs and discussing the interplay between a function and its rate of change. The next two examples illustrate this process in the context of economics applications.

EXAMPLE 8 **Agricultural Exports and Imports** Over the past few decades, the United States has exported more agricultural products than it has imported, maintaining a positive balance of trade in this area. However, the trade balance fluctuated considerably during that period. The graph in Figure 11 approximates the rate of change of the balance of trade over a 15-year period, where $B(t)$ is the balance of trade (in billions of dollars) and t is time (in years).

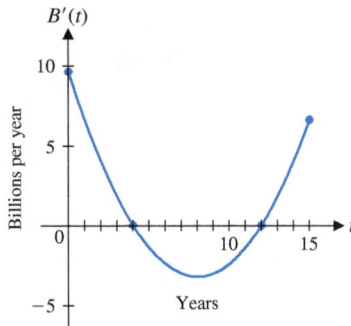

Figure 11 Rate of change of the balance of trade

(A) Write a brief description of the graph of $y = B(t)$, including a discussion of any local extrema.

(B) Sketch a possible graph of $y = B(t)$.

SOLUTION (A) The graph of the derivative $y = B'(t)$ contains the same essential information as a sign chart. That is, we see that $B'(t)$ is positive on $(0, 4)$, 0 at $t = 4$, negative on $(4, 12)$, 0 at $t = 12$, and positive on $(12, 15)$. The trade balance increases for the first 4 years to a local maximum, decreases for the next 8 years to a local minimum, and then increases for the final 3 years.

(B) Without additional information concerning the actual values of $y = B(t)$, we cannot produce an accurate graph. However, we can sketch a possible graph that illustrates the important features, as shown in Figure 12. The absence of a scale on the vertical axis is a consequence of the lack of information about the values of $B(t)$.

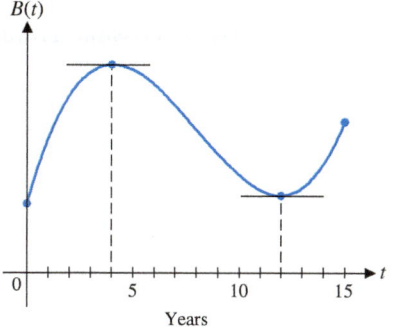

Figure 12 Balance of trade

Matched Problem 8 The graph in Figure 13 approximates the rate of change of the U.S. share of the total world production of motor vehicles over a 20-year period, where $S(t)$ is the U.S. share (as a percentage) and t is time (in years).

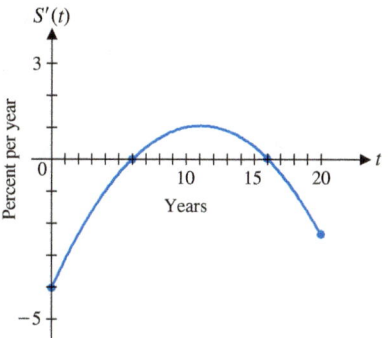

Figure 13

✎ (A) Write a brief description of the graph of $y = S(t)$, including a discussion of any local extrema.

(B) Sketch a possible graph of $y = S(t)$.

EXAMPLE 9 **Revenue Analysis** The graph of the total revenue $R(x)$ (in dollars) from the sale of x bookcases is shown in Figure 14.

✎ (A) Write a brief description of the graph of the marginal revenue function $y = R'(x)$, including a discussion of any x intercepts.

(B) Sketch a possible graph of $y = R'(x)$.

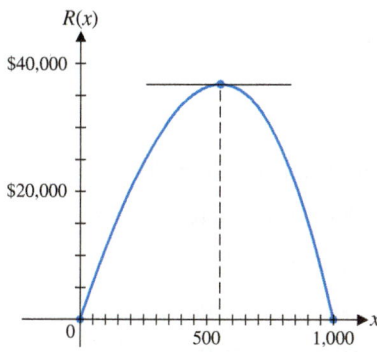

Figure 14 Revenue

SOLUTION   (A) The graph of $y = R(x)$ indicates that $R(x)$ increases on $(0, 550)$, has a local maximum at $x = 550$, and decreases on $(550, 1,000)$. Consequently, the marginal revenue function $R'(x)$ must be positive on $(0, 550)$, 0 at $x = 550$, and negative on $(550, 1,000)$.

(B) A possible graph of $y = R'(x)$ illustrating the information summarized in part (A) is shown in Figure 15.

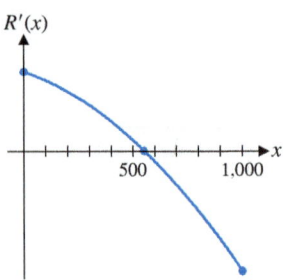

Figure 15 Marginal revenue

Matched Problem 9   The graph of the total revenue $R(x)$ (in dollars) from the sale of x desks is shown in Figure 16.

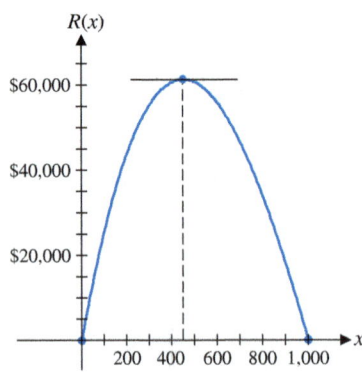

Figure 16

(A) Write a brief description of the graph of the marginal revenue function $y = R'(x)$, including a discussion of any x intercepts.

(B) Sketch a possible graph of $y = R'(x)$.

Comparing Examples 8 and 9, we see that we were able to obtain more information about the function from the graph of its derivative (Example 8) than we were when the process was reversed (Example 9). In the next section, we introduce some ideas that will help us obtain additional information about the derivative from the graph of the function.

Exercises 5-1

A

Problems 1–8 refer to the following graph of $y = f(x)$:

1. Identify the intervals on which $f(x)$ is increasing.

2. Identify the intervals on which $f(x)$ is decreasing.

3. Identify the intervals on which $f'(x) < 0$.

4. Identify the intervals on which $f'(x) > 0$.

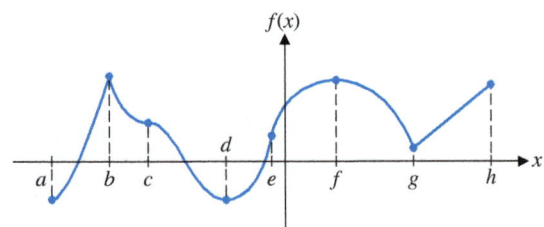

Figure for 1–8

5. Identify the x coordinates of the points where $f'(x) = 0$.

6. Identify the x coordinates of the points where $f'(x)$ does not exist.

7. Identify the x coordinates of the points where $f(x)$ has a local maximum.

8. Identify the x coordinates of the points where $f(x)$ has a local minimum.

In Problems 9 and 10, $f(x)$ is continuous on $(-\infty, \infty)$ and has critical values at $x = a, b, c,$ and d. Use the sign chart for $f'(x)$ to determine whether f has a local maximum, a local minimum, or neither at each critical value.

9.

10.

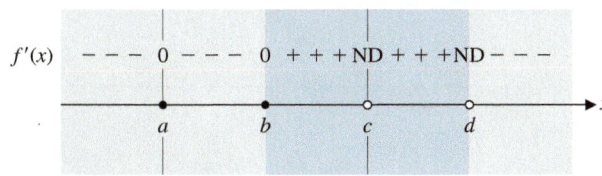

In Problems 11–18, match the graph of f with one of the sign charts a–h in the figure.

11.

12.

13.

14.

15.

16.

17.

18.

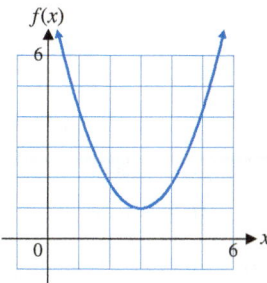

(a)

$f'(x)$ $- - - - - - - -$ 0 $+ + + + + + + +$

x — 3

(b)

$f'(x)$ $- - - - - - - -$ ND $+ + + + + + +$

x — 3

(c)

$f'(x)$ $+ + + + + + + +$ 0 $+ + + + + + + +$

x — 3

(d)

$f'(x)$ $+ + + + + + + +$ ND $+ + + + + + + +$

x — 3

(e)

$f'(x)$ $+ + + + + + + +$ 0 $- - - - - - - -$

x — 3

(f)

$f'(x)$ $+ + + + + + + +$ ND $- - - - - - - -$

x — 3

(g)

$f'(x)$ $- - - - - - - -$ ND $- - - - - - - -$

x — 3

(h)

$f'(x)$ $- - - - - - - -$ 0 $- - - - - - - -$

x — 3

Figure for 11–18

B

In Problems 19–26, find (A) $f'(x)$, (B) the critical values of f, and (C) the partition numbers for f'.

19. $f(x) = x^3 - 12x + 8$

20. $f(x) = x^3 - 27x + 30$

21. $f(x) = (x + 5)^{1/3}$

22. $f(x) = (x - 9)^{2/3}$

23. $f(x) = \dfrac{6}{x + 2}$

24. $f(x) = \dfrac{5}{x - 4}$

25. $f(x) = |x|$

26. $f(x) = |x + 3|$

In Problems 27–40, find the intervals on which f(x) is increasing, the intervals on which f(x) is decreasing, and the local extrema.

27. $f(x) = 2x^2 - 4x$

28. $f(x) = -3x^2 - 12x$

29. $f(x) = -2x^2 - 16x - 25$

30. $f(x) = -3x^2 + 12x - 5$

31. $f(x) = x^3 + 4x - 5$

32. $f(x) = -x^3 - 4x + 8$

33. $f(x) = 2x^3 - 3x^2 - 36x$

34. $f(x) = -2x^3 + 3x^2 + 120x$

35. $f(x) = 3x^4 - 4x^3 + 5$

36. $f(x) = x^4 + 2x^3 + 5$

37. $f(x) = (x - 1)e^{-x}$

38. $f(x) = x \ln x - x$

39. $f(x) = 4x^{1/3} - x^{2/3}$

40. $f(x) = (x^2 - 9)^{2/3}$

In Problems 41–46, use a graphing calculator to approximate the critical values of f(x) to two decimal places. Find the intervals on which f(x) is increasing, the intervals on which f(x) is decreasing, and the local extrema.

41. $f(x) = x^4 - 4x^3 + 9x$

42. $f(x) = x^4 + 5x^3 - 15x$

43. $f(x) = x \ln x - (x - 2)^3$

44. $f(x) = e^{-x} - 3x^2$

45. $f(x) = e^x - 2x^2$

46. $f(x) = \dfrac{\ln x}{x} - 5x + x^2$

In Problems 47–54, find the intervals on which f(x) is increasing and the intervals on which f(x) is decreasing. Then sketch the graph. Add horizontal tangent lines.

47. $f(x) = 4 + 8x - x^2$

48. $f(x) = 2x^2 - 8x + 9$

49. $f(x) = x^3 - 3x + 1$

50. $f(x) = x^3 - 12x + 2$

51. $f(x) = 10 - 12x + 6x^2 - x^3$

52. $f(x) = x^3 + 3x^2 + 3x$

53. $f(x) = x^4 - 18x^2$

54. $f(x) = -x^4 + 50x^2$

In Problems 55–62, f(x) is continuous on $(-\infty, \infty)$. Use the given information to sketch the graph of f.

55.

x	-2	-1	0	1	2
$f(x)$	-1	1	2	3	1

56.

x	-2	-1	0	1	2
$f(x)$	1	3	2	1	-1

57.

x	-2	-1	0	2	4
$f(x)$	2	1	2	1	0

58.

x	-2	-1	0	2	3
$f(x)$	-3	0	2	-1	0

59. $f(-2) = 4, f(0) = 0, f(2) = -4$;

$f'(-2) = 0, f'(0) = 0, f'(2) = 0$;

$f'(x) > 0$ on $(-\infty, -2)$ and $(2, \infty)$;

$f'(x) < 0$ on $(-2, 0)$ and $(0, 2)$

60. $f(-2) = -1, f(0) = 0, f(2) = 1$;

$f'(-2) = 0, f'(2) = 0$;

$f'(x) > 0$ on $(-\infty, -2), (-2, 2),$ and $(2, \infty)$

61. $f(-1) = 2, f(0) = 0, f(1) = -2$;

$f'(-1) = 0, f'(1) = 0, f'(0)$ is not defined;

$f'(x) > 0$ on $(-\infty, -1)$ and $(1, \infty)$;

$f'(x) < 0$ on $(-1, 0)$ and $(0, 1)$

62. $f(-1) = 2, f(0) = 0, f(1) = 2$;

$f'(-1) = 0, f'(1) = 0, f'(0)$ is not defined;

$f'(x) > 0$ on $(-\infty, -1)$ and $(0, 1)$;

$f'(x) < 0$ on $(-1, 0)$ and $(1, \infty)$

Problems 63–68 involve functions f_1–f_6 and their derivatives, g_1–g_6. Use the graphs shown in figures (A) and (B) to match each function f_i with its derivative g_j.

63. f_1 **64.** f_2 **65.** f_3

66. f_4 **67.** f_5 **68.** f_6

Figure (A) for 63–68

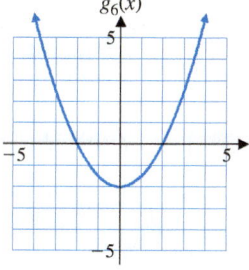

Figure (B) for 63–68

In Problems 69–74, use the given graph of $y = f'(x)$ to find the intervals on which f is increasing, the intervals on which f is decreasing, and the local extrema. Sketch a possible graph of $y = f(x)$.

69.

70.

71.

72.

73.

74.
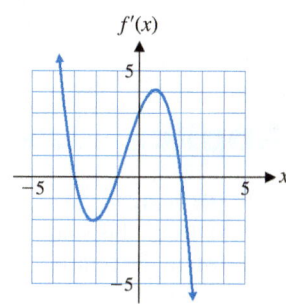

In Problems 75–78, use the given graph of $y = f(x)$ to find the intervals on which $f'(x) > 0$, the intervals on which $f'(x) < 0$, and the values of x for which $f'(x) = 0$. Sketch a possible graph of $y = f'(x)$.

75.

76.

77.

78.

In Problems 79–86, find the critical values, the intervals on which f(x) is increasing, the intervals on which f(x) is decreasing, and the local extrema. Do not graph.

79. $f(x) = x + \dfrac{4}{x}$

80. $f(x) = \dfrac{9}{x} + x$

81. $f(x) = 1 + \dfrac{1}{x} + \dfrac{1}{x^2}$

82. $f(x) = 3 - \dfrac{4}{x} - \dfrac{2}{x^2}$

83. $f(x) = \dfrac{x^2}{x - 2}$

84. $f(x) = \dfrac{x^2}{x + 1}$

85. $f(x) = x^4(x - 6)^2$

86. $f(x) = x^3(x - 5)^2$

87. Let $f(x) = x^3 + kx$, where k is a constant. Discuss the number of local extrema and the shape of the graph of f if
(A) $k > 0$ (B) $k < 0$ (C) $k = 0$

88. Let $f(x) = x^4 + kx^2$, where k is a constant. Discuss the number of local extrema and the shape of the graph of f if
(A) $k > 0$ (B) $k < 0$ (C) $k = 0$

Applications

89. Profit analysis. The graph of the total profit $P(x)$ (in dollars) from the sale of x cordless electric screwdrivers is shown in the figure.

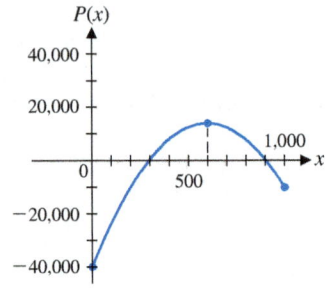

Figure for 89

(A) Write a brief description of the graph of the marginal profit function $y = P'(x)$, including a discussion of any x intercepts.

(B) Sketch a possible graph of $y = P'(x)$.

90. Revenue analysis. The graph of the total revenue $R(x)$ (in dollars) from the sale of x cordless electric screwdrivers is shown in the figure.

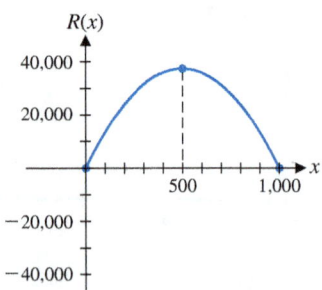

Figure for 90

(A) Write a brief description of the graph of the marginal revenue function $y = R'(x)$, including a discussion of any x intercepts.

(B) Sketch a possible graph of $y = R'(x)$.

91. Price analysis. The figure approximates the rate of change of the price of bacon over a 70-month period, where $B(t)$ is the price of a pound of sliced bacon (in dollars) and t is time (in months).

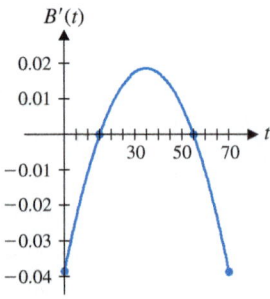

Figure for 91

(A) Write a brief description of the graph of $y = B(t)$, including a discussion of any local extrema.

(B) Sketch a possible graph of $y = B(t)$.

92. Price analysis. The figure approximates the rate of change of the price of eggs over a 70-month period, where $E(t)$ is the price of a dozen eggs (in dollars) and t is time (in months).

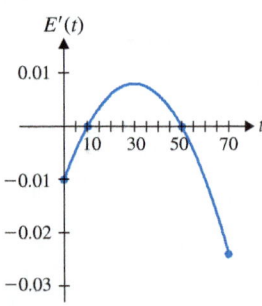

Figure for 92

(A) Write a brief description of the graph of $y = E(t)$, including a discussion of any local extrema.

(B) Sketch a possible graph of $y = E(t)$.

93. Average cost. A manufacturer incurs the following costs in producing x water ski vests in one day, for $0 < x < 150$: fixed costs, \$320; unit production cost, \$20 per vest; equipment maintenance and repairs, $0.05x^2$ dollars. So, the cost of manufacturing x vests in one day is given by

$$C(x) = 0.05x^2 + 20x + 320 \qquad 0 < x < 150$$

(A) What is the average cost $\overline{C}(x)$ per vest if x vests are produced in one day?

(B) Find the critical values of $\overline{C}(x)$, the intervals on which the average cost per vest is decreasing, the intervals on which the average cost per vest is increasing, and the local extrema. Do not graph.

94. Average cost. A manufacturer incurs the following costs in producing x rain jackets in one day for $0 < x < 200$: fixed costs, \$450; unit production cost, \$30 per jacket; equipment maintenance and repairs, $0.08x^2$ dollars.

(A) What is the average cost $\overline{C}(x)$ per jacket if x jackets are produced in one day?

(B) Find the critical values of $\overline{C}(x)$, the intervals on which the average cost per jacket is decreasing, the intervals on which the average cost per jacket is increasing, and the local extrema. Do not graph.

95. Marginal analysis. Show that profit will be increasing over production intervals (a, b) for which marginal revenue is greater than marginal cost. [*Hint:* $P(x) = R(x) - C(x)$]

96. Marginal analysis. Show that profit will be decreasing over production intervals (a, b) for which marginal revenue is less than marginal cost.

97. Medicine. A drug is injected into the bloodstream of a patient through the right arm. The drug concentration in the bloodstream of the left arm t hours after the injection is approximated by

$$C(t) = \frac{0.28t}{t^2 + 4} \qquad 0 < t < 24$$

Find the critical values of $C(t)$, the intervals on which the drug concentration is increasing, the intervals on which concentration of the drug is decreasing, and the local extrema. Do not graph.

98. Medicine. The concentration $C(t)$, in milligrams per cubic centimeter, of a particular drug in a patient's bloodstream is given by

$$C(t) = \frac{0.3t}{t^2 + 6t + 9} \qquad 0 < t < 12$$

where t is the number of hours after the drug is taken orally. Find the critical values of $C(t)$, the intervals on which the drug concentration is increasing, the intervals on which the drug concentration is decreasing, and the local extrema. Do not graph.

Answers to Matched Problems

1. (A) Horizontal tangent line at $x = 3$.

(B) Decreasing on $(-\infty, 3)$; increasing on $(3, \infty)$

(C)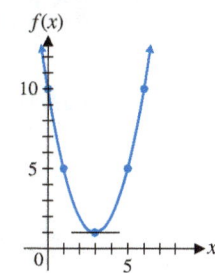

2. Partition number: $x = 0$; critical value: $x = 0$; decreasing for all x

3. Partition number: $x = -1$; critical value: $x = -1$; increasing for all x

4. Partition number: $x = 0$; no critical values; decreasing on $(-\infty, 0)$ and $(0, \infty)$

5. Partition number: $x = 5$; critical value: $x = 5$; increasing on $(0, 5)$; decreasing on $(5, \infty)$

6. Increasing on $(-3, 1)$; decreasing on $(-\infty, -3)$ and $(1, \infty)$; local maximum at $x = 1$; local minimum at $x = -3$

7. (A) Critical values: $x = 2$, $x = 4$

(B) Local maximum at $x = 2$; local minimum at $x = 4$

(C)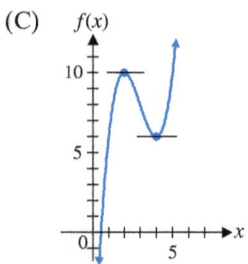

8. (A) The U.S. share of the world market decreases for 6 years to a local minimum, increases for the next 10 years to a local maximum, and then decreases for the final 4 years.

(B)

9. (A) The marginal revenue is positive on $(0, 450)$, 0 at $x = 450$, and negative on $(450, 1,000)$.

(B)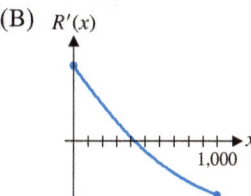

5-2 Second Derivative and Graphs

In Section 5-1, we saw that the derivative can be used when a graph is rising and falling. Now we want to see what the *second derivative* (the derivative of the derivative) can tell us about the shape of a graph.

Using Concavity as a Graphing Tool

Consider the functions

$$f(x) = x^2 \quad \text{and} \quad g(x) = \sqrt{x}$$

for x in the interval $(0, \infty)$. Since

$$f'(x) = 2x > 0 \quad \text{for } 0 < x < \infty$$

and

$$g'(x) = \frac{1}{2\sqrt{x}} > 0 \quad \text{for } 0 < x < \infty$$

both functions are increasing on $(0, \infty)$.

EXPLORE & DISCUSS 1

(A) Discuss the difference in the shapes of the graphs of f and g shown in Figure 1.

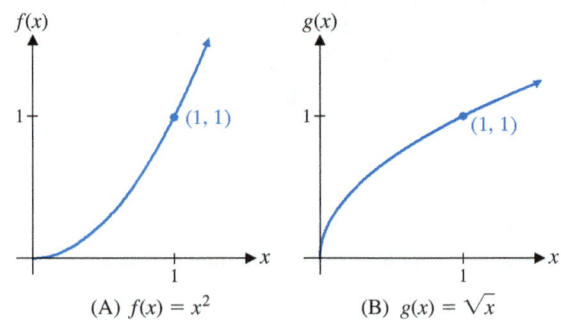

(A) $f(x) = x^2$ (B) $g(x) = \sqrt{x}$

Figure 1

(B) Complete the following table, and discuss the relationship between the values of the derivatives of f and g and the shapes of their graphs:

x	0.25	0.5	0.75	1
$f'(x)$				
$g'(x)$				

We use the term *concave upward* to describe a graph that opens upward and *concave downward* to describe a graph that opens downward. Thus, the graph of f in Figure 1A is concave upward, and the graph of g in Figure 1B is concave downward. Finding a mathematical formulation of concavity will help us sketch and analyze graphs.

We examine the slopes of f and g at various points on their graphs (see Fig. 2) and make two observations about each graph:

1. Looking at the graph of f in Figure 2A, we see that $f'(x)$ (the slope of the tangent line) is *increasing* and that the graph lies *above* each tangent line;

2. Looking at Figure 2B, we see that $g'(x)$ is *decreasing* and that the graph lies *below* each tangent line.

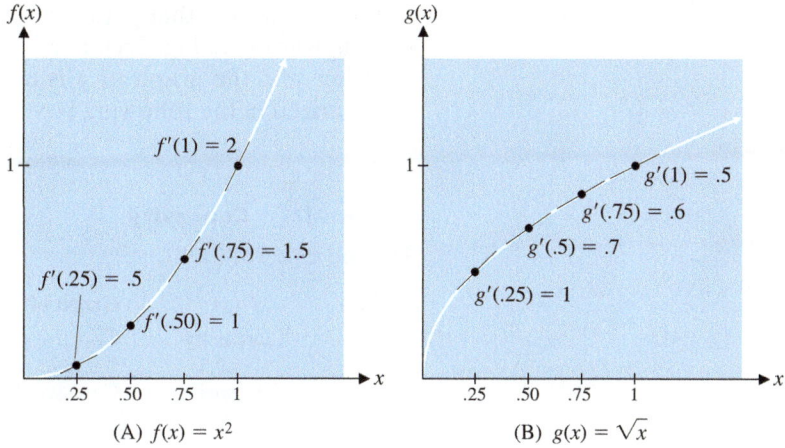

(A) $f(x) = x^2$

(B) $g(x) = \sqrt{x}$

Figure 2

With these ideas in mind, we can state the general definition of concavity.

DEFINITION   Concavity

The graph of a function f is **concave upward** on the interval (a, b) if $f'(x)$ is *increasing* on (a, b) and is **concave downward** on the interval (a, b) if $f'(x)$ is *decreasing* on (a, b).

Geometrically, the graph is concave upward on (a, b) if it lies above its tangent lines in (a, b) and is concave downward on (a, b) if it lies below its tangent lines in (a, b).

How can we determine when $f'(x)$ is increasing or decreasing? In Section 5-1, we used the derivative to determine when a function is increasing or decreasing. To determine when the function $f'(x)$ is increasing or decreasing, we use the derivative of $f'(x)$. The derivative of the derivative of a function is called the *second derivative* of the function. Various notations for the second derivative are given in the following box:

NOTATION   Second Derivative

For $y = f(x)$, the **second derivative** of f, provided that it exists, is

$$f''(x) = \frac{d}{dx}f'(x)$$

Other notations for $f''(x)$ are

$$\frac{d^2y}{dx^2} \quad \text{and} \quad y''$$

Returning to the functions f and g discussed at the beginning of this section, we have

$$f(x) = x^2 \qquad\qquad g(x) = \sqrt{x} = x^{1/2}$$

$$f'(x) = 2x \qquad\qquad g'(x) = \frac{1}{2}x^{-1/2} = \frac{1}{2\sqrt{x}}$$

$$f''(x) = \frac{d}{dx}2x = 2 \qquad g''(x) = \frac{d}{dx}\frac{1}{2}x^{-1/2} = -\frac{1}{4}x^{-3/2} = -\frac{1}{4\sqrt{x^3}}$$

For $x > 0$, we see that $f''(x) > 0$; so, $f'(x)$ is increasing, and the graph of f is concave upward (see Fig. 2A). For $x > 0$, we also see that $g''(x) < 0$; so, $g'(x)$ is decreasing, and the graph of g is concave downward (see Fig. 2B). These ideas are summarized in the following box:

SUMMARY **Concavity**

For the interval (a, b),

$f''(x)$	$f'(x)$	Graph of $y = f(x)$	Examples
$+$	Increasing	Concave upward	
$-$	Decreasing	Concave downward	

CONCEPTUAL INSIGHT

Be careful not to confuse concavity with falling and rising. A graph that is concave upward on an interval may be falling, rising, or both falling and rising on that interval. A similar statement holds for a graph that is concave downward. See Figure 3.

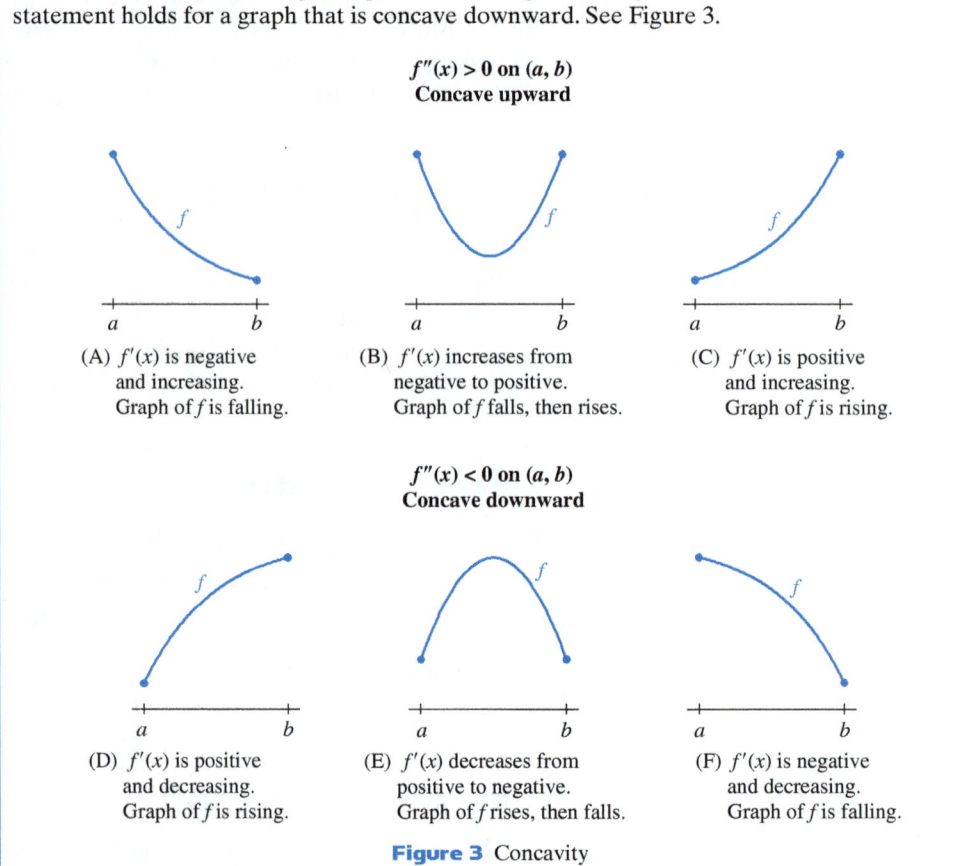

$f''(x) > 0$ on (a, b)
Concave upward

(A) $f'(x)$ is negative and increasing. Graph of f is falling.

(B) $f'(x)$ increases from negative to positive. Graph of f falls, then rises.

(C) $f'(x)$ is positive and increasing. Graph of f is rising.

$f''(x) < 0$ on (a, b)
Concave downward

(D) $f'(x)$ is positive and decreasing. Graph of f is rising.

(E) $f'(x)$ decreases from positive to negative. Graph of f rises, then falls.

(F) $f'(x)$ is negative and decreasing. Graph of f is falling.

Figure 3 Concavity

EXAMPLE 1 **Concavity of Graphs** Determine the intervals on which the graph of each function is concave upward and the intervals on which it is concave downward. Sketch a graph of each function.

(A) $f(x) = e^x$ (B) $g(x) = \ln x$ (C) $h(x) = x^3$

SOLUTION

(A) $f(x) = e^x$

$f'(x) = e^x$

$f''(x) = e^x$

Since $f''(x) > 0$ on $(-\infty, \infty)$, the graph of $f(x) = e^x$ [Fig. 4(A)] is concave upward on $(-\infty, \infty)$.

(B) $g(x) = \ln x$

$g'(x) = \dfrac{1}{x}$

$g''(x) = -\dfrac{1}{x^2}$

The domain of $g(x) = \ln x$ is $(0, \infty)$ and $g''(x) < 0$ on this interval, so the graph of $g(x) = \ln x$ [Fig. 4(B)] is concave downward on $(0, \infty)$.

(C) $h(x) = x^3$

$h'(x) = 3x^2$

$h''(x) = 6x$

Since $h''(x) < 0$ when $x < 0$ and $h''(x) > 0$ when $x > 0$, the graph of $h(x) = x^3$ [Fig. 4(C)] is concave downward on $(-\infty, 0)$ and concave upward on $(0, \infty)$.

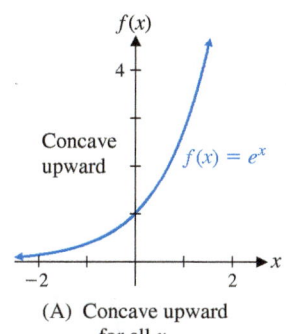

(A) Concave upward for all x.

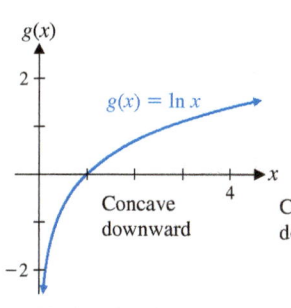

(B) Concave downward for $x > 0$.

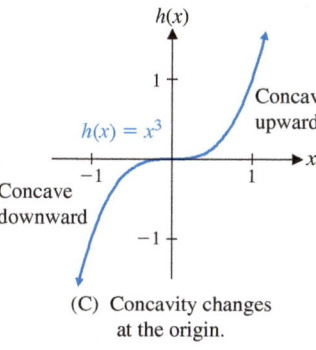

(C) Concavity changes at the origin.

Figure 4

Matched Problem 1   Determine the intervals on which the graph of each function is concave upward and the intervals on which it is cancave downward. Sketch a graph of each function.

(A) $f(x) = -e^{-x}$

(B) $g(x) = \ln \dfrac{1}{x}$

(C) $h(x) = x^{1/3}$

Refer to Example 1. The graphs of $f(x) = e^x$ and $g(x) = \ln x$ never change concavity. But the graph of $h(x) = x^3$ changes concavity at $(0, 0)$. This point is called an *inflection point*.

Finding Inflection Points

An **inflection point** is a point on the graph of the function where the concavity changes (from upward to downward or from downward to upward). For the concavity to change at a point, $f''(x)$ must change sign at that point. But in Section 3-2, we saw that the partition numbers* identify the points where a function can change sign.

THEOREM 1   Inflection Points

If $y = f(x)$ is continuous on (a, b) and has an inflection point at $x = c$, then either $f''(c) = 0$ or $f''(c)$ does not exist.

---

*As we did with the first derivative, we assume that if f'' is discontinuous at c, then $f''(c)$ does not exist.

Note that inflection points can occur only at partition numbers of f'', but not every partition number of f'' produces an inflection point. Two additional requirements must be satisfied for an inflection point to occur:

A partition number c for f'' produces an inflection point for the graph of f only if

1. $f''(x)$ changes sign at c and

2. c is in the domain of f.

Figure 5 illustrates several typical cases.

If $f'(c)$ exists and $f''(x)$ changes sign at $x = c$, then the tangent line at an inflection point $(c, f(c))$ will always lie below the graph on the side that is concave upward and above the graph on the side that is concave downward (see Fig. 5A, B, and C).

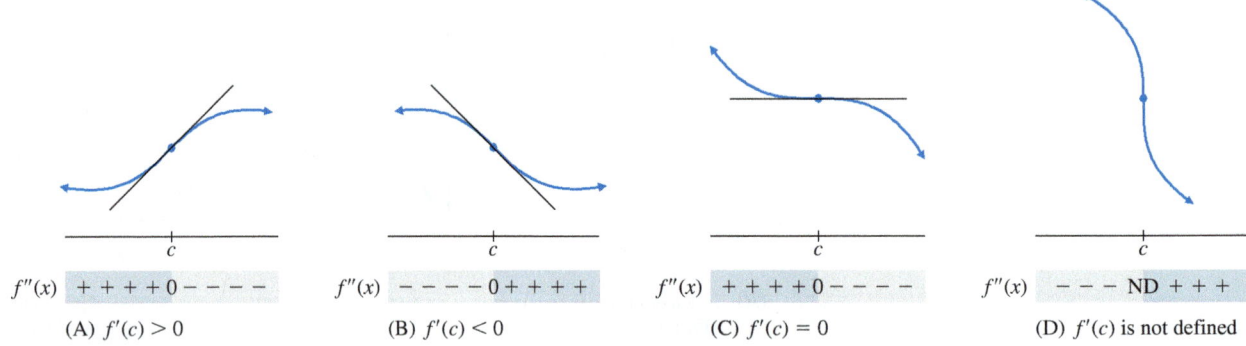

$f''(x)$ $+ + + + 0 - - - -$	$f''(x)$ $- - - - 0 + + + +$	$f''(x)$ $+ + + + 0 - - - -$	$f''(x)$ $- - -$ ND $+ + +$
(A) $f'(c) > 0$	(B) $f'(c) < 0$	(C) $f'(c) = 0$	(D) $f'(c)$ is not defined

Figure 5 Inflection points

EXAMPLE 2   **Locating Inflection Points**   Find the inflection point(s) of

$$f(x) = x^3 - 6x^2 + 9x + 1$$

SOLUTION   Since inflection points occur at values of x where $f''(x)$ changes sign, we construct a sign chart for $f''(x)$.

$$f(x) = x^3 - 6x^2 + 9x + 1$$

$$f'(x) = 3x^2 - 12x + 9$$

$$f''(x) = 6x - 12 = 6(x - 2)$$

The sign chart for $f''(x) = 6(x - 2)$ (partition number is 2) is as follows:

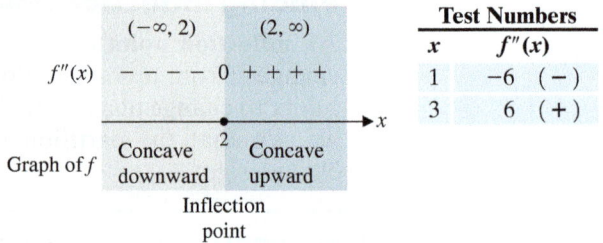

	Test Numbers	
	x	$f''(x)$
	1	-6 $(-)$
	3	6 $(+)$

From the sign chart, we see that the graph of f has an inflection point at $x = 2$. That is, the point

$$(2, f(2)) = (2, 3) \qquad f(2) = 2^3 - 6 \cdot 2^2 + 9 \cdot 2 + 1 = 3$$

is an inflection point on the graph of f.

Matched Problem 2   Find the inflection point(s) of

$$f(x) = x^3 - 9x^2 + 24x - 10$$

EXAMPLE 3 **Locating Inflection Points** Find the inflection point(s) of

$$f(x) = \ln(x^2 - 4x + 5)$$

SOLUTION First we find the domain of f. Since $\ln x$ is defined only for $x > 0$, f is defined only for

$$x^2 - 4x + 5 > 0 \qquad \text{\textit{Use completing the square (Section 2-3).}}$$
$$(x - 2)^2 + 1 > 0 \qquad \text{\textit{True for all x.}}$$

So the domain of f is $(-\infty, \infty)$. Now we find $f''(x)$ and construct a sign chart for it.

$$f(x) = \ln(x^2 - 4x + 5)$$

$$f'(x) = \frac{2x - 4}{x^2 - 4x + 5}$$

$$f''(x) = \frac{(x^2 - 4x + 5)(2x - 4)' - (2x - 4)(x^2 - 4x + 5)'}{(x^2 - 4x + 5)^2}$$

$$= \frac{(x^2 - 4x + 5)2 - (2x - 4)(2x - 4)}{(x^2 - 4x + 5)^2}$$

$$= \frac{2x^2 - 8x + 10 - 4x^2 + 16x - 16}{(x^2 - 4x + 5)^2}$$

$$= \frac{-2x^2 + 8x - 6}{(x^2 - 4x + 5)^2}$$

$$= \frac{-2(x - 1)(x - 3)}{(x^2 - 4x + 5)^2}$$

The partition numbers for $f''(x)$ are $x = 1$ and $x = 3$.
Sign chart for $f''(x)$:

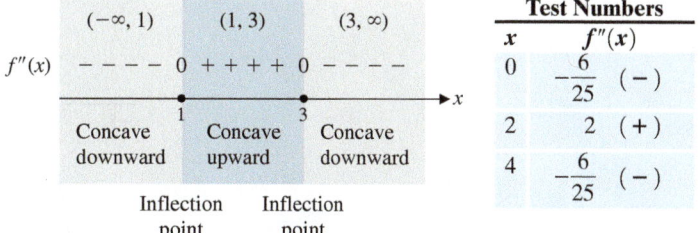

	Test Numbers	
x	$f''(x)$	
0	$-\dfrac{6}{25}$	$(-)$
2	2	$(+)$
4	$-\dfrac{6}{25}$	$(-)$

The sign chart shows that the graph of f has inflection points at $x = 1$ and $x = 3$.

Matched Problem 3 Find the inflection point(s) of

$$f(x) = \ln(x^2 - 2x + 5)$$

CONCEPTUAL INSIGHT

It is important to remember that the partition numbers for f'' are only *candidates* for inflection points. The function f must be defined at $x = c$, and the second derivative must change sign at $x = c$ in order for the graph to have an inflection point at $x = c$. For example, consider

$$f(x) = x^4 \qquad g(x) = \frac{1}{x}$$

$$f'(x) = 4x^3 \qquad g'(x) = -\frac{1}{x^2}$$

$$f''(x) = 12x^2 \qquad g''(x) = \frac{2}{x^3}$$

In each case, $x = 0$ is a partition number for the second derivative, but neither the graph of $f(x)$ nor the graph of $g(x)$ has an inflection point at $x = 0$. Function f does not have an inflection point at $x = 0$ because $f''(x)$ does not change sign at $x = 0$ (see Fig. 6A). Function g does not have an inflection point at $x = 0$ because $g(0)$ is not defined (see Fig. 6B).

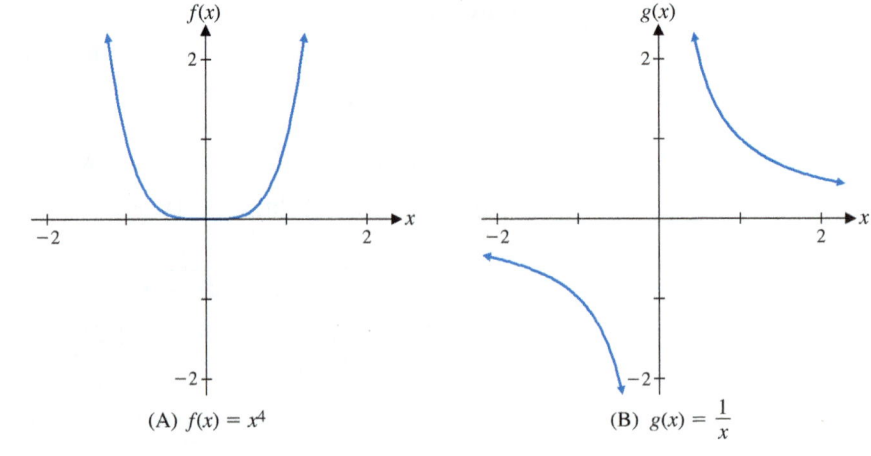

(A) $f(x) = x^4$ (B) $g(x) = \dfrac{1}{x}$

Figure 6

Analyzing Graphs

In the next example, we combine increasing/decreasing properties with concavity properties to analyze the graph of a function.

EXAMPLE 4   **Analyzing a Graph** Figure 7 shows the graph of the derivative of a function f. Use this graph to discuss the graph of f. Include a sketch of a possible graph of f.

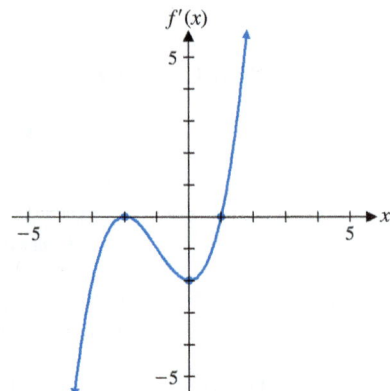

Figure 7

SOLUTION   The sign of the derivative determines where the original function is increasing and decreasing, and the increasing/decreasing properties of the derivative determine the concavity of the original function. The relevant information obtained from the graph of f' is summarized in Table 1, and a possible graph of f is shown in Figure 8.

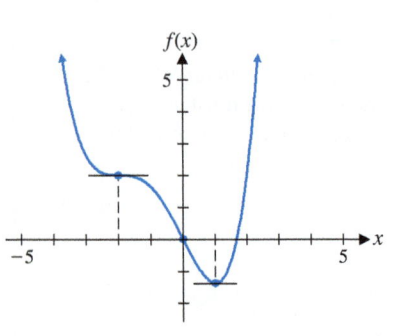

Figure 8

Table 1		
x	$f'(x)$ (Fig. 7)	$f(x)$ (Fig. 8)
$-\infty < x < -2$	Negative and increasing	Decreasing and concave upward
$x = -2$	Local maximum	Inflection point
$-2 < x < 0$	Negative and decreasing	Decreasing and concave downward
$x = 0$	Local minimum	Inflection point
$0 < x < 1$	Negative and increasing	Decreasing and concave upward
$x = 1$	x intercept	Local minimum
$1 < x < \infty$	Positive and increasing	Increasing and concave upward

Matched Problem 4 Figure 9 shows the graph of the derivative of a function f. Use this graph to discuss the graph of f. Include a sketch of a possible graph of f.

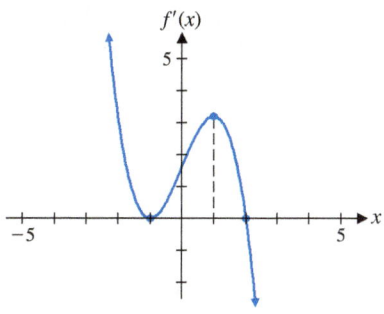

Figure 9

Curve Sketching

Graphing calculators and computers produce the graph of a function by plotting many points. However, important points on a plot many be difficult to identify. Using information gained from the function $f(x)$ and its derivatives, and plotting the important points—intercepts, local extrema, and inflection points—we can sketch by hand a very good representation of the graph of $f(x)$. This graphing process is called **curve sketching**.

PROCEDURE   Graphing Strategy (First Version)*

Step 1 *Analyze f(x).* Find the domain and the intercepts. The x intercepts are the solutions of $f(x) = 0$, and the y intercept is $f(0)$.

Step 2 *Analyze f'(x).* Find the partition numbers for, and critical values of, $f'(x)$. Construct a sign chart for $f'(x)$, determine the intervals on which f is increasing and decreasing, and find local maxima and minima.

Step 3 *Analyze f"(x).* Find the partition numbers for $f''(x)$. Construct a sign chart for $f''(x)$, determine the intervals on which the graph of f is concave upward and concave downward, and find inflection points.

Step 4 *Sketch the graph of f.* Locate intercepts, local maxima and minima, and inflection points. Sketch in what you know from steps 1–3. Plot additional points as needed and complete the sketch.

EXAMPLE 5 **Using the Graphing Strategy** Follow the graphing strategy and analyze the function

$$f(x) = x^4 - 2x^3$$

State all the pertinent information and sketch the graph of f.

SOLUTION **Step 1** *Analyze f(x).* Since f is a polynomial, its domain is $(-\infty, \infty)$.

x intercept: $f(x) = 0$

$$x^4 - 2x^3 = 0$$
$$x^3(x - 2) = 0$$
$$x = 0, 2$$

y intercept: $f(0) = 0$

---

*We will modify this summary in Section 5-4 to include additional information about the graph of f.

Step 2 *Analyze* $f'(x)$. $f'(x) = 4x^3 - 6x^2 = 4x^2(x - \frac{3}{2})$

Critical values of $f(x)$: 0 and $\frac{3}{2}$

Partition numbers for $f'(x)$: 0 and $\frac{3}{2}$

Sign chart for $f'(x)$:

Test Numbers	
x	$f'(x)$
-1	-10 $(-)$
1	-2 $(-)$
2	8 $(+)$

So $f(x)$ is decreasing on $(-\infty, \frac{3}{2})$, is increasing on $(\frac{3}{2}, \infty)$, and has a local minimum at $x = \frac{3}{2}$.

Step 3 *Analyze* $f''(x)$. $f''(x) = 12x^2 - 12x = 12x(x - 1)$

Partition numbers for $f''(x)$: 0 and 1

Sign chart for $f''(x)$:

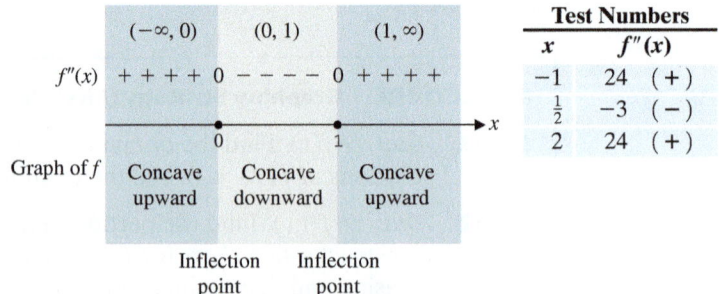

Test Numbers	
x	$f''(x)$
-1	24 $(+)$
$\frac{1}{2}$	-3 $(-)$
2	24 $(+)$

So the graph of f is concave upward on $(-\infty, 0)$ and $(1, \infty)$, is concave downward on $(0, 1)$, and has inflection points at $x = 0$ and $x = 1$.

Step 4 *Sketch the graph of f.*

Test Numbers	
x	$f(x)$
0	0
1	-1
$\frac{3}{2}$	$-\frac{27}{16}$
2	0

Matched Problem 5 Follow the graphing strategy and analyze the function $f(x) = x^4 + 4x^3$. State all the pertinent information and sketch the graph of f.

CONCEPTUAL INSIGHT

Refer to the solution of Example 5. Combining the sign charts for $f'(x)$ and $f''(x)$ (Fig. 10) partitions the real-number line into intervals on which neither $f'(x)$ nor $f''(x)$ changes sign. On each of these intervals, the graph of $f(x)$ must have one of four basic shapes (see also Fig. 3, parts A, C, D, and F on page 286). This reduces sketching the graph of a function to plotting the points identified in the graphing strategy and connecting them with one of the basic shapes.

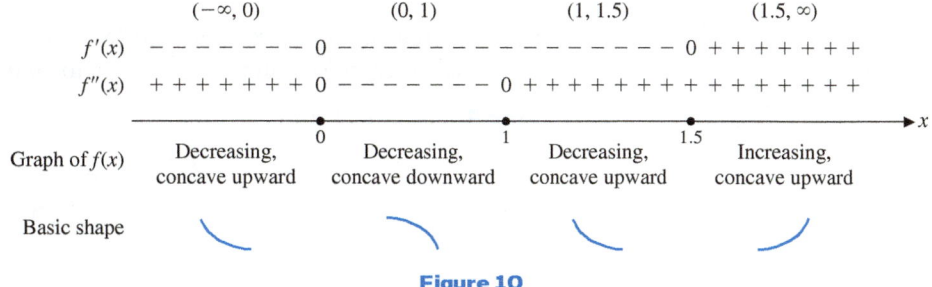

Figure 10

EXAMPLE 6 **Using the Graphing Strategy** Follow the graphing strategy and analyze the function

$$f(x) = 3x^{5/3} - 20x$$

State all the pertinent information and sketch the graph of f. Round any decimal values to two decimal places.

SOLUTION **Step 1** *Analyze* $f(x)$. $f(x) = 3x^{5/3} - 20x$

Since x^p is defined for any x and any positive p, the domain of f is $(-\infty, \infty)$.

x intercepts: Solve $f(x) = 0$

$$3x^{5/3} - 20x = 0$$

$$3x\left(x^{2/3} - \frac{20}{3}\right) = 0 \qquad (a^2 - b^2) = (a - b)(a + b)$$

$$3x\left(x^{1/3} - \sqrt{\frac{20}{3}}\right)\left(x^{1/3} + \sqrt{\frac{20}{3}}\right) = 0$$

The x intercepts of f are

$$x = 0, \quad x = \left(\sqrt{\frac{20}{3}}\right)^3 \approx 17.21, \quad x = \left(-\sqrt{\frac{20}{3}}\right)^3 \approx -17.21$$

y intercept: $f(0) = 0$.

Step 2 Analyze $f'(x)$.

$$f'(x) = 5x^{2/3} - 20$$

$$= 5(x^{2/3} - 4) \qquad \text{Again, } a^2 - b^2 = (a - b)(a + b)$$

$$= 5(x^{1/3} - 2)(x^{1/3} + 2)$$

Critical values of f: $x = 2^3 = 8$ and $x = (-2)^3 = -8$.
Partition numbers for f: $-8, 8$
Sign chart for $f'(x)$:

	$(-\infty, -8)$	$(-8, 8)$	$(8, \infty)$
$f'(x)$	+ + + + 0	– – – – 0	+ + + +

Test Numbers

x	$f'(x)$
-12	6.21 (+)
0	-20 (–)
12	6.21 (+)

Increasing   Decreasing   Increasing

Local maximum   Local minimum

So f is increasing on $(-\infty, -8)$ and $(8, \infty)$ and decreasing on $(-8, 8)$. Therefore, $f(-8)$ is a local maximum, and $f(8)$ is a local minimum.

Step 3 *Analyze $f''(x)$.*

$$f'(x) = 5x^{2/3} - 20$$
$$f''(x) = \frac{10}{3}x^{-1/3} = \frac{10}{3x^{1/3}}$$

Partition number for f'': 0

Sign chart for $f''(x)$:

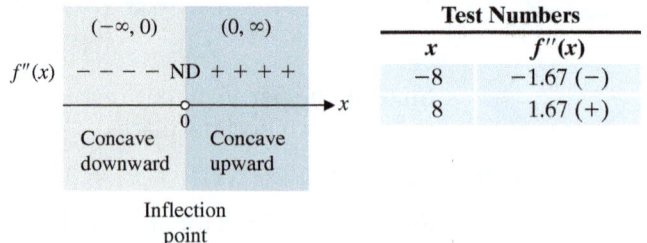

	$(-\infty, 0)$	$(0, \infty)$
$f''(x)$	– – – – ND	+ + + +

Test Numbers

x	$f''(x)$
-8	-1.67 (–)
8	1.67 (+)

Concave downward   Concave upward

Inflection point

So f is concave downward on $(-\infty, 0)$, is concave upward on $(0, \infty)$, and has an inflection point at $x = 0$.

Step 4 *Sketch the graph of f.*

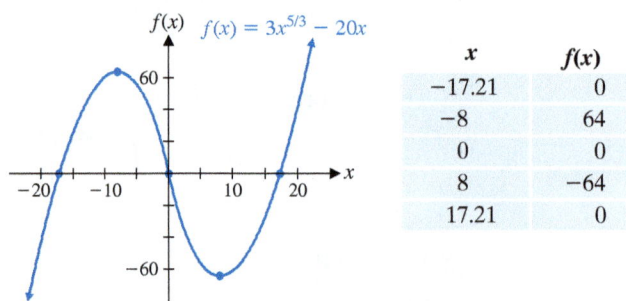

$f(x) = 3x^{5/3} - 20x$

x	$f(x)$
-17.21	0
-8	64
0	0
8	-64
17.21	0

Matched Problem 6   Follow the graphing strategy and analyze the function $f(x) = 3x^{2/3} - x$. State all the pertinent information and sketch the graph of f. Round any decimal values to two decimal places.

Point of Diminishing Returns

If a company decides to increase spending on advertising, it would expect sales to increase. At first, sales will increase at an increasing rate and then increase at a decreasing rate. The value of x where the rate of change of sales goes from increasing to decreasing is called the **point of diminishing returns**. This is also the point where the rate of change has a maximum value. Money spent after this point may increase sales but at a lower rate.

EXAMPLE 7   **Maximum Rate of Change** Currently, a discount appliance store is selling 200 large-screen television sets monthly. If the store invests x thousand in an advertising campaign, the ad company estimates that sales will increase to

$$N(x) = 3x^3 - 0.25x^4 + 200 \qquad 0 \le x \le 9$$

When is the rate of change of sales increasing and when is it decreasing? What is the point of diminishing returns and the maximum rate of change of sales? Graph N and N' on the same coordinate system.

SOLUTION   The rate of change of sales with respect to advertising expenditures is

$$N'(x) = 9x^2 - x^3 = x^2(9 - x)$$

To determine when $N'(x)$ is increasing and decreasing, we find $N''(x)$, the derivative of $N'(x)$:

$$N''(x) = 18x - 3x^2 = 3x(6 - x)$$

The information obtained by analyzing the signs of $N'(x)$ and $N''(x)$ is summarized in Table 2 (sign charts are omitted).

Table 2

x	$N''(x)$	$N'(x)$	$N'(x)$	$N(x)$
$0 < x < 6$	+	+	Increasing	Increasing, concave upward
$x = 6$	0	+	Local maximum	Inflection point
$6 < x < 9$	−	+	Decreasing	Increasing, concave downward

Examining Table 2, we see that $N'(x)$ is increasing on $(0, 6)$ and decreasing on $(6, 9)$. The point of diminishing returns is $x = 6$ and the maximum rate of change is $N'(6) = 108$. Note that $N'(x)$ has a local maximum and $N(x)$ has an inflection point at $x = 6$.

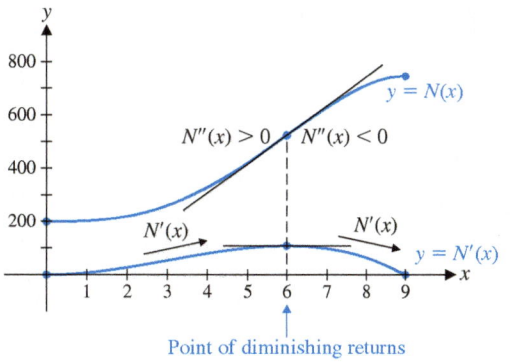

Matched Problem 7   Repeat Example 7 for

$$N(x) = 4x^3 - 0.25x^4 + 500 \qquad 0 \le x \le 12$$

Exercises 5-2

A

1. Use the graph of $y = f(x)$ to identify
   (A) Intervals on which the graph of f is concave upward
   (B) Intervals on which the graph of f is concave downward
   (C) Intervals on which $f''(x) < 0$
   (D) Intervals on which $f''(x) > 0$
   (E) Intervals on which $f'(x)$ is increasing
   (F) Intervals on which $f'(x)$ is decreasing
   (G) The x coordinates of inflection points
   (H) The x coordinates of local extrema for $f'(x)$

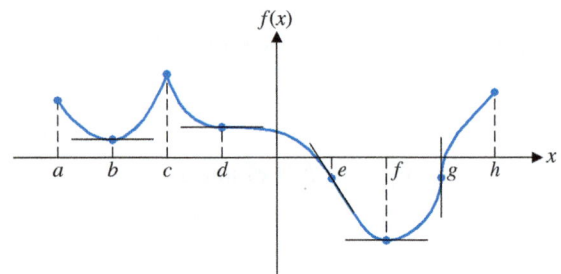

Figure for 1

2. Use the graph of $y = g(x)$ to identify
   (A) Intervals on which the graph of g is concave upward
   (B) Intervals on which the graph of g is concave downward
   (C) Intervals on which $g''(x) < 0$
   (D) Intervals on which $g''(x) > 0$
   (E) Intervals on which $g'(x)$ is increasing
   (F) Intervals on which $g'(x)$ is decreasing
   (G) The x coordinates of inflection points
   (H) The x coordinates of local extrema for $g'(x)$

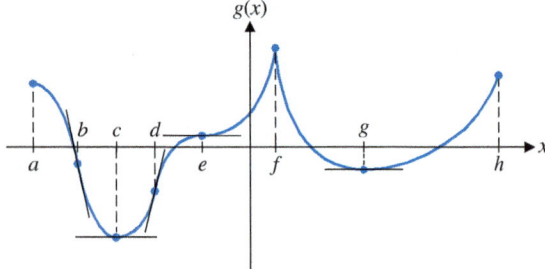

Figure for 2

In Problems 3–6, match the indicated conditions with one of the graphs (A)–(D) shown in the figure.

3. $f'(x) > 0$ and $f''(x) > 0$ on (a, b)

4. $f'(x) > 0$ and $f''(x) < 0$ on (a, b)

5. $f'(x) < 0$ and $f''(x) > 0$ on (a, b)

6. $f'(x) < 0$ and $f''(x) < 0$ on (a, b)

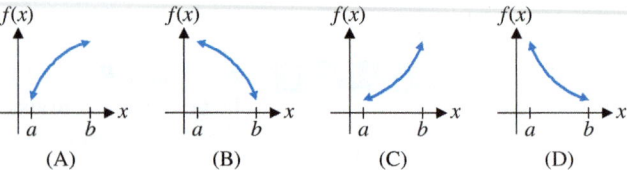

Figure for 3–6

In Problems 7–18, find the indicated derivative for each function.

7. $f''(x)$ for $f(x) = 2x^3 - 4x^2 + 5x - 6$

8. $g''(x)$ for $g(x) = -x^3 + 2x^2 - 3x + 9$

9. $h''(x)$ for $h(x) = 2x^{-1} - 3x^{-2}$

10. $k''(x)$ for $k(x) = -6x^{-2} + 12x^{-3}$

11. d^2y/dx^2 for $y = x^2 - 18x^{1/2}$

12. d^2y/dx^2 for $y = x^3 - 24x^{1/3}$

13. y'' for $y = (x^2 + 9)^4$

14. y'' for $y = (x^2 - 16)^5$

15. $f''(x)$ for $f(x) = e^{-x^2}$

16. $f''(x)$ for $f(x) = xe^{-x}$

17. y'' for $y = \dfrac{\ln x}{x^2}$

18. y'' for $y = x^2 \ln x$

In Problems 19–24, find the x and y coordinates of all inflection points.

19. $f(x) = x^3 + 30x^2$

20. $f(x) = x^3 - 24x^2$

21. $f(x) = x^{5/3} + 2$

22. $f(x) = 5 - x^{4/3}$

23. $f(x) = 1 + x + x^{2/5}$

24. $f(x) = x^{3/5} - 6x + 7$

In Problems 25–34, find the intervals on which the graph of f is concave upward, the intervals on which the graph of f is concave downward, and the inflection points.

25. $f(x) = x^4 + 6x^2$

26. $f(x) = x^4 + 6x$

27. $f(x) = x^3 - 4x^2 + 5x - 2$

28. $f(x) = -x^3 - 5x^2 + 4x - 3$

29. $f(x) = -x^4 + 12x^3 - 12x + 24$

30. $f(x) = x^4 - 2x^3 - 36x + 12$

31. $f(x) = \ln(x^2 - 2x + 10)$

32. $f(x) = \ln(x^2 + 6x + 13)$

33. $f(x) = 8e^x - e^{2x}$

34. $f(x) = e^{3x} - 9e^x$

In Problems 35–42, f(x) is continuous on $(-\infty, \infty)$. *Use the given information to sketch the graph of f.*

35.

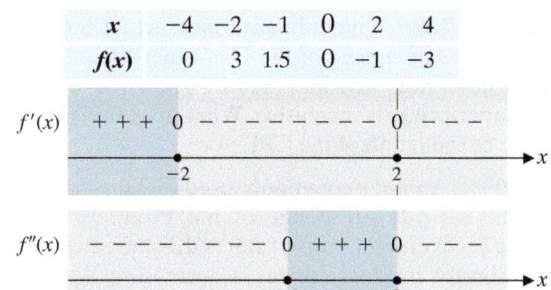

x	-4	-2	-1	0	2	4
$f(x)$	0	3	1.5	0	-1	-3

36.

x	-4	-2	-1	0	2	4
$f(x)$	0	-2	-1	0	1	3

37.

x	-3	0	1	2	4	5
$f(x)$	-4	0	2	1	-1	0

38.

x	-4	-2	0	2	4	6
$f(x)$	0	3	0	-2	0	3

39. $f(0) = 2, f(1) = 0, f(2) = -2;$

$f'(0) = 0, f'(2) = 0;$

$f'(x) > 0$ on $(-\infty, 0)$ and $(2, \infty);$

$f'(x) < 0$ on $(0, 2);$

$f''(1) = 0;$

$f''(x) > 0$ on $(1, \infty);$

$f''(x) < 0$ on $(-\infty, 1)$

40. $f(-2) = -2, f(0) = 1, f(2) = 4;$

$f'(-2) = 0, f'(2) = 0;$

$f'(x) > 0$ on $(-2, 2);$

$f'(x) < 0$ on $(-\infty, -2)$ and $(2, \infty);$

$f''(0) = 0;$

$f''(x) > 0$ on $(-\infty, 0);$

$f''(x) < 0$ on $(0, \infty)$

41. $f(-1) = 0, f(0) = -2, f(1) = 0;$

$f'(0) = 0, f'(-1)$ and $f'(1)$ are not defined;

$f'(x) > 0$ on $(0, 1)$ and $(1, \infty);$

$f'(x) < 0$ on $(-\infty, -1)$ and $(-1, 0);$

$f''(-1)$ and $f''(1)$ are not defined;

$f''(x) > 0$ on $(-1, 1);$

$f''(x) < 0$ on $(-\infty, -1)$ and $(1, \infty)$

42. $f(0) = -2, f(1) = 0, f(2) = 4;$

$f'(0) = 0, f'(2) = 0, f'(1)$ is not defined;

$f'(x) > 0$ on $(0, 1)$ and $(1, 2);$

$f'(x) < 0$ on $(-\infty, 0)$ and $(2, \infty);$

$f''(1)$ is not defined;

$f''(x) > 0$ on $(-\infty, 1);$

$f''(x) < 0$ on $(1, \infty)$

B

In Problems 43–64, summarize the pertinent information obtained by applying the graphing strategy and sketch the graph of $y = f(x)$.

43. $f(x) = (x - 2)(x^2 - 4x - 8)$

44. $f(x) = (x - 3)(x^2 - 6x - 3)$

45. $f(x) = (x + 1)(x^2 - x + 2)$

46. $f(x) = (1 - x)(x^2 + x + 4)$

47. $f(x) = -0.25x^4 + x^3$

48. $f(x) = 0.25x^4 - 2x^3$

49. $f(x) = 16x(x - 1)^3$

50. $f(x) = -4x(x + 2)^3$

51. $f(x) = (x^2 + 3)(9 - x^2)$

52. $f(x) = (x^2 + 3)(x^2 - 1)$

53. $f(x) = (x^2 - 4)^2$

54. $f(x) = (x^2 - 1)(x^2 - 5)$

55. $f(x) = 2x^6 - 3x^5$

56. $f(x) = 3x^5 - 5x^4$

57. $f(x) = 1 - e^{-x}$

58. $f(x) = 2 - 3e^{-2x}$

59. $f(x) = e^{0.5x} + 4e^{-0.5x}$

60. $f(x) = 2e^{0.5x} + e^{-0.5x}$

61. $f(x) = -4 + 2\ln x$

62. $f(x) = 5 - 3 \ln x$

63. $f(x) = \ln(x + 4) - 2$

64. $f(x) = 1 - \ln(x - 3)$

In Problems 65–68, use the graph of $y = f'(x)$ to discuss the graph of $y = f(x)$. Organize your conclusions in a table (see Example 4), and sketch a possible graph of $y = f(x)$.

65.

66.

67.

68.

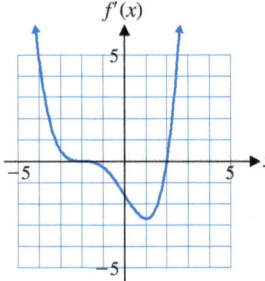

In Problems 69–76, apply steps 1–3 of the graphing strategy to f(x). Use a graphing calculator to approximate (to two decimal places) x intercepts, critical values, and x coordinates of inflection points. Summarize all the pertinent information.

69. $f(x) = x^4 - 5x^3 + 3x^2 + 8x - 5$

70. $f(x) = x^4 + 2x^3 - 5x^2 - 4x + 4$

71. $f(x) = x^4 - 21x^3 + 100x^2 + 20x + 100$

72. $f(x) = x^4 - 12x^3 + 28x^2 + 76x - 50$

73. $f(x) = -x^4 - x^3 + 2x^2 - 2x + 3$

74. $f(x) = -x^4 + x^3 + x^2 + 6$

75. $f(x) = 0.1x^5 + 0.3x^4 - 4x^3 - 5x^2 + 40x + 30$

76. $f(x) = x^5 + 4x^4 - 7x^3 - 20x^2 + 20x - 20$

C

In Problems 77–80, assume that f is a polynomial.

77. Explain how you can locate inflection points for the graph of $y = f(x)$ by examining the graph of $y = f'(x)$.

78. Explain how you can determine where $f'(x)$ is increasing or decreasing by examining the graph of $y = f(x)$.

79. Explain how you can locate local maxima and minima for the graph of $y = f'(x)$ by examining the graph of $y = f(x)$.

80. Explain how you can locate local maxima and minima for the graph of $y = f(x)$ by examining the graph of $y = f'(x)$.

Applications

81. Inflation. One commonly used measure of inflation is the annual rate of change of the Consumer Price Index (CPI). A TV news story says that the rate of change of inflation for consumer prices is increasing. What does this say about the shape of the graph of the CPI?

82. Inflation. Another commonly used measure of inflation is the annual rate of change of the Producer Price Index (PPI). A government report states that the rate of change of inflation for producer prices is decreasing. What does this say about the shape of the graph of the PPI?

83. Cost analysis. A company manufactures a variety of camp stoves at different locations. The total cost $C(x)$ (in dollars) of producing x camp stoves per week at plant A is shown in the figure. Discuss the graph of the marginal cost function $C'(x)$ and interpret the graph of $C'(x)$ in terms of the efficiency of the production process at this plant.

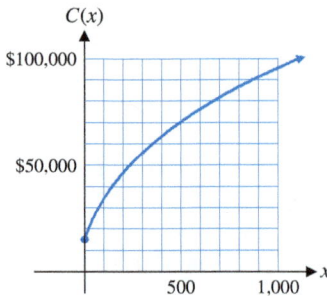

Figure for 83   Production costs at plant A

84. Cost analysis. The company in Problem 83 produces the same camp stove at another plant. The total cost $C(x)$ (in dollars) of producing x camp stoves per week at plant B is shown in the figure. Discuss the graph of the marginal cost function $C'(x)$ and interpret the graph of $C'(x)$ in terms of the efficiency of the production process at plant B. Compare the production processes at the two plants.

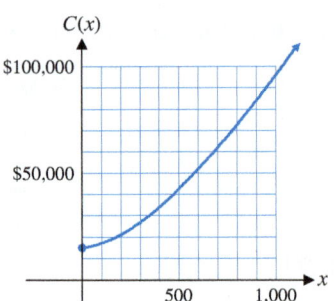

Figure for 84   Production costs at plant B

85. Revenue. The marketing research department of a computer company used a large city to test market the firm's new laptop. The department found that the relationship between price p (dollars per unit) and the demand x (units per week) was given approximately by

$$p = 1,296 - 0.12x^2 \qquad 0 < x < 80$$

So, weekly revenue can be approximated by

$$R(x) = xp = 1{,}296x - 0.12x^3 \qquad 0 < x < 80$$

(A) Find the local extrema for the revenue function.

(B) On which intervals is the graph of the revenue function concave upward? Concave downward?

86. **Profit.** Suppose that the cost equation for the company in Problem 85 is

$$C(x) = 830 + 396x$$

(A) Find the local extrema for the profit function.

(B) On which intervals is the graph of the profit function concave upward? Concave downward?

87. **Revenue.** A dairy is planning to introduce and promote a new line of organic ice cream. After test marketing the new line in a large city, the marketing research department found that the demand in that city is given approximately by

$$p = 10e^{-x} \qquad 0 \le x \le 5$$

where x thousand quarts were sold per week at a price of $\$p$ each.

(A) Find the local extrema for the revenue function.

(B) On which intervals is the graph of the revenue function concave upward? Concave downward?

88. **Revenue.** A national food service runs food concessions for sporting events throughout the country. The company's marketing research department chose a particular football stadium to test market a new jumbo hot dog. It was found that the demand for the new hot dog is given approximately by

$$p = 8 - 2\ln x \qquad 5 \le x \le 50$$

where x is the number of hot dogs (in thousands) that can be sold during one game at a price of $\$p$.

(A) Find the local extrema for the revenue function.

(B) On which intervals is the graph of the revenue function concave upward? Concave downward?

89. **Production: point of diminishing returns.** A T-shirt manufacturer is planning to expand its workforce. It estimates that the number of T-shirts produced by hiring x new workers is given by

$$T(x) = -0.25x^4 + 5x^3 \qquad 0 \le x \le 15$$

When is the rate of change of T-shirt production increasing and when is it decreasing? What is the point of diminishing returns and the maximum rate of change of T-shirt production? Graph T and T' on the same coordinate system.

90. **Production: point of diminishing returns.** A baseball cap manufacturer is planning to expand its workforce. It estimates that the number of baseball caps produced by hiring x new workers is given by

$$T(x) = -0.25x^4 + 6x^3 \qquad 0 \le x \le 18$$

When is the rate of change of baseball cap production increasing and when is it decreasing? What is the point of diminishing returns and the maximum rate of change of

baseball cap production? Graph T and T' on the same coordinate system.

91. **Advertising: point of diminishing returns.** A company estimates that it will sell $N(x)$ units of a product after spending $\$x$ thousand on advertising, as given by

$$N(x) = -0.25x^4 + 23x^3 - 540x^2 + 80{,}000 \qquad 24 \le x \le 45$$

When is the rate of change of sales increasing and when is it decreasing? What is the point of diminishing returns and the maximum rate of change of sales? Graph N and N' on the same coordinate system.

92. **Advertising: point of diminishing returns.** A company estimates that it will sell $N(x)$ units of a product after spending $\$x$ thousand on advertising, as given by

$$N(x) = -0.25x^4 + 13x^3 - 180x^2 + 10{,}000 \qquad 15 \le x \le 24$$

When is the rate of change of sales increasing and when is it decreasing? What is the point of diminishing returns and the maximum rate of change of sales? Graph N and N' on the same coordinate system.

93. **Advertising.** An automobile dealer uses TV advertising to promote car sales. On the basis of past records, the dealer arrived at the following data, where x is the number of ads placed monthly and y is the number of cars sold that month:

Number of Ads	Number of Cars
x	y
10	325
12	339
20	417
30	546
35	615
40	682
50	795

(A) Enter the data in a graphing calculator and find a cubic regression equation for the number of cars sold monthly as a function of the number of ads.

(B) How many ads should the dealer place each month to maximize the rate of change of sales with respect to the number of ads, and how many cars can the dealer expect to sell with this number of ads? Round answers to the nearest integer.

94. **Advertising.** A sporting goods chain places TV ads to promote golf club sales. The marketing director used past records to determine the following data, where x is the number of ads placed monthly and y is the number of golf clubs sold that month.

Number of Ads	Number of Golf Clubs
x	y
10	345
14	488
20	746
30	1,228
40	1,671
50	1,955

(A) Enter the data in a graphing calculator and find a cubic regression equation for the number of golf clubs sold monthly as a function of the number of ads.

(B) How many ads should the store manager place each month to maximize the rate of change of sales with respect to the number of ads, and how many golf clubs can the manager expect to sell with this number of ads? Round answers to the nearest integer.

95. Population growth: bacteria. A drug that stimulates reproduction is introduced into a colony of bacteria. After t minutes, the number of bacteria is given approximately by

$$N(t) = 1{,}000 + 30t^2 - t^3 \qquad 0 \le t \le 20$$

(A) When is the rate of growth, $N'(t)$, increasing? Decreasing?

(B) Find the inflection points for the graph of N.

(C) Sketch the graphs of N and N' on the same coordinate system.

(D) What is the maximum rate of growth?

96. Drug sensitivity. One hour after x milligrams of a particular drug are given to a person, the change in body temperature $T(x)$, in degrees Fahrenheit, is given by

$$T(x) = x^2\left(1 - \frac{x}{9}\right) \qquad 0 \le x \le 6$$

The rate $T'(x)$ at which $T(x)$ changes with respect to the size of the dosage x is called the *sensitivity* of the body to the dosage.

(A) When is $T'(x)$ increasing? Decreasing?

(B) Where does the graph of T have inflection points?

(C) Sketch the graphs of T and T' on the same coordinate system.

(D) What is the maximum value of $T'(x)$?

97. Learning. The time T (in minutes) it takes a person to learn a list of length n is

$$T(n) = 0.08n^3 - 1.2n^2 + 6n \qquad n \ge 0$$

(A) When is the rate of change of T with respect to the length of the list increasing? Decreasing?

(B) Where does the graph of T have inflection points?

(C) Graph T and T' on the same coordinate system.

(D) What is the minimum value of $T'(n)$?

Answers to Matched Problems

1. (A) Concave downward on $(-\infty, \infty)$

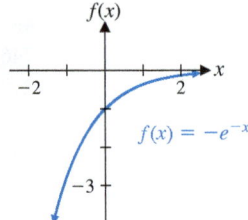

(B) Concave upward on $(0, \infty)$

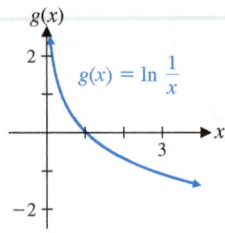

(C) Concave upward on $(-\infty, 0)$ and concave downward on $(0, \infty)$

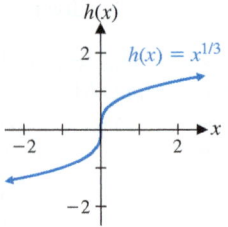

2. The only inflection point is $(3, f(3)) = (3, 8)$.

3. The inflection points are $(-1, f(-1)) = (-1, \ln 8)$ and $(3, f(3)) = (3, \ln 8)$.

4.

x	$f'(x)$	$f(x)$
$-\infty < x < -1$	Positive and decreasing	Increasing and concave downward
$x = -1$	Local minimum	Inflection point
$-1 < x < 1$	Positive and increasing	Increasing and concave upward
$x = 1$	Local maximum	Inflection point
$1 < x < 2$	Positive and decreasing	Increasing and concave downward
$x = 2$	x intercept	Local maximum
$2 < x < \infty$	Negative and decreasing	Decreasing and concave downward

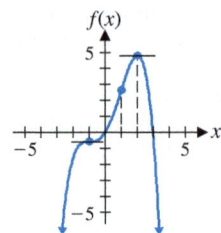

5. x intercepts: $-4, 0$; y intercept: $f(0) = 0$

Decreasing on $(-\infty, -3)$; increasing on $(-3, \infty)$; local minimum at $x = -3$

Concave upward on $(-\infty, -2)$ and $(0, \infty)$; concave downward on $(-2, 0)$

Inflection points at $x = -2$ and $x = 0$

x	$f(x)$
-4	0
-3	-27
-2	-16
0	0

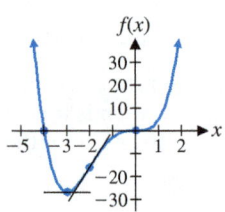

6. x intercepts: $0, 27$; y intercept: $f(0) = 0$

Decreasing on $(-\infty, 0)$ and $(8, \infty)$; increasing on $(0, 8)$; local minimum: $f(0) = 0$; local maximum: $f(8) = 4$

Concave downward on $(-\infty, 0)$ and $(0, \infty)$; no inflection points

x	$f(x)$
0	0
8	4
27	0

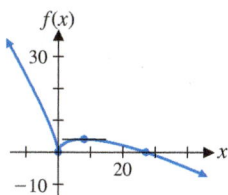

7. $N'(x)$ is increasing on $(0, 8)$ and decreasing on $(8, 12)$. The point of diminishing returns is $x = 8$ and the maximum rate of change is $N'(8) = 256$.

Point of diminishing returns

5-3 L'Hôpital's Rule

Introduction

The ability to evaluate a wide variety of different types of limits is one of the skills that are necessary to apply the techniques of calculus successfully. Limits play a fundamental role in the development of the derivative and are an important graphing tool. In order to deal effectively with graphs, we need to develop some more methods for evaluating limits.

In this section, we discuss a powerful technique for evaluating limits of quotients called *L'Hôpital's rule*. The rule is named after the French mathematician Marquis de L'Hôpital (1661–1704). To use L'Hôpital's rule, it is necessary to be familiar with the limit properties of some basic functions. Figure 1 reviews some limits involving powers of x that were discussed earlier.

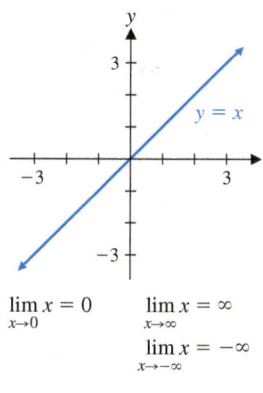

$$\lim_{x \to 0} x = 0 \qquad \lim_{x \to \infty} x = \infty$$
$$\lim_{x \to -\infty} x = -\infty$$

(A) $y = x$

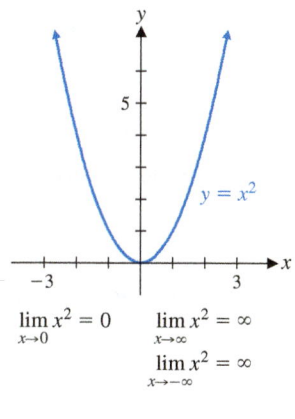

$$\lim_{x \to 0} x^2 = 0 \qquad \lim_{x \to \infty} x^2 = \infty$$
$$\lim_{x \to -\infty} x^2 = \infty$$

(B) $y = x^2$

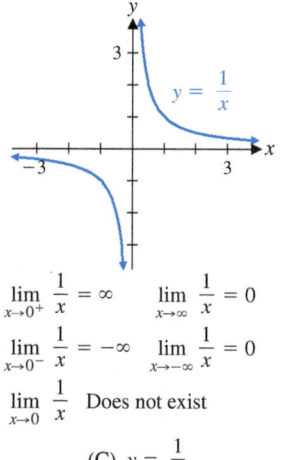

$$\lim_{x \to 0^+} \frac{1}{x} = \infty \qquad \lim_{x \to \infty} \frac{1}{x} = 0$$
$$\lim_{x \to 0^-} \frac{1}{x} = -\infty \qquad \lim_{x \to -\infty} \frac{1}{x} = 0$$
$$\lim_{x \to 0} \frac{1}{x} \text{ Does not exist}$$

(C) $y = \frac{1}{x}$

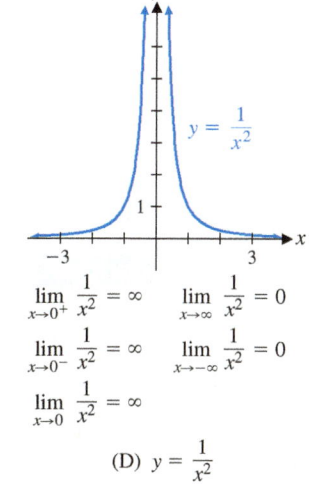

$$\lim_{x \to 0^+} \frac{1}{x^2} = \infty \qquad \lim_{x \to \infty} \frac{1}{x^2} = 0$$
$$\lim_{x \to 0^-} \frac{1}{x^2} = \infty \qquad \lim_{x \to -\infty} \frac{1}{x^2} = 0$$
$$\lim_{x \to 0} \frac{1}{x^2} = \infty$$

(D) $y = \frac{1}{x^2}$

Figure 1 Limits involving powers of x

The limits in Figure 1 are easily extended to functions of the form $f(x) = (x - c)^n$ and $g(x) = 1/(x - c)^n$. In general, if n is an odd integer, then limits involving $(x - c)^n$ or $1/(x - c)^n$ as x approaches c (or $\pm\infty$) behave, respectively, like the limits of x and $1/x$ as x approaches 0 (or $\pm\infty$). If n is an even integer, then limits involving these expressions behave, respectively, like the limits of x^2 and $1/x^2$ as x approaches 0 (or $\pm\infty$).

EXAMPLE 1   **Limits Involving Powers of $x - c$**

(A) $\lim\limits_{x \to 2} \dfrac{5}{(x - 2)^4} = \infty$ Compare with $\lim\limits_{x \to 0} \dfrac{1}{x^2}$ in Figure 1.

(B) $\lim\limits_{x \to -1^-} \dfrac{4}{(x + 1)^3} = -\infty$ Compare with $\lim\limits_{x \to 0^-} \dfrac{1}{x}$ in Figure 1.

(C) $\lim\limits_{x \to \infty} \dfrac{4}{(x - 9)^6} = 0$ Compare with $\lim\limits_{x \to \infty} \dfrac{1}{x^2}$ in Figure 1.

(D) $\lim\limits_{x \to -\infty} 3x^3 = -\infty$ Compare with $\lim\limits_{x \to -\infty} x$ in Figure 1.

Matched Problem 1   Evaluate each limit.

(A) $\lim\limits_{x \to 3^+} \dfrac{7}{(x - 3)^5}$ (B) $\lim\limits_{x \to -4} \dfrac{6}{(x + 4)^6}$

(C) $\lim\limits_{x \to -\infty} \dfrac{3}{(x + 2)^3}$ (D) $\lim\limits_{x \to \infty} 5x^4$

Figure 2 reviews limits of exponential and logarithmic functions.

The limits in Figure 2 also generalize to other simple exponential and logarithmic forms.

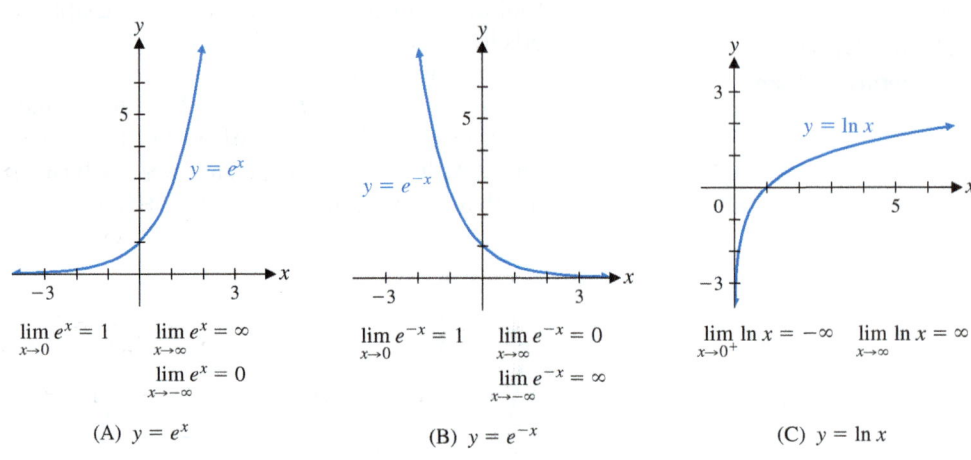

$\lim\limits_{x \to 0} e^x = 1$ $\lim\limits_{x \to \infty} e^x = \infty$

$\lim\limits_{x \to -\infty} e^x = 0$

(A) $y = e^x$

$\lim\limits_{x \to 0} e^{-x} = 1$ $\lim\limits_{x \to \infty} e^{-x} = 0$

$\lim\limits_{x \to -\infty} e^{-x} = \infty$

(B) $y = e^{-x}$

$\lim\limits_{x \to 0^+} \ln x = -\infty$ $\lim\limits_{x \to \infty} \ln x = \infty$

(C) $y = \ln x$

Figure 2 Limits involving exponential and logarithmic functions

EXAMPLE 2   **Limits Involving Exponential and Logarithmic Forms**

(A) $\lim\limits_{x \to \infty} 2e^{3x} = \infty$ Compare with $\lim\limits_{x \to \infty} e^x$ in Figure 2.

(B) $\lim\limits_{x \to \infty} 4e^{-5x} = 0$ Compare with $\lim\limits_{x \to \infty} e^{-x}$ in Figure 2.

(C) $\lim\limits_{x \to \infty} \ln(x + 4) = \infty$ Compare with $\lim\limits_{x \to \infty} \ln x$ in Figure 2.

(D) $\lim\limits_{x \to 2^+} \ln(x - 2) = -\infty$ Compare with $\lim\limits_{x \to 0^+} \ln x$ in Figure 2.

Matched Problem 2 Evaluate each limit.

(A) $\displaystyle\lim_{x \to -\infty} 2e^{-6x}$

(B) $\displaystyle\lim_{x \to -\infty} 3e^{2x}$

(C) $\displaystyle\lim_{x \to -4^+} \ln(x + 4)$

(D) $\displaystyle\lim_{x \to \infty} \ln(x - 10)$

Now that we have reviewed the limit properties of some basic functions, we are ready to consider the main topic of this section: L'Hôpital's rule.

L'Hôpital's Rule and the Indeterminate Form 0/0

Recall that the limit

$$\lim_{x \to c} \frac{f(x)}{g(x)}$$

is a 0/0 indeterminate form if

$$\lim_{x \to c} f(x) = 0 \quad \text{and} \quad \lim_{x \to c} g(x) = 0$$

The quotient property for limits in Section 3-1 does not apply since $\displaystyle\lim_{x \to c} g(x) = 0$.

If we are dealing with a 0/0 indeterminate form, the limit may or may not exist, and we cannot tell which is true without further investigation.

Each of the following is a 0/0 indeterminate form:

$$\lim_{x \to 2} \frac{x^2 - 4}{x - 2} \quad \text{and} \quad \lim_{x \to 1} \frac{e^x - e}{x - 1}$$

The first limit can be evaluated by performing some algebraic simplifications, such as

$$\lim_{x \to 2} \frac{x^2 - 4}{x - 2} = \lim_{x \to 2} \frac{(x - 2)(x + 2)}{x - 2} = \lim_{x \to 2}(x + 2) = 4$$

The second cannot. Instead, we turn to the powerful **L'Hôpital's rule**, which we now state without proof. This rule can be used whenever a limit is a 0/0 indeterminate form.

THEOREM 1 L'Hôpital's Rule for 0/0 Indeterminate Forms: Version 1

For c a real number,

if $\displaystyle\lim_{x \to c} f(x) = 0$ and $\displaystyle\lim_{x \to c} g(x) = 0$, then

$$\lim_{x \to c} \frac{f(x)}{g(x)} = \lim_{x \to c} \frac{f'(x)}{g'(x)}$$

provided that the second limit exists or is $+\infty$ or $-\infty$.

EXAMPLE 3 **L'Hôpital's Rule** Evaluate $\displaystyle\lim_{x \to 1} \frac{e^x - e}{x - 1}$.

SOLUTION **Step 1** *Check to see if L'Hôpital's rule applies:*

$$\lim_{x \to 1}(e^x - e) = e^1 - e = 0 \quad \text{and} \quad \lim_{x \to 1}(x - 1) = 1 - 1 = 0$$

L'Hôpital's rule does apply.

Step 2 *Apply L'Hôpital's rule:*

$$\underset{0/0 \text{ form}}{\lim_{x \to 1} \frac{e^x - e}{x - 1}} = \lim_{x \to 1} \frac{\dfrac{d}{dx}(e^x - e)}{\dfrac{d}{dx}(x - 1)} \qquad \text{Apply L'Hôpital's rule.}$$

$$= \lim_{x \to 1} \frac{e^x}{1} \qquad e^x \text{ is continuous at } x = 1.$$

$$= \frac{e^1}{1} = e$$

Matched Problem 3   Evaluate $\displaystyle\lim_{x \to 4} \frac{e^x - e^4}{x - 4}$.

CONCEPTUAL INSIGHT

In L'Hôpital's rule, the symbol $f'(x)/g'(x)$ represents the derivative of $f(x)$ divided by the derivative of $g(x)$, not the derivative of the quotient $f(x)/g(x)$.

When applying L'Hôpital's rule to a 0/0 indeterminate form, be certain that you differentiate the numerator and denominator separately.

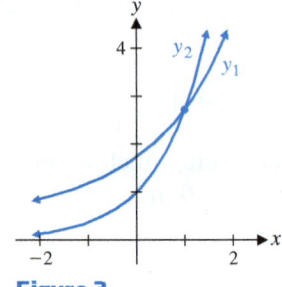

Figure 3

The functions

$$y_1 = \frac{e^x - e}{x - 1} \qquad \text{and} \qquad y_2 = \frac{e^x}{1}$$

of Example 3 are different functions (see Fig. 3), but both functions have the same limit e as x approaches 1. Although y_1 is undefined at $x = 1$, the graph of y_1 provides a check of the answer to Example 3.

EXAMPLE 4   **L'Hôpital's Rule** Evaluate $\displaystyle\lim_{x \to 0} \frac{\ln(1 + x^2)}{x^4}$.

SOLUTION   **Step 1** *Check to see if L'Hôpital's rule applies:*

$$\lim_{x \to 0} \ln(1 + x^2) = \ln 1 = 0 \qquad \text{and} \qquad \lim_{x \to 0} x^4 = 0$$

L'Hôpital's rule does apply.

Step 2 *Apply L'Hôpital's rule:*

$$\underset{0/0 \text{ form}}{\lim_{x \to 0} \frac{\ln(1 + x^2)}{x^4}} = \lim_{x \to 0} \frac{\dfrac{d}{dx}\ln(1 + x^2)}{\dfrac{d}{dx}x^4} \qquad \text{Apply L'Hôpital's rule.}$$

$$\lim_{x \to 0} \frac{\ln(1 + x^2)}{x^4} = \lim_{x \to 0} \frac{\dfrac{2x}{1 + x^2}}{4x^3} \qquad \text{Multiply numerator and denominator by } 1/4x^3.$$

$$= \lim_{x \to 0} \frac{\dfrac{2x}{1 + x^2} \dfrac{1}{4x^3}}{4x^3 \dfrac{1}{4x^3}} \qquad \text{Simplify.}$$

$$= \lim_{x \to 0} \frac{1}{2x^2(1 + x^2)}$$ *Apply Theorem 1 in Section 3-2*
and compare with Fig. 1(D).

$$= \infty$$

Matched Problem 4   Evaluate $\lim_{x \to 1} \dfrac{\ln x}{(x - 1)^3}$.

EXAMPLE 5   **L'Hôpital's Rule May Not Be Applicable**   Evaluate $\lim_{x \to 1} \dfrac{\ln x}{x}$.

SOLUTION   **Step 1** *Check to see if L'Hôpital's rule applies:*

$$\lim_{x \to 1} \ln x = \ln 1 = 0, \quad \text{but} \quad \lim_{x \to 1} x = 1 \neq 0$$

L'Hôpital's rule does not apply.

Step 2 *Evaluate by another method.* The quotient property for limits from Section 3-1 does apply, and we have

$$\lim_{x \to 1} \frac{\ln x}{x} = \frac{\lim_{x \to 1} \ln x}{\lim_{x \to 1} x} = \frac{\ln 1}{1} = \frac{0}{1} = 0$$

Note that applying L'Hôpital's rule would give us an incorrect result:

$$\lim_{x \to 1} \frac{\ln x}{x} \neq \lim_{x \to 1} \frac{\dfrac{d}{dx} \ln x}{\dfrac{d}{dx} x} = \lim_{x \to 1} \frac{1/x}{1} = 1$$

Matched Problem 5   Evaluate $\lim_{x \to 0} \dfrac{x}{e^x}$.

CONCEPTUAL INSIGHT

As Example 5 illustrates, all limits involving quotients are not 0/0 indeterminate forms.

You must always check to see if L'Hôpital's rule applies before you use it.

EXAMPLE 6   **Repeated Application of L'Hôpital's Rule**   Evaluate

$$\lim_{x \to 0} \frac{x^2}{e^x - 1 - x}$$

SOLUTION   **Step 1** *Check to see if L'Hôpital's rule applies:*

$$\lim_{x \to 0} x^2 = 0 \quad \text{and} \quad \lim_{x \to 0} (e^x - 1 - x) = 0$$

L'Hôpital's rule does apply.

Step 2 Apply L'Hôpital's rule:

0/0 form

$$\lim_{x \to 0} \frac{x^2}{e^x - 1 - x} = \lim_{x \to 0} \frac{\dfrac{d}{dx} x^2}{\dfrac{d}{dx}(e^x - 1 - x)} = \lim_{x \to 0} \frac{2x}{e^x - 1}$$

Since $\lim_{x \to 0} 2x = 0$ and $\lim_{x \to 0}(e^x - 1) = 0$, the new limit obtained is also a 0/0 indeterminate form, and L'Hôpital's rule can be applied again.

Step 3 *Apply L'Hôpital's rule again:*

$$
\lim_{x \to 0} \frac{2x}{e^x - 1} = \overbrace{\lim_{x \to 0} \frac{\dfrac{d}{dx} 2x}{\dfrac{d}{dx}(e^x - 1)}}^{0/0 \text{ form}} = \lim_{x \to 0} \frac{2}{e^x} = \frac{2}{e^0} = 2
$$

Therefore,

$$
\lim_{x \to 0} \frac{x^2}{e^x - 1 - x} = \lim_{x \to 0} \frac{2x}{e^x - 1} = \lim_{x \to 0} \frac{2x}{e^x} = 2
$$

Matched Problem 6   Evaluate $\displaystyle\lim_{x \to 0} \frac{e^{2x} - 1 - 2x}{x^2}$

One-Sided Limits and Limits at ∞

In addition to examining the limit as x approaches c, we have discussed one-sided limits and limits at ∞ in Chapter 3. L'Hôpital's rule is valid in these cases also.

THEOREM 2   L'Hôpital's Rule for 0/0 Indeterminate Forms: Version 2 (for one-sided limits and limits at infinity)

The first version of L'Hôpital's rule (Theorem 1) remains valid if the symbol $x \to c$ is replaced everywhere it occurs with one of the following symbols:

$$
x \to c^+ \qquad x \to c^- \qquad x \to \infty \qquad x \to -\infty
$$

For example, if $\displaystyle\lim_{x \to \infty} f(x) = 0$ and $\displaystyle\lim_{x \to \infty} g(x) = 0$, then

$$
\lim_{x \to \infty} \frac{f(x)}{g(x)} = \lim_{x \to \infty} \frac{f'(x)}{g'(x)}
$$

provided that the second limit exists or is $+\infty$ or $-\infty$. Similar rules can be written for $x \to c^+$, $x \to c^-$, and $x \to -\infty$.

EXAMPLE 7   **L'Hôpital's Rule for One-Sided Limits** Evaluate $\displaystyle\lim_{x \to 1^+} \frac{\ln x}{(x - 1)^2}$.

SOLUTION   **Step 1** *Check to see if L'Hôpital's rule applies:*

$$
\lim_{x \to 1^+} \ln x = 0 \qquad \text{and} \qquad \lim_{x \to 1^+} (x - 1)^2 = 0
$$

L'Hôpital's rule does apply.

Step 2 *Apply L'Hôpital's rule:*

$$
\lim_{x \to 1^+} \frac{\ln x}{(x - 1)^2} = \overbrace{\lim_{x \to 1^+} \frac{\dfrac{d}{dx}(\ln x)}{\dfrac{d}{dx}(x - 1)^2}}^{0/0 \text{ form}} \qquad \text{Apply L'Hôpital's rule.}
$$

$$
= \lim_{x \to 1^+} \frac{1/x}{2(x - 1)} \qquad \text{Simplify.}
$$

$$
= \lim_{x \to 1^+} \frac{1}{2x(x - 1)}
$$

$$
= \infty
$$

The limit as $x \to 1^+$ is ∞ because $1/2x(x-1)$ has a vertical asymptote at $x = 1$ (Theorem 1, Section 3-2) and $x(x - 1) > 0$ for $x > 1$.

Matched Problem 7   Evaluate $\displaystyle\lim_{x\to 1^-} \frac{\ln x}{(x - 1)^2}$.

EXAMPLE 8   **L'Hôpital's Rule for Limits at Infinity**   Evaluate $\displaystyle\lim_{x\to\infty} \frac{\ln(1 + e^{-x})}{e^{-x}}$.

SOLUTION   **Step 1**   *Check to see if L'Hôpital's rule applies:*

$$\lim_{x\to\infty} \ln(1 + e^{-x}) = \ln(1 + 0) = \ln 1 = 0 \text{ and } \lim_{x\to\infty} e^{-x} = 0$$

L'Hôpital's rule does apply.

Step 2   *Apply L'Hôpital's rule:*

$$\underset{0/0 \text{ form}}{\lim_{x\to\infty} \frac{\ln(1 + e^{-x})}{e^{-x}}} = \lim_{x\to\infty} \frac{\dfrac{d}{dx}[\ln(1 + e^{-x})]}{\dfrac{d}{dx}e^{-x}} \qquad \text{Apply L'Hôpital's rule.}$$

$$= \lim_{x\to\infty} \frac{-e^{-x}/(1 + e^{-x})}{-e^{-x}} \qquad \text{Multiply numerator and denominator by } -e^x.$$

$$\lim_{x\to\infty} \frac{\ln(1 + e^{-x})}{e^{-x}} = \lim_{x\to\infty} \frac{1}{1 + e^{-x}} \qquad \lim_{x\to\infty} e^{-x} = 0$$

$$= \frac{1}{1 + 0} = 1$$

Matched Problem 8   Evaluate $\displaystyle\lim_{x\to -\infty} \frac{\ln(1 + 2e^{x})}{e^x}$.

L'Hôpital's Rule and the Indeterminate Form ∞/∞

In Section 3-2, we discussed techniques for evaluating limits of rational functions such as

$$\lim_{x\to\infty} \frac{2x^2}{x^3 + 3} \qquad \lim_{x\to\infty} \frac{4x^3}{2x^2 + 5} \qquad \lim_{x\to\infty} \frac{3x^3}{5x^3 + 6} \qquad (1)$$

Each of these limits is an ∞/∞ *indeterminate form*. In general, if $\lim_{x\to c} f(x) = \pm\infty$ and $\lim_{x\to c} g(x) = \pm\infty$, then

$$\lim_{x\to c} \frac{f(x)}{g(x)}$$

is called an ∞/∞ **indeterminate form**. Furthermore, $x \to c$ can be replaced in all three limits above with $x \to c^+$, $x \to c^-$, $x \to \infty$, or $x \to -\infty$. It can be shown that L'Hôpital's rule also applies to these ∞/∞ indeterminate forms.

THEOREM 3   L'Hôpital's Rule for the Indeterminate Form ∞/∞: Version 3
Versions 1 and 2 of L'Hôpital's rule for the indeterminate form 0/0 are also valid if the limit of f and the limit of g are both infinite; that is, both $+\infty$ and $-\infty$ are permissible for either limit.

For example, if $\lim_{x \to c^+} f(x) = \infty$ and $\lim_{x \to c^+} g(x) = -\infty$, then L'Hôpital's rule can be applied to $\lim_{x \to c^+} [f(x)/g(x)]$.

> **EXPLORE & DISCUSS 1**
>
> Evaluate each of the limits in (1) in two ways:
>
> 1. Use Theorem 4 in Section 3-2.
> 2. Use L'Hôpital's rule.
>
> Given a choice, which method would you choose? Why?

EXAMPLE 9   **L'Hôpital's Rule for the Indeterminate Form ∞/∞**   Evaluate $\lim_{x \to \infty} \dfrac{\ln x}{x^2}$.

SOLUTION   **Step 1** *Check to see if L'Hôpital's rule applies:*

$$\lim_{x \to \infty} \ln x = \infty \qquad \text{and} \qquad \lim_{x \to \infty} x^2 = \infty$$

L'Hôpital's rule does apply.

Step 2 *Apply L'Hôpital's rule:*

$$\underset{\infty/\infty \text{ form}}{\lim_{x \to \infty} \frac{\ln x}{x^2}} = \lim_{x \to \infty} \frac{\dfrac{d}{dx}(\ln x)}{\dfrac{d}{dx} x^2} \qquad \text{Apply L'Hôpital's rule.}$$

$$= \lim_{x \to \infty} \frac{1/x}{2x} \qquad \text{Simplify.}$$

$$\lim_{x \to \infty} \frac{\ln x}{x^2} = \lim_{x \to \infty} \frac{1}{2x^2} \qquad \text{See Figure 1(D).}$$

$$= 0$$

Matched Problem 9   Evaluate $\lim_{x \to \infty} \dfrac{\ln x}{x}$.

EXAMPLE 10   **L'Hôpital's Rule for the Indeterminate Form ∞/∞**   Evaluate $\lim_{x \to \infty} \dfrac{e^x}{x^2}$.

SOLUTION   **Step 1** *Check to see if L'Hôpital's rule applies:*

$$\lim_{x \to \infty} e^x = \infty \qquad \text{and} \qquad \lim_{x \to \infty} x^2 = \infty$$

L'Hôpital's rule does apply.

Step 2 *Apply L'Hôpital's rule:*

$$\underset{\infty/\infty \text{ form}}{\lim_{x \to \infty} \frac{e^x}{x^2}} = \lim_{x \to \infty} \frac{\dfrac{d}{dx} e^x}{\dfrac{d}{dx} x^2} = \lim_{x \to \infty} \frac{e^x}{2x}$$

Since $\lim_{x \to \infty} e^x = \infty$ and $\lim_{x \to \infty} 2x = \infty$, this limit is an ∞/∞ indeterminate form and L'Hôpital's rule can be applied again.

Step 3 *Apply L'Hôpital's rule again:*

$$\underset{\infty/\infty \text{ form}}{\lim_{x \to \infty} \frac{e^x}{2x}} = \lim_{x \to \infty} \frac{\dfrac{d}{dx} e^x}{\dfrac{d}{dx} 2x} = \lim_{x \to \infty} \frac{e^x}{2} = \infty$$

Therefore,

$$\lim_{x \to \infty} \frac{e^x}{x^2} = \lim_{x \to \infty} \frac{e^x}{2x} = \lim_{x \to \infty} \frac{e^x}{2} = \infty$$

Matched Problem 10   Evaluate $\lim\limits_{x \to \infty} \dfrac{e^{2x}}{x^2}$.

CONCEPTUAL INSIGHT

The three versions of L'Hôpital's rule cover a multitude of limits—far too many to remember case by case. Instead, we suggest you use the following pattern, common to all versions, as a memory aid:

1. All versions involve three limits: $\lim f(x)/g(x)$, $\lim f(x)$, and $\lim g(x)$.

2. The independent variable x must behave the same way in all three limits. The acceptable behaviors are $x \to c$, $x \to c^+$, $x \to c^-$, $x \to \infty$, or $x \to -\infty$.

3. The form of $\lim f(x)/g(x)$ must be $\dfrac{0}{0}$ or $\dfrac{\pm\infty}{\pm\infty}$ and both $\lim f(x)$ and $\lim g(x)$ must approach 0 or both must approach $\pm\infty$.

Exercises 5-3

A

Use L'Hôpital's rule to find each limit in Problems 1–14.

1. $\lim\limits_{x \to 2} \dfrac{x^4 - 16}{x^3 - 8}$

2. $\lim\limits_{x \to 1} \dfrac{x^6 - 1}{x^5 - 1}$

3. $\lim\limits_{x \to 2} \dfrac{x^2 + x - 6}{x^2 + 6x - 16}$

4. $\lim\limits_{x \to 4} \dfrac{x^2 - 8x + 16}{x^2 - 5x + 4}$

5. $\lim\limits_{x \to 0} \dfrac{e^x - 1}{x}$

6. $\lim\limits_{x \to 1} \dfrac{\ln x}{x - 1}$

7. $\lim\limits_{x \to 0} \dfrac{\ln(1 + 4x)}{x}$

8. $\lim\limits_{x \to 0} \dfrac{e^{2x} - 1}{x}$

9. $\lim\limits_{x \to \infty} \dfrac{2x^2 + 7}{5x^3 + 9}$

10. $\lim\limits_{x \to \infty} \dfrac{3x^4 + 6}{2x^2 + 5}$

11. $\lim\limits_{x \to \infty} \dfrac{e^{3x}}{x}$

12. $\lim\limits_{x \to \infty} \dfrac{x}{e^{4x}}$

13. $\lim\limits_{x \to \infty} \dfrac{x^2}{\ln x}$

14. $\lim\limits_{x \to \infty} \dfrac{\ln x}{x^4}$

✎ *In Problems 15–18, explain why L'Hôpital's rule does not apply. If the limit exists, find it by other means.*

15. $\lim\limits_{x \to 1} \dfrac{x^2 + 5x + 4}{x^3 + 1}$

16. $\lim\limits_{x \to \infty} \dfrac{e^{-x}}{\ln x}$

17. $\lim\limits_{x \to 2} \dfrac{x + 2}{(x - 2)^4}$

18. $\lim\limits_{x \to -3} \dfrac{x^2}{(x + 3)^5}$

In Problems 19–22, the limit can be found in two ways. Use L'Hôpital's rule to find the limit and check your answer using an algebraic simplification.

19. $\lim\limits_{x \to -4} \dfrac{x + 4}{x^2 - 16}$

20. $\lim\limits_{x \to 5} \dfrac{x - 5}{x^2 - 25}$

21. $\lim\limits_{x \to 9} \dfrac{x^2 - x - 72}{x - 9}$

22. $\lim\limits_{x \to 7} \dfrac{x - 7}{x^2 + 6x - 91}$

B

Find each limit in Problems 23–46. Note that L'Hôpital's rule does not apply to every problem, and some problems will require more than one application of L'Hôpital's rule.

23. $\lim\limits_{x \to 0} \dfrac{e^{4x} + 1 - 4x}{x^2}$

24. $\lim\limits_{x \to 0} \dfrac{3x + 1 - e^{3x}}{x^2}$

25. $\lim\limits_{x \to 2} \dfrac{\ln(x - 1)}{x - 1}$

26. $\lim\limits_{x \to -1} \dfrac{\ln(x + 2)}{x + 2}$

27. $\lim\limits_{x \to 0^+} \dfrac{\ln(1 + x^2)}{x^3}$

28. $\lim\limits_{x \to 0^-} \dfrac{\ln(1 + 2x)}{x^2}$

29. $\lim\limits_{x \to 0^+} \dfrac{\ln(1 + \sqrt{x})}{x}$

30. $\lim\limits_{x \to 0^+} \dfrac{\ln(1 + x)}{\sqrt{x}}$

31. $\lim\limits_{x \to -2} \dfrac{x^2 + 2x + 1}{x^2 + x + 1}$

32. $\lim\limits_{x \to 1} \dfrac{2x^3 - 3x^2 + 1}{x^3 - 3x + 2}$

33. $\lim\limits_{x \to -1} \dfrac{x^3 + x^2 - x - 1}{x^3 + 4x^2 + 5x + 2}$

34. $\lim\limits_{x \to 3} \dfrac{x^3 + 3x^2 - x - 3}{x^2 + 6x + 9}$

35. $\lim\limits_{x \to 2^-} \dfrac{x^3 - 12x + 16}{x^3 - 6x^2 + 12x - 8}$

36. $\lim\limits_{x \to 1^+} \dfrac{x^3 + x^2 - x + 1}{x^3 + 3x^2 + 3x - 1}$

37. $\lim\limits_{x \to \infty} \dfrac{3x^2 + 5x}{4x^3 + 7}$

38. $\lim\limits_{x \to \infty} \dfrac{4x^2 + 9x}{5x^2 + 8}$

39. $\lim\limits_{x \to \infty} \dfrac{x^2}{e^{2x}}$

40. $\lim\limits_{x \to \infty} \dfrac{e^{3x}}{x^3}$

53. $\lim\limits_{x \to \infty} \dfrac{\sqrt{1 + x^2}}{x}$

41. $\lim\limits_{x \to \infty} \dfrac{1 + e^{-x}}{1 + x^2}$

42. $\lim\limits_{x \to -\infty} \dfrac{1 + e^{-x}}{1 + x^2}$

54. $\lim\limits_{x \to -\infty} \dfrac{x}{\sqrt{4 + x^2}}$

43. $\lim\limits_{x \to \infty} \dfrac{e^{-x}}{\ln(1 + 4e^{-x})}$

44. $\lim\limits_{x \to \infty} \dfrac{\ln(1 + 2e^{-x})}{\ln(1 + e^{-x})}$

55. $\lim\limits_{x \to -\infty} \dfrac{\sqrt[3]{x^3 + 1}}{x}$

45. $\lim\limits_{x \to 0} \dfrac{e^x - e^{-x} - 2x}{x^3}$

46. $\lim\limits_{x \to 0} \dfrac{e^{2x} - 1 - 2x - 2x^2}{x^3}$

56. $\lim\limits_{x \to \infty} \dfrac{x^2}{\sqrt[3]{(x^3 + 1)^2}}$

C

47. Find $\lim\limits_{x \to 0^+} (x \ln x)$.

[*Hint*: Write $x \ln x = (\ln x)/x^{-1}$.]

48. Find $\lim\limits_{x \to 0^+} (\sqrt{x} \ln x)$.

[*Hint*: Write $\sqrt{x} \ln x = (\ln x)/x^{-1/2}$.]

In Problems 49–52, n is a positive integer. Find each limit.

49. $\lim\limits_{x \to \infty} \dfrac{\ln x}{x^n}$

50. $\lim\limits_{x \to \infty} \dfrac{x^n}{\ln x}$

51. $\lim\limits_{x \to \infty} \dfrac{e^x}{x^n}$

52. $\lim\limits_{x \to \infty} \dfrac{x^n}{e^x}$

In Problems 53–56, show that the repeated application of L'Hôpital's rule does not lead to a solution. Then use algebraic manipulation to evaluate each limit. [Hint: If $x > 0$ and $n > 0$, then $\sqrt[n]{x^n} = x$.]

Answers to Matched Problems

1. (A) ∞
  (B) ∞
  (C) 0
  (D) ∞

2. (A) ∞
  (B) 0
  (C) $-\infty$
  (D) ∞

3. e^4 **4.** ∞ **5.** 0 **6.** 2

7. $-\infty$ **8.** 2 **9.** 0 **10.** ∞

5-4 Curve-Sketching Techniques

- Modifying the Graphing Strategy
- Using the Graphing Strategy
- Modeling Average Cost

When we summarized the graphing strategy in Section 5-2, we omitted one important topic: asymptotes. Polynomial functions do not have any asymptotes. Asymptotes of rational functions were discussed in Section 3-3, but what about all the other functions, such as logarithmic and exponential functions? Since investigating asymptotes always involves limits, we can now use L'Hôpital's rule (Section 5-3) as a tool for finding asymptotes of many different types of functions.

Modifying the Graphing Strategy

The first version of the graphing strategy in Section 5-2 made no mention of asymptotes. Including information about asymptotes produces the following (and final) version of the graphing strategy.

> **PROCEDURE** **Graphing Strategy (Final Version)**
>
> **Step 1** *Analyze $f(x)$.*
>   (A) Find the domain of f.
>   (B) Find the intercepts.
>   (C) Find asymptotes.

Step 2 *Analyze $f'(x)$*. Find the partition numbers for, and critical values of, $f'(x)$. Construct a sign chart for $f'(x)$, determine the intervals on which f is increasing and decreasing, and find local maxima and minima.

Step 3 *Analyze $f''(x)$*. Find the partition numbers of $f''(x)$. Construct a sign chart for $f''(x)$, determine the intervals on which the graph of f is concave upward and concave downward, and find inflection points.

Step 4 *Sketch the graph of f*. Draw asymptotes and locate intercepts, local maxima and minima, and inflection points. Sketch in what you know from steps 1–3. Plot additional points as needed and complete the sketch.

Using the Graphing Strategy

We will illustrate the graphing strategy with several examples. From now on, you should always use the final version of the graphing strategy. If a function does not have any asymptotes, simply state this fact.

EXAMPLE 1 **Using the Graphing Strategy** Use the graphing strategy to analyze the function $f(x) = (x - 1)/(x - 2)$. State all the pertinent information and sketch the graph of f.

SOLUTION **Step 1** *Analyze $f(x)$*. $f(x) = \dfrac{x - 1}{x - 2}$

(A) Domain: All real x, except $x = 2$

(B) y intercept: $f(0) = \dfrac{0 - 1}{0 - 2} = \dfrac{1}{2}$

x intercepts: Since a fraction is 0 when its numerator is 0 and its denominator is not 0, the x intercept is $x = 1$.

(C) Horizontal asymptote: $\dfrac{a_m x^m}{b_n x^n} = \dfrac{x}{x} = 1$

So the line $y = 1$ is a horizontal asymptote.

Vertical asymptote: The denominator is 0 for $x = 2$, and the numerator is not 0 for this value. Therefore, the line $x = 2$ is a vertical asymptote.

Step 2 *Analyze $f'(x)$*. $f'(x) = \dfrac{(x - 2)(1) - (x - 1)(1)}{(x - 2)^2} = \dfrac{-1}{(x - 2)^2}$

Critical values of $f(x)$: None

Partition number for $f'(x)$: $x = 2$

Sign chart for $f'(x)$:

	$(-\infty, 2)$	$(2, \infty)$	Test Numbers	
			x	$f'(x)$
$f'(x)$	$- - - -$ ND $- - - -$		1	-1 $(-)$
			3	-1 $(-)$
	2			
$f(x)$	Decreasing	Decreasing		

So $f(x)$ is decreasing on $(-\infty, 2)$ and $(2, \infty)$. There are no local extrema.

Step 3 *Analyze $f''(x)$*. $f''(x) = \dfrac{2}{(x - 2)^3}$

Partition number for $f''(x)$: $x = 2$

Sign chart for $f''(x)$:

	$(-\infty, 2)$	$(2, \infty)$
$f''(x)$	$-----$ ND	$+ + + +$

Test Numbers	
x	$f''(x)$
1	-2 $(-)$
3	2 $(+)$

Graph of f — Concave downward / Concave upward

The graph of f is concave downward on $(-\infty, 2)$ and concave upward on $(2, \infty)$. Since $f(2)$ is not defined, there is no inflection point at $x = 2$, even though $f''(x)$ changes sign at $x = 2$.

Step 4 *Sketch the graph of f.* Insert intercepts and asymptotes, and plot a few additional points (for functions with asymptotes, plotting additional points is often helpful). Then sketch the graph.

x	$f(x)$
-2	$\frac{3}{4}$
0	$\frac{1}{2}$
1	0
$\frac{3}{2}$	-1
$\frac{5}{2}$	3
3	2
4	$\frac{3}{2}$

Matched Problem 1 Follow the graphing strategy and analyze the function $f(x) = 2x/(1 - x)$. State all the pertinent information and sketch the graph of f.

EXAMPLE 2 **Using the Graphing Strategy** Use the graphing strategy to analyze the function

$$g(x) = \frac{2x - 1}{x^2}$$

State all pertinent information and sketch the graph of g.

SOLUTION **Step 1** *Analyze $g(x)$.*

(A) Domain: All real x, except $x = 0$

(B) x intercept: $x = \dfrac{1}{2} = 0.5$

y intercept: Since 0 is not in the domain of g, there is no y intercept.

(C) Horizontal asymptote: $y = 0$ (the x axis)

Vertical asymptote: The denominator of $g(x)$ is 0 at $x = 0$ and the numerator is not. So the line $x = 0$ (the y axis) is a vertical asymptote.

Step 2 *Analyze $g'(x)$.*

$$g(x) = \frac{2x - 1}{x^2} = \frac{2}{x} - \frac{1}{x^2} = 2x^{-1} - x^{-2}$$

$$g'(x) = -2x^{-2} + 2x^{-3} = -\frac{2}{x^2} + \frac{2}{x^3} = \frac{-2x + 2}{x^3}$$

$$= \frac{2(1 - x)}{x^3}$$

Critical values of $g(x)$: $x = 1$

Partition numbers for $g'(x)$: $x = 0, x = 1$

Sign chart for $g'(x)$:

Function $f(x)$ is decreasing on $(-\infty, 0)$ and $(1, \infty)$, is increasing on $(0, 1)$, and has a local maximum at $x = 1$.

Step 3 *Analyze $g''(x)$.*

$$g'(x) = -2x^{-2} + 2x^{-3}$$

$$g''(x) = 4x^{-3} - 6x^{-4} = \frac{4}{x^3} - \frac{6}{x^4} = \frac{4x - 6}{x^4} = \frac{2(2x - 3)}{x^4}$$

Partition numbers for $g''(x)$: $x = 0, x = \dfrac{3}{2} = 1.5$

Sign chart for $g''(x)$:

Function $g(x)$ is concave downward on $(-\infty, 0)$ and $(0, 1.5)$, is concave upward on $(1.5, \infty)$, and has an inflection point at $x = 1.5$.

Step 4 *Sketch the graph of g.* Plot key points, note that the coordinate axes are asymptotes, and sketch the graph.

x	$g(x)$
-10	-0.21
-1	-3
0.5	0
1	1
1.5	0.89
10	0.19

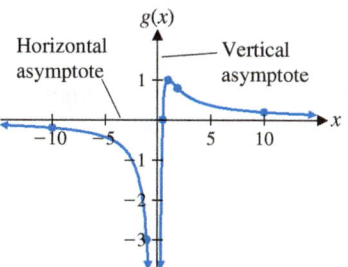

Matched Problem 2 Use the graphing strategy to analyze the function

$$h(x) = \frac{4x + 3}{x^2}$$

State all pertinent information and sketch the graph of h.

EXAMPLE 3 **Graphing Strategy** Follow the steps of the graphing strategy and analyze the function $f(x) = xe^x$. State all the pertinent information and sketch the graph of f.

SOLUTION **Step 1** Analyze $f(x)$: $f(x) = xe^x$.

(A) Domain: All real numbers

(B) y intercept: $f(0) = 0$

x intercept: $xe^x = 0$ for $x = 0$ only, since $e^x > 0$ for all x.

(C) Vertical asymptotes: None

Horizontal asymptotes: We use tables to determine the nature of the graph of f as $x \to \infty$ and $x \to -\infty$:

x	1	5	10	$\to \infty$
$f(x)$	2.72	742.07	220,264.66	$\to \infty$

x	-1	-5	-10	$\to -\infty$
$f(x)$	-0.37	-0.03	$-0.000\,45$	$\to 0$

Step 2 *Analyze $f'(x)$:*

$$f'(x) = x\frac{d}{dx}e^x + e^x\frac{d}{dx}x$$

$$= xe^x + e^x = e^x(x + 1)$$

Critical value of $f(x)$: -1

Partition number for $f'(x)$: -1

Sign chart for $f'(x)$:

	$(-\infty, -1)$	$(-1, \infty)$
$f'(x)$	$----\ 0$	$+++++$
	-1	
$f(x)$	Decreasing	Increasing

Test Numbers	
x	$f'(x)$
-2	$-e^{-2}$ $(-)$
0	1 $(+)$

So $f(x)$ decreases on $(-\infty, -1)$, has a local minimum at $x = -1$, and increases on $(-1, \infty)$.

Step 3 *Analyze $f''(x)$:*

$$f''(x) = e^x\frac{d}{dx}(x + 1) + (x + 1)\frac{d}{dx}e^x$$

$$= e^x + (x + 1)e^x = e^x(x + 2)$$

Sign chart for $f''(x)$ (partition number is -2):

	$(-\infty, -2)$	$(-2, \infty)$
$f''(x)$	$----\ 0$	$+++++$
	-2	
Graph of f	Concave downward	Concave upward
	Inflection point	

Test Numbers	
x	$f''(x)$
-3	$-e^{-3}$ $(-)$
-1	e^{-1} $(+)$

The graph of f is concave downward on $(-\infty, -2)$, has an inflection point at $x = -2$, and is concave upward on $(-2, \infty)$.

Step 4 *Sketch the graph of f, using the information from steps 1 to 3:*

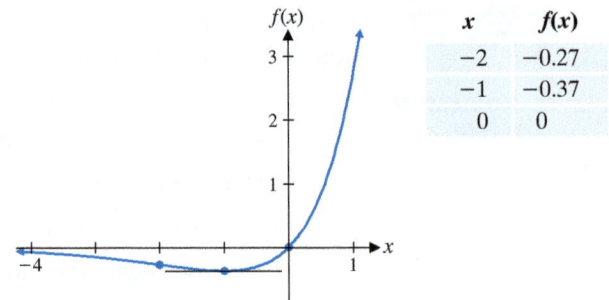

x	$f(x)$
-2	-0.27
-1	-0.37
0	0

Matched Problem 3   Analyze the function $f(x) = xe^{-0.5x}$. State all the pertinent information and sketch the graph of f.

EXPLORE & DISCUSS 1

Refer to the discussion of vertical asymptotes in the solution of Example 3. We used tables of values to estimate limits at infinity and determine horizontal asymptotes. In some cases, the functions involved in these limits can be written in a form that allows us to use L'Hôpital's rule.

$$\underset{-\infty \cdot 0 \text{ form}}{\lim_{x \to -\infty} f(x)} = \lim_{x \to -\infty} xe^x \quad \text{Rewrite as a fraction.}$$

$$\underset{-\infty/\infty \text{ form}}{= \lim_{x \to -\infty} \frac{x}{e^{-x}}} \quad \text{Apply L'Hôpital's rule.}$$

$$= \lim_{x \to -\infty} \frac{1}{-e^{-x}} \quad \text{Simplify.}$$

$$= \lim_{x \to -\infty} (-e^x) \quad \text{Property of } e^x$$

$$= 0$$

Use algebraic manipulation and L'Hôpital's rule to verify the value of each of the following limits:

(A) $\lim\limits_{x \to \infty} xe^{-0.5x} = 0$

(B) $\lim\limits_{x \to 0^+} x^2(\ln x - 0.5) = 0$

(C) $\lim\limits_{x \to 0^+} x \ln x = 0$

EXAMPLE 4   **Graphing Strategy** Let $f(x) = x^2 \ln x - 0.5x^2$. Follow the steps in the graphing strategy and analyze this function. State all the pertinent information and sketch the graph of f.

SOLUTION   *Step 1* *Analyze* $f(x)$: $f(x) = x^2 \ln x - 0.5x^2 = x^2(\ln x - 0.5)$.

(A) Domain: $(0, \infty)$

(B) y intercept: None [$f(0)$ is not defined.]

x intercept: Solve $x^2(\ln x - 0.5) = 0$

$\ln x - 0.5 = 0 \qquad \text{or} \qquad x^2 = 0 \qquad$ *Discard, since 0 is not in the domain of f.*

$\ln x = 0.5$ *In x = a if and only if x = e^a.*

$x = e^{0.5}$ *x intercept*

(C) Asymptotes: None. The following tables suggest the nature of the graph as $x \to 0^+$ and as $x \to \infty$:

x	0.1	0.01	0.001	$\to 0^+$	*See Explore &*
$f(x)$	-0.0280	-0.00051	-0.00007	$\to 0$	*Discuss 1(B).*

x	10	100	1,000	$\to \infty$	
$f(x)$	180	41,000	6,400,000	$\to \infty$	

Step 2 *Analyze $f'(x)$:*

$$f'(x) = x^2 \frac{d}{dx} \ln x + (\ln x) \frac{d}{dx} x^2 - 0.5 \frac{d}{dx} x^2$$

$$= x^2 \frac{1}{x} + (\ln x) 2x - 0.5(2x)$$

$$= x + 2x \ln x - x$$

$$= 2x \ln x$$

Critical value of $f(x)$: 1

Partition number for $f'(x)$: 1

Sign chart for $f'(x)$:

	Test Numbers	
x	$f'(x)$	
0.5	-0.2983	$(-)$
2	0.7726	$(+)$

The function $f(x)$ decreases on $(0, 1)$, has a local minimum at $x = 1$, and increases on $(1, \infty)$.

Step 3 *Analyze $f''(x)$:*

$$f''(x) = 2x \frac{d}{dx} (\ln x) + (\ln x) \frac{d}{dx} (2x)$$

$$= 2x \frac{1}{x} + (\ln x) 2$$

$$= 2 + 2 \ln x = 0$$

$$2 \ln x = -2$$

$$\ln x = -1$$

$$x = e^{-1} \approx 0.3679$$

Sign chart for $f'(x)$ (partition number is e^{-1}):

	Test Numbers	
x	$f'(x)$	
0.2	-1.2189	$(-)$
1	2	$(+)$

The graph of $f(x)$ is concave downward on $(0, e^{-1})$, has an inflection point at $x = e^{-1}$, and is concave upward on (e^{-1}, ∞).

Step 4 *Sketch the graph of f, using the information from steps 1 to 3:*

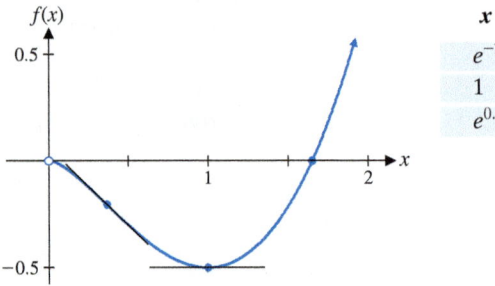

x	$f(x)$
e^{-1}	$-1.5e^{-2}$
1	-0.5
$e^{0.5}$	0

Matched Problem 4   Analyze the function $f(x) = x \ln x$. State all pertinent information and sketch the graph of f.

Modeling Average Cost

EXAMPLE 5   **Average Cost**   Given the cost function $C(x) = 5{,}000 + 0.5x^2$, where x is the number of items produced, use the graphing strategy to analyze the graph of the average cost function. State all the pertinent information and sketch the graph of the average cost function. Find the marginal cost function and graph it on the same set of coordinate axes.

SOLUTION   The average cost function is

$$\overline{C}(x) = \frac{5{,}000 + 0.5x^2}{x} = \frac{5{,}000}{x} + 0.5x$$

Step 1   *Analyze $\overline{C}(x)$.*

(A) Domain: Since negative values of x do not make sense and $\overline{C}(0)$ is not defined, the domain is the set of positive real numbers.

(B) Intercepts: None

(C) Horizontal asymptote: $\dfrac{a_m x^m}{b_n x^n} = \dfrac{0.5x^2}{x} = 0.5x$

So there is no horizontal asymptote.

Vertical asymptote: The line $x = 0$ is a vertical asymptote since the denominator is 0 and the numerator is not 0 for $x = 0$.

Oblique asymptotes: If a graph approaches a line that is neither horizontal nor vertical as x approaches ∞ or $-\infty$, then that line is called an **oblique asymptote**. If x is a large positive number, then $5{,}000/x$ is very small and

$$\overline{C}(x) = \frac{5{,}000}{x} + 0.5x \approx 0.5x$$

That is,

$$\lim_{x \to \infty} [\overline{C}(x) - 0.5x] = \lim_{x \to \infty} \frac{5{,}000}{x} = 0$$

This implies that the graph of $y = \overline{C}(x)$ approaches the line $y = 0.5x$ as x approaches ∞. That line is an oblique asymptote for the graph of $y = \overline{C}(x)$.*

Step 2   *Analyze $\overline{C}'(x)$.*

$$\overline{C}'(x) = -\frac{5{,}000}{x^2} + 0.5$$

$$= \frac{0.5x^2 - 5{,}000}{x^2}$$

$$= \frac{0.5(x - 100)(x + 100)}{x^2}$$

---

*If $f(x) = n(x)/d(x)$ is a rational function for which the degree of $n(x)$ is 1 more than the degree of $d(x)$, then we can use polynomial long division to write $f(x) = mx + b + r(x)/d(x)$, where the degree of $r(x)$ is less than the degree of $d(x)$. The line $y = mx + b$ is then an oblique asymptote for the graph of $y = f(x)$.

Critical value for $\overline{C}(x)$:   100
Partition numbers for $\overline{C}'(x)$:   0 and 100
Sign chart for $\overline{C}'(x)$:

So $\overline{C}(x)$ is decreasing on $(0, 100)$, is increasing on $(100, \infty)$, and has a local minimum at $x = 100$.

Step 3 *Analyze $\overline{C}''(x)$*: $\overline{C}''(x) = \dfrac{10{,}000}{x^3}$.

$\overline{C}''(x)$ is positive for all positive x, so the graph of $y = \overline{C}(x)$ is concave upward on $(0, \infty)$.

Step 4 *Sketch the graph of $\overline{C}$.* The graph of $\overline{C}$ is shown in Figure 1.

Figure 1

The marginal cost function is $C'(x) = x$. The graph of this linear function is also shown in Figure 1.

Figure 1 illustrates an important principle in economics:

The minimum average cost occurs when the average cost is equal to the marginal cost.

Matched Problem 5   Given the cost function $C(x) = 1{,}600 + 0.25x^2$, where x is the number of items produced,

(A) Use the graphing strategy to analyze the graph of the average cost function. State all the pertinent information and sketch the graph of the average cost function. Find the marginal cost function and graph it on the same set of coordinate axes. Include any oblique asymptotes.

(B) Find the minimum average cost.

Exercises 5-4

A

1. Use the graph of *f* in the figure to identify the following:

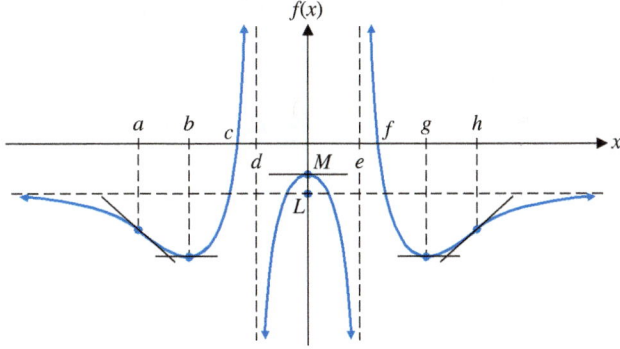

Figure for 1

(A) the intervals on which $f'(x) < 0$

(B) the intervals on which $f'(x) > 0$

(C) the intervals on which $f(x)$ is increasing

(D) the intervals on which $f(x)$ is decreasing

(E) the x coordinate(s) of the point(s) where $f(x)$ has a local maximum

(F) the x coordinate(s) of the point(s) where $f(x)$ has a local minimum

(G) the intervals on which $f''(x) < 0$

(H) the intervals on which $f''(x) > 0$

(I) the intervals on which the graph of f is concave upward

(J) the intervals on which the graph of f is concave downward

(K) the x coordinate(s) of the inflection point(s)

(L) the horizontal asymptote(s)

(M) the vertical asymptote(s)

2. Repeat Problem 1 for the following graph of *f*:

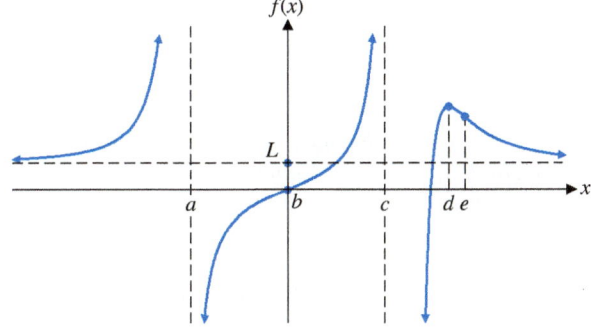

Figure for 2

In Problems 3–10, use the given information to sketch the graph of f. Assume that f is continuous on its domain and that all intercepts are included in the table of values.

3. Domain: All real x; $\lim\limits_{x \to \pm\infty} f(x) = 2$

x	−4	−2	0	2	4
$f(x)$	0	−2	0	−2	0

4. Domain: All real x; $\lim\limits_{x \to -\infty} f(x) = -3$; $\lim\limits_{x \to \infty} f(x) = 3$

x	−2	−1	0	1	2
$f(x)$	0	2	0	−2	0

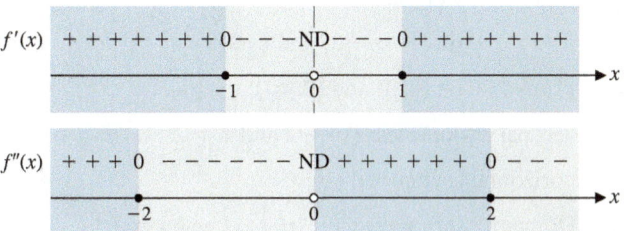

5. Domain: All real x, except $x = -2$;
$\lim\limits_{x \to -2^-} f(x) = \infty$; $\lim\limits_{x \to -2^+} f(x) = -\infty$; $\lim\limits_{x \to \infty} f(x) = 1$

x	−4	0	4	6
$f(x)$	0	0	3	2

6. Domain: All real x, except
$x = 1$; $\lim\limits_{x \to 1^-} f(x) = \infty$; $\lim\limits_{x \to 1^+} f(x) = \infty$; $\lim\limits_{x \to \infty} f(x) = -2$

x	−4	−2	0	2
$f(x)$	0	−2	0	0

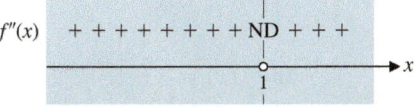

7. Domain: All real x, except $x = -1$;

$f(-3) = 2, f(-2) = 3, f(0) = -1, f(1) = 0$;

$f'(x) > 0$ on $(-\infty, -1)$ and $(-1, \infty)$;

$f''(x) > 0$ on $(-\infty, -1)$; $f''(x) < 0$ on $(-1, \infty)$;

vertical asymptote: $x = -1$;

horizontal asymptote: $y = 1$

8. Domain: All real x, except $x = 1$;

$f(0) = -2, f(2) = 0$;

$f'(x) < 0$ on $(-\infty, 1)$ and $(1, \infty)$;

$f''(x) < 0$ on $(-\infty, 1)$;

$f''(x) > 0$ on $(1, \infty)$;

vertical asymptote: $x = 1$;

horizontal asymptote: $y = -1$

9. Domain: All real x, except $x = -2$ and $x = 2$;

$f(-3) = -1, f(0) = 0, f(3) = 1$;

$f'(x) < 0$ on $(-\infty, -2)$ and $(2, \infty)$;

$f'(x) > 0$ on $(-2, 2)$;

$f''(x) < 0$ on $(-\infty, -2)$ and $(-2, 0)$;

$f''(x) > 0$ on $(0, 2)$ and $(2, \infty)$;

vertical asymptotes: $x = -2$ and $x = 2$;

horizontal asymptote: $y = 0$

10. Domain: All real x, except $x = -1$ and $x = 1$;

$f(-2) = 1, f(0) = 0, f(2) = 1$;

$f'(x) > 0$ on $(-\infty, -1)$ and $(0, 1)$;

$f'(x) < 0$ on $(-1, 0)$ and $(1, \infty)$;

$f''(x) > 0$ on $(-\infty, -1)$, $(-1, 1)$, and $(1, \infty)$;

vertical asymptotes: $x = -1$ and $x = 1$;

horizontal asymptote: $y = 0$

In Problems 11–16, find the domain and intercepts.

11. $f(x) = \sqrt{x + 4}$

12. $f(x) = \sqrt{x + 25}$

13. $f(x) = 75 - 5x$

14. $f(x) = 4x + 52$

15. $f(x) = \dfrac{95}{x - 5}$

16. $f(x) = \dfrac{102}{x + 3}$

In Problems 17–56, summarize the pertinent information obtained by applying the graphing strategy and sketch the graph of $y = f(x)$.

17. $f(x) = \dfrac{x + 3}{x - 3}$

18. $f(x) = \dfrac{2x - 4}{x + 2}$

19. $f(x) = \dfrac{x}{x - 2}$

20. $f(x) = \dfrac{2 + x}{3 - x}$

21. $f(x) = 5 + 5e^{-0.1x}$

22. $f(x) = 3 + 7e^{-0.2x}$

23. $f(x) = 5xe^{-0.2x}$

24. $f(x) = 10xe^{-0.1x}$

25. $f(x) = \ln(1 - x)$

26. $f(x) = \ln(2x + 4)$

27. $f(x) = x - \ln x$

28. $f(x) = \ln(x^2 + 4)$

29. $f(x) = \dfrac{x}{x^2 - 4}$

30. $f(x) = \dfrac{1}{x^2 - 4}$

31. $f(x) = \dfrac{1}{1 + x^2}$

32. $f(x) = \dfrac{x^2}{1 + x^2}$

33. $f(x) = \dfrac{2x}{1 - x^2}$

34. $f(x) = \dfrac{2x}{x^2 - 9}$

35. $f(x) = \dfrac{-5x}{(x - 1)^2}$

36. $f(x) = \dfrac{x}{(x - 2)^2}$

37. $f(x) = \dfrac{x^2 + x - 2}{x^2}$

38. $f(x) = \dfrac{x^2 - 5x - 6}{x^2}$

39. $f(x) = \dfrac{x^2}{x - 1}$

40. $f(x) = \dfrac{x^2}{2 + x}$

41. $f(x) = \dfrac{3x^2 + 2}{x^2 - 9}$

42. $f(x) = \dfrac{2x^2 + 5}{4 - x^2}$

43. $f(x) = \dfrac{x^3}{x - 2}$

44. $f(x) = \dfrac{x^3}{4 - x}$

45. $f(x) = (3 - x)e^x$

46. $f(x) = (x - 2)e^x$

47. $f(x) = e^{-0.5x^2}$

48. $f(x) = e^{-2x^2}$

49. $f(x) = x^2 \ln x$

50. $f(x) = \dfrac{\ln x}{x}$

51. $f(x) = (\ln x)^2$

52. $f(x) = \dfrac{x}{\ln x}$

53. $f(x) = \dfrac{1}{x^2 + 2x - 8}$

54. $f(x) = \dfrac{1}{3 - 2x - x^2}$

55. $f(x) = \dfrac{x^3}{3 - x^2}$

56. $f(x) = \dfrac{x^3}{x^2 - 12}$

In Problems 57–64, show that the line $y = x$ is an oblique asymptote for the graph of $y = f(x)$, summarize all pertinent information obtained by applying the graphing strategy, and sketch the graph of $y = f(x)$.

57. $f(x) = x + \dfrac{4}{x}$

58. $f(x) = x - \dfrac{9}{x}$

59. $f(x) = x - \dfrac{4}{x^2}$

60. $f(x) = x + \dfrac{32}{x^2}$

61. $f(x) = x - \dfrac{9}{x^3}$

62. $f(x) = x + \dfrac{27}{x^3}$

63. $f(x) = x + \dfrac{1}{x} + \dfrac{4}{x^3}$

64. $f(x) = x - \dfrac{16}{x^3}$

In Problems 65–72, summarize all pertinent information obtained by applying the graphing strategy, and sketch the graph of $y = f(x)$. [Note: These rational functions are not reduced to lowest terms.]

65. $f(x) = \dfrac{x^2 + x - 6}{x^2 - 6x + 8}$

66. $f(x) = \dfrac{x^2 + x - 6}{x^2 - x - 12}$

67. $f(x) = \dfrac{2x^2 + x - 15}{x^2 - 9}$

68. $f(x) = \dfrac{2x^2 + 11x + 14}{x^2 - 4}$

69. $f(x) = \dfrac{x^3 - 5x^2 + 6x}{x^2 - x - 2}$

70. $f(x) = \dfrac{x^3 - 5x^2 - 6x}{x^2 + 3x + 2}$

71. $f(x) = \dfrac{x^2 + x - 2}{x^2 - 2x + 1}$

72. $f(x) = \dfrac{x^2 + x - 2}{x^2 + 4x + 4}$

Applications

73. Revenue. The marketing research department for a computer company used a large city to test market the firm's new laptop. The department found that the relationship between price p (dollars per unit) and demand x (units sold per week) was given approximately by

$$p = 1,296 - 0.12x^2 \qquad 0 \le x \le 80$$

So, weekly revenue can be approximated by

$$R(x) = xp = 1,296x - 0.12x^3 \qquad 0 \le x \le 80$$

Graph the revenue function R.

74. Profit. Suppose that the cost function $C(x)$ (in dollars) for the company in Problem 73 is

$$C(x) = 830 + 396x$$

(A) Write an equation for the profit $P(x)$.

(B) Graph the profit function P.

75. Pollution. In Silicon Valley, a number of computer firms were found to be contaminating underground water supplies with toxic chemicals stored in leaking underground containers. A water quality control agency ordered the companies to take immediate corrective action and contribute to a monetary pool for the testing and cleanup of the underground contamination. Suppose that the required monetary pool (in millions of dollars) is given by

$$P(x) = \dfrac{2x}{1 - x} \qquad 0 \le x < 1$$

where x is the percentage (expressed as a decimal fraction) of the total contaminant removed.

(A) Where is $P(x)$ increasing? Decreasing?

(B) Where is the graph of P concave upward? Downward?

(C) Find any horizontal and vertical asymptotes.

(D) Find the x and y intercepts.

(E) Sketch a graph of P.

76. Employee training. A company producing dive watches has established that, on average, a new employee can assemble $N(t)$ dive watches per day after t days of on-the-job training, as given by

$$N(t) = \dfrac{100t}{t + 9} \qquad t \ge 0$$

(A) Where is $N(t)$ increasing? Decreasing?

(B) Where is the graph of N concave upward? Downward?

(C) Find any horizontal and vertical asymptotes.

(D) Find the intercepts.

(E) Sketch a graph of N.

77. Replacement time. An outboard motor has an initial price of $3,200. A service contract costs $300 for the first year and increases $100 per year thereafter. The total cost of the outboard motor (in dollars) after n years is given by

$$C(n) = 3,200 + 250n + 50n^2 \qquad n \ge 1$$

(A) Write an expression for the average cost per year, $\overline{C}(n)$, for n years.

(B) Graph the average cost function found in part (A).

(C) When is the average cost per year minimum? (This time is frequently referred to as the **replacement time** for this piece of equipment.)

78. Construction costs. The management of a manufacturing plant wishes to add a fenced-in rectangular storage yard of 20,000 square feet, using a building as one side of the yard (see the figure). If x is the distance (in feet) from the building to the fence, show that the length of the fence required for the yard is given by

$$L(x) = 2x + \dfrac{20,000}{x} \qquad x > 0$$

Figure for 78

(A) Graph L.

(B) What are the dimensions of the rectangle requiring the least amount of fencing?

79. Average and marginal costs. The total daily cost (in dollars) of producing x mountain bikes is given by

$$C(x) = 1,000 + 5x + 0.1x^2$$

(A) Sketch the graphs of the average cost function and the marginal cost function on the same set of coordinate axes. Include any oblique asymptotes.

(B) Find the minimum average cost.

80. Average and marginal costs. The total daily cost (in dollars) of producing x city bikes is given by

$$C(x) = 500 + 2x + 0.2x^2$$

(A) Sketch the graphs of the average cost function and the marginal cost function on the same set of coordinate axes. Include any oblique asymptotes.

(B) Find the minimum average cost.

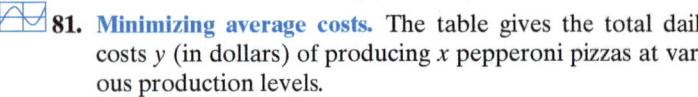 **81. Minimizing average costs.** The table gives the total daily costs y (in dollars) of producing x pepperoni pizzas at various production levels.

Number of Pizzas	Total Costs
x	y
50	395
100	475
150	640
200	910
250	1,140
300	1,450

(A) Enter the data into a graphing calculator and find a quadratic regression equation for the total costs.

(B) Use the regression equation from part (A) to find the minimum average cost (to the nearest cent) and the corresponding production level (to the nearest integer).

82. Minimizing average costs. The table gives the total daily costs y (in dollars) of producing x deluxe pizzas at various production levels.

Number of Pizzas	Total Costs
x	y
50	595
100	755
150	1,110
200	1,380
250	1,875
300	2,410

(A) Enter the data into a graphing calculator and find a quadratic regression equation for the total costs.

(B) Use the regression equation from part (A) to find the minimum average cost (to the nearest cent) and the corresponding production level (to the nearest integer).

83. Medicine. A drug is injected into the bloodstream of a patient through her right arm. The drug concentration in the bloodstream of the left arm t hours after the injection is given by

$$C(t) = \frac{0.14t}{t^2 + 1}$$

Graph C.

84. Physiology. In a study on the speed of muscle contraction in frogs under various loads, researchers found that the speed of contraction decreases with increasing loads. More precisely, they found that the relationship between speed of contraction, S (in centimeters per second), and load w (in grams) is given approximately by

$$S(w) = \frac{26 + 0.06w}{w} \qquad w \geq 5$$

Graph S.

85. Psychology: retention. Each student in a psychology class is given one day to memorize the same list of 30 special characters. The lists are turned in at the end of the day, and for each succeeding day for 30 days, each student is asked to turn in a list of as many of the symbols as can be recalled. Averages are taken, and it is found that

$$N(t) = \frac{5t + 20}{t} \qquad t \geq 1$$

provides a good approximation of the average number $N(t)$ of symbols retained after t days. Graph N.

Answers to Matched Problems

1. Domain: All real x, except $x = 1$
y intercept: $f(0) = 0$; x intercept: 0
Horizontal asymptote: $y = -2$;
vertical asymptote: $x = 1$
Increasing on $(-\infty, 1)$ and $(1, \infty)$
Concave upward on $(-\infty, 1)$; concave downward on $(1, \infty)$

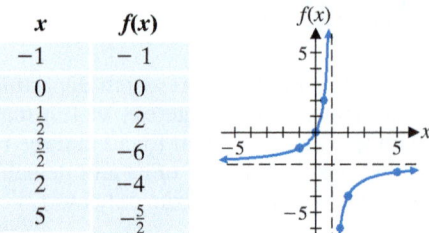

x	$f(x)$
-1	-1
0	0
$\frac{1}{2}$	2
$\frac{3}{2}$	-6
2	-4
5	$-\frac{5}{2}$

2. Domain: All real x, except $x = 0$
x intercept: $= -\dfrac{3}{4} = -0.75$
$h(0)$ is not defined
Vertical asymptote: $x = 0$ (the y axis)
Horizontal asymptote: $y = 0$ (the x axis)
Increasing on $(-1.5, 0)$
Decreasing on $(-\infty, -1.5)$ and $(0, \infty)$
Local minimum at $x = 1.5$
Concave upward on $(-2.25, 0)$ and $(0, \infty)$
Concave downward on $(-\infty, -2.25)$
Inflection point at $x = -2.25$

x	$h(x)$
-10	-0.37
-2.25	-1.19
-1.5	-1.33
-0.75	0
2	2.75
10	0.43

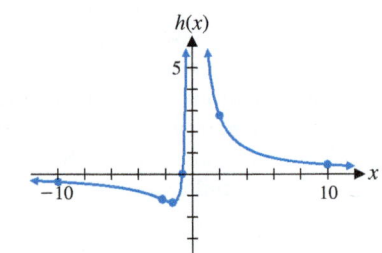

3. Domain: $(-\infty, \infty)$
   y intercept: $f(0) = 0$
   x intercept: $x = 0$
   Horizontal asymptote: $y = 0$ (the x axis)
   Increasing on $(-\infty, 2)$
   Decreasing on $(2, \infty)$
   Local maximum at $x = 2$
   Concave downward on $(-\infty, 4)$
   Concave upward on $(4, \infty)$
   Inflection point at $x = 4$

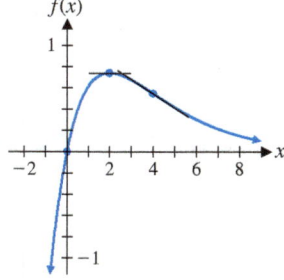

4. Domain: $(0, \infty)$
   y intercept: None [$f(0)$ is not defined]
   x intercept: $x = 1$
   Increasing on (e^{-1}, ∞)
   Decreasing on $(0, e^{-1})$
   Local minimum at $x = e^{-1} \approx 0.368$
   Concave upward on $(0, \infty)$

x	5	10	100	$\to \infty$	
$f(x)$	8.05	23.03	460.52	$\to \infty$	

x	0.1	0.01	0.001	0.000 1	$\to 0$
$f(x)$	-0.23	-0.046	$-0.006\ 9$	$-0.000\ 92$	$\to 0$

5. (A) Domain: $(0, \infty)$
   Intercepts: None
   Vertical asymptote: $x = 0$; oblique asymptote: $y = 0.25x$
   Decreasing on $(0, 80)$; increasing on $(80, \infty)$; local minimum at $x = 80$
   Concave upward on $(0, \infty)$

   (B) Minimum average cost is 40 at $x = 80$.

5-5   Absolute Maxima and Minima

- Absolute Maxima and Minima
- Second Derivative and Extrema

Now we will consider one of the most important applications of the derivative: finding the absolute maximum or minimum value of a function. An economist may be interested in the price or production level of a commodity that will bring a maximum profit; a doctor may be interested in the time it takes for a drug to reach its maximum concentration in the bloodstream after an injection; and a city planner might be interested in the location of heavy industry in a city in order to produce minimum pollution in residential and business areas. In this section, we develop the procedures needed to find the absolute maximum and absolute minimum values of a function.

Absolute Maxima and Minima

Recall that $f(c)$ is a local maximum value if $f(x) \le f(c)$ for x near c and a local minimum value if $f(x) \ge f(c)$ for x near c. Now we are interested in finding the largest and the smallest values of $f(x)$ throughout its domain.

DEFINITION Absolute Maxima and Minima

If $f(c) \geq f(x)$ for all x in the domain of f, then $f(c)$ is called the **absolute maximum value** of f.

If $f(c) \leq f(x)$ for all x in the domain of f, then $f(c)$ is called the **absolute minimum value** of f.

Figure 1 illustrates some typical examples.

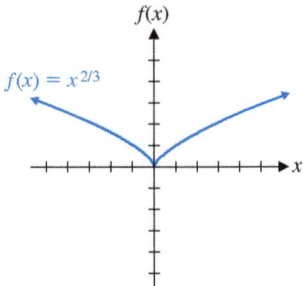

(A) No absolute maximum or minimum
   One local maximum at $x = -2$
   One local minimum at $x = 2$

(B) Absolute maximum at $x = 0$
   No absolute minimum

(C) Absolute minimum at $x = 0$
   No absolute maximum

Figure 1

CONCEPTUAL INSIGHT

If $f(c)$ is the absolute maximum value of a function f, then $f(c)$ is obviously a "value" of f. It is common practice to omit "value" and to refer to $f(c)$ as the **absolute maximum** of f. In either usage, note that c is a value of x in the domain of f where the absolute maximum occurs. It is incorrect to refer to c as the absolute maximum. Collectively, the absolute maximum and the absolute minimum are referred to as **absolute extrema**.

In many applications, the domain of a function is restricted because of practical or physical considerations. If the domain is restricted to some closed interval, as is often the case, then Theorem 1 applies.

THEOREM 1 Extreme Value Theorem

A function f that is continuous on a closed interval $[a, b]$ has both an absolute maximum value and an absolute minimum value on that interval.

It is important to understand that the absolute maximum and minimum values depend on both the function f and the interval $[a, b]$. Figure 2 illustrates four cases.

In all four cases illustrated in Figure 2, the absolute maximum value and absolute minimum value occur at a critical value or an endpoint. This property is generalized in Theorem 2. Note that both the absolute maximum value and the absolute minimum value are unique, but each can occur at more than one point in the interval (Fig. 2D).

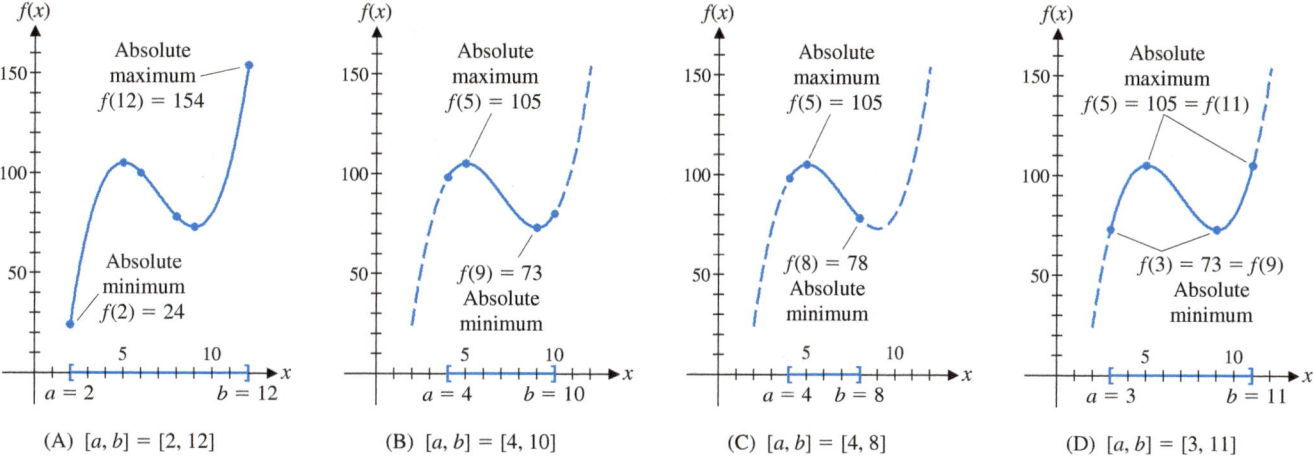

Figure 2 Absolute extrema for $f(x) = x^3 - 21x^2 + 135x - 170$ for various closed intervals

> **THEOREM 2   Locating Absolute Extrema**
>
> Absolute extrema (if they exist) must always occur at critical values or at endpoints.

To find the absolute maximum or minimum value of a continuous function on a closed interval, we simply identify the endpoints and critical values in the interval, evaluate the function at each, and then choose the largest and smallest values out of this group.

> **PROCEDURE Finding Absolute Extrema on a Closed Interval**
>
> **Step 1** Check to make certain that f is continuous over $[a, b]$.
>
> **Step 2** Find the critical values in the interval (a, b).
>
> **Step 3** Evaluate f at the endpoints a and b and at the critical values found in step 2.
>
> **Step 4** The absolute maximum $f(x)$ on $[a, b]$ is the largest value found in step 3.
>
> **Step 5** The absolute minimum $f(x)$ on $[a, b]$ is the smallest value found in step 3.

EXAMPLE 1 **Finding Absolute Extrema** Find the absolute maximum and absolute minimum values of

$$f(x) = x^3 + 3x^2 - 9x - 7$$

on each of the following intervals:

(A) $[-6, 4]$ (B) $[-4, 2]$ (C) $[-2, 2]$

SOLUTION (A) The function is continuous for all values of x.

$$f'(x) = 3x^2 + 6x - 9 = 3(x - 1)(x + 3)$$

So, $x = -3$ and $x = 1$ are critical values in the interval $(-6, 4)$. Evaluate f at the endpoints and critical values ($-6, -3, 1$, and 4), and choose the maximum and minimum from these:

$$f(-6) = -61 \quad \text{Absolute minimum}$$

$$f(-3) = 20$$

$$f(1) = -12$$

$$f(4) = 69 \quad \text{Absolute maximum}$$

(B) Interval: $[-4, 2]$

x	$f(x)$	
-4	13	
-3	20	Absolute maximum
1	-12	Absolute minimum
2	-5	

(C) Interval: $[-2, 2]$

x	$f(x)$	
-2	15	Absolute maximum
1	-12	Absolute minimum
2	-5	

The critical value x = −3 is not included in this table, because it is not in the interval [−2, 2].

Matched Problem 1 Find the absolute maximum and absolute minimum values of

$$f(x) = x^3 - 12x$$

on each of the following intervals:

(A) $[-5, 5]$ (B) $[-3, 3]$ (C) $[-3, 1]$

Now, suppose that we want to find the absolute maximum or minimum value of a function that is continuous on an interval that is not closed. Since Theorem 1 no longer applies, we cannot be certain that the absolute maximum or minimum value exists. Figure 3 illustrates several ways that functions can fail to have absolute extrema.

In general, the best procedure to follow in searching for absolute extrema on an interval that is not of the form $[a, b]$ is to sketch a graph of the function. However, many applications can be solved with a new tool that does not require any graphing.

Second Derivative and Extrema

The second derivative can be used to classify the local extrema of a function. Suppose that f is a function satisfying $f'(c) = 0$ and $f''(c) > 0$. First, note that if $f''(c) > 0$, then it follows from the properties of limits* that $f''(x) > 0$ in some

---

*Actually, we are assuming that $f''(x)$ is continuous in an interval containing c. It is unlikely that we will encounter a function for which $f''(c)$ exists but $f''(x)$ is not continuous in an interval containing c.

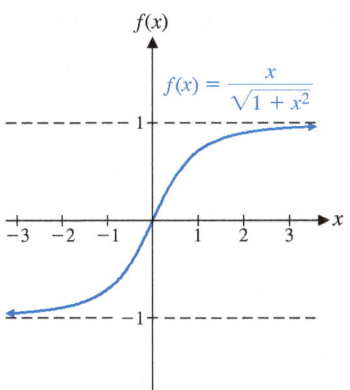

(A) No absolute extrema on $(-\infty, \infty)$:
$-1 < f(x) < 1$ for all x
$[f(x) \neq 1$ or -1 for any $x]$

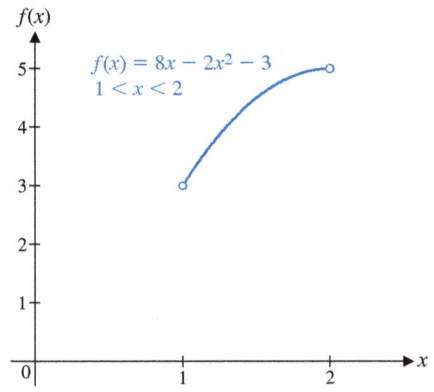

(B) No absolute extrema on $(1, 2)$:
$3 < f(x) < 5$ for $x \in (1, 2)$
$[f(x) \neq 3$ or 5 for any $x \in (1, 2)]$

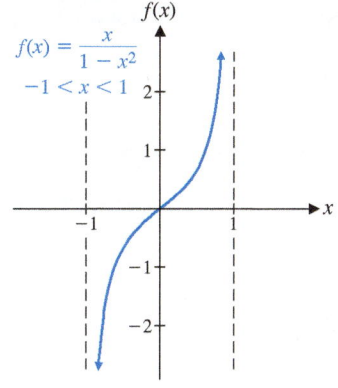

(C) No absolute extrema on $(-1, 1)$:
Graph has vertical asymptotes
at $x = -1$ and $x = 1$

Figure 3 Functions with no absolute extrema

interval (m, n) containing c. Thus, the graph of f must be concave upward in this interval. But this implies that $f'(x)$ is increasing in the interval. Since $f'(c) = 0$, $f'(x)$ must change from negative to positive at $x = c$, and $f(c)$ is a local minimum (see Fig. 4). Reasoning in the same fashion, we conclude that if $f'(c) = 0$ and $f''(c) < 0$, then $f(c)$ is a local maximum. Of course, it is possible that both $f'(c) = 0$ and $f''(c) = 0$. In this case, the second derivative cannot be used to determine the shape of the graph around $x = c$; $f(c)$ may be a local minimum, a local maximum, or neither.

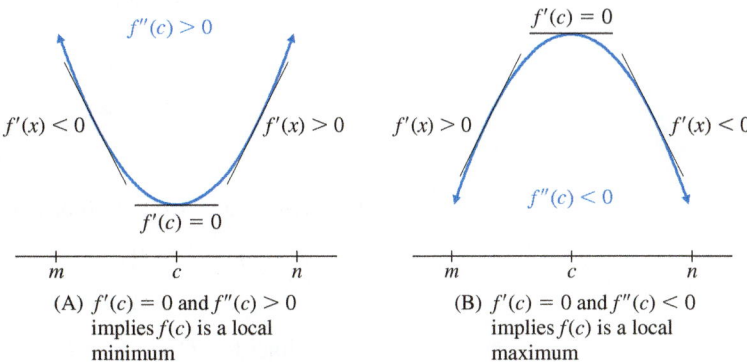

(A) $f'(c) = 0$ and $f''(c) > 0$
implies $f(c)$ is a local
minimum

(B) $f'(c) = 0$ and $f''(c) < 0$
implies $f(c)$ is a local
maximum

Figure 4 Second derivative and local extrema

The sign of the second derivative provides a simple test for identifying local maxima and minima. This test is most useful when we do not want to draw the graph of the function. If we are interested in drawing the graph and have already constructed the sign chart for $f'(x)$, then the first-derivative test can be used to identify the local extrema.

RESULT Second-Derivative Test

Let c be a critical value of $f(x)$.

$f'(c)$	$f''(c)$	Graph of f is:	$f(c)$	Example
0	+	Concave upward	Local minimum	$\smile$
0	−	Concave downward	Local maximum	$\frown$
0	0	?	Test does not apply	

EXAMPLE 2   **Testing Local Extrema** Find the local maxima and minima for each function. Use the second-derivative test when it applies.

(A) $f(x) = x^3 - 6x^2 + 9x + 1$

(B) $f(x) = xe^{-0.2x}$

(C) $f(x) = \frac{1}{6}x^6 - 4x^5 + 25x^4$

SOLUTION   (A) Take first and second derivatives and find critical values:

$$f(x) = x^3 - 6x^2 + 9x + 1$$

$$f'(x) = 3x^2 - 12x + 9 = 3(x - 1)(x - 3)$$

$$f''(x) = 6x - 12 = 6(x - 2)$$

Critical values are $x = 1$ and $x = 3$.

$$f''(1) = -6 < 0 \qquad f \text{ has a local maximum at } x = 1.$$

$$f''(3) = 6 > 0 \qquad f \text{ has a local minimum at } x = 3.$$

(B) $$f(x) = xe^{-0.2x}$$

$$f'(x) = e^{-0.2x} + xe^{-0.2x}(-0.2)$$

$$= e^{-0.2x}(1 - 0.2x)$$

$$f''(x) = e^{-0.2x}(-0.2)(1 - 0.2x) + e^{-0.2x}(-0.2)$$

$$= e^{-0.2x}(0.04x - 0.4)$$

Critical value: $x = 1/0.2 = 5$

$f''(5) = e^{-1}(-0.2) < 0$

f has a local maximum at $x = 5$.

(C) $$f(x) = \frac{1}{6}x^6 - 4x^5 + 25x^4$$

$$f'(x) = x^5 - 20x^4 + 100x^3 = x^3(x - 10)^2$$

$$f''(x) = 5x^4 - 80x^3 + 300x^2$$

Critical values are $x = 0$ and $x = 10$.

$$f''(0) = 0 \qquad \textit{The second-derivative test fails at both critical values, so}$$

$$f''(10) = 0 \qquad \textit{the first-derivative test must be used.}$$

Sign chart for $f'(x) = x^3(x - 10)^2$ (partition numbers are 0 and 10):

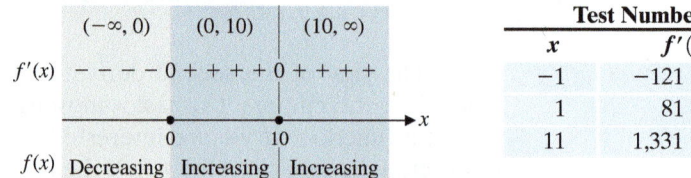

	Test Numbers	
x	$f'(x)$	
-1	-121	$(-)$
1	81	$(+)$
11	$1,331$	$(+)$

From the chart, we see that $f(x)$ has a local minimum at $x = 0$ and does not have a local extremum at $x = 10$.

Matched Problem 2   Find the local maxima and minima for each function. Use the second-derivative test when it applies.

(A) $f(x) = x^3 - 9x^2 + 24x - 10$

(B) $f(x) = e^x - 5x$

(C) $f(x) = 10x^6 - 24x^5 + 15x^4$

CONCEPTUAL INSIGHT

The second-derivative test does not apply if $f''(c) = 0$ or if $f''(c)$ is not defined. As Example 2C illustrates, if $f''(c) = 0$, then $f(c)$ may or may not be a local extremum. Some other method, such as the first-derivative test, must be used when $f''(c) = 0$ or $f''(c)$ does not exist.

The solution of many optimization problems involves searching for an absolute extremum. If the function in question has only one critical value, then the second-derivative test not only classifies the local extremum but also guarantees that the local extremum is, in fact, the absolute extremum.

THEOREM 3   Second-Derivative Test for Absolute Extremum

Let f be continuous on an interval I with only one critical value c in I.

If $f'(c) = 0$ and $f''(c) > 0$, then $f(c)$ is the absolute minimum of f on I.

If $f'(c) = 0$ and $f''(c) < 0$, then $f(c)$ is the absolute maximum of f on I.

Since the second-derivative test cannot be applied when $f''(c) = 0$ or $f''(c)$ does not exist, Theorem 3 makes no mention of these cases.

EXAMPLE 3   **Finding an Absolute Extremum on an Open Interval**   Find the absolute minimum value of each function on $(0, \infty)$.

(A) $f(x) = x + \dfrac{4}{x}$

(B) $f(x) = (\ln x)^2 - 3 \ln x$

SOLUTION   (A) $f(x) = x + \dfrac{4}{x}$

$$f'(x) = 1 - \frac{4}{x^2} = \frac{x^2 - 4}{x^2} = \frac{(x - 2)(x + 2)}{x^2} \qquad \text{Critical values are } x = -2 \text{ and } x = 2.$$

$$f''(x) = \frac{8}{x^3}$$

The only critical value in the interval $(0, \infty)$ is $x = 2$. Since $f''(2) = 1 > 0$, $f(2) = 4$ is the absolute minimum value of f on $(0, \infty)$.

(B) $f(x) = (\ln x)^2 - 3 \ln x$

$$f'(x) = (2 \ln x)\frac{1}{x} - \frac{3}{x} = \frac{2 \ln x - 3}{x} \qquad \text{Critical value is } x = e^{3/2}.$$

$$f''(x) = \frac{x\dfrac{2}{x} - (2 \ln x - 3)}{x^2} = \frac{5 - 2 \ln x}{x^2}$$

The only critical value in the interval $(0, \infty)$ is $x = e^{3/2}$. Since $f''(e^{3/2}) = 2/e^3 > 0$, $f(e^{3/2}) = -2.25$ is the absolute minimum value of f on $(0, \infty)$.

Matched Problem 3 Find the absolute maximum value of each function on $(0, \infty)$.

(A) $f(x) = 12 - x - \dfrac{5}{x}$ (B) $f(x) = 5 \ln x - x$

Exercises 5-5

A

Problems 1–10 refer to the graph of $y = f(x)$ shown here. Find the absolute minimum and the absolute maximum over the indicated interval.

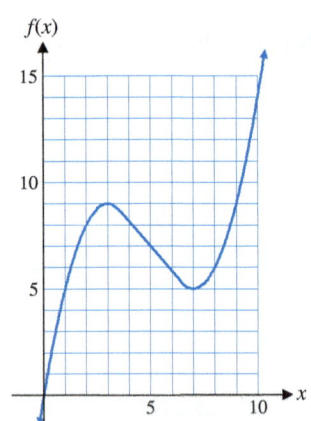

Figure for 1–10

1. $[0, 10]$ **2.** $[2, 8]$ **3.** $[0, 8]$ **4.** $[2, 10]$

5. $[1, 10]$ **6.** $[0, 9]$ **7.** $[1, 9]$ **8.** $[0, 2]$

9. $[2, 5]$ **10.** $[5, 8]$

In Problems 11–16, find the absolute maximum and absolute minimum of each function on the given interval.

11. $f(x) = 4x + 9$ on $[2, 3]$ **12.** $f(x) = -3x + 20$ on $[-2, 6]$

13. $f(x) = e^{-x}$ on $[-1, 1]$ **14.** $f(x) = \ln x$ on $[1, 2]$

15. $f(x) = 9 - x^2$ on $[-4, 4]$

16. $f(x) = x^2 - 6x + 7$ on $[0, 10]$

In Problems 17–32, find the absolute maximum and minimum, if either exists, for each function.

17. $f(x) = x^2 - 2x + 3$ **18.** $f(x) = x^2 + 4x - 3$

19. $f(x) = -x^2 - 6x + 9$ **20.** $f(x) = -x^2 + 2x + 4$

21. $f(x) = x^3 + x$ **22.** $f(x) = -x^3 - 2x$

23. $f(x) = 8x^3 - 2x^4$ **24.** $f(x) = x^4 - 4x^3$

25. $f(x) = x + \dfrac{16}{x}$ **26.** $f(x) = x + \dfrac{25}{x}$

27. $f(x) = \dfrac{x^2}{x^2 + 1}$ **28.** $f(x) = \dfrac{1}{x^2 + 1}$

29. $f(x) = \dfrac{2x}{x^2 + 1}$ **30.** $f(x) = \dfrac{-8x}{x^2 + 4}$

31. $f(x) = \dfrac{x^2 - 1}{x^2 + 1}$ **32.** $f(x) = \dfrac{9 - x^2}{x^2 + 4}$

B

In Problems 33–56, find the indicated extremum of each function on the given interval.

33. Absolute minimum value on $[0, \infty)$ for
$$f(x) = 2x^2 - 8x + 6$$

34. Absolute maximum value on $[0, \infty)$ for
$$f(x) = 6x - x^2 + 4$$

35. Absolute maximum value on $[0, \infty)$ for
$$f(x) = 3x^2 - x^3$$

36. Absolute minimum value on $[0, \infty)$ for
$$f(x) = x^3 - 6x^2$$

37. Absolute minimum value on $[0, \infty)$ for
$$f(x) = (x + 4)(x - 2)^2$$

38. Absolute minimum value on $[0, \infty)$ for
$$f(x) = (2 - x)(x + 1)^2$$

39. Absolute maximum value on $(0, \infty)$ for
$$f(x) = 2x^4 - 8x^3$$

40. Absolute maximum value on $(0, \infty)$ for
$$f(x) = 4x^3 - 8x^4$$

41. Absolute maximum value on $(0, \infty)$ for
$$f(x) = 20 - 3x - \dfrac{12}{x}$$

42. Absolute minimum value on $(0, \infty)$ for
$$f(x) = 4 + x + \dfrac{9}{x}$$

43. Absolute maximum value on $(0, \infty)$ for
$$f(x) = 10 + 2x + \dfrac{64}{x^2}$$

44. Absolute maximum value on $(0, \infty)$ for
$$f(x) = 20 - 4x - \dfrac{250}{x^2}$$

45. Absolute minimum value on $(0, \infty)$ for

$$f(x) = x + \frac{1}{x} + \frac{30}{x^3}$$

46. Absolute minimum value on $(0, \infty)$ for

$$f(x) = 2x + \frac{5}{x} + \frac{4}{x^3}$$

47. Absolute minimum value on $(0, \infty)$ for

$$f(x) = \frac{e^x}{x^2}$$

48. Absolute maximum value on $(0, \infty)$ for

$$f(x) = \frac{x^4}{e^x}$$

49. Absolute maximum value on $(0, \infty)$ for

$$f(x) = \frac{x^3}{e^x}$$

50. Absolute minimum value on $(0, \infty)$ for

$$f(x) = \frac{e^x}{x}$$

51. Absolute maximum value on $(0, \infty)$ for

$$f(x) = 5x - 2x \ln x$$

52. Absolute minimum value on $(0, \infty)$ for

$$f(x) = 4x \ln x - 7x$$

53. Absolute maximum value on $(0, \infty)$ for

$$f(x) = x^2(3 - \ln x)$$

54. Absolute minimum value on $(0, \infty)$ for

$$f(x) = x^3(\ln x - 2)$$

55. Absolute maximum value on $(0, \infty)$ for

$$f(x) = \ln(xe^{-x})$$

56. Absolute maximum value on $(0, \infty)$ for

$$f(x) = \ln(x^2 e^{-x})$$

In Problems 57–62, find the absolute maximum and minimum, if either exists, for each function on the indicated intervals.

57. $f(x) = x^3 - 6x^2 + 9x - 6$
   (A) $[-1, 5]$ (B) $[-1, 3]$ (C) $[2, 5]$

58. $f(x) = 2x^3 - 3x^2 - 12x + 24$
   (A) $[-3, 4]$ (B) $[-2, 3]$ (C) $[-2, 1]$

59. $f(x) = (x - 1)(x - 5)^3 + 1$
   (A) $[0, 3]$ (B) $[1, 7]$ (C) $[3, 6]$

60. $f(x) = x^4 - 8x^2 + 16$
   (A) $[-1, 3]$ (B) $[0, 2]$ (C) $[-3, 4]$

61. $f(x) = x^4 - 4x^3 + 5$
   (A) $[-1, 2]$ (B) $[0, 4]$ (C) $[-1, 1]$

62. $f(x) = x^4 - 18x^2 + 32$
   (A) $[-4, 4]$ (B) $[-1, 1]$ (C) $[1, 3]$

In Problems 63–70, describe the graph of f at the given point relative to the existence of a local maximum or minimum with one of the following phrases: "Local maximum," "Local minimum," "Neither," or "Unable to determine from the given information." Assume that f(x) is continuous on $(-\infty, \infty)$.

63. $(2, f(2))$ if $f'(2) = 0$ and $f''(2) > 0$

64. $(4, f(4))$ if $f'(4) = 1$ and $f''(4) < 0$

65. $(-3, f(-3))$ if $f'(-3) = 0$ and $f''(-3) = 0$

66. $(-1, f(-1))$ if $f'(-1) = 0$ and $f''(-1) < 0$

67. $(6, f(6))$ if $f'(6) = 1$ and $f''(6)$ does not exist

68. $(5, f(5))$ if $f'(5) = 0$ and $f''(5)$ does not exist

69. $(-2, f(-2))$ if $f'(-2) = 0$ and $f''(-2) < 0$

70. $(1, f(1))$ if $f'(1) = 0$ and $f''(1) > 0$

Answers to Matched Problems

1. (A) Absolute maximum: $f(5) = 65$; absolute minimum: $f(-5) = -65$

   (B) Absolute maximum: $f(-2) = 16$; absolute minimum: $f(2) = -16$

   (C) Absolute maximum: $f(-2) = 16$; absolute minimum: $f(1) = -11$

2. (A) $f(2)$ is a local maximum; $f(4)$ is a local minimum.

   (B) $f(\ln 5) = 5 - 5 \ln 5$ is a local minimum.

   (C) $f(0)$ is a local minimum; there is no local extremum at $x = 1$.

3. (A) $f(\sqrt{5}) = 12 - 2\sqrt{5}$ (B) $f(5) = 5 \ln 5 - 5$

5-6 Optimization

- Area and Perimeter
- Maximizing Revenue and Profit
- Inventory Control

Now we can use calculus to solve **optimization problems**—problems that involve finding the absolute maximum value or the absolute minimum value of a function. As you work through this section, note that the statement of the problem does not usually include the function to be optimized. Often, it is your responsibility to find the function and then to find its absolute extremum.

Area and Perimeter

The techniques used to solve optimization problems are best illustrated through examples.

EXAMPLE 1   **Maximizing Area** A homeowner has $320 to spend on building a fence around a rectangular garden. Three sides of the fence will be constructed with wire fencing at a cost of $2 per linear foot. The fourth side will be constructed with wood fencing at a cost of $6 per linear foot. Find the dimensions and the area of the largest garden that can be enclosed with $320 worth of fencing.

SOLUTION   To begin, we draw a figure (Fig. 1), introduce variables, and look for relationships among the variables.

Since we don't know the dimensions of the garden, the lengths of fencing are represented by the variables x and y. The costs of the fencing materials are fixed and are represented by constants.

Now we look for relationships among the variables. The area of the garden is

$$A = xy$$

while the cost of the fencing is

$$C = 2y + 2x + 2y + 6x$$
$$= 8x + 4y$$

Figure 1

The problem states that the homeowner has $320 to spend on fencing. We assume that enclosing the largest area will use all the money available for fencing. The problem has now been reduced to

Maximize $A = xy$ subject to $8x + 4y = 320$

Before we can use calculus to find the maximum area A, we must express A as a function of a single variable. We use the cost equation to eliminate one of the variables in the area expression (we choose to eliminate y—either will work).

$$8x + 4y = 320$$
$$4y = 320 - 8x$$
$$y = 80 - 2x$$
$$A = xy = x(80 - 2x) = 80x - 2x^2$$

Now we consider the permissible values of x. Because x is one of the dimensions of a rectangle, x must satisfy

$$x \geq 0 \quad \text{Length is always nonnegative.}$$

And because $y = 80 - 2x$ is also a dimension of a rectangle, y must satisfy

$$y = 80 - 2x \geq 0 \quad \text{Width is always nonnegative.}$$
$$80 \geq 2x$$
$$40 \geq x \quad \text{or} \quad x \leq 40$$

We summarize the preceding discussion by stating the following model for this optimization problem:

Maximize $A(x) = 80x - 2x^2$ for $0 \leq x \leq 40$

Next, we find any critical values of A:

$$A'(x) = 80 - 4x = 0$$
$$80 = 4x$$
$$x = \frac{80}{4} = 20 \quad \text{Critical value}$$

Table 1

x	$A(x)$
0	0
20	800
40	0

Since $A(x)$ is continuous on $[0, 40]$, the absolute maximum value of A, if it exists, must occur at a critical value or an endpoint. Evaluating A at these values (Table 1), we see that the maximum area is 800 when

$$x = 20 \quad \text{and} \quad y = 80 - 2(20) = 40$$

Finally, we must answer the questions posed in the problem. The dimensions of the garden with the maximum area of 800 square feet are 20 feet by 40 feet, with one 20-foot side of wood fencing.

Matched Problem 1   Repeat Example 1 if the wood fencing costs $8 per linear foot and all other information remains the same.

We summarize the steps in the solution of Example 1 in the following box:

PROCEDURE Strategy for Solving Optimization Problems

Step 1   Introduce variables, look for relationships among the variables, and construct a mathematical model of the form

Maximize (or minimize) $f(x)$ on the interval I

Step 2   Find the critical values of $f(x)$.

Step 3   Use the procedures developed in Section 5-5 to find the absolute maximum (or minimum) value of $f(x)$ on the interval I and the value(s) of x where this occurs.

Step 4   Use the solution to the mathematical model to answer all the questions asked in the problem.

EXAMPLE 2   **Minimizing Perimeter**   Refer to Example 1. The homeowner judges that an area of 800 square feet for the garden is too small and decides to increase the area to 1,250 square feet. What is the minimum cost of building a fence that will enclose a garden with an area of 1,250 square feet? What are the dimensions of this garden? Assume that the cost of fencing remains unchanged.

SOLUTION   Refer to Figure 1 and the solution of Example 1. This time we want to minimize the cost of the fencing that will enclose 1,250 square feet. The problem can be expressed as

$$\text{Minimize} \quad C = 8x + 4y \quad \text{subject to} \quad xy = 1,250$$

Since x and y represent distances, we know that $x \geq 0$ and $y \geq 0$. But neither variable can equal 0 because their product must be 1,250.

$$xy = 1,250 \qquad \qquad \textit{Solve the area equation for y.}$$

$$y = \frac{1,250}{x}$$

$$C(x) = 8x + 4\frac{1,250}{x} \qquad \textit{Substitute for y in the cost equation.}$$

$$= 8x + \frac{5,000}{x} \qquad x > 0$$

The model for this problem is

$$\text{Minimize } C(x) = 8x + \frac{5{,}000}{x} \qquad \text{for } x > 0$$

$$= 8x + 5{,}000x^{-1}$$

$$C'(x) = 8 - 5{,}000x^{-2}$$

$$= 8 - \frac{5{,}000}{x^2} = 0$$

$$8 = \frac{5{,}000}{x^2}$$

$$x^2 = \frac{5{,}000}{8} = 625$$

$$x = \sqrt{625} = 25$$

The negative square root is discarded, since x > 0.

We use the second derivative to determine the behavior at $x = 25$.

$$\boxed{C'(x) = 8 - 5{,}000x^{-2}}$$

$$C''(x) = 0 + 10{,}000x^{-3} = \frac{10{,}000}{x^3}$$

$$C''(25) = \frac{10{,}000}{25^3} = 0.64 > 0$$

The second-derivative test shows that $C(x)$ has a local minimum at $x = 25$, and since $x = 25$ is the only critical value of $x > 0$, then $C(25)$ must be the absolute minimum value of $C(x)$ for $x > 0$. When $x = 25$, the cost is

$$C(25) = 8(25) + \frac{5{,}000}{25} = 200 + 200 = \$400$$

and

$$y = \frac{1{,}250}{25} = 50$$

The minimal cost for enclosing a 1,250-square-foot garden is $400, and the dimensions are 25 feet by 50 feet, with one 25-foot side of wood fencing.

Matched Problem 2   Repeat Example 2 if the homeowner wants to enclose an 1,800-square-foot garden and all other data remain unchanged.

CONCEPTUAL INSIGHT

The restrictions on the variables in the solutions of Examples 1 and 2 are typical of problems involving areas or perimeters (or the cost of the perimeter):

$$8x + 4y = 320 \quad \text{Cost of fencing (Example 1)}$$

$$xy = 1{,}250 \qquad \text{Area of garden (Example 2)}$$

The equation in Example 1 restricts the values of x to

$$0 \le x \le 40 \qquad \text{or} \qquad [0, 40]$$

The endpoints are included in the interval for our convenience (a closed interval is easier to work with than an open one). The area function is defined at each endpoint, so it does no harm to include them.

The equation in Example 2 restricts the values of x to

$$x > 0 \quad \text{or} \quad (0, \infty)$$

Neither endpoint can be included in this interval. We cannot include 0 because the area is not defined when $x = 0$, and we can never include ∞ as an endpoint. Remember, ∞ is not a number; it is a symbol that indicates the interval is unbounded.

Maximizing Revenue and Profit

EXAMPLE 3 **Maximizing Revenue** An office supply company sells x permanent markers per year at $\$p$ per marker. The price–demand equation for these markers is $p = 10 - 0.001x$. What price should the company charge for the markers to maximize revenue? What is the maximum revenue?

SOLUTION

$$\text{Revenue} = \text{price} \times \text{demand}$$

$$R(x) = (10 - 0.001x)x$$

$$= 10x - 0.001x^2$$

Both price and demand must be nonnegative, so

$$x \geq 0 \quad \text{and} \quad p = 10 - 0.001x \geq 0$$

$$10 \geq 0.001x$$

$$10{,}000 \geq x$$

The mathematical model for this problem is

$$\text{Maximize} \quad R(x) = 10x - 0.001x^2 \quad 0 \leq x \leq 10{,}000$$

$$R'(x) = 10 - 0.002x$$

$$10 - 0.002x = 0$$

$$10 = 0.002x$$

$$x = \frac{10}{0.002} = 5{,}000 \qquad \textcolor{teal}{Critical\ value}$$

Use the second-derivative test for absolute extrema:

$$R''(x) = -0.002 < 0 \quad \text{for all } x$$

$$\text{Max} \quad R(x) = R(5{,}000) = \$25{,}000$$

When the demand is $x = 5{,}000$, the price is

$$10 - 0.001(5{,}000) = \$5 \quad \textcolor{teal}{p = 10 - 0.001x}$$

The company will realize a maximum revenue of \$25,000 when the price of a marker is \$5.

Matched Problem 3 An office supply company sells x heavy-duty paper shredders per year at $\$p$ per shredder. The price–demand equation for these shredders is

$$p = 300 - \frac{x}{30}$$

What price should the company charge for the shredders to maximize revenue? What is the maximum revenue?

EXAMPLE 4 **Maximizing Profit** The total annual cost of manufacturing x permanent markers for the office supply company in Example 3 is

$$C(x) = 5{,}000 + 2x$$

What is the company's maximum profit? What should the company charge for each marker, and how many markers should be produced?

SOLUTION Using the revenue model in Example 3, we have

$$\text{Profit} = \text{Revenue} - \text{Cost}$$

$$P(x) = R(x) - C(x)$$

$$= 10x - 0.001x^2 - 5{,}000 - 2x$$

$$= 8x - 0.001x^2 - 5{,}000$$

The mathematical model for profit is

$$\text{Maximize} \quad P(x) = 8x - 0.001x^2 - 5{,}000 \qquad 0 \le x \le 10{,}000$$

The restrictions on x come from the revenue model in Example 3.

$$P'(x) = 8 - 0.002x = 0$$

$$8 = 0.002x$$

$$x = \frac{8}{0.002} = 4{,}000 \quad \textit{Critical value}$$

$$P''(x) = -0.002 < 0 \quad \text{for all } x$$

Since $x = 4{,}000$ is the only critical value and $P''(x) < 0$,

$$\text{Max } P(x) = P(4{,}000) = \$11{,}000$$

Using the price–demand equation from Example 3 with $x = 4{,}000$, we find that

$$p = 10 - 0.001(4{,}000) = \$6 \quad \textit{p = 10 − 0.001x}$$

A maximum profit of \$11,000 is realized when 4,000 markers are manufactured annually and sold for \$6 each.

The results in Examples 3 and 4 are illustrated in Figure 2.

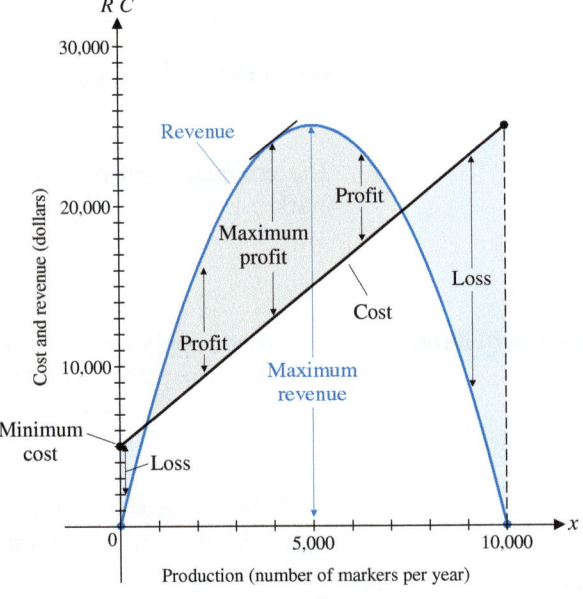

Figure 2

CONCEPTUAL INSIGHT

In Figure 2, notice that the maximum revenue and the maximum profit occur at different production levels. The maximum profit occurs when

$$P'(x) = R'(x) - C'(x) = 0$$

that is, when the marginal revenue is equal to the marginal cost. Notice that the slopes of the revenue function and the cost function are the same at this production level.

Matched Problem 4

The annual cost of manufacturing x paper shredders for the office supply company in Matched Problem 3 is $C(x) = 90,000 + 30x$. What is the company's maximum profit? What should it charge for each shredder, and how many shredders should it produce?

EXAMPLE 5

Maximizing Profit The government decides to tax the company in Example 4 $2 for each marker produced. Taking into account this additional cost, how many markers should the company manufacture each week to maximize its weekly profit? What is the maximum weekly profit? How much should the company charge for the markers to realize the maximum weekly profit?

SOLUTION The tax of $2 per unit changes the company's cost equation:

$$C(x) = \text{original cost} + \text{tax}$$

$$= 5,000 + 2x + 2x$$

$$= 5,000 + 4x$$

The new profit function is

$$P(x) = R(x) - C(x)$$

$$= 10x - 0.001x^2 - 5,000 - 4x$$

$$= 6x - 0.001x^2 - 5,000$$

So, we must solve the following equation:

$$\text{Maximize} \quad P(x) = 6x - 0.001x^2 - 5,000 \quad 0 \le x \le 10,000$$

$$P'(x) = 6 - 0.002x$$

$$6 - 0.002x = 0$$

$$x = 3,000 \qquad \textit{Critical value}$$

$$P''(x) = -0.002 < 0 \quad \text{for all } x$$

$$\text{Max } P(x) = P(3,000) = \$4,000$$

Using the price–demand equation (Example 3) with $x = 3,000$, we find that

$$p = 10 - 0.001(3,000) = \$7 \quad \textit{p = 10 − 0.001x}$$

The company's maximum profit is $4,000 when 3,000 markers are produced and sold weekly at a price of $7.

Even though the tax caused the company's cost to increase by $2 per marker, the price that the company should charge to maximize its profit increases by only $1. The company must absorb the other $1, with a resulting decrease of $7,000 in maximum profit.

Matched Problem 5   The government decides to tax the office supply company in Matched Problem 4 $20 for each shredder produced. Taking into account this additional cost, how many shredders should the company manufacture each week to maximize its weekly profit? What is the maximum weekly profit? How much should the company charge for the shredders to realize the maximum weekly profit?

EXAMPLE 6   **Maximizing Revenue**   When a management training company prices its seminar on management techniques at $400 per person, 1,000 people will attend the seminar. The company estimates that for each $5 reduction in price, an additional 20 people will attend the seminar. How much should the company charge for the seminar in order to maximize its revenue? What is the maximum revenue?

SOLUTION   Let x represent the number of $5 price reductions.

$$400 - 5x = \text{price per customer}$$

$$1,000 + 20x = \text{number of customers}$$

$$\text{Revenue} = (\text{price per customer})(\text{number of customers})$$

$$R(x) = (400 - 5x) \times (1,000 + 20x)$$

Since price cannot be negative, we have

$$400 - 5x \geq 0$$

$$400 \geq 5x$$

$$80 \geq x \quad \text{or} \quad x \leq 80$$

A negative value of x would result in a price increase. Since the problem is stated in terms of price reductions, we must restrict x so that $x \geq 0$. Putting all this together, we have the following model:

$$\text{Maximize} \quad R(x) = (400 - 5x)(1,000 + 20x) \quad \text{for } 0 \leq x \leq 80$$

$$R(x) = 400,000 + 3,000x - 100x^2$$

$$R'(x) = 3,000 - 200x = 0$$

$$3,000 = 200x$$

$$x = 15 \quad \textit{Critical value}$$

Table 2

x	$R(x)$
0	400,000
15	422,500
80	0

Since $R(x)$ is continuous on the interval $[0, 80]$, we can determine the behavior of the graph by constructing a table. Table 2 shows that $R(15) = \$422,500$ is the absolute maximum revenue. The price of attending the seminar at $x = 15$ is $400 - 5(15) = \$325$. The company should charge $325 for the seminar in order to receive a maximum revenue of $422,500.

Matched Problem 6   A walnut grower estimates from past records that if 20 trees are planted per acre, then each tree will average 60 pounds of nuts per year. If, for each additional tree planted per acre, the average yield per tree drops 2 pounds, then how many trees should be planted to maximize the yield per acre? What is the maximum yield?

EXAMPLE 7   **Maximizing Revenue**   After additional analysis, the management training company in Example 6 decides that its estimate of attendance was too high. Its new estimate is that only 10 additional people will attend the seminar for each $5 decrease in price. All other information remains the same. How much should the company charge for the seminar now in order to maximize revenue? What is the new maximum revenue?

SOLUTION   Under the new assumption, the model becomes

$$\text{Maximize} \quad R(x) = (400 - 5x)(1{,}000 + 10x) \qquad 0 \le x \le 80$$

$$= 400{,}000 - 1{,}000x - 50x^2$$

$$R'(x) = -1{,}000 - 100x = 0$$

$$-1{,}000 = 100x$$

$$x = -10 \quad \textit{Critical value}$$

Table 3

x	$R(x)$
0	400,000
80	0

Note that $x = -10$ is not in the interval $[0, 80]$. Since $R(x)$ is continuous on $[0, 80]$, we can use a table to find the absolute maximum revenue. Table 3 shows that the maximum revenue is $R(0) = \$400{,}000$. The company should leave the price at \$400. Any \$5 decreases in price will lower the revenue.

Matched Problem 7   After further analysis, the walnut grower in Matched Problem 6 determines that each additional tree planted will reduce the average yield by 4 pounds. All other information remains the same. How many additional trees per acre should the grower plant now in order to maximize the yield? What is the new maximum yield?

CONCEPTUAL INSIGHT

The solution in Example 7 is called an **endpoint solution** because the optimal value occurs at the endpoint of an interval rather than at a critical value in the interior of the interval. It is always important to verify that the optimal value has been found.

Inventory Control

EXAMPLE 8   **Inventory Control** A multimedia company anticipates that there will be a demand for 20,000 copies of a certain DVD during the next year. It costs the company \$0.50 to store a DVD for one year. Each time it must make additional DVDs, it costs \$200 to set up the equipment. How many DVDs should the company make during each production run to minimize its total storage and setup costs?

SOLUTION   This type of problem is called an **inventory control problem**. One of the basic assumptions made in such problems is that the demand is uniform. For example, if there are 250 working days in a year, then the daily demand would be $20{,}000 \div 250 = 80$ DVDs. The company could decide to produce all 20,000 DVDs at the beginning of the year. This would certainly minimize the setup costs but would result in very large storage costs. At the other extreme, the company could produce 80 DVDs each day. This would minimize the storage costs but would result in very large setup costs. Somewhere between these two extremes is the optimal solution that will minimize the total storage and setup costs. Let

x = number of DVDs manufactured during each production run

y = number of production runs

It is easy to see that the total setup cost for the year is $200y$, but what is the total storage cost? If the demand is uniform, then the number of DVDs in storage between production runs will decrease from x to 0, and the average number in storage each day is $x/2$. This result is illustrated in Figure 3.

Since it costs \$0.50 to store a DVD for one year, the total storage cost is $0.5(x/2) = 0.25x$ and the total cost is

$$\text{total cost} = \text{setup cost} + \text{storage cost}$$

$$C = 200y + 0.25x$$

Figure 3

In order to write the total cost C as a function of one variable, we must find a relationship between x and y. If the company produces x DVDs in each of y production runs, then the total number of DVDs produced is xy.

$$xy = 20{,}000$$

$$y = \frac{20{,}000}{x}$$

Certainly, x must be at least 1 and cannot exceed 20,000. We must solve the following equation:

$$\text{Minimize} \quad C(x) = 200\left(\frac{20{,}000}{x}\right) + 0.25x \quad 1 \le x \le 20{,}000$$

$$C(x) = \frac{4{,}000{,}000}{x} + 0.25x$$

$$C'(x) = -\frac{4{,}000{,}000}{x^2} + 0.25$$

$$-\frac{4{,}000{,}000}{x^2} + 0.25 = 0$$

$$x^2 = \frac{4{,}000{,}000}{0.25}$$

$$x^2 = 16{,}000{,}000 \qquad -4{,}000 \text{ is not a critical value, since}$$

$$x = 4{,}000 \qquad\qquad 1 \le x \le 20{,}000.$$

$$C''(x) = \frac{8{,}000{,}000}{x^3} > 0 \qquad \text{for } x \in (1, 20{,}000)$$

Therefore,

$$\text{Min } C(x) = C(4{,}000) = 2{,}000$$

$$y = \frac{20{,}000}{4{,}000} = 5$$

The company will minimize its total cost by making 4,000 DVDs five times during the year.

Matched Problem 8   Repeat Example 8 if it costs $250 to set up a production run and $0.40 to store a DVD for one year.

Exercises 5-6

Preliminary word problems:

1. Find two numbers whose sum is 15 and whose product is a maximum.

2. Find two numbers whose sum is 21 and whose product is a maximum.

3. Find two numbers whose difference is 15 and whose product is a minimum.

4. Find two numbers whose difference is 21 and whose product is a minimum.

5. Find two positive numbers whose product is 15 and whose sum is a minimum.

6. Find two positive numbers whose product is 21 and whose sum is a minimum.

7. Find the dimensions of a rectangle with an area of 200 square feet that has the minimum perimeter.

8. Find the dimensions of a rectangle with an area of 108 square feet that has the minimum perimeter.

9. Find the dimensions of a rectangle with a perimeter of 148 feet that has the maximum area.

10. Find the dimensions of a rectangle with a perimeter of 76 feet that has the maximum area.

11. Maximum revenue and profit. A company manufactures and sells x videophones per week. The weekly price–demand and cost equations are, respectively,

$$p = 500 - 0.5x \quad \text{and} \quad C(x) = 20,000 + 135x$$

(A) What price should the company charge for the phones, and how many phones should be produced to maximize the weekly revenue? What is the maximum weekly revenue?

(B) What is the maximum weekly profit? How much should the company charge for the phones, and how many phones should be produced to realize the maximum weekly profit?

12. Maximum revenue and profit. A company manufactures and sells x digital cameras per week. The weekly price–demand and cost equations are, respectively,

$$p = 400 - 0.4x \quad \text{and} \quad C(x) = 2,000 + 160x$$

(A) What price should the company charge for the cameras, and how many cameras should be produced to maximize the weekly revenue? What is the maximum revenue?

(B) What is the maximum weekly profit? How much should the company charge for the cameras, and how many cameras should be produced to realize the maximum weekly profit?

13. Maximum revenue and profit. A company manufactures and sells x television sets per month. The monthly cost and price–demand equations are

$$C(x) = 72,000 + 60x$$

$$p = 200 - \frac{x}{30} \qquad 0 \le x \le 6,000$$

(A) Find the maximum revenue.

(B) Find the maximum profit, the production level that will realize the maximum profit, and the price the company should charge for each television set.

(C) If the government decides to tax the company $5 for each set it produces, how many sets should the company manufacture each month to maximize its profit? What is the maximum profit? What should the company charge for each set?

14. Maximum revenue and profit. Repeat Problem 13 for

$$C(x) = 60,000 + 60x$$

$$p = 200 - \frac{x}{50} \qquad 0 \le x \le 10,000$$

15. Maximum profit. The following table contains price–demand and total cost data for the production of extreme-cold sleeping bags, where p is the wholesale price (in dollars) of a sleeping bag for an annual demand of x sleeping bags and C is the total cost (in dollars) of producing x sleeping bags:

(A) Find a quadratic regression equation for the price–demand data, using x as the independent variable.

x	p	C
950	240	130,000
1,200	210	150,000
1,800	160	180,000
2,050	120	190,000

(B) Find a linear regression equation for the cost data, using x as the independent variable.

(C) What is the maximum profit? What is the wholesale price per extreme-cold sleeping bag that should be charged to realize the maximum profit? Round answers to the nearest dollar.

16. Maximum profit. The following table contains price–demand and total cost data for the production of regular sleeping bags, where p is the wholesale price (in dollars) of a sleeping bag for an annual demand of x sleeping bags and C is the total cost (in dollars) of producing x sleeping bags:

x	p	C
2,300	98	145,000
3,300	84	170,000
4,500	67	190,000
5,200	51	210,000

(A) Find a quadratic regression equation for the price–demand data, using x as the independent variable.

(B) Find a linear regression equation for the cost data, using x as the independent variable.

(C) What is the maximum profit? What is the wholesale price per regular sleeping bag that should be charged to realize the maximum profit? Round answers to the nearest dollar.

17. Maximum revenue. A deli sells 640 sandwiches per day at a price of $8 each.

(A) A market survey shows that for every $0.10 reduction in price, 40 more sandwiches will be sold. How much should the deli charge for a sandwich in order to maximize revenue?

(B) A different market survey shows that for every $0.20 reduction in the original $8 price, 15 more sandwiches will be sold. Now how much should the deli charge for a sandwich in order to maximize revenue?

18. Maximum revenue. A university student center sells 1,600 cups of coffee per day at a price of $2.40.

(A) A market survey shows that for every $0.05 reduction in price, 50 more cups of coffee will be sold. How much should the student center charge for a cup of coffee in order to maximize revenue?

(B) A different market survey shows that for every $0.10 reduction in the original $2.40 price, 60 more cups of coffee will be sold. Now how much should the student center charge for a cup of coffee in order to maximize revenue?

19. Car rental. A car rental agency rents 200 cars per day at a rate of $30 per day. For each $1 increase in rate, 5 fewer cars are rented. At what rate should the cars be rented to produce the maximum income? What is the maximum income?

20. Rental income. A 300-room hotel in Las Vegas is filled to capacity every night at $80 a room. For each $1 increase in rent, 3 fewer rooms are rented. If each rented room costs $10 to service per day, how much should the management charge for each room to maximize gross profit? What is the maximum gross profit?

21. Agriculture. A commercial cherry grower estimates from past records that if 30 trees are planted per acre, then each tree will yield an average of 50 pounds of cherries per season. If, for each additional tree planted per acre (up to 20), the average yield per tree is reduced by 1 pound, how many trees should be planted per acre to obtain the maximum yield per acre? What is the maximum yield?

22. Agriculture. A commercial pear grower must decide on the optimum time to have fruit picked and sold. If the pears are picked now, they will bring 30¢ per pound, with each tree yielding an average of 60 pounds of salable pears. If the average yield per tree increases 6 pounds per tree per week for the next 4 weeks, but the price drops 2¢ per pound per week, when should the pears be picked to realize the maximum return per tree? What is the maximum return?

23. Manufacturing. A candy box is to be made out of a piece of cardboard that measures 8 by 12 inches. Squares of equal size will be cut out of each corner, and then the ends and sides will be folded up to form a rectangular box. What size square should be cut from each corner to obtain a maximum volume?

24. Packaging. A parcel delivery service will deliver a package only if the length plus girth (distance around) does not exceed 108 inches.

 (A) Find the dimensions of a rectangular box with square ends that satisfies the delivery service's restriction and has maximum volume. What is the maximum volume?

 (B) Find the dimensions (radius and height) of a cylindrical container that meets the delivery service's requirement and has maximum volume. What is the maximum volume?

Figure for 24

25. Construction costs. A fence is to be built to enclose a rectangular area of 800 square feet. The fence along three sides is to be made of material that costs $6 per foot. The material for the fourth side costs $18 per foot. Find the dimensions of the rectangle that will allow for the most economical fence to be built.

26. Construction costs. The owner of a retail lumber store wants to construct a fence to enclose an outdoor storage area adjacent to the store, using all of the store as part of one side of the area (see the figure). Find the dimensions that will enclose the largest area if

 (A) 240 feet of fencing material are used.

 (B) 400 feet of fencing material are used.

Figure for 26

27. Inventory control. A paint manufacturer has a uniform annual demand for 16,000 cans of automobile primer. It costs $4 to store one can of paint for one year and $500 to set up the plant for production of the primer. How many times a year should the company produce this primer in order to minimize the total storage and setup costs?

28. Inventory control. A pharmacy has a uniform annual demand for 200 bottles of a certain antibiotic. It costs $10 to store one bottle for one year and $40 to place an order. How many times during the year should the pharmacy order the antibiotic in order to minimize the total storage and reorder costs?

29. Inventory control. A publishing company sells 50,000 copies of a certain book each year. It costs the company $1 to store a book for one year. Each time that it prints additional copies, it costs the company $1,000 to set up the presses. How many books should the company produce during each printing in order to minimize its total storage and setup costs?

30. Operational costs. The cost per hour for fuel to run a train is $v^2/4$ dollars, where v is the speed of the train in miles per hour. (Note that the cost goes up as the square of the speed.) Other costs, including labor, are $300 per hour. How fast should the train travel on a 360-mile trip to minimize the total cost for the trip?

31. Construction costs. A freshwater pipeline is to be run from a source on the edge of a lake to a small resort community on an island 5 miles offshore, as indicated in the figure.

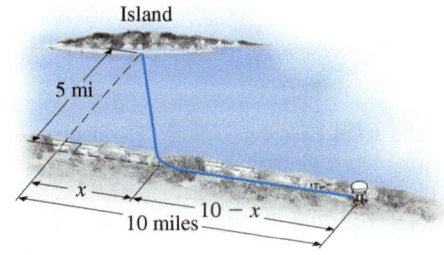

Figure for 31

(A) If it costs 1.4 times as much to lay the pipe in the lake as it does on land, what should x be (in miles) to minimize the total cost of the project?

(B) If it costs only 1.1 times as much to lay the pipe in the lake as it does on land, what should x be to minimize the total cost of the project? [*Note:* Compare with Problem 34.]

32. Drug concentration. The concentration $C(t)$, in milligrams per cubic centimeter, of a particular drug in a patient's bloodstream is given by

$$C(t) = \frac{0.16t}{t^2 + 4t + 4}$$

where t is the number of hours after the drug is taken. How many hours after the drug is taken will the concentration be maximum? What is the maximum concentration?

33. Bacteria control. A lake used for recreational swimming is treated periodically to control harmful bacteria growth. Suppose that t days after a treatment, the concentration of bacteria per cubic centimeter is given by

$$C(t) = 30t^2 - 240t + 500 \qquad 0 \le t \le 8$$

How many days after a treatment will the concentration be minimal? What is the minimum concentration?

34. Bird flights. Some birds tend to avoid flights over large bodies of water during daylight hours. Suppose that an adult bird with this tendency is taken from its nesting area on the edge of a large lake to an island 5 miles offshore and is then released (see the figure).

Figure for 34

(A) If it takes 1.4 times as much energy to fly over water as land, how far up the shore (x, in miles) should the bird head to minimize the total energy expended in returning to the nesting area?

(B) If it takes only 1.1 times as much energy to fly over water as land, how far up the shore should the bird head to minimize the total energy expended in returning to the nesting area? [*Note:* Compare with Problem 31.]

35. Botany. If it is known from past experiments that the height (in feet) of a certain plant after t months is given approximately by

$$H(t) = 4t^{1/2} - 2t \qquad 0 \le t \le 2$$

then how long, on average, will it take a plant to reach its maximum height? What is the maximum height?

36. Pollution. Two heavily industrial areas are located 10 miles apart, as shown in the figure. If the concentration of partic-

ulate matter (in parts per million) decreases as the reciprocal of the square of the distance from the source, and if area A_1 emits eight times the particulate matter as A_2, then the concentration of particulate matter at any point between the two areas is given by

$$C(x) = \frac{8k}{x^2} + \frac{k}{(10 - x)^2} \qquad 0.5 \le x \le 9.5, \quad k > 0$$

How far from A_1 will the concentration of particulate matter between the two areas be at a minimum?

Figure for 36

37. Politics. In a newly incorporated city, it is estimated that the voting population (in thousands) will increase according to

$$N(t) = 30 + 12t^2 - t^3 \qquad 0 \le t \le 8$$

where t is time in years. When will the rate of increase be most rapid?

38. Learning. A large grocery chain found that, on average, a checker can recall $P\%$ of a given price list x hours after starting work, as given approximately by

$$P(x) = 96x - 24x^2 \qquad 0 \le x \le 3$$

At what time x does the checker recall a maximum percentage? What is the maximum?

Answers to Matched Problems

1. The dimensions of the garden with the maximum area of 640 square feet are 16 feet by 40 feet, with one 16-foot side with wood fencing.

2. The minimal cost for enclosing a 1,800-square-foot garden is $480, and the dimensions are 30 feet by 60 feet, with one 30-foot side with wood fencing.

3. The company will realize a maximum revenue of $675,000 when the price of a shredder is $150.

4. A maximum profit of $456,750 is realized when 4,050 shredders are manufactured annually and sold for $165 each.

5. A maximum profit of $378,750 is realized when 3,075 shredders are manufactured annually and sold for $175 each.

6. The maximum yield is 1,250 pounds per acre when 5 additional trees are planted on each acre.

7. The maximum yield is 1,200 pounds when no additional trees are planted.

8. The company should produce 5,000 DVDs four times a year.

Chapter 5   Review

Important Terms, Symbols, and Concepts

Review Exercises

Work through all the problems in this chapter review, and check your answers in the back of the book. Answers to all review problems are there, along with section numbers in italics to indicate where each type of problem is discussed. Where weaknesses show up, review appropriate sections in the text.

A

Problems 1–8 refer to the following graph of $y = f(x)$. Identify the points or intervals on the x axis that produce the indicated behavior.

1. $f(x)$ is increasing.

2. $f'(x) < 0$

3. The graph of f is concave downward.

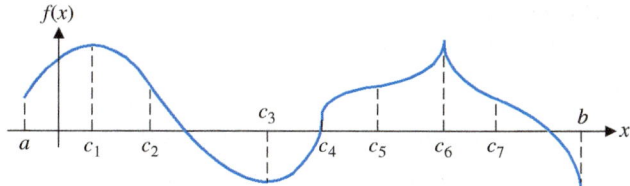

Figure for 1–8

4. Local minima

5. Absolute maxima

6. $f'(x)$ appears to be 0.

7. $f'(x)$ does not exist.

8. Inflection points

In Problems 9 and 10, use the given information to sketch the graph of f. Assume that f is continuous on its domain and that all intercepts are included in the information given.

9. Domain: All real x

x	-3	-2	-1	0	2	3
$f(x)$	0	3	2	0	-3	0

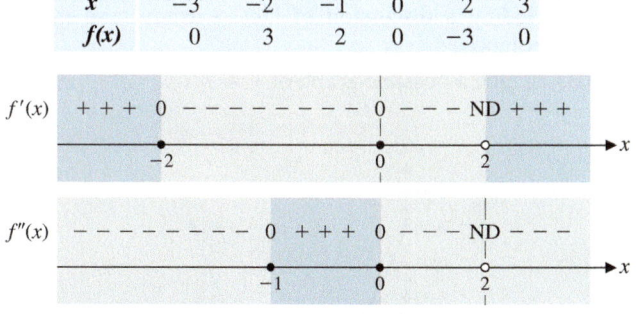

10. Domain: All real x

$$f(-2) = 1, f(0) = 0, f(2) = 1;$$
$$f'(0) = 0; f'(x) < 0 \text{ on } (-\infty, 0);$$
$$f'(x) > 0 \text{ on } (0, \infty);$$
$$f''(-2) = 0, f''(2) = 0;$$
$$f''(x) < 0 \text{ on } (-\infty, -2) \text{ and } (2, \infty);$$
$$f''(x) > 0 \text{ on } (-2, 2);$$
$$\lim_{x \to -\infty} f(x) = 2; \lim_{x \to \infty} f(x) = 2$$

11. Find $f''(x)$ for $f(x) = x^4 + 5x^3$.

12. Find y'' for $y = 3x + \dfrac{4}{x}$.

In Problems 13 and 14, find the domain and intercepts.

13. $f(x) = \dfrac{5 + x}{4 - x}$

14. $f(x) = \ln(x + 2)$

In Problems 15 and 16, find the horizontal and vertical asymptotes.

15. $f(x) = \dfrac{x + 3}{x^2 - 4}$

16. $f(x) = \dfrac{2x - 7}{3x + 10}$

In Problems 17 and 18, find the x and y coordinates of all inflection points.

17. $f(x) = x^4 - 12x^2$

18. $f(x) = (2x + 1)^{1/3} - 6$

In Problems 19 and 20, find (A) $f'(x)$, (B) the critical values of f, and (C) the partition numbers for f'.

19. $f(x) = x^{1/5}$

20. $f(x) = x^{-1/5}$

In Problems 21–30, summarize all the pertinent information obtained by applying the final version of the graphing strategy (Section 5–4) to f, and sketch the graph of f.

21. $f(x) = x^3 - 18x^2 + 81x$ **22.** $f(x) = (x + 4)(x - 2)^2$

23. $f(x) = 8x^3 - 2x^4$ **24.** $f(x) = (x - 1)^3(x + 3)$

25. $f(x) = \dfrac{3x}{x + 2}$ **26.** $f(x) = \dfrac{x^2}{x^2 + 27}$

27. $f(x) = \dfrac{x}{(x + 2)^2}$ **28.** $f(x) = \dfrac{x^3}{x^2 + 3}$

29. $f(x) = 5 - 5e^{-x}$ **30.** $f(x) = x^3 \ln x$

Find each limit in Problems 31–40.

31. $\lim\limits_{x \to 0} \dfrac{e^{3x} - 1}{x}$ **32.** $\lim\limits_{x \to 2} \dfrac{x^2 - 5x + 6}{x^2 + x - 6}$

33. $\lim\limits_{x \to 0^-} \dfrac{\ln(1 + x)}{x^2}$ **34.** $\lim\limits_{x \to 0} \dfrac{\ln(1 + x)}{1 + x}$

35. $\lim\limits_{x \to \infty} \dfrac{e^{4x}}{x^2}$ **36.** $\lim\limits_{x \to 0} \dfrac{e^x + e^{-x} - 2}{x^2}$

37. $\lim\limits_{x \to 0^+} \dfrac{\sqrt{1 + x} - 1}{\sqrt{x}}$ **38.** $\lim\limits_{x \to \infty} \dfrac{\ln x}{x^5}$

39. $\lim\limits_{x \to \infty} \dfrac{\ln(1 + 6x)}{\ln(1 + 3x)}$ **40.** $\lim\limits_{x \to 0} \dfrac{\ln(1 + 6x)}{\ln(1 + 3x)}$

41. Use the graph of $y = f'(x)$ shown here to discuss the graph of $y = f(x)$. Organize your conclusions in a table (see Example 4, Section 5-2). Sketch a possible graph of $y = f(x)$.

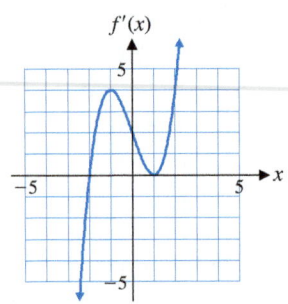

$f'(x)$

Figure for 41 and 42

42. Refer to the above graph of $y = f'(x)$. Which of the following could be the graph of $y = f''(x)$?

(A)

$f''(x)$

(B)

$f''(x)$

(C)

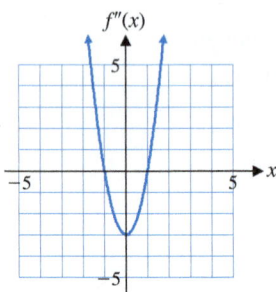

$f''(x)$

43. Use the second-derivative test to find any local extrema for

$$f(x) = x^3 - 6x^2 - 15x + 12$$

44. Find the absolute maximum and absolute minimum, if either exists, for

$$y = f(x) = x^3 - 12x + 12 \qquad -3 \le x \le 5$$

45. Find the absolute minimum, if it exists, for

$$y = f(x) = x^2 + \frac{16}{x^2} \qquad x > 0$$

46. Find the absolute maximum value, if it exists, for

$$f(x) = 11x - 2x \ln x \qquad x > 0$$

47. Find the absolute maximum value, if it exists, for

$$f(x) = 10xe^{-2x} \qquad x > 0$$

48. Let $y = f(x)$ be a polynomial function with local minima at $x = a$ and $x = b$, $a < b$. Must f have at least one local maximum between a and b? Justify your answer.

49. The derivative of $f(x) = x^{-1}$ is $f'(x) = -x^{-2}$. Since $f'(x) < 0$ for $x \ne 0$, is it correct to say that $f(x)$ is decreasing for all x except $x = 0$? Explain.

50. Discuss the difference between a partition number for $f'(x)$ and a critical value of $f(x)$, and illustrate with examples.

51. Find the absolute maximum for $f'(x)$ if

$$f(x) = 6x^2 - x^3 + 8$$

Graph f and f' on the same coordinate system for $0 \le x \le 4$.

52. Find two positive numbers whose product is 400 and whose sum is a minimum. What is the minimum sum?

In Problems 53 and 54, apply the graphing strategy and summarize the pertinent information. Round any approximate values to two decimal places.

53. $f(x) = x^4 + x^3 - 4x^2 - 3x + 4$

54. $f(x) = 0.25x^4 - 5x^3 + 31x^2 - 70x$

55. Find the absolute maximum value, if it exists, for

$$f(x) = 3x - x^2 + e^{-x} x > 0$$

56. Find the absolute maximum value, if it exists, for

$$f(x) = \frac{\ln x}{e^x} x > 0$$

Applications

57. Price analysis. The graph in the figure approximates the rate of change of the price of tomatoes over a 60-month period, where $p(t)$ is the price of a pound of tomatoes and t is time (in months).

(A) Write a brief description of the graph of $y = p(t)$, including a discussion of local extrema and inflection points.

(B) Sketch a possible graph of $y = p(t)$.

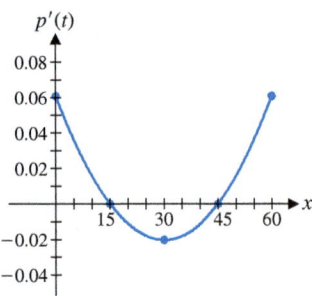

Figure for 57

58. Maximum revenue and profit. A company manufactures and sells x e-book readers per month. The monthly cost and price–demand equations are, respectively,

$$C(x) = 350x + 50,000$$

$$p = 500 - 0.025x \qquad 0 \le x \le 20,000$$

(A) Find the maximum revenue.

(B) How many readers should the company manufacture each month to maximize its profit? What is the maximum monthly profit? How much should the company charge for each reader?

(C) If the government decides to tax the company $20 for each reader it produces, how many readers should the company manufacture each month to maximize its profit? What is the maximum monthly profit? How much should the company charge for each reader?

59. Construction. A fence is to be built to enclose a rectangular area. The fence along three sides is to be made of material that costs $5 per foot. The material for the fourth side costs $15 per foot.

(A) If the area is 5,000 square feet, find the dimensions of the rectangle that will allow for the most economical fence.

(B) If $3,000 is available for the fencing, find the dimensions of the rectangle that will enclose the most area.

60. Rental income. A 200-room hotel in Reno is filled to capacity every night at a rate of $40 per room. For each $1 increase in the nightly rate, 4 fewer rooms are rented. If each rented room costs $8 a day to service, how much should the management charge per room in order to maximize gross profit? What is the maximum gross profit?

61. Inventory control. A computer store sells 7,200 boxes of storage disks annually. It costs the store $0.20 to store a box of disks for one year. Each time it reorders disks, the store must pay a $5.00 service charge for processing the order. How many times during the year should the store order disks to minimize the total storage and reorder costs?

62. Average cost. The total cost of producing x dorm refrigerators per day is given by

$$C(x) = 4,000 + 10x + 0.1x^2$$

Find the minimum average cost. Graph the average cost and the marginal cost functions on the same coordinate system. Include any oblique asymptotes.

63. Average cost. The cost of producing x wheeled picnic coolers is given by

$$C(x) = 200 + 50x - 50 \ln x \qquad x \geq 1$$

Find the minimum average cost.

64. Marginal analysis. The price–demand equation for a GPS device is

$$p(x) = 1,000e^{-0.02x}$$

where x is the monthly demand and p is the price in dollars. Find the production level and price per unit that produce the maximum revenue. What is the maximum revenue?

65. Maximum revenue. Graph the revenue function from Problem 64 for $0 \leq x \leq 100$.

66. Maximum profit. Refer to Problem 64. If the GPS devices cost the store $220 each, find the price (to the nearest cent) that maximizes the profit. What is the maximum profit (to the nearest dollar)?

67. Maximum profit. The data in the table show the daily demand x for cream puffs at a state fair at various price levels p. If it costs $1 to make a cream puff, use logarithmic regression ($p = a + b \ln x$) to find the price (to the nearest cent) that maximizes profit.

Demand x	Price per Cream Puff($) p
3,125	1.99
3,879	1.89
5,263	1.79
5,792	1.69
6,748	1.59
8,120	1.49

68. Construction costs. The ceiling supports in a new discount department store are 12 feet apart. Lights are to be hung from these supports by chains in the shape of a "Y." If the lights are 10 feet below the ceiling, what is the shortest length of chain that can be used to support these lights?

69. Average cost. The table gives the total daily cost y (in dollars) of producing x dozen chocolate chip cookies at various production levels.

Dozens of Cookies x	Total Cost y
50	119
100	187
150	248
200	382
250	505
300	695

(A) Enter the data into a graphing calculator and find a quadratic regression equation for the total cost.

(B) Use the regression equation from part (A) to find the minimum average cost (to the nearest cent) and the corresponding production level (to the nearest integer).

70. Advertising—point of diminishing returns. A company estimates that it will sell $N(x)$ units of a product after spending $\$x$ thousand on advertising, as given by

$$N(x) = -0.25x^4 + 11x^3 - 108x^2 + 3,000$$
$$9 \le x \le 24$$

When is the rate of change of sales increasing and when is it decreasing? What is the point of diminishing returns and the maximum rate of change of sales? Graph N and N' on the same coordinate system.

71. Advertising. A chain of appliance stores uses TV ads to promote its HDTV sales. Analyzing past records produced the data in the following table, where x is the number of ads placed monthly and y is the number of HDTVs sold that month:

Number of Ads	Number of HDTVs
x	y
10	271
20	427
25	526
30	629
45	887
48	917

(A) Enter the data into a graphing calculator, set the calculator to display two decimal places, and find a cubic regression equation for the number of HDTVs sold monthly as a function of the number of ads.

(B) How many ads should be placed each month to maximize the rate of change of sales with respect to the number of ads, and how many HDTVs can be expected to be sold with that number of ads? Round answers to the nearest integer.

72. Bacteria control. If t days after a treatment the bacteria count per cubic centimeter in a body of water is given by

$$C(t) = 20t^2 - 120t + 800 \qquad 0 \le t \le 9$$

then in how many days will the count be a minimum?

73. Politics. In a new suburb, it is estimated that the number of registered voters will grow according to

$$N = 10 + 6t^2 - t^3 \qquad 0 \le t \le 5$$

where t is time in years and N is in thousands. When will the rate of increase be a maximum?

Multivariable Calculus

Introduction

In previous chapters, we have applied the key concepts of calculus, the derivative and the integral, to functions with one independent variable. The graph of such a function is a curve in the plane. In Chapter 6, we extend the key concepts of calculus to functions with two independent variables. Graphs of such functions are surfaces in a three-dimensional coordinate system. We use functions with two independent variables to study how production depends on both labor and capital; how braking distance depends on both the weight and speed of a car; how resistance in a blood vessel depends on both its length and radius. In Section 6-5, we justify the method of least squares and use the method to construct linear models (see, for example, Problem 31 in Section 6-5 on global warming).

6-1   Functions of Several Variables

- Functions of Two or More Independent Variables
- Examples of Functions of Several Variables
- Three-Dimensional Coordinate Systems

Functions of Two or More Independent Variables

In Section 2-1, we introduced the concept of a function with one independent variable. Now we broaden the concept to include functions with more than one independent variable.

A small manufacturing company produces a standard type of surfboard. If fixed costs are \$500 per week and variable costs are \$70 per board produced, the weekly cost function is given by

$$C(x) = 500 + 70x \qquad (1)$$

where x is the number of boards produced per week. The cost function is a function of a single independent variable x. For each value of x from the domain of C, there exists exactly one value of $C(x)$ in the range of C.

Now, suppose that the company decides to add a high-performance competition board to its line. If the fixed costs for the competition board are \$200 per week and the variable costs are \$100 per board, then the cost function (1) must be modified to

$$C(x, y) = 700 + 70x + 100y \qquad (2)$$

where $C(x, y)$ is the cost for a weekly output of x standard boards and y competition boards. Equation (2) is an example of a function with two independent variables x and y. Of course, as the company expands its product line even further, its weekly cost function must be modified to include more and more independent variables, one for each new product produced.

In general, an equation of the form

$$z = f(x, y)$$

describes a **function of two independent variables** if, for each permissible ordered pair (x, y), there is one and only one value of z determined by $f(x, y)$. The variables x and y are **independent variables**, and the variable z is a **dependent variable**. The set of all ordered pairs of permissible values of x and y is the **domain** of the function, and the set of all corresponding values $f(x, y)$ is the **range** of the function. Unless otherwise stated, we will assume that the domain of a function specified by an equation of the form $z = f(x, y)$ is the set of all ordered pairs of real numbers (x, y) such that $f(x, y)$ is also a real number. It should be noted, however, that certain conditions in practical problems often lead to further restrictions on the domain of a function.

We can similarly define functions of three independent variables, $w = f(x, y, z)$; of four independent variables, $u = f(w, x, y, z)$; and so on. In this chapter, we concern ourselves primarily with functions of two independent variables.

EXAMPLE 1   **Evaluating a Function of Two Independent Variables** For the cost function $C(x, y) = 700 + 70x + 100y$ described earlier, find $C(10, 5)$.

SOLUTION
$$C(10, 5) = 700 + 70(10) + 100(5)$$
$$= \$1,900$$

Matched Problem 1   Find $C(20, 10)$ for the cost function in Example 1.

EXAMPLE 2   **Evaluating a Function of Three Independent Variables** For the function $f(x, y, z) = 2x^2 - 3xy + 3z + 1$, find $f(3, 0, -1)$.

SOLUTION
$$f(3, 0, -1) = 2(3)^2 - 3(3)(0) + 3(-1) + 1$$
$$= 18 - 0 - 3 + 1 = 16$$

Matched Problem 2   Find $f(-2, 2, 3)$ for f in Example 2.

EXAMPLE 3   **Revenue, Cost, and Profit Functions** Suppose the surfboard company discussed earlier has determined that the demand equations for its two types of boards are given by

$$p = 210 - 4x + y$$
$$q = 300 + x - 12y$$

where p is the price of the standard board, q is the price of the competition board, x is the weekly demand for standard boards, and y is the weekly demand for competition boards.

(A) Find the weekly revenue function $R(x, y)$, and evaluate $R(20, 10)$.

(B) If the weekly cost function is

$$C(x, y) = 700 + 70x + 100y$$

find the weekly profit function $P(x, y)$ and evaluate $P(20, 10)$.

SOLUTION   (A)

$$\text{Revenue} = \begin{pmatrix} \text{demand for} \\ \text{standard} \\ \text{boards} \end{pmatrix} \times \begin{pmatrix} \text{price of a} \\ \text{standard} \\ \text{board} \end{pmatrix} + \begin{pmatrix} \text{demand for} \\ \text{competition} \\ \text{boards} \end{pmatrix} \times \begin{pmatrix} \text{price of a} \\ \text{competition} \\ \text{board} \end{pmatrix}$$

$$R(x, y) = xp + yq$$
$$= x(210 - 4x + y) + y(300 + x - 12y)$$
$$= 210x + 300y - 4x^2 + 2xy - 12y^2$$
$$R(20, 10) = 210(20) + 300(10) - 4(20)^2 + 2(20)(10) - 12(10)^2$$
$$= \$4,800$$

(B) $\text{Profit} = \text{revenue} - \text{cost}$

$$P(x, y) = R(x, y) - C(x, y)$$
$$= 210x + 300y - 4x^2 + 2xy - 12y^2 - 700 - 70x - 100y$$
$$= 140x + 200y - 4x^2 + 2xy - 12y^2 - 700$$
$$P(20, 10) = 140(20) + 200(10) - 4(20)^2 + 2(20)(10) - 12(10)^2 - 700$$
$$= \$1,700$$

Matched Problem 3   Repeat Example 3 if the demand and cost equations are given by

$$p = 220 - 6x + y$$
$$q = 300 + 3x - 10y$$
$$C(x, y) = 40x + 80y + 1,000$$

Examples of Functions of Several Variables

A number of concepts can be considered as functions of two or more variables.

Area of a rectangle	$A(x, y) = xy$	$A = \text{area}$
Volume of a box	$V(x, y, z) = xyz$	$V = \text{volume}$
Volume of a right circular cylinder	$V(r, h) = \pi r^2 h$	
Simple interest	$A(P, r, t) = P(1 + rt)$	$A = \text{amount}$ $P = \text{principal}$ $r = \text{annual rate}$ $t = \text{time in years}$
Compound interest	$A(P, r, t, n) = P\left(1 + \dfrac{r}{n}\right)^{nt}$	$A = \text{amount}$ $P = \text{principal}$ $r = \text{annual rate}$ $t = \text{time in years}$ $n = \text{number of compound}$ periods per year
IQ	$Q(M, C) = \dfrac{M}{C}(100)$	$Q = \text{IQ} = \text{intelligence}$ quotient $M = \text{MA} = \text{mental age}$ $C = \text{CA}$ $= \text{chronological age}$
Resistance for blood flow in a vessel (Poiseuille's law)	$R(L, r) = k\dfrac{L}{r^4}$	$R = \text{resistance}$ $L = \text{length of vessel}$ $r = \text{radius of vessel}$ $k = \text{constant}$

EXAMPLE 4   **Package Design** A company uses a box with a square base and an open top for a bath assortment (see figure). If x is the length (in inches) of each side of the base and y is the height (in inches), find the total amount of material $M(x, y)$ required to construct one of these boxes, and evaluate $M(5, 10)$.

SOLUTION

$$\text{Area of base} = x^2$$
$$\text{Area of one side} = xy$$
$$\text{Total material} = (\text{area of base}) + 4(\text{area of one side})$$
$$M(x, y) = x^2 + 4xy$$
$$M(5, 10) = (5)^2 + 4(5)(10)$$
$$= 225 \text{ square inches}$$

Matched Problem 4   For the box in Example 4, find the volume $V(x, y)$ and evaluate $V(5, 10)$.

The next example concerns the **Cobb–Douglas production function**

$$f(x, y) = kx^m y^n$$

where k, m, and n are positive constants with $m + n = 1$. Economists use this function to describe the number of units $f(x, y)$ produced from the utilization of x units of labor and y units of capital (for equipment such as tools, machinery, buildings, and so on). Cobb–Douglas production functions are also used to describe the productivity of a single industry, of a group of industries producing the same product, or even of an entire country.

EXAMPLE 5 **Productivity** The productivity of a steel-manufacturing company is given approximately by the function

$$f(x, y) = 10x^{0.2}y^{0.8}$$

with the utilization of x units of labor and y units of capital. If the company uses 3,000 units of labor and 1,000 units of capital, how many units of steel will be produced?

SOLUTION The number of units of steel produced is given by

$$f(3,000, 1,000) = 10(3,000)^{0.2}(1,000)^{0.8} \quad \textit{Use a calculator.}$$

$$\approx 12{,}457 \text{ units}$$

Matched Problem 5 Refer to Example 5. Find the steel production if the company uses 1,000 units of labor and 2,000 units of capital.

Three-Dimensional Coordinate Systems

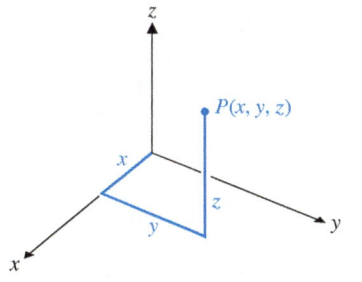

Figure 1 Rectangular coordinate system

We now take a brief look at graphs of functions of two independent variables. Since functions of the form $z = f(x, y)$ involve two independent variables x and y, and one dependent variable z, we need a *three-dimensional coordinate system* for their graphs. A **three-dimensional coordinate system** is formed by three mutually perpendicular number lines intersecting at their origins (see Fig. 1). In such a system, every ordered **triplet of numbers (x, y, z)** can be associated with a unique point, and conversely.

EXAMPLE 6 **Three-Dimensional Coordinates** Locate $(-3, 5, 2)$ in a rectangular coordinate system.

SOLUTION

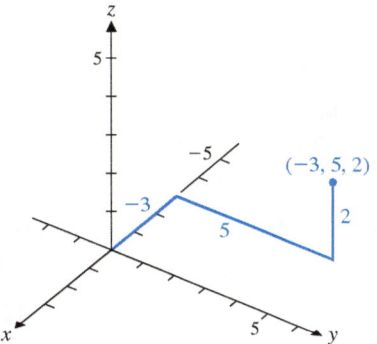

Matched Problem 6 Find the coordinates of the corners $A, C, G,$ and D of the rectangular box shown in the following figure.

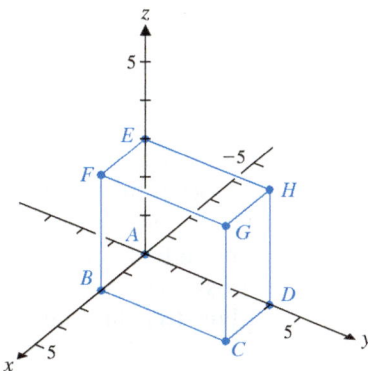

EXPLORE & DISCUSS 1

Imagine that you are facing the front of a classroom whose rectangular walls meet at right angles. Suppose that the point of intersection of the floor, front wall, and left-side wall is the origin of a three-dimensional coordinate system in which every point in the room has nonnegative coordinates. Then the plane $z = 0$ (or, equivalently, the xy plane) can be described as "the floor," and the plane $z = 2$ can be described as "the plane parallel to, but 2 units above, the floor." Give similar descriptions of the following planes:

(A) $x = 0$ (B) $x = 3$ (C) $y = 0$ (D) $y = 4$ (E) $x = -1$

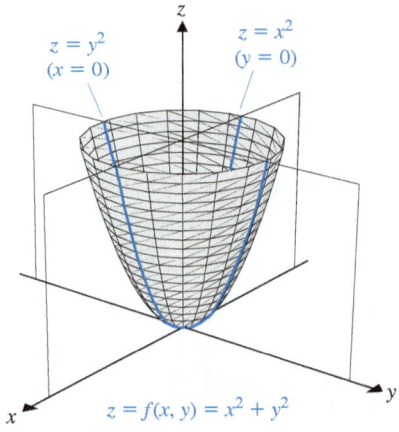

$z = y^2$
$(x = 0)$

$z = x^2$
$(y = 0)$

$z = f(x, y) = x^2 + y^2$

Figure 2 Paraboloid

What does the graph of $z = x^2 + y^2$ look like? If we let $x = 0$ and graph $z = 0^2 + y^2 = y^2$ in the yz plane, we obtain a parabola; if we let $y = 0$ and graph $z = x^2 + 0^2 = x^2$ in the xz plane, we obtain another parabola. The graph of $z = x^2 + y^2$ is either one of these parabolas rotated around the z axis (see Fig. 2). This cup-shaped figure is a *surface* and is called a **paraboloid**.

In general, the graph of any function of the form $z = f(x, y)$ is called a **surface**. The graph of such a function is the graph of all ordered triplets of numbers (x, y, z) that satisfy the equation. Graphing functions of two independent variables is a difficult task, and the general process will not be dealt with in this book. We present only a few simple graphs to suggest extensions of earlier geometric interpretations of the derivative and local maxima and minima to functions of two variables. Note that $z = f(x, y) = x^2 + y^2$ appears (see Fig. 2) to have a local minimum at $(x, y) = (0, 0)$. Figure 3 shows a local maximum at $(x, y) = (0, 0)$.

Figure 4 shows a point at $(x, y) = (0, 0)$, called a **saddle point**, that is neither a local minimum nor a local maximum. Note that if $x = 0$, the saddle point is a local minimum, and if $y = 0$, the saddle point is a local maximum. More will be said about local maxima and minima in Section 6-3.

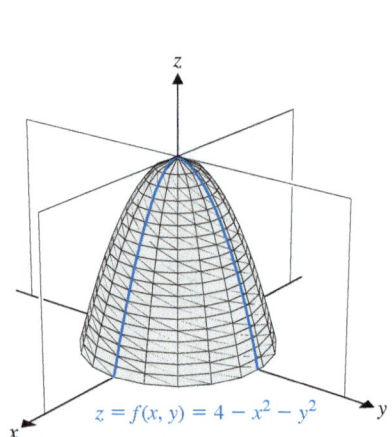

$z = f(x, y) = 4 - x^2 - y^2$

Figure 3 Local maximum:
$f(0, 0) = 4$

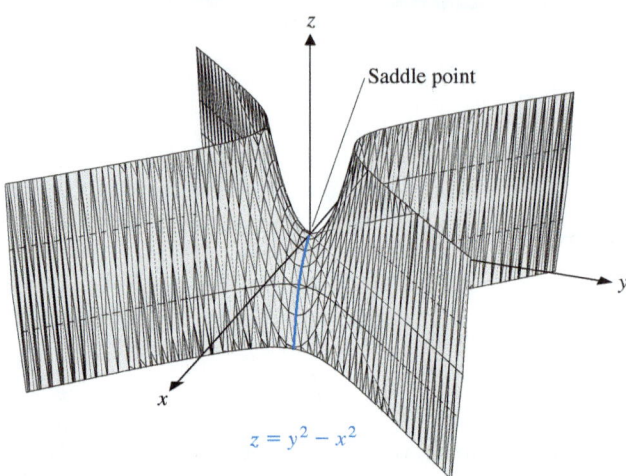

Saddle point

$z = y^2 - x^2$

Figure 4 Saddle point at $(0, 0, 0)$

Some graphing calculators are designed to draw graphs (like those of Figs. 2, 3, and 4) of functions of two independent variables. Others, such as the graphing calculator used for the displays in this book, are designed to draw graphs of functions of one independent variable. When using the latter type of calculator, we can graph cross sections produced by cutting surfaces with planes parallel to the xz plane or yz plane to gain insight into the graph of a function of two independent variables.

EXAMPLE 7 **Graphing Cross Sections**

(A) Describe the cross sections of $f(x, y) = 2x^2 + y^2$ in the planes $y = 0$, $y = 1$, $y = 2$, $y = 3$, and $y = 4$.

(B) Describe the cross sections of $f(x, y) = 2x^2 + y^2$ in the planes $x = 0$, $x = 1$, $x = 2$, $x = 3$, and $x = 4$.

SOLUTION (A) The cross section of $f(x, y) = 2x^2 + y^2$ produced by cutting it with the plane $y = 0$ is the graph of the function $f(x, 0) = 2x^2$ in this plane. We can examine the shape of this cross section by graphing $y_1 = 2x^2$ on a graphing calculator (Fig. 5). Similarly, the graphs of $y_2 = f(x, 1) = 2x^2 + 1$, $y_3 = f(x, 2) = 2x^2 + 4$, $y_4 = f(x, 3) = 2x^2 + 9$, and $y_5 = f(x, 4) = 2x^2 + 16$ show the shapes of the other four cross sections (see Fig. 5). Each of these is a parabola that opens upward. Note the correspondence between the graphs in Figure 5 and the actual cross sections of $f(x, y) = 2x^2 + y^2$ shown in Figure 6.

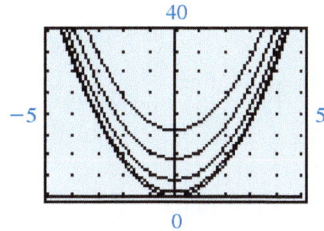

Figure 5

$y_1 = 2x^2$ $y_4 = 2x^2 + 9$

$y_2 = 2x^2 + 1$ $y_5 = 2x^2 + 16$

$y_3 = 2x^2 + 4$

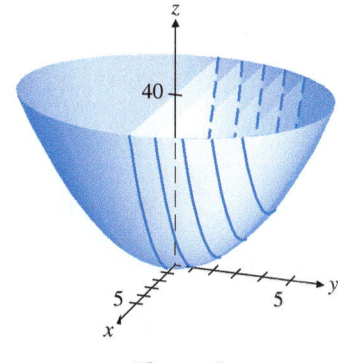

Figure 6

Figure 7

$y_1 = x^2$ $y_4 = 18 + x^2$

$y_2 = 2 + x^2$ $y_5 = 32 + x^2$

$y_3 = 8 + x^2$

(B) The five cross sections are represented by the graphs of the functions $f(0, y) = y^2$, $f(1, y) = 2 + y^2$, $f(2, y) = 8 + y^2$, $f(3, y) = 18 + y^2$, and $f(4, y) = 32 + y^2$. These five functions are graphed in Figure 7. (Note that changing the name of the independent variable from y to x for graphing purposes does not affect the graph displayed.) Each of the five cross sections is a parabola that opens upward.

Matched Problem 7 (A) Describe the cross sections of $g(x, y) = y^2 - x^2$ in the planes $y = 0$, $y = 1$, $y = 2$, $y = 3$, and $y = 4$.

(B) Describe the cross sections of $g(x, y) = y^2 - x^2$ in the planes $x = 0$, $x = 1$, $x = 2$, $x = 3$, and $x = 4$.

CONCEPTUAL INSIGHT

The graph of the *equation*

$$x^2 + y^2 + z^2 = 4 \qquad (3)$$

is the graph of all ordered triplets of numbers (x, y, z) that satisfy the equation. The Pythagorean theorem can be used to show that the distance from the point (x, y, z) to the origin $(0, 0, 0)$ is equal to

$$\sqrt{x^2 + y^2 + z^2}$$

Therefore, the graph of (3) consists of all points that are at a distance 2 from the origin — that is, all points on the sphere of radius 2 and with center at the origin. Recall that a circle in the plane is *not* the graph of a function $y = f(x)$, because it fails the vertical-line test (Section 2-1). Similarly, a sphere is *not* the graph of a *function* $z = f(x, y)$ of two variables.

Exercises 6-1

In Problems 1–10, find the indicated values of the functions

$$f(x, y) = 2x + 7y - 5 \quad \text{and} \quad g(x, y) = \frac{88}{x^2 + 3y}$$

1. $f(4, -1)$　　　　　**2.** $f(0, 10)$

3. $f(8, 0)$　　　　　**4.** $f(5, 6)$

5. $g(1, 7)$　　　　　**6.** $g(-2, 0)$

7. $g(3, -3)$　　　　　**8.** $g(0, 0)$

9. $3f(-2, 2) + 5g(-2, 2)$

10. $2f(10, -4) - 7g(10, -4)$

In Problems 11–14, find the indicated values of

$$f(x, y, z) = 2x - 3y^2 + 5z^3 - 1$$

11. $f(0, 0, 0)$　　　　　**12.** $f(0, 0, 2)$

13. $f(6, -5, 0)$　　　　　**14.** $f(-10, 4, -3)$

In Problems 15–24, find the indicated value of the given function.

15. $P(13, 5)$ for $P(n, r) = \dfrac{n!}{(n - r)!}$

16. $C(13, 5)$ for $C(n, r) = \dfrac{n!}{r!(n - r)!}$

17. $V(4, 12)$ for $V(R, h) = \pi R^2 h$

18. $T(4, 12)$ for $T(R, h) = 2\pi R(R + h)$

19. $S(3, 10)$ for $S(R, h) = \pi R\sqrt{R^2 + h^2}$

20. $W(3, 10)$ for $W(R, h) = \dfrac{1}{3}\pi R^2 h$

21. $A(100, 0.06, 3)$ for $A(P, r, t) = P + Prt$

22. $A(10, 0.04, 3, 2)$ for $A(P, r, t, n) = P\left(1 + \dfrac{r}{n}\right)^{tn}$

23. $P(0.05, 12)$ for $P(r, T) = \displaystyle\int_0^T 4{,}000e^{-rt}\, dt$

24. $F(0.07, 10)$ for $F(r, T) = \displaystyle\int_0^T 4{,}000e^{r(T-t)}\, dt$

In Problems 25–30, find the indicated function f of a single variable.

25. $f(x) = G(x, 0)$ for $G(x, y) = x^2 + 3xy + y^2 - 7$

26. $f(y) = H(0, y)$ for $H(x, y) = x^2 - 5xy - y^2 + 2$

27. $f(y) = K(4, y)$ for $K(x, y) = 10xy + 3x - 2y + 8$

28. $f(x) = L(x, -2)$ for $L(x, y) = 25 - x + 5y - 6xy$

29. $f(y) = M(y, y)$ for $M(x, y) = x^2 y - 3xy^2 + 5$

30. $f(x) = N(x, 2x)$ for $N(x, y) = 3xy + x^2 - y^2 + 1$

31. Let $F(x, y) = 2x + 3y - 6$. Find all values of y such that $F(0, y) = 0$.

32. Let $F(x, y) = 5x - 4y + 12$. Find all values of x such that $F(x, 0) = 0$.

33. Let $F(x, y) = 2xy + 3x - 4y - 1$. Find all values of x such that $F(x, x) = 0$.

34. Let $F(x, y) = xy + 2x^2 + y^2 - 25$. Find all values of y such that $F(y, y) = 0$.

35. Let $F(x, y) = x^2 + e^x y - y^2$. Find all values of x such that $F(x, 2) = 0$.

36. Let $G(a, b, c) = a^3 + b^3 + c^3 - (ab + ac + bc) - 6$. Find all values of b such that $G(2, b, 1) = 0$.

37. For the function $f(x, y) = x^2 + 2y^2$, find

$$\frac{f(x + h, y) - f(x, y)}{h}$$

38. For the function $f(x, y) = x^2 + 2y^2$, find

$$\frac{f(x, y + k) - f(x, y)}{k}$$

39. For the function $f(x, y) = 2xy^2$, find

$$\frac{f(x + h, y) - f(x, y)}{h}$$

40. For the function $f(x, y) = 2xy^2$, find

$$\frac{f(x, y + k) - f(x, y)}{k}$$

41. Find the coordinates of E and F in the figure for Matched Problem 6 on page 353.

42. Find the coordinates of B and H in the figure for Matched Problem 6 on page 353.

In Problems 43–48, use a graphing calculator as necessary to explore the graphs of the indicated cross sections.

43. Let $f(x, y) = x^2$.

(A) Explain why the cross sections of the surface $z = f(x, y)$ produced by cutting it with planes parallel to $y = 0$ are parabolas.

(B) Describe the cross sections of the surface in the planes $x = 0$, $x = 1$, and $x = 2$.

(C) Describe the surface $z = f(x, y)$.

44. Let $f(x, y) = \sqrt{4 - y^2}$.

(A) Explain why the cross sections of the surface $z = f(x, y)$ produced by cutting it with planes parallel to $x = 0$ are semicircles of radius 2.

(B) Describe the cross sections of the surface in the planes $y = 0$, $y = 2$, and $y = 3$.

(C) Describe the surface $z = f(x, y)$.

45. Let $f(x, y) = \sqrt{36 - x^2 - y^2}$.

(A) Describe the cross sections of the surface $z = f(x, y)$ produced by cutting it with the planes $y = 1$, $y = 2$, $y = 3$, $y = 4$, and $y = 5$.

(B) Describe the cross sections of the surface in the planes $x = 0$, $x = 1$, $x = 2$, $x = 3$, $x = 4$, and $x = 5$.

(C) Describe the surface $z = f(x, y)$.

46. Let $f(x, y) = 100 + 10x + 25y - x^2 - 5y^2$.

(A) Describe the cross sections of the surface $z = f(x, y)$ produced by cutting it with the planes $y = 0$, $y = 1$, $y = 2$, and $y = 3$.

(B) Describe the cross sections of the surface in the planes $x = 0$, $x = 1$, $x = 2$, and $x = 3$.

(C) Describe the surface $z = f(x, y)$.

47. Let $f(x, y) = e^{-(x^2+y^2)}$.

(A) Explain why $f(a, b) = f(c, d)$ whenever (a, b) and (c, d) are points on the same circle centered at the origin in the xy plane.

(B) Describe the cross sections of the surface $z = f(x, y)$ produced by cutting it with the planes $x = 0$, $y = 0$, and $x = y$.

(C) Describe the surface $z = f(x, y)$.

48. Let $f(x, y) = 4 - \sqrt{x^2 + y^2}$.

(A) Explain why $f(a, b) = f(c, d)$ whenever (a, b) and (c, d) are points on the same circle with center at the origin in the xy plane.

(B) Describe the cross sections of the surface $z = f(x, y)$ produced by cutting it with the planes $x = 0$, $y = 0$, and $x = y$.

(C) Describe the surface $z = f(x, y)$.

Applications

49. **Cost function.** A small manufacturing company produces two models of a surfboard: a standard model and a competition model. If the standard model is produced at a variable cost of \$210 each and the competition model at a variable cost of \$300 each, and if the total fixed costs per month are \$6,000, then the monthly cost function is given by

$$C(x, y) = 6{,}000 + 210x + 300y$$

where x and y are the numbers of standard and competition models produced per month, respectively. Find $C(20, 10)$, $C(50, 5)$, and $C(30, 30)$.

50. **Advertising and sales.** A company spends \$$x$ thousand per week on online advertising and \$$y$ thousand per week on TV advertising. Its weekly sales are found to be given by

$$S(x, y) = 5x^2y^3$$

Find $S(3, 2)$ and $S(2, 3)$.

51. **Revenue function.** A supermarket sells two brands of coffee: brand A at \$$p$ per pound and brand B at \$$q$ per pound. The daily demand equations for brands A and B are, respectively,

$$x = 200 - 5p + 4q$$
$$y = 300 + 2p - 4q$$

(both in pounds). Find the daily revenue function $R(p, q)$. Evaluate $R(2, 3)$ and $R(3, 2)$.

52. **Revenue, cost, and profit functions.** A company manufactures 10- and 3-speed bicycles. The weekly demand and cost equations are

$$p = 230 - 9x + y$$
$$q = 130 + x - 4y$$
$$C(x, y) = 200 + 80x + 30y$$

where \$$p$ is the price of a 10-speed bicycle, \$$q$ is the price of a 3-speed bicycle, x is the weekly demand for 10-speed bicycles, y is the weekly demand for 3-speed bicycles, and $C(x, y)$ is the cost function. Find the weekly revenue function $R(x, y)$ and the weekly profit function $P(x, y)$. Evaluate $R(10, 15)$ and $P(10, 15)$.

53. **Productivity.** The Cobb–Douglas production function for a petroleum company is given by

$$f(x, y) = 20x^{0.4}y^{0.6}$$

where x is the utilization of labor and y is the utilization of capital. If the company uses 1,250 units of labor and 1,700 units of capital, how many units of petroleum will be produced?

54. **Productivity.** The petroleum company in Problem 53 is taken over by another company that decides to double both the units of labor and the units of capital utilized in the production of petroleum. Use the Cobb–Douglas production function given in Problem 53 to find the amount of petroleum that will be produced by this increased utilization of labor and capital. What is the effect on productivity of doubling both the units of labor and the units of capital?

55. **Future value.** At the end of each year, \$5,000 is invested into an IRA earning 3% compounded annually.

(A) How much will be in the account at the end of 30 years? Use the annuity formula

$$F(P, i, n) = P\frac{(1 + i)^n - 1}{i}$$

where

P = periodic payment
i = rate per period
n = number of payments (periods)
F = FV = future value

(B) Use graphical approximation methods to determine the rate of interest that would produce \$300,000 in the account at the end of 30 years.

56. **Package design.** The packaging department in a company has been asked to design a rectangular box with no top and a partition down the middle (see the figure on the next page). Let x, y, and z be the dimensions of the box (in inches).

Figure for 56

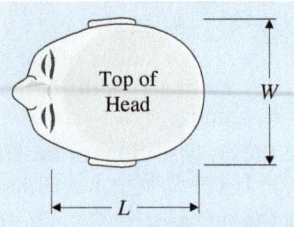

Figure for 59

(A) Find the total amount of material $M(x, y, z)$ used in constructing one of these boxes, and evaluate $M(10, 12, 6)$.

(B) Suppose that the box will have a square base and a volume of 720 cubic inches. Use graphical approximation methods to determine the dimensions that require the least amount of material.

57. Marine biology. Using scuba-diving gear, a marine biologist estimates the time of a dive according to the equation

$$T(V, x) = \frac{33V}{x + 33}$$

where

T = time of dive in minutes

V = volume of air, at sea level pressure, compressed into tanks

x = depth of dive in feet

Find $T(70, 47)$ and $T(60, 27)$.

58. Blood flow. Poiseuille's law states that the resistance R for blood flowing in a blood vessel varies directly as the length L of the vessel and inversely as the fourth power of its radius r. Stated as an equation,

$$R(L, r) = k \frac{L}{r^4} \qquad k \text{ a constant}$$

Find $R(8, 1)$ and $R(4, 0.2)$.

59. Physical anthropology. Anthropologists use an index called the *cephalic index*. The cephalic index C varies directly as the width W of the head and inversely as the length L of the head (both viewed from the top). In terms of an equation,

$$C(W, L) = 100 \frac{W}{L}$$

where

W = width in inches

L = length in inches

Find $C(6, 8)$ and $C(8.1, 9)$.

60. Safety research. Under ideal conditions, if a person driving a car slams on the brakes and skids to a stop, the length of the skid marks (in feet) is given by the formula

$$L(w, v) = kwv^2$$

where

k = constant

w = weight of car in pounds

v = speed of car in miles per hour

For $k = 0.000\ 013\ 3$, find $L(2,000, 40)$ and $L(3,000, 60)$.

61. Psychology. The intelligence quotient (IQ) is defined to be the ratio of mental age (MA), as determined by certain tests, to chronological age (CA), multiplied by 100. Stated as an equation,

$$Q(M, C) = \frac{M}{C} \cdot 100$$

where

$$Q = \text{IQ} \qquad M = \text{MA} \qquad C = \text{CA}$$

Find $Q(12, 10)$ and $Q(10, 12)$.

Answers to Matched Problems

1. $3,100

2. 30

3. (A) $R(x, y) = 220x + 300y - 6x^2 + 4xy - 10y^2$;
   $R(20, 10) = \$4,800$
   (B) $P(x, y) = 180x + 220y - 6x^2 + 4xy - 10y^2 - 1,000; P(20, 10) = \$2,200$

4. $V(x, y) = x^2y; V(5, 10) = 250$ in.^{3}

5. 17,411 units

6. $A\ (0, 0, 0); C(2, 4, 0); G(2, 4, 3); D(0, 4, 0)$

7. (A) Each cross section is a parabola that opens downward.
   (B) Each cross section is a parabola that opens upward.

6-2 Partial Derivatives

- Partial Derivatives
- Second-Order Partial Derivatives

Partial Derivatives

We know how to differentiate many kinds of functions of one independent variable and how to interpret the derivatives that result. What about functions with two or more independent variables? Let's return to the surfboard example considered on page 350.

For the company producing only the standard board, the cost function was

$$C(x) = 500 + 70x$$

Differentiating with respect to x, we obtain the marginal cost function

$$C'(x) = 70$$

Since the marginal cost is constant, \$70 is the change in cost for a 1-unit increase in production at any output level.

For the company producing two types of boards—a standard model and a competition model—the cost function was

$$C(x, y) = 700 + 70x + 100y$$

Now suppose that we differentiate with respect to x, holding y fixed, and denote the resulting function by $C_x(x, y)$; or suppose we differentiate with respect to y, holding x fixed, and denote the resulting function by $C_y(x, y)$. Differentiating in this way, we obtain

$$C_x(x, y) = 70 \qquad C_y(x, y) = 100$$

Each of these functions is called a **partial derivative**, and, in this example, each represents marginal cost. The first is the change in cost due to a 1-unit increase in production of the standard board with the production of the competition model held fixed. The second is the change in cost due to a 1-unit increase in production of the competition board with the production of the standard board held fixed.

In general, if $z = f(x, y)$, then the **partial derivative of f with respect to x**, denoted $\partial z/\partial x$, f_x, or $f_x(x, y)$, is defined by

$$\frac{\partial z}{\partial x} = \lim_{h \to 0} \frac{f(x + h, y) - f(x, y)}{h}$$

provided that the limit exists. We recognize this formula as the ordinary derivative of f with respect to x, holding y constant. We can continue to use all the derivative rules and properties discussed in Chapters 3 to 5 and apply them to partial derivatives.

Similarly, the **partial derivative of f with respect to y**, denoted $\partial z/\partial y$, f_y, or $f_y(x, y)$, is defined by

$$\frac{\partial z}{\partial y} = \lim_{k \to 0} \frac{f(x, y + k) - f(x, y)}{k}$$

which is the ordinary derivative with respect to y, holding x constant.

Parallel definitions and interpretations hold for functions with three or more independent variables.

EXAMPLE 1 **Partial Derivatives** For $z = f(x, y) = 2x^2 - 3x^2y + 5y + 1$, find

(A) $\partial z/\partial x$ (B) $f_x(2, 3)$

SOLUTION (A) $z = 2x^2 - 3x^2y + 5y + 1$

Differentiating with respect to x, holding y constant (that is, treating y as a constant), we obtain

$$\frac{\partial z}{\partial x} = 4x - 6xy$$

Figure 1 $y_1 = -7x^2 + 16$

(B) $f(x, y) = 2x^2 - 3x^2y + 5y + 1$

First, differentiate with respect to x. From part (A), we have

$$f_x(x, y) = 4x - 6xy$$

Then evaluate this equation at $(2, 3)$:

$$f_x(2, 3) = 4(2) - 6(2)(3) = -28$$

In part 1(B), an alternative approach would be to substitute $y = 3$ into $f(x, y)$ and graph the function $f(x, 3) = -7x^2 + 16$, which represents the cross section of the surface $z = f(x, y)$ produced by cutting it with the plane $y = 3$. Then determine the slope of the tangent line when $x = 2$. Again, we conclude that $f_x(2, 3) = -28$ (see Fig. 1).

Matched Problem 1   For f in Example 1, find

(A) $\partial z/\partial y$

(B) $f_y(2, 3)$

EXAMPLE 2   **Partial Derivatives Using the Chain Rule** For $z = f(x, y) = e^{x^2+y^2}$, find

(A) $\partial z/\partial x$

(B) $f_y(2, 1)$

SOLUTION   (A) Using the chain rule [thinking of $z = e^u, u = u(x)$; y is held constant], we obtain

$$\frac{\partial z}{\partial x} = e^{x^2+y^2}\frac{\partial(x^2 + y^2)}{\partial x}$$

$$= 2xe^{x^2+y^2}$$

(B)

$$f_y(x, y) = e^{x^2+y^2}\frac{\partial(x^2 + y^2)}{\partial y} = 2ye^{x^2+y^2}$$

$$f_y(2, 1) = 2(1)e^{(2)^2+(1)^2}$$

$$= 2e^5$$

Matched Problem 2   For $z = f(x, y) = (x^2 + 2xy)^5$, find

(A) $\partial z/\partial y$

(B) $f_x(1, 0)$

EXAMPLE 3   **Profit** The profit function for the surfboard company in Example 3 of Section 6-1 was

$$P(x, y) = 140x + 200y - 4x^2 + 2xy - 12y^2 - 700$$

Find $P_x(15, 10)$ and $P_x(30, 10)$, and interpret the results.

SOLUTION

$$P_x(x, y) = 140 - 8x + 2y$$

$$P_x(15, 10) = 140 - 8(15) + 2(10) = 40$$

$$P_x(30, 10) = 140 - 8(30) + 2(10) = -80$$

At a production level of 15 standard and 10 competition boards per week, increasing the production of standard boards by 1 unit and holding the production of competition boards fixed at 10 will increase profit by approximately \$40. At a production level of 30 standard and 10 competition boards per week, increasing the production of standard boards by 1 unit and holding the production of competition boards fixed at 10 will decrease profit by approximately \$80.

Matched Problem 3 For the profit function in Example 3, find $P_y(25, 10)$ and $P_y(25, 15)$, and interpret the results.

EXAMPLE 4 **Productivity** The productivity of a major computer manufacturer is given approximately by the Cobb–Douglas production function

$$f(x, y) = 15x^{0.4}y^{0.6}$$

with the utilization of x units of labor and y units of capital. The partial derivative $f_x(x, y)$ represents the rate of change of productivity with respect to labor and is called the **marginal productivity of labor**. The partial derivative $f_y(x, y)$ represents the rate of change of productivity with respect to capital and is called the **marginal productivity of capital**. If the company is currently utilizing 4,000 units of labor and 2,500 units of capital, find the marginal productivity of labor and the marginal productivity of capital. For the greatest increase in productivity, should the management of the company encourage increased use of labor or increased use of capital?

SOLUTION

$$f_x(x, y) = 6x^{-0.6}y^{0.6}$$

$$f_x(4,000, 2,500) = 6(4,000)^{-0.6}(2,500)^{0.6}$$

$$\approx 4.53 \qquad \text{Marginal productivity of labor}$$

$$f_y(x, y) = 9x^{0.4}y^{-0.4}$$

$$f_y(4,000, 2,500) = 9(4,000)^{0.4}(2,500)^{-0.4}$$

$$\approx 10.86 \qquad \text{Marginal productivity of capital}$$

At the current level of utilization of 4,000 units of labor and 2,500 units of capital, each 1-unit increase in labor utilization (keeping capital utilization fixed at 2,500 units) will increase production by approximately 4.53 units, and each 1-unit increase in capital utilization (keeping labor utilization fixed at 4,000 units) will increase production by approximately 10.86 units. The management of the company should encourage increased use of capital.

Matched Problem 4 The productivity of an airplane-manufacturing company is given approximately by the Cobb–Douglas production function

$$f(x, y) = 40x^{0.3}y^{0.7}$$

(A) Find $f_x(x, y)$ and $f_y(x, y)$.

(B) If the company is currently using 1,500 units of labor and 4,500 units of capital, find the marginal productivity of labor and the marginal productivity of capital.

(C) For the greatest increase in productivity, should the management of the company encourage increased use of labor or increased use of capital?

Partial derivatives have simple geometric interpretations, as shown in Figure 2. If we hold x fixed at $x = a$, then $f_y(a, y)$ is the slope of the curve obtained by intersecting the surface $z = f(x, y)$ with the plane $x = a$. A similar interpretation is given to $f_x(x, b)$.

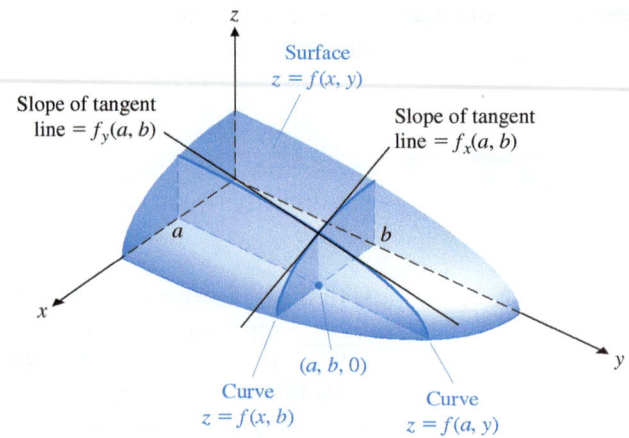

Figure 2

Second-Order Partial Derivatives

The function

$$z = f(x, y) = x^4 y^7$$

has two **first-order partial derivatives**:

$$\frac{\partial z}{\partial x} = f_x = f_x(x, y) = 4x^3 y^7 \qquad \text{and} \qquad \frac{\partial z}{\partial y} = f_y = f_y(x, y) = 7x^4 y^6$$

Each of these partial derivatives, in turn, has two partial derivatives called **second-order partial derivatives** of $z = f(x, y)$. Generalizing the various notations we have for first-order partial derivatives, we write the four second-order partial derivatives of $z = f(x, y) = x^4 y^7$ as

Equivalent notations

$$f_{xx} = f_{xx}(x, y) = \frac{\partial^2 z}{\partial x^2} = \frac{\partial}{\partial x}\left(\frac{\partial z}{\partial x}\right) = \frac{\partial}{\partial x}(4x^3 y^7) = 12x^2 y^7$$

$$f_{xy} = f_{xy}(x, y) = \frac{\partial^2 z}{\partial y\,\partial x} = \frac{\partial}{\partial y}\left(\frac{\partial z}{\partial x}\right) = \frac{\partial}{\partial y}(4x^3 y^7) = 28x^3 y^6$$

$$f_{yx} = f_{yx}(x, y) = \frac{\partial^2 z}{\partial x\,\partial y} = \frac{\partial}{\partial x}\left(\frac{\partial z}{\partial y}\right) = \frac{\partial}{\partial x}(7x^4 y^6) = 28x^3 y^6$$

$$f_{yy} = f_{yy}(x, y) = \frac{\partial^2 z}{\partial y^2} = \frac{\partial}{\partial y}\left(\frac{\partial z}{\partial y}\right) = \frac{\partial}{\partial y}(7x^4 y^6) = 42x^4 y^5$$

In the mixed partial derivative $\partial^2 z/\partial y\,\partial x = f_{xy}$, we started with $z = f(x, y)$ and first differentiated with respect to x (holding y constant). Then we differentiated with respect to y (holding x constant). In the other mixed partial derivative, $\partial^2 z/\partial x\,\partial y = f_{yx}$, the order of differentiation was reversed; however, the final result was the same—that is, $f_{xy} = f_{yx}$. Although it is possible to find functions for which $f_{xy} \neq f_{yx}$, such functions rarely occur in applications involving partial derivatives. For all the functions in this book, we will assume that $f_{xy} = f_{yx}$.

In general, we have the following definitions:

> **DEFINITION Second-Order Partial Derivatives**
>
> If $z = f(x, y)$, then
>
> $$f_{xx} = f_{xx}(x, y) = \frac{\partial^2 z}{\partial x^2} = \frac{\partial}{\partial x}\left(\frac{\partial z}{\partial x}\right)$$
>
> $$f_{xy} = f_{xy}(x, y) = \frac{\partial^2 z}{\partial y\, \partial x} = \frac{\partial}{\partial y}\left(\frac{\partial z}{\partial x}\right)$$
>
> $$f_{yx} = f_{yx}(x, y) = \frac{\partial^2 z}{\partial x\, \partial y} = \frac{\partial}{\partial x}\left(\frac{\partial z}{\partial y}\right)$$
>
> $$f_{yy} = f_{yy}(x, y) = \frac{\partial^2 z}{\partial y^2} = \frac{\partial}{\partial y}\left(\frac{\partial z}{\partial y}\right)$$

EXAMPLE 5 **Second-Order Partial Derivatives** For $z = f(x, y) = 3x^2 - 2xy^3 + 1$, find

(A) $\dfrac{\partial^2 z}{\partial x\, \partial y},\ \dfrac{\partial^2 z}{\partial y\, \partial x}$
(B) $\dfrac{\partial^2 z}{\partial x^2}$
(C) $f_{yx}(2, 1)$

SOLUTION (A) First differentiate with respect to y and then with respect to x:

$$\frac{\partial z}{\partial y} = -6xy^2 \qquad \frac{\partial^2 z}{\partial x\, \partial y} = \frac{\partial}{\partial x}\left(\frac{\partial z}{\partial y}\right) = \frac{\partial}{\partial x}(-6xy^2) = -6y^2$$

Now differentiate with respect to x and then with respect to y:

$$\frac{\partial z}{\partial x} = 6x - 2y^3 \qquad \frac{\partial^2 z}{\partial y\, \partial x} = \frac{\partial}{\partial y}\left(\frac{\partial z}{\partial x}\right) = \frac{\partial}{\partial y}(6x - 2y^3) = -6y^2$$

(B) Differentiate with respect to x twice:

$$\frac{\partial z}{\partial x} = 6x - 2y^3 \qquad \frac{\partial^2 z}{\partial x^2} = \frac{\partial}{\partial x}\left(\frac{\partial z}{\partial x}\right) = 6$$

(C) First find $f_{yx}(x, y)$; then evaluate the resulting equation at $(2, 1)$. Again, remember that f_{yx} signifies differentiation first with respect to y and then with respect to x.

$$f_y(x, y) = -6xy^2 \qquad f_{yx}(x, y) = -6y^2$$

and

$$f_{yx}(2, 1) = -6(1)^2 = -6$$

Matched Problem 5 For $z = f(x, y) = x^3 y - 2y^4 + 3$, find

(A) $\dfrac{\partial^2 z}{\partial y\, \partial x}$
(B) $\dfrac{\partial^2 z}{\partial y^2}$

(C) $f_{xy}(2, 3)$
(D) $f_{yx}(2, 3)$

> **CONCEPTUAL INSIGHT**
>
> Although the mixed second-order partial derivatives f_{xy} and f_{yx} are equal for all functions considered in this book, it is a good idea to compute both of them, as in Example 5(A), as a check on your work. By contrast, the other two second-order partial derivatives, f_{xx} and f_{yy}, are generally not equal to each other. For example, for the function
>
> $$f(x, y) = 3x^2 - 2xy^3 + 1$$
>
> of Example 5,
>
> $$f_{xx} = 6 \quad \text{and} \quad f_{yy} = -12xy$$

Exercises 6-2

A

In Problems 1–18, find the indicated first-order partial derivative for each function $z = f(x, y)$.

1. $f_x(x, y)$ if $f(x, y) = 4x - 3y + 6$

2. $f_x(x, y)$ if $f(x, y) = 7x + 8y - 2$

3. $f_y(x, y)$ if $f(x, y) = x^2 - 3xy + 2y^2$

4. $f_y(x, y)$ if $f(x, y) = 3x^2 + 2xy - 7y^2$

5. $\dfrac{\partial z}{\partial x}$ if $z = x^3 + 4x^2y + 2y^3$

6. $\dfrac{\partial z}{\partial y}$ if $z = 4x^2y - 5xy^2$

7. $\dfrac{\partial z}{\partial y}$ if $z = (5x + 2y)^{10}$

8. $\dfrac{\partial z}{\partial x}$ if $z = (2x - 3y)^8$

In Problems 9–16, find the indicated value.

9. $f_x(1, 3)$ if $f(x, y) = 5x^3y - 4xy^2$

10. $f_x(4, 1)$ if $f(x, y) = x^2y^2 - 5xy^3$

11. $f_y(1, 0)$ if $f(x, y) = 3xe^y$

12. $f_y(2, 4)$ if $f(x, y) = x^4 \ln y$

13. $f_y(2, 1)$ if $f(x, y) = e^{x^2} - 4y$

14. $f_y(3, 3)$ if $f(x, y) = e^{3x} - y^2$

15. $f_x(1, -1)$ if $f(x, y) = \dfrac{2xy}{1 + x^2y^2}$

16. $f_x(-1, 2)$ if $f(x, y) = \dfrac{x^2 - y^2}{1 + x^2}$

🖉 *In Problems 17–22, $M(x,y) = 68 + 0.3x - 0.8y$ gives the mileage (in mpg) of a new car as a function of tire pressure x (in psi) and speed (in mph). Find the indicated quantity (include the appropriate units) and explain what it means.*

17. $M(32, 40)$

18. $M(22, 40)$

19. $M(32, 50)$

20. $M(22, 50)$

21. $M_x(32, 50)$

22. $M_y(32, 50)$

B

In Problems 23–34, find the indicated second-order partial derivative for each function $f(x, y)$.

23. $f_{xx}(x, y)$ if $f(x, y) = 6x - 5y + 3$

24. $f_{yx}(x, y)$ if $f(x, y) = -2x + y + 8$

25. $f_{xy}(x, y)$ if $f(x, y) = 4x^2 + 6y^2 - 10$

26. $f_{yy}(x, y)$ if $f(x, y) = x^2 + 9y^2 - 4$

27. $f_{xy}(x, y)$ if $f(x, y) = e^{xy^2}$

28. $f_{yx}(x, y)$ if $f(x, y) = e^{3x+2y}$

29. $f_{yy}(x, y)$ if $f(x, y) = \dfrac{\ln x}{y}$

30. $f_{xx}(x, y)$ if $f(x, y) = \dfrac{3 \ln x}{y^2}$

31. $f_{xx}(x, y)$ if $f(x, y) = (2x + y)^5$

32. $f_{yx}(x, y)$ if $f(x, y) = (3x - 8y)^6$

33. $f_{xy}(x, y)$ if $f(x, y) = (x^2 + y^4)^{10}$

34. $f_{yy}(x, y)$ if $f(x, y) = (1 + 2xy^2)^8$

In Problems 35–44, find the indicated function or value if $C(x, y) = 3x^2 + 10xy - 8y^2 + 4x - 15y - 120$.

35. $C_x(x, y)$ **36.** $C_y(x, y)$

37. $C_x(3, -2)$ **38.** $C_y(3, -2)$

39. $C_{xx}(x, y)$ **40.** $C_{yy}(x, y)$

41. $C_{xy}(x, y)$ **42.** $C_{yx}(x, y)$

43. $C_{xx}(3, -2)$ **44.** $C_{yy}(3, -2)$

In Problems 45–56, find the indicated function or value if $S(x, y) = x^3 \ln y + 4y^2e^x$.

45. $S_y(x, y)$ **46.** $S_x(x, y)$

47. $S_y(-1, 1)$ **48.** $S_x(-1, 1)$

49. $S_{yx}(x, y)$ **50.** $S_{xy}(x, y)$

51. $S_{yy}(x, y)$ **52.** $S_{xx}(x, y)$

53. $S_{yx}(-1, 1)$ **54.** $S_{xy}(-1, 1)$

55. $S_{yy}(-1, 1)$ **56.** $S_{xx}(-1, 1)$

In Problems 57–62, $S(T, r) = 50(T - 40)(5 - r)$ *gives an ice cream shop's daily sales as a function of temperature T (in °F) and rain r (in inches). Find the indicated quantity (include the appropriate units) and explain what it means.*

57. $S(60, 2)$
58. $S(80, 0)$

59. $S_r(90, 1)$
60. $S_T(90, 1)$

61. $S_{Tr}(90, 1)$
62. $S_{rT}(90, 1)$

63. (A) Let $f(x, y) = y^3 + 4y^2 - 5y + 3$. Show that $\partial f / \partial x = 0$.
   (B) Explain why there are an infinite number of functions $g(x, y)$ such that $\partial g / \partial x = 0$.

64. (A) Find an example of a function $f(x, y)$ such that $\partial f / \partial x = 3$ and $\partial f / \partial y = 2$.
   (B) How many such functions are there? Explain.

In Problems 65–70, find $f_{xx}(x, y), f_{xy}(x, y), f_{yx}(x, y),$ *and* $f_{yy}(x, y)$ *for each function f.*

65. $f(x, y) = x^2y^2 + x^3 + y$

66. $f(x, y) = x^3y^3 + x + y^2$

67. $f(x, y) = \dfrac{x}{y} - \dfrac{y}{x}$

68. $f(x, y) = \dfrac{x^2}{y} - \dfrac{y^2}{x}$

69. $f(x, y) = xe^{xy}$

70. $f(x, y) = x \ln(xy)$

C

71. For
$$P(x, y) = -x^2 + 2xy - 2y^2 - 4x + 12y - 5$$
find all values of x and y such that
$$P_x(x, y) = 0 \quad \text{and} \quad P_y(x, y) = 0$$
simultaneously.

72. For
$$C(x, y) = 2x^2 + 2xy + 3y^2 - 16x - 18y + 54$$
find all values of x and y such that
$$C_x(x, y) = 0 \quad \text{and} \quad C_y(x, y) = 0$$
simultaneously.

73. For
$$F(x, y) = x^3 - 2x^2y^2 - 2x - 4y + 10$$
find all values of x and y such that
$$F_x(x, y) = 0 \quad \text{and} \quad F_y(x, y) = 0$$
simultaneously.

74. For
$$G(x, y) = x^2 \ln y - 3x - 2y + 1$$
find all values of x and y such that
$$G_x(x, y) = 0 \quad \text{and} \quad G_y(x, y) = 0$$
simultaneously

75. Let $f(x, y) = 3x^2 + y^2 - 4x - 6y + 2$.
   (A) Find the minimum value of $f(x, y)$ when $y = 1$.
   (B) Explain why the answer to part (A) is not the minimum value of the function $f(x, y)$.

76. Let $f(x, y) = 5 - 2x + 4y - 3x^2 - y^2$.
   (A) Find the maximum value of $f(x, y)$ when $x = 2$.
   (B) Explain why the answer to part (A) is not the maximum value of the function $f(x, y)$.

77. Let $f(x, y) = 4 - x^4y + 3xy^2 + y^5$.
   (A) Use graphical approximation methods to find c (to three decimal places) such that $f(c, 2)$ is the maximum value of $f(x, y)$ when $y = 2$.
   (B) Find $f_x(c, 2)$ and $f_y(c, 2)$.

78. Let $f(x, y) = e^x + 2e^y + 3xy^2 + 1$.
   (A) Use graphical approximation methods to find d (to three decimal places) such that $f(1, d)$ is the minimum value of $f(x, y)$ when $x = 1$.
   (B) Find $f_x(1, d)$ and $f_y(1, d)$.

In Problems 79 and 80, show that the function f satisfies $f_{xx}(x, y) + f_{yy}(x, y) = 0$.

79. $f(x, y) = \ln(x^2 + y^2)$

80. $f(x, y) = x^3 - 3xy^2$

81. For $f(x, y) = x^2 + 2y^2$, find
   (A) $\displaystyle\lim_{h \to 0} \dfrac{f(x + h, y) - f(x, y)}{h}$
   (B) $\displaystyle\lim_{k \to 0} \dfrac{f(x, y + k) - f(x, y)}{k}$

82. For $f(x, y) = 2xy^2$, find
   (A) $\displaystyle\lim_{h \to 0} \dfrac{f(x + h, y) - f(x, y)}{h}$
   (B) $\displaystyle\lim_{k \to 0} \dfrac{f(x, y + k) - f(x, y)}{k}$

Applications

83. **Profit function.** A firm produces two types of calculators each week, x of type A and y of type B. The weekly revenue and cost functions (in dollars) are
$$R(x, y) = 80x + 90y + 0.04xy - 0.05x^2 - 0.05y^2$$
$$C(x, y) = 8x + 6y + 20,000$$
Find $P_x(1{,}200, 1{,}800)$ and $P_y(1{,}200, 1{,}800)$, and interpret the results.

84. **Advertising and sales.** A company spends $\$x$ per week on online advertising and $\$y$ per week on TV advertising. Its weekly sales were found to be given by
$$S(x, y) = 10x^{0.4}y^{0.8}$$
Find $S_x(3{,}000, 2{,}000)$ and $S_y(3{,}000, 2{,}000)$, and interpret the results.

85. Demand equations. A supermarket sells two brands of coffee: brand A at $\$p$ per pound and brand B at $\$q$ per pound. The daily demands x and y (in pounds) for brands A and B, respectively, are given by

$$x = 200 - 5p + 4q$$

$$y = 300 + 2p - 4q$$

Find $\partial x/\partial p$ and $\partial y/\partial p$, and interpret the results.

86. Revenue and profit functions. A company manufactures 10- and 3-speed bicycles. The weekly demand and cost functions are

$$p = 230 - 9x + y$$

$$q = 130 + x - 4y$$

$$C(x, y) = 200 + 80x + 30y$$

where $\$p$ is the price of a 10-speed bicycle, $\$q$ is the price of a 3-speed bicycle, x is the weekly demand for 10-speed bicycles, y is the weekly demand for 3-speed bicycles, and $C(x, y)$ is the cost function. Find $R_x(10, 5)$ and $P_x(10, 5)$, and interpret the results.

87. Productivity. The productivity of a certain third-world country is given approximately by the function

$$f(x, y) = 10x^{0.75}y^{0.25}$$

with the utilization of x units of labor and y units of capital.

(A) Find $f_x(x, y)$ and $f_y(x, y)$.

(B) If the country is now using 600 units of labor and 100 units of capital, find the marginal productivity of labor and the marginal productivity of capital.

(C) For the greatest increase in the country's productivity, should the government encourage increased use of labor or increased use of capital?

88. Productivity. The productivity of an automobile-manufacturing company is given approximately by the function

$$f(x, y) = 50\sqrt{xy} = 50x^{0.5}y^{0.5}$$

with the utilization of x units of labor and y units of capital.

(A) Find $f_x(x, y)$ and $f_y(x, y)$.

(B) If the company is now using 250 units of labor and 125 units of capital, find the marginal productivity of labor and the marginal productivity of capital.

(C) For the greatest increase in the company's productivity, should the management encourage increased use of labor or increased use of capital?

Problems 89–92 refer to the following: If a decrease in demand for one product results in an increase in demand for another product, the two products are said to be **competitive,** *or* **substitute, products.** *(Real whipping cream and imitation whipping cream are examples of competitive, or substitute, products.) If a decrease in demand for one product results in a decrease in demand for another product, the two products are said to be* **complementary products.** *(Fishing boats and outboard motors are examples of complementary products.) Partial derivatives can be used to test whether two products are competitive, complementary, or neither. We start with demand functions for two*

products such that the demand for either depends on the prices for both:

$$x = f(p, q) \quad \text{Demand function for product A}$$

$$y = g(p, q) \quad \text{Demand function for product B}$$

The variables x and y represent the number of units demanded of products A and B, respectively, at a price p for 1 unit of product A and a price q for 1 unit of product B. Normally, if the price of A increases while the price of B is held constant, then the demand for A will decrease; that is, $f_p(p, q) < 0$. Then, if A and B are competitive products, the demand for B will increase; that is, $g_p(p, q) > 0$. Similarly, if the price of B increases while the price of A is held constant, the demand for B will decrease; that is, $g_q(p, q) < 0$. Then, if A and B are competitive products, the demand for A will increase; that is, $f_q(p, q) > 0$. Reasoning similarly for complementary products, we arrive at the following test:

Test for Competitive and Complementary Products

Partial Derivatives		Products A and B
$f_q(p, q) > 0$ and	$g_p(p, q) > 0$	Competitive (substitute)
$f_q(p, q) < 0$ and	$g_p(p, q) < 0$	Complementary
$f_q(p, q) \geq 0$ and	$g_p(p, q) \leq 0$	Neither
$f_q(p, q) \leq 0$ and	$g_p(p, q) \geq 0$	Neither

Use this test in Problems 89–92 to determine whether the indicated products are competitive, complementary, or neither.

89. Product demand. The weekly demand equations for the sale of butter and margarine in a supermarket are

$$x = f(p, q) = 8{,}000 - 0.09p^2 + 0.08q^2 \quad \text{Butter}$$

$$y = g(p, q) = 15{,}000 + 0.04p^2 - 0.3q^2 \quad \text{Margarine}$$

90. Product demand. The daily demand equations for the sale of brand A coffee and brand B coffee in a supermarket are

$$x = f(p, q) = 200 - 5p + 4q \quad \text{Brand A coffee}$$

$$y = g(p, q) = 300 + 2p - 4q \quad \text{Brand B coffee}$$

91. Product demand. The monthly demand equations for the sale of skis and ski boots in a sporting goods store are

$$x = f(p, q) = 800 - 0.004p^2 - 0.003q^2 \quad \text{Skis}$$

$$y = g(p, q) = 600 - 0.003p^2 - 0.002q^2 \quad \text{Ski boots}$$

92. Product demand. The monthly demand equations for the sale of tennis rackets and tennis balls in a sporting goods store are

$$x = f(p, q) = 500 - 0.5p - q^2 \quad \text{Tennis rackets}$$

$$y = g(p, q) = 10{,}000 - 8p - 100q^2 \quad \text{Tennis balls (cans)}$$

93. Medicine. The following empirical formula relates the surface area A (in square inches) of an average human body to its weight w (in pounds) and its height h (in inches):

$$A = f(w, h) = 15.64w^{0.425}h^{0.725}$$

(A) Find $f_w(w, h)$ and $f_h(w, h)$.

(B) For a 65-pound child who is 57 inches tall, find $f_w(65, 57)$ and $f_h(65, 57)$, and interpret the results.

94. Blood flow. Poiseuille's law states that the resistance R for blood flowing in a blood vessel varies directly as the length L of the vessel and inversely as the fourth power of its radius r. Stated as an equation,

$$R(L, r) = k\frac{L}{r^4} \qquad k \text{ a constant}$$

Find $R_L(4, 0.2)$ and $R_r(4, 0.2)$, and interpret the results.

95. Physical anthropology. Anthropologists use the cephalic index C, which varies directly as the width W of the head and inversely as the length L of the head (both viewed from the top). In terms of an equation,

$$C(W, L) = 100\frac{W}{L}$$

where

W = width in inches L = length in inches

Find $C_W(6, 8)$ and $C_L(6, 8)$, and interpret the results.

Figure for 95

96. Safety research. Under ideal conditions, if a person driving a car slams on the brakes and skids to a stop, the length of the skid marks (in feet) is given by the formula

$$L(w, v) = kwv^2$$

where

k = constant
w = weight of car in pounds
v = speed of car in miles per hour

For $k = 0.000\,013\,3$, find $L_w(2{,}500, 60)$ and $L_v(2{,}500, 60)$, and interpret the results.

Answers to Matched Problems

1. (A) $\partial z/\partial y = -3x^2 + 5$
   (B) $f_y(2, 3) = -7$
2. (A) $10x(x^2 + 2xy)^4$
   (B) 10
3. $P_y(25, 10) = 10$: At a production level of $x = 25$ and $y = 10$, increasing y by 1 unit and holding x fixed at 25 will increase profit by approximately \$10; $P_y(25, 15) = -110$: At a production level of $x = 25$ and $y = 15$, increasing y by 1 unit and holding x fixed at 25 will decrease profit by approximately \$110
4. (A) $f_x(x, y) = 12x^{-0.7}y^{0.7}$; $f_y(x, y) = 28x^{0.3}y^{-0.3}$
   (B) Marginal productivity of labor ≈ 25.89; marginal productivity of capital ≈ 20.14
   (C) Labor
5. (A) $3x^2$ (B) $-24y^2$
   (C) 12 (D) 12

6-3 Maxima and Minima

We are now ready to undertake a brief, but useful, analysis of local maxima and minima for functions of the type $z = f(x, y)$. We will extend the second-derivative test developed for functions of a single independent variable. We assume that all second-order partial derivatives exist for the function f in some circular region in the xy plane. This guarantees that the surface $z = f(x, y)$ has no sharp points, breaks, or ruptures. In other words, we are dealing only with smooth surfaces with no edges (like the edge of a box), breaks (like an earthquake fault), or sharp points (like the bottom point of a golf tee). (See Figure 1.)

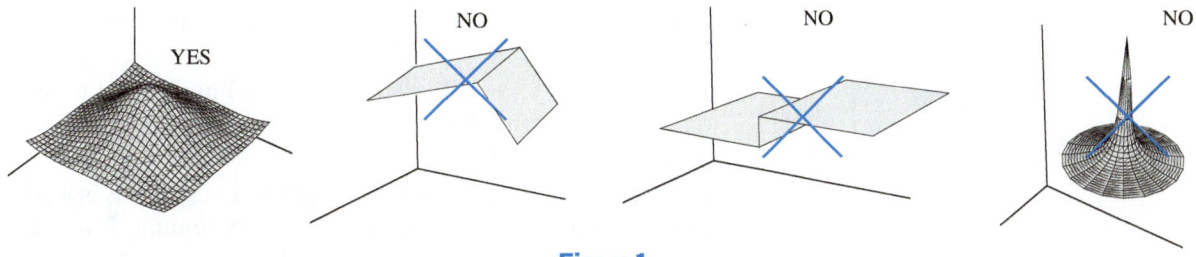

Figure 1

In addition, we will not concern ourselves with boundary points or absolute maxima–minima theory. Despite these restrictions, the procedure we will describe will help us solve a large number of useful problems.

What does it mean for $f(a, b)$ to be a local maximum or a local minimum? We say that $f(a, b)$ **is a local maximum** if there exists a circular region in the domain of f with (a, b) as the center, such that

$$f(a, b) \geq f(x, y)$$

for all (x, y) in the region. Similarly, we say that $f(a, b)$ **is a local minimum** if there exists a circular region in the domain of f with (a, b) as the center, such that

$$f(a, b) \leq f(x, y)$$

for all (x, y) in the region. Figure 2A illustrates a local maximum, Figure 2B a local minimum, and Figure 2C a **saddle point**, which is neither a local maximum nor a local minimum.

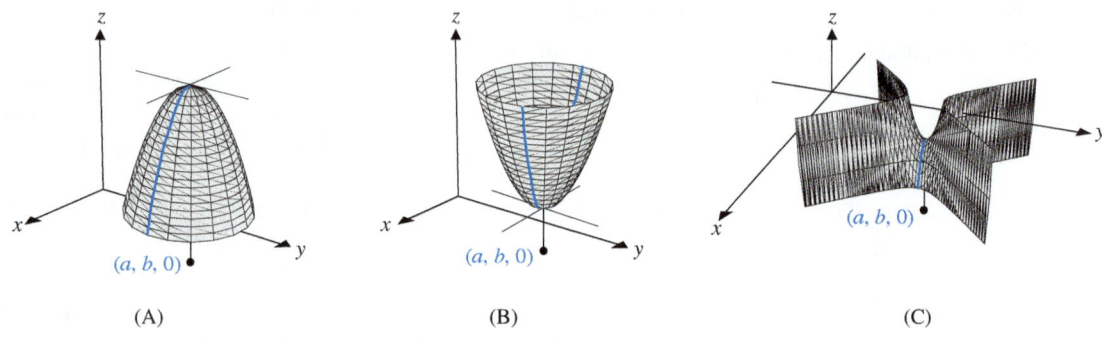

(A)   (B)   (C)

Figure 2

What happens to $f_x(a, b)$ and $f_y(a, b)$ if $f(a, b)$ is a local minimum or a local maximum and the partial derivatives of f exist in a circular region containing (a, b)? Figure 2 suggests that $f_x(a, b) = 0$ and $f_y(a, b) = 0$, since the tangent lines to the given curves are horizontal. Theorem 1 indicates that our intuitive reasoning is correct.

THEOREM 1   Local Extrema and Partial Derivatives

Let $f(a, b)$ be a local extremum (a local maximum or a local minimum) for the function f. If both f_x and f_y exist at (a, b), then

$$f_x(a, b) = 0 \qquad \text{and} \qquad f_y(a, b) = 0 \tag{1}$$

The converse of this theorem is false. If $f_x(a, b) = 0$ and $f_y(a, b) = 0$, then $f(a, b)$ may or may not be a local extremum; for example, the point $(a, b, f(a, b))$ may be a saddle point (see Fig. 2C).

Theorem 1 gives us *necessary* (but not *sufficient*) conditions for $f(a, b)$ to be a local extremum. We find all points (a, b) such that $f_x(a, b) = 0$ and $f_y(a, b) = 0$ and test these further to determine whether $f(a, b)$ is a local extremum or a saddle point. Points (a, b) such that conditions (1) hold are called **critical points**.

EXPLORE & DISCUSS 1

(A) Let $f(x, y) = y^2 + 1$. Explain why $f(x, y)$ has a local minimum at every point on the x axis. Verify that every point on the x axis is a critical point. Explain why the graph of $z = f(x, y)$ could be described as a trough.

(B) Let $g(x, y) = x^3$. Show that every point on the y axis is a critical point. Explain why no point on the y axis is a local extremum. Explain why the graph of $z = g(x, y)$ could be described as a slide.

The next theorem, using second-derivative tests, gives us *sufficient* conditions for a critical point to produce a local extremum or a saddle point.

THEOREM 2   Second-Derivative Test for Local Extrema
If

1. $z = f(x, y)$

2. $f_x(a, b) = 0$ and $f_y(a, b) = 0$ [(a, b) is a critical point]

3. All second-order partial derivatives of f exist in some circular region containing (a, b) as center.

4. $A = f_{xx}(a, b), \quad B = f_{xy}(a, b), \quad C = f_{yy}(a, b)$

Then

Case 1.   If $AC - B^2 > 0$ and $A < 0$, then $f(a, b)$ is a local maximum.

Case 2.   If $AC - B^2 > 0$ and $A > 0$, then $f(a, b)$ is a local minimum.

Case 3.   If $AC - B^2 < 0$, then f has a saddle point at (a, b).

Case 4.   If $AC - B^2 = 0$, the test fails.

CONCEPTUAL INSIGHT

The condition $A = f_{xx}(a, b) < 0$ in case 1 of Theorem 2 is analogous to the condition $f''(c) < 0$ in the second-derivative test for a function of one variable (Section 5-5), which implies that the function is concave downward and therefore has a local maximum. Similarly, the condition $A = f_{xx}(a, b) > 0$ in case 2 is analogous to the condition $f''(c) > 0$ in the earlier second-derivative test, which implies that the function is concave upward and therefore has a local minimum.

To illustrate the use of Theorem 2, we find the local extremum for a very simple function whose solution is almost obvious: $z = f(x, y) = x^2 + y^2 + 2$. From the function f itself and its graph (Fig. 3), it is clear that a local minimum is found at $(0, 0)$. Let us see how Theorem 2 confirms this observation.

Step 1   Find critical points: Find (x, y) such that $f_x(x, y) = 0$ and $f_y(x, y) = 0$ simultaneously:

$$f_x(x, y) = 2x = 0 \qquad f_y(x, y) = 2y = 0$$
$$x = 0 \qquad\qquad y = 0$$

The only critical point is $(a, b) = (0, 0)$.

Step 2   Compute $A = f_{xx}(0, 0)$, $B = f_{xy}(0, 0)$, and $C = f_{yy}(0, 0)$:

$$f_{xx}(x, y) = 2; \qquad \text{so,} \qquad A = f_{xx}(0, 0) = 2$$

$$f_{xy}(x, y) = 0; \qquad \text{so,} \qquad B = f_{xy}(0, 0) = 0$$

$$f_{yy}(x, y) = 2; \qquad \text{so,} \qquad C = f_{yy}(0, 0) = 2$$

Step 3   Evaluate $AC - B^2$ and try to classify the critical point $(0, 0)$ by using Theorem 2:

$$AC - B^2 = (2)(2) - (0)^2 = 4 > 0 \qquad \text{and} \qquad A = 2 > 0$$

Therefore, case 2 in Theorem 2 holds. That is, $f(0, 0) = 2$ is a local minimum.

We will now use Theorem 2 to analyze extrema without the aid of graphs.

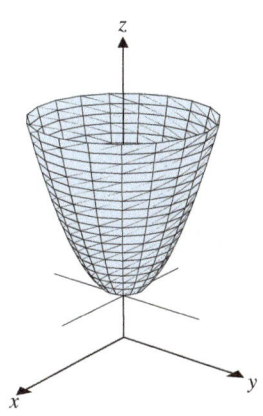

Figure 3

EXAMPLE 1 **Finding Local Extrema** Use Theorem 2 to find local extrema of

$$f(x, y) = -x^2 - y^2 + 6x + 8y - 21$$

SOLUTION **Step 1** Find critical points: Find (x, y) such that $f_x(x, y) = 0$ and $f_y(x, y) = 0$ simultaneously:

$$f_x(x, y) = -2x + 6 = 0 \qquad f_y(x, y) = -2y + 8 = 0$$
$$x = 3 \qquad\qquad\qquad y = 4$$

The only critical point is $(a, b) = (3, 4)$.

Step 2 Compute $A = f_{xx}(3, 4)$, $B = f_{xy}(3, 4)$, and $C = f_{yy}(3, 4)$:

$$f_{xx}(x, y) = -2; \qquad \text{so,} \qquad A = f_{xx}(3, 4) = -2$$
$$f_{xy}(x, y) = 0; \qquad \text{so,} \qquad B = f_{xy}(3, 4) = 0$$
$$f_{yy}(x, y) = -2; \qquad \text{so,} \qquad C = f_{yy}(3, 4) = -2$$

Step 3 Evaluate $AC - B^2$ and try to classify the critical point $(3, 4)$ by using Theorem 2:

$$AC - B^2 = (-2)(-2) - (0)^2 = 4 > 0 \qquad \text{and} \qquad A = -2 < 0$$

Therefore, case 1 in Theorem 2 holds, and $f(3, 4) = 4$ is a local maximum.

Matched Problem 1 Use Theorem 2 to find local extrema of

$$f(x, y) = x^2 + y^2 - 10x - 2y + 36$$

EXAMPLE 2 **Finding Local Extrema: Multiple Critical Points** Use Theorem 2 to find local extrema of

$$f(x, y) = x^3 + y^3 - 6xy$$

SOLUTION **Step 1** Find critical points of $f(x, y) = x^3 + y^3 - 6xy$:

$$f_x(x, y) = 3x^2 - 6y = 0 \qquad\qquad \text{Solve for y.}$$
$$6y = 3x^2$$
$$y = \tfrac{1}{2}x^2 \qquad\qquad\qquad\qquad\qquad\qquad (2)$$
$$f_y(x, y) = 3y^2 - 6x = 0$$
$$3y^2 = 6x \qquad\qquad \text{Use equation (2) to eliminate y.}$$
$$3(\tfrac{1}{2}x^2)^2 = 6x$$
$$\tfrac{3}{4}x^4 = 6x \qquad\qquad \text{Solve for x.}$$
$$3x^4 - 24x = 0$$
$$3x(x^3 - 8) = 0$$
$$x = 0 \qquad \text{or} \qquad x = 2$$
$$y = 0 \qquad\qquad y = \tfrac{1}{2}(2)^2 = 2$$

The critical points are $(0, 0)$ and $(2, 2)$.

Since there are two critical points, steps 2 and 3 must be performed twice.

Test (0, 0) **Step 2** Compute $A = f_{xx}(0, 0)$, $B = f_{xy}(0, 0)$, and $C = f_{yy}(0, 0)$:

$$f_{xx}(x, y) = 6x; \qquad \text{so,} \qquad A = f_{xx}(0, 0) = 0$$

$$f_{xy}(x, y) = -6; \quad \text{so,} \quad B = f_{xy}(0, 0) = -6$$

$$f_{yy}(x, y) = 6y; \quad \text{so,} \quad C = f_{yy}(0, 0) = 0$$

Step 3 Evaluate $AC - B^2$ and try to classify the critical point $(0, 0)$ by using Theorem 2:

$$AC - B^2 = (0)(0) - (-6)^2 = -36 < 0$$

Therefore, case 3 in Theorem 2 applies. That is, f has a saddle point at $(0, 0)$. Now we will consider the second critical point, $(2, 2)$:

Test (2, 2) **Step 2** Compute $A = f_{xx}(2, 2)$, $B = f_{xy}(2, 2)$, and $C = f_{yy}(2, 2)$:

$$f_{xx}(x, y) = 6x; \quad \text{so,} \quad A = f_{xx}(2, 2) = 12$$

$$f_{xy}(x, y) = -6; \quad \text{so,} \quad B = f_{xy}(2, 2) = -6$$

$$f_{yy}(x, y) = 6y; \quad \text{so,} \quad C = f_{yy}(2, 2) = 12$$

Step 3 Evaluate $AC - B^2$ and try to classify the critical point $(2, 2)$ by using Theorem 2:

$$AC - B^2 = (12)(12) - (-6)^2 = 108 > 0 \quad \text{and} \quad A = 12 > 0$$

So, case 2 in Theorem 2 applies, and $f(2, 2) = -8$ is a local minimum.

Our conclusions in Example 2 may be confirmed geometrically by graphing cross sections of the function f. The cross sections of f in the planes $y = 0$, $x = 0$, $y = x$, and $y = -x$ [each of these planes contains $(0, 0)$] are represented by the graphs of the functions $f(x, 0) = x^3, f(0, y) = y^3, f(x, x) = 2x^3 - 6x^2$, and $f(x, -x) = 6x^2$, respectively, as shown in Figure 4A (note that the first two functions have the same graph). The cross sections of f in the planes $y = 2$, $x = 2$, $y = x$, and $y = 4 - x$ [each of these planes contains $(2, 2)$] are represented by the graphs of $f(x, 2) = x^3 - 12x + 8, f(2, y) = y^3 - 12y + 8, f(x, x) = 2x^3 - 6x^2$, and $f(x, 4 - x) = x^3 + (4 - x)^3 + 6x^2 - 24x$, respectively, as shown in Figure 4B (the first two functions have the same graph). Figure 4B illustrates the fact that since f has a local minimum at $(2, 2)$, each of the cross sections of f through $(2, 2)$ has a local minimum of -8 at $(2, 2)$. Figure 4A, by contrast, indicates that some cross sections of f through $(0, 0)$ have a local minimum, some a local maximum, and some neither one, at $(0, 0)$.

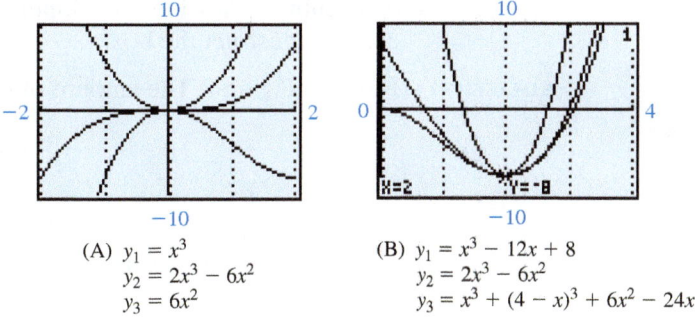

(A) $y_1 = x^3$
$y_2 = 2x^3 - 6x^2$
$y_3 = 6x^2$

(B) $y_1 = x^3 - 12x + 8$
$y_2 = 2x^3 - 6x^2$
$y_3 = x^3 + (4 - x)^3 + 6x^2 - 24x$

Figure 4

Matched Problem 2 Use Theorem 2 to find local extrema for $f(x, y) = x^3 + y^2 - 6xy$.

EXAMPLE 3   **Profit** Suppose that the surfboard company discussed earlier has developed the yearly profit equation

$$P(x, y) = -22x^2 + 22xy - 11y^2 + 110x - 44y - 23$$

where x is the number (in thousands) of standard surfboards produced per year, y is the number (in thousands) of competition surfboards produced per year, and P is profit (in thousands of dollars). How many of each type of board should be produced per year to realize a maximum profit? What is the maximum profit?

SOLUTION   **Step 1** Find critical points:

$$P_x(x, y) = -44x + 22y + 110 = 0$$

$$P_y(x, y) = 22x - 22y - 44 = 0$$

Solving this system, we obtain $(3, 1)$ as the only critical point.

Step 2 Compute $A = P_{xx}(3, 1)$, $B = P_{xy}(3, 1)$, and $C = P_{yy}(3, 1)$:

$$P_{xx}(x, y) = -44; \qquad \text{so,} \qquad A = P_{xx}(3, 1) = -44$$

$$P_{xy}(x, y) = 22; \qquad \text{so,} \qquad B = P_{xy}(3, 1) = 22$$

$$P_{yy}(x, y) = -22; \qquad \text{so,} \qquad C = P_{yy}(3, 1) = -22$$

Step 3 Evaluate $AC - B^2$ and try to classify the critical point $(3, 1)$ by using Theorem 2:

$$AC - B^2 = (-44)(-22) - 22^2 = 484 > 0 \qquad \text{and} \qquad A = -44 < 0$$

Therefore, case 1 in Theorem 2 applies. That is, $P(3, 1) = 120$ is a local maximum. A maximum profit of $120,000 is obtained by producing and selling 3,000 standard boards and 1,000 competition boards per year.

Matched Problem 3   Repeat Example 3 with

$$P(x, y) = -66x^2 + 132xy - 99y^2 + 132x - 66y - 19$$

EXAMPLE 4   **Package Design** The packaging department in a company is to design a rectangular box with no top and a partition down the middle. The box must have a volume of 48 cubic inches. Find the dimensions that will minimize the amount of material used to construct the box.

SOLUTION   Refer to Figure 5. The amount of material used in constructing this box is

$$\begin{array}{ccc} & \text{Front,} & \text{Sides,} \\ \text{Base} & \text{back} & \text{partition} \end{array}$$

$$M = xy + 2xz + 3yz \tag{3}$$

The volume of the box is

$$V = xyz = 48 \tag{4}$$

Since Theorem 2 applies only to functions with two independent variables, we must use equation (4) to eliminate one of the variables in equation (3):

$$M = xy + 2xz + 3yz \qquad \text{Substitute } z = 48/xy.$$

$$= xy + 2x\left(\frac{48}{xy}\right) + 3y\left(\frac{48}{xy}\right)$$

$$= xy + \frac{96}{y} + \frac{144}{x}$$

Figure 5

So, we must find the minimum value of

$$M(x, y) = xy + \frac{96}{y} + \frac{144}{x} \qquad x > 0 \qquad \text{and} \qquad y > 0$$

Step 1 Find critical points:

$$M_x(x, y) = y - \frac{144}{x^2} = 0$$

$$y = \frac{144}{x^2} \qquad\qquad (5)$$

$$M_y(x, y) = x - \frac{96}{y^2} = 0$$

$$x = \frac{96}{y^2} \qquad\qquad \textit{Solve for } y^2.$$

$$y^2 = \frac{96}{x} \qquad\qquad \textit{Use equation (5) to eliminate } y \textit{ and solve for } x.$$

$$\left(\frac{144}{x^2}\right)^2 = \frac{96}{x}$$

$$\frac{20{,}736}{x^4} = \frac{96}{x} \qquad\qquad \textit{Multiply both sides by } x^4/96 \\ \textit{(recall that } x > 0).$$

$$x^3 = \frac{20{,}736}{96} = 216$$

$$x = 6 \qquad\qquad \textit{Use equation (5) to find } y.$$

$$y = \frac{144}{36} = 4$$

Therefore, $(6, 4)$ is the only critical point.

Step 2 Compute $A = M_{xx}(6, 4)$, $B = M_{xy}(6, 4)$, and $C = M_{yy}(6, 4)$:

$$M_{xx}(x, y) = \frac{288}{x^3}; \qquad \text{so,} \qquad A = M_{xx}(6, 4) = \tfrac{288}{216} = \tfrac{4}{3}$$

$$M_{xy}(x, y) = 1; \qquad \text{so,} \qquad B = M_{xy}(6, 4) = 1$$

$$M_{yy}(x, y) = \frac{192}{y^3}; \qquad \text{so,} \qquad C = M_{yy}(6, 4) = \tfrac{192}{64} = 3$$

Step 3 Evaluate $AC - B^2$ and try to classify the critical point $(6, 4)$ by using Theorem 2:

$$AC - B^2 = \left(\tfrac{4}{3}\right)(3) - (1)^2 = 3 > 0 \qquad \text{and} \qquad A = \tfrac{4}{3} > 0$$

Case 2 in Theorem 2 applies, and $M(x, y)$ has a local minimum at $(6, 4)$. If $x = 6$ and $y = 4$, then

$$z = \frac{48}{xy} = \frac{48}{(6)(4)} = 2$$

The dimensions that will require the minimum amount of material are 6 inches by 4 inches by 2 inches (see Fig. 6).

2 inches

4 inches

6 inches

Figure 6

Matched Problem 4 If the box in Example 4 must have a volume of 384 cubic inches, find the dimensions that will require the least amount of material.

Exercises 6-3

A

🖎 *In Problems 1–4, find $f_x(x, y)$ and $f_y(x, y)$, and explain, using Theorem 1, why $f(x, y)$ has no local extrema.*

1. $f(x, y) = 4x + 5y - 6$

2. $f(x, y) = 10 - 2x - 3y + x^2$

3. $f(x, y) = 3.7 - 1.2x + 6.8y + 0.2y^3 + x^4$

4. $f(x, y) = x^3 - y^2 + 7x + 3y + 1$

Use Theorem 2 to find local extrema in Problems 5–24.

5. $f(x, y) = 6 - x^2 - 4x - y^2$

6. $f(x, y) = 3 - x^2 - y^2 + 6y$

7. $f(x, y) = x^2 + y^2 + 2x - 6y + 14$

8. $f(x, y) = x^2 + y^2 - 4x + 6y + 23$

B

9. $f(x, y) = xy + 2x - 3y - 2$

10. $f(x, y) = x^2 - y^2 + 2x + 6y - 4$

11. $f(x, y) = -3x^2 + 2xy - 2y^2 + 14x + 2y + 10$

12. $f(x, y) = -x^2 + xy - 2y^2 + x + 10y - 5$

13. $f(x, y) = 2x^2 - 2xy + 3y^2 - 4x - 8y + 20$

14. $f(x, y) = 2x^2 - xy + y^2 - x - 5y + 8$

C

15. $f(x, y) = e^{xy}$

16. $f(x, y) = x^2y - xy^2$

17. $f(x, y) = x^3 + y^3 - 3xy$

18. $f(x, y) = 2y^3 - 6xy - x^2$

19. $f(x, y) = 2x^4 + y^2 - 12xy$

20. $f(x, y) = 16xy - x^4 - 2y^2$

21. $f(x, y) = x^3 - 3xy^2 + 6y^2$

22. $f(x, y) = 2x^2 - 2x^2y + 6y^3$

 23. $f(x, y) = y^3 + 2x^2y^2 - 3x - 2y + 8$

 24. $f(x, y) = x \ln y + x^2 - 4x - 5y + 3$

🖎 25. Explain why $f(x, y) = x^2$ has an infinite number of local extrema.

🖎 26. (A) Find the local extrema of the functions $f(x, y) = x + y$, $g(x, y) = x^2 + y^2$, and $h(x, y) = x^3 + y^3$.

   (B) Discuss the local extrema of the function $k(x, y) = x^n + y^n$, where n is a positive integer.

27. (A) Show that $(0, 0)$ is a critical point of the function $f(x, y) = x^4e^y + x^2y^4 + 1$, but that the second-derivative test for local extrema fails.

   (B) Use cross sections, as in Example 2, to decide whether f has a local maximum, a local minimum, or a saddle point at $(0, 0)$.

28. (A) Show that $(0, 0)$ is a critical point of the function $g(x, y) = e^{xy^2} + x^2y^3 + 2$, but that the second-derivative test for local extrema fails.

   (B) Use cross sections, as in Example 2, to decide whether g has a local maximum, a local minimum, or a saddle point at $(0, 0)$.

Applications

29. **Product mix for maximum profit.** A firm produces two types of earphones per year: x thousand of type A and y thousand of type B. If the revenue and cost equations for the year are (in millions of dollars)

$$R(x, y) = 2x + 3y$$

$$C(x, y) = x^2 - 2xy + 2y^2 + 6x - 9y + 5$$

determine how many of each type of earphone should be produced per year to maximize profit. What is the maximum profit?

30. **Automation–labor mix for minimum cost.** The annual labor and automated equipment cost (in millions of dollars) for a company's production of HDTV's is given by

$$C(x, y) = 2x^2 + 2xy + 3y^2 - 16x - 18y + 54$$

where x is the amount spent per year on labor and y is the amount spent per year on automated equipment (both in millions of dollars). Determine how much should be spent on each per year to minimize this cost. What is the minimum cost?

31. **Maximizing profit.** A store sells two brands of camping chairs. The store pays $60 for each brand A chair and $80 for each brand B chair. The research department has estimated the weekly demand equations for these two competitive products to be

$x = 260 - 3p + q$ *Demand equation for brand A*

$y = 180 + p - 2q$ *Demand equation for brand B*

where p is the selling price for brand A and q is the selling price for brand B.

   (A) Determine the demands x and y when $p = \$100$ and $q = \$120$; when $p = \$110$ and $q = \$110$.

   (B) How should the store price each chair to maximize weekly profits? What is the maximum weekly profit? [*Hint:* $C = 60x + 80y$, $R = px + qy$, and $P = R - C$.]

32. **Maximizing profit.** A store sells two brands of laptop sleeves. The store pays $25 for each brand A sleeve and $30 for each brand B sleeve. A consulting firm has estimated the daily demand equations for these two competitive products to be

$x = 130 - 4p + q$ *Demand equation for brand A*

$y = 115 + 2p - 3q$ *Demand equation for brand B*

where p is the selling price for brand A and q is the selling price for brand B.

(A) Determine the demands x and y when $p = \$40$ and $q = \$50$; when $p = \$45$ and $q = \$55$.

(B) How should the store price each brand of sleeve to maximize daily profits? What is the maximum daily profit? [*Hint:* $C = 25x + 30y$, $R = px + qy$, and $P = R - C$.]

33. Minimizing cost. A satellite TV station is to be located at $P(x, y)$ so that the sum of the squares of the distances from P to the three towns A, B, and C is a minimum (see the figure). Find the coordinates of P, the location that will minimize the cost of providing satellite TV for all three towns.

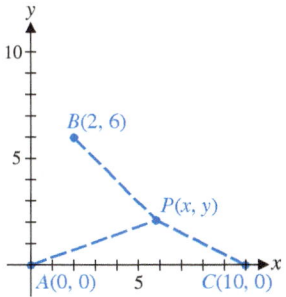

Figure for 33

34. Minimizing cost. Repeat Problem 33, replacing the coordinates of B with $B(6, 9)$ and the coordinates of C with $C(9, 0)$.

35. Minimum material. A rectangular box with no top and two parallel partitions (see the figure) must hold a volume of 64 cubic inches. Find the dimensions that will require the least amount of material.

Figure for 35

36. Minimum material. A rectangular box with no top and two intersecting partitions (see the figure) must hold a volume of 72 cubic inches. Find the dimensions that will require the least amount of material.

Figure for 36

37. Maximum volume. A mailing service states that a rectangular package cannot have the sum of its length and girth exceed 120 inches (see the figure). What are the dimensions of the largest (in volume) mailing carton that can be constructed to meet these restrictions?

Figure for 37

38. Maximum shipping volume. A shipping box is to be reinforced with steel bands in all three directions, as shown in the figure. A total of 150 inches of steel tape is to be used, with 6 inches of waste because of a 2-inch overlap in each direction. Find the dimensions of the box with maximum volume that can be taped as described.

Figure for 38

Answers to Matched Problems

1. $f(5, 1) = 10$ is a local minimum

2. f has a saddle point at $(0, 0)$; $f(6, 18) = -108$ is a local minimum

3. Local maximum for $x = 2$ and $y = 1$; $P(2, 1) = 80$; a maximum profit of \$80,000 is obtained by producing and selling 2,000 standard boards and 1,000 competition boards

4. 12 in. by 8 in. by 4 in.

6-4   Maxima and Minima Using Lagrange Multipliers

- Functions of Two Independent Variables
- Functions of Three Independent Variables

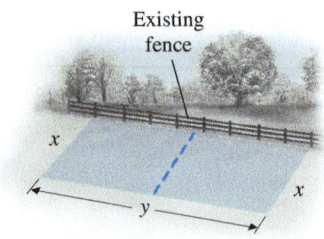

Figure 1

Functions of Two Independent Variables

We now consider a powerful method of solving a certain class of maxima–minima problems. Joseph Louis Lagrange (1736–1813), an eighteenth-century French mathematician, discovered this method, called the **method of Lagrange multipliers**. We introduce the method through an example.

A rancher wants to construct two feeding pens of the same size along an existing fence (see Fig. 1). If the rancher has 720 feet of fencing materials available, how long should x and y be in order to obtain the maximum total area? What is the maximum area?

The total area is given by

$$f(x, y) = xy$$

which can be made as large as we like, provided that there are no restrictions on x and y. But there are restrictions on x and y, since we have only 720 feet of fencing. The variables x and y must be chosen so that

$$3x + y = 720$$

This restriction on x and y, called a **constraint**, leads to the following maxima–minima problem:

$$\text{Maximize} \quad f(x, y) = xy \tag{1}$$

$$\text{subject to} \quad 3x + y = 720, \quad \text{or} \quad 3x + y - 720 = 0 \tag{2}$$

This problem is one of a general class of problems of the form

$$\text{Maximize (or minimize)} \quad z = f(x, y) \tag{3}$$

$$\text{subject to} \quad g(x, y) = 0 \tag{4}$$

Of course, we could try to solve equation (4) for y in terms of x, or for x in terms of y, then substitute the result into equation (3), and use methods developed in Section 5-5 for functions of a single variable. But what if equation (4) is more complicated than equation (2), and solving for one variable in terms of the other is either very difficult or impossible? In the method of Lagrange multipliers, we will work with $g(x, y)$ directly and avoid solving equation (4) for one variable in terms of the other. In addition, the method generalizes to functions of arbitrarily many variables subject to one or more constraints.

Now to the method: We form a new function F, using functions f and g in equations (3) and (4), as follows:

$$F(x, y, \lambda) = f(x, y) + \lambda g(x, y) \tag{5}$$

Here, λ (the Greek lowercase letter lambda) is called a **Lagrange multiplier**. Theorem 1 gives the basis for the method.

THEOREM 1   Method of Lagrange Multipliers for Functions of Two Variables
Any local maxima or minima of the function $z = f(x, y)$ subject to the constraint $g(x, y) = 0$ will be among those points (x_0, y_0) for which (x_0, y_0, λ_0) is a solution of the system

$$F_x(x, y, \lambda) = 0$$

$$F_y(x, y, \lambda) = 0$$

$$F_\lambda(x, y, \lambda) = 0$$

where $F(x, y, \lambda) = f(x, y) + \lambda g(x, y)$, provided that all the partial derivatives exist.

We now use the method of Lagrange multipliers to solve the fence problem.

Step 1 Formulate the problem in the form of equations (3) and (4):

$$\text{Maximize} \quad f(x, y) = xy$$

$$\text{subject to} \quad g(x, y) = 3x + y - 720 = 0$$

Step 2 Form the function F, introducing the Lagrange multiplier λ:

$$F(x, y, \lambda) = f(x, y) + \lambda g(x, y)$$

$$= xy + \lambda(3x + y - 720)$$

Step 3 Solve the system $F_x = 0$, $F_y = 0$, $F_\lambda = 0$ (the solutions are **critical points** of F):

$$F_x = y + 3\lambda = 0$$

$$F_y = x + \lambda = 0$$

$$F_\lambda = 3x + y - 720 = 0$$

From the first two equations, we see that

$$y = -3\lambda$$

$$x = -\lambda$$

Substitute these values for x and y into the third equation and solve for λ:

$$-3\lambda - 3\lambda = 720$$

$$-6\lambda = 720$$

$$\lambda = -120$$

So,

$$y = -3(-120) = 360 \text{ feet}$$

$$x = -(-120) = 120 \text{ feet}$$

and $(x_0, y_0, \lambda_0) = (120, 360, -120)$ is the only critical point of F.

Step 4 According to Theorem 1, if the function $f(x, y)$, subject to the constraint $g(x, y) = 0$, has a local maximum or minimum, that maximum or minimum must occur at $x = 120$, $y = 360$. Although it is possible to develop a test similar to Theorem 2 in Section 6-3 to determine the nature of this local extremum, we will not do so. [Note that Theorem 2 cannot be applied to $f(x, y)$ at $(120, 360)$, since this point is not a critical point of the unconstrained function $f(x, y)$.] We simply assume that the maximum value of $f(x, y)$ must occur for $x = 120$, $y = 360$.

$$\text{Max } f(x, y) = f(120, 360)$$

$$= (120)(360) = 43{,}200 \text{ square feet}$$

The key steps in applying the method of Lagrange multipliers are as follows:

PROCEDURE   Method of Lagrange Multipliers: Key Steps

Step 1 Write the problem in the form

$$\text{Maximize (or minimize)} \quad z = f(x, y)$$

$$\text{subject to} \qquad\qquad g(x, y) = 0$$

Step 2 Form the function F:

$$F(x, y, \lambda) = f(x, y) + \lambda g(x, y)$$

Step 3 Find the critical points of F; that is, solve the system

$$F_x(x, y, \lambda) = 0$$
$$F_y(x, y, \lambda) = 0$$
$$F_\lambda(x, y, \lambda) = 0$$

Step 4 If (x_0, y_0, λ_0) is the only critical point of F, we assume that (x_0, y_0) will always produce the solution to the problems we consider. If F has more than one critical point, we evaluate $z = f(x, y)$ at (x_0, y_0) for each critical point (x_0, y_0, λ_0) of F. For the problems we consider, we assume that the largest of these values is the maximum value of $f(x, y)$, subject to the constraint $g(x, y) = 0$, and the smallest is the minimum value of $f(x, y)$, subject to the constraint $g(x, y) = 0$.

EXAMPLE 1   **Minimization Subject to a Constraint** Minimize $f(x, y) = x^2 + y^2$ subject to $x + y = 10$.

SOLUTION   **Step 1**   Minimize $f(x, y) = x^2 + y^2$

subject to $g(x, y) = x + y - 10 = 0$

Step 2 $F(x, y, \lambda) = x^2 + y^2 + \lambda(x + y - 10)$

Step 3 $F_x = 2x + \lambda = 0$

$F_y = 2y + \lambda = 0$

$F_\lambda = x + y - 10 = 0$

From the first two equations, $x = -\lambda/2$ and $y = -\lambda/2$. Substituting these values into the third equation, we obtain

$$-\frac{\lambda}{2} - \frac{\lambda}{2} = 10$$

$$-\lambda = 10$$

$$\lambda = -10$$

The only critical point is $(x_0, y_0, \lambda_0) = (5, 5, -10)$.

Step 4 Since $(5, 5, -10)$ is the only critical point of F, we conclude that (see step 4 in the box)

$$\text{Min } f(x, y) = f(5, 5) = (5)^2 + (5)^2 = 50$$

Since $g(x, y)$ in Example 1 has a relatively simple form, an alternative to the method of Lagrange multipliers is to solve $g(x, y) = 0$ for y and then substitute into $f(x, y)$ to obtain the function $h(x) = f(x, 10 - x) = x^2 + (10 - x)^2$ in the single variable x. Then we minimize h (see Fig. 2). From Figure 2, we conclude that Min $f(x, y) = f(5, 5) = 50$. This technique depends on being able to solve the constraint for one of the two variables and so is not always available as an alternative to the method of Lagrange multipliers.

Figure 2 $h(x) = x^2 + (10 - x)^2$

Matched Problem 1   Maximize $f(x, y) = 25 - x^2 - y^2$ subject to $x + y = 4$.

Figures 3 and 4 illustrate the results obtained in Example 1 and Matched Problem 1, respectively.

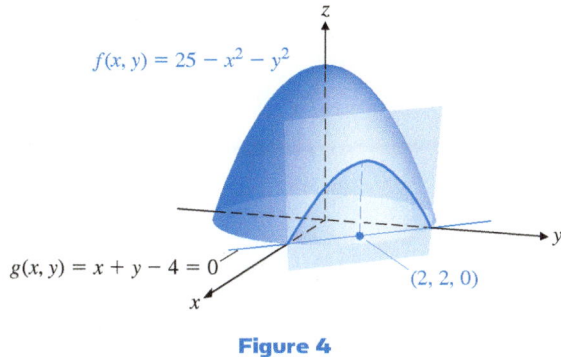

Figure 4

Figure 3

Consider the problem of minimizing $f(x, y) = 3x^2 + 5y^2$ subject to the constraint $g(x, y) = 2x + 3y - 6 = 0$.

(A) Compute the value of $f(x, y)$ when x and y are integers, $0 \le x \le 3$, $0 \le y \le 2$. Record your answers in the empty boxes next to the points (x, y) in Figure 5.

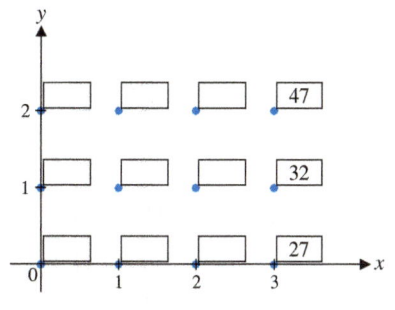

Figure 5

(B) Graph the constraint $g(x, y) = 0$.

(C) Estimate the minimum value of f on the basis of your graph and the computations from part (A).

(D) Use the method of Lagrange multipliers to solve the minimization problem.

EXAMPLE 2

Productivity The Cobb–Douglas production function for a new product is given by

$$N(x, y) = 16x^{0.25}y^{0.75}$$

where x is the number of units of labor and y is the number of units of capital required to produce $N(x, y)$ units of the product. Each unit of labor costs \$50 and each unit of capital costs \$100. If \$500,000 has been budgeted for the production of this product, how should that amount be allocated between labor and capital in order to maximize production? What is the maximum number of units that can be produced?

SOLUTION

The total cost of using x units of labor and y units of capital is $50x + 100y$. Thus, the constraint imposed by the \$500,000 budget is

$$50x + 100y = 500,000$$

Step 1   Maximize $N(x, y) = 16x^{0.25}y^{0.75}$

subject to $g(x, y) = 50x + 100y - 500,000 = 0$

Step 2 $$F(x, y, \lambda) = 16x^{0.25}y^{0.75} + \lambda(50x + 100y - 500{,}000)$$

Step 3 $$F_x = 4x^{-0.75}y^{0.75} + 50\lambda = 0$$

$$F_y = 12x^{0.25}y^{-0.25} + 100\lambda = 0$$

$$F_\lambda = 50x + 100y - 500{,}000 = 0$$

From the first two equations,

$$\lambda = -\tfrac{2}{25}x^{-0.75}y^{0.75} \qquad \text{and} \qquad \lambda = -\tfrac{3}{25}x^{0.25}y^{-0.25}$$

Therefore,

$$-\tfrac{2}{25}x^{-0.75}y^{0.75} = -\tfrac{3}{25}x^{0.25}y^{-0.25} \quad \textcolor{blue}{\text{Multiply both sides by } x^{0.75}\,y^{0.25}.}$$

$$-\tfrac{2}{25}y = -\tfrac{3}{25}x \qquad\qquad \textcolor{blue}{(\text{We can assume that } x \neq 0 \text{ and } y \neq 0.)}$$

$$y = \tfrac{3}{2}x$$

Now substitute for y in the third equation and solve for x:

$$50x + 100\left(\tfrac{3}{2}x\right) - 500{,}000 = 0$$

$$200x = 500{,}000$$

$$x = 2{,}500$$

So,

$$y = \tfrac{3}{2}(2{,}500) = 3{,}750$$

and

$$\lambda = -\tfrac{2}{25}(2{,}500)^{-0.75}(3{,}750)^{0.75} \approx -0.1084$$

The only critical point of F is $(2{,}500, 3{,}750, -0.1084)$.

Step 4   Since F has only one critical point, we conclude that maximum productivity occurs when 2,500 units of labor and 3,750 units of capital are used (see step 4 in the method of Lagrange multipliers).

$$\text{Max } N(x, y) = N(2{,}500, 3{,}750)$$

$$= 16(2{,}500)^{0.25}(3{,}750)^{0.75}$$

$$\approx 54{,}216 \text{ units}$$

The negative of the value of the Lagrange multiplier found in step 3 is called the **marginal productivity of money** and gives the approximate increase in production for each additional dollar spent on production. In Example 2, increasing the production budget from \$500,000 to \$600,000 would result in an approximate increase in production of

$$0.1084(100{,}000) = 10{,}840 \text{ units}$$

Note that simplifying the constraint equation

$$50x + 100y - 500{,}000 = 0$$

to

$$x + 2y - 10{,}000 = 0$$

before forming the function $F(x, y, \lambda)$ would make it difficult to interpret $-\lambda$ correctly. **In marginal productivity problems, the constraint equation should not be simplified.**

Matched Problem 2 The Cobb–Douglas production function for a new product is given by

$$N(x, y) = 20x^{0.5}y^{0.5}$$

where x is the number of units of labor and y is the number of units of capital required to produce $N(x, y)$ units of the product. Each unit of labor costs \$40 and each unit of capital costs \$120.

(A) If \$300,000 has been budgeted for the production of this product, how should that amount be allocated in order to maximize production? What is the maximum production?

(B) Find the marginal productivity of money in this case, and estimate the increase in production if an additional \$40,000 is budgeted for production.

EXPLORE & DISCUSS 2

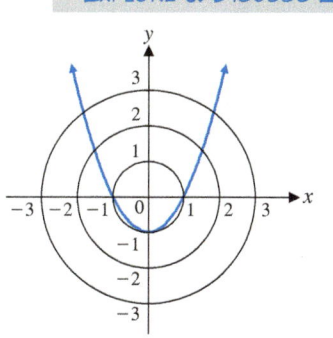

Figure 6

Consider the problem of maximizing $f(x, y) = 4 - x^2 - y^2$ subject to the constraint $g(x, y) = y - x^2 + 1 = 0$.

(A) Explain why $f(x, y) = 3$ whenever (x, y) is a point on the circle of radius 1 centered at the origin. What is the value of $f(x, y)$ when (x, y) is a point on the circle of radius 2 centered at the origin? On the circle of radius 3 centered at the origin? (See Fig. 6.)

(B) Explain why some points on the parabola $y - x^2 + 1 = 0$ lie inside the circle $x^2 + y^2 = 1$.

(C) In light of part (B), would you guess that the maximum value of $f(x, y)$ subject to the constraint is greater than 3? Explain.

(D) Use Lagrange multipliers to solve the maximization problem.

Functions of Three Independent Variables

The method of Lagrange multipliers can be extended to functions with arbitrarily many independent variables with one or more constraints. We now state a theorem for functions with three independent variables and one constraint, and we consider an example that demonstrates the advantage of the method of Lagrange multipliers over the method used in Section 6-3.

THEOREM 2   Method of Lagrange Multipliers for Functions of Three Variables

Any local maxima or minima of the function $w = f(x, y, z)$, subject to the constraint $g(x, y, z) = 0$, will be among the set of points (x_0, y_0, z_0) for which $(x_0, y_0, z_0, \lambda_0)$ is a solution of the system

$$F_x(x, y, z, \lambda) = 0$$
$$F_y(x, y, z, \lambda) = 0$$
$$F_z(x, y, z, \lambda) = 0$$
$$F_\lambda(x, y, z, \lambda) = 0$$

where $F(x, y, z, \lambda) = f(x, y, z) + \lambda g(x, y, z)$, provided that all the partial derivatives exist.

EXAMPLE 3 **Package Design** A rectangular box with an open top and one partition is to be constructed from 162 square inches of cardboard (Fig. 7). Find the dimensions that result in a box with the largest possible volume.

SOLUTION We must maximize

$$V(x, y, z) = xyz$$

subject to the constraint that the amount of material used is 162 square inches. So $x, y,$ and z must satisfy

$$xy + 2xz + 3yz = 162$$

Figure 7

Step 1   Maximize $V(x, y, z) = xyz$
subject to $g(x, y, z) = xy + 2xz + 3yz - 162 = 0$

Step 2 $F(x, y, z, \lambda) = xyz + \lambda(xy + 2xz + 3yz - 162)$

Step 3

$$F_x = yz + \lambda(y + 2z) = 0$$
$$F_y = xz + \lambda(x + 3z) = 0$$
$$F_z = xy + \lambda(2x + 3y) = 0$$
$$F_\lambda = xy + 2xz + 3yz - 162 = 0$$

From the first two equations, we can write

$$\lambda = \frac{-yz}{y + 2z} \qquad \lambda = \frac{-xz}{x + 3z}$$

Eliminating λ, we have

$$\frac{-yz}{y + 2z} = \frac{-xz}{x + 3z}$$

$$-xyz - 3yz^2 = -xyz - 2xz^2$$

$$3yz^2 = 2xz^2 \qquad \textit{We can assume that } z \neq 0.$$

$$3y = 2x$$

$$x = \tfrac{3}{2}y$$

From the second and third equations,

$$\lambda = \frac{-xz}{x + 3z} \qquad \lambda = \frac{-xy}{2x + 3y}$$

Eliminating λ, we have

$$\frac{-xz}{x + 3z} = \frac{-xy}{2x + 3y}$$

$$-2x^2z - 3xyz = -x^2y - 3xyz$$

$$2x^2z = x^2y \qquad \textit{We can assume that } x \neq 0.$$

$$2z = y$$

$$z = \tfrac{1}{2}y$$

Substituting $x = \tfrac{3}{2}y$ and $z = \tfrac{1}{2}y$ into the fourth equation, we have

$$\left(\tfrac{3}{2}y\right)y + 2\left(\tfrac{3}{2}y\right)\left(\tfrac{1}{2}y\right) + 3y\left(\tfrac{1}{2}y\right) - 162 = 0$$

$$\tfrac{3}{2}y^2 + \tfrac{3}{2}y^2 + \tfrac{3}{2}y^2 = 162$$

$$y^2 = 36 \qquad \textit{We can assume that } y > 0.$$

$$y = 6$$

$$x = \tfrac{3}{2}(6) = 9 \qquad \textit{Using } x = \tfrac{3}{2}y$$

$$z = \tfrac{1}{2}(6) = 3 \qquad \textit{Using } z = \tfrac{1}{2}y$$

and finally,

Figure 8

$$\lambda = \frac{-(6)(3)}{6 + 2(3)} = -\frac{3}{2} \quad \text{Using } \lambda = \frac{-yz}{y + 2z}$$

The only critical point of F with x, y, and z all positive is $(9, 6, 3, -\frac{3}{2})$.

Step 4 The box with the maximum volume has dimensions 9 inches by 6 inches by 3 inches (see Fig. 8).

Matched Problem 3 A box of the same type as in Example 3 is to be constructed from 288 square inches of cardboard. Find the dimensions that result in a box with the largest possible volume.

CONCEPTUAL INSIGHT

An alternative to the method of Lagrange multipliers would be to solve Example 3 by means of Theorem 2 (the second-derivative test for local extrema) in Section 6-3. That approach involves solving the material constraint for one of the variables, say, z:

$$z = \frac{162 - xy}{2x + 3y}$$

Then we would eliminate z in the volume function to obtain a function of two variables:

$$V(x, y) = xy\frac{162 - xy}{2x + 3y}$$

The method of Lagrange multipliers allows us to avoid the formidable tasks of calculating the partial derivatives of V and finding the critical points of V in order to apply Theorem 2.

Exercises 6-4

A

Use the method of Lagrange multipliers in Problems 1–4.

1. Maximize $f(x, y) = 2xy$

   subject to $x + y = 6$

2. Minimize $f(x, y) = 6xy$

   subject to $y - x = 6$

3. Minimize $f(x, y) = x^2 + y^2$

   subject to $3x + 4y = 25$

4. Maximize $f(x, y) = 25 - x^2 - y^2$

   subject to $2x + y = 10$

B

In Problems 5 and 6, use Theorem 1 to explain why no maxima or minima exist.

5. Minimize $f(x, y) = 4y - 3x$

   subject to $2x + 5y = 3$

6. Maximize $f(x, y) = 6x + 5y + 24$

   subject to $3x + 2y = 4$

Use the method of Lagrange multipliers in Problems 7–16.

7. Find the maximum and minimum of $f(x, y) = 2xy$ subject to $x^2 + y^2 = 18$.

8. Find the maximum and minimum of $f(x, y) = x^2 - y^2$ subject to $x^2 + y^2 = 25$.

9. Maximize the product of two numbers if their sum must be 10.

10. Minimize the product of two numbers if their difference must be 10.

C

11. Minimize $f(x, y, z) = x^2 + y^2 + z^2$

   subject to $2x - y + 3z = -28$

12. Maximize $f(x, y, z) = xyz$

   subject to $2x + y + 2z = 120$

13. Maximize and minimize $f(x, y, z) = x + y + z$

   subject to $x^2 + y^2 + z^2 = 12$

14. Maximize and minimize $f(x, y, z) = 2x + 4y + 4z$

   subject to $x^2 + y^2 + z^2 = 9$

15. Maximize $f(x, y) = y + xy^2$

subject to $x + y^2 = 1$

16. Maximize and minimize $f(x, y) = x + e^y$

subject to $x^2 + y^2 = 1$

In Problems 17 and 18, use Theorem 1 to explain why no maxima or minima exist.

17. Maximize $f(x, y) = e^x + 3e^y$

subject to $x - 2y = 6$

18. Minimize $f(x, y) = x^3 + 2y^3$

subject to $6x - 2y = 1$

19. Consider the problem of maximizing $f(x, y)$ subject to $g(x, y) = 0$, where $g(x, y) = y - 5$. Explain how the maximization problem can be solved without using the method of Lagrange multipliers.

20. Consider the problem of minimizing $f(x, y)$ subject to $g(x, y) = 0$, where $g(x, y) = 4x - y + 3$. Explain how the minimization problem can be solved without using the method of Lagrange multipliers.

21. Consider the problem of maximizing $f(x, y) = e^{-(x^2+y^2)}$ subject to the constraint $g(x, y) = x^2 + y - 1 = 0$.

(A) Solve the constraint equation for y, and then substitute into $f(x, y)$ to obtain a function $h(x)$ of the single variable x. Solve the original maximization problem by maximizing h (round answers to three decimal places).

(B) Confirm your answer by the method of Lagrange multipliers.

22. Consider the problem of minimizing

$$f(x, y) = x^2 + 2y^2$$

subject to the constraint $g(x, y) = ye^{x^2} - 1 = 0$.

(A) Solve the constraint equation for y, and then substitute into $f(x, y)$ to obtain a function $h(x)$ of the single variable x. Solve the original minimization problem by minimizing h (round answers to three decimal places).

(B) Confirm your answer by the method of Lagrange multipliers.

Applications

23. Budgeting for least cost. A manufacturing company produces two models of an HDTV per week, x units of model A and y units of model B at a cost (in dollars) of

$$C(x, y) = 6x^2 + 12y^2$$

If it is necessary (because of shipping considerations) that

$$x + y = 90$$

how many of each type of set should be manufactured per week to minimize cost? What is the minimum cost?

24. Budgeting for maximum production. A manufacturing firm has budgeted $60,000 per month for labor and materi-

als. If $x thousand is spent on labor and $y thousand is spent on materials, and if the monthly output (in units) is given by

$$N(x, y) = 4xy - 8x$$

then how should the $60,000 be allocated to labor and materials in order to maximize N? What is the maximum N?

25. Productivity. A consulting firm for a manufacturing company arrived at the following Cobb–Douglas production function for a particular product:

$$N(x, y) = 50x^{0.8}y^{0.2}$$

In this equation, x is the number of units of labor and y is the number of units of capital required to produce $N(x, y)$ units of the product. Each unit of labor costs $40 and each unit of capital costs $80.

(A) If $400,000 is budgeted for production of the product, determine how that amount should be allocated to maximize production, and find the maximum production.

(B) Find the marginal productivity of money in this case, and estimate the increase in production if an additional $50,000 is budgeted for the production of the product.

26. Productivity. The research department of a manufacturing company arrived at the following Cobb–Douglas production function for a particular product:

$$N(x, y) = 10x^{0.6}y^{0.4}$$

In this equation, x is the number of units of labor and y is the number of units of capital required to produce $N(x, y)$ units of the product. Each unit of labor costs $30 and each unit of capital costs $60.

(A) If $300,000 is budgeted for production of the product, determine how that amount should be allocated to maximize production, and find the maximum production.

(B) Find the marginal productivity of money in this case, and estimate the increase in production if an additional $80,000 is budgeted for the production of the product.

27. Maximum volume. A rectangular box with no top and two intersecting partitions is to be constructed from 192 square inches of cardboard (see the figure). Find the dimensions that will maximize the volume.

Figure for 27

28. Maximum volume. A mailing service states that a rectangular package shall have the sum of its length and girth not to exceed 120 inches (see the figure). What are the dimensions of the largest (in volume) mailing carton that can be constructed to meet these restrictions?

Figure for 28

29. Agriculture. Three pens of the same size are to be built along an existing fence (see the figure). If 400 feet of fencing is available, what length should x and y be to produce the maximum total area? What is the maximum area?

Figure for 29

30. Diet and minimum cost. A group of guinea pigs is to receive 25,600 calories per week. Two available foods produce $200xy$ calories for a mixture of x kilograms of type M food and y kilograms of type N food. If type M costs \$1 per kilogram and type N costs \$2 per kilogram, how much of each type of food should be used to minimize weekly food costs? What is the minimum cost?

Note: $x \geq 0$, $y \geq 0$

Answers to Matched Problems

1. Max $f(x, y) = f(2, 2) = 17$ (see Fig. 4)

2. (A) 3,750 units of labor and 1,250 units of capital;
Max $N(x, y) = N(3,750, 1,250) \approx 43,301$ units

(B) Marginal productivity of money ≈ 0.1443; increase in production $\approx 5,774$ units

3. 12 in. by 8 in. by 4 in.

6-5 Method of Least Squares

- Least Squares Approximation
- Applications

Least Squares Approximation

Regression analysis is the process of fitting an elementary function to a set of data points by the **method of least squares**. The mechanics of using regression techniques were introduced in Chapter 1. Now, using the optimization techniques of Section 6-3, we can develop and explain the mathematical foundation of the method of least squares. We begin with **linear regression**, the process of finding the equation of the line that is the "best" approximation to a set of data points.

Suppose that a manufacturer wants to approximate the cost function for a product. The value of the cost function has been determined for certain levels of production, as listed in Table 1. Although these points do not all lie on a line (see Fig. 1),

Table 1

Number of Units x (hundreds)	Cost y (thousand $)
2	4
5	6
6	7
9	8

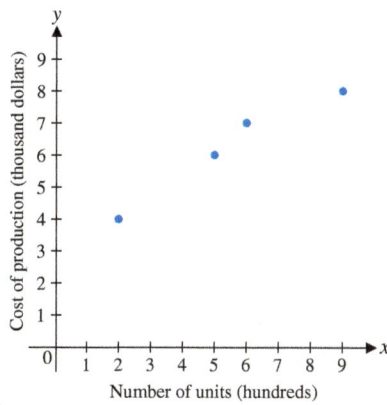

Figure 1

they are very close to being linear. The manufacturer would like to approximate the cost function by a linear function—that is, determine values a and b so that the line

$$y = ax + b$$

is, in some sense, the "best" approximation to the cost function.

What do we mean by "best"? Since the line $y = ax + b$ will not go through all four points, it is reasonable to examine the differences between the y coordinates of the points listed in the table and the y coordinates of the corresponding points on the line. Each of these differences is called the **residual** at that point (see Fig. 2). For example, at $x = 2$, the point from Table 1 is $(2, 4)$ and the point on the line is $(2, 2a + b)$, so the residual is

$$4 - (2a + b) = 4 - 2a - b$$

All the residuals are listed in Table 2.

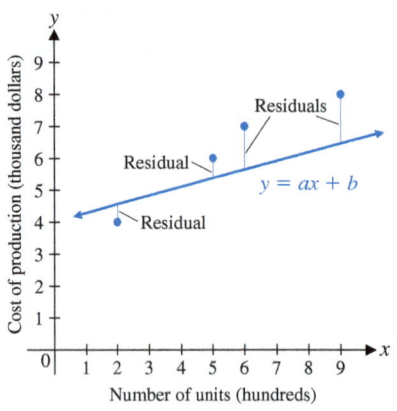

Figure 2

Table 2

x	y	$ax + b$	Residual
2	4	$2a + b$	$4 - 2a - b$
5	6	$5a + b$	$6 - 5a - b$
6	7	$6a + b$	$7 - 6a - b$
9	8	$9a + b$	$8 - 9a - b$

Our criterion for the "best" approximation is the following: Determine the values of a and b that *minimize the sum of the squares* of the residuals. The resulting line is called the **least squares line**, or the **regression line**. To this end, we minimize

$$F(a, b) = (4 - 2a - b)^2 + (6 - 5a - b)^2 + (7 - 6a - b)^2 + (8 - 9a - b)^2$$

Step 1 Find critical points:

$$
\begin{aligned}
F_a(a, b) &= 2(4 - 2a - b)(-2) + 2(6 - 5a - b)(-5) \\
&\quad + 2(7 - 6a - b)(-6) + 2(8 - 9a - b)(-9) \\
&= -304 + 292a + 44b = 0 \\
F_b(a, b) &= 2(4 - 2a - b)(-1) + 2(6 - 5a - b)(-1) \\
&\quad + 2(7 - 6a - b)(-1) + 2(8 - 9a - b)(-1) \\
&= -50 + 44a + 8b = 0
\end{aligned}
$$

After dividing each equation by 2, we solve the system

$$
\begin{aligned}
146a + 22b &= 152 \\
22a + 4b &= 25
\end{aligned}
$$

obtaining $(a, b) = (0.58, 3.06)$ as the only critical point.

Step 2 Compute $A = F_{aa}(a, b)$, $B = F_{ab}(a, b)$, and $C = F_{bb}(a, b)$:

$$F_{aa}(a, b) = 292; \quad \text{so,} \quad A = F_{aa}(0.58, 3.06) = 292$$

$$F_{ab}(a, b) = 44; \quad \text{so,} \quad B = F_{ab}(0.58, 3.06) = 44$$

$$F_{bb}(a, b) = 8; \quad \text{so,} \quad C = F_{bb}(0.58, 3.06) = 8$$

Step 3 Evaluate $AC - B^2$ and try to classify the critical point (a, b) by using Theorem 2 in Section 6-3:

$$AC - B^2 = (292)(8) - (44)^2 = 400 > 0 \qquad \text{and} \qquad A = 292 > 0$$

Therefore, case 2 in Theorem 2 applies, and $F(a, b)$ has a local minimum at the critical point $(0.58, 3.06)$.

So, the least squares line for the given data is

$$y = 0.58x + 3.06 \qquad \textcolor{blue}{\textit{Least squares line}}$$

The sum of the squares of the residuals is minimized for this choice of a and b (see Fig. 3).

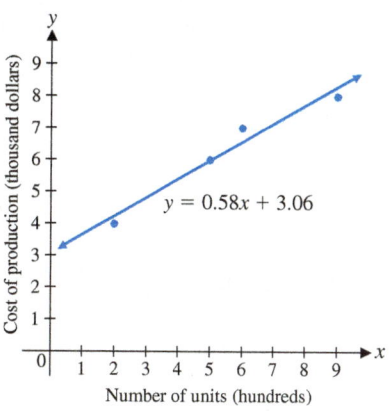

Figure 3

This linear function can now be used by the manufacturer to estimate any of the quantities normally associated with the cost function—such as costs, marginal costs, average costs, and so on. For example, the cost of producing 2,000 units is approximately

$$y = (0.58)(20) + 3.06 = 14.66, \qquad \text{or} \qquad \$14,660$$

The marginal cost function is

$$\frac{dy}{dx} = 0.58$$

The average cost function is

$$\bar{y} = \frac{0.58x + 3.06}{x}$$

In general, if we are given a set of n points $(x_1, y_1), (x_2, y_2), \ldots, (x_n, y_n)$, we want to determine the line $y = ax + b$ for which the sum of the squares of the residuals is minimized. Using summation notation, we find that the sum of the squares of the residuals is given by

$$F(a, b) = \sum_{k=1}^{n} (y_k - ax_k - b)^2$$

Note that in this expression the variables are a and b, and the x_k and y_k are all known values. To minimize $F(a, b)$, we thus compute the partial derivatives with respect to a and b and set them equal to 0:

$$F_a(a, b) = \sum_{k=1}^{n} 2(y_k - ax_k - b)(-x_k) = 0$$

$$F_b(a, b) = \sum_{k=1}^{n} 2(y_k - ax_k - b)(-1) = 0$$

Dividing each equation by 2 and simplifying, we see that the coefficients a and b of the least squares line $y = ax + b$ must satisfy the following system of *normal equations:*

$$\left(\sum_{k=1}^{n} x_k^2\right)a + \left(\sum_{k=1}^{n} x_k\right)b = \sum_{k=1}^{n} x_k y_k$$

$$\left(\sum_{k=1}^{n} x_k\right)a + nb = \sum_{k=1}^{n} y_k$$

Solving this system for a and b produces the formulas given in Theorem 1.

THEOREM 1   Least Squares Approximation

For a set of n points $(x_1, y_1), (x_2, y_2), \ldots, (x_n, y_n)$, the coefficients of the least squares line $y = ax + b$ are the solutions of the system of **normal equations**

$$\left(\sum_{k=1}^{n} x_k^2\right)a + \left(\sum_{k=1}^{n} x_k\right)b = \sum_{k=1}^{n} x_k y_k \tag{1}$$

$$\left(\sum_{k=1}^{n} x_k\right)a + nb = \sum_{k=1}^{n} y_k$$

and are given by the formulas

$$a = \frac{n\left(\sum_{k=1}^{n} x_k y_k\right) - \left(\sum_{k=1}^{n} x_k\right)\left(\sum_{k=1}^{n} y_k\right)}{n\left(\sum_{k=1}^{n} x_k^2\right) - \left(\sum_{k=1}^{n} x_k\right)^2} \tag{2}$$

$$b = \frac{\sum_{k=1}^{n} y_k - a\left(\sum_{k=1}^{n} x_k\right)}{n} \tag{3}$$

Now we return to the data in Table 1 and tabulate the sums required for the normal equations and their solution in Table 3.

The normal equations (1) are then

Table 3

	x_k	y_k	$x_k y_k$	x_k^2
	2	4	8	4
	5	6	30	25
	6	7	42	36
	9	8	72	81
Totals	22	25	152	146

$$146a + 22b = 152$$

$$22a + 4b = 25$$

The solution of the normal equations given by equations (2) and (3) is

$$a = \frac{4(152) - (22)(25)}{4(146) - (22)^2} = 0.58$$

$$b = \frac{25 - 0.58(22)}{4} = 3.06$$

Compare these results with step 1 on page 486. Note that Table 3 provides a convenient format for the computation of step 1.

 Many graphing calculators have a linear regression feature that solves the system of normal equations obtained by setting the partial derivatives of the sum of squares of the residuals equal to 0. Therefore, in practice, we simply enter the given data points and use the linear regression feature to determine the line $y = ax + b$ that best fits the data (see Fig. 4). There is no need to compute partial derivatives or even to tabulate sums (as in Table 3).

(A) (B) (C) $y_1 = 0.58x + 3.06$

Figure 4

EXPLORE & DISCUSS 1

(A) Plot the four points $(0, 0), (0, 1), (10, 0),$ and $(10, 1)$. Which line would you guess "best" fits these four points? Use formulas (2) and (3) to test your conjecture.

(B) Plot the four points $(0, 0), (0, 10), (1, 0)$ and $(1, 10)$. Which line would you guess "best" fits these four points? Use formulas (2) and (3) to test your conjecture.

(C) If either of your conjectures was wrong, explain how your reasoning was mistaken.

CONCEPTUAL INSIGHT

Formula (2) for a is undefined if the denominator equals 0. When can this happen? Suppose $n = 3$. Then

$$n\left(\sum_{k=1}^{n} x_k^2\right) - \left(\sum_{k=1}^{n} x_k\right)^2$$

$$= 3\left(x_1^2 + x_2^2 + x_3^2\right) - (x_1 + x_2 + x_3)^2$$
$$= 3\left(x_1^2 + x_2^2 + x_3^2\right) - \left(x_1^2 + x_2^2 + x_3^2 + 2x_1x_2 + 2x_1x_3 + 2x_2x_3\right)$$
$$= 2\left(x_1^2 + x_2^2 + x_3^2\right) - (2x_1x_2 + 2x_1x_3 + 2x_2x_3)$$
$$= \left(x_1^2 + x_2^2\right) + \left(x_1^2 + x_3^2\right) + \left(x_2^2 + x_3^2\right) - (2x_1x_2 + 2x_1x_3 + 2x_2x_3)$$
$$= \left(x_1^2 - 2x_1x_2 + x_2^2\right) + \left(x_1^2 - 2x_1x_3 + x_3^2\right) + \left(x_2^2 - 2x_2x_3 + x_3^2\right)$$
$$= (x_1 - x_2)^2 + (x_1 - x_3)^2 + (x_2 - x_3)^2$$

and the last expression is equal to 0 if and only if $x_1 = x_2 = x_3$ (i.e., if and only if the three points all lie on the same vertical line). A similar algebraic manipulation works for any integer $n > 1$, showing that, in formula (2) for a, the denominator equals 0 if and only if all n points lie on the same vertical line.

The method of least squares can also be applied to find the quadratic equation $y = ax^2 + bx + c$ that best fits a set of data points. In this case, the sum of the squares of the residuals is a function of three variables:

$$F(a, b, c) = \sum_{k=1}^{n} \left(y_k - ax_k^2 - bx_k - c\right)^2$$

There are now three partial derivatives to compute and set equal to 0:

$$F_a(a, b, c) = \sum_{k=1}^{n} 2\left(y_k - ax_k^2 - bx_k - c\right)\left(-x_k^2\right) = 0$$

$$F_b(a, b, c) = \sum_{k=1}^{n} 2\left(y_k - ax_k^2 - bx_k - c\right)(-x_k) = 0$$

$$F_c(a, b, c) = \sum_{k=1}^{n} 2\left(y_k - ax_k^2 - bx_k - c\right)(-1) = 0$$

The resulting set of three linear equations in the three variables a, b, and c is called the *set of normal equations for quadratic regression.*

 A quadratic regression feature on a calculator is designed to solve such normal equations after the given set of points has been entered. Figure 5 illustrates the computation for the data of Table 1.

(A)   (B)   (C) $y_1 = -0.0417x^2 + 1.0383x + 2.06$

Figure 5

EXPLORE & DISCUSS 2

(A) Use the graphs in Figures 4 and 5 to predict which technique, linear regression or quadratic regression, yields the smaller sum of squares of the residuals for the data of Table 1. Explain.

(B) Confirm your prediction by computing the sum of squares of the residuals in each case.

The method of least squares can also be applied to other regression equations—for example, cubic, quartic, logarithmic, exponential, and power regression models. Details are explored in some of the exercises at the end of this section.

Applications

EXAMPLE 1   **Exam Scores** Table 4 lists the midterm and final examination scores of 10 students in a calculus course.

Table 4

Midterm	Final	Midterm	Final
49	61	78	77
53	47	83	81
67	72	85	79
71	76	91	93
74	68	99	99

(A) Use formulas (1), (2), and (3) to find the normal equations and the least squares line for the data given in Table 4.

 (B) Use the linear regression feature on a graphing calculator to find and graph the least squares line.

(C) Use the least squares line to predict the final examination score of a student who scored 95 on the midterm examination.

SOLUTION   (A) Table 5 shows a convenient way to compute all the sums in the formulas for a and b.

Table 5

x_k	y_k	$x_k y_k$	x_k^2
49	61	2,989	2,401
53	47	2,491	2,809
67	72	4,824	4,489
71	76	5,396	5,041
74	68	5,032	5,476
78	77	6,006	6,084
83	81	6,723	6,889
85	79	6,715	7,225
91	93	8,463	8,281
99	99	9,801	9,801
Totals 750	753	58,440	58,496

From the last line in Table 5, we have

$$\sum_{k=1}^{10} x_k = 750 \qquad \sum_{k=1}^{10} y_k = 753 \qquad \sum_{k=1}^{10} x_k y_k = 58{,}440 \qquad \sum_{k=1}^{10} x_k^2 = 58{,}496$$

and the normal equations are

$$58{,}496a + 750b = 58{,}440$$
$$750a + 10b = 753$$

Using formulas (2) and (3), we obtain

$$a = \frac{10(58{,}440) - (750)(753)}{10(58{,}496) - (750)^2} = \frac{19{,}650}{22{,}460} \approx 0.875$$

$$b = \frac{753 - 0.875(750)}{10} = 9.675$$

The least squares line is given (approximately) by

$$y = 0.875x + 9.675$$

(B) We enter the data and use the linear regression feature, as shown in Figure 6. [The discrepancy between values of a and b in the preceding calculations and those in Figure 6B is due to rounding in part (A).]

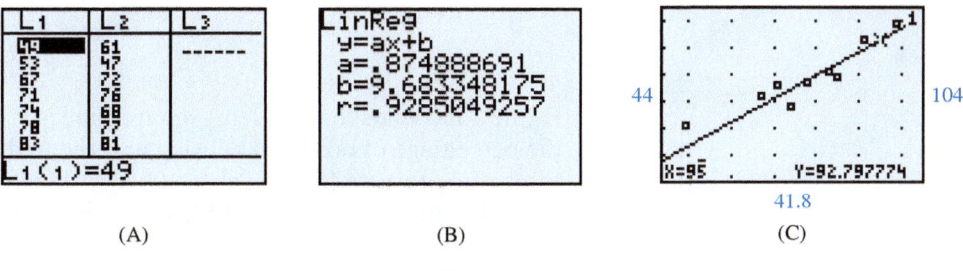

(A)　　　　　　　(B)　　　　　　　(C)

Figure 6

(C) If $x = 95$, then $y = 0.875(95) + 9.675 \approx 92.8$ is the predicted score on the final exam. This is also indicated in Figure 6C. If we assume that the exam score must be an integer, then we would predict a score of 93.

Matched Problem 1 Repeat Example 1 for the scores listed in Table 6.

Table 6

Midterm	Final	Midterm	Final
54	50	84	80
60	66	88	95
75	80	89	85
76	68	97	94
78	71	99	86

EXAMPLE 2 **Energy Consumption** The use of fuel oil for home heating in the United States has declined steadily for several decades. Table 7 lists the percentage of occupied housing units in the United States that were heated by fuel oil for various years between 1960 and 2000. Use the data in the table and linear regression to estimate the percentage of occupied housing units in the United States that were heated by fuel oil in the year 2005.

Table 7 Occupied Housing Units Heated by Fuel Oil

Year	Percent	Year	Percent
1960	32.4	1990	13.2
1970	26.0	2000	9.0
1980	18.1		

SOLUTION We enter the data, with $x = 0$ representing 1960, $x = 10$ representing 1970, and so on, and use linear regression as shown in Figure 7.

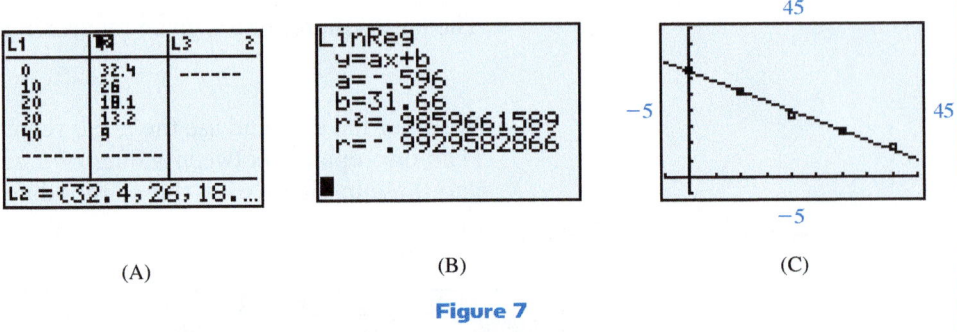

(A)   (B)   (C)

Figure 7

Figure 7 indicates that the least squares line is $y = -0.596x + 31.66$. To estimate the percentage of occupied housing units heated by fuel oil in the year 2005 (corresponding to $x = 45$), we substitute $x = 45$ in the equation of the least squares line: $-0.596(45) + 31.66 = 4.84$. The estimated percentage for 2005 is 4.84%.

Matched Problem 2 In 1950, coal was still a major source of fuel for home energy consumption, and the percentage of occupied housing units heated by fuel oil was only 22.1%. Add the data for 1950 to the data for Example 2, and compute the new least squares line and the new estimate for the percentage of occupied housing units heated by fuel oil in the year 2005. Discuss the discrepancy between the two estimates. (As in Example 2, let $x = 0$ represent 1960.)

Exercises 6-5

A

In Problems 1–6, find the least squares line. Graph the data and the least squares line.

1.

x	y
1	1
2	3
3	4
4	3

2.

x	y
1	−2
2	−1
3	3
4	5

3.

x	y
1	8
2	5
3	4
4	0

4.

x	y
1	20
2	14
3	11
4	3

5.

x	y
1	3
2	4
3	5
4	6

6.

x	y
1	2
2	3
3	3
4	2

B

In Problems 7–14, find the least squares line and use it to estimate y for the indicated value of x. Round answers to two decimal places.

7.

x	y
1	3
2	1
2	2
3	0

Estimate y when x = 2.5.

8.

x	y
1	0
3	1
3	6
3	4

Estimate y when x = 3.

9.

x	y
0	10
5	22
10	31
15	46
20	51

Estimate y when x = 25.

10.

x	y
−5	60
0	50
5	30
10	20
15	15

Estimate y when x = 20.

11.

x	y
−1	14
1	12
3	8
5	6
7	5

Estimate y when x = 2.

12.

x	y
2	−4
6	0
10	8
14	12
18	14

Estimate y when x = 15.

13.

x	y	x	y
0.5	25	9.5	12
2	22	11	11
3.5	21	12.5	8
5	21	14	5
6.5	18	15.5	1

Estimate y when x = 8.

14.

x	y	x	y
0	−15	12	11
2	−9	14	13
4	−7	16	19
6	−7	18	25
8	−1	20	33

Estimate y when x = 10.

C

15. To find the coefficients of the parabola

$$y = ax^2 + bx + c$$

that is the "best" fit to the points (1, 2), (2, 1), (3, 1), and (4, 3), minimize the sum of the squares of the residuals

$$F(a, b, c) = (a + b + c − 2)^2$$
$$+ (4a + 2b + c − 1)^2$$
$$+ (9a + 3b + c − 1)^2$$
$$+ (16a + 4b + c − 3)^2$$

by solving the system of normal equations

$$F_a(a, b, c) = 0 \qquad F_b(a, b, c) = 0 \qquad F_c(a, b, c) = 0$$

for a, b, and c. Graph the points and the parabola.

16. Repeat Problem 15 for the points (−1, −2), (0, 1), (1, 2), and (2, 0).

Problems 17 and 18 refer to the system of normal equations and the formulas for a and b given on page 488.

17. Verify formulas (2) and (3) by solving the system of normal equations (1) for a and b.

18. If

$$\bar{x} = \frac{1}{n}\sum_{k=1}^{n} x_k \qquad \text{and} \qquad \bar{y} = \frac{1}{n}\sum_{k=1}^{n} y_k$$

are the averages of the x and y coordinates, respectively, show that the point $(\bar{x}, \bar{y})$ satisfies the equation of the least squares line, y = ax + b.

19. (A) Suppose that n = 5 and the x coordinates of the data points $(x_1, y_1), (x_2, y_2), \ldots, (x_n, y_n)$ are −2, −1, 0, 1, 2. Show that system (1) in the text implies that

$$a = \frac{\sum x_k y_k}{\sum x_k^2}$$

and that b is equal to the average of the values of y_k.

(B) Show that the conclusion of part (A) holds whenever the average of the x coordinates of the data points is 0.

20. (A) Give an example of a set of six data points such that half of the points lie above the least squares line and half lie below.

(B) Give an example of a set of six data points such that just one of the points lies above the least squares line and five lie below.

21. (A) Find the linear and quadratic functions that best fit the data points (0, 1.3), (1, 0.6), (2, 1.5), (3, 3.6), and (4, 7.4). Round coefficients to two decimal places.

(B) Which of the two functions best fits the data? Explain.

22. (A) Find the linear, quadratic, and logarithmic functions that best fit the data points (1, 3.2), (2, 4.2), (3, 4.7), (4, 5.0), and (5, 5.3). (Round coefficients to two decimal places.)

(B) Which of the three functions best fits the data? Explain.

23. Describe the normal equations for cubic regression. How many equations are there? What are the variables? What techniques could be used to solve the equations?

24. Describe the normal equations for quartic regression. How many equations are there? What are the variables? What techniques could be used to solve the equations?

Applications

25. Crime rate. Data on U.S. property crimes (in number of crimes per 100,000 population) are given in the table for the years 2001 through 2006.

U.S. Property Crime Rates	
Year	**Rate**
2001	3,658
2002	3,631
2003	3,591
2004	3,514
2005	3,432
2006	3,335

(A) Find the least squares line for the data, using $x = 0$ for 2000.

(B) Use the least squares line to predict the property crime rate in 2016.

26. Cable TV revenue. Data for cable TV revenue are given in the table for the years 2003 through 2007.

Cable TV Revenue	
Year	**Thousands of dollars**
2003	53,991
2004	59,428
2005	65,041
2006	71,843
2007	78,886

(A) Find the least squares line for the data, using $x = 0$ for 2000.

(B) Use the least squares line to predict cable TV revenue in 2017.

27. Maximizing profit. The market research department for a drugstore chain chose two summer resort areas to test market a new sunscreen lotion packaged in 4-ounce plastic bottles. After a summer of varying the selling price and recording the monthly demand, the research department arrived at the following demand table, where y is the number of bottles purchased per month (in thousands) at x dollars per bottle:

x	y
5.0	2.0
5.5	1.8
6.0	1.4
6.5	1.2
7.0	1.1

(A) Use the method of least squares to find a demand equation.

(B) If each bottle of sunscreen costs the drugstore chain $4, how should the sunscreen be priced to achieve a maximum monthly profit? [*Hint:* Use the result of part (A), with $C = 4y$, $R = xy$, and $P = R - C$.]

28. Maximizing profit. A market research consultant for a supermarket chain chose a large city to test market a new brand of mixed nuts packaged in 8-ounce cans. After a year of varying the selling price and recording the monthly demand, the consultant arrived at the following demand table, where y is the number of cans purchased per month (in thousands) at x dollars per can:

x	y
4.0	4.2
4.5	3.5
5.0	2.7
5.5	1.5
6.0	0.7

(A) Use the method of least squares to find a demand equation.

(B) If each can of nuts costs the supermarket chain $3, how should the nuts be priced to achieve a maximum monthly profit?

29. Olympic Games. The table gives the winning heights in the pole vault in the Olympic Games from 1980 to 2008.

Olympic Pole Vault Winning Height	
Year	**Height (ft)**
1980	18.96
1984	18.85
1988	19.35
1992	19.02
1996	19.42
2000	19.35
2004	19.52
2008	19.56

(A) Use a graphing calculator to find the least squares line for the data, letting $x = 0$ for 1980.

(B) Estimate the winning height in the pole vault in the Olympic Games of 2020.

30. Biology. In biology, there is an approximate rule, called the *bioclimatic rule for temperate climates*. This rule states that in spring and early summer, periodic phenomena such as the blossoming of flowers, the appearance of insects, and the ripening of fruit usually come about 4 days later for each 500 feet of altitude. Stated as a formula, the rule becomes

$$d = 8h \qquad 0 \le h \le 4$$

where d is the change in days and h is the altitude (in thousands of feet). To test this rule, an experiment was set up to record the difference in blossoming times of the same type of apple tree at different altitudes. A summary of the results is given in the following table:

h	d
0	0
1	7
2	18
3	28
4	33

(A) Use the method of least squares to find a linear equation relating h and d. Does the bioclimatic rule $d = 8h$ appear to be approximately correct?

(B) How much longer will it take this type of apple tree to blossom at 3.5 thousand feet than at sea level? [Use the linear equation found in part (A).]

 31. Global warming. Average global temperatures from 1885 to 2005 are given in the table.

Average Global Temperatures

Year	°F	Year	°F
1885	56.65	1955	57.06
1895	56.64	1965	57.05
1905	56.52	1975	57.04
1915	56.57	1985	57.36
1925	56.74	1995	57.64
1935	57.00	2005	58.59
1945	57.13		

(A) Find the least squares line for the data, using $x = 0$ for 1885.

(B) Use the least squares line to estimate the average global temperature in 2085.

32. Air pollution. Data for emissions of sulfur dioxide in the United States from selected years from 1972 to 2007 are given in the table.

Emissions of Sulfur Dioxide in the United States

Year	Million Short Tons	Year	Million Short Tons
1972	30.4	1992	22.1
1977	28.6	1997	18.8
1982	23.2	2002	15.4
1987	22.2	2007	12.4

(A) Find the least squares line for the data, using $x = 0$ for 1972.

(B) Use the least squares line to estimate the emissions of sulfur dioxide in the United States in 2017.

Answers to Matched Problems

1. (A) $y = 0.85x + 9.47$

(B)

(C) 90.3

2. $y = -0.37x + 25.75$; 9.1%

6-6 Double Integrals over Rectangular Regions

- Introduction
- Definition of the Double Integral
- Average Value over Rectangular Regions
- Volume and Double Integrals

Introduction

We have generalized the concept of differentiation to functions with two or more independent variables. How can we do the same with integration, and how can we interpret the results? Let's look first at the operation of antidifferentiation. We can antidifferentiate a function of two or more variables with respect to one of the variables by treating all the other variables as though they were constants. Thus, this operation is the reverse operation of partial differentiation, just as ordinary antidifferentiation is the reverse operation of ordinary differentiation. We write $\int f(x, y)\, dx$ to indicate that we are to antidifferentiate $f(x, y)$ with respect to x, holding y fixed; we write $\int f(x, y)\, dy$ to indicate that we are to antidifferentiate $f(x, y)$ with respect to y, holding x fixed.

EXAMPLE 1   **Partial Antidifferentiation** Evaluate:

(A) $\displaystyle\int (6xy^2 + 3x^2)\, dy$ (B) $\displaystyle\int (6xy^2 + 3x^2)\, dx$

SOLUTION   (A) Treating x as a constant and using the properties of antidifferentiation from Section 6-1, we have

$$\int (6xy^2 + 3x^2)\, dy = \int 6xy^2\, dy + \int 3x^2\, dy$$

$$= 6x \int y^2\, dy + 3x^2 \int dy$$

$$= 6x \left(\frac{y^3}{3}\right) + 3x^2(y) + C(x)$$

$$= 2xy^3 + 3x^2y + C(x)$$

The dy tells us that we are looking for the antiderivative of $6xy^2 + 3x^2$ with respect to y only, holding x constant.

Note that the constant of integration can be *any function of x alone* since for any such function,

$$\frac{\partial}{\partial y} C(x) = 0$$

Check:

We can verify that our answer is correct by using partial differentiation:

$$\frac{\partial}{\partial y}[2xy^3 + 3x^2y + C(x)] = 6xy^2 + 3x^2 + 0$$

$$= 6xy^2 + 3x^2$$

(B) We treat y as a constant:

$$\int (6xy^2 + 3x^2)\, dx = \int 6xy^2\, dx + \int 3x^2\, dx$$

$$= 6y^2 \int x\, dx + 3 \int x^2\, dx$$

$$= 6y^2 \left(\frac{x^2}{2}\right) + 3\left(\frac{x^3}{3}\right) + E(y)$$

$$= 3x^2y^2 + x^3 + E(y)$$

The antiderivative contains an arbitrary function $E(y)$ of y alone.

Check:

$$\frac{\partial}{\partial x}[3x^2y^2 + x^3 + E(y)] = 6xy^2 + 3x^2 + 0$$

$$= 6xy^2 + 3x^2$$

Matched Problem 1   Evaluate   (A) $\displaystyle\int (4xy + 12x^2y^3)\, dy$ (B) $\displaystyle\int (4xy + 12x^2y^3)\, dx$

Now that we have extended the concept of antidifferentiation to functions with two variables, we also can evaluate definite integrals of the form

$$\int_a^b f(x, y)\, dx \quad \text{or} \quad \int_c^d f(x, y)\, dy$$

EXAMPLE 2   **Evaluating a Partial Antiderivative** Evaluate, substituting the limits of integration in y if dy is used and in x if dx is used:

(A) $\displaystyle\int_0^2 (6xy^2 + 3x^2)\,dy$ (B) $\displaystyle\int_0^1 (6xy^2 + 3x^2)\,dx$

SOLUTION   (A) From Example 1A, we know that

$$\int (6xy^2 + 3x^2)\,dy = 2xy^3 + 3x^2y + C(x)$$

According to properties of the definite integral for a function of one variable, we can use any antiderivative to evaluate the definite integral. Thus, choosing $C(x) = 0$, we have

$$\int_0^2 (6xy^2 + 3x^2)\,dy = (2xy^3 + 3x^2y)\big|_{y=0}^{y=2}$$

$$= [2x(2)^3 + 3x^2(2)] - [2x(0)^3 + 3x^2(0)]$$

$$= 16x + 6x^2$$

(B) From Example 1B, we know that

$$\int (6xy^2 + 3x^2)\,dx = 3x^2y^2 + x^3 + E(y)$$

Choosing $E(y) = 0$, we have

$$\int_0^1 (6xy^2 + 3x^2)\,dx = (3x^2y^2 + x^3)\big|_{x=0}^{x=1}$$

$$= [3y^2(1)^2 + (1)^3] - [3y^2(0)^2 + (0)^3]$$

$$= 3y^2 + 1$$

Matched Problem 2   Evaluate:

(A) $\displaystyle\int_0^1 (4xy + 12x^2y^3)\,dy$ (B) $\displaystyle\int_0^3 (4xy + 12x^2y^3)\,dx$

Integrating and evaluating a definite integral with integrand $f(x, y)$ with respect to y produces a function of x alone (or a constant). Likewise, integrating and evaluating a definite integral with integrand $f(x, y)$ with respect to x produces a function of y alone (or a constant). Each of these results, involving at most one variable, can now be used as an integrand in a second definite integral.

EXAMPLE 3   **Evaluating Integrals** Evaluate:

(A) $\displaystyle\int_0^1 \left[\int_0^2 (6xy^2 + 3x^2)\,dy \right] dx$ (B) $\displaystyle\int_0^2 \left[\int_0^1 (6xy^2 + 3x^2)\,dx \right] dy$

SOLUTION   (A) Example 2A showed that

$$\int_0^2 (6xy^2 + 3x^2)\,dy = 16x + 6x^2$$

Therefore,

$$\int_0^1 \left[\int_0^2 (6xy^2 + 3x^2)\,dy \right] dx = \int_0^1 (16x + 6x^2)\,dx$$

$$= (8x^2 + 2x^3)\big|_{x=0}^{x=1}$$

$$= [8(1)^2 + 2(1)^3] - [8(0)^2 + 2(0)^3] = 10$$

(B) Example 2B showed that

$$\int_0^1 (6xy^2 + 3x^2)\, dx = 3y^2 + 1$$

Therefore,

$$\int_0^2 \left[\int_0^1 (6xy^2 + 3x^2)\, dx \right] dy = \int_0^2 (3y^2 + 1)\, dy$$

$$= (y^3 + y)\big|_{y=0}^{y=2}$$

$$= [(2)^3 + 2] - [(0)^3 + 0] = 10$$

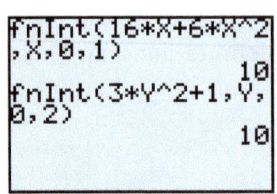

Figure 1

A numerical integration command can be used as an alternative to the fundamental theorem of calculus to evaluate the last integrals in Examples 3A and 3B, $\int_0^1 (16x + 6x^2)\, dx$ and $\int_0^2 (3y^2 + 1)\, dy$, since the integrand in each case is a function of a single variable (see Fig. 1).

Matched Problem 3   Evaluate

(A) $\displaystyle\int_0^3 \left[\int_0^1 (4xy + 12x^2 y^3)\, dy \right] dx$ (B) $\displaystyle\int_0^1 \left[\int_0^3 (4xy + 12x^2 y^3)\, dx \right] dy$

Definition of the Double Integral

Notice that the answers in Examples 3A and 3B are identical. This is not an accident. In fact, it is this property that enables us to define the *double integral*, as follows:

DEFINITION Double Integral

The **double integral** of a function $f(x, y)$ over a rectangle

$$R = \{(x, y)\,|\,a \le x \le b,\ c \le y \le d\}$$

is

$$\iint\limits_R f(x, y)\, dA = \int_a^b \left[\int_c^d f(x, y)\, dy \right] dx$$

$$= \int_c^d \left[\int_a^b f(x, y)\, dx \right] dy$$

In the double integral $\iint_R f(x, y)\, dA, f(x, y)$ is called the **integrand**, and R is called the **region of integration**. The expression dA indicates that this is an integral over a two-dimensional region. The integrals

$$\int_a^b \left[\int_c^d f(x, y)\, dy \right] dx \qquad \text{and} \qquad \int_c^d \left[\int_a^b f(x, y)\, dx \right] dy$$

are referred to as **iterated integrals** (the brackets are often omitted), and the order in which dx and dy are written indicates the order of integration. This is not the most general definition of the double integral over a rectangular region; however, it is equivalent to the general definition for all the functions we will consider.

EXAMPLE 4   **Evaluating a Double Integral** Evaluate

$$\iint_R (x + y)\, dA \quad \text{over} \quad R = \{(x, y)|1 \le x \le 3, \ -1 \le y \le 2\}$$

SOLUTION   Region R is illustrated in Figure 2. We can choose either order of iteration. As a check, we will evaluate the integral both ways:

$$\iint_R (x + y)\, dA = \int_1^3 \int_{-1}^2 (x + y)\, dy\, dx$$

$$= \int_1^3 \left[\left(xy + \frac{y^2}{2} \right) \Big|_{y=-1}^{y=2} \right] dx$$

$$= \int_1^3 \left[(2x + 2) - \left(-x + \tfrac{1}{2} \right) \right] dx$$

$$= \int_1^3 \left(3x + \tfrac{3}{2} \right) dx$$

$$= \left(\tfrac{3}{2}x^2 + \tfrac{3}{2}x \right) \Big|_{x=1}^{x=3}$$

$$= \left(\tfrac{27}{2} + \tfrac{9}{2} \right) - \left(\tfrac{3}{2} + \tfrac{3}{2} \right) = 18 - 3 = 15$$

$$\iint_R (x + y)\, dA = \int_{-1}^2 \int_1^3 (x + y)\, dx\, dy$$

$$= \int_{-1}^2 \left[\left(\frac{x^2}{2} + xy \right) \Big|_{x=1}^{x=3} \right] dy$$

$$= \int_{-1}^2 \left[\left(\tfrac{9}{2} + 3y \right) - \left(\tfrac{1}{2} + y \right) \right] dy$$

$$= \int_{-1}^2 (4 + 2y)\, dy$$

$$= (4y + y^2) \Big|_{y=-1}^{y=2}$$

$$= (8 + 4) - (-4 + 1) = 12 - (-3) = 15$$

Figure 2

Matched Problem 4   Evaluate

$$\iint_R (2x - y)\, dA \quad \text{over} \quad R = \{(x, y)|-1 \le x \le 5, \ 2 \le y \le 4\}$$

both ways.

EXAMPLE 5   **Double Integral of an Exponential Function** Evaluate

$$\iint_R 2xe^{x^2+y}\, dA \quad \text{over} \quad R = \{(x, y)|0 \le x \le 1, \ -1 \le y \le 1\}$$

SOLUTION   Region R is illustrated in Figure 3.

$$\iint_R 2xe^{x^2+y}\, dA = \int_{-1}^1 \int_0^1 2xe^{x^2+y}\, dx\, dy$$

$$= \int_{-1}^1 \left[(e^{x^2+y}) \Big|_{x=0}^{x=1} \right] dy$$

$$= \int_{-1}^1 (e^{1+y} - e^y)\, dy$$

Figure 3

$$= (e^{1+y} - e^y)\big|_{y=-1}^{y=1}$$

$$= (e^2 - e) - (e^0 - e^{-1})$$

$$= e^2 - e - 1 + e^{-1}$$

Matched Problem 5   Evaluate $\displaystyle\iint_R \frac{x}{y^2}e^{x/y}\, dA$ over $R = \{(x, y)|0 \le x \le 1,\ 1 \le y \le 2\}$.

Average Value over Rectangular Regions

In Section 7-5, the average value of a function $f(x)$ over an interval $[a, b]$ is defined as

$$\frac{1}{b-a}\int_a^b f(x)\, dx$$

This definition is easily extended to functions of two variables over rectangular regions as follows (notice that the denominator $(b - a)(d - c)$ is simply the area of the rectangle R):

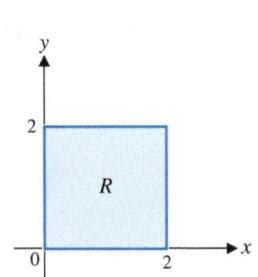

> **DEFINITION**   **Average Value over Rectangular Regions**
>
> The **average value** of the function $f(x, y)$ over the rectangle
>
> $$R = \{(x, y)|a \le x \le b,\ c \le y \le d\}$$
>
> is
>
> $$\frac{1}{(b - a)(d - c)}\iint_R f(x, y)\, dA$$

EXAMPLE 6   **Average Value**   Find the average value of $f(x, y) = 4 - \frac{1}{2}x - \frac{1}{2}y$ over the rectangle $R = \{(x, y)|0 \le x \le 2,\ 0 \le y \le 2\}$.

SOLUTION   Region R is illustrated in Figure 4. We have

$$\frac{1}{(b - a)(d - c)}\iint_R f(x, y)\, dA = \frac{1}{(2 - 0)(2 - 0)}\iint_R \left(4 - \frac{1}{2}x - \frac{1}{2}y\right) dA$$

$$= \frac{1}{4}\int_0^2 \int_0^2 \left(4 - \frac{1}{2}x - \frac{1}{2}y\right) dy\, dx$$

$$= \frac{1}{4}\int_0^2 \left[\left(4y - \frac{1}{2}xy - \frac{1}{4}y^2\right)\Big|_{y=0}^{y=2}\right] dx$$

$$= \frac{1}{4}\int_0^2 (7 - x)\, dx$$

$$= \frac{1}{4}\left(7x - \frac{1}{2}x^2\right)\big|_{x=0}^{x=2}$$

$$= \frac{1}{4}(12) = 3$$

Figure 5 illustrates the surface $z = f(x, y)$, and our calculations show that 3 is the average of the z values over the region R.

Figure 4

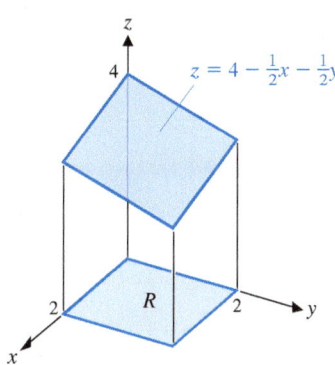

Figure 5

Matched Problem 6   Find the average value of $f(x, y) = x + 2y$ over the rectangle

$$R = \{(x, y)|0 \le x \le 2, \quad 0 \le y \le 1\}$$

EXPLORE & DISCUSS 1

(A) Which of the functions $f(x, y) = 4 - x^2 - y^2$ and $g(x, y) = 4 - x - y$ would you guess has the greater average value over the rectangle $R = \{(x, y)|0 \le x \le 1, \quad 0 \le y \le 1\}$? Explain.

(B) Use double integrals to check the correctness of your guess in part (A).

Volume and Double Integrals

One application of the definite integral of a function with one variable is the calculation of areas, so it is not surprising that the definite integral of a function of two variables can be used to calculate volumes of solids.

> **THEOREM 1   Volume under a Surface**
> If $f(x, y) \ge 0$ over a rectangle $R = \{(x, y)|a \le x \le b, \quad c \le y \le d\}$, then the volume of the solid formed by graphing f over the rectangle R is given by
>
> $$V = \iint\limits_{R} f(x, y) \, dA$$

EXAMPLE 7   **Volume**   Find the volume of the solid under the graph of $f(x, y) = 1 + x^2 + y^2$ over the rectangle $R = \{(x, y)|0 \le x \le 1, \quad 0 \le y \le 1\}$.

SOLUTION   Figure 6 shows the region R, and Figure 7 illustrates the volume under consideration.

$$V = \iint\limits_{R} (1 + x^2 + y^2) \, dA$$

$$= \int_0^1 \int_0^1 (1 + x^2 + y^2) \, dx \, dy$$

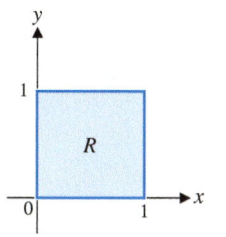

Figure 6

$$= \int_0^1 \left[\left(x + \tfrac{1}{3}x^3 + xy^2 \right) \Big|_{x=0}^{\,x=1} \right] dy$$

$$= \int_0^1 \left(\tfrac{4}{3} + y^2 \right) dy$$

$$= \left(\tfrac{4}{3}y + \tfrac{1}{3}y^3 \right)\big|_{y=0}^{y=1} = \tfrac{5}{3} \text{ cubic units}$$

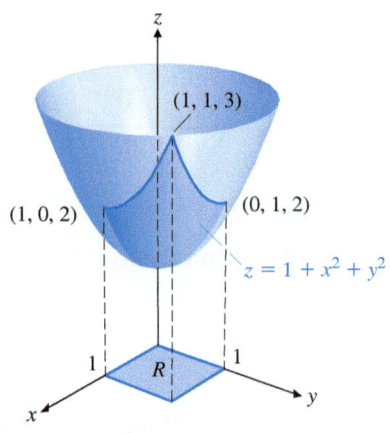

Figure 7

Matched Problem 7   Find the volume of the solid under the graph of $f(x, y) = 1 + x + y$ over the rectangle $R = \{(x, y) | 0 \le x \le 1, \ 0 \le y \le 2\}$.

CONCEPTUAL INSIGHT

Double integrals can be defined over regions that are more general than rectangles. For example, let $R > 0$. Then the function $f(x, y) = \sqrt{R^2 - (x^2 + y^2)}$ can be integrated over the circular region $C = \{(x, y) | x^2 + y^2 \le R^2\}$. In fact, it can be shown that

$$\iint\limits_C \sqrt{R^2 - (x^2 + y^2)} \, dx \, dy = \frac{2\pi R^3}{3}$$

Because $x^2 + y^2 + z^2 = R^2$ is the equation of a sphere of radius R centered at the origin, the double integral over C represents the volume of the upper hemisphere. Therefore, the volume of a sphere of radius R is given by

$$V = \frac{4\pi R^3}{3} \qquad \text{Volume of sphere of radius } R$$

Double integrals can also be used to obtain volume formulas for other geometric figures (see Table 1, Appendix C).

Exercises 6-6

A

In Problems 1–8, find each antiderivative. Then use the antiderivative to evaluate the definite integral.

1. (A) $\displaystyle \int 12x^2 y^3 \, dy$ (B) $\displaystyle \int_0^1 12x^2 y^3 \, dy$

2. (A) $\displaystyle \int 12x^2 y^3 \, dx$ (B) $\displaystyle \int_{-1}^2 12x^2 y^3 \, dx$

3. (A) $\displaystyle \int (4x + 6y + 5) \, dx$

     (B) $\displaystyle \int_{-2}^3 (4x + 6y + 5) \, dx$

4. (A) $\displaystyle \int (4x + 6y + 5) \, dy$

     (B) $\displaystyle \int_1^4 (4x + 6y + 5) \, dy$

5. (A) $\int \dfrac{x}{\sqrt{y + x^2}}\, dx$ **(B)** $\int_0^2 \dfrac{x}{\sqrt{y + x^2}}\, dx$

6. (A) $\int \dfrac{x}{\sqrt{y + x^2}}\, dy$ **(B)** $\int_1^5 \dfrac{x}{\sqrt{y + x^2}}\, dy$

7. (A) $\int \dfrac{\ln x}{xy}\, dy$ **(B)** $\int_1^{e^2} \dfrac{\ln x}{xy}\, dy$

8. (A) $\int \dfrac{\ln x}{xy}\, dx$ **(B)** $\int_1^e \dfrac{\ln x}{xy}\, dx$

B

In Problems 9–16, evaluate each iterated integral. (See the indicated problem for the evaluation of the inner integral.)

9. $\displaystyle\int_{-1}^{2}\int_{0}^{1} 12x^2 y^3 \, dy\, dx$

(See Problem 1.)

10. $\displaystyle\int_{0}^{1}\int_{-1}^{2} 12x^2 y^3 \, dx\, dy$

(See Problem 2.)

11. $\displaystyle\int_{1}^{4}\int_{-2}^{3} (4x + 6y + 5) \, dx\, dy$

(See Problem 3.)

12. $\displaystyle\int_{-2}^{3}\int_{1}^{4} (4x + 6y + 5) \, dy\, dx$

(See Problem 4.)

13. $\displaystyle\int_{1}^{5}\int_{0}^{2} \dfrac{x}{\sqrt{y + x^2}} \, dx\, dy$

(See Problem 5.)

14. $\displaystyle\int_{0}^{2}\int_{1}^{5} \dfrac{x}{\sqrt{y + x^2}} \, dy\, dx$

(See Problem 6.)

15. $\displaystyle\int_{1}^{e}\int_{1}^{e^2} \dfrac{\ln x}{xy} \, dy\, dx$

(See Problem 7.)

16. $\displaystyle\int_{1}^{e^2}\int_{1}^{e} \dfrac{\ln x}{xy} \, dx\, dy$

(See Problem 8.)

Use both orders of iteration to evaluate each double integral in Problems 17–20.

17. $\displaystyle\iint_R xy\, dA; R = \{(x, y)|0 \le x \le 2, \ 0 \le y \le 4\}$

18. $\displaystyle\iint_R \sqrt{xy}\, dA; R = \{(x, y)|1 \le x \le 4, \ 1 \le y \le 9\}$

19. $\displaystyle\iint_R (x + y)^5\, dA; R = \{(x, y)|-1 \le x \le 1, \ 1 \le y \le 2\}$

20. $\displaystyle\iint_R xe^y\, dA; R = \{(x, y)|-2 \le x \le 3, \ 0 \le y \le 2\}$

In Problems 21–24, find the average value of each function over the given rectangle.

21. $f(x, y) = (x + y)^2$;

$R = \{(x, y)|1 \le x \le 5, \ -1 \le y \le 1\}$

22. $f(x, y) = x^2 + y^2$;

$R = \{(x, y)|-1 \le x \le 2, \ 1 \le y \le 4\}$

23. $f(x, y) = x/y; R = \{(x, y)|1 \le x \le 4, \ 2 \le y \le 7\}$

24. $f(x, y) = x^2 y^3$;

$R = \{(x, y)|-1 \le x \le 1, \ 0 \le y \le 2\}$

In Problems 25–28, find the volume of the solid under the graph of each function over the given rectangle.

25. $f(x, y) = 2 - x^2 - y^2$;

$R = \{(x, y)|0 \le x \le 1, \ 0 \le y \le 1\}$

26. $f(x, y) = 5 - x$;

$R = \{(x, y)|0 \le x \le 5, \ 0 \le y \le 5\}$

27. $f(x, y) = 4 - y^2$;

$R = \{(x, y)|0 \le x \le 2, \ 0 \le y \le 2\}$

28. $f(x, y) = e^{-x-y}$;

$R = \{(x, y)|0 \le x \le 1, \ 0 \le y \le 1\}$

C

Evaluate each double integral in Problems 29–32. Select the order of integration carefully; each problem is easy to do one way and difficult the other.

29. $\displaystyle\iint_R xe^{xy}\, dA; R = \{(x, y)|0 \le x \le 1, \ 1 \le y \le 2\}$

30. $\displaystyle\iint_R xye^{x^2 y}\, dA; R = \{(x, y)|0 \le x \le 1, \ 1 \le y \le 2\}$

31. $\displaystyle\iint_R \dfrac{2y + 3xy^2}{1 + x^2}\, dA;$

$R = \{(x, y)|0 \le x \le 1, \ -1 \le y \le 1\}$

32. $\displaystyle\iint_R \dfrac{2x + 2y}{1 + 4y + y^2}\, dA;$

$R = \{(x, y)|1 \le x \le 3, \ 0 \le y \le 1\}$

33. Show that $\int_0^2\int_0^2 (1 - y)\, dx\, dy = 0$. Does the double integral represent the volume of a solid? Explain.

34. (A) Find the average values of the functions $f(x, y) = x + y$, $g(x, y) = x^2 + y^2$, and $h(x, y) = x^3 + y^3$ over the rectangle

$R = \{(x, y)|0 \le x \le 1, \ 0 \le y \le 1\}$

(B) Does the average value of $k(x, y) = x^n + y^n$ over the rectangle

$R_1 = \{(x, y)|0 \le x \le 1, \ 0 \le y \le 1\}$

increase or decrease as n increases? Explain.

(C) Does the average value of $k(x, y) = x^n + y^n$ over the rectangle

$R_2 = \{(x, y)|0 \le x \le 2, \ 0 \le y \le 2\}$

increase or decrease as n increases? Explain.

35. Let $f(x, y) = x^3 + y^2 - e^{-x} - 1$.

   (A) Find the average value of $f(x, y)$ over the rectangle
   $$R = \{(x, y)|-2 \le x \le 2, \quad -2 \le y \le 2\}.$$

   (B) Graph the set of all points (x, y) in R for which $f(x, y) = 0$.

   (C) For which points (x, y) in R is $f(x, y)$ greater than 0? Less than 0? Explain.

36. Find the dimensions of the square S centered at the origin for which the average value of $f(x, y) = x^2 e^y$ over S is equal to 100.

Applications

37. Multiplier principle. Suppose that Congress enacts a one-time-only 10% tax rebate that is expected to infuse $\$y$ billion, $5 \le y \le 7$, into the economy. If every person and every corporation is expected to spend a proportion $x, 0.6 \le x \le 0.8$, of each dollar received, then, by the **multiplier principle** in economics, the total amount of spending S (in billions of dollars) generated by this tax rebate is given by

$$S(x, y) = \frac{y}{1 - x}$$

What is the average total amount of spending for the indicated ranges of the values of x and y? Set up a double integral and evaluate it.

38. Multiplier principle. Repeat Problem 37 if $6 \le y \le 10$ and $0.7 \le x \le 0.9$.

39. Cobb–Douglas production function. If an industry invests x thousand labor-hours, $10 \le x \le 20$, and $\$y$ million, $1 \le y \le 2$, in the production of N thousand units of a certain item, then N is given by

$$N(x, y) = x^{0.75} y^{0.25}$$

What is the average number of units produced for the indicated ranges of x and y? Set up a double integral and evaluate it.

40. Cobb–Douglas production function. Repeat Problem 39 for

$$N(x, y) = x^{0.5} y^{0.5}$$

where $10 \le x \le 30$ and $1 \le y \le 3$.

41. Population distribution. In order to study the population distribution of a certain species of insect, a biologist has constructed an artificial habitat in the shape of a rectangle 16 feet long and 12 feet wide. The only food available to the insects in this habitat is located at its center. The biologist has determined that the concentration C of insects per square foot at a point d units from the food supply (see the figure) is given approximately by

$$C = 10 - \tfrac{1}{10}d^2$$

What is the average concentration of insects throughout the habitat? Express C as a function of x and y, set up a double integral, and evaluate it.

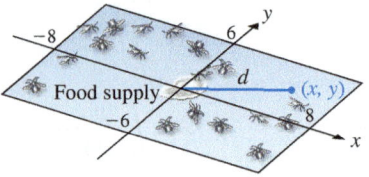

Figure for 41

42. Population distribution. Repeat Problem 41 for a square habitat that measures 12 feet on each side, where the insect concentration is given by

$$C = 8 - \tfrac{1}{10}d^2$$

43. Pollution. A heavy industrial plant located in the center of a small town emits particulate matter into the atmosphere. Suppose that the concentration of particulate matter (in parts per million) at a point d miles from the plant (see the figure) is given by

$$C = 100 - 15d^2$$

If the boundaries of the town form a rectangle 4 miles long and 2 miles wide, what is the average concentration of particulate matter throughout the town? Express C as a function of x and y, set up a double integral, and evaluate it.

Figure for 43

44. Pollution. Repeat Problem 43 if the boundaries of the town form a rectangle 8 miles long and 4 miles wide and the concentration of particulate matter is given by

$$C = 100 - 3d^2$$

45. Safety research. Under ideal conditions, if a person driving a car slams on the brakes and skids to a stop, the length of the skid marks (in feet) is given by the formula

$$L = 0.000\,013\,3xy^2$$

where x is the weight of the car (in pounds) and y is the speed of the car (in miles per hour). What is the average length of the skid marks for cars weighing between 2,000 and 3,000 pounds and traveling at speeds between 50 and 60 miles per hour? Set up a double integral and evaluate it.

46. Safety research. Repeat Problem 45 for cars weighing between 2,000 and 2,500 pounds and traveling at speeds between 40 and 50 miles per hour.

47. Psychology. The intelligence quotient Q for a person with mental age x and chronological age y is given by

$$Q(x, y) = 100\frac{x}{y}$$

In a group of sixth-graders, the mental age varies between 8 and 16 years and the chronological age varies between 10 and 12 years. What is the average intelligence quotient for this group? Set up a double integral and evaluate it.

48. Psychology. Repeat Problem 47 for a group with mental ages between 6 and 14 years and chronological ages between 8 and 10 years.

Answers to Matched Problems

1. (A) $2xy^2 + 3x^2y^4 + C(x)$ (B) $2x^2y + 4x^3y^3 + E(y)$
2. (A) $2x + 3x^2$ (B) $18y + 108y^3$
3. (A) 36   (B) 36
4. 12
5. $e - 2e^{1/2} + 1$
6. 2
7. 5 cubic units

6-7   Double Integrals over More General Regions

- Regular Regions
- Double Integrals over Regular Regions
- Reversing the Order of Integration
- Volume and Double Integrals

In this section, we extend the concept of double integration discussed in Section 6-6 to nonrectangular regions. We begin with an example and some new terminology.

Regular Regions

Let R be the region graphed in Figure 1. We can describe R with the following inequalities:

$$R = \{(x, y)|x \le y \le 6x - x^2, \ 0 \le x \le 5\}$$

The region R can be viewed as a union of vertical line segments. For each x in the interval $[0, 5]$, the line segment from the point $(x, g(x))$ to the point $(x, f(x))$ lies in the region R. Any region that can be covered by vertical line segments in this manner is called a *regular x region*.

Now consider the region S in Figure 2. It can be described with the following inequalities:

$$S = \{(x, y)|y^2 \le x \le y + 2, \ -1 \le y \le 2\}$$

The region S can be viewed as a union of horizontal line segments going from the graph of $h(y) = y^2$ to the graph of $k(y) = y + 2$ on the interval $[-1, 2]$. Regions that can be described in this manner are called *regular y regions*.

In general, *regular regions* are defined as follows:

Figure 1

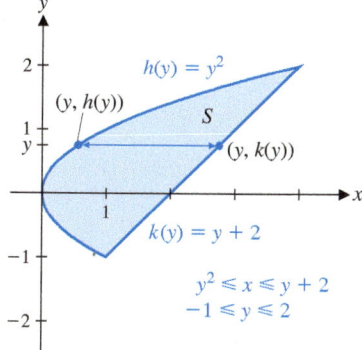

Figure 2

DEFINITION Regular Regions

A region R in the xy plane is a **regular x region** if there exist functions $f(x)$ and $g(x)$ and numbers a and b such that

$$R = \{(x, y)|g(x) \le y \le f(x), \ a \le x \le b\}$$

A region R is a **regular y region** if there exist functions $h(y)$ and $k(y)$ and numbers c and d such that

$$R = \{(x, y)|h(y) \le x \le k(y), \ c \le y \le d\}$$

See Figure 3 for a geometric interpretation.

CONCEPTUAL INSIGHT

If, for some region R, there is a horizontal line that has a nonempty intersection I with R, and if I is neither a closed interval nor a point, then R is *not* a regular y region. Similarly, if, for some region R, there is a vertical line that has a nonempty intersection I with R, and if I is neither a closed interval nor a point, then R is *not* a regular x region (see Fig. 3).

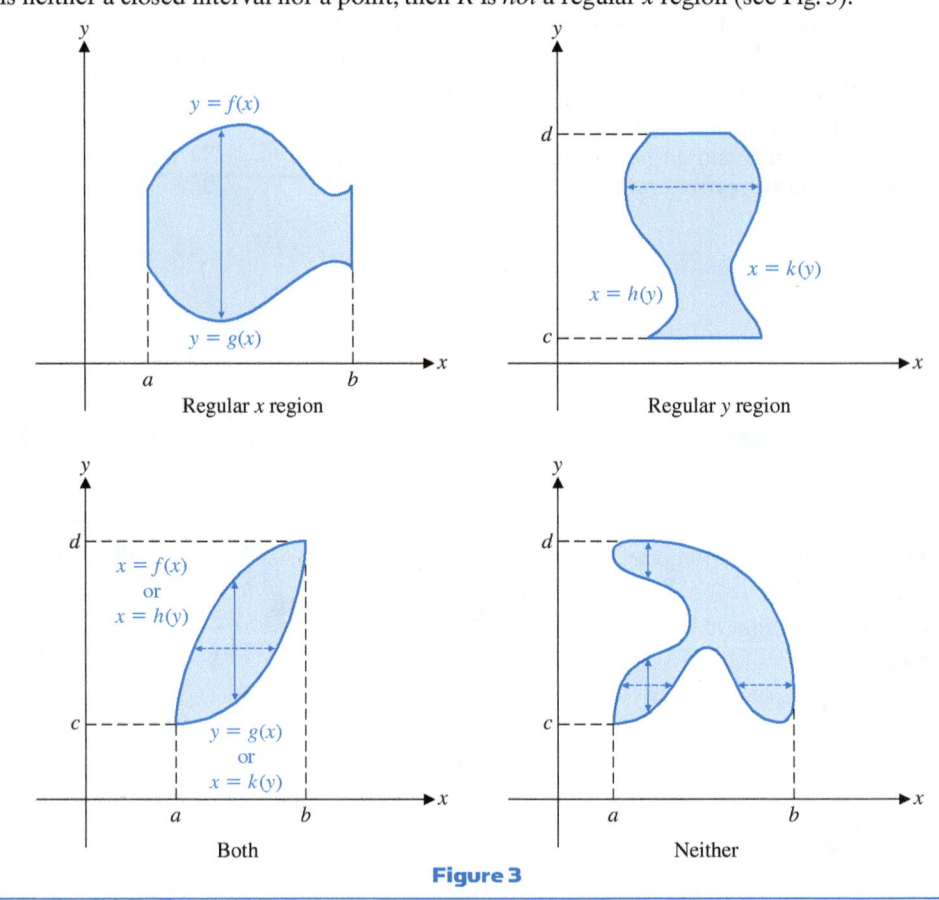

Figure 3

EXAMPLE 1   **Describing a Regular x Region** The region R is bounded by the graphs of $y = 4 - x^2$ and $y = x - 2$, $x \geq 0$, and the y axis. Graph R and use set notation and double inequalities to describe R as a regular x region.

SOLUTION   As the solid line in the following figure indicates, R can be covered by vertical line segments that go from the graph of $y = x - 2$ to the graph of $y = 4 - x^2$. So, R is a regular x region. In terms of set notation and double inequalities, we can write

$$R = \{(x, y) \mid x - 2 \leq y \leq 4 - x^2, \ 0 \leq x \leq 2\}$$

CONCEPTUAL INSIGHT

The region R of Example 1 is also a regular y region, since $R = \{(x, y) | 0 \leq x \leq k(y),$ $-2 \leq y \leq 4\}$, where

$$k(y) = \begin{cases} 2 + y & \text{if } -2 \leq y \leq 0 \\ \sqrt{4 - y} & \text{if } 0 \leq y \leq 4 \end{cases}$$

But because $k(y)$ is piecewise defined, this description is more complicated than the description of R in Example 1 as a regular x region.

Matched Problem 1 Describe the region R bounded by the graphs of $x = 6 - y$ and $x = y^2$, $y \geq 0$, and the x axis as a regular y region.

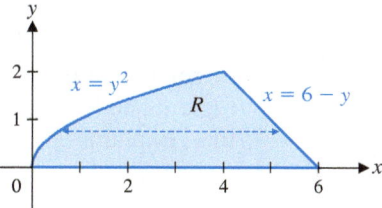

EXAMPLE 2 **Describing Regular Regions** The region R is bounded by the graphs of $x + y^2 = 9$ and $x + 3y = 9$. Graph R and describe R as a regular x region, a regular y region, both, or neither. Represent R in set notation and with double inequalities.

SOLUTION

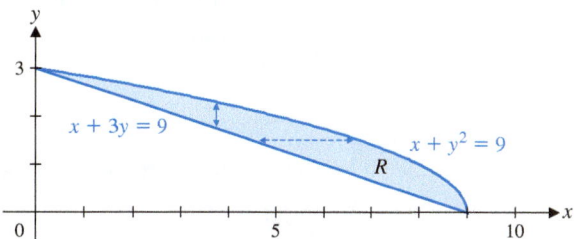

Region R can be covered by vertical line segments that go from the graph of $x + 3y = 9$ to the graph of $x + y^2 = 9$. Thus, R is a regular x region. In order to describe R with inequalities, we must solve each equation for y in terms of x:

$$\begin{array}{ll} x + 3y = 9 & x + y^2 = 9 \\ 3y = 9 - x & y^2 = 9 - x \\ y = 3 - \tfrac{1}{3}x & y = \sqrt{9 - x} \end{array}$$
We use the positive square root, since the graph is in the first quadrant.

So,

$$R = \{(x, y) | 3 - \tfrac{1}{3}x \leq y \leq \sqrt{9 - x}, \ 0 \leq x \leq 9\}$$

Since region R also can be covered by horizontal line segments (see the dashed line in the preceding figure) that go from the graph of $x + 3y = 9$ to the graph of $x + y^2 = 9$, it is a regular y region. Now we must solve each equation for x in terms of y:

$$\begin{array}{ll} x + 3y = 9 & x + y^2 = 9 \\ x = 9 - 3y & x = 9 - y^2 \end{array}$$

Therefore,

$$R = \{(x, y) | 9 - 3y \leq x \leq 9 - y^2, \ 0 \leq y \leq 3\}$$

Matched Problem 2   Repeat Example 2 for the region bounded by the graphs of $2y - x = 4$ and $y^2 - x = 4$, as shown in the following figure:

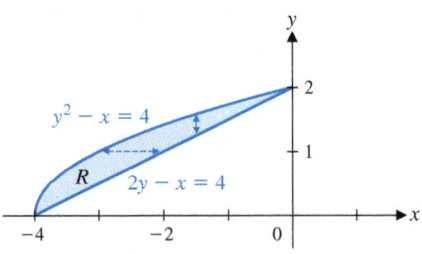

Explore & Discuss 1

A E I O U

Consider the vowels A, E, I, O, U, written in block letters as shown in the margin, to be regions of the plane. One of the vowels is a regular x region, but not a regular y region; one is a regular y region, but not a regular x region; one is both; two are neither. Explain.

Double Integrals over Regular Regions

Now we want to extend the definition of double integration to include regular x regions and regular y regions. The order of integration now depends on the nature of the region R. If R is a regular x region, we integrate with respect to y first, while if R is a regular y region, we integrate with respect to x first.

> Note that the variable limits of integration (when present) are always on the inner integral, and the constant limits of integration are always on the outer integral.

DEFINITION   Double Integration over Regular Regions

Regular x Region

If $R = \{(x, y) | g(x) \le y \le f(x), \quad a \le x \le b\}$, then

$$\iint\limits_{R} F(x, y) \, dA = \int_a^b \left[\int_{g(x)}^{f(x)} F(x, y) \, dy \right] dx$$

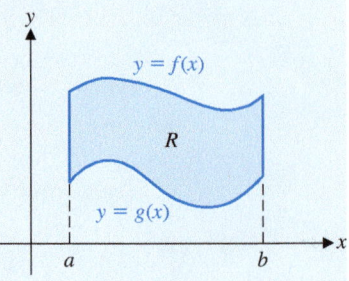

Regular y Region

If $R = \{(x, y) \mid h(y) \le x \le k(y), \quad c \le y \le d\}$, then

$$\iint\limits_{R} F(x, y) \, dA = \int_c^d \left[\int_{h(y)}^{k(y)} F(x, y) \, dx \right] dy$$

EXAMPLE 3   **Evaluating a Double Integral** Evaluate $\iint\limits_{R} 2xy \, dA$, where R is the region bounded by the graphs of $y = -x$ and $y = x^2$, $x \geq 0$, and the graph of $x = 1$.

SOLUTION   From the graph, we can see that R is a regular x region described by

$$R = \{(x, y)\,|\, -x \leq y \leq x^2, \ 0 \leq x \leq 1\}$$

$$\iint\limits_{R} 2xy \, dA = \int_0^1 \left[\int_{-x}^{x^2} 2xy \, dy \right] dx$$

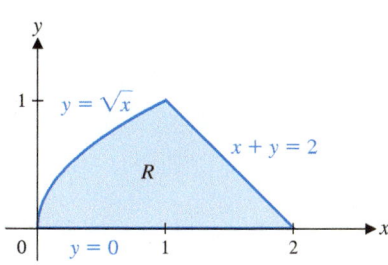

$$= \int_0^1 \left[xy^2 \Big|_{y=-x}^{y=x^2} \right] dx$$

$$= \int_0^1 \left[x(x^2)^2 - x(-x)^2 \right] dx$$

$$= \int_0^1 (x^5 - x^3) \, dx$$

$$= \left(\frac{x^6}{6} - \frac{x^4}{4} \right) \Big|_{x=0}^{x=1}$$

$$= \left(\tfrac{1}{6} - \tfrac{1}{4} \right) - (0 - 0) = -\tfrac{1}{12}$$

Matched Problem 3   Evaluate $\iint\limits_{R} 3xy^2 \, dA$, where R is the region in Example 3.

EXAMPLE 4   **Evaluating a Double Integral** Evaluate $\iint\limits_{R} (2x + y) \, dA$, where R is the region bounded by the graphs of $y = \sqrt{x}$, $x + y = 2$, and $y = 0$.

SOLUTION   From the graph, we can see that R is a regular y region. After solving each equation for x, we can write

$$R = \{(x, y)\,|\, y^2 \leq x \leq 2 - y, \ 0 \leq y \leq 1\}$$

$$\iint\limits_{R} (2x + y) \, dA = \int_0^1 \left[\int_{y^2}^{2-y} (2x + y) \, dx \right] dy$$

$$= \int_0^1 \left[(x^2 + yx) \Big|_{x=y^2}^{x=2-y} \right] dy$$

$$= \int_0^1 \{ [(2 - y)^2 + y(2 - y)] - [(y^2)^2 + y(y^2)] \} \, dy$$

$$= \int_0^1 (4 - 2y - y^3 - y^4) \, dy$$

$$= \left(4y - y^2 - \tfrac{1}{4} y^4 - \tfrac{1}{5} y^5 \right) \Big|_{y=0}^{y=1}$$

$$= \left(4 - 1 - \tfrac{1}{4} - \tfrac{1}{5} \right) - 0 = \tfrac{51}{20}$$

Matched Problem 4   Evaluate $\iint\limits_{R} (y - 4x)\,dA$, where R is the region in Example 4.

EXAMPLE 5   **Evaluating a Double Integral**   The region R is bounded by the graphs of $y = \sqrt{x}$ and $y = \frac{1}{2}x$. Evaluate $\iint\limits_{R} 4xy^3\,dA$ two different ways.

SOLUTION   Region R is both a regular x region and a regular y region:
$$R = \{(x, y)|\tfrac{1}{2}x \le y \le \sqrt{x},\ \ 0 \le x \le 4\} \quad \text{Regular } x \text{ region}$$
$$R = \{(x, y)|y^2 \le x \le 2y,\ \ 0 \le y \le 2\} \quad \text{Regular } y \text{ region}$$

Using the first representation (a regular x region), we obtain

$$\iint\limits_{R} 4xy^3\,dA = \int_0^4 \left[\int_{x/2}^{\sqrt{x}} 4xy^3\,dy\right]dx$$

$$= \int_0^4 \left[xy^4\,\Big|_{y=x/2}^{y=\sqrt{x}}\right]dx$$

$$= \int_0^4 [x(\sqrt{x})^4 - x(\tfrac{1}{2}x)^4]\,dx$$

$$= \int_0^4 \left(x^3 - \tfrac{1}{16}x^5\right)dx$$

$$= \left(\tfrac{1}{4}x^4 - \tfrac{1}{96}x^6\right)\Big|_{x=0}^{x=4}$$

$$= \left(64 - \tfrac{128}{3}\right) - 0 = \tfrac{64}{3}$$

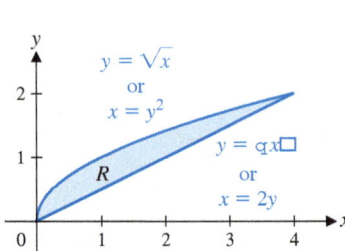

Using the second representation (a regular y region), we obtain

$$\iint\limits_{R} 4xy^3\,dA = \int_0^2 \left[\int_{y^2}^{2y} 4xy^3\,dx\right]dy$$

$$= \int_0^2 \left[2x^2y^3\,\Big|_{x=y^2}^{x=2y}\right]dy$$

$$= \int_0^2 [2(2y)^2y^3 - 2(y^2)^2y^3]\,dy$$

$$= \int_0^2 (8y^5 - 2y^7)\,dy$$

$$= \left(\tfrac{4}{3}y^6 - \tfrac{1}{4}y^8\right)\Big|_{y=0}^{y=2}$$

$$= \left(\tfrac{256}{3} - 64\right) - 0 = \tfrac{64}{3}$$

Matched Problem 5   The region R is bounded by the graphs of $y = x$ and $y = \frac{1}{2}x^2$. Evaluate $\iint\limits_{R} 4xy^3\,dA$ two different ways.

Reversing the Order of Integration

Example 5 shows that

$$\iint\limits_{R} 4xy^3 \, dA = \int_0^4 \left[\int_{x/2}^{\sqrt{x}} 4xy^3 \, dy \right] dx = \int_0^2 \left[\int_{y^2}^{2y} 4xy^3 \, dx \right] dy$$

In general, if R is both a regular x region and a regular y region, then the two iterated integrals are equal. In rectangular regions, reversing the order of integration in an iterated integral was a simple matter. As Example 5 illustrates, the process is more complicated in nonrectangular regions. The next example illustrates how to start with an iterated integral and reverse the order of integration. Since we are interested in the reversal process and not in the value of either integral, the integrand will not be specified.

EXAMPLE 6   **Reversing the Order of Integration**   Reverse the order of integration in

$$\int_1^3 \left[\int_0^{x-1} f(x, y) \, dy \right] dx$$

SOLUTION   The order of integration indicates that the region of integration is a regular x region:

$$R = \{(x, y) | 0 \le y \le x - 1, \ 1 \le x \le 3\}$$

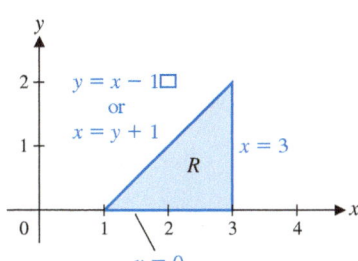

Graph region R to determine whether it is also a regular y region. The graph shows that R is also a regular y region, and we can write

$$R = \{(x, y) | y + 1 \le x \le 3, \ 0 \le y \le 2\}$$

$$\int_1^3 \left[\int_0^{x-1} f(x, y) \, dy \right] dx = \int_0^2 \left[\int_{y+1}^3 f(x, y) \, dx \right] dy$$

Matched Problem 6   Reverse the order of integration in $\int_2^4 \left[\int_0^{4-x} f(x, y) \, dy \right] dx$.

EXPLORE & DISCUSS 2   Explain the difficulty in evaluating $\int_0^2 \int_{x^2}^4 xe^{y^2} dy \, dx$ and how it can be overcome by reversing the order of integration.

Volume and Double Integrals

In Section 6-6, we used the double integral to calculate the volume of a solid with a rectangular base. In general, if a solid can be described by the graph of a positive function $f(x, y)$ over a regular region R (not necessarily a rectangle), then the double integral of the function f over the region R still represents the volume of the corresponding solid.

EXAMPLE 7 **Volume** The region R is bounded by the graphs of $x + y = 1$, $y = 0$, and $x = 0$. Find the volume of the solid under the graph of $z = 1 - x - y$ over the region R.

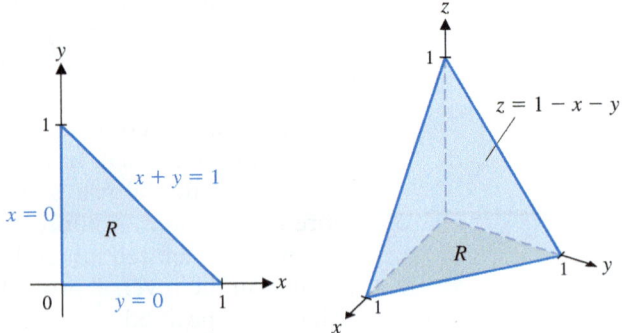

SOLUTION The graph of R shows that R is both a regular x region and a regular y region. We choose to use the regular x region:

$$R = \{(x, y) | 0 \le y \le 1 - x, \ 0 \le x \le 1\}$$

The volume of the solid is

$$V = \iint\limits_{R} (1 - x - y) \, dA = \int_0^1 \left[\int_0^{1-x} (1 - x - y) \, dy \right] dx$$

$$= \int_0^1 \left[\left(y - xy - \tfrac{1}{2}y^2 \right) \Big|_{y=0}^{y=1-x} \right] dx$$

$$= \int_0^1 \left[(1 - x) - x(1 - x) - \tfrac{1}{2}(1 - x)^2 \right] dx$$

$$= \int_0^1 \left(\tfrac{1}{2} - x + \tfrac{1}{2}x^2 \right) dx$$

$$= \left(\tfrac{1}{2}x - \tfrac{1}{2}x^2 + \tfrac{1}{6}x^3 \right) \Big|_{x=0}^{x=1}$$

$$= \left(\tfrac{1}{2} - \tfrac{1}{2} + \tfrac{1}{6} \right) - 0 = \tfrac{1}{6}$$

Matched Problem 7 The region R is bounded by the graphs of $y + 2x = 2$, $y = 0$, and $x = 0$. Find the volume of the solid under the graph of $z = 2 - 2x - y$ over the region R. [*Hint*: Sketch the region first; the solid does not have to be sketched.]

Exercises 6-7

A

In Problems 1–6, graph the region R bounded by the graphs of the equations. Use set notation and double inequalities to describe R as a regular x region and a regular y region in Problems 1 and 2, and as a regular x region or a regular y region, whichever is simpler, in Problems 3–6.

1. $y = 4 - x^2$, $y = 0$, $0 \le x \le 2$

2. $y = x^2$, $y = 9$, $0 \le x \le 3$

3. $y = x^3$, $y = 12 - 2x$, $x = 0$

4. $y = 5 - x$, $y = 1 + x$, $y = 0$

5. $y^2 = 2x$, $y = x - 4$

6. $y = 4 + 3x - x^2$, $x + y = 4$

Evaluate each integral in Problems 7–10.

7. $\int_0^1 \int_0^x (x + y) \, dy \, dx$

8. $\int_0^2 \int_0^y xy \, dx \, dy$

9. $\int_0^1 \int_{y^3}^{\sqrt{y}} (2x + y) \, dx \, dy$

10. $\int_1^4 \int_x^{x^2} (x^2 + 2y) \, dy \, dx$

B

✎ *In Problems 11–14, give a verbal description of the region R and determine whether R is a regular x region, a regular y region, both, or neither.*

11. $R = \{(x, y) \,||x| \le 2, \ |y| \le 3\}$

12. $R = \{(x, y) \,|1 \le x^2 + y^2 \le 4\}$

13. $R = \{(x, y) \,|x^2 + y^2 \ge 1, \ |x| \le 2, \ 0 \le y \le 2\}$

14. $R = \{(x, y) \,||x| + |y| \le 1\}$

In Problems 15–20, use the description of the region R to evaluate the indicated integral.

15. $\iint\limits_R (x^2 + y^2) \, dA;$

$R = \{(x, y) \,|0 \le y \le 2x, \ 0 \le x \le 2\}$

16. $\iint\limits_R 2x^2 y \, dA;$

$R = \{(x, y) \,|0 \le y \le 9 - x^2, \ -3 \le x \le 3\}$

17. $\iint\limits_R (x + y - 2)^3 \, dA;$

$R = \{(x, y) \,|0 \le x \le y + 2, \ 0 \le y \le 1\}$

18. $\iint\limits_R (2x + 3y) \, dA;$

$R = \{(x, y) \,|y^2 - 4 \le x \le 4 - 2y, \ 0 \le y \le 2\}$

19. $\iint\limits_R e^{x+y} \, dA;$

$R = \{(x, y) \,|-x \le y \le x, \ 0 \le x \le 2\}$

20. $\iint\limits_R \dfrac{x}{\sqrt{x^2 + y^2}} \, dA;$

$R = \{(x, y) \,|0 \le x \le \sqrt{4y - y^2}, \ 0 \le y \le 2\}$

In Problems 21–26, graph the region R bounded by the graphs of the indicated equations. Describe R in set notation with double inequalities, and evaluate the indicated integral.

21. $y = x + 1, y = 0, x = 0, x = 1;$

$\iint\limits_R \sqrt{1 + x + y} \, dA$

22. $y = x^2, \ y = \sqrt{x} \, ; \ \iint\limits_R 12xy \, dA$

23. $y = 4x - x^2, \ y = 0; \ \iint\limits_R \sqrt{y + x^2} \, dA$

24. $x = 1 + 3y, \ x = 1 - y, \ y = 1;$

$\iint\limits_R (x + y + 1)^3 \, dA$

25. $y = 1 - \sqrt{x}, y = 1 + \sqrt{x}, x = 4;$

$\iint\limits_R x(y - 1)^2 \, dA$

26. $y = \tfrac{1}{2}x, y = 6 - x, \ y = 1; \ \iint\limits_R \dfrac{1}{x + y} \, dA$

In Problems 27–32, evaluate each integral. Graph the region of integration, reverse the order of integration, and then evaluate the integral with the order reversed.

27. $\int_0^3 \int_0^{3-x} (x + 2y) \, dy \, dx$

28. $\int_0^2 \int_0^y (y - x)^4 \, dx \, dy$

29. $\int_0^1 \int_0^{1-x^2} x\sqrt{y} \, dy \, dx$

30. $\int_0^2 \int_{x^3}^{4x} (1 + 2y) \, dy \, dx$

31. $\int_0^4 \int_{x/4}^{\sqrt{x}/2} x \, dy \, dx$

32. $\int_0^4 \int_{y^2/4}^{2\sqrt{y}} (1 + 2xy) \, dx \, dy$

In Problems 33–36, find the volume of the solid under the graph of f (x, y) over the region R bounded by the graphs of the indicated equations. Sketch the region R; the solid does not have to be sketched.

33. $f(x, y) = 4 - x - y$; R is the region bounded by the graphs of $x + y = 4, y = 0, x = 0$

34. $f(x, y) = (x - y)^2$; R is the region bounded by the graphs of $y = x, y = 2, x = 0$

35. $f(x, y) = 4$; R is the region bounded by the graphs of $y = 1 - x^2$ and $y = 0$ for $0 \le x \le 1$

36. $f(x, y) = 4xy$; R is the region bounded by the graphs of $y = \sqrt{1 - x^2}$ and $y = 0$ for $0 \le x \le 1$

C

In Problems 37–40, reverse the order of integration for each integral. Evaluate the integral with the order reversed. Do not attempt to evaluate the integral in the original form.

37. $\int_0^2 \int_{x^2}^4 \dfrac{4x}{1 + y^2} \, dy \, dx$

38. $\int_0^1 \int_y^1 \sqrt{1 - x^2} \, dx \, dy$

39. $\int_0^1 \int_{y^2}^1 4ye^{x^2} \, dx \, dy$

40. $\int_0^4 \int_{\sqrt{x}}^2 \sqrt{3x + y^2} \, dy \, dx$

In Problems 41–46, use a graphing calculator to graph the region R bounded by the graphs of the indicated equations. Use approximation techniques to find intersection points correct to two decimal places. Describe R in set notation with double inequalities, and evaluate the indicated integral correct to two decimal places.

41. $y = 1 + \sqrt{x}, \quad y = x^2, \quad x = 0; \ \iint\limits_R x \, dA$

42. $y = 1 + \sqrt[3]{x}, y = x, x = 0; \ \iint\limits_R x \, dA$

43. $y = \sqrt[3]{x}, y = 1 - x, y = 0; \ \iint\limits_R 24xy \, dA$

44. $y = x^3, y = 1 - x, y = 0; \ \iint\limits_R 48xy \, dA$

45. $y = e^{-x}, \quad y = 3 - x; \quad \iint\limits_R 4y \, dA$

46. $y = e^x, \quad y = 2 + x; \quad \iint\limits_R 8y \, dA$

3. $\frac{13}{40}$

4. $-\frac{77}{20}$

5. $\frac{16}{3}$

6. $\int_0^2 \int_2^{4-y} f(x, y) \, dx \, dy$

7. $\frac{2}{3}$

Answers to Matched Problems

1. $R = \{(x, y) | y^2 \le x \le 6 - y, \ 0 \le y \le 2\}$

2. R is both a regular x region and a regular y region:

$R = \{(x, y) \mid \frac{1}{2}x + 2 \le y \le \sqrt{x + 4}, \ -4 \le x \le 0\}$

$R = \{(x, y) | y^2 - 4 \le x \le 2y - 4, \ 0 \le y \le 2\}$

Chapter 6   Review

Important Terms, Symbols, and Concepts

6-1 Functions of Several Variables

- An equation of the form $z = f(x, y)$ describes a **function of two independent variables** if, for each permissible ordered pair (x, y), there is one and only one value of z determined by $f(x, y)$. The variables x and y are **independent variables**, and z is a **dependent variable**. The set of all ordered pairs of permissible values of x and y is the **domain** of the function, and the set of all corresponding values $f(x, y)$ is the **range**. Functions of more than two independent variables are defined similarly.
- The graph of $z = f(x, y)$ consists of all triples (x, y, z) in a **three-dimensional coordinate system** that satisfy the equation. The graphs of the functions $z = f(x, y) = x^2 + y^2$ and $z = g(x, y) = x^2 - y^2$, for example, are **surfaces**; the first has a local minimum, and the second has a **saddle point**, at $(0, 0)$.

6-2 Partial Derivatives

- If $z = f(x, y)$, then the **partial derivative of f with respect to x**, denoted as $\partial z/\partial x$, f_x, or $f_x(x, y)$, is

$$\frac{\partial z}{\partial x} = \lim_{h \to 0} \frac{f(x + h, y) - f(x, y)}{h}$$

Similarly, the **partial derivative of f with respect to y**, denoted as $\partial z/\partial y$, f_y, or $f_y(x, y)$, is

$$\frac{\partial z}{\partial y} = \lim_{k \to 0} \frac{f(x, y + k) - f(x, y)}{k}$$

The partial derivatives $\partial z/\partial x$ and $\partial z/\partial y$ are said to be **first-order partial derivatives**.
- There are four **second-order partial derivatives** of $z = f(x, y)$:

$$f_{xx} = f_{xx}(x, y) = \frac{\partial^2 z}{\partial x^2} = \frac{\partial}{\partial x}\left(\frac{\partial z}{\partial x}\right)$$

$$f_{xy} = f_{xy}(x, y) = \frac{\partial^2 z}{\partial y \, \partial x} = \frac{\partial}{\partial y}\left(\frac{\partial z}{\partial x}\right)$$

$$f_{yx} = f_{yx}(x, y) = \frac{\partial^2 z}{\partial x \, \partial y} = \frac{\partial}{\partial x}\left(\frac{\partial z}{\partial y}\right)$$

$$f_{yy} = f_{yy}(x, y) = \frac{\partial^2 z}{\partial y^2} = \frac{\partial}{\partial y}\left(\frac{\partial z}{\partial y}\right)$$

6-3 Maxima and Minima

- If $f(a, b) \ge f(x, y)$ for all (x, y) in a circular region in the domain of f with (a, b) as center, then $f(a, b)$ is a **local maximum**. If $f(a, b) \le f(x, y)$ for all (x, y) in such a region, then $f(a, b)$ is a **local minimum**.
- If a function $f(x, y)$ has a local maximum or minimum at the point (a, b), and f_x and f_y exist at (a, b), then both first-order partial derivatives equal 0 at (a, b) [Theorem 1, p. 468].
- The second-derivative test for local extrema (Theorem 2, p. 469) gives conditions on the first- and second-order partial derivatives of $f(x, y)$, which guarantee that $f(a, b)$ is a local maximum, local minimum, or saddle point.

6-4 Maxima and Minima Using Lagrange Multipliers

- The **method of Lagrange multipliers** can be used to find local extrema of a function $z = f(x, y)$ subject to the constraint $g(x, y) = 0$. A procedure that lists the key steps in the method is given on page 377.

Ex. 1, p. 378
Ex. 2, p. 379
Ex. 3, p. 381

- The method of Lagrange multipliers can be extended to functions with arbitrarily many independent variables with one or more constraints (see Theorem 1, p. 376, and Theorem 2, p. 381, for the method when there are two and three independent variables, respectively).

6-5 Method of Least Squares

- **Linear regression** is the process of fitting a line $y = ax + b$ to a set of data points $(x_1, y_1), (x_2, y_2), \ldots, (x_n, y_n)$ by using the **method of least squares**.

Ex. 1, p. 390

- We minimize $F(a, b) = \sum_{k=1}^{n} (y_k - ax_k - b)^2$, the **sum of the squares of the residuals**, by computing the first-order partial derivatives of F and setting them equal to 0. Solving for a and b gives the formulas

$$a = \frac{n\left(\sum_{k=1}^{n} x_k y_k\right) - \left(\sum_{k=1}^{n} x_k\right)\left(\sum_{k=1}^{n} y_k\right)}{n\left(\sum_{k=1}^{n} x_k^2\right) - \left(\sum_{k=1}^{n} x_k\right)^2}$$

$$b = \frac{\sum_{k=1}^{n} y_k - a\left(\sum_{k=1}^{n} x_k\right)}{n}$$

- Graphing calculators have built-in routines to calculate linear—as well as quadratic, cubic, quartic, logarithmic, exponential, power, and trigonometric—regression equations.

Ex. 2, p. 392

6-6 Double Integrals over Rectangular Regions

- The **double integral** of a function $f(x, y)$ over a rectangle

$$R = \{(x, y) | a \leq x \leq b, \ c \leq y \leq d\}$$

is

Ex. 1, p. 396
Ex. 2, p. 397
Ex. 3, p. 397
Ex. 4, p. 399
Ex. 5, p. 399

$$\iint\limits_{R} f(x, y) \, dA = \int_a^b \left[\int_c^d f(x, y) \, dy \right] dx$$

$$= \int_c^d \left[\int_a^b f(x, y) \, dx \right] dy$$

- In the double integral $\iint_R f(x, y) \, dA$, $f(x, y)$ is called the **integrand** and R is called the **region of integration**. The expression dA indicates that this is an integral over a two-dimensional region. The integrals

$$\int_a^b \left[\int_c^d f(x, y) \, dy \right] dx \qquad \text{and} \qquad \int_c^d \left[\int_a^b f(x, y) \, dx \right] dy$$

are referred to as **iterated integrals** (the brackets are often omitted), and the order in which dx and dy are written indicates the order of integration.

- The **average value** of the function $f(x, y)$ over the rectangle

Ex. 6, p. 400

$$R = \{(x, y) | a \leq x \leq b, \ c \leq y \leq d\}$$

is

$$\frac{1}{(b - a)(d - c)} \iint\limits_{R} f(x, y) \, dA$$

- If $f(x, y) \geq 0$ over a rectangle $R = \{(x, y) | a \leq x \leq b, c \leq y \leq d\}$, then the volume of the solid formed by graphing f over the rectangle R is given by

Ex. 7, p. 401

$$V = \iint\limits_{R} f(x, y) \, dA$$

6-7 Double Integrals over More General Regions

- A region R in the xy plane is a **regular x region** if there exist functions $f(x)$ and $g(x)$ and numbers a and b such that

$$R = \{(x, y)\,|\, g(x) \le y \le f(x), \quad a \le x \le b\}$$

Ex. 1, p. 406

- A region R in the xy plane is a **regular y region** if there exist functions $h(y)$ and $k(y)$ and numbers c and d such that

$$R = \{(x, y)\,|\, h(y) \le x \le k(y), \quad c \le y \le d\}$$

Ex. 2, p. 407

- The double integral of a function $F(x, y)$ over a regular x region $R = \{(x, y)\,|\, g(x) \le y \le f(x), \ a \le x \le b\}$ is

$$\iint\limits_{R} F(x, y)\, dA = \int_a^b \left[\int_{g(x)}^{f(x)} F(x, y)\, dy \right] dx$$

Ex. 3, p. 409
Ex. 4, p. 409
Ex. 5, p. 410
Ex. 6, p. 411

- The double integral of a function $F(x, y)$ over a regular y region $R = \{(x, y)\,|\, h(y) \le x \le k(y), \ c \le y \le d\}$ is

$$\iint\limits_{R} F(x, y)\, dA = \int_c^d \left[\int_{h(y)}^{k(y)} F(x, y)\, dx \right] dy$$

Ex. 7, p. 412

Review Exercises

Work through all the problems in this chapter review and check your answers in the back of the book. Answers to all review problems are there, along with section numbers in italics to indicate where each type of problem is discussed. Where weaknesses show up, review appropriate sections of the text.

A

1. For $f(x, y) = 2{,}000 + 40x + 70y$, find $f(5, 10), f_x(x, y)$, and $f_y(x, y)$.

2. For $z = x^3 y^2$, find $\partial^2 z / \partial x^2$ and $\partial^2 z / \partial x\, \partial y$.

3. Evaluate $\int (6xy^2 + 4y)\, dy$.

4. Evaluate $\int (6xy^2 + 4y)\, dx$.

5. Evaluate $\int_0^1 \int_0^1 4xy\, dy\, dx$.

 6. For $f(x, y) = 6 + 5x - 2y + 3x^2 + x^3$, find $f_x(x, y)$, and $f_y(x, y)$, and explain why $f(x, y)$ has no local extrema.

B

7. For $f(x, y) = 3x^2 - 2xy + y^2 - 2x + 3y - 7$, find $f(2, 3)$, $f_y(x, y)$, and $f_y(2, 3)$.

8. For $f(x, y) = -4x^2 + 4xy - 3y^2 + 4x + 10y + 81$, find $[f_{xx}(2, 3)][f_{yy}(2, 3)] - [f_{xy}(2, 3)]^2$.

9. If $f(x, y) = x + 3y$ and $g(x, y) = x^2 + y^2 - 10$, find the critical points of $F(x, y, \lambda) = f(x, y) + \lambda g(x, y)$.

10. Use the least squares line for the data in the following table to estimate y when $x = 10$.

x	y
2	12
4	10
6	7
8	3

11. For $R = \{(x, y)\,|\,-1 \le x \le 1, \ 1 \le y \le 2\}$, evaluate the following in two ways:

$$\iint\limits_{R} (4x + 6y)\, dA$$

12. For $R = \{(x, y)\,|\,\sqrt{y} \le x \le 1, \ 0 \le y \le 1\}$, evaluate

$$\iint\limits_{R} (6x + y)\, dA$$

C

13. For $f(x, y) = e^{x^2 + 2y}$, find f_x, f_y, and f_{xy}.

14. For $f(x, y) = (x^2 + y^2)^5$, find f_x and f_{xy}.

15. Find all critical points and test for extrema for

$$f(x, y) = x^3 - 12x + y^2 - 6y$$

16. Use Lagrange multipliers to maximize $f(x, y) = xy$ subject to $2x + 3y = 24$.

17. Use Lagrange multipliers to minimize $f(x, y, z) = x^2 + y^2 + z^2$ subject to $2x + y + 2z = 9$.

18. Find the least squares line for the data in the following table.

x	y	x	y
10	50	60	80
20	45	70	85
30	50	80	90
40	55	90	90
50	65	100	110

19. Find the average value of $f(x, y) = x^{2/3} y^{1/3}$ over the rectangle

$$R = \{(x, y)\,|\,-8 \le x \le 8, \ 0 \le y \le 27\}$$

20. Find the volume of the solid under the graph of $z = 3x^2 + 3y^2$ over the rectangle

$$R = \{(x, y)\,|\, 0 \le x \le 1, \ -1 \le y \le 1\}$$

21. Without doing any computation, predict the average value of $f(x, y) = x + y$ over the rectangle $R = \{(x, y)\,|\,-10 \le x \le 10, \ -10 \le y \le 10\}$. Then check the correctness of your prediction by evaluating a double integral.

22. (A) Find the dimensions of the square S centered at the origin such that the average value of

$$f(x, y) = \frac{e^x}{y + 10}$$

over S is equal to 5.

 (B) Is there a square centered at the origin over which

$$f(x, y) = \frac{e^x}{y + 10}$$

has average value 0.05? Explain.

 23. Explain why the function $f(x, y) = 4x^3 - 5y^3$, subject to the constraint $3x + 2y = 7$, has no maxima or minima.

24. Find the volume of the solid under the graph of $F(x, y) = 60x^2y$ over the region R bounded by the graph of $x + y = 1$ and the coordinate axes.

Applications

25. Maximizing profit. A company produces x units of product A and y units of product B (both in hundreds per month). The monthly profit equation (in thousands of dollars) is given by

$$P(x, y) = -4x^2 + 4xy - 3y^2 + 4x + 10y + 81$$

 (A) Find $P_x(1, 3)$ and interpret the results.

(B) How many of each product should be produced each month to maximize profit? What is the maximum profit?

26. Minimizing material. A rectangular box with no top and six compartments (see the figure) is to have a volume of 96 cubic inches. Find the dimensions that will require the least amount of material.

Figure for 26

27. Profit. A company's annual profits (in millions of dollars) over a 5-year period are given in the following table. Use the least squares line to estimate the profit for the sixth year.

Year	Profit
1	2
2	2.5
3	3.1
4	4.2
5	4.3

28. Productivity. The Cobb–Douglas production function for a product is

$$N(x, y) = 10x^{0.8}y^{0.2}$$

where x is the number of units of labor and y is the number of units of capital required to produce N units of the product.

(A) Find the marginal productivity of labor and the marginal productivity of capital at $x = 40$ and $y = 50$. For the greatest increase in productivity, should management encourage increased use of labor or increased use of capital?

(B) If each unit of labor costs $100, each unit of capital costs $50, and $10,000 is budgeted for production of this product, use the method of Lagrange multipliers to determine the allocations of labor and capital that will maximize the number of units produced and find the maximum production. Find the marginal productivity of money and approximate the increase in production that would result from an increase of $2,000 in the amount budgeted for production.

(C) If $50 \le x \le 100$ and $20 \le y \le 40$, find the average number of units produced. Set up a double integral, and evaluate it.

29. Marine biology. The function used for timing dives with scuba gear is

$$T(V, x) = \frac{33V}{x + 33}$$

where T is the time of the dive in minutes, V is the volume of air (in cubic feet, at sea-level pressure) compressed into tanks, and x is the depth of the dive in feet. Find $T_x(70, 17)$ and interpret the results.

30. Pollution. A heavy industrial plant located in the center of a small town emits particulate matter into the atmosphere. Suppose that the concentration of particulate matter (in parts per million) at a point d miles from the plant is given by

$$C = 100 - 24d^2$$

If the boundaries of the town form a square 4 miles long and 4 miles wide, what is the average concentration of particulate matter throughout the town? Express C as a function of x and y, and set up a double integral and evaluate it.

31. Sociology. A sociologist found that the number n of long-distance telephone calls between two cities during a given period varied (approximately) jointly as the populations P_1 and P_2 of the two cities and varied inversely as the distance d between the cities. An equation for a period of 1 week is

$$n(P_1, P_2, d) = 0.001\frac{P_1P_2}{d}$$

Find $n(100,000, 50,000, 100)$.

32. Education. At the beginning of the semester, students in a foreign language course take a proficiency exam. The same exam is given at the end of the semester. The results for 5 students are shown in the following table. Use the least squares line to estimate the second exam score of a student who scored 40 on the first exam.

First Exam	Second Exam
30	60
50	75
60	80
70	85
90	90

33. Population density. The following table gives the U.S. population per square mile for the years 1900–2000:

U.S. Population Density

Year	Population (per Square Mile)	Year	Population (per Square Mile)
1900	25.6	1960	50.6
1910	31.0	1970	57.4
1920	35.6	1980	64.0
1930	41.2	1990	70.4
1940	44.2	2000	77.8
1950	50.7		

(A) Find the least squares line for the data, using $x = 0$ for 1900.

(B) Use the least squares line to estimate the population density in the United States in the year 2020.

(C) Now use quadratic regression and exponential regression to obtain the estimate of part (B).

34. Life expectancy. The following table gives life expectancies for males and females in a sample of Central and South American countries:

Life Expectancies for Central and South American Countries

Males	Females	Males	Females
62.30	67.50	70.15	74.10
68.05	75.05	62.93	66.58
72.40	77.04	68.43	74.88
63.39	67.59	66.68	72.80
55.11	59.43		

(A) Find the least squares line for the data.

(B) Use the least squares line to estimate the life expectancy of a female in a Central or South American country in which the life expectancy for males is 60 years.

(C) Now use quadratic regression and logarithmic regression to obtain the estimate of part (B).

Integration

Introduction

In Chapters 3, 4, and 5, we studied the *derivative* and its applications. In Chapter 7, we introduce the *integral*, the second key concept of calculus. The integral can be used to calculate areas, volumes, the index of income concentration, and consumers' surplus. At first glance, the integral may appear to be unrelated to the derivative. There is, however, a close connection between these two concepts, which is made precise by the *fundamental theorem of calculus* (Section 7-5). We consider many applications of integrals and differential equations in Chapter 7. See, for example, Problem 67 in Section 7-3, which explores how the age of an archeological site or artifact can be estimated.

7-1 Antiderivatives and Indefinite Integrals

- Antiderivatives
- Indefinite Integrals: Formulas and Properties
- Applications

Many operations in mathematics have reverses—addition and subtraction, multiplication and division, powers and roots. We now know how to find the derivatives of many functions. The reverse operation, *antidifferentiation* (the reconstruction of a function from its derivative), will receive our attention in this and the next two sections.

Antiderivatives

A function F is an **antiderivative** of a function f if $F'(x) = f(x)$.

The function $F(x) = \dfrac{x^3}{3}$ is an antiderivative of the function $f(x) = x^2$ because

$$\frac{d}{dx}\left(\frac{x^3}{3}\right) = x^2$$

However, $F(x)$ is not the only antiderivative of x^2. Note also that

$$\frac{d}{dx}\left(\frac{x^3}{3} + 2\right) = x^2 \qquad \frac{d}{dx}\left(\frac{x^3}{3} - \pi\right) = x^2 \qquad \frac{d}{dx}\left(\frac{x^3}{3} + \sqrt{5}\right) = x^2$$

Therefore,

$$\frac{x^3}{3} + 2 \qquad \frac{x^3}{3} - \pi \qquad \frac{x^3}{3} + \sqrt{5}$$

are also antiderivatives of x^2 because each has x^2 as a derivative. In fact, it appears that

$$\frac{x^3}{3} + C \qquad \text{for any real number } C$$

is an antiderivative of x^2 because

$$\frac{d}{dx}\left(\frac{x^3}{3} + C\right) = x^2$$

Antidifferentiation of a given function does not give a unique function, but an entire family of functions.

Does the expression

$$\frac{x^3}{3} + C \qquad \text{with } C \text{ any real number}$$

include all antiderivatives of x^2? Theorem 1 (stated without proof) indicates that the answer is yes.

THEOREM 1 Antiderivatives

If the derivatives of two functions are equal on an open interval (a, b), then the functions differ by at most a constant. Symbolically, if F and G are differentiable functions on the interval (a, b) and $F'(x) = G'(x)$ for all x in (a, b), then $F(x) = G(x) + k$ for some constant k.

CONCEPTUAL INSIGHT

Suppose that $F(x)$ is an antiderivative of $f(x)$. If $G(x)$ is any other antiderivative of $f(x)$, then by Theorem 1, the graph of $G(x)$ is a vertical translation of the graph of $F(x)$ (see Section 2-2).

EXAMPLE 1   **A Family of Antiderivatives** Note that

$$\frac{d}{dx}\left(\frac{x^2}{2}\right) = x$$

(A) Find all antiderivatives of $f(x) = x$.

(B) Graph the antiderivative of $f(x) = x$ that passes through the point $(0, 0)$; through the point $(0, 1)$; through the point $(0, 2)$.

(C) How are the graphs of the three antiderivatives in part (B) related?

SOLUTION   (A) By Theorem 1, any antiderivative of $f(x)$ has the form

$$F(x) = \frac{x^2}{2} + k$$

where k is a real number.

(B) Because $F(0) = (0^2/2) + k = k$, the functions

$$F_0(x) = \frac{x^2}{2}, \quad F_1(x) = \frac{x^2}{2} + 1, \quad \text{and} \quad F_2(x) = \frac{x^2}{2} + 2$$

pass through the points $(0, 0)$, $(0, 1)$, and $(0, 2)$, respectively (see Fig. 1).

(C) The graphs of the three antiderivatives are vertical translations of each other.

Figure 1

Matched Problem 1   Note that

$$\frac{d}{dx}(x^3) = 3x^2$$

(A) Find all antiderivatives of $f(x) = 3x^2$.

(B) Graph the antiderivative of $f(x) = 3x^2$ that passes through the point $(0, 0)$; through the point $(0, 1)$; through the point $(0, 2)$.

(C) How are the graphs of the three antiderivatives in part (B) related?

Indefinite Integrals: Formulas and Properties

Theorem 1 states that if the derivatives of two functions are equal, then the functions differ by at most a constant. We use the symbol

$$\int f(x)\, dx$$

called the **indefinite integral**, to represent the family of all antiderivatives of $f(x)$, and we write

$$\int f(x)\, dx = F(x) + C \quad \text{if} \quad F'(x) = f(x)$$

The symbol $\int$ is called an **integral sign**, and the function $f(x)$ is called the **integrand**. The symbol dx indicates that the antidifferentiation is performed with respect to the variable x. (We will have more to say about the symbols $\int$ and dx later in the chapter.) The arbitrary constant C is called the **constant of integration**. Referring to the preceding discussion, we can write

$$\int x^2\, dx = \frac{x^3}{3} + C \quad \text{since} \quad \frac{d}{dx}\left(\frac{x^3}{3} + C\right) = x^2$$

Of course, variables other than x can be used in indefinite integrals. For example,

$$\int t^2 \, dt = \frac{t^3}{3} + C \qquad \text{since} \qquad \frac{d}{dt}\left(\frac{t^3}{3} + C\right) = t^2$$

or

$$\int u^2 \, du = \frac{u^3}{3} + C \qquad \text{since} \qquad \frac{d}{du}\left(\frac{u^3}{3} + C\right) = u^2$$

The fact that indefinite integration and differentiation are reverse operations, except for the addition of the constant of integration, can be expressed symbolically as

$$\frac{d}{dx}\left[\int f(x) \, dx\right] = f(x) \qquad \text{\textit{The derivative of the indefinite integral of f(x) is f(x).}}$$

and

$$\int F'(x) \, dx = F(x) + C \qquad \text{\textit{The indefinite integral of the derivative of F(x) is F(x) + C.}}$$

We can develop formulas for the indefinite integrals of certain basic functions from the formulas for derivatives in Chapters 3 and 4.

> **FORMULAS Indefinite Integrals of Basic Functions**
>
> For C a constant,
>
> **1.** $\displaystyle\int x^n \, dx = \frac{x^{n+1}}{n+1} + C, \qquad n \neq -1$
>
> **2.** $\displaystyle\int e^x \, dx = e^x + C$
>
> **3.** $\displaystyle\int \frac{1}{x} \, dx = \ln|x| + C, \qquad x \neq 0$

To justify each formula, show that the derivative of the right-hand side is the integrand of the left-hand side (see Problems 77–80 in Exercise 7-1). Note that formula 1 does not give the antiderivative of x^{-1} (because $x^{n+1}/(n+1)$ is undefined when $n = -1$), but formula 3 does.

EXPLORE & DISCUSS 1

Formulas 1, 2, and 3 do *not* provide a formula for the indefinite integral of the function $\ln x$. Show that if $x > 0$, then

$$\int \ln x \, dx = x \ln x - x + C$$

by differentiating the right-hand side.

We can obtain properties of the indefinite integral from derivative properties that were established in Chapter 3.

> **PROPERTIES Indefinite Integrals**
>
> For k a constant,
>
> **4.** $\displaystyle\int k f(x) \, dx = k \int f(x) \, dx$
>
> **5.** $\displaystyle\int [f(x) \pm g(x)] \, dx = \int f(x) \, dx \pm \int g(x) \, dx$

Property 4 states that

> **The indefinite integral of a constant times a function is the constant times the indefinite integral of the function.**

Property 5 states that

> **The indefinite integral of the sum of two functions is the sum of the indefinite integrals, and the indefinite integral of the difference of two functions is the difference of the indefinite integrals.**

To establish property 4, let F be a function such that $F'(x) = f(x)$. Then

$$k \int f(x)\, dx = k \int F'(x)\, dx = k[F(x) + C_1] = kF(x) + kC_1$$

and since $[kF(x)]' = kF'(x) = kf(x)$, we have

$$\int kf(x)\, dx = \int kF'(x)\, dx = kF(x) + C_2$$

But $kF(x) + kC_1$ and $kF(x) + C_2$ describe the same set of functions, because C_1 and C_2 are arbitrary real numbers. Property 4 is established. Property 5 can be established in a similar manner (see Problems 81 and 82 in Exercise 7-1).

⚠ **CAUTION** Property 4 states that **a constant factor can be moved across an integral sign. A variable factor cannot be moved across an integral sign:**

CONSTANT FACTOR	*VARIABLE FACTOR*
$\displaystyle\int 5x^{1/2}\, dx = 5 \int x^{1/2}\, dx$	$\displaystyle\int xx^{1/2}\, dx \neq x \int x^{1/2}\, dx$

Indefinite integral formulas and properties can be used together to find indefinite integrals for many frequently encountered functions. If $n = 0$, then formula 1 gives

$$\int dx = x + C$$

Therefore, by property 4,

$$\int k\, dx = k(x + C) = kx + kC$$

Because kC is a constant, we replace it with a single symbol that denotes an arbitrary constant (usually C), and write

$$\int k\, dx = kx + C$$

In words,

> **The indefinite integral of a constant function with value k is $kx + C$.**

Similarly, using property 5 and then formulas 2 and 3, we obtain

$$\int \left(e^x + \frac{1}{x} \right) dx = \int e^x\, dx + \int \frac{1}{x}\, dx$$

$$= e^x + C_1 + \ln|x| + C_2$$

Because $C_1 + C_2$ is a constant, we replace it with the symbol C and write

$$\int \left(e^x + \frac{1}{x} \right) dx = e^x + \ln|x| + C$$

EXAMPLE 2 **Using Indefinite Integral Properties and Formulas**

(A) $\displaystyle\int 5\,dx = 5x + C$

(B) $\displaystyle\int 9e^x\,dx = 9\int e^x\,dx = 9e^x + C$

(C) $\displaystyle\int 5t^7\,dt = 5\int t^7\,dt = 5\frac{t^8}{8} + C = \frac{5}{8}t^8 + C$

(D) $\displaystyle\int (4x^3 + 2x - 1)\,dx = \int 4x^3\,dx + \int 2x\,dx - \int dx$

$\qquad\qquad\qquad = 4\int x^3\,dx + 2\int x\,dx - \int dx$

$\qquad\qquad\qquad = \dfrac{4x^4}{4} + \dfrac{2x^2}{2} - x + C$

$\qquad\qquad\qquad = x^4 + x^2 - x + C$

Property 4 can be extended to the sum and difference of an arbitrary number of functions.

(E) $\displaystyle\int \left(2e^x + \frac{3}{x}\right)dx = 2\int e^x\,dx + 3\int \frac{1}{x}\,dx$

$\qquad\qquad\qquad = 2e^x + 3\ln|x| + C$

To check any of the results in Example 2, we differentiate the final result to obtain the integrand in the original indefinite integral. When you evaluate an indefinite integral, do not forget to include the arbitrary constant C.

Matched Problem 2 Find each indefinite integral:

(A) $\displaystyle\int 2\,dx$
(B) $\displaystyle\int 16e^t\,dt$

(C) $\displaystyle\int 3x^4\,dx$
(D) $\displaystyle\int (2x^5 - 3x^2 + 1)\,dx$

(E) $\displaystyle\int \left(\frac{5}{x} - 4e^x\right)dx$

EXAMPLE 3 **Using Indefinite Integral Properties and Formulas**

(A) $\displaystyle\int \frac{4}{x^3}\,dx = \int 4x^{-3}\,dx = \frac{4x^{-3+1}}{-3+1} + C = -2x^{-2} + C$

(B) $\displaystyle\int 5\sqrt[3]{u^2}\,du = 5\int u^{2/3}\,du = 5\frac{u^{(2/3)+1}}{\frac{2}{3}+1} + C$

$\qquad\qquad\qquad = 5\dfrac{u^{5/3}}{\frac{5}{3}} + C = 3u^{5/3} + C$

(C) $\displaystyle\int \frac{x^3 - 3}{x^2}\,dx = \int \left(\frac{x^3}{x^2} - \frac{3}{x^2}\right)dx$

$\qquad\qquad\qquad = \int (x - 3x^{-2})\,dx$

$\qquad\qquad\qquad = \int x\,dx - 3\int x^{-2}\,dx$

$\qquad\qquad\qquad = \dfrac{x^{1+1}}{1+1} - 3\dfrac{x^{-2+1}}{-2+1} + C$

$\qquad\qquad\qquad = \tfrac{1}{2}x^2 + 3x^{-1} + C$

(D) $\displaystyle\int\left(\frac{2}{\sqrt[3]{x}} - 6\sqrt{x}\right)dx = \int(2x^{-1/3} - 6x^{1/2})\,dx$

$$= 2\int x^{-1/3}\,dx - 6\int x^{1/2}\,dx$$

$$= 2\frac{x^{(-1/3)+1}}{-\frac{1}{3}+1} - 6\frac{x^{(1/2)+1}}{\frac{1}{2}+1} + C$$

$$= 2\frac{x^{2/3}}{\frac{2}{3}} - 6\frac{x^{3/2}}{\frac{3}{2}} + C$$

$$= 3x^{2/3} - 4x^{3/2} + C$$

(E) $\displaystyle\int x(x^2 + 2)\,dx = \int(x^3 + 2x)\,dx = \frac{x^4}{4} + x^2 + C$

Matched Problem 3   Find each indefinite integral:

(A) $\displaystyle\int\left(2x^{2/3} - \frac{3}{x^4}\right)dx$

(B) $\displaystyle\int 4\sqrt[5]{w^3}\,dw$

(C) $\displaystyle\int\frac{x^4 - 8x^3}{x^2}\,dx$

(D) $\displaystyle\int\left(8\sqrt[3]{x} - \frac{6}{\sqrt{x}}\right)dx$

(E) $\displaystyle\int(x^2 - 2)(x + 3)\,dx$

⚠ **CAUTION**

1. Note from Example 3(E) that

$$\int x(x^2 + 2)\,dx \neq \frac{x^2}{2}\left(\frac{x^3}{3} + 2x\right) + C$$

In general, the **indefinite integral of a product is not the product of the indefinite integrals.** (This is expected because the derivative of a product is not the product of the derivatives.)

2.

$$\int e^x\,dx \neq \frac{e^{x+1}}{x+1} + C$$

The power rule applies only to power functions of the form x^n, where the exponent n is a real constant not equal to -1 and the base x is the variable. The function e^x is an exponential function with variable exponent x and constant base e. The correct form is

$$\int e^x\,dx = e^x + C$$

3. Not all elementary functions have elementary antiderivatives. It is impossible, for example, to give a formula for the antiderivative of $f(x) = e^{x^2}$ in terms of elementary functions. Nevertheless, finding such a formula, when it exists, can markedly simplify the solution of certain problems.

Applications

Let's consider some applications of the indefinite integral.

EXAMPLE 4   **Curves** Find the equation of the curve that passes through $(2, 5)$ if the slope of the curve is given by $dy/dx = 2x$ at any point x.

SOLUTION   We want to find a function $y = f(x)$ such that

$$\frac{dy}{dx} = 2x \tag{1}$$

and

$$y = 5 \quad \text{when} \quad x = 2 \tag{2}$$

If $dy/dx = 2x$, then

$$y = \int 2x \, dx \tag{3}$$

$$= x^2 + C$$

Since $y = 5$ when $x = 2$, we determine the *particular value of C* so that

$$5 = 2^2 + C$$

So $C = 1$, and

$$y = x^2 + 1$$

is the *particular antiderivative* out of all those possible from equation (3) that satisfies both equations (1) and (2) (see Fig. 2).

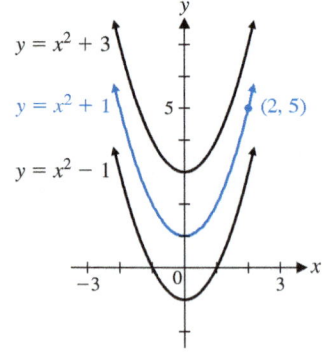

$y = x^2 + 3$
$y = x^2 + 1$
$y = x^2 - 1$
$(2, 5)$

Figure 2 $y = x^2 + C$

Matched Problem 4   Find the equation of the curve that passes through $(2, 6)$ if the slope of the curve is given by $dy/dx = 3x^2$ at any point x.

In certain situations, it is easier to determine the rate at which something happens than to determine how much of it has happened in a given length of time (for example, population growth rates, business growth rates, the rate of healing of a wound, rates of learning or forgetting). If a rate function (derivative) is given and we know the value of the dependent variable for a given value of the independent variable, then we can often find the original function by integration.

EXAMPLE 5   **Cost Function** If the marginal cost of producing x units of a commodity is given by

$$C'(x) = 0.3x^2 + 2x$$

and the fixed cost is $2,000, find the cost function $C(x)$ and the cost of producing 20 units.

SOLUTION   Recall that marginal cost is the derivative of the cost function and that fixed cost is cost at a zero production level. So we want to find $C(x)$, given

$$C'(x) = 0.3x^2 + 2x \qquad C(0) = 2,000$$

We find the indefinite integral of $0.3x^2 + 2x$ and determine the arbitrary integration constant using $C(0) = 2,000$:

$$C'(x) = 0.3x^2 + 2x$$

$$C(x) = \int (0.3x^2 + 2x) \, dx$$

$$= 0.1x^3 + x^2 + K \qquad \textit{Since C represents the cost, we use}$$
$$\textit{K for the constant of integration.}$$

But

$$C(0) = (0.1)0^3 + 0^2 + K = 2,000$$

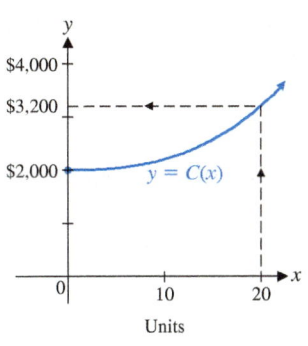

Figure 3

So $K = 2,000$, and the cost function is

$$C(x) = 0.1x^3 + x^2 + 2,000$$

We now find $C(20)$, the cost of producing 20 units:

$$C(20) = (0.1)20^3 + 20^2 + 2,000$$

$$= \$3,200$$

See Figure 3 for a geometric representation.

Matched Problem 5   Find the revenue function $R(x)$ when the marginal revenue is

$$R'(x) = 400 - 0.4x$$

and no revenue results at a zero production level. What is the revenue at a production level of 1,000 units?

EXAMPLE 6   **Advertising**   A satellite radio station is launching an aggressive advertising campaign in order to increase the number of daily listeners. The station currently has 27,000 daily listeners, and management expects the number of daily listeners, $S(t)$, to grow at the rate of

$$S'(t) = 60t^{1/2}$$

listeners per day, where t is the number of days since the campaign began. How long should the campaign last if the station wants the number of daily listeners to grow to 41,000?

SOLUTION   We must solve the equation $S(t) = 41,000$ for t, given that

$$S'(t) = 60t^{1/2} \quad \text{and} \quad S(0) = 27,000$$

First, we use integration to find $S(t)$:

$$\int S(t) = 60t^{1/2} \, dt$$

$$= 60\frac{t^{3/2}}{\frac{3}{2}} + C$$

$$= 40t^{3/2} + C$$

Since

$$S(0) = 40(0)^{3/2} + C = 27,000$$

we have $C = 27,000$ and

$$S(t) = 40t^{3/2} + 27,000$$

Now we solve the equation $S(t) = 41,000$ for t:

$$40t^{3/2} + 27,000 = 41,000$$

$$40t^{3/2} = 14,000$$

$$t^{3/2} = 350$$

$$t = 350^{2/3} \qquad \text{Use a calculator.}$$

$$= 49.664\ 419\ldots$$

The advertising campaign should last approximately 50 days.

Matched Problem 6   There are 64,000 subscribers to an online fashion magazine. Due to competition from a new magazine, the number $C(t)$ of subscribers is expected to decrease at the rate of

$$C'(t) = -600t^{1/3}$$

subscribers per month, where t is the time in months since the new magazine began publication. How long will it take until the number of subscribers to the online fashion magazine drops to 46,000?

Exercises 7-1

A

In Problems 1–16, find each indefinite integral. Check by differentiating.

1. $\displaystyle\int 7\,dx$ **2.** $\displaystyle\int 10\,dx$

3. $\displaystyle\int 8x\,dx$ **4.** $\displaystyle\int 14x\,dx$

5. $\displaystyle\int 9x^2\,dx$ **6.** $\displaystyle\int 15x^2\,dx$

7. $\displaystyle\int x^5\,dx$ **8.** $\displaystyle\int x^8\,dx$

9. $\displaystyle\int x^{-3}\,dx$ **10.** $\displaystyle\int x^{-4}\,dx$

11. $\displaystyle\int 10x^{3/2}\,dx$ **12.** $\displaystyle\int 8x^{1/3}\,dx$

13. $\displaystyle\int \frac{3}{z}\,dz$ **14.** $\displaystyle\int \frac{7}{z}\,dz$

15. $\displaystyle\int 16e^u\,du$ **16.** $\displaystyle\int 5e^u\,du$

In Problems 17–24, find all the antiderivatives for each derivative.

17. $\dfrac{dy}{dx} = 200x^4$ **18.** $\dfrac{dx}{dt} = 42t^5$

19. $\dfrac{dP}{dx} = 24 - 6x$ **20.** $\dfrac{dy}{dx} = 3x^2 - 4x^3$

21. $\dfrac{dy}{dx} = e^x + 3$ **22.** $\dfrac{dy}{dx} = x - e^x$

23. $\dfrac{dx}{dt} = 5t^{-1} + 1$ **24.** $\dfrac{du}{dv} = \dfrac{4}{v} + \dfrac{v}{4}$

B

In Problems 25–30, discuss the validity of each statement. If the statement is always true, explain why. If not, give a counterexample.

25. The constant function $f(x) = \pi$ is an antiderivative of the constant function $k(x) = 0$.

26. The constant function $k(x) = 0$ is an antiderivative of the constant function $f(x) = \pi$.

27. If n is an integer, then $x^{n+1}/(n + 1)$ is an antiderivative of x^n.

28. The constant function $k(x) = 0$ is an antiderivative of itself.

29. The function $h(x) = 5e^x$ is an antiderivative of itself.

30. The constant function $g(x) = 5e^\pi$ is an antiderivative of itself.

In Problems 31–34, could the three graphs in each figure be antiderivatives of the same function? Explain.

31.   **32.**

33.   **34.**

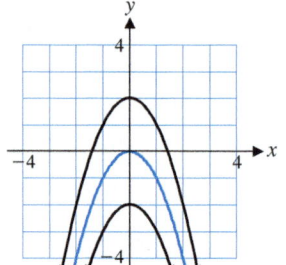

In Problems 35–50, find each indefinite integral. (Check by differentiation.)

35. $\displaystyle\int 5x(1 - x)\,dx$ **36.** $\displaystyle\int x^2(1 + x^3)\,dx$

37. $\displaystyle\int \frac{du}{\sqrt{u}}$ **38.** $\displaystyle\int \frac{dt}{\sqrt[3]{t}}$

39. $\displaystyle\int \frac{dx}{4x^3}$ **40.** $\displaystyle\int \frac{6\,dm}{m^2}$

41. $\displaystyle\int \frac{4 + u}{u}\, du$

42. $\displaystyle\int \frac{1 - y^2}{3y}\, dy$

43. $\displaystyle\int (5e^z + 4)\, dz$

44. $\displaystyle\int \frac{e^t - t}{2}\, dt$

45. $\displaystyle\int \left(3x^2 - \frac{2}{x^2}\right) dx$

46. $\displaystyle\int \left(4x^3 + \frac{2}{x^3}\right) dx$

47. $\displaystyle\int \left(3\sqrt{x} + \frac{2}{\sqrt{x}}\right) dx$

48. $\displaystyle\int \left(\frac{2}{\sqrt[3]{x}} - \sqrt[3]{x^2}\right) dx$

49. $\displaystyle\int \frac{e^x - 3x}{4}\, dx$

50. $\displaystyle\int \frac{e^x - 3x^2}{2}\, dx$

In Problems 51–58, find the particular antiderivative of each derivative that satisfies the given condition.

51. $C'(x) = 6x^2 - 4x; C(0) = 3{,}000$

52. $R'(x) = 600 - 0.6x; R(0) = 0$

53. $\dfrac{dx}{dt} = \dfrac{20}{\sqrt{t}}; x(1) = 40$

54. $\dfrac{dR}{dt} = \dfrac{100}{t^2}; R(1) = 400$

55. $\dfrac{dy}{dx} = 2x^{-2} + 3x^{-1} - 1; y(1) = 0$

56. $\dfrac{dy}{dx} = 3x^{-1} + x^{-2}; y(1) = 1$

57. $\dfrac{dx}{dt} = 4e^t - 2; x(0) = 1$

58. $\dfrac{dy}{dt} = 5e^t - 4; y(0) = -1$

59. Find the equation of the curve that passes through $(2, 3)$ if its slope is given by
$$\frac{dy}{dx} = 4x - 3$$
for each x.

60. Find the equation of the curve that passes through $(1, 3)$ if its slope is given by
$$\frac{dy}{dx} = 12x^2 - 12x$$
for each x.

C

In Problems 61–66, find each indefinite integral.

61. $\displaystyle\int \frac{2x^4 - x}{x^3}\, dx$

62. $\displaystyle\int \frac{x^{-1} - x^4}{x^2}\, dx$

63. $\displaystyle\int \frac{x^5 - 2x}{x^4}\, dx$

64. $\displaystyle\int \frac{1 - 3x^4}{x^2}\, dx$

65. $\displaystyle\int \frac{x^2 e^x - 2x}{x^2}\, dx$

66. $\displaystyle\int \frac{1 - xe^x}{x}\, dx$

In Problems 67–72, find the particular antiderivative of each derivative that satisfies the given condition.

67. $\dfrac{dM}{dt} = \dfrac{t^2 - 1}{t^2}; M(4) = 5$

68. $\dfrac{dR}{dx} = \dfrac{1 - x^4}{x^3}; R(1) = 4$

69. $\dfrac{dy}{dx} = \dfrac{5x + 2}{\sqrt[3]{x}}; y(1) = 0$

70. $\dfrac{dx}{dt} = \dfrac{\sqrt{t^3} - t}{\sqrt{t^3}}; x(9) = 4$

71. $p'(x) = -\dfrac{10}{x^2}; p(1) = 20$

72. $p'(x) = \dfrac{10}{x^3}; p(1) = 15$

In Problems 73–76, find the derivative or indefinite integral as indicated.

73. $\dfrac{d}{dx}\left(\displaystyle\int x^3\, dx\right)$

74. $\dfrac{d}{dt}\left(\displaystyle\int \frac{\ln t}{t}\, dt\right)$

75. $\displaystyle\int \frac{d}{dx}(x^4 + 3x^2 + 1)\, dx$

76. $\displaystyle\int \frac{d}{du}(e^{u^2})\, du$

77. Use differentiation to justify the formula
$$\int x^n\, dx = \frac{x^{n+1}}{n + 1} + C$$
provided that $n \neq -1$.

78. Use differentiation to justify the formula
$$\int e^x\, dx = e^x + C$$

79. Assuming that $x > 0$, use differentiation to justify the formula
$$\int \frac{1}{x}\, dx = \ln|x| + C$$

80. Assuming that $x < 0$, use differentiation to justify the formula
$$\int \frac{1}{x}\, dx = \ln|x| + C$$

81. Show that the indefinite integral of the sum of two functions is the sum of the indefinite integrals.

[*Hint:* Assume that $\int f(x)\, dx = F(x) + C_1$ and $\int g(x)\, dx = G(x) + C_2$. Using differentiation, show that $F(x) + C_1 + G(x) + C_2$ is the indefinite integral of the function $s(x) = f(x) + g(x)$.]

82. Show that the indefinite integral of the difference of two functions is the difference of the indefinite integrals.

Applications

83. Cost function. The marginal average cost of producing x sports watches is given by

$$\overline{C}'(x) = -\frac{1,000}{x^2} \qquad \overline{C}(100) = 25$$

where $\overline{C}(x)$ is the average cost in dollars. Find the average cost function and the cost function. What are the fixed costs?

84. Renewable energy. In 2007, U.S. consumption of renewable energy was 6.8 quadrillion Btu (or 6.8×10^{15} Btu). Since the 1960s, consumption has been growing at a rate (in quadrillion Btu's per year) given by

$$f'(t) = 0.004t + 0.062$$

where t is years after 1960. Find $f(t)$ and estimate U.S. consumption of renewable energy in 2020.

85. Production costs. The graph of the marginal cost function from the production of x thousand bottles of sunscreen per month [where cost $C(x)$ is in thousands of dollars per month] is given in the figure.

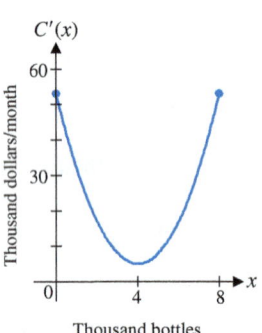

Figure for 85

(A) Using the graph shown, describe the shape of the graph of the cost function $C(x)$ as x increases from 0 to 8,000 bottles per month.

(B) Given the equation of the marginal cost function,

$$C'(x) = 3x^2 - 24x + 53$$

find the cost function if monthly fixed costs at 0 output are $30,000. What is the cost of manufacturing 4,000 bottles per month? 8,000 bottles per month?

(C) Graph the cost function for $0 \le x \le 8$. [Check the shape of the graph relative to the analysis in part (A).]

(D) Why do you think that the graph of the cost function is steeper at both ends than in the middle?

86. Revenue. The graph of the marginal revenue function from the sale of x sports watches is given in the figure.

(A) Using the graph shown, describe the shape of the graph of the revenue function $R(x)$ as x increases from 0 to 1,000.

(B) Find the equation of the marginal revenue function (the linear function shown in the figure).

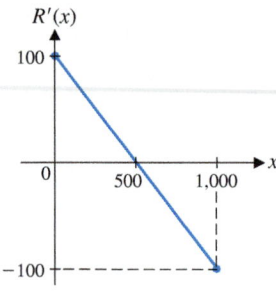

Figure for 86

(C) Find the equation of the revenue function that satisfies $R(0) = 0$. Graph the revenue function over the interval $[0, 1,000]$. [Check the shape of the graph relative to the analysis in part (A).]

(D) Find the price–demand equation and determine the price when the demand is 700 units.

87. Sales analysis. Monthly sales of an SUV model are expected to decline at the rate of

$$S'(t) = -24t^{1/3}$$

SUVs per month, where t is time in months and $S(t)$ is the number of SUVs sold each month. The company plans to stop manufacturing this model when monthly sales reach 300 SUVs. If monthly sales now ($t = 0$) are 1,200 SUVs, find $S(t)$. How long will the company continue to manufacture this model?

88. Sales analysis. The rate of change of the monthly sales of a newly released football game is given by

$$S'(t) = 500t^{1/4} \qquad S(0) = 0$$

where t is the number of months since the game was released and $S(t)$ is the number of games sold each month. Find $S(t)$. When will monthly sales reach 20,000 games?

89. Sales analysis. Repeat Problem 87 if $S'(t) = -24t^{1/3} - 70$ and all other information remains the same. Use a graphing calculator to approximate the solution of the equation $S(t) = 800$ to two decimal places.

90. Sales analysis. Repeat Problem 88 if $S'(t) = 500t^{1/4} + 300$ and all other information remains the same. Use a graphing calculator to approximate the solution of the equation $S(t) = 20,000$ to two decimal places.

91. Labor costs. A defense contractor is starting production on a new missile control system. On the basis of data collected during the assembly of the first 16 control systems, the production manager obtained the following function describing the rate of labor use:

$$g(x) = 2,400x^{-1/2}$$

$g(x)$ is the number of labor-hours required to assemble the xth unit of the control system. For example, after assembly of 16 units, the rate of assembly is 600 labor-hours per unit, and after assembly of 25 units, the rate of assembly is 480 labor-hours per unit. The more units assembled, the more efficient the process. If 19,200 labor-hours are required to assemble of the first 16 units, how many labor-hours $L(x)$ will be required to assemble the first x units? The first 25 units?

92. Labor costs. If the rate of labor use in Problem 91 is

$$g(x) = 2{,}000x^{-1/3}$$

and if the first 8 control units require 12,000 labor-hours, how many labor-hours, $L(x)$, will be required for the first x control units? The first 27 control units?

93. Weight–height. For an average person, the rate of change of weight W (in pounds) with respect to height h (in inches) is given approximately by

$$\frac{dW}{dh} = 0.0015h^2$$

Find $W(h)$ if $W(60) = 108$ pounds. Find the weight of an average person who is 5 feet, 10 inches, tall.

94. Wound healing. The area A of a healing wound changes at a rate given approximately by

$$\frac{dA}{dt} = -4t^{-3} \qquad 1 \le t \le 10$$

where t is time in days and $A(1) = 2$ square centimeters. What will the area of the wound be in 10 days?

95. Urban growth. The rate of growth of the population $N(t)$ of a new city t years after its incorporation is estimated to be

$$\frac{dN}{dt} = 400 + 600\sqrt{t} \qquad 0 \le t \le 9$$

If the population was 5,000 at the time of incorporation, find the population 9 years later.

96. Learning. A college language class was chosen for an experiment in learning. Using a list of 50 words, the experiment involved measuring the rate of vocabulary memorization at different times during a continuous 5-hour study session. It was found that the average rate of learning for the entire class was inversely proportional to the time spent studying and was given approximately by

$$V'(t) = \frac{15}{t} \qquad 1 \le t \le 5$$

If the average number of words memorized after 1 hour of study was 15 words, what was the average number of words memorized after t hours of study for $1 \le t \le 5$? After 4 hours of study? Round answer to the nearest whole number.

Answers to Matched Problems

1. (A) $x^3 + C$

(B)

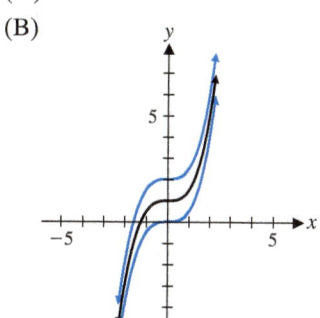

(C) The graphs are vertical translations of each other.

2. (A) $2x + C$

(B) $16e^t + C$

(C) $\frac{3}{5}x^5 + C$

(D) $\frac{1}{3}x^6 - x^3 + x + C$

(E) $5\ln|x| - 4e^x + C$

3. (A) $\frac{6}{5}x^{5/3} + x^{-3} + C$

(B) $\frac{5}{2}w^{8/5} + C$

(C) $\frac{1}{3}x^3 - 4x^2 + C$

(D) $6x^{4/3} - 12x^{1/2} + C$

(E) $\frac{1}{4}x^4 + x^3 - x^2 - 6x + C$

4. $y = x^3 - 2$

5. $R(x) = 400x - 0.2x^2$; $R(1{,}000) = \$200{,}000$

6. $t = (40)^{3/4} \approx 16$ mo

7-2 Integration by Substitution

- Reversing the Chain Rule
- Integration by Substitution
- Additional Substitution Techniques
- Application

Many of the indefinite integral formulas introduced in the preceding section are based on corresponding derivative formulas studied earlier. We now consider indefinite integral formulas and procedures based on the chain rule for differentiation.

Reversing the Chain Rule

Recall the chain rule:

$$\frac{d}{dx}f[g(x)] = f'[g(x)]g'(x)$$

The expression on the right is formed from the expression on the left by taking the derivative of the outside function f and multiplying it by the derivative of the inside function g. If we recognize an integrand as a chain-rule form $f'[g(x)]g'(x)$, we can easily find an antiderivative and its indefinite integral:

$$\int f'[g(x)]g'(x)\,dx = f[g(x)] + C \tag{1}$$

We are interested in finding the indefinite integral

$$\int 3x^2 e^{x^3-1}\,dx \tag{2}$$

The integrand appears to be the chain-rule form $e^{g(x)}g'(x)$, which is the derivative of $e^{g(x)}$. Since

$$\frac{d}{dx}e^{x^3-1} = 3x^2 e^{x^3-1}$$

it follows that

$$\int 3x^2 e^{x^3-1}\,dx = e^{x^3-1} + C \tag{3}$$

How does the following indefinite integral differ from integral (2)?

$$\int x^2 e^{x^3-1}\,dx \tag{4}$$

It is missing the constant factor 3. That is, $x^2 e^{x^3-1}$ is within a constant factor of being the derivative of e^{x^3-1}. But because a constant factor can be moved across the integral sign, this causes us little trouble in finding the indefinite integral of $x^2 e^{x^3-1}$. We introduce the constant factor 3 and at the same time multiply by $\frac{1}{3}$ and move the $\frac{1}{3}$ factor outside the integral sign. This is equivalent to multiplying the integrand in integral (4) by 1:

$$\int x^2 e^{x^3-1}\,dx = \int \frac{3}{3} x^2 e^{x^3-1}\,dx \tag{5}$$

$$= \frac{1}{3}\int 3x^2 e^{x^3-1}\,dx = \frac{1}{3}e^{x^3-1} + C$$

The derivative of the rightmost side of equation (5) is the integrand of the indefinite integral (4). Check this.

How does the following indefinite integral differ from integral (2)?

$$\int 3x e^{x^3-1}\,dx \tag{6}$$

It is missing a variable factor x. This is more serious. As tempting as it might be, we *cannot* adjust integral (6) by introducing the variable factor x and moving $1/x$ outside the integral sign, as we did with the constant 3 in equation (5).

⚠️ **CAUTION** A constant factor can be moved across an integral sign, but a variable factor cannot.

There is nothing wrong with educated guessing when you are looking for an antiderivative of a given function. You have only to check the result by differentiation. If you are right, you go on your way; if you are wrong, you simply try another approach.

In Section 4-4, we saw that the chain rule extends the derivative formulas for x^n, e^x, and $\ln x$ to derivative formulas for $[f(x)]^n$, $e^{f(x)}$, and $\ln [f(x)]$. The chain rule can also be used to extend the indefinite integral formulas discussed in Section 7-1. Some general formulas are summarized in the following box:

FORMULAS **General Indefinite Integral Formulas**

1. $\displaystyle\int [f(x)]^n f'(x)\, dx = \frac{[f(x)]^{n+1}}{n+1} + C, n \neq -1$

2. $\displaystyle\int e^{f(x)} f'(x)\, dx = e^{f(x)} + C$

3. $\displaystyle\int \frac{1}{f(x)} f'(x)\, dx = \ln|f(x)| + C$

We can verify each formula by using the chain rule to show that the derivative of the function on the right is the integrand on the left. For example,

$$\frac{d}{dx}[e^{f(x)} + C] = e^{f(x)} f'(x)$$

verifies formula 2.

EXAMPLE 1 **Reversing the Chain Rule**

(A) $\displaystyle\int (3x+4)^{10}(3)\, dx = \frac{(3x+4)^{11}}{11} + C$ Formula 1 with $f(x) = 3x + 4$ and $f'(x) = 3$

Check:

$$\frac{d}{dx}\frac{(3x+4)^{11}}{11} = 11\frac{(3x+4)^{10}}{11}\frac{d}{dx}(3x+4) = (3x+4)^{10}(3)$$

(B) $\displaystyle\int e^{x^2}(2x)\, dx = e^{x^2} + C$ Formula 2 with $f(x) = x^2$ and $f'(x) = 2x$

Check:

$$\frac{d}{dx}e^{x^2} = e^{x^2}\frac{d}{dx}x^2 = e^{x^2}(2x)$$

(C) $\displaystyle\int \frac{1}{1+x^3}3x^2\, dx = \ln|1 + x^3| + C$ Formula 3 with $f(x) = 1 + x^3$ and $f'(x) = 3x^2$

Check:

$$\frac{d}{dx}\ln|1 + x^3| = \frac{1}{1+x^3}\frac{d}{dx}(1 + x^3) = \frac{1}{1+x^3}3x^2$$

Matched Problem 1 Find each indefinite integral.

(A) $\displaystyle\int (2x^3 - 3)^{20}(6x^2)\, dx$ (B) $\displaystyle\int e^{5x}(5)\, dx$

(C) $\displaystyle\int \frac{1}{4+x^2}2x\, dx$

Integration by Substitution

The key step in using formulas 1, 2, and 3 is recognizing the form of the integrand. Some people find it difficult to identify $f(x)$ and $f'(x)$ in these formulas and prefer to use a *substitution* to simplify the integrand. The *method of substitution*, which we now discuss, becomes increasingly useful as one progresses in studies of integration.

We start by introducing the idea of the *differential*. We represented the derivative by the symbol dy/dx taken as a whole. We now define dy and dx as two separate quantities with the property that their ratio is still equal to $f'(x)$:

> **DEFINITION Differentials**
>
> If $y = f(x)$ defines a differentiable function, then
>
> 1. The **differential** dx of the independent variable x is an arbitrary real number.
> 2. The **differential** dy of the dependent variable y is defined as the product of $f'(x)$ and dx:
> $$dy = f'(x)\, dx$$

Differentials involve mathematical subtleties that are treated carefully in advanced mathematics courses. Here, we are interested in them mainly as a bookkeeping device to aid in the process of finding indefinite integrals. We can always check an indefinite integral by differentiating.

EXAMPLE 2 **Differentials**

(A) If $y = f(x) = x^2$, then
$$dy = f'(x)\, dx = 2x\, dx$$

(B) If $u = g(x) = e^{3x}$, then
$$du = g'(x)\, dx = 3e^{3x}\, dx$$

(C) If $w = h(t) = \ln(4 + 5t)$, then
$$dw = h'(t)\, dt = \frac{5}{4 + 5t}\, dt$$

Matched Problem 2

(A) Find dy for $y = f(x) = x^3$.

(B) Find du for $u = h(x) = \ln(2 + x^2)$.

(C) Find dv for $v = g(t) = e^{-5t}$.

The **method of substitution** is developed through Examples 3–6.

EXAMPLE 3 **Using Substitution** Find $\int (x^2 + 2x + 5)^5 (2x + 2)\, dx$.

SOLUTION If
$$u = x^2 + 2x + 5$$

then the differential of u is
$$du = (2x + 2)\, dx$$

Notice that du is one of the factors in the integrand. Substitute u for $x^2 + 2x + 5$ and du for $(2x + 2)\, dx$ to obtain

$$\int (x^2 + 2x + 5)^5 (2x + 2)\, dx = \int u^5\, du$$

$$= \frac{u^6}{6} + C$$

$$= \frac{1}{6}(x^2 + 2x + 5)^6 + C \quad \text{Since } u = x^2 + 2x + 5$$

Check:

$$\frac{d}{dx}\frac{1}{6}(x^2 + 2x + 5)^6 = \frac{1}{6}(6)(x^2 + 2x + 5)^5\frac{d}{dx}(x^2 + 2x + 5)$$

$$= (x^2 + 2x + 5)^5(2x + 2)$$

Matched Problem 3 Find $\int (x^2 - 3x + 7)^4(2x - 3)\, dx$ by substitution.

The substitution method is also called the **change-of-variable method** since u replaces the variable x in the process. Substituting $u = f(x)$ and $du = f'(x)\, dx$ in formulas 1, 2, and 3 produces the general indefinite integral formulas 4, 5, and 6:

FORMULAS General Indefinite Integral Formulas

4. $\int u^n\, du = \dfrac{u^{n+1}}{n + 1} + C, \qquad n \neq -1$

5. $\int e^u\, du = e^u + C$

6. $\int \dfrac{1}{u}\, du = \ln|u| + C$

These formulas are valid if u is an independent variable, or if u is a function of another variable and du is the differential of u with respect to that variable.

The substitution method for evaluating certain indefinite integrals is outlined as follows:

PROCEDURE Integration by Substitution

Step 1 Select a substitution that appears to simplify the integrand. In particular, try to select u so that du is a factor in the integrand.

Step 2 Express the integrand entirely in terms of u and du, completely eliminating the original variable and its differential.

Step 3 Evaluate the new integral if possible.

Step 4 Express the antiderivative found in step 3 in terms of the original variable.

EXAMPLE 4 **Using Substitution** Use a substitution to find each indefinite integral.

(A) $\int (3x + 4)^6(3)\, dx$

(B) $\int e^{t^2}(2t)\, dt$

SOLUTION (A) If we let $u = 3x + 4$, then $du = 3\,dx$, and

$$\int (3x + 4)^6 (3)\,dx = \int u^6\,du \qquad \textit{Use formula 4.}$$

$$= \frac{u^7}{7} + C$$

$$= \frac{(3x + 4)^7}{7} + C \qquad \textit{Since } u = 3x + 4$$

Check:

$$\frac{d}{dx}\frac{(3x + 4)^7}{7} = \frac{7(3x + 4)^6}{7}\frac{d}{dx}(3x + 4) = (3x + 4)^6(3)$$

(B) If we let $u = t^2$, then $du = 2t\,dt$, and

$$\int e^{t^2}(2t)\,dt = \int e^u\,du \qquad \textit{Use formula 5.}$$

$$= e^u + C$$

$$= e^{t^2} + C \qquad \textit{Since } u = t^2$$

Check:

$$\frac{d}{dt}e^{t^2} = e^{t^2}\frac{d}{dt}t^2 = e^{t^2}(2t)$$

Matched Problem 4 Use a substitution to find each indefinite integral.

(A) $\displaystyle\int (2x^3 - 3)^4 (6x^2)\,dx$ (B) $\displaystyle\int e^{5w}(5)\,dw$

Additional Substitution Techniques

In order to use the substitution method, **the integrand must be expressed entirely in terms of u and du.** In some cases, the integrand must be modified before making a substitution and using one of the integration formulas. Example 5 illustrates this process.

EXAMPLE 5 **Substitution Techniques** Integrate.

(A) $\displaystyle\int \frac{1}{4x + 7}\,dx$ (B) $\displaystyle\int te^{-t^2}\,dt$

(C) $\displaystyle\int 4x^2\sqrt{x^3 + 5}\,dx$

SOLUTION (A) If $u = 4x + 7$, then $du = 4\,dx$. We are missing a factor of 4 in the integrand to match formula 6 exactly. Recalling that a constant factor can be moved across an integral sign, we proceed as follows:

$$\int \frac{1}{4x + 7}\,dx = \int \frac{1}{4x + 7}\frac{4}{4}\,dx$$

$$= \frac{1}{4}\int \frac{1}{4x + 7}4\,dx \quad \textit{Substitute } u = 4x + 7 \textit{ and } du = 4\,dx.$$

$$= \frac{1}{4}\int \frac{1}{u}\,du \qquad \textit{Use formula 6.}$$

$$= \tfrac{1}{4}\ln|u| + C$$

$$= \tfrac{1}{4}\ln|4x + 7| + C \quad \textit{Since } u = 4x + 7$$

Check:

$$\frac{d}{dx}\frac{1}{4}\ln|4x + 7| = \frac{1}{4}\frac{1}{4x + 7}\frac{d}{dx}(4x + 7) = \frac{1}{4}\frac{1}{4x + 7}4 = \frac{1}{4x + 7}$$

(B) If $u = -t^2$, then $du = -2t\,dt$. Proceed as in part (A):

$$\int te^{-t^2}\,dt = \int e^{-t^2}\frac{-2}{-2}t\,dt$$

$$= -\frac{1}{2}\int e^{-t^2}(-2t)\,dt \quad \textit{Substitute } u = -t^2 \textit{ and } du = -2t\,dt.$$

$$= -\frac{1}{2}\int e^u\,du \qquad\qquad \textit{Use formula 5.}$$

$$= -\tfrac{1}{2}e^u + C$$

$$= -\tfrac{1}{2}e^{-t^2} + C \qquad\qquad \textit{Since } u = -t^2$$

Check:

$$\frac{d}{dt}(-\tfrac{1}{2}e^{-t^2}) = -\tfrac{1}{2}e^{-t^2}\frac{d}{dt}(-t^2) = -\tfrac{1}{2}e^{-t^2}(-2t) = te^{-t^2}$$

(C) $\displaystyle\int 4x^2\sqrt{x^3 + 5}\,dx = 4\int \sqrt{x^3 + 5}(x^2)\,dx \quad$ Move the 4 across the integral sign and proceed as before.

$$= 4\int \sqrt{x^3 + 5}\frac{3}{3}(x^2)\,dx$$

$$= \frac{4}{3}\int \sqrt{x^3 + 5}(3x^2)\,dx \quad \textit{Substitute } u = x^3 + 5 \textit{ and } du = 3x^2\,dx.$$

$$= \frac{4}{3}\int \sqrt{u}\,du$$

$$= \frac{4}{3}\int u^{1/2}\,du \qquad\qquad \textit{Use formula 4.}$$

$$= \frac{4}{3}\frac{u^{3/2}}{\frac{3}{2}} + C$$

$$= \tfrac{8}{9}u^{3/2} + C$$

$$= \tfrac{8}{9}(x^3 + 5)^{3/2} + C \qquad \textit{Since } u = x^3 + 5$$

Check:

$$\frac{d}{dx}\left[\tfrac{8}{9}(x^3 + 5)^{3/2}\right] = \tfrac{4}{3}(x^3 + 5)^{1/2}\frac{d}{dx}(x^3 + 5)$$

$$= \tfrac{4}{3}(x^3 + 5)^{1/2}(3x^2) = 4x^2\sqrt{x^3 + 5}$$

Matched Problem 5   Integrate.

(A) $\displaystyle\int e^{-3x}\,dx$ (B) $\displaystyle\int \frac{x}{x^2 - 9}\,dx$

(C) $\displaystyle\int 5t^2(t^3 + 4)^{-2}\,dt$

Even if it is not possible to find a substitution that makes an integrand match one of the integration formulas exactly, a substitution may simplify the integrand sufficiently so that other techniques can be used.

EXAMPLE 6   **Substitution Techniques**   Find $\int \dfrac{x}{\sqrt{x+2}}\,dx$.

SOLUTION   Proceeding as before, if we let $u = x + 2$, then $du = dx$ and

$$\int \frac{x}{\sqrt{x+2}}\,dx = \int \frac{x}{\sqrt{u}}\,du$$

Notice that this substitution is not complete because we have not expressed the integrand entirely in terms of u and du. As we noted earlier, only a constant factor can be moved across an integral sign, so we cannot move x outside the integral sign. Instead, we must return to the original substitution, solve for x in terms of u, and use the resulting equation to complete the substitution:

$$u = x + 2 \qquad \textit{Solve for x in terms of u.}$$

$$u - 2 = x \qquad \textit{Substitute this expression for x.}$$

Thus,

$$\int \frac{x}{\sqrt{x+2}}\,dx = \int \frac{u-2}{\sqrt{u}}\,du \qquad \textit{Simplify the integrand.}$$

$$= \int \frac{u-2}{u^{1/2}}\,du$$

$$= \int (u^{1/2} - 2u^{-1/2})\,du$$

$$\boxed{= \int u^{1/2}\,du - 2\int u^{-1/2}\,du}$$

$$= \frac{u^{3/2}}{\frac{3}{2}} - 2\frac{u^{1/2}}{\frac{1}{2}} + C$$

$$= \tfrac{2}{3}(x+2)^{3/2} - 4(x+2)^{1/2} + C \qquad \textit{Since u = x + 2}$$

Check:

$$\frac{d}{dx}\left[\tfrac{2}{3}(x+2)^{3/2} - 4(x+2)^{1/2}\right] = (x+2)^{1/2} - 2(x+2)^{-1/2}$$

$$= \frac{x+2}{(x+2)^{1/2}} - \frac{2}{(x+2)^{1/2}}$$

$$= \frac{x}{(x+2)^{1/2}}$$

Matched Problem 6   Find $\int x\sqrt{x+1}\,dx$.

We can find the indefinite integral of some functions in more than one way. For example, we can use substitution to find

$$\int x(1+x^2)^2\,dx$$

by letting $u = 1 + x^2$. As a second approach, we can expand the integrand, obtaining

$$\int (x + 2x^3 + x^5)\,dx$$

for which we can easily calculate an antiderivative. In such a case, choose the approach that you prefer.

There are also some functions for which substitution is not an effective approach to finding the indefinite integral. For example, substitution is not helpful in finding

$$\int e^{x^2} \, dx \quad \text{or} \quad \int \ln x \, dx$$

Application

EXAMPLE 7   **Price–Demand** The market research department of a supermarket chain has determined that, for one store, the marginal price $p'(x)$ at x tubes per week for a certain brand of toothpaste is given by

$$p'(x) = -0.015e^{-0.01x}$$

Find the price–demand equation if the weekly demand is 50 tubes when the price of a tube is \$4.35. Find the weekly demand when the price of a tube is \$3.89.

SOLUTION

$$p(x) = \int -0.015e^{-0.01x} \, dx$$

$$= -0.015 \int e^{-0.01x} \, dx$$

$$= -0.015 \int e^{-0.01x} \frac{-0.01}{-0.01} \, dx$$

$$= \frac{-0.015}{-0.01} \int e^{-0.01x}(-0.01) \, dx \qquad \text{Substitute } u = -0.01x \text{ and } du = -0.01 \, dx.$$

$$= 1.5 \int e^{u} \, du$$

$$= 1.5e^{u} + C$$

$$= 1.5e^{-0.01x} + C \qquad \text{Since } u = -0.01x$$

We find C by noting that

$$p(50) = 1.5e^{-0.01(50)} + C = \$4.35$$

$$C = \$4.35 - 1.5e^{-0.5} \qquad \text{Use a calculator.}$$

$$= \$4.35 - 0.91$$

$$= \$3.44$$

So,

$$p(x) = 1.5e^{-0.01x} + 3.44$$

To find the demand when the price is \$3.89, we solve $p(x) = \$3.89$ for x:

$$1.5e^{-0.01x} + 3.44 = 3.89$$

$$1.5e^{-0.01x} = 0.45$$

$$e^{-0.01x} = 0.3$$

$$-0.01x = \ln 0.3$$

$$x = -100 \ln 0.3 \approx 120 \text{ tubes}$$

Matched Problem 7 The marginal price $p'(x)$ at a supply level of x tubes per week for a certain brand of toothpaste is given by

$$p'(x) = 0.001e^{0.01x}$$

Find the price–supply equation if the supplier is willing to supply 100 tubes per week at a price of $3.65 each. How many tubes would the supplier be willing to supply at a price of $3.98 each?

We conclude with two final cautions. The first was stated earlier, but it is worth repeating.

⚠️ **CAUTION**

1. A variable cannot be moved across an integral sign.
2. An integral must be expressed entirely in terms of u and du before applying integration formulas 4, 5, and 6.

Exercises 7-2

A

In Problems 1–36, find each indefinite integral and check the result by differentiating.

1. $\int (3x + 5)^2 (3)\,dx$

2. $\int (6x - 1)^3 (6)\,dx$

3. $\int (x^2 - 1)^5 (2x)\,dx$

4. $\int (x^6 + 1)^4 (6x^5)\,dx$

5. $\int (5x^3 + 1)^{-3}(15x^2)\,dx$

6. $\int (4x^2 - 3)^{-6}(8x)\,dx$

7. $\int e^{5x}(5)\,dx$

8. $\int e^{x^3}(3x^2)\,dx$

9. $\int \frac{1}{1 + x^2}(2x)\,dx$

10. $\int \frac{1}{5x - 7}(5)\,dx$

11. $\int \sqrt{1 + x^4}\,(4x^3)\,dx$

12. $\int (x^2 + 9)^{-1/2}(2x)\,dx$

B

13. $\int (x + 3)^{10}\,dx$

14. $\int (x - 3)^{-4}\,dx$

15. $\int (6t - 7)^{-2}\,dt$

16. $\int (5t + 1)^3\,dt$

17. $\int (t^2 + 1)^5\,t\,dt$

18. $\int (t^3 + 4)^{-2}\,t^2\,dt$

19. $\int xe^{x^2}\,dx$

20. $\int e^{-0.01x}\,dx$

21. $\int \frac{1}{5x + 4}\,dx$

22. $\int \frac{x}{1 + x^2}\,dx$

23. $\int e^{1-t}\,dt$

24. $\int \frac{3}{2 - t}\,dt$

25. $\int \frac{t}{(3t^2 + 1)^4}\,dt$

26. $\int \frac{t^2}{(t^3 - 2)^5}\,dt$

27. $\int x\sqrt{x + 4}\,dx$

28. $\int x\sqrt{x - 9}\,dx$

29. $\int \frac{x}{\sqrt{x - 3}}\,dx$

30. $\int \frac{x}{\sqrt{x + 5}}\,dx$

31. $\int x(x - 4)^9\,dx$

32. $\int x(x + 6)^8\,dx$

33. $\int e^{2x}(1 + e^{2x})^3\,dx$

34. $\int e^{-x}(1 - e^{-x})^4\,dx$

35. $\int \frac{1 + x}{4 + 2x + x^2}\,dx$

36. $\int \frac{x^2 - 1}{x^3 - 3x + 7}\,dx$

In Problems 37–42, the indefinite integral can be found in more than one way. First use the substitution method to find the indefinite integral. Then find it without using substitution. Check that your answers are equivalent.

37. $\int 5(5x + 3)\,dx$

38. $\int -7(4 - 7x)\,dx$

39. $\int 2x(x^2 - 1)\,dx$

40. $\int 3x^2(x^3 + 1)\,dx$

41. $\int 5x^4(x^5)^4\,dx$

42. $\int 8x^7(x^8)^3\,dx$

〰️ *In Problems 43–48, suppose that the indicated "solutions" were given to you by a student whom you are tutoring.*

(A) *How would you have the student check each solution?*

(B) *Is the solution right or wrong? If the solution is wrong, explain what is wrong and how it can be corrected.*

(C) *Show a correct solution for each incorrect solution, and check the result by differentiation.*

43. $\int \frac{1}{2x - 3}\,dx = \ln|2x - 3| + C$

44. $\int \frac{x}{x^2 + 5}\,dx = \ln|x^2 + 5| + C$

45. $\displaystyle\int x^3 e^{x^4}\,dx = e^{x^4} + C$

46. $\displaystyle\int e^{4x-5}\,dx = e^{4x-5} + C$

47. $\displaystyle\int 2(x^2 - 2)^2\,dx = \dfrac{(x^2 - 2)^2}{3x} + C$

48. $\displaystyle\int (-10x)(x^2 - 3)^{-4}\,dx = (x^2 - 3)^{-5} + C$

C

In Problems 49–60, find each indefinite integral and check the result by differentiating.

49. $\displaystyle\int x\sqrt{3x^2 + 7}\,dx$

50. $\displaystyle\int x^2\sqrt{2x^3 + 1}\,dx$

51. $\displaystyle\int x(x^3 + 2)^2\,dx$

52. $\displaystyle\int x(x^2 + 2)^2\,dx$

53. $\displaystyle\int x^2(x^3 + 2)^2\,dx$

54. $\displaystyle\int (x^2 + 2)^2\,dx$

55. $\displaystyle\int \dfrac{x^3}{\sqrt{2x^4 + 3}}\,dx$

56. $\displaystyle\int \dfrac{x^2}{\sqrt{4x^3 - 1}}\,dx$

57. $\displaystyle\int \dfrac{(\ln x)^3}{x}\,dx$

58. $\displaystyle\int \dfrac{e^x}{1 + e^x}\,dx$

59. $\displaystyle\int \dfrac{1}{x^2}e^{-1/x}\,dx$

60. $\displaystyle\int \dfrac{1}{x\ln x}\,dx$

In Problems 61–66, find the antiderivative of each derivative.

61. $\dfrac{dx}{dt} = 7t^2(t^3 + 5)^6$

62. $\dfrac{dm}{dn} = 10n(n^2 - 8)^7$

63. $\dfrac{dy}{dt} = \dfrac{3t}{\sqrt{t^2 - 4}}$

64. $\dfrac{dy}{dx} = \dfrac{5x^2}{(x^3 - 7)^4}$

65. $\dfrac{dp}{dx} = \dfrac{e^x + e^{-x}}{(e^x - e^{-x})^2}$

66. $\dfrac{dm}{dt} = \dfrac{\ln(t - 5)}{t - 5}$

Applications

67. Price–demand equation. The marginal price for a weekly demand of x bottles of shampoo in a drugstore is given by

$$p'(x) = \dfrac{-6{,}000}{(3x + 50)^2}$$

Find the price–demand equation if the weekly demand is 150 when the price of a bottle of shampoo is \$8. What is the weekly demand when the price is \$6.50?

68. Price–supply equation. The marginal price at a supply level of x bottles of shampoo per week is given by

$$p'(x) = \dfrac{300}{(3x + 25)^2}$$

Find the price–supply equation if the distributor of the shampoo is willing to supply 75 bottles a week at a price of \$5.00 per bottle. How many bottles would the supplier be willing to supply at a price of \$5.15 per bottle?

69. Cost function. The weekly marginal cost of producing x pairs of tennis shoes is given by

$$C'(x) = 12 + \dfrac{500}{x + 1}$$

where $C(x)$ is cost in dollars. If the fixed costs are \$2,000 per week, find the cost function. What is the average cost per pair of shoes if 1,000 pairs of shoes are produced each week?

70. Revenue function. The weekly marginal revenue from the sale of x pairs of tennis shoes is given by

$$R'(x) = 40 - 0.02x + \dfrac{200}{x + 1} \qquad R(0) = 0$$

where $R(x)$ is revenue in dollars. Find the revenue function. Find the revenue from the sale of 1,000 pairs of shoes.

71. Marketing. An automobile company is ready to introduce a new line of hybrid cars through a national sales campaign. After test marketing the line in a carefully selected city, the marketing research department estimates that sales (in millions of dollars) will increase at the monthly rate of

$$S'(t) = 10 - 10e^{-0.1t} \qquad 0 \le t \le 24$$

t months after the campaign has started.

(A) What will be the total sales $S(t)$ t months after the beginning of the national campaign if we assume no sales at the beginning of the campaign?

(B) What are the estimated total sales for the first 12 months of the campaign?

(C) When will the estimated total sales reach \$100 million? Use a graphing calculator to approximate the answer to two decimal places.

72. Marketing. Repeat Problem 71 if the monthly rate of increase in sales is found to be approximated by

$$S'(t) = 20 - 20e^{-0.05t} \qquad 0 \le t \le 24$$

73. Oil production. Using production and geological data, the management of an oil company estimates that oil will be pumped from a field producing at a rate given by

$$R(t) = \dfrac{100}{t + 1} + 5 \qquad 0 \le t \le 20$$

where $R(t)$ is the rate of production (in thousands of barrels per year) t years after pumping begins. How many barrels of oil $Q(t)$ will the field produce in the first t years if $Q(0) = 0$? How many barrels will be produced in the first 9 years?

74. Oil production. Assume that the rate in Problem 73 is found to be

$$R(t) = \dfrac{120t}{t^2 + 1} + 3 \qquad 0 \le t \le 20$$

(A) When is the rate of production greatest?

(B) How many barrels of oil $Q(t)$ will the field produce in the first t years if $Q(0) = 0$? How many barrels will be produced in the first 5 years?

(C) How long (to the nearest tenth of a year) will it take to produce a total of a quarter of a million barrels of oil?

75. Biology. A yeast culture is growing at the rate of $w'(t) = 0.2e^{0.1t}$ grams per hour. If the starting culture weighs 2 grams, what will be the weight of the culture $W(t)$ after t hours? After 8 hours?

76. Medicine. The rate of healing for a skin wound (in square centimeters per day) is approximated by $A'(t) = -0.9e^{-0.1t}$. If the initial wound has an area of 9 square centimeters, what will its area $A(t)$ be after t days? After 5 days?

 77. Pollution. A contaminated lake is treated with a bactericide. The rate of increase in harmful bacteria t days after the treatment is given by

$$\frac{dN}{dt} = -\frac{2{,}000t}{1 + t^2} \qquad 0 \le t \le 10$$

where $N(t)$ is the number of bacteria per milliliter of water. Since dN/dt is negative, the count of harmful bacteria is decreasing.

(A) Find the minimum value of dN/dt.

(B) If the initial count was 5,000 bacteria per milliliter, find $N(t)$ and then find the bacteria count after 10 days.

(C) When (to two decimal places) is the bacteria count 1,000 bacteria per milliliter?

78. Pollution. An oil tanker aground on a reef is losing oil and producing an oil slick that is radiating outward at a rate given approximately by

$$\frac{dR}{dt} = \frac{60}{\sqrt{t + 9}} \qquad t \ge 0$$

where R is the radius (in feet) of the circular slick after t minutes. Find the radius of the slick after 16 minutes if the radius is 0 when $t = 0$.

79. Learning. An average student enrolled in an advanced typing class progressed at a rate of $N'(t) = 6e^{-0.1t}$ words per minute per week t weeks after enrolling in a 15-week course. If, at the beginning of the course, a student could type 40 words per minute, how many words per minute $N(t)$ would the student be expected to type t weeks into the course? After completing the course?

80. Learning. An average student enrolled in a stenotyping class progressed at a rate of $N'(t) = 12e^{-0.06t}$ words per minute per week t weeks after enrolling in a 15-week course. If, at the beginning of the course, a student could stenotype at zero words per minute, how many words per minute $N(t)$ would the student be expected to handle t weeks into the course? After completing the course?

81. College enrollment. The projected rate of increase in enrollment at a new college is estimated by

$$\frac{dE}{dt} = 5{,}000(t + 1)^{-3/2} \qquad t \ge 0$$

where $E(t)$ is the projected enrollment in t years. If enrollment is 2,000 now $(t = 0)$, find the projected enrollment 15 years from now.

Answers to Matched Problems

1. (A) $\frac{1}{21}(2x^3 - 3)^{21} + C$
   (B) $e^{5x} + C$
   (C) $\ln|4 + x^2| + C$ or $\ln(4 + x^2) + C$, since $4 + x^2 > 0$

2. (A) $dy = 3x^2\,dx$
   (B) $du = \dfrac{2x}{2 + x^2}dx$
   (C) $dv = -5e^{-5t}\,dt$

3. $\frac{1}{5}(x^2 - 3x + 7)^5 + C$

4. (A) $\frac{1}{5}(2x^3 - 3)^5 + C$
   (B) $e^{5w} + C$

5. (A) $-\frac{1}{3}e^{-3x} + C$
   (B) $\frac{1}{2}\ln|x^2 - 9| + C$
   (C) $-\frac{5}{3}(t^3 + 4)^{-1} + C$

6. $\frac{2}{5}(x + 1)^{5/2} - \frac{2}{3}(x + 1)^{3/2} + C$

7. $p(x) = 0.1e^{0.01x} + 3.38$; 179 tubes

7-3 Differential Equations; Growth and Decay

- Differential Equations and Slope Fields
- Continuous Compound Interest Revisited
- Exponential Growth Law
- Population Growth, Radioactive Decay, and Learning
- Comparison of Exponential Growth Phenomena

In the preceding section, we considered equations of the form

$$\frac{dy}{dx} = 6x^2 - 4x \qquad p'(x) = -400e^{-0.04x}$$

These are examples of *differential equations*. In general, an equation is a **differential equation** if it involves an unknown function and one or more of its derivatives. Other examples of differential equations are

$$\frac{dy}{dx} = ky \qquad y'' - xy' + x^2 = 5 \qquad \frac{dy}{dx} = 2xy$$

The first and third equations are called **first-order** equations because each involves a first derivative but no higher derivative. The second equation is called a **second-**

order equation because it involves a second derivative and no higher derivatives. Finding solutions of different types of differential equations (functions that satisfy the equation) is the subject matter of entire books and courses on this topic. Here, we consider only a few special first-order equations that have immediate and significant applications.

We start by looking at some first-order equations geometrically, in terms of *slope fields*. We then consider continuous compound interest as modeled by a first-order differential equation. From this treatment, we can generalize our approach to a wide variety of other types of growth phenomena.

Differential Equations and Slope Fields

We introduce the concept of *slope field* through an example. Consider the first-order differential equation

$$\frac{dy}{dx} = 0.2y \tag{1}$$

A function f is a solution of equation (1) if $y = f(x)$ satisfies equation (1) for all values of x in the domain of f. Geometrically interpreted, equation (1) gives us the slope of a solution curve that passes through the point (x, y). For example, if $y = f(x)$ is a solution of equation (1) that passes through the point $(0, 2)$, then the slope of f at $(0, 2)$ is given by

$$\frac{dy}{dx} = 0.2(2) = 0.4$$

We indicate this relationship by drawing a short segment of the tangent line at the point $(0, 2)$, as shown in Figure 1A. The procedure is repeated for points $(-3, 1)$ and $(2, 3)$. Assuming that the graph of f passes through all three points, we sketch an approximate graph of f in Figure 1B.

(A) (B)

Figure 1

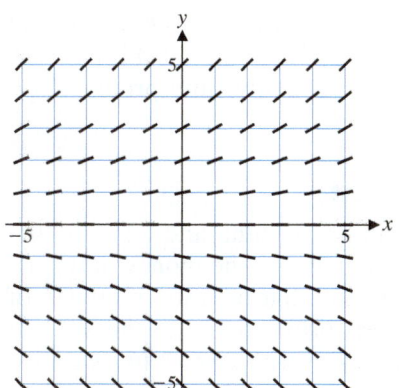

Figure 2

If we continue the process of drawing tangent line segments at each point grid in Figure 1—a task easily handled by computers, but not by hand—we obtain a *slope field*. A slope field for differential equation (1), drawn by a computer, is shown in Figure 2. In general, a **slope field** for a first-order differential equation is obtained by drawing tangent line segments determined by the equation at each point in a grid.

EXPLORE & DISCUSS 1

(A) In Figure 1A (or a copy), draw tangent line segments for a solution curve of differential equation (1) that passes through $(-3, -1)$, $(0, -2)$, and $(2, -3)$.

(B) In Figure 1B (or a copy), sketch an approximate graph of the solution curve that passes through the three points given in part (A). Repeat the tangent line segments first.

(C) What type of function, of all the elementary functions discussed in the first two chapters, appears to be a solution of differential equation (1)?

In Explore & Discuss 1, if you guessed that all solutions of equation (1) are exponential functions, you are to be congratulated. We now show that

$$y = Ce^{0.2x} \tag{2}$$

is a solution of equation (1) for any real number C. We substitute $y = Ce^{0.2x}$ into equation (1) to see if the left side is equal to the right side for all real x:

$$\frac{dy}{dx} = 0.2y$$

Left side: $\dfrac{dy}{dx} = \dfrac{d}{dx}(Ce^{0.2x}) = 0.2Ce^{0.2x}$

Right side: $0.2y = 0.2Ce^{0.2x}$

So equation (2) is a solution of equation (1) for C any real number. Which values of C will produce solution curves that pass through $(0, 2)$ and $(0, -2)$, respectively? Substituting the coordinates of each point into equation (2) and solving for C, we obtain

$$y = 2e^{0.2x} \qquad \text{and} \qquad y = -2e^{0.2x} \tag{3}$$

The graphs of equations (3) are shown in Figure 3, and they confirm the results shown in Figure 1B.

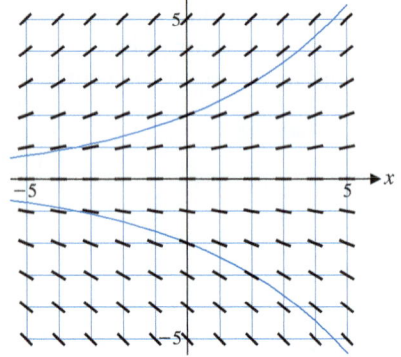

Figure 3

CONCEPTUAL INSIGHT

For a complicated first-order differential equation, say,

$$\frac{dy}{dx} = \frac{3 + \sqrt{xy}}{x^2 - 5y^4}$$

it may be impossible to find a formula analogous to (2) for its solutions. Nevertheless, it is routine to evaluate the right-hand side at each point in a grid. The resulting slope field provides a graphical representation of the solutions of the differential equation.

Drawing slope fields by hand is not a task for human beings: A 20-by-20 grid would require drawing 400 tangent line segments! Repetitive tasks of this type are what computers are for. A few problems in Exercise 6-3 involve interpreting slope fields, not drawing them.

Continuous Compound Interest Revisited

Let P be the initial amount of money deposited in an account, and let A be the amount in the account at any time t. Instead of assuming that the money in the account earns a particular rate of interest, suppose we say that the rate of growth of the amount of money in the account at any time t is proportional to the amount present at that time. Since dA/dt is the rate of growth of A with respect to t, we have

$$\frac{dA}{dt} = rA \qquad A(0) = P \qquad A, P > 0 \tag{4}$$

where r is an appropriate constant. We would like to find a function $A = A(t)$ that satisfies these conditions. Multiplying both sides of equation (4) by $1/A$, we obtain

$$\frac{1}{A}\frac{dA}{dt} = r$$

Now we integrate each side with respect to t:

$$\int \frac{1}{A}\frac{dA}{dt}\,dt = \int r\,dt \qquad \frac{dA}{dt}dt = A'(t)dt = dA$$

$$\int \frac{1}{A}\,dA = \int r\,dt$$

$$\ln|A| = rt + C \qquad |A| = A, \text{ since } A > 0$$

$$\ln A = rt + C$$

We convert this last equation into the equivalent exponential form

$$A = e^{rt+C} \qquad \text{Definition of logarithmic function:}$$
$$y = \ln x \text{ if and only if } x = e^y$$

$$= e^C e^{rt} \qquad \text{Property of exponents: } b^m b^n = b^{m+n}$$

Since $A(0) = P$, we evaluate $A(t) = e^C e^{rt}$ at $t = 0$ and set the result equal to P:

$$A(0) = e^C e^0 = e^C = P$$

Hence, $e^C = P$, and we can rewrite $A = e^C e^{rt}$ in the form

$$A = Pe^{rt}$$

This is the same continuous compound interest formula obtained in Section 5-1, where the principal P is invested at an annual nominal rate r compounded continuously for t years.

Exponential Growth Law

In general, if the rate of change of a quantity Q with respect to time is proportional to the amount of Q present and $Q(0) = Q_0$, then, proceeding in exactly the same way as we just did, we obtain the following theorem:

THEOREM 1   Exponential Growth Law

If $\dfrac{dQ}{dt} = rQ$ and $Q(0) = Q_0$, then $Q = Q_0 e^{rt}$,

where

$Q_0 =$ amount of Q at $t = 0$
$r =$ relative growth rate (expressed as a decimal)
$t =$ time
$Q =$ quantity at time t

The constant r in the exponential growth law is called the **relative growth rate**. If the relative growth rate is $r = 0.02$, then the quantity Q is growing at a rate $dQ/dt = 0.02Q$ (that is, 2% of the quantity Q per unit of time t). Note the distinction between the relative growth rate r and the rate of growth dQ/dt of the quantity Q. If $r < 0$, then $dQ/dt < 0$ and Q is decreasing. This type of growth is called **exponential decay**.

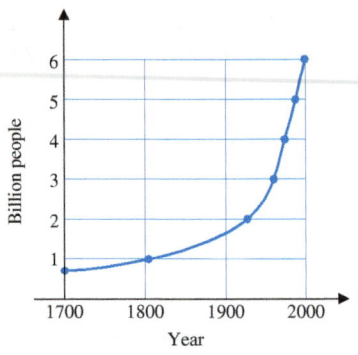

Figure 4 World population growth

Once we know that the rate of growth is proportional to the amount present, we recognize exponential growth and can use Theorem 1 without solving the differential equation each time. The exponential growth law applies not only to money invested at interest compounded continuously, but also to many other types of problems — population growth, radioactive decay, the depletion of a natural resource, and so on.

Population Growth, Radioactive Decay, and Learning

The world population passed 1 billion in 1804, 2 billion in 1927, 3 billion in 1960, 4 billion in 1974, 5 billion in 1987, and 6 billion in 1999, as illustrated in Figure 4. **Population growth** over certain periods often can be approximated by the exponential growth law of Theorem 1.

EXAMPLE 1   **Population Growth** India had a population of about 1.2 billion in 2010 ($t = 0$). Let P represent the population (in billions) t years after 2010, and assume a growth rate of 1.5% compounded continuously.

(A) Find an equation that represents India's population growth after 2010, assuming that the 1.5% growth rate continues.

(B) What is the estimated population (to the nearest tenth of a billion) of India in the year 2030?

(C) Graph the equation found in part (A) from 2000 to 2030.

SOLUTION   (A) The exponential growth law applies, and we have

$$\frac{dP}{dt} = 0.015P \qquad P(0) = 1.2$$

Therefore,

$$P = 1.2e^{0.015t} \tag{5}$$

(B) Using equation (5), we can estimate the population in India in 2030 ($t = 20$):

$$P = 1.2e^{0.015(20)} = 1.6 \text{ billion people}$$

(C) The graph is shown in Figure 5.

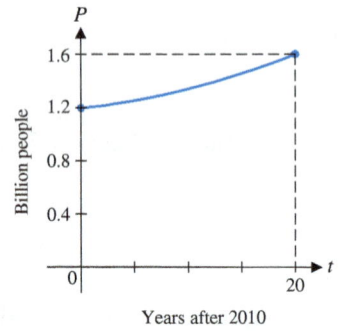

Figure 5 Population of India

Matched Problem 1   Assuming the same continuous compound growth rate as in Example 1, what will India's population be (to the nearest tenth of a billion) in the year 2020?

EXAMPLE 2   **Population Growth** If the exponential growth law applies to Canada's population growth, at what continuous compound growth rate will the population double over the next 100 years?

SOLUTION   We must find r, given that $P = 2P_0$ and $t = 100$:

$$P = P_0e^{rt}$$
$$2P_0 = P_0e^{100r}$$
$$2 = e^{100r} \qquad \text{Take the natural logarithm}$$
$$100r = \ln 2 \qquad \text{of both sides and reverse}$$
$$\qquad\qquad \text{the equation.}$$
$$r = \frac{\ln 2}{100}$$
$$\approx 0.0069 \quad \text{or} \quad 0.69\%$$

Matched Problem 2   If the exponential growth law applies to population growth in Nigeria, find the doubling time (to the nearest year) of the population if it grows at 2.1% per year compounded continuously.

We now turn to another type of exponential growth: **radioactive decay**. In 1946, Willard Libby (who later received a Nobel Prize in chemistry) found that as long as a plant or animal is alive, radioactive carbon-14 is maintained at a constant level in its tissues. Once the plant or animal is dead, however, the radioactive carbon-14 diminishes by radioactive decay at a rate proportional to the amount present.

$$\frac{dQ}{dt} = rQ \qquad Q(0) = Q_0$$

This is another example of the exponential growth law. The continuous compound rate of decay for radioactive carbon-14 is 0.000 123 8, so $r = -0.000\ 123\ 8$, since decay implies a negative continuous compound growth rate.

EXAMPLE 3   **Archaeology**   A human bone fragment was found at an archaeological site in Africa. If 10% of the original amount of radioactive carbon-14 was present, estimate the age of the bone (to the nearest 100 years).

SOLUTION   By the exponential growth law for

$$\frac{dQ}{dt} = -0.000\ 123\ 8Q \qquad Q(0) = Q_0$$

we have

$$Q = Q_0 e^{-0.0001238t}$$

We must find t so that $Q = 0.1Q_0$ (since the amount of carbon-14 present now is 10% of the amount Q_0 present at the death of the person).

$$0.1Q_0 = Q_0 e^{-0.0001238t}$$

$$0.1 = e^{-0.0001238t}$$

$$\ln 0.1 = \ln e^{-0.0001238t}$$

$$t = \frac{\ln 0.1}{-0.000\ 123\ 8} \approx 18{,}600 \text{ years}$$

Figure 6 $y_1 = e^{-0.0001238x}$; $y_2 = 0.1$

See Figure 6 for a graphical solution to Example 3.

Matched Problem 3   Estimate the age of the bone in Example 3 (to the nearest 100 years) if 50% of the original amount of carbon-14 is present.

In learning certain skills, such as typing and swimming, one often assumes that there is a maximum skill attainable—say, M—and the rate of improvement is proportional to the difference between what has been achieved y and the maximum attainable M. Mathematically,

$$\frac{dy}{dt} = k(M - y) \qquad y(0) = 0$$

We solve this type of problem with the same technique used to obtain the exponential growth law. First, multiply both sides of the first equation by $1/(M - y)$ to get

$$\frac{1}{M - y}\frac{dy}{dt} = k$$

and then integrate each side with respect to t:

$$\int \frac{1}{M-y} \frac{dy}{dt} dt = \int k\, dt$$

$$-\int \frac{1}{M-y}\left(-\frac{dy}{dt}\right) dt = \int k\, dt \qquad \text{Substitute } u = M - y \text{ and}$$

$$-\int \frac{1}{u} du = \int k\, dt \qquad du = -dy = -\frac{dy}{dt} dt.$$

$$-\ln|u| = kt + C \qquad \text{Substitute } M - y, \text{ which is } > 0, \text{ for } u.$$

$$-\ln(M-y) = kt + C \qquad \text{Multiply both sides by } -1.$$

$$\ln(M-y) = -kt - C$$

Change this last equation to an equivalent exponential form:

$$M - y = e^{-kt - C}$$

$$M - y = e^{-C} e^{-kt}$$

$$y = M - e^{-C} e^{-kt}$$

Now, $y(0) = 0$; hence,

$$y(0) = M - e^{-C} e^{0} = 0$$

Solving for e^{-C}, we obtain

$$e^{-C} = M$$

and our final solution is

$$y = M - Me^{-kt} = M(1 - e^{-kt})$$

EXAMPLE 4 **Learning** For a particular person learning to swim, the distance y (in feet) that the person is able to swim in 1 minute after t hours of practice is given approximately by

$$y = 50(1 - e^{-0.04t})$$

What is the rate of improvement (to two decimal places) after 10 hours of practice?

SOLUTION

$$y = 50 - 50e^{-0.04t}$$

$$y'(t) = 2e^{-0.04t}$$

$$y'(10) = 2e^{-0.04(10)} \approx 1.34 \text{ feet per hour of practice}$$

Matched Problem 4 In Example 4, what is the rate of improvement (to two decimal places) after 50 hours of practice?

Comparison of Exponential Growth Phenomena

The graphs and equations given in Table 1 compare several widely used growth models. These models are divided into two groups: unlimited growth and limited growth. Following each equation and graph is a short (and necessarily incomplete) list of areas in which the models are used.

Table 1   Exponential Growth

Description	Model	Solution	Graph	Uses
Unlimited growth: Rate of growth is proportional to the amount present	$\dfrac{dy}{dt} = ky$ $k, t > 0$ $y(0) = c$	$y = ce^{kt}$		• Short-term population growth (people, bacteria, etc.) • Growth of money at continuous compound interest • Price–supply curves
Exponential decay: Rate of growth is proportional to the amount present	$\dfrac{dy}{dt} = -ky$ $k, t > 0$ $y(0) = c$	$y = ce^{-kt}$		• Depletion of natural resources • Radioactive decay • Absorption of light in water • Price–demand curves • Atmospheric pressure (t is altitude)
Limited growth: Rate of growth is proportional to the difference between the amount present and a fixed limit	$\dfrac{dy}{dt} = k(M - y)$ $k, t > 0$ $y(0) = 0$	$y = M(1 - e^{-kt})$		• Sales fads (for example, skateboards) • Depreciation of equipment • Company growth • Learning
Logistic growth: Rate of growth is proportional to the amount present and to the difference between the amount present and a fixed limit	$\dfrac{dy}{dt} = ky(M - y)$ $k, t > 0$ $y(0) = \dfrac{M}{1 + c}$	$y = \dfrac{M}{1 + ce^{-kMt}}$		• Long-term population growth • Epidemics • Sales of new products • Spread of a rumor • Company growth

Exercises 7-3

A

In Problems 1–12, find the general or particular solution, as indicated, for each differential equation.

1. $\dfrac{dy}{dx} = 6x$

2. $\dfrac{dy}{dx} = 3x^{-2}$

3. $\dfrac{dy}{dx} = \dfrac{7}{x}$

4. $\dfrac{dy}{dx} = e^{0.1x}$

5. $\dfrac{dy}{dx} = e^{0.02x}$

6. $\dfrac{dy}{dx} = 8x^{-1}$

7. $\dfrac{dy}{dx} = x^2 - x;\ y(0) = 0$

8. $\dfrac{dy}{dx} = \sqrt{x};\ y(0) = 0$

9. $\dfrac{dy}{dx} = -2xe^{-x^2};\ y(0) = 3$

10. $\dfrac{dy}{dx} = e^{x-3};\ y(3) = -5$

11. $\dfrac{dy}{dx} = \dfrac{2}{1 + x};\ y(0) = 5$

12. $\dfrac{dy}{dx} = \dfrac{1}{4(3 - x)};\ y(0) = 1$

B

Problems 13–18 refer to the following slope fields:

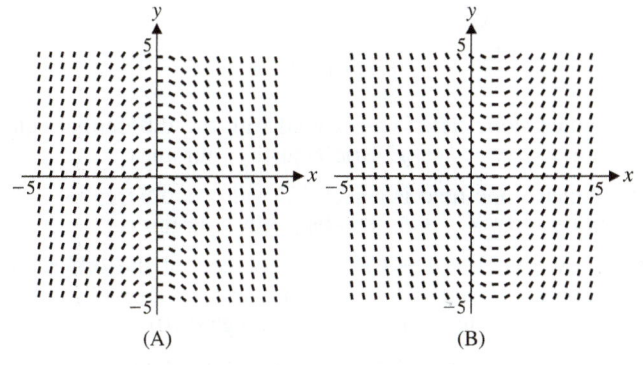

(A)   (B)

Figure for 13–18

13. Which slope field is associated with the differential equation $dy/dx = x - 1$? Briefly justify your answer.

14. Which slope field is associated with the differential equation $dy/dx = -x$? Briefly justify your answer.

15. Solve the differential equation $dy/dx = x - 1$ and find the particular solution that passes through $(0, -2)$.

16. Solve the differential equation $dy/dx = -x$ and find the particular solution that passes through $(0, 3)$.

17. Graph the particular solution found in Problem 15 in the appropriate figure A or B (or a copy).

18. Graph the particular solution found in Problem 16 in the appropriate figure A or B (or a copy).

In Problems 19–26, find the general or particular solution, as indicated, for each differential equation.

19. $\dfrac{dy}{dt} = 2y$

20. $\dfrac{dy}{dt} = -3y$

21. $\dfrac{dy}{dx} = -0.5y; \ y(0) = 100$

22. $\dfrac{dy}{dx} = 0.1y; \ y(0) = -2.5$

23. $\dfrac{dx}{dt} = -5x$

24. $\dfrac{dx}{dt} = 4t$

25. $\dfrac{dx}{dt} = -5t$

26. $\dfrac{dx}{dt} = 4x$

Problems 27–34 refer to the following slope fields:

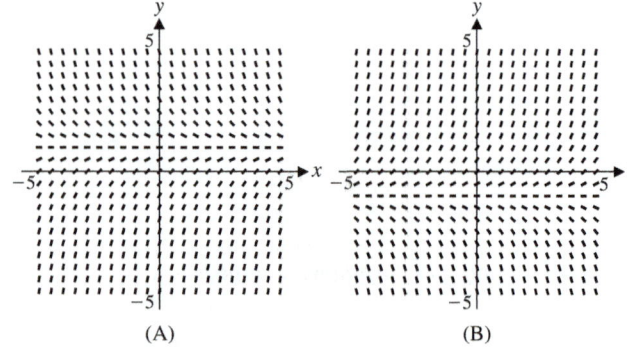

(A)   (B)

Figure for 27–34

27. Which slope field is associated with the differential equation $dy/dx = 1 - y$? Briefly justify your answer.

28. Which slope field is associated with the differential equation $dy/dx = y + 1$? Briefly justify your answer.

29. Show that $y = 1 - Ce^{-x}$ is a solution of the differential equation $dy/dx = 1 - y$ for any real number C. Find the particular solution that passes through $(0, 0)$.

30. Show that $y = Ce^x - 1$ is a solution of the differential equation $dy/dx = y + 1$ for any real number C. Find the particular solution that passes through $(0, 0)$.

31. Graph the particular solution found in Problem 29 in the appropriate figure, Figure A or Figure B (or a copy).

32. Graph the particular solution found in Problem 30 in the appropriate figure A or B (or a copy).

33. Use a graphing calculator to graph $y = 1 - Ce^{-x}$ for $C = -2, -1, 1,$ and $2,$ for $-5 \le x \le 5, -5 \le y \le 5,$ all in the same viewing window. Observe how the solution curves go with the flow of the tangent line segments in the corresponding slope field shown in figure A or figure B.

34. Use a graphing calculator to graph $y = Ce^x - 1$ for $C = -2, -1, 1,$ and $2,$ for $-5 \le x \le 5, -5 \le y \le 5,$ all in the same viewing window. Observe how the solution curves go with the flow of the tangent line segments in the corresponding slope field shown in figure A or figure B.

35. Show that $y = \sqrt{C - x^2}$ is a solution of the differential equation $dy/dx = -x/y$ for any positive real number C. Find the particular solution that passes through $(3, 4)$.

36. Show that $y = \sqrt{x^2 + C}$ is a solution of the differential equation $dy/dx = x/y$ for any real number C. Find the particular solution that passes through $(-6, 7)$.

37. Show that $y = Cx$ is a solution of the differential equation $dy/dx = y/x$ for any real number C. Find the particular solution that passes through $(-8, 24)$.

38. Show that $y = C/x$ is a solution of the differential equation $dy/dx = -y/x$ for any real number C. Find the particular solution that passes through $(2, 5)$.

39. Show that $y = 1/(1 + ce^{-t})$ is a solution of the differential equation $dy/dt = y(1 - y)$ for any real number c. Find the particular solution that passes through $(0, -1)$.

40. Show that $y = 2/(1 + ce^{-6t})$ is a solution of the differential equation $dy/dt = 3y(2 - y)$ for any real number c. Find the particular solution that passes through $(0, 1)$.

In Problems 41–48, use a graphing calculator to graph the given examples of the various cases in Table 1 on page 449.

41. Unlimited growth:

$y = 1,000e^{0.08t}$
$0 \le t \le 15$
$0 \le y \le 3,500$

42. Unlimited growth:

$y = 5,250e^{0.12t}$
$0 \le t \le 10$
$0 \le y \le 20,000$

43. Exponential decay:

$p = 100e^{-0.05x}$
$0 \le x \le 30$
$0 \le p \le 100$

44. Exponential decay:

$p = 1,000e^{-0.08x}$
$0 \le x \le 40$
$0 \le p \le 1,000$

45. Limited growth:

$N = 100(1 - e^{-0.05t})$
$0 \le t \le 100$
$0 \le N \le 100$

46. Limited growth:

$N = 1,000(1 - e^{-0.07t})$
$0 \le t \le 70$
$0 \le N \le 1,000$

47. Logistic growth:

$N = \dfrac{1,000}{1 + 999e^{-0.4t}}$
$0 \le t \le 40$
$0 \le N \le 1,000$

48. Logistic growth:

$N = \dfrac{400}{1 + 99e^{-0.4t}}$
$0 \le t \le 30$
$0 \le N \le 400$

49. Show that the rate of logistic growth, $dy/dt = ky(M - y)$, has its maximum value when $y = M/2$.

50. Find the value of t for which the logistic function

$$y = \dfrac{M}{1 + ce^{-kMt}}$$

is equal to $M/2$.

51. Let $Q(t)$ denote the population of the world at time t. In 1999, the world population was 6.0 billion and increasing at 1.3% per year; in 2009, it was 6.8 billion and increasing at 1.2% per year. In which year, 1999 or 2009, was dQ/dt (the rate of growth of Q with respect to t) greater? Explain.

52. Refer to Problem 51. Explain why the world population function $Q(t)$ does not satisfy an exponential growth law.

Applications

53. Continuous compound interest. Find the amount A in an account after t years if

$$\dfrac{dA}{dt} = 0.03A \quad \text{and} \quad A(0) = 1,000$$

54. Continuous compound interest. Find the amount A in an account after t years if

$$\dfrac{dA}{dt} = 0.02A \quad \text{and} \quad A(0) = 5,250$$

55. Continuous compound interest. Find the amount A in an account after t years if

$$\dfrac{dA}{dt} = rA \quad A(0) = 8,000 \quad A(2) = 8,260.14$$

56. Continuous compound interest. Find the amount A in an account after t years if

$$\dfrac{dA}{dt} = rA \quad A(0) = 5,000 \quad A(5) = 5,581.39$$

57. Price–demand. The marginal price dp/dx at x units of demand per week is proportional to the price p. There is no weekly demand at a price of $100 per unit $[p(0) = 100]$, and there is a weekly demand of 5 units at a price of $77.88 per unit $[p(5) = 77.88]$.

(A) Find the price–demand equation.

(B) At a demand of 10 units per week, what is the price?

(C) Graph the price–demand equation for $0 \le x \le 25$.

58. Price–supply. The marginal price dp/dx at x units of supply per day is proportional to the price p. There is no supply at a price of $10 per unit $[p(0) = 10]$, and there is a daily supply of 50 units at a price of $12.84 per unit $[p(50) = 12.84]$.

(A) Find the price–supply equation.

(B) At a supply of 100 units per day, what is the price?

(C) Graph the price–supply equation for $0 \le x \le 250$.

59. Advertising. A company is trying to expose a new product to as many people as possible through TV ads. Suppose that the rate of exposure to new people is proportional to the number of those who have not seen the product out of L possible viewers. No one is aware of the product at the start of the campaign, and after 10 days, 40% of L are aware of the product. Mathematically,

$$\dfrac{dN}{dt} = k(L - N) \quad N(0) = 0 \quad N(10) = 0.4L$$

(A) Solve the differential equation.

(B) What percent of L will have been exposed after 5 days of the campaign?

(C) How many days will it take to expose 80% of L?

(D) Graph the solution found in part (A) for $0 \le t \le 90$.

60. Advertising. Suppose that the differential equation for Problem 59 is

$$\dfrac{dN}{dt} = k(L - N) \quad N(0) = 0 \quad N(10) = 0.1L$$

(A) Explain what the equation $N(10) = 0.1L$ means.

(B) Solve the differential equation.

(C) How many days will it take to expose 50% of L?

(D) Graph the solution found in part (B) for $0 \le t \le 300$.

61. Biology. For relatively clear bodies of water, the intensity of light is reduced according to

$$\dfrac{dI}{dx} = -kI \quad I(0) = I_0$$

where I is the intensity of light at x feet below the surface. For the Sargasso Sea off the West Indies, $k = 0.00942$. Find I in terms of x, and find the depth at which the light is reduced to half of that at the surface.

62. Blood pressure. Under certain assumptions, the blood pressure P in the largest artery in the human body (the aorta) changes between beats with respect to time t according to

$$\dfrac{dP}{dt} = -aP \quad P(0) = P_0$$

where a is a constant. Find $P = P(t)$ that satisfies both conditions.

63. Drug concentration. A single injection of a drug is administered to a patient. The amount Q in the body then decreases at a rate proportional to the amount present. For a particular drug, the rate is 4% per hour. Thus,

$$\dfrac{dQ}{dt} = -0.04Q \quad Q(0) = Q_0$$

where t is time in hours.

(A) If the initial injection is 3 milliliters [$Q(0) = 3$], find $Q = Q(t)$ satisfying both conditions.

(B) How many milliliters (to two decimal places) are in the body after 10 hours?

(C) How many hours (to two decimal places) will it take for only 1 milliliter of the drug to be left in the body?

(D) Graph the solution found in part (A).

64. Simple epidemic. A community of 1,000 people is homogeneously mixed. One person who has just returned from another community has influenza. Assume that the home community has not had influenza shots and all are susceptible. One mathematical model assumes that influenza tends to spread at a rate in direct proportion to the number N who have the disease and to the number $1,000 - N$ who have not yet contracted the disease. Mathematically,

$$\frac{dN}{dt} = kN(1,000 - N) \qquad N(0) = 1$$

where N is the number of people who have contracted influenza after t days. For $k = 0.0004$, $N(t)$ is given by

$$N(t) = \frac{1,000}{1 + 999e^{-0.4t}}$$

(A) How many people have contracted influenza after 10 days? After 20 days?

(B) How many days will it take until half the community has contracted influenza?

(C) Find $\lim_{t \to \infty} N(t)$.

(D) Graph $N = N(t)$ for $0 \le t \le 30$.

65. Nuclear accident. One of the dangerous radioactive isotopes detected after the Chernobyl nuclear disaster in 1986 was cesium-137. If 93.3% of the cesium-137 emitted during the disaster was still present 3 years later, find the continuous compound rate of decay of this isotope.

66. Insecticides. Many countries have banned the use of the insecticide DDT because of its long-term adverse effects. Five years after a particular country stopped using DDT, the amount of DDT in the ecosystem had declined to 75% of the amount present at the time of the ban. Find the continuous compound rate of decay of DDT.

67. Archaeology. A skull found in an ancient tomb has 5% of the original amount of radioactive carbon-14 present. Estimate the age of the skull. (See Example 3.)

68. Learning. For a person learning to type, the number N of words per minute that the person could type after t hours of practice was given by

$$N = 100(1 - e^{-0.02t})$$

What is the rate of improvement after 10 hours of practice? After 40 hours of practice?

69. Small-group analysis. In a study on small-group dynamics, sociologists found that when 10 members of a discussion group were ranked according to the number of times each participated, the number $N(k)$ of times that the kth-ranked person participated was given by

$$N(k) = N_1 e^{-0.11(k-1)} \qquad 1 \le k \le 10$$

where N_1 is the number of times that the first-ranked person participated in the discussion. If $N_1 = 180$, in a discussion group of 10 people, estimate how many times the sixth-ranked person participated. How about the 10th-ranked person?

70. Perception. The Weber–Fechner law concerns a person's sensed perception of various strengths of stimulation involving weights, sound, light, shock, taste, and so on. One form of the law states that the rate of change of sensed sensation S with respect to stimulus R is inversely proportional to the strength of the stimulus R. So

$$\frac{dS}{dR} = \frac{k}{R}$$

where k is a constant. If we let R_0 be the threshold level at which the stimulus R can be detected (the least amount of sound, light, weight, and so on, that can be detected), then

$$S(R_0) = 0$$

Find a function S in terms of R that satisfies these conditions.

71. Rumor propagation. Sociologists have found that a rumor tends to spread at a rate in direct proportion to the number x who have heard it and to the number $P - x$ who have not, where P is the total population. If a resident of a 400-student dormitory hears a rumor that there is a case of TB on campus, then $P = 400$ and

$$\frac{dx}{dt} = 0.001x(400 - x) \qquad x(0) = 1$$

where t is time (in minutes). From these conditions, it can be shown that

$$x(t) = \frac{400}{1 + 399e^{-0.4t}}$$

(A) How many people have heard the rumor after 5 minutes? after 20 minutes?

(B) Find $\lim_{t \to \infty} x(t)$.

(C) Graph $x = x(t)$ for $0 \le t \le 30$.

72. Rumor propagation. In Problem 71, how long (to the nearest minute) will it take for half of the group of 400 to have heard the rumor?

Answers to Matched Problems

1. 1.4 billion people   2. 33 yr

3. 5,600 yr   4. 0.27 ft/hr

7-4   The Definite Integral

- Approximating Areas by Left and Right Sums
- The Definite Integral as a Limit of Sums
- Properties of the Definite Integral

The first three sections of this chapter focused on the *indefinite integral*. In this section, we introduce the *definite integral*. The definite integral is used to compute areas, probabilities, average values of functions, future values of continuous income streams, and many other quantities. Initially, the concept of the definite integral may seem unrelated to the notion of the indefinite integral. There is, however, a close connection between the two integrals. The fundamental theorem of calculus, discussed in Section 7-5, makes that connection precise.

Approximating Areas by Left and Right Sums

How do we find the shaded area in Figure 1? That is, how do we find the area bounded by the graph of $f(x) = 0.25x^2 + 1$, the x axis, and the vertical lines $x = 1$ and $x = 5$? [This cumbersome description is usually shortened to "the area under the graph of $f(x) = 0.25x^2 + 1$ from $x = 1$ to $x = 5$."] Our standard geometric area formulas do not apply directly, but the formula for the area of a rectangle can be used indirectly. To see how, we look at a method of approximating the area under the graph by using rectangles. This method will give us any accuracy desired, which is quite different from finding the area exactly. Our first area approximation is made by dividing the interval $[1, 5]$ on the x axis into four equal parts, each of length

$$\Delta x = \frac{5 - 1}{4} = 1^*$$

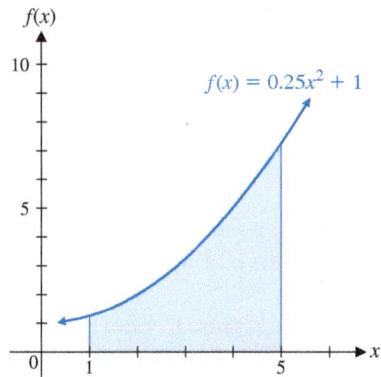

Figure 1 What is the shaded area?

We then place a **left rectangle** on each subinterval, that is, a rectangle whose base is the subinterval and whose height is the value of the function at the left endpoint of the subinterval (see Fig. 2).

Summing the areas of the left rectangles in Figure 2 results in a **left sum** of four rectangles, denoted by L_4, as follows:

$$L_4 = f(1) \cdot 1 + f(2) \cdot 1 + f(3) \cdot 1 + f(4) \cdot 1$$
$$= 1.25 + 2.00 + 3.25 + 5 = 11.5$$

From Figure 3, since $f(x)$ is increasing, we see that the left sum L_4 underestimates the area, and we can write

$$11.5 = L_4 < \text{Area}$$

Figure 2 Left rectangles

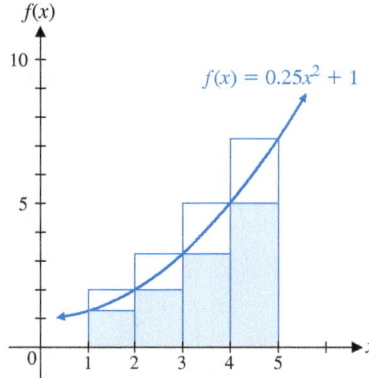

Figure 3 Left and right rectangles

*It is customary to denote the length of the subintervals by Δx, which is read "delta x," since Δ is the Greek capital letter delta.

EXPLORE & DISCUSS 1

If $f(x)$ were decreasing over the interval $[1, 5]$, would the left sum L_4 over- or under-estimate the actual area under the curve? Explain.

Similarly, we use the right endpoint of each subinterval to find the height of the **right rectangle** placed on top of it. Superimposing right rectangles on top of Figure 2, we get Figure 3.

Summing the areas of the right rectangles in Figure 3 results in a **right sum** of four rectangles, denoted by R_4, as follows (compare R_4 with L_4 and note that R_4 can be obtained from L_4 by deleting one rectangular area and adding one more):

$$R_4 = f(2) \cdot 1 + f(3) \cdot 1 + f(4) \cdot 1 + f(5) \cdot 1$$
$$= 2.00 + 3.25 + 5.00 + 7.25 = 17.5 \cdot$$

From Figure 3, since $f(x)$ is increasing, we see that the right sum R_4 overestimates the area, and we conclude that the actual area is between 11.5 and 17.5. That is,

$$11.5 = L_4 < \text{Area} < R_4 = 17.5$$

EXPLORE & DISCUSS 2

If $f(x)$ in Figure 3 were decreasing over the interval $[1, 5]$, would the right sum R_4 overestimate or underestimate the actual area under the curve? Explain.

The first approximation of the area under the curve in Figure 1 is fairly coarse, but the method outlined can be continued with increasingly accurate results by dividing the interval $[1, 5]$ into more and more subintervals of equal horizontal length. Of course, this is not a job for hand calculation, but a job that computers are designed to do.* Figure 4 shows left- and right-rectangle approximations for 16 equal subdivisions.

For this case,

$$\Delta x = \frac{5 - 1}{16} = 0.25$$

$$L_{16} = f(1) \cdot \Delta x + f(1.25) \cdot \Delta x + \cdots + f(4.75) \cdot \Delta x$$
$$= 13.59$$

$$R_{16} = f(1.25) \cdot \Delta x + f(1.50) \cdot \Delta x + \cdots + f(5) \cdot \Delta x$$
$$= 15.09$$

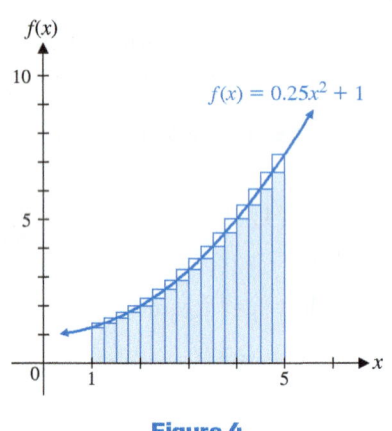

$f(x) = 0.25x^2 + 1$

Figure 4

Thus, we now know that the area under the curve is between 13.59 and 15.09. That is,

$$13.59 = L_{16} < \text{Area} < R_{16} = 15.09$$

For 100 equal subdivisions, computer calculations give us

$$14.214 = L_{100} < \text{Area} < R_{100} = 14.454$$

The **error in an approximation** is the absolute value of the difference between the approximation and the actual value. In general, neither the actual value nor the error in an approximation is known. However, it is often possible to calculate an **error bound**—a positive number such that the error is guaranteed to be less than or equal to that number.

The error in the approximation of the area under the graph of f from $x = 1$ to $x = 5$ by the left sum L_{16} (or the right sum R_{16}) is less than the sum of the areas of the small rectangles in Figure 4. By stacking those rectangles (see Fig. 5), we see that

$$\text{Error} = |\text{Area} - L_{16}| < |f(5) - f(1)| \cdot \Delta x = 1.5$$

Therefore, 1.5 is an error bound for the approximation of the area under f by L_{16}. We can apply the same stacking argument to any positive function that is increasing on $[a, b]$ or decreasing on $[a, b]$, to obtain the error bound in Theorem 1.

---

*The computer software that accompanies this book will perform these calculations (see the preface).

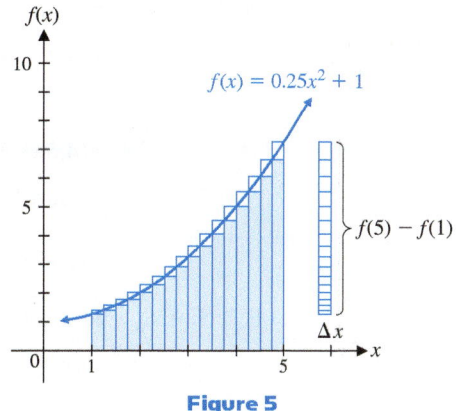

Figure 5

THEOREM 1   Error Bounds for Approximations of Area by Left or Right Sums
If $f(x) > 0$ and is either increasing on $[a, b]$ or decreasing on $[a, b]$, then

$$|f(b) - f(a)| \cdot \frac{b - a}{n}$$

is an error bound for the approximation of the area between the graph of f and the x axis, from $x = a$ to $x = b$, by L_n or R_n.

Because the error bound of Theorem 1 approaches 0 as $n \to \infty$, it can be shown that left and right sums, for certain functions, approach the same limit as $n \to \infty$.

THEOREM 2   Limits of Left and Right Sums
If $f(x) > 0$ and is either increasing on $[a, b]$ or decreasing on $[a, b]$, then its left and right sums approach the same real number as $n \to \infty$.

The number approached as $n \to \infty$ by the left and right sums in Theorem 2 is the area between the graph of f and the x axis from $x = a$ to $x = b$.

EXAMPLE 1   **Approximating Areas** Given the function $f(x) = 9 - 0.25x^2$, we want to approximate the area under $y = f(x)$ from $x = 2$ to $x = 5$.

(A) Graph the function over the interval $[0, 6]$. Then draw left and right rectangles for the interval $[2, 5]$ with $n = 6$.

(B) Calculate L_6, R_6, and error bounds for each.

(C) How large should n be in order for the approximation of the area by L_n or R_n to be within 0.05 of the true value?

SOLUTION   (A) $\Delta x = 0.5$:

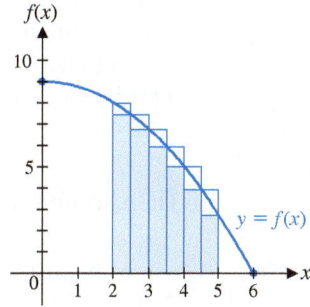

(B) $L_6 = f(2) \cdot \Delta x + f(2.5) \cdot \Delta x + f(3) \cdot \Delta x + f(3.5) \cdot \Delta x + f(4) \cdot \Delta x$
$+ f(4.5) \cdot \Delta x = 18.53$

$R_6 = f(2.5) \cdot \Delta x + f(3) \cdot \Delta x + f(3.5) \cdot \Delta x + f(4) \cdot \Delta x$
$+ f(4.5) \cdot \Delta x + f(5) \cdot \Delta x = 15.91$

The error bound for L_6 and R_6 is

$$\text{error} \leq |f(5) - f(2)|\frac{5 - 2}{6} = |2.75 - 8|(0.5) = 2.625$$

(C)   For L_n and R_n, find n such that error ≤ 0.05:

$$|f(b) - f(a)|\frac{b - a}{n} \leq 0.05$$

$$|2.75 - 8|\frac{3}{n} \leq 0.05$$

$$|-5.25|\frac{3}{n} \leq 0.05$$

$$15.75 \leq 0.05n$$

$$n \geq \frac{15.75}{0.05} = 315$$

Matched Problem 1   Given the function $f(x) = 8 - 0.5x^2$, we want to approximate the area under $y = f(x)$ from $x = 1$ to $x = 3$.

(A) Graph the function over the interval $[0, 4]$. Then draw left and right rectangles for the interval $[1, 3]$ with $n = 4$.

(B) Calculate L_4, R_4, and error bounds for each.

(C) How large should n be in order for the approximation of the area by L_n or R_n to be within 0.5 of the true value?

CONCEPTUAL INSIGHT

Note from Example 1(C) that a relatively large value of n ($n = 315$) is required to approximate the area by L_n or R_n to within 0.05. In other words, 315 rectangles must be used, and 315 terms must be summed, to guarantee that the error does not exceed 0.05. We can obtain a more efficient approximation of the area (fewer terms are summed to achieve a given accuracy) by replacing rectangles with trapezoids. The resulting **trapezoidal rule**, and other methods for approximating areas, are discussed in Group Activity 1 in this book's Web site.

The Definite Integral as a Limit of Sums

Left and right sums are special cases of more general sums, called *Riemann sums* [named after the German mathematician Georg Riemann (1826–1866)], that are used to approximate areas by means of rectangles.

Let f be a function defined on the interval $[a, b]$. We partition $[a, b]$ into n subintervals of equal length $\Delta x = (b - a)/n$ with endpoints

$$a = x_0 < x_1 < x_2 < \cdots < x_n = b$$

Then, using **summation notation** (see Appendix B-1), we have

Left sum: $L_n = f(x_0)\Delta x + f(x_1)\Delta x + \cdots + f(x_{n-1})\Delta x = \sum_{k=1}^{n} f(x_{k-1})\Delta x$

Right sum: $R_n = f(x_1)\Delta x + f(x_2)\Delta x + \cdots + f(x_n)\Delta x = \sum_{k=1}^{n} f(x_k)\Delta x$

Riemann sum: $S_n = f(c_1)\Delta x + f(c_2)\Delta x + \cdots + f(c_n)\Delta x = \sum_{k=1}^{n} f(c_k)\Delta x$

In a **Riemann sum**,* each c_k is required to belong to the subinterval $[x_{k-1}, x_k]$. Left and right sums are the special cases of Riemann sums in which c_k is the left endpoint or right endpoint, respectively, of the subinterval. If $f(x) > 0$, then each term of a Riemann sum S_n represents the area of a rectangle having height $f(c_k)$ and width Δx (see Fig. 6). If $f(x)$ has both positive and negative values, then some terms of S_n represent areas of rectangles, and others represent the negatives of areas of rectangles, depending on the sign of $f(c_k)$ (see Fig. 7).

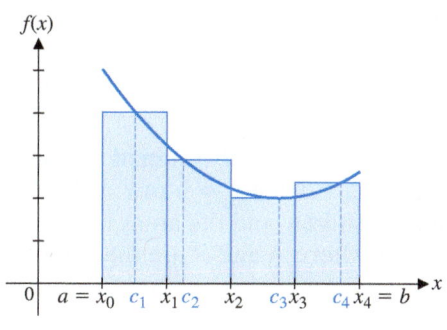

Figure 6   **Figure 7**

EXAMPLE 2 **Riemann Sums** Consider the function $f(x) = 15 - x^2$ on $[1, 5]$. Partition the interval $[1, 5]$ into four subintervals of equal length. For each subinterval $[x_{k-1}, x_k]$, let c_k be the midpoint. Calculate the corresponding Riemann sum S_4. (Riemann sums for which the c_k are the midpoints of the subintervals are called **midpoint sums**.)

SOLUTION
$$\Delta x = \frac{5 - 1}{4} = 1$$
$$S_4 = f(c_1) \cdot \Delta x + f(c_2) \cdot \Delta x + f(c_3) \cdot \Delta x + f(c_4) \cdot \Delta x$$
$$= f(1.5) \cdot 1 + f(2.5) \cdot 1 + f(3.5) \cdot 1 + f(4.5) \cdot 1$$
$$= 12.75 + 8.75 + 2.75 - 5.25 = 19$$

Matched Problem 2 Consider the function $f(x) = x^2 - 2x - 10$ on $[2, 8]$. Partition the interval $[2, 8]$ into three subintervals of equal length. For each subinterval $[x_{k-1}, x_k]$, let c_k be the midpoint. Calculate the corresponding Riemann sum S_3.

By analyzing properties of a continuous function on a closed interval, it can be shown that the conclusion of Theorem 2 is valid if f is continuous. In that case, not just left and right sums, but Riemann sums, have the same limit as $n \to \infty$.

THEOREM 3 Limit of Riemann Sums
If f is a continuous function on $[a, b]$, then the Riemann sums for f on $[a, b]$ approach a real number limit I as $n \to \infty$.†

---

*The term *Riemann sum* is often applied to more general sums in which the subintervals $[x_{k-1}, x_k]$ are not required to have the same length. Such sums are not considered in this book.

†The precise meaning of this limit statement is as follows: For each $e > 0$, there exists some $d > 0$ such that $|S_n - I| < e$ whenever S_n is a Riemann sum for f on $[a, b]$ for which $\Delta x < d$.

DEFINITION Definite Integral

Let f be a continuous function on $[a, b]$. The limit I of Riemann sums for f on $[a, b]$, guaranteed to exist by Theorem 2, is called the **definite integral** of f from a to b and is denoted as

$$\int_a^b f(x)\, dx$$

The **integrand** is $f(x)$, the **lower limit of integration** is a, and the **upper limit of integration** is b.

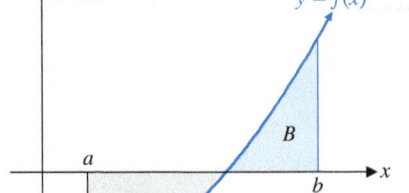

Figure 8 $\int_a^b f(x)\, dx = -A + B$

Because area is a positive quantity, the definite integral has the following geometric interpretation:

$$\int_a^b f(x)\, dx$$

represents the cumulative sum of the signed areas between the graph of f and the x axis from $x = a$ to $x = b$, where the areas above the x axis are counted positively and the areas below the x axis are counted negatively (see Fig. 8, where A and B are the actual areas of the indicated regions).

EXAMPLE 3   **Definite Integrals** Calculate the definite integrals by referring to Figure 9.

(A) $\displaystyle\int_a^b f(x)\, dx$

(B) $\displaystyle\int_a^c f(x)\, dx$

(C) $\displaystyle\int_b^c f(x)\, dx$

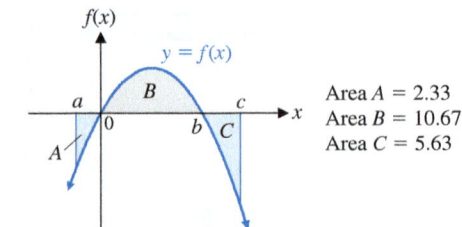

Area $A = 2.33$
Area $B = 10.67$
Area $C = 5.63$

Figure 9

SOLUTION   (A) $\displaystyle\int_a^b f(x)\, dx = -2.33 + 10.67 = 8.34$

(B) $\displaystyle\int_a^c f(x)\, dx = -2.33 + 10.67 - 5.63 = 2.71$

(C) $\displaystyle\int_b^c f(x)\, dx = -5.63$

Matched Problem 3   Referring to the figure for Example 3, calculate the definite integrals.

(A) $\displaystyle\int_a^0 f(x)\, dx$ (B) $\displaystyle\int_0^c f(x)\, dx$ (C) $\displaystyle\int_0^b f(x)\, dx$

Properties of the Definite Integral

Because the definite integral is defined as the limit of Riemann sums, many properties of sums are also properties of the definite integral. Note that Properties 3 and 4 are similar to the indefinite integral properties given in Section 7-1.

PROPERTIES   Properties of Definite Integrals

1. $\displaystyle\int_a^a f(x)\,dx = 0$

2. $\displaystyle\int_a^b f(x)\,dx = -\int_b^a f(x)\,dx$

3. $\displaystyle\int_a^b kf(x)\,dx = k\int_a^b f(x)\,dx,\, k \text{ a constant}$

4. $\displaystyle\int_a^b [f(x) \pm g(x)]\,dx = \int_a^b f(x)\,dx \pm \int_a^b g(x)\,dx$

5. $\displaystyle\int_a^b f(x)\,dx = \int_a^c f(x)\,dx + \int_c^b f(x)\,dx$

EXAMPLE 4   **Using Properties of the Definite Integral** If

$$\int_0^2 x\,dx = 2 \qquad \int_0^2 x^2\,dx = \frac{8}{3} \qquad \int_2^3 x^2\,dx = \frac{19}{3}$$

then

(A) $\displaystyle\int_0^2 12x^2\,dx = 12\int_0^2 x^2\,dx = 12\left(\frac{8}{3}\right) = 32$

(B) $\displaystyle\int_0^2 (2x - 6x^2)\,dx = 2\int_0^2 x\,dx - 6\int_0^2 x^2\,dx = 2(2) - 6\left(\frac{8}{3}\right) = -12$

(C) $\displaystyle\int_3^2 x^2\,dx = -\int_2^3 x^2\,dx = -\frac{19}{3}$

(D) $\displaystyle\int_5^5 3x^2\,dx = 0$

(E) $\displaystyle\int_0^3 3x^2\,dx = 3\int_0^2 x^2\,dx + 3\int_2^3 x^2\,dx = 3\left(\frac{8}{3}\right) + 3\left(\frac{19}{3}\right) = 27$

Matched Problem 4   Using the same integral values given in Example 4, find

(A) $\displaystyle\int_2^3 6x^2\,dx$ (B) $\displaystyle\int_0^2 (9x^2 - 4x)\,dx$ (C) $\displaystyle\int_2^0 3x\,dx$

(D) $\displaystyle\int_{-2}^{-2} 3x\,dx$ (E) $\displaystyle\int_0^3 12x^2\,dx$

Exercises 7-4

A

Problems 1–4 refer to the rectangles A, B, C, D, and E in the following figure.

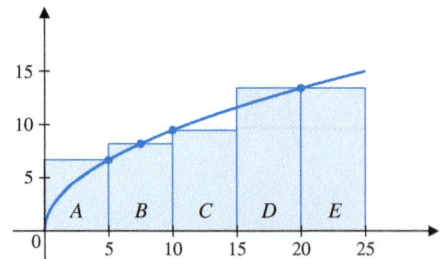

1. Which rectangles are left rectangles?

2. Which rectangles are right rectangles?

3. Which rectangles are neither left nor right rectangles?

4. Which rectangles are both left and right rectangles?

Problems 5–8 refer to the rectangles F, G, H, I, and J in the following figure.

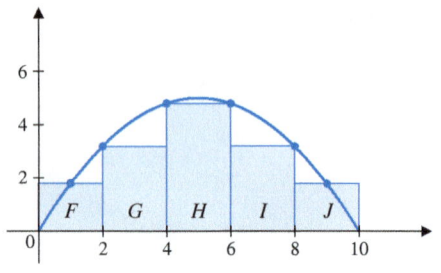

Figure for 5–8

5. Which rectangles are right rectangles?

6. Which rectangles are left rectangles?

7. Which rectangles are both left and right rectangles?

8. Which rectangles are neither left nor right rectangles?

Problems 9–16 involve estimating the area under the curves in Figures A–D from $x = 1$ to $x = 4$. For each figure, divide the interval $[1, 4]$ into three equal subintervals.

9. Draw in left and right rectangles for Figures A and B.

10. Draw in left and right rectangles for Figures C and D.

11. Using the results of Problem 9, compute L_3 and R_3 for Figure A and for Figure B.

12. Using the results of Problem 10, compute L_3 and R_3 for Figure C and for Figure D.

13. Replace the question marks with L_3 and R_3 as appropriate. Explain your choice.

$$? \le \int_1^4 f(x)\,dx \le ? \qquad ? \le \int_1^4 g(x)\,dx \le ?$$

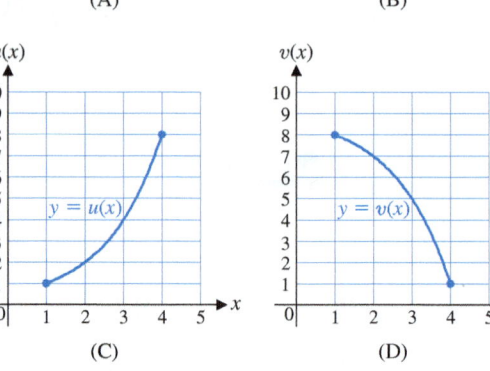

Figure for 9–16

14. Replace the question marks with L_3 and R_3 as appropriate. Explain your choice.

$$? \le \int_1^4 u(x)\,dx \le ? \qquad ? \le \int_1^4 v(x)\,dx \le ?$$

15. Compute error bounds for L_3 and R_3 found in Problem 11 for both figures.

16. Compute error bounds for L_3 and R_3 found in Problem 12 for both figures.

In Problems 17–20, calculate the indicated Riemann sum S_n for the function $f(x) = 25 - 3x^2$.

17. Partition $[-2, 8]$ into five subintervals of equal length, and for each subinterval $[x_{k-1}, x_k]$, let $c_k = (x_{k-1} + x_k)/2$.

18. Partition $[0, 12]$ into four subintervals of equal length, and for each subinterval $[x_{k-1}, x_k]$, let $c_k = (x_{k-1} + 2x_k)/3$.

19. Partition $[0, 12]$ into four subintervals of equal length, and for each subinterval $[x_{k-1}, x_k]$, let $c_k = (2x_{k-1} + x_k)/3$.

20. Partition $[-5, 5]$ into five subintervals of equal length, and for each subinterval $[x_{k-1}, x_k]$, let $c_k = (x_{k-1} + x_k)/2$.

In Problems 21–24, calculate the indicated Riemann sum S_n for the function $f(x) = x^2 - 5x - 6$.

21. Partition $[0, 3]$ into three subintervals of equal length, and let $c_1 = 0.7$, $c_2 = 1.8$, and $c_3 = 2.4$.

22. Partition $[0, 3]$ into three subintervals of equal length, and let $c_1 = 0.2$, $c_2 = 1.5$, and $c_3 = 2.8$.

23. Partition $[1, 7]$ into six subintervals of equal length, and let $c_1 = 1, c_2 = 3, c_3 = 3, c_4 = 5, c_5 = 5$, and $c_6 = 7$.

24. Partition $[1, 7]$ into six subintervals of equal length, and let $c_1 = 2, c_2 = 2, c_3 = 4, c_4 = 4, c_5 = 6$, and $c_6 = 6$.

In Problems 25–36, calculate the definite integral by referring to the figure with the indicated areas.

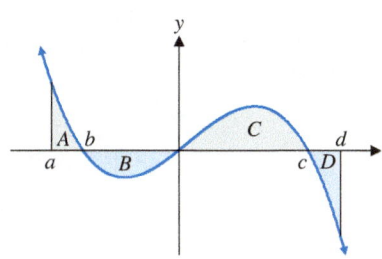

Area $A = 1.408$
Area $B = 2.475$
Area $C = 5.333$
Area $D = 1.792$

Figure for 25–36

25. $\displaystyle\int_b^0 f(x)\, dx$

26. $\displaystyle\int_0^c f(x)\, dx$

27. $\displaystyle\int_a^c f(x)\, dx$

28. $\displaystyle\int_b^d f(x)\, dx$

29. $\displaystyle\int_a^d f(x)\, dx$

30. $\displaystyle\int_0^d f(x)\, dx$

31. $\displaystyle\int_c^0 f(x)\, dx$

32. $\displaystyle\int_d^a f(x)\, dx$

33. $\displaystyle\int_0^a f(x)\, dx$

34. $\displaystyle\int_c^a f(x)\, dx$

35. $\displaystyle\int_d^b f(x)\, dx$

36. $\displaystyle\int_c^b f(x)\, dx$

In Problems 37–48, calculate the definite integral, given that

$$\int_1^4 x\, dx = 7.5 \qquad \int_1^4 x^2\, dx = 21 \qquad \int_4^5 x^2\, dx = \frac{61}{3}$$

37. $\displaystyle\int_1^4 2x\, dx$

38. $\displaystyle\int_1^4 3x^2\, dx$

39. $\displaystyle\int_1^4 (5x + x^2)\, dx$

40. $\displaystyle\int_1^4 (7x - 2x^2)\, dx$

41. $\displaystyle\int_1^4 (x^2 - 10x)\, dx$

42. $\displaystyle\int_1^4 (4x^2 - 9x)\, dx$

43. $\displaystyle\int_1^5 6x^2\, dx$

44. $\displaystyle\int_1^5 -4x^2\, dx$

45. $\displaystyle\int_4^4 (7x - 2)^2\, dx$

46. $\displaystyle\int_5^5 (10 - 7x + x^2)\, dx$

47. $\displaystyle\int_5^4 9x^2\, dx$

48. $\displaystyle\int_4^1 x(1 - x)\, dx$

B

In Problems 49–54, discuss the validity of each statement. If the statement is always true, explain why. If it is not always true, give a counterexample.

49. If $\int_a^b f(x)\, dx = 0$, then $f(x) = 0$ for all x in $[a, b]$.

50. If $f(x) = 0$ for all x in $[a, b]$, then $\int_a^b f(x)\, dx = 0$.

51. If $f(x) = 2x$ on $[0, 10]$, then there is a positive integer n for which the left sum L_n equals the exact area under the graph of f from $x = 0$ to $x = 10$.

52. If $f(x) = 2x$ on $[0, 10]$ and n is a positive integer, then there is some Riemann sum S_n that equals the exact area under the graph of f from $x = 0$ to $x = 10$.

53. If the area under the graph of f on $[a, b]$ is equal to both the left sum L_n and the right sum R_n for some positive integer n, then f is constant on $[a, b]$.

54. If f is a decreasing function on $[a, b]$, then the area under the graph of f is greater than the left sum L_n and less than the right sum R_n, for any positive integer n.

Problems 55 and 56 refer to the following figure showing two parcels of land along a river:

Figure for 55 and 56

55. You want to purchase both parcels of land shown in the figure and make a quick check on their combined area. There is no equation for the river frontage, so you use the average of the left and right sums of rectangles covering the area. The 1,000-foot baseline is divided into 10 equal parts. At the end of each subinterval, a measurement is made from the baseline to the river, and the results are tabulated. Let x be the distance from the left end of the baseline and let $h(x)$ be the distance from the baseline to the river at x. Use L_{10} to estimate the combined area of both parcels, and calculate an error bound for this estimate. How many subdivisions of the baseline would be required so that the error incurred in using L_n would not exceed 2,500 square feet?

x	0	100	200	300	400	500
$h(x)$	0	183	235	245	260	286

x	600	700	800	900	1,000
$h(x)$	322	388	453	489	500

56. Refer to Problem 55. Use R_{10} to estimate the combined area of both parcels, and calculate an error bound for this estimate. How many subdivisions of the baseline would be required so that the error incurred in using R_n would not exceed 1,000 square feet?

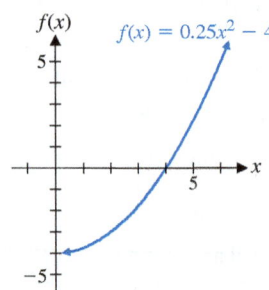

Problems 57 and 58 refer to the following figure:

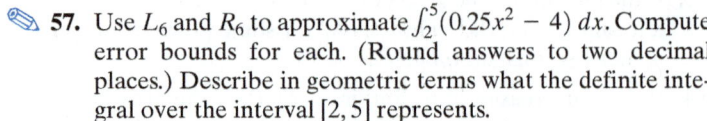

Figure for 57 and 58

57. Use L_6 and R_6 to approximate $\int_2^5 (0.25x^2 - 4)\,dx$. Compute error bounds for each. (Round answers to two decimal places.) Describe in geometric terms what the definite integral over the interval $[2, 5]$ represents.

58. Use L_5 and R_5 to approximate $\int_1^6 (0.25x^2 - 4)\,dx$. Compute error bounds for each. (Round answers to two decimal places.) Describe in geometric terms what the definite integral over the interval $[1, 6]$ represents.

For Problems 59–62, use a graphing calculator to determine the intervals on which each function is increasing or decreasing.

59. $f(x) = e^{-x^2}$

60. $f(x) = \dfrac{3}{1 + 2e^{-x}}$

61. $f(x) = x^4 - 2x^2 + 3$

62. $f(x) = e^{x^2}$

In Problems 63–66, the left sum L_n or the right sum R_n is used to approximate the definite integral to the indicated accuracy. How large must n be chosen in each case? (Each function is increasing over the indicated interval.)

63. $\displaystyle\int_1^3 \ln x\,dx = R_n \pm 0.1$

64. $\displaystyle\int_0^{10} \ln(x^2 + 1)\,dx = L_n \pm 0.5$

65. $\displaystyle\int_1^3 x^x\,dx = L_n \pm 0.5$ 66. $\displaystyle\int_1^4 x^x\,dx = R_n \pm 0.5$

Applications

67. **Employee training.** A company producing electric motors has established that, on the average, a new employee can assemble $N(t)$ components per day after t days of on-the-job training, as shown in the following table (a new employee's productivity increases continuously with time on the job):

t	0	20	40	60	80	100	120
N(t)	10	51	68	76	81	84	86

Use left and right sums to estimate the area under the graph of $N(t)$ from $t = 0$ to $t = 60$. Use three subintervals of equal length for each. Calculate an error bound for each estimate.

68. **Employee training.** For a new employee in Problem 67, use left and right sums to estimate the area under the graph of $N(t)$ from $t = 20$ to $t = 100$. Use four equal subintervals for each. Replace the question marks with the values of L_4 or R_4 as appropriate:

$$? \le \int_{20}^{100} N(t)\,dt \le ?$$

69. **Medicine.** The rate of healing, $A'(t)$ (in square centimeters per day), for a certain type of skin wound is given approximately by the following table:

t	0	1	2	3	4	5
A'(t)	0.90	0.81	0.74	0.67	0.60	0.55

t	6	7	8	9	10
A'(t)	0.49	0.45	0.40	0.36	0.33

(A) Use left and right sums over five equal subintervals to approximate the area under the graph of $A'(t)$ from $t = 0$ to $t = 5$.

(B) Replace the question marks with values of L_5 and R_5 as appropriate:

$$? \le \int_0^5 A'(t)\,dt \le ?$$

70. **Medicine.** Refer to Problem 69. Use left and right sums over five equal subintervals to approximate the area under the graph of $A'(t)$ from $t = 5$ to $t = 10$. Calculate an error bound for this estimate.

71. **Learning.** A psychologist found that, on average, the rate of learning a list of special symbols in a code $N'(x)$ after x days of practice was given approximately by the following table values:

x	0	2	4	6	8	10	12
N'(x)	29	26	23	21	19	17	15

Use left and right sums over three equal subintervals to approximate the area under the graph of $N'(x)$ from $x = 6$ to $x = 12$. Calculate an error bound for this estimate.

72. **Learning.** For the data in Problem 71, use left and right sums over three equal subintervals to approximate the area under the graph of $N'(x)$ from $x = 0$ to $x = 6$. Replace the question marks with values of L_3 and R_3 as appropriate:

$$? \le \int_0^6 N'(x)\,dx \le ?$$

Answers to Matched Problems

1. (A) $\Delta x = 0.5$:

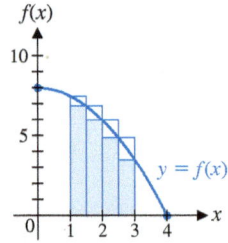

(B) $L_4 = 12.625$, $R_4 = 10.625$; error for L_4 and $R_4 = 2$

(C) $n > 16$ for L_n and R_n

2. $S_3 = 46$

3. (A) -2.33 (B) 5.04 (C) 10.67

4. (A) 38 (B) 16 (C) -6

(D) 0 (E) 108

7-5   The Fundamental Theorem of Calculus

- Introduction to the Fundamental Theorem
- Evaluating Definite Integrals
- Recognizing a Definite Integral: Average Value

The definite integral of a function f on an interval $[a, b]$ is a number, the area (if $f(x) > 0$) between the graph of f and the x axis from $x = a$ to $x = b$. The indefinite integral of a function is a family of antiderivatives. In this section, we explain the connection between these two integrals, a connection made precise by the fundamental theorem of calculus.

Introduction to the Fundamental Theorem

Suppose that the daily cost function for a small manufacturing firm is given (in dollars) by

$$C(x) = 180x + 200 \qquad 0 \le x \le 20$$

Then the marginal cost function is given (in dollars per unit) by

$$C'(x) = 180$$

What is the change in cost as production is increased from $x = 5$ units to $x = 10$ units? That change is equal to

$$C(10) - C(5) = (180 \cdot 10 + 200) - (180 \cdot 5 + 200)$$
$$= 180(10 - 5)$$
$$= \$900$$

Notice that $180(10 - 5)$ is equal to the area between the graph of $C'(x)$ and the x axis from $x = 5$ to $x = 10$. Therefore,

$$C(10) - C(5) = \int_5^{10} 180 \, dx$$

In other words, the change in cost from $x = 5$ to $x = 10$ is equal to the area between the marginal cost function and the x axis from $x = 5$ to $x = 10$ (see Fig. 1).

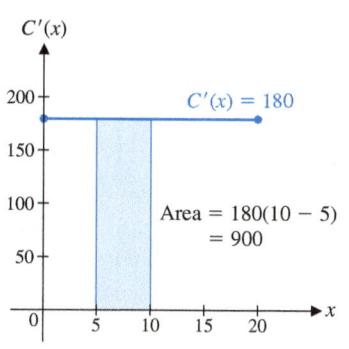

Figure 1

CONCEPTUAL INSIGHT

Consider the formula for the slope of a line:

$$m = \frac{y_2 - y_1}{x_2 - x_1}$$

Multiplying both sides of this equation by $x_2 - x_1$ gives

$$y_2 - y_1 = m(x_2 - x_1)$$

The right-hand side, $m(x_2 - x_1)$, is equal to the area of a rectangle of height m and width $x_2 - x_1$. So the change in y coordinates is equal to the area under the constant function with value m from $x = x_1$ to $x = x_2$.

EXAMPLE 1   **Change in Cost vs Area under Marginal Cost**   The daily cost function for a company (in dollars) is given by

$$C(x) = -5x^2 + 210x + 400 \qquad 0 \leq x \leq 20$$

(A) Graph $C(x)$ for $0 \leq x \leq 20$, calculate the change in cost from $x = 5$ to $x = 10$, and indicate that change in cost on the graph.

(B) Graph the marginal cost function $C'(x)$ for $0 \leq x \leq 20$, and use geometric formulas (see Appendix C) to calculate the area between $C'(x)$ and the x axis from $x = 5$ to $x = 10$.

(C) Compare the results of the calculations in parts (A) and (B).

SOLUTION   (A) $C(10) - C(5) = 2,000 - 1,325 = 675$, and this change in cost is indicated in Figure 2A.

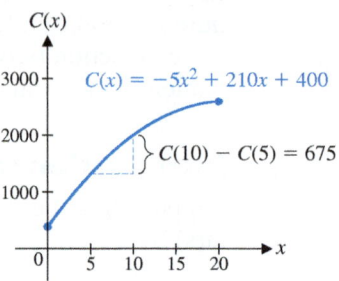

Figure 2(A)

(B) $C'(x) = -10x + 210$, so the area between $C'(x)$ and the x axis from $x = 5$ to $x = 10$ (see Fig. 2B) is the area of a trapezoid (geometric formulas are given in Appendix C):

$$\text{Area} = \frac{C'(5) + C'(10)}{2}(10 - 5) = \frac{160 + 110}{2}(5) = 675$$

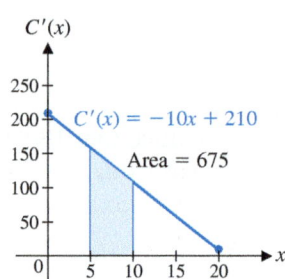

Figure 2(B)

(C) The change in cost from $x = 5$ to $x = 10$ is equal to the area between the marginal cost function and the x axis from $x = 5$ to $x = 10$.

Matched Problem 1   Repeat Example 1 for the daily cost function

$$C(x) = -7.5x^2 + 305x + 625$$

The connection illustrated in Example 1, between the change in a function from $x = a$ to $x = b$ and the area under the derivative of the function, provides the link between antiderivatives (or indefinite integrals) and the definite integral. This link is known as the fundamental theorem of calculus. (See Problems 59 and 60 in Exercise 7-5 for an outline of its proof.)

> **THEOREM 1   Fundamental Theorem of Calculus**
> If f is a continuous function on $[a, b]$, and F is any antiderivative of f, then
>
> $$\int_a^b f(x)\, dx = F(b) - F(a)$$

CONCEPTUAL INSIGHT

Because a definite integral is the limit of Riemann sums, we expect that it would be diffi-cult to calculate definite integrals exactly. The fundamental theorem, however, gives us an easy method for evaluating definite integrals, *provided that we can find an antiderivative* $F(x)$ of $f(x)$: Simply calculate the difference $F(b) - F(a)$. But what if we are unable to find an antiderivative of $f(x)$? In that case, we must resort to left sums, right sums, or other approximation methods to approximate the definite integral. However, it is often useful to remember that such an approximation is also an estimate of the change $F(b) - F(a)$.

Evaluating Definite Integrals

By the fundamental theorem, we can evaluate $\int_a^b f(x)\, dx$ easily and exactly when-ever we can find an antiderivative $F(x)$ of $f(x)$. We simply calculate the difference $F(b) - F(a)$.

Now you know why we studied techniques of indefinite integration before this section—so that we would have methods of finding antiderivatives of large classes of elementary functions for use with the fundamental theorem. It is important to re-member that

> **Any antiderivative of $f(x)$ can be used in the fundamental theorem. One generally chooses the simplest antiderivative by letting $C = 0$, since any other value of C will drop out in computing the difference $F(b) - F(a)$.**

In evaluating definite integrals by the fundamental theorem, it is convenient to use the notation $F(x)\big|_a^b$, which represents the change in $F(x)$ from $x = a$ to $x = b$, as an intermediate step in the calculation. This technique is illustrated in the follow-ing examples.

EXAMPLE 2   **Evaluating Definite Integrals** Evaluate $\displaystyle\int_1^2 \left(2x + 3e^x - \frac{4}{x}\right) dx$.

SOLUTION

$$\int_1^2 \left(2x + 3e^x - \frac{4}{x}\right) dx = 2\int_1^2 x\, dx + 3\int_1^2 e^x\, dx - 4\int_1^2 \frac{1}{x}\, dx$$

$$= 2\frac{x^2}{2}\bigg|_1^2 + 3e^x\bigg|_1^2 - 4\ln|x|\bigg|_1^2$$

$$= (2^2 - 1^2) + (3e^2 - 3e^1) - (4\ln 2 - 4\ln 1)$$

$$= 3 + 3e^2 - 3e - 4\ln 2 \approx 14.24$$

Matched Problem 2   Evaluate $\displaystyle\int_1^3 \left(4x - 2e^x + \frac{5}{x}\right) dx$.

The evaluation of a definite integral is a two-step process: First, find an antiderivative. Then find the change in that antiderivative. If *substitution techniques* are required to find the antiderivative, there are two different ways to proceed. The next example illustrates both methods.

EXAMPLE 3 **Definite Integrals and Substitution Techniques** Evaluate

$$\int_0^5 \frac{x}{x^2 + 10}\,dx$$

SOLUTION We solve this problem using substitution in two different ways.
Method 1. Use substitution in an indefinite integral to find an antiderivative as a function of x. Then evaluate the definite integral.

$$\int \frac{x}{x^2+10}\,dx = \frac{1}{2}\int \frac{1}{x^2+10}2x\,dx \quad \text{Substitute } u = x^2 + 10 \text{ and } du = 2x\,dx.$$
$$= \frac{1}{2}\int \frac{1}{u}\,du$$
$$= \tfrac{1}{2}\ln|u| + C$$
$$= \tfrac{1}{2}\ln(x^2+10) + C \quad \text{Since } u = x^2 + 10 > 0$$

We choose $C = 0$ and use the antiderivative $\tfrac{1}{2}\ln(x^2+10)$ to evaluate the definite integral.

$$\int_0^5 \frac{x}{x^2+10}\,dx = \frac{1}{2}\ln(x^2+10)\Big|_0^5$$
$$= \tfrac{1}{2}\ln 35 - \tfrac{1}{2}\ln 10 \approx 0.626$$

Method 2. Substitute directly into the definite integral, changing both the variable of integration and the limits of integration. In the definite integral

$$\int_0^5 \frac{x}{x^2+10}\,dx$$

the upper limit is $x = 5$ and the lower limit is $x = 0$. When we make the substitution $u = x^2 + 10$ in this definite integral, we must change the limits of integration to the corresponding values of u:

$$x = 5 \quad \text{implies} \quad u = 5^2 + 10 = 35 \quad \text{New upper limit}$$
$$x = 0 \quad \text{implies} \quad u = 0^2 + 10 = 10 \quad \text{New lower limit}$$

We have

$$\int_0^5 \frac{x}{x^2+10}\,dx = \frac{1}{2}\int_0^5 \frac{1}{x^2+10}2x\,dx$$
$$= \frac{1}{2}\int_{10}^{35} \frac{1}{u}\,du$$
$$= \frac{1}{2}\left(\ln|u|\Big|_{10}^{35}\right)$$
$$= \tfrac{1}{2}(\ln 35 - \ln 10) \approx 0.626$$

Matched Problem 3 Use both methods described in Example 3 to evaluate $\int_0^1 \frac{1}{2x+4}\,dx$.

EXAMPLE 4 **Definite Integrals and Substitution** Use method 2 described in Example 3 to evaluate

$$\int_{-4}^1 \sqrt{5-t}\,dt$$

SOLUTION   If $u = 5 - t$, then $du = -dt$, and

$t = 1$	implies	$u = 5 - 1 = 4$	New upper limit
$t = -4$	implies	$u = 5 - (-4) = 9$	New lower limit

Notice that the lower limit for u is larger than the upper limit. Be careful not to reverse these two values when substituting into the definite integral:

$$\int_{-4}^{1} \sqrt{5 - t} \, dt = -\int_{-4}^{1} \sqrt{5 - t} \, (-dt)$$

$$= -\int_{9}^{4} \sqrt{u} \, du$$

$$= -\int_{9}^{4} u^{1/2} \, du$$

$$= -\left(\frac{u^{3/2}}{\frac{3}{2}} \bigg|_{9}^{4} \right)$$

$$= -\left[\tfrac{2}{3}(4)^{3/2} - \tfrac{2}{3}(9)^{3/2} \right]$$

$$= -\left[\tfrac{16}{3} - \tfrac{54}{3} \right] = \tfrac{38}{3} \approx 12.667$$

Matched Problem 4   Use method 2 described in Example 3 to evaluate $\int_{2}^{5} \dfrac{1}{\sqrt{6 - t}} \, dt$.

EXAMPLE 5   **Change in Profit**   A company manufactures x HDTVs per month. The monthly marginal profit (in dollars) is given by

$$P'(x) = 165 - 0.1x \qquad 0 \le x \le 4{,}000$$

The company is currently manufacturing 1,500 HDTVs per month, but is planning to increase production. Find the change in the monthly profit if monthly production is increased to 1,600 HDTVs.

SOLUTION

$$P(1{,}600) - P(1{,}500) = \int_{1{,}500}^{1{,}600} (165 - 0.1x) \, dx$$

$$= (165x - 0.05x^2)\big|_{1{,}500}^{1{,}600}$$

$$= \left[165(1{,}600) - 0.05(1{,}600)^2 \right]$$

$$\quad - \left[165(1{,}500) - 0.05(1{,}500)^2 \right]$$

$$= 136{,}000 - 135{,}000$$

$$= 1{,}000$$

Increasing monthly production from 1,500 units to 1,600 units will increase the monthly profit by $1,000.

Matched Problem 5   Repeat Example 5 if

$$P'(x) = 300 - 0.2x \qquad 0 \le x \le 3{,}000$$

and monthly production is increased from 1,400 to 1,500 HDTVs.

EXAMPLE 6   **Useful Life**   An amusement company maintains records for each video game installed in an arcade. Suppose that $C(t)$ and $R(t)$ represent the total accumulated costs and revenues (in thousands of dollars), respectively, t years after a particular game has been installed. Suppose also that

$$C'(t) = 2 \qquad R'(t) = 9e^{-0.5t}$$

The value of t for which $C'(t) = R'(t)$ is called the **useful life** of the game.

(A) Find the useful life of the game, to the nearest year.

(B) Find the total profit accumulated during the useful life of the game.

SOLUTION (A) $R'(t) = C'(t)$

$$9e^{-0.5t} = 2$$

$$e^{-0.5t} = \tfrac{2}{9} \qquad \textcolor{blue}{\textit{Convert to equivalent logarithmic form.}}$$

$$-0.5t = \ln\tfrac{2}{9}$$

$$t = -2\ln\tfrac{2}{9} \approx 3 \text{ years}$$

Thus, the game has a useful life of 3 years. This is illustrated graphically in Figure 3.

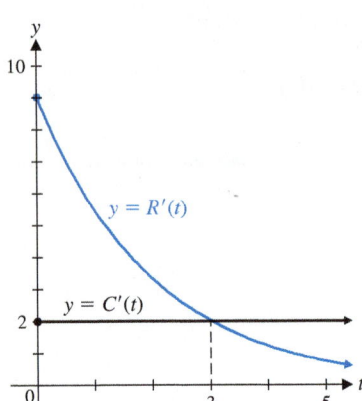

Figure 3 Useful life

(B) The total profit accumulated during the useful life of the game is

$$P(3) - P(0) = \int_0^3 P'(t)\, dt$$

$$= \int_0^3 [R'(t) - C'(t)]\, dt$$

$$= \int_0^3 (9e^{-0.5t} - 2)\, dt$$

$$= \left(\frac{9}{-0.5}e^{-0.5t} - 2t\right)\Big|_0^3 \qquad \textit{Recall: } \int e^{ax}\, dx = \frac{1}{a}e^{ax} + C$$

$$= (-18e^{-0.5t} - 2t)|_0^3$$

$$= (-18e^{-1.5} - 6) - (-18e^0 - 0)$$

$$= 12 - 18e^{-1.5} \approx 7.984 \quad \text{or} \quad \$7,984$$

Matched Problem 6 Repeat Example 6 if $C'(t) = 1$ and $R'(t) = 7.5e^{-0.5t}$.

EXAMPLE 7 **Numerical Integration on a Graphing Calculator** Evaluate $\int_{-1}^{2} e^{-x^2}\, dx$ to three decimal places.

SOLUTION The integrand e^{-x^2} does not have an elementary antiderivative, so we are unable to use the fundamental theorem to evaluate the definite integral. Instead, we use a numerical integration routine that has been preprogrammed into a graphing calculator. (Consult your user's manual for specific details.) Such a routine is an approximation algorithm, more powerful than the left-sum and right-sum methods discussed in Section 7-4. From Figure 4,

$$\int_{-1}^{2} e^{-x^2}\, dx = 1.629$$

Figure 4

Matched Problem 7 Evaluate $\int_{1.5}^{4.3} \frac{x}{\ln x}\, dx$ to three decimal places.

Recognizing a Definite Integral: Average Value

Recall that the derivative of a function f was defined in Section 3-3 by

$$f'(x) = \lim_{h \to 0} \frac{f(x + h) - f(x)}{h}$$

This form is generally not easy to compute directly but is easy to recognize in certain practical problems (slope, instantaneous velocity, rates of change, and so on). Once we know that we are dealing with a derivative, we proceed to try to compute the derivative with the use of derivative formulas and rules.

Similarly, evaluating a definite integral with the use of the definition

$$\int_a^b f(x)\, dx = \lim_{n \to \infty} [f(c_1)\Delta x_1 + f(c_2)\Delta x_2 + \cdots + f(c_n)\Delta x_n] \tag{1}$$

is generally not easy, but the form on the right occurs naturally in many practical problems. We can use the fundamental theorem to evaluate the definite integral (once it is recognized) if an antiderivative can be found; otherwise, we will approximate it with a rectangle sum. We will now illustrate these points by finding the *average value* of a continuous function.

Suppose that the temperature F (in degrees Fahrenheit) in the middle of a small shallow lake from 8 AM ($t = 0$) to 6 PM ($t = 10$) during the month of May is given approximately as shown in Figure 5.

How can we compute the average temperature from 8 AM to 6 PM? We know that the average of a finite number of values $a_1, a_2, \ldots, a_n$ is given by

$$\text{average} = \frac{a_1 + a_2 + \cdots + a_n}{n}$$

But how can we handle a continuous function with infinitely many values? It would seem reasonable to divide the time interval $[0, 10]$ into n equal subintervals, compute the temperature at a point in each subinterval, and then use the average of the temperatures as an approximation of the average value of the continuous function $F = F(t)$ over $[0, 10]$. We would expect the approximations to improve as n increases. In fact, we would define the limit of the average of n values as $n \to \infty$ as the *average value of F over* $[0, 10]$ if the limit exists. This is exactly what we will do:

$$\binom{\text{average temperature}}{\text{for } n \text{ values}} = \frac{1}{n}[F(t_1) + F(t_2) + \cdots + F(t_n)] \tag{2}$$

Here t_k is a point in the kth subinterval. We will call the limit of equation (2) as $n \to \infty$ the *average temperature over the time interval* $[0, 10]$.

Form (2) resembles form (1), but we are missing the Δt_k. We take care of this by multiplying equation (2) by $(b - a)/(b - a)$, which will change the form of equation (2) without changing its value:

$$\frac{b - a}{b - a} \cdot \frac{1}{n}[F(t_1) + F(t_2) + \cdots + F(t_n)] = \frac{1}{b - a} \cdot \frac{b - a}{n}[F(t_1) + F(t_2) + \cdots + F(t_n)]$$

$$= \frac{1}{b - a}\left[F(t_1)\frac{b - a}{n} + F(t_2)\frac{b - a}{n} + \cdots + F(t_n)\frac{b - a}{n}\right]$$

$$= \frac{1}{b - a}[F(t_1)\Delta t + F(t_2)\Delta t + \cdots + F(t_n)\Delta t]$$

Therefore,

$$\binom{\text{average temperature}}{\text{over } [a, b] = [0, 10]} = \lim_{n \to \infty}\left\{\frac{1}{b - a}[F(t_1)\Delta t + F(t_2)\Delta t + \cdots + F(t_n)\Delta t]\right\}$$

$$= \frac{1}{b - a}\left\{\lim_{n \to \infty}[F(t_1)\Delta t + F(t_2)\Delta t + \cdots + F(t_n)\Delta t]\right\}$$

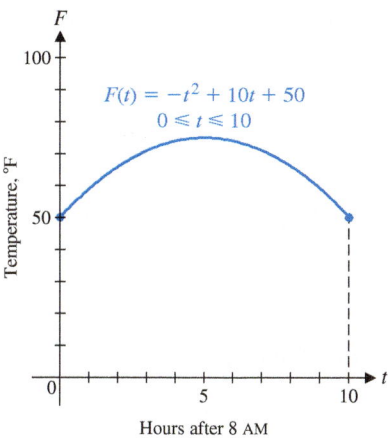

$F(t) = -t^2 + 10t + 50$
$0 \le t \le 10$

Temperature, °F

Hours after 8 AM

Figure 5

The limit inside the braces is of form (1)—that is, a definite integral. So

$$\begin{pmatrix} \text{average temperature} \\ \text{over } [a, b] = [0, 10] \end{pmatrix} = \frac{1}{b - a} \int_a^b F(t) \, dt$$

$$= \frac{1}{10 - 0} \int_0^{10} (-t^2 + 10t + 50) \, dt$$

We now use the fundamental theorem to evaluate the definite integral.

$$= \frac{1}{10} \left(-\frac{t^3}{3} + 5t^2 + 50t \right) \Big|_0^{10}$$

$$= \frac{200}{3} \approx 67°\text{F}$$

Proceeding as before for an arbitrary continuous function f over an interval $[a, b]$, we obtain the following general formula:

DEFINITION **Average Value of a Continuous Function f over $[a, b]$**

$$\frac{1}{b - a} \int_a^b f(x) \, dx$$

EXPLORE & DISCUSS 1

In Figure 6, the rectangle shown has the same area as the area under the graph of $y = f(x)$ from $x = a$ to $x = b$. Explain how the average value of $f(x)$ over the interval $[a, b]$ is related to the height of the rectangle.

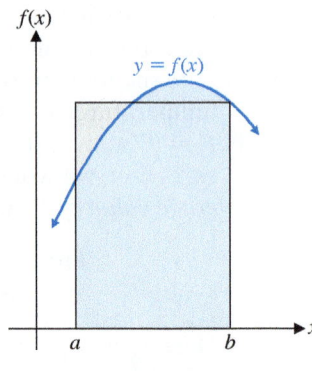

Figure 6

EXAMPLE 8 **Average Value of a Function** Find the average value of $f(x) = x - 3x^2$ over the interval $[-1, 2]$.

SOLUTION

$$\frac{1}{b - a} \int_a^b f(x) \, dx = \frac{1}{2 - (-1)} \int_{-1}^2 (x - 3x^2) \, dx$$

$$= \frac{1}{3} \left(\frac{x^2}{2} - x^3 \right) \Big|_{-1}^2 = -\frac{5}{2}$$

Matched Problem 8 Find the average value of $g(t) = 6t^2 - 2t$ over the interval $[-2, 3]$.

EXAMPLE 9   **Average Price** Given the demand function

$$p = D(x) = 100e^{-0.05x}$$

find the average price (in dollars) over the demand interval $[40, 60]$.

SOLUTION   Average price $= \dfrac{1}{b-a} \displaystyle\int_a^b D(x)\,dx$

$$= \frac{1}{60-40} \int_{40}^{60} 100e^{-0.05x}\,dx$$

$$= \frac{100}{20} \int_{40}^{60} e^{-0.05x}\,dx \qquad \text{Use } \int e^{ax}\,dx = \frac{1}{a}e^{ax}, a \neq 0.$$

$$= -\frac{5}{0.05} e^{-0.05x} \Big|_{40}^{60}$$

$$= 100(e^{-2} - e^{-3}) \approx \$8.55$$

Matched Problem 9   Given the supply equation

$$p = S(x) = 10e^{0.05x}$$

find the average price (in dollars) over the supply interval $[20, 30]$.

Exercises 7-5

A

In Problems 1–4,

(A) *Calculate the change in $F(x)$ from $x = 10$ to $x = 15$.*

(B) *Graph $F'(x)$ and use geometric formulas (see Appendix C) to calculate the area between the graph of $F'(x)$ and the x axis from $x = 10$ to $x = 15$.*

(C) *Verify that your answers to (A) and (B) are equal, as is guaranteed by the fundamental theorem of calculus.*

1. $F(x) = 3x^2 + 160$

2. $F(x) = 9x + 120$

3. $F(x) = -x^2 + 42x + 240$

4. $F(x) = x^2 + 30x + 210$

Evaluate the integrals in Problems 5–24.

5. $\displaystyle\int_0^{10} 4\,dx$

6. $\displaystyle\int_0^8 9x\,dx$

7. $\displaystyle\int_0^6 x^2\,dx$

8. $\displaystyle\int_0^4 x^3\,dx$

9. $\displaystyle\int_1^4 (5x + 3)\,dx$

10. $\displaystyle\int_2^5 (2x - 1)\,dx$

11. $\displaystyle\int_0^1 e^x\,dx$

12. $\displaystyle\int_0^2 4e^x\,dx$

13. $\displaystyle\int_1^2 \frac{1}{x}\,dx$

14. $\displaystyle\int_1^5 \frac{2}{x}\,dx$

15. $\displaystyle\int_{-2}^2 (x^3 + 7x)\,dx$

16. $\displaystyle\int_0^8 (0.25x - 1)\,dx$

17. $\displaystyle\int_2^5 (2x + 9)\,dx$

18. $\displaystyle\int_1^4 (6x - 5)\,dx$

19. $\displaystyle\int_5^2 (2x + 9)\,dx$

20. $\displaystyle\int_4^1 (6x - 5)\,dx$

21. $\displaystyle\int_2^3 (6 - x^3)\,dx$

22. $\displaystyle\int_6^9 (5 - x^2)\,dx$

23. $\displaystyle\int_6^6 (x^2 - 5x + 1)^{10}\,dx$

24. $\displaystyle\int_{-3}^{-3} (x^2 + 4x + 2)^8\,dx$

B

Evaluate the integrals in Problems 25–40.

25. $\displaystyle\int_1^2 (2x^{-2} - 3)\,dx$

26. $\displaystyle\int_1^2 (5 - 16x^{-3})\,dx$

27. $\displaystyle\int_1^4 3\sqrt{x}\,dx$

28. $\displaystyle\int_4^{25} \frac{2}{\sqrt{x}}\,dx$

29. $\displaystyle\int_2^3 12(x^2 - 4)^5 x\,dx$

30. $\displaystyle\int_0^1 32(x^2 + 1)^7 x\,dx$

31. $\int_3^9 \frac{1}{x-1}\,dx$

32. $\int_2^8 \frac{1}{x+1}\,dx$

33. $\int_{-5}^{10} e^{-0.05x}\,dx$

34. $\int_{-10}^{25} e^{-0.01x}\,dx$

35. $\int_1^e \frac{\ln t}{t}\,dt$

36. $\int_e^{e^2} \frac{(\ln t)^2}{t}\,dt$

37. $\int_0^1 xe^{-x^2}\,dx$

38. $\int_0^1 xe^{x^2}\,dx$

39. $\int_1^1 e^{x^2}\,dx$

40. $\int_{-1}^{-1} e^{-x^2}\,dx$

In Problems 41–48,

(A) *Find the average value of each function over the indicated interval.*

 (B) *Use a graphing calculator to graph the function and its average value over the indicated interval in the same viewing window.*

41. $f(x) = 500 - 50x;\ [0, 10]$

42. $g(x) = 2x + 7;\ [0, 5]$

43. $f(t) = 3t^2 - 2t;\ [-1, 2]$

44. $g(t) = 4t - 3t^2;\ [-2, 2]$

45. $f(x) = \sqrt[3]{x};\ [1, 8]$

46. $g(x) = \sqrt{x+1};\ [3, 8]$

47. $f(x) = 4e^{-0.2x};\ [0, 10]$

48. $f(x) = 64e^{0.08x};\ [0, 10]$

C

Evaluate the integrals in Problems 49–54.

49. $\int_2^3 x\sqrt{2x^2 - 3}\,dx$

50. $\int_0^1 x\sqrt{3x^2 + 2}\,dx$

51. $\int_0^1 \frac{x-1}{x^2 - 2x + 3}\,dx$

52. $\int_1^2 \frac{x+1}{2x^2 + 4x + 4}\,dx$

53. $\int_{-1}^1 \frac{e^{-x} - e^x}{(e^{-x} + e^x)^2}\,dx$

54. $\int_6^7 \frac{\ln(t-5)}{t-5}\,dt$

 Use a numerical integration routine to evaluate each definite integral in Problems 55–58 (to three decimal places).

55. $\int_{1.7}^{3.5} x \ln x\,dx$

56. $\int_{-1}^1 e^{x^2}\,dx$

57. $\int_{-2}^2 \frac{1}{1 + x^2}\,dx$

58. $\int_0^3 \sqrt{9 - x^2}\,dx$

59. The **mean value theorem** states that if $F(x)$ is a differentiable function on the interval $[a, b]$, then there exists some number c between a and b such that

$$F'(c) = \frac{F(b) - F(a)}{b - a}$$

Explain why the mean value theorem implies that if a car averages 60 miles per hour in some 10-minute interval, then the car's instantaneous velocity is 60 miles per hour at least once in that interval.

60. The fundamental theorem of calculus can be proved by showing that, for every positive integer n, there is a Riemann sum for f on $[a, b]$ that is equal to $F(b) - F(a)$. By the mean value theorem (see Problem 59), within each subinterval $[x_{k-1}, x_k]$ that belongs to a partition of $[a, b]$, there is some c_k such that

$$f(c_k) = F'(c_k) = \frac{F(x_k) - F(x_{k-1})}{x_k - x_{k-1}}$$

Multiplying by the denominator $x_k - x_{k-1}$, we get

$$f(c_k)(x_k - x_{k-1}) = F(x_k) - F(x_{k-1})$$

Show that the Riemann sum

$$S_n = \sum_{k=1}^n f(c_k)(x_k - x_{k-1})$$

is equal to $F(b) - F(a)$.

Applications

61. Cost. A company manufactures mountain bikes. The research department produced the marginal cost function

$$C'(x) = 500 - \frac{x}{3} \qquad 0 \le x \le 900$$

where $C'(x)$ is in dollars and x is the number of bikes produced per month. Compute the increase in cost going from a production level of 300 bikes per month to 900 bikes per month. Set up a definite integral and evaluate it.

62. Cost. Referring to Problem 61, compute the increase in cost going from a production level of 0 bikes per month to 600 bikes per month. Set up a definite integral and evaluate it.

63. Salvage value. A new piece of industrial equipment will depreciate in value, rapidly at first and then less rapidly as time goes on. Suppose that the rate (in dollars per year) at which the book value of a new milling machine changes is given approximately by

$$V'(t) = f(t) = 500(t - 12) \qquad 0 \le t \le 10$$

where $V(t)$ is the value of the machine after t years. What is the total loss in value of the machine in the first 5 years? In the second 5 years? Set up appropriate integrals and solve.

64. Maintenance costs. Maintenance costs for an apartment house generally increase as the building gets older. From past records, the rate of increase in maintenance costs (in dollars per year) for a particular apartment complex is given approximately by

$$M'(x) = f(x) = 90x^2 + 5,000$$

where x is the age of the apartment complex in years and $M(x)$ is the total (accumulated) cost of maintenance for x years. Write a definite integral that will give the total maintenance costs from the end of the second year to the end of the seventh year, and evaluate the integral.

 65. Employee training. A company producing computer components has established that, on the average, a new employee can assemble $N(t)$ components per day after t days of on-the-job training, as indicated in the following table (a new employee's productivity usually increases with time on the job, up to a leveling-off point):

t	0	20	40	60	80	100	120
$N(t)$	10	51	68	76	81	84	85

(A) Find a quadratic regression equation for the data, and graph it and the data set in the same viewing window.

(B) Use the regression equation and a numerical integration routine on a graphing calculator to approximate the number of units assembled by a new employee during the first 100 days on the job.

 66. Employee training. Refer to Problem 65.

(A) Find a cubic regression equation for the data, and graph it and the data set in the same viewing window.

(B) Use the regression equation and a numerical integration routine on a graphing calculator to approximate the number of units assembled by a new employee during the second 60 days on the job.

67. Useful life. The total accumulated costs $C(t)$ and revenues $R(t)$ (in thousands of dollars), respectively, for a photocopying machine satisfy

$$C'(t) = \tfrac{1}{11}t \quad \text{and} \quad R'(t) = 5te^{-t^2}$$

where t is time in years. Find the useful life of the machine, to the nearest year. What is the total profit accumulated during the useful life of the machine?

68. Useful life. The total accumulated costs $C(t)$ and revenues $R(t)$ (in thousands of dollars), respectively, for a coal mine satisfy

$$C'(t) = 3 \quad \text{and} \quad R'(t) = 15e^{-0.1t}$$

where t is the number of years that the mine has been in operation. Find the useful life of the mine, to the nearest year. What is the total profit accumulated during the useful life of the mine?

69. Average cost. The total cost (in dollars) of manufacturing x auto body frames is $C(x) = 60,000 + 300x$.

(A) Find the average cost per unit if 500 frames are produced. [*Hint*: Recall that $\overline{C}(x)$ is the average cost per unit.]

(B) Find the average value of the cost function over the interval $[0, 500]$.

(C) Discuss the difference between parts (A) and (B).

 70. Average cost. The total cost (in dollars) of printing x dictionaries is $C(x) = 20,000 + 10x$.

(A) Find the average cost per unit if 1,000 dictionaries are produced.

(B) Find the average value of the cost function over the interval $[0, 1,000]$.

(C) Discuss the difference between parts (A) and (B).

 71. Cost. The marginal cost at various levels of output per month for a company that manufactures sunglasses is shown in the following table, with the output x given in thousands of units per month and the total cost $C(x)$ given in thousands of dollars per month:

x	0	1	2	3	4	5	6	7	8
$C'(x)$	58	30	18	9	5	7	17	33	51

(A) Find a quadratic regression equation for the data, and graph it and the data set in the same viewing window.

(B) Use the regression equation and a numerical integration routine on a graphing calculator to approximate (to the nearest dollar) the increased cost in going from a production level of 2 thousand sunglasses per month to 8 thousand sunglasses per month.

72. Cost. Refer to Problem 71.

(A) Find a cubic regression equation for the data, and graph it and the data set in the same viewing window.

(B) Use the regression equation and a numerical integration routine on a graphing calculator to approximate (to the nearest dollar) the increased cost in going from a production level of 1 thousand sunglasses per month to 7 thousand sunglasses per month.

73. Supply function. Given the supply function

$$p = S(x) = 10(e^{0.02x} - 1)$$

find the average price (in dollars) over the supply interval $[20, 30]$.

74. Demand function. Given the demand function

$$p = D(x) = \frac{1,000}{x}$$

find the average price (in dollars) over the demand interval $[400, 600]$.

75. Labor costs and learning. A defense contractor is starting production on a new missile control system. On the basis of data collected during assembly of the first 16 control systems, the production manager obtained the following function for the rate of labor use:

$$g(x) = 2,400x^{-1/2}$$

$g(x)$ is the number of labor-hours required to assemble the xth unit of a control system. Approximately how many labor-hours will be required to assemble the 17th through the 25th control units? [*Hint*: Let $a = 16$ and $b = 25$.]

76. Labor costs and learning. If the rate of labor use in Problem 75 is

$$g(x) = 2,000x^{-1/3}$$

then approximately how many labor-hours will be required to assemble the 9th through the 27th control units? [*Hint*: Let $a = 8$ and $b = 27$.]

77. Inventory. A store orders 600 units of a product every 3 months. If the product is steadily depleted to 0 by the end of each 3 months, the inventory on hand I at any time t during the year is shown in the following figure:

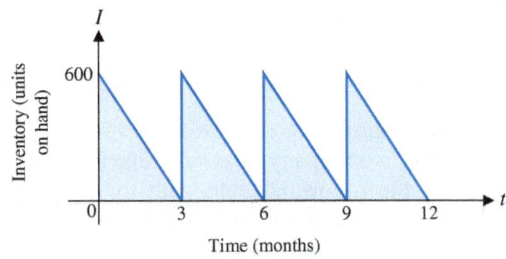

Figure for 77

(A) Write an inventory function (assume that it is continuous) for the first 3 months. [The graph is a straight line joining $(0, 600)$ and $(3, 0)$.]

(B) What is the average number of units on hand for a 3-month period?

78. Repeat Problem 77 with an order of 1,200 units every 4 months.

79. Oil production. Using production and geological data, the management of an oil company estimates that oil will be pumped from a producing field at a rate given by

$$R(t) = \frac{100}{t + 1} + 5 \qquad 0 \le t \le 20$$

where $R(t)$ is the rate of production (in thousands of barrels per year) t years after pumping begins. Approximately how many barrels of oil will the field produce during the first 10 years of production? From the end of the 10th year to the end of the 20th year of production?

80. Oil production. In Problem 79, if the rate is found to be

$$R(t) = \frac{120t}{t^2 + 1} + 3 \qquad 0 \le t \le 20$$

then approximately how many barrels of oil will the field produce during the first 5 years of production? The second 5 years of production?

81. Biology. A yeast culture weighing 2 grams is expected to grow at the rate of $W'(t) = 0.2e^{0.1t}$ grams per hour at a higher controlled temperature. How much will the weight of the culture increase during the first 8 hours of growth? How much will the weight of the culture increase from the end of the 8th hour to the end of the 16th hour of growth?

82. Medicine. The rate of healing of a skin wound (in square centimeters per day) is given approximately by $A'(t) = -0.9e^{-0.1t}$. The initial wound has an area of 9 square centimeters. How much will the area change during the first 5 days? The second 5 days?

83. Temperature. If the temperature $C(t)$ in an aquarium changes according to

$$C(t) = t^3 - 2t + 10 \qquad 0 \le t \le 2$$

(in degrees Celsius) over a 2-hour period, what is the average temperature over this period?

84. Medicine. A drug is injected into the bloodstream of a patient through her right arm. The drug concentration in the bloodstream of the left arm t hours after the injection is given by

$$C(t) = \frac{0.14t}{t^2 + 1}$$

What is the average drug concentration in the bloodstream of the left arm during the first hour after the injection? During the first 2 hours after the injection?

85. Politics. Public awareness of a congressional candidate before and after a successful campaign was approximated by

$$P(t) = \frac{8.4t}{t^2 + 49} + 0.1 \qquad 0 \le t \le 24$$

where t is time in months after the campaign started and $P(t)$ is the fraction of the number of people in the congressional district who could recall the candidate's name. What is the average fraction of the number of people who could recall the candidate's name during the first 7 months of the campaign? During the first 2 years of the campaign?

86. Population composition. The number of children in a large city was found to increase and then decrease rather drastically. If the number of children over a 6-year period was given by

$$N(t) = -\tfrac{1}{4}t^2 + t + 4 \qquad 0 \le t \le 6$$

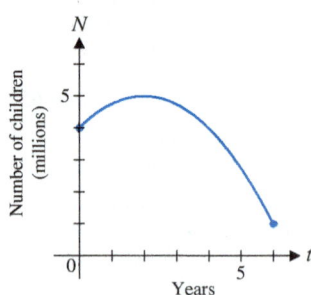

Figure for 86

what was the average number of children in the city over the 6-year period? [Assume that $N = N(t)$ is continuous.]

Answers to Matched Problems

1. (A)

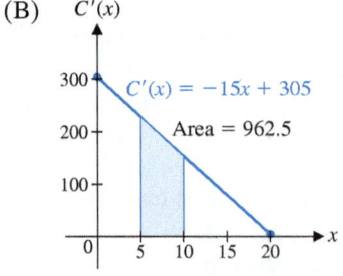

$$C(x) = -7.5x^2 + 305x + 625$$

$$C(10) - C(5) = 962.5$$

(B)

$C'(x)$

$$C'(x) = -15x + 305$$

Area = 962.5

(C) The change in cost from $x = 5$ to $x = 10$ is equal to the area between the marginal cost function and the x axis from $x = 5$ to $x = 10$.

2. $16 + 2e - 2e^3 + 5 \ln 3 \approx -13.241$

3. $\frac{1}{2}(\ln 6 - \ln 4) \approx 0.203$

4. 2   **5.** $1,000

6. (A) $-2 \ln \frac{2}{15} \approx 4$ yr

(B) $11 - 15e^{-2} \approx 8.970$ or $8,970

7. 8.017   **8.** 13   **9.** $35.27

Chapter 7 Review

7-1 Antiderivatives and Indefinite Integrals

- A function F is an **antiderivative** of a function f if $F'(x) = f(x)$.

- If F and G are both antiderivatives of f, then F and G differ by a constant; that is, $F(x) = G(x) + k$ for some constant k.
- We use the symbol $\int f(x)\, dx$, called an **indefinite integral,** to represent the family of all antiderivatives of f, and we write

$$\int f(x)\, dx = F(x) + C$$

The symbol $\int$ is called an **integral sign,** $f(x)$ is the **integrand,** and C is the **constant of integration.**

- Indefinite integrals of basic functions are given by the formulas on page 422.

- Properties of indefinite integrals are given on page 422; in particular, a constant factor can be moved across an integral sign. However, a variable factor *cannot* be moved across an integral sign.

7-2 Integration by Substitution

- The **method of substitution** (also called the **change-of-variable method**) is a technique for finding indefinite integrals. It is based on the following formula, which is obtained by reversing the chain rule:

$$\int f'[g(x)]g'(x)\, dx = f[g(x)] + C$$

- This formula implies the general indefinite integral formulas on page 435.

- When using the method of substitution, it is helpful to use differentials as a bookkeeping device:

  1. The **differential dx** of the independent variable x is an arbitrary real number.

  2. The **differential dy** of the dependent variable y is defined by $dy = f'(x)\, dx$.

- Guidelines for using the substitution method are given by the procedure on page 435.

7-3 Differential Equations; Growth and Decay

- An equation is a **differential equation** if it involves an unknown function and one or more of its derivatives.
- The equation

$$\frac{dy}{dx} = 3x(1 + xy^2)$$

is a **first-order** differential equation because it involves the first derivative of the unknown function y but no second or higher order derivative.

- A **slope field** can be constructed for the preceding differential equation by drawing a tangent line segment with slope $3x(1 + xy^2)$ at each point (x, y) of a grid. The slope field gives a graphical representation of the functions that are solutions of the differential equation.

- The differential equation

$$\frac{dQ}{dt} = rQ$$

(in words, the rate at which the unknown function Q increases is proportional to Q) is called the **exponential growth law.** The constant r is called the **relative growth rate.** The solutions of the exponential growth law are the functions

$$Q(t) = Q_0 e^{rt}$$

where Q_0 denotes $Q(0)$, the amount present at time $t = 0$. These functions can be used to solve problems in population growth, continuous compound interest, radioactive decay, blood pressure, and light absorption.

- Table 1 on page 449 gives the solutions of other first-order differential equations that can be used to model the limited or logistic growth of epidemics, sales, and corporations.

7-4 The Definite Integral

- If the function f is positive on $[a, b]$, then the area between the graph of f and the x axis from $x = a$ to $x = b$ can be approximated by partitioning $[a, b]$ into n subintervals $[x_{k-1}, x_k]$ of equal length $\Delta x = (b - a)/n$ and summing the areas of n rectangles. This can be done using **left sums, right sums,** or, more generally, **Riemann sums:**

Ex. 1, p. 455
Ex. 2, p. 457

Left sum: $\qquad L_n = \sum_{k=1}^{n} f(x_{k-1})\Delta x$

Right sum: $\qquad R_n = \sum_{k=1}^{n} f(x_k)\Delta x$

Riemann sum: $\quad S_n = \sum_{k=1}^{n} f(c_k)\Delta x$

  In a Riemann sum, each c_k is required to belong to the subinterval $[x_{k-1}, x_k]$. Left sums and right sums are the special cases of Riemann sums in which c_k is the left endpoint and right endpoint, respectively, of the subinterval.
- The **error in an approximation** is the absolute value of the difference between the approximation and the actual value. An **error bound** is a positive number such that the error is guaranteed to be less than or equal to that number.
- Theorem 1 on page 455 gives error bounds for the approximation of the area between the graph of a positive function f and the x axis from $x = a$ to $x = b$, by left sums or right sums, if f is either increasing or decreasing.
- If $f(x) > 0$ and is either increasing on $[a, b]$ or decreasing on $[a, b]$, then the left and right sums of $f(x)$ approach the same real number as $n \to \infty$ (Theorem 2, page 455).
- If f is a continuous function on $[a, b]$, then the Riemann sums for f on $[a, b]$ approach a real-number limit I as $n \to \infty$ (Theorem 3, page 457).
- Let f be a continuous function on $[a, b]$. Then the limit I of Riemann sums for f on $[a, b]$, guaranteed to exist by Theorem 3, is called the **definite integral** of f from a to b and is denoted

$$\int_a^b f(x)\, dx$$

  The **integrand** is $f(x)$, the **lower limit of integration** is a, and the **upper limit of integration** is b.
- Geometrically, the definite integral

Ex. 3, p. 458

$$\int_a^b f(x)\, dx$$

  represents the cumulative sum of the signed areas between the graph of f and the x axis from $x = a$ to $x = b$.
- Properties of the definite integral are given on page 389.

Ex. 4, p. 459

7-5 The Fundamental Theorem of Calculus

- If f is a continuous function on $[a, b]$ and F is any antiderivative of f, then

Ex. 1, p. 464
Ex. 2, p. 465

$$\int_a^b f(x)\, dx = F(b) - F(a)$$

  This is the fundamental theorem of calculus (see page 465).

Ex. 3, p. 466
Ex. 4, p. 466
Ex. 5, p. 467
Ex. 6, p. 467
Ex. 7, p. 468

- The fundamental theorem gives an easy and exact method for evaluating definite integrals, provided that we can find an antiderivative $F(x)$ of $f(x)$. In practice, we first find an antiderivative $F(x)$ (when possible), using techniques for computing indefinite integrals. Then we calculate the difference $F(b) - F(a)$. If it is impossible to find an antiderivative, we must resort to left or right sums, or other approximation methods, to evaluate the definite integral. Graphing calculators have a built-in numerical approximation routine, more powerful than left- or right-sum methods, for this purpose.
- If f is a continuous function on $[a, b]$, then the **average value** of f over $[a, b]$ is defined to be

Ex. 8, p. 470

$$\frac{1}{b - a} \int_a^b f(x)\, dx$$

Review Exercises

Work through all the problems in this chapter review and check your answers in the back of the book. Answers to all review problems are there, along with section numbers in italics to indicate where each type of problem is discussed. Where weaknesses show up, review appropriate sections of the text.

A

Find each integral in Problems 1–6.

1. $\int (6x + 3)\, dx$

2. $\int_{10}^{20} 5\, dx$

3. $\int_0^9 (4 - t^2)\, dt$

4. $\int (1 - t^2)^3 t\, dt$

5. $\int \frac{1 + u^4}{u}\, du$

6. $\int_0^1 xe^{-2x^2}\, dx$

In Problems 7 and 8, find the derivative or indefinite integral as indicated.

7. $\dfrac{d}{dx}\left(\int e^{-x^2}\, dx \right)$

8. $\int \dfrac{d}{dx}\left(\sqrt{4 + 5x} \right) dx$

9. Find a function $y = f(x)$ that satisfies both conditions:

$$\frac{dy}{dx} = 3x^2 - 2 \qquad f(0) = 4$$

10. Find all antiderivatives of

(A) $\dfrac{dy}{dx} = 8x^3 - 4x - 1$ (B) $\dfrac{dx}{dt} = e^t - 4t^{-1}$

11. Approximate $\int_1^5 (x^2 + 1)\, dx$, using a right sum with $n = 2$. Calculate an error bound for this approximation.

12. Evaluate the integral in Problem 11, using the fundamental theorem of calculus, and calculate the actual error $|I - R_2|$ produced by using R_2.

13. Use the following table of values and a left sum with $n = 4$ to approximate $\int_1^{17} f(x)\, dx$:

x	1	5	9	13	17
$f(x)$	1.2	3.4	2.6	0.5	0.1

14. Find the average value of $f(x) = 6x^2 + 2x$ over the interval $[-1, 2]$.

15. Describe a rectangle that has the same area as the area under the graph of $f(x) = 6x^2 + 2x$ from $x = -1$ to $x = 2$ (see Problem 14).

In Problems 16 and 17, calculate the indicated Riemann sum S_n for the function $f(x) = 100 - x^2$.

16. Partition $[3, 11]$ into four subintervals of equal length, and for each subinterval $[x_{i-1}, x_i]$, let $c_i = (x_{i-1} + x_i)/2$.

17. Partition $[-5, 5]$ into five subintervals of equal length and let $c_1 = -4$, $c_2 = -1$, $c_3 = 1$, $c_4 = 2$, and $c_5 = 5$.

B

Use the graph and actual areas of the indicated regions in the figure to evaluate the integrals in Problems 18–25:

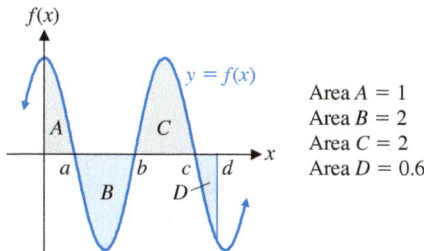

Figure for 18–25

18. $\int_a^b 5f(x)\, dx$

19. $\int_b^c \dfrac{f(x)}{5}\, dx$

20. $\int_b^d f(x)\, dx$

21. $\int_a^c f(x)\, dx$

22. $\int_0^d f(x)\, dx$

23. $\int_b^a f(x)\, dx$

24. $\int_c^b f(x)\, dx$

25. $\int_d^0 f(x)\, dx$

Problems 26–31 refer to the slope field shown in the figure:

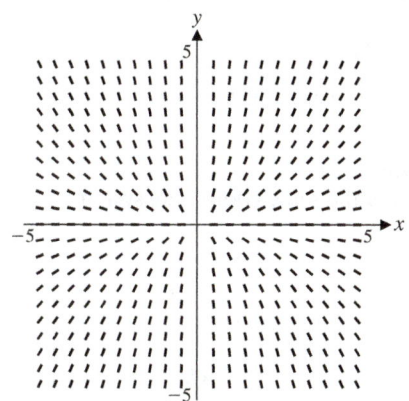

Figure for 26–31

26. (A) For $dy/dx = (2y)/x$, what is the slope of a solution curve at $(2, 1)$? At $(-2, -1)$?

(B) For $dy/dx = (2x)/y$, what is the slope of a solution curve at $(2, 1)$? At $(-2, -1)$?

27. Is the slope field shown in the figure for $dy/dx = (2x)/y$ or for $dy/dx = (2y)/x$? Explain.

28. Show that $y = Cx^2$ is a solution of $dy/dx = (2y)/x$ for any real number C.

29. Referring to Problem 28, find the particular solution of $dy/dx = (2y)/x$ that passes through $(2, 1)$. Through $(-2, -1)$.

30. Graph the two particular solutions found in Problem 29 in the slope field shown (or a copy).

 31. Use a graphing calculator to graph, in the same viewing window, graphs of $y = Cx^2$ for $C = -2, -1, 1,$ and 2 for $-5 \le x \le 5$ and $-5 \le y \le 5$.

Find each integral in Problems 32–42.

32. $\displaystyle\int_{-1}^{1} \sqrt{1 + x} \, dx$

33. $\displaystyle\int_{-1}^{0} x^2(x^3 + 2)^{-2} \, dx$

34. $\displaystyle\int 5e^{-t} \, dt$

35. $\displaystyle\int_{1}^{e} \frac{1 + t^2}{t} \, dt$

36. $\displaystyle\int xe^{3x^2} \, dx$

37. $\displaystyle\int_{-3}^{1} \frac{1}{\sqrt{2 - x}} \, dx$

38. $\displaystyle\int_{0}^{3} \frac{x}{1 + x^2} \, dx$

39. $\displaystyle\int_{0}^{3} \frac{x}{(1 + x^2)^2} \, dx$

40. $\displaystyle\int x^3(2x^4 + 5)^5 \, dx$

41. $\displaystyle\int \frac{e^{-x}}{e^{-x} + 3} \, dx$

42. $\displaystyle\int \frac{e^x}{(e^x + 2)^2} \, dx$

43. Find a function $y = f(x)$ that satisfies both conditions:

$$\frac{dy}{dx} = 3x^{-1} - x^{-2} \qquad f(1) = 5$$

44. Find the equation of the curve that passes through $(2, 10)$ if its slope is given by

$$\frac{dy}{dx} = 6x + 1$$

for each x.

45. (A) Find the average value of $f(x) = 3\sqrt{x}$ over the interval $[1, 9]$.

(B) Graph $f(x) = 3\sqrt{x}$ and its average over the interval $[1, 9]$ in the same coordinate system.

C

Find each integral in Problems 46–50.

46. $\displaystyle\int \frac{(\ln x)^2}{x} \, dx$

47. $\displaystyle\int x(x^3 - 1)^2 \, dx$

48. $\displaystyle\int \frac{x}{\sqrt{6 - x}} \, dx$

49. $\displaystyle\int_{0}^{7} x\sqrt{16 - x} \, dx$

50. $\displaystyle\int_{1}^{1} (x + 1)^9 dx$

51. Find a function $y = f(x)$ that satisfies both conditions:

$$\frac{dy}{dx} = 9x^2 e^{x^3} \qquad f(0) = 2$$

52. Solve the differential equation

$$\frac{dN}{dt} = 0.06N \qquad N(0) = 800 \qquad N > 0$$

Graph Problems 53–56 on a graphing calculator, and identify each curve as unlimited growth, exponential decay, limited growth, or logistic growth:

53. $N = 50(1 - e^{-0.07t}); 0 \le t \le 80, 0 \le N \le 60$

54. $p = 500e^{-0.03x}; 0 \le x \le 100, 0 \le p \le 500$

55. $A = 200e^{0.08t}; 0 \le t \le 20, 0 \le A \le 1{,}000$

56. $N = \dfrac{100}{1 + 9e^{-0.3t}}; 0 \le t \le 25, 0 \le N \le 100$

Use a numerical integration routine to evaluate each definite integral in Problems 57–59 (to three decimal places).

57. $\displaystyle\int_{-0.5}^{0.6} \frac{1}{\sqrt{1 - x^2}} \, dx$

58. $\displaystyle\int_{-2}^{3} x^2 e^x \, dx$

59. $\displaystyle\int_{0.5}^{2.5} \frac{\ln x}{x^2} \, dx$

Applications

60. Cost. A company manufactures downhill skis. The research department produced the marginal cost graph shown in the accompanying figure, where $C'(x)$ is in dollars and x is the number of pairs of skis produced per week. Estimate the increase in cost going from a production level of 200 to 600 pairs of skis per week. Use left and right sums over two equal subintervals. Replace the question marks with the values of L_2 and R_2 as appropriate:

$$? \le \int_{200}^{600} C'(x) \, dx \le ?$$

Figure for 60

61. Cost. Assuming that the marginal cost function in Problem 60 is linear, find its equation and write a definite integral that represents the increase in costs going from a production level of 200 to 600 pairs of skis per week. Evaluate the definite integral.

62. Profit and production. The weekly marginal profit for an output of x units is given approximately by

$$P'(x) = 150 - \frac{x}{10} \qquad 0 \le x \le 40$$

What is the total change in profit for a change in production from 10 units per week to 40 units? Set up a definite integral and evaluate it.

63. Profit function. If the marginal profit for producing x units per day is given by

$$P'(x) = 100 - 0.02x \qquad P(0) = 0$$

where $P(x)$ is the profit in dollars, find the profit function P and the profit on 10 units of production per day.

64. Resource depletion. An oil well starts out producing oil at the rate of 60,000 barrels of oil per year, but the production rate is expected to decrease by 4,000 barrels per year. Thus, if $P(t)$ is the total production (in thousands of barrels) in t years, then

$$P'(t) = f(t) = 60 - 4t \qquad 0 \le t \le 15$$

Write a definite integral that will give the total production after 15 years of operation, and evaluate the integral.

65. Inventory. Suppose that the inventory of a certain item t months after the first of the year is given approximately by

$$I(t) = 10 + 36t - 3t^2 \qquad 0 \le t \le 12$$

What is the average inventory for the second quarter of the year?

66. Price–supply. Given the price–supply function

$$p = S(x) = 8(e^{0.05x} - 1)$$

find the average price (in dollars) over the supply interval $[40, 50]$.

67. Useful life. The total accumulated costs C(t) and revenues R(t) (in thousands of dollars), respectively, for a coal mine satisfy

$$C'(t) = 3 \qquad \text{and} \qquad R'(t) = 20e^{-0.1t}$$

where t is the number of years that the mine has been in operation. Find the useful life of the mine, to the nearest year. What is the total profit accumulated during the useful life of the mine?

68. Marketing. The market research department for an automobile company estimates that sales (in millions of dollars) of a new electric car will increase at the monthly rate of

$$S'(t) = 4e^{-0.08t} \qquad 0 \le t \le 24$$

t months after the introduction of the car. What will be the total sales S(t) t months after the car is introduced if we assume that there were 0 sales at the time the car entered the marketplace? What are the estimated total sales during the first 12 months after the introduction of the car? How long will it take for the total sales to reach $40 million?

69. Wound healing. The area of a healing skin wound changes at a rate given approximately by

$$\frac{dA}{dt} = -5t^{-2} \qquad 1 \le t \le 5$$

where t is time in days and $A(1) = 5$ square centimeters. What will be the area of the wound in 5 days?

70. Pollution. An environmental protection agency estimates that the rate of seepage of toxic chemicals from a waste dump (in gallons per year) is given by

$$R(t) = \frac{1,000}{(1 + t)^2}$$

where t is the time in years since the discovery of the seepage. Find the total amount of toxic chemicals that seep from the dump during the first 4 years of its discovery.

71. Population. The population of Mexico was 111 million in 2009 and was growing at a rate of 1.13% per year, compounded continuously.

(A) Assuming that the population continues to grow at this rate, estimate the population of Mexico in the year 2025.

(B) At the growth rate indicated, how long will it take the population of Mexico to double?

72. Archaeology. The continuous compound rate of decay for carbon-14 is $r = -0.000\ 123\ 8$. A piece of animal bone found at an archaeological site contains 4% of the original amount of carbon-14. Estimate the age of the bone.

73. Learning. An average student enrolled in a typing class progressed at a rate of $N'(t) = 7e^{-0.1t}$ words per minute t weeks after enrolling in a 15-week course. If a student could type 25 words per minute at the beginning of the course, how many words per minute $N(t)$ would the student be expected to type t weeks into the course? After completing the course?

Additional Integration Topics

Introduction

In Chapter 8 we explore additional applications and techniques of integration. We use the integral to find probabilities and to calculate several quantities that are important in business and economics: the total income and future value produced by a continuous income stream, consumers' and producers' surplus, and the Gini index of income concentration. The Gini index is a single number that measures the equality of a country's income distribution (see Problems 87 and 88, for example, in Section 8-1).

8-1   Area Between Curves

- Area Between Two Curves
- Application: Income Distribution

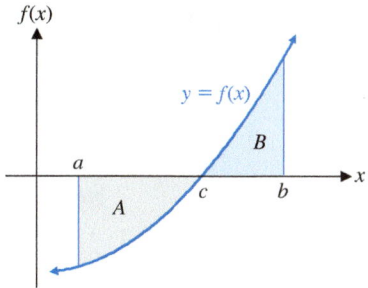

Figure 1 $\int_a^b f(x)\, dx = -A + B$

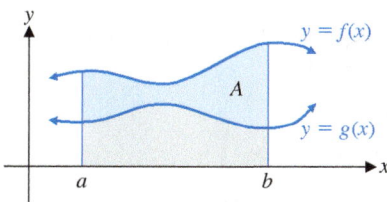

Figure 2

In Chapter 7, we found that the definite integral $\int_a^b f(x)\, dx$ represents the sum of the signed areas between the graph of $y = f(x)$ and the x axis from $x = a$ to $x = b$, where the areas above the x axis are counted positively and the areas below the x axis are counted negatively (see Fig. 1). In this section, we are interested in using the definite integral to find the actual area between a curve and the x axis or the actual area between two curves. These areas are always nonnegative quantities—**area measure is never negative**.

Area Between Two Curves

Consider the area bounded by $y = f(x)$ and $y = g(x)$, where $f(x) \geq g(x) \geq 0$, for $a \leq x \leq b$, as shown in Figure 2.

$$\begin{pmatrix} \text{Area } A \text{ between} \\ f(x) \text{ and } g(x) \end{pmatrix} = \begin{pmatrix} \text{area} \\ \text{under } f(x) \end{pmatrix} - \begin{pmatrix} \text{area} \\ \text{under } g(x) \end{pmatrix} \quad \begin{array}{l} \textit{Areas are from } x = a \textit{ to} \\ x = b \textit{ above the } x \textit{ axis.} \end{array}$$

$$= \int_a^b f(x)\, dx - \int_a^b g(x)\, dx \quad \begin{array}{l} \textit{Use definite integral} \\ \textit{property 4 (Section 6-4).} \end{array}$$

$$= \int_a^b [f(x) - g(x)]\, dx$$

It can be shown that the preceding result does not require $f(x)$ or $g(x)$ to remain positive over the interval $[a, b]$. A more general result is stated in the following box:

THEOREM 1   Area Between Two Curves

If f and g are continuous and $f(x) \geq g(x)$ over the interval $[a, b]$, then the area bounded by $y = f(x)$ and $y = g(x)$ for $a \leq x \leq b$ is given exactly by

$$A = \int_a^b [f(x) - g(x)]\, dx$$

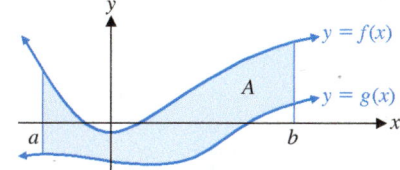

CONCEPTUAL INSIGHT

Theorem 1 requires the graph of f to be *above* (or equal to) the graph of g throughout $[a, b]$, but f and g can be either positive, negative, or 0. In Section 7-4, we considered the special cases of Theorem 1 in which (1) f is positive and g is the zero function on $[a, b]$; and (2) f is the zero function and g is negative on $[a, b]$:

Special case 1. If f is continuous and positive over $[a, b]$, then the area bounded by the graph of f and the x axis for $a \leq x \leq b$ is given exactly by

$$\int_a^b f(x)\, dx$$

Special case 2. If g is continuous and negative over $[a, b]$, then the area bounded by the graph of g and the x axis for $a \leq x \leq b$ is given exactly by

$$\int_a^b -g(x)\, dx$$

EXAMPLE 1   **Area Between a Curve and the x Axis** Find the area bounded by $f(x) = 6x - x^2$ and $y = 0$ for $1 \le x \le 4$.

SOLUTION   We sketch a graph of the region first (Fig. 3). The solution of every area problem should begin with a sketch. Since $f(x) \ge 0$ on $[1, 4]$,

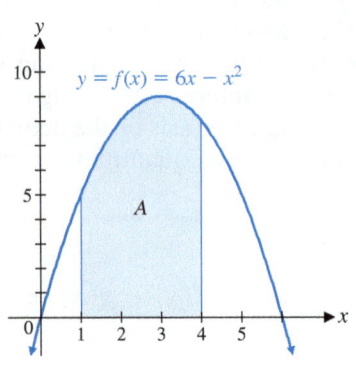

$y = f(x) = 6x - x^2$

A

Figure 3

$$A = \int_1^4 (6x - x^2)\, dx = \left(3x^2 - \frac{x^3}{3} \right)\Bigg|_1^4$$

$$= \left[3(4)^2 - \frac{(4)^3}{3} \right] - \left[3(1)^2 - \frac{(1)^3}{3} \right]$$

$$= 48 - \tfrac{64}{3} - 3 + \tfrac{1}{3}$$

$$= 48 - 21 - 3$$

$$= 24$$

Matched Problem 1   Find the area bounded by $f(x) = x^2 + 1$ and $y = 0$ for $-1 \le x \le 3$.

EXAMPLE 2   **Area Between a Curve and the x Axis** Find the area between the graph of $f(x) = x^2 - 2x$ and the x axis over the indicated intervals:

(A) $[1, 2]$ (B) $[-1, 1]$

SOLUTION   We begin by sketching the graph of f, as shown in Figure 4.

(A) From the graph, we see that $f(x) \le 0$ for $1 \le x \le 2$, so we integrate $-f(x)$:

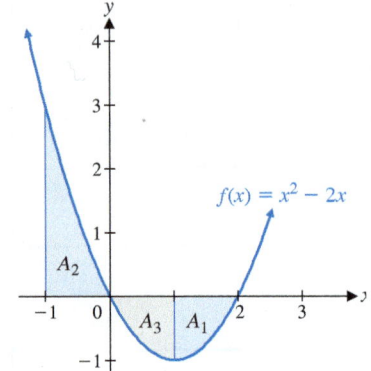

$f(x) = x^2 - 2x$

A_2

A_3 A_1

Figure 4

$$A_1 = \int_1^2 [-f(x)]\, dx$$

$$= \int_1^2 (2x - x^2)\, dx$$

$$= \left(x^2 - \frac{x^3}{3} \right)\Bigg|_1^2$$

$$= \left[(2)^2 - \frac{(2)^3}{3} \right] - \left[(1)^2 - \frac{(1)^3}{3} \right]$$

$$= 4 - \tfrac{8}{3} - 1 + \tfrac{1}{3} \quad = \tfrac{2}{3} \approx 0.667$$

(B) Since the graph shows that $f(x) \ge 0$ on $[-1, 0]$ and $f(x) \le 0$ on $[0, 1]$, the computation of this area will require two integrals:

$$A = A_2 + A_3$$

$$= \int_{-1}^0 f(x)\, dx + \int_0^1 [-f(x)]\, dx$$

$$= \int_{-1}^0 (x^2 - 2x)\, dx + \int_0^1 (2x - x^2)\, dx$$

$$= \left(\frac{x^3}{3} - x^2 \right)\Bigg|_{-1}^0 + \left(x^2 - \frac{x^3}{3} \right)\Bigg|_0^1$$

$$= \tfrac{4}{3} + \tfrac{2}{3} \quad = 2$$

Matched Problem 2   Find the area between the graph of $f(x) = x^2 - 9$ and the x axis over the indicated intervals:

(A) $[0, 2]$ (B) $[2, 4]$

EXAMPLE 3   **Area Between Two Curves** Find the area bounded by the graphs of $f(x) = \frac{1}{2}x + 3$, $g(x) = -x^2 + 1$, $x = -2$, and $x = 1$.

SOLUTION   We first sketch the area (Fig. 5) and then set up and evaluate an appropriate definite integral. We observe from the graph that $f(x) \geq g(x)$ for $-2 \leq x \leq 1$, so

$$A = \int_{-2}^{1} [f(x) - g(x)]\, dx = \int_{-2}^{1} \left[\left(\frac{x}{2} + 3 \right) - (-x^2 + 1) \right] dx$$

$$= \int_{-2}^{1} \left(x^2 + \frac{x}{2} + 2 \right) dx$$

$$= \left(\frac{x^3}{3} + \frac{x^2}{4} + 2x \right) \Big|_{-2}^{1}$$

$$= \left(\frac{1}{3} + \frac{1}{4} + 2 \right) - \left(\frac{-8}{3} + \frac{4}{4} - 4 \right) = \frac{33}{4} = 8.25$$

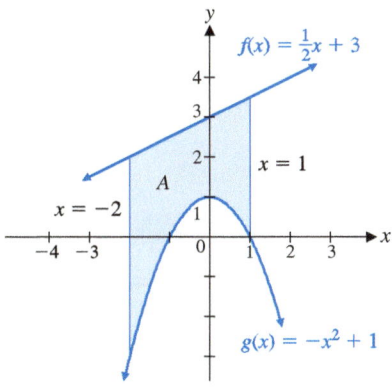

Figure 5

Matched Problem 3   Find the area bounded by $f(x) = x^2 - 1$, $g(x) = -\frac{1}{2}x - 3$, $x = -1$, and $x = 2$.

EXAMPLE 4   **Area Between Two Curves** Find the area bounded by $f(x) = 5 - x^2$ and $g(x) = 2 - 2x$.

SOLUTION   First, graph f and g on the same coordinate system, as shown in Figure 6. Since the statement of the problem does not include any limits on the values of x, we must determine the appropriate values from the graph. The graph of f is a parabola and the graph of g is a line. The area bounded by these two graphs extends from the intersection point on the left to the intersection point on the right. To find these intersection points, we solve the equation $f(x) = g(x)$ for x:

$$f(x) = g(x)$$
$$5 - x^2 = 2 - 2x$$
$$x^2 - 2x - 3 = 0$$
$$x = -1, 3$$

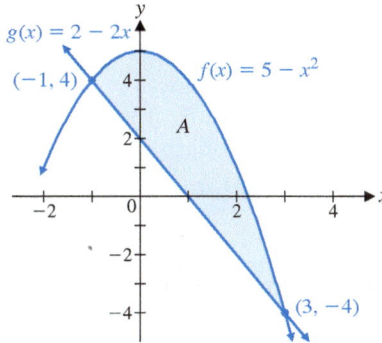

Figure 6

You should check these values in the original equations. (Note that the area between the graphs for $x < -1$ is unbounded on the left, and the area between the graphs for $x > 3$ is unbounded on the right.) Figure 6 shows that $f(x) \geq g(x)$ over the interval $[-1, 3]$, so we have

$$A = \int_{-1}^{3} [f(x) - g(x)]\, dx = \int_{-1}^{3} [5 - x^2 - (2 - 2x)]\, dx$$

$$= \int_{-1}^{3} (3 + 2x - x^2)\, dx$$

$$= \left(3x + x^2 - \frac{x^3}{3} \right) \Big|_{-1}^{3}$$

$$= \left[3(3) + (3)^2 - \frac{(3)^3}{3} \right] - \left[3(-1) + (-1)^2 - \frac{(-1)^3}{3} \right] = \frac{32}{3} \approx 10.667$$

Matched Problem 4 Find the area bounded by $f(x) = 6 - x^2$ and $g(x) = x$.

EXAMPLE 5 **Area Between Two Curves** Find the area bounded by $f(x) = x^2 - x$ and $g(x) = 2x$ for $-2 \le x \le 3$.

SOLUTION The graphs of f and g are shown in Figure 7. Examining the graph, we see that $f(x) \ge g(x)$ on the interval $[-2, 0]$, but $g(x) \ge f(x)$ on the interval $[0, 3]$. Thus, two integrals are required to compute this area:

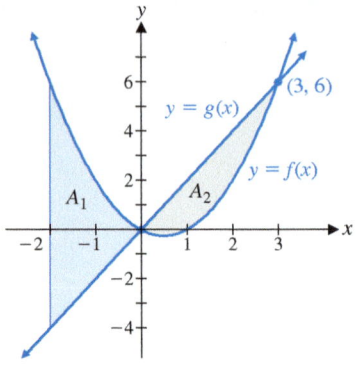

Figure 7

$$A_1 = \int_{-2}^{0} [f(x) - g(x)] \, dx \qquad f(x) \ge g(x) \text{ on } [-2, 0]$$

$$= \int_{-2}^{0} [x^2 - x - 2x] \, dx$$

$$= \int_{-2}^{0} (x^2 - 3x) \, dx$$

$$= \left(\frac{x^3}{3} - \frac{3}{2}x^2 \right) \Big|_{-2}^{0}$$

$$= (0) - \left[\frac{(-2)^3}{3} - \frac{3}{2}(-2)^2 \right] = \frac{26}{3} \approx 8.667$$

$$A_2 = \int_{0}^{3} [g(x) - f(x)] \, dx \qquad g(x) \ge f(x) \text{ on } [0, 3]$$

$$= \int_{0}^{3} [2x - (x^2 - x)] \, dx$$

$$= \int_{0}^{3} (3x - x^2) \, dx$$

$$= \left(\frac{3}{2}x^2 - \frac{x^3}{3} \right) \Big|_{0}^{3}$$

$$= \left[\frac{3}{2}(3)^2 - \frac{(3)^3}{3} \right] - (0) = \frac{9}{2} = 4.5$$

The total area between the two graphs is

$$A = A_1 + A_2 = \tfrac{26}{3} + \tfrac{9}{2} = \tfrac{79}{6} \approx 13.167$$

Matched Problem 5 Find the area bounded by $f(x) = 2x^2$ and $g(x) = 4 - 2x$ for $-2 \le x \le 2$.

EXAMPLE 6 **Computing Areas with a Numerical Integration Routine** Find the area (to three decimal places) bounded by $f(x) = e^{-x^2}$ and $g(x) = x^2 - 1$.

SOLUTION First, we use a graphing calculator to graph the functions f and g and find their intersection points (see Fig. 8A). We see that the graph of f is bell shaped and the graph of g is a parabola. We note that $f(x) \ge g(x)$ on the interval $[-1.131, 1.131]$ and compute the area A by a numerical integration routine (see Fig. 8B):

$$A = \int_{-1.131}^{1.131} \left[e^{-x^2} - (x^2 - 1) \right] dx = 2.876$$

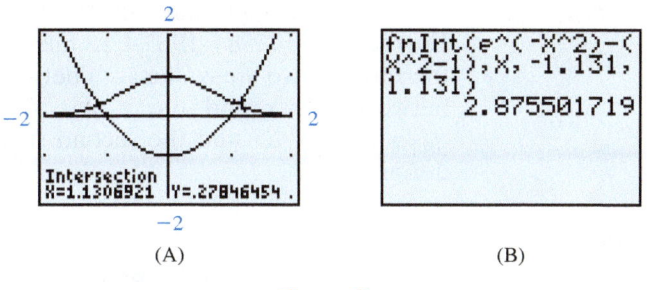

(A)   (B)

Figure 8

Matched Problem 6   Find the area (to three decimal places) bounded by the graphs of $f(x) = x^2 \ln x$ and $g(x) = 3x - 3$.

Application: Income Distribution

The U.S. Census Bureau compiles and analyzes a great deal of data having to do with the distribution of income among families in the United States. For 2006, the Bureau reported that the lowest 20% of families received 3% of all family income and the top 20% received 50%. Table 1 and Figure 9 give a detailed picture of the distribution of family income in 2006.

The graph of $y = f(x)$ in Figure 9 is called a **Lorenz curve** and is generally found by using *regression analysis,* a technique for fitting a function to a data set over a given interval. The variable **x represents the cumulative percentage of families at or below a given income level,** and **y represents the cumulative percentage of total family income received.** For example, data point (0.40, 0.12) in Table 1 indicates that the bottom 40% of families (those with incomes under \$38,000) received 12% of the total income for all families in 2006, data point (0.60, 0.27) indicates that the bottom 60% of families received 27% of the total income for all families that year, and so on.

Table 1 Family Income Distribution in the United States, 2006		
Income Level	x	y
Under \$20,000	0.20	0.03
Under \$38,000	0.40	0.12
Under \$60,000	0.60	0.27
Under \$97,000	0.80	0.49

Source: U.S. Census Bureau

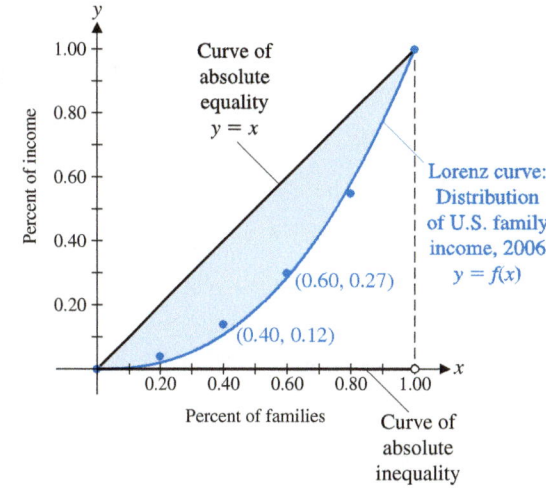

Figure 9 Lorenz curve

Absolute equality of income would occur if the area between the Lorenz curve and $y = x$ were 0. In this case, the Lorenz curve would be $y = x$ and all families would receive equal shares of the total income. That is, 5% of the families would receive 5% of the income, 20% of the families would receive 20% of the income, 65% of the families would receive 65% of the income, and so on. The maximum possible area between a Lorenz curve and $y = x$ is $\frac{1}{2}$, the area of the triangle below $y = x$. In this case, we would have **absolute inequality**: All the income would be in the hands of one family and the rest would have none. In actuality, Lorenz curves lie between these two extremes. But as the shaded area increases, the greater is the inequality of income distribution.

We use a single number, the **Gini index** [named after the Italian sociologist Corrado Gini (1884–1965)], to measure income concentration. The Gini index is the ratio of two areas: the area between $y = x$ and the Lorenz curve, and the area between $y = x$ and the x axis, from $x = 0$ to $x = 1$. The first area equals $\int_0^1 [x - f(x)]\,dx$ and the second (triangular) area equals $\frac{1}{2}$, giving the following definition:

DEFINITION   Gini Index of Income Concentration

If $y = f(x)$ is the equation of a Lorenz curve, then

$$\text{Gini index} = 2 \int_0^1 [x - f(x)]\,dx$$

The Gini index is always a number between 0 and 1:

A Gini index of 0 indicates absolute equality—all people share equally in the income. A Gini index of 1 indicates absolute inequality—one person has all the income and the rest have none.

The closer the index is to 0, the closer the income is to being equally distributed. The closer the index is to 1, the closer the income is to being concentrated in a few hands. The Gini index of income concentration is used to compare income distributions at various points in time, between different groups of people, before and after taxes are paid, between different countries, and so on.

EXAMPLE 7   **Distribution of Income**   The Lorenz curve for the distribution of income in a certain country in 2010 is given by $f(x) = x^{2.6}$. Economists predict that the Lorenz curve for the country in the year 2025 will be given by $g(x) = x^{1.8}$. Find the Gini index of income concentration for each curve, and interpret the results.

SOLUTION   The Lorenz curves are shown in Figure 10.

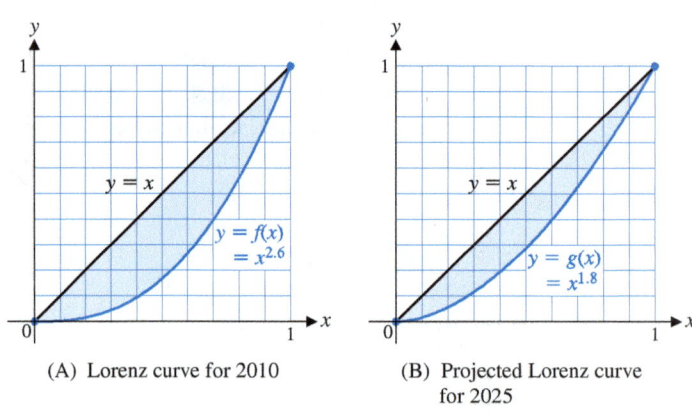

(A) Lorenz curve for 2010   (B) Projected Lorenz curve for 2025

Figure 10

The Gini index in 2010 is (see Fig. 10A)

$$2 \int_0^1 [x - f(x)]\,dx = 2 \int_0^1 [x - x^{2.6}]\,dx = 2\left(\frac{1}{2}x^2 - \frac{1}{3.6}x^{3.6} \right)\Big|_0^1$$

$$= 2\left(\frac{1}{2} - \frac{1}{3.6} \right) \approx 0.444$$

The projected Gini index in 2025 is (see Fig. 10B)

$$2 \int_0^1 [x - g(x)] \, dx = 2 \int_0^1 [x - x^{1.8}] \, dx = 2\left(\frac{1}{2}x^2 - \frac{1}{2.8}x^{2.8}\right)\Big|_0^1$$

$$= 2\left(\frac{1}{2} - \frac{1}{2.8}\right) \approx 0.286$$

If this projection is correct, the Gini index will decrease, and income will be more equally distributed in the year 2025 than in 2010.

Matched Problem 7 Repeat Example 7 if the projected Lorenz curve in the year 2025 is given by $g(x) = x^{3.8}$.

EXPLORE & DISCUSS 1 Do you agree or disagree with each of the following statements (explain your answers by referring to the data in Table 2):

(A) In countries with a low Gini index, there is little incentive for individuals to strive for success, and therefore productivity is low.

(B) In countries with a high Gini index, it is almost impossible to rise out of poverty, and therefore productivity is low.

Table 2

Country	Gini Index	Per Capita Gross Domestic Product
Brazil	0.57	$8,402
China	0.47	6,757
France	0.33	30,386
Germany	0.28	29,461
Japan	0.25	31,267
Jordan	0.39	5,530
Mexico	0.46	10,751
Russia	0.40	10,845
Sweden	0.25	32,525
United States	0.41	41,890

Source: The World Bank

Exercises 8-1

A

Problems 1–6 refer to Figures A–D on page 488. Set up definite integrals in Problems 1–4 that represent the indicated shaded area.

1. Shaded area in Figure B

2. Shaded area in Figure A

3. Shaded area in Figure C

4. Shaded area in Figure D

5. Explain why $\int_a^b h(x) \, dx$ does not represent the area between the graph of $y = h(x)$ and the x axis from $x = a$ to $x = b$ in Figure C.

6. Explain why $\int_a^b [-h(x)] \, dx$ represents the area between the graph of $y = h(x)$ and the x axis from $x = a$ to $x = b$ in Figure C.

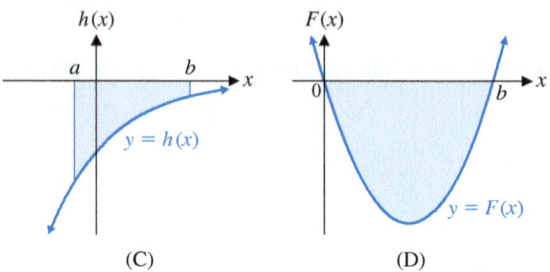

(A)   (B)

(C)   (D)

Figures for 1–6

In Problems 7–20, find the area bounded by the graphs of the indicated equations over the given interval. Compute answers to three decimal places.

7. $y = x + 4; y = 0; 0 \le x \le 4$

8. $y = -x + 10; y = 0; -2 \le x \le 2$

9. $y = x^2 - 20; y = 0; -3 \le x \le 0$

10. $y = x^2 + 2; y = 0; 0 \le x \le 3$

11. $y = -x^2 + 10; y = 0; -3 \le x \le 3$

12. $y = -2x^2; y = 0; -6 \le x \le 0$

13. $y = x^3 + 1; y = 0; 0 \le x \le 2$

14. $y = -x^3 + 3; y = 0; -2 \le x \le 1$

15. $y = x(1 - x); y = 0; -1 \le x \le 0$

16. $y = -x(3 - x); y = 0; 1 \le x \le 2$

17. $y = -e^x; y = 0; -1 \le x \le 1$

18. $y = e^x; y = 0; 0 \le x \le 1$

19. $y = \dfrac{1}{x}; y = 0; 1 \le x \le e$

20. $y = -\dfrac{1}{x}; y = 0; -1 \le x \le -\dfrac{1}{e}$

B

In Problems 21–26, use a definite integral to find the area bounded by the graphs of the indicated equations over the given interval. Then check your answer by finding the area without using a definite integral. [Hint: Partition the region into triangles and/or rectangles].

21. $y = x; y = -2; 0 \le x \le 3$

22. $y = -x; y = 8; 0 \le x \le 5$

23. $y = 2x + 3; y = x - 1; 4 \le x \le 6$

24. $y = 3x + 6; y = \dfrac{1}{2}x + 1; 8 \le x \le 10$

25. $y = -4x + 1; y = \dfrac{3}{2}x + 1; 0 \le x \le 4$

26. $y = -2x - 3; y = x - 3; 0 \le x \le 20$

Problems 27–36 refer to Figures A and B. Set up definite integrals in Problems 27–34 that represent the indicated shaded areas over the given intervals.

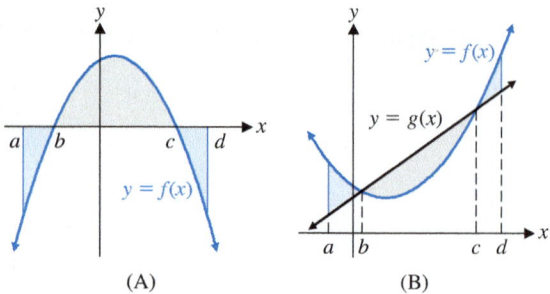

(A)   (B)

Figures for 27–34

27. Over interval $[a, b]$ in Figure A

28. Over interval $[c, d]$ in Figure A

29. Over interval $[b, d]$ in Figure A

30. Over interval $[a, c]$ in Figure A

31. Over interval $[c, d]$ in Figure B

32. Over interval $[a, b]$ in Figure B

33. Over interval $[a, c]$ in Figure B

34. Over interval $[b, d]$ in Figure B

35. Referring to Figure B, explain how you would use definite integrals and the functions f and g to find the area bounded by the two functions from $x = a$ to $x = d$.

36. Referring to Figure A, explain how you would use definite integrals to find the area between the graph of $y = f(x)$ and the x axis from $x = a$ to $x = d$.

In Problems 37–52, find the area bounded by the graphs of the indicated equations over the given intervals (when stated). Compute answers to three decimal places.

37. $y = -x; y = 0; -2 \le x \le 1$

38. $y = -x + 1; y = 0; -1 \le x \le 2$

39. $y = x^2 - 4; y = 0; 0 \le x \le 3$

40. $y = 4 - x^2; y = 0; 0 \le x \le 4$

41. $y = x^2 - 3x; y = 0; -2 \le x \le 2$

42. $y = -x^2 - 2x; y = 0; -2 \le x \le 1$

43. $y = -2x + 8; y = 12; -1 \le x \le 2$

44. $y = 2x + 6; y = 3; -1 \le x \le 2$

45. $y = 3x^2; y = 12$

46. $y = x^2; y = 9$

47. $y = 4 - x^2; y = -5$

48. $y = x^2 - 1; y = 3$

49. $y = x^2 + 1; y = 2x - 2; -1 \leq x \leq 2$

50. $y = x^2 - 1; y = x - 2; -2 \leq x \leq 1$

51. $y = e^{0.5x}; y = -\dfrac{1}{x}; 1 \leq x \leq 2$

52. $y = \dfrac{1}{x}; y = -e^x; 0.5 \leq x \leq 1$

In Problems 53–58, set up a definite integral that represents the area bounded by the graphs of the indicated equations over the given interval. Find the areas to three decimal places. [Hint: A circle of radius r, with center at the origin, has equation $x^2 + y^2 = r^2$ and area πr^2].

53. $y = \sqrt{9 - x^2}; y = 0; -3 \leq x \leq 3$

54. $y = \sqrt{25 - x^2}; y = 0; -5 \leq x \leq 5$

55. $y = -\sqrt{16 - x^2}; y = 0; 0 \leq x \leq 4$

56. $y = -\sqrt{36 - x^2}; y = 0; -6 \leq x \leq 0$

57. $y = -\sqrt{4 - x^2}; y = \sqrt{4 - x^2}; -2 \leq x \leq 2$

58. $y = -\sqrt{100 - x^2}; y = \sqrt{100 - x^2}; -10 \leq x \leq 10$

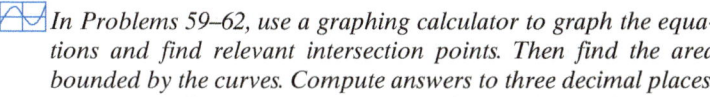 *In Problems 59–62, use a graphing calculator to graph the equations and find relevant intersection points. Then find the area bounded by the curves. Compute answers to three decimal places.*

59. $y = 3 - 5x - 2x^2; y = 2x^2 + 3x - 2$

60. $y = 3 - 2x^2; y = 2x^2 - 4x$

61. $y = -0.5x + 2.25; y = \dfrac{1}{x}$

62. $y = x - 4.25; y = -\dfrac{1}{x}$

C

In Problems 63–68, find the area bounded by the graphs of the indicated equations over the given intervals (when stated). Compute answers to three decimal places.

63. $y = e^x; y = e^{-x}; 0 \leq x \leq 4$

64. $y = e^x; y = -e^{-x}; 1 \leq x \leq 2$

65. $y = x^3; y = 4x$

66. $y = x^3 + 1; y = x + 1$

67. $y = x^3 - 3x^2 - 9x + 12; y = x + 12$

68. $y = x^3 - 6x^2 + 9x; y = x$

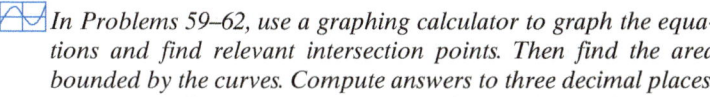 *In Problems 69–74, use a graphing calculator to graph the equations and find relevant intersection points. Then find the area bounded by the curves. Compute answers to three decimal places.*

69. $y = x^3 - x^2 + 2; y = -x^3 + 8x - 2$

70. $y = 2x^3 + 2x^2 - x; y = -2x^3 - 2x^2 + 2x$

71. $y = e^{-x}; y = 3 - 2x$

72. $y = 2 - (x + 1)^2; y = e^{x+1}$

73. $y = e^x; y = 5x - x^3$

74. $y = 2 - e^x; y = x^3 + 3x^2$

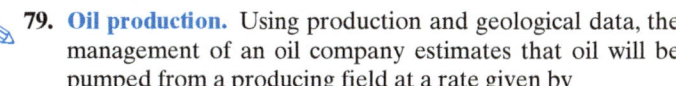 *In Problems 75–78, use a numerical integration routine on a graphing calculator to find the area bounded by the graphs of the indicated equations over the given interval (when stated). Compute answers to three decimal places.*

75. $y = e^{-x}; y = \sqrt{\ln x}; 2 \leq x \leq 5$

76. $y = x^2 + 3x + 1; y = e^{e^x}; -3 \leq x \leq 0$

77. $y = e^{x^2}; y = x + 2$

78. $y = \ln(\ln x); y = 0.01x$

Applications

In the applications that follow, it is helpful to sketch graphs to get a clearer understanding of each problem and to interpret results. A graphing calculator will prove useful if you have one, but it is not necessary.

79. Oil production. Using production and geological data, the management of an oil company estimates that oil will be pumped from a producing field at a rate given by

$$R(t) = \frac{100}{t + 10} + 10 \qquad 0 \leq t \leq 15$$

where $R(t)$ is the rate of production (in thousands of barrels per year) t years after pumping begins. Find the area between the graph of R and the t axis over the interval $[5, 10]$ and interpret the results.

80. Oil production. In Problem 79, if the rate is found to be

$$R(t) = \frac{100t}{t^2 + 25} + 4 \qquad 0 \leq t \leq 25$$

then find the area between the graph of R and the t axis over the interval $[5, 15]$ and interpret the results.

81. Useful life. An amusement company maintains records for each video game it installs in an arcade. Suppose that $C(t)$ and $R(t)$ represent the total accumulated costs and revenues (in thousands of dollars), respectively, t years after a particular game has been installed. If

$$C'(t) = 2 \qquad \text{and} \qquad R'(t) = 9e^{-0.3t}$$

then find the area between the graphs of C' and R' over the interval on the t axis from 0 to the useful life of the game and interpret the results.

82. Useful life. Repeat Problem 81 if

$$C'(t) = 2t \qquad \text{and} \qquad R'(t) = 5te^{-0.1t^2}$$

83. Income distribution. In a study on the effects of World War II on the U.S. economy, an economist used data from the U.S. Census Bureau to produce the following Lorenz curves for the distribution of U.S. income in 1935 and in 1947:

$$f(x) = x^{2.4} \qquad \text{Lorenz curve for 1935}$$

$$g(x) = x^{1.6} \qquad \text{Lorenz curve for 1947}$$

Find the Gini index of income concentration for each Lorenz curve and interpret the results.

84. Income distribution. Using data from the U.S. Census Bureau, an economist produced the following Lorenz curves for the distribution of U.S. income in 1962 and in 1972:

$$f(x) = \tfrac{3}{10}x + \tfrac{7}{10}x^2 \quad \text{Lorenz curve for 1962}$$

$$g(x) = \tfrac{1}{2}x + \tfrac{1}{2}x^2 \quad \text{Lorenz curve for 1972}$$

Find the Gini index of income concentration for each Lorenz curve and interpret the results.

85. Distribution of wealth. Lorenz curves also can provide a relative measure of the distribution of a country's total assets. Using data in a report by the U.S. Congressional Joint Economic Committee, an economist produced the following Lorenz curves for the distribution of total U.S. assets in 1963 and in 1983:

$$f(x) = x^{10} \quad \text{Lorenz curve for 1963}$$

$$g(x) = x^{12} \quad \text{Lorenz curve for 1983}$$

Find the Gini index of income concentration for each Lorenz curve and interpret the results.

86. Income distribution. The government of a small country is planning sweeping changes in the tax structure in order to provide a more equitable distribution of income. The Lorenz curves for the current income distribution and for the projected income distribution after enactment of the tax changes are as follows:

$$f(x) = x^{2.3} \quad \text{Current Lorenz curve}$$

$$g(x) = 0.4x + 0.6x^2 \quad \substack{\text{Projected Lorenz curve after} \\ \text{changes in tax laws}}$$

Find the Gini index of income concentration for each Lorenz curve. Will the proposed changes provide a more equitable income distribution? Explain.

87. Distribution of wealth. The data in the following table describe the distribution of wealth in a country:

x	0	0.20	0.40	0.60	0.80	1
y	0	0.12	0.31	0.54	0.78	1

(A) Use quadratic regression to find the equation of a Lorenz curve for the data.

(B) Use the regression equation and a numerical integration routine to approximate the Gini index of income concentration.

88. Distribution of wealth. Refer to Problem 87.

(A) Use cubic regression to find the equation of a Lorenz curve for the data.

(B) Use the cubic regression equation you found in Part (A) and a numerical integration routine to approximate the Gini index of income concentration.

89. Biology. A yeast culture is growing at a rate of $W'(t) = 0.3e^{0.1t}$ grams per hour. Find the area between the graph of W' and the t axis over the interval $[0, 10]$ and interpret the results.

90. Natural resource depletion. The instantaneous rate of change in demand for U.S. lumber since 1970 ($t = 0$), in billions of cubic feet per year, is given by

$$Q'(t) = 12 + 0.006t^2 \quad 0 \le t \le 50$$

Find the area between the graph of Q' and the t axis over the interval $[15, 20]$, and interpret the results.

91. Learning. A college language class was chosen for a learning experiment. Using a list of 50 words, the experiment measured the rate of vocabulary memorization at different times during a continuous 5-hour study session. The average rate of learning for the entire class was inversely proportional to the time spent studying and was given approximately by

$$V'(t) = \frac{15}{t} \quad 1 \le t \le 5$$

Find the area between the graph of V' and the t axis over the interval $[2, 4]$, and interpret the results.

92. Learning. Repeat Problem 91 if $V'(t) = 13/t^{1/2}$ and the interval is changed to $[1, 4]$.

Answers to Matched Problems

1. $A = \int_{-1}^{3} (x^2 + 1)\, dx = \tfrac{40}{3} \approx 13.333$

2. (A) $A = \int_{0}^{2} (9 - x^2)\, dx = \tfrac{46}{3} \approx 15.333$

   (B) $A = \int_{2}^{3} (9 - x^2)\, dx + \int_{3}^{4} (x^2 - 9)\, dx = 6$

3. $A = \int_{-1}^{2}\left[(x^2 - 1) - \left(-\dfrac{x}{2} - 3\right)\right] dx = \dfrac{39}{4} = 9.75$

4. $A = \int_{-3}^{2}[(6 - x^2) - x]\, dx = \tfrac{125}{6} \approx 20.833$

5. $A = \int_{-2}^{1}[(4 - 2x) - 2x^2]\, dx$
   $+ \int_{1}^{2}[2x^2 - (4 - 2x)]\, dx = \tfrac{38}{3} \approx 12.667$

6. 0.443

7. Gini index of income concentration ≈ 0.583; income will be less equally distributed in 2025.

8-2   Applications in Business and Economics

- Probability Density Functions
- Continuous Income Stream
- Future Value of a Continuous Income Stream
- Consumers' and Producers' Surplus

This section contains important applications of the definite integral to business and economics. Included are three independent topics: probability density functions, continuous income streams, and consumers' and producers' surplus. Any of the three may be covered in any order as time and interests dictate.

Probability Density Functions

We now take a brief, informal look at the use of the definite integral to determine probabilities. A more formal treatment of the subject requires the use of the special "improper" integral form $\int_{-\infty}^{\infty} f(x)\, dx$, which we will not discuss.

Suppose that an experiment is designed in such a way that any real number x on the interval $[c, d]$ is a possible outcome. For example, x may represent an IQ score, the height of a person in inches, or the life of a lightbulb in hours. Technically, we refer to x as a *continuous random variable*.

In certain situations, we can find a function f with x as an independent variable such that the function f can be used to determine the probability that the outcome x of an experiment will be in the interval $[c, d]$. Such a function, called a **probability density function**, must satisfy the following three conditions (see Fig. 1):

1. $f(x) \geq 0$ for all real x.
2. The area under the graph of $f(x)$ over the interval $(-\infty, \infty)$ is exactly 1.
3. If $[c, d]$ is a subinterval of $(-\infty, \infty)$, then

$$\text{Probability } (c \leq x \leq d) = \int_{c}^{d} f(x)\, dx$$

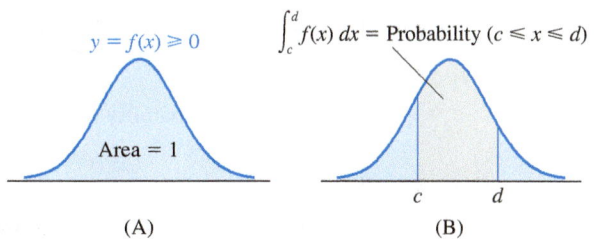

Figure 1 Probability density function

EXAMPLE 1   **Duration of Telephone Calls**   Suppose that the length of telephone calls (in minutes) is a continuous random variable with the probability density function shown in Figure 2:

$$f(t) = \begin{cases} \frac{1}{4} e^{-t/4} & \text{if } t \geq 0 \\ 0 & \text{otherwise} \end{cases}$$

(A) Determine the probability that a call selected at random will last between 2 and 3 minutes.

(B) Find b (to two decimal places) so that the probability of a call selected at random lasting between 2 and b minutes is .5.

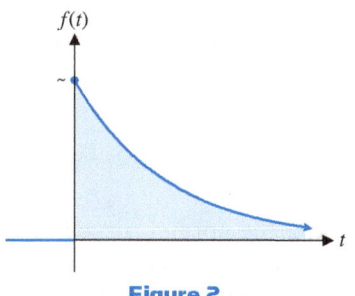

Figure 2

SOLUTION   (A) $\text{Probability } (2 \leq t \leq 3) = \int_{2}^{3} \frac{1}{4} e^{-t/4}\, dt$

$$= (-e^{-t/4})\big|_{2}^{3}$$

$$= -e^{-3/4} + e^{-1/2} \approx .13$$

(B) We want to find b such that Probability $(2 \leq t \leq b) = .5$.

$$\int_2^b \tfrac{1}{4} e^{-t/4} \, dt = .5$$

$$-e^{-b/4} + e^{-1/2} = .5 \qquad\qquad \textit{Solve for b.}$$

$$e^{-b/4} = e^{-.5} - .5$$

$$-\frac{b}{4} = \ln(e^{-.5} - .5)$$

$$b = 8.96 \text{ minutes}$$

So the probability of a call selected at random lasting from 2 to 8.96 minutes is .5.

Matched Problem 1

(A) In Example 1, find the probability that a call selected at random will last 4 minutes or less.

(B) Find b (to two decimal places) so that the probability of a call selected at random lasting b minutes or less is .9

CONCEPTUAL INSIGHT

The probability that a phone call in Example 1 lasts exactly 2 minutes (not 1.999 minutes, not 1.999 999 minutes) is given by

$$\text{Probability } (2 \leq t \leq 2) = \int_2^2 \tfrac{1}{4} e^{-t/4} \, dt \quad \textit{Use Property 1, Section 6-4}$$

$$= 0$$

In fact, for any *continuous* random variable x with probability density function $f(x)$, the probability that x is exactly equal to a constant c is equal to 0:

$$\text{Probability } (c \leq x \leq c) = \int_c^c f(x) \, dx \quad \textit{Use Property 1, Section 6-4}$$

$$= 0$$

In this respect, a *continuous* random variable differs from a *discrete* random variable. If x, for example, is the discrete random variable that represents the number of dots that appear on the top face when a fair die is rolled, then

$$\text{Probability } (2 \leq x \leq 2) = \tfrac{1}{6}$$

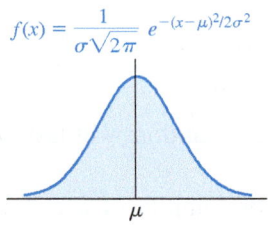

$$f(x) = \frac{1}{\sigma\sqrt{2\pi}} \, e^{-(x-\mu)^2/2\sigma^2}$$

Figure 3 Normal curve

One of the most important probability density functions, the **normal probability density function**, is defined as follows and graphed in Figure 3:

$$f(x) = \frac{1}{\sigma\sqrt{2\pi}} e^{-(x-\mu)^2/2\sigma^2} \quad \begin{array}{l} \boldsymbol{\mu} \textit{ is the mean.} \\ \boldsymbol{\sigma} \textit{ is the standard deviation.} \end{array}$$

It can be shown (but not easily) that the area under the normal curve in Figure 3 over the interval $(-\infty, \infty)$ is exactly 1. Since $\int e^{-x^2} \, dx$ is nonintegrable in terms of elementary functions (that is, the antiderivative cannot be expressed as a finite combination of simple functions), probabilities such as

$$\text{Probability } (c \leq x \leq d) = \frac{1}{\sigma\sqrt{2\pi}} \int_c^d e^{-(x-\mu)^2/2\sigma^2} \, dx$$

can be determined by making an appropriate substitution in the integrand and then using a table of areas under the standard normal curve (that is, the normal curve with $\mu = 0$ and $\sigma = 1$). As an alternative to a table, calculators and computers can be used to compute areas under normal curves.

Continuous Income Stream

We start with a simple example having an obvious solution and generalize the concept to examples having less obvious solutions.

Suppose that an aunt has established a trust that pays you $2,000 a year for 10 years. What is the total amount you will receive from the trust by the end of the 10th year? Since there are 10 payments of $2,000 each, you will receive

$$10 \times \$2,000 = \$20,000$$

We now look at the same problem from a different point of view. Let's assume that the income stream is continuous at a rate of $2,000 per year. In Figure 4, the area under the graph of $f(t) = 2,000$ from 0 to t represents the income accumulated t years after the start. For example, for $t = \frac{1}{4}$ year, the income would be $\frac{1}{4}(2,000) = \$500$; for $t = \frac{1}{2}$ year, the income would be $\frac{1}{2}(2,000) = \$1,000$; for $t = 1$ year, the income would be $1(2,000) = \$2,000$; for $t = 5.3$ years, the income would be $5.3(2,000) = \$10,600$; and for $t = 10$ years, the income would be $10(2,000) = \$20,000$. The total income over a 10-year period—that is, the area under the graph of $f(t) = 2,000$ from 0 to 10—is also given by the definite integral

$$\int_0^{10} 2,000 \, dt = 2,000t\Big|_0^{10} = 2,000(10) - 2,000(0) = \$20,000$$

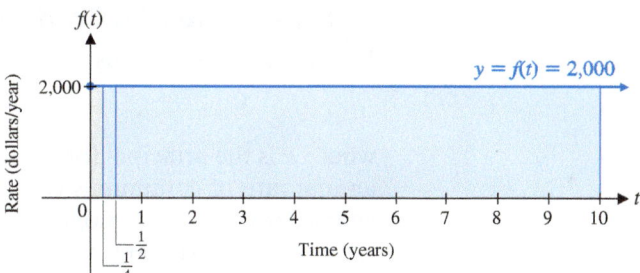

Figure 4 Continuous income stream

EXAMPLE 2   **Continuous Income Stream** The rate of change of the income produced by a vending machine is given by

$$f(t) = 5,000e^{0.04t}$$

where t is time in years since the installation of the machine. Find the total income produced by the machine during the first 5 years of operation.

SOLUTION   The area under the graph of the rate-of-change function from 0 to 5 represents the total change in income over the first 5 years (Fig. 5), and is given by a definite integral:

$$\text{Total income} = \int_0^5 5,000e^{0.04t} \, dt$$

$$= 125,000e^{0.04t}\Big|_0^5$$

$$= 125,000e^{0.04(5)} - 125,000e^{0.04(0)}$$

$$= 152,675 - 125,000$$

$$= \$27,675 \qquad \text{Rounded to the nearest dollar}$$

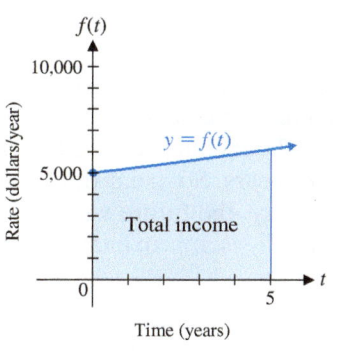

Figure 5 Continuous income stream

The vending machine produces a total income of $27,675 during the first 5 years of operation.

Matched Problem 2 Referring to Example 2, find the total income produced (to the nearest dollar) during the second 5 years of operation.

In reality, income from a vending machine is not usually received as a single payment at the end of each year, even though the rate is given as a yearly rate. Income is usually collected on a daily or weekly basis. In problems of this type, it is convenient to assume that income is actually received in a **continuous stream**; that is, we assume that the rate at which income is received is a continuous function of time. The rate of change is called the **rate of flow** of the continuous income stream. In general, we have the following definition:

DEFINITION Total Income for a Continuous Income Stream

If $f(t)$ is the rate of flow of a continuous income stream, then the **total income** produced during the period from $t = a$ to $t = b$ is

$$\text{Total income} = \int_a^b f(t)\, dt$$

Future Value of a Continuous Income Stream

In Section 4-1, we discussed the continuous compound interest formula

$$A = Pe^{rt}$$

where P is the principal (or present value), A is the amount (or future value), r is the annual rate of continuous compounding (expressed as a decimal), and t is time in years. For example, if money is worth 12% compounded continuously, then the future value of a \$10,000 investment in 5 years is (to the nearest dollar)

$$A = 10{,}000e^{0.12(5)} = \$18{,}221$$

We want to apply the future value concept to the income produced by a continuous income stream. Suppose that $f(t)$ is the rate of flow of a continuous income stream, and the income produced by this continuous income stream is invested as soon as it is received at a rate r, compounded continuously. We already know how to find the total income produced after T years, but how can we find the total of the income produced and the interest earned by this income? Since the income is received in a continuous flow, we cannot just use the formula $A = Pe^{rt}$. This formula is valid only for a single deposit P, not for a continuous flow of income. Instead, we use a Riemann sum approach that will allow us to apply the formula $A = Pe^{rt}$ repeatedly. To begin, we divide the time interval $[0, T]$ into n equal subintervals of length Δt and choose an arbitrary point c_k in each subinterval, as shown in Figure 6.

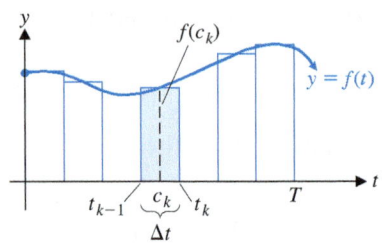

Figure 6

The total income produced during the period from $t = t_{k-1}$ to $t = t_k$ is equal to the area under the graph of $f(t)$ over this subinterval and is approximately equal to $f(c_k)\, \Delta t$, the area of the shaded rectangle in Figure 6. The income received during this period will earn interest for approximately $T - c_k$ years. So, from the future-value formula $A = Pe^{rt}$ with $P = f(c_k)\, \Delta t$ and $t = T - c_k$, the future value of the income produced during the period from $t = t_{k-1}$ to $t = t_k$ is approximately equal to

$$f(c_k)\, \Delta t\, e^{(T - c_k)r}$$

The total of these approximate future values over n subintervals is then

$$f(c_1)\,\Delta t\,e^{(T-c_1)r} + f(c_2)\,\Delta t\,e^{(T-c_2)r} + \cdots + f(c_n)\,\Delta t\,e^{(T-c_n)r} = \sum_{k=1}^{n} f(c_k)e^{r(T-c_k)}\,\Delta t$$

This equation has the form of a Riemann sum, the limit of which is a definite integral. (See the definition of the definite integral in Section 7-4.) Therefore, the *future value FV* of the income produced by the continuous income stream is given by

$$FV = \int_0^T f(t)e^{r(T-t)}\,dt$$

Since r and T are constants, we also can write

$$FV = \int_0^T f(t)e^{rT}e^{-rt}\,dt = e^{rT}\int_0^T f(t)e^{-rt}\,dt \qquad (1)$$

This last form is preferable, since the integral is usually easier to evaluate than the first form.

DEFINITION Future Value of a Continuous Income Stream

If $f(t)$ is the rate of flow of a continuous income stream, $0 \le t \le T$, and if the income is continuously invested at a rate r, compounded continuously, then the **future value FV** at the end of T years is given by

$$FV = \int_0^T f(t)e^{r(T-t)}\,dt = e^{rT}\int_0^T f(t)e^{-rt}\,dt$$

The future value of a continuous income stream is the total value of all money produced by the continuous income stream (income and interest) at the end of T years.

We return to the trust that your aunt set up for you. Suppose that the $2,000 per year you receive from the trust is invested as soon as it is received at 8%, compounded continuously. We consider the trust income to be a continuous income stream with a flow rate of $2,000 per year. What is its future value (to the nearest dollar) by the end of the 10th year? Using the definite integral for future value from the preceding box, we have

$$FV = e^{rT}\int_0^T f(t)e^{-rt}\,dt$$

$$FV = e^{0.08(10)}\int_0^{10} 2{,}000e^{-0.08t}\,dt \qquad r = 0.08,\ T = 10,\ f(t) = 2{,}000$$

$$= 2{,}000e^{0.8}\int_0^{10} e^{-0.08t}\,dt$$

$$= 2{,}000e^{0.8}\left[\frac{e^{-0.08t}}{-0.08}\right]\Big|_0^{10}$$

$$= 2{,}000e^{0.8}[-12.5e^{-0.8} + 12.5] = \$30{,}639$$

At the end of 10 years, you will have received $30,639, including interest. How much is interest? Since you received $20,000 in income from the trust, the interest is the difference between the future value and income. So,

$$\$30{,}639 - \$20{,}000 = \$10{,}639$$

is the interest earned by the income received from the trust over the 10-year period.

EXAMPLE 3   **Future Value of a Continuous Income Stream** Using the continuous income rate of flow for the vending machine in Example 2, namely,

$$f(t) = 5{,}000e^{0.04t}$$

find the future value of this income stream at 12%, compounded continuously for 5 years, and find the total interest earned. Compute answers to the nearest dollar.

SOLUTION   Using the formula

$$FV = e^{rT} \int_0^T f(t)e^{-rt}\, dt$$

with $r = 0.12$, $T = 5$, and $f(t) = 5{,}000e^{0.04t}$, we have

$$FV = e^{0.12(5)} \int_0^5 5{,}000e^{0.04t}e^{-0.12t}\, dt$$

$$= 5{,}000e^{0.6} \int_0^5 e^{-0.08t}\, dt$$

$$= 5{,}000e^{0.6} \left(\frac{e^{-0.08t}}{-0.08}\right)\Big|_0^5$$

$$= 5{,}000e^{0.6}(-12.5e^{-0.4} + 12.5)$$

$$= \$37{,}545 \quad \text{\textit{Rounded to the nearest dollar}}$$

The future value of the income stream at 12% compounded continuously at the end of 5 years is \$37,545.

In Example 2, we saw that the total income produced by this vending machine over a 5-year period was \$27,675. The difference between future value and income is interest. So,

$$\$37{,}545 - \$27{,}675 = \$9{,}870$$

is the interest earned by the income produced by the vending machine during the 5-year period.

Matched Problem 3   Repeat Example 3 if the interest rate is 9%, compounded continuously.

Consumers' and Producers' Surplus

Let $p = D(x)$ be the price–demand equation for a product, where x is the number of units of the product that consumers will purchase at a price of \$$p$ per unit. Suppose that $\bar{p}$, is the current price and $\bar{x}$ is the number of units that can be sold at that price. Then the price–demand curve in Figure 7 shows that if the price is higher than $\bar{p}$, the demand x is less than $\bar{x}$, but some consumers are still willing to pay the higher price. Consumers who are willing to pay more than $\bar{p}$, but who are still able to buy the product at $\bar{p}$, have saved money. We want to determine the total amount saved by all the consumers who are willing to pay a price higher than $\bar{p}$ for the product.

To do this, consider the interval $[c_k, c_k + \Delta x]$, where $c_k + \Delta x < \bar{x}$. If the price remained constant over that interval, the savings on each unit would be the difference between $D(c_k)$, the price consumers are willing to pay, and $\bar{p}$, the price they actually pay. Since Δx represents the number of units purchased by consumers over the interval, the total savings to consumers over this interval is approximately equal to

$$[D(c_k) - \bar{p}]\, \Delta x \quad \text{\textit{(savings per unit)} $\times$ \textit{(number of units)}}$$

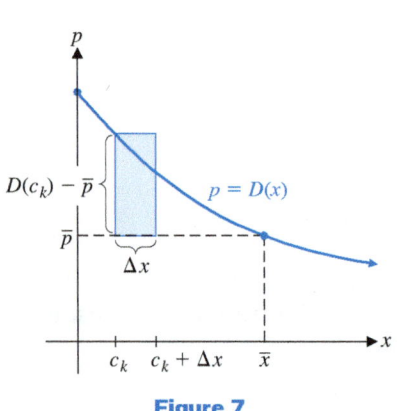

Figure 7

which is the area of the shaded rectangle shown in Figure 7. If we divide the interval $[0, \bar{x}]$ into n equal subintervals, then the total savings to consumers is approximately equal to

$$[D(c_1) - \bar{p}] \, \Delta x + [D(c_2) - \bar{p}] \, \Delta x + \cdots + [D(c_n) - \bar{p}] \, \Delta x = \sum_{k=1}^{n} [D(c_k) - \bar{p}] \, \Delta x$$

which we recognize as a Riemann sum for the integral

$$\int_0^{\bar{x}} [D(x) - \bar{p}] \, dx$$

We define the *consumers' surplus* to be this integral.

Figure 8

DEFINITION   Consumers' Surplus

If $(\bar{x}, \bar{p})$ is a point on the graph of the price–demand equation $p = D(x)$ for a particular product, then the **consumers' surplus CS** at a price level of $\bar{p}$ is

$$CS = \int_0^{\bar{x}} [D(x) - \bar{p}] \, dx$$

which is the area between $p = \bar{p}$ and $p = D(x)$ from $x = 0$ to $x = \bar{x}$, as shown in Figure 8.

The consumers' surplus represents the total savings to consumers who are willing to pay more than $\bar{p}$ for the product but are still able to buy the product for $\bar{p}$.

EXAMPLE 4   **Consumers' Surplus**   Find the consumers' surplus at a price level of \$8 for the price–demand equation

$$p = D(x) = 20 - 0.05x$$

SOLUTION   **Step 1**   Find $\bar{x}$, the demand when the price is $\bar{p} = 8$:

$$\bar{p} = 20 - 0.05\bar{x}$$
$$8 = 20 - 0.05\bar{x}$$
$$0.05\bar{x} = 12$$
$$\bar{x} = 240$$

Figure 9

Step 2   Sketch a graph, as shown in Figure 9.

Step 3   Find the consumers' surplus (the shaded area in the graph):

$$CS = \int_0^{\bar{x}} [D(x) - \bar{p}] \, dx$$
$$= \int_0^{240} (20 - 0.05x - 8) \, dx$$
$$= \int_0^{240} (12 - 0.05x) \, dx$$
$$= (12x - 0.025x^2)\Big|_0^{240}$$
$$= 2{,}880 - 1{,}440 = \$1{,}440$$

The total savings to consumers who are willing to pay a higher price for the product is \$1,440.

Matched Problem 4   Repeat Example 4 for a price level of \$4.

If $p = S(x)$ is the price–supply equation for a product, $\overline{p}$ is the current price, and $\overline{x}$ is the current supply, then some suppliers are still willing to supply some units at a lower price than $\overline{p}$. The additional money that these suppliers gain from the higher price is called the *producers' surplus* and can be expressed in terms of a definite integral (proceeding as we did for the consumers' surplus).

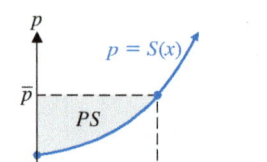

Figure 10

> ### DEFINITION Producers' Surplus
>
> If $(\overline{x}, \overline{p})$ is a point on the graph of the price–supply equation $p = S(x)$, then the **producers' surplus PS** at a price level of $\overline{p}$ is
>
> $$PS = \int_0^{\overline{x}} [\overline{p} - S(x)] \, dx$$
>
> which is the area between $p = \overline{p}$ and $p = S(x)$ from $x = 0$ to $x = \overline{x}$, as shown in Figure 10.
>
> The producers' surplus represents the total gain to producers who are willing to supply units at a lower price than $\overline{p}$ but are still able to supply units at $\overline{p}$.

EXAMPLE 5   **Producers' Surplus** Find the producers' surplus at a price level of $20 for the price–supply equation

$$p = S(x) = 2 + 0.0002x^2$$

SOLUTION   **Step 1** Find $\overline{x}$, the supply when the price is $\overline{p} = 20$:

$$\overline{p} = 2 + 0.0002\overline{x}^2$$
$$20 = 2 + 0.0002\overline{x}^2$$
$$0.0002\overline{x}^2 = 18$$
$$\overline{x}^2 = 90,000$$
$$\overline{x} = 300 \qquad \text{There is only one solution, since } \overline{x} \geq 0.$$

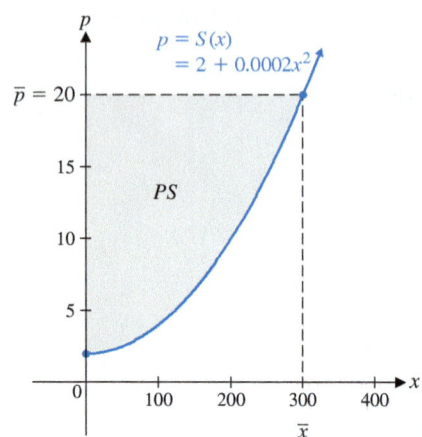

Figure 11

Step 2 Sketch a graph, as shown in Figure 11.

Step 3 Find the producers' surplus (the shaded area in the graph):

$$PS = \int_0^{\overline{x}} [\overline{p} - S(x)] \, dx = \int_0^{300} [20 - (2 + 0.0002x^2)] \, dx$$

$$= \int_0^{300} (18 - 0.0002x^2) \, dx = \left(18x - 0.0002\frac{x^3}{3} \right) \Bigg|_0^{300}$$

$$= 5,400 - 1,800 = \$3,600$$

The total gain to producers who are willing to supply units at a lower price is $3,600.

Matched Problem 5   Repeat Example 5 for a price level of $4.

In a free competitive market, the price of a product is determined by the relationship between supply and demand. If $p = D(x)$ and $p = S(x)$ are the price–demand and price–supply equations, respectively, for a product and if $(\overline{x}, \overline{p})$ is the point of intersection of these equations, then $\overline{p}$ is called the **equilibrium price** and $\overline{x}$ is called the **equilibrium quantity**. If the price stabilizes at the equilibrium price $\overline{p}$, then this is the price level that will determine both the consumers' surplus and the producers' surplus.

EXAMPLE 6   **Equilibrium Price and Consumers' and Producers' Surplus**   Find the equilibrium price and then find the consumers' surplus and producers' surplus at the equilibrium price level, if

$$p = D(x) = 20 - 0.05x \quad \text{and} \quad p = S(x) = 2 + 0.0002x^2$$

SOLUTION   **Step 1**   Find the equilibrium point. Set $D(x)$ equal to $S(x)$ and solve:

$$D(x) = S(x)$$
$$20 - 0.05x = 2 + 0.0002x^2$$
$$0.0002x^2 + 0.05x - 18 = 0$$
$$x^2 + 250x - 90,000 = 0$$
$$x = 200, -450$$

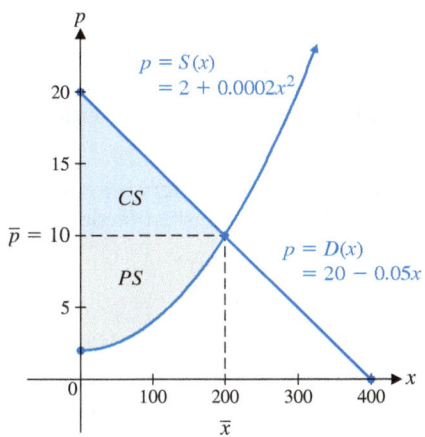

Figure 12

Since x cannot be negative, the only solution is $x = 200$. The equilibrium price can be determined by using $D(x)$ or $S(x)$. We will use both to check our work:

$$\overline{p} = D(200) \qquad\qquad \overline{p} = S(200)$$
$$= 20 - 0.05(200) = 10 \qquad = 2 + 0.0002(200)^2 = 10$$

The equilibrium price is $\overline{p} = 10$, and the equilibrium quantity is $\overline{x} = 200$.

Step 2   Sketch a graph, as shown in Figure 12.

Step 3   Find the consumers' surplus:

$$CS = \int_0^{\overline{x}} [D(x) - \overline{p}]\, dx = \int_0^{200} (20 - 0.05x - 10)\, dx$$
$$= \int_0^{200} (10 - 0.05x)\, dx$$
$$= (10x - 0.025x^2)\big|_0^{200}$$
$$= 2,000 - 1,000 = \$1,000$$

Step 4   Find the producers' surplus:

$$PS = \int_0^{\overline{x}} [\overline{p} - S(x)]\, dx$$
$$= \int_0^{200} [10 - (2 + 0.0002x^2)]\, dx$$
$$= \int_0^{200} (8 - 0.0002x^2)\, dx$$
$$= \left(8x - 0.0002\frac{x^3}{3}\right)\Big|_0^{200}$$
$$= 1,600 - \frac{1,600}{3} \approx \$1,067 \qquad \text{Rounded to the nearest dollar}$$

(A)

```
fnInt(20-.05*X-1
0,X,0,200)
            1000
fnInt(10-(2+.000
2*X^2),X,0,200)
      1066.666667
```

(B)

Figure 13

A graphing calculator offers an alternative approach to finding the equilibrium point for Example 6 (Fig. 13A). A numerical integration command can then be used to find the consumers' and producers' surplus (Fig. 13B).

Matched Problem 6   Repeat Example 6 for

$$p = D(x) = 25 - 0.001x^2 \quad \text{and} \quad p = S(x) = 5 + 0.1x$$

Exercises 8-2

A

In Problems 1–10, evaluate each definite integral to two decimal places.

1. $\displaystyle\int_0^1 e^{-2t}\, dt$

2. $\displaystyle\int_1^3 e^{-t}\, dt$

3. $\displaystyle\int_0^2 e^{4(2-t)}\, dt$

4. $\displaystyle\int_0^1 e^{3(1-t)}\, dt$

5. $\displaystyle\int_0^8 e^{0.06(8-t)}\, dt$

6. $\displaystyle\int_1^{10} e^{0.07(10-t)}\, dt$

7. $\displaystyle\int_0^{20} e^{0.08t}\, e^{0.12(20-t)}\, dt$

8. $\displaystyle\int_0^{15} e^{0.05t}\, e^{0.06(15-t)}\, dt$

9. $\displaystyle\int_0^{30} 500\, e^{0.02t}\, e^{0.09(30-t)}\, dt$

10. $\displaystyle\int_0^{25} 900\, e^{0.03t}\, e^{0.04(25-t)}\, dt$

B

In Problems 11 and 12, explain which of (A), (B), and (C) are equal before evaluating the expressions. Then evaluate each expression to two decimal places.

11. (A) $\displaystyle\int_0^8 e^{0.07(8-t)}\, dt$

   (B) $\displaystyle\int_0^8 (e^{0.56} - e^{0.07t})\, dt$

   (C) $e^{0.56}\displaystyle\int_0^8 e^{-0.07t}\, dt$

12. (A) $\displaystyle\int_0^{10} 2{,}000 e^{0.05t} e^{0.12(10-t)}\, dt$

   (B) $2{,}000 e^{1.2}\displaystyle\int_0^{10} e^{-0.07t}\, dt$

   (C) $2{,}000 e^{0.05}\displaystyle\int_0^{10} e^{0.12(10-t)}\, dt$

Applications

Unless stated to the contrary, compute all monetary answers to the nearest dollar.

13. The life expectancy (in years) of a microwave oven is a continuous random variable with probability density function

$$f(x) = \begin{cases} 2/(x+2)^2 & \text{if } x \ge 0 \\ 0 & \text{otherwise} \end{cases}$$

   (A) Find the probability that a randomly selected microwave oven lasts at most 6 years.

   (B) Find the probability that a randomly selected microwave oven lasts from 6 to 12 years.

   (C) Graph $y = f(x)$ for $[0, 12]$ and show the shaded region for part (A).

14. The shelf life (in years) of a laser pointer battery is a continuous random variable with probability density function

$$f(x) = \begin{cases} 1/(x+1)^2 & \text{if } x \ge 0 \\ 0 & \text{otherwise} \end{cases}$$

   (A) Find the probability that a randomly selected laser pointer battery has a shelf life of 3 years or less.

   (B) Find the probability that a randomly selected laser pointer battery has a shelf life of from 3 to 9 years.

   (C) Graph $y = f(x)$ for $[0, 10]$ and show the shaded region for part (A).

15. In Problem 13, find d so that the probability of a randomly selected microwave oven lasting d years or less is .8.

16. In Problem 14, find d so that the probability of a randomly selected laser pointer battery lasting d years or less is .5.

17. A manufacturer guarantees a product for 1 year. The time to failure of the product after it is sold is given by the probability density function

$$f(t) = \begin{cases} .01e^{-.01t} & \text{if } t \ge 0 \\ 0 & \text{otherwise} \end{cases}$$

   where t is time in months. What is the probability that a buyer chosen at random will have a product failure

   (A) During the warranty period?

   (B) During the second year after purchase?

18. In a certain city, the daily use of water (in hundreds of gallons) per household is a continuous random variable with probability density function

$$f(x) = \begin{cases} .15e^{-.15x} & \text{if } x \ge 0 \\ 0 & \text{otherwise} \end{cases}$$

   Find the probability that a household chosen at random will use

   (A) At most 400 gallons of water per day

   (B) Between 300 and 600 gallons of water per day

19. In Problem 17, what is the probability that the product will last at least 1 year? [*Hint:* Recall that the total area under the probability density function curve is 1.]

20. In Problem 18, what is the probability that a household will use more than 400 gallons of water per day? [See the hint in Problem 19.]

21. Find the total income produced by a continuous income stream in the first 5 years if the rate of flow is $f(t) = 2,500$.

22. Find the total income produced by a continuous income stream in the first 10 years if the rate of flow is $f(t) = 3,000$.

23. Interpret the results of Problem 21 with both a graph and a description of the graph.

24. Interpret the results of Problem 22 with both a graph and a description of the graph.

25. Find the total income produced by a continuous income stream in the first 3 years if the rate of flow is $f(t) = 400e^{0.05t}$.

26. Find the total income produced by a continuous income stream in the first 2 years if the rate of flow is $f(t) = 600e^{0.06t}$.

27. Interpret the results of Problem 25 with both a graph and a description of the graph.

28. Interpret the results of Problem 26 with both a graph and a description of the graph.

29. Starting at age 25, you deposit $2,000 a year into an IRA account. Treat the yearly deposits into the account as a continuous income stream. If money in the account earns 5%, compounded continuously, how much will be in the account 40 years later, when you retire at age 65? How much of the final amount is interest?

30. Suppose in Problem 29 that you start the IRA deposits at age 30, but the account earns 6%, compounded continuously. Treat the yearly deposits into the account as a continuous income stream. How much will be in the account 35 years later when you retire at age 65? How much of the final amount is interest?

31. Find the future value at 3.25% interest, compounded continuously for 4 years, of the continuous income stream with rate of flow $f(t) = 1,650e^{-0.02t}$.

32. Find the future value, at 2.95% interest, compounded continuously for 6 years, of the continuous income stream with rate of flow $f(t) = 2,000e^{0.06t}$.

33. Compute the interest earned in Problem 31.

34. Compute the interest earned in Problem 32.

35. An investor is presented with a choice of two investments: an established clothing store and a new computer store. Each choice requires the same initial investment and each produces a continuous income stream of 4%, compounded continuously. The rate of flow of income from the clothing store is $f(t) = 12,000$, and the rate of flow of income from the computer store is expected to be $g(t) = 10,000e^{0.05t}$. Compare the future values of these investments to determine which is the better choice over the next 5 years.

36. Refer to Problem 35. Which investment is the better choice over the next 10 years?

37. An investor has $10,000 to invest in either a bond that matures in 5 years or a business that will produce a continuous stream of income over the next 5 years with rate of flow $f(t) = 2,150$. If both the bond and the continuous income stream earn 3.75%, compounded continuously, which is the better investment?

38. Refer to Problem 37. Which is the better investment if the rate of the income from the business is $f(t) = 2,250$?

39. A business is planning to purchase a piece of equipment that will produce a continuous stream of income for 8 years with rate of flow $f(t) = 9,000$. If the continuous income stream earns 6.95%, compounded continuously, what single deposit into an account earning the same interest rate will produce the same future value as the continuous income stream? (This deposit is called the **present value** of the continuous income stream.)

40. Refer to Problem 39. Find the present value of a continuous income stream at 7.65%, compounded continuously for 12 years, if the rate of flow is $f(t) = 1,000e^{0.03t}$.

41. Find the future value at a rate r, compounded continuously for T years, of a continuous income stream with rate of flow $f(t) = k$, where k is a constant.

42. Find the future value at a rate r, compounded continuously for T years, of a continuous income stream with rate of flow $f(t) = ke^{ct}$, where c and k are constants, $c \neq r$.

43. Find the consumers' surplus at a price level of $\bar{p} = \$150$ for the price–demand equation

$$p = D(x) = 400 - 0.05x$$

44. Find the consumers' surplus at a price level of $\bar{p} = \$120$ for the price–demand equation

$$p = D(x) = 200 - 0.02x$$

45. Interpret the results of Problem 43 with both a graph and a description of the graph.

46. Interpret the results of Problem 44 with both a graph and a description of the graph.

47. Find the producers' surplus at a price level of $\bar{p} = \$67$ for the price–supply equation

$$p = S(x) = 10 + 0.1x + 0.0003x^2$$

48. Find the producers' surplus at a price level of $\bar{p} = \$55$ for the price–supply equation

$$p = S(x) = 15 + 0.1x + 0.003x^2$$

49. Interpret the results of Problem 47 with both a graph and a description of the graph.

50. Interpret the results of Problem 48 with both a graph and a description of the graph.

In Problems 51–58, find the consumers' surplus and the producers' surplus at the equilibrium price level for the given price–demand and price–supply equations. Include a graph that identifies the consumers' surplus and the producers' surplus. Round all values to the nearest integer.

51. $p = D(x) = 50 - 0.1x; p = S(x) = 11 + 0.05x$

52. $p = D(x) = 25 - 0.004x^2; p = S(x) = 5 + 0.004x^2$

53. $p = D(x) = 80e^{-0.001x}; p = S(x) = 30e^{0.001x}$

54. $p = D(x) = 185e^{-0.005x}; p = S(x) = 25e^{0.005x}$

55. $p = D(x) = 80 - 0.04x; p = S(x) = 30e^{0.001x}$

56. $p = D(x) = 190 - 0.2x; p = S(x) = 25e^{0.005x}$

57. $p = D(x) = 80e^{-0.001x}; p = S(x) = 15 + 0.0001x^2$

58. $p = D(x) = 185e^{-0.005x}; p = S(x) = 20 + 0.002x^2$

59. The following tables give price–demand and price–supply data for the sale of soybeans at a grain market, where x is the number of bushels of soybeans (in thousands of bushels) and p is the price per bushel (in dollars):

Tables for 59–60

Price–Demand		Price–Supply	
x	$p = D(x)$	x	$p = S(x)$
0	6.70	0	6.43
10	6.59	10	6.45
20	6.52	20	6.48
30	6.47	30	6.53
40	6.45	40	6.62

Use quadratic regression to model the price–demand data and linear regression to model the price–supply data.

(A) Find the equilibrium quantity (to three decimal places) and equilibrium price (to the nearest cent).

(B) Use a numerical integration routine to find the consumers' surplus and producers' surplus at the equilibrium price level.

60. Repeat Problem 59, using quadratic regression to model both sets of data.

Answers to Matched Problems

1. (A) .63   (B) 9.21 min

2. $33,803

3. FV = $34,691; interest = $7,016

4. $2,560

5. $133

6. $\bar{p}$ = 15; CS = $667; PS = $500

8-3   Integration by Parts

In Section 7-1, we promised to return later to the indefinite integral

$$\int \ln x \, dx$$

since none of the integration techniques considered up to that time could be used to find an antiderivative for ln x. We now develop a very useful technique, called *integration by parts,* that will enable us to find not only the preceding integral, but also many others, including integrals such as

$$\int x \ln x \, dx \quad \text{and} \quad \int xe^x \, dx$$

The method of integration by parts is based on the product formula for derivatives. If f and g are differentiable functions, then

$$\frac{d}{dx}[f(x)g(x)] = f(x)g'(x) + g(x)f'(x)$$

which can be written in the equivalent form

$$f(x)g'(x) = \frac{d}{dx}[f(x)g(x)] - g(x)f'(x)$$

Integrating both sides, we obtain

$$\int f(x)g'(x) \, dx = \int \frac{d}{dx}[f(x)g(x)] \, dx - \int g(x)f'(x) \, dx$$

The first integral to the right of the equal sign is $f(x)g(x) + C$. Why? We will leave out the constant of integration for now, since we can add it after integrating the second integral to the right of the equal sign. So,

$$\int f(x)g'(x) \, dx = f(x)g(x) - \int g(x)f'(x) \, dx$$

This equation can be transformed into a more convenient form by letting $u = f(x)$ and $v = g(x)$; then $du = f'(x) \, dx$ and $dv = g'(x) \, dx$. Making these substitutions, we obtain the **integration-by-parts formula:**

Integration-by-Parts Formula

$$\int u \, dv = uv - \int v \, du$$

This formula can be very useful when the integral on the left is difficult or impossible to integrate with standard formulas. If u and dv are chosen with care—this is the crucial part of the process—then the integral on the right side may be easier to integrate than the one on the left. The formula provides us with another tool that is helpful in many, but not all, cases. We are able to easily check the results by differentiating to get the original integrand, a good habit to develop.

EXAMPLE 1   **Integration by Parts** Find $\int xe^x \, dx$, using integration by parts, and check the result.

SOLUTION   First, write the integration-by-parts formula:

$$\int u \, dv = uv - \int v \, du \tag{1}$$

Now try to identify u and dv in $\int xe^x \, dx$ so that $\int v \, du$ on the right side of (1) is easier to integrate than $\int u \, dv = \int xe^x \, dx$ on the left side. There are essentially two reasonable choices in selecting u and dv in $\int xe^x \, dx$:

$$
\begin{array}{cc}
\text{Choice 1} & \text{Choice 2} \\
u \quad dv & u \quad dv \\
\int x \, e^x \, dx & \int e^x \, x \, dx
\end{array}
$$

We pursue choice 1 and leave choice 2 for you to explore (see Explore & Discuss 1 following this example).

From choice 1, $u = x$ and $dv = e^x \, dx$. Looking at formula (1), we need du and v to complete the right side. Let

$$u = x \qquad dv = e^x \, dx$$

Then,

$$du = dx \qquad \int dv = \int e^x \, dx$$

$$v = e^x$$

Any constant may be added to v, but we will always choose 0 for simplicity. The general arbitrary constant of integration will be added at the end of the process. Substituting these results into formula (1), we obtain

$$\int u \, dv = uv - \int v \, du$$

$$\int xe^x \, dx = xe^x - \int e^x \, dx \qquad \text{The right integral is easy to integrate.}$$

$$= xe^x - e^x + C \qquad \text{Now add the arbitrary constant C.}$$

Check:

$$\frac{d}{dx}(xe^x - e^x + C) = xe^x + e^x - e^x = xe^x$$

Pursue choice 2 in Example 1, using the integration-by-parts formula, and explain why this choice does not work out.

Matched Problem 1   Find $\int xe^{2x}\,dx$.

EXAMPLE 2   **Integration by Parts**   Find $\int x \ln x\,dx$.

SOLUTION   As before, we have essentially two choices in choosing u and dv:

$$
\begin{array}{cc}
\text{Choice 1} & \text{Choice 2} \\
\overset{u}{\underbrace{}}\ \overset{dv}{\underbrace{}} & \overset{u}{\underbrace{}}\ \overset{dv}{\underbrace{}} \\
\int x\ \ln x\,dx & \int \ln x\ x\,dx
\end{array}
$$

Choice 1 is rejected since we do not yet know how to find an antiderivative of $\ln x$. So we move to choice 2 and choose $u = \ln x$ and $dv = x\,dx$. Then we proceed as in Example 1. Let

$$u = \ln x \qquad dv = x\,dx$$

Then,

$$du = \frac{1}{x}dx \qquad \int dv = \int x\,dx$$

$$v = \frac{x^2}{2}$$

Substitute these results into the integration-by-parts formula:

$$\int u\,dv = uv - \int v\,du$$

$$\int x \ln x\,dx = (\ln x)\left(\frac{x^2}{2}\right) - \int\left(\frac{x^2}{2}\right)\left(\frac{1}{x}\right)dx$$

$$= \frac{x^2}{2}\ln x - \int \frac{x}{2}dx \qquad \textcolor{blue}{\textit{An easy integral to evaluate}}$$

$$= \frac{x^2}{2}\ln x - \frac{x^2}{4} + C$$

Check:

$$\frac{d}{dx}\left(\frac{x^2}{2}\ln x - \frac{x^2}{4} + C\right) = x \ln x + \left(\frac{x^2}{2}\cdot\frac{1}{x}\right) - \frac{x}{2} = x \ln x$$

Matched Problem 2   Find $\int x \ln 2x\,dx$.

CONCEPTUAL INSIGHT

As you may have discovered in Explore & Discuss 1, some choices for u and dv will lead to integrals that are more complicated than the original integral. This does not mean that there is an error in either the calculations or the integration-by-parts formula. It simply means that the particular choice of u and dv does not change the problem into one we can solve. When this happens, we must look for a different choice of u and dv. In some problems, it is possible that no choice will work.

Guidelines for selecting u and dv for integration by parts are summarized in the following box:

SUMMARY Integration by Parts: Selection of u and dv

For $\int u\ dv = uv - \int v\ du$,

1. The product $u\ dv$ must equal the original integrand.
2. It must be possible to integrate dv (preferably by using standard formulas or simple substitutions).
3. The new integral $\int v\ du$ should not be more complicated than the original integral $\int u\ dv$.
4. For integrals involving $x^p e^{ax}$, try

$$u = x^p \quad \text{and} \quad dv = e^{ax}\ dx$$

5. For integrals involving $x^p (\ln x)^q$, try

$$u = (\ln x)^q \quad \text{and} \quad dv = x^p\ dx$$

In some cases, repeated use of the integration-by-parts formula will lead to the evaluation of the original integral. The next example provides an illustration of such a case.

EXAMPLE 3 **Repeated Use of Integration by Parts** Find $\int x^2 e^{-x}\ dx$.

SOLUTION Following suggestion 4 in the box, we choose

$$u = x^2 \qquad dv = e^{-x}\ dx$$

Then,

$$du = 2x\ dx \qquad v = -e^{-x}$$

and

$$x^2 e^{-x}\ dx = x^2(-e^{-x}) - (-e^{-x})2x\ dx$$
$$= -x^2 e^{-x} + 2\ xe^{-x}\ dx \tag{2}$$

The new integral is not one we can evaluate by standard formulas, but it is simpler than the original integral. Applying the integration-by-parts formula to it will produce an even simpler integral. For the integral $\int xe^{-x}\ dx$, we choose

$$u = x \qquad dv = e^{-x}\ dx$$

Then,

$$du = dx \qquad v = -e^{-x}$$

and

$$\int xe^{-x}\ dx = x(-e^{-x}) - \int (-e^{-x})\ dx$$
$$= -xe^{-x} + \int e^{-x}\ dx$$
$$= -xe^{-x} - e^{-x} \qquad \text{\color{blue}Choose 0 for the constant.} \tag{3}$$

Substituting equation (3) into equation (2), we have

$$\int x^2 e^{-x}\ dx = -x^2 e^{-x} + 2(-xe^{-x} - e^{-x}) + C \qquad \text{\color{blue}Add an arbitrary constant here.}$$

$$= -x^2 e^{-x} - 2xe^{-x} - 2e^{-x} + C$$

Check:

$$\frac{d}{dx}(-x^2e^{-x} - 2xe^{-x} - 2e^{-x} + C) = x^2e^{-x} - 2xe^{-x} + 2xe^{-x} - 2e^{-x} + 2e^{-x}$$

$$= x^2e^{-x}$$

Matched Problem 3 Find $\int x^2 e^{2x}\, dx$.

EXAMPLE 4 **Using Integration by Parts** Find $\int_1^e \ln x\, dx$ and interpret the result geometrically.

SOLUTION First, we find $\int \ln x\, dx$. Then we return to the definite integral. Following suggestion 5 in the box (with $p = 0$), we choose

$$u = \ln x \qquad dv = dx$$

Then,

$$du = \frac{1}{x}dx \qquad v = x$$

$$\int \ln x\, dx = (\ln x)(x) - \int (x)\frac{1}{x}dx$$

$$= x \ln x - x + C$$

This is the important result we mentioned at the beginning of this section. Now we have

$$\int_1^e \ln x\, dx = (x \ln x - x)\Big|_1^e$$

$$= (e \ln e - e) - (1 \ln 1 - 1)$$

$$= (e - e) - (0 - 1)$$

$$= 1$$

The integral represents the area under the curve $y = \ln x$ from $x = 1$ to $x = e$, as shown in Figure 1.

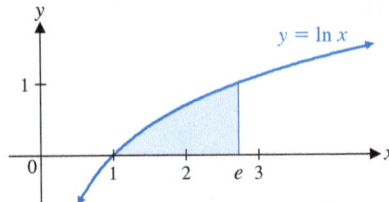

Figure 1

Matched Problem 4 Find $\int_1^2 \ln 3x\, dx$.

EXPLORE & DISCUSS 2 Try using the integration-by-parts formula on $\int e^{x^2}\, dx$, and explain why it does not work.

Exercises 8-3

A

In Problems 1–4, integrate by parts. Assume that $x > 0$ whenever the natural logarithm function is involved.

1. $\int xe^{3x}\, dx$

2. $\int xe^{4x}\, dx$

3. $\int x^2 \ln x\, dx$

4. $\int x^3 \ln x\, dx$

B

5. If you want to use integration by parts to find $\int (x + 1)^5(x + 2)\, dx$, which is the better choice for u: $u = (x + 1)^5$ or $u = x + 2$? Explain your choice and then integrate.

6. If you want to use integration by parts to find $\int (5x - 7)(x - 1)^4\, dx$, which is the better choice for u: $u = 5x - 7$ or $u = (x - 1)^4$? Explain your choice and then integrate.

Problems 7–20 are mixed—some require integration by parts, and others can be solved with techniques considered earlier. Integrate as indicated, assuming $x > 0$ whenever the natural logarithm function is involved.

7. $\displaystyle\int xe^{-x}\,dx$

8. $\displaystyle\int (x-1)e^{-x}\,dx$

9. $\displaystyle\int xe^{x^2}\,dx$

10. $\displaystyle\int xe^{-x^2}\,dx$

11. $\displaystyle\int_0^1 (x-3)e^x\,dx$

12. $\displaystyle\int_0^1 (x+1)e^x\,dx$

13. $\displaystyle\int_1^3 \ln 2x\,dx$

14. $\displaystyle\int_1^2 \ln\left(\frac{x}{2}\right)dx$

15. $\displaystyle\int \frac{2x}{x^2+1}\,dx$

16. $\displaystyle\int \frac{x^2}{x^3+5}\,dx$

17. $\displaystyle\int \frac{\ln x}{x}\,dx$

18. $\displaystyle\int \frac{e^x}{e^x+1}\,dx$

19. $\displaystyle\int \sqrt{x}\ln x\,dx$

20. $\displaystyle\int \frac{\ln x}{\sqrt{x}}\,dx$

In Problems 21–24, the integral can be found in more than one way. First use integration by parts, then use a method that does not involve integration by parts. Which method do you prefer?

21. $\displaystyle\int (x-3)(x+1)^2\,dx$

22. $\displaystyle\int (x+2)(x-1)^2\,dx$

23. $\displaystyle\int (2x+1)(x-2)^2\,dx$

24. $\displaystyle\int (5x-1)(x+2)^2\,dx$

In Problems 25–28, illustrate each integral graphically and describe what the integral represents in terms of areas.

25. Problem 11

26. Problem 12

27. Problem 13

28. Problem 14

C

Problems 29–50 are mixed—some may require use of the integration-by-parts formula along with techniques we have considered earlier; others may require repeated use of the integration-by-parts formula. Assume that $g(x) > 0$ whenever $\ln g(x)$ is involved.

29. $\displaystyle\int x^2 e^x\,dx$

30. $\displaystyle\int x^3 e^x\,dx$

31. $\displaystyle\int xe^{ax}\,dx, a \neq 0$

32. $\displaystyle\int \ln(ax)\,dx, a > 0$

33. $\displaystyle\int_1^e \frac{\ln x}{x^2}\,dx$

34. $\displaystyle\int_1^2 x^3 e^{x^2}\,dx$

35. $\displaystyle\int_0^2 \ln(x+4)\,dx$

36. $\displaystyle\int_0^2 \ln(4-x)\,dx$

37. $\displaystyle\int xe^{x-2}\,dx$

38. $\displaystyle\int xe^{x+1}\,dx$

39. $\displaystyle\int x\ln(1+x^2)\,dx$

40. $\displaystyle\int x\ln(1+x)\,dx$

41. $\displaystyle\int e^x \ln(1+e^x)\,dx$

42. $\displaystyle\int \frac{\ln(1+\sqrt{x})}{\sqrt{x}}\,dx$

43. $\displaystyle\int (\ln x)^2\,dx$

44. $\displaystyle\int x(\ln x)^2\,dx$

45. $\displaystyle\int (\ln x)^3\,dx$

46. $\displaystyle\int x(\ln x)^3\,dx$

47. $\displaystyle\int_1^e \ln(x^2)\,dx$

48. $\displaystyle\int_1^e \ln(x^4)\,dx$

49. $\displaystyle\int_0^1 \ln(e^{x^2})\,dx$

50. $\displaystyle\int_1^2 \ln(xe^x)\,dx$

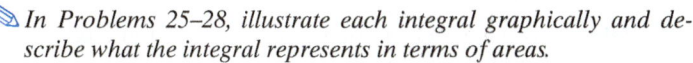 *In Problems 51–54, use a graphing calculator to graph each equation over the indicated interval and find the area between the curve and the x axis over that interval. Find answers to two decimal places.*

51. $y = x - 2 - \ln x; 1 \le x \le 4$

52. $y = 6 - x^2 - \ln x; 1 \le x \le 4$

53. $y = 5 - xe^x; 0 \le x \le 3$

54. $y = xe^x + x - 6; 0 \le x \le 3$

Applications

55. Profit. If the marginal profit (in millions of dollars per year) is given by

$$P'(t) = 2t - te^{-t}$$

use an appropriate definite integral to find the total profit (to the nearest million dollars) earned over the first 5 years of operation.

56. Production. An oil field is estimated to produce oil at a rate of $R(t)$ thousand barrels per month t months from now, as given by

$$R(t) = 10te^{-0.1t}$$

Use an appropriate definite integral to find the total production (to the nearest thousand barrels) in the first year of operation.

57. Profit. Interpret the results of Problem 55 with both a graph and a description of the graph.

58. Production. Interpret the results of Problem 56 with both a graph and a description of the graph.

59. Continuous income stream. Find the future value at 3.95%, compounded continuously, for 5 years of a continuous income stream with a rate of flow of

$$f(t) = 1,000 - 200t$$

60. Continuous income stream. Find the interest earned at 4.15%, compounded continuously, for 4 years for a continuous income stream with a rate of flow of

$$f(t) = 1,000 - 250t$$

61. Income distribution. Find the Gini index of income concentration for the Lorenz curve with equation

$$y = xe^{x-1}$$

62. Income distribution. Find the Gini index of income concentration for the Lorenz curve with equation

$$y = x^2 e^{x-1}$$

63. Income distribution. Interpret the results of Problem 61 with both a graph and a description of the graph.

64. Income distribution. Interpret the results of Problem 62 with both a graph and a description of the graph.

65. Sales analysis. Monthly sales of a particular personal computer are expected to decline at the rate of

$$S'(t) = -4te^{0.1t}$$

computers per month, where t is time in months and $S(t)$ is the number of computers sold each month. The company plans to stop manufacturing this computer when monthly sales reach 800 computers. If monthly sales now ($t = 0$) are 2,000 computers, find $S(t)$. How long, to the nearest month, will the company continue to manufacture the computer?

66. Sales analysis. The rate of change of the monthly sales of a new basketball game is given by

$$S'(t) = 350 \ln(t + 1) \qquad S(0) = 0$$

where t is the number of months since the game was released and $S(t)$ is the number of games sold each month. Find $S(t)$. When, to the nearest month, will monthly sales reach 15,000 games?

67. Consumers' surplus. Find the consumers' surplus (to the nearest dollar) at a price level of $\bar{p} = \$2.089$ for the price–demand equation

$$p = D(x) = 9 - \ln(x + 4)$$

Use $\bar{x}$ computed to the nearest higher unit.

68. Producers' surplus. Find the producers' surplus (to the nearest dollar) at a price level of $\bar{p} = \$26$ for the price–supply equation

$$p = S(x) = 5 \ln(x + 1)$$

Use $\bar{x}$ computed to the nearest higher unit.

69. Consumers' surplus. Interpret the results of Problem 67 with both a graph and a description of the graph.

70. Producers' surplus. Interpret the results of Problem 68 with both a graph and a description of the graph.

71. Pollution. The concentration of particulate matter (in parts per million) t hours after a factory ceases operation for the day is given by

$$C(t) = \frac{20 \ln(t + 1)}{(t + 1)^2}$$

Find the average concentration for the period from $t = 0$ to $t = 5$.

72. Medicine. After a person takes a pill, the drug contained in the pill is assimilated into the bloodstream. The rate of assimilation t minutes after taking the pill is

$$R(t) = te^{-0.2t}$$

Find the total amount of the drug that is assimilated into the bloodstream during the first 10 minutes after the pill is taken.

73. Learning. A student enrolled in an advanced typing class progressed at a rate of

$$N'(t) = (t + 6)e^{-0.25t}$$

words per minute per week t weeks after enrolling in a 15-week course. If a student could type 40 words per minute at the beginning of the course, then how many words per minute $N(t)$ would the student be expected to type t weeks into the course? How long, to the nearest week, should it take the student to achieve the 70-word-per-minute level? How many words per minute should the student be able to type by the end of the course?

74. Learning. A student enrolled in a stenotyping class progressed at a rate of

$$N'(t) = (t + 10)e^{-0.1t}$$

words per minute per week t weeks after enrolling in a 15-week course. If a student had no knowledge of stenotyping (that is, if the student could stenotype at 0 words per minute) at the beginning of the course, then how many words per minute $N(t)$ would the student be expected to handle t weeks into the course? How long, to the nearest week, should it take the student to achieve 90 words per minute? How many words per minute should the student be able to handle by the end of the course?

75. Politics. The number of voters (in thousands) in a certain city is given by

$$N(t) = 20 + 4t - 5te^{-0.1t}$$

where t is time in years. Find the average number of voters during the period from $t = 0$ to $t = 5$.

Answers to Matched Problems

1. $\dfrac{x}{2}e^{2x} - \dfrac{1}{4}e^{2x} + C$

2. $\dfrac{x^2}{2}\ln 2x - \dfrac{x^2}{4} + C$

3. $\dfrac{x^2}{2}e^{2x} - \dfrac{x}{2}e^{2x} + \dfrac{1}{4}e^{2x} + C$

4. $2 \ln 6 - \ln 3 - 1 \approx 1.4849$

8-4   Integration Using Tables

- Using a Table of Integrals
- Substitution and Integral Tables
- Reduction Formulas
- Application

A **table of integrals** is a list of integration formulas used to evaluate integrals. Table II of Appendix C on pages 630–632 contains a list of integral formulas. Some of these formulas can be derived with the integration techniques discussed earlier, while others require techniques we have not considered. However, it is possible to verify each formula by differentiating the right side.

Using a Table of Integrals

The formulas in Table II on pages 630–632 are organized by categories, such as "Integrals Involving $a + bu$," "Integrals Involving $\sqrt{u^2 - a^2}$," and so on. The variable u is the variable of integration. All other symbols represent constants. To use a table to evaluate an integral, you must first find the category that most closely agrees with the form of the integrand and then find a formula in that category that you can make to match the integrand exactly by assigning values to the constants in the formula.

EXAMPLE 1   **Integration Using Tables**   Use Table II to find

$$\int \frac{x}{(5 + 2x)(4 - 3x)}\,dx$$

SOLUTION   Since the integrand

$$f(x) = \frac{x}{(5 + 2x)(4 - 3x)}$$

is a rational function involving terms of the form $a + bu$ and $c + du$, we examine formulas 15 to 20 in Table II on page 630 to see if any of the integrands in these formulas can be made to match $f(x)$ exactly. Comparing the integrand in formula 16 with $f(x)$, we see that this integrand will match $f(x)$ if we let $a = 5, b = 2, c = 4$, and $d = -3$. Letting $u = x$ and substituting for a, b, c, and d in formula 16, we have

$$\int \frac{u}{(a + bu)(c + du)}\,du = \frac{1}{ad - bc}\left(\frac{a}{b}\ln|a + bu| - \frac{c}{d}\ln|c + du|\right) \quad \text{Formula 16}$$

$$\int \frac{x}{\underset{a\quad\ b\quad c\quad d}{(5 + 2x)(4 - 3x)}}\,dx = \frac{1}{\underset{a\cdot d - b\cdot c = 5\cdot(-3) - 2\cdot 4 = -23}{5\cdot(-3) - 2\cdot 4}}\left(\frac{5}{2}\ln|5 + 2x| - \frac{4}{-3}\ln|4 - 3x|\right) + C$$

$$= -\tfrac{5}{46}\ln|5 + 2x| - \tfrac{4}{69}\ln|4 - 3x| + C$$

Notice that the constant of integration, C, is not included in any of the formulas in Table II. However, you must still include C in all antiderivatives.

Matched Problem 1   Use Table II to find $\displaystyle\int \frac{1}{(5 + 3x)^2(1 + x)}\,dx$.

EXAMPLE 2   **Integration Using Tables**   Evaluate $\displaystyle\int_3^4 \frac{1}{x\sqrt{25 - x^2}}\,dx$.

SOLUTION   First, we use Table II to find

$$\int \frac{1}{x\sqrt{25 - x^2}}\,dx$$

Since the integrand involves the expression $\sqrt{25 - x^2}$, we examine formulas 29 to 31 in Table II and select formula 29 with $a^2 = 25$ and $a = 5$:

$$\int \frac{1}{u\sqrt{a^2 - u^2}}\, du = -\frac{1}{a}\ln\left|\frac{a + \sqrt{a^2 - u^2}}{u}\right| \qquad \text{Formula 29}$$

$$\int \frac{1}{x\sqrt{25 - x^2}}\, dx = -\frac{1}{5}\ln\left|\frac{5 + \sqrt{25 - x^2}}{x}\right| + C$$

So

$$\int_3^4 \frac{1}{x\sqrt{25 - x^2}}\, dx = -\frac{1}{5}\ln\left|\frac{5 + \sqrt{25 - x^2}}{x}\right|\Bigg|_3^4$$

$$= -\frac{1}{5}\ln\left|\frac{5 + 3}{4}\right| + \frac{1}{5}\ln\left|\frac{5 + 4}{3}\right|$$

$$= -\tfrac{1}{5}\ln 2 + \tfrac{1}{5}\ln 3 = \tfrac{1}{5}\ln 1.5 \approx 0.0811$$

Matched Problem 2   Evaluate $\displaystyle\int_6^8 \frac{1}{x^2\sqrt{100 - x^2}}\, dx$.

Substitution and Integral Tables

As Examples 1 and 2 illustrate, if the integral we want to evaluate can be made to match one in the table exactly, then evaluating the indefinite integral consists simply of substituting the correct values of the constants into the formula. But what happens if we cannot match an integral with one of the formulas in the table? In many cases, a substitution will change the given integral into one that corresponds to a table entry.

EXAMPLE 3   **Integration Using Substitution and Tables** Find $\displaystyle\int \frac{x^2}{\sqrt{16x^2 - 25}}\, dx$.

SOLUTION   In order to relate this integral to one of the formulas involving $\sqrt{u^2 - a^2}$ (formulas 40 to 45 in Table II), we observe that if $u = 4x$, then

$$u^2 = 16x^2 \qquad \text{and} \qquad \sqrt{16x^2 - 25} = \sqrt{u^2 - 25}$$

So, we will use the substitution $u = 4x$ to change this integral into one that appears in the table:

$$\int \frac{x^2}{\sqrt{16x^2 - 25}}\, dx = \frac{1}{4}\int \frac{\frac{1}{16}u^2}{\sqrt{u^2 - 25}}\, du \qquad \text{\color{blue}Substitution:}$$
$$\qquad\qquad\qquad\qquad\qquad \text{\color{blue}$u = 4x$, $du = 4\,dx$, $x = \tfrac{1}{4}u$}$$

$$= \frac{1}{64}\int \frac{u^2}{\sqrt{u^2 - 25}}\, du$$

This last integral can be evaluated with the aid of formula 44 in Table II with $a = 5$:

$$\int \frac{u^2}{\sqrt{u^2 - a^2}}\, du = \frac{1}{2}\left(u\sqrt{u^2 - a^2} + a^2\ln\left|u + \sqrt{u^2 - a^2}\right|\right) \qquad \text{Formula 44}$$

$$\int \frac{x^2}{\sqrt{16x^2 - 25}}\, dx = \frac{1}{64}\int \frac{u^2}{\sqrt{u^2 - 25}}\, du \qquad \text{\color{blue}Use formula 44 with $a = 5$.}$$

$$= \tfrac{1}{128}\left(u\sqrt{u^2 - 25} + 25\ln\left|u + \sqrt{u^2 - 25}\right|\right) + C \quad \text{\color{blue}Substitute $u = 4x$.}$$

$$= \tfrac{1}{128}\left(4x\sqrt{16x^2 - 25} + 25\ln\left|4x + \sqrt{16x^2 - 25}\right|\right) + C$$

Matched Problem 3   Find $\int \sqrt{9x^2 - 16} \, dx$.

EXAMPLE 4   **Integration Using Substitution and Tables** Find $\int \dfrac{x}{\sqrt{x^4 + 1}} \, dx$.

SOLUTION   None of the formulas in Table II involve fourth powers; however, if we let $u = x^2$, then

$$\sqrt{x^4 + 1} = \sqrt{u^2 + 1}$$

and this form does appear in formulas 32 to 39. Thus, we substitute $u = x^2$:

$$\int \frac{1}{\sqrt{x^4 + 1}} x \, dx = \frac{1}{2} \int \frac{1}{\sqrt{u^2 + 1}} \, du \qquad \begin{array}{l} \text{Substitution:} \\ u = x^2, \, du = 2x \, dx \end{array}$$

We recognize the last integral as formula 36 with $a = 1$:

$$\int \frac{1}{\sqrt{u^2 + a^2}} \, du = \ln|u + \sqrt{u^2 + a^2}| \qquad \text{Formula 36}$$

$$\int \frac{x}{\sqrt{x^4 + 1}} \, dx = \frac{1}{2} \int \frac{1}{\sqrt{u^2 + 1}} \, du \qquad \text{Use formula 36 with } a = 1.$$

$$= \tfrac{1}{2} \ln|u + \sqrt{u^2 + 1}| + C \qquad \text{Substitute } u = x^2.$$

$$= \tfrac{1}{2} \ln|x^2 + \sqrt{x^4 + 1}| + C$$

Matched Problem 4   Find $\int x\sqrt{x^4 + 1} \, dx$.

Reduction Formulas

EXAMPLE 5   **Using Reduction Formulas** Use Table II to find $\int x^2 e^{3x} \, dx$.

SOLUTION   Since the integrand involves the function e^{3x}, we examine formulas 46–48 and conclude that formula 47 can be used for this problem. Letting $u = x$, $n = 2$, and $a = 3$ in formula 47, we have

$$\int u^n e^{au} \, du = \frac{u^n e^{au}}{a} - \frac{n}{a} \int u^{n-1} e^{au} \, du \qquad \text{Formula 47}$$

$$\int x^2 e^{3x} \, dx = \frac{x^2 e^{3x}}{3} - \frac{2}{3} \int x e^{3x} \, dx$$

Notice that the expression on the right still contains an integral, but the exponent of x has been reduced by 1. Formulas of this type are called **reduction formulas** and are designed to be applied repeatedly until an integral that can be evaluated is obtained. Applying formula 47 to $\int x e^{3x} \, dx$ with $n = 1$, we have

$$\int x^2 e^{3x} \, dx = \frac{x^2 e^{3x}}{3} - \frac{2}{3} \left(\frac{x e^{3x}}{3} - \frac{1}{3} \int e^{3x} \, dx \right)$$

$$= \frac{x^2 e^{3x}}{3} - \frac{2x e^{3x}}{9} + \frac{2}{9} \int e^{3x} \, dx$$

This last expression contains an integral that is easy to evaluate:

$$\int e^{3x} \, dx = \tfrac{1}{3} e^{3x}$$

After making a final substitution and adding a constant of integration, we have

$$\int x^2 e^{3x} \, dx = \frac{x^2 e^{3x}}{3} - \frac{2x e^{3x}}{9} + \frac{2}{27} e^{3x} + C$$

Matched Problem 5   Use Table II to find $\int (\ln x)^2\, dx$.

Application

EXAMPLE 6   **Producers' Surplus** Find the producers' surplus at a price level of $20 for the price–supply equation

$$p = S(x) = \frac{5x}{500 - x}$$

SOLUTION   **Step 1** Find $\bar{x}$, the supply when the price is $\bar{p} = 20$:

$$\bar{p} = \frac{5\bar{x}}{500 - \bar{x}}$$

$$20 = \frac{5\bar{x}}{500 - \bar{x}}$$

$$10{,}000 - 20\bar{x} = 5\bar{x}$$

$$10{,}000 = 25\bar{x}$$

$$\bar{x} = 400$$

Step 2 Sketch a graph, as shown in Figure 1.

Figure 1

Step 3 Find the producers' surplus (the shaded area of the graph):

$$PS = \int_0^{\bar{x}} [\bar{p} - S(x)]\, dx$$

$$= \int_0^{400} \left(20 - \frac{5x}{500 - x} \right) dx$$

$$= \int_0^{400} \frac{10{,}000 - 25x}{500 - x}\, dx$$

Use formula 20 with $a = 10{,}000$, $b = -25$, $c = 500$, and $d = -1$:

$$\int \frac{a + bu}{c + du}\, du = \frac{bu}{d} + \frac{ad - bc}{d^2} \ln|c + du| \qquad \text{Formula 20}$$

$$PS = \left. (25x + 2{,}500 \ln|500 - x|) \right|_0^{400}$$

$$= 10{,}000 + 2{,}500 \ln|100| - 2{,}500 \ln|500|$$

$$\approx \$5{,}976$$

Matched Problem 6   Find the consumers' surplus at a price level of $10 for the price–demand equation

$$p = D(x) = \frac{20x - 8{,}000}{x - 500}$$

Exercises 8-4

A

Use Table II on pages 630–632 to find each indefinite integral in Problems 1–14.

1. $\int \dfrac{1}{x(1 + x)} dx$

2. $\int \dfrac{1}{x^2(1 + x)} dx$

3. $\int \dfrac{1}{(3 + x)^2(5 + 2x)} dx$

4. $\int \dfrac{x}{(5 + 2x)^2(2 + x)} dx$

5. $\int \dfrac{x}{\sqrt{16 + x}} dx$

6. $\int \dfrac{1}{x\sqrt{16 + x}} dx$

7. $\int \dfrac{1}{x\sqrt{1 - x^2}} dx$

8. $\int \dfrac{\sqrt{9 - x^2}}{x} dx$

9. $\int \dfrac{1}{x\sqrt{x^2 + 4}} dx$

10. $\int \dfrac{1}{x^2\sqrt{x^2 - 16}} dx$

11. $\int x^2 \ln x \, dx$

12. $\int x^3 \ln x \, dx$

13. $\int \dfrac{1}{1 + e^x} dx$

14. $\int \dfrac{1}{5 + 2e^{3x}} dx$

Evaluate each definite integral in Problems 15–20. Use Table II on pages 630–632 to find the antiderivative.

15. $\int_1^3 \dfrac{x^2}{3 + x} dx$

16. $\int_2^6 \dfrac{x}{(6 + x)^2} dx$

17. $\int_0^7 \dfrac{1}{(3 + x)(1 + x)} dx$

18. $\int_0^7 \dfrac{x}{(3 + x)(1 + x)} dx$

19. $\int_0^4 \dfrac{1}{\sqrt{x^2 + 9}} dx$

20. $\int_4^5 \sqrt{x^2 - 16} \, dx$

B

In Problems 21–32, use substitution techniques and Table II to find each indefinite integral.

21. $\int \dfrac{\sqrt{4x^2 + 1}}{x^2} dx$

22. $\int x^2\sqrt{9x^2 - 1} \, dx$

23. $\int \dfrac{x}{\sqrt{x^4 - 16}} dx$

24. $\int x\sqrt{x^4 - 16} \, dx$

25. $\int x^2\sqrt{x^6 + 4} \, dx$

26. $\int \dfrac{x^2}{\sqrt{x^6 + 4}} dx$

27. $\int \dfrac{1}{x^3\sqrt{4 - x^4}} dx$

28. $\int \dfrac{\sqrt{x^4 + 4}}{x} dx$

29. $\int \dfrac{e^x}{(2 + e^x)(3 + 4e^x)} dx$

30. $\int \dfrac{e^x}{(4 + e^x)^2(2 + e^x)} dx$

31. $\int \dfrac{\ln x}{x\sqrt{4 + \ln x}} dx$

32. $\int \dfrac{1}{x \ln x\sqrt{4 + \ln x}} dx$

C

In Problems 33–38, use Table II to find each indefinite integral.

33. $\int x^2 e^{5x} \, dx$

34. $\int x^2 e^{-4x} \, dx$

35. $\int x^3 e^{-x} \, dx$

36. $\int x^3 e^{2x} \, dx$

37. $\int (\ln x)^3 \, dx$

38. $\int (\ln x)^4 \, dx$

Problems 39–46 are mixed—some require the use of Table II, and others can be solved with techniques considered earlier.

39. $\int_3^5 x\sqrt{x^2 - 9} \, dx$

40. $\int_3^5 x^2\sqrt{x^2 - 9} \, dx$

41. $\int_2^4 \dfrac{1}{x^2 - 1} dx$

42. $\int_2^4 \dfrac{x}{(x^2 - 1)^2} dx$

43. $\int \dfrac{\ln x}{x^2} dx$

44. $\int \dfrac{(\ln x)^2}{x} dx$

45. $\int \dfrac{x}{\sqrt{x^2 - 1}} dx$

46. $\int \dfrac{x^2}{\sqrt{x^2 - 1}} dx$

In Problems 47–50, find the area bounded by the graphs of $y = f(x)$ and $y = g(x)$ to two decimal places. Use a graphing calculator to approximate intersection points to two decimal places.

47. $f(x) = \dfrac{10}{\sqrt{x^2 + 1}}; g(x) = x^2 + 3x$

48. $f(x) = \sqrt{1 + x^2}; g(x) = 5x - x^2$

49. $f(x) = x\sqrt{4 + x}; g(x) = 1 + x$

50. $f(x) = \dfrac{x}{\sqrt{x + 4}}; g(x) = x - 2$

Applications

Use Table II to evaluate all integrals involved in any solutions of Problems 51–74.

51. Consumers' surplus. Find the consumers' surplus at a price level of $\overline{p} = \$15$ for the price–demand equation

$$p = D(x) = \dfrac{7,500 - 30x}{300 - x}$$

52. Producers' surplus. Find the producers' surplus at a price level of $\overline{p} = \$20$ for the price–supply equation

$$p = S(x) = \dfrac{10x}{300 - x}$$

53. Consumers' surplus. Graph the price–demand equation and the price-level equation $\overline{p} = 15$ of Problem 51 in the same coordinate system. What region represents the consumers' surplus?

54. Producers' surplus. Graph the price–supply equation and the price-level equation $\overline{p} = 20$ of Problem 52 in the same coordinate system. What region represents the producers' surplus?

55. Cost. A company manufactures downhill skis. It has fixed costs of $25,000 and a marginal cost given by

$$C'(x) = \dfrac{250 + 10x}{1 + 0.05x}$$

where $C(x)$ is the total cost at an output of x pairs of skis. Find the cost function $C(x)$ and determine the production level (to the nearest unit) that produces a cost of $150,000. What is the cost (to the nearest dollar) for a production level of 850 pairs of skis?

56. Cost. A company manufactures a portable DVD player. It has fixed costs of $11,000 per week and a marginal cost given by

$$C'(x) = \dfrac{65 + 20x}{1 + 0.4x}$$

where $C(x)$ is the total cost per week at an output of x players per week. Find the cost function $C(x)$ and determine the production level (to the nearest unit) that produces a cost of $52,000 per week. What is the cost (to the nearest dollar) for a production level of 700 players per week?

57. Continuous income stream. Find the future value at 4.4%, compounded continuously, for 10 years for the continuous income stream with rate of flow $f(t) = 50t^2$.

58. Continuous income stream. Find the interest earned at 3.7%, compounded continuously, for 5 years for the continuous income stream with rate of flow $f(t) = 200t$.

59. Income distribution. Find the Gini index of income concentration for the Lorenz curve with equation

$$y = \tfrac{1}{2}x\sqrt{1 + 3x}$$

60. Income distribution. Find the Gini index of income concentration for the Lorenz curve with equation

$$y = \tfrac{1}{2}x^2\sqrt{1 + 3x}$$

61. Income distribution. Graph $y = x$ and the Lorenz curve of Problem 59 over the interval $[0, 1]$. Discuss the effect of the area bounded by $y = x$ and the Lorenz curve getting smaller relative to the equitable distribution of income.

62. Income distribution. Graph $y = x$ and the Lorenz curve of Problem 60 over the interval $[0, 1]$. Discuss the effect of the area bounded by $y = x$ and the Lorenz curve getting larger relative to the equitable distribution of income.

63. Marketing. After test marketing a new high-fiber cereal, the market research department of a major food producer estimates that monthly sales (in millions of dollars) will grow at the monthly rate of

$$S'(t) = \dfrac{t^2}{(1 + t)^2}$$

t months after the cereal is introduced. If we assume 0 sales at the time the cereal is introduced, find $S(t)$, the total sales, t months after the cereal is introduced. Find the total sales during the first 2 years that the cereal is on the market.

64. Average price. At a discount department store, the price–demand equation for premium motor oil is given by

$$p = D(x) = \dfrac{50}{\sqrt{100 + 6x}}$$

where x is the number of cans of oil that can be sold at a price of $\$p$. Find the average price over the demand interval $[50, 250]$.

65. Marketing. For the cereal of Problem 63, show the sales over the first 2 years geometrically, and describe the geometric representation.

66. Price–demand. For the motor oil of Problem 64, graph the price–demand equation and the line representing the average price in the same coordinate system over the interval [50, 250]. Describe how the areas under the two curves over the interval [50, 250] are related.

 67. Profit. The marginal profit for a small car agency that sells x cars per week is given by

$$P'(x) = x\sqrt{2 + 3x}$$

where $P(x)$ is the profit in dollars. The agency's profit on the sale of only 1 car per week is $-\$2,000$. Find the profit function and the number of cars that must be sold (to the nearest unit) to produce a profit of $\$13,000$ per week. How much weekly profit (to the nearest dollar) will the agency have if 80 cars are sold per week?

 68. Revenue. The marginal revenue for a company that manufactures and sells x graphing calculators per week is given by

$$R'(x) = \frac{x}{\sqrt{1 + 2x}} \qquad R(0) = 0$$

where $R(x)$ is the revenue in dollars. Find the revenue function and the number of calculators that must be sold (to the nearest unit) to produce $\$10,000$ in revenue per week. How much weekly revenue (to the nearest dollar) will the company have if 1,000 calculators are sold per week?

69. Pollution. An oil tanker is producing an oil slick that is radiating outward at a rate given approximately by

$$\frac{dR}{dt} = \frac{100}{\sqrt{t^2 + 9}} \qquad t \geq 0$$

where R is the radius (in feet) of the circular slick after t minutes. Find the radius of the slick after 4 minutes if the radius is 0 when $t = 0$.

70. Pollution. The concentration of particulate matter (in parts per million) during a 24-hour period is given approximately by

$$C(t) = t\sqrt{24 - t} \qquad 0 \leq t \leq 24$$

where t is time in hours. Find the average concentration during the period from $t = 0$ to $t = 24$.

71. Learning. A person learns N items at a rate given approximately by

$$N'(t) = \frac{60}{\sqrt{t^2 + 25}} \qquad t \geq 0$$

where t is the number of hours of continuous study. Determine the total number of items learned in the first 12 hours of continuous study.

72. Politics. The number of voters (in thousands) in a metropolitan area is given approximately by

$$f(t) = \frac{500}{2 + 3e^{-t}} \qquad t \geq 0$$

where t is time in years. Find the average number of voters during the period from $t = 0$ to $t = 10$.

 73. Learning. Interpret Problem 71 geometrically. Describe the geometric interpretation.

74. Politics. For the voters of Problem 72, graph $y = f(t)$ and the line representing the average number of voters over the interval [0, 10] in the same coordinate system. Describe how the areas under the two curves over the interval [0, 10] are related.

Answers to Matched Problems

1. $\frac{1}{2}\left(\dfrac{1}{5 + 3x}\right) + \dfrac{1}{4}\ln\left|\dfrac{1 + x}{5 + 3x}\right| + C$

2. $\frac{7}{1,200} \approx 0.0058$

3. $\frac{1}{6}(3x\sqrt{9x^2 - 16} - 16\ln|3x + \sqrt{9x^2 - 16}|) + C$

4. $\frac{1}{4}(x^2\sqrt{x^4 + 1} + \ln|x^2 + \sqrt{x^4 + 1}|) + C$

5. $x(\ln x)^2 - 2x\ln x + 2x + C$

6. $3,000 + 2,000\ln 200 - 2,000\ln 500 \approx \$1,167$

Chapter 8   Review

Important Terms, Symbols, and Concepts

8-1 Area Between Curves

EXAMPLES

- If f and g are continuous and $f(x) \geq g(x)$ over the interval $[a, b]$, then the area bounded by $y = f(x)$ and $y = g(x)$ for $a \leq x \leq b$ is given exactly by

$$A = \int_a^b [f(x) - g(x)]\,dx$$

- A graphical representation of the distribution of income among a population can be found by plotting data points (x, y), where **x represents the cumulative percentage of families at or below a given income level** and **y represents the cumulative percentage of total family income received.** Regression analysis can be used to find a particular function $y = f(x)$, called a **Lorenz curve,** that best fits the data.

- A single number, the **Gini index,** measures income concentration:

$$\text{Gini index} = 2\int_0^1 [x - f(x)]\,dx$$

A Gini index of 0 indicates **absolute equality:** All families share equally in the income. A Gini index of 1 indicates **absolute inequality:** One family has all of the income and the rest have none.

Ex. 1, p. 482
Ex. 2, p. 482
Ex. 3, p. 483
Ex. 4, p. 483
Ex. 5, p. 484
Ex. 6, p. 484

Ex. 7, p. 486

8-2 Applications in Business and Economics

- *Probability Density Functions.* If any real number x in an interval is a possible outcome of an experiment, then x is said to be a **continuous random variable.** The probability distribution of a continuous random variable is described by a **probability density function** f that satisfies the following conditions:
  1. $f(x) \geq 0$ for all real x.
  2. The area under the graph of $f(x)$ over the interval $(-\infty, \infty)$ is exactly 1.
  3. If $[c, d]$ is a subinterval of $(-\infty, \infty)$, then

$$\text{Probability} \quad (c \leq x \leq d) = \int_c^d f(x)\, dx$$

- *Continuous Income Stream* If the rate at which income is received—its **rate of flow**—is a continuous function $f(t)$ of time, then the income is said to be a **continuous income stream.** The **total income** produced by a continuous income stream from $t = a$ to $t = b$ is

$$\text{Total income} = \int_a^b f(t)\, dt$$

The **future value** of a continuous income stream that is invested at rate r, compounded continuously, for $0 \leq t \leq T$, is

$$FV = \int_0^T f(t)e^{r(T-t)}\, dt$$

- *Consumers' and Producers' Surplus* If $(\bar{x}, \bar{p})$ is a point on the graph of a price–demand equation $p = D(x)$, then the **consumers' surplus** at a price level of $\bar{p}$ is

$$CS = \int_0^{\bar{x}} [D(x) - \bar{p}]\, dx$$

The consumers' surplus represents the total savings to consumers who are willing to pay more than $\bar{p}$ but are still able to buy the product for $\bar{p}$.

Similarly, for a point $(\bar{x}, \bar{p})$ on the graph of a price–supply equation $p = S(x)$, the **producers' surplus** at a price level of $\bar{p}$ is

$$PS = \int_0^{\bar{x}} [\bar{p} - S(x)]\, dx$$

The producers' surplus represents the total gain to producers who are willing to supply units at a lower price $\bar{p}$, but are still able to supply units at $\bar{p}$.

If $(\bar{x}, \bar{p})$ is the intersection point of a price–demand equation $p = D(x)$ and a price–supply equation $p = S(x)$, then $\bar{p}$ is called the **equilibrium price** and $\bar{x}$ is called the **equilibrium quantity**.

8-3 Integration by Parts

- Some indefinite integrals, but not all, can be found by means of the **integration-by-parts formula**

$$\int u\, dv = uv - \int v\, du$$

- Select u and dv with the help of the guidelines in the summary on page 505.

8-4 Integration Using Tables

- A **table of integrals** is a list of integration formulas that can be used to find indefinite or definite integrals of frequently encountered functions. Such a list appears in Table II of Appendix C on pages 630–632.

Review Exercises

Work through all the problems in this chapter review and check your answers in the back of the book. Answers to all review problems are there, along with section numbers in italics to indicate where each type of problem is discussed. Where weaknesses show up, review appropriate sections of the text.

Compute all numerical answers to three decimal places unless directed otherwise.

A

In Problems 1–3, set up definite integrals that represent the shaded areas in the figure over the indicated intervals.

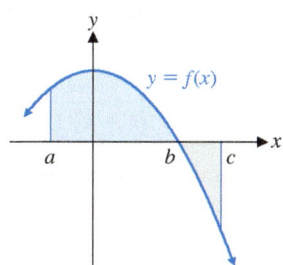

Figure for 1–3

1. Interval $[a, b]$

2. Interval $[b, c]$

3. Interval $[a, c]$

4. Sketch a graph of the area between the graphs of $y = \ln x$ and $y = 0$ over the interval $[0.5, e]$ and find the area.

In Problems 5–10, evaluate each integral.

5. $\int xe^{4x}\, dx$

6. $\int x \ln x\, dx$

7. $\int \dfrac{\ln x}{x}\, dx$

8. $\int \dfrac{x}{1 + x^2}\, dx$

9. $\int \dfrac{1}{x(1 + x)^2}\, dx$

10. $\int \dfrac{1}{x^2\sqrt{1 + x}}\, dx$

In Problems 11–16, find the area bounded by the graphs of the indicated equations over the given interval.

11. $y = 5 - 2x - 6x^2;\ y = 0, 1 \le x \le 2$

12. $y = 5x + 7;\ y = 12, -3 \le x \le 1$

13. $y = -x + 2;\ y = x^2 + 3, -1 \le x \le 4$

14. $y = \dfrac{1}{x};\ y = -e^{-x}, 1 \le x \le 2$

15. $y = x;\ y = -x^3, -2 \le x \le 2$

16. $y = x^2;\ y = -x^4; -2 \le x \le 2$

B

In Problems 17–20, set up definite integrals that represent the shaded areas in the figure over the indicated intervals.

17. Interval $[a, b]$

18. Interval $[b, c]$

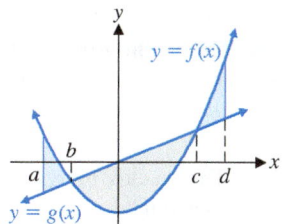

Figure for 17–20

19. Interval $[b, d]$

20. Interval $[a, d]$

21. Sketch a graph of the area bounded by the graphs of $y = x^2 - 6x + 9$ and $y = 9 - x$ and find the area.

In Problems 22–27, evaluate each integral.

22. $\int_0^1 xe^x\, dx$

23. $\int_0^3 \dfrac{x^2}{\sqrt{x^2 + 16}}\, dx$

24. $\int \sqrt{9x^2 - 49}\, dx$

25. $\int te^{-0.5t}\, dt$

26. $\int x^2 \ln x\, dx$

27. $\int \dfrac{1}{1 + 2e^x}\, dx$

28. Sketch a graph of the area bounded by the indicated graphs, and find the area. In part (B), approximate intersection points and area to two decimal places.
    (A) $y = x^3 - 6x^2 + 9x;\ y = x$
     (B) $y = x^3 - 6x^2 + 9x;\ y = x + 1$

C

In Problems 29–36, evaluate each integral.

29. $\int \dfrac{(\ln x)^2}{x}\, dx$

30. $\int x(\ln x)^2\, dx$

31. $\int \dfrac{x}{\sqrt{x^2 - 36}}\, dx$

32. $\int \dfrac{x}{\sqrt{x^4 - 36}}\, dx$

33. $\int_0^4 x \ln(10 - x)\, dx$

34. $\int (\ln x)^2\, dx$

35. $\int xe^{-2x^2}\, dx$

36. $\int x^2 e^{-2x}\, dx$

37. Use a numerical integration routine on a graphing calculator to find the area in the first quadrant that is below the graph of

$$y = \dfrac{6}{2 + 5e^{-x}}$$

and above the graph of $y = 0.2x + 1.6$.

Applications

38. **Product warranty.** A manufacturer warrants a product for parts and labor for 1 year and for parts only for a second year. The time to a failure of the product after it is sold is given by the probability density function

$$f(t) = \begin{cases} 0.21e^{-0.21t} & \text{if } t \ge 0 \\ 0 & \text{otherwise} \end{cases}$$

What is the probability that a buyer chosen at random will have a product failure

(A) During the first year of warranty?

(B) During the second year of warranty?

39. Product warranty. Graph the probability density function for Problem 38 over the interval [0, 3], interpret part (B) of Problem 38 geometrically and describe the geometric representation.

40. Revenue function. The weekly marginal revenue from the sale of x hair dryers is given by

$$R'(x) = 65 - 6\ln(x + 1) \qquad R(0) = 0$$

where $R(x)$ is the revenue in dollars. Find the revenue function and the production level (to the nearest unit) for a revenue of $20,000 per week. What is the weekly revenue (to the nearest dollar) at a production level of 1,000 hair dryers per week?

41. Continuous income stream. The rate of flow (in dollars per year) of a continuous income stream for a 5-year period is given by

$$f(t) = 2{,}500e^{0.05t} \qquad 0 \le t \le 5$$

(A) Graph $y = f(t)$ over $[0, 5]$ and shade the area that represents the total income received from the end of the first year to the end of the fourth year.

(B) Find the total income received, to the nearest dollar, from the end of the first year to the end of the fourth year.

42. Future value of a continuous income stream. The continuous income stream in Problem 41 is invested at 4%, compounded continuously.

(A) Find the future value (to the nearest dollar) at the end of the 5-year period.

(B) Find the interest earned (to the nearest dollar) during the 5-year period.

43. Income distribution. An economist produced the following Lorenz curves for the current income distribution and the projected income distribution 10 years from now in a certain country:

$$f(x) = 0.1x + 0.9x^2 \qquad \textit{Current Lorenz curve}$$

$$g(x) = x^{1.5} \qquad \textit{Projected Lorenz curve}$$

(A) Graph $y = x$ and the current Lorenz curve on one set of coordinate axes for [0, 1] and graph $y = x$ and the projected Lorenz curve on another set of coordinate axes over the same interval.

(B) Looking at the areas bounded by the Lorenz curves and $y = x$, can you say that the income will be more or less equitably distributed 10 years from now?

(C) Compute the Gini index of income concentration (to one decimal place) for the current and projected curves. What can you say about the distribution of income 10 years from now? Is it more equitable or less?

44. Consumers' and producers' surplus. Find the consumers' surplus and the producers' surplus at the equilibrium price level for each pair of price–demand and price–supply equations. Include a graph that identifies the consumers' surplus and the producers' surplus. Round all values to the nearest integer.

(A) $p = D(x) = 70 - 0.2x$;

$p = S(x) = 13 + 0.0012x^2$

(B) $p = D(x) = 70 - 0.2x$; $p = S(x) = 13e^{0.006x}$

45. Producers' surplus. The accompanying table gives price–supply data for the sale of hogs at a livestock market, where x is the number of pounds (in thousands) and p is the price per pound (in cents):

Price–Supply	
x	$p = S(x)$
0	43.50
10	46.74
20	50.05
30	54.72
40	59.18

(A) Using quadratic regression to model the data, find the demand at a price of 52.50 cents per pound.

(B) Use a numerical integration routine to find the producers' surplus (to the nearest dollar) at a price level of 52.50 cents per pound.

46. Drug assimilation. The rate at which the body eliminates a certain drug (in milliliters per hour) is given by

$$R(t) = \frac{60t}{(t + 1)^2(t + 2)}$$

where t is the number of hours since the drug was administered. How much of the drug is eliminated during the first hour after it was administered? During the fourth hour?

47. With the aid of a graphing calculator, illustrate Problem 46 geometrically.

48. Medicine. For a particular doctor, the length of time (in hours) spent with a patient per office visit has the probability density function

$$f(t) = \begin{cases} \dfrac{\frac{4}{3}}{(t + 1)^2} & \text{if } 0 \le t \le 3 \\ 0 & \text{otherwise} \end{cases}$$

(A) What is the probability that this doctor will spend less than 1 hour with a randomly selected patient?

(B) What is the probability that this doctor will spend more than 1 hour with a randomly selected patient?

49. Medicine. Illustrate part (B) in Problem 48 geometrically. Describe the geometric interpretation.

50. Politics. The rate of change of the voting population of a city with respect to time t (in years) is estimated to be

$$N'(t) = \frac{100t}{(1 + t^2)^2}$$

where $N(t)$ is in thousands. If $N(0)$ is the current voting population, how much will this population increase during the next 3 years?

51. Psychology. Rats were trained to go through a maze by rewarding them with a food pellet upon successful completion of the run. After the seventh successful run, the probability density function for length of time (in minutes) until success on the eighth trial was given by

$$f(t) = \begin{cases} .5e^{-.5t} & \text{if } t \ge 0 \\ 0 & \text{otherwise} \end{cases}$$

What is the probability that a rat selected at random after seven successful runs will take 2 or more minutes to complete the eighth run successfully? [Recall that the area under a probability density function curve from $-\infty$ to ∞ is 1.]

CHAPTER 8 Appendix

Applications of Integral Calculus: Gini Index and Streams of Income

1   The Gini Index of Concentration for the Distribution of Income or Wealth

In the last few years, Americans have become more aware of the growing income disparity. We hear that a small percentage, namely, the upper 1% of Americans earn a disproportionately high proportion of the total income in the US. Whereas the lower 99% of Americans earn a much smaller relative proportion of the total income. Since 2005 pre-tax household income earned by the top 1% has exceeded 15% of the total income.

The **Gini Index** is a statistic that provides a useful measure of the inequality in the distribution of income or other resources, such as land and taxes, or wealth in general. It takes into account all percentiles at once, rather than, say, just the 99th percentile. It was introduced by economists C. Gini and M. O. Lorenz in the early 1900's, and is based on the **Lorenz curve** associated with the distribution of income or wealth. Although the discussion that follows is about the Gini Index for income distribution, it applies to the distribution of wealth or other economic or physical resources as well.

Government reports indicate that income inequality has been increasing since the 1970s, whereas it had been declining for the preceding two decades before 1970. Furthermore, as of 2006, the US had one of the highest levels of income inequality, as measured by the Gini Index among developed countries.

1.1 Lorenz Curves

The **Lorenz curve of income distribution** is defined as follows. Consider the population of all household income for a particular country and a specific year that are listed from the lowest income to the highest. For each fraction or percentage x between zero and one, $I(x)$ is the annual household income level at which the fraction of households with income equal or lower than $I(x)$ is equal to x. In other words, $I(x)$ is the $100x^{th}$ percentile of the population of household income. That is

$$I(x) = \begin{array}{l}\text{The household income for which the fraction of households} \\ \text{with income equal or lower than } I(x) \text{ is equal to } x.\end{array}$$

$$= \text{The household income whose percentile is equal to } x.$$

For example,

- $I(.25)$ is the lower quartile or, equivalently, the 25th percentile of household income;
- $I(.50)$ is the median or, equivalently, the 50th percentile of household income;
- $I(.75)$ is the upper quartile or, equivalently, the 75th percentile of household income.

The Lorenz curve of income distribution is the function of the variable x defined by

$$L(x) = \frac{\text{Total income of all households with an income percentile less or equal to } I(x)}{\text{Grand total of all household income}}$$

$$= \text{Proportion of total income earned by those households with income percentile less than } I(x). \tag{1}$$

For example,

- $L(.25)$ is the proportion of total income earned by the lower quartile of the population;
- $L(.50)$ is the proportion of total income earned by the lower half of the population;
- $L(.75)$ is the proportion of total income earned by the lower three quarters of the population.
- $L(x)$ is the proportion of total income earned by the bottom proportion x of the population, for all $0 \leq x \leq 1$.

It is easy to see that the Lorenz curve $y = L(x)$ is a polygonal curve or a continuous curve, depending on whether the distribution of household income is discrete or continuous, respectively. We shall assume that it is a continuous function. It has the following properties:

It is concave up and typically resembles curves like $y = x^2$.;

- It lies below the graph of the line $y = x$;
- It rises from $f(0) = 0$ at $x = 0$ to $f(1) = 1$ at $x = 1$.

Furthermore, the area under the Lorenz curve is less or equal to 0.5. In the sequel, we work with a continuous distribution of income or wealth in order to illustrate an application of integral calculus that helps to understand income inequality or wealth inequality.

1.2 The Gini Index

The Gini Index of income inequality for a Lorenz curve

The Gini Index of income inequality or Gini Index for a Lorenz curve $L(x)$ is twice the area between the line $y = x$ of absolute equality, and the Lorenz curve $y = L(x)$, as shown in the graphic example below. That is

$$\textbf{The Gini Index of Income Inequality} = 2\left[\int_0^1 x\,dx - \int_0^1 L(x)\,dx \right]$$

$$= 2\int_0^1 (x - L(x))\,dx.$$

The Gini index is a measure of income inequality based on the Lorenz curve. Note that the closer to zero the Gini Index is, the more equitable is the distribution. On the other hand, the closer to 1 is the Gini Index, the less equitable is the distribution. We now present two formulas that are helpful in computing the Gini Index and discuss several examples concerning the distribution of income or wealth for the US and other European countries. The interested reader will find lots of articles or references on the Gini Index to explore on the Internet, starting with Wikipedia, where there are links to free software that will compute the Gini Index of any data set.

Area between absolute equality and $f(x) = x^{2.6}$

THEOREM 1

Let $a, b \geq 1$.

(i) The Gini Index for a Lorenz curve $y = L(x) = x^a$ is equal to

$$\frac{a - 1}{a + 1} \tag{2}$$

(ii) The Gini Index for a Lorenz curve $y = L(x) = px^a + qx^b$, a mixture of powers of x, where $p + q = 1$, is equal to

$$\frac{a - 1}{a + 1} p + \frac{b - 1}{b + 1} q. \tag{3}$$

Proof

(i)

$$2 \int_0^1 (x - x^a) \, dx = 2 \left[\frac{x^2}{2} - \frac{x^{a-1}}{a = 1} \right]_{x=0}^{x=1}$$

$$= 2 \left[\frac{1}{2} - \frac{1}{a + 1} \right]$$

$$= 1 - \frac{2}{a + 1} = \frac{a - 1}{a + 1}.$$

(ii) Apply (i) twice.

$$2 \int_0^1 (x - (px^a + qx^b)) \, dx = 2 \int_0^1 (px + qx - (px^a + qx^b)) \, dx$$

$$= 2 \int_0^1 [p(x - x^a) + q(x - x^b)] \, dx$$

$$= 2p \int_0^1 (x - x^a) \, dx + 2q \int_0^1 (x - x^b) \, dx$$

$$= \frac{a - 1}{a = 1} p + \frac{b - 1}{b + 1} q, \text{ by (i)}.$$

Remarks. Clearly, the larger the exponent in a Lorenz curve, which is a power of x as in (1), the more concave the curve, and the larger is the Gini Index. Whereas the closer to 1 is the exponent, the closer is the Lorenz curve to the line $y = x$, and the smaller is the Gini Index. This can easily be seen from the equation $(a - 1)/(a + 1) = 1 - 2/(a + 1)$ for the Gini Index, which increases to 1 as a increases indefinitely and decreases to 0 as a decreases to 1. The same applies to the dependence of the Gini Index in (2) on the exponents a and b, where the Lorenz curve is a mixture of two powers of x.

1.3 Examples of the Gini Index

EXAMPLE 1 The Lorenz for the income distribution of the US for the year 1993 was

$$L(x) = x^{2.4}.$$

The Gini Index is

$$2\int_0^1 (x - x^{2.4})\, dx = 2\left[\frac{1}{2}x^2 - \frac{1}{3.4}x^{3.4}\right]_{x=0}^1 = 2\left[\frac{1}{2} - \frac{1}{3.4}\right] \approx 0.412,$$

or

$$\frac{a-1}{a+1} = \frac{2.4-1}{2.4+1} = \frac{1.4}{3.4} \approx 0.412, \text{ using the formula (2)}$$

EXAMPLE 2 The Lorenz Curve for household income and the Gini Index for Sweden in 1990 were:

$$L(x) = x^{2.1}. \text{ The Gini Index} = \frac{a-1}{a+1} = \frac{11}{31} \approx 0.367, \text{ using (2)},$$

which is lower than the index for the US in example 1.

EXAMPLE 3 The Gini index can be used for wealth distribution instead of household income. The Lorenz curve for wealth distribution in the US for the year 2000 was

$$y = L(x) = x^{6.9}.$$

$$\text{The Gini Index} = 2\int_0^1 (x - x^{6.9})\, dx = 2\left[\frac{1}{2}x^2 - \frac{1}{7.9.}x^{7.9}\right]_{x=0}^1$$

$$= 2\left[\frac{1}{2} - \frac{1}{7.9}\right] \approx 0.747,$$

or

$$\frac{a-1}{a+1} = \frac{5.9}{7.9} = \frac{59}{79} \approx 0.747, \text{ using (2)}.$$

EXAMPLE 4 The Lorenz curve for household wealth in Great Britain in 1990 was

$$L(x) = x^{15.3}.$$

The Gini Index $= \frac{15.3-1}{15.3+1} = \frac{143}{163} \approx 0.877$. The index is much higher than that for US wealth in example 3.

EXAMPLE 5 The following table shows Gini indices for the US for a range of years.

Income Distribution	Lorenz Curve $L(x)$	The Gini Index
US 1929	$x^{3.2}$	$\frac{22}{42} \approx 0.524$
US 1935	$x^{2.4}$	$\frac{14}{34} \approx 0.412$
US 1947	$x^{1.6}$	$\frac{6}{26} \approx 0.231$
US 1993	$x^{2.4}$	$\frac{14}{34} \approx 0.412$
US 2005	$x^{2.6}$	$\frac{16}{36} \approx 0.444$
US 2009	$x^{2.7}$	$\frac{17}{37} \approx 0.459$

EXAMPLE 6 Next we have examples of Gini indices where the Lorenz curve is a polynomial.

Income Distribution	Lorenz Curve $L(x)$	The Gini Index
1962	$\frac{3}{10}x + \frac{7}{10}x^2$	0.233
1972	$\frac{1}{2}x + \frac{1}{2}x^2$	0.167

Both Lorenz curves are mixtures of x and x^2, and we use the equation (3) in theorem 1.

The Gini Index for 1962 is

$$= \frac{a-1}{a+1}p + \frac{b-1}{b+1}q$$

$$= \frac{1-1}{1+1}\cdot\frac{3}{10} + \frac{2-1}{2+1}\cdot\frac{7}{10} \approx 0.233$$

The Gini Index for 1972 is

$$= \frac{1-1}{1+1}\cdot\frac{1}{2} + \frac{2-1}{2+1}\cdot\frac{1}{2} \approx 0.167$$

2 Continuous Streams of Income

In this supplement to Sections 8.2 and 9.1 we focus on the present value of a stream of income.

It is sometimes useful to model payment made in the form of a continuous stream. Instead a discrete sequence of payments, a continuous flow is assumed. The **flow rate** is a continuous function $f(t)$ given in dollars per year, where t is time in years. This flow rate may be constant or may vary with time t. This models the situation where many payments are made spread out over an interval of time. Accordingly, continuous compound interest will be assumed, at some constant annual rate r. In practice this rate is pegged to some bond instrument, such as a treasury bill and may be adjusted by other factors.

One might think of water flowing into a tub at rate variable $f(t)$. Except here the substance flowing in is money, which accumulates interest compounded continuously.

If $f(t) = k$ is a constant rate (say, in dollars per year), then the total income over an interval of b years is simply bk dollars, just interpret as area of rectangle with base b and height k.

2.1 Total Income Produced by a Stream

In general, the **total income over b years** (over $[0, b]$) is the definite integral of the flow rate over the interval of payment:

$$\int_0^b f(t)\, dt$$

This is of course the area under the curve as x goes from 0 to b.

EXAMPLE Income is produced by a machine is modeled as continuous stream at the flow rate $f(t) = 5000e^{.04t}$. This says that the income starts out at the rate of 5000$ per year, and grows exponentially at the relative rate of 4%. The total income over the first 25 years is

$$\int_0^{25} 5000e^{.04t}dt = 5000\cdot 25e^{.04t}\Big|_0^{25}$$

$$= 125000(e-1) = 2.15 \times 10^5 \text{ dollars}$$

$$= \$215,000.$$

2.2 Present Value of a Stream of Income

2.2.1 Present Value

We review the idea of present value of an amount of money in the future under continuous interest, as mentioned in Section 4.1 (exercises).

An amount F of money at t years in the future satisfies $F = Pe^{rt}$ where the **present value** (i.e. value at $t = 0$) is P. While P can be thought of as the "principal" it is sometimes only theoretical amount at $t = 0$, which may not be actually deposited into an account. The interest rate r is known as the **relative growth rate.** This formula is presented as the limiting case of compound interest in section 4.1 and as an example of exponential growth in section 7.3. Solving for P yields

$$P = F/e^{rt} = Fe^{-rt}$$

This formula gives the amount P that would have to be invested to accumulate to the amount F, t years in the future. The present value is thus considered to be financially equivalent to the future value. Note that the present value P is always smaller than the future value. For example 100 dollars 10 years in the future at interest rate 4% has a present value of $100e^{-.04(10)}$, about 67 dollars. To say the same thing, a 67 dollar initial amount will accumulate to about 100 dollars in 10 years. These two values (67 now, or 100 in 10 years) are viewed as being equivalent under the interest rate assumed.

The present value gives a way of evaluating and comparing income from future sources. This encodes the idea of the *time value of money,* a fundamental concept of financial theory. Restating:

> P is the value now (i.e. at $t = 0$) that is financially equivalent to a single future amount F at time t, assuming a continuous compound interest rate of r

The fixed rate r is a standardized ambient market rate (often associated to treasury bill or another bond rate). Note e^{-rt} is a discount (shrinkage) factor — it is positive and less than one.

EXAMPLE

Suppose you are promised 1 million dollars in 100 years, assuming 5% continuous interest. The present value is $P = 1 \cdot e^{-.05(100)} = .00674$ million dollars $= \$6740$. The discount factor is about 0.67%. Equivalently, this amount will grow to a cool million in 100 years at 5% continuous interest.

Assuming 10% interest instead, we get a much smaller amount $P = 10^6 \cdot e^{-.1(100)} = \$45.40$.

Exercise 1

Would you rather be paid a lump sum of 1 million dollars in ten years or 100,000 in 2 years? Assume 5% continuous compound interest. The solution is obtained by computing the present value of each alternative.

2.2.2 The Present Value (PV) of an Income Stream

Assume a flow rate $f(t)$ and continuous interest rate r. Then the **present value of the income stream** that lasts b years is

$$PV = \int_0^b f(t)\, e^{-rt}\, dt \tag{4}$$

Idea: over small interval of length Δt with right endpoint t_k, the PV of the chunk of income $f(t_k)\, \Delta t$ is $f(t_k)\, e^{-rt_k}\, \Delta t$. Forming the Riemann sum $\sum_{k=1}^n f(t_k)\, e^{-rt_k} \Delta t$ and taking $\lim_{\Delta t \to 0}$ results in the present value (PV) integral.

2.2.3 The Future Value (FV) of an Income Stream

Assume a flow rate $f(t)$ and continuous interest rate r. Then the **future value of the income stream** that lasts b years is

$$FV = PV \cdot e^{rb} \tag{5}$$

This just the Pe^{rt} formula applied to the PV of the stream at terminal time $t = b$.[1]

EXAMPLE

(constant flow rate) Suppose the flow rate is a constant 2000 dollars per year over a period of 10 years, with 8% continuous interest. Then the total amount paid is $2000 \cdot 10 = 20{,}000$. This is computable as $\int_0^{10} f(t)\, dt = \int_0^{10} 2000\, dt = 2000t \big|_0^{10} = 2000 \cdot 10$.

The PV is using (4)

$$\int_0^{10} e^{-rt} f(t)\, dt = \int_0^{10} 2000 e^{-.08t}\, dt$$

$$= -\frac{2000}{.08} e^{-.08t} \bigg|_0^{10}$$

$$= \frac{2000}{.08} (1 - e^{-.08(10)}) = 13{,}767 \ dollars.$$

The FV at $b = 10$ is $13767 \cdot e^{.08(10)} = \$30639$ using (5)

So the total interest paid after 10 years is $30369 - 20000 = \$10369$.

EXAMPLE

(non-constant flow rate) Suppose the flow rate is $2000 e^{.07t}$ dollars per year over a period of 10 years, with 8% continuous interest. Then the total amount paid is

$$\int_0^{10} f(t)\, dt = \int_0^{10} 2000 e^{.07t}\, dt$$

$$= \frac{2000}{.07} e^{.07t} \bigg|_0^{10}$$

$$= \frac{2000}{.07} (e^{.7} - 1) = \$28{,}964.$$

The PV is (using (4))

$$\int_0^{10} e^{-rt} f(t)\, dt = \int_0^{10} 2000 e^{-.08t} e^{.07t}\, dt$$

$$= \int_0^{10} 2000 e^{-.01t}\, dt = -\frac{2000}{.01} e^{-.01t} \bigg|_0^{10}$$

$$= \frac{2000}{.01} (1 - e^{-.01(10)})$$

$$= 19033 \ dollars$$

The FV at $b = 10$ years is $19033 \cdot e^{.08(10)} = 42359$ using (5)

The total interest paid after 10 years is the difference between FV and total amount paid: $42359 - 28964 = 13\,395$ dollars.

---

[1] Also:

$$FV = PV \cdot e^{rb} = \left(\int_0^b f(t)\, e^{-rt} dt \right) \cdot e^{rb}$$

$$= \int_0^b f(t) \cdot e^{rb} e^{-rt} dt = \int_0^b f(t) \cdot e^{r(b-t)}\, dt$$

giving the FV formula (as in Section 8.2 with $b = T$). But it is better just to compute the PV first, and then multiply by e^{rb}) as is done here.

EXAMPLE (another non-constant flow rate) Suppose the flow rate is $2000e^{.08t}$ dollars per year over a period of 10 years, with 8% continuous interest. Then the total amount paid is

$$\int_0^{10} f(t)dt = \int_0^{10} 2000e^{.08t}dt$$

$$= \frac{2000}{.08}e^{.08t}\Big|_0^{10}$$

$$= \frac{2000}{.08}(e^{.8} - 1) = \$30,639$$

The PV is by (4):

$$\int_0^{10} e^{-rt}f(t)dt = \int_0^{10} 2000e^{-.08t}e^{.08t}\,dt$$

$$= \int_0^{10} 2000dt$$

$$= 20,000 \text{ dollars}$$

The FV at $b = 10$ years is $20000 \cdot e^{.08(10)} = 44511$ using (5).

So the total interest paid after 10 years is the difference between FV and total amount paid: $\$44511 - 30639 = 13395$ dollars.

Exercise 2 (arbitrary constant flow rate) *Suppose the flow rate is Q dollars per year over a period of b years, with continuous interest rate r. Show that the PV of the stream is*

$$\frac{Q}{r}(1 - e^{-rb}) \tag{6}$$

Proof: The PV is $\int_0^b e^{-rt}Qdt = \frac{Q}{-r}e^{-rt}\big|_0^b = \frac{Q}{r}(1 - e^{-rb})$ dollars, using (4).

On the other hand, the FV at the terminus $t = b$ years is using (5)

$$\frac{Q}{r}(1 - e^{-rb}) \cdot e^{rb} = \frac{Q}{r}(e^{rb} - 1)$$

What is the approximate value of the PV (6) if b is very large? The limiting value is studied next.

2.3 Present Value of Perpetual Income Stream

The duration of the stream of income can be made indefinite by taking the time limit to infinity. The resulting present value of the perpetual stream, if it exists, is known as Capital Value (CV) in Section 8.2. It is defined to be

$$CV = \int_0^\infty f(t)e^{-rt}dt$$

THEOREM (arbitrary constant flow rate):
Suppose the flow rate is Q dollars per year at continuous interest rate r. The PV
($= $ CV) of the perpetual stream is Q/r:
Proof: Using (4)

$$\int_0^\infty Qe^{-rt}dt = \lim_{b\to\infty} Q \int_0^b e^{-rt}dt$$

$$= \lim_{b\to\infty} Q \frac{1}{-r}e^{-rt}\Big|_0^b$$

$$= \lim_{b \to \infty} \frac{Q}{r}(1 - e^{-rb}) \text{ (as in (6))}$$

$$= Q/r \text{ (since } \lim_{b \to \infty} \frac{1}{e^{rb}} = 0).$$

This can be applied to a company earning at flat rate of Q dollars per year forever. The CV $= Q/r$ is the present value of the company (its stock).

EXAMPLE

(non-constant flow rate) Suppose a company produces earnings at the rate $13e^{.07t}$ million dollars per year, with 8% continuous interest.

The CV is by (4)

$$\int_0^\infty e^{-rt} f(t)\, dt = \int_0^\infty 13e^{-.08t} e^{.07t}\, dt$$

$$= \int_0^\infty 13e^{-.01t}\, dt$$

$$= \lim_{b \to \infty} -\frac{13}{.01} e^{-.01t} \Big|_0^b$$

$$= \lim_{b \to \infty} \frac{13}{.01}(1 - e^{-.01b}) = \frac{13}{.01}$$

$= 1300$ million dollars $= 1.3$ billion dollars. Here $g = .07$ is the relative growth rate of the company. This generalizes to

Exercise 3   *Assume* r > g *with* f (t) = Qe^{gt} *and continuous interest rate* r *and constant* Q > 0. *Then show that the capital value is*

$$\frac{Q}{r - g}$$

2.3.1 Price to Earnings Ratio

The theoretical price of a company is the value of the company, interpreted as the net present value of future earnings. It is "theoretical" since the value depends on several assumptions, and the actual market price may differ. This can be interpreted as the capital value, where the income stream is precisely the profit stream of the company. Here the earnings are represented by the stream of income at current flow rate $f(0)$ and the price is the present value of the stream, i.e. the CV. The theoretical price to earnings ratio (PE) is the ratio CV/$f(0)$ with continuous interest rate r. We have two interesting cases

1. The flow rate is a constant $Q > 0$; and using the theorem above the PE ratio is

$$\text{CV}/f(0) = \frac{Q}{r}\frac{1}{Q} = \frac{1}{r}.$$

2. The flow rate is $f(t) = Qe^{gt}$, as above. Here g is the continuous relative growth rate of the stock, which may be negative. The price to earnings ratio is, by the exercise above, is

$$\text{CV}/f(0) = \frac{Q}{r - g}\frac{1}{Q} = \frac{1}{r - g} \text{ if } r > g.$$

Note that this reduces to case 1 when $g = 0$.

Exercise 4   *Suppose that a company has flat earnings at 1 million dollars annually, at 8% continuous compound interest. Find its CV and theoretical PE ratio.* _____

Exercise 5   *Suppose that a company has earnings right now of 1 million dollars per year and is growing continuously at the rate of 5% per year, assuming 8% continuous compound interest. Find its CV and theoretical PE ratio.* _____

Exercise 6   *What happens to the CV if the condition* $r > g$ *is not satisfied?* _____

Probability
and Calculus

Introduction

In this chapter, we discuss this topic in much more detail. In the first section we introduce improper integrals, a new integral form that is fundamental to the work that follows. After discussing basic concepts in the next two sections, we consider several commonly used probability distributions, culminating with the very important normal distribution.

9-1   Improper Integrals

- Improper Integrals
- Application: Capital Value

We are now going to consider an integral form that has wide application in probability studies as well as other areas. Earlier, when we introduced the idea of a definite integral,

$$\int_a^b f(x)\,dx \tag{1}$$

we required f to be continuous over a closed interval $[a, b]$. Now we are going to extend the meaning of (1) so that the interval $[a, b]$ may become infinite in length.

EXPLORE & DISCUSS 1

For each of the following functions, find the area under the graph from $x = 1$ to $x = b$, $b > 1$ (see Fig. 1). Discuss the behavior of this area as b assumes larger and larger values.

(A) $f(x) = \dfrac{1}{x^{5/4}} = x^{-5/4}$ (B) $g(x) = \dfrac{1}{x^{4/5}} = x^{-4/5}$

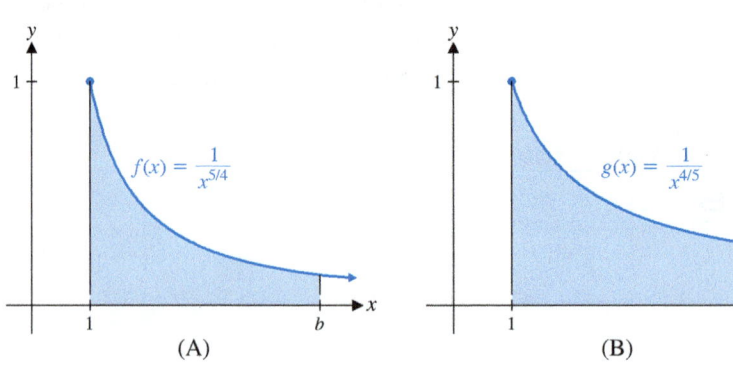

(A)   (B)

Figure 1

Let us investigate a particular example that will motivate several general definitions. What would be a reasonable interpretation for the following expression?

$$\int_1^\infty \frac{dx}{x^2}$$

Sketching a graph of $f(x) = 1/x^2$, $x \geq 1$ (see Fig. 2), we note that for any fixed $b > 1$, $\int_1^b f(x)\,dx$ is the area between the graph of $y = 1/x^2$, the x axis, $x = 1$, and $x = b$.

Let us see what happens when we let $b \to \infty$; that is, when we compute the following limit:

$$\lim_{b \to \infty} \int_1^b \frac{dx}{x^2} = \lim_{b \to \infty}\left[(-x^{-1})\Big|_1^b\right]$$

$$= \lim_{b \to \infty}\left(-\frac{1}{b} + 1\right) = 1$$

Did you expect this result? No matter how large b is taken, the area under the graph from $x = 1$ to $x = b$ never exceeds 1, and in the limit it *is* 1 (see Fig. 3). This suggests that we write

$$\int_1^\infty \frac{dx}{x^2} = \lim_{b \to \infty} \int_1^b \frac{dx}{x^2} = 1$$

Figure 2

y

$f(x) = \dfrac{1}{x^2}$

Area $= \displaystyle\int_1^b \frac{dx}{x^2}$

Figure 3

y

$f(x) = \dfrac{1}{x^2}$

Area $= \displaystyle\int_1^\infty \frac{dx}{x^2} = 1$

This integral is an example of an *improper integral.* In general, the forms

$$\int_{-\infty}^{b} f(x)\, dx \qquad \int_{a}^{\infty} f(x)\, dx \qquad \int_{-\infty}^{\infty} f(x)\, dx$$

where f is continuous over the indicated interval, are called **improper integrals.** (These integrals are "improper" because the interval of integration is unbounded, as indicated by the use of ∞ for one or both limits of integration. There are other types of improper integrals involving certain types of points of discontinuity within the interval of integration, but these will not be considered here.) Each type of improper integral above is formally defined in the following box:

DEFINITION Improper Integrals

If f is continuous over the indicated interval and the limit exists, then

1. $\displaystyle \int_{a}^{\infty} f(x)\, <dx = \lim_{b \to \infty} \int_{a}^{b} f(x)\, dx$

2. $\displaystyle \int_{-\infty}^{\infty} f(x)\, dx = \lim_{a \to -\infty} \int_{a}^{b} f(x)\, dx$

3. $\displaystyle \int_{-\infty}^{\infty} f(x)\, dx = \int_{-\infty}^{c} f(x)\, dx + \int_{c}^{\infty} f(x)\, dx$

where c is any point on $(-\infty, \infty)$, provided both improper integrals on the right exist.

If the indicated limit exists, then the improper integral is said to exist or to **converge;** if the limit does not exist, then the improper integral is said not to exist or to **diverge** (and no value is assigned to it).

EXAMPLE 1 **Evaluating an Improper Integral** Evaluate the following, if it converges:

$$\int_{2}^{\infty} \frac{dx}{x}$$

SOLUTION
$$\int_{2}^{\infty} \frac{dx}{x} = \lim_{b \to \infty} \int_{2}^{b} \frac{dx}{x}$$

$$= \lim_{b \to \infty} \left[(\ln x) \Big|_{2}^{b} \right]$$

$$= \lim_{b \to \infty} (\ln b - \ln 2)$$

Since $\ln b \to \infty$ as $b \to \infty$, the limit does not exist. Hence, the improper integral diverges.

Matched Problem 1 Evaluate the following, if it converges:

$$\int_{3}^{\infty} \frac{dx}{(x - 1)^{2}}.$$

EXAMPLE 2 **Evaluating an Improper Integral** Evaluate the following, if it converges:

$$\int_{-\infty}^{2} e^{x}\, dx$$

SOLUTION
$$\int_{-\infty}^{2} e^{x}\, dx = \lim_{a \to -\infty} \int_{a}^{2} e^{x}\, dx$$

$$= \lim_{a \to -\infty} \left(e^{x} \Big|_{a}^{2} \right)$$

$$= \lim_{a \to -\infty} (e^{2} - e^{a}) = e^{2} - 0 = e^{2} \quad \text{The integral converges.}$$

Matched Problem 2   Evaluate the following, if it converges:

$$\int_{-\infty}^{-1} x^{-2}\, dx.$$

EXAMPLE 3   **Evaluating an Improper Integral**   Evaluate the following, if it converges:

$$\int_{-\infty}^{\infty} \frac{2x}{(1 + x^2)^2}\, dx$$

SOLUTION $\displaystyle \int_{-\infty}^{\infty} \frac{2x}{(1 + x^2)^2}\, dx = \int_{-\infty}^{0} (1 + x^2)^{-2} 2x\, dx + \int_{0}^{\infty} (1 + x^2)^{-2} 2x\, dx$

$$= \lim_{a \to -\infty} \int_{a}^{0} (1 + x^2)^{-2} 2x\, dx + \lim_{b \to \infty} \int_{0}^{b} (1 + x^2)^{-2} 2x\, dx$$

$$= \lim_{a \to -\infty} \left[\frac{(1 + x^2)^{-1}}{-1} \Big|_{a}^{0} \right] + \lim_{b \to \infty} \left[\frac{(1 + x^2)^{-1}}{-1} \Big|_{0}^{b} \right]$$

$$= \lim_{a \to -\infty} \left[-1 + \frac{1}{1 + a^2} \right] + \lim_{b \to \infty} \left[-\frac{1}{1 + b^2} + 1 \right]$$

$$= -1 + 1 = 0 \quad \text{The integral converges.}$$

Matched Problem 3   Evaluate the following, if it converges:

$$\int_{-\infty}^{\infty} \frac{dx}{e^x}.$$

CONCEPTUAL INSIGHT

It is impossible, in general, solely by inspecting the graph of the function, to determine whether an improper integral converges or diverges. For example, the functions

$$f(x) = \frac{1}{x} \quad \text{and} \quad g(x) = \frac{1}{x^{1.001}}$$

have graphs that are nearly indistinguishable on $[1, \infty)$. But by evaluating the appropriate limits, it can be shown that $\int_{0}^{\infty} f(x)\, dx$ diverges and $\int_{0}^{\infty} g(x)\, dx$ converges.

EXPLORE & DISCUSS 2   Let

$$f(x) = \begin{cases} 2 - x & \text{if } 0 \leq x \leq 1 \\ 0 & \text{otherwise} \end{cases}$$

(A) Graph $y = f(x)$ and explain why f is not continuous at $x = 0$ and at $x = 1$.

(B) Find $\displaystyle \int_{0}^{1} f(x)\, dx.$

(C) Discuss possible interpretations of the improper integral $\displaystyle \int_{-\infty}^{\infty} f(x)\, dx.$

In probability studies, we frequently encounter functions that are 0 over one or more intervals on the real axis. An improper integral involving such a function can be simplified by deleting the intervals where the function is 0 and considering the integral over the remaining intervals. The next example illustrates this process.

EXAMPLE 4 **Evaluating an Improper Integral** Evaluate $\int_{-\infty}^{\infty} f(x)\, dx$ for

$$f(x) = \begin{cases} \dfrac{4}{(x+2)^2} & \text{if } x \geq 0 \\ 0 & \text{otherwise} \end{cases}$$

SOLUTION The function f is discontinuous at $x = 0$ (see Fig. 4). However, since $f(x) = 0$ for $x < 0$, we can proceed as follows:

$$\int_{-\infty}^{\infty} f(x)\, dx = \int_{0}^{\infty} f(x)\, dx \qquad \textit{f(x) = 0 for x < 0.}$$

$$= \int_{0}^{\infty} \frac{4}{(x+2)^2}\, dx$$

$$= \lim_{b \to \infty} \int_{0}^{b} \frac{4}{(x+2)^2}\, dx$$

$$= \lim_{b \to \infty} \left[-4(x+2)^{-1} \Big|_{0}^{b} \right]$$

$$= \lim_{b \to \infty} \left(\frac{-4}{b+2} + 2 \right)$$

$$= 2 \qquad \textit{The integral converges.}$$

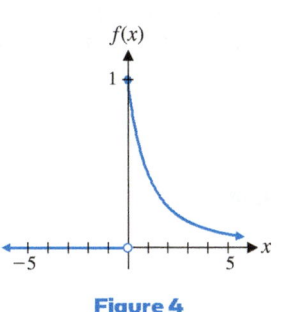

Figure 4

Matched Problem 4 Evaluate $\int_{-\infty}^{\infty} f(x)\, dx$ for $f(x) = \begin{cases} 4x - x^2 & \text{if } 0 \leq x \leq 4 \\ 0 & \text{otherwise.} \end{cases}$

EXAMPLE 5 **Oil Production** It is estimated that an oil well will produce oil at a rate of $R(t)$ million barrels per year t years from now, as given by

$$R(t) = 8e^{-0.05t} - 8e^{-0.1t}$$

Estimate the total amount of oil that will be produced by this well.

SOLUTION The total amount of oil produced in T years of operation is $\int_{0}^{T} R(t)\, dt$ (see Fig. 5). At some point in the future, the annual production rate will become so low that it will no longer be economically feasible to operate the well. However, since we do not know when this will occur, it is convenient to assume that the well is operated indefinitely so we can use an improper integral. Thus, the total amount of oil produced is approximately

$$\int_{0}^{\infty} R(t)\, dt = \lim_{T \to \infty} \int_{0}^{T} R(t)\, dt$$

$$= \lim_{T \to \infty} \int_{0}^{T} (8e^{-0.05t} - 8e^{-0.1t})\, dt$$

$$= \lim_{T \to \infty} \left[(-160e^{-0.05t} + 80e^{-0.1t}) \Big|_{0}^{T} \right]$$

$$= \lim_{T \to \infty} (-160e^{-0.05T} + 80e^{-0.1T} + 80)$$

$$= 80 \quad \text{or} \quad 80{,}000{,}000 \text{ barrels}$$

The total production over any finite time period $[0, T]$ is less than 80,000,000 barrels, and the larger T is, the closer the total production will be to 80,000,000 barrels. Figure 6 displays the total production over $[0, T]$ for values of $T(= X)$ from 50 to 110 years.

Figure 5

Figure 6 Total production over $[0, x]$

Matched Problem 5 The annual production rate (in millions of barrels) for an oil well is given by

$$R(t) = 4e^{-0.1t} - e^{-0.2t}$$

Assuming that the well is operated indefinitely, find the total production.

Application: Capital Value

Recall that if money is invested at a rate r compounded continuously for T years, then the present value, PV, and the future value, FV, are related by the continuous compound interest formula

$$FV = PVe^{rT}$$

which also can be written as

$$PV = e^{-rT}FV$$

This relationship is valid for both single deposits and amounts generated by continuous income streams. Thus, if $f(t)$ is a continuous income stream, the future value is given by

$$FV = e^{rT}\int_0^T f(t)e^{-rt}\, dt \quad \text{Future value}$$

and, using the compound interest formula, the present value is given by

$$PV = e^{-rT}FV = e^{-rT}e^{rT}\int_0^T f(t)e^{-rt}\, dt$$

Since $e^{-rT}e^{rT} = 1$, we have

$$PV = \int_0^T f(t)e^{-rt}\, dt \quad \text{Present value}$$

If we let T approach ∞ in this formula for present value, we obtain an improper integral that represents the *capital value* of the income stream.

DEFINITION   Capital Value of a Perpetual Income Stream

A continuous income stream is called **perpetual** if it never stops producing income. The **capital value, CV,** of a perpetual income stream $f(t)$ at a rate r compounded continuously is the present value over the time interval $[0, \infty)$. That is,

$$CV = \int_0^\infty f(t)e^{-rt}\, dt$$

Capital value provides a method for expressing the worth (in terms of today's dollars) of an investment that will produce income for an indefinite period of time.

EXAMPLE 6   **Capital Value** A family has leased the oil rights of a property to a petroleum company in return for a perpetual annual payment of $1,200. Find the capital value of this lease at 5% compounded continuously.

SOLUTION   The annual payments from the oil company produce a continuous income stream with rate of flow $f(t) = 1,200$ that continues indefinitely. (It is common practice to treat a sequence of equal periodic payments as a continuous income stream with a constant rate of flow, even if the income is received only at the end of each period.) Thus, the capital value is

$$CV = \int_0^\infty f(t)e^{-rt}\, dt$$

$$CV = \int_0^\infty 1,200e^{-0.05t}\, dt$$

$$= \lim_{T\to\infty} \int_0^T 1,200e^{-0.05t}\, dt$$

$$= \lim_{T\to\infty} \left[-24,000e^{-0.05t} \Big|_0^T \right]$$

$$= \lim_{T\to\infty} (-24,000e^{-0.05T} + 24,000) = \$24,000$$

Matched Problem 6   Repeat Example 6 if the interest rate is 6% compounded continuously.

Exercise 9-1

A

In Problems 1–16, find the value of each improper integral that converges.

1. $\displaystyle\int_1^\infty \frac{dx}{x^3}$

2. $\displaystyle\int_1^\infty x\,dx$

3. $\displaystyle\int_1^\infty x^2\,dx$

4. $\displaystyle\int_1^\infty \frac{dx}{x^4}$

5. $\displaystyle\int_{-\infty}^0 e^{x/2}\,dx$

6. $\displaystyle\int_{-\infty}^{-1} \frac{dx}{x}$

7. $\displaystyle\int_4^\infty \frac{dx}{100\sqrt{x}}$

8. $\displaystyle\int_1^\infty e^{-2x}\,dx$

9. $\displaystyle\int_9^\infty \frac{dx}{x\sqrt{x}}$

10. $\displaystyle\int_8^\infty \frac{dx}{\sqrt[3]{x}}$

11. $\displaystyle\int_0^\infty \frac{dx}{(x+1)^{2/3}}$

12. $\displaystyle\int_0^\infty \frac{dx}{\sqrt{x+1}}$

13. $\displaystyle\int_1^\infty \frac{dx}{x^{0.99}}$

14. $\displaystyle\int_1^\infty \frac{dx}{x^{1.01}}$

15. $\displaystyle 0.3\int_0^\infty e^{-0.3x}\,dx$

16. $\displaystyle 0.01\int_0^\infty e^{-0.1x}\,dx$

In Problems 17–22, graph $y = f(x)$ and find the value of $\int_{-\infty}^\infty f(x)\,dx$ if it converges.

17. $f(x) = \begin{cases} 1 + x^2 & \text{if } 0 \le x \le 2 \\ 0 & \text{otherwise} \end{cases}$

18. $f(x) = \begin{cases} 2 + x & \text{if } 0 \le x \le 3 \\ 0 & \text{otherwise} \end{cases}$

19. $f(x) = \begin{cases} e^{-0.1x} & \text{if } x \ge 0 \\ 0 & \text{otherwise} \end{cases}$

20. $f(x) = \begin{cases} 2 - e^{-0.2x} & \text{if } x \ge 0 \\ 0 & \text{otherwise} \end{cases}$

21. $f(x) = \begin{cases} 4/(x+2) & \text{if } x \ge 2 \\ 0 & \text{otherwise} \end{cases}$

22. $f(x) = \begin{cases} 4/(x+1)^2 & \text{if } x \ge 1 \\ 0 & \text{otherwise} \end{cases}$

In Problems 23–26, discuss the validity of each statement. If the statement is always true, explain why. If not, give a counterexample.

23. If f is a continuous increasing function on $[0, \infty)$ such that $f(x) > 0$ for all x, then $\int_0^\infty f(x)\,dx$ diverges.

24. If f is a continuous decreasing function on $[0, \infty)$ such that $f(x) > 0$ for all x, then $\int_0^\infty f(x)\,dx$ converges.

25. If f is a continuous function on $[1, \infty)$ such that $f(x) > 0$ for all x and $f(x) \ge 10$ for $1 \le x \le 1,000$, then $\int_1^\infty f(x)\,dx$ diverges.

26. If f is a continuous decreasing function on $[1, \infty)$ such that $0 < f(x) \le 0.001$ for all x, then $\int_1^\infty f(x)\,dx$ converges.

In Problems 27–36, find $F(b)$, use a graphing utility to graph $F(b)$, and use the graph to estimate $\lim\limits_{b\to\infty} F(b)$ or, in Problems 31 and 32, $\lim\limits_{b\to-\infty} F(b)$.

27. From Problem 1, $F(b) = \displaystyle\int_1^b \frac{dx}{x^3}$

28. From Problem 2, $F(b) = \displaystyle\int_1^b x\,dx$

29. From Problem 3, $F(b) = \displaystyle\int_1^b x^2\,dx$

30. From Problem 4, $F(b) = \displaystyle\int_1^b \frac{dx}{x^4}$

31. From Problem 5, $F(b) = \displaystyle\int_b^0 e^{x/2}\,dx$

32. From Problem 6, $F(b) = \displaystyle\int_b^{-1} \frac{dx}{x}$

33. From Problem 7, $F(b) = \displaystyle\int_4^b \frac{dx}{100\sqrt{x}}$

34. From Problem 8, $F(b) = \displaystyle\int_1^b e^{-2x}\,dx$

35. From Problem 9, $F(b) = \displaystyle\int_9^b \frac{dx}{x\sqrt{x}}$

36. From Problem 10, $F(b) = \displaystyle\int_8^b \frac{dx}{\sqrt[3]{x}}$

37. If f is continuous on $[0, \infty)$ and $\int_0^\infty f(x)\,dx$ converges, does $\int_1^\infty f(x)\,dx$ converge? Explain.

38. If f is continuous on $[0, \infty)$ and $\int_0^\infty f(x)\,dx$ diverges, does $\int_1^\infty f(x)\,dx$ diverge? Explain.

C

In Problems 39–48, find the value of each improper integral that converges.

39. $\displaystyle\int_0^\infty \frac{1}{k} e^{-x/k}\,dx, k > 0$

40. $\displaystyle\int_1^\infty \frac{1}{x^p}\,dx, p > 1$

41. $\displaystyle\int_{-\infty}^\infty \frac{x}{1+x^2}\,dx$

42. $\displaystyle\int_{-\infty}^\infty xe^{-x^2}\,dx$

43. $\displaystyle\int_0^\infty (e^{-x} - e^{-2x})\,dx$

44. $\displaystyle\int_0^\infty (e^{-x} + 2e^x)\,dx$

45. $\displaystyle\int_{-\infty}^0 \frac{1}{\sqrt{1-x}}\,dx$

46. $\displaystyle\int_{-\infty}^0 \frac{1}{\sqrt[m]{(1-x)^4}}\,dx$

47. $\displaystyle\int_1^\infty \frac{\ln x}{x}\,dx$

48. $\displaystyle\int_0^\infty \frac{e^{-x}}{1+e^{-x}}\,dx$

Applications

49. Capital value. The perpetual annual rent for a property is $6,000. Find the capital value at 5% compounded continuously.

50. Capital value. The perpetual annual rent for a property is $10,000. Find the capital value at 4% compounded continuously.

51. Capital value. A trust fund produces a perpetual stream of income with rate of flow

$$f(t) = 1,500e^{0.04t}$$

Find the capital value at 7% compounded continuously.

52. Capital value. A trust fund produces a perpetual stream of income with rate of flow

$$f(t) = 1,000e^{0.02t}$$

Find the capital value at 6% compounded continuously.

53. Capital value. Refer to Problem 49. Discuss the effect on the capital value if the interest rate is increased to 6%, and then if decreased to 4%. Interpret these results.

54. Capital value. Refer to Problem 50. Discuss the effect on the capital value if the interest rate is increased to 5%, and then if decreased to 2%. Interpret these results.

55. Production. The rate of production of a natural gas well (in billions of cubic feet per year) is given by (see figure below)

$$R(t) = 3e^{-0.2t} - 3e^{-0.4t}$$

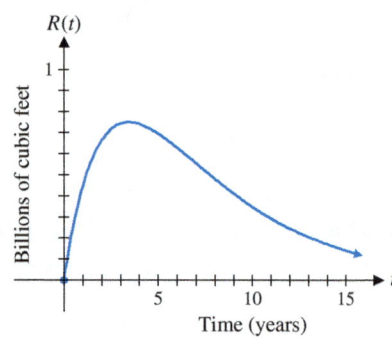

(A) Assuming that the well is operated indefinitely, find the total production.

(B) When will the output from the well reach 50% of the total production? Round answer to two decimal places.

56. Production. The rate of production of an oil well (in millions of barrels per year) is given by

$$R(t) = \frac{1,000t}{(50 + t^2)^2}$$

(A) Assuming that the well is operated indefinitely, find the total production.

(B) When will the output from the well reach 50% of the total production? Round answer to two decimal places.

57. Pollution. It has been estimated that the rate of seepage of toxic chemicals from a waste dump is $R(t)$ gallons per year t years from now, where

$$R(t) = \frac{500}{(1 + t)^2}$$

Assuming that this seepage continues indefinitely, find the total amount of toxic chemicals that seep from the dump.

58. Drug assimilation. When a person takes a drug, the body does not assimilate all of the drug. One way to determine the amount of the drug assimilated is to measure the rate at which the drug is eliminated from the body. If the rate of elimination of the drug (in milliliters per minute) is given by

$$R(t) = 3e^{-0.1t} - 3e^{-0.3t}$$

where t is the time in minutes since the drug was administered, how much of the drug is eliminated from the body?

59. Immigration. In order to control rising population, a country is planning to institute a new policy for immigration. Analysts estimate that the rate of immigration into the country t years after the policy is instituted (in millions of immigrants per year) will be given by

$$R(t) = \frac{400}{(5 + t)^3}$$

Find the total number of immigrants that will enter the country under this policy, assuming that the policy is applied indefinitely.

Answers to Matched Problems

1. $\frac{1}{2}$ **2.** 1

3. Diverges   **4.** $\frac{32}{3}$

5. 35,000,000 barrels   **6.** $20,000

9-2   Continuous Random Variables

- Continuous Random Variables
- Probability Density Functions
- Cumulative Distribution Functions

Continuous Random Variables

Probability theory is concerned with determining the long-run frequency of the outcomes of an experiment when the outcome each time the experiment is performed is uncertain. For example, when a fair coin is tossed, it is uncertain whether it will turn up heads or tails. However, probability theory enables us to predict that the coin will turn up heads approximately 50% of the time.

Table 1 Number of Heads in the Toss of Three Coins

Sample Space S	Number of Heads $X(e_j)$
e_1: TTT	0
e_2: TTH	1
e_3: THT	1
e_4: HTT	1
e_5: THH	2
e_6: HTH	2
e_7: HHT	2
e_8: HHH	3

Assigning a value to each outcome of an experiment defines a function called a *random variable.* For example, consider the experiment of tossing three coins and the function X whose values are the number of heads. The possible outcomes of the experiment and the corresponding values of X are shown in Table 1. Since the set of possible outcomes for this experiment is a finite set, X is called a **discrete random variable.**

Things are more complicated if an experiment has an infinite number of outcomes. For example, consider the experiment of turning on light bulbs and letting them remain lit until they burn out. One bulb may be defective and not light at all. Another may still be lit 50 or 100 years from now. Let X be the function whose values (in hours) are the life expectancies of the light bulbs. Theoretically, there is no reason to exclude any of the values in the interval $[0, \infty)$ as a possibility for the life expectancy of the bulb. Thus, we assume that the range of X is $[0, \infty)$, and we call X a *continuous random variable.*

> **DEFINITION Continuous Random Variable**
>
> A **continuous random variable** X is a function that assigns a numeric value to each outcome of an experiment. The set of possible values of X is an interval of real numbers. This interval may be open or closed, and it may be bounded or unbounded.

The term *continuous* refers to the fact that the values of the random variable form a continuous set of real numbers, such as $[0, \infty)$, rather than a discrete set, such as $\{0, 1, 2, 3\}$ or $\{2, 4, 8, \ldots\}$.

EXPLORE & DISCUSS 1

For each of the following experiments, determine whether the indicated random variable is discrete or continuous.

(A) Two dice are rolled, and X is the sum of the dots on the upturned faces of the dice.

(B) A dart is thrown at a circular dart board, and X is the distance from the dart to the center of the board.

(C) The drive-up window of a bank is open 10 hours a day, and X is the number of customers served in one day.

(D) The drive-up window of a bank is open 10 hours a day, and X is the time (in minutes) each customer must wait for service.

Table 2 Probability Distribution

Number of Heads x	0	1	2	3
Probability $p(x)$	$\frac{1}{8}$	$\frac{3}{8}$	$\frac{3}{8}$	$\frac{1}{8}$

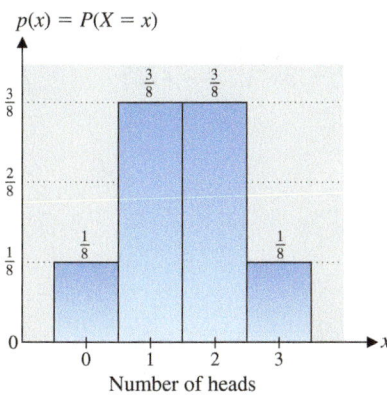

Figure 1 Histogram for a probability distribution

Probability Density Functions

If X is a discrete random variable, then the probability that X lies in a given interval can be computed by addition. For example, consider the experiment of tossing three coins, and let X represent the number of heads. The probability distribution for X is shown in Table 2 and illustrated in the histogram in Figure 1. We denote the probability that X is between 0 and 2, inclusively, by $P(0 \le X \le 2)$ and compute it by addition:

$$P(0 \le X \le 2) = P(X = 0) + P(X = 1) + P(X = 2)$$
$$= \tfrac{1}{8} + \tfrac{3}{8} + \tfrac{3}{8} = \tfrac{7}{8}$$

If X is a continuous random variable, the same approach will not work. Since X can now assume any real number in the interval $[0, 2]$, it is impossible to write $P(0 \le X \le 2)$ as a finite or even an infinite sum. (What would be the second term in such a sum? That is, what is the "next" real number after 0? Think about this.) Instead, we introduce a new type of function, called a *probability density function,* and use integrals involving this function to compute the probability that a continuous random variable X lies in a given interval. For example, we might use a probability density function to find the probability that the actual amount of soda in a 12-ounce can is between 11.9 and 12.1 ounces, or that the speed of a car involved in an accident was between 60 and 65 miles per hour.

For convenience in stating definitions and formulas, we will assume that the value of a continuous random variable can be any real number; that is, the range is $(-\infty, \infty)$.

DEFINITION Probability Density Function

The function $f(x)$ is a probability density function for a continuous random variable X if

1. $f(x) \geq 0$ for all $x \in (-\infty, \infty)$

2. $\displaystyle\int_{-\infty}^{\infty} f(x)\, dx = 1$

3. The probability that X lies in the interval $[c, d]$ is given by

$$P(c \leq X \leq d) = \int_{c}^{d} f(x)\, dx$$

$\displaystyle\int_{-\infty}^{\infty} f(x)\, dx = 1$ $\displaystyle\int_{c}^{d} f(x)\, dx = \text{Probability}(c \leq X \leq d)$

$y = f(x) \geq 0$

Area $= 1$

Range of $X = (-\infty, \infty) = $ Domain of f

EXAMPLE 1   **Using a Probability Density Function** Let

$$f(x) = \begin{cases} 12x^2 - 12x^3 & \text{if } 0 \leq x \leq 1 \\ 0 & \text{otherwise} \end{cases}$$

Graph f and verify that f satisfies the first two conditions for a probability density function. Then find each of the following probabilities and illustrate graphically:

(A) $P(.4 \leq X \leq .7)$ (B) $P(X \leq .5)$

(C) $P(X \geq .6)$ (D) $P(X = .3)$

SOLUTION   The graph of f (Fig. 2) shows that $f(x) \geq 0$ for all x. Also,

$$\int_{-\infty}^{\infty} f(x)\, dx = \int_{0}^{1} (12x^2 - 12x^3)\, dx = (4x^3 - 3x^4)\Big|_{0}^{1} = 1$$

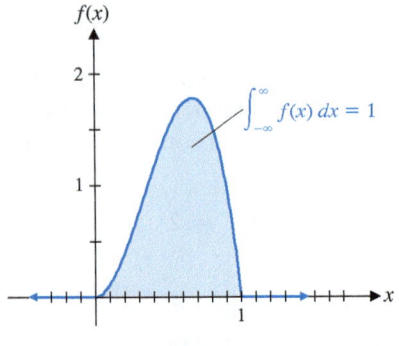

$\displaystyle\int_{-\infty}^{\infty} f(x)\, dx = 1$

Figure 2

(A) $P(.4 \leq X \leq .7) = \displaystyle\int_{.4}^{.7} f(x)\, dx$

$$= \int_{.4}^{.7} (12x^2 - 12x^3)\, dx = (4x^3 - 3x^4)\Big|_{.4}^{.7} = .4725$$

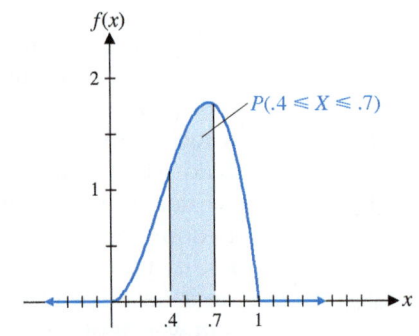

$P(.4 \leq X \leq .7)$

(B) $P(X \le .5) = \int_0^{.5} f(x)\, dx$ Note that $f(x) = 0$ for $x < 0$.

$$= \int_0^{.5} (12x^2 - 12x^3)\, dx = (4x^3 - 3x^4)\Big|_0^{.5} = .3125$$

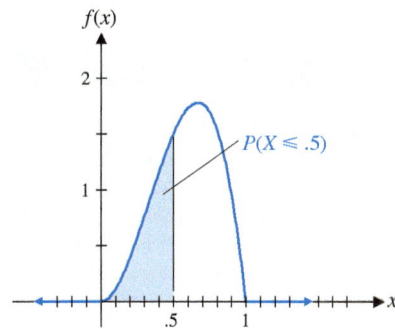

(C) $P(X \ge .6) = \int_{.6}^1 f(x)\, dx$ Note that $f(x) = 0$ for $f(x) = 0$ for $x < 0$.

$$= \int_{.6}^1 (12x^2 - 12x^3)\, dx = (4x^3 - 3x^4)\Big|_{.6}^1 = .5248$$

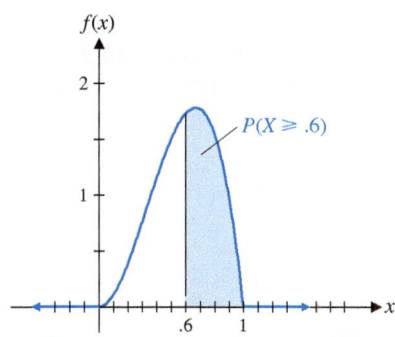

(D) $P(X = .3) = \int_{.3}^{.3} f(x)\, dx = 0$ Definite integral property 1

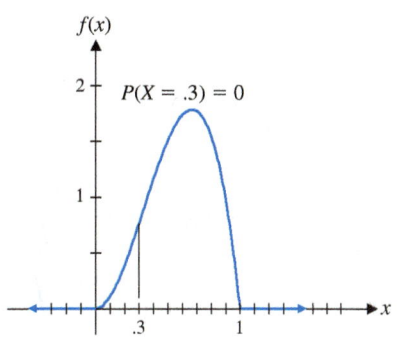

Matched Problem 1   Let

$$f(x) = \begin{cases} 6x - 6x^2 & \text{if } 0 \le x \le 1 \\ 0 & \text{otherwise} \end{cases}$$

Graph f and verify that f satisfies the first two conditions for a probability density function. Then find each of the following probabilities and illustrate graphically:

(A) $P(.3 \le X \le .8)$

(B) $P(X \le .7)$

(C) $P(X \ge .4)$

(D) $P(X = .5)$

CONCEPTUAL INSIGHT

There is a fundamental difference between discrete and continuous random variables. In the discrete case, there is a *probability distribution* $P(x)$ that gives the probability of each possible value of the random variable. The probability, for example, that a roll of a fair die will give a 5, is equal to $\frac{1}{6}$. In the continuous case, the probability that an outcome is exactly 5 (not 5.01, not 4.999) is equal to 0. For in the continuous case it is the integral of the *probability density function* $f(x)$ that gives the probability that the outcome lies in a certain interval:

$$P(X = 5) = P(5 \le X \le 5) = \int_5^5 f(x)\, dx = 0$$

If X is a continuous random variable and c is any real number, then the probability that the outcome is exactly c is

$$P(X = c) = P(c \le X \le c) = \int_c^c f(x)\, dx = 0$$

The fact that $P(X = c) = 0$ also implies that excluding either end point from an interval does not change the probability that the random variable lies in that interval; that is,

$$P(a < X < b) = P(a < X \le b) = P(a \le X < b) = P(a \le X \le b)$$

$$= \int_a^b f(x)\, dx$$

EXAMPLE 2   **Using a Probability Density Function** Use the probability density function in Example 1 to compute $P(.1 < X \le .2)$ and $P(X > .9)$.

SOLUTION $P(.1 < X \le .2) = \displaystyle\int_{.1}^{.2} f(x)\, dx$

$f(x) = \begin{cases} 12x^2 - 12x^3 & \text{if } 0 \le x \le 1 \\ 0 & \text{otherwise} \end{cases}$

$$= \int_{.1}^{.2} (12x^2 - 12x^3)\, dx$$

$$= (4x^3 - 3x^4)\Big|_{.1}^{.2}$$

$$= .0272 - .0037 = .0235$$

$$P(X > .9) = \int_{.9}^{\infty} f(x)\, dx$$

$$= \int_{.9}^{1} (12x^2 - 12x^3)\, dx$$

$$= (4x^3 - 3x^4)\Big|_{.9}^{1}$$

$$= 1 - .9477 = .0523$$

Matched Problem 2   Use the probability density function in Matched Problem 1 to compute the probabilities $P(.2 \le X < .4)$ and $P(X < .8)$.

EXAMPLE 3

Shelf Life The shelf life (in months) of a certain drug is a continuous random variable with probability density function (see Fig. 3)

$$f(x) = \begin{cases} 50/(x + 50)^2 & \text{if } x \geq 0 \\ 0 & \text{otherwise} \end{cases}$$

Find the probability that the drug has a shelf life of

(A) Between 10 and 20 months

(B) At most 30 months

(C) More than 25 months

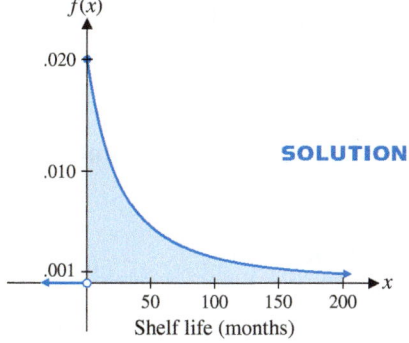

Figure 3

SOLUTION

(A) $P(10 \leq X \leq 20) = \int_{10}^{20} f(x)\, dx = \int_{10}^{20} \frac{50}{(x + 50)^2}\, dx = \frac{-50}{x + 50}\Big|_{10}^{20}$

$\qquad = (-\frac{50}{70}) - (-\frac{50}{60}) = \frac{5}{42}$

(B) $P(X \leq 30) = \int_{-\infty}^{30} f(x)\, dx = \int_{0}^{30} \frac{50}{(x + 50)^2}\, dx = \frac{-50}{x + 50}\Big|_{0}^{30}$

$\qquad = (-\frac{50}{80}) - (-1) = \frac{3}{8}$

(C) $P(X > 25) = \int_{25}^{\infty} f(x)\, dx = \int_{25}^{\infty} \frac{50}{(x + 50)^2}\, dx$

This improper integral can be evaluated directly using the techniques discussed in Section 9-1. However, there is another method that does not involve evaluation of any improper integrals. Since f is a probability density function, we can write

$$1 = \int_{-\infty}^{\infty} f(x)\, dx$$

$$= \int_{0}^{\infty} \frac{50}{(x + 50)^2}\, dx \qquad \int_{a}^{b} f(x)\, dx = \int_{a}^{c} f(x)\, dx + \int_{c}^{b} f(x)\, dx$$

$$= \int_{0}^{25} \frac{50}{(x + 50)^2}\, dx + \int_{25}^{\infty} \frac{50}{(x + 50)^2}\, dx$$

Solving this last equation for $\int_{25}^{\infty} [50/(x + 50)^2]\, dx$, we have

$$\int_{25}^{\infty} \frac{50}{(x + 50)^2}\, dx = 1 - \int_{0}^{25} \frac{50}{(x + 50)^2}\, dx$$

$$= 1 - \frac{-50}{x + 50}\Big|_{0}^{25}$$

$$= 1 - (\frac{-50}{75}) - 1 = \frac{2}{3}$$

Thus, $P(X > 25) = \frac{2}{3}$.

Matched Problem 3

In Example 3, find the probability that the drug has a shelf life of

(A) Between 50 and 100 months

(B) At most 20 months

(C) More than 10 months

Cumulative Distribution Functions

EXPLORE & DISCUSS 2

Let

$$f(x) = \begin{cases} \frac{1}{2} & \text{if } 0 \leq x \leq 2 \\ 0 & \text{otherwise} \end{cases} \qquad \text{and} \qquad F(x) = \begin{cases} 0 & \text{if } x < 0 \\ \frac{1}{2} x & \text{if } 0 \leq x \leq 2 \\ 1 & \text{if } x > 2 \end{cases}$$

(A) Graph each function.

(B) Is f a probability density function? Is F?

(C) Is there any relationship between f and F?

(D) Evaluate $\displaystyle\int_{.5}^{1.5} f(x)\,dx$.

(E) Can you use F to evaluate the integral in part (D)?

Each time we compute the probability for a continuous random variable, we must find the antiderivative of the probability density function. This antiderivative is used so often that it is convenient to give it a name.

> **DEFINITION**   **Cumulative Distribution Function**
>
> If f is a probability density function, then the associated **cumulative distribution function F** is defined by
>
> $$F(x) = P(X \leq x) = \int_{-\infty}^{x} f(t)\,dt$$
>
> Furthermore,
>
> $$P(c \leq X \leq d) = F(d) - F(c)$$
>
> Figure 4 gives a geometric interpretation of these ideas.

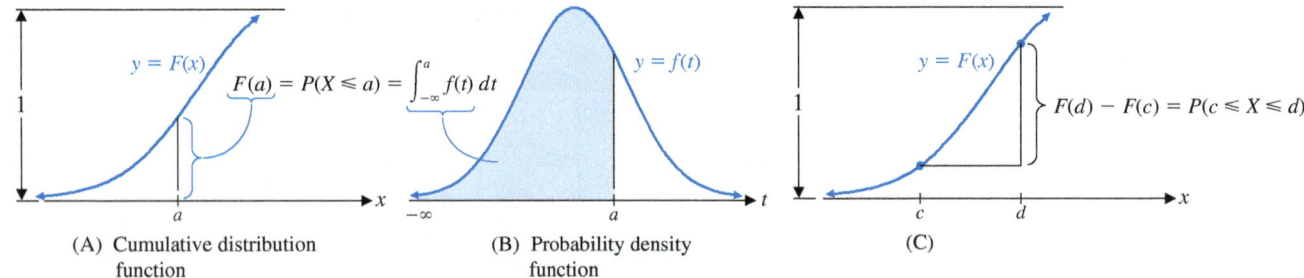

(A) Cumulative distribution   (B) Probability density   (C)
       function   function

Figure 4

Notice that $F(x) = \int_{-\infty}^{x} f(t)\,dt$ is a function of x, the upper limit of integration, not t, the variable in the integrand. We state some important properties of cumulative distribution functions in the next box. These properties follow directly from the fact that $F(x)$ can be interpreted geometrically as the area under the graph of $y = f(t)$ from $-\infty$ to x (see Fig. 4B).

> **PROPERTIES**   **Cumulative Distribution Functions**
>
> If f is a probability density function and
>
> $$F(x) = \int_{-\infty}^{x} f(t)\,dt$$
>
> is the associated cumulative distribution function, then
>
> 1. $F'(x) = f(x)$ wherever f is continuous
> 2. $0 \leq F(x) \leq 1, \quad -\infty < x < \infty$
> 3. $F(x)$ is nondecreasing on $(-\infty, \infty)$*
>
> ――――――――
> *A function $F(x)$ is nondecreasing on (a, b) if $F(x_1) \leq F(x_2)$ for $a < x_1 < x_2 < b$.

EXAMPLE 4   **Using a Cumulative Distribution Function** Find the cumulative distribution function for the probability density function in Example 1, and use it to compute $P(.1 \leq X \leq .9)$.

SOLUTION If $x < 0$, then

$$F(x) = \int_{-\infty}^{x} f(t)\, dt \quad f(x) = \begin{cases} 12x^2 - 12x^3 & \text{if } 0 \le x \le 1 \\ 0 & \text{otherwise} \end{cases}$$

$$= \int_{-\infty}^{x} 0\, dt = 0$$

If $0 \le x \le 1$, then

$$F(x) = \int_{-\infty}^{x} f(t)\, dt = \int_{-\infty}^{0} f(t)\, dt + \int_{0}^{x} f(t)\, dt$$

$$= 0 + \int_{0}^{x} (12t^2 - 12t^3)\, dt = (4t^3 - 3t^4)\Big|_{0}^{x}$$

$$= 4x^3 - 3x^4$$

If $x > 1$, then

$$F(x) = \int_{-\infty}^{x} f(t)\, dt = \int_{-\infty}^{0} f(t)\, dt + \int_{0}^{1} f(t)\, dt + \int_{1}^{x} f(t)\, dt$$

$$= 0 + 1 + 0 = 1$$

Thus,

$$F(x) = \begin{cases} 0 & \text{if } x < 0 \\ 4x^3 - 3x^4 & \text{if } 0 \le x \le 1 \\ 1 & \text{if } x > 1 \end{cases}$$

And

$$P(.1 \le X \le .9) = F(.9) - F(.1) = .9477 - .0037 = .944$$

See Figure 5.

(A)

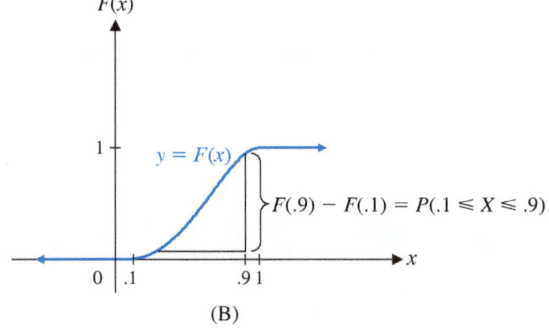

(B)

Figure 5

Matched Problem 4 Find the cumulative distribution function for the probability density function in Matched Problem 1, and use it to compute $P(.3 \le X \le .7)$.

EXAMPLE 5 **Shelf Life** Returning to the discussion of the shelf life of a drug in Example 3, suppose a pharmacist wants to be 95% certain that the drug is still good when it is sold. How long is it safe to leave the drug on the shelf?

SOLUTION Let x be the number of months the drug has been on the shelf when it is sold. The probability that the shelf life of the drug is less than the number of months it has been sitting on the shelf is $P(0 \le X \le x)$. The pharmacist wants this probability to be .05. Thus, we must solve the equation $P(0 \le X \le x) = .05$ for x. First, we will find the cumulative distribution function F. For $x < 0$, we see that $F(x) = 0$. For $x \ge 0$,

$$F(x) = \int_{0}^{x} \frac{50}{(50 + t)^2}\, dt = \frac{-50}{50 + t}\Big|_{0}^{x} = \frac{-50}{50 + x} - (-1) = 1 - \frac{50}{50 + x}$$

$$= \frac{x}{50 + x}$$

Thus,

$$F(x) = \begin{cases} 0 & \text{if } x < 0 \\ x/(50 + x) & \text{if } x \geq 0 \end{cases}$$

Now, to solve the equation $P(0 \leq X \leq x) = .05$, we solve

$$F(x) - F(0) = .05 \quad F(0) = 0$$

$$\frac{x}{50 + x} = .05$$

$$x = 2.5 + .05x$$

$$.95x = 2.5$$

$$x \approx 2.6$$

If the drug is sold during the first 2.6 months it is on the shelf, then the probability that it is still good is .95.

Figure 6 shows an alternative method for solving Example 5 on a graphing utility.

Figure 6

Matched Problem 5   Repeat Example 5 if the pharmacist wants the probability that the drug is still good to be .99.

Exercise 9-2

A

In Problems 1 and 2, graph f, and show that f satisfies the first two conditions for a probability density function.

1. $f(x) = \begin{cases} \frac{1}{8}x & \text{if } 0 \leq x \leq 4 \\ 0 & \text{otherwise} \end{cases}$

2. $f(x) = \begin{cases} \frac{1}{9}x^2 & \text{if } 0 \leq x \leq 3 \\ 0 & \text{otherwise} \end{cases}$

In Problems 3–6, is f a probability density function? Explain.

3. $f(x) = \begin{cases} 2 + 2x & \text{if } -1 \leq x \leq 0 \\ 0 & \text{otherwise} \end{cases}$

4. $f(x) = \begin{cases} e^{-x} & \text{if } 0 \leq x \leq 4 \\ 0 & \text{otherwise} \end{cases}$

5. $f(x) = \begin{cases} 0.2 & \text{if } 1 \leq x \leq 5 \\ 0 & \text{otherwise} \end{cases}$

6. $f(x) = \begin{cases} 3x^2 - \dfrac{1}{2} & \text{if } -1 \leq x \leq 1 \\ 0 & \text{otherwise} \end{cases}$

7. Use the function in Problem 1 to find the indicated probabilities. Illustrate each probability with a graph.

(A) $P(1 \leq X \leq 3)$

(B) $P(X \leq 2)$

(C) $P(X > 3)$

8. Use the function in Problem 2 to find the indicated probabilities. Illustrate each probability with a graph.

(A) $P(1 \leq X \leq 2)$

(B) $P(X \geq 1)$

(C) $P(X < 2)$

9. Use the function in Problem 1 to find the indicated probabilities.

(A) $P(X = 1)$

(B) $P(X > 5)$

(C) $P(X < 5)$

10. Use the function in Problem 2 to find the indicated probabilities.

(A) $P(X = 2)$

(B) $P(X > 4)$

(C) $P(X < 4)$

11. Find and graph the cumulative distribution function associated with the function in Problem 1.

12. Find and graph the cumulative distribution function associated with the function in Problem 2.

13. Use the cumulative distribution function from Problem 11 to find the indicated probabilities.

(A) $P(2 \leq X \leq 4)$

(B) $P(0 < X < 2)$

14. Use the cumulative distribution function from Problem 12 to find the indicated probabilities.

(A) $P(0 \leq X \leq 1)$

(B) $P(2 < X < 3)$

15. Use the cumulative distribution function from Problem 11 to find the value of x that satisfies each equation.

(A) $P(0 \leq X \leq x) = \frac{1}{4}$

(B) $P(0 \leq X \leq x) = \frac{1}{9}$

16. Use the cumulative distribution function from Problem 12 to find the value of x that satisfies each equation.

(A) $P(0 \leq X \leq x) = \frac{1}{8}$

(B) $P(0 \leq X \leq x) = \frac{1}{64}$

B

In Problems 17 and 18, graph f, and show that f satisfies the first two conditions for a probability density function.

17. $f(x) = \begin{cases} 2/(1 + x)^3 & \text{if } x \geq 0 \\ 0 & \text{otherwise} \end{cases}$

18. $f(x) = \begin{cases} 2/(2 + x)^2 & \text{if } x \geq 0 \\ 0 & \text{otherwise} \end{cases}$

19. Use the function in Problem 17 to find the indicated probabilities.

(A) $P(1 \leq X \leq 4)$

(B) $P(X > 3)$

(C) $P(X \leq 2)$

20. Use the function in Problem 18 to find the indicated probabilities.

(A) $P(2 \leq X \leq 8)$

(B) $P(X \geq 3)$

(C) $P(X < 1)$

21. Find and graph the cumulative distribution function associated with the function in Problem 17.

22. Find and graph the cumulative distribution function associated with the function in Problem 18.

23. Use the cumulative distribution function from Problem 21 to find the value of x that satisfies each equation.

(A) $P(0 \leq X \leq x) = \frac{3}{4}$

(B) $P(X \geq x) = \frac{1}{16}$

24. Use the cumulative distribution function from Problem 22 to find the value of x that satisfies each equation.

(A) $P(0 \leq X \leq x) = \frac{3}{4}$ (B) $P(X > x) = \frac{1}{3}$

In Problems 25–28, find the associated cumulative distribution function. Graph both functions (on separate sets of axes).

25. $f(x) = \begin{cases} \frac{3}{2} x - \frac{3}{4} x^2 & \text{if } 0 \leq x \leq 2 \\ 0 & \text{otherwise} \end{cases}$

26. $f(x) = \begin{cases} \frac{3}{4} - \frac{3}{4} x^2 & \text{if } -1 \leq x \leq 1 \\ 0 & \text{otherwise} \end{cases}$

27. $f(x) = \begin{cases} \frac{1}{2} + \frac{1}{2} x^3 & \text{if } -1 \leq x \leq 1 \\ 0 & \text{otherwise} \end{cases}$

28. $f(x) = \begin{cases} \frac{3}{4} - \frac{3}{8}\sqrt{x} & \text{if } 0 \leq x \leq 4 \\ 0 & \text{otherwise} \end{cases}$

In Problems 29–32, use a graphing utility to approximate (to two decimal places) the value of x that satisfies the given equation for the indicated cumulative distribution function F(x).

29. $P(0 \leq X \leq x) = .2$ for $F(x)$ from Problem 25

30. $P(-1 \leq X \leq x) = .4$ for $F(x)$ from Problem 26

31. $P(-1 \leq X \leq x) = .6$ for $F(x)$ from Problem 27

32. $P(0 \leq X \leq x) = .7$ for $F(x)$ from Problem 28

For each function f in Problems 33–36, find a constant k so that kf is a probability density function, or explain why no such constant exists.

33. $f(x) = \begin{cases} e^{-x/3} & \text{if } x \geq 0 \\ 0 & \text{otherwise} \end{cases}$

34. $f(x) = \begin{cases} e^{-x/3} & \text{if } x \leq 0 \\ 0 & \text{otherwise} \end{cases}$

35. $f(x) = \begin{cases} 1/x^5 & \text{if } x \geq 1 \\ 0 & \text{otherwise} \end{cases}$

36. $f(x) = \begin{cases} x^3 & \text{if } 0 \leq x \leq 4 \\ 0 & \text{otherwise} \end{cases}$

37. Can the function

$$F(x) = \begin{cases} 2 - 2x & \text{if } 0 \leq x \leq 1 \\ 0 & \text{otherwise} \end{cases}$$

be a cumulative distribution function? Explain.

38. Can the function

$$F(x) = \begin{cases} 0.5 - 0.5x^3 & \text{if } -1 \leq x \leq 1 \\ 0 & \text{otherwise} \end{cases}$$

be a cumulative distribution function? Explain.

C

In Problems 39 and 40, use a graphing utility to graph f and compute the integrals $\int_{-N}^{N} f(x)\, dx$ for N = 1, 2, and 3. Discuss the evidence for concluding that f is a probability density function.

39. $f(x) = \dfrac{1}{\sqrt{2\pi}} e^{-0.5x^2}$ **40.** $f(x) = \dfrac{2}{\sqrt{2\pi}} e^{-2(x-1.5)^2}$

In Problems 41–44, $F(x)$ is the cumulative distribution function for a continuous random variable X. Find the probability density function $f(x)$ associated with each $F(x)$.

41. $F(x) = \begin{cases} 0 & \text{if } x < 0 \\ x^2 & \text{if } 0 \le x \le 1 \\ 1 & \text{if } x > 1 \end{cases}$

42. $F(x) = \begin{cases} 0 & \text{if } x < 1 \\ \frac{1}{2}x - \frac{1}{2} & \text{if } 1 \le x \le 3 \\ 1 & \text{if } x > 3 \end{cases}$

43. $F(x) = \begin{cases} 0 & \text{if } x < 0 \\ 6x^2 - 8x^3 + 3x^4 & \text{if } 0 \le x \le 1 \\ 1 & \text{if } x > 1 \end{cases}$

44. $F(x) = \begin{cases} 1 - (1/x^3) & \text{if } x \ge 1 \\ 0 & \text{otherwise} \end{cases}$

In Problems 45 and 46, find the associated cumulative distribution function.

45. $f(x) = \begin{cases} x & \text{if } 0 \le x \le 1 \\ 2 - x & \text{if } 1 < x \le 2 \\ 0 & \text{otherwise} \end{cases}$

46. $f(x) = \begin{cases} \frac{1}{4} & \text{if } 0 \le x \le 1 \\ \frac{1}{2} & \text{if } 1 < x \le 2 \\ \frac{1}{4} & \text{if } 2 < x \le 3 \\ 0 & \text{otherwise} \end{cases}$

Applications

47. **Electricity consumption.** The daily demand for electricity (in millions of kilowatt-hours) in a large city is a continuous random variable with probability density function

$$f(x) = \begin{cases} .2 - .02x & \text{if } 0 \le x \le 10 \\ 0 & \text{otherwise} \end{cases}$$

(A) Evaluate $\int_2^6 f(x)\, dx$ and interpret the results.

(B) What is the probability that the daily demand for electricity is less than 8 million kilowatt-hours?

(C) What is the probability that 5 million kilowatt-hours will not be sufficient to meet the daily demand?

48. **Gasoline consumption.** The daily demand for gasoline (in millions of gallons) in a large city is a continuous random variable with probability density function

$$f(x) = \begin{cases} .4 - .08x & \text{if } 0 \le x \le 5 \\ 0 & \text{otherwise} \end{cases}$$

(A) Evaluate $\int_1^4 f(x)\, dx$ and interpret the results.

(B) What is the probability that the daily demand is less than 2 million gallons?

(C) What is the probability that 3 million gallons will not be sufficient to meet the daily demand?

49. **Time-sharing.** In a computer time-sharing network, the time it takes (in seconds) to respond to a user's request is a continuous random variable with probability density function given by

$$f(x) = \begin{cases} \frac{1}{10}e^{-x/10} & \text{if } x \ge 0 \\ 0 & \text{otherwise} \end{cases}$$

(A) Evaluate $\int_5^{10} f(x)\, dx$ and interpret the results.

(B) What is the probability that the computer responds within 1 second?

(C) What is the probability that a user must wait more than 4 seconds for a response?

50. **Waiting time.** The time (in minutes) a customer must wait in line at a bank is a continuous random variable with probability density function given by

$$f(x) = \begin{cases} \frac{1}{2}e^{-x/2} & \text{if } x \ge 0 \\ 0 & \text{otherwise} \end{cases}$$

(A) Evaluate $\int_2^6 f(x)\, dx$ and interpret the results.

(B) What is the probability that a customer waits less than 3 minutes?

(C) What is the probability that a customer waits more than 5 minutes?

51. **Demand.** The weekly demand for hamburger (in thousands of pounds) for a chain of supermarkets is a continuous random variable with probability density function given by

$$f(x) = \begin{cases} .003x\sqrt{100 - x^2} & \text{if } 0 \le x \le 10 \\ 0 & \text{otherwise} \end{cases}$$

(A) What is the probability that more than 4,000 pounds of hamburger are demanded?

(B) The manager of the meat department orders 8,000 pounds of hamburger. What is the probability that the demand will not exceed this amount?

(C) The manager wants the probability that the demand does not exceed the amount ordered to be .9. How much hamburger meat should be ordered?

52. **Demand.** The demand for a weekly sports magazine (in thousands of copies) in a certain city is a continuous random variable with probability density function given by

$$f(x) = \begin{cases} \frac{3}{125} x\sqrt{25 - x^2} & \text{if } 0 \le x \le 5 \\ 0 & \text{otherwise} \end{cases}$$

(A) The magazine's distributor in this city orders 3,000 copies of the magazine. What is the probability that the demand exceeds this number?

(B) What is the probability that the demand does not exceed 4,000 copies?

(C) If the distributor wants to be 95% certain that the demand does not exceed the number ordered, how many copies should be ordered?

53. **Life expectancy.** The life expectancy (in minutes) of a certain microscopic organism is a continuous random variable with probability density function given by

$$f(x) = \begin{cases} \frac{1}{5,000}(10x^3 - x^4) & \text{if } 0 \le x \le 10 \\ 0 & \text{otherwise} \end{cases}$$

(A) What is the probability that an organism lives for at least 7 minutes?

(B) What is the probability that an organism lives for at most 5 minutes?

54. Life expectancy. The life expectancy (in months) of plants of a certain species is a continuous random variable with probability density function given by

$$f(x) = \begin{cases} \frac{1}{36}(6x - x^2) & \text{if } 0 \le x \le 6 \\ 0 & \text{otherwise} \end{cases}$$

(A) What is the probability that one of these plants survives for at least 4 months?

(B) What is the probability that one of these plants survives for at most 5 months?

55. Shelf life. The shelf life (in days) of a perishable drug is a continuous random variable with probability density function given by

$$f(x) = \begin{cases} 800x/(400 + x^2)^2 & \text{if } x \ge 0 \\ 0 & \text{otherwise} \end{cases}$$

(A) What is the probability that the drug has a shelf life of at most 20 days?

(B) What is the probability that the shelf life exceeds 15 days?

(C) If the user wants the probability that the drug is still good to be .8, when is the last time it should be used?

56. Shelf life. Repeat Problem 55 if

$$f(x) = \begin{cases} 200x/(100 + x^2)^2 & \text{if } x \ge 0 \\ 0 & \text{otherwise} \end{cases}$$

57. Learning. The number of words per minute a beginner can type after 1 week of practice is a continuous random variable with probability density function given by

$$f(x) = \begin{cases} \frac{1}{20}e^{-x/20} & \text{if } x \ge 0 \\ 0 & \text{otherwise} \end{cases}$$

(A) What is the probability that a beginner can type at least 30 words per minute after 1 week of practice?

(B) What is the probability that a beginner can type at least 80 words per minute after 1 week of practice?

58. Learning. The number of hours it takes a chimpanzee to learn a new task is a continuous random variable with probability density function given by

$$f(x) = \begin{cases} \frac{4}{9}x^2 - \frac{4}{27}x^3 & \text{if } 0 \le x \le 3 \\ 0 & \text{otherwise} \end{cases}$$

(A) What is the probability that the chimpanzee learns the task in the first hour?

(B) What is the probability that the chimpanzee does not learn the task in the first 2 hours?

Answers to Matched Problems

1.

(A) .68

(B) .784

(C) .648

(D) 0

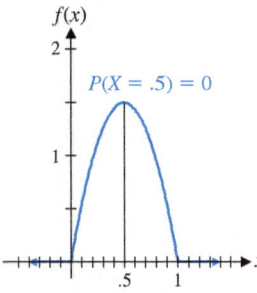

2. .248; .896

3. (A) $\frac{1}{6}$ (B) $\frac{2}{7}$ (C) $\frac{5}{6}$

4. $F(x) = \begin{cases} 0 & \text{if } x < 0 \\ 3x^2 - 2x^3 & \text{if } 0 \le x \le 1; \ P(.3 \le X \le .7) = .568 \\ 1 & \text{if } x > 1 \end{cases}$

5. Approx. $\frac{1}{2}$ month, or 15 days

9-3   Expected Value, Standard Deviation, and Median

- Expected Value (or Mean) and Standard Deviation
- Alternative Formula for Variance
- Median

Expected Value (or Mean) and Standard Deviation

Statisticians use a measure of central tendency called the mean and a measure of dispersion called the standard deviation to help describe sets of data. These same terms are used to describe both discrete and continuous random variables. For example, if X is the discrete random variable representing the number of heads when three coins are tossed (Table 1), then the mean and standard deviation are computed as follows:

$$\text{Mean} = 0 \cdot \tfrac{1}{8} + 1 \cdot \tfrac{3}{8} + 2 \cdot \tfrac{3}{8} + 3 \cdot \tfrac{1}{8} = \tfrac{3}{2}$$

Table 1

x	0	1	2	3
$p(x)$	$\tfrac{1}{8}$	$\tfrac{3}{8}$	$\tfrac{3}{8}$	$\tfrac{1}{8}$

$$\text{Variance} = (0 - \tfrac{3}{2})^2 \cdot \tfrac{1}{8} + (1 - \tfrac{3}{2})^2 \cdot \tfrac{3}{8} + (2 - \tfrac{3}{2})^2 \cdot \tfrac{3}{8} + (3 - \tfrac{3}{2})^2 \cdot \tfrac{1}{8}$$

$$= \tfrac{3}{4}$$

$$\text{standard deviation} = \sqrt{\text{Variance}} = \sqrt{\tfrac{3}{4}} \approx .8660$$

Notice that these quantities were computed using finite sums and the probabilities in Table 1.

The extension of these ideas to continuous random variables makes use of integration and the probability density function.

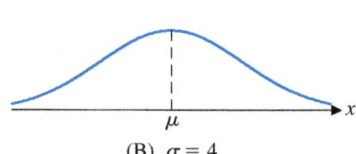

Figure 1

> **DEFINITION   Expected Value (or Mean) and Standard Deviation for a Continuous Random Variable**
>
> Let $f(x)$ be the probability density function for a continuous random variable X. The **expected value, or mean, of X** is
>
> $$\mu = E(X) = \int_{-\infty}^{\infty} x f(x)\, dx$$
>
> The **variance** is
>
> $$V(X) = \int_{-\infty}^{\infty} (x - \mu)^2 f(x)\, dx$$
>
> and the **standard deviation** is
>
> $$\sigma = \sqrt{V(X)}$$

Just as in the discrete case, the mean of a continuous random variable is a measure of central tendency for the variable, and the standard deviation is a measure of the dispersion of the variable about the mean. This is illustrated in Figure 1. The probability density function in Figure 1A has a standard deviation of 1. Most of the area under the curve is near the mean. In Figure 1B, the standard deviation is four times as large, and the area under the graph is much more spread out.

EXAMPLE 1   **Computing Expected Value (Mean), Variance, and Standard Deviation** Find the expected value (mean), variance, and standard deviation for

$$f(x) = \begin{cases} 12x^2 - 12x^3 & \text{if } 0 \le x \le 1 \\ 0 & \text{otherwise} \end{cases}$$

SOLUTION
$$\mu = E(X) = \int_{-\infty}^{\infty} x f(x) = \int_0^1 x(12x^2 - 12x^3)\, dx$$

$$= \int_0^1 (12x^3 - 12x^4)\, dx$$

$$= (3x^4 - \tfrac{12}{5}x^5) \Big|_0^1 = \tfrac{3}{5} \quad \text{Expected value (mean)}$$

$$V(X) = \int_{-\infty}^{\infty} (x - \mu)^2 f(x)\, dx = \int_0^1 (x - \tfrac{3}{5})^2 (12x^2 - 12x^3)\, dx$$

$$= \int_0^1 (x^2 - \tfrac{6}{5}x + \tfrac{9}{25})(12x^2 - 12x^3)\, dx$$

$$= \int_0^1 (\tfrac{108}{25}x^2 - \tfrac{468}{25}x^3 + \tfrac{132}{5}x^4 - 12x^5)\, dx$$

$$= (\tfrac{36}{25}x^3 - \tfrac{117}{25}x^4 + \tfrac{132}{25}x^5 - 2x^6)\Big|_0^1 = \tfrac{1}{25} \qquad \text{Variance}$$

$$\sigma = \sqrt{V(X)} = \sqrt{\tfrac{1}{25}} = \tfrac{1}{5} \qquad\qquad \text{Standard deviation}$$

Matched Problem 1   Find the expected value (mean), variance, and standard deviation for

$$f(x) = \begin{cases} 6x - 6x^2 & \text{if } 0 \le x \le 1 \\ 0 & \text{otherwise} \end{cases}$$

EXPLORE & DISCUSS 1   Consider the probability density function

$$f(x) = \begin{cases} \tfrac{1}{2}x & \text{if } 0 \le x \le 2 \\ 0 & \text{otherwise} \end{cases}$$

(A) Find the mean μ.

(B) The vertical line $x = \mu$ divides the area under the graph of $y = f(x)$ into two regions. Do these regions have equal areas? If not, find a number m with the property that the vertical line $x = m$ divides the area under the graph of $y = f(x)$ into two regions with equal areas.

The graph of the probability density function considered in Example 1 is shown in Figure 2A, with the areas to the left and right of the mean indicated (computations omitted). As you discovered in the preceding Explore–Discuss 1, the vertical line through the mean does not divide the region under the graph of f into two regions with equal areas. The value of x that accomplishes this is the *median*, which we will discuss later in this section. However, the mean does have a geometric interpretation. If the shaded region in Figure 2A were drawn on a piece of wood of uniform thickness and the wood cut around the outside of the region, then the resulting object would balance on a wedge placed at the mean $\mu = .6$ (see Fig. 2B). For this reason, the mean m is often referred to as a *measure of central tendency*.

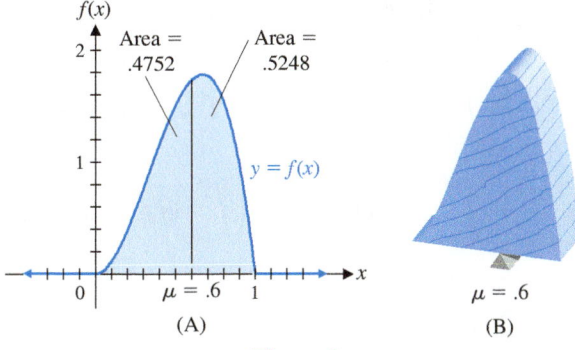

(A)   (B)

Figure 2

EXAMPLE 2   **Life Expectancy** The life expectancy (in hours) for a particular brand of light bulbs is a continuous random variable with probability density function

$$f(x) = \begin{cases} \dfrac{1}{100} - \dfrac{1}{20{,}000}\, x & \text{if } 0 \le x \le 200 \\ 0 & \text{otherwise} \end{cases}$$

(A) What is the average life expectancy of one of these light bulbs?

(B) What is the probability that a bulb will last longer than this average?

SOLUTION   (A) Since the value of this random variable is the number of hours a bulb lasts, the average life expectancy is just the expected value of the random variable. Thus,

$$E(X) = \int_{-\infty}^{\infty} xf(x)\,dx = \int_{0}^{200} x(\tfrac{1}{100} - \tfrac{1}{20,000}\,x)\,dx$$

$$= \int_{0}^{200} (\tfrac{1}{100}\,x - \tfrac{1}{20,000}\,x^2)\,dx = (\tfrac{1}{200}\,x^2 - \tfrac{1}{60,000}\,x^3)\Big|_{0}^{200}$$

$$= \tfrac{200}{3} \approx 66.67 \text{ hours}$$

(B) The probability that a bulb lasts longer than $\tfrac{200}{3}$ hours is

$$P(X > \tfrac{200}{3}) = \int_{200/3}^{\infty} f(x)\,dx = \int_{200/3}^{200} (\tfrac{1}{100} - \tfrac{1}{20,000}\,x)\,dx$$

$$= (\tfrac{1}{100}\,x - \tfrac{1}{40,000}\,x^2)\Big|_{200/3}^{200}$$

$$= 1 - \tfrac{5}{9} = \tfrac{4}{9} \approx 0.44$$

Matched Problem 2   Repeat Example 2 if the probability density function is

$$f(x) = \begin{cases} \tfrac{1}{200} - \tfrac{1}{90,000}\,x & \text{if } 0 \le x \le 300 \\ 0 & \text{otherwise} \end{cases}$$

Alternative Formula for Variance

The term $(x - \mu)^2$ in the formula for $V(X)$ introduces some complicated algebraic manipulations in the evaluation of the integral. We can use the properties of the definite integral to simplify this formula. Thus,

$$V(X) = \int_{-\infty}^{\infty} (x - \mu)^2 f(x)\,dx$$

$$= \int_{-\infty}^{\infty} (x^2 - 2x\mu + \mu^2)f(x)\,dx \qquad \text{Algebra}$$

$$= \int_{-\infty}^{\infty} [x^2 f(x) - 2x\mu f(x) + \mu^2 f(x)]\,dx \qquad \text{Algebra}$$

$$= \int_{-\infty}^{\infty} x^2 f(x)\,dx - \int_{-\infty}^{\infty} 2x\mu f(x)\,dx + \int_{-\infty}^{\infty} \mu^2 f(x)\,dx \qquad \text{Definite integral property}$$

$$= \int_{-\infty}^{\infty} x^2 f(x)\,dx - 2\mu \int_{-\infty}^{\infty} xf(x)\,dx + \mu^2 \int_{-\infty}^{\infty} f(x)\,dx \qquad \text{Definite integral property}$$

$$= \int_{-\infty}^{\infty} x^2 f(x)\,dx - 2\mu(\mu) + \mu^2(1) \qquad \int_{-\infty}^{\infty} xf(x)\,dx = \mu, \ \int_{-\infty}^{\infty} xf(x)\,dx = 1$$

$$= \int_{-\infty}^{\infty} x^2 f(x)\,dx - \mu^2$$

In general, it will be easier to evaluate $\int_{-\infty}^{\infty} x^2 f(x)\,dx$ than to evaluate $\int_{-\infty}^{\infty} (x - \mu)^2 f(x)\,dx$.

> **THEOREM 1   Alternative Formula for Variance**
>
> $$V(X) = \int_{-\infty}^{\infty} x^2 f(x)\,dx - \mu^2$$

EXAMPLE 3   **Computing Variance** Use the alternative formula for variance (Theorem 1) to compute the variance in Example 1.

SOLUTION   From Example 1, we have $\mu = \int_{-\infty}^{\infty} xf(x)\,dx = \frac{3}{5}$. Thus,

$$\int_{-\infty}^{\infty} x^2 f(x)\,dx = \int_0^1 x^2(12x^2 - 12x^3)\,dx \qquad f(x) = \begin{cases} 12x^2 - 12x^3 & \text{if } 0 \le x \le 1 \\ 0 & \text{otherwise} \end{cases}$$

$$= \int_0^1 (12x^4 - 12x^5)\,dx$$

$$= \left(\tfrac{12}{5}x^5 - \tfrac{12}{6}x^6\right)\Big|_0^1 = \tfrac{2}{5} = .4$$

$$V(X) = \int_{-\infty}^{\infty} x^2 f(x)\,dx - \mu^2 = \tfrac{2}{5} - (\tfrac{3}{5})^2 = \tfrac{1}{25} = .04$$

Matched Problem 3   Use the alternative formula for variance (Theorem 1) to compute the variance in Matched Problem 1.

EXAMPLE 4   **Computing Expected Value (Mean), Variance, and Standard Deviation** Find the expected value (mean), variance, and standard deviation for

$$f(x) = \begin{cases} 3/x^4 & \text{if } x \ge 1 \\ 0 & \text{otherwise} \end{cases}$$

SOLUTION

$$\mu = \int_{-\infty}^{\infty} xf(x)\,dx = \int_1^{\infty} x\,\frac{3}{x^4}\,dx = \lim_{R \to \infty} \int_1^R \frac{3}{x^3}\,dx$$

$$= \lim_{R \to \infty}\left[-\frac{3}{2}\left(\frac{1}{x^2}\right)\right]\Big|_1^R = \lim_{R \to \infty}\left[-\frac{3}{2}\left(\frac{1}{R^2}\right) + \frac{3}{2}\right] = \frac{3}{2} = 1.5$$

$$\int_{-\infty}^{\infty} x^2 f(x)\,dx = \int_1^{\infty} x^2\,\frac{3}{x^4}\,dx = \lim_{R \to \infty} \int_1^R \frac{3}{x^2}\,dx = \lim_{R \to \infty}\left(-\frac{3}{x}\right)\Big|_1^R$$

$$= \lim_{R \to \infty}\left(-\frac{3}{R} + 3\right) = 3$$

$$V(X) = \int_{-\infty}^{\infty} x^2 f(x)\,dx - \mu^2 = 3 - \left(\frac{3}{2}\right)^2 = \frac{3}{4} = .75$$

$$\sigma = \sqrt{V(X)} = \sqrt{\frac{3}{4}} = \frac{\sqrt{3}}{2} \approx .8660$$

Matched Problem 4   Find the expected value (mean), variance, and standard deviation for

$$f(x) = \begin{cases} 24/x^4 & \text{if } x \ge 2 \\ 0 & \text{otherwise} \end{cases}$$

Median

Another measurement often used to describe the properties of a random variable is the median. A **median** is a value of the random variable that divides the area under the graph of the probability density function into two equal parts (see Fig. 3). If m is a median, then m must satisfy

$$P(X \le m) = \tfrac{1}{2}$$

Generally, this equation is solved by first finding the cumulative distribution function.

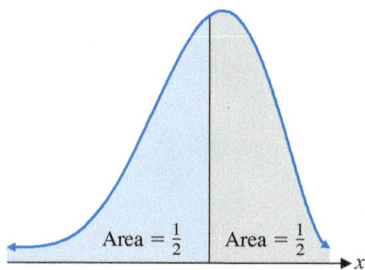

Area $= \frac{1}{2}$ Area $= \frac{1}{2}$

Figure 3 Large Calculator

EXAMPLE 5   **Finding the Median** Find the median of the continuous random variable with probability density function

$$f(x) = \begin{cases} 3/x^4 & \text{if } x \geq 1 \\ 0 & \text{otherwise} \end{cases}$$

SOLUTION   **Step 1.** **Find the cumulative distribution function:** For $x < 1$, we have $F(x) = 0$. If $x \geq 1$, then

$$F(x) = \int_{-\infty}^{x} f(t)\, dt = \int_{1}^{x} \frac{3}{t^4}\, dt = -\frac{1}{t^3}\Big|_{1}^{x} = -\frac{1}{x^3} + 1 = 1 - \frac{1}{x^3}$$

Step 2. **Solve the equation** $P(X \leq m) = \frac{1}{2}$ **for** m:

$$F(m) = P(X \leq m)$$
$$1 - \frac{1}{m^3} = \frac{1}{2}$$
$$\frac{1}{2} = \frac{1}{m^3}$$
$$m^3 = 2$$
$$m = \sqrt[n]{2}$$

Thus, the median is $\sqrt[3]{2} \approx 1.26$.

Figure 4 shows an alternative method for solving Example 5 on a graphing utility.

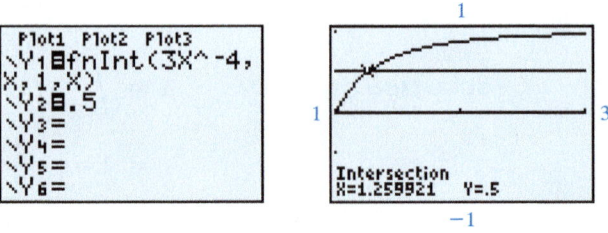

Figure 4

Matched Problem 5   Find the median of the continuous random variable with probability density function

$$f(x) = \begin{cases} 24/x^4 & \text{if } x \geq 2 \\ 0 & \text{otherwise} \end{cases}$$

EXPLORE & DISCUSS 2   Consider the probability density function f defined by the graph in Figure 5. Does a median exist? Is it unique? Explain.

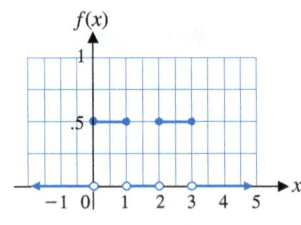

Figure 5

EXAMPLE 6   **Life Expectancy** In Example 2, find the median life expectancy of a lightbulb.

SOLUTION   **Step 1.** **Find the cumulative distribution function:** If $x < 0$, we have $F(x) = 0$. If $0 \le x \le 200$, then

$$F(x) = \int_{-\infty}^{x} f(t)\, dt \qquad\qquad f(x) = \begin{cases} \frac{1}{100} - \frac{1}{20{,}000}\, x & \text{if } 0 \le x \le 200 \\ 0 & \text{otherwise} \end{cases}$$

$$= \int_{0}^{x} \left(\tfrac{1}{100} - \tfrac{1}{20{,}000}\, t \right) dt$$

$$= \left(\tfrac{1}{100} t - \tfrac{1}{40{,}000} t^2 \right)\Big|_{0}^{x}$$

$$= \tfrac{1}{100}\, x - \tfrac{1}{40{,}000}\, x^2$$

If $x > 200$, then

$$F(x) = \int_{-\infty}^{x} f(t)\, dt = \int_{-\infty}^{0} f(t)\, dt + \int_{0}^{200} f(t)\, dt + \int_{200}^{x} f(t)\, dt$$

$$= 0 + 1 + 0 = 1$$

Thus,

$$F(x) = \begin{cases} 0 & \text{if } x < 0 \\ \tfrac{1}{100}\, x - \tfrac{1}{40{,}000}\, x^2 & \text{if } 0 \le x \le 200 \\ 1 & \text{if } x > 200 \end{cases}$$

Step 2. **Solve the equation** $P(X \le m) = \frac{1}{2}$ **for** m:

$$F(m) = P(X \le m) = \tfrac{1}{2}$$

$$\tfrac{1}{100} m - \tfrac{1}{40{,}000} m^2 = \tfrac{1}{2}$$

$$m^2 - 400m + 20{,}000 = 0 \qquad \textit{The solution must occur for } 0 \le m \le 200.$$

This quadratic equation has two solutions, $200 + 100\sqrt{2}$ and $200 - 100\sqrt{2}$. Since m must lie in the interval $[0, 200]$, the second solution is the correct answer. Thus, the median life expectancy is $200 - 100\sqrt{2} \approx 58.58$ hours.

Matched Problem 6   In Matched Problem 2, find the median life expectancy of a lightbulb.

Figure 6 shows the mean and the median for two of the probability density functions discussed in this section (see Examples 2, 4, 5, and 6). As we noted earlier, the mean and the median generally are not equal.

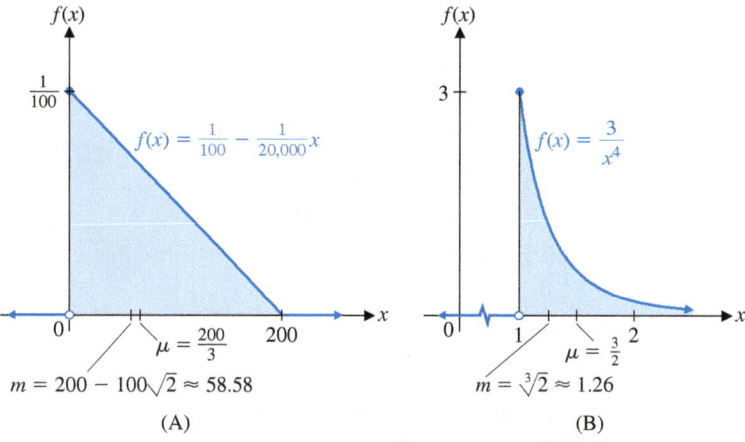

(A)   (B)

Figure 6

CONCEPTUAL INSIGHT

Although the median is smaller than the mean for the probability distributions of Figure 6, for other distributions the median is larger than the mean. To obtain examples of such distributions, simply reflect the graphs of Figure 6 in the y axis.

Exercise 9-3

A

In Problems 1–6, find the mean, variance, and standard deviation.

1. $f(x) = \begin{cases} 2 & \text{if } 0 \le x \le 0.5 \\ 0 & \text{otherwise} \end{cases}$

2. $f(x) = \begin{cases} \frac{1}{4} & \text{if } -1 \le x \le 3 \\ 0 & \text{otherwise} \end{cases}$

3. $f(x) = \begin{cases} \frac{1}{8}x & \text{if } 0 \le x \le 4 \\ 0 & \text{otherwise} \end{cases}$

4. $f(x) = \begin{cases} 18x & \text{if } 0 \le x \le \frac{1}{3} \\ 0 & \text{otherwise} \end{cases}$

5. $f(x) = \begin{cases} 4 - 2x & \text{if } 1 \le x \le 2 \\ 0 & \text{otherwise} \end{cases}$

6. $f(x) = \begin{cases} 2 - \frac{1}{2}x & \text{if } 2 \le x \le 4 \\ 0 & \text{otherwise} \end{cases}$

In Problems 7–12, find the median.

7. $f(x) = \begin{cases} 0.2 & \text{if } -2 \le x \le 3 \\ 0 & \text{otherwise} \end{cases}$

8. $f(x) = \begin{cases} 10 & \text{if } 2.5 \le x \le 2.6 \\ 0 & \text{otherwise} \end{cases}$

9. $f(x) = \begin{cases} \frac{1}{6}x & \text{if } 2 \le x \le 4 \\ 0 & \text{otherwise} \end{cases}$

10. $f(x) = \begin{cases} \frac{1}{4}x & \text{if } 1 \le x \le 3 \\ 0 & \text{otherwise} \end{cases}$

11. $f(x) = \begin{cases} \frac{1}{2} - \frac{1}{8}x & \text{if } 0 \le x \le 4 \\ 0 & \text{otherwise} \end{cases}$

12. $f(x) = \begin{cases} 2 - 2x & \text{if } 0 \le x \le 1 \\ 0 & \text{otherwise} \end{cases}$

B

Problems 13 and 14 refer to random variables X_1 and X_2 with probability density functions f_1 and f_2, respectively, where

$$f_1(x) = \begin{cases} 0.5 & \text{if } 0 \le x \le 2 \\ 0 & \text{otherwise} \end{cases}$$

$$f_2(x) = \begin{cases} 0.25 & \text{if } -2 \le x \le 2 \\ 0 & \text{otherwise} \end{cases}$$

13. Explain how you can predict from the graphs of f_1 and f_2 which random variable, X_1 or X_2, has the greater mean. Check your prediction by computing the mean of each.

14. Explain how you can predict from the graphs of f_1 and f_2 which random variable, X_1 or X_2, has the greater variance. Check your prediction by computing the variance of each.

Problems 15 and 16 refer to random variables X_1 and X_2 with probability density functions g_1 and g_2, respectively, where

$$g_1(x) = \begin{cases} 0.5 & \text{if } 0 \le x < 1 \\ 0.25 & \text{if } 1 \le x \le 3 \\ 0 & \text{otherwise} \end{cases}$$

$$g_2(x) = \begin{cases} 0.5 & \text{if } 0 \le x < 1 \\ 0.125 & \text{if } 1 \le x \le 5 \\ 0 & \text{otherwise} \end{cases}$$

15. Explain how you can predict from the graphs of g_1 and g_2 which random variable, X_1 or X_2, has the greater median. Check your prediction by computing the median of each.

16. Explain how you can predict from the graphs of g_1 and g_2 which random variable, X_1 or X_2, has the greater mean. Check your prediction by computing the mean of each.

In Problems 17–20, find the mean, variance, and standard deviation.

17. $f(x) = \begin{cases} 4/x^5 & \text{if } x \ge 1 \\ 0 & \text{otherwise} \end{cases}$

18. $f(x) = \begin{cases} 5/x^6 & \text{if } x \ge 1 \\ 0 & \text{otherwise} \end{cases}$

19. $f(x) = \begin{cases} 64/x^5 & \text{if } x \ge 2 \\ 0 & \text{otherwise} \end{cases}$

20. $f(x) = \begin{cases} 81/x^4 & \text{if } x \ge 3 \\ 0 & \text{otherwise} \end{cases}$

In Problems 21 and 22, use a graphing utility to approximate the mean, variance, and standard deviation by replacing integrals over $(-\infty, \infty)$ by integrals over $[-10, 10]$.

21. $f(x) = \dfrac{2}{\sqrt{2\pi}} e^{-2(x-1)^2}$

22. $f(x) = \dfrac{1}{1.4\sqrt{2\pi}} e^{-(x+2.1)^2/3.92}$

In Problems 23–30, find the median.

23. $f(x) = \begin{cases} 1/x & \text{if } 1 \le x \le e \\ 0 & \text{otherwise} \end{cases}$

24. $f(x) = \begin{cases} 1/(2x) & \text{if } 1 \le x \le e^2 \\ 0 & \text{otherwise} \end{cases}$

25. $f(x) = \begin{cases} 4/(2 + x)^2 & \text{if } 0 \le x \le 2 \\ 0 & \text{otherwise} \end{cases}$

26. $f(x) = \begin{cases} 2/(1 + x)^2 & \text{if } 0 \le x \le 1 \\ 0 & \text{otherwise} \end{cases}$

27. $f(x) = \begin{cases} 1/(1 + x)^2 & \text{if } x \ge 0 \\ 0 & \text{otherwise} \end{cases}$

28. $f(x) = \begin{cases} 3/(3 + x)^2 & \text{if } x \ge 0 \\ 0 & \text{otherwise} \end{cases}$

29. $f(x) = \begin{cases} 2e^{-2x} & \text{if } x \ge 0 \\ 0 & \text{otherwise} \end{cases}$

30. $f(x) = \begin{cases} e^{-x} & \text{if } x \ge 0 \\ 0 & \text{otherwise} \end{cases}$

In Problems 31 and 32, use the graph of the cumulative distribution function F to find the median. (Assume each portion of the graph of F is a straight line segment.) Is the median unique? Explain.

31.

32.

In Problems 33 and 34, use the graph of the probability density function f to find the cumulative distribution function F. (Assume each portion of the graph of f is a straight line segment.) Find the median. Is the median unique? Explain.

33.

34.

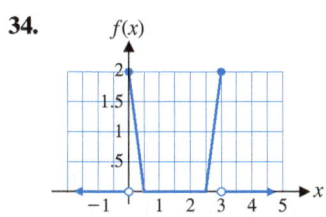

C

In Problems 35 and 36, $f(x)$ is a continuous probability density function with mean μ and standard deviation σ; a and b are constants. Evaluate each integral, expressing the result in terms of a, b, μ, and σ.

35. $\int_{-\infty}^{\infty} (ax + b)f(x)\, dx$

36. $\int_{-\infty}^{\infty} (x - a)^2 f(x)\, dx$

The **quartile points** for a probability density function are the values x_1, x_2, x_3 that divide the area under the graph of the function into four equal parts. Find the quartile points for the probability density functions in Problems 37–40.

37. $f(x) = \begin{cases} \frac{1}{2}x & \text{if } 0 \le x \le 2 \\ 0 & \text{otherwise} \end{cases}$

38. $f(x) = \begin{cases} 3x^2 & \text{if } 0 \le x \le 1 \\ 0 & \text{otherwise} \end{cases}$

39. $f(x) = \begin{cases} 3/(3 + x)^2 & \text{if } x \ge 0 \\ 0 & \text{otherwise} \end{cases}$

40. $f(x) = \begin{cases} 1/(1 + x)^2 & \text{if } x \ge 0 \\ 0 & \text{otherwise} \end{cases}$

In Problems 41–44, use a graphing utility to approximate the median of the indicated probability density function f (to two decimal places).

41. $f(x) = \begin{cases} 4x - 4x^3 & \text{if } 0 \le x \le 1 \\ 0 & \text{otherwise} \end{cases}$

42. $f(x) = \begin{cases} 3x - 3x^5 & \text{if } 0 \le x \le 1 \\ 0 & \text{otherwise} \end{cases}$

43. $f(x) = \begin{cases} \dfrac{1}{2x^2} + \dfrac{3}{2}x^4 & \text{if } x \ge 1 \\ 0 & \text{otherwise} \end{cases}$

44. $f(x) = \begin{cases} \dfrac{2}{3x^3} + \dfrac{2}{x^4} & \text{if } x \ge 1 \\ 0 & \text{otherwise} \end{cases}$

Applications

45. Profit. A building contractor's profit (in thousands of dollars) on each unit in a subdivision is a continuous random variable with probability density function given by

$$f(x) = \begin{cases} 0.08(20 - x) & \text{if } 15 \le x \le 20 \\ 0 & \text{otherwise} \end{cases}$$

(A) Find the contractor's expected profit.

(B) Find the median profit.

46. Electricity consumption. The daily consumption of electricity (in millions of kilowatt-hours) in a large city is a continuous random variable with probability density function

$$f(x) = \begin{cases} .2 - .02x & \text{if } 0 \le x \le 10 \\ 0 & \text{otherwise} \end{cases}$$

(A) Find the expected daily consumption of electricity.

(B) Find the median daily consumption of electricity.

47. Waiting time. The time (in minutes) a customer must wait in line at a bank is a continuous random variable with probability density function given by

$$f(x) = \begin{cases} \frac{1}{3}e^{-x/3} & \text{if } x \ge 0 \\ 0 & \text{otherwise} \end{cases}$$

Find the median waiting time.

48. Product life. The life expectancy (in years) of an automobile battery is a continuous random variable with probability density function given by

$$f(x) = \begin{cases} \frac{1}{2}e^{-x/2} & \text{if } x \ge 0 \\ 0 & \text{otherwise} \end{cases}$$

Find the median life expectancy.

49. Water consumption. The daily consumption of water (in millions of gallons) in a small city is a continuous random variable with probability density function given by

$$f(x) = \begin{cases} 1/(1 + x^2)^{3/2} & \text{if } x \ge 0 \\ 0 & \text{otherwise} \end{cases}$$

Find the expected daily consumption.

50. Gasoline consumption. The daily consumption of gasoline (in millions of gallons) in a large city is a continuous random variable with probability density function

$$f(x) = \begin{cases} 4/(4 + x^2)^{3/2} & \text{if } x \ge 0 \\ 0 & \text{otherwise} \end{cases}$$

Find the expected daily consumption of gasoline.

51. Life expectancy. The life expectancy of a certain microscopic organism (in minutes) is a continuous random variable with probability density function given by

$$f(x) = \begin{cases} \frac{1}{5,000}(10x^3 - x^4) & \text{if } 0 \le x \le 10 \\ 0 & \text{otherwise} \end{cases}$$

Find the mean life expectancy of one of these organisms.

52. Life expectancy. The life expectancy (in months) of plants of a certain species is a continuous random variable with probability density function given by

$$f(x) = \begin{cases} \frac{1}{36}(6x - x^2) & \text{if } 0 \le x \le 6 \\ 0 & \text{otherwise} \end{cases}$$

Find the mean life expectancy of one of these plants.

53. Shelf life. The shelf life (in days) of a perishable drug is a continuous random variable with probability density function given by

$$f(x) = \begin{cases} 800x/(400 + x^2)^2 & \text{if } x \ge 0 \\ 0 & \text{otherwise} \end{cases}$$

Find the median shelf life.

54. Shelf life. Repeat Problem 53 if

$$f(x) = \begin{cases} 200x/(100 + x^2)^2 & \text{if } x \ge 0 \\ 0 & \text{otherwise} \end{cases}$$

55. Learning. The number of hours it takes a chimpanzee to learn a new task is a continuous random variable with probability density function given by

$$f(x) = \begin{cases} \frac{4}{9}x^2 - \frac{4}{27}x^3 & \text{if } 0 \le x \le 3 \\ 0 & \text{otherwise} \end{cases}$$

What is the expected number of hours it will take a chimpanzee to learn the task?

56. Voter turnout. The number of registered voters (in thousands) who vote in an off-year election in a small town is a continuous random variable with probability density function given by

$$f(x) = \begin{cases} 1 - \frac{1}{8}x & \text{if } 4 \le x \le 8 \\ 0 & \text{otherwise} \end{cases}$$

(A) Find the expected voter turnout.

(B) Find the median voter turnout.

Answers to Matched Problems

1. $\mu = \frac{1}{2}; V(X) = \frac{1}{20}; \sigma \approx .2236$

2. (A) 125 hr (B) $\frac{133}{288}$

3. $\frac{1}{20}$

4. $\mu = 3; V(X) = 3; \sigma = \sqrt{3} \approx 1.732$

5. $m = \sqrt[3]{16} = 2\sqrt[3]{2} \approx 2.52$

6. $m = 450 - 150\sqrt{5} \approx 114.59$ hr

9-4 Special Probability Distributions

- Uniform Distributions
- Exponential Distributions
- Normal Distributions

Now that we have developed some general properties of probability density functions, we can examine some specific functions that are used extensively in applications.

Uniform Distributions

EXPLORE & DISCUSS 1

Consider a random variable X with probability density function

$$f(x) = \begin{cases} \frac{1}{10} & \text{if } 0 \le x \le 10 \\ 0 & \text{otherwise} \end{cases}$$

Graph f and use the graph to find the indicated probabilities.

(A) Find $P(0 \le X \le 1)$, and $P(2 \le X \le 3)$, $P(6 \le X \le 7)$, $P(9 \le X \le 10)$.

(B) If $a \ge 0$, $n \ge 0$, and $a + n \le 10$, find $P(a \le X \le a + n)$.

(C) If I is an interval of length n that lies in the interval $[0, 10]$, what is the probability that the random variable X lies in I? Does this probability depend on the location of the interval I? Explain.

Simple probability density functions like the example given in the preceding Explore–Discuss 1 turn out to be quite useful. The next example will lead to some general observations about functions of this form.

EXAMPLE 1 **Waiting Time** A bus arrives every 30 minutes at a particular bus stop. If an individual arrives at the bus stop at a random time (that is, with no knowledge of the bus schedule), then the random variable X, representing the time this individual must spend waiting for the next bus, is said to be *uniformly distributed* on the interval $[0, 30]$. This means that the probability that X lies in a small interval of fixed length is independent of the location of that interval within $[0, 30]$. Thus, the probability of waiting between 0 and 5 minutes is the same as the probability of waiting between 5 and 10 minutes or the probability of waiting between 18 and 23 minutes. It can be shown that the probability density function for this uniformly distributed random variable is

$$f(x) = \begin{cases} \frac{1}{30} & \text{if } 0 \le x \le 30 \\ 0 & \text{otherwise} \end{cases}$$

Thus,

$$P(0 \le X \le 5) = \int_0^5 \frac{1}{30}\, dx = \frac{x}{30}\Big|_0^5 = \frac{5}{30} - \frac{0}{30} = \frac{1}{6}$$

$$P(5 \le X \le 10) = \int_5^{10} \frac{1}{30}\, dx = \frac{x}{30}\Big|_5^{10} = \frac{10}{30} - \frac{5}{30} = \frac{1}{6}$$

$$P(18 \le X \le 23) = \int_{18}^{23} \frac{1}{30}\, dx = \frac{x}{30}\Big|_{18}^{23} = \frac{23}{30} - \frac{18}{30} = \frac{1}{6}$$

Figure 1

Each of these probabilities is represented as an area under the graph of f in Figure 1.

Matched Problem 1 Use the probability density function given in Example 1 to find the probability that the individual waits:

(A) Between 0 and 10 minutes

(B) Between 10 and 20 minutes

(C) Between 17 and 27 minutes

In general, if the outcomes of an experiment lie in an interval $[a, b]$ and if the probability of the outcome lying in a small interval of fixed length is independent of the location of this small interval within $[a, b]$, then we say that the continuous random variable for this experiment is **uniformly distributed** on the interval $[a, b]$. The **uniform probability density function** is

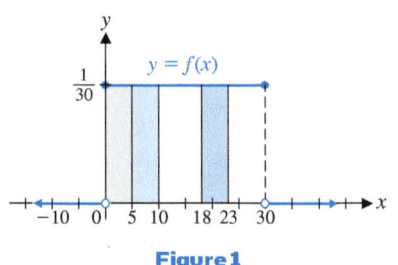

$$f(x) = \begin{cases} \dfrac{1}{b - a} & \text{if } a \le x \le b \\ 0 & \text{otherwise} \end{cases}$$

Figure 2

See Figure 2. Since $f(x) \ge 0$ and

$$\int_{-\infty}^{\infty} f(x)\, dx = \int_a^b \frac{1}{b - a}\, dx = \frac{x}{b - a}\Big|_a^b = \frac{b}{b - a} - \frac{a}{b - a} = 1$$

f satisfies the necessary conditions for a probability density function.

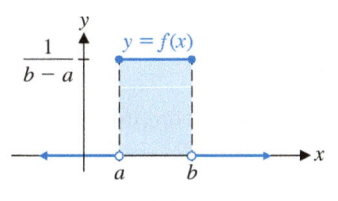

If F is the associated cumulative distribution function, then for $x < a$, $F(x) = 0$. For $a \le x \le b$, we have

$$F(x) = \int_{-\infty}^{x} f(t)\, dt = \int_{a}^{x} \frac{1}{b-a}\, dt = \frac{t}{b-a}\Big|_{a}^{x}$$

$$= \frac{x}{b-a} - \frac{a}{b-a} = \frac{x-a}{b-a}$$

For $x > b$, $F(x) = 1$.

Using the techniques discussed in the preceding section, it can be shown that (see Problems 49–52 in Exercise 9-4)

$$\mu = m = \frac{1}{2}(a+b) \qquad \text{and} \qquad \sigma = \frac{1}{\sqrt{12}}(b-a)$$

These properties are summarized in the next box.

SUMMARY **Uniform Probability Density Function**

$$f(x) = \begin{cases} \dfrac{1}{b-a} & \text{if } a \le x \le b \\ 0 & \text{otherwise} \end{cases} \qquad F(x) = \begin{cases} 0 & \text{if } x < a \\ \dfrac{x-a}{b-a} & \text{if } a \le x \le b \\ 1 & \text{if } x > b \end{cases}$$

Mean $\mu = \frac{1}{2}(a+b)$ **Median** $m = \frac{1}{2}(a+b)$

Standard deviation $\sigma = \dfrac{1}{\sqrt{12}}(b-a)$

EXAMPLE 2 **Electrical Current** Standard electrical current is uniformly distributed between 110 and 120 volts. What is the probability that the current is between 113 and 118 volts?

SOLUTION Since we are told that the current is uniformly distributed on the interval [110, 120], we choose the uniform probability density function

$$f(x) = \begin{cases} \frac{1}{10} & \text{if } 110 \le x \le 120 \\ 0 & \text{otherwise} \end{cases}$$

Then

$$P(113 \le X \le 118) = \int_{113}^{118} \frac{1}{10}\, dx = \frac{x}{10}\Big|_{113}^{118} = \frac{118}{10} - \frac{113}{10} = \frac{1}{2}$$

Matched Problem 2 In Example 2, what is the probability that the current is at least 116 volts?

Exponential Distributions

EXPLORE & DISCUSS 1 Determine a relationship between the positive constants a and b that will make the following function a probability density function.

$$f(x) = \begin{cases} ae^{-bx} & \text{if } x \ge 0 \\ 0 & \text{otherwise} \end{cases}$$

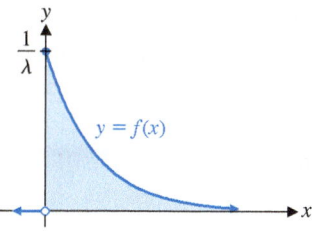

A continuous random variable has an **exponential distribution** and is referred to as an **exponential random variable** if its probability density function is the **exponential probability density function,**

$$f(x) = \begin{cases} (1/\lambda)e^{-x/\lambda} & \text{if } x \geq 0 \\ 0 & \text{otherwise} \end{cases}$$

where λ is a positive constant. Exponential random variables are used in a variety of applications, including studies of the length of telephone conversations, the time customers spend waiting in line at a bank, and the life expectancy of machine parts.

Since $f(x) \geq 0$ and

$$\int_{-\infty}^{\infty} f(x)\, dx = \int_{0}^{\infty} \frac{1}{\lambda} e^{-x/\lambda}\, dx$$

$$= \lim_{R \to \infty} \int_{0}^{R} \frac{1}{\lambda} e^{-x/\lambda}\, dx$$

$$= \lim_{R \to \infty} (-e^{-x/\lambda}) \Big|_{0}^{R}$$

$$= \lim_{R \to \infty} (-e^{-R/\lambda} + 1) = 1$$

f satisfies the conditions for a probability density function. If F is the cumulative distribution function, we see that $F(x) = 0$ for $x < 0$. For $x \geq 0$, we have

$$F(x) = \int_{-\infty}^{x} f(t)\, dt = \int_{0}^{x} \frac{1}{\lambda} e^{-t/\lambda}\, dt$$

$$= -e^{-t/\lambda} \Big|_{0}^{x} = 1 - e^{-x/\lambda}$$

To find the median, we solve

$$F(m) = P(X \leq m) = \frac{1}{2}$$

$$1 - e^{-m/\lambda} = \frac{1}{2}$$

$$\frac{1}{2} = e^{-m/\lambda}$$

$$\ln \frac{1}{2} = -\frac{m}{\lambda}$$

$$m = -\lambda \ln \frac{1}{2} = \lambda \ln 2 \quad \text{Note:} \quad \ln \tfrac{1}{2} = -\ln 2$$

Integration by parts can be used to show that $\mu = \lambda$ and $\sigma = \lambda$. The calculations are not included here. These results are summarized in the next box.

SUMMARY   **Exponential Probability Density Function**

$$f(x) = \begin{cases} (1/\lambda)e^{-x/\lambda} & \text{if } x \geq 0 \\ 0 & \text{otherwise} \end{cases} \qquad F(x) = \begin{cases} 1 - e^{-x/\lambda} & \text{if } x \geq 0 \\ 0 & \text{otherwise} \end{cases}$$

Mean $\mu = \lambda$ **Median** $m = \lambda \ln 2$
Standard deviation $\sigma = \lambda$

EXAMPLE 3   **Arrival Rates**   The length of time between calls received by the switchboard in a large legal firm is an exponential random variable. The average length of time between calls is 20 seconds. If a call has just been received, what is the probability that no calls are received in the next 30 seconds?

SOLUTION   Let X be the random variable that represents the length of time between calls received (in seconds). Since the average length of time is 20 seconds, we have $\mu = \lambda = 20$. Thus, the probability density function for X is

$$f(x) = \begin{cases} \frac{1}{20}\, e^{-x/20} & \text{if } x \geq 0 \\ 0 & \text{otherwise} \end{cases}$$

and

$$P(X \geq 30) = \int_{30}^{\infty} \tfrac{1}{20} e^{-x/20}\, dx \qquad\qquad \int_{0}^{30} f(x)\, dx + \int_{30}^{\infty} f(x)\, dx = 1$$

$$= 1 - \int_{0}^{30} \tfrac{1}{20}\, e^{-x/20}\, dx$$

$$= 1 - \left. \left(-e^{-x/20}\right) \right|_{0}^{30} = e^{-1.5} \approx .223$$

Matched Problem 3   In Example 3, if a call has just been received, what is the probability that no calls are received in the next 10 seconds?

Normal Distributions

We now consider the most important of all the probability density functions, the *normal probability density function*. This function is at the heart of a great deal of statistical theory, and it is also a useful tool in its own right for solving problems.

A continuous random variable X has a **normal distribution** and is referred to as a **normal random variable** if its probability density function is the **normal probability density function**

$$f(x) = \frac{1}{\pi\sqrt{2\pi}}\, e^{-(x-\mu)^2/(2\sigma^2)} \tag{1}$$

where μ is any constant and σ is any positive constant. It can be shown, but not easily, that

$$\int_{-\infty}^{\infty} f(x)\, dx = 1$$

and

$$E(X) = \int_{-\infty}^{\infty} xf(x)\, dx = \mu \qquad V(X) = \int_{-\infty}^{\infty} (x-\mu)^2 f(x)\, dx = \sigma^2$$

Thus, μ is the mean of the normal probability density function and s is the standard deviation. The graph of $f(x)$ is always a bell-shaped curve called a **normal curve.** Figure 3 illustrates three normal curves for different values of μ and σ.

CONCEPTUAL INSIGHT

Equation (1) for the normal probability density function is fairly complicated and involves both $\pi \approx 3.1416$ and $e \approx 2.7183$. Given values of μ and σ, however, the function is completely specified, and we could plot points or use a graphing calculator to produce its graph. Substituting $x + h$ for x in equation (1) produces an equation of the same form but with a different value of μ. Therefore, any horizontal translation of a normal curve is also a normal curve.

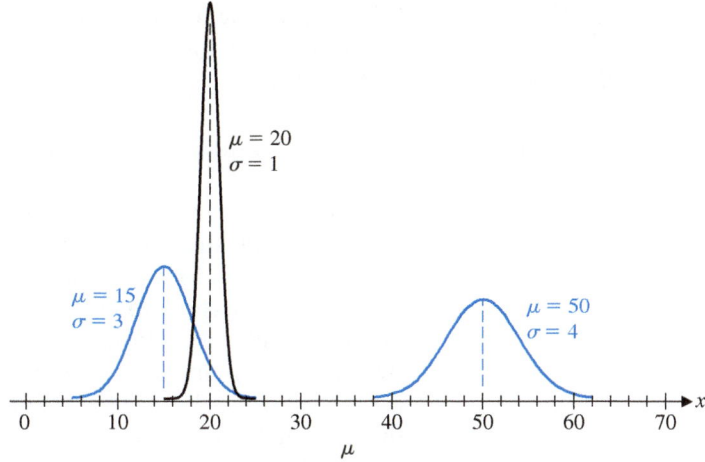

Figure 3 Normal probability distributions

The standard deviation measures the dispersion of the normal probability density function about the mean. A small standard deviation indicates a tight clustering about the mean and thus a tall, narrow curve; a large standard deviation indicates a large deviation from the mean and thus a broad, flat curve. Notice that each of the normal curves in Figure 3 is symmetric about a vertical line through the mean. This is true for any normal curve. Thus, the line $x = \mu$ divides the region under a normal curve into two regions with equal areas. Since the total area under a normal curve is always 1, the area of each of these regions is .5. This implies that the median of a normal random variable is always equal to the mean (see Fig. 4).

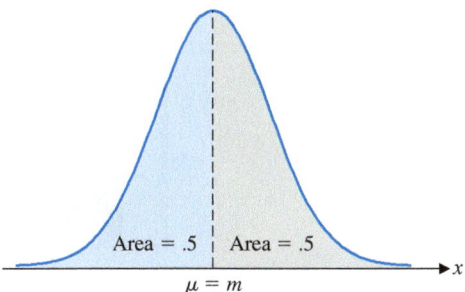

Figure 4 The mean and median of a normal random variable

The properties of the normal probability density function are summarized in the next box for ease of reference.

SUMMARY   Normal Probability Density Function

$$f(x) = \frac{1}{\sigma\sqrt{2p}} e^{-(x-\mu)^2/(2\sigma^2)} \quad \sigma > 0$$

Mean μ

Median μ

Standard deviation σ

The graph of $f(x)$ is symmetric with respect to the line $x = \mu$.

The cumulative distribution function for a normal random variable is given formally by

$$F(x) = \frac{1}{s\sqrt{2p}} \int_{-\infty}^{x} e^{-(t-\mu)^2/(2\sigma^2)} \, dt$$

Figure 5

It is not possible to express $F(x)$ as a finite combination of the functions with which we are familiar. Furthermore, we cannot use antidifferentiation to evaluate probabilities such as

$$P(c \le X \le d) = \frac{1}{\sigma\sqrt{2\pi}} \int_c^d e^{-(x-\mu)^2/(2\sigma^2)} \, dx$$

However, numerical integration can be used to approximate probabilities of this type, as illustrated in Figure 5, where we use a graphing utility to approximate $P(0 \le X \le 1.5)$ for the normal probability density function with $\mu = 0$ and $\sigma = 1$.

It is also possible to use a table to approximate probabilities involving the normal probability density function. Remarkably, the area under a normal curve between a mean μ and a given number of standard deviations to the right (or left) of μ is the same, regardless of the shape of the normal curve. For example, the area under the normal curve with $\mu = 3$, $\sigma = 5$ from $\mu = 3$ to $\mu + 1.5\sigma = 10.5$ is equal to the area under the normal curve with $\mu = 15$, $\sigma = 2$ from $\mu = 15$ to $\mu = 1.5\sigma = 18$ (see Fig. 6 below, noting that the shaded regions have the same areas, or equivalently, the same numbers of pixels). Therefore, such areas for any normal curve can be easily determined from the areas for the **standard normal curve,** that is, the normal curve with mean 0 and standard deviation 1. In fact, if z represents the number of standard deviations that a measurement x is from a mean μ, then the area under a normal curve from μ to $\mu + z\sigma$ equals the area under the standard normal curve from 0 to z (see Fig. 7). Table III in Appendix C lists those areas for the standard normal curve.

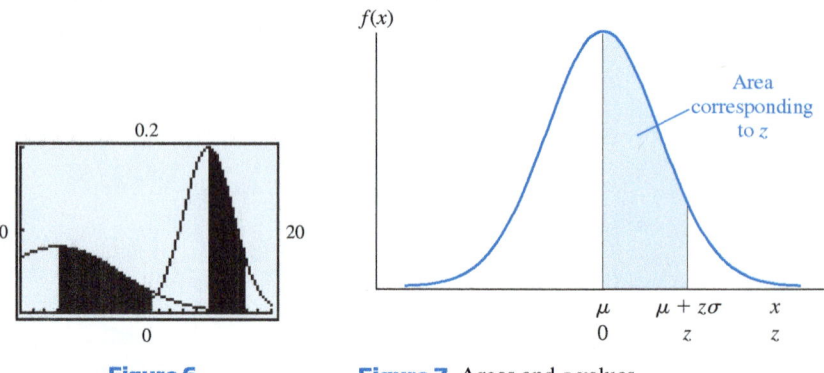

Figure 6   **Figure 7** Areas and z values

In the discussion that follows we will concentrate on using Table III in Appendix C to approximate areas under the normal curve. If desired, numerical integration on a graphing utility can be used to obtain the same results.

EXAMPLE 4   **Finding Probabilities for a Normal Distribution** A manufacturing process produces lightbulbs with life expectancies that are normally distributed with a mean of 500 hours and a standard deviation of 100 hours. What percentage of the lightbulbs can be expected to last between 500 and 670 hours?

SOLUTION   To answer this question, we first determine how many standard deviations 670 is from 500, the mean. This is easily done by dividing the distance between 500 and 670 by 100, the standard deviation. Thus,

$$z = \frac{670 - 500}{100} = \frac{170}{100} = 1.70$$

That is, 670 is 1.7 standard deviations from 500, the mean. Referring to Table III in Appendix C, we see that .4554 corresponds to $z = 1.70$. Also, because the total area under a normal curve is 1, we conclude that 45.54% of the lightbulbs produced will last between 500 and 670 hours (see Fig. 8).

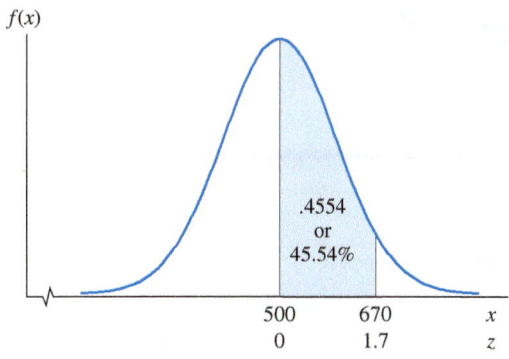

Figure 8 Lightbulb life expectancy—positive z

Matched Problem 4   In Example 4, what percentage of the lightbulbs can be expected to last between 500 and 750 hours?

In general, to find how many standard deviations a measurement x is from a mean m, first determine the distance between x and μ, and then divide by σ:

$$z = \frac{\text{Distance between } x \text{ and } \mu}{\text{Standard deviation}} = \frac{x - \mu}{\mu}$$

EXAMPLE 5   **Finding Probabilities for a Normal Distribution** From all lightbulbs produced (see Example 4), what is the probability of a lightbulb chosen at random lasting between 380 and 500 hours?

SOLUTION   To answer this, we first find z:

$$z = \frac{x - \mu}{\sigma} = \frac{380 - 500}{100} = -1.20$$

It is usually a good idea to draw a rough sketch of a normal curve and insert relevant data (see Fig. 9).

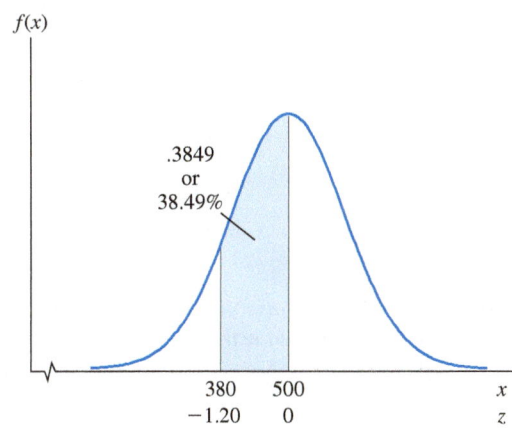

Figure 9 Lightbulb life expectancy—negative z

Table III in Appendix C does not include negative values for z, but because normal curves are symmetrical with respect to a vertical line through the mean, we simply use the absolute value (positive value) of z for the table. Thus, the area corresponding to $z = -1.20$ is the same as the area corresponding to $z = 1.20$, which is .3849, so the probability of a lightbulb chosen at random lasting between 380 and 500 hours is .3849.

Matched Problem 5   In Example 5, what is the probability of a lightbulb chosen at random lasting between 400 and 500 hours?

Exercises 9-4

A

In Problems 1–4, find the probability density function f and the associated cumulative distribution function F for the continuous random variable X if

1. X is uniformly distributed on $[0, 2]$.

2. X is uniformly distributed on $[3, 6]$.

3. X is an exponential random variable with $\lambda = \frac{1}{2}$.

4. X is an exponential random variable with $\lambda = \frac{1}{4}$.

In Problems 5–8, find the mean, median, and standard deviation of the continuous random variable X if

5. X is uniformly distributed on $[1, 5]$.

6. X is uniformly distributed on $[2, 8]$.

7. X is an exponential random variable with $\lambda = 5$.

8. X is an exponential random variable with $\lambda = 3$.

In Problems 9–16, use Table III in Appendix C to find the area under the standard normal curve from 0 to the indicated value of z.

9. 1.5	10. 0.8
11. 20.72	12. 3.18
13. 2.01	14. 22.39
15. 21.93	16. 1.67

In Problems 17–24, given a normal distribution with mean 100 and standard deviation 10, use Table III in Appendix C to find the area under this normal curve from the mean to the indicated measurement.

17. 120	18. 135
19. 95	20. 113
21. 106	22. 87
23. 78	24. 99

B

Problems 25 and 26 refer to random variables X_1 and X_2 with probability density functions f_1 and f_2, respectively, where

$$f_1(x) = \frac{1}{\sqrt{2\pi}} e^{-x^2/2}$$

$$f_2(x) = \begin{cases} 0.4 & \text{if } -1.25 \le x \le 1.25 \\ 0 & \text{otherwise} \end{cases}$$

25. Explain how you can predict from the graphs of f_1 and f_2 which random variable, X_1 or X_2, has the greater standard deviation. Check your prediction by finding the standard deviation of each.

26. Explain how you can predict from the graphs of f_1 and f_2 which random variable, X_1 or X_2, has the greater median. Check your prediction by finding the median of each.

Problems 27 and 28 refer to random variables X_1 and X_2 with probability density functions g_1 and g_2, respectively, where

$$g_1(x) = \frac{1}{\sqrt{2\pi}} e^{-(x-1)^2/2}$$

$$g_2(x) = \begin{cases} e^{-x} & \text{if } x \ge 0 \\ 0 & \text{otherwise} \end{cases}$$

27. Explain how you can predict from the graphs of g_1 and g_2 which random variable, X_1 or X_2, has the greater median. Check your prediction by finding the median of each.

28. Explain how you can predict from the graphs of g_1 and g_2 which random variable, X_1 or X_2, has the greater mean. Check your prediction by finding the mean of each.

In Problems 29–32, X is a continuous random variable with mean μ. Find μ, and then find $P(X \le \mu)$ if

29. X is uniformly distributed on $[0, 4]$.

30. X is uniformly distributed on $[0, 10]$.

31. X is an exponential random variable with $\lambda = 1$.

32. X is an exponential random variable with $\lambda = 2$.

In Problems 33–36, X is a continuous random variable with mean m and standard deviation s. Find m and s, and then find $P(\mu - \sigma \le X \le \mu + \sigma)$ if

33. X is uniformly distributed on $[-5, 5]$.

34. X is uniformly distributed on $[-2, 2]$.

35. X is an exponential random variable with $m = 6 \ln 2$.

36. X is an exponential random variable with $m = 4 \ln 2$.

Given a normal random variable X with mean 70 and standard deviation 8, find the indicated probabilities in Problems 37–44.

37. $P(60 \le X \le 80)$	38. $P(50 \le X \le 90)$
39. $P(62 \le X \le 74)$	40. $P(66 \le X \le 78)$
41. $P(X \ge 88)$	42. $P(X \ge 90)$
43. $P(X \le 60)$	44. $P(X \le 56)$

Problems 45–48 refer to the normal random variable X with mean μ and standard deviation s. Use a graphing utility with a built-in numerical integration routine to compute the indicated probabilities.

45. Approximate $P(\mu - \sigma \le X \le \mu + \sigma)$ to four decimal places for $\mu = 0$ and

   (A) $\sigma = .5$

   (B) $\sigma = 1$

   (C) $\sigma = 2$

Write a brief verbal explanation of the result indicated by these calculations.

46. Approximate $P(\mu - 2\sigma \leq X \leq \mu + 2\sigma)$ to four decimal places for $\mu = 0$ and

(A) $\sigma = .5$

(B) $\sigma = 1$

(C) $\sigma = 2$

Write a brief verbal explanation of the result indicated by these calculations.

47. Approximate $P(\mu - 3\sigma \leq X \leq \mu + 3\sigma)$ to four decimal places for $\mu = 0$ and

(A) $\sigma = .5$

(B) $\sigma = 1$

(C) $\sigma = 2$

Write a brief verbal explanation of the result indicated by these calculations.

48. In Problems 45–47, if μ is changed from 0 to 5, will any of the probabilities change? Explain.

C

Problems 49–52 refer to the uniformly distributed random variable X with probability density function

$$f(x) = \begin{cases} \dfrac{1}{b-a} & \text{if } a \geq x \leq b \\ 0 & \text{otherwise} \end{cases}$$

49. Show that $\mu = (a + b)/2$.

50. Show that $m = (a + b)/2$.

51. Show that $\int_{-\infty}^{\infty} x^2 f(x)\, dx = (b^2 + ab + a^2)/3$.

52. Show that $V(X) = (b - a)^2/12$.

53. Use numerical integration on a graphing utility to approximate the mean, median, and standard deviation of the probability density function

$$f(x) = \begin{cases} 5e^{-5x} & \text{if } x \geq 0 \\ 0 & \text{otherwise} \end{cases}$$

Explain why it is reasonable to replace integrals over the interval $(-\infty, \infty)$ by integrals over the interval $[0, 10]$. Verify that your answers agree with the results expected for exponential probability density functions.

54. Use numerical integration on a graphing utility to approximate the mean, median, and standard deviation of the probability density function

$$f(x) = \begin{cases} 0.36e^{-0.36x} & \text{if } x \geq 0 \\ 0 & \text{otherwise} \end{cases}$$

Explain why it is reasonable to replace integrals over the interval $(-\infty, \infty)$ by integrals over the interval $[0, 100]$. Verify that your answers agree with the results expected for exponential probability density functions.

Problems 55–58 refer to a random variable X with probability density function

$$f(x) = \begin{cases} p/x^{p+1} & \text{if } x \geq 1 \\ 0 & \text{otherwise} \end{cases}$$

*where p is a positive constant. This random variable is often referred to as a **Pareto* random variable.***

55. Verify that f satisfies the necessary conditions for a probability density function and find the mean, if it exists. What restrictions must you place on p to ensure that the mean exists?

56. Find the standard deviation, if it exists. What restrictions must you place on p to ensure that the standard deviation exists?

57. Find the median.

58. The mean and the median for a uniformly distributed random variable are equal, as are the mean and standard deviation for an exponentially distributed random variable. Use a graph to illustrate the relationship between the mean, median, and standard deviation for a Pareto random variable. Approximate abscissas of intersection points to two decimal places.

Applications

59. Waiting time. The time (in minutes) applicants must wait to receive a driver's examination is uniformly distributed on the interval $[0, 40]$. What is the probability that an applicant must wait more than 25 minutes?

60. Waiting time. The time (in minutes) passengers must wait for a commuter plane in a large airport is uniformly distributed on the interval $[0, 60]$. What is the probability that a passenger waits less than 20 minutes?

61. Communications. The length of time for telephone conversations (in minutes) is exponentially distributed. The average (mean) length of a conversation is 3 minutes. What is the probability that a conversation lasts less than 2 minutes?

62. Waiting time. The waiting time (in minutes) for customers at a drive-in bank is an exponential random variable. The average (mean) time a customer waits is 4 minutes. What is the probability that a customer waits more than 5 minutes?

63. Service time. The time between failures of a photo zcopier is an exponential random variable. Half the copiers require service during the first 2 years of operation. What is the probability that a copier requires service during the first year of operation?

64. Component failure. The life expectancy (in years) of a component in a computer is an exponential random variable. Half the components fail in the first 3 years. The company that manufactures the component offers a 1-year warranty. What is the probability that a component will fail during the warranty period?

---

*Named after Vilfredo Pareto (1848–1923), an Italian sociologist and economist. Although best known for his theories on political behavior, Pareto also developed new applications of mathematics to economics.

65. Sales. The annual sales for salespeople at a business machine company are normally distributed. The average (mean) annual sales are $200,000 with a standard deviation of $20,000. What percentage of the salespeople would be expected to make annual sales of $240,000 or more?

66. Guarantees. The life expectancy of a car battery is normally distributed. The average (mean) lifetime is 170 weeks with a standard deviation of 10 weeks. If the company guarantees the battery for 3 years, what percentage of the batteries sold would be expected to be returned before the end of the warranty period?

67. Quality control. A manufacturing process produces a precision part whose length is a normal random variable with mean 100 millimeters and standard deviation 2 millimeters. All parts deviating by more than 5 millimeters from the mean must be rejected. On average, what percentage of the parts must be rejected?

68. Quality control. *An* automated manufacturing process produces a component whose width is a normal random variable with mean 7.55 centimeters and standard deviation 0.02 centimeter. All components deviating by more than 0.05 centimeter from the mean must be rejected. On average, what percentage of the parts must be rejected?

69. Survival time. The time of death (in years) after patients have contracted a certain disease is exponentially distributed. The probability that a patient dies within 1 year is .3.

(A) What is the expected time of death?

(B) What is the probability that a patient survives longer than the expected time of death?

70. Survival time. Repeat Problem 69 if the probability that a patient dies within 1 year is .5.

71. Medicine. The healing time of a certain type of incision is normally distributed with an average healing time of 240 hours and a standard deviation of 20 hours. What percentage of the people having this incision would heal in 8 days or less?

72. Agriculture. The height of a hay crop is normally distributed with an average height of 38 inches and a standard de-

viation of 1.5 inches. What percentage of the crop will be 40 inches or more?

73. Learning. The time (in minutes) it takes an adult to memorize a sequence of random digits is an exponential random variable. The average time is 2 minutes. What is the probability that it takes an adult over 5 minutes to memorize the digits?

74. Psychology. The time (in seconds) it takes rats to find their way through a maze is exponentially distributed. The average time is 30 seconds. What is the probability that it takes a rat over 1 minute to find a path through the maze?

75. Grading on a curve. An instructor grades on a curve by assuming the grades on a test are normally distributed. If the average grade is 70 and the standard deviation is 8, find the test scores for each grade interval if the instructor wishes to assign grades as follows: 10% A's, 20% B's, 40% C's, 20% D's, and 10% F's. Round answers to one decimal place.

76. Psychology. A test devised to measure aggressive–passive personalities was standardized on a large group of people. The scores were normally distributed with a mean of 50 and a standard deviation of 10. If we want to designate the highest 10% as aggressive, the next 20% as moderately aggressive, the middle 40% as average, the next 20% as moderately passive, and the lowest 10% as passive, what range of scores will be covered by these five designations? Round answers to one decimal place.

Answers to Matched Problems

1. (A) $\frac{1}{3}$ (B) $\frac{1}{3}$ (C) $\frac{1}{3}$

2. $\frac{2}{5}$

3. $e^{-0.5} \approx .607$

4. 49.38%

5. .3413

Chapter 9   Review

Important Terms, Symbols, and Concepts

9-1 Improper Integrams

If f is continuous over the indicated interval and the limit exists, then the improper integral on the left in each of the following equations is defined by

- $$\int_a^\infty f(x)\,dx = \lim_{b \to \infty} \int_a^b f(x)\,dx$$

- $$\int_{-\infty}^b f(x)\,dx = \lim_{a \to -\infty} \int_a^b f(x)\,dx$$

- $$\int_{-\infty}^\infty f(x)\,dx = \int_{-\infty}^c f(x)\,dx + \int_c^\infty f(x)\,dx$$

where c is any point on $(-\infty, \infty)$, provided both integrals on the right exist.

The improper integral **converges** if the limit exists; otherwise it **diverges** (and no value is assigned to it).

A continuous income stream is **perpetual** if it never stops producing income. The **capital value**, CV, of a perpetual income stream $f(t)$, at a rate r compounded continuously, is defined by

$$CV = \int_0^\infty f(t)e^{-rt}\,dt$$

Capital value provides a method for expressing the worth (in terms of today's dollars) of an investment that will produce income indefinitely.

9-2 Continuous Random Variables

A **continuous random variable** X is a function that assigns a numeric value to each outcome of an experiment. The set of possible values of X is an interval of real numbers. This interval may be open or closed, bounded or unbounded.

The function $f(x)$ is a **probability density function** for a continuous random variable X if:

- $f(x) \geq 0$ for all x in $(-\infty, \infty)$

- $\displaystyle\int_{-\infty}^\infty f(x)\,d = 1$

- The probability that X lies in the interval $[c, d]$ is:

$$P(c \leq X \leq d) = \int_c^d f(x)\,dx$$

If f is a probability density function, then the associated **cumulative distribution function** F is:

$$F(x) = P(X \leq x) = \int_{-\infty}^x f(t)\,dt$$

The cumulative distribution function F has the properties

- $F'(x) = f(x)$ wherever f is continuous
- $0 \leq F(x) \leq 1$ for x in $(-\infty, \infty)$
- $F(x)$ is nondecreasing on $(-\infty, \infty)$

9-3 Expected Value, Standard Deviation, and Median

Let $f(x)$ be the probability density function for a continuous random variable X. The **expected value, or mean, of** X is

$$\mu = E(X) = \int_{-\infty}^\infty xf(x)\,dx$$

The **variance** is:

$$V(X) = \int_{-\infty}^\infty (x - \mu)^2 f(x)\,dx = \int_{-\infty}^\infty x^2 f(x)\,dx - \mu^2$$

and the **standard deviation** is $\sigma = \sqrt{V(X)}$

The **median** is a value of the random variable that divides the area under the graph of the probability density function into two equal parts, each of area $\frac{1}{2}$.

9-4 Special Probability Distributions

A continuous random variable has a **uniform distribution** on $[a, b]$, and a **uniform probability density function,** if

$$f(x) = \begin{cases} \dfrac{1}{b - a} & \text{if } a \leq x \leq b \\ 0 & \text{otherwise} \end{cases}$$

A continuous random variable has an **exponential distribution,** and an **exponential probability density function,** if

$$f(x) = \begin{cases} (1/\lambda)e^{-x/\lambda} & \text{if } x \geq 0 \\ 0 & \text{otherwise} \end{cases}$$

A continuous random variable has a **normal distribution,** and a **normal probability density function,** if

$$f(x) = \frac{1}{\sigma\sqrt{2\pi}}\, e^{-(x-\mu)^2/(2\sigma^2)}$$

where μ (the mean) is any constant and σ (the standard deviation) is any positive constant. The graph of a normal probability density function is called a **normal curve.** The **standard normal curve** has mean 0 and standard deviation 1. The area under any normal curve from μ to $x = \mu + z\sigma$ is equal to the area under the standard normal curve from 0 to z, where $z = \dfrac{x - \mu}{\sigma}$.

Review Exercises

Work through all the problems in this chapter review and check your answers in the back of this supplement. Answers to all review problems are there along with section numbers in italics to indicate where each type of problem is discussed. Where weaknesses show up, review appropriate sections in the text.

A

In Problems 1–3, evaluate each improper integral, if it converges.

1. $\int_0^\infty e^{-2x}\,dx$

2. $\int_0^\infty \frac{1}{x+1}\,dx$

3. $\int_1^\infty \frac{16}{x^3}\,dx$

Problems 4–7 refer to the continuous random variable X with probability density function

$$f(x) = \begin{cases} 1 - \frac{1}{2}x & \text{if } 0 \le x \le 2 \\ 0 & \text{otherwise} \end{cases}$$

4. Find $P(0 \le X \le 1)$ and illustrate with a graph.

5. Find the mean, variance, and standard deviation.

6. Find and graph the associated cumulative distribution function.

7. Find the median.

8. If X is a normal random variable with a mean of 100 and a standard deviation of 10, find $P(100 \le X \le 118)$.

In Problems 9 and 10, find the probability density function f and the associated cumulative distribution function F for the continuous random variable X if

9. X is uniformly distributed on $[5, 15]$.

10. X is an exponential random variable with $\lambda = \frac{1}{5}$.

In Problems 11 and 12, is f a probability density function? Explain.

11. $f(x) = e^{-x^2}$

12. $f(x) = \begin{cases} 0.75(x^2 - 4x + 3) & \text{if } 0 \le x \le 4 \\ 0 & \text{otherwise} \end{cases}$

B

Problems 13–16 refer to the continuous random variable X with probability density function

$$f(x) = \begin{cases} \frac{5}{2}x^{-7/2} & \text{if } x \ge 1 \\ 0 & \text{otherwise} \end{cases}$$

13. Find $P(1 \le X \le 4)$ and illustrate with a graph.

14. Find the mean, variance, and standard deviation.

15. Find and graph the associated cumulative distribution function.

16. Find the median.

Problems 17–20 refer to an exponentially distributed random variable X.

17. If $P(4 \le X) = e^{-2}$, find the probability density function.

18. Find $P(0 \le X \le 2)$.

19. Find the associated cumulative distribution function.

20. Find the mean, standard deviation, and median.

21. Given a normal distribution with mean 50 and standard deviation 6, find the area under the normal curve:

(A) Between 41 and 62

(B) From 59 on

22. Given a normal random variable X with mean 82 and standard deviation 8, find:

(A) $P(84 \le X \le 94)$

(B) $P(X \ge 60)$

In Problems 23 and 24, evaluate each improper integral, if it converges.

23. $\int_{-\infty}^0 e^x\,dx$

24. $\int_0^\infty \frac{1}{(x+3)^2}\,dx$

25. If f is continuous on $[-1, \infty)$ and $\int_{-1}^\infty f(x)\,dx$ converges, does $\int_1^\infty f(x)\,dx$ converge? Explain.

In Problems 26 and 27, find a constant k so that kf is a probability density function, or explain why no such k exists.

26. $f(x) = \begin{cases} e^{-10x} & \text{if } x \ge 0 \\ 0 & \text{otherwise} \end{cases}$

27. $f(x) = \begin{cases} e^{10x} & \text{if } x \ge 0 \\ 0 & \text{otherwise} \end{cases}$

28. If X is an exponentially distributed random variable with median $m = 3 \ln 2$ and mean μ, find $P(X \le \mu)$.

C

29. Find the mean and the median of the continuous random variable with probability density function

$$f(x) = \begin{cases} 50/(x+5)^3 & \text{if } x \ge 0 \\ 0 & \text{otherwise} \end{cases}$$

30. Use a graphing utility to approximate (to two decimal places) the median of the continuous random variable with probability density function

$$f(x) = \begin{cases} \frac{.8}{x^2} + \frac{.8}{x^5} & \text{if } x \ge 1 \\ 0 & \text{otherwise} \end{cases}$$

31. Find $\int_{-\infty}^\infty \frac{e^x}{(1+e^x)^2}\,dx$, if it converges.

32. If f is a continuous probability density function with mean μ and standard deviation σ and a, b, and c are constants, evaluate the following integral. Express the result in terms of μ, σ, a, b, and c.

$$\int_{-\infty}^{\infty} (ax^2 + bx + c)f(x)\, dx$$

Problems 33–36 refer to random variables X_1 and X_2 with probability density functions f_1 and f_2, respectively, where

$$f_1(x) = \begin{cases} 0.25xe^{-x/2} & \text{if } x \geq 0 \\ 0 & \text{otherwise} \end{cases}$$

$$f_2(x) = \begin{cases} 0.0625x^2e^{-x/2} & \text{if } x \geq 0 \\ 0 & \text{otherwise} \end{cases}$$

33. Explain how you can predict from the graphs of f_1 and f_2 which random variable, X_1 or X_2, has the greater mean.

34. Explain how you can predict from the graphs of f_1 and f_2 which random variable, X_1 or X_2, has the greater variance.

35. Check your prediction in Problem 33 by using numerical integration to approximate the means of X_1 and X_2.

36. Check your prediction in Problem 34 by using numerical integration to approximate the variances of X_1 and X_2.

Applications

37. Production. The rate of production of an oil well (in millions of barrels per year) is given by (see the figure below)

$$R(t) = 12e^{-0.3t} - 12e^{-0.6t}$$

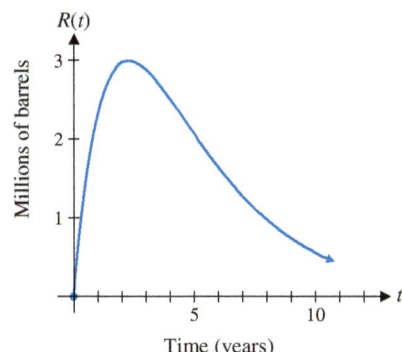

Time (years)

(A) Assuming that the well is operated indefinitely, find the total production.

(B) When will the output from the well reach 50% of the total production? Round answer to two decimal places.

36. Demand. The manager of a movie theater has determined that the weekly demand for popcorn (in pounds) is a continuous random variable with probability density function

$$f(x) = \begin{cases} .02(1 - .01x) & \text{if } 0 \leq x \leq 100 \\ 0 & \text{otherwise} \end{cases}$$

(A) Evaluate $\int_{40}^{100} f(x)\, dx$ and interpret.

(B) If the manager has 50 pounds of popcorn on hand at the beginning of the week, what is the probability that this will be enough to meet the weekly demand?

(C) If the manager wants the probability that the supply on hand exceeds the weekly demand to be .96, how much popcorn must be on hand at the beginning of the week?

39. Capital value. The perpetual annual rent for a property is $2,400. Find the capital value at 6% compounded continuously.

40. Credit applications. The percentage of applications (expressed as a decimal between 0 and 1) for a national credit card processed on the same day the applications are received is a continuous random variable with probability density function

$$f(x) = \begin{cases} 6x(1 - x) & \text{if } 0 \leq x \leq 1 \\ 0 & \text{otherwise} \end{cases}$$

(A) What is the probability that at least 20% of the applications received are processed the same day they arrive?

(B) What is the expected percentage of applications processed the same day they arrive?

(C) What is the median percentage of applications processed the same day they arrive?

41. Computer failure. A computer manufacturer has determined that the time between failures for its computer is an exponentially distributed random variable with a mean failure time of 4,000 hours. Suppose a particular computer has just been repaired.

(A) What is the probability that the computer operates for the next 4,000 hours without failure?

(B) What is the probability that the computer fails in the next 1,000 hours?

42. Radial tire failure. The life expectancy (in miles) of a certain brand of radial tires is a normal random variable with a mean of 35,000 miles and a standard deviation of 5,000 miles. What is the probability that a tire fails during the first 25,000 miles of use?

43. Personnel screening. The scores on a screening test for new technicians are normally distributed with mean 100 and standard deviation 10. Find the approximate percentage of applicants taking the test who score

(A) Between 92 and 108

(B) 115 or higher

44. Medicine. The shelf life (in months) of a certain drug is a continuous random variable with probability density function

$$f(x) = \begin{cases} 10/(x + 10)^2 & \text{if } x \geq 0 \\ 0 & \text{otherwise} \end{cases}$$

(A) Evaluate $\int_2^8 f(x)\, dx$ and interpret.

(B) What is the probability that the drug is still usable after 5 months?

(C) What is the median shelf life?

45. Life expectancy. The life expectancy (in months) after dogs have contracted a certain disease is an exponentially distributed random variable. The probability of surviving more than 1 month is e^{-2}. After contracting the disease,

(A) What is the probability of the dog surviving more than 2 months?

(B) What is the mean life expectancy?

46. Drug assimilation. The rate at which the body eliminates a drug (in milliliters per hour) is given by

$$R(t) = 15e^{-0.2t} - 15e^{-0.3t}$$

where t is the number of hours since the drug was administered.

(A) What is the total amount of the drug that is eliminated by the body?

(B) How long will it take for 50% of the total amount to be eliminated? Round answer to two decimal places.

47. Testing. The IQ scores for 6-year-old children in a certain area are normally distributed with a mean of 108 and a standard deviation of 12. What percentage of the children can be expected to have IQ scores of 135 or more?

48. Politics. The rate of change of the voting population of a city with respect to time t (in years) is estimated to be

$$N'(t) = \frac{100t}{(1 + t^2)^2}$$

where $N(t)$ is in thousands. If $N(0)$ is the current voting population, how much will this population increase during the next 3 years? If the population continues to grow at this rate indefinitely, what is the total increase in the voting population?

APPENDIX A Basic Algebra Review

Appendix A reviews some important basic algebra concepts usually studied in earlier courses. The material may be studied systematically before beginning the rest of the book or reviewed as needed.

A-1 Real Numbers

- Set of Real Numbers
- Real Number Line
- Basic Real Number Properties
- Further Properties
- Fraction Properties

The rules for manipulating and reasoning with symbols in algebra depend, in large measure, on properties of the real numbers. In this section we look at some of the important properties of this number system. To make our discussions here and elsewhere in the book clearer and more precise, we occasionally make use of simple *set* concepts and notation.

Set of Real Numbers

Informally, a **real number** is any number that has a decimal representation. Table 1 describes the set of real numbers and some of its important subsets. Figure 1 illustrates how these sets of numbers are related.

The set of integers contains all the natural numbers and something else—their negatives and 0. The set of rational numbers contains all the integers and something else—noninteger ratios of integers. And the set of real numbers contains all the rational numbers and something else—irrational numbers.

Table 1 Set of Real Numbers

Symbol	Name	Description	Examples
N	Natural numbers	Counting numbers (also called positive integers)	$1, 2, 3, \ldots$
Z	Integers	Natural numbers, their negatives, and 0	$\ldots, -2, -1, 0, 1, 2, \ldots$
Q	Rational numbers	Numbers that can be represented as a/b, where a and b are integers and $b \neq 0$; decimal representations are repeating or terminating	$-4, 0, 1, 25, \frac{-3}{5}, \frac{2}{3}, 3.67, -0.33\overline{3}, 5.272\,7\overline{27}$*
I	Irrational numbers	Numbers that can be represented as nonrepeating and nonterminating decimal numbers	$\sqrt{2}, \pi, \sqrt[3]{7}, 1.414\,213\ldots, 2.718\,281\,82\ldots$
R	Real numbers	Rational and irrational numbers	

*The overbar indicates that the number (or block of numbers) repeats indefinitely. The space after every third digit is used to help keep track of the number of decimal places.

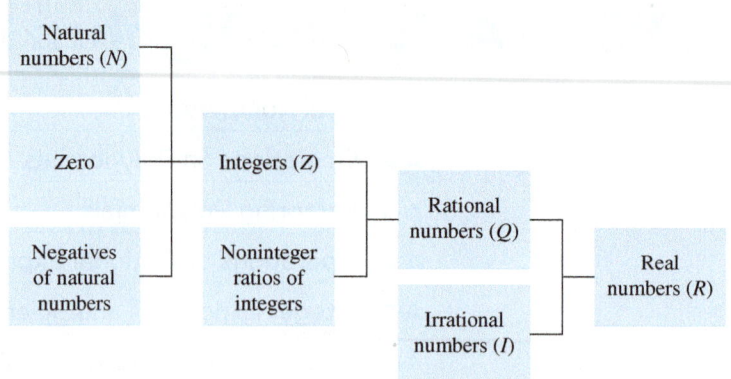

Figure 1 Real numbers and important subsets

Real Number Line

A one-to-one correspondence exists between the set of real numbers and the set of points on a line. That is, each real number corresponds to exactly one point, and each point corresponds to exactly one real number. A line with a real number associated with each point, and vice versa, as shown in Figure 2, is called a **real number line**, or simply a **real line**. Each number associated with a point is called the coordinate of the point.

Figure 2 Real number line

The point with coordinate 0 is called the **origin**. The arrow on the right end of the line indicates a positive direction. The coordinates of all points to the right of the origin are called positive **real numbers**, and those to the left of the origin are called **negative real numbers**. The real number 0 is neither positive nor negative.

Basic Real Number Properties

We now take a look at some of the basic properties of the real number system that enable us to convert algebraic expressions into *equivalent forms*.

SUMMARY **Basic Properties of the Set of Real Numbers**

Let a, b, and c be arbitrary elements in the set of real numbers R.

Addition Properties

Associative: $(a + b) + c = a + (b + c)$

Commutative: $a + b = b + a$

Identity: 0 is the additive identity; that is, $0 + a = a + 0 = a$ for all a in R, and 0 is the only element in R with this property.

Inverse: For each a in R, $-a$, is its unique additive inverse; that is, $a + (-a) = (-a) + a = 0$ and $-a$ is the only element in R relative to a with this property.

Multiplication Properties

Associative: $(ab)c = a(bc)$

Commutative: $ab = ba$

Identity: 1 is the multiplicative identity; that is, $(1)a = a(1) = a$ for all a in R, and 1 is the only element in R with this property.

Inverse: For each a in R, $a \neq 0$, $1/a$ is its unique multiplicative inverse; that is, $a(1/a) = (1/a)a = 1$, and $1/a$ is the only element in R relative to a with this property.

Distributive Properties

$$a(b + c) = ab + ac \qquad (a + b)c = ac + bc$$

You are already familiar with the **commutative properties** for addition and multiplication. They indicate that the order in which the addition or multiplication of two numbers is performed does not matter. For example,

$$7 + 2 = 2 + 7 \quad \text{and} \quad 3 \cdot 5 = 5 \cdot 3$$

Is there a commutative property relative to subtraction or division? That is, does $a - b = b - a$ or does $a \div b = b \div a$ for all real numbers a and b (division by 0 excluded)? The answer is no, since, for example,

$$8 - 6 \neq 6 - 8 \quad \text{and} \quad 10 \div 5 \neq 5 \div 10$$

When computing

$$3 + 2 + 6 \quad \text{or} \quad 3 \cdot 2 \cdot 6$$

why don't we need parentheses to indicate which two numbers are to be added or multiplied first? The answer is to be found in the **associative properties**. These properties allow us to write

$$(3 + 2) + 6 = 3 + (2 + 6) \quad \text{and} \quad (3 \cdot 2) \cdot 6 = 3 \cdot (2 \cdot 6)$$

so it does not matter how we group numbers relative to either operation. Is there an associative property for subtraction or division? The answer is no, since, for example,

$$(12 - 6) - 2 \neq 12 - (6 - 2) \quad \text{and} \quad (12 \div 6) \div 2 \neq 12 \div (6 \div 2)$$

Evaluate each side of each equation to see why.

What number added to a given number will give that number back again? What number times a given number will give that number back again? The answers are 0 and 1, respectively. Because of this, 0 and 1 are called the **identity elements** for the real numbers. Hence, for any real numbers a and b,

$$0 + 5 = 5 \quad \text{and} \quad (a + b) + 0 = a + b$$
$$1 \cdot 4 = 4 \quad \text{and} \quad (a + b) \cdot 1 = a + b$$

We now consider **inverses**. For each real number a, there is a unique real number $-a$ such that $a + (-a) = 0$. The number $-a$ is called the **additive inverse** of a, or the **negative** of a. For example, the additive inverse (or negative) of 7 is -7, since $7 + (-7) = 0$. The additive inverse (or negative) of -7 is $-(-7) = 7$, since $-7 + [-(-7)] = 0$.

CONCEPTUAL INSIGHT

Do not confuse negation with the sign of a number. If a is a real number, $-a$ is the negative of a and may be positive or negative. Specifically, if a is negative, then $-a$ is positive and if a is positive, then $-a$ is negative.

For each nonzero real number a, there is a unique real number $1/a$ such that $a(1/a) = 1$. The number $1/a$ is called the **multiplicative inverse** of a, or the **reciprocal** of a. For example, the multiplicative inverse (or reciprocal) of 4 is $\frac{1}{4}$, since $4\left(\frac{1}{4}\right) = 1$. (Also note that 4 is the multiplicative inverse of $\frac{1}{4}$.) The number 0 has no multiplicative inverse.

We now turn to the **distributive properties**, which involve both multiplication and addition. Consider the following two computations:

$$5(3 + 4) = 5 \cdot 7 = 35 \quad 5 \cdot 3 + 5 \cdot 4 = 15 + 20 = 35$$

Thus,

$$5(3 + 4) = 5 \cdot 3 + 5 \cdot 4$$

and we say that multiplication by 5 *distributes* over the sum $(3 + 4)$. In general, **multiplication distributes over addition** in the real number system. Two more illustrations are

$$9(m + n) = 9m + 9n \quad (7 + 2)u = 7u + 2u$$

EXAMPLE 1 **Real Number Properties** State the real number property that justifies the indicated statement.

Statement	Property Illustrated
(A) $x(y + z) = (y + z)x$	Commutative $(\cdot)$
(B) $5(2y) = (5 \cdot 2)y$	Associative $(\cdot)$
(C) $2 + (y + 7) = 2 + (7 + y)$	Commutative $(+)$
(D) $4z + 6z = (4 + 6)z$	Distributive
(E) If $m + n = 0$, then $n = -m$.	Inverse $(+)$

Matched Problem 1 State the real number property that justifies the indicated statement.

(A) $8 + (3 + y) = (8 + 3) + y$

(B) $(x + y) + z = z + (x + y)$

(C) $(a + b)(x + y) = a(x + y) + b(x + y)$

(D) $5xy + 0 = 5xy$

(E) If $xy = 1, x \neq 0$, then $y = 1/x$.

Further Properties

Subtraction and *division* can be defined in terms of addition and multiplication, respectively:

DEFINITION **Subtraction and Division**

For all real numbers a and b,

Subtraction: $\quad a - b = a + (-b)$ $\quad\quad 7 - (-5) = 7 + [-(-5)]$
$$= 7 + 5 = 12$$

Division: $\quad\quad a \div b = a\left(\dfrac{1}{b}\right), b \neq 0 \quad\quad 9 \div 4 = 9\left(\dfrac{1}{4}\right) = \dfrac{9}{4}$

To subtract b from a, add the negative (the additive inverse) of b to a. To divide a by b, multiply a by the reciprocal (the multiplicative inverse) of b. Note that division by 0 is not defined, since 0 does not have a reciprocal. **0 can never be used as a divisor!**

The following properties of negatives can be proved using the preceding assumed properties and definitions.

THEOREM 1 **Negative Properties**

For all real numbers a and b,

1. $-(-a) = a$

2. $(-a)b = -(ab)$
$\quad\quad = a(-b) = -ab$

3. $(-a)(-b) = ab$

4. $(-1)a = -a$

5. $\dfrac{-a}{b} = -\dfrac{a}{b} = \dfrac{a}{-b}, b \neq 0$

6. $\dfrac{-a}{-b} = -\dfrac{-a}{b} = -\dfrac{a}{-b} = \dfrac{a}{b}, b \neq 0$

We now state two important properties involving 0.

THEOREM 2 **Zero Properties**

For all real numbers a and b,

1. $a \cdot 0 = 0$ $\quad 0 \cdot 0 = 0 \quad (-35)(0) = 0$

2. $ab = 0$ $\quad$ if and only if $\quad a = 0 \quad$ or $\quad b = 0$
$\quad\quad$ If $(3x + 2)(x - 7) = 0$, then either $3x + 2 = 0$ or $x - 7 = 0$.

Fraction Properties

Recall that the quotient $a \div b$ $(b \neq 0)$ written in the form a/b is called a **fraction**. The quantity a is called the **numerator**, and the quantity b is called the **denominator**.

THEOREM 3   Fraction Properties

For all real numbers a, b, c, d, and k (division by 0 excluded):

1. $\dfrac{a}{b} = \dfrac{c}{d}$ if and only if $ad = bc$ $\dfrac{4}{6} = \dfrac{6}{9}$ since $4 \cdot 9 = 6 \cdot 6$

2. $\dfrac{ka}{kb} = \dfrac{a}{b}$
   $\dfrac{7 \cdot 3}{7 \cdot 5} = \dfrac{3}{5}$

3. $\dfrac{a}{b} \cdot \dfrac{c}{d} = \dfrac{ac}{bd}$
   $\dfrac{3}{5} \cdot \dfrac{7}{8} = \dfrac{3 \cdot 7}{5 \cdot 8}$

4. $\dfrac{a}{b} \div \dfrac{c}{d} = \dfrac{a}{b} \cdot \dfrac{d}{c}$
   $\dfrac{2}{3} \div \dfrac{5}{7} = \dfrac{2}{3} \cdot \dfrac{7}{5}$

5. $\dfrac{a}{b} + \dfrac{c}{b} = \dfrac{a + c}{b}$
   $\dfrac{3}{6} + \dfrac{5}{6} = \dfrac{3 + 5}{6}$

6. $\dfrac{a}{b} - \dfrac{c}{b} = \dfrac{a - c}{b}$
   $\dfrac{7}{8} - \dfrac{3}{8} = \dfrac{7 - 3}{8}$

7. $\dfrac{a}{b} + \dfrac{c}{d} = \dfrac{ad + bc}{bd}$
   $\dfrac{2}{3} + \dfrac{3}{5} = \dfrac{2 \cdot 5 + 3 \cdot 3}{3 \cdot 5}$

Exercises A-1

All variables represent real numbers.

A

In Problems 1–6, replace each question mark with an appropriate expression that will illustrate the use of the indicated real number property.

1. Commutative property $(\cdot)$: $uv = ?$

2. Commutative property $(+)$: $x + 7 = ?$

3. Associative property $(+)$: $3 + (7 + y) = ?$

4. Associative property $(\cdot)$: $x(yz) = ?$

5. Identity property $(\cdot)$: $1(u + v) = ?$

6. Identity property $(+)$: $0 + 9m = ?$

In Problems 7–26, indicate true (T) or false (F).

7. $5(8m) = (5 \cdot 8)m$

8. $a + cb = a + bc$

9. $5x + 7x = (5 + 7)x$

10. $uv(w + x) = uvw + uvx$

11. $-2(-a)(2x - y) = 2a(-4x + y)$

12. $8 \div (-5) = 8\left(\frac{1}{-5}\right)$

13. $(x + 3) + 2x = 2x + (x + 3)$

14. $\dfrac{x}{3y} \div \dfrac{5y}{x} = \dfrac{15y^2}{x^2}$

15. $\dfrac{2x}{-(x + 3)} = -\dfrac{2x}{x + 3}$

16. $-\dfrac{2x}{-(x - 3)} = \dfrac{2x}{x - 3}$

17. $(-3)\left(\frac{1}{-3}\right) = 1$

18. $(-0.5) + (0.5) = 0$

19. $-x^2y^2 = (-1)x^2y^2$

20. $[-(x + 2)](-x) = (x + 2)x$

21. $\dfrac{a}{b} + \dfrac{c}{d} = \dfrac{a + c}{b + d}$

22. $\dfrac{k}{k + b} = \dfrac{1}{1 + b}$

23. $(x + 8)(x + 6) = (x + 8)x + (x + 8)6$

24. $u(u - 2v) + v(u - 2v) = (u + v)(u - 2v)$

25. If $(x - 2)(2x + 3) = 0$, then either $x - 2 = 0$ or $2x + 3 = 0$.

26. If either $x - 2 = 0$ or $2x + 3 = 0$, then $(x - 2)(2x + 3) = 0$.

B

27. If $uv = 1$, does either u or v have to be 1? Explain.

28. If $uv = 0$, does either u or v have to be 0? Explain.

29. Indicate whether the following are true (T) or false (F):

    (A) All integers are natural numbers.

    (B) All rational numbers are real numbers.

    (C) All natural numbers are rational numbers.

30. Indicate whether the following are true (T) or false (F):

    (A) All natural numbers are integers.

    (B) All real numbers are irrational.

    (C) All rational numbers are real numbers.

31. Give an example of a real number that is not a rational number.

32. Give an example of a rational number that is not an integer.

33. Given the sets of numbers N (natural numbers), Z (integers), Q (rational numbers), and R (real numbers), indicate to which set(s) each of the following numbers belongs:

(A) 8   (B) $\sqrt{2}$ (C) -1.414 (D) $\frac{-5}{2}$

34. Given the sets of numbers N, Z, Q, and R (see Problem 33), indicate to which set(s) each of the following numbers belongs:

(A) -3 (B) 3.14   (C) π (D) $\frac{2}{3}$

35. Indicate true (T) or false (F), and for each false statement find real number replacements for a, b, and c that will provide a counterexample. For all real numbers a, b, and c,

(A) $a(b - c) = ab - c$

(B) $(a - b) - c = a - (b - c)$

(C) $a(bc) = (ab)c$

(D) $(a \div b) \div c = a \div (b \div c)$

36. Indicate true (T) or false (F), and for each false statement find real number replacements for a and b that will provide a counterexample. For all real numbers a and b,

(A) $a + b = b + a$

(B) $a - b = b - a$

(C) $ab = ba$

(D) $a \div b = b \div a$

C

37. If $c = 0.151\,515\ldots$, then $100c = 15.151\,5\ldots$ and

$$100c - c = 15.151\,5\ldots - 0.151\,515\ldots$$
$$99c = 15$$
$$c = \tfrac{15}{99} = \tfrac{5}{33}$$

Proceeding similarly, convert the repeating decimal $0.090\,909\ldots$ into a fraction. (All repeating decimals are rational numbers, and all rational numbers have repeating decimal representations.)

38. Repeat Problem 37 for $0.181\,818\ldots$.

Use a calculator to express each number in Problems 39 and 40 as a decimal to the capacity of your calculator. Observe the repeating decimal representation of the rational numbers and the nonrepeating decimal representation of the irrational numbers.

39. (A) $\frac{13}{6}$ (B) $\sqrt{21}$ (C) $\frac{7}{16}$ (D) $\frac{29}{111}$

40. (A) $\frac{8}{9}$ (B) $\frac{3}{11}$ (C) $\sqrt{5}$ (D) $\frac{11}{8}$

Answers to Matched Problems

1. (A) Associative $(+)$
(B) Commutative $(+)$
(C) Distributive
(D) Identity $(+)$
(E) Inverse $(\cdot)$

A-2   Operations on Polynomials

- Natural Number Exponents
- Polynomials
- Combining Like Terms
- Addition and Subtraction
- Multiplication
- Combined Operations

This section covers basic operations on *polynomials*. Our discussion starts with a brief review of natural number exponents. Integer and rational exponents and their properties will be discussed in detail in subsequent sections. (Natural numbers, integers, and rational numbers are important parts of the real number system; see Table 1 and Figure 1 in Appendix A-1.)

Natural Number Exponents

We define a **natural number exponent** as follows:

DEFINITION Natural Number Exponent

For n a natural number and b any real number,

$$b^n = b \cdot b \cdot \cdots \cdot b \qquad n \text{ factors of } b$$
$$3^5 = 3 \cdot 3 \cdot 3 \cdot 3 \cdot 3 \qquad 5 \text{ factors of } 3$$

where n is called the exponent and b is called the **base**.

Along with this definition, we state the **first property of exponents**:

> **THEOREM 1** **First Property of Exponents**
> For any natural numbers m and n, and any real number b:
> $$b^m b^n = b^{m+n} \quad (2t^4)(5t^3) \boxed{= 2 \cdot 5 t^{4+3}} = 10t^7$$

Polynomials

Algebraic expressions are formed by using constants and variables and the algebraic operations of addition, subtraction, multiplication, division, raising to powers, and taking roots. Special types of algebraic expressions are called *polynomials*. A **polynomial in one variable** x is constructed by adding or subtracting constants and terms of the form ax^n, where a is a real number and n is a natural number. A **polynomial in two variables** x and y is constructed by adding and subtracting constants and terms of the form $ax^m y^n$, where a is a real number and m and n are natural numbers. Polynomials in three and more variables are defined in a similar manner.

Polynomials		Not Polynomials	
8	0	$\dfrac{1}{x}$	$\dfrac{x - y}{x^2 + y^2}$
$3x^3 - 6x + 7$	$6x + 3$		
$2x^2 - 7xy - 8y^2$	$9y^3 + 4y^2 - y + 4$	$\sqrt{x^3 - 2x}$	$2x^{-2} - 3x^{-1}$
$2x - 3y + 2$	$u^5 - 3u^3v^2 + 2uv^4 - v^4$		

Polynomial forms are encountered frequently in mathematics. For the efficient study of polynomials, it is useful to classify them according to their *degree*. If a term in a polynomial has only one variable as a factor, then the **degree of the term** is the power of the variable. If two or more variables are present in a term as factors, then the **degree of the term** is the sum of the powers of the variables. The **degree of a polynomial** is the degree of the nonzero term with the highest degree in the polynomial. Any nonzero constant is defined to be a **polynomial of degree 0**. The number 0 is also a polynomial but is not assigned a degree.

EXAMPLE 1 **Degree**

(A) The degree of the first term in $5x^3 + \sqrt{3}x - \frac{1}{2}$ is 3, the degree of the second term is 1, the degree of the third term is 0, and the degree of the whole polynomial is 3 (the same as the degree of the term with the highest degree).

(B) The degree of the first term in $8u^3v^2 - \sqrt{7}uv^2$ is 5, the degree of the second term is 3, and the degree of the whole polynomial is 5.

Matched Problem 1

(A) Given the polynomial $6x^5 + 7x^3 - 2$, what is the degree of the first term? The second term? The third term? The whole polynomial?

(B) Given the polynomial $2u^4v^2 - 5uv^3$, what is the degree of the first term? The second term? The whole polynomial?

In addition to classifying polynomials by degree, we also call a single-term polynomial a **monomial**, a two-term polynomial a **binomial**, and a three-term polynomial a **trinomial**.

Combining Like Terms

The concept of *coefficient* plays a central role in the process of combining *like terms*. A constant in a term of a polynomial, including the sign that precedes it, is called the **numerical coefficient**, or simply, the **coefficient**, of the term. If a constant does not appear, or only a $+$ sign appears, the coefficient is understood to be 1. If only a $-$ sign appears, the coefficient is understood to be -1. Given the polynomial

$$5x^4 - x^3 - 3x^2 + x - 7 \quad \boxed{= 5x^4 + (-1)x^3 + (-3)x^2 + 1x + (-7)}$$

the coefficient of the first term is 5, the coefficient of the second term is -1, the coefficient of the third term is -3, the coefficient of the fourth term is 1, and the coefficient of the fifth term is -7.

The following distributive properties are fundamental to the process of combining *like terms*.

THEOREM 2   Distributive Properties of Real Numbers

1. $a(b + c) = (b + c)a = ab + ac$
2. $a(b - c) = (b - c)a = ab - ac$
3. $a(b + c + \cdots + f) = ab + ac + \cdots + af$

Two terms in a polynomial are called **like terms** if they have exactly the same variable factors to the same powers. The numerical coefficients may or may not be the same. Since constant terms involve no variables, all constant terms are like terms. If a polynomial contains two or more like terms, these terms can be combined into a single term by making use of distributive properties. The following example illustrates the reasoning behind the process:

$$3x^2y - 5xy^2 + x^2y - 2x^2y \quad \boxed{\begin{aligned} &= 3x^2y + x^2y - 2x^2y - 5xy^2 \\ &= (3x^2y + 1x^2y - 2x^2y) - 5xy^2 \\ &= (3 + 1 - 2)x^2y - 5xy^2 \end{aligned}} \quad \text{\textit{Note the use of distributive properties.}}$$
$$= 2x^2y - 5xy^2$$

Free use is made of the real number properties discussed in Appendix A-1.

How can we simplify expressions such as $4(x - 2y) - 3(2x - 7y)$? We clear the expression of parentheses using distributive properties, and combine like terms:

$$4(x - 2y) - 3(2x - 7y) = 4x - 8y - 6x + 21y$$
$$= -2x + 13y$$

EXAMPLE 2   **Removing Parentheses**   Remove parentheses and simplify:

(A) $2(3x^2 - 2x + 5) + (x^2 + 3x - 7) \quad \boxed{= 2(3x^2 - 2x + 5) + 1(x^2 + 3x - 7)}$
$$= 6x^2 - 4x + 10 + x^2 + 3x - 7$$
$$= 7x^2 - x + 3$$

(B) $(x^3 - 2x - 6) - (2x^3 - x^2 + 2x - 3)$

$\boxed{= 1(x^3 - 2x - 6) + (-1)(2x^3 - x^2 + 2x - 3)}$ \quad \text{\textit{Be careful with the sign here}}
$$= x^3 - 2x - 6 - 2x^3 + x^2 - 2x + 3$$
$$= -x^3 + x^2 - 4x - 3$$

(C) $[3x^2 - (2x + 1)] - (x^2 - 1) = [3x^2 - 2x - 1] - (x^2 - 1)$ \quad \text{\textit{Remove inner}}
$$= 3x^2 - 2x - 1 - x^2 + 1 \qquad \text{\textit{parentheses}}$$
$$= 2x^2 - 2x \qquad\qquad\qquad \text{\textit{first.}}$$

Matched Problem 2   Remove parentheses and simplify:

(A) $3(u^2 - 2v^2) + (u^2 + 5v^2)$
(B) $(m^3 - 3m^2 + m - 1) - (2m^3 - m + 3)$
(C) $(x^3 - 2) - [2x^3 - (3x + 4)]$

Addition and Subtraction

Addition and subtraction of polynomials can be thought of in terms of removing parentheses and combining like terms, as illustrated in Example 2. Horizontal and vertical arrangements are illustrated in the next two examples. You should be able to work either way, letting the situation dictate your choice.

EXAMPLE 3 **Adding Polynomials** Add horizontally and vertically:

$$x^4 - 3x^3 + x^2, \quad -x^3 - 2x^2 + 3x, \quad \text{and} \quad 3x^2 - 4x - 5$$

SOLUTION Add horizontally:

$$(x^4 - 3x^3 + x^2) + (-x^3 - 2x^2 + 3x) + (3x^2 - 4x - 5)$$
$$= x^4 - 3x^3 + x^2 - x^3 - 2x^2 + 3x + 3x^2 - 4x - 5$$
$$= x^4 - 4x^3 + 2x^2 - x - 5$$

Or vertically, by lining up like terms and adding their coefficients:

$$
\begin{array}{r}
x^4 - 3x^3 + x^2 \phantom{{}- 4x - 5} \\
- x^3 - 2x^2 + 3x \phantom{{}- 5} \\
3x^2 - 4x - 5 \\
\hline
x^4 - 4x^3 + 2x^2 - x - 5
\end{array}
$$

Matched Problem 3 Add horizontally and vertically:

$$3x^4 - 2x^3 - 4x^2, \quad x^3 - 2x^2 - 5x, \quad \text{and} \quad x^2 + 7x - 2$$

EXAMPLE 4 **Subtracting Polynomials** Subtract $4x^2 - 3x + 5$ from $x^2 - 8$, both horizontally and vertically.

SOLUTION
$$(x^2 - 8) - (4x^2 - 3x + 5) \qquad \text{or} \qquad$$
$$= x^2 - 8 - 4x^2 + 3x - 5$$
$$= -3x^2 + 3x - 13$$

$$
\begin{array}{r}
x^2 \phantom{{}+ 3x} - 8 \\
-4x^2 + 3x - 5 \quad \leftarrow \textit{Change signs and add.} \\
\hline
-3x^2 + 3x - 13
\end{array}
$$

Matched Problem 4 Subtract $2x^2 - 5x + 4$ from $5x^2 - 6$, both horizontally and vertically.

Multiplication

Multiplication of algebraic expressions involves the extensive use of distributive properties for real numbers, as well as other real number properties.

EXAMPLE 5 **Multiplying Polynomials** Multiply: $(2x - 3)(3x^2 - 2x + 3)$

SOLUTION $(2x - 3)(3x^2 - 2x + 3) \;\boxed{= 2x(3x^2 - 2x + 3) - 3(3x^2 - 2x + 3)}$

$$= 6x^3 - 4x^2 + 6x - 9x^2 + 6x - 9$$
$$= 6x^3 - 13x^2 + 12x - 9$$

Or, using a vertical arrangement,

$$
\begin{array}{r}
3x^2 - 2x + 3 \\
2x - 3 \\
\hline
6x^3 - 4x^2 + 6x \phantom{{}- 9} \\
- 9x^2 + 6x - 9 \\
\hline
6x^3 - 13x^2 + 12x - 9
\end{array}
$$

Matched Problem 5 Multiply: $(2x - 3)(2x^2 + 3x - 2)$

Thus, to multiply two polynomials, multiply each term of one by each term of the other, and combine like terms.

Products of binomial factors occur frequently, so it is useful to develop procedures that will enable us to write down their products by inspection. To find the product $(2x - 1)(3x + 2)$, we proceed as follows:

$$(2x - 1)(3x + 2) \quad = 6x^2 + 4x - 3x - 2 \qquad \text{The inner and outer products}$$
$$\qquad\qquad\qquad\qquad\qquad\qquad\qquad\text{are like terms, so combine into}$$
$$= 6x^2 + x - 2 \qquad\qquad\quad \text{a single term.}$$

To speed the process, we do the step in the dashed box mentally.

Products of certain binomial factors occur so frequently that it is useful to learn formulas for their products. The following formulas are easily verified by multiplying the factors on the left.

THEOREM 3   Special Products

1. $(a - b)(a + b) = a^2 - b^2$
2. $(a + b)^2 = a^2 + 2ab + b^2$
3. $(a - b)^2 = a^2 - 2ab + b^2$

EXAMPLE 6 **Special Products** Multiply mentally, where possible.

(A) $(2x - 3y)(5x + 2y)$ (B) $(3a - 2b)(3a + 2b)$

(C) $(5x - 3)^2$ (D) $(m + 2n)^3$

SOLUTION (A) $(2x - 3y)(5x + 2y) = 10x^2 + 4xy - 15xy - 6y^2$

$$= 10x^2 - 11xy - 6y^2$$

(B) $(3a - 2b)(3a + 2b) = (3a)^2 - (2b)^2$

$$= 9a^2 - 4b^2$$

(C) $(5x - 3)^2 = (5x)^2 - 2(5x)(3) + 3^2$

$$= 25x^2 - 30x + 9$$

(D) $(m + 2n)^3 = (m + 2n)^2(m + 2n)$

$$= (m^2 + 4mn + 4n^2)(m + 2n)$$
$$= m^2(m + 2n) + 4mn(m + 2n) + 4n^2(m + 2n)$$
$$= m^3 + 2m^2n + 4m^2n + 8mn^2 + 4mn^2 + 8n^3$$
$$= m^3 + 6m^2n + 12mn^2 + 8n^3$$

Matched Problem 6 Multiply mentally, where possible.

(A) $(4u - 3v)(2u + v)$

(B) $(2xy + 3)(2xy - 3)$

(C) $(m + 4n)(m - 4n)$

(D) $(2u - 3v)^2$

(E) $(2x - y)^3$

Combined Operations

We complete this section by considering several examples that use all the operations just discussed. Note that in simplifying, we usually remove grouping symbols starting from the inside. That is, we remove parentheses ( ) first, then brackets [ ], and finally braces { }, if present. Also,

DEFINITION Order of Operations

Multiplication and division precede addition and subtraction, and taking powers precedes multiplication and division.

$$2 \cdot 3 + 4 = 6 + 4 = 10, \quad \text{not} \quad 2 \cdot 7 = 14$$

$$\frac{10^2}{2} = \frac{100}{2} = 50, \quad \text{not} \quad 5^2 = 25$$

EXAMPLE 7 **Combined Operations** Perform the indicated operations and simplify:

(A) $3x - \{5 - 3[x - x(3 - x)]\} = 3x - \{5 - 3[x - 3x + x^2]\}$
$$= 3x - \{5 - 3x + 9x - 3x^2\}$$
$$= 3x - 5 + 3x - 9x + 3x^2$$
$$= 3x^2 - 3x - 5$$

(B) $(x - 2y)(2x + 3y) - (2x + y)^2 = 2x^2 - xy - 6y^2 - (4x^2 + 4xy + y^2)$
$$= 2x^2 - xy - 6y^2 - 4x^2 - 4xy - y^2$$
$$= -2x^2 - 5xy - 7y^2$$

Matched Problem 7 Perform the indicated operations and simplify:

(A) $2t - \{7 - 2[t - t(4 + t)]\}$

(B) $(u - 3v)^2 - (2u - v)(2u + v)$

Exercises A-2

A

Problems 1–8 refer to the following polynomials:

(A) $2x - 3$ (B) $2x^2 - x + 2$ (C) $x^3 + 2x^2 - x + 3$

1. What is the degree of (C)?

2. What is the degree of (A)?

3. Add (B) and (C).

4. Add (A) and (B).

5. Subtract (B) from (C).

6. Subtract (A) from (B).

7. Multiply (B) and (C).

8. Multiply (A) and (C).

In Problems 9–30, perform the indicated operations and simplify.

9. $2(u - 1) - (3u + 2) - 2(2u - 3)$

10. $2(x - 1) + 3(2x - 3) - (4x - 5)$

11. $4a - 2a[5 - 3(a + 2)]$

12. $2y - 3y[4 - 2(y - 1)]$

13. $(a + b)(a - b)$

14. $(m - n)(m + n)$

15. $(3x - 5)(2x + 1)$

16. $(4t - 3)(t - 2)$

17. $(2x - 3y)(x + 2y)$

18. $(3x + 2y)(x - 3y)$

19. $(3y + 2)(3y - 2)$

20. $(2m - 7)(2m + 7)$

21. $-(2x - 3)^2$

22. $-(5 - 3x)^2$

23. $(4m + 3n)(4m - 3n)$

24. $(3x - 2y)(3x + 2y)$

25. $(3u + 4v)^2$

26. $(4x - y)^2$

27. $(a - b)(a^2 + ab + b^2)$

28. $(a + b)(a^2 - ab + b^2)$

29. $[(x - y) + 3z][(x - y) - 3z]$

30. $[a - (2b - c)][a + (2b - c)]$

B

In Problems 31–44, perform the indicated operations and simplify.

31. $m - \{m - [m - (m - 1)]\}$

32. $2x - 3\{x + 2[x - (x + 5)] + 1\}$

33. $(x^2 - 2xy + y^2)(x^2 + 2xy + y^2)$

34. $(3x - 2y)^2(2x + 5y)$

35. $(5a - 2b)^2 - (2b + 5a)^2$

36. $(2x - 1)^2 - (3x + 2)(3x - 2)$

37. $(m - 2)^2 - (m - 2)(m + 2)$

38. $(x - 3)(x + 3) - (x - 3)^2$

39. $(x - 2y)(2x + y) - (x + 2y)(2x - y)$

40. $(3m + n)(m - 3n) - (m + 3n)(3m - n)$

41. $(u + v)^3$

42. $(x - y)^3$

43. $(x - 2y)^3$

44. $(2m - n)^3$

45. Subtract the sum of the last two polynomials from the sum of the first two: $2x^2 - 4xy + y^2, 3xy - y^2,$ $x^2 - 2xy - y^2, -x^2 + 3xy - 2y^2$

46. Subtract the sum of the first two polynomials from the sum of the last two: $3m^2 - 2m + 5, 4m^2 - m, 3m^2 - 3m - 2,$ $m^3 + m^2 + 2$

C

In Problems 47–50, perform the indicated operations and simplify.

47. $[(2x - 1)^2 - x(3x + 1)]^2$

48. $[5x(3x + 1) - 5(2x - 1)^2]^2$

49. $2\{(x - 3)(x^2 - 2x + 1) - x[3 - x(x - 2)]\}$

50. $-3x\{x[x - x(2 - x)] - (x + 2)(x^2 - 3)\}$

51. If you are given two polynomials, one of degree m and the other of degree n, where m is greater than n, what is the degree of their product?

52. What is the degree of the sum of the two polynomials in Problem 51?

53. How does the answer to Problem 51 change if the two polynomials can have the same degree?

54. How does the answer to Problem 52 change if the two polynomials can have the same degree?

55. Show by example that, in general, $(a + b)^2 \neq a^2 + b^2$. Discuss possible conditions on a and b that would make this a valid equation.

56. Show by example that, in general, $(a - b)^2 \neq a^2 - b^2$. Discuss possible conditions on a and b that would make this a valid equation.

Applications

57. **Investment.** You have $10,000 to invest, part at 9% and the rest at 12%. If x is the amount invested at 9%, write an algebraic expression that represents the total annual income from both investments. Simplify the expression.

58. **Investment.** A person has $100,000 to invest. If x are invested in a money market account yielding 7% and twice that amount in certificates of deposit yielding 9%, and if the rest is invested in high-grade bonds yielding 11%, write an algebraic expression that represents the total annual income from all three investments. Simplify the expression.

59. **Gross receipts.** Four thousand tickets are to be sold for a musical show. If x tickets are to be sold for $20 each and three times that number for $30 each, and if the rest are sold for $50 each, write an algebraic expression that represents the gross receipts from ticket sales, assuming all tickets are sold. Simplify the expression.

60. **Gross receipts.** Six thousand tickets are to be sold for a concert, some for $20 each and the rest for $35 each. If x is the number of $20 tickets sold, write an algebraic expression that represents the gross receipts from ticket sales, assuming all tickets are sold. Simplify the expression.

61. **Nutrition.** Food mix A contains 2% fat, and food mix B contains 6% fat. A 10-kilogram diet mix of foods A and B is formed. If x kilograms of food A are used, write an algebraic expression that represents the total number of kilograms of fat in the final food mix. Simplify the expression.

62. **Nutrition.** Each ounce of food M contains 8 units of calcium, and each ounce of food N contains 5 units of calcium. A 160-ounce diet mix is formed using foods M and N. If x is the number of ounces of food M used, write an algebraic expression that represents the total number of units of calcium in the diet mix. Simplify the expression.

Answers to Matched Problems

1. (A) $5, 3, 0, 5$ (B) $6, 4, 6$

2. (A) $4u^2 - v^2$
   (B) $-m^3 - 3m^2 + 2m - 4$
   (C) $-x^3 + 3x + 2$

3. $3x^4 - x^3 - 5x^2 + 2x - 2$

4. $3x^2 + 5x - 10$

5. $4x^3 - 13x + 6$

6. (A) $8u^2 - 2uv - 3v^2$
   (B) $4x^2y^2 - 9$
   (C) $m^2 - 16n^2$
   (D) $4u^2 - 12uv + 9v^2$
   (E) $8x^3 - 12x^2y + 6xy^2 - y^3$

7. (A) $-2t^2 - 4t - 7$ (B) $-3u^2 - 6uv + 10v^2$

A-3 Factoring Polynomials

A polynomial is written in factored form if it is written as the product of two or more polynomials. The following polynomials are written in factored form:

$$4x^2y - 6xy^2 = 2xy(2x - 3y) \qquad 2x^3 - 8x = 2x(x - 2)(x + 2)$$
$$x^2 - x - 6 = (x - 3)(x + 2) \qquad 5m^2 + 20 = 5(m^2 + 4)$$

Unless stated to the contrary, we will limit our discussion of factoring polynomials to polynomials with integer coefficients.

A polynomial with integer coefficients is said to be **factored completely** if each factor cannot be expressed as the product of two or more polynomials with integer coefficients, other than itself or 1. All the polynomials above, as we will see by the conclusion of this section, are factored completely.

Writing polynomials in completely factored form is often a difficult task. But accomplishing it can lead to the simplification of certain algebraic expressions and to the solution of certain types of equations and inequalities. The distributive properties for real numbers are central to the factoring process.

Common Factors

Generally, a first step in any factoring procedure is to factor out all factors common to all terms.

EXAMPLE 1   **Common Factors**   Factor out all factors common to all terms.

(A) $3x^3y - 6x^2y^2 - 3xy^3$

(B) $3y(2y + 5) + 2(2y + 5)$

SOLUTION   (A) $3x^3y - 6x^2y^2 - 3xy^3 \boxed{= (3xy)x^2 - (3xy)2xy - (3xy)y^2}$

$$= 3xy(x^2 - 2xy - y^2)$$

(B) $3y(2y + 5) + 2(2y + 5) \boxed{= 3y(2y + 5) + 2(2y + 5)}$

$$= (3y + 2)(2y + 5)$$

Matched Problem 1   Factor out all factors common to all terms.

(A) $2x^3y - 8x^2y^2 - 6xy^3$ (B) $2x(3x - 2) - 7(3x - 2)$

Factoring by Grouping

Occasionally, polynomials can be factored by grouping terms in such a way that we obtain results that look like Example 1B. We can then complete the factoring following the steps used in that example. This process will prove useful in the next subsection, where an efficient method is developed for factoring a second-degree polynomial as the product of two first-degree polynomials, if such factors exist.

EXAMPLE 2   **Factoring by Grouping**   Factor by grouping.

(A) $3x^2 - 3x - x + 1$

(B) $4x^2 - 2xy - 6xy + 3y^2$

(C) $y^2 + xz + xy + yz$

SOLUTION   (A) $3x^2 - 3x - x + 1$

$= (3x^2 - 3x) - (x - 1)$ *Group the first two and the last two terms.*

$= 3x(x - 1) - (x - 1)$ *Factor out any common factors from each*

$= (x - 1)(3x - 1)$ *group. The common factor $(x - 1)$ can be taken out, and the factoring is complete.*

(B) $4x^2 - 2xy - 6xy + 3y^2 = (4x^2 - 2xy) - (6xy - 3y^2)$
$$= 2x(2x - y) - 3y(2x - y)$$
$$= (2x - y)(2x - 3y)$$

(C) If, as in parts (A) and (B), we group the first two terms and the last two terms of $y^2 + xz + xy + yz$, no common factor can be taken out of each group to complete the factoring. However, if the two middle terms are reversed, we can proceed as before:

$$y^2 + xz + xy + yz = y^2 + xy + xz + yz$$
$$= (y^2 + xy) + (xz + yz)$$
$$= y(y + x) + z(x + y)$$
$$= y(x + y) + z(x + y)$$
$$= (x + y)(y + z)$$

Matched Problem 2 Factor by grouping.

(A) $6x^2 + 2x + 9x + 3$ (B) $2u^2 + 6uv - 3uv - 9v^2$

(C) $ac + bd + bc + ad$

Factoring Second-Degree Polynomials

We now turn our attention to factoring second-degree polynomials of the form

$$2x^2 - 5x - 3 \quad \text{and} \quad 2x^2 + 3xy - 2y^2$$

into the product of two first-degree polynomials with integer coefficients. Since many second-degree polynomials with integer coefficients cannot be factored in this way, it would be useful to know ahead of time that the factors we are seeking actually exist. The factoring approach we use, involving the *ac test*, determines at the beginning whether first-degree factors with integer coefficients do exist. Then, if they exist, the test provides a simple method for finding them.

THEOREM 1 *ac* **Test for Factorability**

If in polynomials of the form

$$ax^2 + bx + c \quad \text{or} \quad ax^2 + bxy + cy^2 \qquad (1)$$

the product ac has two integer factors p and q whose sum is the coefficient b of the middle term; that is, if integers p and q exist so that

$$pq = ac \quad \text{and} \quad p + q = b \qquad (2)$$

then the polynomials have first-degree factors with integer coefficients. If no integers p and q exist that satisfy equations (2), then the polynomials in equations (1) will not have first-degree factors with integer coefficients.

If integers p and q exist that satisfy equations (2) in the *ac* test, the factoring always can be completed as follows: Using $b = p + q$, split the middle terms in equations (1) to obtain

$$ax^2 + bx + c = ax^2 + px + qx + c$$
$$ax^2 + bxy + cy^2 = ax^2 + pxy + qxy + cy^2$$

Complete the factoring by grouping the first two terms and the last two terms as in Example 2. This process always works, and it does not matter if the two middle terms on the right are interchanged.

Several examples should make the process clear. After a little practice, you will perform many of the steps mentally and will find the process fast and efficient.

EXAMPLE 3 **Factoring Second-Degree Polynomials** Factor, if possible, using integer coefficients.

(A) $4x^2 - 4x - 3$

(B) $2x^2 - 3x - 4$

(C) $6x^2 - 25xy + 4y^2$

SOLUTION (A) $4x^2 - 4x - 3$

Step 1 Use the *ac* test to test for factorability. Comparing $4x^2 - 4x - 3$ with $ax^2 + bx + c$, we see that $a = 4, b = -4$, and $c = -3$. Multiply a and c to obtain

$$ac = (4)(-3) = -12$$

pq

(1)(−12) *All factor pairs*
(−1)(12) *of −12 = ac*
(2)(−6)
(−2)(6)
(3)(−4)
(−3)(4)

List all pairs of integers whose product is -12, as shown in the margin. These are called **factor pairs** of -12. Then try to find a factor pair that sums to $b = -4$, the coefficient of the middle term in $4x^2 - 4x - 3$. (In practice, this part of Step 1 is often done mentally and can be done rather quickly.) Notice that the factor pair 2 and -6 sums to -4. By the *ac* test, $4x^2 - 4x - 3$ has first-degree factors with integer coefficients.

Step 2 Split the middle term, using $b = p + q$, and complete the factoring by grouping. Using $-4 = 2 + (-6)$, we split the middle term in $4x^2 - 4x - 3$ and complete the factoring by grouping:

$$4x^2 - 4x - 3 = 4x^2 + 2x - 6x - 3$$
$$= (4x^2 + 2x) - (6x + 3)$$
$$= 2x(2x + 1) - 3(2x + 1)$$
$$= (2x + 1)(2x - 3)$$

The result can be checked by multiplying the two factors to obtain the original polynomial.

(B) $2x^2 - 3x - 4$

Step 1 Use the *ac* test to test for factorability:

$$ac = (2)(-4) = -8$$

pq

(−1)(8) *All factor pairs of*
(1)(−8) *−8 = ac*
(−2)(4)
(2)(−4)

Does -8 have a factor pair whose sum is -3? None of the factor pairs listed in the margin sums to $-3 = b$, the coefficient of the middle term in $2x^2 - 3x - 4$. According to the *ac* test, we can conclude that $2x^2 - 3x - 4$ does not have first-degree factors with integer coefficients, and we say that the polynomial is **not factorable**.

(C) $6x^2 - 25xy + 4y^2$

Step 1 Use the *ac* test to test for factorability:

$$ac = (6)(4) = 24$$

Mentally checking through the factor pairs of 24, keeping in mind that their sum must be $-25 = b$, we see that if $p = -1$ and $q = -24$, then

$$pq = (-1)(-24) = 24 = ac$$

and

$$p + q = (-1) + (-24) = -25 = b$$

So the polynomial is factorable.

Step 2 Split the middle term, using $b = p + q$, and complete the factoring by grouping. Using $-25 = (-1) + (-24)$, we split the middle term in $6x^2 - 25xy + 4y^2$ and complete the factoring by grouping:

$$6x^2 - 25xy + 4y^2 = 6x^2 - xy - 24xy + 4y^2$$
$$= (6x^2 - xy) - (24xy - 4y^2)$$
$$= x(6x - y) - 4y(6x - y)$$
$$= (6x - y)(x - 4y)$$

The check is left to the reader.

Matched Problem 3   Factor, if possible, using integer coefficients.

(A) $2x^2 + 11x - 6$ (B) $4x^2 + 11x - 6$ (C) $6x^2 + 5xy - 4y^2$

Special Factoring Formulas

The factoring formulas listed in the following box will enable us to factor certain polynomial forms that occur frequently. These formulas can be established by multiplying the factors on the right.

THEOREM 2   Special Factoring Formulas

Perfect square:	1. $u^2 + 2uv + v^2 = (u + v)^2$
Perfect square:	2. $u^2 - 2uv + v^2 = (u - v)^2$
Difference of squares:	3. $u^2 - v^2 = (u - v)(u + v)$
Difference of cubes:	4. $u^3 - v^3 = (u - v)(u^2 + uv + v^2)$
Sum of cubes:	5. $u^3 + v^3 = (u + v)(u^2 - uv + v^2)$

CAUTION Notice that $u^2 + v^2$ is not included in the list of special factoring formulas. In fact,

$$u^2 + v^2 \neq (au + bv)(cu + dv)$$

for any choice of real number coefficients a, b, c, and d.

EXAMPLE 4   **Factoring** Factor completely.

(A) $4m^2 - 12mn + 9n^2$ (B) $x^2 - 16y^2$ (C) $z^3 - 1$
(D) $m^3 + n^3$ (E) $a^2 - 4(b + 2)^2$

SOLUTION   (A) $4m^2 - 12mn + 9n^2 = (2m - 3n)^2$

(B) $x^2 - 16y^2 = x^2 - (4y)^2 = (x - 4y)(x + 4y)$

(C) $z^3 - 1 = (z - 1)(z^2 + z + 1)$ Use the ac test to verify that $z^2 + z + 1$ cannot be factored.

(D) $m^3 + n^3 = (m + n)(m^2 - mn + n^2)$ Use the ac test to verify that $m^2 - mn + n^2$ cannot be factored.

(E) $a^2 - 4(b + 2)^2 = [a - 2(b + 2)][a + 2(b + 2)]$

Matched Problem 4   Factor completely:

(A) $x^2 + 6xy + 9y^2$ (B) $9x^2 - 4y^2$ (C) $8m^3 - 1$
(D) $x^3 + y^3z^3$ (E) $9(m - 3)^2 - 4n^2$

Combined Factoring Techniques

We complete this section by considering several factoring problems that involve combinations of the preceding techniques.

> **PROCEDURE** **Factoring Polynomials**
> **Step 1** Take out any factors common to all terms.
> **Step 2** Use any of the special formulas listed in Theorem 2 that are applicable.
> **Step 3** Apply the *ac* test to any remaining second-degree polynomial factors.
>
> **Note:** It may be necessary to perform some of these steps more than once. Furthermore, the order of applying these steps can vary.

EXAMPLE 5   **Combined Factoring Techniques** Factor completely.
(A) $3x^3 - 48x$
(B) $3u^4 - 3u^3v - 9u^2v^2$
(C) $3m^4 - 24mn^3$
(D) $3x^4 - 5x^2 + 2$

SOLUTION
(A) $3x^3 - 48x = 3x(x^2 - 16) = 3x(x - 4)(x + 4)$
(B) $3u^4 - 3u^3v - 9u^2v^2 = 3u^2(u^2 - uv - 3v^2)$
(C) $3m^4 - 24mn^3 = 3m(m^3 - 8n^3) = 3m(m - 2n)(m^2 + 2mn + 4n^2)$
(D) $3x^4 - 5x^2 + 2 = (3x^2 - 2)(x^2 - 1) = (3x^2 - 2)(x - 1)(x + 1)$

Matched Problem 5   Factor completely.
(A) $18x^3 - 8x$
(B) $4m^3n - 2m^2n^2 + 2mn^3$
(C) $2t^4 - 16t$
(D) $2y^4 - 5y^2 - 12$

Exercises A-3

A

In Problems 1–8, factor out all factors common to all terms.

1. $6m^4 - 9m^3 - 3m^2$
2. $6x^4 - 8x^3 - 2x^2$
3. $8u^3v - 6u^2v^2 + 4uv^3$
4. $10x^3y + 20x^2y^2 - 15xy^3$
5. $7m(2m - 3) + 5(2m - 3)$
6. $5x(x + 1) - 3(x + 1)$
7. $4ab(2c + d) - (2c + d)$
8. $12a(b - 2c) - 15b(b - 2c)$

In Problems 9–18, factor by grouping.

9. $2x^2 - x + 4x - 2$
10. $x^2 - 3x + 2x - 6$
11. $3y^2 - 3y + 2y - 2$
12. $2x^2 - x + 6x - 3$
13. $2x^2 + 8x - x - 4$
14. $6x^2 + 9x - 2x - 3$
15. $wy - wz + xy - xz$
16. $ac + ad + bc + bd$
17. $am - 3bm + 2na - 6bn$
18. $ab + 6 + 2a + 3b$

B

In Problems 19–56, factor completely. If a polynomial cannot be factored, say so.

19. $3y^2 - y - 2$
20. $2x^2 + 5x - 3$
21. $u^2 - 2uv - 15v^2$
22. $x^2 - 4xy - 12y^2$
23. $m^2 - 6m - 3$
24. $x^2 + x - 4$
25. $w^2x^2 - y^2$
26. $25m^2 - 16n^2$
27. $9m^2 - 6mn + n^2$
28. $x^2 + 10xy + 25y^2$
29. $y^2 + 16$
30. $u^2 + 81$
31. $4z^2 - 28z + 48$
32. $6x^2 + 48x + 72$
33. $2x^4 - 24x^3 + 40x^2$
34. $2y^3 - 22y^2 + 48y$
35. $4xy^2 - 12xy + 9x$
36. $16x^2y - 8xy + y$
37. $6m^2 - mn - 12n^2$
38. $6s^2 + 7st - 3t^2$
39. $4u^3v - uv^3$
40. $x^3y - 9xy^3$
41. $2x^3 - 2x^2 + 8x$
42. $3m^3 - 6m^2 + 15m$
43. $8x^3 - 27y^3$
44. $5x^3 + 40y^3$
45. $x^4y + 8xy$
46. $8a^3 - 1$

C

47. $(x + 2)^2 - 9y^2$
48. $(a - b)^2 - 4(c - d)^2$
49. $5u^2 + 4uv - 2v^2$
50. $3x^2 - 2xy - 4y^2$
51. $6(x - y)^2 + 23(x - y) - 4$
52. $4(A + B)^2 - 5(A + B) - 6$
53. $y^4 - 3y^2 - 4$
54. $m^4 - n^4$
55. $15y(x - y)^3 + 12x(x - y)^2$
56. $15x^2(3x - 1)^4 + 60x^3(3x - 1)^3$

In Problems 57–60, discuss the validity of each statement. If the statement is true, explain why. If not, give a counterexample.

57. If n is a positive integer greater than 1, then $u^n - v^n$ can be factored.

58. If m and n are positive integers and $m \neq n$, then $u^m - v^n$ is not factorable.

59. If n is a positive integer greater than 1, then $u^n + v^n$ can be factored.

60. If k is a positive integer, then $u^{2k+1} + v^{2k+1}$ can be factored.

Answers to Matched Problems

1. (A) $2xy(x^2 - 4xy - 3y^2)$ (B) $(2x - 7)(3x - 2)$

2. (A) $(3x + 1)(2x + 3)$

  (B) $(u + 3v)(2u - 3v)$

  (C) $(a + b)(c + d)$

3. (A) $(2x - 1)(x + 6)$

  (B) Not factorable

  (C) $(3x + 4y)(2x - y)$

4. (A) $(x + 3y)^2$

  (B) $(3x - 2y)(3x + 2y)$

  (C) $(2m - 1)(4m^2 + 2m + 1)$

  (D) $(x + yz)(x^2 - xyz + y^2z^2)$

  (E) $[3(m - 3) - 2n][3(m - 3) + 2n]$

5. (A) $2x(3x - 2)(3x + 2)$

  (B) $2mn(2m^2 - mn + n^2)$

  (C) $2t(t - 2)(t^2 + 2t + 4)$

  (D) $(2y^2 + 3)(y - 2)(y + 2)$

A-4 Operations on Rational Expressions

- Reducing to Lowest Terms
- Multiplication and Division
- Addition and Subtraction
- Compound Fractions

We now turn our attention to fractional forms. A quotient of two algebraic expressions (division by 0 excluded) is called a **fractional expression**. If both the numerator and the denominator are polynomials, the fractional expression is called a **rational expression**. Some examples of rational expressions are

$$\frac{1}{x^3 + 2x} \qquad \frac{5}{x} \qquad \frac{x + 7}{3x^2 - 5x + 1} \qquad \frac{x^2 - 2x + 4}{1}$$

In this section, we discuss basic operations on rational expressions. Since variables represent real numbers in the rational expressions we will consider, the properties of real number fractions summarized in Appendix A-1 will play a central role.

> **AGREEMENT Variable Restriction**
> Even though not always explicitly stated, we always assume that variables are restricted so that division by 0 is excluded.

For example, given the rational expression

$$\frac{2x + 5}{x(x + 2)(x - 3)}$$

the variable x is understood to be restricted from being 0, -2, or 3, since these values would cause the denominator to be 0.

Reducing to Lowest Terms

Central to the process of reducing rational expressions to *lowest terms* is the *fundamental property of fractions*, which we restate here for convenient reference:

> **THEOREM 1 Fundamental Property of Fractions**
> If a, b, and k are real numbers with $b, k \neq 0$, then
>
> $$\frac{ka}{kb} = \frac{a}{b} \qquad \frac{5 \cdot 2}{5 \cdot 7} = \frac{2}{7} \qquad \frac{x(x + 4)}{2(x + 4)} = \frac{x}{2}, \quad x \neq -4$$

Using this property from left to right to eliminate all common factors from the numerator and the denominator of a given fraction is referred to as **reducing a fraction to lowest terms**. We are actually dividing the numerator and denominator by the same nonzero common factor.

Using the property from right to left—that is, multiplying the numerator and denominator by the same nonzero factor—is referred to as **raising a fraction to higher terms**. We will use the property in both directions in the material that follows.

EXAMPLE 1   **Reducing to Lowest Terms**   Reduce each rational expression to lowest terms.

(A) $\dfrac{6x^2 + x - 1}{2x^2 - x - 1} = \dfrac{(2x + 1)(3x - 1)}{(2x + 1)(x - 1)}$ *Factor numerator and denominator completely.*

$\qquad\qquad\quad = \dfrac{3x - 1}{x - 1}$ *Divide numerator and denominator by the common factor $(2x + 1)$.*

(B) $\dfrac{x^4 - 8x}{3x^3 - 2x^2 - 8x} = \dfrac{x(x - 2)(x^2 + 2x + 4)}{x(x - 2)(3x + 4)}$

$\qquad\qquad\qquad = \dfrac{x^2 + 2x + 4}{3x + 4}$

Matched Problem 1   Reduce each rational expression to lowest terms.

(A) $\dfrac{x^2 - 6x + 9}{x^2 - 9}$ (B) $\dfrac{x^3 - 1}{x^2 - 1}$

CONCEPTUAL INSIGHT

Using Theorem 1 to divide the numerator and denominator of a fraction by a common factor is often referred to as **canceling**. This operation can be denoted by drawing a slanted line through each common factor and writing any remaining factors above or below the common factor. Canceling is often incorrectly applied to individual terms in the numerator or denominator, instead of to common factors. For example,

$$\frac{14 - 5}{2} = \frac{9}{2}$$ *Theorem 1 does not apply. There are no common factors in the numerator.*

$$\frac{14 - 5}{2} \neq \frac{\overset{7}{\cancel{14}} - 5}{\underset{1}{\cancel{2}}} = 2$$ *Incorrect use of Theorem 1. To cancel 2 in the denominator, 2 must be a factor of each term in the numerator.*

Multiplication and Division

Since we are restricting variable replacements to real numbers, multiplication and division of rational expressions follow the rules for multiplying and dividing real number fractions summarized in Appendix A-1.

THEOREM 2   Multiplication and Division

If $a, b, c,$ and d are real numbers, then

1. $\dfrac{a}{b} \cdot \dfrac{c}{d} = \dfrac{ac}{bd}$, $b, d \neq 0$ $\dfrac{3}{5} \cdot \dfrac{x}{x + 5} = \dfrac{3x}{5(x + 5)}$

2. $\dfrac{a}{b} \div \dfrac{c}{d} = \dfrac{a}{b} \cdot \dfrac{d}{c}$, $b, c, d \neq 0$ $\dfrac{3}{5} \div \dfrac{x}{x + 5} = \dfrac{3}{5} \cdot \dfrac{x + 5}{x}$

EXAMPLE 2 **Multiplication and Division** Perform the indicated operations and reduce to lowest terms.

(A) $\dfrac{10x^3y}{3xy + 9y} \cdot \dfrac{x^2 - 9}{4x^2 - 12x}$ *Factor numerators and denominators. Then divide any numerator and any denominator with a like common factor.*

$$= \dfrac{\overset{5x^2}{\cancel{10x^3y}}}{\underset{3\cdot 1}{\cancel{3y(x + 3)}}} \cdot \dfrac{\overset{1\cdot 1}{\cancel{(x - 3)(x + 3)}}}{\underset{2\cdot 1}{\cancel{4x(x - 3)}}}$$

$$= \dfrac{5x^2}{6}$$

(B) $\dfrac{4 - 2x}{4} \div (x - 2) = \dfrac{\overset{1}{\cancel{2}(2 - x)}}{\underset{2}{\cancel{4}}} \cdot \dfrac{1}{x - 2}$ $x - 2 = \dfrac{x - 2}{1}$

$$= \dfrac{2 - x}{2(x - 2)} = \dfrac{\overset{-1}{\cancel{-(x - 2)}}}{\underset{1}{\cancel{2(x - 2)}}}$$ $b - a = -(a - b)$, a useful change in some problems

$$= -\dfrac{1}{2}$$

Matched Problem 2 Perform the indicated operations and reduce to lowest terms.

(A) $\dfrac{12x^2y^3}{2xy^2 + 6xy} \cdot \dfrac{y^2 + 6y + 9}{3y^3 + 9y^2}$ (B) $(4 - x) \div \dfrac{x^2 - 16}{5}$

Addition and Subtraction

Again, because we are restricting variable replacements to real numbers, addition and subtraction of rational expressions follow the rules for adding and subtracting real number fractions.

THEOREM 3 **Addition and Subtraction**

For $a, b,$ and c real numbers,

1. $\dfrac{a}{b} + \dfrac{c}{b} = \dfrac{a + c}{b}$, $b \neq 0$ $\dfrac{x}{x + 5} + \dfrac{8}{x + 5} = \dfrac{x + 8}{x + 5}$

2. $\dfrac{a}{b} - \dfrac{c}{b} = \dfrac{a - c}{b}$, $b \neq 0$ $\dfrac{x}{3x^2y^2} - \dfrac{x + 7}{3x^2y^2} = \dfrac{x - (x + 7)}{3x^2y^2}$

We add rational expressions with the same denominators by adding or subtracting their numerators and placing the result over the common denominator. If the denominators are not the same, we raise the fractions to higher terms, using the fundamental property of fractions to obtain common denominators, and then proceed as described.

Even though any common denominator will do, our work will be simplified if the *least common denominator (LCD)* is used. Often, the LCD is obvious, but if it is not, the steps in the next box describe how to find it.

> **PROCEDURE** **Least Common Denominator**
> The least common denominator (LCD) of two or more rational expressions is found as follows:
>
> 1. Factor each denominator completely, including integer factors.
> 2. Identify each different factor from all the denominators.
> 3. Form a product using each different factor to the highest power that occurs in any one denominator. This product is the LCD.

EXAMPLE 3 **Addition and Subtraction** Combine into a single fraction and reduce to lowest terms.

(A) $\dfrac{3}{10} + \dfrac{5}{6} - \dfrac{11}{45}$ (B) $\dfrac{4}{9x} - \dfrac{5x}{6y^2} + 1$ (C) $\dfrac{1}{x-1} - \dfrac{1}{x} - \dfrac{2}{x^2-1}$

SOLUTION (A) To find the LCD, factor each denominator completely:

$$\left.\begin{array}{l} 10 = 2 \cdot 5 \\ 6 = 2 \cdot 3 \\ 45 = 3^2 \cdot 5 \end{array}\right\} \quad LCD = 2 \cdot 3^2 \cdot 5 = 90$$

Now use the fundamental property of fractions to make each denominator 90:

$$\frac{3}{10} + \frac{5}{6} - \frac{11}{45} = \frac{9 \cdot 3}{9 \cdot 10} + \frac{15 \cdot 5}{15 \cdot 6} - \frac{2 \cdot 11}{2 \cdot 45}$$

$$= \boxed{\frac{27}{90} + \frac{75}{90} - \frac{22}{90}}$$

$$= \frac{27 + 75 - 22}{90} = \frac{80}{90} = \frac{8}{9}$$

(B) $\left.\begin{array}{l} 9x = 3^2 x \\ 6y^2 = 2 \cdot 3y^2 \end{array}\right\} \quad LCD = 2 \cdot 3^2 xy^2 = 18xy^2$

$$\frac{4}{9x} - \frac{5x}{6y^2} + 1 = \frac{2y^2 \cdot 4}{2y^2 \cdot 9x} - \frac{3x \cdot 5x}{3x \cdot 6y^2} + \frac{18xy^2}{18xy^2}$$

$$= \frac{8y^2 - 15x^2 + 18xy^2}{18xy^2}$$

(C) $\dfrac{1}{x-1} - \dfrac{1}{x} - \dfrac{2}{x^2-1}$

$$= \frac{1}{x-1} - \frac{1}{x} - \frac{2}{(x-1)(x+1)} \qquad LCD = x(x-1)(x+1)$$

$$= \frac{x(x+1) - (x-1)(x+1) - 2x}{x(x-1)(x+1)}$$

$$= \frac{x^2 + x - x^2 + 1 - 2x}{x(x-1)(x+1)}$$

$$= \frac{1-x}{x(x-1)(x+1)}$$

$$= \frac{\overset{-1}{\cancel{-(x-1)}}}{x\underset{1}{\cancel{(x-1)}}(x+1)} = \frac{-1}{x(x+1)}$$

Matched Problem 3   Combine into a single fraction and reduce to lowest terms.

(A) $\dfrac{5}{28} - \dfrac{1}{10} + \dfrac{6}{35}$

(B) $\dfrac{1}{4x^2} - \dfrac{2x+1}{3x^3} + \dfrac{3}{12x}$

(C) $\dfrac{2}{x^2 - 4x + 4} + \dfrac{1}{x} - \dfrac{1}{x-2}$

Compound Fractions

A fractional expression with fractions in its numerator, denominator, or both is called a **compound fraction**. It is often necessary to represent a compound fraction as a **simple fraction**—that is (in all cases we will consider), as the quotient of two polynomials. The process does not involve any new concepts. It is a matter of applying old concepts and processes in the correct sequence.

EXAMPLE 4   **Simplifying Compound Fractions** Express as a simple fraction reduced to lowest terms:

(A) $\dfrac{\dfrac{1}{5+h} - \dfrac{1}{5}}{h}$

(B) $\dfrac{\dfrac{y}{x^2} - \dfrac{x}{y^2}}{\dfrac{y}{x} - \dfrac{x}{y}}$

SOLUTION   We will simplify the expressions in parts (A) and (B) using two different methods—each is suited to the particular type of problem.

(A) We simplify this expression by combining the numerator into a single fraction and using division of rational forms.

$$\dfrac{\dfrac{1}{5+h} - \dfrac{1}{5}}{h} = \left[\dfrac{1}{5+h} - \dfrac{1}{5}\right] \div \dfrac{h}{1}$$

$$= \dfrac{5 - 5 - h}{5(5+h)} \cdot \dfrac{1}{h}$$

$$= \dfrac{-h}{5(5+h)h} = \dfrac{-1}{5(5+h)}$$

(B) The method used here makes effective use of the fundamental property of fractions in the form

$$\dfrac{a}{b} = \dfrac{ka}{kb} \qquad b, k \neq 0$$

Multiply the numerator and denominator by the LCD of all fractions in the numerator and denominator—in this case, $x^2 y^2$:

$$\dfrac{x^2 y^2 \left(\dfrac{y}{x^2} - \dfrac{x}{y^2}\right)}{x^2 y^2 \left(\dfrac{y}{x} - \dfrac{x}{y}\right)} = \dfrac{x^2 y^2 \dfrac{y}{x^2} - x^2 y^2 \dfrac{x}{y^2}}{x^2 y^2 \dfrac{y}{x} - x^2 y^2 \dfrac{x}{y}} = \dfrac{y^3 - x^3}{xy^3 - x^3 y}$$

$$= \dfrac{\overset{1}{\cancel{(y-x)}}(y^2 + xy + x^2)}{xy\underset{1}{\cancel{(y-x)}}(y+x)}$$

$$= \dfrac{y^2 + xy + x^2}{xy(y+x)} \quad \text{or} \quad \dfrac{x^2 + xy + y^2}{xy(x+y)}$$

Matched Problem 4   Express as a simple fraction reduced to lowest terms:

(A) $\dfrac{\dfrac{1}{2+h} - \dfrac{1}{2}}{h}$

(B) $\dfrac{\dfrac{a}{b} - \dfrac{b}{a}}{\dfrac{a}{b} + 2 + \dfrac{b}{a}}$

Exercises A-4

A

In Problems 1–18, perform the indicated operations and reduce answers to lowest terms.

1. $\dfrac{d^5}{3a} \div \left(\dfrac{d^2}{6a^2} \cdot \dfrac{a}{4d^3}\right)$

2. $\left(\dfrac{d^5}{3a} \div \dfrac{d^2}{6a^2}\right) \cdot \dfrac{a}{4d^3}$

3. $\dfrac{x^2}{12} + \dfrac{x}{18} - \dfrac{1}{30}$

4. $\dfrac{2y}{18} - \dfrac{-1}{28} - \dfrac{y}{42}$

5. $\dfrac{4m-3}{18m^3} + \dfrac{3}{4m} - \dfrac{2m-1}{6m^2}$

6. $\dfrac{3x+8}{4x^2} - \dfrac{2x-1}{x^3} - \dfrac{5}{8x}$

7. $\dfrac{x^2-9}{x^2-3x} \div (x^2 - x - 12)$

8. $\dfrac{2x^2+7x+3}{4x^2-1} \div (x+3)$

9. $\dfrac{2}{x} - \dfrac{1}{x-3}$

10. $\dfrac{5}{m-2} - \dfrac{3}{2m+1}$

11. $\dfrac{2}{(x+1)^2} - \dfrac{5}{x^2-x-2}$

12. $\dfrac{3}{x^2-5x+6} - \dfrac{5}{(x-2)^2}$

13. $\dfrac{x+1}{x-1} - 1$

14. $m - 3 - \dfrac{m-1}{m-2}$

15. $\dfrac{3}{a-1} - \dfrac{2}{1-a}$

16. $\dfrac{5}{x-3} - \dfrac{2}{3-x}$

17. $\dfrac{2x}{x^2-16} - \dfrac{x-4}{x^2+4x}$

18. $\dfrac{m+2}{m^2-2m} - \dfrac{m}{m^2-4}$

B

In Problems 19–30, perform the indicated operations and reduce answers to lowest terms. Represent any compound fractions as simple fractions reduced to lowest terms.

19. $\dfrac{x^2}{x^2+2x+1} + \dfrac{x-1}{3x+3} - \dfrac{1}{6}$

20. $\dfrac{y}{y^2-y-2} - \dfrac{1}{y^2+5y-14} - \dfrac{2}{y^2+8y+7}$

21. $\dfrac{1-\dfrac{x}{y}}{2-\dfrac{y}{x}}$

22. $\dfrac{2}{5-\dfrac{3}{4x+1}}$

23. $\dfrac{c+2}{5c-5} - \dfrac{c-2}{3c-3} + \dfrac{c}{1-c}$

24. $\dfrac{x+7}{ax-bx} + \dfrac{y+9}{by-ay}$

25. $\dfrac{1+\dfrac{3}{x}}{x-\dfrac{9}{x}}$

26. $\dfrac{1-\dfrac{y^2}{x^2}}{1-\dfrac{y}{x}}$

27. $\dfrac{\dfrac{1}{2(x+h)} - \dfrac{1}{2x}}{h}$

28. $\dfrac{\dfrac{1}{x+h} - \dfrac{1}{x}}{h}$

29. $\dfrac{\dfrac{x}{y} - 2 + \dfrac{y}{x}}{\dfrac{x}{y} - \dfrac{y}{x}}$

30. $\dfrac{1+\dfrac{2}{x} - \dfrac{15}{x^2}}{1+\dfrac{4}{x} - \dfrac{5}{x^2}}$

In Problems 31–38, imagine that the indicated "solutions" were given to you by a student whom you were tutoring in this class.

(A) Is the solution correct? If the solution is incorrect, explain what is wrong and how it can be corrected.

(B) Show a correct solution for each incorrect solution.

31. $\dfrac{x^2+4x+3}{x+3} = \dfrac{x^2+4x}{x} = x + 4$

32. $\dfrac{x^2-3x-4}{x-4} = \dfrac{x^2-3x}{x} = x - 3$

33. $\dfrac{(x+h)^2 - x^2}{h} = (x+1)^2 - x^2 = 2x + 1$

34. $\dfrac{(x+h)^3 - x^3}{h} = (x+1)^3 - x^3 = 3x^2 + 3x + 1$

35. $\dfrac{x^2-3x}{x^2-2x-3} + x - 3 = \dfrac{x^2-3x+x-3}{x^2-2x-3} = 1$

36. $\dfrac{2}{x-1} - \dfrac{x+3}{x^2-1} = \dfrac{2x+2-x-3}{x^2-1} = \dfrac{1}{x+1}$

37. $\dfrac{2x^2}{x^2-4} - \dfrac{x}{x-2} = \dfrac{2x^2-x^2-2x}{x^2-4} = \dfrac{x}{x+2}$

38. $x + \dfrac{x-2}{x^2-3x+2} = \dfrac{x+x-2}{x^2-3x+2} = \dfrac{2}{x-2}$

C

Represent the compound fractions in Problems 39–42 as simple fractions reduced to lowest terms.

39. $\dfrac{\dfrac{1}{3(x+h)^2} - \dfrac{1}{3x^2}}{h}$

40. $\dfrac{\dfrac{1}{(x+h)^2} - \dfrac{1}{x^2}}{h}$

41. $x - \dfrac{2}{1 - \dfrac{1}{x}}$

42. $2 - \dfrac{1}{1 - \dfrac{2}{a+2}}$

Answers to Matched Problems

1. (A) $\dfrac{x-3}{x+3}$ (B) $\dfrac{x^2+x+1}{x+1}$

2. (A) $2x$ (B) $\dfrac{-5}{x+4}$

3. (A) $\dfrac{1}{4}$ (B) $\dfrac{3x^2-5x-4}{12x^3}$ (C) $\dfrac{4}{x(x-2)^2}$

4. (A) $\dfrac{-1}{2(2+h)}$ (B) $\dfrac{a-b}{a+b}$

A-5 Integer Exponents and Scientific Notation

- Integer Exponents
- Scientific Notation

We now review basic operations on integer exponents and scientific notation.

Integer Exponents

> **DEFINITION Integer Exponents**
>
> For n an integer and a a real number:
>
> **1.** For n a positive integer,
> $$a^n = a \cdot a \cdot \,\cdots\, \cdot a \qquad n \text{ factors of } a \qquad 5^4 = 5 \cdot 5 \cdot 5 \cdot 5$$
>
> **2.** For $n = 0$,
> $$a^0 = 1 \qquad a \neq 0 \qquad 12^0 = 1$$
> $$0^0 \text{ is not defined.}$$
>
> **3.** For n a negative integer,
> $$a^n = \frac{1}{a^{-n}} \qquad a \neq 0 \qquad a^{-3} = \frac{1}{a^{-(-3)}} = \frac{1}{a^3}$$
>
> [If n is negative, then $(-n)$ is positive.]
> **Note:** It can be shown that for *all* integers n,
> $$a^{-n} = \frac{1}{a^n} \quad \text{and} \quad a^n = \frac{1}{a^{-n}} \qquad a \neq 0 \qquad a^5 = \frac{1}{a^{-5}}, \quad a^{-5} = \frac{1}{a^5}$$

The following properties are very useful in working with integer exponents.

> **THEOREM 1 Exponent Properties**
> For n and m integers and a and b real numbers,
>
> **1.** $a^m a^n = a^{m+n}$ $\qquad\qquad a^8 a^{-3} = a^{8+(-3)} = a^5$
>
> **2.** $(a^n)^m = a^{mn}$ $\qquad\qquad (a^{-2})^3 = a^{3(-2)} = a^{-6}$
>
> **3.** $(ab)^m = a^m b^m$ $\qquad\qquad (ab)^{-2} = a^{-2}b^{-2}$
>
> **4.** $\left(\dfrac{a}{b}\right)^m = \dfrac{a^m}{b^m} \qquad b \neq 0 \qquad \left(\dfrac{a}{b}\right)^5 = \dfrac{a^5}{b^5}$
>
> **5.** $\dfrac{a^m}{a^n} = a^{m-n} = \dfrac{1}{a^{n-m}} \qquad a \neq 0 \qquad \dfrac{a^{-3}}{a^7} = \dfrac{1}{a^{7-(-3)}} = \dfrac{1}{a^{10}}$

Exponents are frequently encountered in algebraic applications. You should sharpen your skills in using exponents by reviewing the preceding basic definitions and properties and the examples that follow.

EXAMPLE 1   **Simplifying Exponent Forms** Simplify, and express the answers using positive exponents only.

(A) $(2x^3)(3x^5) \boxed{= 2 \cdot 3x^{3+5}} = 6x^8$

(B) $x^5x^{-9} = x^{-4} = \dfrac{1}{x^4}$

(C) $\dfrac{x^5}{x^7} \boxed{= x^{5-7}} = x^{-2} = \dfrac{1}{x^2}$ or $\dfrac{x^5}{x^7} \boxed{= \dfrac{1}{x^{7-5}}} = \dfrac{1}{x^2}$

(D) $\dfrac{x^{-3}}{y^{-4}} = \dfrac{y^4}{x^3}$

(E) $(u^{-3}v^2)^{-2} \boxed{= (u^{-3})^{-2}(v^2)^{-2}} = u^6v^{-4} = \dfrac{u^6}{v^4}$

(F) $\left(\dfrac{y^{-5}}{y^{-2}}\right)^{-2} \boxed{= \dfrac{(y^{-5})^{-2}}{(y^{-2})^{-2}}} = \dfrac{y^{10}}{y^4} = y^6$

(G) $\dfrac{4m^{-3}n^{-5}}{6m^{-4}n^3} \boxed{= \dfrac{2m^{-3-(-4)}}{3n^{3-(-5)}}} = \dfrac{2m}{3n^8}$

Matched Problem 1   Simplify, and express the answers using positive exponents only.

(A) $(3y^4)(2y^3)$ (B) m^2m^{-6} (C) $(u^3v^{-2})^{-2}$

(C) $\left(\dfrac{y^{-6}}{y^{-2}}\right)^{-1}$ (D) $\dfrac{8x^{-2}y^{-4}}{6x^{-5}y^2}$

EXAMPLE 2   **Converting to a Simple Fraction** Write $\dfrac{1-x}{x^{-1}-1}$ as a simple fraction with positive exponents.

SOLUTION   First note that

$$\dfrac{1-x}{x^{-1}-1} \neq \dfrac{x(1-x)}{-1} \qquad \text{A common error}$$

The original expression is a complex fraction, and we proceed to simplify it as follows:

$$\dfrac{1-x}{x^{-1}-1} = \dfrac{1-x}{\dfrac{1}{x}-1} \qquad \text{Multiply numerator and denominator by } x \text{ to clear internal fractions.}$$

$$= \dfrac{x(1-x)}{x\left(\dfrac{1}{x}-1\right)}$$

$$= \dfrac{x(1-x)}{1-x} = x$$

Matched Problem 2   Write $\dfrac{1+x^{-1}}{1-x^{-2}}$ as a simple fraction with positive exponents.

Scientific Notation

In the real world, one often encounters very large and very small numbers. For example,

- The public debt in the United States in 2008, to the nearest billion dollars, was

$$\$10,025,000,000,000$$

- The world population in the year 2025, to the nearest million, is projected to be

$$7,947,000,000$$

- The sound intensity of a normal conversation is

$$0.000\ 000\ 000\ 316 \text{ watt per square centimeter*}$$

It is generally troublesome to write and work with numbers of this type in standard decimal form. The first and last example cannot even be entered into many calculators as they are written. But with exponents defined for all integers, we can now express any finite decimal form as the product of a number between 1 and 10 and an integer power of 10, that is, in the form

$$a \times 10^n \qquad 1 \le a < 10, \quad a \text{ in decimal form}, \quad n \text{ an integer}$$

A number expressed in this form is said to be in **scientific notation**. The following are some examples of numbers in standard decimal notation and in scientific notation:

DECIMAL AND SCIENTIFIC NOTATION

$$7 = 7 \times 10^0 \qquad\qquad 0.5 = 5 \times 10^{-1}$$
$$67 = 6.7 \times 10 \qquad\qquad 0.45 = 4.5 \times 10^{-1}$$
$$580 = 5.8 \times 10^2 \qquad\qquad 0.0032 = 3.2 \times 10^{-3}$$
$$43,000 = 4.3 \times 10^4 \qquad\qquad 0.000\ 045 = 4.5 \times 10^{-5}$$
$$73,400,000 = 7.34 \times 10^7 \qquad 0.000\ 000\ 391 = 3.91 \times 10^{-7}$$

Note that the power of 10 used corresponds to the number of places we move the decimal to form a number between 1 and 10. The power is positive if the decimal is moved to the left and negative if it is moved to the right. Positive exponents are associated with numbers greater than or equal to 10; negative exponents are associated with positive numbers less than 1; and a zero exponent is associated with a number that is 1 or greater, but less than 10.

EXAMPLE 3 **Scientific Notation**

(A) Write each number in scientific notation:

$$7,320,000 \qquad \text{and} \qquad 0.000\ 000\ 54$$

(B) Write each number in standard decimal form:

$$4.32 \times 10^6 \qquad \text{and} \qquad 4.32 \times 10^{-5}$$

SOLUTION

(A)
$$7,320,000 = 7.320\ 000. \times 10^6 = 7.32 \times 10^6$$
$$\text{6 places left}$$
$$\text{Positive exponent}$$

$$0.000\ 000\ 54 = 0.000\ 000\ 5.4 \times 10^{-7} = 5.4 \times 10^{-7}$$
$$\text{7 places right}$$
$$\text{Negative exponent}$$

---

*We write 0.000 000 000 316 in place of 0.000000000316, because it is then easier to keep track of the number of decimal places. We follow this convention when there are more than five decimal places to the right of the decimal.

(B) $4.32 \times 10^6 = 4,320,000$ $\qquad$ $4.32 \times 10^{-5} = \dfrac{4.32}{10^5} = 0.000\ 043\ 2$

6 places right $\qquad\qquad$ 5 places left

Positive exponent 6 $\qquad\qquad$ Negative exponent −5

Matched Problem 3 (A) Write each number in scientific notation: 47,100; 2,443,000,000; 1.45

(B) Write each number in standard decimal form: 3.07×10^8; 5.98×10^{-6}

Exercises A-5

A

In Problems 1–14, simplify and express answers using positive exponents only. Variables are restricted to avoid division by 0.

1. $2x^{-9}$ $\qquad$ **2.** $3y^{-5}$ $\qquad$ **3.** $\dfrac{3}{2w^{-7}}$

4. $\dfrac{5}{4x^{-9}}$ $\qquad$ **5.** $2x^{-8}x^5$ $\qquad$ **6.** $3c^{-9}c^4$

7. $\dfrac{w^{-8}}{w^{-3}}$ $\qquad$ **8.** $\dfrac{m^{-11}}{m^{-5}}$ $\qquad$ **9.** $(2a^{-3})^2$

10. $7d^{-4}d^4$ $\qquad$ **11.** $(a^{-3})^2$ $\qquad$ **12.** $(5b^{-2})^2$

13. $(2x^4)^{-3}$ $\qquad$ **14.** $(a^{-3}b^4)^{-3}$

In Problems 15–20, write each number in scientific notation.

15. 82,300,000,000 $\qquad$ **16.** 5,380,000

17. 0.783 $\qquad$ **18.** 0.019

19. 0.000 034 $\qquad$ **20.** 0.000 000 007 832

In Problems 21–28, write each number in standard decimal notation.

21. 4×10^4 $\qquad$ **22.** 9×10^6

23. 7×10^{-3} $\qquad$ **24.** 2×10^{-5}

25. 6.171×10^7 $\qquad$ **26.** 3.044×10^3

27. 8.08×10^{-4} $\qquad$ **28.** 1.13×10^{-2}

B

In Problems 29–38, simplify and express answers using positive exponents only.

29. $(22 + 31)^0$ $\qquad$ **30.** $(2x^3y^4)^0$

31. $\dfrac{10^{-3} \cdot 10^4}{10^{-11} \cdot 10^{-2}}$ $\qquad$ **32.** $\dfrac{10^{-17} \cdot 10^{-5}}{10^{-3} \cdot 10^{-14}}$

33. $(5x^2y^{-3})^{-2}$ $\qquad$ **34.** $(2m^{-3}n^2)^{-3}$

35. $\left(\dfrac{-5}{2x^3}\right)^{-2}$ $\qquad$ **36.** $\left(\dfrac{2a}{3b^2}\right)^{-3}$

37. $\dfrac{8x^{-3}y^{-1}}{6x^2y^{-4}}$ $\qquad$ **38.** $\dfrac{9m^{-4}n^3}{12m^{-1}n^{-1}}$

In Problems 39–42, write each expression in the form $ax^p + bx^q$ or $ax^p + bx^q + cx^r$, where a, b, and c are real numbers and p, q, and r are integers. For example,

$$\dfrac{2x^4 - 3x^2 + 1}{2x^3} = \boxed{\dfrac{2x^4}{2x^3} - \dfrac{3x^2}{2x^3} + \dfrac{1}{2x^3}} = x - \dfrac{3}{2}x^{-1} + \dfrac{1}{2}x^{-3}$$

39. $\dfrac{7x^5 - x^2}{4x^5}$ $\qquad$ **40.** $\dfrac{5x^3 - 2}{3x^2}$

41. $\dfrac{5x^4 - 3x^2 + 8}{2x^2}$ $\qquad$ **42.** $\dfrac{2x^3 - 3x^2 + x}{2x^2}$

Write each expression in Problems 43–46 with positive exponents only, and as a single fraction reduced to lowest terms.

43. $\dfrac{3x^2(x - 1)^2 - 2x^3(x - 1)}{(x - 1)^4}$

44. $\dfrac{5x^4(x + 3)^2 - 2x^5(x + 3)}{(x + 3)^4}$

45. $2x^{-2}(x - 1) - 2x^{-3}(x - 1)^2$

46. $2x(x + 3)^{-1} - x^2(x + 3)^{-2}$

In Problems 47–50, convert each number to scientific notation and simplify. Express the answer in both scientific notation and in standard decimal form.

47. $\dfrac{9,600,000,000}{(1,600,000)(0.000\ 000\ 25)}$

48. $\dfrac{(60,000)(0.000\ 003)}{(0.0004)(1,500,000)}$

49. $\dfrac{(1,250,000)(0.000\ 38)}{0.0152}$

50. $\dfrac{(0.000\ 000\ 82)(230,000)}{(625,000)(0.0082)}$

51. What is the result of entering 2^{3^2} on a calculator?

52. Refer to Problem 51. What is the difference between $2^{(3^2)}$ and $(2^3)^2$? Which agrees with the value of 2^{3^2} obtained with a calculator?

53. If $n = 0$, then property 1 in Theorem 1 implies that $a^m a^0 = a^{m+0} = a^m$. Explain how this helps motivate the definition of a^0.

54. If $m = -n$, then property 1 in Theorem 1 implies that $a^{-n}a^n = a^0 = 1$. Explain how this helps motivate the definition of a^{-n}.

C

Write the fractions in Problems 55–58 as simple fractions reduced to lowest terms.

55. $\dfrac{u + v}{u^{-1} + v^{-1}}$

56. $\dfrac{x^{-2} - y^{-2}}{x^{-1} + y^{-1}}$

57. $\dfrac{b^{-2} - c^{-2}}{b^{-3} - c^{-3}}$

58. $\dfrac{xy^{-2} - yx^{-2}}{y^{-1} - x^{-1}}$

Applications

Problems 59 and 60 refer to Table 1.

Table 1 U.S. Public Debt, Interest on Debt, and Population

Year	Public Debt ($)	Interest on Debt ($)	Population
2000	5,674,000,000,000	362,000,000,000	281,000,000
2008	10,025,000,000,000	451,000,000,000	304,000,000

59. Public debt. Carry out the following computations using scientific notation, and write final answers in standard decimal form.

(A) What was the per capita debt in 2008 (to the nearest dollar)?

(B) What was the per capita interest paid on the debt in 2008 (to the nearest dollar)?

(C) What was the percentage interest paid on the debt in 2008 (to two decimal places)?

60. Public debt. Carry out the following computations using scientific notation, and write final answers in standard decimal form.

(A) What was the per capita debt in 2000 (to the nearest dollar)?

(B) What was the per capita interest paid on the debt in 2000 (to the nearest dollar)?

(C) What was the percentage interest paid on the debt in 2000 (to two decimal places)?

Air pollution. *Air quality standards establish maximum amounts of pollutants considered acceptable in the air. The amounts are frequently given in parts per million (ppm). A standard of 30 ppm also can be expressed as follows:*

$$30 \text{ ppm} = \frac{30}{1,000,000} = \frac{3 \times 10}{10^6}$$

$$= 3 \times 10^{-5} = 0.000\ 03 = 0.003\%$$

In Problems 61 and 62, express the given standard:

(A) In scientific notation

(B) In standard decimal notation

(C) As a percent

61. 9 ppm, the standard for carbon monoxide, when averaged over a period of 8 hours

62. 0.03 ppm, the standard for sulfur oxides, when averaged over a year

63. Crime. In 2008, the United States had a violent crime rate of 466 per 100,000 people and a population of 304 million people. How many violent crimes occurred that year? Compute the answer using scientific notation and convert the answer to standard decimal form (to the nearest thousand).

64. Population density. The United States had a 2008 population of 304 million people and a land area of 3,539,000 square miles. What was the population density? Compute the answer using scientific notation and convert the answer to standard decimal form (to one decimal place).

Answers to Matched Problems

1. (A) $6y^7$ (B) $\dfrac{1}{m^4}$ (C) $\dfrac{v^4}{u^6}$ (D) y^4 (E) $\dfrac{4x^3}{3y^6}$

2. $\dfrac{x}{x - 1}$

3. (A) 4.7×10^4; 2.443×10^9; 1.45×10^0
   (B) 307,000,000; 0.000 005 98

A-6 Rational Exponents and Radicals

- *n*th Roots of Real Numbers
- Rational Exponents and Radicals
- Properties of Radicals

Square roots may now be generalized to *n*th roots, and the meaning of exponent may be generalized to include all rational numbers.

*n*th Roots of Real Numbers

Consider a square of side r with area 36 square inches. We can write

$$r^2 = 36$$

and conclude that side r is a number whose square is 36. We say that r is a **square root** of b if $r^2 = b$. Similarly, we say that r is a **cube root** of b if $r^3 = b$. And, in general,

DEFINITION *n*th **Root**

For any natural number n,

$$r \text{ is an } \textbf{\textit{n}th root} \text{ of } b \text{ if } r^n = b$$

So 4 is a square root of 16, since $4^2 = 16$; -2 is a cube root of -8, since $(-2)^3 = -8$. Since $(-4)^2 = 16$, we see that -4 is also a square root of 16. It can be shown that any positive number has two real square roots, two real 4th roots, and, in general, two real nth roots if n is even. Negative numbers have no real square roots, no real 4th roots, and, in general, no real nth roots if n is even. The reason is that no real number raised to an even power can be negative. For odd roots, the situation is simpler. Every real number has exactly one real cube root, one real 5th root, and, in general, one real nth root if n is odd.

Additional roots can be considered in the *complex number system*. In this book, we restrict our interest to *real roots of real numbers*, and *root* will always be interpreted to mean "real root."

Rational Exponents and Radicals

We now turn to the question of what symbols to use to represent nth roots. For n a natural number greater than 1, we use

$$b^{1/n} \qquad \text{or} \qquad \sqrt[n]{b}$$

to represent a **real *n*th root of *b*.** The exponent form is motivated by the fact that $(b^{1/n})^n = b$ if exponent laws are to continue to hold for rational exponents. The other form is called an **nth root radical.** In the expression below, the symbol $\sqrt{}$ is called a **radical,** n is the **index** of the radical, and b is the **radicand:**

When the index is 2, it is usually omitted. That is, when dealing with square roots, we simply use $\sqrt{b}$ rather than $\sqrt[2]{b}$. If there are two real nth roots, both $b^{1/n}$ and $\sqrt[n]{b}$ denote the positive root, called the **principal *n*th root.**

EXAMPLE 1   **Finding *n*th Roots**   Evaluate each of the following:

(A) $4^{1/2}$ and $\sqrt{4}$ (B) $-4^{1/2}$ and $-\sqrt{4}$ (C) $(-4)^{1/2}$ and $\sqrt{-4}$

(D) $8^{1/3}$ and $\sqrt[3]{8}$ (E) $(-8)^{1/3}$ and $\sqrt[3]{-8}$ (F) $-8^{1/3}$ and $-\sqrt[3]{8}$

SOLUTION   (A) $4^{1/2} = \sqrt{4} = 2$ $(\sqrt{4} \neq \pm2)$ (B) $-4^{1/2} = -\sqrt{4} = -2$

(C) $(-4)^{1/2}$ and $\sqrt{-4}$ are not real numbers

(D) $8^{1/3} = \sqrt[3]{8} = 2$ (E) $(-8)^{1/3} = \sqrt[3]{-8} = -2$

(F) $-8^{1/3} = -\sqrt[3]{8} = -2$

Matched Problem 1   Evaluate each of the following:

(A) $16^{1/2}$ (B) $-\sqrt{16}$ (C) $\sqrt[3]{-27}$ (D) $(-9)^{1/2}$ (E) $(\sqrt[4]{81})^3$

CAUTION   The symbol $\sqrt{4}$ represents the single number 2, not ±2. Do not confuse $\sqrt{4}$ with the solutions of the equation $x^2 = 4$, which are usually written in the form $x = \pm\sqrt{4} = \pm2$.

We now define b^r for any rational number $r = m/n$.

DEFINITION Rational Exponents

If m and n are natural numbers without common prime factors, b is a real number, and b is nonnegative when n is even, then

$$b^{m/n} = \begin{cases} (b^{1/n})^m = (\sqrt[n]{b})^m \\ (b^m)^{1/n} = \sqrt[n]{b^m} \end{cases} \quad \begin{aligned} 8^{2/3} &= (8^{1/3})^2 = (\sqrt[3]{8})^2 = 2^2 = 4 \\ 8^{2/3} &= (8^2)^{1/3} = \sqrt[3]{8^2} = \sqrt[3]{64} = 4 \end{aligned}$$

and

$$b^{-m/n} = \frac{1}{b^{m/n}} \quad b \neq 0 \qquad 8^{-2/3} = \frac{1}{8^{2/3}} = \frac{1}{4}$$

Note that the two definitions of $b^{m/n}$ are equivalent under the indicated restrictions on m, n, and b.

CONCEPTUAL INSIGHT

All the properties for integer exponents listed in Theorem 1 in Section A-5 also hold for rational exponents, provided that b is nonnegative when n is even. This restriction on b is necessary to avoid nonreal results. For example,

$$(-4)^{3/2} = \sqrt{(-4)^3} = \sqrt{-64} \qquad \text{Not a real number}$$

To avoid nonreal results, all variables in the remainder of this discussion represent positive real numbers.

EXAMPLE 2 **From Rational Exponent Form to Radical Form and Vice Versa** Change rational exponent form to radical form.

(A) $x^{1/7} = \sqrt[7]{x}$

(B) $(3u^2v^3)^{3/5} = \sqrt[5]{(3u^2v^3)^3}$ or $(\sqrt[5]{3u^2v^3})^3$ The first is usually preferred.

(C) $y^{-2/3} = \frac{1}{y^{2/3}} = \frac{1}{\sqrt[3]{y^2}}$ or $\sqrt[3]{y^{-2}}$ or $\sqrt[3]{\frac{1}{y^2}}$

Change radical form to rational exponent form.

(D) $\sqrt[5]{6} = 6^{1/5}$ (E) $-\sqrt[3]{x^2} = -x^{2/3}$

(F) $\sqrt{x^2 + y^2} = (x^2 + y^2)^{1/2}$ Note that $(x^2 + y^2)^{1/2} \neq x + y$. Why?

Matched Problem 2   Convert to radical form.

(A) $u^{1/5}$ (B) $(6x^2y^5)^{2/9}$ (C) $(3xy)^{-3/5}$

Convert to rational exponent form.

(D) $\sqrt[4]{9u}$ (E) $-\sqrt[7]{(2x)^4}$ (F) $\sqrt[3]{x^3 + y^3}$

EXAMPLE 3 **Working with Rational Exponents** Simplify each and express answers using positive exponents only. If rational exponents appear in final answers, convert to radical form.

(A) $(3x^{1/3})(2x^{1/2}) = 6x^{1/3+1/2} = 6x^{5/6} = 6\sqrt[6]{x^5}$

(B) $(-8)^{5/3} = [(-8)^{1/3}]^5 = (-2)^5 = -32$

(C) $(2x^{1/3}y^{-2/3})^3 = 8xy^{-2} = \frac{8x}{y^2}$

(D) $\left(\frac{4x^{1/3}}{x^{1/2}}\right)^{1/2} = \frac{4^{1/2}x^{1/6}}{x^{1/4}} = \frac{2}{x^{1/4-1/6}} = \frac{2}{x^{1/12}} = \frac{2}{\sqrt[12]{x}}$

Matched Problem 3 Simplify each and express answers using positive exponents only. If rational exponents appear in final answers, convert to radical form.

(A) $9^{3/2}$ (B) $(-27)^{4/3}$ (C) $(5y^{1/4})(2y^{1/3})$ (D) $(2x^{-3/4}y^{1/4})^4$

(E) $\left(\dfrac{8x^{1/2}}{x^{2/3}}\right)^{1/3}$

EXAMPLE 4 **Working with Rational Exponents** Multiply, and express answers using positive exponents only.

(A) $3y^{2/3}(2y^{1/3} - y^2)$ (B) $(2u^{1/2} + v^{1/2})(u^{1/2} - 3v^{1/2})$

SOLUTION (A) $3y^{2/3}(2y^{1/3} - y^2) \boxed{= 6y^{2/3+1/3} - 3y^{2/3+2}}$

$$= 6y - 3y^{8/3}$$

(B) $(2u^{1/2} + v^{1/2})(u^{1/2} - 3v^{1/2}) = 2u - 5u^{1/2}v^{1/2} - 3v$

Matched Problem 4 Multiply, and express answers using positive exponents only.

(A) $2c^{1/4}(5c^3 - c^{3/4})$ (B) $(7x^{1/2} - y^{1/2})(2x^{1/2} + 3y^{1/2})$

EXAMPLE 5 **Working with Rational Exponents** Write the following expression in the form $ax^p + bx^q$, where a and b are real numbers and p and q are rational numbers:

$$\frac{2\sqrt{x} - 3\sqrt[3]{x^2}}{2\sqrt[3]{x}}$$

SOLUTION $\dfrac{2\sqrt{x} - 3\sqrt[3]{x^2}}{2\sqrt[3]{x}} = \dfrac{2x^{1/2} - 3x^{2/3}}{2x^{1/3}}$ *Change to rational exponent form.*

$$= \frac{2x^{1/2}}{2x^{1/3}} - \frac{3x^{2/3}}{2x^{1/3}}$$ *Separate into two fractions.*

$$= x^{1/6} - 1.5x^{1/3}$$

Matched Problem 5 Write the following expression in the form $ax^p + bx^q$, where a and b are real numbers and p and q are rational numbers:

$$\frac{5\sqrt[3]{x} - 4\sqrt{x}}{2\sqrt{x^3}}$$

Properties of Radicals

Changing or simplifying radical expressions is aided by several properties of radicals that follow directly from the properties of exponents considered earlier.

THEOREM 1   Properties of Radicals

If c, n, and m are natural numbers greater than or equal to 2, and if x and y are positive real numbers, then

1. $\sqrt[n]{x^n} = x$ $\sqrt[3]{x^3} = x$

2. $\sqrt[n]{xy} = \sqrt[n]{x}\sqrt[n]{y}$ $\sqrt[5]{xy} = \sqrt[5]{x}\,\sqrt[5]{y}$.

3. $\sqrt[n]{\dfrac{x}{y}} = \dfrac{\sqrt[n]{x}}{\sqrt[n]{y}}$ $\sqrt[4]{\dfrac{x}{y}} = \dfrac{\sqrt[4]{x}}{\sqrt[4]{y}}$

EXAMPLE 6 **Applying Properties of Radicals** Simplify using properties of radicals.

(A) $\sqrt[4]{(3x^4y^3)^4}$ (B) $\sqrt[4]{8}\sqrt[4]{2}$ (C) $\sqrt[3]{\dfrac{xy}{27}}$

SOLUTION (A) $\sqrt[4]{(3x^4y^3)^4} = 3x^4y^3$ *Property 1*

(B) $\sqrt[4]{8}\sqrt[4]{2} = \sqrt[4]{16} = \sqrt[4]{2^4} = 2$ *Properties 2 and 1*

(C) $\sqrt[3]{\dfrac{xy}{27}} = \dfrac{\sqrt[3]{xy}}{\sqrt[3]{27}} = \dfrac{\sqrt[3]{xy}}{3}$ or $\dfrac{1}{3}\sqrt[3]{xy}$ *Properties 3 and 1*

Matched Problem 6 Simplify using properties of radicals.

(A) $\sqrt[7]{(x^3 + y^3)^7}$ (B) $\sqrt[3]{8y^3}$ (C) $\dfrac{\sqrt[3]{16x^4y}}{\sqrt[3]{2xy}}$

What is the best form for a radical expression? There are many answers, depending on what use we wish to make of the expression. In deriving certain formulas, it is sometimes useful to clear either a denominator or a numerator of radicals. The process is referred to as **rationalizing** the denominator or numerator. Examples 7 and 8 illustrate the rationalizing process.

EXAMPLE 7 **Rationalizing Denominators** Rationalize each denominator.

(A) $\dfrac{6x}{\sqrt{2x}}$ (B) $\dfrac{6}{\sqrt{7} - \sqrt{5}}$ (C) $\dfrac{x - 4}{\sqrt{x} + 2}$

SOLUTION (A) $\dfrac{6x}{\sqrt{2x}} = \dfrac{6x}{\sqrt{2x}} \cdot \dfrac{\sqrt{2x}}{\sqrt{2x}} = \dfrac{6x\sqrt{2x}}{2x} = 3\sqrt{2x}$

(B) $\dfrac{6}{\sqrt{7} - \sqrt{5}} = \dfrac{6}{\sqrt{7} - \sqrt{5}} \cdot \dfrac{\sqrt{7} + \sqrt{5}}{\sqrt{7} + \sqrt{5}}$

$= \dfrac{6(\sqrt{7} + \sqrt{5})}{2} = 3(\sqrt{7} + \sqrt{5})$

(C) $\dfrac{x - 4}{\sqrt{x} + 2} = \dfrac{x - 4}{\sqrt{x} + 2} \cdot \dfrac{\sqrt{x} - 2}{\sqrt{x} - 2}$

$= \dfrac{(x - 4)(\sqrt{x} - 2)}{x - 4} = \sqrt{x} - 2$

Matched Problem 7 Rationalize each denominator.

(A) $\dfrac{12ab^2}{\sqrt{3ab}}$ (B) $\dfrac{9}{\sqrt{6} + \sqrt{3}}$ (C) $\dfrac{x^2 - y^2}{\sqrt{x} - \sqrt{y}}$

EXAMPLE 8 **Rationalizing Numerators** Rationalize each numerator.

(A) $\dfrac{\sqrt{2}}{2\sqrt{3}}$ (B) $\dfrac{3 + \sqrt{m}}{9 - m}$ (C) $\dfrac{\sqrt{2 + h} - \sqrt{2}}{h}$

SOLUTION (A) $\dfrac{\sqrt{2}}{2\sqrt{3}} = \dfrac{\sqrt{2}}{2\sqrt{3}} \cdot \dfrac{\sqrt{2}}{\sqrt{2}} = \dfrac{2}{2\sqrt{6}} = \dfrac{1}{\sqrt{6}}$

(B) $\dfrac{3 + \sqrt{m}}{9 - m} = \dfrac{3 + \sqrt{m}}{9 - m} \cdot \dfrac{3 - \sqrt{m}}{3 - \sqrt{m}} = \dfrac{9 - m}{(9 - m)(3 - \sqrt{m})} = \dfrac{1}{3 - \sqrt{m}}$

(C) $\dfrac{\sqrt{2 + h} - \sqrt{2}}{h} = \dfrac{\sqrt{2 + h} - \sqrt{2}}{h} \cdot \dfrac{\sqrt{2 + h} + \sqrt{2}}{\sqrt{2 + h} + \sqrt{2}}$

$= \dfrac{h}{h(\sqrt{2 + h} + \sqrt{2})} = \dfrac{1}{\sqrt{2 + h} + \sqrt{2}}$

Matched Problem 8   Rationalize each numerator.

(A) $\dfrac{\sqrt{3}}{3\sqrt{2}}$ (B) $\dfrac{2 - \sqrt{n}}{4 - n}$ (C) $\dfrac{\sqrt{3 + h} - \sqrt{3}}{h}$

Exercises A-6

A

Change each expression in Problems 1–6 to radical form. Do not simplify.

1. $6x^{3/5}$ **2.** $7y^{2/5}$ **3.** $(32x^2y^3)^{3/5}$

4. $(7x^2y)^{5/7}$ **5.** $(x^2 + y^2)^{1/2}$ **6.** $x^{1/2} + y^{1/2}$

Change each expression in Problems 7–12 to rational exponent form. Do not simplify.

7. $5\sqrt[4]{x^3}$ **8.** $7m\sqrt[5]{n^2}$ **9.** $\sqrt[5]{(2x^2y)^3}$

10. $\sqrt[7]{(8x^4y)^3}$ **11.** $\sqrt[3]{x} + \sqrt[3]{y}$ **12.** $\sqrt[3]{x^2 + y^3}$

In Problems 13–24, find rational number representations for each, if they exist.

13. $25^{1/2}$ **14.** $64^{1/3}$ **15.** $16^{3/2}$

16. $16^{3/4}$ **17.** $-49^{1/2}$ **18.** $(-49)^{1/2}$

19. $-64^{2/3}$ **20.** $(-64)^{2/3}$ **21.** $\left(\frac{4}{25}\right)^{3/2}$

22. $\left(\frac{8}{27}\right)^{2/3}$ **23.** $9^{-3/2}$ **24.** $8^{-2/3}$

In Problems 25–34, simplify each expression and write answers using positive exponents only. All variables represent positive real numbers.

25. $x^{4/5}x^{-2/5}$ **26.** $y^{-3/7}y^{4/7}$ **27.** $\dfrac{m^{2/3}}{m^{-1/3}}$

28. $\dfrac{x^{1/4}}{x^{3/4}}$ **29.** $(8x^3y^{-6})^{1/3}$ **30.** $(4u^{-2}v^4)^{1/2}$

31. $\left(\dfrac{4x^{-2}}{y^4}\right)^{-1/2}$ **32.** $\left(\dfrac{w^4}{9x^{-2}}\right)^{-1/2}$ **33.** $\dfrac{(8x)^{-1/3}}{12x^{1/4}}$

34. $\dfrac{6a^{3/4}}{15a^{-1/3}}$

Simplify each expression in Problems 35–40 using properties of radicals. All variables represent positive real numbers.

35. $\sqrt[5]{(2x + 3)^5}$ **36.** $\sqrt[3]{(7 + 2y)^3}$

37. $\sqrt{6x}\sqrt{15x^3}\sqrt{30x^7}$ **38.** $\sqrt[5]{16a^4}\,\sqrt[5]{4a^2}\,\sqrt[5]{8a^3}$

39. $\dfrac{\sqrt{6x}\sqrt{10}}{\sqrt{15x}}$ **40.** $\dfrac{\sqrt{8}\sqrt{12y}}{\sqrt{6y}}$

B

In Problems 41–48, multiply, and express answers using positive exponents only.

41. $3x^{3/4}(4x^{1/4} - 2x^8)$

42. $2m^{1/3}(3m^{2/3} - m^6)$

43. $(3u^{1/2} - v^{1/2})(u^{1/2} - 4v^{1/2})$

44. $(a^{1/2} + 2b^{1/2})(a^{1/2} - 3b^{1/2})$

45. $(6m^{1/2} + n^{-1/2})(6m - n^{-1/2})$

46. $(2x - 3y^{1/3})(2x^{1/3} + 1)$

47. $(3x^{1/2} - y^{1/2})^2$

48. $(x^{1/2} + 2y^{1/2})^2$

Write each expression in Problems 49–54 in the form $ax^p + bx^q$, where a and b are real numbers and p and q are rational numbers.

49. $\dfrac{\sqrt[3]{x^2} + 2}{2\sqrt[3]{x}}$ **50.** $\dfrac{12\sqrt{x} - 3}{4\sqrt{x}}$ **51.** $\dfrac{2\sqrt[4]{x^3} + \sqrt[3]{x}}{3x}$

52. $\dfrac{3\sqrt[3]{x^2} + \sqrt{x}}{5x}$ **53.** $\dfrac{2\sqrt[3]{x} - \sqrt{x}}{4\sqrt{x}}$ **54.** $\dfrac{x^2 - 4\sqrt{x}}{2\sqrt[3]{x}}$

Rationalize the denominators in Problems 55–60.

55. $\dfrac{12mn^2}{\sqrt{3mn}}$ **56.** $\dfrac{14x^2}{\sqrt{7x}}$ **57.** $\dfrac{2(x + 3)}{\sqrt{x - 2}}$

58. $\dfrac{3(x + 1)}{\sqrt{x + 4}}$ **59.** $\dfrac{7(x - y)^2}{\sqrt{x} - \sqrt{y}}$ **60.** $\dfrac{3a - 3b}{\sqrt{a} + \sqrt{b}}$

Rationalize the numerators in Problems 61–66.

61. $\dfrac{\sqrt{5xy}}{5x^2y^2}$ **62.** $\dfrac{\sqrt{3mn}}{3mn}$

63. $\dfrac{\sqrt{x + h} - \sqrt{x}}{h}$ **64.** $\dfrac{\sqrt{2(a + h)} - \sqrt{2a}}{h}$

65. $\dfrac{\sqrt{t} - \sqrt{x}}{t^2 - x^2}$ **66.** $\dfrac{\sqrt{x} - \sqrt{y}}{\sqrt{x} + \sqrt{y}}$

Problems 67–70 illustrate common errors involving rational exponents. In each case, find numerical examples that show that the left side is not always equal to the right side.

67. $(x + y)^{1/2} \neq x^{1/2} + y^{1/2}$ **68.** $(x^3 + y^3)^{1/3} \neq x + y$

69. $(x + y)^{1/3} \neq \dfrac{1}{(x + y)^3}$ **70.** $(x + y)^{-1/2} \neq \dfrac{1}{(x + y)^2}$

C

In Problems 71–82, discuss the validity of each statement. If the statement is true, explain why. If not, give a counterexample.

71. $\sqrt{x^2} = x$ for all real numbers x

72. $\sqrt{x^2} = |x|$ for all real numbers x

73. $\sqrt[3]{x^3} = |x|$ for all real numbers x

74. $\sqrt[3]{x^3} = x$ for all real numbers x

75. If $r < 0$, then r has no cube roots.

76. If $r < 0$, then r has no square roots.

77. If $r > 0$, then r has two square roots.

78. If $r > 0$, then r has three cube roots.

79. The fourth roots of 100 are $\sqrt{10}$ and $-\sqrt{10}$.

80. The square roots of $2\sqrt{6} - 5$ are $\sqrt{3} - \sqrt{2}$ and $\sqrt{2} - \sqrt{3}$.

81. $\sqrt{355 - 60\sqrt{35}} = 5\sqrt{7} - 6\sqrt{5}$

82. $\sqrt[3]{7 - 5\sqrt{2}} = 1 - \sqrt{2}$

In Problems 83–88, simplify by writing each expression as a simple or single fraction reduced to lowest terms and without negative exponents.

83. $-\frac{1}{2}(x - 2)(x + 3)^{-3/2} + (x + 3)^{-1/2}$

84. $2(x - 2)^{-1/2} - \frac{1}{2}(2x + 3)(x - 2)^{-3/2}$

85. $\dfrac{(x - 1)^{1/2} - x\left(\frac{1}{2}\right)(x - 1)^{-1/2}}{x - 1}$

86. $\dfrac{(2x - 1)^{1/2} - (x + 2)\left(\frac{1}{2}\right)(2x - 1)^{-1/2}(2)}{2x - 1}$

87. $\dfrac{(x + 2)^{2/3} - x\left(\frac{2}{3}\right)(x + 2)^{-1/3}}{(x + 2)^{4/3}}$

88. $\dfrac{2(3x - 1)^{1/3} - (2x + 1)\left(\frac{1}{3}\right)(3x - 1)^{-2/3}(3)}{(3x - 1)^{2/3}}$

In Problems 89–94, evaluate using a calculator. (Refer to the instruction book for your calculator to see how exponential forms are evaluated.)

89. $22^{3/2}$

90. $15^{5/4}$

91. $827^{-3/8}$

92. $103^{-3/4}$

93. $37.09^{7/3}$

94. $2.876^{8/5}$

In Problems 95 and 96, evaluate each expression on a calculator and determine which pairs have the same value. Verify these results algebraically.

95. (A) $\sqrt{3} + \sqrt{5}$ (B) $\sqrt{2 + \sqrt{3}} + \sqrt{2 - \sqrt{3}}$

(C) $1 + \sqrt{3}$ (D) $\sqrt[3]{10 + 6\sqrt{3}}$

(E) $\sqrt{8 + \sqrt{60}}$ (F) $\sqrt{6}$

96. (A) $2\sqrt[3]{2} + \sqrt{5}$ (B) $\sqrt{8}$

(C) $\sqrt{3} + \sqrt{7}$ (D) $\sqrt{3 + \sqrt{8}} + \sqrt{3 - \sqrt{8}}$

(E) $\sqrt{10 + \sqrt{84}}$ (F) $1 + \sqrt{5}$

Answers to Matched Problems

1. (A) 4   (B) -4

(C) -3 (D) Not a real number (E) 27

2. (A) $\sqrt[5]{u}$ (B) $\sqrt[9]{(6x^2y^5)^2}$ or $\left(\sqrt[9]{6x^2y^5}\right)^2$

(C) $1/\sqrt[5]{(3xy)^3}$ (D) $(9u)^{1/4}$ (E) $-(2x)^{4/7}$

(F) $(x^3 + y^3)^{1/3}$ (not $x + y$)

3. (A) 27   (B) 81

(C) $10y^{7/12} = 10\sqrt[12]{y^7}$ (D) $16y/x^3$

(E) $2/x^{1/18} = 2/\sqrt[18]{x}$

4. (A) $10c^{13/4} - 2c$ (B) $14x + 19x^{1/2}y^{1/2} - 3y$

5. $2.5x^{-7/6} - 2x^{-1}$

6. (A) $x^3 + y^3$ (B) $2y$ (C) $2x$

7. (A) $4b\sqrt{3ab}$

(B) $3(\sqrt{6} - \sqrt{3})$

(C) $(x + y)(\sqrt{x} + \sqrt{y})$

8. (A) $\dfrac{1}{\sqrt{6}}$ (B) $\dfrac{1}{2 + \sqrt{n}}$ (C) $\dfrac{1}{\sqrt{3 + h} + \sqrt{3}}$

A-7   Quadratic Equations

- Solution by Square Root
- Solution by Factoring
- Quadratic Formula
- Quadratic Formula and Factoring
- Application: Supply and Demand

In this section we consider equations involving second-degree polynomials.

DEFINITION Quadratic Equation

A **quadratic equation** in one variable is any equation that can be written in the form

$$ax^2 + bx + c = 0 \qquad a \neq 0 \qquad \text{Standard form}$$

where x is a variable and a, b, and c are constants.

The equations

$$5x^2 - 3x + 7 = 0 \quad \text{and} \quad 18 = 32t^2 - 12t$$

are both quadratic equations, since they are either in the standard form or can be transformed into this form.

We restrict our review to finding real solutions to quadratic equations.

Solution by Square Root

The easiest type of quadratic equation to solve is the special form where the first-degree term is missing:

$$ax^2 + c = 0 \qquad a \neq 0$$

The method of solution of this special form makes direct use of the square-root property:

THEOREM 1 Square-Root Property

If $a^2 = b$, then $a = \pm\sqrt{b}$.

EXAMPLE 1 **Square-Root Method** Use the square-root property to solve each equation.

(A) $x^2 - 7 = 0$ (B) $2x^2 - 10 = 0$

(C) $3x^2 + 27 = 0$ (D) $(x - 8)^2 = 9$

SOLUTION (A) $x^2 - 7 = 0$

$$x^2 = 7 \qquad \textit{What real number squared is 7?}$$
$$x = \pm\sqrt{7} \qquad \textit{Short for } \sqrt{7} \text{ and } -\sqrt{7}$$

(B) $2x^2 - 10 = 0$

$$2x^2 = 10$$
$$x^2 = 5 \qquad \textit{What real number squared is 5?}$$
$$x = \pm\sqrt{5}$$

(C) $3x^2 + 27 = 0$

$$3x^2 = -27$$
$$x^2 = -9 \qquad \textit{What real number squared is } -9?$$

No real solution, since no real number squared is negative.

(D) $(x - 8)^2 = 9$

$$x - 8 = \pm\sqrt{9}$$
$$x - 8 = \pm 3$$
$$x = 8 \pm 3 = 5 \quad \text{or} \quad 11$$

Matched Problem 1 Use the square-root property to solve each equation.

(A) $x^2 - 6 = 0$ (B) $3x^2 - 12 = 0$

(C) $x^2 + 4 = 0$ (D) $(x + 5)^2 = 1$

Solution by Factoring

If the left side of a quadratic equation when written in standard form can be factored, the equation can be solved very quickly. The method of solution by factoring rests on a basic property of real numbers, first mentioned in Section A-1.

CONCEPTUAL INSIGHT

Theorem 2 in Section A-1 states that if a and b are real numbers, then $ab = 0$ if and only if $a = 0$ or $b = 0$. To see that this property is useful for solving quadratic equations, consider the following:

$$x^2 - 4x + 3 = 0 \tag{1}$$
$$(x - 1)(x - 3) = 0$$
$$x - 1 = 0 \quad \text{or} \quad x - 3 = 0$$
$$x = 1 \quad \text{or} \quad x = 3$$

You should check these solutions in equation (1).

If one side of the equation is not 0, then this method cannot be used. For example, consider

$$x^2 - 4x + 3 = 8 \tag{2}$$
$$(x - 1)(x - 3) = 8$$
$$x - 1 \neq 8 \quad \text{or} \quad x - 3 \neq 8 \qquad ab = 8 \text{ does not imply}$$
$$\qquad\qquad\qquad\qquad\qquad\qquad \text{that } a = 8 \text{ or } b = 8.$$
$$x = 9 \quad \text{or} \quad x = 11$$

Verify that neither $x = 9$ nor $x = 11$ is a solution for equation (2).

EXAMPLE 2

Factoring Method Solve by factoring using integer coefficients, if possible.

(A) $3x^2 - 6x - 24 = 0$ (B) $3y^2 = 2y$ (C) $x^2 - 2x - 1 = 0$

SOLUTION

(A) $3x^2 - 6x - 24 = 0$ *Divide both sides by 3, since 3 is a factor of each coefficient.*

$\qquad x^2 - 2x - 8 = 0$ *Factor the left side, if possible.*

$\qquad (x - 4)(x + 2) = 0$

$\qquad x - 4 = 0 \quad \text{or} \quad x + 2 = 0$

$\qquad\qquad x = 4 \quad \text{or} \qquad\quad x = -2$

(B) $\qquad\qquad\qquad 3y^2 = 2y$

$\qquad\qquad\qquad 3y^2 - 2y = 0$ *We lose the solution $y = 0$ if both sides are divided*

$\qquad\qquad\qquad y(3y - 2) = 0$ *by y ($3y^2 = 2y$ and $3y = 2$ are not equivalent).*

$\qquad y = 0 \quad \text{or} \quad 3y - 2 = 0$

$\qquad\qquad\qquad\qquad 3y = 2$

$\qquad\qquad\qquad\qquad y = \frac{2}{3}$

(C) $x^2 - 2x - 1 = 0$

This equation cannot be factored using integer coefficients. We will solve this type of equation by another method, considered below.

Matched Problem 2

Solve by factoring using integer coefficients, if possible.

(A) $2x^2 + 4x - 30 = 0$ (B) $2x^2 = 3x$ (C) $2x^2 - 8x + 3 = 0$

$\qquad\qquad\qquad\qquad\qquad\qquad\quad x^2 = 25$

Note that an equation such as   can be solved by either the square-root or the factoring method, and the results are the same (as they should be). Solve this equation both ways and compare.

Also, note that the factoring method can be extended to higher-degree polynomial equations. Consider the following:

$$x^3 - x = 0$$
$$x(x^2 - 1) = 0$$
$$x(x - 1)(x + 1) = 0$$
$$x = 0 \quad \text{or} \quad x - 1 = 0 \quad \text{or} \quad x + 1 = 0$$

Solution: $x = 0, 1, -1$

Check these solutions in the original equation.

The factoring and square-root methods are fast and easy to use when they apply. However, there are quadratic equations that look simple but cannot be solved by either method. For example, as was noted in Example 2C, the polynomial in

$$x^2 - 2x - 1 = 0$$

cannot be factored using integer coefficients. This brings us to the well-known and widely used *quadratic formula*.

Quadratic Formula

There is a method called *completing the square* that will work for all quadratic equations. After briefly reviewing this method, we will use it to develop the quadratic formula, which can be used to solve any quadratic equation.

The method of **completing the square** is based on the process of transforming a quadratic equation in standard form,

$$ax^2 + bx + c = 0$$

into the form

$$(x + A)^2 = B$$

where A and B are constants. Then, this last equation can be solved easily (if it has a real solution) by the square-root method discussed above.

Consider the equation from Example 2C:

$$x^2 - 2x - 1 = 0 \qquad (3)$$

Since the left side does not factor using integer coefficients, we add 1 to each side to remove the constant term from the left side:

$$x^2 - 2x = 1 \qquad (4)$$

Now we try to find a number that we can add to each side to make the left side a square of a first-degree polynomial. Note the following square of a binomial:

$$(x + m)^2 = x^2 + 2mx + m^2$$

We see that the third term on the right is the square of one-half the coefficient of x in the second term on the right. To complete the square in equation (4), we add the square of one-half the coefficient of x, $\left(-\frac{2}{2}\right)^2 = 1$, to each side. (This rule works only when the coefficient of x^2 is 1, that is, $a = 1$.) Thus,

$$x^2 - 2x + \mathbf{1} = 1 + \mathbf{1}$$

The left side is the square of $x - 1$, and we write

$$(x - 1)^2 = 2$$

What number squared is 2?

$$x - 1 = \pm\sqrt{2}$$
$$x = 1 \pm \sqrt{2}$$

And equation (3) is solved!

Let us try the method on the general quadratic equation

$$ax^2 + bx + c = 0 \qquad a \neq 0 \qquad (5)$$

and solve it once and for all for x in terms of the coefficients a, b, and c. We start by multiplying both sides of equation (5) by $1/a$ to obtain

$$x^2 + \frac{b}{a}x + \frac{c}{a} = 0$$

Add $-c/a$ to both sides:

$$x^2 + \frac{b}{a}x = -\frac{c}{a}$$

Now we complete the square on the left side by adding the square of one-half the coefficient of x, that is, $(b/2a)^2 = b^2/4a^2$, to each side:

$$x^2 + \frac{b}{a}x + \frac{b^2}{4a^2} = \frac{b^2}{4a^2} - \frac{c}{a}$$

Writing the left side as a square and combining the right side into a single fraction, we obtain

$$\left(x + \frac{b}{2a}\right)^2 = \frac{b^2 - 4ac}{4a^2}$$

Now we solve by the square-root method:

$$x + \frac{b}{2a} = \pm\sqrt{\frac{b^2 - 4ac}{4a^2}}$$

$$x = -\frac{b}{2a} \pm \frac{\sqrt{b^2 - 4ac}}{2a} \qquad \text{Since } \pm\sqrt{4a^2} = \pm 2a \text{ for any real number } a$$

When this is written as a single fraction, it becomes the **quadratic formula**:

Quadratic Formula

If $ax^2 + bx + c = 0$, $a \neq 0$, then

$$x = \frac{-b \pm \sqrt{b^2 - 4ac}}{2a}$$

This formula is generally used to solve quadratic equations when the square-root or factoring methods do not work. The quantity $b^2 - 4ac$ under the radical is called the **discriminant**, and it gives us the useful information about solutions listed in Table 1.

Table 1

$b^2 - 4ac$	$ax^2 + bx + c = 0$
Positive	Two real solutions
Zero	One real solution
Negative	No real solutions

EXAMPLE 3 **Quadratic Formula Method** Solve $x^2 - 2x - 1 = 0$ using the quadratic formula.

SOLUTION $x^2 - 2x - 1 = 0$

$$x = \frac{-b \pm \sqrt{b^2 - 4ac}}{2a} \qquad a = 1, b = -2, c = -1$$

$$= \frac{-(-2) \pm \sqrt{(-2)^2 - 4(1)(-1)}}{2(1)}$$

$$= \frac{2 \pm \sqrt{8}}{2} = \frac{2 \pm 2\sqrt{2}}{2} = 1 \pm \sqrt{2} \approx -0.414 \quad \text{or} \quad 2.414$$

CHECK $x^2 - 2x - 1 = 0$
When $x = 1 + \sqrt{2}$,

$$(1 + \sqrt{2})^2 - 2(1 + \sqrt{2}) - 1 = 1 + 2\sqrt{2} + 2 - 2 - 2\sqrt{2} - 1 = 0$$

When $x = 1 - \sqrt{2}$,

$$(1 - \sqrt{2})^2 - 2(1 - \sqrt{2}) - 1 = 1 - 2\sqrt{2} + 2 - 2 + 2\sqrt{2} - 1 = 0$$

Matched Problem 3 Solve $2x^2 - 4x - 3 = 0$ using the quadratic formula.

If we try to solve $x^2 - 6x + 11 = 0$ using the quadratic formula, we obtain

$$x = \frac{6 \pm \sqrt{-8}}{2}$$

which is not a real number. (Why?)

Quadratic Formula and Factoring

As in Section A-3, we restrict our interest in factoring to polynomials with integer coefficients. If a polynomial cannot be factored as a product of lower-degree polynomials with integer coefficients, we say that the polynomial is **not factorable in the integers**.

How can you factor the quadratic polynomial $x^2 - 13x - 2,310$? We start by solving the corresponding quadratic equation using the quadratic formula:

$$x^2 - 13x - 2,310 = 0$$

$$x = \frac{-(-13) \pm \sqrt{(-13)^2 - 4(1)(-2,310)}}{2}$$

$$x = \frac{-(-13) \pm \sqrt{9,409}}{2}$$

$$= \frac{13 \pm 97}{2} = 55 \quad \text{or} \quad -42$$

Now we write

$$x^2 - 13x - 2,310 = [x - 55][x - (-42)] = (x - 55)(x + 42)$$

Multiplying the two factors on the right produces the second-degree polynomial on the left.

What is behind this procedure? The following two theorems justify and generalize the process:

THEOREM 2 Factorability Theorem

A second-degree polynomial, $ax^2 + bx + c$, with integer coefficients can be expressed as the product of two first-degree polynomials with integer coefficients if and only if $\sqrt{b^2 - 4ac}$ is an integer.

THEOREM 3 Factor Theorem

If r_1 and r_2 are solutions to the second-degree equation $ax^2 + bx + c = 0$, then

$$ax^2 + bx + c = a(x - r_1)(x - r_2)$$

EXAMPLE 4 **Factoring with the Aid of the Discriminant** Factor, if possible, using integer coefficients.

(A) $4x^2 - 65x + 264$ (B) $2x^2 - 33x - 306$

SOLUTION (A) $4x^2 - 65x + 264$

Step 1 Test for factorability:

$$\sqrt{b^2 - 4ac} = \sqrt{(-65)^2 - 4(4)(264)} = 1$$

Since the result is an integer, the polynomial has first-degree factors with integer coefficients.

Step 2 Factor, using the factor theorem. Find the solutions to the corresponding quadratic equation using the quadratic formula:

$$4x^2 - 65x + 264 = 0 \quad \text{From step 1}$$

$$x = \frac{-(-65) \pm 1}{2 \cdot 4} = \frac{33}{4} \quad \text{or} \quad 8$$

Thus,

$$4x^2 - 65x + 264 = 4\left(x - \frac{33}{4}\right)(x - 8)$$

$$= (4x - 33)(x - 8)$$

(B) $2x^2 - 33x - 306$

Step 1 Test for factorability:

$$\sqrt{b^2 - 4ac} = \sqrt{(-33)^2 - 4(2)(-306)} = \sqrt{3{,}537}$$

Since $\sqrt{3{,}537}$ is not an integer, the polynomial is not factorable in the integers.

Matched Problem 4 Factor, if possible, using integer coefficients.

(A) $3x^2 - 28x - 464$ (B) $9x^2 + 320x - 144$

Application: Supply and Demand

Supply-and-demand analysis is a very important part of business and economics. In general, producers are willing to supply more of an item as the price of an item increases and less of an item as the price decreases. Similarly, buyers are willing to buy less of an item as the price increases, and more of an item as the price decreases. We have a dynamic situation where the price, supply, and demand fluctuate until a price is reached at which the supply is equal to the demand. In economic theory, this point is called the **equilibrium point**. If the price increases from this point, the supply will increase and the demand will decrease; if the price decreases from this point, the supply will decrease and the demand will increase.

EXAMPLE 5 **Supply and Demand** At a large summer beach resort, the weekly supply-and-demand equations for folding beach chairs are

$$p = \frac{x}{140} + \frac{3}{4} \quad \text{Supply equation}$$

$$p = \frac{5{,}670}{x} \quad \text{Demand equation}$$

The supply equation indicates that the supplier is willing to sell x units at a price of p dollars per unit. The demand equation indicates that consumers are willing to buy x units at a price of p dollars per unit. How many units are required for supply to equal demand? At what price will supply equal demand?

SOLUTION Set the right side of the supply equation equal to the right side of the demand equation and solve for x.

$$\frac{x}{140} + \frac{3}{4} = \frac{5{,}670}{x} \qquad \text{Multiply by 140x, the LCD.}$$

$$x^2 + 105x = 793{,}800 \qquad \text{Write in standard form.}$$

$$x^2 + 105x - 793{,}800 = 0 \qquad \text{Use the quadratic formula.}$$

$$x = \frac{-105 \pm \sqrt{105^2 - 4(1)(-793{,}800)}}{2}$$

$$x = 840 \text{ units}$$

The negative root is discarded since a negative number of units cannot be produced or sold. Substitute $x = 840$ back into either the supply equation or the demand equation to find the equilibrium price (we use the demand equation).

$$p = \frac{5{,}670}{x} = \frac{5{,}670}{840} = \$6.75$$

At a price of $6.75 the supplier is willing to supply 840 chairs and consumers are willing to buy 840 chairs during a week.

Matched Problem 5 Repeat Example 5 if near the end of summer, the supply-and-demand equations are

$$p = \frac{x}{80} - \frac{1}{20} \qquad \textit{Supply equation}$$

$$p = \frac{1{,}264}{x} \qquad \textit{Demand equation}$$

Exercises A-7

Find only real solutions in the problems below. If there are no real solutions, say so.

A

Solve Problems 1–4 by the square-root method.

1. $2x^2 - 22 = 0$ **2.** $3m^2 - 21 = 0$

3. $(3x - 1)^2 = 25$ **4.** $(2x + 1)^2 = 16$

Solve Problems 5–8 by factoring.

5. $2u^2 - 8u - 24 = 0$ **6.** $3x^2 - 18x + 15 = 0$

7. $x^2 = 2x$ **8.** $n^2 = 3n$

Solve Problems 9–12 by using the quadratic formula.

9. $x^2 - 6x - 3 = 0$ **10.** $m^2 + 8m + 3 = 0$

11. $3u^2 + 12u + 6 = 0$ **12.** $2x^2 - 20x - 6 = 0$

B

Solve Problems 13–30 by using any method.

13. $\dfrac{2x^2}{3} = 5x$ **14.** $x^2 = -\dfrac{3}{4}x$

15. $4u^2 - 9 = 0$ **16.** $9y^2 - 25 = 0$

17. $8x^2 + 20x = 12$ **18.** $9x^2 - 6 = 15x$

19. $x^2 = 1 - x$ **20.** $m^2 = 1 - 3m$

21. $2x^2 = 6x - 3$ **22.** $2x^2 = 4x - 1$

23. $y^2 - 4y = -8$ **24.** $x^2 - 2x = -3$

25. $(2x + 3)^2 = 11$ **26.** $(5x - 2)^2 = 7$

27. $\dfrac{3}{p} = p$ **28.** $x - \dfrac{7}{x} = 0$

29. $2 - \dfrac{2}{m^2} = \dfrac{3}{m}$ **30.** $2 + \dfrac{5}{u} = \dfrac{3}{u^2}$

In Problems 31–38, factor, if possible, as the product of two first-degree polynomials with integer coefficients. Use the quadratic formula and the factor theorem.

31. $x^2 + 40x - 84$ **32.** $x^2 - 28x - 128$

33. $x^2 - 32x + 144$ **34.** $x^2 + 52x + 208$

35. $2x^2 + 15x - 108$ **36.** $3x^2 - 32x - 140$

37. $4x^2 + 241x - 434$ **38.** $6x^2 - 427x - 360$

C

39. Solve $A = P(1 + r)^2$ for r in terms of A and P; that is, isolate r on the left side of the equation (with coefficient 1) and end up with an algebraic expression on the right side involving A and P but not r. Write the answer using positive square roots only.

40. Solve $x^2 + 3mx - 3n = 0$ for x in terms of m and n.

41. Consider the quadratic equation

$$x^2 + 4x + c = 0$$

where c is a real number. Discuss the relationship between the values of c and the three types of roots listed in Table 1 on page 608.

42. Consider the quadratic equation

$$x^2 - 2x + c = 0$$

where c is a real number. Discuss the relationship between the values of c and the three types of roots listed in Table 1 on page 608.

Applications

43. Supply and demand. A company wholesales shampoo in a particular city. Their marketing research department established the following weekly supply-and-demand equations:

$$p = \frac{x}{450} + \frac{1}{2} \qquad \text{Supply equation}$$

$$p = \frac{6,300}{x} \qquad \text{Demand equation}$$

How many units are required for supply to equal demand? At what price per bottle will supply equal demand?

44. Supply and demand. An importer sells an automatic camera to outlets in a large city. During the summer, the weekly supply-and-demand equations are

$$p = \frac{x}{6} + 9 \qquad \text{Supply equation}$$

$$p = \frac{24,840}{x} \qquad \text{Demand equation}$$

How many units are required for supply to equal demand? At what price will supply equal demand?

45. Interest rate. If P dollars are invested at $100r$ percent compounded annually, at the end of 2 years it will grow to $A = P(1 + r)^2$. At what interest rate will \$484 grow to \$625 in 2 years? (*Note:* If $A = 625$ and $P = 484$ find r.)

46. Interest rate. Using the formula in Problem 45, determine the interest rate that will make \$1,000 grow to \$1,210 in 2 years.

47. Ecology. To measure the velocity v (in feet per second) of a stream, we position a hollow L-shaped tube with one end under the water pointing upstream and the other end pointing straight up a couple of feet out of the water. The water will then be pushed up the tube a certain distance h (in feet) above the surface of the stream. Physicists have shown that $v^2 = 64h$. Approximately how fast is a stream flowing if $h = 1$ foot? If $h = 0.5$ foot?

48. Safety research. It is of considerable importance to know the least number of feet d in which a car can be stopped, including reaction time of the driver, at various speeds v (in miles per hour). Safety research has produced the formula $d = 0.044v^2 + 1.1v$. If it took a car 550 feet to stop, estimate the car's speed at the moment the stopping process was started.

Answers to Matched Problems

1. (A) $\pm\sqrt{6}$ (B) ± 2

   (C) No real solution   (D) $-6, -4$

2. (A) $-5, 3$ (B) $0, \frac{3}{2}$

   (C) Cannot be factored using integer coefficients

3. $(2 \pm \sqrt{10})/2$

4. (A) Cannot be factored using integer coefficients

   (B) $(9x - 4)(x + 36)$

5. 320 chairs at \$3.95 each

APPENDIX B Special Topics

B-1 Sequences, Series, and Summation Notation

- Sequences
- Series and Summation Notation

If someone asked you to list all natural numbers that are perfect squares, you might begin by writing

$$1, 4, 9, 16, 25, 36$$

But you would soon realize that it is impossible to actually list all the perfect squares, since there are an infinite number of them. However, you could represent this collection of numbers in several different ways. One common method is to write

$$1, 4, 9, \ldots, n^2, \ldots \quad n \in N$$

where N is the set of natural numbers. A list of numbers such as this is generally called a *sequence*.

Sequences

Consider the function f given by

$$f(n) = 2n + 1 \tag{1}$$

where the domain of f is the set of natural numbers N. Note that

$$f(1) = 3, \quad f(2) = 5, \quad f(3) = 7, \quad \ldots$$

The function f is an example of a sequence. In general, a **sequence** is a function with domain a set of successive integers. Instead of the standard function notation used in equation (1), sequences are usually defined in terms of a special notation.

The range value $f(n)$ is usually symbolized more compactly with a symbol such as a_n. Thus, in place of equation (1), we write

$$a_n = 2n + 1$$

and the domain is understood to be the set of natural numbers unless something is said to the contrary or the context indicates otherwise. The elements in the range are called **terms of the sequence**; a_1 is the first term, a_2 is the second term, and a_n is the **nth term**, or **general term**.

$$a_1 = 2(1) + 1 = 3 \quad \text{First term}$$
$$a_2 = 2(2) + 1 = 5 \quad \text{Second term}$$
$$a_3 = 2(3) + 1 = 7 \quad \text{Third term}$$
$$\vdots$$
$$a_n = 2n + 1 \quad \text{General term}$$

The ordered list of elements

$$3, 5, 7, \ldots, 2n + 1, \ldots$$

obtained by writing the terms of the sequence in their natural order with respect to the domain values is often informally referred to as a sequence. A sequence also may be represented in the abbreviated form $\{a_n\}$, where a symbol for the nth term is written within braces. For example, we could refer to the sequence $3, 5, 7, \ldots, 2n + 1, \ldots$ as the sequence $\{2n + 1\}$.

If the domain of a sequence is a finite set of successive integers, then the sequence is called a **finite sequence**. If the domain is an infinite set of successive integers, then the sequence is called an **infinite sequence.** The sequence $\{2n + 1\}$ discussed above is an infinite sequence.

EXAMPLE 1   **Writing the Terms of a Sequence** Write the first four terms of each sequence:

(A) $a_n = 3n - 2$ (B) $\left\{ \dfrac{(-1)^n}{n} \right\}$

SOLUTION   (A) $1, 4, 7, 10$ (B) $-1, \dfrac{1}{2}, \dfrac{-1}{3}, \dfrac{1}{4}$

Matched Problem 1   Write the first four terms of each sequence:

(A) $a_n = -n + 3$ (B) $\left\{ \dfrac{(-1)^n}{2^n} \right\}$

Now that we have seen how to use the general term to find the first few terms in a sequence, we consider the reverse problem. That is, can a sequence be defined just by listing the first three or four terms of the sequence? And can we then use these initial terms to find a formula for the nth term? In general, without other information, the answer to the first question is no. Many different sequences may start off with the same terms. Simply listing the first three terms (or any other finite number of terms) does not specify a particular sequence.

What about the second question? That is, given a few terms, can we find the general formula for at least one sequence whose first few terms agree with the given terms? The answer to this question is a qualified yes. If we can observe a simple pattern in the given terms, we usually can construct a general term that will produce that pattern. The next example illustrates this approach.

EXAMPLE 2   **Finding the General Term of a Sequence** Find the general term of a sequence whose first four terms are

(A) $3, 4, 5, 6, \ldots$ (B) $5, -25, 125, -625, \ldots$

SOLUTION   (A) Since these terms are consecutive integers, one solution is $a_n = n, n \geq 3$. If we want the domain of the sequence to be all natural numbers, another solution is $b_n = n + 2$.

(B) Each of these terms can be written as the product of a power of 5 and a power of -1:

$$5 = (-1)^0 5^1 = a_1$$
$$-25 = (-1)^1 5^2 = a_2$$
$$125 = (-1)^2 5^3 = a_3$$
$$-625 = (-1)^3 5^4 = a_4$$

If we choose the domain to be all natural numbers, a solution is

$$a_n = (-1)^{n-1} 5^n$$

Matched Problem 2   Find the general term of a sequence whose first four terms are
(A) $3, 6, 9, 12, \ldots$ (B) $1, -2, 4, -8, \ldots$

In general, there is usually more than one way of representing the nth term of a given sequence (see the solution of Example 2A). However, unless something is stated to the contrary, we assume that the domain of the sequence is the set of natural numbers N.

Series and Summation Notation

If $a_1, a_2, a_3, \ldots, a_n, \ldots$ is a sequence, the expression

$$a_1 + a_2 + a_3 + \cdots + a_n + \cdots$$

is called a **series**. If the sequence is finite, the corresponding series is a **finite series**. If the sequence is infinite, the corresponding series is an **infinite series**. We consider only finite series in this section. For example,

$$1, 3, 5, 7, 9 \qquad \textit{Finite sequence}$$
$$1 + 3 + 5 + 7 + 9 \qquad \textit{Finite series}$$

Notice that we can easily evaluate this series by adding the five terms:

$$1 + 3 + 5 + 7 + 9 = 25$$

Series are often represented in a compact form called **summation notation**. Consider the following examples:

$$\sum_{k=3}^{6} k^2 = 3^2 + 4^2 + 5^2 + 6^2$$
$$= 9 + 16 + 25 + 36 = 86$$

$$\sum_{k=0}^{2} (4k + 1) = (4 \cdot 0 + 1) + (4 \cdot 1 + 1) + (4 \cdot 2 + 1)$$
$$= 1 + 5 + 9 = 15$$

In each case, the terms of the series on the right are obtained from the expression on the left by successively replacing the **summing index k** with integers, starting with the number indicated below the **summation sign Σ** and ending with the number that appears above Σ. The summing index may be represented by letters other than k and may start at any integer and end at any integer greater than or equal to the starting integer. If we are given the finite sequence

$$\frac{1}{2}, \frac{1}{4}, \frac{1}{8}, \ldots, \frac{1}{2^n}$$

the corresponding series is

$$\frac{1}{2} + \frac{1}{4} + \frac{1}{8} + \cdots + \frac{1}{2^n} = \sum_{j=1}^{n} \frac{1}{2^j}$$

where we have used j for the summing index.

EXAMPLE 3   **Summation Notation**   Write

$$\sum_{k=1}^{5} \frac{k}{k^2 + 1}$$

without summation notation. Do not evaluate the sum.

SOLUTION
$$\sum_{k=1}^{5} \frac{k}{k^2 + 1} = \frac{1}{1^2 + 1} + \frac{2}{2^2 + 1} + \frac{3}{3^2 + 1} + \frac{4}{4^2 + 1} + \frac{5}{5^2 + 1}$$
$$= \frac{1}{2} + \frac{2}{5} + \frac{3}{10} + \frac{4}{17} + \frac{5}{26}$$

Matched Problem 3   Write

$$\sum_{k=1}^{5} \frac{k + 1}{k}$$

without summation notation. Do not evaluate the sum.

If the terms of a series are alternately positive and negative, we call the series an **alternating series**. The next example deals with the representation of such a series.

EXAMPLE 4   **Summation Notation** Write the alternating series

$$\frac{1}{2} - \frac{1}{4} + \frac{1}{6} - \frac{1}{8} + \frac{1}{10} - \frac{1}{12}$$

using summation notation with

(A) The summing index k starting at 1

(B) The summing index j starting at 0

SOLUTION   (A) $(-1)^{k+1}$ provides the alternation of sign, and $1/(2k)$ provides the other part of each term. So, we can write

$$\frac{1}{2} - \frac{1}{4} + \frac{1}{6} - \frac{1}{8} + \frac{1}{10} - \frac{1}{12} = \sum_{k=1}^{6} \frac{(-1)^{k+1}}{2k}$$

(B) $(-1)^{j}$ provides the alternation of sign, and $1/[2(j+1)]$ provides the other part of each term. So, we can write

$$\frac{1}{2} - \frac{1}{4} + \frac{1}{6} - \frac{1}{8} + \frac{1}{10} - \frac{1}{12} = \sum_{j=0}^{5} \frac{(-1)^{j}}{2(j+1)}$$

Matched Problem 4   Write the alternating series

$$1 - \frac{1}{3} + \frac{1}{9} - \frac{1}{27} + \frac{1}{81}$$

using summation notation with

(A) The summing index k starting at 1

(B) The summing index j starting at 0

Summation notation provides a compact notation for the sum of any list of numbers, even if the numbers are not generated by a formula. For example, suppose that the results of an examination taken by a class of 10 students are given in the following list:

$$87, 77, 95, 83, 86, 73, 95, 68, 75, 86$$

If we let $a_1, a_2, a_3, \ldots, a_{10}$ represent these 10 scores, then the average test score is given by

$$\frac{1}{10}\sum_{k=1}^{10} a_k = \frac{1}{10}(87 + 77 + 95 + 83 + 86 + 73 + 95 + 68 + 75 + 86)$$

$$= \frac{1}{10}(825) = 82.5$$

More generally, in statistics, the **arithmetic mean** $\bar{a}$ of a list of n numbers $a_1, a_2, \ldots, a_n$ is defined as

$$\bar{a} = \frac{1}{n}\sum_{k=1}^{n} a_k$$

EXAMPLE 5   **Arithmetic Mean** Find the arithmetic mean of $3, 5, 4, 7, 4, 2, 3,$ and 6.

SOLUTION $\bar{a} = \dfrac{1}{8}\sum_{k=1}^{8} a_k = \dfrac{1}{8}(3 + 5 + 4 + 7 + 4 + 2 + 3 + 6) = \dfrac{1}{8}(34) = 4.25$

Matched Problem 5   Find the arithmetic mean of $9, 3, 8, 4, 3,$ and 6.

Exercises B-1

A

Write the first four terms for each sequence in Problems 1–6.

1. $a_n = 2n + 3$

2. $a_n = 4n - 3$

3. $a_n = \dfrac{n + 2}{n + 1}$

4. $a_n = \dfrac{2n + 1}{2n}$

5. $a_n = (-3)^{n+1}$

6. $a_n = \left(-\dfrac{1}{4}\right)^{n-1}$

7. Write the 10th term of the sequence in Problem 1.

8. Write the 15th term of the sequence in Problem 2.

9. Write the 99th term of the sequence in Problem 3.

10. Write the 200th term of the sequence in Problem 4.

In Problems 11–16, write each series in expanded form without summation notation, and evaluate.

11. $\displaystyle\sum_{k=1}^{6} k$

12. $\displaystyle\sum_{k=1}^{5} k^2$

13. $\displaystyle\sum_{k=4}^{7} (2k - 3)$

14. $\displaystyle\sum_{k=0}^{4} (-2)^k$

15. $\displaystyle\sum_{k=0}^{3} \dfrac{1}{10^k}$

16. $\displaystyle\sum_{k=1}^{4} \dfrac{1}{2^k}$

Find the arithmetic mean of each list of numbers in Problems 17–20.

17. 5, 4, 2, 1, and 6

18. 7, 9, 9, 2, and 4

19. 96, 65, 82, 74, 91, 88, 87, 91, 77, and 74

20. 100, 62, 95, 91, 82, 87, 70, 75, 87, and 82

B

Write the first five terms of each sequence in Problems 21–26.

21. $a_n = \dfrac{(-1)^{n+1}}{2^n}$

22. $a_n = (-1)^n (n - 1)^2$

23. $a_n = n[1 + (-1)^n]$

24. $a_n = \dfrac{1 - (-1)^n}{n}$

25. $a_n = \left(-\dfrac{3}{2}\right)^{n-1}$

26. $a_n = \left(-\dfrac{1}{2}\right)^{n+1}$

In Problems 27–42, find the general term of a sequence whose first four terms agree with the given terms.

27. $-2, -1, 0, 1, \ldots$

28. $4, 5, 6, 7, \ldots$

29. $4, 8, 12, 16, \ldots$

30. $-3, -6, -9, -12, \ldots$

31. $\frac{1}{2}, \frac{3}{4}, \frac{5}{6}, \frac{7}{8}, \ldots$

32. $\frac{1}{2}, \frac{2}{3}, \frac{3}{4}, \frac{4}{5}, \ldots$

33. $1, -2, 3, -4, \ldots$

34. $-2, 4, -8, 16, \ldots$

35. $1, -3, 5, -7, \ldots$

36. $3, -6, 9, -12, \ldots$

37. $1, \frac{2}{5}, \frac{4}{25}, \frac{8}{125}, \ldots$

38. $\frac{4}{3}, \frac{16}{9}, \frac{64}{27}, \frac{256}{81}, \ldots$

39. $x, x^2, x^3, x^4, \ldots$

40. $1, 2x, 3x^2, 4x^3, \ldots$

41. $x, -x^3, x^5, -x^7, \ldots$

42. $x, \dfrac{x^2}{2}, \dfrac{x^3}{3}, \dfrac{x^4}{4}, \ldots$

Write each series in Problems 43–50 in expanded form without summation notation. Do not evaluate.

43. $\displaystyle\sum_{k=1}^{5} (-1)^{k+1} (2k - 1)^2$

44. $\displaystyle\sum_{k=1}^{4} \dfrac{(-2)^{k+1}}{2k + 1}$

45. $\displaystyle\sum_{k=2}^{5} \dfrac{2^k}{2k + 3}$

46. $\displaystyle\sum_{k=3}^{7} \dfrac{(-1)^k}{k^2 - k}$

47. $\displaystyle\sum_{k=1}^{5} x^{k-1}$

48. $\displaystyle\sum_{k=1}^{3} \dfrac{1}{k} x^{k+1}$

49. $\displaystyle\sum_{k=0}^{4} \dfrac{(-1)^k x^{2k+1}}{2k + 1}$

50. $\displaystyle\sum_{k=0}^{4} \dfrac{(-1)^k x^{2k}}{2k + 2}$

Write each series in Problems 51–54 using summation notation with

(A) The summing index k starting at k = 1

(B) The summing index j starting at j = 0

51. $2 + 3 + 4 + 5 + 6$

52. $1^2 + 2^2 + 3^2 + 4^2$

53. $1 - \frac{1}{2} + \frac{1}{3} - \frac{1}{4}$

54. $1 - \frac{1}{3} + \frac{1}{5} - \frac{1}{7} + \frac{1}{9}$

Write each series in Problems 55–58 using summation notation with the summing index k starting at k = 1.

55. $2 + \dfrac{3}{2} + \dfrac{4}{3} + \cdots + \dfrac{n + 1}{n}$

56. $1 + \dfrac{1}{2^2} + \dfrac{1}{3^2} + \cdots + \dfrac{1}{n^2}$

57. $\dfrac{1}{2} - \dfrac{1}{4} + \dfrac{1}{8} - \cdots + \dfrac{(-1)^{n+1}}{2^n}$

58. $1 - 4 + 9 - \cdots + (-1)^{n+1} n^2$

C

✎ *In Problems 59–62, discuss the validity of each statement. If the statement is true, explain why. If not, give a counterexample.*

59. For each positive integer n, the sum of the series
$1 + \dfrac{1}{2} + \dfrac{1}{3} + \cdots + \dfrac{1}{n}$ is less than 4.

60. For each positive integer n, the sum of the series
$\dfrac{1}{2} + \dfrac{1}{4} + \dfrac{1}{8} + \cdots + \dfrac{1}{2^n}$ is less than 1.

61. For each positive integer n, the sum of the series
$\dfrac{1}{2} - \dfrac{1}{4} + \dfrac{1}{8} - \cdots + \dfrac{(-1)^{n+1}}{2^n}$ is greater than or equal to $\dfrac{1}{4}$.

62. For each positive integer n, the sum of the series
$1 - \dfrac{1}{2} + \dfrac{1}{3} - \dfrac{1}{4} + \cdots + \dfrac{(-1)^{n+1}}{n}$ is greater than or equal to $\dfrac{1}{2}$.

*Some sequences are defined by a **recursion formula**—that is, a formula that defines each term of the sequence in terms of one or more of the preceding terms. For example, if $\{a_n\}$ is defined by*

$$a_1 = 1 \quad and \quad a_n = 2a_{n-1} + 1 \quad for \ n \geq 2$$

then

$$a_2 = 2a_1 + 1 = 2 \cdot 1 + 1 = 3$$
$$a_3 = 2a_2 + 1 = 2 \cdot 3 + 1 = 7$$
$$a_4 = 2a_3 + 1 = 2 \cdot 7 + 1 = 15$$

and so on. In Problems 63–66, write the first five terms of each sequence.

63. $a_1 = 2$ and $a_n = 3a_{n-1} + 2$ for $n \geq 2$

64. $a_1 = 3$ and $a_n = 2a_{n-1} - 2$ for $n \geq 2$

65. $a_1 = 1$ and $a_n = 2a_{n-1}$ for $n \geq 2$

66. $a_1 = 1$ and $a_n = -\frac{1}{3}a_{n-1}$ for $n \geq 2$

If A is a positive real number, the terms of the sequence defined by

$$a_1 = \frac{A}{2} \quad and \quad a_n = \frac{1}{2}\left(a_{n-1} + \frac{A}{a_{n-1}}\right) \quad for \ n \geq 2$$

can be used to approximate $\sqrt{A}$ to any decimal place accuracy desired. In Problems 67 and 68, compute the first four terms of this sequence for the indicated value of A, and compare the fourth term with the value of $\sqrt{A}$ obtained from a calculator.

67. $A = 2$

68. $A = 6$

69. The sequence defined recursively by $a_1 = 1$, $a_2 = 1$, $a_n = a_{n-1} + a_{n-2}$ for $n \geq 3$ is called the *Fibonacci sequence*. Find the first ten terms of the Fibonacci sequence.

70. The sequence defined by $b_n = \dfrac{\sqrt{5}}{5}\left(\dfrac{1 + \sqrt{5}}{2}\right)^n$ is related to the Fibonacci sequence. Find the first ten terms (to three decimal places) of the sequence $\{b_n\}$ and describe the relationship.

Answers to Matched Problems

1. (A) $2, 1, 0, -1$ (B) $\frac{-1}{2}, \frac{1}{4}, \frac{-1}{8}, \frac{1}{16}$

2. (A) $a_n = 3n$ (B) $a_n = (-2)^{n-1}$

3. $2 + \frac{3}{2} + \frac{4}{3} + \frac{5}{4} + \frac{6}{5}$

4. (A) $\displaystyle\sum_{k=1}^{5} \frac{(-1)^{k-1}}{3^{k-1}}$ (B) $\displaystyle\sum_{j=0}^{4} \frac{(-1)^j}{3^j}$

5. 5.5

B-2   Arithmetic and Geometric Sequences

- Arithmetic and Geometric Sequences
- nth-Term Formulas
- Sum Formulas for Finite Arithmetic Series
- Sum Formulas for Finite Geometric Series
- Sum Formula for Infinite Geometric Series
- Applications

For most sequences, it is difficult to sum an arbitrary number of terms of the sequence without adding term by term. But particular types of sequences—*arithmetic sequences* and *geometric sequences*—have certain properties that lead to convenient and useful formulas for the sums of the corresponding *arithmetic series* and *geometric series*.

Arithmetic and Geometric Sequences

The sequence $5, 7, 9, 11, 13, \ldots, 5 + 2(n - 1), \ldots$, where each term after the first is obtained by adding 2 to the preceding term, is an example of an arithmetic sequence. The sequence $5, 10, 20, 40, 80, \ldots, 5(2)^{n-1}, \ldots$, where each term after the first is obtained by multiplying the preceding term by 2, is an example of a geometric sequence.

> **DEFINITION Arithmetic Sequence**
>
> A sequence of numbers
>
> $$a_1, a_2, a_3, \ldots, a_n, \ldots$$
>
> is called an **arithmetic sequence** if there is a constant d, called the **common difference**, such that
>
> $$a_n - a_{n-1} = d$$
>
> That is,
>
> $$a_n = a_{n-1} + d \qquad \text{for every } n > 1$$

DEFINITION Geometric Sequence

A sequence of numbers

$$a_1, a_2, a_3, \ldots, a_n, \ldots$$

is called a **geometric sequence** if there exists a nonzero constant r, called a **common ratio**, such that

$$\frac{a_n}{a_{n-1}} = r$$

That is,

$$a_n = ra_{n-1} \quad \text{for every } n > 1$$

EXAMPLE 1 **Recognizing Arithmetic and Geometric Sequences** Which of the following can be the first four terms of an arithmetic sequence? Of a geometric sequence?

(A) $1, 2, 3, 5, \ldots$ 　　　　　　　　(B) $-1, 3, -9, 27, \ldots$

(C) $3, 3, 3, 3, \ldots$ 　　　　　　　　(D) $10, 8.5, 7, 5.5, \ldots$

SOLUTION (A) Since $2 - 1 \neq 5 - 3$, there is no common difference, so the sequence is not an arithmetic sequence. Since $2/1 \neq 3/2$, there is no common ratio, so the sequence is not geometric either.

(B) The sequence is geometric with common ratio -3. It is not arithmetic.

(C) The sequence is arithmetic with common difference 0, and is also geometric with common ratio 1.

(D) The sequence is arithmetic with common difference -1.5. It is not geometric.

Matched Problem 1 Which of the following can be the first four terms of an arithmetic sequence? Of a geometric sequence?

(A) $8, 2, 0.5, 0.125, \ldots$ 　　　(B) $-7, -2, 3, 8, \ldots$ 　　　(C) $1, 5, 25, 100, \ldots$

nth-Term Formulas

If $\{a_n\}$ is an arithmetic sequence with common difference d, then

$$a_2 = a_1 + d$$
$$a_3 = a_2 + d = a_1 + 2d$$
$$a_4 = a_3 + d = a_1 + 3d$$

This suggests that

THEOREM 1 nth Term of an Arithmetic Sequence

$$a_n = a_1 + (n - 1)d \quad \text{for all } n > 1 \tag{1}$$

Similarly, if $\{a_n\}$ is a geometric sequence with common ratio r, then

$$a_2 = a_1 r$$
$$a_3 = a_2 r = a_1 r^2$$
$$a_4 = a_3 r = a_1 r^3$$

This suggests that

> **THEOREM 2** *n*th Term of a Geometric Sequence
>
> $$a_n = a_1 r^{n-1} \quad \text{for all } n > 1 \tag{2}$$

EXAMPLE 2 **Finding Terms in Arithmetic and Geometric Sequences**

(A) If the 1st and 10th terms of an arithmetic sequence are 3 and 30, respectively, find the 40th term of the sequence.

(B) If the 1st and 10th terms of a geometric sequence are 3 and 30, find the 40th term to three decimal places.

SOLUTION (A) First use formula (1) with $a_1 = 3$ and $a_{10} = 30$ to find d:

$$a_n = a_1 + (n - 1)d$$
$$a_{10} = a_1 + (10 - 1)d$$
$$30 = 3 + 9d$$
$$d = 3$$

Now find a_{40}:

$$a_{40} = 3 + 39 \cdot 3 = 120$$

(B) First use formula (2) with $a_1 = 3$ and $a_{10} = 30$ to find r:

$$a_n = a_1 r^{n-1}$$
$$a_{10} = a_1 r^{10-1}$$
$$30 = 3r^9$$
$$r^9 = 10$$
$$r = 10^{1/9}$$

Now find a_{40}:

$$a_{40} = 3(10^{1/9})^{39} = 3(10^{39/9}) = 64{,}633.041$$

Matched Problem 2 (A) If the 1st and 15th terms of an arithmetic sequence are -5 and 23, respectively, find the 73rd term of the sequence.

(B) Find the 8th term of the geometric sequence

$$\frac{1}{64}, \frac{-1}{32}, \frac{1}{16}, \dots$$

Sum Formulas for Finite Arithmetic Series

If $a_1, a_2, a_3, \dots, a_n$ is a finite arithmetic sequence, then the corresponding series $a_1 + a_2 + a_3 + \cdots + a_n$ is called a *finite arithmetic series*. We will derive two simple and very useful formulas for the sum of a finite arithmetic series. Let d be the common difference of the arithmetic sequence $a_1, a_2, a_3, \dots, a_n$ and let S_n denote the sum of the series $a_1 + a_2 + a_3 + \cdots + a_n$. Then

$$S_n = a_1 + (a_1 + d) + \cdots + [a_1 + (n - 2)d] + [a_1 + (n - 1)d]$$

Reversing the order of the sum, we obtain

$$S_n = [a_1 + (n - 1)d] + [a_1 + (n - 2)d] + \cdots + (a_1 + d) + a_1$$

Something interesting happens if we combine these last two equations by addition (adding corresponding terms on the right sides):

$$2S_n = [2a_1 + (n - 1)d] + [2a_1 + (n - 1)d] + \cdots + [2a_1 + (n - 1)d] + [2a_1 + (n - 1)d]$$

All the terms on the right side are the same, and there are n of them. Thus,

$$2S_n = n[2a_1 + (n - 1)d]$$

and we have the following general formula:

THEOREM 3   Sum of a Finite Arithmetic Series: First Form

$$S_n = \frac{n}{2}[2a_1 + (n - 1)d] \tag{3}$$

Replacing

$$[a_1 + (n - 1)d] \quad \text{in} \quad \frac{n}{2}[a_1 + a_1 + (n - 1)d]$$

by a_n from equation (1), we obtain a second useful formula for the sum:

THEOREM 4   Sum of a Finite Arithmetic Series: Second Form

$$S_n = \frac{n}{2}(a_1 + a_n) \tag{4}$$

EXAMPLE 3   **Finding a Sum** Find the sum of the first 30 terms in the arithmetic sequence:

$$3, 8, 13, 18, \ldots$$

SOLUTION   Use formula (3) with $n = 30$, $a_1 = 3$, and $d = 5$:

$$S_{30} = \frac{30}{2}[2 \cdot 3 + (30 - 1)5] = 2,265$$

Matched Problem 3   Find the sum of the first 40 terms in the arithmetic sequence:

$$15, 13, 11, 9, \ldots$$

EXAMPLE 4   **Finding a Sum** Find the sum of all the even numbers between 31 and 87.

SOLUTION   First, find n using equation (1):

$$a_n = a_1 + (n - 1)d$$
$$86 = 32 + (n - 1)2$$
$$n = 28$$

Now find S_{28} using formula (4):

$$S_n = \frac{n}{2}(a_1 + a_n)$$

$$S_{28} = \frac{28}{2}(32 + 86) = 1,652$$

Matched Problem 4   Find the sum of all the odd numbers between 24 and 208.

Sum Formulas for Finite Geometric Series

If $a_1, a_2, a_3, \ldots, a_n$ is a finite geometric sequence, then the corresponding series $a_1 + a_2 + a_3 + \cdots + a_n$ is called a *finite geometric series*. As with arithmetic series, we can derive two simple and very useful formulas for the sum of a finite geometric series. Let r be the common ratio of the geometric sequence $a_1, a_2, a_3, \ldots, a_n$ and let S_n denote the sum of the series $a_1 + a_2 + a_3 + \cdots + a_n$. Then

$$S_n = a_1 + a_1 r + a_1 r^2 + \cdots + a_1 r^{n-2} + a_1 r^{n-1}$$

If we multiply both sides by r, we obtain

$$rS_n = a_1 r + a_1 r^2 + a_1 r^3 + \cdots + a_1 r^{n-1} + a_1 r^n$$

Now combine these last two equations by subtraction to obtain

$$rS_n - S_n = (a_1 r + a_1 r^2 + a_1 r^3 + \cdots + a_1 r^{n-1} + a_1 r^n) - (a_1 + a_1 r + a_1 r^2 + \cdots + a_1 r^{n-2} + a_1 r^{n-1})$$
$$(r-1)S_n = a_1 r^n - a_1$$

Notice how many terms drop out on the right side. Solving for S_n, we have

---

THEOREM 5   Sum of a Finite Geometric Series: First Form

$$S_n = \frac{a_1(r^n - 1)}{r - 1} \qquad r \neq 1 \tag{5}$$

Since $a_n = a_1 r^{n-1}$, or $ra_n = a_1 r^n$, formula (5) also can be written in the form

---

THEOREM 6   Sum of a Finite Geometric Series: Second Form

$$S_n = \frac{ra_n - a_1}{r - 1} \qquad r \neq 1 \tag{6}$$

EXAMPLE 5   **Finding a Sum**   Find the sum (to 2 decimal places) of the first ten terms of the geometric sequence:

$$1, 1.05, 1.05^2, \ldots$$

SOLUTION   Use formula (5) with $a_1 = 1$, $r = 1.05$, and $n = 10$:

$$S_n = \frac{a_1(r^n - 1)}{r - 1}$$
$$S_{10} = \frac{1(1.05^{10} - 1)}{1.05 - 1}$$
$$\approx \frac{0.6289}{0.05} \approx 12.58$$

Matched Problem 5   Find the sum of the first eight terms of the geometric sequence:

$$100, 100(1.08), 100(1.08)^2, \ldots$$

Sum Formula for Infinite Geometric Series

Given a geometric series, what happens to the sum S_n of the first n terms as n increases without stopping? To answer this question, let us write formula (5) in the form

$$S_n = \frac{a_1 r^n}{r - 1} - \frac{a_1}{r - 1}$$

It is possible to show that if $-1 < r < 1$, then r^n will approach 0 as n increases. The first term above will approach 0 and S_n can be made as close as we please to the second term, $-a_1/(r - 1)$ [which can be written as $a_1/(1 - r)$], by taking n sufficiently large. So, if the common ratio r is between -1 and 1, we conclude that the sum of an infinite geometric series is

> **THEOREM 7 Sum of an Infinite Geometric Series**
>
> $$S_\infty = \frac{a_1}{1 - r} \qquad -1 < r < 1 \qquad (7)$$

If $r \leq -1$ or $r \geq 1$, then an infinite geometric series has no sum.

Applications

EXAMPLE 6 **Loan Repayment** A person borrows \$3,600 and agrees to repay the loan in monthly installments over 3 years. The agreement is to pay 1% of the unpaid balance each month for using the money and \$100 each month to reduce the loan. What is the total cost of the loan over the 3 years?

SOLUTION Let us look at the problem relative to a time line:

The total cost of the loan is

$$1 + 2 + \cdots + 34 + 35 + 36$$

The terms form a finite arithmetic series with $n = 36$, $a_1 = 1$, and $a_{36} = 36$, so we can use formula (4):

$$S_n = \frac{n}{2}(a_1 + a_n)$$

$$S_{36} = \frac{36}{2}(1 + 36) = \$666$$

We conclude that the total cost of the loan over 3 years is \$666.

Matched Problem 6 Repeat Example 6 with a loan of \$6,000 over 5 years.

EXAMPLE 7 **Economy Stimulation** The government has decided on a tax rebate program to stimulate the economy. Suppose that you receive \$1,200 and you spend 80% of this, and each of the people who receive what you spend also spend 80% of what they receive, and this process continues without end. According to the **multiplier principle** in economics, the effect of your \$1,200 tax rebate on the economy is multiplied many times. What is the total amount spent if the process continues as indicated?

SOLUTION We need to find the sum of an infinite geometric series with the first amount spent being $a_1 = (0.8)(\$1,200) = \$960$ and $r = 0.8$. Using formula (7), we obtain

$$S_\infty = \frac{a_1}{1 - r}$$

$$= \frac{\$960}{1 - 0.8} = \$4,800$$

Assuming the process continues as indicated, we would expect the \$1,200 tax rebate to result in about \$4,800 of spending.

Matched Problem 7 Repeat Example 7 with a tax rebate of \$2,000.

Exercises B-2

A

In Problems 1 and 2, determine whether the indicated sequence can be the first three terms of an arithmetic or geometric sequence, and, if so, find the common difference or common ratio and the next two terms of the sequence.

1. (A) $-11, -16, -21, \ldots$ (B) $2, -4, 8, \ldots$

   (C) $1, 4, 9, \ldots$ (D) $\frac{1}{2}, \frac{1}{6}, \frac{1}{18}, \ldots$

2. (A) $5, 20, 100, \ldots$ (B) $-5, -5, -5, \ldots$

   (C) $7, 6.5, 6, \ldots$ (D) $512, 256, 128, \ldots$

B

In Problems 3–8, determine whether the finite series is arithmetic, geometric, both, or neither. If the series is arithmetic or geometric, find its sum.

3. $\sum_{k=1}^{101} (-1)^{k+1}$ 4. $\sum_{k=1}^{200} 3$

5. $1 + \frac{1}{2} + \frac{1}{3} + \cdots + \frac{1}{50}$

6. $3 - 9 + 27 - \cdots - 3^{20}$

7. $5 + 4.9 + 4.8 + \cdots + 0.1$

8. $1 - \frac{1}{4} + \frac{1}{9} - \cdots - \frac{1}{100^2}$

Let $a_1, a_2, a_3, \ldots, a_n, \ldots$ be an arithmetic sequence. In Problems 9–14, find the indicated quantities.

9. $a_1 = 7; d = 4; a_2 = ?; a_3 = ?$

10. $a_1 = -2; d = -3; a_2 = ?; a_3 = ?$

11. $a_1 = 2; d = 4; a_{21} = ?; S_{31} = ?$

12. $a_1 = 8; d = -10; a_{15} = ?; S_{23} = ?$

13. $a_1 = 18; a_{20} = 75; S_{20} = ?$

14. $a_1 = 203; a_{30} = 261; S_{30} = ?$

Let $a_1, a_2, a_3, \ldots, a_n, \ldots$ be a geometric sequence. In Problems 15–24, find the indicated quantities.

15. $a_1 = 3; r = -2; a_2 = ?; a_3 = ?; a_4 = ?$

16. $a_1 = 32; r = -\frac{1}{2}; a_2 = ?; a_3 = ?; a_4 = ?$

17. $a_1 = 1; a_7 = 729; r = -3; S_7 = ?$

18. $a_1 = 3; a_7 = 2,187; r = 3; S_7 = ?$

19. $a_1 = 100; r = 1.08; a_{10} = ?$

20. $a_1 = 240; r = 1.06; a_{12} = ?$

21. $a_1 = 100; a_9 = 200; r = ?$

22. $a_1 = 100; a_{10} = 300; r = ?$

23. $a_1 = 500; r = 0.6; S_{10} = ?; S_\infty = ?$

24. $a_1 = 8,000; r = 0.4; S_{10} = ?; S_\infty = ?$

25. $S_{41} = \sum_{k=1}^{41} (3k + 3) = ?$ 26. $S_{50} = \sum_{k=1}^{50} (2k - 3) = ?$

27. $S_8 = \sum_{k=1}^{8} (-2)^{k-1} = ?$ 28. $S_8 = \sum_{k=1}^{8} 2^k = ?$

29. Find the sum of all the odd integers between 12 and 68.

30. Find the sum of all the even integers between 23 and 97.

31. Find the sum of each infinite geometric sequence (if it exists).

   (A) $2, 4, 8, \ldots$ (B) $2, -\frac{1}{2}, \frac{1}{8}, \ldots$

32. Repeat Problem 31 for:

   (A) $16, 4, 1, \ldots$ (B) $1, -3, 9, \ldots$

C

33. Find $f(1) + f(2) + f(3) + \cdots + f(50)$ if $f(x) = 2x - 3$.

34. Find $g(1) + g(2) + g(3) + \cdots + g(100)$ if $g(t) = 18 - 3t$.

35. Find $f(1) + f(2) + \cdots + f(10)$ if $f(x) = \left(\frac{1}{2}\right)^x$.

36. Find $g(1) + g(2) + \cdots + g(10)$ if $g(x) = 2^x$.

37. Show that the sum of the first n odd positive integers is n^2, using appropriate formulas from this section.

38. Show that the sum of the first n even positive integers is $n + n^2$, using formulas in this section.

39. If $r = 1$, neither the first form nor the second form for the sum of a finite geometric series is valid. Find a formula for the sum of a finite geometric series if $r = 1$.

40. If all of the terms of an infinite geometric series are less than 1, could the sum be greater than 1,000? Explain.

41. Does there exist a finite arithmetic series with $a_1 = 1$ and $a_n = 1.1$ that has sum equal to 100? Explain.

42. Does there exist a finite arithmetic series with $a_1 = 1$ and $a_n = 1.1$ that has sum equal to 105? Explain.

43. Does there exist an infinite geometric series with $a_1 = 10$ that has sum equal to 6? Explain.

44. Does there exist an infinite geometric series with $a_1 = 10$ that has sum equal to 5? Explain.

Applications

45. **Loan repayment.** If you borrow $4,800 and repay the loan by paying $200 per month to reduce the loan and 1% of the unpaid balance each month for the use of the money, what is the total cost of the loan over 24 months?

46. **Loan repayment.** If you borrow $5,400 and repay the loan by paying $300 per month to reduce the loan and 1.5% of the unpaid balance each month for the use of the money, what is the total cost of the loan over 18 months?

47. **Economy stimulation.** The government, through a subsidy program, distributes $5,000,000. If we assume that each person or agency spends 70% of what is received, and 70% of

this is spent, and so on, how much total increase in spending results from this government action? (Let $a_1 = \$3,500,000$.)

48. **Economy stimulation.** Due to reduced taxes, a person has an extra \$1,200 in spendable income. If we assume that the person spends 65% of this on consumer goods, and the producers of these goods in turn spend 65% on consumer goods, and that this process continues indefinitely, what is the total amount spent (to the nearest dollar) on consumer goods?

49. **Compound interest.** If \$1,000 is invested at 5% compounded annually, the amount A present after n years forms a geometric sequence with common ratio $1 + 0.05 = 1.05$. Use a geometric sequence formula to find the amount A in the account (to the nearest cent) after 10 years. After 20 years. (*Hint*: Use a time line.)

50. **Compound interest.** If \$$P$ is invested at $100r$% compounded annually, the amount A present after n years forms a geometric sequence with common ratio $1 + r$. Write a formula for the amount present after n years. (*Hint*: Use a time line.)

B-3 Binomial Theorem

- Factorial
- Development of the Binomial Theorem

The binomial form

$$(a + b)^n$$

where n is a natural number, appears more frequently than you might expect. The coefficients in the expansion play an important role in probability studies. The *binomial formula,* which we will derive informally, enables us to expand $(a + b)^n$ directly for n any natural number. Since the formula involves *factorials*, we digress for a moment here to introduce this important concept.

Factorial

For n a natural number, **n factorial**, denoted by $n!$, is the product of the first n natural numbers. **Zero factorial** is defined to be 1. That is,

> **DEFINITION** *n* **Factorial**
>
> $$n! = n \cdot (n - 1) \cdot \cdots \cdot 2 \cdot 1$$
> $$1! = 1$$
> $$0! = 1$$

It is also useful to note that $n!$ can be defined recursively.

> **DEFINITION** *n* **Factorial—Recursive Definition**
>
> $$n! = n \cdot (n - 1)! \qquad n \geq 1$$

EXAMPLE 1 **Factorial Forms** Evaluate.

(A) $5! = 5 \cdot 4 \cdot 3 \cdot 2 \cdot 1 = 120$

(B) $\dfrac{8!}{7!} = \dfrac{8 \cdot \cancel{7!}}{\cancel{7!}} = 8$

(C) $\dfrac{10!}{7!} = \dfrac{10 \cdot 9 \cdot 8 \cdot \cancel{7!}}{\cancel{7!}} = 720$

Matched Problem 1   Evaluate.

(A) $4!$ (B) $\dfrac{7!}{6!}$ (C) $\dfrac{8!}{5!}$

The following formula involving factorials has applications in many areas of mathematics and statistics. We will use this formula to provide a more concise form for the expressions encountered later in this discussion.

THEOREM 1   For n and r integers satisfying $0 \le r \le n$,

$$C_{n,r} = \frac{n!}{r!(n-r)!}$$

EXAMPLE 2   Evaluating $C_{n,r}$

(A) $C_{9,2} = \dfrac{9!}{2!(9-2)!} = \dfrac{9!}{2!7!} = \dfrac{9 \cdot 8 \cdot \cancel{7!}}{2 \cdot \cancel{7!}} = 36$

(B) $C_{5,5} = \dfrac{5!}{5!(5-5)!} = \dfrac{5!}{5!0!} = \dfrac{5!}{5!} = 1$

Matched Problem 2   Find

(A) $C_{5,2}$ (B) $C_{6,0}$

Development of the Binomial Theorem

Let us expand $(a + b)^n$ for several values of n to see if we can observe a pattern that leads to a general formula for the expansion for any natural number n:

$$(a + b)^1 = a + b$$
$$(a + b)^2 = a^2 + 2ab + b^2$$
$$(a + b)^3 = a^3 + 3a^2b + 3ab^2 + b^3$$
$$(a + b)^4 = a^4 + 4a^3b + 6a^2b^2 + 4ab^3 + b^4$$
$$(a + b)^5 = a^5 + 5a^4b + 10a^3b^2 + 10a^2b^3 + 5ab^4 + b^5$$

CONCEPTUAL INSIGHT

1. The expansion of $(a + b)^n$ has $(n + 1)$ terms.
2. The power of a decreases by 1 for each term as we move from left to right.
3. The power of b increases by 1 for each term as we move from left to right.
4. In each term, the sum of the powers of a and b always equals n.
5. Starting with a given term, we can get the coefficient of the next term by multiplying the coefficient of the given term by the exponent of a and dividing by the number that represents the position of the term in the series of terms. For example, in the expansion of $(a + b)^4$ above, the coefficient of the third term is found from the second term by multiplying 4 and 3, and then dividing by 2 [that is, the coefficient of the third term $= (4 \cdot 3)/2 = 6$].

We now postulate these same properties for the general case:

$$(a + b)^n = a^n + \frac{n}{1}a^{n-1}b + \frac{n(n-1)}{1 \cdot 2}a^{n-2}b^2 + \frac{n(n-1)(n-2)}{1 \cdot 2 \cdot 3}a^{n-3}b^3 + \cdots + b^n$$

$$= \frac{n!}{0!(n-0)!}a^n + \frac{n!}{1!(n-1)!}a^{n-1}b + \frac{n!}{2!(n-2)!}a^{n-2}b^2 + \frac{n!}{3!(n-3)!}a^{n-3}b^3 + \cdots + \frac{n!}{n!(n-n)!}b^n$$

$$= C_{n,0}a^n + C_{n,1}a^{n-1}b + C_{n,2}a^{n-2}b^2 + C_{n,3}a^{n-3}b^3 + \cdots + C_{n,n}b^n$$

And we are led to the formula in the binomial theorem:

> **THEOREM 2 Binomial Theorem**
> For all natural numbers n,
> $$(a + b)^n = C_{n,0}a^n + C_{n,1}a^{n-1}b + C_{n,2}a^{n-2}b^2 + C_{n,3}a^{n-3}b^3 + \cdots + C_{n,n}b^n$$

EXAMPLE 3 **Using the Binomial Theorem** Use the binomial theorem to expand $(u + v)^6$.

SOLUTION
$$(u + v)^6 = C_{6,0}u^6 + C_{6,1}u^5v + C_{6,2}u^4v^2 + C_{6,3}u^3v^3 + C_{6,4}u^2v^4 + C_{6,5}uv^5 + C_{6,6}v^6$$
$$= u^6 + 6u^5v + 15u^4v^2 + 20u^3v^3 + 15u^2v^4 + 6uv^5 + v^6$$

Matched Problem 3 Use the binomial theorem to expand $(x + 2)^5$.

EXAMPLE 4 **Using the Binomial Theorem** Use the binomial theorem to find the sixth term in the expansion of $(x - 1)^{18}$.

SOLUTION
$$\text{Sixth term} = C_{18,5}x^{13}(-1)^5 = \frac{18!}{5!(18 - 5)!}x^{13}(-1)$$
$$= -8{,}568x^{13}$$

Matched Problem 4 Use the binomial theorem to find the fourth term in the expansion of $(x - 2)^{20}$.

Exercises B-3

A

In Problems 1–20, evaluate each expression.

1. $6!$ **2.** $7!$ **3.** $\dfrac{10!}{9!}$

4. $\dfrac{20!}{19!}$ **5.** $\dfrac{12!}{9!}$ **6.** $\dfrac{10!}{6!}$

7. $\dfrac{5!}{2!3!}$ **8.** $\dfrac{7!}{3!4!}$ **9.** $\dfrac{6!}{5!(6 - 5)!}$

10. $\dfrac{7!}{4!(7 - 4)!}$ **11.** $\dfrac{20!}{3!17!}$ **12.** $\dfrac{52!}{50!2!}$

B

13. $C_{5,3}$ **14.** $C_{7,3}$ **15.** $C_{6,5}$ **16.** $C_{7,4}$

17. $C_{5,0}$ **18.** $C_{5,5}$ **19.** $C_{18,15}$ **20.** $C_{18,3}$

Expand each expression in Problems 21–26 using the binomial theorem.

21. $(a + b)^4$ **22.** $(m + n)^5$

23. $(x - 1)^6$ **24.** $(u - 2)^5$

25. $(2a - b)^5$ **26.** $(x - 2y)^5$

Find the indicated term in each expansion in Problems 27–32.

27. $(x - 1)^{18}$; 5th term   **28.** $(x - 3)^{20}$; 3rd term

29. $(p + q)^{15}$; 7th term   **30.** $(p + q)^{15}$; 13th term

31. $(2x + y)^{12}$; 11th term   **32.** $(2x + y)^{12}$; 3rd term

C

33. Show that $C_{n,0} = C_{n,n}$ for $n \geq 0$.

34. Show that $C_{n,r} = C_{n,n-r}$ for $n \geq r \geq 0$.

35. The triangle next is called **Pascal's triangle**. Can you guess what the next two rows at the bottom are? Compare these numbers with the coefficients of binomial expansions.

$$
\begin{array}{ccccccccc}
 & & & & 1 & & & & \\
 & & & 1 & & 1 & & & \\
 & & 1 & & 2 & & 1 & & \\
 & 1 & & 3 & & 3 & & 1 & \\
1 & & 4 & & 6 & & 4 & & 1
\end{array}
$$

36. Explain why the sum of the entries in each row of Pascal's triangle is a power of 2. (*Hint:* Let $a = b = 1$ in the binomial theorem.)

37. Explain why the alternating sum of the entries in each row of Pascal's triangle (e.g., $1 - 4 + 6 - 4 + 1$) is equal to 0.

38. Show that $C_{n,r} = \frac{n - r + 1}{r}C_{n,r-1}$ for $n \geq r \geq 1$.

39. Show that $C_{n,r-1} + C_{n,r} = C_{n+1,r}$ for $n \geq r \geq 1$.

Answers to Matched Problems

1. (A) 24   (B) 7   (C) 336

2. (A) 10   (B) 1

3. $x^5 + 10x^4 + 40x^3 + 80x^2 + 80x + 32$

4. $-9{,}120x^{17}$

APPENDIX C Tables

Table I Basic Geometric Formulas

1. Similar Triangles

   (A) Two triangles are similar if two angles of one triangle have the same measure as two angles of the other.

   (B) If two triangles are similar, their corresponding sides are proportional:

$$\frac{a}{a'} = \frac{b}{b'} = \frac{c}{c'}$$

2. Pythagorean Theorem

$$c^2 = a^2 + b^2$$

3. Rectangle

   $A = ab$ Area

   $P = 2a + 2b$ Perimeter

4. Parallelogram

   $h = $ height

   $A = ah = ab \sin \theta$ Area

   $P = 2a + 2b$ Perimeter

5. Triangle

   $h = $ height

   $A = \frac{1}{2}hc$ Area

   $P = a + b + c$ Perimeter

   $s = \frac{1}{2}(a + b + c)$ Semiperimeter

   $A = \sqrt{s(s - a)(s - b)(s - c)}$ Area: Heron's formula

6. Trapezoid

Base a is parallel to base b.

   $h = $ height

   $A = \frac{1}{2}(a + b)h$ Area

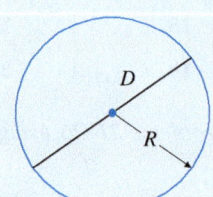

7. Circle

   $R = $ radius

   $D = $ diameter

   $D = 2R$

   $A = \pi R^2 = \frac{1}{4}\pi D^2$ Area

   $C = 2\pi R = \pi D$ Circumference

   $\dfrac{C}{D} = \pi$ For all circles

   $\pi \approx 3.14159$

8. Rectangular Solid

   $V = abc$ Volume

   $T = 2ab + 2ac + 2bc$ Total surface area

Table I   Basic Geometric Formulas   (continued)

9. Right Circular Cylinder

R = radius of base
h = height
$V = \pi R^2 h$ Volume
$S = 2\pi R h$ Lateral surface area
$T = 2\pi R(R + h)$ Total surface area

10. Right Circular Cone

R = radius of base
h = height
s = slant height
$V = \frac{1}{3}\pi R^2 h$ Volume
$S = \pi R s = \pi R \sqrt{R^2 + h^2}$ Lateral surface area
$T = \pi R(R + s) = \pi R(R + \sqrt{R^2 + h^2})$ Total surface area

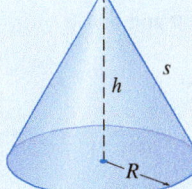

11. Sphere

R = radius
D = diameter
$D = 2R$
$V = \frac{4}{3}\pi R^3 = \frac{1}{6}\pi D^3$ Volume
$S = 4\pi R^2 = \pi D^2$ Surface area

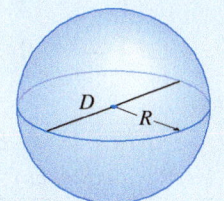

Table II   Integration Formulas

[*Note:* The constant of integration is omitted for each integral, but must be included in any particular application of a formula. The variable u is the variable of integration; all other symbols represent constants.]

Integrals Involving u^n

1. $\displaystyle\int u^n\,du = \frac{u^{n+1}}{n+1}, \quad n \neq -1$

2. $\displaystyle\int u^{-1}\,du = \int \frac{1}{u}\,du = \ln|u|$

Integrals Involving $a + bu, a \neq 0$ and $b \neq 0$

3. $\displaystyle\int \frac{1}{a+bu}\,du = \frac{1}{b}\ln|a+bu|$

4. $\displaystyle\int \frac{u}{a+bu}\,du = \frac{u}{b} - \frac{a}{b^2}\ln|a+bu|$

5. $\displaystyle\int \frac{u^2}{a+bu}\,du = \frac{(a+bu)^2}{2b^3} - \frac{2a(a+bu)}{b^3} + \frac{a^2}{b^3}\ln|a+bu|$

6. $\displaystyle\int \frac{u}{(a+bu)^2}\,du = \frac{1}{b^2}\left(\ln|a+bu| + \frac{a}{a+bu}\right)$

7. $\displaystyle\int \frac{u^2}{(a+bu)^2}\,du = \frac{(a+bu)}{b^3} - \frac{a^2}{b^3(a+bu)} - \frac{2a}{b^3}\ln|a+bu|$

8. $\displaystyle\int u(a+bu)^n\,du = \frac{(a+bu)^{n+2}}{(n+2)b^2} - \frac{a(a+bu)^{n+1}}{(n+1)b^2}, \quad n \neq -1, -2$

9. $\displaystyle\int \frac{1}{u(a+bu)}\,du = \frac{1}{a}\ln\left|\frac{u}{a+bu}\right|$

10. $\displaystyle\int \frac{1}{u^2(a+bu)}\,du = -\frac{1}{au} + \frac{b}{a^2}\ln\left|\frac{a+bu}{u}\right|$

11. $\displaystyle\int \frac{1}{u(a+bu)^2}\,du = \frac{1}{a(a+bu)} + \frac{1}{a^2}\ln\left|\frac{u}{a+bu}\right|$

12. $\displaystyle\int \frac{1}{u^2(a+bu)^2}\,du = -\frac{a+2bu}{a^2u(a+bu)} + \frac{2b}{a^3}\ln\left|\frac{a+bu}{u}\right|$

Integrals Involving $a^2 - u^2, a > 0$

13. $\displaystyle\int \frac{1}{u^2 - a^2}\,du = \frac{1}{2a}\ln\left|\frac{u-a}{u+a}\right|$

14. $\displaystyle\int \frac{1}{a^2 - u^2}\,du = \frac{1}{2a}\ln\left|\frac{u+a}{u-a}\right|$

Integrals Involving $(a + bu)$ and $(c + du), b \neq 0, d \neq 0$, and $ad - bc \neq 0$

15. $\displaystyle\int \frac{1}{(a+bu)(c+du)}\,du = \frac{1}{ad-bc}\ln\left|\frac{c+du}{a+bu}\right|$

16. $\displaystyle\int \frac{u}{(a+bu)(c+du)}\,du = \frac{1}{ad-bc}\left(\frac{a}{b}\ln|a+bu| - \frac{c}{d}\ln|c+du|\right)$

17. $\displaystyle\int \frac{u^2}{(a+bu)(c+du)}\,du = \frac{1}{bd}u - \frac{1}{ad-bc}\left(\frac{a^2}{b^2}\ln|a+bu| - \frac{c^2}{d^2}\ln|c+du|\right)$

18. $\displaystyle\int \frac{1}{(a+bu)^2(c+du)}\,du = \frac{1}{ad-bc}\frac{1}{a+bu} + \frac{d}{(ad-bc)^2}\ln\left|\frac{c+du}{a+bu}\right|$

19. $\displaystyle\int \frac{u}{(a+bu)^2(c+du)}\,du = -\frac{a}{b(ad-bc)}\frac{1}{a+bu} - \frac{c}{(ad-bc)^2}\ln\left|\frac{c+du}{a+bu}\right|$

20. $\displaystyle\int \frac{a+bu}{c+du}\,du = \frac{bu}{d} + \frac{ad-bc}{d^2}\ln|c+du|$

Table II   (continued)

Integrals Involving $\sqrt{a + bu}, a \neq 0$ and $b \neq 0$

21. $\displaystyle \int \sqrt{a + bu}\, du = \frac{2\sqrt{(a + bu)^3}}{3b}$

22. $\displaystyle \int u\sqrt{a + bu}\, du = \frac{2(3bu - 2a)}{15b^2}\sqrt{(a + bu)^3}$

23. $\displaystyle \int u^2\sqrt{a + bu}\, du = \frac{2(15b^2u^2 - 12abu + 8a^2)}{105b^3}\sqrt{(a + bu)^3}$

24. $\displaystyle \int \frac{1}{\sqrt{a + bu}}\, du = \frac{2\sqrt{a + bu}}{b}$

25. $\displaystyle \int \frac{u}{\sqrt{a + bu}}\, du = \frac{2(bu - 2a)}{3b^2}\sqrt{a + bu}$

26. $\displaystyle \int \frac{u^2}{\sqrt{a + bu}}\, du = \frac{2(3b^2u^2 - 4abu + 8a^2)}{15b^3}\sqrt{a + bu}$

27. $\displaystyle \int \frac{1}{u\sqrt{a + bu}}\, du = \frac{1}{\sqrt{a}}\ln\left|\frac{\sqrt{a + bu} - \sqrt{a}}{\sqrt{a + bu} + \sqrt{a}}\right|,\quad a > 0$

28. $\displaystyle \int \frac{1}{u^2\sqrt{a + bu}}\, du = -\frac{\sqrt{a + bu}}{au} - \frac{b}{2a\sqrt{a}}\ln\left|\frac{\sqrt{a + bu} - \sqrt{a}}{\sqrt{a + bu} + \sqrt{a}}\right|,\quad a > 0$

Integrals Involving $\sqrt{a^2 - u^2}, a > 0$

29. $\displaystyle \int \frac{1}{u\sqrt{a^2 - u^2}}\, du = -\frac{1}{a}\ln\left|\frac{a + \sqrt{a^2 - u^2}}{u}\right|$

30. $\displaystyle \int \frac{1}{u^2\sqrt{a^2 - u^2}}\, du = -\frac{\sqrt{a^2 - u^2}}{a^2 u}$

31. $\displaystyle \int \frac{\sqrt{a^2 - u^2}}{u}\, du = \sqrt{a^2 - u^2} - a\ln\left|\frac{a + \sqrt{a^2 - u^2}}{u}\right|$

Integrals Involving $\sqrt{u^2 + a^2}, a > 0$

32. $\displaystyle \int \sqrt{u^2 + a^2}\, du = \frac{1}{2}\left(u\sqrt{u^2 + a^2} + a^2\ln|u + \sqrt{u^2 + a^2}|\right)$

33. $\displaystyle \int u^2\sqrt{u^2 + a^2}\, du = \frac{1}{8}\left[u(2u^2 + a^2)\sqrt{u^2 + a^2} - a^4\ln|u + \sqrt{u^2 + a^2}|\right]$

34. $\displaystyle \int \frac{\sqrt{u^2 + a^2}}{u}\, du = \sqrt{u^2 + a^2} - a\ln\left|\frac{a + \sqrt{u^2 + a^2}}{u}\right|$

35. $\displaystyle \int \frac{\sqrt{u^2 + a^2}}{u^2}\, du = -\frac{\sqrt{u^2 + a^2}}{u} + \ln|u + \sqrt{u^2 + a^2}|$

36. $\displaystyle \int \frac{1}{\sqrt{u^2 + a^2}}\, du = \ln|u + \sqrt{u^2 + a^2}|$

37. $\displaystyle \int \frac{1}{u\sqrt{u^2 + a^2}}\, du = \frac{1}{a}\ln\left|\frac{u}{a + \sqrt{u^2 + a^2}}\right|$

38. $\displaystyle \int \frac{u^2}{\sqrt{u^2 + a^2}}\, du = \frac{1}{2}\left(u\sqrt{u^2 + a^2} - a^2\ln|u + \sqrt{u^2 + a^2}|\right)$

39. $\displaystyle \int \frac{1}{u^2\sqrt{u^2 + a^2}}\, du = -\frac{\sqrt{u^2 + a^2}}{a^2 u}$

(continued)

Table II (continued)

Integrals Involving $\sqrt{u^2 - a^2}, a > 0$

40. $\displaystyle\int \sqrt{u^2 - a^2} \, du = \frac{1}{2}\left(u\sqrt{u^2 - a^2} - a^2 \ln|u + \sqrt{u^2 - a^2}|\right)$

41. $\displaystyle\int u^2\sqrt{u^2 - a^2} \, du = \frac{1}{8}\left[u(2u^2 - a^2)\sqrt{u^2 - a^2} - a^4 \ln|u + \sqrt{u^2 - a^2}|\right]$

42. $\displaystyle\int \frac{\sqrt{u^2 - a^2}}{u^2} \, du = -\frac{\sqrt{u^2 - a^2}}{u} + \ln|u + \sqrt{u^2 - a^2}|$

43. $\displaystyle\int \frac{1}{\sqrt{u^2 - a^2}} \, du = \ln|u + \sqrt{u^2 - a^2}|$

44. $\displaystyle\int \frac{u^2}{\sqrt{u^2 - a^2}} \, du = \frac{1}{2}\left(u\sqrt{u^2 - a^2} + a^2 \ln|u + \sqrt{u^2 - a^2}|\right)$

45. $\displaystyle\int \frac{1}{u^2\sqrt{u^2 - a^2}} \, du = \frac{\sqrt{u^2 - a^2}}{a^2 u}$

Integrals Involving $e^{au}, a \neq 0$

46. $\displaystyle\int e^{au} \, du = \frac{e^{au}}{a}$

47. $\displaystyle\int u^n e^{au} \, du = \frac{u^n e^{au}}{a} - \frac{n}{a}\int u^{n-1} e^{au} \, du$

48. $\displaystyle\int \frac{1}{c + de^{au}} \, du = \frac{u}{c} - \frac{1}{ac}\ln|c + de^{au}|, \quad c \neq 0$

Integrals Involving ln u

49. $\displaystyle\int \ln u \, du = u \ln u - u$

50. $\displaystyle\int \frac{\ln u}{u} \, du = \frac{1}{2}(\ln u)^2$

51. $\displaystyle\int u^n \ln u \, du = \frac{u^{n+1}}{n+1}\ln u - \frac{u^{n+1}}{(n+1)^2}, \quad n \neq -1$

52. $\displaystyle\int (\ln u)^n \, du = u(\ln u)^n - n\int (\ln u)^{n-1} \, du$

Integrals Involving Trigonometric Functions of $au, a \neq 0$

53. $\displaystyle\int \sin au \, du = -\frac{1}{a}\cos au$

54. $\displaystyle\int \cos au \, du = \frac{1}{a}\sin au$

55. $\displaystyle\int \tan au \, du = -\frac{1}{a}\ln|\cos au|$

56. $\displaystyle\int \cot au \, du = \frac{1}{a}\ln|\sin au|$

57. $\displaystyle\int \sec au \, du = \frac{1}{a}\ln|\sec au + \tan au|$

58. $\displaystyle\int \csc au \, du = \frac{1}{a}\ln|\csc au - \cot au|$

59. $\displaystyle\int (\sin au)^2 \, du = \frac{u}{2} - \frac{1}{4a}\sin 2au$

60. $\displaystyle\int (\cos au)^2 \, du = \frac{u}{2} + \frac{1}{4a}\sin 2au$

61. $\displaystyle\int (\sin au)^n \, du = -\frac{1}{an}(\sin au)^{n-1}\cos au + \frac{n-1}{n}\int (\sin au)^{n-2} \, du, \quad n \neq 0$

62. $\displaystyle\int (\cos au)^n \, du = \frac{1}{an}\sin au(\cos au)^{n-1} + \frac{n-1}{n}\int (\cos au)^{n-2} \, du, \quad n \neq 0$

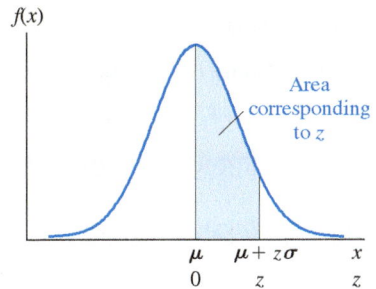

f(x)

Area corresponding to z

μ $\mu + z\sigma$ x
0   z   z

Table III   Area Under the Standard Normal Curve

(Table Entries Represent the Area Under the Standard Normal Curve from 0 to z, z $ 0)

z	.00	.01	.02	.03	.04	.05	.06	.07	.08	.09
0.0	0.0000	0.0040	0.0080	0.0120	0.0160	0.0199	0.0239	0.0279	0.0319	0.0359
0.1	0.0398	0.0438	0.0478	0.0517	0.0557	0.0596	0.0636	0.0675	0.0714	0.0753
0.2	0.0793	0.0832	0.0871	0.0910	0.0948	0.0987	0.1026	0.1064	0.1103	0.1141
0.3	0.1179	0.1217	0.1255	0.1293	0.1331	0.1368	0.1406	0.1443	0.1480	0.1517
0.4	0.1554	0.1591	0.1628	0.1664	0.1700	0.1736	0.1772	0.1808	0.1844	0.1879
0.5	0.1915	0.1950	0.1985	0.2019	0.2054	0.2088	0.2123	0.2157	0.2190	0.2224
0.6	0.2257	0.2291	0.2324	0.2357	0.2389	0.2422	0.2454	0.2486	0.2517	0.2549
0.7	0.2580	0.2611	0.2642	0.2673	0.2704	0.2734	0.2764	0.2794	0.2823	0.2852
0.8	0.2881	0.2910	0.2939	0.2967	0.2995	0.3023	0.3051	0.3078	0.3106	0.3133
0.9	0.3159	0.3186	0.3212	0.3238	0.3264	0.3289	0.3315	0.3340	0.3365	0.3389
1.0	0.3413	0.3438	0.3461	0.3485	0.3508	0.3531	0.3554	0.3577	0.3599	0.3621
1.1	0.3643	0.3665	0.3686	0.3708	0.3729	0.3749	0.3770	0.3790	0.3810	0.3830
1.2	0.3849	0.3869	0.3888	0.3907	0.3925	0.3944	0.3962	0.3980	0.3997	0.4015
1.3	0.4032	0.4049	0.4066	0.4082	0.4099	0.4115	0.4131	0.4147	0.4162	0.4177
1.4	0.4192	0.4207	0.4222	0.4236	0.4251	0.4265	0.4279	0.4292	0.4306	0.4319
1.5	0.4332	0.4345	0.4357	0.4370	0.4382	0.4394	0.4406	0.4418	0.4429	0.4441
1.6	0.4452	0.4463	0.4474	0.4484	0.4495	0.4505	0.4515	0.4525	0.4535	0.4545
1.7	0.4554	0.4564	0.4573	0.4582	0.4591	0.4599	0.4608	0.4616	0.4625	0.4633
1.8	0.4641	0.4649	0.4656	0.4664	0.4671	0.4678	0.4686	0.4693	0.4699	0.4706
1.9	0.4713	0.4719	0.4726	0.4732	0.4738	0.4744	0.4750	0.4756	0.4761	0.4767
2.0	0.4772	0.4778	0.4783	0.4788	0.4793	0.4798	0.4803	0.4808	0.4812	0.4817
2.1	0.4821	0.4826	0.4830	0.4834	0.4838	0.4842	0.4846	0.4850	0.4854	0.4857

Table III Area Under the Standard Normal Curve (continued)

(Table Entries Represent the Area Under the Standard Normal Curve from 0 to z, z $ 0)

z	.00	.01	.02	.03	.04	.05	.06	.07	.08	.09
2.2	0.4861	0.4864	0.4868	0.4871	0.4875	0.4878	0.4881	0.4884	0.4887	0.4890
2.3	0.4893	0.4896	0.4898	0.4901	0.4904	0.4906	0.4909	0.4911	0.4913	0.4916
2.4	0.4918	0.4920	0.4922	0.4925	0.4927	0.4929	0.4931	0.4932	0.4934	0.4936
2.5	0.4938	0.4940	0.4941	0.4943	0.4945	0.4946	0.4948	0.4949	0.4951	0.4952
2.6	0.4953	0.4955	0.4956	0.4957	0.4959	0.4960	0.4961	0.4962	0.4963	0.4964
2.7	0.4965	0.4966	0.4967	0.4968	0.4969	0.4970	0.4971	0.4972	0.4973	0.4974
2.8	0.4974	0.4975	0.4976	0.4977	0.4977	0.4978	0.4979	0.4979	0.4980	0.4981
2.9	0.4981	0.4982	0.4982	0.4983	0.4984	0.4984	0.4985	0.4985	0.4986	0.4986
3.0	0.4987	0.4987	0.4987	0.4988	0.4988	0.4989	0.4989	0.4989	0.4990	0.4990
3.1	0.4990	0.4991	0.4991	0.4991	0.4992	0.4992	0.4992	0.4992	0.4993	0.4993
3.2	0.4993	0.4993	0.4994	0.4994	0.4994	0.4994	0.4994	0.4995	0.4995	0.4995
3.3	0.4995	0.4995	0.4995	0.4996	0.4996	0.4996	0.4996	0.4996	0.4996	0.4997
3.4	0.4997	0.4997	0.4997	0.4997	0.4997	0.4997	0.4997	0.4997	0.4997	0.4998
3.5	0.4998	0.4998	0.4998	0.4998	0.4998	0.4998	0.4998	0.4998	0.4998	0.4998
3.6	0.4998	0.4998	0.4999	0.4999	0.4999	0.4999	0.4999	0.4999	0.4999	0.4999
3.7	0.4999	0.4999	0.4999	0.4999	0.4999	0.4999	0.4999	0.4999	0.4999	0.4999
3.8	0.4999	0.4999	0.4999	0.4999	0.4999	0.4999	0.4999	0.4999	0.4999	0.4999
3.9	0.5000	0.5000	0.5000	0.5000	0.5000	0.5000	0.5000	0.5000	0.5000	0.5000

ANSWERS

Diagnostic Algebra Test

1. (A) $(y + z)x$ (B) $(2 + x) + y$ (C) $2x + 3x$ *(A-1)*   **2.** $x^3 - 3x^2 + 4x + 8$ *(A-2)*   **3.** $x^3 + 3x^2 - 2x + 12$ *(A-2)*
4. $-3x^5 + 2x^3 - 24x^2 + 16$ *(A-2)*   **5.** (A) 1   (B) 1   (C) 2   (D) 3 *(A-2)*   **6.** (A) 3   (B) 1   (C) -3 (D) 1 *(A-2)*
7. $14x^2 - 30x$ *(A-2)*   **8.** $6x^2 - 5xy - 4y^2$ *(A-2)*   **9.** $4a^2 - 12ab + 9b^2$ *(A-2)*   **10.** $4xy - 2y^2$ *(A-2)*
11. $9x^6 - 12x^3y + 4y^2$ *(A-2)*   **12.** $x^3 - 6x^2y + 12xy^2 - 8y^3$ *(A-2)*   **13.** (A) 4.065×10^{12} (B) 7.3×10^{-3} *(A-5)*
14. (A) 255,000,000   (B) 0.000 406 *(A-5)*   **15.** (A) T   (B) F *(A-1)*   **16.** 0 and -3 are two examples of infinitely many. *(A-1)*
17. $6x^5y^{15}$ *(A-5)*   **18.** $3u^4/v^2$ *(A-5)*   **19.** 6×10^2 *(A-5)*   **20.** x^6/y^4 *(A-5)*   **21.** $u^{7/3}$ *(A-6)*   **22.** $3a^2/b$ *(A-6)*   **23.** $\frac{5}{9}$ *(A-5)*
24. $x + 2x^{1/2}y^{1/2} + y$ *(A-6)*   **25.** $6x + 7x^{1/2}y^{1/2} - 3y$ *(A-6)*   **26.** $(3x - 1)(4x + 3)$ *(A-3)*   **27.** $(4x - 3y)(2x - 3y)$ *(A-3)*
28. Not factorable relative to the integers *(A-3)*   **29.** $3n(2n - 5)(n + 1)$ *(A-3)*   **30.** $(x - y)(7x - y)$ *(A-3)*
31. $3x(2x + 1)(2 - x)$ *(A-3)*   **32.** $\dfrac{12a^3b - 40b^2 - 5a}{30a^3b^2}$ *(A-4)*   **33.** $\dfrac{7x - 4}{6x(x - 4)}$ *(A-4)*   **34.** $\dfrac{-8(x + 2)}{x(x - 4)(x + 4)}$ *(A-4)*
35. $2x + y$ *(A-4)*   **36.** $\dfrac{-1}{7(7 + h)}$ *(A-4)*   **37.** $\dfrac{xy}{y - x}$ *(A-6)*
38. (A) Subtraction   (B) Commutative (+)   (C) Distributive   (D) Associative ($\cdot$)   (E) Negatives   (F) Identity (+) *(A-1)*
39. $6x^{2/5} - 7(x - 1)^{3/4}$ *(A-6)*   **40.** $2\sqrt{x} - 3\sqrt[3]{x^2}$ *(A-6)*   **41.** $2 - \frac{3}{2}x^{-1/2}$ *(A-6)*   **42.** $\sqrt{3x}$ *(A-6)*   **43.** $\sqrt{x} + \sqrt{5}$ *(A-6)*
44. $\dfrac{1}{\sqrt{x - 5}}$ *(A-6)*   **45.** $\dfrac{1}{\sqrt{u + h} + \sqrt{u}}$ *(A-6)*   **46.** $x = 0, 5$ *(A-7)*   **47.** $x = \pm\sqrt{7}$ *(A-7)*   **48.** $x = -4, 5$ *(A-7)*   **49.** $x = 1, \frac{1}{6}$ *(A-7)*

Chapter 1

Exercises 1–1

1. $m = 5$ **3.** $x < -\dfrac{7}{2}$ **5.** $x \le 4$ **7.** $x < -3$ or $(-\infty, -3)$ **9.** $-1 \le x \le 2$

or $[-1, 2]$ **11.** $x = -\dfrac{3}{2}$ **13.** $y < -\dfrac{15}{2}$ **15.** $u = -\dfrac{3}{4}$ **17.** $x = 10$ **19.** $y \ge 3$ **21.** $x = 36$

23. $m < \dfrac{36}{7}$ **25.** $3 \le x < 7$ or $[3, 7)$ **27.** $-20 \le C \le 20$ or $[-20, 20]$

29. $y = \dfrac{3}{4}x - 3$ **31.** $y = -(A/B)x + (C/B) = (-Ax + C)/B$ **33.** $C = \dfrac{5}{9}(F - 32)$

35. $-2 < x \le 1$ or $(-2, 1]$ **37.** (A) and (C): $a > 0$ and $b > 0$, or $a < 0$ and $b < 0$ (B) and (D): $a > 0$
and $b < 0$, or $a < 0$ and $b > 0$ **39.** Negative   **41.** True   **43.** False   **45.** True   **47.** 4,500 \$35 tickets and 5,000 \$55 tickets
49. \$180,000 in Fund A and \$320,000 in Fund B   **51.** \$14,943   **53.** (A) \$420   (B) \$55   **55.** 34 rounds   **57.** \$32,000   **59.** 5,851
books   **61.** (B) 6,180 books   (C) At least \$11.50   **63.** 5,000   **65.** 12.6 yr   **67.** 66.7% to 100%

Exercises 1–2

1. (D)   **3.** (C)   **5.**   **7.**

9. Slope = 5; y intercept = -7 **11.** Slope = $-\dfrac{5}{2}$; y intercept = -9 **13.** Slope = $\dfrac{1}{4}$; y intercept = $\dfrac{2}{3}$ **15.** $y = 2x + 1$

17. $y = -\dfrac{1}{3}x + 6$ **19.** x intercept: -1; y intercept: -2; $y = -2x - 2$ **21.** x intercept: -3; y intercept: 1; $y = x/3 + 1$

23.   **25.**   **27.**   **29.** -4 **31.** $-\dfrac{3}{5}$ **33.** 2

35.   **37.**   **39.** 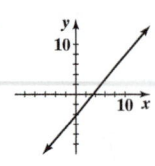   (B) x intercept: 3.5; y intercept: -4.2

(C)

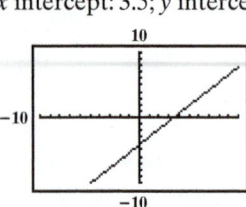

(D) x intercept: 3.5; y intercept: -4.2

41. $x = 4, y = -3$ **43.** $x = -1.5, y = -3.5$ **45.** $y = -4x + 5$

47. $y = \dfrac{3}{2}x + 1$ **49.** $y = 4.6$ **51.** (A) $m = \dfrac{2}{3}$ (B) $-2x + 3y = 11$ (C) $y = \dfrac{2}{3}x + \dfrac{11}{3}$ **53.** (A) $m = -\dfrac{5}{4}$

(B) $5x + 4y = -14$ (C) $y = -\dfrac{5}{4}x - \dfrac{7}{2}$ **55.** (A) Not defined   (B) $x = 5$ (C) None   **57.** (A) $m = 0$ (B) $y = 5$ (C) $y = 5$

59. The graphs have the same
y intercept, $(0, 2)$.

61. $C = 124 + 0.12x$; 1,050 donuts

63. (A) $C = 75x + 1,647$
(B)

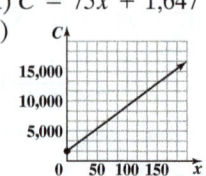

(C) The y intercept, \$1,647, is the
fixed cost and the slope, \$75, is
the cost per club.

65. (A) $R = 1.4C - 7$
(B) \$137

67. (A) $V = -7,500t + 157,000$
(B) \$112,000
(C) During the 12th year
(D)

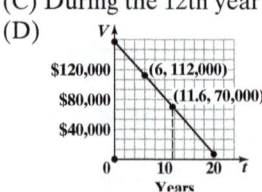

69. (A) $T = -1.84x + 212$
(B) $205.56°F$
(C) 6,522 ft
(D)

71. (A) $T = 70 - 3.6A$
(B) 10,000 ft

73. (A) $N = -0.0075t + 2.76$
(B) 2.46 persons

75. (A) $f = -0.5t + 21$
(B) 2022

77. (A) $p = 0.000225x + 0.5925$
(B) $p = -0.0009x + 9.39$
(C) $(7,820, 2.352)$
(D)

79. (A) $s = \dfrac{2}{5}w$

(B) 8 in
(C) 9 lb

81. (A) $y = 0.5x + 9$
(B) $y = 0.3x + 9$
(C)

(D) A: 39%; B: 27%

Exercises 1-3

1. (A) $w = 49 + 1.7h$ (B) The rate of change of weight with respect to height is 1.7 kg/in.   (C) 55.8 kg   (D) 5′6.5″
3. (A) $p = 0.445\overline{d} + 14.7$ (B) The rate of change of pressure with respect to depth is $0.4\overline{45}$ lb/in.^{2} per foot.   (C) 37 lb/in.^{2}
(D) 99 ft   **5.** (A) $a = 2,880 - 24t$ (B) -24 ft/sec   (C) 24 ft/sec   **7.** $s = 0.6t + 331$; the rate of change of the speed of sound with
respect to temperature is 0.6 m/s per °C.

9. (A)

(B) The rate of change of fossil fuel production is −0.3% per year.
(C) 74.1% of total production
(D) 2034

11. (A)

(B) 2023

13. (A)

(B) 16,200,000
(C) Undergraduate enrollment is increasing at a rate of 170,000 students per year.

15. (A)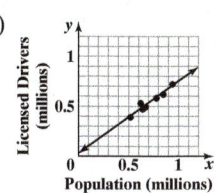

(B) 1,110,000
(C) 986,000

17. (A)

(B) $650 billion

19. (A)

(B) 2°F
(C) 22.75%

21. (A) The rate of change of height with respect to Dbh is 1.37 ft/in. (B) Height increases by approximately 1.37 ft. (C) 18 ft
(D) 20 in **23.** (A) The monthly price is increasing at a rate of $1.84 per year. (B) $67.22 **25.** (A) Male enrollment is increasing at a rate of 50,000 students per year; female enrollment is increasing at a rate of 170,000 students per year. (B) Male: 7.7 million; female: 12.2 million (C) 2023 **27.** Men: $y = -0.094x + 50.107$; women: $y = -0.076x + 55.512$; no **29.** Supply: $y = 0.2x + 0.87$; demand: $y = -0.15x + 3.5$; equilibrium price $= \$2.37$

Chapter 1 Review Exercises

1. $x = 2.8$ *(1-1)* **2.** $x = 2$ *(1-1)* **3.** $y = 1.8 - 0.4x$ *(1-1)* **4.** $x = \frac{4}{3}y + \frac{7}{3}$ *(1-1)*

5. $y < \frac{13}{4}$ or $\left(-\infty, \frac{13}{4}\right)$ ⟵————————➤ y *(1-1)*   **6.** $1 \le x < 3$ or $[1, 3)$ ————————➤ x *(1-1)*
$\frac{13}{4}$ 1 3

7. $x \ge \frac{9}{2}$ or $\left[\frac{9}{2}, \infty\right)$ ————————➤ x *(1-1)* **8.**   *(1-2)*
$\frac{9}{2}$

9. $2x + 3y = 12$ *(1-2)* **10.** x intercept $= 9$; y intercept $= -6$; slope $= \frac{2}{3}$ *(1-2)*

11. $y = -\frac{2}{3}x + 6$ *(1-2)* **12.** Vertical line: $x = -6$; horizontal line: $y = 5$ *(1-2)*

13. (A) $y = -\frac{2}{3}x$ (B) $y = 3$ *(1-2)* **14.** (A) $3x + 2y = 1$ (B) $y = 5$ (C) $x = -2$ *(1-2)* **15.** $x = \frac{25}{2}$ *(1-1)*

16. $u = 36$ *(1-1)* **17.** $x = \frac{30}{11}$ *(1-1)* **18.** $x = 21$ *(1-1)* **19.** $x = 4$ *(1-1)* **20.** $x < 4$ or $(-\infty, 4)$ *(1-1)* ⟵————————➤ x
                                                                                                                                              4

21. $x \ge 1$ or $[1, \infty)$ *(1-1)* ————————➤ x **22.** $x < -\frac{143}{17}$ or $\left(-\infty, -\frac{143}{17}\right)$ ⟵————————➤ x *(1-1)*
                                        1 $-\frac{143}{17}$

23. $1 < x \le 4$ or $(1, 4]$ ————————➤ x *(1-1)* **24.** $\frac{3}{8} \le x \le \frac{7}{8}$ or $\left[\frac{3}{8}, \frac{7}{8}\right]$ ————————➤ x *(1-1)*
        1 4 $\frac{3}{8}$ $\frac{7}{8}$

25. **(A)** *(1-2)* **26.** The graph of $x = -3$ is a vertical line with x intercept -3, and the graph of $y = 2$ is a horizontal line with y intercept 2. *(1-2)*

27. (A) An oblique line through the origin with slope $-3/4$ (B) A vertical line with x intercept $-4/3$ (C) The x axis   (D) An oblique line with x intercept 12 and y intercept 9 *(1-2)*   **28.** $\dfrac{2A - bh}{h}$ *(1-1)*   **29.** $\dfrac{S - P}{St}$ *(1-1)*   **30.** $a < 0$ and b any real number *(1-1)*   **31.** Less than *(1-1)*   **32.** The graphs appear to be perpendicular to each other. (It can be shown that if the slopes of two slant lines are the negative reciprocals of each other, then the two lines are perpendicular.) *(1-2)*   **33.** \$75,000 *(1-1)*   **34.** 9,375 CDs *(1-1)*   **35.** (A) $m = 132 - 0.6x$ (B) $M = 187 - 0.85x$ (C) Between 120 and 170 beats per minute   (D) Between 102 and 144.5 beats per minute *(1-3)*   **36.** (A) $V = 224,000 - 15,500t$ (B) \$38,000 *(1-2)*   **37.** (A) $R = 1.6C$ (B) \$192   (C) \$110   (D) The slope is 1.6. This is the rate of change of retail price with respect to cost. *(1-2)*   **38.** \$400; \$800 *(1-1)*   **39.** Demand: $p = 5.24 - 0.00125x$; 1,560 bottles *(1-2)*   **40.** (A)   (B) $-30°F$ (C) 45% *(1-3)*

41. (A) The dropout rate is decreasing at a rate of 0.164 percentage points per year.   (B)   (C) 2016 *(1-3)*

42. (A) The CPI is increasing at a rate of 4.66 units per year.   (B) 264.5 *(1-3)*   **43.** (A) The rate of change of tree height with respect to Dbh is 0.74.   (B) Tree height increases by about 0.74 ft.   (C) 21 ft   (D) 16 in. *(1-3)*

Chapter 2

Exercises 2-1

1.   **3.**   **5.**   **7.**

9. A function   **11.** Not a function
13. A function   **15.** A function   **17.** Not a function
19. A function   **21.** Linear   **23.** Neither
25. Constant   **27.** Linear

29.   **31.**   **33.**   **35.**   **37.**

39. $y = 0$ **41.** $y = 4$ **43.** $x = -5, 0, 4$ **45.** $x = -6$ **47.** 7   **49.** 48   **51.** 25   **53.** 22   **55.** All real numbers
57. All real numbers except -4 **59.** $x \leq 7$ **61.** A function with domain R **63.** Not a function; for example, when $x = 1$, $y = \pm 3$
65. A function with domain all real numbers except $x = 4$ **67.** Not a function; for example, when $x = 4$, $y = \pm 3$
69. $4x^2 - 1$ **71.** $x^2 - 6x + 8$ **73.** $15 + 8h + h^2$ **75.** $8h + h^2$ **77.** (A) $4x + 4h - 3$ (B) $4h$ (C) 4
79. (A) $4x^2 + 8xh + 4h^2 - 7x - 7h + 6$ (B) $8xh + 4h^2 - 7h$ (C) $8x + 4h - 7$ **81.** (A) $20x + 20h - x^2 - 2xh - h^2$
(B) $20h - 2xh - h^2$ (C) $20 - 2x - h$ **83.** $P(w) = 2w + \dfrac{50}{w}$, $w > 0$ **85.** $A(l) = l(50 - l), 0 < l < 50$

87. (A)

(B) \$54, \$42

89. (A) $R(x) = (75 - 3x)x$, $1 \leq x \leq 20$

(B)

x	$R(x)$
1	72
4	252
8	408
12	468
16	432
20	300

(C)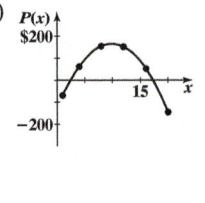

91. (A) $P(x) = 59x - 3x^2 - 125$, $1 \leq x \leq 20$

(B)

x	$P(x)$
1	-69
4	63
8	155
12	151
16	51
20	-145

(C)

93. (A) $V(x) = x(8 - 2x)(12 - 2x)$

(B) $0 < x < 4$

(C)

x	$V(x)$
1	60
2	64
3	36

(D)

95. (A) The graph indicates that there is a value of x near 2 that will produce a volume of 65.

(B) The table shows $x = 1.9$ to one decimal place:

x	1.8	1.9	2
$V(x)$	66.5	65.4	64

(C) $x = 1.93$ to two decimal places

97. $v = \dfrac{75 - w}{15 + w}$; 1.9032 cm/sec

Exercises 2-2

1. Domain: all real numbers; range: $[3, \infty)$ **3.** Domain: all real numbers; range: $(-\infty, 0]$ **5.** Domain: $[0, \infty)$; range: $[-10, \infty)$

7. Domain: all real numbers; range: all real numbers

9.

11.

13.

15.

17.

19.

21. The graph of $g(x) = -|x + 3|$ is the graph of $y = |x|$ reflected in the x axis and shifted 3 units to the left.

23. The graph of $f(x) = (x - 4)^2 - 3$ is the graph of $y = x^2$ shifted 4 units to the right and 3 units down.

25. The graph of $f(x) = 7 - \sqrt{x}$ is the graph of $y = \sqrt{x}$ reflected in the x axis and shifted 7 units up.

27. The graph of $h(x) = -3|x|$ is the graph of $y = |x|$ reflected in the x axis and vertically expanded by a factor of 3.

29. The graph of the basic function $y = x^2$ is shifted 2 units to the left and 3 units down. Equation: $y = (x + 2)^2 - 3$.

31. The graph of the basic function $y = x^2$ is reflected in the x axis and shifted 3 units to the right and 2 units up. Equation: $y = 2 - (x - 3)^2$.

33. The graph of the basic function $y = \sqrt{x}$ is reflected in the x axis and shifted 4 units up. Equation: $y = 4 - \sqrt{x}$.

35. The graph of the basic function $y = x^3$ is shifted 2 units to the left and 1 unit down. Equation: $y = (x + 2)^3 - 1$.

37. $g(x) = \sqrt{x - 2} - 3$

39. $g(x) = -|x + 3|$

41. $g(x) = -(x - 2)^3 - 1$

43.

45.

47.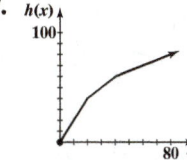

49. The graph of the basic function $y = |x|$ is reflected in the x axis and vertically contracted by a factor of 0.5. Equation: $y = -0.5|x|$.

51. The graph of the basic function $y = x^2$ is reflected in the x axis and vertically expanded by a factor of 2. Equation: $y = -2x^2$.

53. The graph of the basic function $y = \sqrt[3]{x}$ is reflected in the x axis and vertically expanded by a factor of 3. Equation: $y = -3\sqrt[3]{x}$.

55. Reversing the order does not change the result.

57. Reversing the order can change the result.

59. Reversing the order can change the result.

61. (A) The graph of the basic function $y = \sqrt{x}$ is reflected in the x axis, vertically expanded by a factor of 4, and shifted up 115 units.

(B)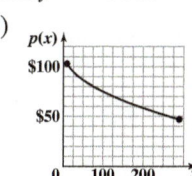

63. (A) The graph of the basic function is vertically contracted by a factor of 0.000 48 and shifted right 500 units and up 60,000 units.

(B)

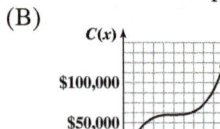

65. (A) $S(x) = \begin{cases} 8.5 + 0.065x & \text{if } 0 \leq x \leq 700 \\ -9 + 0.09x & \text{if } x > 700 \end{cases}$

(B)

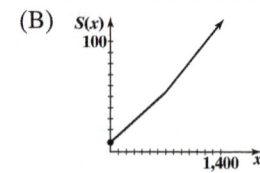

67. (A) $T(x) = \begin{cases} 0.035x & \text{if } 0 \leq x \leq 30,000 \\ 0.0625x - 825 & \text{if } 30,000 < x \leq 60,000 \\ 0.0645x - 945 & \text{if } x > 60,000 \end{cases}$

(B)

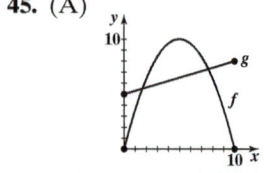

(C) $1,675; $3,570

69. (A) The graph of the basic function is vertically expanded by a factor of 5.5 and shifted down 220 units.

(B)

71. (A) The graph of the basic function is vertically expanded by a factor of 7.08.

(B)

Exercises 2-3

1. $(x - 2)^2 - 1$ **3.** $-(x - 3)^2 + 5$ **5.** The graph of $f(x)$ is the graph of $y = x^2$ shifted right 2 units and down 1 unit.

7. The graph of $m(x)$ is the graph of $y = x^2$ reflected in the x axis, then shifted right 3 units and up 5 units.

9. (A) m (B) g (C) f (D) n

11. (A) x intercepts: 1, 3; y intercept: -3 (B) Vertex: $(2, 1)$ (C) Maximum: 1 (D) Range: $y \leq 1$ or $(-\infty, 1]$

13. (A) x intercepts: $-3, -1$; y intercept: 3 (B) Vertex: $(-2, -1)$ (C) Minimum: -1 (D) Range: $y \geq -1$ or $[-1, \infty)$

15. (A) x intercepts: $3 \pm \sqrt{2}$; y intercept: -7 (B) Vertex: $(3, 2)$ (C) Maximum: 2 (D) Range: $y \leq 2$ or $(-\infty, 2]$

17. (A) x intercepts: $-1 \pm \sqrt{2}$; y intercept: -1 (B) Vertex: $(-1, -2)$ (C) Minimum: -2 (D) Range: $y \geq -2$ or $[-2, \infty)$

19. $y = -[x - (-2)]^2 + 5$ or $y = -(x + 2)^2 + 5$ **21.** $y = (x - 1)^2 - 3$

23. Vertex form: $(x - 4)^2 - 4$ (A) x intercepts: 2, 6; y intercept: 12 (B) Vertex: $(4, -4)$ (C) Minimum: -4
(D) Range: $y \geq -4$ or $[-4, \infty)$

25. Vertex form: $-4(x - 2)^2 + 1$ (A) x intercepts: 1.5, 2.5; y intercept: -15 (B) Vertex: $(2, 1)$ (C) Maximum: 1
(D) Range: $y \leq 1$ or $(-\infty, 1]$

27. Vertex form: $0.5(x - 2)^2 + 3$ (A) x intercepts: none; y intercept: 5 (B) Vertex: $(2, 3)$ (C) Minimum: 3
(D) Range: $y \geq 3$ or $[3, \infty)$

29. (A) $-4.87, 8.21$ (B) $-3.44, 6.78$ (C) No solution **31.** 651.0417

33. $g(x) = 0.25(x - 3)^2 - 9.25$ (A) x intercepts: $-3.08, 9.08$; y intercept: -7 (B) Vertex: $(3, -9.25)$
(C) Minimum: -9.25 (D) Range: $y \geq -9.25$ or $[-9.25, \infty)$

35. $f(x) = -0.12(x - 4)^2 + 3.12$ (A) x intercepts: $-1.1, 9.1$; y intercept: 1.2 (B) Vertex: $(4, 3.12)$ (C) Maximum: 3.12
(D) Range: $y \leq 3.12$ or $(-\infty, 3.12]$

37. $x = -5.37, 0.37$ **39.** $-1.37 < x < 2.16$ **41.** $x \leq -0.74$ or $x \geq 4.19$

43. Axis: $x = 2$; vertex: $(2, 4)$; range: $y \geq 4$ or $[4, \infty)$; no x intercepts

45. (A)

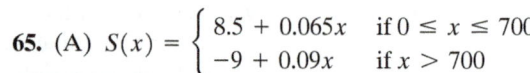

(B) 1.64, 7.61
(C) $1.64 < x < 7.61$
(D) $0 \leq x < 1.64$ or $7.61 < x \leq 10$

47. (A)

(B) 1.10, 5.57
(C) $1.10 < x < 5.57$
(D) $0 \leq x < 1.10$ or $5.57 < x \leq 8$

57. (A)

x	28	30	32	34	36
Mileage	45	52	55	51	47
$f(x)$	45.3	51.8	54.2	52.4	46.5

(B)

(C) $f(31) = 53.50$ thousand miles; $f(35) = 49.95$ thousand miles

61. (A)

(B) 12.5 (12,500,000 chips); $468,750,000
(C) $37.50

63. (A)

(B) 2,415,000 chips and 17,251,000 chips
(C) Loss: $1 \leq x < 2.415$ or $17.251 < x \leq 20$; profit: $2.415 < x < 17.251$

65. (A) $P(x) = 59x - 3x^2 - 125$

(C) Intercepts and break-even points: 2,415,000 chips and 17,251,000 chips

(E) Maximum profit is $165,083,000 at a production level of 9,833,000 chips. This is much smaller than the maximum revenue of $468,750,000.

67. $x = 0.14$ cm

69. 10.6 mph

Exercises 2–4

1. (A) 1 (B) −3/5 (C) 3 **3.** (A) 2 (B) 3,−3 (C) −9 **5.** (A) 3 (B) −3, 2, 5 (C) 30 **7.** (A) 2 (B) −4/3, 9/2 (C) −36
9. (A) 6 (B) None (C) 1 **11.** (A) 4 (B) Negative **13.** (A) 5 (B) Negative **15.** (A) 1 (B) Negative
17. (A) 6 (B) Positive **19.** 10 **21.** 1

23. (A) x intercept: −2; y intercept: −1
(B) Domain: all real numbers except 2
(C) Vertical asymptote: $x = 2$; horizontal asymptote: $y = 1$
(D)   (E)

25. (A) x intercept: 0; y intercept: 0
(B) Domain: all real numbers except −2
(C) Vertical asymptote: $x = -2$; horizontal asymptote: $y = 3$
(D)   (E)

27. (A) x intercept: 2; y intercept: −1
(B) Domain: all real numbers except 4
(C) Vertical asymptote: $x = 4$; horizontal asymptote: $y = -2$
(D)   (E)

29. (A)   (B)

31. (A)   (B)

33. $y = \dfrac{5}{6}$ **35.** $y = \dfrac{1}{4}$ **37.** $y = 0$ **39.** None **41.** $x = -1, x = 1, x = -3, x = 3$ **43.** $x = 5$ **45.** $x = -6, x = 6$

47. (A) x intercept: 0; y intercept: 0
(B) Vertical asymptotes: $x = -2, x = 3$; horizontal asymptote: $y = 2$
(C)   (D)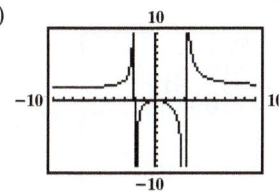

49. (A) x intercept: $\pm\sqrt{3}$; y intercept: $-\dfrac{2}{3}$
(B) Vertical asymptotes: $x = -3, x = 3$; horizontal asymptote: $y = -2$
(C)   (D)

51. (A) x intercept: 0; y intercept: 0   (B) Vertical asymptotes: $x = -3$, $x = 2$; horizontal asymptote: $y = 0$

(C)   (D)   **53.** $f(x) = x^2 - x - 2$ **55.** $f(x) = 4x - x^3$

57. (A) $C(x) = 180x + 200$ (B) $\overline{C}(x) = \dfrac{180x + 200}{x}$ **59.** (A) $\overline{C}(n) = \dfrac{2{,}500 + 175n + 25n^2}{n}$ (C) 10 yr; \$675.00 per year

(C)   (D) \$180 per board   (B)   (D) 10 yr; \$675.00 per year

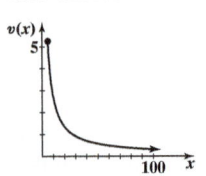

61. (A) $\overline{C}(x) = \dfrac{0.00048(x - 500)^3 + 60{,}000}{x}$

(B)

63. (A)

```
CubicReg
 y=ax³+bx²+cx+d
 a=-2.296296E-4
 b=.0083968254
 c=-.191005291
 d=17.83968254
```

(B) 8.9 lb

65. (A) 0.06 cm/sec

(B)

(C) 750 cases per month; \$90 per case

67. (A)

```
CubicReg
 y=ax³+bx²+cx+d
 a=1.3333333E-4
 b=-.0134285714
 c=.3238095238
 d=8.514285714
```

(B) 8.4

Exercises 2-5

1. (A) k **3.**   **5.**   **7.**   **9.**   **11.** The graph of g is the graph of f reflected in the x axis.
(B) g
(C) h **13.** The graph of g is the graph of f shifted 1 unit to the left.
(D) f

15. The graph of g is the graph of f shifted 1 unit up.

17. The graph of g is the graph of f vertically expanded by a factor of 2 and shifted to the left 2 units.

19. (A)   (B)   (C)   (D)   **21.**   **23.**

            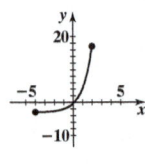

25.   **27.** $a = 1, -1$ **29.** $x = 1$ **39.**   **41.**   **43.** \$16,064.07

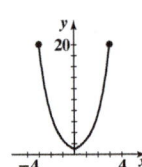   **31.** $x = -1, 6$ **33.** $x = 3$ **45.** (A) \$2,633.56   (B) \$7,079.54

35. $x = 3$ **37.** $x = -3, 0$ **47.** \$10,706

   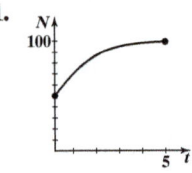   **49.** (A) \$10,553.57   (B) \$10,507.42

(C) \$10,525.03

51. N approaches 2 as t increases without bound.

53. (A) \$10,411,000

```
ExpReg
 y=a*b^x
 a=692.8121333
 b=1.094533635
```

(B) The model gives an average salary of \$1,710,000 in 2000. Inclusion of the data for 2000 gives an annual salary of \$10,721,000 in 2020.

55. (A) 10%   (B) 1%

57. (A) $P = 6.8e^{0.0114t}$
    (B) 2020: 7.7 billion; 2030: 8.6 billion

59. (A) 28,652,000

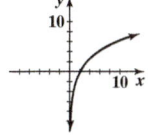

Exercises 2–6

1. $27 = 3^3$ **3.** $10^0 = 1$ **5.** $8 = 4^{3/2}$ **7.** $\log_7 49 = 2$ **9.** $\log_4 8 = \frac{3}{2}$ **11.** $\log_b A = u$ **13.** 0   **15.** 1   **17.** -3 **19.** 3

21. $\log_b P - \log_b Q$ **23.** $5\log_b L$ **25.** q^p **27.** $x = 9$ **29.** $y = 2$ **31.** $b = 10$ **33.** $x = 2$ **35.** False   **37.** True   **39.** True

41. False   **43.** $x = 2$ **45.** $x = 8$ **47.** $x = 7$ **49.** No solution

51.

53. The graph of $y = \log_2(x - 2)$ is the graph of $y = \log_2 x$ shifted to the right 2 units.

55. Domain: $(-1, \infty)$; range: all real numbers

57. (A) 3.547 43   (B) $-2.160\,32$ (C) 5.626 29   (D) $-3.197\,04$

59. (A) 13.4431   (B) 0.0089   (C) 16.0595   (D) 0.1514

61. 1.0792   **63.** 1.4595   **65.** 18.3559

67. Increasing: $(0, \infty)$

69. Decreasing: $(0, 1]$
Increasing: $[1, \infty)$

71. Increasing: $(-2, \infty)$

73. Increasing: $(0, \infty)$

75. Because $b^0 = 1$ for any permissible base $b (b > 0, b \neq 1)$.

77. $x > \sqrt{x} > \ln x$ for $1 < x \leq 16$

79. 4 yr   **81.** 9.87 yr; 9.80 yr

83. 7.51 yr

85. (A) 5,373

(B) 7,220

89. 160.1 bushels/acre

91. 887 yr

Chapter 2 Review Exercises

1.   *(2-1)*   **2.**   *(2-1)*   **3.**   *(2-1)*

4. (A) Not a function   (B) A function
    (C) A function   (D) Not a function *(2-1)*

5. (A) -2 (B) -8 (C) 0   (D) Not defined *(2-1)*

6. $v = \ln u$ *(2-6)*   **7.** $y = \log x$ *(2-6)*

8. $M = e^N$ *(2-6)*   **9.** $u = 10^v$ *(2-6)*

10. $x = 9$ *(2-6)*   **11.** $x = 6$ *(2-6)*   **12.** $x = 4$ *(2-6)*

13. $x = 2.157$ *(2-6)*   **14.** $x = 13.128$ *(2-6)*

15. $x = 1{,}273.503$ *(2-6)*   **16.** $x = 0.318$ *(2-6)*

17. (A) $y = 4$ (B) $x = 0$ (C) $y = 1$ (D) $x = -1$ or 1   (E) $y = -2$ (F) $x = -5$ or 5 *(2-1)*

18. (A)   (B)   (C)   (D)   *(2-2)*

19. $f(x) = -(x - 2)^2 + 4$. The graph of $f(x)$ is the graph of $y = x^2$ reflected in the x axis, then shifted right 2 units and up 4 units. *(2-2)*

20. (A) g (B) m (C) n (D) f *(2-2, 2-3)*   **21.** (A) x intercepts: -4, 0; y intercept: 0   (B) Vertex: $(-2, -4)$

    (C) Minimum: -4 (D) Range: $y \geq -4$ or $[-4, \infty)$ *(2-3)*   **22.** Quadratic *(2-3)*   **23.** Linear *(2-1)*   **24.** None *(2-1, 2-3)*

25. Constant *(2-1)*   **26.** $x = 8$ *(2-6)*   **27.** $x = 3$ *(2-6)*   **28.** $x = 3$ *(2-5)*   **29.** $x = -1, 3$ *(2-5)*   **30.** $x = 0, \frac{3}{2}$ *(2-5)*

31. $x = -2$ *(2-6)*   **32.** $x = \frac{1}{2}$ *(2-6)*   **33.** $x = 27$ *(2-6)*   **34.** $x = 13.3113$ *(2-6)*   **35.** $x = 158.7552$ *(2-6)*

36. $x = 0.0097$ *(2-6)*   **37.** $x = 1.4359$ *(2-6)*   **38.** $x = 1.4650$ *(2-6)*   **39.** $x = 92.1034$ *(2-6)*   **40.** $x = 9.0065$ *(2-6)*

41. $x = 2.1081$ *(2-6)*   **42.** (A) All real numbers except $x = -2$ and 3   (B) $x < 5$ *(2-1)*

43. Vertex form: $4\left(x + \frac{1}{2}\right)^2 - 4$; x intercepts: $-\frac{3}{2}$ and $\frac{1}{2}$; y intercept: -3; vertex: $\left(-\frac{1}{2}, -4\right)$; minimum: -4;

    range: $y \geq -4$ or $[-4, \infty)$ *(2-3)*   **44.** $(-1.54, -0.79)$; $(0.69, 0.99)$ *(2-5, 2-6)*

45. *(2-1)*

47. 6 *(2-1)*   **48.** −19 *(2-1)*   **49.** $10x − 4$ *(2-1)*   **50.** $21 − 5x$ *(2-1)*

51. (A) −1   (B) $−1 − 2h$ (C) $−2h$ (D) −2 *(2-1)*

52. (A) $a^2 − 3a + 1$ (B) $a^2 + 2ah + h^2 − 3a − 3h + 1$
(C) $2ah + h^2 − 3h$ (D) $2a + h − 3$ *(2-1)*

53. The graph of function m is the graph of $y = |x|$ reflected in the x axis and shifted to the right 4 units. *(2-2)*

54. The graph of function g is the graph of $y = x^3$ vertically contracted by a factor of 0.3 and shifted up 3 units. *(2-2)*

46. *(2-1)*

55. The graph of $y = x^2$ is vertically expanded by a factor of 2, reflected in the x axis, and shifted to the left 3 units. Equation: $y = −2(x + 3)^2$ *(2-2)*

56. $f(x) = 2\sqrt{x + 3} − 1$ *(2-2)*   **57.** $y = 0$ *(2-4)*   **58.** $y = \dfrac{3}{4}$ *(2-4)*   **59.** None *(2-4)*   **60.** $x = −10, x = 10$ *(2-4)*

61. $x = −2$ *(2-4)*   **62.** True *(2-3)*   **63.** False *(2-3)*   **64.** False *(2-3)*   **65.** True *(2-4)*

66. True *(2-5)*   **67.** True *(2-3)*   **68.** *(2-2)*   **69.** *(2-2)*

70. $y = −(x − 4)^2 + 3$ *(2-2, 2-3)*

71. $f(x) = −0.4(x − 4)^2 + 7.6$ (A) x intercepts: −0.4, 8.4; y intercept: 1.2
(B) Vertex: $(4.0, 7.6)$ (C) Maximum: 7.6   (D) Range: $y \le 7.6$ or $(−\infty, 7.6]$ *(2-3)*

72.

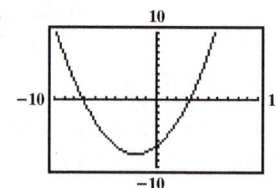

(A) x intercepts: −0.4, 8.4; y intercept: 1.2
(B) Vertex: $(4.0, 7.6)$ (C) Maximum: 7.6
(D) Range: $y \le 7.6$ or $(−\infty, 7.6]$ *(2-3)*

73. $\log 10^\pi = \pi$ and $10^{\log\sqrt{2}} = \sqrt{2}$; $\ln e^\pi = \pi$ and $e^{\ln\sqrt{2}} = \sqrt{2}$ *(2-6)*

74. $x = 2$ *(2-6)*   **75.** $x = 2$ *(2-6)*   **76.** $x = 1$ *(2-6)*   **77.** $x = 300$ *(2-6)*

78. $y = ce^{−5t}$ *(2-6)*

79. If $\log_1 x = y$, then $1^y = x$; that is, $1 = x$ for all positive real numbers x, which is not possible. *(2-6)*

80. The graph of $y = \sqrt[3]{x}$ is vertically expanded by a factor of 2, reflected in the x axis, and shifted 1 unit left and 1 unit down. Equation: $y = −2\sqrt[3]{x + 1} − 1$. *(2-2)*

81. $G(x) = 0.3(x + 2)^2 − 8.1$ (A) x intercepts: −7.2, 3.2; y intercept: −6.9
(B) Vertex: $(−2, −8.1)$ (C) Minimum: −8.1   (D) Range: $y \ge −8.1$ or $[−8.1, \infty)$
(2-3)

82.

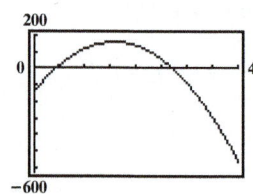

(A) x intercepts: −7.2, 3.2; y intercept: −6.9   (B) Vertex: $(−2, −8.1)$ (C) Minimum: −8.1
(D) Range: $y \ge −8.1$ or $[−8.1, \infty)$ *(2-3)*

83. (A) $S(x) = \begin{cases} 3 & \text{if } 0 \le x \le 20 \\ 0.057x + 1.86 & \text{if } 20 < x \le 200 \\ 0.0346x + 6.34 & \text{if } 200 < x \le 1{,}000 \\ 0.0217x + 19.24 & \text{if } x > 1{,}000 \end{cases}$

(B) *(2-2)*

84. \$6,521.89 *(2-5)*   **85.** \$6,362.50 *(2-5)*   **86.** 201 months ($\approx$16.7 years) *(2-5)*   **87.** 9.38 yr *(2-5)*

88. (A)

(B) $R = C$ for $x = 4.686$ thousand units (4,686 units) and for $x = 27.314$ thousand units (27,314 units); $R < C$ for $1 \le x < 4.686$ or $27.314 < x \le 40$; $R > C$ for $4.686 < x < 27.314$.
(C) Maximum revenue is 500 thousand dollars (\$500,000). This occurs at an output of 20 thousand units (20,000 units). At this output, the wholesale price is $p(20) = \$25$. *(2-3)*

89. (A) $P(x) = R(x) − C(x) = x(50 − 1.25x) − (160 + 10x)$

(B) $P = 0$ for $x = 4.686$ thousand units (4,686 units) and for $x = 27.314$ thousand units (27,314 units); $P < 0$ for $1 \le x < 4.686$ or $27.314 < x \le 40$; $P > 0$ for $4.686 < x < 27.314$.
(C) Maximum profit is 160 thousand dollars (\$160,000). This occurs at an output of 16 thousand units (16,000 units). At this output, the wholesale price is $p(16) = \$30$. *(2-3)*

90. (A) $A(x) = -\frac{3}{2}x^2 + 420x$

(B) Domain: $0 \leq x \leq 280$

(C)

(D) There are two solutions to the equation
$A(x) = 25,000$, one near 90 and another near 190.

(E) 86 ft; 194 ft

(F) Maximum combined area is 29,400 ft^2. This occurs for
$x = 140$ ft and $y = 105$ ft. *(2-3)*

91. (A) 2,833 sets   (B) 4,836

(D) Equilibrium price: \$131.59; equilibrium quantity: 3,587
cookware sets *(2-3)*

92. (A) 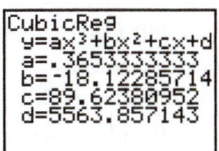   (B) 1,806 *(2-4)*

93. (A) $N = 2^{2t}$ or $N = 4^t$ (B) 15 days *(2-5)*   **94.** $k = 0.009\,42$; 489 ft *(2-6)*
95. (A) 5,569,000 *(2-6)*   **96.** 23.1 yr *(2-5)*   **97.** (A) \$931 billion   (B) 2018 *(2-5)*

Chapter 3

Exercises 3-1

1. 2 **3.** 1.25 **5.** (A) 2 (B) 2 (C) 2 (D) 2 **7.** (A) 1 (B) 2 (C) Does not exist (D) 2 (E) No **9.** 2 **11.** 0.5
13. (A) 1 (B) 2 (C) Does not exist (D) Does not exist (E) No **15.** (A) 1 (B) 1 (C) 1 (D) 3 (E) Yes
17. (A) −2 (B) −2 (C) −2 (D) 1 (E) Yes **19.** (A) 2 (B) 2 (C) 2 (D) Does not exist (E) Yes

21. 12 **23.** 1 **25.** −4 **27.** −1.5 **29.** 3 **31.** 15 **33.** −6 **35.** $\frac{7}{5}$ **37.** 3

39.   **41.**   **43.** (A) 1 (B) 1 (C) 1 (D) 1

45. (A) 2 (B) 1 (C) Does not exist (D) Does not exist
47. (A) −6 (B) Does not exist (C) 6
49. (A) 1 (B) −1 (C) Does not exist (D) Does not exist
51. (A) Does not exist (B) $\frac{1}{2}$ (C) $\frac{1}{4}$ **53.** (A) −5 (B) −3 (C) 0

55. (A) 0 (B) −1 (C) Does not exist **57.** (A) 1 (B) $\frac{1}{3}$ (C) $\frac{3}{4}$ **59.** False **61.** True **63.** False **65.** 3 **67.** 4

69. (A) $\lim_{x\to 1^-} f(x) = 2$ (B) $\lim_{x\to 1^-} f(x) = 3$ (C) $m = 1.5$ (D) The graph in (A) is broken when it jumps from
$\lim_{x\to 1^+} f(x) = 3$ $\lim_{x\to 1^+} f(x) = 2$ (1, 2) up to (1, 3). The graph in (B) is also
broken when it jumps down from (1, 3) to (1, 2).
The graph in (C) is one continuous piece, with
no breaks or jumps.

71. $2a$ **73.** $1/(2\sqrt{a})$

75. (A) $F(x) = \begin{cases} 0.99 & \text{if } 0 < x \leq 20 \\ 0.07x - 0.41 & \text{if } x \geq 20 \end{cases}$

(B)

(C) All 3 limits are 0.99.

79. (A) $D(x) = \begin{cases} x & \text{if } 0 \leq x < 300 \\ 0.97x & \text{if } 300 \leq x < 1,000 \\ 0.95x & \text{if } 1,000 \leq x < 3,000 \\ 0.93x & \text{if } 3,000 \leq x < 5,000 \\ 0.9x & \text{if } x \geq 5,000 \end{cases}$

(B) $\lim_{x\to 1,000} D(x)$ does not exist because
$\lim_{x\to 1,000^-} D(x) = 970$ and $\lim_{x\to 1,000^+} D(x) = 950$;
$\lim_{x\to 3,000} D(x)$ does not exist because
$\lim_{x\to 3,000^-} D(x) = 2,850$ and $\lim_{x\to 3,000^+} D(x) = 2,790$

81. $F(x) = \begin{cases} 20x & \text{if } 0 < x \le 4{,}000 \\ 80{,}000 & \text{if } x \ge 4{,}000 \end{cases}$

$\lim\limits_{x \to 4{,}000} F(x) = 80{,}000; \quad \lim\limits_{x \to 8{,}000} F(x) = 80{,}000$

83. $\lim\limits_{x \to 5} f(x)$ does not exist; $\quad \lim\limits_{x \to 10} f(x) = 0;$

$\lim\limits_{x \to 5} g(x) = 0; \quad \lim\limits_{x \to 10} g(x) = 1$

Exercises 3-2

1. -2 **3.** $-\infty$ **5.** Does not exist **7.** 0 **9.** (A) $-\infty$ (B) ∞ (C) Does not exist
11. (A) ∞ (B) ∞ (C) ∞ **13.** (A) 3 (B) 3 (C) 3 **15.** (A) $-\infty$ (B) ∞ (C) Does not exist
17. (A) ∞ (B) $-\infty$ **19.** (A) $-\infty$ (B) $-\infty$ **21.** $\lim\limits_{x \to -3^-} f(x) = -\infty; \lim\limits_{x \to -3^+} f(x) = \infty; x = -3$ is a vertical asymptote
23. $\lim\limits_{x \to -2^-} h(x) = \infty; \lim\limits_{x \to -2^+} h(x) = -\infty; \lim\limits_{x \to 2^-} h(x) = -\infty; \lim\limits_{x \to 2^+} h(x) = \infty; x = -2$ and $x = 2$ are vertical asymptotes
25. No zeros of denominator; no vertical asymptotes
27. $\lim\limits_{x \to 1^-} H(x) = -\infty; \lim\limits_{x \to 1^+} H(x) = \infty; \lim\limits_{x \to 3} H(x) = 2; x = 1$ is a vertical asymptote
29. $\lim\limits_{x \to 0} T(x) = -\infty; \lim\limits_{x \to 4} T(x) = \infty; x = 0$ and $x = 4$ are vertical asymptotes

31. (A) $\dfrac{47}{41} \approx 1.146$ (B) $\dfrac{407}{491} \approx 0.829$ (C) $\dfrac{4}{5} = 0.8$ **33.** (A) $\dfrac{2{,}011}{138} \approx 14.572$ (B) $\dfrac{12{,}511}{348} \approx 35.951$ (C) ∞

35. (A) $-\dfrac{8{,}568}{46{,}653} \approx -0.184$ (B) $-\dfrac{143{,}136}{1{,}492{,}989} \approx 0.096$ (C) 0

37. (A) $-\dfrac{7{,}010}{996} \approx -7.038$ (B) $-\dfrac{56{,}010}{7{,}996} \approx -7.005$ (C) -7

39. Horizontal asymptote: $y = 2$; vertical asymptote: $x = -2$
41. Horizontal asymptote: $y = 1$; vertical asymptotes: $x = -1$ and $x = 1$
43. No horizontal asymptotes; no vertical asymptotes **45.** Horizontal asymptote: $y = 0$; no vertical asymptotes
47. No horizontal asymptotes; vertical asymptote: $x = 3$
49. Horizontal asymptote: $y = 2$; vertical asymptotes: $x = -1$ and $x = 2$
51. Horizontal asymptote: $y = 2$; vertical asymptote: $x = -1$
53. False **55.** False **57.** True **59.** If $n \ge 1$ and $a_n > 0$, then the limit is ∞. If $n \ge 1$ and $a_n < 0$, then the limit is $-\infty$.
61. $\lim\limits_{x \to \infty} f(x) = \infty; \lim\limits_{x \to -\infty} f(x) = \infty$ **63.** $\lim\limits_{x \to \infty} f(x) = \infty; \lim\limits_{x \to -\infty} f(x) = -\infty$ **65.** $\lim\limits_{x \to \infty} f(x) = \infty; \lim\limits_{x \to -\infty} f(x) = -\infty$
67. $\lim\limits_{x \to \infty} f(x) = -\infty; \lim\limits_{x \to -\infty} f(x) = -\infty$

69. (A) $C(x) = 180x + 200$

(B) $\overline{C}(x) = \dfrac{180x + 200}{x}$

(C)

(D) $180 per board

71. (A) $C_e(x) = 950 + 56x; \overline{C}_e(x) = \dfrac{950}{x} + 56$

(B) $C_c(x) = 900 + 66x; \overline{C}_c(x) = \dfrac{900}{x} + 66$

(C) At $x = 5$ years
(D) At $x = 5$ years
(E) $\lim\limits_{x \to \infty} \overline{C}_e(x) = 56; \lim\limits_{x \to \infty} \overline{C}_c(x) = 66$

73. The long-term drug concentration is 5 mg/ml.
75. (A) $18 million
(B) $38 million
(C) $\lim\limits_{x \to 1^-} P(x) = \infty$

77. (C) $V_{\max} = 4, K_M = 20$

(D) $v(s) = \dfrac{4s}{20 + s}$

(E) $v = \dfrac{12}{7}$ when $s = 15$; $s = 60$ when $v = 3$

79. (A) $C_{\max} = 18, M = 150$

(B) $C(T) = \dfrac{18T}{150 + T}$
(C) $C = 14.4$ when $T = 600$ K; $T = 300$ K when $C = 12$

Exercises 3-3

1. f is continuous at $x = 1$, since $\lim_{x\to 1} f(x) = f(1)$.

3. f is discontinuous at $x = 1$, since $\lim_{x\to 1} f(x) \neq f(1)$.

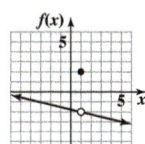

5. f is discontinuous at $x = 1$, since $\lim_{x\to 1} f(x)$ does not exist.

7. 1.9 **9.** 0.9 **11.** (A) 2 (B) 1 (C) Does not exist (D) 1 (E) No **13.** (A) 1 (B) 1 (C) 1 (D) 3 (E) No
15. 0.9 **17.** 2.05 **19.** (A) 1 (B) 1 (C) 1 (D) 3 (E) No **21.** (A) 2 (B) −1 (C) Does not exist (D) 2 (E) No

23. All x **25.** All x, except $x = -2$ **27.** All x, except $x = -4$ and $x = 1$ **29.** All x **31.** All x, except $x = \pm\frac{3}{2}$

33. (A)

(B) 1 (C) 2 (D) No (E) All integers
35. $-3 < x < 4$; $(-3, 4)$ **37.** $x < 3$ or $x > 7$; $(-\infty, 3) \cup (7, \infty)$
39. $x < -2$ or $0 < x < 2$; $(-\infty, -2) \cup (0, 2)$ **41.** $-5 < x < 0$ or $x > 3$; $(-5, 0) \cup (3, \infty)$
43. (A) $(-4, -2) \cup (0, 2) \cup (4, \infty)$ (B) $(-\infty, -4) \cup (-2, 0) \cup (2, 4)$
45. (A) $(-\infty, -2.5308) \cup (-0.7198, \infty)$ (B) $(-2.5308, -0.7198)$
47. (A) $(-\infty, -2.1451) \cup (-1, -0.5240) \cup (1, 2.6691)$
(B) $(-2.1451, -1) \cup (-0.5240, 1) \cup (2.6691, \infty)$

49. $[6, \infty)$ **51.** $(-\infty, \infty)$ **53.** $(-\infty, -3] \cup [3, \infty)$ **55.** $(-\infty, \infty)$

57. Since $\lim_{x\to 1^-} f(x) = 2$ and $\lim_{x\to 1^+} f(x) = 4$, $\lim_{x\to 1} f(x)$ does not exist and f is not continuous at $x = 1$.

59. This function is continuous for all x.

61. Since $\lim_{x\to 0} f(x) = 0$ and $f(0) = 1$, $\lim_{x\to 0} f(x) \neq f(0)$ and f is not continuous at $x = 0$.

63. (A) Yes (B) No (C) Yes (D) No (E) Yes **65.** True **67.** False **69.** True
71. x intercepts: $x = -5, 2$ **73.** x intercepts: $x = -6, -1, 4$ **75.** No, but this does not contradict Theorem 2, since f is discontinuous at $x = 1$.

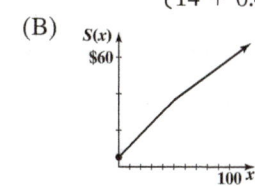

77. (A)
$$P(x) = \begin{cases} 0.44 & \text{if } 0 < x \le 1 \\ 0.61 & \text{if } 1 < x \le 2 \\ 0.78 & \text{if } 2 < x \le 3 \\ 0.95 & \text{if } 3 < x \le 3.5 \end{cases}$$

(B)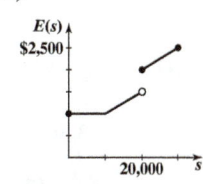

(C) Yes; no

81. (A) $S(x) = \begin{cases} 5 + 0.63x & \text{if } 0 \le x \le 50 \\ 14 + 0.45x & \text{if } 50 < x \end{cases}$

(B)

(C) Yes

83. (A)

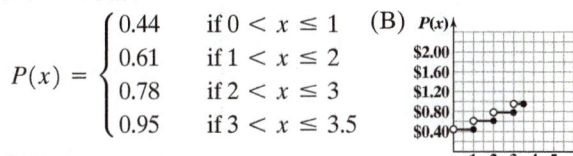

(B) $\lim_{s\to 10,000} E(s) = \$1{,}000$; $E(10{,}000) = \$1{,}000$
(C) $\lim_{s\to 20,000} E(s)$ does not exist; $E(20{,}000) = \$2{,}000$
(D) Yes; no

85. (A) t_2, t_3, t_4, t_6, t_7 (B) $\lim_{t\to t_5} N(t) = 7$; $N(t_5) = 7$ (C) $\lim_{t\to t_3} N(t)$ does not exist; $N(t_3) = 4$

Exercises 3-4

1. (A) -3; slope of the secant line through $(1, f(1))$ and $(2, f(2))$
(B) $-2 - h$; slope of the secant line through $(1, f(1))$ and $(1 + h, f(1 + h))$
(C) -2; slope of the tangent line at $(1, f(1))$
3. (A) 15 (B) 15 (C) $6 + 3h$ (D) 6 (E) 6 (F) 6 (G) $y = 6x - 3$
5. $f'(x) = 0$; $f'(1) = 0$, $f'(2) = 0$, $f'(3) = 0$ **7.** $f'(x) = 3$; $f'(1) = 3$, $f'(2) = 3$, $f'(3) = 3$
9. $f'(x) = -6x$; $f'(1) = -6$, $f'(2) = -12$, $f'(3) = -18$ **11.** $f'(x) = 2x + 6$; $f'(1) = 8$, $f'(2) = 10$, $f'(3) = 12$
13. $f'(x) = 4x - 7$; $f'(1) = -3$, $f'(2) = 1$, $f'(3) = 5$ **15.** $f'(x) = -2x + 4$; $f'(1) = 2$, $f'(2) = 0$, $f'(3) = -2$

17. $f'(x) = 6x^2; f'(1) = 6, f'(2) = 24, f'(3) = 54$ **19.** $f'(x) = -\dfrac{4}{x^2}; f'(1) = -4, f'(2) = -1, f'(3) = -\dfrac{4}{9}$

21. $f'(x) = \dfrac{3}{2\sqrt{x}}; f'(1) = \dfrac{3}{2}, f'(2) = \dfrac{3}{2\sqrt{2}}$ or $\dfrac{3\sqrt{2}}{4}, f'(3) = \dfrac{3}{2\sqrt{3}}$ or $\dfrac{\sqrt{3}}{2}$

23. $f'(x) = \dfrac{5}{\sqrt{x+5}}; f'(1) = \dfrac{5}{\sqrt{6}}$ or $\dfrac{5\sqrt{6}}{6}, f'(2) = \dfrac{5}{\sqrt{7}}$ or $\dfrac{5\sqrt{7}}{7}, f'(3) = \dfrac{5}{2\sqrt{2}}$ or $\dfrac{5\sqrt{2}}{4}$

25. $f'(x) = \dfrac{6}{(x+2)^2}; f'(1) = \dfrac{2}{3}, f'(2) = \dfrac{3}{8}, f'(3) = \dfrac{6}{25}$

27. (A) 5   (B) $3 + h$ (C) 3   (D) $y = 3x - 1$ **29.** (A) 5 m/s   (B) $3 + h$ m/s   (C) 3 m/s

31. Yes   **33.** No   **35.** Yes   **37.** Yes

39. (A) $f'(x) = 2x - 4$ (B) $-4, 0, 4$

(C)

41. $v = f'(x) = 8x - 2$; 6 ft/s, 22 ft/s, 38 ft/s

43. (A) The graphs of g and h are vertical translations of the graph of f. All three functions should have the same derivative.

(B) $2x$

45. True

47. False

49. False

51. f is nondifferentiable at $x = 1$

53. f is differentiable for all real numbers

55. No   **57.** No   **59.** $f'(0) = 0$ **61.** 6 s; 192 ft/s

63. (A) \$8.75

(B) $R'(x) = 60 - 0.05x$

(C) $R(1,000) = 35,000; R'(1,000) = 10$; At a production level of 1,000 car seats, the revenue is \$35,000 and is increasing at the rate of \$10 per seat.

65. (A) $S'(t) = 1/\sqrt{t + 10}$

(B) $S(15) = 10; S'(15) = 0.2$. After 15 months, the total sales are \$10 million and are increasing at the rate of \$0.2 million, or \$200,000, per month.

(C) The estimated total sales are \$10.2 million after 16 months and \$10.4 million after 17 months.

67. (A) $p'(t) = 328t + 161$

(B) $p(10) = 30,336; p'(10) = 413.4$; In 2015, 30,336 metric tons of tungsten are consumed and this quantity is increasing at the rate of 3,441 metric tons per year.

69. (A)

(B) $R(20) = 1,403.5$ billion kilowatts, $R'(20) = -10.7$ billion kilowatts. In 2020, 1,403.5 billion kilowatts will be sold and the amount sold is decreasing at the rate of 10.7 billion kilowatts per year.

71. (A) $P'(t) = 12 - 2t$

(B) $P(3) = 107; P'(3) = 6$. After 3 hours, the ozone level is 107 ppb and is increasing at the rate of 6 ppb per hour.

Exercises 3-5

1. 0   **3.** $9x^8$ **5.** $3x^2$ **7.** $-4x^{-5}$ **9.** $\dfrac{8}{3}x^{5/3}$ **11.** $-\dfrac{10}{x^{11}}$ **13.** $10x$ **15.** $2.8x^6$ **17.** $\dfrac{x^2}{6}$ **19.** 12   **21.** 2   **23.** 9   **25.** 2   **27.** $4t - 3$

29. $-10x^{-3} - 9x^{-2}$ **31.** $1.5u^{-0.7} - 8.8u^{1.2}$ **33.** $0.5 - 3.3t^2$ **35.** $-\dfrac{8}{5}x^{-5}$ **37.** $3x + \dfrac{14}{5}x^{-3}$ **39.** $-\dfrac{20}{9}w^{-5} + \dfrac{5}{3}w^{-2/3}$

41. $2u^{-1/3} - \dfrac{5}{3}u^{-2/3}$ **43.** $-\dfrac{9}{5}t^{-8/5} + 3t^{-3/2}$ **45.** $-\dfrac{1}{3}x^{-4/3}$ **47.** $-0.6x^{-3/2} + 6.4x^{-3} + 1$

49. (A) $f'(x) = 6 - 2x$ (B) $f'(2) = 2; f'(4) = -2$ (C) $y = 2x + 4; y = -2x + 16$ (D) $x = 3$

51. (A) $f'(x) = 12x^3 - 12x$ (B) $f'(2) = 72; f'(4) = 720$ (C) $y = 72x - 127; y = 720x - 2,215$ (D) $x = -1, 0, 1$

53. (A) $v = f'(x) = 176 - 32x$ (B) $f'(0) = 176$ ft/s; $f'(3) = 80$ ft/s   (C) 5.5 s

55. (A) $v = f'(x) = 3x^2 - 18x + 15$ (B) $f'(0) = 15$ ft/s; $f'(3) = -12$ ft/s   (C) $x = 1$ s, $x = 5$ s

57. $f'(x) = 2x - 3 - 2x^{-1/2} = 2x - 3 - \dfrac{2}{x^{1/2}}; x = 2.1777$ **59.** $f'(x) = 4\sqrt[3]{x} - 3x - 3; x = -2.9018$

61. $f'(x) = 0.2x^3 + 0.3x^2 - 3x - 1.6; x = -4.4607, -0.5159, 3.4765$

63. $f'(x) = 0.8x^3 - 9.36x^2 + 32.5x - 28.25; x = 1.3050$

69. $8x - 4$ **71.** $-20x^{-2}$ **73.** $-\dfrac{1}{4}x^{-2} + \dfrac{2}{3}x^{-3}$ **75.** True   **77.** False   **79.** True

81. (A) $S'(t) = 0.09t^2 + t + 2$

(B) $S(5) = 29.25, S'(5) = 9.25$. After 5 months, sales are \$29.25 million and are increasing at the rate of \$9.25 million per month.

(C) $S(10) = 103, S'(10) = 21$. After 10 months, sales are \$103 million and are increasing at the rate of \$21 million per month.

83. (A) $N'(x) = 3,780/x^2$
  (B) $N'(10) = 37.8$. At the \$10,000 level of advertising, sales are increasing at the rate of 37.8 boats per \$1,000 spent on advertising.
  $N'(20) = 9.45$. At the \$20,000 level of advertising, sales are increasing at the rate of 9.45 boats per \$1,000 spent on advertising.

85. (A)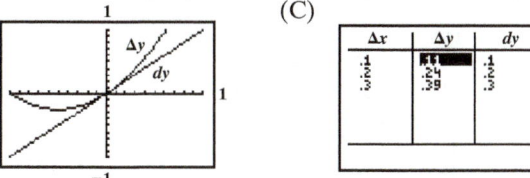
  (B) In 2016, 55.7% of male high-school graduates enroll in college and the percentage is decreasing at the rate of 1.9% per year.
87. (A) -1.37 beats/min
  (B) -0.58 beat/min
89. (A) 25 items/h
  (B) 8.33 items/h

Exercises 3-6

1. $\Delta x = 3$; $\Delta y = 45$; $\Delta y/\Delta x = 15$ **3.** 12   **5.** 12   **7.** $dy = (24x - 3x^2)\,dx$ **9.** $dy = \left(2x - \dfrac{x^2}{3}\right)dx$ **11.** $dy = -\dfrac{295}{x^{3/2}}\,dx$

13. (A) $12 + 3\Delta x$ (B) 12   **15.** $dy = (8x + 4)\,dx$ **17.** $dy = (1 - 9x^{-2})\,dx$ **19.** $dy = 1.4$; $\Delta y = 1.44$

21. $dy = -3$; $\Delta y = -3\frac{1}{3}$ **23.** 120 in.^{3}

25. (A) $\Delta y = \Delta x + (\Delta x)^2$; $dy = \Delta x$
  (B)
  (C)

27. (A) $\Delta y = -\Delta x + (\Delta x)^2 + (\Delta x)^3$; $dy = -\Delta x$
  (B)
  (C)

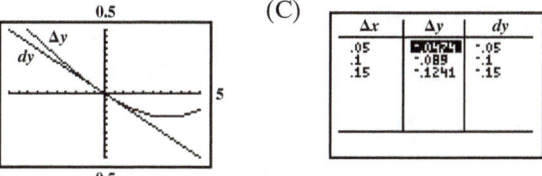

29. True   **31.** False   **33.** $dy = \left(\dfrac{2}{3}x^{-1/3} - \dfrac{10}{3}x^{2/3}\right)dx$ **35.** $dy = 3.9$; $\Delta y = 3.83$ **37.** 40-unit increase; 20-unit increase

39. $-\$2.50$; \$1.25   **41.** -1.37/min; -0.58/min   **43.** 1.26 mm^2 **45.** 3 wpm
47. (A) 2,100 increase   (B) 4,800 increase   (C) 2,100 increase

Exercises 3-7

1. $C'(x) = 0.8$ **3.** $C'(x) = 4.6 - 0.02x$ **5.** $R'(x) = 4 - 0.02x$ **7.** $R'(x) = 12 - 0.08x$ **9.** $P'(x) = 3.2 - 0.02x$

11. $P'(x) = 7.4 - 0.06x$ **13.** $\overline{C}(x) = 1.1 + \dfrac{145}{x}$ **15.** $\overline{C}'(x) = -\dfrac{145}{x^2}$ **17.** $P(x) = 3.9x - 0.02x^2 - 145$

19. $\overline{P}(x) = 3.9 - 0.02x - \dfrac{145}{x}$ **21.** True   **23.** False   **25.** (A) \$29.50   (B) \$30

27. (A) \$420
  (B) $\overline{C}'(500) = -0.24$. At a production level of 500 frames, average cost is decreasing at the rate of 24¢ per frame.
  (C) Approximately \$419.76
29. (A) \$14.70   (B) \$15
31. (A) $P'(450) = 0.5$. At a production level of 450 cassettes, profit is increasing at the rate of 50¢ per cassette.
  (B) $P'(750) = -2.5$. At a production level of 750 cassettes, profit is decreasing at the rate of \$2.50 per cassette.
33. (A) \$13.50
  (B) $\overline{P}'(50) = \$0.27$. At a production level of 50 mowers, the average profit per mower is increasing at the rate of \$0.27 per mower.
  (C) Approximately \$13.77
35. (A) $p = 100 - 0.025x$, domain: $0 \le x \le 4,000$
  (B) $R(x) = 100x - 0.025x^2$, domain: $0 \le x \le 4,000$
  (C) $R'(1,600) = 20$. At a production level of 1,600 radios, revenue is increasing at the rate of \$20 per radio.
  (D) $R'(2,500) = -25$. At a production level of 2,500 radios, revenue is decreasing at the rate of \$25 per radio.
37. (A) $p = 200 - \frac{1}{30}x$, domain: $0 \le x \le 6,000$
  (B) $C'(x) = 60$
  (C) $R(x) = 200x - (x^2/30)$, domain: $0 \le x \le 6,000$
  (D) $R'(x) = 200 - (x/15)$
  (E) $R'(1,500) = 100$. At a production level of 1,500 saws, revenue is increasing at the rate of \$100 per saw.
    $R'(4,500) = -100$. At a production level of 4,500 saws, revenue is decreasing at the rate of \$100 per saw.

(F) Break-even points: (600, 108,000) and (3,600, 288,000)

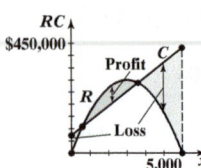

39. (A) $p = 20 - 0.02x$, domain: $0 \leq x \leq 1,000$
(B) $R(x) = 20x - 0.02x^2$, domain: $0 \leq x \leq 1,000$
(C) $C(x) = 4x + 1,400$

41. (A) $x = 500$
(B) $P(x) = 176x - 0.2x^2 - 21,900$
(C) $x = 440$
(D) Break-even points: (150, 25,500) and (730, 39,420); x intercepts for $P(x)$: $x = 150$ and $x = 730$

45. (A)

(G) $P(x) = -(x^2/30) + 140x - 72,000$
(H) $P'(x) = -(x/15) + 140$
(I) $P'(1,500) = 40$. At a production level of 1,500 saws, profit is increasing at the rate of $40 per saw.
$P'(3,000) = -60$. At a production level of 3,000 saws, profit is decreasing at the rate of $60 per saw.

(D) Break-even points: (100, 1,800) and (700, 4,200)
(E) $P(x) = 16x - 0.02x^2 - 1,400$
(F) $P'(250) = 6$. At a production level of 250 toasters, profit is increasing at the rate of $6 per toaster. $P'(475) = -3$. At a production level of 475 toasters, profit is decreasing at the rate of $3 per toaster.

43. (A) $R(x) = 20x - x^{3/2}$
(B) Break-even points: (44, 588), (258, 1,016)

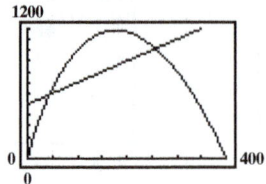

(B) Fixed costs $\approx$ $721,680
Variable costs $\approx$ $121

(C) (713, 807,703), (5,423, 1,376,227)
(D) $254 \leq p \leq $1,133

Chapter 3 Review Exercises

1. (A) 16 (B) 8 (C) 8 (D) 4 (E) 4 (F) 4 *(3-2)* 2. $f'(x) = -3$ *(3-2)* 3. (A) 22 (B) 8 (C) 2 (D) -5 *(3-1)*
4. 1.5 *(3-1)* 5. 3.5 *(3-1)* 6. 3.75 *(3-1)* 7. 3.75 *(3-1)* 8. (A) 1 (B) 1 (C) 1 (D) 1 *(3-1)*
9. (A) 2 (B) 3 (C) Does not exist (D) 3 *(3-1)* 10. (A) 4 (B) 4 (C) 4 (D) Does not exist *(3-1)*
11. (A) Does not exist (B) 3 (C) No *(3-3)* 12. (A) 2 (B) Not defined (C) No *(3-3)*
13. (A) 1 (B) 1 (C) Yes *(3-3)* 14. 5 *(3-2)* 15. 5 *(3-2)* 16. ∞ *(3-2)* 17. $-\infty$ *(3-2)* 18. 0 *(3-1)* 19. 0 *(3-1)* 20. 0 *(3-1)*
21. Vertical asymptote: $x = 2$ *(3-3)* 22. Horizontal asymptote: $y = 5$ *(3-2)* 23. $x = 2$ 24. $f'(x) = 10x$ *(3-4)*

25. (A) -3 (B) 6 (C) -2 (D) 3 (E) -11 *(3-5)* 26. $x^2 - 10x$ *(3-5)* 27. $x^{-1/2} - 3 = \dfrac{1}{x^{1/2}} - 3$ *(3-5)* 28. 0 *(3-5)*

29. $-\dfrac{3}{2}x^{-2} + \dfrac{15}{4}x^2 = \dfrac{-3}{2x^2} + \dfrac{15x^2}{4}$ *(3-5)* 30. $-2x^{-5} + x^3 = \dfrac{-2}{x^5} + x^3$ *(3-5)* 31. $f'(x) = 12x^3 + 9x^2 - 2$ *(3-5)*

32. $\Delta x = 2, \Delta y = 10, \Delta y/\Delta x = 5$ *(3-6)* 33. 5 *(3-6)* 34. 6 *(3-6)* 35. $\Delta y = 0.64; dy = 0.6$ *(3-6)*

36. (A) 4 (B) 6 (C) Does not exist (D) 6 (E) No *(3-3)* 37. (A) 3 (B) 3 (C) 3 (D) 3 (E) Yes *(3-3)*
38. (A) $(8, \infty)$ (B) $[0, 8]$ *(3-3)* 39. $(-3, 4)$ *(3-3)* 40. $(-3, 0) \cup (5, \infty)$ *(3-3)* 41. $(-2.3429, -0.4707) \cup (1.8136, \infty)$ *(3-3)*

42. (A) 3 (B) $2 + 0.5h$ (C) 2 *(3-4)* 43. $-x^{-4} + 10x^{-3}$ *(3-4)* 44. $\dfrac{3}{4}x^{-1/2} - \dfrac{5}{6}x^{-3/2} = \dfrac{3}{4\sqrt{x}} - \dfrac{5}{6\sqrt{x^3}}$ *(3-5)*

45. $0.6x^{-2/3} - 0.3x^{-4/3} = \dfrac{0.6}{x^{2/3}} - \dfrac{0.3}{x^{4/3}}$ *(3-4)* 46. $-\dfrac{3}{5}(-3)x^{-4} = \dfrac{9}{5x^4}$ *(3-5)* 47. (A) $m = f'(1) = 2$ (B) $y = 2x + 3$ *(3-4, 3-5)*
48. $x = 5$ *(3-4)* 49. $x = -5, x = 3$ *(3-5)* 50. $x = -1.3401, 0.5771, 2.2630$ *(3-4)* 51. ± 2.4824 *(3-5)*
52. (A) $v = f'(x) = 16x - 4$ (B) 44 ft/sec *(3-5)* 53. (A) $v = f'(x) = -10x + 16$ (B) $x = 1.6$ sec *(3-5)*

54. (A) The graph of g is the graph of f shifted 4 units to the right, and the graph of h is the graph of f shifted 3 units to the left:

(B) The graph of g' is the graph of f' shifted 4 units to the right, and the graph of h' is the graph of f' shifted 3 units to the left:

55. $(-\infty, \infty)$ *(3-3)* **56.** $(-\infty, 2) \cup (2, \infty)$ *(3-3)* **57.** $(-\infty, -4) \cup (-4, 1) \cup (1, \infty)$ *(3-3)* **58.** $(-\infty, \infty)$ *(3-3)*

59. $[-2, 2]$ *(3-3)* **60.**   (A) -1 (B) Does not exist   (C) $-\dfrac{2}{3}$ *(3-1)* **61.**   (A) $\dfrac{1}{2}$ (B) 0 (C) Does not exist *(3-1)*

62. (A) -1 (B) 1 (C) Does not exist *(3-1)* **63.**   (A) $-\dfrac{1}{6}$ (B) Does not exist   (C) $-\dfrac{1}{3}$ *(3-1)*

64. (A) 0 (B) -1 (C) Does not exist *(3-1)* **65.**   (A) $\dfrac{2}{3}$ (B) $\dfrac{2}{3}$ (C) Does not exist *(3-2)*

66. (A) ∞ (B) $-\infty$ (C) ∞ *(3-3)* **67.**   (A) 0 (B) 0 (C) Does not exist *(3-2)* **68.** 4 *(3-1)* **69.** $\dfrac{-1}{(x+2)^2}$ *(3-1)*

70. $2x - 1$ *(3-4)* **71.** $1/(2\sqrt{x})$ *(3-4)* **72.** Yes *(3-4)* **73.** No *(3-4)* **74.** No *(3-4)* **75.** No *(3-4)* **76.** Yes *(3-4)* **77.** Yes *(3-4)*

78. Horizontal asymptote: $y = 5$; vertical asymptote: $x = 7$ *(3-2)*

79. Horizontal asymptote: $y = 0$; vertical asymptote: $x = 4$ *(3-2)*

80. No horizontal asymptotes; vertical asymptote: $x = 3$ *(3-2)*

81. Horizontal asymptotes: $y = 1$; vertical asymptotes: $x = -2, x = 1$ *(3-2)*

82. Horizontal asymptote: $y = 1$; vertical asymptotes: $x = -1, x = 1$ *(3-2)*

83. The domain of $f'(x)$ is all real numbers except $x = 0$. At $x = 0$, the graph of $y = f(x)$ is smooth, but it has a vertical tangent. *(3-4)*

84. (A) $\lim_{x \to 1^-} f(x) = 1; \lim_{x \to 1^+} f(x) = -1$ (B) $\lim_{x \to 1^-} f(x) = -1; \lim_{x \to 1^+} f(x) = 1$ (C) $m = 1$

(D) The graphs in (A) and (B) have discontinuities at $x = 1$; the graph in (C) does not. *(3-2)*

85. (A) 1 (B) -1 (C) Does not exist   (D) No *(3-4)*

86. (A) $S(x) = \begin{cases} 7.47 + 0.4x & \text{if } 0 \le x \le 90 \\ 24.786 + 0.2076x & \text{if } 90 < x \end{cases}$

(B)

(C) Yes *(3-2)*

87. (A) $\$179.90$ (B) $\$180$ *(3-7)*

88. (A) $C(100) = 9{,}500; C'(100) = 50$. At a production level of 100 bicycles, the total cost is $\$9{,}500$, and cost is increasing at the rate of $\$50$ per bicycle.

(B) $\overline{C}(100) = 95; \overline{C}'(100) = -0.45$. At a production level of 100 bicycles, the average cost is $\$95$, and average cost is decreasing at a rate of $\$0.45$ per bicycle. *(3-7)*

89. The approximate cost of producing the 201st printer is greater than that of the 601st printer. Since these marginal costs are decreasing, the manufacturing process is becoming more efficient. *(3-7)*

90. (A) $C'(x) = 2; \overline{C}(x) = 2 + \dfrac{9{,}000}{x}; \overline{C}'(x) = \dfrac{-9{,}000}{x^2}$

(B) $R(x) = xp = 25x - 0.01x^2; R'(x) = 25 - 0.02x; \overline{R}(x) = 25 - 0.01x; \overline{R}'(x) = -0.01$

(C) $P(x) = R(x) - C(x) = 23x - 0.01x^2 - 9{,}000; P'(x) = 23 - 0.02x;$

$\overline{P}(x) = 23 - 0.01x - \dfrac{9{,}000}{x}; \overline{P}'(x) = -0.01 + \dfrac{9{,}000}{x^2}$

(D) $(500, 10{,}000)$ and $(1{,}800, 12{,}600)$

(E) $P'(1{,}000) = 3$. Profit is increasing at the rate of $\$3$ per umbrella.
$P'(1{,}150) = 0$. Profit is flat.
$P'(1{,}400) = -5$. Profit is decreasing at the rate of $\$5$ per umbrella.

(F)

91. (A) 8 (B) 20 *(3-5)*

92. $N(9) = 27; N'(t) = 3.5$; After 9 months, 27,000 pools have been sold and the total sales are increasing at the rate of 3,500 pools per month. *(3-5)*

93. (A)

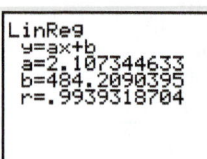

```
CubicReg
y=ax³+bx²+cx+d
a=.001225
b=-.0819285714
c=1.564642857
d=12.08428571
```

(B) $N(50) = 38.6$; $N'(50) = 2.6$. In 2010, natural gas consumption is 38.6 trillion cubic feet and is increasing at the rate of 2.6 trillion cubic feet per year *(3-4)*

94. (A)

```
LinReg
y=ax+b
a=-.0384180791
b=13.59887006
r=-.9897782666
```

(B) Fixed costs: \$484.21; variable costs per kringle: \$2.11

```
LinReg
y=ax+b
a=2.107344633
b=484.2090395
r=.9939318704
```

(C) $(51, 591.15)$, $(248, 1,007.62)$　(D) $\$4.07 < p < \$11.64$ *(3-7)*

95. $C'(10) = -1$; $C'(100) = -0.001$ *(3-5)*

96. $F(4) = 98.16$; $F'(4) = -0.32$; After 4 hours the patient's temperature is 98.16°F and is decreasing at the rate of 0.32°F per hour. *(3-5)*

97. (A) 10 items/h　(B) 5 items/h *(3-5)*

98. (A)

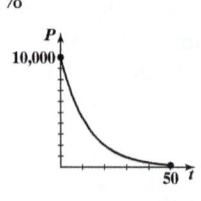

(B) $C(T) = \dfrac{12T}{150 + T}$

(C) $C = 9.6$ at $T = 600$ K, $T = 750$ K when $C = 10$ *(3-3)*

Chapter 4

Exercises 4-1

1. \$1,221.40; \$1,648.72; \$2,225.54　　**5.** 11.55　**7.** 10.99　**9.** 0.14

3.

$A = 6{,}000e^{0.1t}$

11.

n	$[1 + (1/n)]^n$
10	2.593 74
100	2.704 81
1,000	2.716 92
10,000	2.718 15
100,000	2.718 27
1,000,000	2.718 28
10,000,000	2.718 28
↓	↓
∞	$e = 2.718\ 281\ 828\ 459\ldots$

13. $\lim_{n\to\infty}(1 + n)^{1/n} = 1$

15.

$y_3 = \left(1 + \dfrac{1}{n}\right)^{n+1}$

$y_1 = \left(1 + \dfrac{1}{n}\right)^{n}$

17. (A) \$15,143.71　(B) 14.16 yr

19. \$11,890.41

21. 8.11%

23. (A)

P, 10,000, 50 t

(B) $\lim_{t\to\infty}10{,}000e^{-0.08t} = 0$

25. 17.33 yr　**27.** 8.66%　**29.** 7.3 yr

31. (A) 　$A = Pe^{rt}$

$2P = Pe^{rt}$

$2 = e^{rt}$

$rt = \ln 2$

$t = \dfrac{\ln 2}{r}$

(B) t, 35, $t = \dfrac{\ln 2}{r}$, 0.3 r

Although r could be any positive number, the restrictions on r are reasonable in the sense that most investments would be expected to earn a return of between 2% and 30%.

(C) The doubling times (in years) are 13.86, 6.93, 4.62, 3.47, 2.77, and 2.31, respectively.

33. $t = -(\ln 0.5)/0.000\ 433\ 2 \approx 1{,}600$ yr　**35.** $r = (\ln 0.5)/30 \approx -0.0231$　**37.** 53.3 yr　**39.** 1.39%

Exercises 4-2

1. $5e^x + 3$　**3.** $-\dfrac{2}{x} + 2x$　**5.** $3x^2 - 6e^x$　**7.** $e^x + 1 - \dfrac{1}{x}$　**9.** $\dfrac{3}{x}$　**11.** $5 - \dfrac{5}{x}$　**13.** $\dfrac{2}{x} + 4e^x$　**15.** $f'(x) = \dfrac{1}{x}$; $y = x + 2$

17. $f'(x) = 3e^x$; $y = 3x + 3$　**19.** $f'(x) = \dfrac{3}{x}$; $y = \dfrac{3x}{e}$　**21.** $f'(x) = e^x$; $y = ex + 2$　**23.** Yes; yes　**25.** No; no

27. $f(x) = 10x + \ln 10 + \ln x$; $f'(x) = 10 + \dfrac{1}{x}$　**29.** $f(x) = \ln 4 - 3\ln x$; $f'(x) = -\dfrac{3}{x}$　**31.** $\dfrac{1}{x \ln 2}$　**33.** $3^x \ln 3$

35. $2 - \dfrac{1}{x \ln 10}$　**37.** $1 + 10^x \ln 10$　**39.** $\dfrac{3}{x} + \dfrac{2}{x \ln 3}$　**41.** $2^x \ln 2$　**43.** $(-0.82, 0.44)$, $(1.43, 4.18)$, $(8.61, 5503.66)$

45. $(0.49, 0.49)$ **47.** $(3.65, 1.30), (332, 105.11, 12.71)$ **51.** \$28,447/yr; \$18,664/yr; \$11,021/yr

53. $A'(t) = 5{,}000(\ln 4)4^t$; $A'(1) = 27{,}726$ bacteria/hr (rate of change at the end of the first hour); $A'(5) = 7{,}097{,}827$ bacteria/hr (rate of change at the end of the fifth hour)

55. At the 40-lb weight level, blood pressure would increase at the rate of 0.44 mm of mercury per pound of weight gain. At the 90-lb weight level, blood pressure would increase at the rate of 0.19 mm of mercury per pound of weight gain.

57. $dR/dS = k/S$

Exercises 4-3

1. $2x^3(2x) + (x^2 - 2)(6x^2) = 10x^4 - 12x^2$ **3.** $(x - 3)(2) + (2x - 1)(1) = 4x - 7$

5. $\dfrac{(x - 3)(1) - x(1)}{(x - 3)^2} = \dfrac{-3}{(x - 3)^2}$ **7.** $\dfrac{(x - 2)(2) - (2x + 3)(1)}{(x - 2)^2} = \dfrac{-7}{(x - 2)^2}$

9. $3xe^x + 3e^x = 3(x + 1)e^x$ **11.** $x^3\left(\dfrac{1}{x}\right) + 3x^2 \ln x = x^2(1 + 3 \ln x)$

13. $(x^2 + 1)(2) + (2x - 3)(2x) = 6x^2 - 6x + 2$ **15.** $(0.4x + 2)(0.5) + (0.5x - 5)(0.4) = 0.4x - 1$

17. $\dfrac{(2x - 3)(2x) - (x^2 + 1)(2)}{(2x - 3)^2} = \dfrac{2x^2 - 6x - 2}{(2x - 3)^2}$ **19.** $(x^2 + 2)2x + (x^2 - 3)2x = 4x^3 - 2x$

21. $\dfrac{(x^2 - 3)2x - (x^2 + 2)2x}{(x^2 - 3)^2} = \dfrac{-10x}{(x^2 - 3)^2}$ **23.** $\dfrac{(x^2 + 1)e^x - e^x(2x)}{(x^2 + 1)^2} = \dfrac{(x - 1)^2 e^x}{(x^2 + 1)^2}$

25. $\dfrac{(1 + x)\left(\dfrac{1}{x}\right) - \ln x}{(1 + x)^2} = \dfrac{1 + x - x \ln x}{x(1 + x)^2}$ **27.** $xf'(x) + f(x)$ **29.** $x^3 f'(x) + 3x^2 f(x)$ **31.** $\dfrac{x^2 f'(x) - 2xf(x)}{x^4}$

33. $\dfrac{f(x) - xf'(x)}{[f(x)]^2}$ **35.** $e^x f'(x) + f(x)e^x = e^x[f'(x) + f(x)]$ **37.** $\dfrac{f(x)\left(\dfrac{1}{x}\right) - (\ln x)f'(x)}{f(x)^2} = \dfrac{f(x) - (x \ln x)f'(x)}{x f(x)^2}$

39. $(2x + 1)(2x - 3) + (x^2 - 3x)(2) = 6x^2 - 10x - 3$ **41.** $(2.5t - t^2)(4) + (4t + 1.4)(2.5 - 2t) = -12t^2 + 17.2t + 3.5$

43. $\dfrac{(x^2 + 2x)(5) - (5x - 3)(2x + 2)}{(x^2 + 2x)^2} = \dfrac{-5x^2 + 6x + 6}{(x^2 + 2x)^2}$ **45.** $\dfrac{(w^2 - 1)(2w - 3) - (w^2 - 3w + 1)(2w)}{(w^2 - 1)^2} = \dfrac{3w^2 - 4w + 3}{(w^2 - 1)^2}$

47. $(1 + x - x^2)e^x + e^x(1 - 2x) = (2 - x - x^2)e^x$ **49.** $f'(x) = (1 + 3x)(-2) + (5 - 2x)(3)$; $y = -11x + 29$

51. $f'(x) = \dfrac{(3x - 4)(1) - (x - 8)(3)}{(3x - 4)^2}$; $y = 5x - 13$ **53.** $f'(x) = \dfrac{2^x - x(2^x \ln 2)}{2^{2x}}$; $y = \left(\dfrac{1 - 2 \ln 2}{4}\right)x + \ln 2$

55. $f'(x) = (2x - 15)(2x) + (x^2 + 18)(2) = 6(x - 2)(x - 3)$; $x = 2, x = 3$

57. $f'(x) = \dfrac{(x^2 + 1)(1) - x(2x)}{(x^2 + 1)^2} = \dfrac{1 - x^2}{(x^2 + 1)^2}$; $x = -1, x = 1$ **59.** $7x^6 - 3x^2$ **61.** $-27x^{-4} = -\dfrac{27}{x^4}$

63. $(w + 1)2^w \ln 2 + 2^w = [(w + 1) \ln 2 + 1]2^w$ **65.** $\dfrac{(4x^2 + 5x - 1)(6x - 2) - (3x^2 - 2x + 3)(8x + 5)}{(4x^2 + 5x - 1)^2} = \dfrac{23x^2 - 30x - 13}{(4x^2 + 5x - 1)^2}$

67. $9x^{1/3}(3x^2) + (x^3 + 5)(3x^{-2/3}) = \dfrac{30x^3 + 15}{x^{2/3}}$ **69.** $\dfrac{(1 + x^2)\dfrac{1}{x \ln 2} - 2x \log_2 x}{(1 + x^2)^2} = \dfrac{1 + x^2 - 2x^2 \ln x}{x(1 + x^2)^2 \ln 2}$

71. $\dfrac{(x^2 - 3)(2x^{-2/3}) - 6x^{1/3}(2x)}{(x^2 - 3)^2} = \dfrac{-10x^2 - 6}{(x^2 - 3)^2 x^{2/3}}$ **73.** $g'(t) = \dfrac{(3t^2 - 1)(0.2) - (0.2t)(6t)}{(3t^2 - 1)^2} = \dfrac{-0.6t^2 - 0.2}{(3t^2 - 1)^2}$

75. $(20x)\dfrac{1}{x \ln 10} + 20 \log x = \dfrac{20(1 + \ln x)}{\ln 10}$ **77.** $x^{-2/3}(3x^2 - 4x) + (x^3 - 2x^2)\left(-\frac{2}{3}x^{-5/3}\right) = -\frac{8}{3}x^{1/3} + \frac{7}{3}x^{4/3}$

79. $\dfrac{(x^2 + 1)[(2x^2 - 1)(2x) + (x^2 + 3)(4x)] - (2x^2 - 1)(x^2 + 3)(2x)}{(x^2 + 1)^2} = \dfrac{4x^5 + 8x^3 + 16x}{(x^2 + 1)^2}$

81. $\dfrac{e^t(1 + \ln t) - (t \ln t)e^t}{e^{2t}} = \dfrac{1 + \ln t - t \ln t}{e^t}$

83. (A) $S'(t) = \dfrac{(t^2 + 50)(180t) - 90t^2(2t)}{(t^2 + 50)^2} = \dfrac{9{,}000t}{(t^2 + 50)^2}$

(B) $S(10) = 60$; $S'(10) = 4$. After 10 months, the total sales are 60,000 DVDs and sales are increasing at the rate of 4,000 DVDs per month.

(C) Approximately 64,000 DVDs

85. (A) $\dfrac{dx}{dp} = \dfrac{(0.1p + 1)(0) - 4,000(0.1)}{(0.1p + 1)^2} = \dfrac{-400}{(0.1p + 1)^2}$

(B) $x = 800$; $dx/dp = -16$. At a price level of \$40, the demand is 800 DVD players per week and demand is decreasing at the rate of 16 players per dollar.

(C) Approximately 784 DVD players

87. (A) $C'(t) = \dfrac{(t^2 + 1)(0.14) - 0.14t(2t)}{(t^2 + 1)^2} = \dfrac{0.14 - 0.14t^2}{(t^2 + 1)^2}$

(B) $C'(0.5) = 0.0672$. After 0.5 h, concentration is increasing at the rate of 0.0672 mg/cm³ per hour.
$C'(3) = -0.0112$. After 3 h, concentration is decreasing at the rate of 0.0112 mg/cm³ per hour.

89. (A) $N'(x) = \dfrac{(x + 32)(100) - (100x + 200)}{(x + 32)^2} = \dfrac{3,000}{(x + 32)^2}$

(B) $N'(4) = 2.31$; $N'(68) = 0.30$

Exercises 4-4

1. $(3x^2 + 2)^3$ **3.** e^{-x^2} **5.** $y = u^4; u = 3x^2 - x + 5$ **7.** $y = e^u; u = 1 + x + x^2$ **9.** 3 **11.** $(-4x)$

13. $2x$ **15.** $4x^3$ **17.** $2(x + 3)$ **19.** $6(2x + 5)^2$ **21.** $-8(5 - 2x)^3$ **23.** $5(4 + 0.2x)^4(0.2) = (4 + 0.2x)^4$ **25.** $30x(3x^2 + 5)^4$

27. $5e^x$ **29.** $5e^{5x}$ **31.** $-18e^{-6x}$ **33.** $(2x - 5)^{-1/2} = \dfrac{1}{(2x - 5)^{1/2}}$ **35.** $-8x^3(x^4 + 1)^{-3} = \dfrac{-8x^3}{(x^4 + 1)^3}$ **37.** $-\dfrac{2}{x}$ **39.** $\dfrac{6x}{1 + x^2}$

41. $\dfrac{3(1 + \ln x)^2}{x}$ **43.** $f'(x) = 6(2x - 1)^2; y = 6x - 5; x = \frac{1}{2}$ **45.** $f'(x) = 2(4x - 3)^{-1/2} = \dfrac{2}{(4x - 3)^{1/2}}; y = \dfrac{2}{3}x + 1$; none

47. $f'(x) = 10(x - 2)e^{x^2 - 4x + 1}; y = -20ex + 5e; x = 2$ **49.** $12(x^2 - 2)^3(2x) = 24x(x^2 - 2)^3$

51. $-6(t^2 + 3t)^{-4}(2t + 3) = \dfrac{-6(2t + 3)}{(t^2 + 3t)^4}$ **53.** $\dfrac{1}{2}(w^2 + 8)^{-1/2}(2w) = \dfrac{w}{\sqrt{w^2 + 8}}$ **55.** $12xe^{3x} + 4e^{3x} = 4(3x + 1)e^{3x}$

57. $\dfrac{x^3\left(\dfrac{1}{1 + x}\right) - 3x^2 \ln(1 + x)}{x^6} = \dfrac{x - 3(1 + x)\ln(1 + x)}{x^4(1 + x)}$ **59.** $6te^{3(t^2 + 1)}$ **61.** $\dfrac{3x}{x^2 + 3}$

63. $-5(w^3 + 4)^{-6}(3w^2) = \dfrac{-15w^2}{(w^3 + 4)^6}$ **65.** $f'(x) = (4 - x)^3 - 3x(4 - x)^2 = 4(4 - x)^2(1 - x); y = -16x + 48$

67. $f'(x) = \dfrac{(2x - 5)^3 - 6x(2x - 5)^2}{(2x - 5)^6} = \dfrac{-4x - 5}{(2x - 5)^4}; y = -17x + 54$ **69.** $f'(x) = \dfrac{1}{2x\sqrt{\ln x}}; y = \dfrac{x}{2e} + \dfrac{1}{2}$

71. $f'(x) = 2x(x - 5)^3 + 3x^2(x - 5)^2 = 5x(x - 5)^2(x - 2); x = 0, 2, 5$

73. $f'(x) = \dfrac{(2x + 5)^2 - 4x(2x + 5)}{(2x + 5)^4} = \dfrac{5 - 2x}{(2x + 5)^3}; x = \frac{5}{2}$

75. $f'(x) = (x^2 - 8x + 20)^{-1/2}(x - 4) = \dfrac{x - 4}{(x^2 - 8x + 20)^{1/2}}; x = 4$ **77.** No; yes

79. $18x^2(x^2 + 1)^2 + 3(x^2 + 1)^3 = 3(x^2 + 1)^2(7x^2 + 1)$ **81.** $\dfrac{24x^5(x^3 - 7)^3 - (x^3 - 7)^4 6x^2}{4x^6} = \dfrac{3(x^3 - 7)^3(3x^3 + 7)}{2x^4}$

83. $\dfrac{1}{\ln 2}\left(\dfrac{6x}{3x^2 - 1}\right)$ **85.** $(2x + 1)(10^{x^2 + x})(\ln 10)$ **87.** $\dfrac{12x^2 + 5}{(4x^3 + 5x + 7)\ln 3}$ **89.** $2^{x^3 - x^2 + 4x + 1}(3x^2 - 2x + 4)\ln 2$

91. (A) $C'(x) = (2x + 16)^{-1/2} = \dfrac{1}{(2x + 16)^{1/2}}$

(B) $C'(24) = \frac{1}{8}$, or \$12.50. At a production level of 24 cell phones, total cost is increasing at the rate of \$12.50 per cell phone and the cost of producing the 25th cell phone is approximately \$12.50.
$C'(42) = \frac{1}{10}$, or \$10.00. At a production level of 42 cell phones, total cost is increasing at the rate of \$10.00 per cell phone and the cost of producing the 43rd cell phone is approximately \$10.00.

93. (A) $\dfrac{dx}{dp} = 40(p + 25)^{-1/2} = \dfrac{40}{(p + 25)^{1/2}}$

(B) $x = 400$ and $dx/dp = 4$. At a price of \$75, the supply is 400 bicycle helmets per week and supply is increasing at the rate of 4 bicycle helmets per dollar.

95. (A) After 1 hr, the concentration is decreasing at the rate of 1.60 mg/mL per hour; after 4 hr, the concentration is decreasing at the rate of 0.08 mg/mL per hour.

(B)

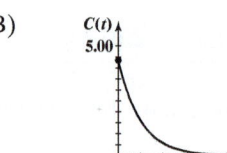

97. 2.27 mm of mercury/yr; 0.81 mm of mercury/yr; 0.41 mm of mercury/yr

99. (A) $f'(n) = n(n-2)^{-1/2} + 2(n-2)^{1/2}$

    (B) $f'(11) = \frac{29}{3} \approx 9.67$. When the list contains 11 items, the learning time is increasing at the rate of 9.67 min per item.

      $f'(27) = \frac{77}{5} = 15.4$. When the list contains 27 items, the learning time is increasing at the rate of 15.4 min per item.

Exercises 4-5

1. $y' = -\dfrac{3}{5}$ **3.** $y' = \dfrac{3x}{2}$ **5.** $y' = 10x; 10$ **7.** $y' = \dfrac{2x}{3y^2}; \dfrac{4}{3}$ **9.** $y' = -\dfrac{3}{2y+2}; -\dfrac{3}{4}$ **11.** $y' = -\dfrac{y}{x}; -\dfrac{3}{2}$

13. $y' = -\dfrac{2y}{2x+1}; 4$ **15.** $y' = \dfrac{6-2y}{x}; -1$ **17.** $y' = \dfrac{2x}{e^y - 2y}; 2$ **19.** $y' = \dfrac{3x^2 y}{y+1}; \dfrac{3}{2}$ **21.** $y' = \dfrac{6x^2 y - y \ln y}{x + 2y}; 2$

23. $x' = \dfrac{2tx - 3t^2}{2x - t^2}; 8$ **25.** $y'|_{(1.6, 1.8)} = -\dfrac{3}{4}; y'|_{(1.6, 0.2)} = \dfrac{3}{4}$ **27.** $y = -x + 5$ **29.** $y = \frac{2}{5}x - \frac{12}{5}; y = \frac{3}{5}x + \frac{12}{5}$

31. $y' = -\dfrac{1}{x}$ **33.** $y' = \dfrac{1}{3(1+y)^2 + 1}; \dfrac{1}{13}$ **35.** $y' = \dfrac{3(x-2y)^2}{6(x-2y)^2 + 4y}; \dfrac{3}{10}$ **37.** $y' = \dfrac{3x^2(7+y^2)^{1/2}}{y}; 16$

39. $y' = \dfrac{y}{2xy^2 - x}; 1$ **41.** $y = 0.63x + 1.04$ **43.** $p' = \dfrac{1}{2p - 2}$ **45.** $p' = -\dfrac{\sqrt{10{,}000 - p^2}}{p}$

47. $\dfrac{dL}{dV} = \dfrac{-(L+m)}{V+n}$ **49.** $\dfrac{dT}{dv} = \dfrac{2}{k}\sqrt{T}$ **51.** $\dfrac{dv}{dT} = \dfrac{k}{2\sqrt{T}}$

Exercises 4-6

1. 30   **3.** $-\frac{16}{3}$ **5.** $-\frac{16}{7}$ **7.** Decreasing at 9 units/sec   **9.** Approx. -3.03 ft/sec   **11.** $dA/dt \approx 126$ ft^2/sec

13. 3,768 cm^3/min   **15.** 6 lb/in.^{2}/hr   **17.** $-\frac{9}{4}$ ft/sec   **19.** $\frac{20}{3}$ ft/sec   **21.** 0.0214 ft/sec; 0.0135 ft/sec; yes, at $t = 0.000\ 19$ sec

23. 3.835 units/sec   **25.** (A) $dC/dt = \$15{,}000/$wk   (B) $dR/dt = -\$50{,}000/$wk   (C) $dP/dt = -\$65{,}000/$wk

27. $ds/dt = \$2{,}207/$wk   **29.** (A) $dx/dt = -12.73$ units/month   (B) $dp/dt = \$1.53/$month   **31.** Approximately 100 ft^3/min

Exercises 4-7

1. $\dfrac{35 - 0.8x}{35x - 0.4x^2}$ **3.** $-\dfrac{4e^{-x}}{7 + 4e^x}$ **5.** $\dfrac{5}{x(12 + 5 \ln x)}$ **7.** 0   **9.** -0.017 **11.** -0.034 **13.** 1.013   **15.** 0.405   **17.** 11.8%

19. 5.4%   **21.** -14.7% **23.** -431.6% **25.** $E(p) = \dfrac{450p}{25{,}000 - 450p}$ **27.** $E(p) = \dfrac{8p^2}{4{,}800 - 4p^2}$ **29.** $E(p) = \dfrac{0.6pe^p}{98 - 0.6e^p}$

31. (A) Inelastic   (B) Unit elasticity   (C) Elastic   **33.** (A) Inelastic   (B) Unit elasticity   (C) Elastic

35. (A) $x = 6{,}000 - 200p$ $0 \le p \le 30$
    (B) $E(p) = \dfrac{p}{30 - p}$
    (C) $E(10) = 0.5$; 5% decrease
    (D) $E(25) = 5$; 50% decrease
    (E) $E(15) = 1$; 10% decrease

37. (A) $x = 3{,}000 - 50p$ $0 \le p \le 60$
    (B) $R(p) = 3{,}000p - 50p^2$
    (C) $E(p) = \dfrac{p}{60 - p}$
    (D) Elastic on $(30, 60)$; inelastic on $(0, 30)$
    (E) Increasing on $(0, 30)$; decreasing on $(30, 60)$
    (F) Decrease
    (G) Increase

39. Elastic on $(3.5, 7)$; inelastic on $(0, 3.5)$

41. Elastic on $(25/\sqrt{3}, 25)$; inelastic on $(0, 25/\sqrt{3})$

43. Elastic on $(48, 72)$; inelastic on $(0, 48)$

45. Elastic on $(25, 25\sqrt{2})$; inelastic on $(0, 25)$

47. $R(p) = 20p(10 - p)$ **49.** $R(p) = 40p(p - 15)^2$ **51.** $R(p) = 30p - 10p\sqrt{p}$

53. $\dfrac{3}{2}$ **55.** $\dfrac{1}{2}$

57. Elastic on $(0, 300)$; inelastic on $(300, 600)$

59. Elastic on $(0, 10\sqrt{3})$; inelastic on $(10\sqrt{3}, 30)$

61. k

63. \$75 per day

65. Increase   **67.** Decrease   **69.** \$3.75   **71.** $p(t) = \dfrac{31}{0.31t + 18.5}$ **73.** -0.08

Chapter 4 Review Exercises

1. $3,136.62; $4,919.21; $12,099.29 *(4-1)*   **2.** $\dfrac{2}{x} + 3e^x$ *(4-2)*   **3.** $2e^{2x-3}$ *(4-4)*   **4.** $\dfrac{2}{2x + 7}$ *(4-4)*

5. (A) $y = \ln(3 + e^x)$ (B) $\dfrac{dy}{dx} = \dfrac{e^x}{3 + e^x}$ *(4-4)*   **6.** $y' = \dfrac{9x^2}{4y}; \dfrac{9}{8}$ *(4-5)*   **7.** $dy/dt = 216$ *(4-6)*

8. (A) $x = 1,000 - 25p$ (B) $\dfrac{p}{40 - p}$ (C) 0.6; demand is inelastic and insensitive to small changes in price.

(D) $1,000p - 25p^2$ (E) Revenue increases *(4-7)*

9. -10 *(4-2)*   **10.** $\displaystyle\lim_{n \to \infty} \left(1 + \dfrac{2}{n}\right)^n = e^2 \approx 7.389\ 06$ *(4-1)*   **11.** $\dfrac{7[(\ln z)^6 + 1]}{z}$ *(4-4)*   **12.** $x^5(1 + 6 \ln x)$ *(4-3)*   **13.** $\dfrac{e^x(x - 6)}{x^7}$ *(4-3)*

14. $\dfrac{6x^2 - 3}{2x^3 - 3x}$ *(4-4)*   **15.** $(3x^2 - 2x)e^{x^3 - x^2}$ *(4-4)*   **16.** $\dfrac{1 - 2x \ln 5x}{xe^{2x}}$ *(4-4)*   **17.** $y = -x + 2; y = -ex + 1$ *(4-4)*

18. $y' = \dfrac{3y - 2x}{8y - 3x}; \dfrac{8}{19}$ *(4-5)*   **19.** $x' = \dfrac{4tx}{3x^2 - 2t^2}; -4$ *(4-5)*   **20.** $y' = \dfrac{1}{e^y + 2y}; 1$ *(4-5)*   **21.** $y' = \dfrac{2xy}{1 + 2y^2}; \dfrac{2}{3}$ *(4-5)*

22. $dy/dt = -2$ units/sec *(4-6)*   **23.** 0.27 ft/sec *(4-6)*   **24.** $dR/dt = 1/\pi \approx 0.318$ in./min *(4-6)*

25. Elastic for $5 < p < 15$; inelastic for $0 < p < 5$ *(4-7)*

26.

27. (A) $y = [\ln(4 - e^x)]^3$ (B) $\dfrac{dy}{dx} = \dfrac{-3e^x[\ln(4 - e^x)]^2}{4 - e^x}$ *(4-4)*

28. $2x(5^{x^2-1})(\ln 5)$ *(4-4)*   **29.** $\left(\dfrac{1}{\ln 5}\right)\dfrac{2x - 1}{x^2 - x}$ *(4-4)*

30. $\dfrac{2x + 1}{2(x^2 + x)\sqrt{\ln(x^2 + x)}}$ *(4-4)*   **31.** $y' = \dfrac{2x - e^{xy}y}{xe^{xy} - 1}; 0$ *(4-5)*

32. The rate of increase of area is proportional to the radius R, so the rate is smallest when $R = 0$, and has no largest value. *(4-6)*

33. Yes, for $-\sqrt{3}/3 < x < \sqrt{3}/3$ *(4-6)*   **34.** (A) 15 yr   (B) 13.9 yr *(4-1)*

35. $A'(t) = 10e^{0.1t}; A'(1) = $11.05/yr; A'(10) = $27.18/yr$ *(4-1)*   **36.** $R'(x) = (1,000 - 20x)e^{-0.02x}$ *(4-4)*

37. $p' = \dfrac{-(5,000 - 2p^3)^{1/2}}{3p^2}$ *(4-5)*   **38.** $dR/dt = $2,242/day$ *(4-6)*   **39.** Decrease price *(4-7)*   **40.** 0.02125 *(4-7)*

41. -1.111 mg/mL per hour; -0.335 mg/mL per hour *(4-4)*   **42.** $dR/dt = -3/(2\pi)$; approx. 0.477 mm/day *(4-6)*

43. (A) Increasing at the rate of 2.68 units/day at the end of 1 day of training; increasing at the rate of 0.54 unit/day after 5 days of training

(B) 7 days *(4-4)*

44. $dT/dt = -1/27 \approx -0.037$ min/operation hour *(4-6)*

Chapter 5

Exercises 5-1

1. $(a, b); (d, f); (g, h)$ **3.** $(b, c); (c, d); (f, g)$ **5.** c, d, f **7.** b, f

9. Local maximum at $x = a$; local minimum at $x = c$; no local extrema at $x = b$ and $x = d$ **11.** e **13.** d **15.** f

17. c **19.** (A) $f'(x) = 3x^2 - 12$ (B) $-2, 2$ (C) $-2, 2$ **21.** (A) $f'(x) = \dfrac{1}{3}(x + 5)^{-2/3}$ (B) -5 (C) -5

23. (A) $f'(x) = -\dfrac{6}{(x + 2)^2}$ (B) None   (C) -2 **25.** (A) $f'(x) = \begin{cases} -1 \text{ if } x < 0 \\ 1 \text{ if } x > 0 \end{cases}$ (B) 0   (C) 0

27. Decreasing on $(-\infty, 1)$; increasing on $(1, \infty)$; $f(1) = -2$ is a local minimum

29. Increasing on $(-\infty, -4)$; decreasing on $(-4, \infty)$; $f(-4) = 7$ is a local maximum

31. Increasing for all x; no local extrema

33. Increasing on $(-\infty, -2)$ and $(3, \infty)$; decreasing on $(-2, 3)$; $f(-2) = 44$ is a local maximum, $f(3) = -81$ is a local minimum

35. Decreasing on $(-\infty, 1)$; increasing on $(1, \infty)$; $f(1) = 4$ is a local minimum

37. Increasing on $(-\infty, 2)$; decreasing on $(2, \infty)$; $f(2) = e^{-2} \approx 0.135$ is a local maximum

39. Increasing on $(-\infty, 8)$; decreasing on $(8, \infty)$; $f(8) = 4$ is a local maximum

41. Critical values: $x = -0.77, 1.08, 2.69$; decreasing on $(-\infty, -0.77)$ and $(1.08, 2.69)$; increasing on $(-0.77, 1.08)$ and $(2.69, \infty)$; local minima at $x = -0.77$ and $x = 2.69$; local maximum at $x = 1.08$

43. Critical values: $x = 1.34, 2.82$; decreasing on $(0, 1.34)$ and $(2.82, \infty)$; increasing on $(1.34, 2.82)$; local minimum at $x = 1.34$; and local maximum at $x = 2.82$

45. Critical values: 0.36, 2.15; increasing on $(-\infty, 0.36)$ and $(2.15, \infty)$; decreasing on $(0.36, 2.15)$; local maximum at $x = 0.36$; local minimum at $x = 2.15$

47. Increasing on $(-\infty, 4)$
Decreasing on $(4, \infty)$
Horizontal tangent at $x = 4$

49. Increasing on $(-\infty, -1), (1, \infty)$
Decreasing on $(-1, 1)$
Horizontal tangents at $x = -1, 1$

51. Decreasing for all x
Horizontal tangent at $x = 2$

53. Decreasing on $(-\infty, -3)$ and $(0, 3)$;
increasing on $(-3, 0)$ and $(3, \infty)$
Horizontal tangents at $x = -3, 0, 3$

55.

57.

59.

61.

63. g_4
65. g_6
67. g_2

69. Increasing on $(-1, 2)$; decreasing on
$(-\infty, -1)$ and $(2, \infty)$; local minimum
at $x = -1$; local maximum at $x = 2$

71. Increasing on $(-1, 2)$ and $(2, \infty)$;
decreasing on $(-\infty, -1)$; local
minimum at $x = -1$

73. Increasing on $(-2, 0)$ and $(3, \infty)$;
decreasing on $(-\infty, -2)$ and $(0, 3)$;
local minima at $x = -2$ and $x = 3$;
local maximum at $x = 0$

75. $f'(x) > 0$ on $(-\infty, -1)$
and $(3, \infty)$; $f'(x) < 0$ on
$(-1, 3)$; $f'(x) = 0$ at
$x = -1$ and $x = 3$

77. $f'(x) > 0$ on $(-2, 1)$ and
$(3, \infty)$; $f'(x) < 0$ on $(-\infty, -2)$
and $(1, 3)$; $f'(x) = 0$ at
$x = -2$, $x = 1$, and $x = 3$

79. Critical values: $x = -2$, $x = 2$; increasing on $(-\infty, -2)$ and $(2, \infty)$; decreasing on $(-2, 0)$ and $(0, 2)$; local maximum at $x = -2$; local minimum at $x = 2$

81. Critical value: $x = -2$; increasing on $(-2, 0)$; decreasing on $(-\infty, -2)$ and $(0, \infty)$; local minimum at $x = -2$

83. Critical values: $x = 0$, $x = 4$; increasing on $(-\infty, 0)$ and $(4, \infty)$; decreasing on $(0, 2)$ and $(2, 4)$; local maximum at $x = 0$; local minimum at $x = 4$

85. Critical values: $x = 0$, $x = 4$, $x = 6$; increasing on $(0, 4)$ and $(6, \infty)$; decreasing on $(-\infty, 0)$ and $(4, 6)$; local maximum at $x = 4$; local minima at $x = 0$ and $x = 6$

87. (A) There are no critical values and no local extrema. The function is increasing for all x.
(B) There are two critical values, $x = \pm\sqrt{-k/3}$. The function increases on $(-\infty, -\sqrt{-k/3})$ to a local maximum at
$x = -\sqrt{-k/3}$, decreases on $(-\sqrt{-k/3}, \sqrt{-k/3})$ to a local minimum at $x = \sqrt{-k/3}$, and increases on $(\sqrt{-k/3}, \infty)$.
(C) The only critical value is $x = 0$. There are no local extrema. The function is increasing for all x.

89. (A) The marginal profit is positive on $(0, 600)$, 0 at
$x = 600$, and negative on $(600, 1,000)$.
(B)

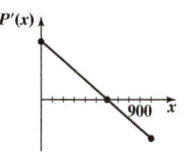

91. (A) The price decreases for the first 15 months to a local
minimum, increases for the next 40 months to a local
maximum, and then decreases for the remaining 15 months.
(B)

93. (A) $\overline{C}(x) = 0.05x + 20 + \dfrac{320}{x}$
(B) Critical value: $x = 80$; decreasing for $0 < x < 80$; increasing for $80 < x < 150$; local minimum at $x = 80$

95. $P(x)$ is increasing over (a, b) if $P'(x) = R'(x) - C'(x) > 0$ over (a, b); that is, if $R'(x) > C'(x)$ over (a, b).

97. Critical value: $t = 2$; increasing on $(0, 2)$; decreasing on $(2, 24)$; $C(2) = 0.07$ is a local maximum.

Exercises 5-2

1. (A) $(a, c), (c, d), (e, g)$ (B) $(d, e), (g, h)$ (C) $(d, e), (g, h)$ (D) $(a, c), (c, d), (e, g)$ (E) $(a, c), (c, d), (e, g)$ (F) $(d, e), (g, h)$
(G) d, e, g (H) d, e, g **3.** (C) **5.** (D) **7.** $12x - 8$ **9.** $4x^{-3} - 18x^{-4}$

11. $2 + \dfrac{9}{2}x^{-3/2}$ **13.** $8(x^2 + 9)^3 + 48x^2(x^2 + 9)^2 = 8(x^2 + 9)^2(7x^2 + 9)$

15. $-2e^{-x^2} + 4x^2e^{-x^2} = 2e^{-x^2}(2x^2 - 1)$ **17.** $\dfrac{x^3\left(-\dfrac{2}{x}\right) - (1 - 2\ln x)3x^2}{x^6} = \dfrac{6\ln x - 5}{x^4}$ **19.** $(-10, 2{,}000)$ **21.** $(0, 2)$ **23.** None

25. Concave upward for all x; no inflection points
27. Concave downward on $(-\infty, \frac{4}{3})$; concave upward on $(\frac{4}{3}, \infty)$; inflection point at $x = \frac{4}{3}$
29. Concave downward on $(-\infty, 0)$ and $(6, \infty)$; concave upward on $(0, 6)$; inflection points at $x = 0$ and $x = 6$
31. Concave upward on $(-2, 4)$; concave downward on $(-\infty, -2)$ and $(4, \infty)$; inflection points at $x = -2$ and $x = 4$
33. Concave upward on $(-\infty, \ln 2)$; concave downward on $(\ln 2, \infty)$; inflection point at $x = \ln 2$

35.

37.

39.

41.

43. Domain: All real numbers
y intercept: 16; x intercepts: $2 - 2\sqrt{3}, 2, 2 + 2\sqrt{3}$
Increasing on $(-\infty, 0)$ and $(4, \infty)$
Decreasing on $(0, 4)$
Local maximum at $x = 0$, local minimum at $x = 4$
Concave downward on $(-\infty, 2)$
Concave upward on $(2, \infty)$
Inflection point at $x = 2$

45. Domain: All real numbers
y intercept: 2; x intercept: -1
Increasing on $(-\infty, \infty)$
Concave downward on $(-\infty, 0)$
Concave upward on $(0, \infty)$
Inflection point at $x = 0$

47. Domain: All real numbers
y intercept: 0; x intercepts: 0, 4
Increasing on $(-\infty, 3)$
Decreasing on $(3, \infty)$
Local maximum at $x = 3$
Concave upward on $(0, 2)$
Concave downward on $(-\infty, 0)$ and $(2, \infty)$
Inflection points at $x = 0$ and $x = 2$

49. Domain: All real numbers
y intercept: 0; x intercepts: 0, 1
Increasing on $(0.25, \infty)$
Decreasing on $(-\infty, 0.25)$
Local minimum at $x = 0.25$
Concave upward on $(-\infty, 0.5)$ and $(1, \infty)$
Concave downward on $(0.5, 1)$
Inflection points at $x = 0.5$ and $x = 1$

51. Domain: All real numbers
y intercept: 27; x intercepts: $-3, 3$
Increasing on $(-\infty, -\sqrt{3})$ and $(0, \sqrt{3})$
Decreasing on $(-\sqrt{3}, 0)$ and $(\sqrt{3}, \infty)$
Local maxima at $x = -\sqrt{3}$ and $x = \sqrt{3}$
Local minimum at $x = 0$
Concave upward on $(-1, 1)$
Concave downward on $(-\infty, -1)$ and $(1, \infty)$
Inflection points at $x = -1$ and $x = 1$

53. Domain: All real numbers
y intercept: 16; x intercepts: $-2, 2$
Decreasing on $(-\infty, -2)$ and $(0, 2)$
Increasing on $(-2, 0)$ and $(2, \infty)$
Local minima at $x = -2$ and $x = 2$
Local maximum at $x = 0$
Concave upward on $(-\infty, -2\sqrt{3}/3)$ and $(2\sqrt{3}/3, \infty)$
Concave downward on $(-2\sqrt{3}/3, 2\sqrt{3}/3)$
Inflection points at $x = -2\sqrt{3}/3$ and $x = 2\sqrt{3}/3$

55. Domain: All real numbers
y intercept: 0; x intercepts: 0, 1.5
Decreasing on $(-\infty, 0)$ and $(0, 1.25)$
Increasing on $(1.25, \infty)$
Local minimum at $x = 1.25$
Concave upward on $(-\infty, 0)$ and $(1, \infty)$
Concave downward on $(0, 1)$
Inflection points at $x = 0$ and $x = 1$

57. Domain: All real numbers
y intercept: 0; x intercept: 0
Increasing on $(-\infty, \infty)$
Concave downward on $(-\infty, \infty)$

59. Domain: All real numbers
y intercept: 5
Decreasing on $(-\infty, \ln 4)$
Increasing on $(\ln 4, \infty)$
Local minimum at $x = \ln 4$
Concave upward on $(-\infty, \infty)$

61. Domain: $(0, \infty)$
x intercept: e^2
Increasing on $(-\infty, \infty)$
Concave downward on $(-\infty, \infty)$

63. Domain: $(-4, \infty)$
x intercept: $e^2 - 4$
Increasing on $(-4, \infty)$
Concave downward on $(-4, \infty)$

65.

x	$f'(x)$	$f(x)$
$-\infty < x < -1$	Positive and decreasing	Increasing and concave downward
$x = -1$	x intercept	Local maximum
$-1 < x < 0$	Negative and decreasing	Decreasing and concave downward
$x = 0$	Local minimum	Inflection point
$0 < x < 2$	Negative and increasing	Decreasing and concave upward
$x = 2$	Local maximum	Inflection point
$2 < x < \infty$	Negative and decreasing	Decreasing and concave downward

67.

x	$f'(x)$	$f(x)$
$-\infty < x < -2$	Negative and increasing	Decreasing and concave upward
$x = -2$	Local maximum	Inflection point
$-2 < x < 0$	Negative and decreasing	Decreasing and concave downward
$x = 0$	Local minimum	Inflection point
$0 < x < 2$	Negative and increasing	Decreasing and concave upward
$x = 2$	Local maximum	Inflection point
$2 < x < \infty$	Negative and decreasing	Decreasing and concave downward

69. Domain: All real numbers
x intercepts: $-1.18, 0.61, 1.87, 3.71$
y intercept: -5
Decreasing on $(-\infty, -0.53)$ and $(1.24, 3.04)$
Increasing on $(-0.53, 1.24)$ and $(3.04, \infty)$
Local minima at $x = -0.53$ and $x = 3.04$
Local maximum at $x = 1.24$
Concave upward on $(-\infty, 0.22)$ and $(2.28, \infty)$
Concave downward on $(0.22, 2.28)$
Inflection points at $x = 0.22$ and $x = 2.28$

71. Domain: All real numbers
y intercept: 100; x intercepts: 8.01, 13.36
Increasing on $(-0.10, 4.57)$ and $(11.28, \infty)$
Decreasing on $(-\infty, -0.10)$ and $(4.57, 11.28)$
Local maximum at $x = 4.57$
Local minima at $x = -0.10$ and $x = 11.28$
Concave upward on $(-\infty, 1.95)$ and $(8.55, \infty)$
Concave downward on $(1.95, 8.55)$
Inflection points at $x = 1.95$ and $x = 8.55$

73. Domain: All real numbers
x intercepts: $-2.40, 1.16$; y intercept: 3
Increasing on $(-\infty, -1.58)$
Decreasing on $(-1.58, \infty)$
Local maximum at $x = -1.58$
Concave downward on $(-\infty, -0.88)$ and $(0.38, \infty)$
Concave upward on $(-0.88, 0.38)$
Inflection points at $x = -0.88$ and $x = 0.38$

75. Domain: All real numbers
x intercepts: $-6.68, -3.64, -0.72$; y intercept: 30
Decreasing on $(-5.59, -2.27)$ and $(1.65, 3.82)$
Increasing on $(-\infty, -5.59)$, $(-2.27, 1.65)$, and $(3.82, \infty)$
Local minima at $x = -2.27$ and $x = 3.82$
Local maxima at $x = -5.59$ and $x = 1.65$
Concave upward on $(-4.31, -0.40)$ and $(2.91, \infty)$
Concave downward on $(-\infty, -4.31)$ and $(-0.40, 2.91)$
Inflection points at $x = -4.31$, $x = -0.40$, and $x = 2.91$

77. If $f'(x)$ has a local extremum at $x = c$, then $f'(x)$ must change from increasing to decreasing or from decreasing to increasing at $x = c$. The graph of $y = f(x)$ must change concavity at $x = c$, and there must be an inflection point at $x = c$.

79. If there is an inflection point on the graph of $y = f(x)$ at $x = c$, then $f(x)$ must change concavity at $x = c$. Consequently, $f'(x)$ must change from increasing to decreasing or from decreasing to increasing at $x = c$, and $x = c$ is a local extremum for $f'(x)$.

81. The graph of the CPI is concave upward.

83. The graph of $y = C'(x)$ is positive and decreasing. Since marginal costs are decreasing, the production process is becoming more efficient as production increases.

85. (A) Local maximum at $x = 60$
(B) Concave downward on the whole interval $(0, 80)$

87. (A) Local maximum at $x = 1$
(B) Concave downward on $(-\infty, 2)$;
concave upward on $(2, \infty)$

89. Increasing on $(0, 10)$; decreasing on $(10, 15)$; point of diminishing returns is $x = 10$, max $T'(x) = T'(10) = 500$

91. Increasing on $(24, 36)$; decreasing on $(36, 45)$; point of diminishing returns is $x = 36$, max $N'(x) = N'(36) = 3,888$

93. (A)

(B) 32 ads to sell 574 cars per month

95. (A) Increasing on $(0, 10)$;
decreasing on $(10, 20)$
(B) Inflection point at $t = 10$
(C)

(D) $N'(10) = 300$

97. (A) Increasing on $(5, \infty)$;
decreasing on $(0, 5)$
(B) Inflection point at $n = 5$
(C)

(D) $T'(5) = 0$

Exercises 5-3

1. $\frac{8}{3}$ **3.** $\frac{1}{2}$ **5.** 1 **7.** 4 **9.** 0 **11.** ∞ **13.** ∞ **15.** 5 **17.** ∞ **19.** $-\frac{1}{8}$ **21.** 17 **23.** 8 **25.** 0 **27.** ∞ **29.** ∞ **31.** $\frac{1}{3}$
33. -2 **35.** $-\infty$ **37.** 0 **39.** 0 **41.** 0 **43.** $\frac{1}{4}$ **45.** $\frac{1}{3}$ **47.** 0 **49.** 0 **51.** ∞ **53.** 1 **55.** 1

Exercises 5-4

1. (A) $(-\infty, b), (0, e), (e, g)$ (B) $(b, d), (d, 0), (g, \infty)$ (C) $(b, d), (d, 0), (g, \infty)$ (D) $(-\infty, b), (0, e), (e, g)$ (E) $x = 0$
(F) $x = b, x = g$ (G) $(-\infty, a), (d, e), (h, \infty)$ (H) $(a, d), (e, h)$ (I) $(a, d), (e, h)$ (J) $(-\infty, a), (d, e), (h, \infty)$
(K) $x = a, x = h$ (L) $y = L$ (M) $x = d, x = e$

3.

5.

7.

9.

11. Domain: $[-4, \infty)$ y intercept: 2; x intercept: -4 **13.** Domain: All real numbers y intercept: 75; x intercept: 15
15. Domain: All real numbers, except 5 y intercept: -19

17. Domain: All real numbers, except 3
y intercept: -1; x intercept: -3
Horizontal asymptote: $y = 1$
Vertical asymptote: $x = 3$
Decreasing on $(-\infty, 3)$ and $(3, \infty)$
Concave upward on $(3, \infty)$
Concave downward on $(-\infty, 3)$

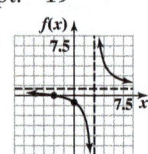

19. Domain: All real numbers, except 2
y intercept: 0; x intercept: 0
Horizontal asymptote: $y = 1$
Vertical asymptote: $x = 2$
Decreasing on $(-\infty, 2)$ and $(2, \infty)$
Concave downward on $(-\infty, 2)$
Concave upward on $(2, \infty)$

21. Domain: $(-\infty, \infty)$
y intercept: 10
Horizontal asymptote: $y = 5$
Decreasing on $(-\infty, \infty)$
Concave upward on $(-\infty, \infty)$

23. Domain: $(-\infty, \infty)$
y intercept: 0; x intercept: 0
Horizontal asymptote: $y = 0$
Increasing on $(-\infty, 5)$
Decreasing on $(5, \infty)$
Local maximum at $x = 5$
Concave upward on $(10, \infty)$
Concave downward on $(-\infty, 10)$
Inflection point at $x = 10$

25. Domain: $(-\infty, 1)$
y intercept: 0; x intercept: 0
Vertical asymptote: $x = 1$
Decreasing on $(-\infty, 1)$
Concave downward on $(-\infty, 1)$

27. Domain: $(0, \infty)$
Vertical asymptote: $x = 0$
Increasing on $(1, \infty)$
Decreasing on $(0, 1)$
Local minimum at $x = 1$
Concave upward on $(0, \infty)$

29. Domain: All real numbers, except ± 2
y intercept: 0; x intercept: 0
Horizontal asymptote: $y = 0$
Vertical asymptotes: $x = -2$, $x = 2$
Decreasing on $(-\infty, -2)$, $(-2, 2)$,
and $(2, \infty)$
Concave upward on $(-2, 0)$ and $(2, \infty)$
Concave downward on $(-\infty, -2)$ and $(0, 2)$
Inflection point at $x = 0$

31. Domain: All real numbers
y intercept: 1
Horizontal asymptote: $y = 0$
Increasing on $(-\infty, 0)$
Decreasing on $(0, \infty)$
Local maximum at $x = 0$
Concave upward on $(-\infty, -\sqrt{3}/3)$ and $(\sqrt{3}/3, \infty)$
Concave downward on $(-\sqrt{3}/3, \sqrt{3}/3)$
Inflection points at $x = -\sqrt{3}/3$ and $x = \sqrt{3}/3$

33. Domain: All real numbers except -1 and 1
y intercept: 0; x intercept: 0
Horizontal asymptote: $y = 0$
Vertical asymptote: $x = -1$ and $x = 1$
Increasing on $(-\infty, -1)$, $(-1, 1)$, and $(1, \infty)$
Concave upward on $(-\infty, -1)$ and $(0, 1)$
Concave downward on $(-1, 0)$ and $(1, \infty)$
Inflection point at $x = 0$

35. Domain: All real numbers except 1
y intercept: 0; x intercept: 0
Horizontal asymptote: $y = 0$
Vertical asymptote: $x = 1$
Increasing on $(-\infty, -1)$ and $(1, \infty)$
Decreasing on $(-1, 1)$
Local maximum at $x = -1$
Concave upward on $(-\infty, -2)$
Concave downward on $(-2, 1)$ and $(1, \infty)$
Inflection point at $x = -2$

37. Domain: All real numbers except 0
Horizontal asymptote: $y = 1$
Vertical asymptote: $x = 0$
Increasing on $(0, 4)$
Decreasing on $(-\infty, 0)$ and $(4, \infty)$
Local maximum at $x = 4$
Concave upward on $(6, \infty)$
Concave downward on $(-\infty, 0)$ and $(0, 6)$
Inflection point at $x = 6$

39. Domain: All real numbers except 1
y intercept: 0; x intercept: 0
Vertical asymptote: $x = 1$
Increasing on $(-\infty, 0)$ and $(2, \infty)$
Decreasing on $(0, 1)$ and $(1, 2)$
Local maximum at $x = 0$
Local minimum at $x = 2$
Concave upward on $(1, \infty)$
Concave downward on $(-\infty, 1)$

41. Domain: All real numbers except -3, 3
y intercept: $-\frac{2}{9}$
Horizontal asymptote: $y = 3$
Vertical asymptotes: $x = -3$, $x = 3$
Increasing on $(-\infty, -3)$ and $(-3, 0)$
Decreasing on $(0, 3)$ and $(3, \infty)$
Local maximum at $x = 0$
Concave upward on $(-\infty, -3)$ and $(3, \infty)$
Concave downward on $(-3, 3)$

43. Domain: All real numbers except 2
y intercept: 0; x intercept: 0
Vertical asymptote: $x = 2$
Increasing on $(3, \infty)$
Decreasing on $(-\infty, 2)$ and $(2, 3)$
Local minimum at $x = 3$
Concave upward on $(-\infty, 0)$ and $(2, \infty)$
Concave downward on $(0, 2)$
Inflection point at $x = 0$

45. Domain: All real numbers
y intercept: 3; x intercept: 3
Horizontal asymptote: $y = 0$
Increasing on $(-\infty, 2)$
Decreasing on $(2, \infty)$
Local maximum at $x = 2$
Concave upward on $(-\infty, 1)$
Concave downward on $(1, \infty)$
Inflection point at $x = 1$

47. Domain: $(-\infty, \infty)$
y intercept: 1
Horizontal asymptote: $y = 0$
Increasing on $(-\infty, 0)$
Decreasing on $(0, \infty)$
Local maximum at $x = 0$
Concave upward on $(-\infty, -1)$ and $(1, \infty)$
Concave downward on $(-1, 1)$
Inflection points at $x = -1$ and $x = 1$

49. Domain: $(0, \infty)$
x intercept: 1
Increasing on $(e^{-1/2}, \infty)$
Decreasing on $(0, e^{-1/2})$
Local minimum at $x = e^{-1/2}$
Concave upward on $(e^{-3/2}, \infty)$
Concave downward on $(0, e^{-3/2})$
Inflection point at $x = e^{-3/2}$

51. Domain: $(0, \infty)$
x intercept: 1
Vertical asymptote: $x = 0$
Increasing on $(1, \infty)$
Decreasing on $(0, 1)$
Local minimum at $x = 1$
Concave upward on $(0, e)$
Concave downward on (e, ∞)
Inflection point at $x = e$

53. Domain: All real numbers except $-4, 2$
y intercept: $-1/8$
Horizontal asymptote: $y = 0$
Vertical asymptote: $x = -4, x = 2$
Increasing on $(-\infty, -4)$ and $(-4, -1)$
Decreasing on $(-1, 2)$ and $(2, \infty)$
Local maximum at $x = -1$
Concave upward on $(-\infty, -4)$ and $(2, \infty)$
Concave downward on $(-4, 2)$

55. Domain: All real numbers except $-\sqrt{3}, \sqrt{3}$
y intercept: 0; x intercept: 0
Vertical asymptote: $x = -\sqrt{3}, x = \sqrt{3}$
Increasing on $(-3, -\sqrt{3})$, $(-\sqrt{3}, \sqrt{3})$, and $(\sqrt{3}, 3)$
Decreasing on $(-\infty, -3)$ and $(3, \infty)$
Local maximum at $x = 3$
Local minimum at $x = -3$
Concave upward on $(-\infty, -\sqrt{3})$ and $(0, \sqrt{3})$
Concave downward on $(-\sqrt{3}, 0)$ and $(\sqrt{3}, \infty)$
Inflection point at $x = 0$

57. Domain: All real numbers except 0
Vertical asymptote: $x = 0$
Oblique asymptote: $y = x$
Increasing on $(-\infty, -2)$ and $(2, \infty)$
Decreasing on $(-2, 0)$ and $(0, 2)$
Local maximum at $x = -2$
Local minimum at $x = 2$
Concave upward on $(0, \infty)$
Concave downward on $(-\infty, 0)$

59. Domain: All real numbers except 0
x intercept: $\sqrt[3]{4}$
Vertical asymptote: $x = 0$
Oblique asymptote: $y = x$
Increasing on $(-\infty, -2)$ and $(0, \infty)$
Local maximum at $x = -2$
Decreasing on $(-2, 0)$
Concave downward on $(-\infty, 0)$ and $(0, \infty)$

61. Domain: All real numbers except 0
x intercepts: $-\sqrt{3}, \sqrt{3}$
Vertical asymptote: $x = 0$
Oblique asymptote: $y = x$
Increasing on $(-\infty, 0)$ and $(0, \infty)$
Concave upward on $(-\infty, 0)$
Concave downward on $(0, \infty)$

63. Domain: All real numbers except 0
Vertical asymptote: $x = 0$
Oblique asymptote: $y = x$
Increasing on $(-\infty, -2)$ and $(2, \infty)$
Decreasing on $(-2, 0)$ and $(0, 2)$
Local maximum at $x = -2$
Local minimum at $x = 2$
Concave upward on $(0, \infty)$
Concave downward on $(-\infty, 0)$

65. Domain: All real numbers except $2, 4$
y intercept: $-3/4$; x intercept: -3
Vertical asymptote: $x = 4$
Horizontal asymptote: $y = 1$
Decreasing on $(-\infty, 2)$, $(2, 4)$, and $(4, \infty)$
Concave upward on $(4, \infty)$
Concave downward on $(-\infty, 2)$ and $(2, 4)$

67. Domain: All real numbers except $-3, 3$
y intercept: $5/3$; x intercept: 2.5
Vertical asymptote: $x = 3$
Horizontal asymptote: $y = 2$
Decreasing on $(-\infty, -3)$, $(-3, 3)$, and $(3, \infty)$
Concave upward on $(3, \infty)$
Concave downward on $(-\infty, -3)$ and $(-3, 3)$

69. Domain: All real numbers except $-1, 2$
y intercept: 0; x intercepts: $0, 3$
Vertical asymptote: $x = -1$
Increasing on $(-\infty, -3)$, $(1, 2)$, and $(2, \infty)$
Decreasing on $(-3, -1)$ and $(-1, 1)$
Local maximum at $x = -3$
Local minimum at $x = 1$
Concave upward on $(-1, 2)$ and $(2, \infty)$
Concave downward on $(-\infty, -1)$

71. Domain: All real numbers except 1
y intercept: -2; x intercept: -2
Vertical asymptote: $x = 1$
Horizontal asymptote: $y = 1$
Decreasing on $(-\infty, 1)$ and $(1, \infty)$
Concave upward on $(1, \infty)$
Concave downward on $(-\infty, 1)$

73.

75. (A) Increasing on $(0, 1)$
(B) Concave upward on $(0, 1)$
(C) $x = 1$ is a vertical asymptote
(D) The origin is both an x and a y intercept
(E)

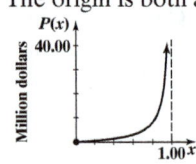

77. (A) $\overline{C}(n) = \dfrac{3,200}{n} + 250 + 50n$

(B)

(C) 8 yr

79. (A)

(B) 25 at $x = 100$

81. (A)

(B) Minimum average cost is $4.35 when 177 pizzas are produced daily.

83.

85.

Exercises 5-5

1. Min $f(x) = f(0) = 0$; Max $f(x) = f(10) = 14$ **3.** Min $f(x) = f(0) = 0$; Max $f(x) = f(3) = 9$
5. Min $f(x) = f(1) = f(7) = 5$; Max $f(x) = f(10) = 14$ **7.** Min $f(x) = f(1) = f(7) = 5$; Max $f(x) = f(3) = f(9) = 9$
9. Min $f(x) = f(5) = 7$; Max $f(x) = f(3) = 9$ **11.** Max $f(x) = f(3) = 21$; Min $f(x) = f(2) = 17$
13. Max $f(x) = f(-1) = e \approx 2.718$; Min $f(x) = e^{-1} \approx 0.368$ **15.** Max $f(x) = f(0) = 9$; Min $f(x) = f(\pm 4) = -7$
17. Min $f(x) = f(1) = 2$; no maximum   **19.** Max $f(x) = f(-3) = 18$; no minimum   **21.** No absolute extrema
23. Max $f(x) = f(3) = 54$; no minimum   **25.** No absolute extrema   **27.** Min $f(x) = f(0) = 0$; no maximum
29. Max $f(x) = f(1) = 1$; Min $f(x) = f(-1) = -1$ **31.** Min $f(x) = f(0) = -1$, no maximum   **33.** Min $f(x) = f(2) = -2$
35. Max $f(x) = f(2) = 4$ **37.** Min $f(x) = f(2) = 0$ **39.** No maximum   **41.** Max $f(x) = f(2) = 8$

43. Min $f(x) = f(4) = 22$ **45.** Min $f(x) = f(\sqrt{10}) = 14/\sqrt{10}$ **47.** Min $f(x) = f(2) = \dfrac{e^2}{4} \approx 1.847$

49. Min $f(x) = f(3) = \dfrac{27}{e^3} \approx 1.344$ **51.** Max $f(x) = f(e^{1.5}) = 2e^{1.5} \approx 8.963$

53. Min $f(x) = f(e^{2.5}) = \dfrac{e^5}{2} \approx 74.207$ **55.** Max $f(x) = f(1) = -1$

57. (A) Max $f(x) = f(5) = 14$; Min $f(x) = f(-1) = -22$ (B) Max $f(x) = f(1) = -2$; Min $f(x) = f(-1) = -22$
   (C) Max $f(x) = f(5) = 14$; Min $f(x) = f(3) = -6$
59. (A) Max $f(x) = f(0) = 126$; Min $f(x) = f(2) = -26$ (B) Max $f(x) = f(7) = 49$; Min $f(x) = f(2) = -26$
   (C) Max $f(x) = f(6) = 6$; Min $f(x) = f(3) = -15$
61. (A) Max $f(x) = f(-1) = 10$; Min $f(x) = f(2) = -11$ (B) Max $f(x) = f(0) = f(4) = 5$; Min $f(x) = f(3) = -22$
   (C) Max $f(x) = f(-1) = 10$; Min $f(x) = f(1) = 2$
63. Local minimum   **65.** Unable to determine   **67.** Neither   **69.** Local maximum

Exercises 5-6

1. 7.5 and 7.5   **3.** 7.5 and -7.5 **5.** $\sqrt{15}$ and $\sqrt{15}$ **7.** $10\sqrt{2}$ ft by $10\sqrt{2}$ ft   **9.** 37 ft by 37 ft
11. (A) Maximum revenue is $125,000 when 500 phones are produced and sold for $250 each.
   (B) Maximum profit is $46,612.50 when 365 phones are produced and sold for $317.50 each.
13. (A) Max $R(x) = R(3,000) = \$300,000$
   (B) Maximum profit is $75,000 when 2,100 sets are manufactured and sold for $130 each.
   (C) Maximum profit is $64,687.50 when 2,025 sets are manufactured and sold for $132.50 each.
15. (A)   (B)   (C) The maximum profit is $118,996 when the price per sleeping bag is $195.

17. (A) $4.80   (B) $8   **19.** $35; $6,125   **21.** 40 trees; 1,600 lb   **23.** $(10 - 2\sqrt{7})/3 = 1.57$ in. squares
25. 20 ft by 40 ft (with the expensive side being one of the short sides)   **27.** 8 production runs per year
29. 10,000 books in 5 printings   **31.** (A) $x = 5.1$ mi   (B) $x = 10$ mi   **33.** 4 days; 20 bacteria/cm^3
35. 1 month; 2 ft   **37.** 4 yr from now

Chapter 5 Review Exercises

1. $(a, c_1), (c_3, c_6)$ (5-1, 5-2)   **2.** $(c_1, c_3), (c_6, b)$ (5-1, 5-2)   **3.** $(a, c_2), (c_4, c_5), (c_7, b)$ (5-1, 5-2)   **4.** c_3 (5-1)
5. c_1, c_6 (5-1)   **6.** c_1, c_3, c_5 (5-1)   **7.** c_4, c_6 (5-1)   **8.** c_2, c_4, c_5, c_7 (5-2)   **11.** $f''(x) = 12x^2 + 30x$ (5-2)
9. (5-2)   **10.** (5-2)   **12.** $y'' = 8/x^3$ (5-2)
13. Domain: All real numbers, except 4
   y intercept: $\frac{5}{4}$; x intercept: -5 (5-2)
14. Domain: $(-2, \infty)$
   y intercept: $\ln 2$; x intercept: -1 (5-2)

15. Horizontal asymptote: $y = 0$;
Vertical asymptotes: $x = -2$, $x = 2$ *(5-4)*

16. Horizontal asymptote: $y = \frac{2}{3}$;
Vertical asymptote: $x = -\frac{10}{3}$ *(5-4)*

17. $(-\sqrt{2}, -20)$, $(\sqrt{2}, -20)$ *(5-2)*   **18.** $(-\frac{1}{2}, -6)$ *(5-2)*   **19.** (A) $f'(x) = \frac{1}{5}x^{-1/5}$ (B) 0   (C) 0 *(5-1)*

20. (A) $f'(x) = -\frac{1}{5}x^{-6/5}$ (B) None   (C) 0 *(5-1)*

21. Domain: All real numbers
y intercept: 0; x intercepts: 0, 9
Increasing on $(-\infty, 3)$ and $(9, \infty)$
Decreasing on $(3, 9)$
Local maximum at $x = 3$
Local minimum at $x = 9$
Concave upward on $(6, \infty)$
Concave downward on $(-\infty, 6)$
Inflection point at $x = 6$ *(5-4)*

22. Domain: All real numbers
y intercept: 16; x intercepts: -4, 2
Increasing on $(-\infty, -2)$ and $(2, \infty)$
Decreasing on $(-2, 2)$
Local maximum at $x = -2$
Local minimum at $x = 2$
Concave upward on $(0, \infty)$
Concave downward on $(-\infty, 0)$
Inflection point at $x = 0$ *(5-4)*

23. Domain: All real numbers
y intercept: 0; x intercepts: 0, 4
Increasing on $(-\infty, 3)$
Decreasing on $(3, \infty)$
Local maximum at $x = 3$
Concave upward on $(0, 2)$
Concave downward on $(-\infty, 0)$ and $(2, \infty)$
Inflection points at $x = 0$ and $x = 2$ *(5-4)*

24. Domain: all real numbers
y intercept: -3; x intercepts: -3, 1
No vertical or horizontal asymptotes
Increasing on $(-2, \infty)$
Decreasing on $(-\infty, -2)$
Local minimum at $x = -2$
Concave upward on $(-\infty, -1)$ and $(1, \infty)$
Concave downward on $(-1, 1)$
Inflection points at $x = -1$ and $x = 1$ *(5-4)*

25. Domain: All real numbers, except -2
y intercept: 0; x intercept: 0
Horizontal asymptote: $y = 3$
Vertical asymptote: $x = -2$
Increasing on $(-\infty, -2)$ and $(-2, \infty)$
Concave upward on $(-\infty, -2)$
Concave downward on $(-2, \infty)$ *(5-4)*

26. Domain: All real numbers
y intercept: 0; x intercept: 0
Horizontal asymptote: $y = 1$
Increasing on $(0, \infty)$
Decreasing on $(-\infty, 0)$
Local minimum at $x = 0$
Concave upward on $(-3, 3)$
Concave downward on $(-\infty, -3)$ and $(3, \infty)$
Inflection points at $x = -3$, 3 *(5-4)*

27. Domain: All real numbers except $x = -2$
y intercept: 0; x intercept: 0
Horizontal asymptote: $y = 0$
Vertical asymptote: $x = -2$
Increasing on $(-2, 2)$
Decreasing on $(-\infty, -2)$ and $(2, \infty)$
Local maximum at $x = 2$
Concave upward on $(4, \infty)$
Concave downward on $(-\infty, -2)$ and $(-2, 4)$
Inflection point at $x = 4$ *(5-4)*

28. Domain: All real numbers
y intercept: 0; x intercept: 0
Increasing on $(-\infty, \infty)$
Concave upward on $(-\infty, -3)$ and $(0, 3)$
Concave downward on $(-3, 0)$ and $(3, \infty)$
Inflection points at $x = -3, 0, 3$ *(5-4)*

29. Domain: All real numbers
y intercept: 0; x intercept: 0
Horizontal asymptote: $y = 5$
Increasing on $(-\infty, \infty)$
Concave downward on $(-\infty, \infty)$

30. Domain: $(0, \infty)$
x intercept: 1
Increasing on $(e^{-1/3}, \infty)$
Decreasing on $(0, e^{-1/3})$
Local minimum at $x = e^{-1/3}$
Concave upward on $(e^{-5/6}, \infty)$
Concave downward on $(0, e^{-5/6})$
Inflection point at $x = e^{-5/6}$

31. 3 *(5-3)*   **32.** $-\frac{1}{5}$ *(5-3)*   **33.** $-\infty$ *(5-3)*   **34.** 0 *(5-3)*   **35.** 0 *(5-3)*
36. 1 *(5-3)*   **37.** 0 *(5-3)*   **38.** 0 *(5-3)*   **39.** 1 *(5-3)*   **40.** 2 *(5-3)*

41.

x	f'(x)	f(x)
$-\infty < x < -2$	Negative and increasing	Decreasing and concave upward
$x = -2$	x intercept	Local minimum
$-2 < x < -1$	Positive and increasing	Increasing and concave upward
$x = -1$	Local maximum	Inflection point
$-1 < x < 1$	Positive and decreasing	Increasing and concave downward
$x = 1$	Local minimum	Inflection point
$1 < x < \infty$	Positive and increasing	Increasing and concave upward

42. (C) *(5-2)*   **43.** Local maximum at $x = -1$; local minimum at $x = 5$ *(5-5)*
44. Min $f(x) = f(2) = -4$; Max $f(x) = f(5) = 77$ *(5-5)*   **45.** Min $f(x) = f(2) = 8$ *(5-5)*
46. Max $f(x) = f(e^{4.5}) = 2e^{4.5} \approx 180.03$ *(5-5)*   **47.** Max $f(x) = f(0.5) = 5e^{-1} \approx 1.84$ *(5-5)*
48. Yes. Since f is continuous on $[a, b]$, f has an absolute maximum on $[a, b]$. But each endpoint is a local minimum; hence, the absolute maximum must occur between a and b. *(5-5)*
49. No, increasing/decreasing properties apply to intervals in the domain of f. It is correct to say that $f(x)$ is decreasing on $(-\infty, 0)$ and $(0, \infty)$. *(5-1)*
50. A critical value for $f(x)$ is a partition number for $f'(x)$ that is also in the domain of f. For example, if $f(x) = x^{-1}$, then 0 is a partition number for $f'(x) = -x^{-2}$, but 0 is not a critical value for $f(x)$ since 0 is not in the domain of f. *(5-1)*

51. Max $f'(x) = f'(2) = 12$ *(5-2, 5-5)*

52. Each number is 20; minimum sum is 40 *(5-6)*
53. Domain: All real numbers
x intercepts: 0.79, 1.64; y intercept: 4
Increasing on $(-1.68, -0.35)$ and $(1.28, \infty)$
Decreasing on $(-\infty, -1.68)$ and $(-0.35, 1.28)$
Local minima at $x = -1.68$ and $x = 1.28$
Local maximum at $x = -0.35$
Concave downward on $(-1.10, 0.60)$
Concave upward on $(-\infty, -1.10)$ and $(0.60, \infty)$
Inflection points at $x = -1.10$ and $x = 0.60$ *(5-4)*

54. Domain: All real numbers
x intercepts; 0, 11.10; y intercept: 0
Increasing on $(1.87, 4.19)$ and $(8.94, \infty)$
Decreasing on $(-\infty, 1.87)$ and $(4.19, 8.94)$
Local maximum at $x = 4.19$
Local minima at $x = 1.87$ and $x = 8.94$
Concave upward on $(-\infty, 2.92)$ and $(7.08, \infty)$
Concave downward on $(2.92, 7.08)$
Inflection points at $x = 2.92$ and $x = 7.08$ *(5-4)*

55. Max $f(x) = f(1.373) = 2.487$ *(5-5)*
56. Max $f(x) = f(1.763) = 0.097$ *(5-5)*

57. (A) For the first 15 months, the graph of the price is increasing and concave downward, with a local maximum at $t = 15$. For the next 15 months, the graph of the price is decreasing and concave downward, with an inflection point at $t = 30$. For the next 15 months, the graph of the price is decreasing and concave upward, with a local minimum at $t = 45$. For the remaining 15 months, the graph of the price is increasing and concave upward.

(B)   *(5-2)*

58. (A) Max $R(x) = R(10,000) = \$2,500,000$
(B) Maximum profit is $175,000 when 3,000 readers are manufactured and sold for $425 each.
(C) Maximum profit is $119,000 when 2,600 readers are manufactured and sold for $435 each. *(5-6)*
59. (A) The expensive side is 50 ft; the other side is 100 ft.
(B) The expensive side is 75 ft; the other side is 150 ft. *(5-6)*
60. $49; $6,724 *(5-6)*   **61.** 12 orders/yr *(5-6)*   **62.** Min $\overline{C}(x) = \overline{C}(200) = 50$
63. Min $\overline{C}(x) = \overline{C}(e^5) \approx \$49.66$ *(5-4)*   **64.** A maximum revenue of $18,394 is realized at a production level of 50 units at $367.88 each. *(5-6)*

65.

66. $549.15; $9,864 *(5-6)*
67. $1.52 *(5-6)*
68. 20.39 feet *(5-6)*
69. (A)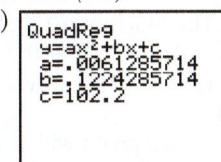

(B) Min $\overline{C}(x) = \overline{C}(129) = \$1.71$ *(5-4)*

70. Increasing on $(0, 18)$; decreasing on $(18, 24)$; point of diminishing returns is $x = 18$, max $N'(x) = N'(18) = 972$ *(5-2)*

71. (A)

```
CubicReg
y=ax³+bx²+cx+d
a=-.01
b=.83
c=-2.3
d=221
```

(B) 28 ads to sell 588 refrigerators per month *(5-2)*

72. 3 days *(5-1)*

73. 2 yr from now *(5-1)*

Chapter 6

Exercises 6-1

1. −4   **3.** 11   **5.** 4   **7.** Not defined   **9.** 59   **11.** −1   **13.** −64   **15.** 154,440   **17.** $192\pi \approx 603.2$ **19.** $3\pi\sqrt{109} \approx 98.4$

21. 118   **23.** 36,095.07   **25.** $f(x) = x^2 - 7$ **27.** $f(y) = 38y + 20$ **29.** $f(y) = -2y^3 + 5$ **31.** $y = 2$ **33.** $x = -\dfrac{1}{2}, 1$

35. −1.926, 0.599   **37.** $2x + h$ **39.** $2y^2$ **41.** $E(0, 0, 3); F(2, 0, 3)$

43. (A) In the plane $y = c$, c any constant, $z = x^2$.
   (B) The y axis; the line parallel to the y axis and passing through the point $(1, 0, 1)$; the line parallel to the y axis and passing through the point $(2, 0, 4)$
   (C) A parabolic "trough" lying on top of the y axis
45. (A) Upper semicircles whose centers lie on the y axis
   (B) Upper semicircles whose centers lie on the x axis
   (C) The upper hemisphere of radius 6 with center at the origin
47. (A) $a^2 + b^2$ and $c^2 + d^2$ both equal the square of the radius of the circle.
   (B) Bell-shaped curves with maximum values of 1 at the origin
   (C) A bell, with maximum value 1 at the origin, extending infinitely far in all directions.
49. $13,200; $18,000; $21,300   **51.** $R(p, q) = -5p^2 + 6pq - 4q^2 + 200p + 300q$; $R(2, 3) = $1,280$; $R(3, 2) = $1,175$
53. 30,065 units   **55.** (A) $237,877.08   (B) 4.4%   **57.** $T(70, 47) \approx 29$ min; $T(60, 27) = 33$ min
59. $C(6, 8) = 75$; $C(8.1, 9) = 90$ **61.** $Q(12, 10) = 120$; $Q(10, 12) \approx 83$

Exercises 6-2

1. $f_x(x, y) = 4$ **3.** $f_y(x, y) = -3x + 4y$ **5.** $\dfrac{\partial z}{\partial x} = 3x^2 + 8xy$ **7.** $\dfrac{\partial z}{\partial y} = 20(5x + 2y)^9$ **9.** 9   **11.** 3   **13.** −4   **15.** 0

17. 45.6 mpg; mileage is 45.6 mpg at a tire pressure of 32 psi and a speed of 40 mph
19. 37.6 mpg; mileage is 37.6 mpg at a tire pressure of 32 psi and a speed of 50 mph
21. 0.3 mpg per psi; mileage increases at a rate of 0.3 mpg per psi of tire pressure
23. $f_{xx}(x, y) = 0$ **25.** $f_{xy}(x, y) = 0$ **27.** $f_{xy}(x, y) = y^2 e^{xy^2}(2xy) + e^{xy^2}(2y) = 2y(1 + xy^2)e^{xy^2}$ **29.** $f_{yy}(x, y) = \dfrac{2\ln x}{y^3}$

31. $f_{xx}(x, y) = 80(2x + y)^3$ **33.** $f_{xy}(x, y) = 720xy^3(x^2 + y^4)^8$ **35.** $C_x(x, y) = 6x + 10y + 4$ **37.** 2   **39.** $C_{xx}(x, y) = 6$

41. $C_{xy}(x, y) = 10$ **43.** 6   **45.** $S_y(x, y) = \dfrac{x^3}{y} + 8ye^x$ **47.** $-1 + 8e^{-1} \approx 1.943$ **49.** $S_{yx}(x, y) = \dfrac{3x^2}{y} + 8ye^x$

51. $S_{yy}(x, y) = -\dfrac{x^3}{y^2} + 8e^x$ **53.** $3 + 8e^{-1} \approx 5.943$ **55.** $1 + 8e^{-1} \approx 3.943$

57. $3,000; daily sales are $3,000 when the temperature is 60° and the rainfall is 2 in.
59. −2,500 $/in.; daily sales decrease at a rate of $2,500 per inch of rain when the temperature is 90° and rainfall is 1 in.
61. − 50 $/in. per °F; S_r decreases at a rate of 50 $/in. per degree of temperature
65. $f_{xx}(x, y) = 2y^2 + 6x$; $f_{xy}(x, y) = 4xy = f_{yx}(x, y)$; $f_{yy}(x, y) = 2x^2$
67. $f_{xx}(x, y) = -2y/x^3$; $f_{xy}(x, y) = (-1/y^2) + (1/x^2) = f_{yx}(x, y)$; $f_{yy}(x, y) = 2x/y^3$
69. $f_{xx}(x, y) = (2y + xy^2)e^{xy}$; $f_{xy}(x, y) = (2x + x^2y)e^{xy} = f_{yx}(x, y)$; $f_{yy}(x, y) = x^3e^{xy}$
71. $x = 2$ and $y = 4$
73. $x = 1.200$ and $y = -0.695$
75. (A) $-\frac{13}{3}$ (B) The function $f(0, y)$, for example, has values less than $-\frac{3}{13}$.
77. (A) $c = 1.145$ (B) $f_x(c, 2) = 0$; $f_y(c, 2) = 92.021$
79. $f_{xx}(x, y) + f_{yy}(x, y) = (2y^2 - 2x^2)/(x^2 + y^2)^2 + (2x^2 - 2y^2)/(x^2 + y^2)^2 = 0$
81. (A) $2x$ (B) $4y$
83. $P_x(1,200, 1,800) = 24$; profit will increase approx. $24 per unit increase in production of type A calculators at the $(1,200, 1,800)$ output level; $P_y(1,200, 1,800) = -48$; profit will decrease approx. $48 per unit increase in production of type B calculators at the $(1,200, 1,800)$ output level

85. $\partial x/\partial p = -5$: a \$1 increase in the price of brand A will decrease the demand for brand A by 5 lb at any price level (p, q);
$\partial y/\partial p = 2$: a \$1 increase in the price of brand A will increase the demand for brand B by 2 lb at any price level (p, q)

87. (A) $f_x(x, y) = 7.5x^{-0.25}y^{0.25}$; $f_y(x, y) = 2.5x^{0.75}y^{-0.75}$
(B) Marginal productivity of labor = $f_x(600, 100) \approx 4.79$; Marginal productivity of capital = $f_y(600, 100) \approx 9.58$
(C) Capital

89. Competitive   **91.** Complementary

93. (A) $f_w(w, h) = 6.65w^{-0.575}h^{0.725}$; $f_h(w, h) = 11.34w^{0.425}h^{-0.275}$
(B) $f_w(65, 57) = 11.31$: for a 65-lb child 57 in. tall, the rate of change in surface area is 11.31 in.^{2} for each pound gained in weight (height is held fixed); $f_h(65, 57) = 21.99$: for a child 57 in. tall, the rate of change in surface area is 21.99 in.^{2} for each inch gained in height (weight is held fixed)

95. $C_W(6, 8) = 12.5$: index increases approx. 12.5 units for a 1-in. increase in width of head (length held fixed) when $W = 6$ and $L = 8$; $C_L(6, 8) = -9.38$: index decreases approx. 9.38 units for a 1-in. increase in length (width held fixed) when $W = 6$ and $L = 8$.

Exercises 6-3

1. $f_x(x, y) = 4$; $f_y(x, y) = 5$; the functions $f_x(x, y)$ and $f_y(x, y)$ never have the value 0.
3. $f_x(x, y) = -1.2 + 4x^3$; $f_y(x, y) = 6.8 + 0.6y^2$; the function $f_y(x, y)$ never has the value 0.
5. $f(-2, 0) = 10$ is a local maximum.   **7.** $f(-1, 3) = 4$ is a local minimum.
9. f has a saddle point at $(3, -2)$.   **11.** $f(3, 2) = 33$ is a local maximum.
13. $f(2, 2) = 8$ is a local minimum.   **15.** f has a saddle point at $(0, 0)$.
17. f has a saddle point at $(0, 0)$; $f(1, 1) = -1$ is a local minimum.
19. f has a saddle point at $(0, 0)$; $f(3, 18) = -162$ and $f(-3, -18) = -162$ are local minima.
21. The test fails at $(0, 0)$; f has saddle points at $(2, 2)$ and $(2, -2)$.   **23.** f has a saddle point at $(0.614, -1.105)$.
25. $f(x, y)$ is nonnegative and equals 0 when $x = 0$, so f has a local minimum at each point of the y axis.
27. (B) Local minimum   **29.** 2,000 type A and 4,000 type B; Max $P = P(2, 4) = \$15$ million
31. (A) When $p = \$110$ and $q = \$120$, $x = 80$ and $y = 40$; when $p = \$110$ and $q = \$110$, $x = 40$ and $y = 70$
(B) A maximum weekly profit of \$4,800 is realized for $p = \$100$ and $q = \$120$.
33. $P(x, y) = P(4, 2)$
35. 8 in. by 4 in. by 2 in.
37. 20 in. by 20 in. by 40 in.

Exercises 6-4

1. Max $f(x, y) = f(3, 3) = 18$ **3.** Min $f(x, y) = f(3, 4) = 25$
5. $F_x = -3 + 2\lambda = 0$ and $F_y = 4 + 5\lambda = 0$ have no simultaneous solution.
7. Max $f(x, y) = f(3, 3) = f(-3, -3) = 18$; Min $f(x, y) = f(3, -3) = f(-3, 3) = -18$
9. Maximum product is 25 when each number is 5.   **11.** Min $f(x, y, z) = f(-4, 2, -6) = 56$
13. Max $f(x, y, z) = f(2, 2, 2) = 6$; Min $f(x, y, z) = f(-2, -2, -2) = -6$
15. Max $f(x, y) = f(0.217, 0.885) = 1.055$ **17.** $F_x = e^x + \lambda = 0$ and $F_y = 3e^y - 2\lambda = 0$ have no simultaneous solution.
19. Maximize $f(x, 5)$, a function of just one independent variable.
21. (A) Max $f(x, y) = f(0.707, 0.5) = f(-0.707, 0.5) = 0.47$
23. 60 of model A and 30 of model B will yield a minimum cost of \$32,400 per week.
25. (A) 8,000 units of labor and 1,000 units of capital; Max $N(x, y) = N(8,000, 1,000) \approx 263,902$ units
(B) Marginal productivity of money ≈ 0.6598; increase in production $\approx 32,990$ units
27. 8 in. by 8 in. by $\frac{8}{3}$ in.   **29.** $x = 50$ ft and $y = 200$ ft; maximum area is 10,000 ft^2

Exercises 6-5

1. $y = 0.7x + 1$ **3.** $y = -2.5x + 10.5$ **5.** $y = x + 2$ **7.** $y = -1.5x + 4.5$; $y = 0.75$ when $x = 2.5$
9. $y = 2.12x + 10.8$; $y = 63.8$ when $x = 25$
11. $y = -1.2x + 12.6$; $y = 10.2$ when $x = 2$
13. $y = -1.53x + 26.67$; $y = 14.4$ when $x = 8$

15. $y = 0.75x^2 - 3.45x + 4.75$ **21.** (A) $y = 1.52x - 0.16$; $y = 0.73x^2 - 1.39x + 1.30$ (B) The quadratic function

23. The normal equations form a system of 4 linear equations in the 4 variables a, b, c, and d, which can be solved by Gauss–Jordan elimination.
25. (A) $y = -65.4x + 3755.7$ (B) 2,709 crimes per 100,000 population
27. (A) $y = -0.48x + 4.38$ (B) \$6.56 per bottle
29. (A) $y = 0.0237x + 18.92$ (B) 19.87 ft

31. (A) $y = 0.0121x + 56.35$ (B) 58.77°F

Exercises 6-6

1. (A) $3x^2y^4 + C(x)$ (B) $3x^2$ **3.** (A) $2x^2 + 6xy + 5x + E(y)$ (B) $35 + 30y$

5. (A) $\sqrt{y + x^2} + E(y)$ (B) $\sqrt{y + 4} - \sqrt{y}$ **7.** (A) $\dfrac{\ln x \ln y}{x} + C(x)$ (B) $\dfrac{2 \ln x}{x}$

9. 9 **11.** 330 **13.** $(56 - 20\sqrt{5})/3$ **15.** 1 **17.** 16 **19.** 49 **21.** $\frac{1}{8}\int_1^5 \int_{-1}^1 (x + y)^2 \, dy \, dx = \frac{32}{3}$

23. $\frac{1}{15}\int_1^4 \int_2^7 (x/y) \, dy \, dx = \frac{1}{2}\ln\frac{7}{2} \approx 0.6264$ **25.** $\frac{4}{3}$ cubic units **27.** $\frac{32}{3}$ cubic units **29.** $\int_0^1 \int_1^2 xe^{xy} \, dy \, dx = \frac{1}{2} + \frac{1}{2}e^2 - e$

31. $\int_0^1 \int_{-1}^1 \dfrac{2y + 3xy^2}{1 + x^2} \, dy \, dx = \ln 2$

35. (A) $\frac{1}{3} + \frac{1}{4}e^{-2} - \frac{1}{4}e^2$

(B)

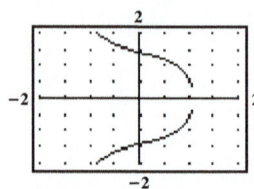

(C) Points to the right of the graph in part (B) are greater than 0; points to the left of the graph are less than 0.

37. $\dfrac{1}{0.4}\int_{0.6}^{0.8} \int_5^7 \dfrac{y}{1 - x} \, dy \, dx = 30 \ln 2 \approx \$20.8$ billion

39. $\frac{1}{10}\int_{10}^{20} \int_1^2 x^{0.75}y^{0.25} \, dy \, dx = \frac{8}{175}(2^{1.25} - 1)(20^{1.75} - 10^{1.75}) \approx 8.375$ or 8,375 units

41. $\frac{1}{192}\int_{-8}^8 \int_{-6}^6 [10 - \frac{1}{10}(x^2 + y^2)] \, dy \, dx = \frac{20}{3}$ insects/ft^2

43. $\frac{1}{8}\int_{-2}^2 \int_{-1}^1 [100 - 15(x^2 + y^2)] \, dy \, dx = 75$ ppm

45. $\frac{1}{10,000}\int_{2,000}^{3,000} \int_{50}^{60} 0.000\,013\,3xy^2 \, dy \, dx \approx 100.86$ ft

47. $\frac{1}{16}\int_8^{16} \int_{10}^{12} 100\dfrac{x}{y} \, dy \, dx = 600 \ln 1.2 \approx 109.4$

Exercises 6-7

1. $R = \{(x, y) | 0 \le y \le 4 - x^2, 0 \le x \le 2\}$
$R = \{(x, y) | 0 \le x \le \sqrt{4 - y}, 0 \le y \le 4\}$

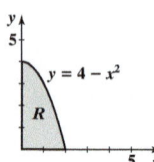

3. R is a regular x region:
$R = \{(x, y) | x^3 \le y \le 12 - 2x, 0 \le x \le 2\}$

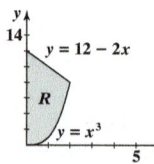

5. R is a regular y region:
$R = \{(x, y) | \frac{1}{2}y^2 \le x \le y + 4, -2 \le y \le 4\}$

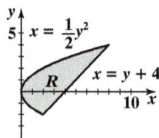

7. $\frac{1}{2}$ **9.** $\frac{39}{70}$

11. R consists of the points on or inside the rectangle with corners $(\pm 2, \pm 3)$; both

13. R is the arch-shaped region consisting of the points on or inside the rectangle with corners $(\pm 2, 0)$ and $(\pm 2, 2)$ that are not inside the circle of radius 1 centered at the origin; regular x region

15. $\frac{56}{3}$ **17.** $-\frac{3}{4}$ **19.** $\frac{1}{2}e^4 - \frac{5}{2}$

21. $R = \{(x, y) | 0 \le y \le x + 1, \; 0 \le x \le 1\}$
$\int_0^1 \int_0^{x+1} \sqrt{1 + x + y} \, dy \, dx = (68 - 24\sqrt{2})/15$

23. $R = \{(x, y) | 0 \le y \le 4x - x^2, 0 \le x \le 4\}$
$\int_0^4 \int_0^{4x-x^2} \sqrt{y + x^2} \, dy \, dx = \frac{128}{5}$

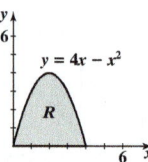

25. $R = \{(x, y) | 1 - \sqrt{x} \le y \le 1 + \sqrt{x}, 0 \le x \le 4\}$
$\int_0^4 \int_{1-\sqrt{x}}^{1+\sqrt{x}} x(y - 1)^2 \, dy \, dx = \frac{512}{21}$

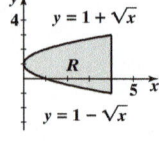

27. $\int_0^3 \int_0^{3-y} (x + 2y) \, dx \, dy = \frac{27}{2}$

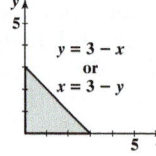

29. $\int_0^1 \int_0^{\sqrt{1-y}} x\sqrt{y}\, dx\, dy = \frac{2}{15}$

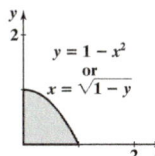

31. $\int_0^1 \int_{4y^2}^{4y} x\, dx\, dy = \frac{16}{15}$

33. $\int_0^4 \int_0^{4-x} (4 - x - y)\, dy\, dx = \frac{32}{3}$

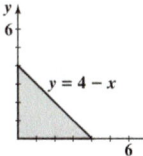

35. $\int_0^1 \int_0^{1-x^2} 4\, dy\, dx = \frac{8}{3}$

37. $\int_0^4 \int_0^{\sqrt{y}} \dfrac{4x}{1 + y^2}\, dx\, dy = \ln 17$

39. $\int_0^1 \int_0^{\sqrt{x}} 4ye^{x^2}\, dy\, dx = e - 1$

41. $R = \{(x, y)\,|\,x^2 \le y \le 1 + \sqrt{x}, 0 \le x \le 1.49\}$

$\int_0^{1.49} \int_{x^2}^{1+\sqrt{x}} x\, dy\, dx \approx 0.96$

43. $R = \{(x, y)\,|\,y^3 \le x \le 1 - y, 0 \le y \le 0.68\}$

$\int_0^{0.68} \int_{y^3}^{1-y} 24xy\, dx\, dy \approx 0.83$

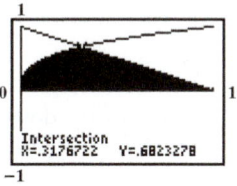

45. $R = \{(x, y)\,|\,e^{-x} \le y \le 3 - x,\quad -1.51 \le x \le 2.95\}$ Regular x region

$R = \{(x, y)\,|\,-\ln y \le x \le 3 - y,\quad 0.05 \le y \le 4.51\}$ Regular y region

$\int_{-1.51}^{2.95} \int_{e^{-x}}^{3-x} 4y\, dy\, dx = \int_{0.05}^{4.51} \int_{-\ln y}^{3-y} 4y\, dx\, dy \approx 40.67$

Chapter 6 Review Exercises

1. $f(5, 10) = 2{,}900; f_x(x, y) = 40; f_y(x, y) = 70$ *(8-1, 8-2)*   **2.** $\partial^2 z/\partial x^2 = 6xy^2; \partial^2 z/\partial x\, \partial y = 6x^2 y$ *(8-2)*

3. $2xy^3 + 2y^2 + C(x)$ *(8-6)*   **4.** $3x^2 y^2 + 4xy + E(y)$ *(8-6)*   **5.** 1 *(8-6)*

6. $f_x(x, y) = 5 + 6x + 3x^2; f_y(x, y) = -2$; the function $f_y(x, y)$ never has the value 0. *(8-3)*

7. $f(2, 3) = 7; f_y(x, y) = -2x + 2y + 3; f_y(2, 3) = 5$ *(8-1, 8-2)*   **8.** $(-8)(-6) - (4)^2 = 32$ *(8-2)*

9. $(1, 3, -\frac{1}{2}), (-1, -3, \frac{1}{2})$ *(8-4)*   **10.** $y = -1.5x + 15.5; y = 0.5$ when $x = 10$ *(8-5)*   **11.** 18 *(8-6)*   **12.** $\frac{8}{5}$ *(8-7)*

13. $f_x(x, y) = 2xe^{x^2+2y}; f_y(x, y) = 2e^{x^2+2y}; f_{xy}(x, y) = 4xe^{x^2+2y}$ *(8-2)*

14. $f_x(x, y) = 10x(x^2 + y^2)^4; f_{xy}(x, y) = 80xy(x^2 + y^2)^3$ *(8-2)*

15. $f(2, 3) = -25$ is a local minimum; f has a saddle point at $(-2, 3)$. *(8-3)*   **16.** Max $f(x, y) = f(6, 4) = 24$ *(8-4)*

17. Min $f(x, y, z) = f(2, 1, 2) = 9$ *(8-4)*   **18.** $y = \frac{116}{165}x + \frac{100}{3}$ *(8-5)*   **19.** $\frac{27}{5}$ *(8-6)*   **20.** 4 cubic units *(8-6)*

21. 0 *(8-6)*   **22.** (A) 12.56   (B) No *(8-6)*

23. $F_x = 12x^2 + 3\lambda = 0, F_y = -15y^2 + 2\lambda = 0,$ and $F_\lambda = 3x + 2y - 7 = 0$ have no simultaneous solution. *(8-4)*   **24.** 1 *(8-7)*

25. (A) $P_x(1, 3) = 8$; profit will increase \$8,000 for a 100-unit increase in product A if the production of product B is held fixed at an output level of $(1, 3)$.

(B) For 200 units of A and 300 units of B, $P(2, 3) = \$100$ thousand is a local maximum. *(8-2, 8-3)*

26. 8 in. by 6 in. by 2 in. *(8-3)*

27. $y = 0.63x + 1.33$; profit in sixth year is \$5.11 million *(8-4)*

28. (A) Marginal productivity of labor ≈ 8.37; marginal productivity of capital ≈ 1.67; management should encourage increased use of labor.

(B) 80 units of labor and 40 units of capital; Max $N(x, y) = N(80, 40) \approx 696$ units; marginal productivity of money ≈ 0.0696; increase in production ≈ 139 units

(C) $\dfrac{1}{1,000} \int_{50}^{100} \int_{20}^{40} 10x^{0.8}y^{0.2}\, dy\, dx = \dfrac{(40^{1.2} - 20^{1.2})(100^{1.8} - 50^{1.8})}{216} = 621$ items *(8-4)*

29. $T_x(70, 17) = -0.924$ min/ft increase in depth when $V = 70$ ft^3 and $x = 17$ ft *(8-2)*

30. $\frac{1}{16}\int_{-2}^{2}\int_{-2}^{2}[100 - 24(x^2 + y^2)]\, dy\, dx = 36$ ppm *(8-6)*   **31.** 50,000 *(8-1)*

32. $y = \frac{1}{2}x + 48$; $y = 68$ when $x = 40$ *(8-5)*

33. (A) $y = 0.4933x + 25.20$ **(B)** 84.40 people/mi^2 **(C)** 89.30 people/mi^2; 97.70 people/mi^2 *(8-5)*

34. (A) $y = 1.069x + 0.522$ **(B)** 64.68 yr   **(C)** 64.78 yr; 64.80 yr *(8-5)*

Chapter 7

Exercises 7-1

1. $7x + C$ **3.** $4x^2 + C$ **5.** $3x^3 + C$ **7.** $(x^6/6) + C$ **9.** $(-x^{-2}/2) + C$ **11.** $4x^{5/2} + C$ **13.** $3\ln|z| + C$ **15.** $16e^u + C$
17. $y = 40x^5 + C$ **19.** $P = 24x - 3x^2 + C$ **21.** $y = e^x + 3x + C$ **23.** $x = 5\ln|t| + t + C$ **25.** True   **27.** False   **29.** True
31. No, since one graph cannot be obtained from another by a vertical translation.
33. Yes, since one graph can be obtained from another by a vertical translation.
35. $(5x^2/2) - (5x^3/3) + C$ **37.** $2\sqrt{u} + C$ **39.** $-(x^{-2}/8) + C$ **41.** $4\ln|u| + u + C$ **43.** $5e^z + 4z + C$
45. $x^3 + 2x^{-1} + C$ **47.** $2x^{3/2} + 4x^{1/2} + C$ **49.** $(e^x/4) - (3x^2/8) + C$ **51.** $C(x) = 2x^3 - 2x^2 + 3,000$
53. $x = 40\sqrt{t}$ **55.** $y = -2x^{-1} + 3\ln|x| - x + 3$ **57.** $x = 4e^t - 2t - 3$ **59.** $y = 2x^2 - 3x + 1$ **61.** $x^2 + x^{-1} + C$
63. $\frac{1}{2}x^2 + x^{-2} + C$ **65.** $e^x - 2\ln|x| + C$ **67.** $M = t + t^{-1} + \frac{3}{4}$ **69.** $y = 3x^{5/3} + 3x^{2/3} - 6$

71. $p(x) = 10x^{-1} + 10$ **73.** x^3 **75.** $x^4 + 3x^2 + C$ **83.** $\overline{C}(x) = 15 + \dfrac{1,000}{x}$; $C(x) = 15x + 1,000$; $C(0) = \$1,000$

85. (A) The cost function increases from 0 to 8, is concave downward from 0 to 4, and is concave upward from 4 to 8. There is an inflection point at $x = 4$.
(B) $C(x) = x^3 - 12x^2 + 53x + 30$; $C(4) = \$114,000$; $C(8) = \$198,000$
(C)

(D) Manufacturing plants are often inefficient at low and high levels of production.

87. $S(t) = 1,200 - 18t^{4/3}$; $50^{3/4} \approx 19$ mo
89. $S(t) = 1,200 - 18t^{4/3} - 70t$; $t \approx 4.05$ mo
91. $L(x) = 4,800x^{1/2}$; 24,000 labor-hours
93. $W(h) = 0.0005h^3$; 171.5 lb
95. 19,400

Exercises 7-2

1. $\frac{1}{3}(3x + 5)^3 + C$ **3.** $\frac{1}{6}(x^2 - 1)^6 + C$ **5.** $-\frac{1}{2}(5x^3 + 1)^{-2} + C$ **7.** $e^{5x} + C$ **9.** $\ln|1 + x^2| + C$ **11.** $\frac{2}{3}(1 + x^4)^{3/2} + C$
13. $\frac{1}{11}(x + 3)^{11} + C$ **15.** $-\frac{1}{6}(6t - 7)^{-1} + C$ **17.** $\frac{1}{12}(t^2 + 1)^6 + C$ **19.** $\frac{1}{2}e^{x^2} + C$ **21.** $\frac{1}{5}\ln|5x + 4| + C$
23. $-e^{1-t} + C$ **25.** $-\frac{1}{18}(3t^2 + 1)^{-3} + C$ **27.** $\frac{2}{5}(x + 4)^{5/2} - \frac{8}{3}(x + 4)^{3/2} + C$ **29.** $\frac{2}{3}(x - 3)^{3/2} + 6(x - 3)^{1/2} + C$
31. $\frac{1}{11}(x - 4)^{11} + \frac{2}{5}(x - 4)^{10} + C$ **33.** $\frac{1}{8}(1 + e^{2x})^4 + C$ **35.** $\frac{1}{2}\ln|4 + 2x + x^2| + C$ **37.** $\frac{1}{2}(5x + 3)^2 + C$
39. $\frac{1}{2}(x^2 - 1)^2 + C$ **41.** $\frac{1}{5}(x^5)^5 + C$ **49.** $\frac{1}{9}(3x^2 + 7)^{3/2} + C$ **51.** $\frac{1}{8}x^8 + \frac{4}{5}x^5 + 2x^2 + C$ **53.** $\frac{1}{9}(x^3 + 2)^3 + C$
55. $\frac{1}{4}(2x^4 + 3)^{1/2} + C$ **57.** $\frac{1}{4}(\ln x)^4 + C$ **59.** $e^{-1/x} + C$ **61.** $x = \frac{1}{3}(t^3 + 5)^7 + C$ **63.** $y = 3(t^2 - 4)^{1/2} + C$
65. $p = -(e^x - e^{-x})^{-1} + C$ **67.** $p(x) = \dfrac{2,000}{3x + 50} + 4$; 250 bottles   **69.** $C(x) = 12x + 500\ln(x + 1) + 2,000$; $\overline{C}(1,000) = \$17.45$
71. (A) $S(t) = 10t + 100e^{-0.1t} - 100$, $0 \le t \le 24$ **(B)** \$50 million   **(C)** 18.41 mo
73. $Q(t) = 100\ln(t + 1) + 5t$, $0 \le t \le 20$; 275 thousand barrels   **75.** $W(t) = 2e^{0.1t}$; 4.45 g
77. (A) $-1,000$ bacteria/mL per day   **(B)** $N(t) = 5,000 - 1,000\ln(1 + t^2)$; 385 bacteria/mL   **(C)** 7.32 days
79. $N(t) = 100 - 60e^{-0.1t}$, $0 \le t \le 15$; 87 words/min
81. $E(t) = 12,000 - 10,000(t + 1)^{-1/2}$; 9,500 students

Exercises 7-3

1. $y = 3x^2 + C$ **3.** $y = 7\ln|x| + C$ **5.** $y = 50e^{0.02x} + C$ **7.** $y = \dfrac{x^3}{3} - \dfrac{x^2}{2}$ **9.** $y = e^{-x^2} + 2$
11. $y = 2\ln|1 + x| + 5$
13. Figure B. When $x = 1$, the slope $dy/dx = 1 - 1 = 0$ for any y. When $x = 0$, the slope $dy/dx = 0 - 1 = -1$ for any y. Both are consistent with the slope field shown in Figure B.
15. $y = \dfrac{x^2}{2} - x + C$; $y = \dfrac{x^2}{2} - x - 2$

17.

19. $y = Ce^{2t}$ **21.** $y = 100e^{-0.5x}$ **23.** $x = Ce^{-5t}$ **25.** $x = -(5t^2/2) + C$

27. Figure A. When $y = 1$, the slope $dy/dx = 1 - 1 = 0$ for any x. When $y = 2$, the slope $dy/dx = 1 - 2 = -1$ for any x. Both are consistent with the slope field shown in Figure A.

29. $y = 1 - e^{-x}$

31.

33.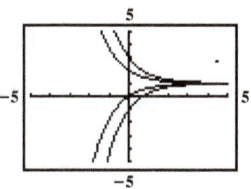

35. $y = \sqrt{25 - x^2}$ **37.** $y = -3x$ **39.** $y = 1/(1 - 2e^{-t})$

41.

43.

45.

47.

49. Apply the second-derivative test to $f(y) = ky(M - y)$. **51.** 2009 **53.** $A = 1,000e^{0.03t}$ **55.** $A = 8,000e^{0.016t}$

57. (A) $p(x) = 100e^{-0.05x}$
(B) $60.65 per unit
(C)

59. (A) $N = L(1 - e^{-0.051t})$
(B) 22.5%
(C) 32 days
(D)

61. $I = I_0 e^{-0.00942x}$; $x \approx 74$ ft
63. (A) $Q = 3e^{-0.04t}$ (B) $Q(10) = 2.01$ mL
(C) 27.47 hr
(D)

65. 0.023 117 **67.** Approx. 24,200 yr **69.** 104 times; 67 times

71. (A) 7 people; 353 people (B) 400
(C)

Exercises 7-4

1. C, E **3.** B **5.** H, I **7.** H

9.

11. Figure A: $L_3 = 13$, $R_3 = 20$; Figure B: $L_3 = 14$, $R_3 = 7$

13. $L_3 \le \int_1^4 f(x)\, dx \le R_3$; $R_3 \le \int_1^4 g(x)\, dx \le L_3$; since $f(x)$ is increasing, L_3 underestimates the area and R_3 overestimates the area; since $g(x)$ is decreasing, the reverse is true.

15. In both figures, the error bound for L_3 and R_3 is 7.

17. $S_5 = -260$ **19.** $S_4 = -1,194$ **21.** $S_3 = -33.01$ **23.** $S_6 = -38$ **25.** -2.475

27. 4.266 **29.** 2.474 **31.** -5.333 **33.** 1.067 **35.** -1.066 **37.** 15 **39.** 58.5

41. -54 **43.** 248 **45.** 0 **47.** -183 **49.** False **51.** False **53.** False

55. $L_{10} = 286,100$ ft^2; error bound is 50,000 ft^2; $n \ge 200$

57. $L_6 = -3.53$, $R_6 = -0.91$; error bound for L_6 and R_6 is 2.63. Geometrically, the definite integral over the interval $[2, 5]$ is the sum of the areas between the curve and the x axis from $x = 2$ to $x = 5$, with the areas below the x axis counted negatively and those above the x axis counted positively.

59. Increasing on $(-\infty, 0]$; decreasing on $[0, \infty)$

61. Increasing on $[-1, 0]$ and $[1, \infty)$; decreasing on $(-\infty, -1]$ and $[0, 1]$ **63.** $n \ge 22$ **65.** $n \ge 104$

67. $L_3 = 2,580$, $R_3 = 3,900$; error bound for L_3 and R_3 is 1,320

69. (A) $L_5 = 3.72$; $R_5 = 3.37$ (B) $R_5 = 3.37 \le \int_0^5 A'(t)\, dt \le 3.72 = L_5$

71. $L_3 = 114$, $R_3 = 102$; error bound for L_3 and R_3 is 12

Exercises 7-5

1. (A) $F(15) - F(10) = 375$
(B)
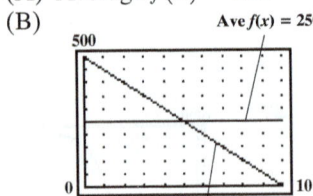

3. (A) $F(15) - F(10) = 85$
(B)
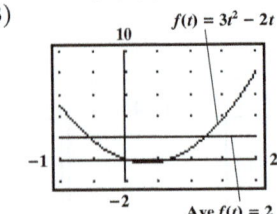

5. 40   **7.** 72   **9.** 46.5   **11.** $e - 1 \approx 1.718$
13. $\ln 2 \approx 0.693$ **15.** 0   **17.** 48   **19.** -48
21. -10.25 **23.** 0   **25.** -2 **27.** 14
29. $5^6 = 15{,}625$ **31.** $\ln 4 \approx 1.386$
33. $20(e^{0.25} - e^{-0.5}) \approx 13.550$ **35.** $\frac{1}{2}$
37. $\frac{1}{2}(1 - e^{-1}) \approx 0.316$ **39.** 0

41. (A) Average $f(x) = 250$
(B)
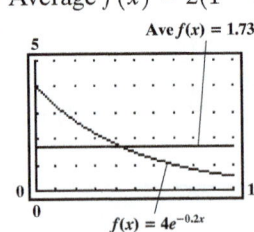

43. (A) Average $f(t) = 2$
(B)

45. (A) Average $f(x) = \frac{45}{28} \approx 1.61$
(B)
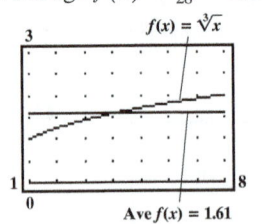

47. (A) Average $f(x) = 2(1 - e^{-2}) \approx 1.73$
(B)

49. $\frac{1}{6}(15^{3/2} - 5^{3/2}) \approx 7.819$ **51.** $\frac{1}{2}(\ln 2 - \ln 3) \approx -0.203$ **53.** 0

55. 4.566   **57.** 2.214   **61.** $\int_{300}^{900}\left(500 - \frac{x}{3}\right) dx = \$180{,}000$

63. $\int_0^5 500(t - 12)\, dt = -\$23{,}750;\ \int_5^{10} 500(t - 12)\, dt = -\$11{,}250$

65. (A)

(B) 6,505

67. Useful life $= \sqrt{\ln 55} \approx 2$ yr; total
profit $= \frac{51}{22} - \frac{5}{2}e^{-4} \approx 2.272$ or \$2,272
69. (A) \$420   (B) \$135,000

71. (A)

(B) \$100,505   **73.** $50e^{0.6} - 50e^{0.4} - 10 \approx \$6.51$
75. 4,800 labor-hours
77. (A) $I = -200t + 600$
(B) $\frac{1}{3}\int_0^3 (-200t + 600)\, dt = 300$
79. $100 \ln 11 + 50 \approx 290$ thousand barrels;
$100 \ln 21 - 100 \ln 11 + 50 \approx 115$ thousand barrels

81. $2e^{0.8} - 2 \approx 2.45$ g; $2e^{1.6} - 2e^{0.8} \approx 5.45$ g   **83.** $10°C$
85. $0.6 \ln 2 + 0.1 \approx 0.516$; $(4.2 \ln 625 + 2.4 - 4.2 \ln 49)/24 \approx 0.546$

Chapter 7 Review Exercises

1. $3x^2 + 3x + C$ *(6-1)*   **2.** 50 *(6-5)*   **3.** -207 *(6-5)*   **4.** $-\frac{1}{8}(1 - t^2)^4 + C$ *(6-2)*   **5.** $\ln|u| + \frac{1}{4}u^4 + C$ *(6-1)*
6. 0.216 *(6-5)*   **7.** e^{-x^2} *(6-1)*   **8.** $\sqrt{4 + 5x} + C$ *(6-1)*   **9.** $y = f(x) = x^3 - 2x + 4$ *(6-3)*

10. (A) $2x^4 - 2x^2 - x + C$ (B) $e^t - 4\ln|t| + C$ *(6-1)*   **11.** $R_2 = 72$; error bound for R_2 is 48 *(6-4)*

12. $\int_1^5 (x^2 + 1)\, dx = \dfrac{136}{3} \approx 45.33$; actual error is $\dfrac{80}{3} \approx 26.67$ *(6-5)*   **13.** $L_4 = 30.8$ *(6-4)*

14. 7 *(6-5)*   **15.** Width $= 2 - (-1) = 3$; height $=$ average $f(x) = 7$ *(6-5)*   **16.** $S_4 = 368$ *(6-4)*   **17.** $S_5 = 906$ *(6-4)*
18. -10 *(6-4)*   **19.** 0.4 *(6-4)*   **20.** 1.4 *(6-4)*   **21.** 0 *(6-4)*   **22.** 0.4 *(6-4)*   **23.** 2 *(6-4)*   **24.** -2 *(6-4)*
25. -0.4 *(6-4)*   **26.** (A) 1;1   (B) 4;4 *(6-3)*
27. $dy/dx = (2y)/x$; the slopes computed in Problem 26A are compatible with the slope field shown. *(6-3)*
29. $y = \frac{1}{4}x^2$; $y = -\frac{1}{4}x^2$ *(6-3)*

30. *(6-3)*

31. *(6-3)*

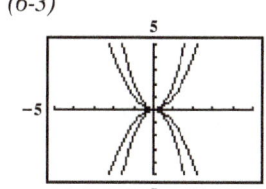

32. $\frac{2}{3}(2)^{3/2} \approx 1.886$ *(6-5)*   **33.** $\frac{1}{6} \approx 0.167$ *(6-5)*

34. $-5e^{-t} + C$ *(6-1)*   **35.** $\frac{1}{2}(1 + e^2)$ *(6-1)*

36. $\frac{1}{6}e^{3x^2} + C$ *(6-2)*   **37.** $2(\sqrt{5} - 1) \approx 2.472$ *(6-5)*

38. $\frac{1}{2}\ln 10 \approx 1.151$ *(6-5)*   **39.** 0.45 *(6-5)*

40. $\frac{1}{48}(2x^4 + 5)^6 + C$ *(6-2)*   **41.** $-\ln(e^{-x} + 3) + C$ *(6-2)*

42. $-(e^x + 2)^{-1} + C$ *(6-2)*

43. $y = f(x) = 3\ln|x| + x^{-1} + 4$ *(6-2, 6-3)*   **44.** $y = 3x^2 + x - 4$ *(6-3)*

45. (A) Average $f(x) = 6.5$

(B)

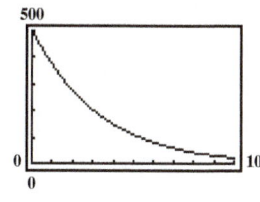

46. $\frac{1}{3}(\ln x)^3 + C$ *(6-2)*

47. $\frac{1}{8}x^8 - \frac{2}{5}x^5 + \frac{1}{2}x^2 + C$ *(6-2)*

48. $\frac{2}{3}(6 - x)^{3/2} - 12(6 - x)^{1/2} + C$ *(6-2)*

49. $\frac{1,234}{15} \approx 82.267$ *(6-5)*

50. 0 *(6-5)*

51. $y = 3e^{x^3} - 1$ *(6-3)*

52. $N = 800e^{0.06t}$ *(6-3)*

(6-5)

53. Limited growth

(6-3)

54. Exponential decay

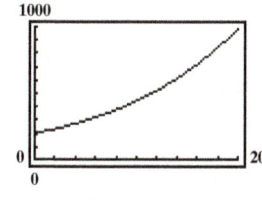

(6-3)

55. Unlimited growth

(6-3)

56. Logistic growth

(6-3)

57. 1.167 *(6-5)*   **58.** 99.074 *(6-5)*   **59.** -0.153 *(6-5)*   **60.** $L_2 = \$180,000$; $R_2 = \$140,000$; $\$140,000 \le \int_{200}^{600} C'(x)\,dx \le \$180,000$ *(6-4)*

61. $\int_{200}^{600} \left(600 - \frac{x}{2}\right) dx = \$160,000$ *(6-5)*   **62.** $\int_{10}^{40} \left(150 - \frac{x}{10}\right) dx = \$4,425$ *(6-5)*   **63.** $P(x) = 100x - 0.01x^2$; $P(10) = \$999$ *(6-3)*

64. $\int_0^{15}(60 - 4t)\,dt = 450$ thousand barrels *(6-5)*   **65.** 109 items *(6-5)*   **66.** $16e^{2.5} - 16e^2 - 8 \approx \$68.70$ *(6-5)*

67. Useful life $= 10\ln\frac{20}{3} \approx 19$ yr; total profit $= 143 - 200e^{-1.9} \approx 113.086$ or $\$113,086$ *(6-5)*

68. $S(t) = 50 - 50e^{-0.08t}$; $50 - 50e^{-0.96} \approx \$31$ million; $-(\ln 0.2)/0.08 \approx 20$ mo *(6-3)*   **69.** 1 cm^2 *(6-3)*   **70.** 800 gal *(6-5)*

71. (A) 133 million   (B) About 61 years *(6-3)*   **72.** $\dfrac{-\ln 0.04}{0.000\,123\,8} \approx 26,000$ yr *(6-3)*

73. $N(t) = 95 - 70e^{-0.1t}$; $N(15) \approx 79$ words/min *(6-3)*

Chapter 8

Exercises 8-1

1. $\int_a^b g(x)\,dx$ **3.** $\int_a^b [-h(x)]\,dx$

5. Since the shaded region in Figure C is below the x axis, $h(x) \le 0$; so, $\int_a^b h(x)\,dx$ represents the negative of the area.

7. 24 **9.** 51 **11.** 42 **13.** 6 **15.** 0.833 **17.** 2.350 **19.** 1 **21.** 10.5 **23.** 18 **25.** 44

27. $\int_a^b [-f(x)]\,dx$ **29.** $\int_b^c f(x)\,dx + \int_c^d [-f(x)]\,dx$ **31.** $\int_c^d [f(x) - g(x)]\,dx$

33. $\int_a^b [f(x) - g(x)]\,dx + \int_b^c [g(x) - f(x)]\,dx$

35. Find the intersection points by solving $f(x) = g(x)$ on the interval $[a, d]$ to determine b and c. Then observe that $f(x) \ge g(x)$

over $[a, b]$, $g(x) \ge f(x)$ over $[b, c]$, and $f(x) \ge g(x)$ over $[c, d]$.

Area $= \int_a^b [f(x) - g(x)]\,dx + \int_b^c [g(x) - f(x)]\,dx + \int_c^d [f(x) - g(x)]\,dx$.

37. 2.5 **39.** 7.667 **41.** 12 **43.** 15 **45.** 32 **47.** 36 **49.** 9 **51.** 2.832

53. $\int_{-3}^3 \sqrt{9 - x^2}\,dx$; 14.137 **55.** $\int_0^4 \sqrt{16 - x^2}\,dx$; 12.566 **57.** $\int_{-2}^2 \sqrt{4 - x^2}\,dx$; 12.566

59. 18 **61.** 1.858 **63.** 52.616 **65.** 8 **67.** 101.75 **69.** 17.979 **71.** 5.113 **73.** 8.290 **75.** 3.166 **77.** 1.385

79. Total production from the end of the fifth year to the end of the 10th year is $50 + 100\ln 20 - 100\ln 15 \approx 79$ thousand barrels.

81. Total profit over the 5-yr useful life of the game is $20 - 30e^{-1.5} \approx 13.306$, or $\$13,306$.

83. 1935: 0.412; 1947: 0.231; income was more equally distributed in 1947.
85. 1963: 0.818; 1983: 0.846; total assets were less equally distributed in 1983.
87. (A) $f(x) = 0.3125x^2 + 0.7175x - 0.015$ (B) 0.104
89. Total weight gain during the first 10 hr is $3e - 3 \approx 5.15$ g.
91. Average number of words learned from $t = 2$ hr to $t = 4$ hr is $15 \ln 4 - 15 \ln 2 \approx 10$.

Exercises 8-2

1. 0.43   **3.** 744.99   **5.** 10.27   **7.** 151.75   **9.** 93,268.66   **11.** (A) 10.72   (B) 3.28   (C) 10.72
13. (A) .75   (B) .11

(C)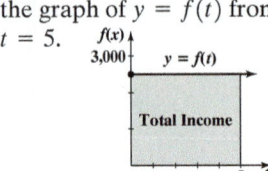

15. 8 yr   **17.** (A) .11   B) .10
19. $P(t \geq 12) = 1 - P(0 \leq t \leq 12) = .89$
21. \$12,500
23. If $f(t)$ is the rate of flow of a continuous income stream, then the total income produced from 0 to 5 yr is the area under the graph of $y = f(t)$ from $t = 0$ to $t = 5$.

25. $8,000(e^{0.15} - 1) \approx \$1,295$
27. If $f(t)$ is the rate of flow of a continuous income stream, then the total income produced from 0 to 3 yr is the area under the graph of $y = f(t)$ from $t = 0$ to $t = 3$.

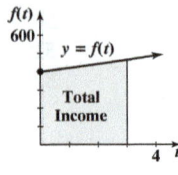

29. \$255,562; \$175,562   **31.** \$6,779.52   **33.** \$436.62
35. Clothing store: \$66,420.83; computer store: \$62,622.66; the clothing store is the better investment.
37. Bond: \$12,062.30 business: \$11,823.87; the bond is the better investment.   **39.** \$55,230
41. $\dfrac{k}{r}(e^{rT} - 1)$ **43.** \$625,000
45. The shaded area is the consumers' surplus and represents the total savings to consumers who are willing to pay more than \$150 for a product but are still able to buy the product for \$150.

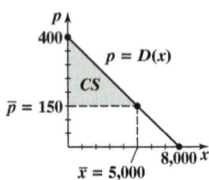

47. \$9,900
49. The area of the region PS is the producers' surplus and represents the total gain to producers who are willing to supply units at a lower price than \$67 but are still able to supply the product at \$67.

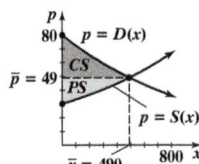

51. $CS = \$3,380;\ PS = \$1,690$

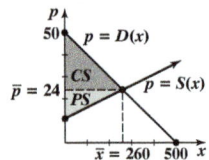

53. $CS = \$6,980;\ PS = \$5,041$

55. $CS = \$7,810;\ PS = \$8,336$

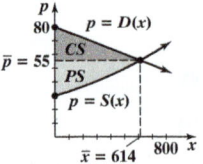

57. $CS = \$8,544;\ PS = \$11,507$

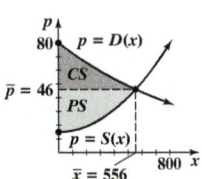

59. (A) $\bar{x} = 21.457;\ \bar{p} = \$6.51$
(B) $CS = 1.774$ or \$1,774; $PS = 1.087$ or \$1,087

Exercises 8-3

1. $\frac{1}{3}xe^{3x} - \frac{1}{9}e^{3x} + C$ **3.** $\frac{x^3}{3}\ln x - \frac{x^3}{9} + C$ **5.** $u = x + 2;\ \dfrac{(x+2)(x+1)^6}{6} - \dfrac{(x+1)^7}{42} + C$ **7.** $-xe^{-x} - e^{-x} + C$
9. $\frac{1}{2}e^{x^2} + C$ **11.** $(xe^x - 4e^x)\big|_0^1 = -3e + 4 \approx -4.1548$ **13.** $(x \ln 2x - x)\big|_1^3 = (3 \ln 6 - 3) - (\ln 2 - 1) \approx 2.6821$
15. $\ln(x^2 + 1) + C$ **17.** $(\ln x)^2/2 + C$ **19.** $\frac{2}{3}x^{3/2}\ln x - \frac{4}{9}x^{3/2} + C$
21. $\dfrac{(x-3)(x+1)^3}{3} - \dfrac{(x+1)^4}{12} + C$ or $\dfrac{x^4}{4} - \dfrac{x^3}{3} - \dfrac{5x^2}{2} - 3x + C$

23. $\dfrac{(2x + 1)(x - 2)^3}{3} - \dfrac{(x - 2)^4}{6} + C$ or $\dfrac{x^4}{2} - \dfrac{7x^3}{3} + 2x^2 + 4x + C$

25. The integral represents the negative of the area between the graph of $y = (x - 3)e^x$ and the x axis from $x = 0$ to $x = 1$.

27. The integral represents the area between the graph of $y = \ln 2x$ and the x axis from $x = 1$ to $x = 3$.

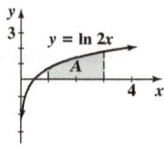

29. $(x^2 - 2x + 2)e^x + C$ **31.** $\dfrac{xe^{ax}}{a} - \dfrac{e^{ax}}{a^2} + C$ **33.** $\left(-\dfrac{\ln x}{x} - \dfrac{1}{x} \right)\Big|_1^e = -\dfrac{2}{e} + 1 \approx 0.2642$

35. $6 \ln 6 - 4 \ln 4 - 2 \approx 3.205$ **37.** $xe^{x-2} - e^{x-2} + C$ **39.** $\frac{1}{2}(1 + x^2) \ln(1 + x^2) - \frac{1}{2}(1 + x^2) + C$

41. $(1 + e^x) \ln(1 + e^x) - (1 + e^x) + C$ **43.** $x(\ln x)^2 - 2x \ln x + 2x + C$ **45.** $x(\ln x)^3 - 3x(\ln x)^2 + 6x \ln x - 6x + C$

47. 2   **49.** $\dfrac{1}{3}$ **51.** 1.56   **53.** 34.98   **55.** $\int_0^5 (2t - te^{-t})\, dt = \$24$ million

57. The total profit for the first 5 yr (in millions of dollars) is the same as the area under the marginal profit function, $P'(t) = 2t - te^{-t}$, from $t = 0$ to $t = 5$.

59. $2,854.88   **61.** 0.264

63. The area bounded by $y = x$ and the Lorenz curve $y = xe^{x-1}$, divided by the area under the graph of $y = x$ from $x = 0$ to $x = 1$, is the Gini index of income concentration. The closer this index is to 0, the more equally distributed the income; the closer the index is to 1, the more concentrated the income in a few hands.

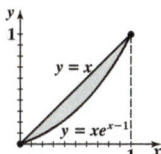

65. $S(t) = 1,600 + 400e^{0.1t} - 40te^{0.1t}$; 15 mo   **67.** $977

69. The area bounded by the price–demand equation, $p = 9 - \ln(x + 4)$, and the price equation, $y = \bar{p} = 2.089$, from $x = 0$ to $x = \bar{x} = 1,000$, represents the consumers' surplus. This is the amount consumers who are willing to pay more than $2.089 save.

71. 2.1388 ppm

73. $N(t) = -4te^{-0.25t} - 40e^{-0.25t} + 80$; 8 wk; 78 words/min

75. 20,980

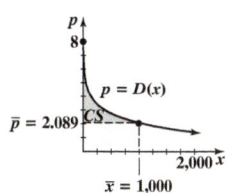

Exercises 8-4

1. $\ln \left| \dfrac{x}{1 + x} \right| + C$ **3.** $\dfrac{1}{3 + x} + 2 \ln \left| \dfrac{5 + 2x}{3 + x} \right| + C$ **5.** $\dfrac{2(x - 32)}{3} \sqrt{16 + x} + C$ **7.** $-\ln \left| \dfrac{1 + \sqrt{1 - x^2}}{x} \right| + C$

9. $\dfrac{1}{2} \ln \left| \dfrac{x}{2 + \sqrt{x^2 + 4}} \right| + C$ **11.** $\frac{1}{3}x^3 \ln x - \frac{1}{9}x^3 + C$ **13.** $x - \ln|1 + e^x| + C$ **15.** $9 \ln \frac{3}{2} - 2 \approx 1.6492$

17. $\frac{1}{2}\ln \frac{12}{5} \approx 0.4377$ **19.** $\ln 3 \approx 1.0986$ **21.** $-\dfrac{\sqrt{4x^2 + 1}}{x} + 2 \ln|2x + \sqrt{4x^2 + 1}| + C$ **23.** $\frac{1}{2}\ln|x^2 + \sqrt{x^4 - 16}| + C$

25. $\frac{1}{6}(x^3 \sqrt{x^6 + 4} + 4 \ln|x^3 + \sqrt{x^6 + 4}|) + C$ **27.** $-\dfrac{\sqrt{4 - x^4}}{8x^2} + C$ **29.** $\dfrac{1}{5} \ln \left| \dfrac{3 + 4e^x}{2 + e^x} \right| + C$ **31.** $\frac{2}{3}(\ln x - 8)\sqrt{4 + \ln x} + C$

33. $\frac{1}{5}x^2 e^{5x} - \frac{2}{25}xe^{5x} + \frac{2}{125}e^{5x} + C$ **35.** $-x^3 e^{-x} - 3x^2 e^{-x} - 6xe^{-x} - 6e^{-x} + C$ **37.** $x(\ln x)^3 - 3x(\ln x)^2 + 6x \ln x - 6x + C$

39. $\frac{64}{3}$ **41.** $\frac{1}{2}\ln \frac{9}{5} \approx 0.2939$ **43.** $\dfrac{-1 - \ln x}{x} + C$ **45.** $\sqrt{x^2 - 1} + C$ **47.** 31.38   **49.** 5.48   **51.** $3,000 + 1,500 \ln \frac{1}{3} \approx \$1,352$

53.

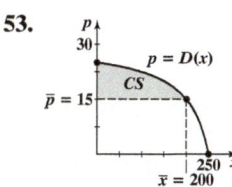

55. $C(x) = 200x + 1,000 \ln(1 + 0.05x) + 25,000$; 608; $198,773

57. $18,673.95   **59.** 0.1407

61. As the area bounded by the two curves gets smaller, the Lorenz curve approaches $y = x$ and the distribution of income approaches perfect equality—all persons share equally in the income available.

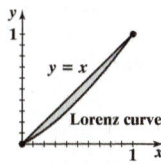

63. $S(t) = 1 + t - \dfrac{1}{1 + t} - 2\ln|1 + t|$; $24.96 - 2\ln 25 \approx \$18.5$ million

65. The total sales (in millions of dollars) over the first 2 yr (24 mo) is the area under the graph of $y = S'(t)$ from $t = 0$ to $t = 24$.

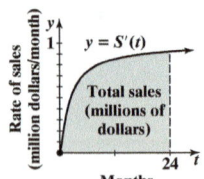

67. $P(x) = \dfrac{2(9x - 4)}{135}(2 + 3x)^{3/2} - 2{,}000.83$; 54; $\$37{,}932$

69. $100\ln 3 \approx 110$ ft

71. $60\ln 5 \approx 97$ items

73. The area under the graph of $y = N'(t)$ from $t = 0$ to $t = 12$ represents the total number of items learned in that time interval.

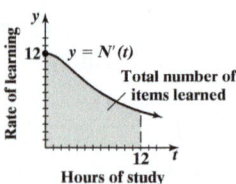

Chapter 8 Review Exercises

1. $\int_a^b f(x)\,dx$ *(7-1)*   **2.** $\int_b^c [-f(x)]\,dx$ *(7-1)*   **3.** $\int_a^b f(x)\,dx + \int_b^c [-f(x)]\,dx$ *(7-1)*

4. Area = 1.153 *(7-1)*

5. $\frac{1}{4}xe^{4x} - \frac{1}{16}e^{4x} + C$ *(7-3, 7-4)*   **6.** $\frac{1}{2}x^2\ln x - \frac{1}{4}x^2 + C$ *(7-3, 7-4)*

7. $\dfrac{(\ln x)^2}{2} + C$ *(6-2)*   **8.** $\dfrac{\ln(1 + x^2)}{2} + C$ *(7-2)*   **9.** $\dfrac{1}{1 + x} + \ln\left|\dfrac{x}{1 + x}\right| + C$ *(7-4)*

10. $-\dfrac{\sqrt{1 + x}}{x} - \frac{1}{2}\ln\left|\dfrac{\sqrt{1 + x} - 1}{\sqrt{1 + x} + 1}\right| + C$ *(7-4)*   **11.** 12 *(7-1)*   **12.** 40 *(7-1)*   **13.** 34.167 *(7-1)*

14. 0.926 *(7-1)*   **15.** 12 *(7-1)*   **16.** 18.133 *(7-1)*   **17.** $\int_a^b [f(x) - g(x)]\,dx$ *(7-1)*

18. $\int_b^c [g(x) - f(x)]\,dx$ *(7-1)*   **19.** $\int_b^c [g(x) - f(x)]\,dx + \int_c^d [f(x) - g(x)]\,dx$ *(7-1)*

20. $\int_a^b [f(x) - g(x)]\,dx + \int_b^c [g(x) - f(x)]\,dx + \int_c^d [f(x) - g(x)]\,dx$ *(7-1)*

21. Area = 20.833 *(7-1)*

22. 1 *(7-3, 7-4)*   **23.** $\frac{15}{2} - 8\ln 8 + 8\ln 4 \approx 1.955$ *(7-4)*

24. $\frac{1}{6}(3x\sqrt{9x^2 - 49} - 49\ln|3x + \sqrt{9x^2 - 49}|) + C$ *(7-4)*

25. $-2te^{-0.5t} - 4e^{-0.5t} + C$ *(7-3, 7-4)*   **26.** $\frac{1}{3}x^3\ln x - \frac{1}{9}x^3 + C$ *(7-3, 7-4)*

27. $x - \ln|1 + 2e^x| + C$ *(7-4)*

28. (A) Area = 8   (B) Area = 8.38 *(7-1)*

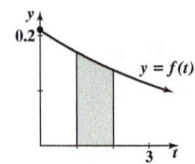

29. $\frac{1}{3}(\ln x)^3 + C$ *(6-2)*

30. $\frac{1}{2}x^2(\ln x)^2 - \frac{1}{2}x^2\ln x + \frac{1}{4}x^2 + C$ *(7-3, 7-4)*

31. $\sqrt{x^2 - 36} + C$ *(6-2)*

32. $\frac{1}{2}\ln|x^2 + \sqrt{x^4 - 36}| + C$ *(7-4)*

33. $50\ln 10 - 42\ln 6 - 24 \approx 15.875$ *(7-3, 7-4)*

34. $x(\ln x)^2 - 2x\ln x + 2x + C$ *(7-3, 7-4)*

35. $-\frac{1}{4}e^{-2x^2} + C$ *(6-2)*

36. $-\frac{1}{2}x^2e^{-2x} - \frac{1}{2}xe^{-2x} - \frac{1}{4}e^{-2x} + C$ *(7-3, 7-4)*

37. 1.703 *(7-1)*   **38.** (A) .189   (B) .154 *(7-2)*

39. The probability that the product will fail during the second year of warranty is the area under the probability density function $y = f(t)$ from $t = 1$ to $t = 2$. *(7-2)*

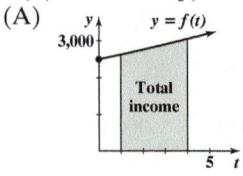

40. $R(x) = 65x - 6[(x + 1)\ln(x + 1) - x]$; 618/wk; $\$29{,}506$ *(7-3)*

41. (A)

(B) $\$8{,}507$ *(7-2)*

42. (A) $\$15{,}655.66$ (B) $\$1{,}454.39$ *(7-2)*

43. (A)

(B) More equitably distributed, since the area bounded by the two curves will have decreased.

(C) Current = 0.3; projected = 0.2; income will be more equitably distributed 10 years from now. *(7-1)*

44. (A) $CS = \$2,250$;   (B) $CS = \$2,890$;
    $PS = \$2,700$ $PS = \$2,278$ *(7-2)*

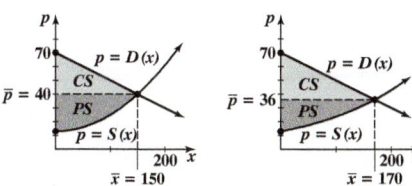

$\bar{x} = 150$ $\bar{x} = 170$

45. (A) 25.403 or 25,403 lb
    (B) $PS = 121.6$ or $\$1,216$ *(7-2)*
46. 4.522 mL; 1.899 mL *(6-5, 7-4)*
47.

     (6-5, 7-1)

48. .667; .333 *(7-2)*

49. The probability that the doctor will
spend more than an hour with a
randomly selected patient is the area
under the probability density
function $y = f(t)$ from $t = 1$ to
$t = 3$. *(7-2)*

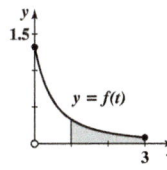

50. 45 thousand *(6-5, 7-1)*   **51.** .368 *(7-2)*

Chapter 9
Exercises 9-1

1. $\displaystyle\int_1^{\infty} \frac{dx}{x^3}$

Note that $\displaystyle\int_1^b \frac{1}{x^3}\,dx = -\frac{1}{2x^2}\Big|_1^b = -\frac{1}{2b^2} + \frac{1}{2}$ and $\displaystyle\lim_{b\to\infty}\left(-\frac{1}{2b^2} + \frac{1}{2}\right) = \frac{1}{2}$. Therefore $\displaystyle\int_1^{\infty} \frac{dx}{x^3} = \frac{1}{2}$.

3. $\displaystyle\int_1^{\infty} x^2\,dx$

Note that $\displaystyle\int_1^b x^2\,dx = \frac{x^3}{3}\Big|_1^b = \frac{b^3}{3} - \frac{1}{3}$ and $\displaystyle\lim_{b\to\infty}\left(\frac{b^3}{3} - \frac{1}{3}\right) = \infty$. Therefore $\displaystyle\int_1^{\infty} x^2\,dx$ diverges.

5. $\displaystyle\int_{-\infty}^{0} e^{x/2}\,dx$

Note that $\displaystyle\int_b^0 e^{x/2}\,dx = 2e^{x/2}\Big|_b^0 = 2 - 2e^{b/2}$ and $\displaystyle\lim_{b\to\infty}(2 - 2e^{b/2}) = 2$. Therefore $\displaystyle\int_{-\infty}^{0} e^{x/2}\,dx = 2$.

7. $\displaystyle\int_4^{\infty} \frac{dx}{100\sqrt{x}}$

Note that $\displaystyle\int_4^b \frac{1}{100\sqrt{x}}\,dx = \frac{1}{50}x^{1/2}\Big|_4^b = \frac{b^{1/2}}{50} - \frac{2}{50}$ and $\displaystyle\lim_{b\to\infty}\frac{b^{1/2}}{50} - \frac{2}{50} = \infty$. Therefore $\displaystyle\int_4^{\infty} \frac{dx}{100\sqrt{x}}$ diverges.

9. $\displaystyle\int_9^{\infty} \frac{dx}{x\sqrt{x}}$

Note that $\displaystyle\int_9^b \frac{dx}{x\sqrt{x}} = \int_9^b x^{-3/2}\,dx = -2x^{-1/2}\Big|_9^b = -\frac{2}{b^{1/2}} + \frac{2}{3}$ and $\displaystyle\lim_{b\to\infty}\left(-\frac{2}{b^{1/2}} + \frac{2}{3}\right) = \frac{2}{3}$. Therefore $\displaystyle\int_9^{\infty} \frac{dx}{x\sqrt{x}} = \frac{2}{3}$.

11. $\displaystyle\int_0^{\infty} \frac{dx}{(x+1)^{2/3}} = \lim_{b\to\infty}\int_0^b \frac{dx}{(x+1)^{2/3}}$

$\displaystyle\qquad = \lim_{b\to\infty}\left(3(x+1)^{1/3}\right)\Big|_0^b$ [Note: If $u = x + 1$, then $du = dx$.]

$\displaystyle\qquad = 3\lim_{b\to\infty}(x+1)^{1/3}\Big|_0^b = 3\lim_{b\to\infty}\left((b+1)^{1/3}\right) - 1$

Since $(b+1)^{1/3} \to \infty$ as $b \to \infty$, the limit does not exist; hence, the improper integral *diverges*.

13. $\displaystyle\int_1^{\infty} \frac{dx}{x^{0.99}} = \lim_{b\to\infty}\int_1^b \frac{dx}{x^{0.99}} = \lim_{b\to\infty}\frac{x^{0.01}}{0.01}\Big|_1^b$

$\displaystyle\qquad = 100\lim_{b\to\infty}(x^{0.01})\Big|_1^b = 100\lim_{b\to\infty}(b^{0.01} - 1)$

Since $b^{0.01} \to \infty$ as $b \to \infty$, the limit does not exist; hence, the improper integral *diverges*.

15. $0.3 \displaystyle\int_0^\infty e^{-0.3x}\, dx = 0.3 \lim_{b\to\infty} \int_{-\infty}^0 e^{-0.3x}\, dx = 0.3 \lim_{b\to\infty} \dfrac{e^{-0.3x}}{-0.3}\Big|_0^b$

$$= -\lim_{b\to\infty} (e^{-0.3x})\Big|_0^b = -\lim_{b\to\infty}(e^{-0.3b} - 1)$$

$$= -\lim_{b\to\infty}\left(\dfrac{1}{e^{0.3b}} - 1\right)$$

$$= -(0-1)\left[\text{Note:}\,\dfrac{1}{e^{0.3b}} \to 0 \text{ as } b \to \infty\right].$$

$$= 1$$

Thus, the improper integral *converges*.

17. The graph of f is shown at the right. Since $f(x) = 0$ for $x < 0$ and $x > 2$,

$$\int_{-\infty}^\infty f(x)\, dx = \int_0^2 f(x)\, dx \qquad (f(x) = 0 \text{ for } x < 0 \text{ and } x > 2)$$

$$= \int_0^2 (1 + x^2)\, dx$$

$$= \left[\left(x + \dfrac{1}{3}x^3\right)\Big|_0^2\right]$$

$$= \dfrac{14}{3}$$

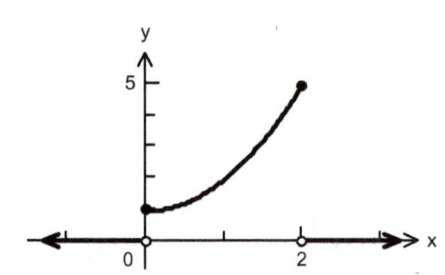

The integral *converges* and has value $\dfrac{14}{3}$.

19. The graph of f is shown at the right. Since $f(x) = 0$ for $x < 0$,

$$\int_{-\infty}^\infty f(x)\,dx = \int_0^\infty f(x) \qquad (f(x) = 0 \text{ for } x < 0)$$

$$= \lim_{b\to\infty} \int_0^b e^{-0.1x}\, dx$$

$$= \lim_{b\to\infty} \left[-10e^{-0.1x}\,\Big|_0^b\right]$$

$$= \lim_{b\to\infty} \left[-10e^{-0.1b} + 10\right] = 10$$

The integral *converges* and has value 10.

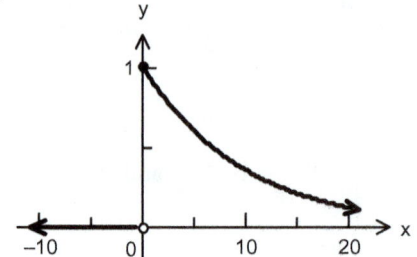

21. The graph of f is shown at the right. Since $f(x) = 0$ for $x < 2$,

$$\int_{-\infty}^\infty f(x)\, dx = \int_2^\infty f(x)\,dx \qquad (f(x) = 0 \text{ for } x < 2)$$

$$= \lim_{b\to\infty} \int_2^b \dfrac{4}{(x + 2)}\, dx$$

$$= \lim_{b\to\infty} \left[4 \ln (x + 2)\,\Big|_2^b\right]$$

$$= \lim_{b\to\infty} \left[4 \ln(b + 2) - 4 \ln 4\right]$$

Since $4 \ln(b + 2) \to \infty$ as $b \to \infty$, the integral *diverges*.

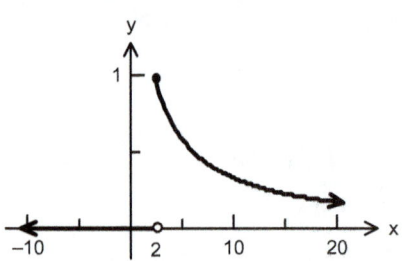

23. Always true. If f is positive, continuous and increasing on $[0, \infty)$, then $f(x) > f(0) > 0$ for all $x > 0$ and

$$\int_0^\infty f(x)\, dx = \lim_{b\to\infty} \int_0^b f(x)\, dx > \lim_{b\to\infty} \int_0^b f(0)dx$$

$$= \lim_{b\to\infty} f(0) \int_0^b dx$$

$$= f(0)\, \lim_{b\to\infty} x\,\Big|_0^b$$

$$= f(0)\, \lim_{b\to\infty} b \to \infty$$

$$\text{as } b \to \infty$$

25. False. Let $f(x) = \begin{cases} 10 & \text{if } 1 \geq x \geq 1000 \\ 10e^{-(x-1000)} & \text{if } x > 1000 \end{cases}$

Then $f(x) > 0$ for all $x \in [1, \infty)$, f is continuous, $f(x) \geq 10$ on $[1, 1000]$, and

$$\int_1^\infty f(x)\,dx = \int_1^{1000} f(x)\,dx + \int_{1000}^\infty f(x)\,dx$$

$$= \int_1^{1000} 10\,dx + \lim_{b \to \infty} \int_{1000}^b 10e^{-(x-1000)}\,dx$$

$$= \Big[10x\Big]_1^{1000} + \lim_{b \to \infty} \Big[-10e^{-(x-1000)}\Big]_{1000}^b$$

$$= 10{,}000 - 10 + \lim_{b \to \infty} \Big[10 - e^{-(b-1000)}\Big] = 10{,}000.$$

The improper integral converges.

27. $F(b) = \displaystyle\int_1^b \frac{dx}{x^3} = -\frac{1}{2b^2} + \frac{1}{2}$

$\displaystyle\lim_{b \to \infty} F(b) = \lim_{b \to \infty} \left(-\frac{1}{2b^2} + \frac{1}{2}\right) = \frac{1}{2}.$

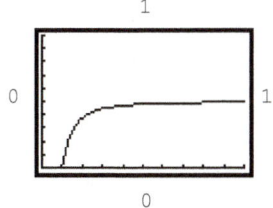

29. $F(b) = \displaystyle\int_1^b x^2\,dx = \frac{b^3}{3} - \frac{1}{3}$

$\displaystyle\lim_{b \to \infty} F(b) = \lim_{b \to \infty} \left(\frac{b^3}{3} - \frac{1}{3}\right) = \infty.$

31. $F(b) = \displaystyle\int_0^b e^{x/2}dx = 2 - 2e^{b/2}$

$\displaystyle\lim_{b \to \infty} F(b) = \lim_{b \to \infty} (2 - 2e^{b/2}) = 2.$

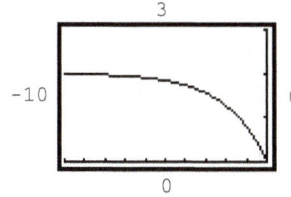

33. $F(b) = \displaystyle\int_4^b \frac{dx}{100\sqrt{x}} = \frac{b^{1/2}}{50} - \frac{1}{25}$

$\displaystyle\lim_{b \to \infty} F(b) = \lim_{b \to \infty} \left(\frac{b^{1/2}}{50} - \frac{1}{25}\right) = \infty$

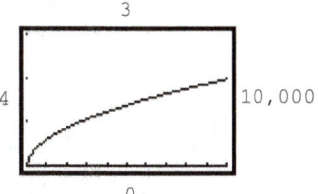

35. $F(b) = \displaystyle\int_9^b \frac{dx}{x\sqrt{x}} = -\frac{2}{b^{1/2}} + \frac{2}{3}$

$\displaystyle\lim_{b \to \infty} F(b) = \lim_{b \to \infty} \left(-\frac{2}{b^{1/2}} + \frac{2}{3}\right) = \frac{2}{3}.$

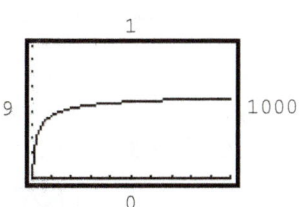

37. f is continuous on $[0, \infty)$ and $\displaystyle\int_0^\infty f(x)\,dx = L$ converges. Since

$$L = \int_0^\infty f(x)\,dx = \lim_{b \to \infty} \int_0^b f(x)\,dx$$

$$= \lim_{b \to \infty}\left[\int_0^1 f(x)\,dx + \int_1^b f(x)\,dx\right]$$

$$= \int_0^1 f(x)\,dx + \lim_{b \to \infty} \int_1^b f(x)\,dx,$$

it follows that

$\lim\limits_{b\to\infty}\displaystyle\int_1^b f(x)\,dx = L - \int_0^1 f(x)\,dx$ exists.

Thus, $\displaystyle\int_1^\infty f(x)\,dx$ converges.

39. $\displaystyle\int_0^\infty \frac{1}{k}e^{-x/k}\,dx = \frac{1}{k}\lim_{b\to\infty}\int_0^b e^{-x/k}\,dx = \frac{1}{k}\lim_{b\to\infty}\left.\frac{e^{-x/k}}{-\dfrac{1}{k}}\right|_0^b = -\lim_{b\to\infty}\left. e^{-x/k}\right|_0^b$

$$= -\lim_{b\to\infty}(e^{-b/k}-1) = -\lim_{b\to\infty}\left(\frac{1}{e^{b/k}}-1\right) = -(0-1) = 1$$

Thus, the improper integral *converges*.

41. $\displaystyle\int_{-\infty}^\infty \frac{x}{1+x^2}\,dx = \int_{-\infty}^c \frac{x}{1+x^2}\,dx + \int_c^\infty \frac{x}{1+x^2}\,dx$

where c is a real number, using $\underline{1}$(c). Let $c = 0$.

Consider $\displaystyle\int_0^\infty \frac{x}{1+x^2}\,dx = \lim_{b\to\infty}\int_0^b \frac{x}{1+x^2}\,dx$

$\displaystyle\int \frac{x}{1+x^2}\,dx = \frac{1}{2}\int\frac{du}{u} = \frac{1}{2}\ln u + C = \frac{1}{2}\ln(1+x^2)+C$

[Let $u = 1+x^2$. Then $du = 2x\,dx$.]

Thus, $\displaystyle\lim_{b\to\infty}\int_0^b \frac{x}{1+x^2}dx = \lim_{b\to\infty}\frac{1}{2}\ln(1+x^2)\Big|_0^b$

$$= \lim_{b\to\infty}\frac{1}{2}\ln(1+b^2) - \frac{1}{2}\ln 1$$

$$= \frac{1}{2}\lim_{b\to\infty}\ln(1+b^2)$$

Since $\ln(1+b^2)\to\infty$ as $b\to\infty$, this improper integral *diverges*. Therefore, $\displaystyle\int_{-\infty}^\infty \frac{x}{1+x^2}\,dx$ diverges.

43. $\displaystyle\int_0^\infty (e^{-x}-e^{-2x})\,dx = \lim_{b\to\infty}\int_0^b (e^{-x}-e^{-2x})\,dx = \lim_{b\to\infty}\left(-e^{-x}-\frac{e^{-2x}}{-2}\right)\Big|_0^b$

$$= \lim_{b\to\infty}\left[-e^{-b}+\frac{e^{-2b}}{2}-\left(-1+\frac{1}{2}\right)\right]$$

$$= \frac{1}{2}+\lim_{b\to\infty}\left(-e^{-b}+\frac{e^{-2b}}{2}\right) = \frac{1}{2}+0 = \frac{1}{2}$$

Thus, the improper integral *converges*.

45. $\displaystyle\int_{-\infty}^0 \frac{dx}{\sqrt{1-x}} = \lim_{a\to-\infty}\int_a^0 \frac{dx}{\sqrt{1-x}}$ (using $\underline{1}$(b))

$\displaystyle\int \frac{dx}{\sqrt{1-x}} = \int -\frac{du}{u^{1/2}} = -\int u^{-1/2}\,du$ Substitution: $u = 1-x$

$\qquad\qquad = -2u^{1/2}+C = -2\sqrt{1-x}+C$ $du = -dx$

$\qquad\qquad\qquad\qquad\qquad\qquad\qquad\qquad\qquad -du = dx$

Therefore, $\displaystyle\lim_{a\to-\infty}\int_a^0 \frac{dx}{\sqrt{1-x}} = \lim_{a\to-\infty}(-2\sqrt{1-x})\big|_a^0 = \lim_{a\to-\infty}(-2+2\sqrt{1-a})$

Since $2\sqrt{1-a}\to\infty$ as $a\to-\infty$, the limit does not exist; hence, the improper integral *diverges*.

47. $\displaystyle\int_1^\infty \frac{\ln x}{x}\,dx = \lim_{b\to\infty}\int_1^b \frac{\ln x}{x}\,dx$ Substitution: $u = \ln x$

$\qquad = \lim_{b\to\infty}\int_0^{\ln b} u\,du$ $du = \frac{1}{x}\,dx$

$\qquad = \lim_{b\to\infty}\frac{u^2}{2}\Big|_0^{\ln b} = \lim_{b\to\infty}\frac{(\ln b)^2}{2}$ Limits: $x = 1$ implies $u = 0$

$\qquad\qquad\qquad\qquad\qquad\qquad\qquad\qquad x = b$ implies $u = \ln b$

Since $\ln b\to\infty$ as $b\to\infty$, the limit does not exist; hence, the improper integral *diverges*.

49. $CV = \displaystyle\int_0^\infty 6000e^{-0.05t}\,dt = \lim_{T\to\infty} \int_0^T 6000e^{-0.05t}\,dt$

$$= \lim_{T\to\infty} \frac{6000}{-0.05}e^{-0.05t}\Big]_0^T = -120{,}000 \cdot \lim_{T\to\infty}[e^{-0.05T}-1]$$

$$= \$120{,}000$$

51. $f(t) = 1500e^{0.04t}$

$$CV = \int_0^\infty 1500e^{0.04t}\cdot e^{0.07t}\,dt = \lim_{T\to\infty}\int_0^T 1500e^{-0.03t}\,dt$$

$$= \lim_{T\to\infty}\left[\frac{1500}{-0.03}e^{-0.03t}\Big|_0^T\right]$$

$$= -50{,}000\lim_{T\to\infty}[e^{-0.03t}-1]$$

$$= \$50{,}000$$

53. At an interest rate of 6%,

$$CV = \int_0^\infty 6000e^{-0.06t}\,dt = \lim_{T\to\infty}\int_0^T 6000e^{-0.06t}\,dt$$

$$= \lim_{T\to\infty}[-100{,}000e^{-0.06t}|_0^T]$$

$$= \lim_{T\to\infty}[-100{,}000e^{-0.06t}+100{,}000]$$

$$= \$100{,}000$$

Increasing the interest rate to 6% decreases the capital value to $100,000. At an interest rate of 4%,

$$CV = \int_0^\infty 6000e^{-0.04t}\,dt = \lim_{T\to\infty}\int_0^T 6000e^{-0.04t}\,dt$$

$$= \lim_{T\to\infty}[-150{,}000e^{-0.04t}|_0^T]$$

$$= \lim_{T\to\infty}[-150{,}000e^{-0.04t}+150{,}000]$$

$$= \$150{,}000$$

Decreasing the interest rate to 4% increases the capital value to $150,000. In general, increasing the interest rate decreases the amount of capital required to establish the income stream.

55. $R(t) = 3e^{-0.2t} - 3e^{-0.4t}$

(A) Total production $= \displaystyle\int_0^\infty R(t)\,dt = \lim_{T\to\infty}\int_0^T R(t)\,dt$

$$= \lim_{T\to\infty}\int_0^T (3e^{-0.2t}-3e^{-0.4t})\,dt$$

$$= \lim_{T\to\infty}[(-15e^{-0.2t}+7.5e^{-0.4t})|_0^T]$$

$$= \lim_{T\to\infty}[-15e^{-0.2T}+7.5e^{-0.4T}+7.5]$$

$$= 7.5 \text{ billion ft}^3$$

(B) 50% of 7.5 billion ft^3 is 3.75 billion ft^3. Thus, we need to solve

$$\int_0^T R(t)\,dt = \int_0^T (3e^{-0.2t}-3e^{-0.4t})\,dt = 3.75$$

for T.

Now,

$$\int_0^T (3e^{-0.2t}-3e^{-0.4t})\,dt = [(-15e^{-0.2t}+7.5e^{-0.4t})|_0^T]$$

$$= -15e^{-0.2T}+7.5e^{-0.4t}+7.5 = 3.75$$

and

$$-15\,e^{-0.2T}+7.5e^{-0.4t} = -3.75$$

Using a graphing utility, we find that $T \approx 6.14$ years.

57. Total seepage $= \displaystyle\int_0^\infty \frac{500}{(1+t)^2}\,dt = \lim_{b\to\infty} 500 \int_0^b \frac{1}{(1+t)^2}\,dt$

$\qquad\qquad\qquad = 500 \lim_{b\to\infty} \int_0^b \frac{1}{(1+t)^2}\,dt$

$\qquad\qquad\qquad = 500 \lim_{b\to\infty} \int_1^{1+b} u^{-2}\,du$

$\qquad\qquad\qquad = 500 \lim_{b\to\infty} \left(-\frac{1}{u}\right)\Big|_1^{1+b}$

$\qquad\qquad\qquad = 500 \lim_{b\to\infty} \left(-\frac{1}{1+b} + 1\right) = 500$ gallons

Substitution: $u = 1 + t$
$\qquad\qquad\qquad du = dt$

Limits: $t = 0$ implies $u = 1$
$\qquad\qquad t = b$ implies $u = 1 + b$

59. Total $= \displaystyle\int_0^\infty R(t)\,dt = \lim_{T\to\infty} \int_0^T R(t)\,dt = \lim_{T\to\infty} \int_0^T \frac{400}{(5+t)^3}\,dt$

$\qquad\qquad\qquad = \lim_{T\to\infty} \left[\frac{-200}{(5+t)^2}\Big|_0^T\right]$

$\qquad\qquad\qquad = \lim_{T\to\infty} \left[\frac{-200}{(5+T)^2} + 8\right] = 8$

Thus, 8 million immigrants will enter the country under this policy.

Exercise 9-2

1. The graph of $f(x)$ is shown at the right. From the graph we see that $f(x) \geq 0$ for $x \in (-\infty, \infty)$.

$\displaystyle\int_{-\infty}^\infty f(x)\,dx = \int_0^4 \frac{1}{8}x\,dx$

$\qquad\qquad = \frac{1}{8}\cdot\frac{x^2}{2}\Big|_0^4$

$\qquad\qquad = \frac{1}{16}(16 - 0)$

$\qquad\qquad = 1$

x	$f(x)$
0	0
1	$\frac{1}{8}$
4	$\frac{1}{2}$
$x > 4$	0

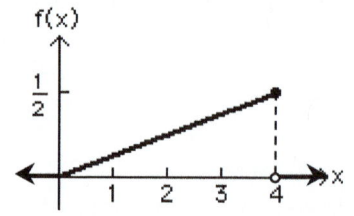

3. $f(x) = \begin{cases} 2 + 2x & \text{if } -1 \leq x \leq 0 \\ 0 & \text{otherwise} \end{cases}$

For $f(x)$ to be a probability density function, it has to satisfy the following two conditions:

(i) $f(x) \leq 0$ for all x

(ii) $\displaystyle\int_{-1}^0 f(x)\,dx = 1$

Note that for $-1 \leq x \leq 0$, $f(x) = 2 + 2x \geq 0$ and hence (i) is satisfied.

$\displaystyle\int_{-1}^0 (2 + 2x)\,dx = (2x + x^2)\big|_{-1}^0 = -(2(-1) + (-1)^2) = -(-2 + 1) = 1.$

Therefore, $f(x)$ given above is a probability density function.

5. $f(x) = \begin{cases} 0.2 & \text{if } \leq x \leq 5 \\ 0 & \text{otherwise)} \end{cases}$

Note that (i) of Problem 3 is satisfied, however,

$\displaystyle\int_1^5 f(x)\,dx = \int_1^5 0.2\,dx = 0.2x\big|_1^5 = 0.2(5) - 0.2(1) = 1 - 0.2 = 0.8$

and hence (ii) of Problem 3 is not satisfied. Thus, $f(x)$ given above is not a probability density function.

7. (A) Using 2(c),

$$P(1 < X < 3) = \int_1^3 \frac{1}{8}x\, dx = \frac{1}{8} \cdot \frac{x^2}{2}\Big|_1^3$$

$$= \frac{1}{16}(9-1) = \frac{8}{16} = \frac{1}{2}.$$

The graph is shown at the right.

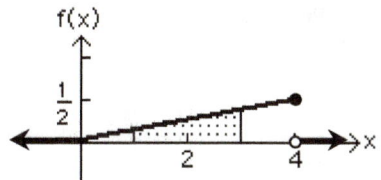

(B) $P(X \le 2) = \int_{-\infty}^2 \frac{1}{8}x\, dx = \int_0^2 \frac{1}{8}x\, dx$ [since $f(x) = 0$ for $x \le 0$]

$$= \frac{1}{8} \cdot \frac{x^2}{2}\Big|_0^2 = \frac{1}{16}(4) = \frac{1}{4}.$$

The graph is shown at the right.

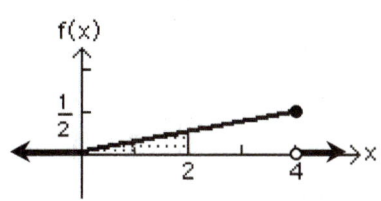

(C) $P(X > 3) = \int_3^\infty \frac{1}{8}x\, dx = \int_3^4 \frac{1}{8}x\, dx$ [since $f(x) = 0$ for $x > 4$]

$$= \frac{1}{8} \cdot \frac{x^2}{2}\Big|_3^4 = \frac{1}{16}(16-9) = \frac{7}{16}.$$

The graph is shown at the right.

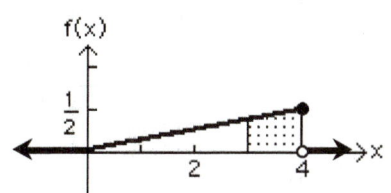

9. (A) $P(X = 1) = \int_1^1 f(x)dx = 0$

(B) $P(X > 5) = \int_5^\infty f(x)dx = \int_5^\infty 0\, dx = 0\ [f(x) = 0$ for $x > 4]$

(C) $P(X < 5) = \int_{-\infty}^5 f(x)dx$

$$= \int_0^4 \frac{1}{8}x\, dx\ [f(x) = 0 \text{ when } x < 0 \text{ and when } x > 4]$$

$$= \frac{1}{8} \cdot \frac{x^2}{2}\Big|_0^4 = \frac{1}{16}(16) = 1.$$

11. If $x < 0$, then

$$F(x) = \int_{-\infty}^x f(t)dt = \int_{-\infty}^x 0\, dt = 0$$

If $0 \le x \le 4$, then

$$F(x) = \int_{-\infty}^x f(t)\, dt = \int_{-\infty}^0 f(t)\, dt + \int_0^x f(t)\, dt = 0 + \int_0^x \frac{1}{8}t\, dt$$

$$= \frac{1}{16}t^2\Big|_0^x = \frac{1}{16}x^2.$$

If $x > 4$, then

$$F(x) = \int_{-\infty}^x f(t)\, dt = \int_{-\infty}^0 f(t)\, dt + \int_0^4 f(t)\, dt + \int_4^x f(t)\, dt$$

$$= 0 + \int_0^4 \frac{1}{8}t\, dt + 0 = \frac{1}{16}t^2\Big|_0^4 = \frac{1}{16}(16-0) = 1.$$

Thus, the cumulative probability distribution function is:

$$F(x) = \begin{cases} 0 & x < 0 \\ \frac{1}{16}x^2 & 0 \le x \le 4 \\ 1 & x > 4 \end{cases}$$

The graph of $F(x)$ is shown at the right.

13. Using 3 and the cumulative probability distribution function F from Problem 11:

(A) $P(2 \le X \le 4) = F(4) - F(2)$

$$= \frac{1}{16}(4)^2 - \frac{1}{16}(2)^2$$

$$= 1 - \frac{1}{4} = \frac{3}{4}.$$

(B) $P(0 < X < 2) = F(2) - F(0)$

$$= \frac{1}{16}(2)^2 - 0$$

$$= \frac{1}{4}.$$

15. From Problem 11:

(A) $P(0 \le X \le x) = F(x) - F(0) = \frac{1}{16} x^2 - 0 = \frac{1}{16} x^2$

Now $\frac{1}{16} x^2 = \frac{1}{4}$

$x^2 = 4$

Since $x \ge 0$, the solution is $x = 2$.

(B) $P(0 \le X \le x) = F(x) - F(0) = \frac{1}{9}$ implies $\frac{1}{16} x^2 = \frac{1}{9}$

$x^2 = \frac{16}{9}$

Since $x \ge 0$, the solution is $x = \frac{4}{3}$.

17. From the graph of $f(x)$ shown below, we see that $f(x) \ge 0$ for $x \in (-\infty, \infty)$.

x	$f(x)$
0	2
2	$\frac{2}{27}$
4	$\frac{2}{125}$
$x < 0$	0

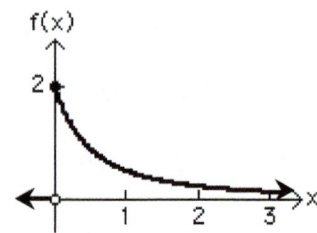

Also,

$$\int_{-\infty}^{\infty} f(x)\, dx = \int_0^{\infty} \frac{2}{(1 + x)^3}\, dx = \lim_{R \to \infty} \int_0^R \frac{2}{(1 + x)^3}\, dx$$

$$= \lim_{R \to \infty} \left(-\frac{1}{(1+x)^2} \right) \Big|_0^R = -\lim_{R \to \infty} \left(\frac{1}{(1 + R)^2} - 1 \right) = -(0 - 1) = 1.$$

19. (A) $P(1 \le X \le 4) = \int_1^4 \frac{2}{(1 + x)^3} dx = \frac{-1}{(1 + x)^2} \Big|_1^4 = \frac{-1}{(1 + 4)^2} - \frac{-1}{(1 + 1)^2}$

$= \frac{-1}{25} + \frac{1}{4} = \frac{21}{100} = .21$

(B) $P(X > 3) = \int_3^{\infty} \frac{2}{(1 + x)^3} dx$

Now, $1 = \int_{-\infty}^{\infty} \frac{2}{(1 + x)^3} dx = \int_{-\infty}^3 \frac{2}{(1 + x)^3} dx + \int_3^{\infty} \frac{2}{(1 + x)^3} dx,$

so $\int_3^{\infty} \frac{2}{(1 + x)^3} dx = 1 - \int_{-\infty}^3 \frac{2}{(1 + x)^3} dx = 1 - \frac{2}{(1 + x)^3} \int_{-\infty}^x dx,$

since $f(x) = 0$ for $x < 0$. Thus,

$\int_3^{\infty} \frac{2}{(1 + x)^3} dx = 1 - \left(\frac{-1}{(1 + x)^2} \right) \Big|_0^3 = 1 - \left(-\frac{1}{16} + 1 \right) = \frac{1}{16}.$

(C) $P(X \le 2) = \int_{-\infty}^2 \frac{2}{(1 + x)^3} dx = \int_0^2 \frac{2}{(1 + x)^3} dx = \frac{-1}{(1 + x)^2} \Big|_0^2 = -\frac{1}{9} + 1 = \frac{8}{9}$

21. If $x < 0$, then

$$F(x) = \int_{-\infty}^x f(t)\, dt = \int_{-\infty}^x 0\, dt = 0.$$

If $x \ge 0$, then

$$F(x) = \int_{-\infty}^x f(t)\, dt = \int_{-\infty}^0 f(t)\, dt + \int_0^x f(t)\, dt = 0 + \int_0^x \frac{2}{(1 + t)^3}\, dt$$

$$= \left(-\frac{1}{(1 + t)^2} \right) \Big|_0^x = -\frac{1}{(1 + x)^2} + 1 \quad \text{or} \quad 1 - \frac{1}{(1 + x)^2}.$$

$$F(x) = \begin{cases} 0 & x < 0 \\ 1 - \dfrac{1}{(1+x)^2} & x \ge 0 \end{cases}$$

The graph of $F(x)$ is shown at the right.

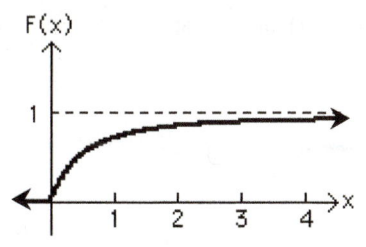

23. From Problem 21,

$$F(x) = \begin{cases} 0 & x < 0 \\ 1 - \dfrac{1}{(1+x)^2} & x \ge 0 \end{cases}$$

and $P(0 \le X \le x) = F(x) - F(0) = 1 - \dfrac{1}{(1+x)^2}$.

(A) Set $P(0 \le X \le x) = \dfrac{3}{4}$ and solve for x.

$$1 - \frac{1}{(1+x)^2} = \frac{3}{4}$$

$$\frac{1}{(1+x)^2} = \frac{1}{4}$$

$$(1+x)^2 = 4$$

Since $x \ge 0$, we have $1 + x = 2$, so $x = 1$.

(B) $1 = P(X \le x) + P(X > x)$. Therefore,

$$P(X > x) = 1 - P(X \le x) = 1 - F(x) = 1 - \left(1 - \frac{1}{(1+x)^2}\right) = \frac{1}{(1+x)^2}.$$

Set $P(X > x) = \dfrac{1}{16}$ and solve for x.

$$\frac{1}{(1+x)^2} = \frac{1}{16}$$

$$(1+x)^2 = 16$$

$$1 + x = 4$$

$$x = 3$$

25. $f(x) = \begin{cases} \dfrac{3}{2}x - \dfrac{3}{4}x^2 & 0 \le x \le 2 \\ 0 & \text{otherwise} \end{cases}$

If $x < 0$, then

$$F(x) = \int_{-\infty}^{x} f(t)dt = \int_{-\infty}^{x} 0 \, dt = 0.$$

If $0 \le x \le 2$, then

$$F(x) = \int_{-\infty}^{x} f(t)\,dt = \int_{-\infty}^{0} f(t)\,dt + \int_{0}^{x} f(t)\,dt = 0 + \int_{0}^{x} \left(\frac{3}{2}t - \frac{3}{4}t^2\right)dt$$

$$= \left(\frac{3}{2}\cdot\frac{t^2}{2} - \frac{3}{4}\cdot\frac{t^3}{3}\right)\Big|_{0}^{x} = \left(\frac{3}{4}t^2 - \frac{1}{4}t^3\right)\Big|_{0}^{x} = \frac{3}{4}x^2 - \frac{1}{4}x^3.$$

If $x > 2$, then

$$F(x) = \int_{-\infty}^{x} f(t)\,dt = \int_{-\infty}^{x} 0 \, dt + \int_{0}^{2} f(t)\,dt + \int_{2}^{x} f(t)\,dt$$

$$= 0 + \int_{0}^{2} \left(\frac{3}{2}t - \frac{3}{4}t^2\right)dt + 0 = \left(\frac{3}{4}t^2 - \frac{1}{4}t^3\right)\Big|_{0}^{2} = \frac{3}{4}(2)^2 - \frac{1}{4}(2)^3 = 1$$

Thus, the cumulative probability distribution function is:

$$F(x) = \begin{cases} 0 & x < 0 \\ \dfrac{3}{4}x^2 - \dfrac{1}{4}x^3 & 0 \le x \le 2 \\ 1 & x > 2 \end{cases}$$

The graphs of $F(x)$ and $f(x)$ are as follows:

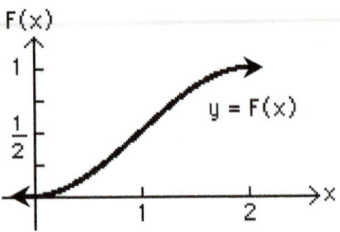

27. $f(x) = \begin{cases} \dfrac{1}{2} + \dfrac{1}{2}x^3 & -1 \le x \le 1 \\ 0 & \text{otherwise} \end{cases}$

If $x < -1$, then

$$F(x) = \int_{-\infty}^{x} f(t)\, dt = \int_{-\infty}^{x} 0\, dt = 0$$

If $-1 \le x \le 1$, then

$$F(x) = \int_{-\infty}^{x} f(t)\, dt = \int_{-\infty}^{-1} f(t)\, dt + \int_{-1}^{x} f(t)\, dt = 0 + \int_{-1}^{x} \left(\frac{1}{2} + \frac{1}{2}t^3 \right) dt$$

$$= \left(\frac{1}{2}t + \frac{1}{8}t^4 \right)\Big|_{-1}^{x} = \frac{1}{2}x + \frac{1}{8}x^4 - \left(-\frac{1}{2} + \frac{1}{8} \right)$$

$$= \frac{3}{8} + \frac{1}{2}x + \frac{1}{8}x^4$$

If $x > 1$, then

$$F(x) = \int_{-\infty}^{x} f(t)\, dt = \int_{-\infty}^{-1} f(t)\, dt + \int_{-1}^{1} f(t)\, dt + \int_{1}^{x} f(t)\, dt$$

$$= 0 + \int_{-1}^{1} \left(\frac{1}{2} + \frac{1}{2}t^3 \right) dt + 0$$

$$= \left(\frac{1}{2}t + \frac{1}{8}t^4 \right)\Big|_{-1}^{1} = \left(\frac{1}{2} + \frac{1}{8} \right) - \left(-\frac{1}{2} + \frac{1}{8} \right)$$

$$= 1$$

Thus, the cumulative probability distribution function is:

$$F(x) = \begin{cases} 0 & x < -1 \\ \dfrac{3}{8} + \dfrac{1}{2}x + \dfrac{1}{8}x^4 & -1 \le x \le 1 \\ 1 & x > 1 \end{cases}$$

The graphs of $f(x)$ and $F(x)$ are:

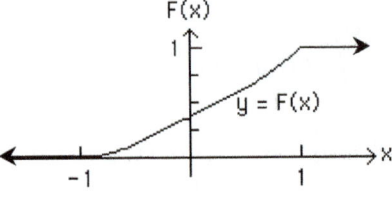

29. The graphs of

$F(x) =$

$$\begin{cases} 0 & x < 0 \\ \dfrac{3}{4}x^2 - \dfrac{1}{4}x^3 & 0 \le x \le 2 \\ 1 & x > 2 \end{cases}$$

and $y = 0.2$ are shown at the right.

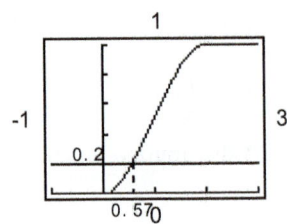

$F(x) = 0.2$ at $x \approx 0.5$

31. The graphs of

$$F(x) = \begin{cases} 0 & x < -1 \\ \dfrac{3}{8} + \dfrac{1}{2}x + \dfrac{1}{8}x^4 & -1 \le x \le 1 \\ 1 & x > 1 \end{cases} \text{ and } y = 0.6 \text{ are shown below.}$$

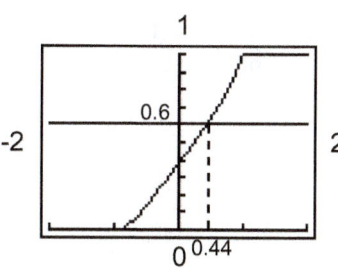

$$F(x) = 0.6 \text{ at } x \approx 0.44$$

33. $f(x) = \begin{cases} e^{-x/3} & \text{if } x \ge 0 \\ 0 & \text{otherwise} \end{cases}$

Note that for $k > 0$, condition (i) of Problem 3 is satisfied. To check the second condition, we observe that

$$\int_0^b e^{-x/3}\, dx = -3e^{-x/3}\Big|_0^b = -3e^{-b/3} + 3 \text{ and}$$

$$\lim_{b \to \infty}(-3e^{-b/3} + 3) = 3, \text{ therefore } \int_0^\infty e^{-x/3}dx = 3.$$

Now, we have to choose k so that $\displaystyle\int_0^\infty kf(x) = 1$

or $k = \dfrac{1}{\displaystyle\int_0^\infty f(x)dx} = \dfrac{1}{3}.$

35. $f(x) = \begin{cases} \dfrac{1}{x^5} & \text{if } x \ge 1 \\ 0 & \text{otherwise} \end{cases}$

Note that for $k > 0$, $kf(x) \ge 0$ for all x.

$$\int_1^b \dfrac{1}{x^5}\, dx = -\dfrac{1}{4}x^{-4}\Big|_1^b = -\dfrac{1}{4b^4} + \dfrac{1}{4} \text{ and}$$

$$\lim_{b \to \infty}\left(-\dfrac{1}{4b^4} + \dfrac{1}{4}\right) = 1, \text{ thus } \int_1^\infty \dfrac{1}{x^5}\, dx = \dfrac{1}{4}.$$

$$\int_1^\infty kf(x)\, dx = 1 \text{ implies that } k = \dfrac{1}{\displaystyle\int_1^\infty \dfrac{1}{x^5}\, dx} = \dfrac{1}{\dfrac{1}{4}} = 4$$

37. No! For example, F does not satisfy (b) and (c) of **4**; $F(x) > 1$ on $\left[0, \dfrac{1}{2}\right]$ and F is *decreasing* on $[0, 1]$.

39. $f(x) = \dfrac{1}{\sqrt{2\pi}}e^{-0.5x^2}$. The graph of f is shown at the right.

$$\int_{-1}^1 f(x)\, dx \approx 0.683; \int_{-2}^2 f(x)\, dx \approx 0.954;$$

$$\int_{-3}^3 f(x)\, dx = 0.997$$

Based on these calculations, it appears that

$$\int_{-\infty}^\infty f(x)dx = 1$$

41. The relationship between $f(x)$ and $F(x)$ is $F'(x) = f(x)$. Thus, if

$$F(x) = \begin{cases} 0 & x < 0 \\ x^2 & 0 \le x \le 1 \\ 1 & x > 1 \end{cases}$$

then

$$f(x) = \begin{cases} 0 & x < 0 \\ 2x & 0 \le x < 1 \\ 0 & x > 1 \end{cases} \quad \text{or} \quad f(x) = \begin{cases} 2x & 0 \le x < 1 \\ 0 & \text{otherwise)} \end{cases}$$

43. $F(x) = \begin{cases} 0 & x < 0 \\ 6x^2 - 8x^3 + 3x^4 & 0 \le x \le 1 \\ 1 & x > 1 \end{cases}$

Thus,

$$f(x) = \begin{cases} 0 & x < 0 \\ 12x - 24x^2 + 12x^3 & 0 \le x \le 1 \\ 0 & \text{otherwise} \end{cases} \quad [\underline{\text{Note:}} f(x) = F'(x).]$$

or

$$f(x) = \begin{cases} 12x(1 - 2x + x^2) & 0 \le x \le 1 \\ 0 & \text{otherwise} \end{cases}$$

45. Given $f(x) = \begin{cases} x & 0 \le x \le 1 \\ 2 - x & 1 < x \ge 2 \\ 0 & \text{otherwise} \end{cases}$

If $x < 0$, then

$$F(x) = \int_{-\infty}^{x} f(t)\, dt = \int_{-\infty}^{x} 0\, dt = 0.$$

If $0 \le x \le 1$, then

$$F(x) = \int_{-\infty}^{x} f(t)\, dt = \int_{-\infty}^{0} f(t)\, dt + \int_{0}^{x} f(t)\, dt = 0 + \int_{0}^{x} t\, dt = \frac{1}{2} t^2 \Big|_0^x = \frac{1}{2} x^2.$$

If $1 < x \le 2$, then

$$F(x) = \int_{-\infty}^{x} f(t)\, dt = \int_{-\infty}^{0} f(t)\, dt + \int_{0}^{1} f(t)\, dt + \int_{1}^{x} f(t)\, dt$$

$$= 0 + \int_{0}^{1} t\, dt + \int_{1}^{x} (2 - t)\, dt = \frac{1}{2} t^2 \Big|_0^1 + \left(2t - \frac{1}{2} t^2 \right)\Big|_1^x$$

$$= \frac{1}{2} + \left(2x - \frac{1}{2} x^2 \right) - \left(2 - \frac{1}{2} \right) = \frac{1}{2} + 2x - \frac{1}{2} x^2 - \frac{3}{2} = 2x - \frac{1}{2} x^2 - 1.$$

If $x > 2$, then

$$F(x) = \int_{-\infty}^{x} f(t)\, dt = \int_{-\infty}^{0} f(t)\, dt + \int_{0}^{1} f(t)\, dt + \int_{1}^{2} f(t)\, dt + \int_{2}^{x} f(t)\, dt$$

$$= 0 + \int_{0}^{1} t\, dt + \int_{1}^{2} (2 - t)\, dt + 0 = \frac{1}{2} t^2 \Big|_0^1 + \left(2t - \frac{1}{2} t^2 \right)\Big|_1^2$$

$$= \frac{1}{2} + \left[\left(4 - \frac{1}{2} \cdot 4 \right) - \left(2 - \frac{1}{2} \right) \right] = 1.$$

Thus, $F(x)$ is given by:

$$F(x) = \begin{cases} 0 & x < 0 \\ \dfrac{1}{2} x^2 & 0 \le x < 1 \\ 2x - \dfrac{1}{2} x^2 - 1 & 1 \le x \le 2 \\ 1 & x > 2 \end{cases}$$

47. $f(x) = \begin{cases} .2 - 0.02x & 0 \le x \le 10 \\ 0 & \text{otherwise} \end{cases}$

(A) $\displaystyle\int_2^6 f(x)\, dx = \int_2^6 (0.2 - 0.02x)\, dx$

$$= [(0.2x - 0.01x^2)\big|_2^6] = 0.48$$

The probability that the daily demand for electricity is between 2 million and 6 million kilowatt hours is 0.48.

(B) $\displaystyle P(X \le 8) = \int_{-\infty}^8 f(x)\, dx = \int_0^8 f(x)\, dx = \int_0^8 (0.2 - 0.02x)\, dx$

$$= \left(0.2x - \frac{0.02x^2}{2} \right)\Big|_0^8 = 0.2(8) - 0.01(8)^2 = 1.6 - 0.64 = 0.96$$

(C) $\displaystyle P(X > 5) = \int_5^\infty f(x)\, dx = \int_5^{10} f(x)\, dx = \int_5^{10} (0.2 - 0.02x)\, dx$

$$= \left(0.2x - \frac{0.02x^2}{2} \right)\Big|_5^{10}$$

$$= [0.2(10) - 0.01(10)^2] - [0.2(5) - 0.01(5)^2]$$

$$= 2 - 1 - (1 - 0.25) = 0.25$$

49. (A) $\displaystyle\int_5^{10} f(x)\, dx = \int_5^{10} \frac{1}{10} e^{-x/10}\, dx$

$$= [- e^{-x/10}\big|_5^{10}] = e^{-1/2} - e^{-1} \cdot 0.239$$

The probability that it takes the computer more than 5 seconds and less than 10 seconds to respond is 0.239.

(B) $\displaystyle P(0 \le X \le 1) = \int_0^1 f(x)\, dx = \int_0^1 \frac{1}{10} e^{-x/10}\, dx$

$$= \frac{1}{10} \int_0^1 e^{-x/10}\, dx = \frac{1}{10} \cdot \frac{e^{-x/10}}{-\frac{1}{10}}\Big|_0^1$$

$$= -(e^{-1/10} - 1) = 1 - e^{-1/10} \approx 0.0952$$

(C) $\displaystyle P(X > 4) = \int_4^\infty \frac{1}{10} e^{-x/10}\, dx = \lim_{R \to \infty} \frac{1}{10} \int_4^R e^{-x/10}\, dx = \lim_{R \to \infty} \left(e^{-x/10}\Big|_4^R \right)$

$$= -\lim_{R \to \infty} (e^{-R/10} - e^{-4/10}) = -(0 - e^{-2/5} = e^{-2/5}) \approx 0.6703$$

51. (A) $\displaystyle P(X > 4) = \int_4^{10} f(x)\, dx \qquad [\underline{\text{Note}}: 4 \text{ stands for 4000.}]$

$$= \int_4^{10} 0.003x\sqrt{100 - x^2}\, dx$$

$$= \frac{-0.003}{2} \int_4^{10} (100 - x^2)^{1/2}(-2x)\, dx \quad [\text{Note: } d(100 - x^2) = -2x\, dx.]$$

$$= -0.0015 \cdot \frac{2}{3}(100 - x^2)^{(3/2)}\Big|_4^{10} = \frac{1}{1000}(100 - x^2)^{3/2}\Big|_4^{10}$$

$$= -\frac{1}{1000}(0 - (84)^{3/2})$$

$$= \frac{(84)^{3/2}}{1000} \approx 0.7699$$

(B) $\displaystyle P(0 \le x \le 8) = \int_0^8 f(x)\, dx = \int_0^8 0.003x\sqrt{100 - x^2}\, dx$

$$= -\frac{1}{1000}(100 - x^2)^{3/2}\Big|_0^8 \quad [\text{refer to part (A)}]$$

$$= -\frac{1}{1000}[(100 - 64)^{3/2} - (100)^{3/2}]$$

$$= -\frac{1}{1000}(36^{3/2} - 100^{3/2}) = \frac{-1}{1000}(216 - 1000) = 0.784$$

(C) We must solve the following for x:

$$\int_0^x f(t)\, dt = 0.9$$

$$\int_0^x 0.003\sqrt{100 - t^2}\, dt = 0.9$$

$$-\frac{1}{1000}(100 - t^2)^{3/2}\Big|_0^x = 0.9 \qquad \text{[refer to part (A)]}$$

$$(100 - t^2)^{3/2}\Big|_0^x = -900$$

$$(100 - x^2)^{3/2} - (100)^{3/2} = -900$$

$$(100 - x^2)^{3/2} - 1000 = -900$$

$$(100 - x^2)^{3/2} = 100$$

$$100 - x^2 = 100^{2/3}$$

$$x^2 = 100 - 100^{2/3}$$

$$x = \sqrt{100 - 100t^{2/3}} \approx 8.858 \quad \text{or} \quad 8858 \text{ pounds}$$

53. (A) $P(7 \le X) = \displaystyle\int_7^\infty f(x)\, dx = \int_7^{10} f(x)\, dx + \int_{10}^\infty f(x)\, dx$

$$= \int_7^{10} \frac{1}{5000}(10x^3 - x^4)\, dx + \int_{10}^\infty 0\, dx = \frac{1}{5000}\left(\frac{10}{4}x^4 - \frac{x^5}{5}\right)\Big|_7^{10} + 0$$

$$= \frac{1}{5000}\left(\left[\frac{5}{2}(10)^4 - \frac{1}{5}(10)^5\right] - \left[\frac{5}{2}(7)^4 - \frac{1}{5}(7)^5\right]\right) \approx 0.47178$$

(B) $P(X \le 5) = \displaystyle\int_{-\infty}^5 f(x)\, dx = \int_{-\infty}^0 f(x)\, dx + \int_0^5 f(x)\, dx$

$$= \int_{-\infty}^0 0\, dx + \int_0^5 \frac{1}{5000}(10x^3 - x^4)\, dx = 0 + \frac{1}{5000}\left(\frac{5}{2}x^4 - \frac{1}{5}x^5\right)\Big|_0^5$$

$$= \frac{1}{5000}\left[\frac{5}{2}(5)^4 - \frac{1}{5}(5)^5\right] = \frac{3}{16} = 0.1875$$

55. (A) $P(X \le 20) = \displaystyle\int_{-\infty}^{20} f(x)\, dx = \int_0^{20} \frac{800x}{(400 + x^2)^2}\, dx = 400\int_0^{20} \frac{2x}{(400 + x^2)^2}\, dx$

$$= \frac{-400}{400 + x^2}\Big|_0^{20} = \frac{-400}{800} + \frac{400}{400} = 0.5$$

(B) $P(X > 15) = \displaystyle\int_{15}^\infty f(x)\, dx = 1 - \int_{-\infty}^{15} f(x)\, dx = 1 - \int_0^{15} \frac{800x}{(400 + x^2)^2}\, dx$

$$= 1 - \left(\frac{-400}{400 + x^2}\right)\Big|_0^{15} = 1 + \frac{400}{625} - 1 = 0.64$$

(C) We must solve the following for x:

$$\int_0^x f(t)\, dt = 0.8$$

$$\int_0^x \frac{800t}{(400 + t^2)^2}\, dt = 0.8$$

$$\frac{-400}{400 + t^2}\Big|_0^x = 0.8$$

$$\frac{-400}{400 + x^2} + 1 = 0.8$$

$$\frac{-400}{400 + x^2} = -0.2$$

$$-400 = -80 - 0.2x^2$$

$$0.2x^2 = 320$$

$$x^2 = 1600$$

$$x = 40 \text{ days}$$

57. (A) $P(X \geq 30) = \int_{30}^{\infty} f(x)\, dx = 1 - \int_{-\infty}^{30} f(x)\, dx = 1 - \int_{0}^{30} \frac{1}{20} e^{-x/20}\, dx$

$= 1 + e^{-x/20}\big|_{0}^{30} = 1 + e^{-30/20} - 1 = e^{-3/2} \approx 0.223$

(B) $P(X \geq 80) = \int_{80}^{\infty} f(x)\, dx = 1 - \int_{-\infty}^{80} f(x)\, dx = 1 - \int_{0}^{80} \frac{1}{20} e^{-x/20}\, dx$

$= 1 + e^{-x/20}\big|_{0}^{80} = 1 + e - 1 = e^{-4} \approx 0.018$

Exercise 9-3

1. $f(x) = \begin{cases} 2 & \text{if } 0 \leq x \leq 0.5 \\ 0 & \text{otherwise} \end{cases}$

$\mu = E(x) = \int_{0}^{0.5} x f(x)\, dx = \int_{0}^{0.5} 2x\, dx = x^2\big|_{0}^{0.5} = 0.25 = \frac{1}{4}$

$V(x) = \int_{0}^{0.5} (x - 0.25)^2\, (2)\, dx = \int_{0}^{0.5} [2x^2 - x + 0.125]\, dx$

$= \left[\frac{2}{3} x^3 - \frac{1}{2} x^2 + 0.125x \right]_{0}^{0.5}$

$= \frac{2}{3}(0.5)^3 - \frac{1}{2}(0.5)^2 + 0.125(0.5)$

$= \frac{1}{48}$

$\sigma = \sqrt{V(X)} = \sqrt{\frac{1}{48}} = 0.144.$

Note: If the density is constant over an interval (a, b) and zero elsewhere, then the mean is the midpoint of the interval $\frac{a+b}{2}$ and the variance is $\frac{(b-a)^2}{12}$ and standard deviation is $\frac{b-a}{2\sqrt{3}}$. Use these to check mean, variance, and standard deviation for this problem.

3. $f(x) = \begin{cases} \dfrac{1}{8} x & \text{if } 0 \leq x \leq 4 \\ 0 & \text{otherwise} \end{cases}$

$\mu = \int_{0}^{4} x\left(\frac{1}{8}x\right) dx = \int_{0}^{4} \frac{1}{8} x^2\, dx = \frac{1}{24} x^3\big|_{0}^{4} = \frac{4^3}{24} = \frac{8}{3}$

$V(x) = \int_{0}^{4} \left(x - \frac{8}{3}\right)^2 \left(\frac{1}{8}x\right) dx = \int_{0}^{4} \frac{1}{8} x\left(x^2 - \frac{16}{3} x + \frac{64}{9}\right) dx$

$= \frac{1}{8} \int_{0}^{4} \left(x^3 - \frac{16}{3} x^2 + \frac{64}{9} x\right) dx = \frac{1}{8} \left[\frac{1}{4} x^4 - \frac{16}{9} x^3 + \frac{32}{9} x^2 \right]_{0}^{4}$

$= \frac{1}{8} \left[\frac{4^4}{4} - \frac{16(4^3)}{9} + \frac{32(4^2)}{9} \right] = \frac{1}{8} \left[\frac{64}{9} \right] = \frac{8}{9}$

$\sigma = \sqrt{V(X)} = \sqrt{\frac{8}{9}} = 0.943.$

5. $\mu = E(X) = \int_{-\infty}^{\infty} x f(x)\, dx = \int_{1}^{2} x(4 - 2x)\, dx = \int_{1}^{2} (4x - 2x^2)\, dx = \left(2x^2 - \frac{2}{3} x^3 \right)\Big|_{1}^{2}$

$= 8 - \frac{16}{3} - \left(2 - \frac{2}{3} \right) = \frac{4}{3} \approx 1.333$

Using 2,

$$V(X) = \int_{-\infty}^{\infty} x^2 f(x)\,dx - \mu^2 = \int_{1}^{2} x^2(4 - 2x)\,dx - \frac{16}{9} = \int_{1}^{2}(4x^2 - 2x^3)dx - \frac{16}{9}$$

$$= \left(\frac{4}{3}x^3 - \frac{1}{2}x^4\right)\Big|_{1}^{2} - \frac{16}{9} = \frac{32}{3} - 8 - \left(\frac{4}{3} - \frac{1}{2}\right) - \frac{16}{9}$$

$$= \frac{1}{18}$$

Thus, $\sigma = \sqrt{V(X)} = \sqrt{\frac{1}{18}} \approx 0.236$.

7. $f(x) = \begin{cases} 0.2 & \text{if } -2 \le x \le 3 \\ 0 & \text{otherwise} \end{cases}$

Since the density is constant over $[-2, 3]$ and zero elsewhere, the median will be the same as the mean which

is $\dfrac{a+b}{2} = \dfrac{-2+3}{2} = \dfrac{1}{2}$. You can find the median directly as follows:

Let m be the median, then $\displaystyle\int_{-2}^{m} f(x)\,dx = \int_{3}^{m} f(x)\,dx = \frac{1}{2}$.

So we need to solve $\displaystyle\int_{-2}^{m} 0.2\,dx = \frac{1}{2}$ or $\displaystyle\int_{3}^{m} 0.2\,dx = \frac{1}{2}$ to find m.

$$\int_{m}^{3} 0.2\,dx = 0.2x\,\Big|_{m}^{3} = 0.6 - 0.2m = \frac{1}{2} \text{ or } 0.2m = 0.1 \text{ or } m = \frac{0.1}{0.2} = 0.5.$$

9. Step 1:   Find the cumulative probability distribution function.

If $x < 2$, then $F(x) = 0$. If $2 \le x \le 4$, then

$$F(x) = \int_{-\infty}^{x} f(t)\,dt = \int_{2}^{x} \frac{1}{6}t\,dt = \frac{1}{12}t^2\,\Big|_{2}^{x} = \frac{x^2}{12} - \frac{1}{3}.$$

If $x > 4$, then

$$F(x) = \int_{-\infty}^{x} f(t)\,dt = \int_{-\infty}^{2} f(t)\,dt + \int_{2}^{4} f(t)\,dt + \int_{4}^{x} f(t)\,dt$$

$$= 0 + \int_{2}^{4} \frac{1}{6}t\,dt + 0 = \frac{1}{12}t^2\,\Big|_{2}^{4} = 1.$$

Thus, $F(x) = \begin{cases} 0 & \text{if } x < 2 \\ \dfrac{x^2}{12} - \dfrac{1}{3} & \text{if } 2 \le x \le 4 \\ 1 & \text{if } x > 4 \end{cases}$

Step 2   Solve the equation $P(X \le m) = \dfrac{1}{2}$ for m.

$$F(m) = P(X \le m) = \frac{1}{2}$$

$$\frac{m^2}{12} - \frac{1}{3} = \frac{1}{2}$$

$$\frac{m^2}{12} = \frac{5}{6}$$

$$m^2 = 10$$

$$m = \sqrt{10} \approx 3.162$$

11.   Step 1:   Find the cumulative probability distribution function.

If $x < 0$, then $F(x) = 0$. If $0 \le x \le 4$, then

$$F(x) = \int_{-\infty}^{x} f(t)\, dt = \int_{0}^{x} \left(\frac{1}{2} - \frac{1}{8} t \right) dt = \left(\frac{1}{2} t - \frac{1}{16} t^2 \right) \Big|_{0}^{x} = \frac{1}{2} x - \frac{1}{16} x^2.$$

If $x > 4$, then

$$F(x) = \int_{-\infty}^{x} f(t)\, dt = \int_{-\infty}^{0} f(t)\, dt + \int_{0}^{4} f(t)\, dt + \int_{4}^{x} f(t)\, dt$$

$$= 0 + \int_{0}^{4} \left(\frac{1}{2} - \frac{1}{8} t \right) dt + 0$$

$$= \left(\frac{1}{2} t - \frac{1}{16} t^2 \right) \Big|_{0}^{4} = 2 - 1 = 1$$

Thus, $F(x) = \begin{cases} 0 & \text{if } x < 0 \\ \dfrac{1}{2} x - \dfrac{1}{16} x^2 & \text{if } 0 \le x \le 4 \\ 1 & \text{if } x > 4 \end{cases}$

Step 2:   Solve the equation $P(X \le m) = \dfrac{1}{2}$ for m.

$$F(m) = P(X \le m) = \frac{1}{2}$$

$$\frac{1}{2} m - \frac{1}{16} m^2 = \frac{1}{2}$$

$$m^2 - 8m + 8 = 0$$

Now, the roots of the quadratic equation are

$$\frac{8 \pm \sqrt{64 - 32}}{2} = 4 \pm 2\sqrt{2}.$$

Since m must lie in the interval $[0, 4]$, $m = 4 - 2\sqrt{2} \approx 1.172$.

13. $f_1(x) = \begin{cases} 0.5 & \text{if } 0 \le x \le 2 \\ 0 & \text{otherwise} \end{cases}$

$f_2(x) = \begin{cases} 0.25 & \text{if } -2 \le x \le 2 \\ 0 & \text{otherwise} \end{cases}$

The mean of X_i, $i = 1, 2$, is the "balance point" of the region bounded by the graph of f_i and the x axis (see the discussion following explore–discuss 1). From this, we conclude that $\mu_1 = 1$ and $\mu_2 = 0$.

$$\mu_1 = \int_{-\infty}^{\infty} x f_1(x)\, dx = \int_{0}^{2} x(0.5)\, dx = \frac{1}{4} x^2 \Big|_{0}^{2} = 1$$

$$\mu_2 = \int_{-\infty}^{\infty} x f_2(x)\, dx = \int_{-2}^{2} x(0.25)\, dx = \frac{1}{8} x^2 \Big|_{-2}^{2} = \frac{1}{2} - \frac{1}{2} = 0$$

15. $g_1(x) = \begin{cases} 0.5 & \text{if } 0 \le x < 1 \\ 0.25 & \text{if } 1 \le x \le 3 \\ 0 & \text{otherwise} \end{cases}$

$g_2(x) = \begin{cases} 0.5 & \text{if } 0 \le x < 1 \\ 0.125 & \text{if } 1 \le x \le 5 \\ 0 & \text{otherwise} \end{cases}$

The median of $X_i, i = 1, 2$, is the value of the random variable that divides the area under the graph of $g_i, i = 1, 2$, into two equal parts. From the two graphs, it is easy to see that the medians $m_1 = m_2 = 1$.

Since $\displaystyle\int_0^1 g_1(x)\,dx = \int_0^1 g_2(x)\,dx = \int_0^1 \frac{1}{2}\,dx = \frac{1}{2}x\,\Big|_0^1 = \frac{1}{2}$ it follows that X_1 and X_2 each have median 1.

17. $\mu = E(X) = \displaystyle\int_{-\infty}^{\infty} xf(x)\,dx = \int_1^{\infty} x \cdot \frac{4}{x^5}\,dx = \int_1^{\infty} \frac{4}{x^4}\,dx = \lim_{R \to \infty} \int_1^R \frac{4}{x^4}\,dx$

$$= \lim_{R \to \infty} \left(-\frac{4}{3x^3} \right)\Big|_1^R = \lim_{R \to \infty} \left(-\frac{4}{3R^3} + \frac{4}{3} \right) = \frac{4}{3}$$

$V(X) = \displaystyle\int_{-\infty}^{\infty} x^2 f(x)\,dx - \mu^2 = \int_1^{\infty} x^2 \cdot \frac{4}{x^5}\,dx - \frac{16}{9} = \int_1^{\infty} \frac{4}{x^3}\,dx - \frac{16}{9}$

$$= \lim_{R \to \infty} \int_1^R \frac{4}{x^3}\,dx - \frac{16}{9} = \lim_{R \to \infty} \left(-\frac{2}{x^2} \right)\Big|_1^R - \frac{16}{9}$$

$$= \lim_{R \to \infty} \left(-\frac{2}{R^2} + 2 \right) - \frac{16}{9} = \frac{2}{9}$$

$\sigma = \sqrt{V(X)} = \sqrt{\dfrac{2}{9}} = \dfrac{\sqrt{2}}{3} \approx 0.471$

19. $\mu = E(X) = \displaystyle\int_{-\infty}^{\infty} xf(x)\,dx = \int_2^{\infty} x \cdot \frac{64}{x^5}\,dx = \int_2^{\infty} \frac{64}{x^4}\,dx = \lim_{R \to \infty} \int_2^R \frac{64}{x^4}\,dx$

$$= \lim_{R \to \infty} \left(\frac{-64}{3x^3} \right)\Big|_2^R = \lim_{R \to \infty} \left(\frac{64}{3R^3} + \frac{64}{24} \right) = \frac{8}{3} \approx 2.667$$

$V(X) = \displaystyle\int_{-\infty}^{\infty} x^2 f(x)\,dx - \mu^2 = \int_2^{\infty} x^2 \cdot \frac{64}{x^5}\,dx - \frac{64}{9} = \int_2^{\infty} \frac{64}{x^3}\,dx - \frac{64}{9}$

$$= \lim_{R \to \infty} \int_2^R \frac{64}{x^3}\,dx - \frac{64}{9} = \lim_{R \to \infty} \left(\frac{-32}{x^2} \right)\Big|_2^R - \frac{64}{9}$$

$$= \lim_{R \to \infty} \left(\frac{-32}{R^2} + 8 \right) - \frac{64}{9} = \frac{8}{9}$$

$\sigma = \sqrt{V(X)} = \sqrt{\dfrac{8}{9}} = \dfrac{2\sqrt{2}}{3} \approx 0.943$

21. $f(x) = \dfrac{2}{\sqrt{2\pi}}\, e^{-2(x-1)^2}$

$$\mu = \int_{-\infty}^{\infty} xf(x)\,dx \approx \int_{-10}^{10} xf(x)\,dx \approx 1$$

$$V(X) = \int_{-\infty}^{\infty} x^2 f(x)\,dx - \mu^2 \approx \int_{-10}^{10} x^2 f(x)\,dx - 1 \approx 1.25 - 1 = 0.25$$

$$\sigma = \sqrt{0.25} = 0.50$$

23. <u>Step 1</u>: Find the cumulative probability distribution function.

If $x < 1$, then $F(x) = 0$. If $1 \leq x \leq e$, then

$$F(x) = \int_{-\infty}^{x} f(t)\, dt = \int_{-\infty}^{1} f(t)\, dt + \int_{1}^{x} f(t)\, dt$$

$$= 0 + \int_{1}^{x} \frac{1}{t}\, dt = \ln t \, \Big|_{1}^{x} = \ln x - \ln 1 = \ln x.$$

If $x > e$, then

$$F(x) = \int_{-\infty}^{x} f(t)\, dt = \int_{-\infty}^{1} f(t)\, dt = \int_{1}^{e} f(t)\, dt = \int_{e}^{x} f(t)\, dt$$

$$= 0 + \int_{1}^{e} \frac{1}{t}\, dt + 0 = \ln t \, \Big|_{1}^{e} = \ln e - \ln 1 = 1.$$

Thus, $F(x) = \begin{cases} 0 & x < 1 \\ \ln x & 1 \leq x \leq e \\ 1 & x > e \end{cases}$

<u>Step 2</u>: Solve the equation $P(X \leq m) = \dfrac{1}{2}$ for m.

$$F(m) = P(X \leq m) = \frac{1}{2}$$

$$\ln m = \frac{1}{2} \quad (1 \leq x \leq e)$$

Thus, the median $m = e^{1/2} \approx 1.649$.

25. <u>Step 1</u>: Find the cumulative probability distribution function.

If $x < 0$, then $F(x) = 0$. If $0 \leq x \leq 2$, then

$$F(x) = \int_{-\infty}^{x} f(t)\, dt = \int_{2}^{x} \frac{4}{(2+t)^2}\, dt = \frac{-4}{2+t} \, \Big|_{0}^{x} = \frac{-4}{2+x} + 2.$$

If $x > 2$, then

$$F(x) = \int_{-\infty}^{x} f(t)\, dt = \int_{-\infty}^{0} f(t)\, dt + \int_{0}^{2} f(t)\, dt + \int_{2}^{x} f(t)\, dt$$

$$= 0 + \int_{0}^{2} \frac{4}{(2+t)^2}\, dt + 0 = \frac{-4}{2+t} \, \Big|_{0}^{2} = -1 + 2 = 1$$

Thus, $F(x) = \begin{cases} 0 & \text{if } x < 0 \\ \dfrac{-4}{2+x} + 2 & \text{if } 0 \leq x \leq 2 \\ 1 & \text{if } x > 2 \end{cases}$

<u>Step 2</u>: Solve the equation $P(X \leq m) = \dfrac{1}{2}$ for m.

$$F(m) = P(X \leq m) = \frac{1}{2}$$

$$\frac{-4}{2+m} + 2 = \frac{1}{2}$$

$$\frac{-4}{2+m} = \frac{-3}{2}$$

$$-6 - 3m = -8$$

$$m = \frac{2}{3}$$

27. <u>Step 1:</u> Find $F(x)$.

If $x < 0$, then $F(x) = 0$. If $x \geq 0$, then

$$F(x) = \int_{-\infty}^{x} f(t)\, dt = \int_{-\infty}^{0} f(t)\, dt + \int_{0}^{x} f(t)\, dt = 0 + \int_{0}^{x} \frac{1}{(1+t)^2}\, dt$$

$$= -\frac{1}{(1+t)}\bigg|_{0}^{x} = -\left(\frac{1}{1+x}\right) = 1 - \frac{1}{1+x} = \frac{x}{1+x}.$$

Thus, $F(x) = \begin{cases} 0 & x < 0 \\ \dfrac{x}{1+x} & x \geq 0 \end{cases}$

<u>Step 2:</u> Solve $P(X \leq m) = \dfrac{1}{2}$ for m.

$$F(m) = P(X \leq m) = \frac{1}{2}$$

$$\frac{m}{1+m} = \frac{1}{2} \quad (x \geq 0)$$

$$2m = 1 + m$$

Thus, the median $m = 1$.

29. <u>Step 1:</u> Find $F(x)$.

If $x < 0$, then $F(x) = 0$. If $x \geq 0$, then

$$F(x) = \int_{-\infty}^{x} f(t)\, dt = \int_{-\infty}^{0} f(t)\, dt + \int_{0}^{x} f(t)\, dt = 0 + \int_{0}^{x} 2e^{-2t}\, dt$$

$$= -e^{-2t}\big|_{0}^{x} = -e^{-2x} + 1$$

Thus, $F(x) = \begin{cases} 0 & \text{if } x < 0 \\ 1 - e^{-2x} & \text{if } x \geq 0 \end{cases}$

<u>Step 2:</u> Solve $P(X \leq m) = \dfrac{1}{2}$ for m.

$$F(m) = P(X \leq m) = \frac{1}{2}$$

$$1 - e^{-2m} = \frac{1}{2}$$

$$e^{-2m} = \frac{1}{2}$$

$$-2m = \ln \frac{1}{2} = -\ln 2$$

$$m = \frac{\ln 2}{2} \approx 0.347$$

31. $F(x) = \dfrac{1}{2}$ for any x satisfying $0.5 \leq x \leq 2.5$. The median is *not* unique.

33. From the graph,

$$f(x) = \begin{cases} 0 & \text{if } x < 0 \\ 1 - x & \text{if } 0 \leq x \leq 1 \\ 0 & \text{if } 1 < x < 2 \\ x - 2 & \text{if } 2 \leq x \leq 3 \\ 0 & \text{if } x > 3 \end{cases}$$

Now $F(x) = \displaystyle\int_{-\infty}^{x} f(t)\, dt$ and:

for $-\infty < x < 0$, $F(x) = \displaystyle\int_{-\infty}^{x} f(t)\, dt = \int_{-\infty}^{x} 0\, dt = 0$; for $0 \leq x < 1$,

$$F(x) = \int_{-\infty}^{x} f(t)\,dt = \int_{-\infty}^{0} 0\,dt + \int_{0}^{x} (1-t)\,dt = \left[\left(t - \frac{1}{2}t^2\right)\Big|_{0}^{x}\right] = x - \frac{1}{2}x^2;$$

for $1 \le x \le 2$, $F(x) = \int_{-\infty}^{x} f(t)\,dt = \int_{-\infty}^{0} 0\,dt + \int_{0}^{1} (1-t)\,dt + \int_{1}^{x} 0\,dt = \left[\left(t - \frac{1}{2}t^2\right)\Big|_{0}^{1}\right] = \frac{1}{2}$; for $2 \le x \le 3$,

$$F(x) = \int_{-\infty}^{x} f(t)\,dt = \int_{-\infty}^{0} 0\,dt + \int_{0}^{1} (1-t)\,dt + \int_{1}^{2} 0\,dt + \int_{2}^{x} (t-2)\,dt$$

$$= \frac{1}{2} + \left[\left(\frac{1}{2}t^2 - 2t\right)\Big|_{2}^{x}\right]$$

$$= \frac{1}{2} + \frac{1}{2}x^2 - 2x + 2$$

$$= \frac{1}{2}x^2 - 2x + \frac{5}{2};$$

For $x > 3$

$$F(x) = \int_{-\infty}^{x} f(t)\,dt = \int_{-\infty}^{0} 0\,dt + \int_{0}^{1} (1-t)\,dt + \int_{1}^{2} 0\,dt + \int_{2}^{3} (t-2)\,dt + \int_{3}^{x} 0\,dt$$

$$= 0 + \frac{1}{2} + 0 + \frac{1}{2} = 1.$$

Thus,

$$F(x) = \begin{cases} 0 & \text{if } x < 0 \\ x - \dfrac{1}{2}x^2 & \text{if } 0 \le x < 1 \\ \dfrac{1}{2} & \text{if } 1 \le x \le 2 \\ \dfrac{1}{2}x^2 - 2x + \dfrac{5}{2} & \text{if } 2 < x \le 3 \\ 1 & \text{if } x > 3 \end{cases}$$

Now, $F(x) = \dfrac{1}{2}$ for any x satisfying $1 \le x \le 2$. The median is *not* unique.

35. Since f is a probability density function,

$$\int_{-\infty}^{\infty} f(x)\,dx = 1 \text{ and } \int_{-\infty}^{\infty} xf(x)\,dx = \mu, \text{ the mean.}$$

Now,

$$\int_{-\infty}^{\infty} (ax+b)f(x)\,dx = \int_{-\infty}^{\infty} axf(x)\,dx + \int_{-\infty}^{\infty} bf(x)\,dx$$

$$= a\int_{-\infty}^{\infty} xf(x)\,dx + b\int_{-\infty}^{\infty} f(x)\,dx = a\mu + b.$$

37. <u>Step 1:</u> Find $F(x)$.

If $x < 0$, then $F(x) = 0$. If $0 \le x \le 2$, then

$$F(x) = \int_{-\infty}^{x} f(t)\,dt = \int_{-\infty}^{0} f(t)\,dt + \int_{0}^{x} f(t)\,dt$$

$$= 0 + \frac{1}{2}\int_{0}^{x} t\,dt = \frac{1}{4}t^2\Big|_{0}^{x} = \frac{1}{4}x^2.$$

If $x > 2$, then

$$F(x) = \int_{-\infty}^{x} f(t)\,dt = \int_{-\infty}^{0} f(t)\,dt + \int_{0}^{2} f(t)\,dt + \int_{2}^{x} f(t)\,dt$$

$$= 0 + \frac{1}{2}\int_{0}^{2} t\,dt + 0 = \frac{1}{4}t^2\Big|_{0}^{2} = 1.$$

Thus,

$$F(x) = \begin{cases} 0 & x < 0 \\ \dfrac{1}{4}x^2 & 0 \le x \le 2 \\ 1 & x > 2 \end{cases}$$

Step 2:   In order to find the quartile point x_1, we solve the following for x_1:

$$F(x_1) = P(X \le x_1) = \frac{1}{4}$$

$$\frac{1}{4}x_1^2 = \frac{1}{4}$$

$$x_1^2 = 1$$

$$x_1 = 1$$

For the quartile point x_2 (or m), we solve the following for x_2:

$$F(x_2) = P(X \le x_2) = \frac{1}{2}$$

$$\frac{1}{4}x_2^2 = \frac{1}{2}$$

$$x_2^2 = 2$$

$$x_2 = \sqrt{2} \approx 1.414$$

For the quartile point x_3, we solve the following for x_3:

$$F(x_3) = P(X \le x_3) = \frac{3}{4}$$

$$\frac{1}{4}x_3^2 = \frac{3}{4}$$

$$x_3^2 = 3$$

$$x_3 = \sqrt{3} \approx 1.732$$

39.   Step 1:   Find $F(x)$.

If $x < 0$, then $F(x) = 0$. If $x \ge 0$, then

$$F(x) = \int_{-\infty}^{x} f(t)\, dt = \int_{-\infty}^{0} f(t)\, dt + \int_{0}^{x} f(t)\, dt = 0 + \int_{0}^{x} \frac{3}{(3+t)^2}\, dt$$

$$= \frac{-3}{3+t}\bigg|_{0}^{x} = \frac{-3}{3+x} + 1.$$

Thus, $F(x) = \begin{cases} 0 & \text{if } x < 0 \\ 1 - \dfrac{3}{3+x} & \text{if } x \ge 0 \end{cases}$

Step 2:   For the quartile point x_1, we solve:

$$F(x_1) = P(X \le x_1) = \frac{1}{4}$$

$$1 - \frac{3}{3+x_1} = \frac{1}{4}$$

$$-\frac{3}{3+x_1} = -\frac{3}{4}$$

$$3 + x_1 = 4$$

$$x_1 = 1$$

For the quartile point x_2, we solve:

$$F(x_2) = P(X \leq x_2) = \frac{1}{2}$$

$$1 - \frac{3}{3 + x_2} = \frac{1}{2}$$

$$-\frac{3}{3 + x_2} = -\frac{1}{2}$$

$$3 + x_2 = 6$$

$$x_2 = 3$$

For the quartile point x_3, we solve:

$$F(x_3) = P(X \leq x_3) = \frac{3}{4}$$

$$1 - \frac{3}{3 + x_3} = \frac{3}{4}$$

$$-\frac{3}{3 + x_3} = -\frac{1}{4}$$

$$3 + x_3 = 12$$

$$x_3 = 9$$

41. Given $f(x) = \begin{cases} 4x - 4x^3 & 0 \leq x \leq 1 \\ 0 & \text{otherwise} \end{cases}$

Step 1. Find $F(x)$.

If $x < 0$, then $F(x) = 0$. If $0 \leq x \leq 1$, then

$$F(x) = \int_{-\infty}^{x} f(t)\, dt = \int_{-\infty}^{0} f(t)\, dt + \int_{0}^{x} f(t)\, dt$$

$$= 0 + \int_{0}^{x} (4t - 4t^3)\, dt = (2t^2 - t^4) \Big|_{0}^{x}$$

$$= 2x^2 - x^4$$

If $x > 1$, then

$$F(x) = \int_{-\infty}^{x} f(t)\, dt = \int_{-\infty}^{0} f(t)\, dt + \int_{0}^{1} f(t)\, dt + \int_{1}^{x} f(t)\, dt$$

$$= 0 + \int_{0}^{1} (4t - 4t^3)\, dt + 0 = (2t^2 - t^4) \Big|_{0}^{1} = 1$$

Thus, $F(x) = \begin{cases} 0 & x < 0 \\ 2x^2 - x^4 & 0 \leq x \leq 1 \\ 1 & x > 1 \end{cases}$

Step 2. Solve the equation $F(m) = \dfrac{1}{2}$ for m.

The graphs of F and $y = \dfrac{1}{2}$ are shown at the right.

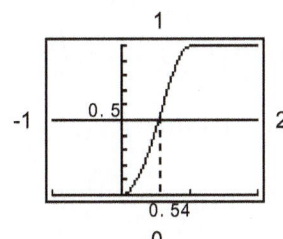

Thus, the median $m \approx 0.54$.

43. $f(x) = \begin{cases} \dfrac{1}{2x^2} + \dfrac{3}{2x^4} & \text{if } x \geq 1 \\ 0 & \text{if } x < 1 \end{cases}$

Step 1:   Find $F(x)$.

If $x < 1$, then $F(x) = 0$. If $x \geq 1$, then

$$F(x) = \int_{-\infty}^{x} f(t)\, dt = \int_{-\infty}^{1} 0\, dt + \int_{1}^{x} \left(\frac{1}{2t^2} + \frac{3}{2t^4} \right) dt$$

$$= \left[\left(-\frac{1}{2t} - \frac{3}{6t^3} \right) \Big|_{1}^{x} \right]$$

$$= 1 - \frac{1}{2x} - \frac{1}{2x^3}$$

Thus, $F(x) = \begin{cases} 0 & \text{if } x < 1 \\ 1 - \dfrac{1}{2x} - \dfrac{1}{2x^3} & \text{if } x \geq 1 \end{cases}$

Step 2.   Solve the equation $F(m) = \dfrac{1}{2}$ for m. The graphs of F and $y = \dfrac{1}{2}$ are shown below.

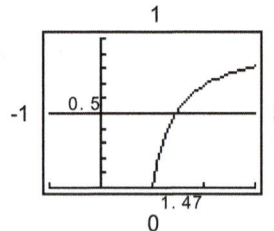

Thus, $m \approx 1.47$.

45.   (A) The contractor's expected profit is given by:

$$E(x) = \int_{-\infty}^{\infty} xf(x)\, dx = \int_{15}^{20} x(0.08)(20 - x)\, dx = 0.08 \int_{15}^{20} (20x - x^2)\, dx$$

$$= 0.08 \left[10x^2 - \frac{1}{3}x^3 \right]_{15}^{20}$$

$$= 0.08 \left(10(20)^2 - \frac{(20)^3}{3} - \left[10(15)^2 - \frac{(15)^3}{3} \right] \right)$$

$$= 0.08(1333.33 - 1125) = 16.667 \text{ or } \$16{,}667.$$

(B) Step 1:   Find $F(x)$.

If $x < 15$, then $F(x) = 0$. If $15 \leq x \leq 20$, then

$$F(x) = \int_{-\infty}^{x} f(t)\, dt = \int_{-\infty}^{15} f(t)\, dt + \int_{15}^{x} f(t)\, dt$$

$$= 0 + 0.08 \int_{15}^{x} (20 - t)\, dt$$

$$= 0.08 \left[20t - \frac{1}{2}t^2 \right]_{15}^{x}$$

$$= 0.08 \left[20x \frac{1}{2}x^2 - (300 - 112.50) \right]$$

$$= 0.08 \left[20x \frac{1}{2}x^2 - 187.5 \right]$$

If $x > 20$, then

$$F(x) = \int_{-\infty}^{x} f(t)\, dt = \int_{-\infty}^{15} f(t)\, dt + \int_{15}^{20} f(t)\, dt + \int_{20}^{x} f(t)\, dt$$

$$= 0 + 0.08 \int_{15}^{20} (20 - x)\, dx + 0$$

$$= 0.08 \left[20x - \frac{1}{2} x^2 \right]_{15}^{20}$$

$$= 0.08 \left[400 - 200 - (300 - 112.5) \right] = 1.$$

Step 2: To find the median profit, x_m, we solve the following:

$$F(x_m) = P(X \le x_m) = \frac{1}{2}$$

$$0.08 \left[20x_m - \frac{1}{2} x_m^2 - 187.5 \right] = \frac{1}{2}$$

$$20x_m - \frac{1}{2} x_m^2 - 187.5 = 6.25$$

$$x_m^2 - 40x_m + 387.5 = 0$$

$$x_m = \frac{40 \pm \sqrt{(40)^2 - 4(387.5)}}{2} \quad \text{(quadratic formula)}$$

$$= \frac{40 \pm \sqrt{50}}{2} = \frac{40 \pm 5\sqrt{2}}{2}$$

Since x_m lies in the interval $[15, 20]$,

$$x_m = \frac{40 - 5\sqrt{2}}{2} = 20 - \frac{5}{2}\sqrt{2} \approx 16.464 \text{ or } \$16,464$$

47. Step 1: Find $F(x)$.

If $x < 0$, then $F(x) = 0$. If $x \ge 0$, then

$$F(x) = \int_{-\infty}^{x} f(t)\, dt = \int_{-\infty}^{0} f(t)\, dt + \int_{0}^{x} f(t)\, dt = 0 + \int_{0}^{x} \frac{1}{3} e^{-t/3}\, dt$$

$$= -e^{-t/3} \Big|_{0}^{x} = -e^{-x/3} + 1.$$

Thus, $F(x) = \begin{cases} 0 & \text{if } x < 0 \\ 1 - e^{-x/3} & \text{if } x \ge 0 \end{cases}$

Step 2: Solve $P(X \le m) = \dfrac{1}{2}$ for m.

$$F(m) = P(X \le m) = \frac{1}{2}$$

$$1 - e^{-m/3} = \frac{1}{2}$$

$$e^{-m/3} = \frac{1}{2}$$

$$\frac{-m}{3} = \ln \frac{1}{2} = -\ln 2$$

$$m = 3 \ln 2 \approx 2.079 \text{ minutes}$$

49. The expected daily consumption is given by:

$$E(X) = \int_{-\infty}^{\infty} xf(x)\,dx = \int_{-\infty}^{\infty} \frac{x}{(1+x^2)^{3/2}}\,dx$$

$$= \lim_{R \to \infty} \int_0^R \frac{x}{(1+x^2)^{3/2}}\,dx$$

$$\left[\underline{\text{Note}}: \text{If } u = 1 + x^2, \text{ then } \frac{du}{2} = x\,dx \text{ and } \int \frac{x}{(1+x^2)^{3/2}}\,dx = \frac{1}{2}\int \frac{du}{u^{3/2}} = \frac{1}{2} \cdot \frac{-2}{u^{1/2}} = \frac{1}{(1+x^2)^{1/2}}.\right]$$

$$= -\lim_{R \to \infty} \frac{1}{(1+x^2)^{1/2}}\Big|_0^R = -\lim_{R \to \infty}\left(\frac{1}{(1+R^2)^{1/2}} - \frac{1}{1}\right) = 1 \text{ or 1 million gallons}$$

51. Mean life expectancy is given by:

$$E(X) = \mu = \int_{-\infty}^{\infty} xf(x)\,dx = \frac{1}{5000}\int_0^{10} x(10x^3 - x^4)\,dx = \frac{1}{5000}\left(10 \cdot \frac{x^5}{5} - \frac{x^6}{6}\right)\Big|_0^{10}$$

$$= \frac{1}{5000}\left(2 \cdot 10^5 - \frac{1}{6} \cdot 10^6 - 0\right) = \frac{1}{5000}\left(\frac{1}{3} \cdot 10^5\right) = \frac{20}{3} \approx 6.7 \text{ minutes}$$

53. Step 1: Find $F(x)$.

If $x < 0$, $F(x) = 0$. If $x \geq 0$, then

$$F(x) = \int_{-\infty}^x f(t)\,dt = \int_{-\infty}^0 f(t)\,dt + \int_0^x f(t)\,dt = 0 + \int_0^x \frac{800t}{(400+t^2)^2}\,dt$$

$$= \frac{-400}{400+t^2}\Big|_0^x = \frac{-400}{400+x^2} + 1.$$

Thus, $F(x) = \begin{cases} 0 & \text{if } x < 0 \\ 1 - \dfrac{400}{400+x^2} & \text{if } x \geq 0 \end{cases}$

Step 2: Solve $P(X \leq m) = \dfrac{1}{2}$ for m.

$$1 - \frac{400}{400+m^2} = \frac{1}{2}$$

$$\frac{-400}{400+m^2} = \frac{-1}{2}$$

$$400 + m^2 = 800$$

$$m^2 = 400$$

$$m = 20 \text{ days}$$

55. The expected number of hours to learn the task is given by:

$$E(X) = \mu = \int_{-\infty}^{\infty} xf(x)\,dx = \int_0^3 x\left(\frac{4}{9}x^2 - \frac{4}{27}x^3\right)dx = \left(\frac{4}{9} \cdot \frac{x^4}{4} - \frac{4}{27} \cdot \frac{x^5}{5}\right)\Big|_0^3$$

$$= \frac{1}{9}(3^4) - \frac{4}{3^3(5)}(3^5) = 9 - \frac{36}{5} = \frac{9}{5} = 1.8 \text{ hours}$$

EXERCISE 9-4

1. Using 1(a) with $[a, b] = [0, 2]$, we have:

$$f(x) = \begin{cases} \dfrac{1}{2-0} & 0 \leq x \leq 2 \\ 0 & \text{otherwise} \end{cases} = \begin{cases} \dfrac{1}{2} & 0 \leq x \leq 2 \\ 0 & \text{otherwise} \end{cases}$$

Using 1(b) with $[a, b] = [0, 2]$, we have:

$$F(x) = \begin{cases} 0 & x < 0 \\ \dfrac{x - 0}{2 - 0} & 0 \le x \le 2 \\ 1 & x > 2 \end{cases} = \begin{cases} 0 & x < 0 \\ \dfrac{x}{2} & 0 \le x \le 2 \\ 1 & x > 2 \end{cases}$$

3. Using 2(a) with $\lambda = \dfrac{1}{2}$, we have:

$$f(x) = \begin{cases} \dfrac{1}{1/2}\, e^{-x/(1/2)} & x \ge 0 \\ 0 & \text{otherwise} \end{cases} = \begin{cases} 2e^{-2x} & x \ge 0 \\ 0 & \text{otherwise} \end{cases}$$

Using 2(b) with $\lambda = \dfrac{1}{2}$, we have:

$$F(x) = \begin{cases} 1 - e^{-x/(1/2)} & x \ge 0 \\ 0 & \text{otherwise} \end{cases} = \begin{cases} 1 - e^{-2x} & x \ge 0 \\ 0 & \text{otherwise} \end{cases}$$

5. Using 1(c), (d), (e), with $[a, b] = [1, 5]$, we have:

Mean: $\mu = \dfrac{1}{2}(a + b) = \dfrac{1}{2}(1 + 5) = 3$

Median: $m = \dfrac{1}{2}(a + b) = \dfrac{1}{2}(1 + 6) = 3$

Standard deviation: $\sigma = \dfrac{1}{\sqrt{12}}(b - a) = \dfrac{1}{\sqrt{12}}(5 - 1) = \dfrac{4}{\sqrt{12}} = \dfrac{2}{\sqrt{3}} \approx 1.155$

7. Using 2(c), (d), (e), with $\lambda = 5$, we have:

Mean: $\mu = 5$

Median: $m = 5 \ln 2 \approx 3.466$

Standard deviation: $\sigma = 5$

9. From Table I, the area under the standard normal curve from 0 to 1.5 is 0.4332.

11. The area under the standard normal curve between -0.72 and 0 is the same as the area between 0 and 0.72 (due to the symmetry of the curve). From Table I, the area under the standard normal curve between 0 and 0.72 is 0.2642.

13. From Table I, the area under the standard normal curve from 0 to 2.01 is 0.4778.

15. The area under the standard normal curve from -1.93 to 0 is the same as the area from 0 to 1.93 which from Table I is 0.4732.

17. The area under a normal curve with mean 100 and standard deviation 10 from 100 to 120 is the same as the area under the standard normal curve from 0 to $\dfrac{120 - 100}{10} = 2$. From Table I, this area is 0.4772.

19. For 95, the area is from $\dfrac{95 - 100}{10} = -\dfrac{1}{2}$ to 0 which is the same as from 0 to $\dfrac{1}{2}$ under the standard normal curve. From Table I for $\dfrac{1}{2} = 0.5$, this area is 0.1915.

21. For 106, the area is from 0 to $\dfrac{106 - 100}{10} = 0.6$ under the standard normal curve. From Table I for 0.6, this area is 0.2257.

23. For 78, the area is from $\dfrac{78 - 100}{10} = -2.2$ to 0 which is the same as from 0 to 2.2 under the standard normal curve. From Table I for 2.2, this area is 0.4861.

25. $f_1(x) = \dfrac{1}{\sqrt{2\pi}} e^{-x^2/2}$

$f_2(x) = \begin{cases} 0.4 & \text{if } -1.25 \le x \le 1.25 \\ 0 & \text{otherwise} \end{cases}$

From the graphs, it appears that X_1 has the greater standard deviation.

$$\mu_1 = \int_{-\infty}^{\infty} x f_1(x)\, dx = 0$$

$$V_1(X) = \int_{-\infty}^{\infty} (x-0)^2 f_1(x)\, dx = \int_{-\infty}^{\infty} x^2 f_1(x)\, dx = 1;\; \sigma_1 = 1$$

$$\mu_2 = \int_{-\infty}^{\infty} x f_2(x)\, dx = \int_{-1.25}^{1.25} 0.4x\, dx = 0$$

$$V_2(X) = \int_{-\infty}^{\infty} (x-0)^2 f(x)\, dx = \int_{-1.25}^{1.25} 0.4x^2\, dx = 0.4\,\frac{x^3}{3}\bigg|_{-1.25}^{1.25} \approx 0.5208$$

$$\sigma_2 = \sqrt{0.5208} = 0.722$$

27. $g_1(x) = \dfrac{1}{\sqrt{2\pi}} e^{-(x-1)^2/2}$

$g_2(x) = \begin{cases} e^{-x} & \text{if } x \ge 0 \\ 0 & \text{otherwise} \end{cases}$

By symmetry, X_1 has median 1. From the graph of g_2, it appears that X_2 has median less than 1.

$$F_2(x) = \int_{-\infty}^{x} g_2(t)\, dt = \int_{0}^{x} e^{-t}\, dt = -e^{-t}\,\big|_0^x = 1 - e^{-x}$$

Now solve the equation $F_2(m) = \dfrac{1}{2}$:

$$1 - e^{-m} = \frac{1}{2}$$

$$e^{-m} = \frac{1}{2}$$

$$-m = \ln(1/2)$$

$$m = 0.693$$

29. X is uniformly distributed on $[0, 4]$. The probability density function is

$$f(x) = \begin{cases} \dfrac{1}{4} & 0 \le x \le 4 \\ 0 & \text{otherwise} \end{cases}$$

and the mean is

$$\mu = \frac{1}{2}(0+4) = 2.$$

Thus, $P(X \le 2) = \displaystyle\int_0^2 f(x)\, dx = \int_0^2 \frac{1}{4}\, dx = \frac{1}{4}x\, \bigg|_0^2 = \frac{1}{2}.$

31. X is an exponential random variable with $\mu = 1$. Then

$$f(x) = \begin{cases} e^{-x} & x \ge 0 \\ 0 & \text{otherwise} \end{cases}$$

and the mean is $\mu = 1$.

Thus, $P(X \le 1) = \displaystyle\int_0^1 e^{-x}\, dx = -e^{-x}\,\bigg|_0^1 = -e^{-1} + 1 = 1 - e^{-1} \approx 0.632.$

33. X is uniformly distributed on $[-5, 5]$. The mean is

$$\mu = \frac{1}{2}(-5 + 5) = 0$$

and the standard deviation is

$$\sigma = \frac{1}{\sqrt{12}}[5 - (-5)] = \frac{10}{\sqrt{12}} = \frac{5}{\sqrt{3}} \approx 2.887.$$

Also,

$$f(x) = \begin{cases} \dfrac{1}{10} & -5 \le x \le 5 \\ 0 & \text{otherwise} \end{cases}$$

Thus, $P(\mu - \sigma \le X \le \mu + \sigma) = P\left(\dfrac{-5}{\sqrt{3}} \le X \le \dfrac{5}{\sqrt{3}}\right) = \displaystyle\int_{-5\sqrt{3}}^{5\sqrt{3}} f(x)\, dx = \int_{-5\sqrt{3}}^{5\sqrt{3}} \frac{1}{10}\, dx$

$$= \frac{1}{10}x \Big|_{-5\sqrt{3}}^{5\sqrt{3}} = \frac{1}{10}\left(\frac{5}{\sqrt{3}}\right) - \frac{1}{10}\left(\frac{-5}{\sqrt{3}}\right)$$

$$= \frac{1}{\sqrt{3}} \approx 0.577.$$

35. X is an exponential random variable with median $m = 6 \ln 2$. Since $m = \lambda \ln 2$, we have $\lambda = 6$. Thus, mean $\mu = 6$ and standard deviation $\sigma = 6$. Also,

$$f(x) = \begin{cases} \dfrac{1}{6}e^{-x/6} & x \ge 0 \\ 0 & \text{otherwise} \end{cases}$$

Now, $P(\mu - \sigma \le X \le \mu + \sigma) = P(0 \le X \le 12) = \displaystyle\int_0^{12} f(x)\, dx = \int_0^{12} \frac{1}{6}e^{-x/6}\, dx = -e^{-x/6}\big|_0^{12} = -e^{-2} + 1 \approx 0.865.$

37. $\mu = 70$, $\sigma = 8$

z (for $x = 60$) $= \dfrac{60 - 70}{8} = -1.25$

z (for $x = 80$) $= \dfrac{80 - 70}{8} = 1.25$

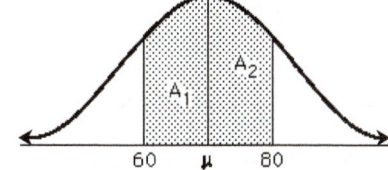

Area $A_1 = 0.3944$. Area $A_2 = 0.3944$.
Total area $= A = A_1 + A_2 = 0.7888$.

39. $\mu = 70$, $\sigma = 8$

z (for $x = 62$) $= \dfrac{62 - 70}{8} = -1.00$

(for $x = 74$) $= \dfrac{74 - 70}{8} = 0.5$

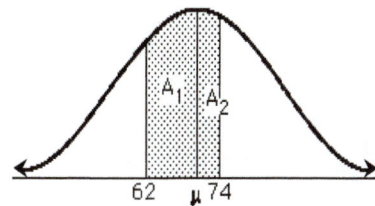

Area $A_1 = 0.3413$. Area $A_2 = 0.1915$. Total area $= A = A_1 + A_2 = 0.5328$.

41. $\mu = 70$, $\sigma = 8$

z (for $x = 88$) $= \dfrac{88 - 70}{8} = 2.25$

Required area $= 0.5 - $ (area corresponding to $z = 2.25$)

$$= 0.5 - 0.4878$$

$$= 0.0122$$

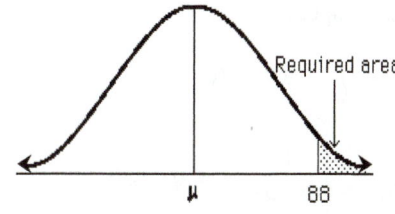

43. $\mu = 70, \sigma = 8$

$$z \text{ (for } x = 60) = \frac{60 - 70}{8} = -1.25$$

Required area $= 0.5 -$ (area corresponding to $z = 1.25$)

$$= 0.5 - 0.3944$$

$$= 0.1056$$

Required area

60   μ

45. Normal probability density function with $\mu = 0$:

$$f(x) = \frac{1}{\sigma \sqrt{2\pi}} e^{-x^2/2\sigma^2}, \sigma > 0.$$

(A) $\sigma = 0.5$

$$P(-0.5 \le X \le 0.5) = \frac{1}{0.5\sqrt{2\pi}} \int_{-0.5}^{0.5} e^{-x^2/0.5} \, dx$$

Using a graphing utility with a numerical integration routine, we find that

$$P(-0.5 \le X \le 0.5) \approx 0.6827$$

(B) $\sigma = 1$

$$P(-1 \le X \le 1) = \frac{1}{\sqrt{2\pi}} \int_{-1}^{1} e^{-x^2/0.5} \, dx \approx 0.6827$$

(C) $\sigma = 2$

$$P(-2 \le X \le 2) = \frac{1}{2\sqrt{2\pi}} \int_{-2}^{2} e^{-x^2/8} - x2/8 \, dx \approx 0.6827$$

Interpretation: the probability of being within one standard deviation of the mean is always 0.6827.

47. Normal probability density function with $\mu = 0$:

$$f(x) = \frac{1}{\sigma \sqrt{2\pi}} e^{-x^2/2\sigma^2}, \sigma > 0.$$

(A) $\sigma = 0.5$

$$P(-1.5 \le X \le 1.5) = \frac{1}{0.5\sqrt{2\pi}} \int_{-1.5}^{1.5} e^{-x^2/0.5} \, dx$$

Using a graphing utility with a numerical integration routine, we find that

$$P(-1.5 \le X \le 1.5) \approx 0.9973$$

(B) $\sigma = 1$

$$P(-3 \le X \le 3) = \frac{1}{\sqrt{2\pi}} \int_{-3}^{3} e^{-x^2/2} \, dx \approx 0.9973$$

(C) $\sigma = 2$

$$P(-6 \le X \le 6) = \frac{1}{2\sqrt{2\pi}} \int_{-6}^{6} e^{-x^2/8} \, dx \approx 0.9973$$

Interpretation: the probability of being within 3 standard deviations of the mean is always 0.9973.

49. From Section 9-3,

$$\mu = \int_{-\infty}^{\infty} xf(x) \, dx,$$

where

$$f(x) = \begin{cases} \dfrac{1}{b-a} & a \le qx \le b \\ 0 & \text{otherwise} \end{cases}$$

Thus,

$$\mu = \int_a^b x\left(\frac{1}{b-a}\right) dx = \frac{1}{b-a} \int_a^b x\, dx = \frac{1}{b-a} \cdot \frac{x^2}{2}\Big|_a^b = \frac{1}{b-a}\left(\frac{b^2}{2} - \frac{a^2}{2}\right)$$

$$= \frac{1}{b-a} \cdot \frac{1}{2}(b-a)(b+a) = \frac{a+b}{2}.$$

51. $$\int_{-\infty}^{\infty} x^2 f(x)\, dx = \int_a^b x^2\left(\frac{1}{b-a}\right) dx = \frac{1}{b-a} \cdot \frac{x^3}{3}\Big|_a^b = \frac{1}{3} \cdot \frac{1}{b-a}(b^3 - a^3)$$

$$= \frac{1}{3} \cdot \frac{1}{b-a}(b-a)(b^2 + ab + a^2) = \frac{1}{3}(b^2 + ab + a^2)$$

53. $f(x) = \begin{cases} 5e^{-5x} & \text{if } x \geq 0 \\ 0 & \text{otherwise} \end{cases}$

Mean: $\mu = \int_{-\infty}^{\infty} xf(x)\, dx = \int_0^{\infty} 5xe^{-5x}\, dx$ since $f(x) = 0$ for $x < 0$.

Also, $10 f(10) \approx 9.644 \times 10^{-21} \approx 0$. Therefore, it is reasonable to replace the integral by

$$\int_0^{10} 5xe^{-5x}\, dx = 0.2$$

Median: $F(x) = \int_{-\infty}^{x} f(t)dt = \int_0^x 5e^{-5x}\, dt, x \geq 0$

$$= -e^{-5t}\big|_0^x 1 - e^{-5x}$$

Now solve the equation $F(m) = \dfrac{1}{2}$:

$$1 - e^{-5m} = \frac{1}{2}$$

$$e^{-5m} = \frac{1}{2}$$

$$-5m = \ln(1/2)$$

$$m = 0.139$$

Standard Deviation:

$$V(X) = \int_{-\infty}^{\infty} x^2 f(x)\, dx - (0.2)^2 = \int_0^{\infty} 5x^2 e^{-5x} dx - 0.04$$

since $f(x) = 0$ for $x < 0$

Also, $100 f(10) \cdot 9.644 \;?\; 10^{20} \approx 0$. Therefore, it is reasonable to integrate over the interval $[0, 10]$:

$$\int_0^{10} 5x^2 e^{-5x}\, dx - 0.04 = 0.08 - 0.04 = 0.04;$$

$$\sigma = \sqrt{0.04} = 0.2$$

55. $f(x) = \begin{cases} \dfrac{p}{x^{p+1}} & \text{if } x \geq 1 \\ 0 & \text{otherwise} \end{cases} \quad p > 0$

(A) Clearly, $f(x) \geq 0$ for all x.

(B) $\displaystyle\int_{-\infty}^{\infty} f(x)\, dx = \int_1^{\infty} \frac{p}{x^{p+1}}\, dx = \lim_{b \to \infty} \int_1^b \frac{p}{x^{p+1}}\, dx$

$$= \lim_{b \to \infty}\left[-\frac{1}{x^p}\Big|_1^b\right]$$

$$= \lim_{b \to \infty}\left[1 - \frac{1}{b^p}\right] = 1$$

Thus, f is a probability density function.

The mean μ of f is given by:

$$\mu = \int_{-\infty}^{\infty} x f(x)\, dx = \int_{-\infty}^{\infty} x \left(\frac{p}{x^{p+1}} \right) dx$$

$$= \int_{1}^{\infty} \frac{p}{x^p}\, dx = \lim_{b \to \infty} \int_{1}^{b} \frac{p}{x^p}\, dx$$

$$= \lim_{b \to \infty} \left[\frac{px^{-p+1}}{-p+1} \,\Big|_{1}^{b} \right], p \neq 1$$

$$= \lim_{b \to \infty} \left[\frac{pb^{-p+1}}{1-p} - \frac{p}{1-p} \right]$$

$$= \frac{p}{p-1} \text{ provided } p > 1;$$

The limit does not exist if $p < 1$.
If $p = 1$, then

$$\int_{1}^{\infty} \frac{1}{x}\, dx = \lim_{b \to \infty} \int_{1}^{b} \frac{1}{x}\, dx$$

$$= \lim_{b \to \infty} [\ln b]$$

does not exist. Thus,

$$\mu = \begin{cases} \dfrac{p}{p-1} & \text{if } p > 1 \\ \text{does not exist} & \text{if } 0 < p \leq 1 \end{cases}$$

57. The median m is given by

$$P(X \leq m) = \frac{1}{2}$$

Clearly, $m > 1$ since $f(x) = 0$ for $x < 1$. Now,

$$P(X \leq m) = \int_{-\infty}^{m} f(x)\, dx = \int_{1}^{m} \frac{p}{x^{p+1}}\, dx$$

$$= \frac{-1}{x^p} \,\Big|_{1}^{m} = 1 - \frac{1}{m^p}$$

and

$$1 - \frac{1}{m^p} = \frac{1}{2}$$

$$\frac{1}{m^p} = \frac{1}{2}$$

$$m^p = 2$$

$$m = \sqrt[p]{2}$$

59. Using the uniform cumulative probability distribution function:

$$P(25 \leq X \leq 40) = F(40) - F(25)$$

$$= \frac{40 - 0}{40 - 0} - \frac{25 - 0}{40 - 0} \quad \text{[Note: } a = 0, b = 40.]$$

$$= 1 - \frac{25}{40} = \frac{15}{40} = \frac{3}{8} = 0.375$$

61. We are given that $\mu = 3$. Also $\lambda = \mu$. Thus, $\lambda = 3$, and we have:

$$P(0 \leq X \leq 2) = F(2) - F(0) = (1 - e^{-2/3}) - (1 - e^0)$$

$$= 1 - e^{-2/3} \approx 0.487$$

63. X is an exponential random variable and the median $m = 2$. Since $m = \lambda \ln 2$, we have $2 = \lambda \ln 2$ or $\lambda = \dfrac{2}{\ln 2}$. Now $P(X \le 1) = F(1)$, where F is the cumulative probability distribution:

$$F(x) = \begin{cases} 1 - e^{-x/\lambda} & x \ge 0 \\ 0 & \text{otherwise} \end{cases}$$

Setting $\lambda = \dfrac{2}{\ln 2}$, we have

$$F(x) = \begin{cases} 1 - e^{-(x\ln 2)/2} & x \ge 0 \\ 0 & \text{otherwise} \end{cases}$$

Thus, $P(X \le 1) = 1 - e^{-(\ln 2)/2} = 1 - e^{-(1/2)\ln 2}$

$$= 1 - e^{\ln(2)^{-1/2}}$$

$$= 1 - 2^{-1/2}$$

$$= 1 - \frac{1}{\sqrt{2}} \approx 0.293.$$

65. $\mu = 200{,}000$, $\sigma = 20{,}000$, $x \ge 240{,}000$

$$z \text{ (for } x = 240{,}000) = \frac{240{,}000 - 200{,}000}{20{,}000} = 2.0$$

Fraction of the salesmen who would be expected to make annual sales of $240,000 or more

$= \text{Area } A_1$

$= 0.5 - (\text{area between ? and } 240{,}000)$

$= 0.5 - 0.4772$

$= 0.0228$

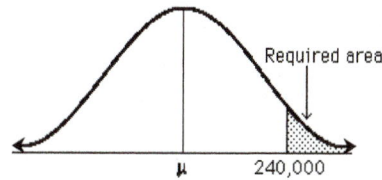

Thus, the percentage of salesmen expected to make annual sales of $240,000 or more is 2.28%.

67. $x = 105$, $x = 95$, $\mu = 100$, $\sigma = 2$

$$z \text{ (for } x = 105) = \frac{105 - 100}{2} = 2.5$$

$$z \text{ (for } x = 95) = \frac{95 - 100}{2} = -2.5$$

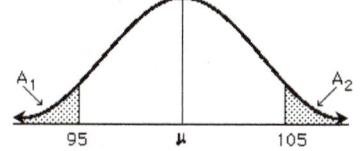

Fraction of parts to be rejected $= \text{Area } A_1 + A_2$

$$= 1 - 2(\text{area corresponding to } z = 2.5)$$

$$= 1 - 2(0.4938)$$

$$= 0.0124$$

Thus, the percentage of parts to be rejected is 1.24%.

69. (A) We are given that

$$P(0 \le X \le 1) = 0.3 = F(1) - F(0) = 1 - e^{-1/\lambda} - (1 - e^0).$$

Thus,

$$1 - e^{-1/\lambda} = 0.3$$

$$e^{-1/\lambda} = 0.7$$

$$-\frac{1}{\lambda} = \ln(0.7)$$

$$\lambda = -\frac{1}{\ln(0.7)}$$

But $E(X) = \mu = \lambda = -\dfrac{1}{\ln(0.7)} \approx 2.8$ years.

(B) $P(X \geq 2.8) = \int_{2.8}^{\infty} f(x)\, dx = \lim_{R \to \infty} \int_{2.8}^{R} \frac{1}{\lambda} e^{-x/\lambda}\, dx$

$$= \lim_{R \to \infty} \int_{2.8}^{R} \frac{1}{2.8} e^{-x/2.8}\, dx$$

$$= \lim_{R \to \infty} (-e^{-R/2.8} + e^{-2.8/2.8})$$

$$= e^{-1} \approx 0.368$$

71. $\mu = 240,\ \sigma = 20$

8 days $= 192$ hours $= x$

z (for $x = 192$) $= \dfrac{192 - 240}{20} = -2.4$

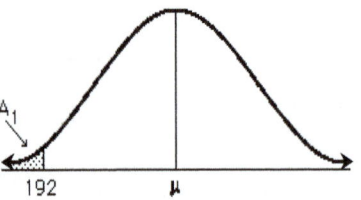

Fraction of people having this incision who would heal in 192 hours or
less $=$ Area A_1

$\quad = 0.5 - (\text{area corresponding to } z = 2.4)$

$\quad = 0.5 - 0.4918$

$\quad = 0.0082$

Thus, the percentage of people who would heal in 8 days or less is 0.82%.

73. X is an exponential random variable with mean $\mu = 2$ (minutes). Thus, $\lambda = 2$ and

$$P(X \geq 5) = \int_{5}^{\infty} f(x)\, dx = \int_{5}^{\infty} \frac{1}{2} e^{-x/2}\, dx = \lim_{R \to \infty} \int_{5}^{R} \frac{1}{2} e^{-x/2}\, dx = \lim_{R \to \infty} (-e^{-x/2}) \Big|_{5}^{R}$$

$$= \lim_{R \to \infty} (-e^{-R/5} + e^{-5/2}) = e^{-5/2} = e^{-2.5} \approx 0.082.$$

75. $\mu = 70,\ \sigma = 8$

We compute $x_1, x_2, x_3,$ and x_4 corresponding to $z_1, z_2, z_3,$ and z_4, respectively. The area between μ and x_3 is 0.2.
Hence, from the table, $z_3 = 0.52$ (approximately). Thus, we have:

$0.52 = \dfrac{x_3 - 70}{8}$

$x_3 - 70 = 4.16 \quad \left[\underline{\text{Note:}}\ z = \dfrac{x - \mu}{\sigma}.\right] \approx 4.2$

and $x_3 = 74.2$. Also, $x_2 = 70 - 4.2 = 65.8$
The area between μ and x_4 is 0.4. Hence, from the table, $z_4 = 1.28$
(approximately). Therefore:

$1.28 = \dfrac{x_4 - 70}{8}$

$\quad x_4 - 70 = 10.24 \approx 10.2$

and $x_4 = 70 + 10.2 = 80.2$.

   Also, $x_1 = 70 - 10.2 = 59.8$. Thus, we have $x_1 = 59.8,\ x_2 = 65.8,\ x_3 = 74.2,\ x_4 = 80.2$. So,
A's $= 80.2$ or greater, B's $= 74.2$ to 80.2, C's $= 65.8$ to 74.2, D's $= 59.8$ to 65.8, and F's $= 59.8$ or lower.

Chapter 9 Review

1. $\displaystyle\int_{0}^{\infty} e^{-2x}\, dx = \lim_{b \to \infty} \int_{0}^{b} e^{-2x}\, dx = \lim_{b \to \infty} \frac{e^{-2x}}{-2} \Big|_{0}^{b} = \lim_{b \to \infty} \left(\frac{-e^{-2b}}{2} + \frac{1}{2} \right) = \frac{1}{2}$

Thus, the improper integral converges.   (9-1)

2. $\displaystyle\int_{0}^{\infty} \frac{1}{x + 1}\, dx = \lim_{b \to \infty} \int_{0}^{b} \frac{1}{x + 1}\, dx$

$$= \lim_{b \to \infty} \int_{1}^{b+1} \frac{1}{u}\, du = \lim_{b \to \infty} \ln u \Big|_{1}^{b+1}$$

$$= \lim_{b \to \infty} \ln|b + 1|$$

$u = x + 1$
$du = dx$
$u = b + 1$ when $x = b$
$u = 1$ when $b = 0$

This limit does not exist. Thus, the improper integral diverges.   (9-1)

3. $\displaystyle\int_{1}^{\infty}\frac{16dx}{x^3}=16\lim_{b\to\infty}\int_{1}^{b}x^{-3}\,dx=16\lim_{b\to\infty}\frac{x^{-2}}{-2}\bigg|_{1}^{b}=16\lim_{b\to\infty}\left(\frac{-b^{-2}}{2}+\frac{1}{2}\right)=16\left(\frac{1}{2}\right)=8$ (9-1)

4. $\displaystyle P(0\le X\le1)=\int_{0}^{1}\left(1-\frac{1}{2}x\right)dx$

$\displaystyle\qquad\qquad\quad=\left(x-\frac{1}{4}x^2\right)\bigg|_{0}^{1}$

$\displaystyle\qquad\qquad\quad=1-\frac{1}{4}=0.75$

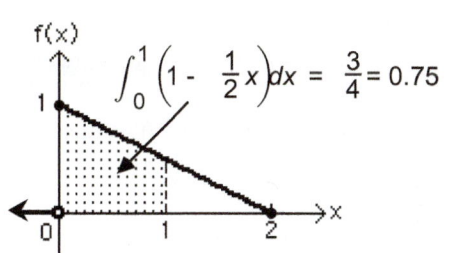

$\displaystyle\int_{0}^{1}\left(1-\frac{1}{2}x\right)dx=\frac{3}{4}=0.75$

The graph of the function is given at the right.   (9-2)

5. $\displaystyle\mu=E(X)=\int_{-\infty}^{\infty}xf(x)\,dx=\int_{0}^{2}x\left(1-\frac{1}{2}x\right)dx=\int_{0}^{2}\left(x-\frac{1}{2}x^2\right)dx$

$\displaystyle\qquad\qquad=\left(\frac{x^2}{2}-\frac{1}{6}x^3\right)\bigg|_{0}^{2}=2-\frac{8}{6}=\frac{2}{3}\approx0.6667$

$\displaystyle V(X)=\int_{-\infty}^{\infty}x^2f(x)\,dx-\mu^2=\int_{0}^{2}x^2\left(1-\frac{1}{2}x\right)dx-\left(\frac{2}{3}\right)^2$

$\displaystyle\qquad\quad=\int_{0}^{2}\left(x^2-\frac{1}{2}x^3\right)dx-\frac{4}{9}=\left(\frac{x^3}{3}-\frac{1}{8}x^4\right)\bigg|_{0}^{2}-\frac{4}{9}$

$\displaystyle\qquad\quad=\frac{8}{3}-\frac{16}{8}-\frac{4}{9}=\frac{16}{72}=\frac{2}{9}\approx0.2222$

$\displaystyle\sigma=\sqrt{V(X)}=\sqrt{\frac{2}{9}}=\frac{\sqrt{2}}{3}\approx0.4714$ (9-3)

6. When $x<0$, then $F(x)=0$. When $0\le x\le2$,

$\displaystyle F(x)=\int_{-\infty}^{x}f(t)\,dt=\int_{-\infty}^{0}f(t)\,dt+\int_{0}^{x}f(t)\,dt=0+\int_{0}^{x}\left(1-\frac{1}{2}t\right)dt$

$\displaystyle\qquad\quad=\left(t-\frac{1}{4}t^2\right)\bigg|_{0}^{x}=x-\frac{1}{4}x^2.$

When $x>2$,

$\displaystyle F(x)=\int_{-\infty}^{x}f(t)\,dt=\int_{-\infty}^{0}f(t)\,dt+\int_{0}^{2}f(t)\,dt+\int_{2}^{x}f(t)\,dt$

$\displaystyle\qquad\quad=0+\int_{0}^{2}\left(1-\frac{1}{2}t\right)dt+0=\left(t-\frac{1}{4}t^2\right)\bigg|_{0}^{2}=2-\frac{1}{4}\cdot4=1.$

Thus,

$$F(x)=\begin{cases}0 & x<0\\[4pt]x-\dfrac{1}{4}x^2 & 0\le x\le2\\[6pt]1 & x>2\end{cases}$$

The graph of $F(x)$ is shown at the right.

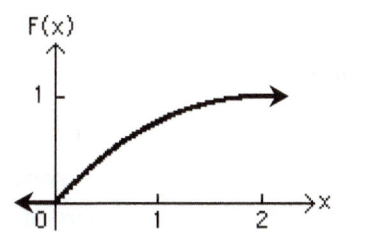

     (9-2)

7. We must solve the following for m:

$\displaystyle F(m)=P(X\le m)=\frac{1}{2}$

$\displaystyle m-\frac{1}{4}m^2=\frac{1}{2}$ (refer to Problem 6)

$m^2-4m+2=0$

$\displaystyle m=\frac{4\pm\sqrt{16-4(2)}}{2}=\frac{4\pm\sqrt{8}}{2}$

$m=2\pm\sqrt{2}$

Thus, the median is $x=2-\sqrt{2}\approx0.5858$.   (9-3)

8. $\mu = 100, \sigma = 10, x = 118.$

$$z = \frac{118 - 100}{10} = 1.8$$

$P(100 \le X \le 118) = P(0 \le Z \le 1.8) = $ area of region over the interval $[0, 1.8]$
$$= 0.4641 \text{ (from Table III).}$$ (9-4)

9.   The uniform probability density function on $[5, 15]$ is given by:

$$f(x) = \begin{cases} \dfrac{1}{15 - 5} & \text{if } 5 \le x \le 15 \\ 0 & \text{otherwise} \end{cases}$$

Thus,

$$f(x) = \begin{cases} \dfrac{1}{10} & \text{if } 5 \le x \le 15 \\ 0 & \text{otherwise} \end{cases}$$

The cumulative distribution function is given by:

$$F(x) = \begin{cases} 0 & \text{if } x < 5 \\ \dfrac{1}{10}(x - 5) & \text{if } 5 \le x \le 15 \\ 1 & \text{if } x > 15 \end{cases}$$ (9-4)

10.   The exponential probability density function with $\lambda = \dfrac{1}{5} = 0.2$ is given by:

$$f(x) = \begin{cases} \dfrac{1}{0.2} e^{-x/0.2} & \text{if } x \ge 0 \\ 0 & \text{otherwise} \end{cases}$$

Thus,

$$f(x) = \begin{cases} 5e^{-5x} & \text{if } x \ge 0 \\ 0 & \text{otherwise} \end{cases}$$

The cumulative distribution function is given by:

$$F(x) = \begin{cases} 0 & \text{if } x < 0 \\ 1 - e^{-5x} & \text{if } x \ge 0 \end{cases}$$ (9-4)

11. $f(x) = e^{-x^2}$

$$\int_{-\infty}^{\infty} f(x)\, dx > \int_{-10}^{10} e^{-x^2}\, dx > 1.772;$$

f is not a probability density function since

$$\int_{-\infty}^{\infty} f(x)\, dx \ne 1$$ (9-2)

12. $f(x) = \begin{cases} 0.75(x^2 - 4x + 3) & \text{if } 0 \le x \le 4 \\ 0 & \text{otherwise} \end{cases}$

f is not a probability density function
since $f(x) < 0$ on $(1, 3)$ (9-2)

13. $P(1 \le X \le 4) = \displaystyle\int_1^4 \frac{5}{2} x^{-7/2}\, dx$

$$= \frac{5}{2}\left(-\frac{2}{5}\right) x^{-5/2} \Big|_1^4$$

$$= -[(4)^{-5/2} (1)^{-5/2}]$$

$$= -\left(\frac{1}{32} - 1\right) = \frac{31}{32}$$

$$\approx 0.9688$$

The graph is shown at the right.

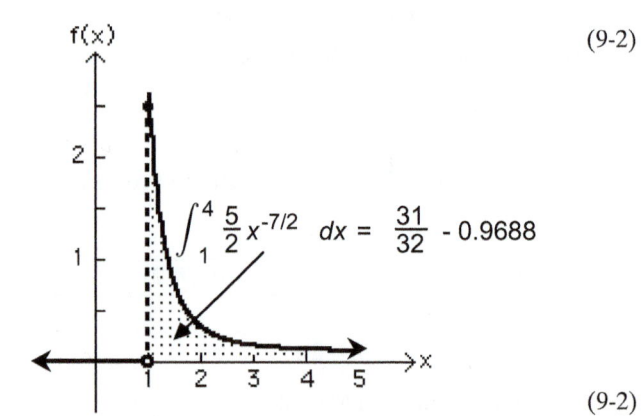

$f(x)$

$$\int_1^4 \frac{5}{2} x^{-7/2}\, dx = \frac{31}{32} - 0.9688$$

(9-2)

14. $\mu = E(X) = \int_{-\infty}^{\infty} xf(x)\,dx = \int_{1}^{\infty} x \cdot \frac{5}{2} x^{-7/2}\,dx$

$$= \lim_{R \to \infty} \int_{1}^{R} \frac{5}{2} x^{-7/2}\,dx = \lim_{R \to \infty} \left(\frac{5}{2} \left(-\frac{2}{3} \right) x^{-3/2} \Big|_{1}^{R} \right)$$

$$= -\frac{5}{3} \lim_{R \to \infty} (R^{-3/2} - 1^{-3/2}) = \frac{5}{3}(-1) = \frac{5}{3} \approx 1.667$$

$$V(X) = \int_{-\infty}^{\infty} x^2 f(x)\,dx - \mu^2 = \int_{1}^{\infty} x^2 \cdot \frac{5}{2} x^{-7/2}\,dx - \left(\frac{5}{3} \right)^2$$

$$= \lim_{R \to \infty} \int_{1}^{R} \frac{5}{2} x^{-3/2}\,dx - \frac{25}{9} = \lim_{R \to \infty} \left(\frac{5}{2}(-2)x^{-1/2} \Big|_{1}^{R} \right) - \frac{25}{9}$$

$$= -5 \lim_{R \to \infty} [R^{-1/2} - (1)^{-7/2}] - \frac{25}{9} = -5(-1) - \frac{25}{9} = 5 - \frac{25}{9} = \frac{20}{9}$$

$$\approx 2.2222$$

$$\sigma = \sqrt{V(X)} = \sqrt{\frac{20}{9}} = \frac{2}{3}\sqrt{5} \approx 1.4907 \qquad (9\text{-}3)$$

15. When $x < 1$, $F(x) = 0$. When $x \geq 1$,

$$F(x) = \int_{-\infty}^{x} f(t)\,dt = \int_{-\infty}^{1} f(t)\,dt + \int_{1}^{x} f(t)\,dt = 0 + \int_{1}^{x} \frac{5}{2} t^{-7/2}\,dt$$

$$= \frac{5}{2} \left(-\frac{2}{5} \right) t^{-5/2} \Big|_{1}^{x} = -(x^{-5/2} - 1^{-5/2}) = 1 - x^{-5/2}.$$

Thus,

$$F(x) = \begin{cases} 1 - x^{-5/2} & x \geq 1 \\ 0 & \text{otherwise} \end{cases}$$

The graph of $F(x)$ is shown at the right.

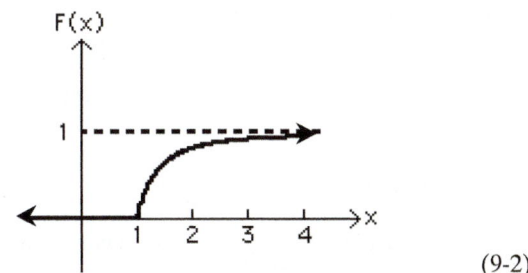

$(9\text{-}2)$

16. We must solve the following for m:

$$F(m) = P(X \leq m) = \frac{1}{2}$$

$$1\, m^{-5/2} = \frac{1}{2} \qquad \text{(refer to Problem 15)}$$

$$m^{-5/2} = \frac{1}{2}$$

$$\frac{1}{m^{5/2}} = \frac{1}{2}$$

$$m^{5/2} = 2 \qquad \text{(square both sides)}$$

$$m^5 = 4$$

$$m = (4)^{1/5}$$

$$m = 2^{2/5} = 1.32 \qquad (9\text{-}3)$$

17. $P(4 \leq X) = \int_{4}^{\infty} f(x)\,dx = e^{-2} \qquad \text{(Solve for } \lambda.\text{)}$

$$= \int_{4}^{\infty} \frac{1}{\lambda} e^{-x/\lambda}\,dx = \lim_{R \to \infty} \frac{1}{\lambda} \int_{4}^{R} e^{-x/\lambda}\,dx = e^{-2}$$

$$= \lim_{R \to \infty} \frac{1}{\lambda} \left. -\lambda e^{-x/\lambda} \right|_{4}^{R} = -\lim_{R \to \infty} (e^{-R/\lambda} - e^{-4/\lambda}) = e^{-2}$$

$$= e^{-4/\lambda} = e^{-2}$$

Thus, we have $-\dfrac{4}{\lambda} = -2$

$$\lambda = 2.$$

The probability density function, with $\lambda = 2$, is:

$$f(x) = \begin{cases} \dfrac{1}{\lambda}e^{-x/\lambda} & x \geq 0 \\ 0 & \text{otherwise} \end{cases} = \begin{cases} \dfrac{1}{2}e^{-x/2} & x \geq 0 \\ 0 & \text{otherwise} \end{cases}$$

(9-4)

18. $P(0 \leq X \leq 2) = \displaystyle\int_0^2 f(x)\,dx = \int_0^2 \dfrac{1}{2}e^{-x/2}\,dx$ (refer to Problem 17)

$$= \dfrac{1}{2}(-2e^{-x/2})\big|_0^2 = -(e^{-2/2} - e^0)$$

$$= -(e^{-1} - 1) \text{ or } (1 - e^{-1}) \approx 0.6321$$

(9-2)

19. $F(x) = \begin{cases} 1-e^{-x/\lambda} & x \geq 0 \\ 0 & \text{otherwise} \end{cases} = \begin{cases} 1-e^{-x/2} & x \geq 0 \\ 0 & \text{otherwise} \end{cases}$ [Note: $\lambda = 2$.]

(9-4)

20. $\mu = \lambda = 2$
$\sigma = \lambda = 2$
$m = \lambda \ln 2 = 2 \ln 2 \approx 1.3863$

(9-4)

21. (A) $\mu = 50, \sigma = 6$

$$z \text{ (for } x = 41) = \dfrac{41 - 50}{6} = -1.5$$

$$z \text{ (for } x = 62) = \dfrac{62 - 50}{6} = 2.0$$

Required area $= A_1 + A_2$
= (area corresponding to $z = 1.5$)
+ (area corresponding to $z = 2$)
$= 0.4332 + 0.4772$
$= 0.9104$

(B) $z \text{ (for } x = 59) = \dfrac{59 - 50}{6} = 1.5$

Required area $= 0.5 - \text{area corresponding to } z = 1.5)$
$= 0.5 - 0.4332$
$= 0.0668$

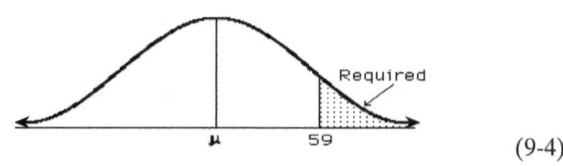

(9-4)

22. Given $\mu = 82$ and $\sigma = 8$.

(A) We first find the number of standard deviations that 84 and 94 are from the mean.

$$\text{For } x = 84: z = \dfrac{84 - 82}{8} = \dfrac{2}{8} = 0.25$$

$$\text{For } x = 94: z = \dfrac{94 - 82}{8} = \dfrac{12}{8} = 1.5$$

Now, $P(84 \leq X \leq 94) = P(0.25 \leq z \leq 1.5)$
$= 0.4332 - 0.0987$
$= 0.3345$

(B) For $x = 60: z = \dfrac{60 - 82}{8} = -\dfrac{22}{8} = -2.75$

$P(X \geq 60) = P(z \geq -2.75) = 0.4970 + 0.5000$
$= 0.9970$

(9-4)

23. $\displaystyle\int_{-\infty}^0 e^x\,dx = \lim_{a \to -\infty}\int_a^0 e^x\,dx = \lim_{a \to -\infty} e^x\big|_a^0 = \lim_{a \to -\infty}(1 - e^a) = 1$, since $e^a \to 0$ as $a \to -\infty$

(9-1)

24. $\displaystyle\int_0^\infty \dfrac{1}{(x+3)^2}\,dx = \lim_{b \to \infty}\int_0^b \dfrac{1}{(x+3)^2}\,dx$ Substitution: $u = x + 3$
$\qquad\qquad\qquad\qquad\qquad du = dx$

$$= \lim_{b \to \infty}\dfrac{-1}{(x+3)}\bigg|_0^b = \lim_{b \to \infty}\left(\dfrac{-1}{b+3} + \dfrac{1}{3}\right)$$

$$= \dfrac{1}{3}$$

Thus, the improper integral converges.

(9-1)

25. Yes. Since $\displaystyle\int_{-1}^{\infty} f(x)\,dx = \int_{-1}^{1} f(x)\,dx + \int_{1}^{\infty} f(x)\,dx$ and $\displaystyle\int_{-1}^{\infty} f(x)\,dx = L$ exists, it follows

that $\displaystyle\int_{1}^{\infty} f(x)\,dx = \int_{1}^{\infty} f(x)\,dx = \int_{-1}^{\infty} f(x)\,dx - \int_{-1}^{1} f(x)\,dx = L - \int_{-1}^{1} f(x)\,dx$ exists.   (9-1)

26. $f(x) = \begin{cases} e^{-10x} & \text{if } x \geq 0 \\ 0 & \text{otherwise} \end{cases}$

$f(x) \geq 0$ on $(-\infty, \infty)$ and

$$\int_{-\infty}^{\infty} f(x)\,dx = \int_{-\infty}^{0} f(x)\,dx + \int_{0}^{\infty} f(x)\,dx$$

$$= \int_{0}^{\infty} e^{-10x}\,dx = \lim_{b \to \infty} \int_{0}^{b} e^{-10x}\,dx$$

$$= \lim_{b \to \infty} \left[-\frac{1}{10} e^{-10x} \right]_{0}^{b}$$

$$= \lim_{b \to \infty} \left[\frac{1}{10} - \frac{1}{10} e^{-10b} \right] = \frac{1}{10}$$

Therefore, let $k = 10$; $f(x) = 10e^{-10x}$ is a probability density function.   (9-2)

27. $f(x) = \begin{cases} e^{10x} & \text{if } x \geq 0 \\ 0 & \text{otherwise} \end{cases}$

Since $\displaystyle\int_{-\infty}^{\infty} f(x)\,dx = \int_{-\infty}^{0} f(x)\,dx + \int_{0}^{\infty} f(x)\,dx$

$$= \int_{0}^{\infty} e^{10x}\,dx = \lim_{b \to \infty} \int_{0}^{b} e^{10x}\,dx$$

$$= \lim_{b \to \infty} \left[\frac{1}{10} e^{10x} \right]_{0}^{b}$$

$$= \lim_{b \to \infty} \left[\frac{1}{10} e^{10b} - \frac{1}{10} \right] \text{ diverges, no constant } k \text{ exists.}$$ (9-2)

28. X is an exponentially distributed random variable with median $m = 3 \ln 2$. It follows that the mean $\mu = 3$ and the probability density function is:

$$f(x) = \begin{cases} (1/3)e^{-x/3} & \text{if } x \geq 0 \\ 0 & \text{otherwise} \end{cases}$$

Now, $\displaystyle P(\overline{X}) \leq 3 = \int_{-\infty}^{3} f(x)\,dx = \int_{-\infty}^{0} f(x)\,dx + \int_{0}^{3} f(x)\,dx$

$$= \int_{0}^{3} \frac{1}{3} e^{-x/3}\,dx$$

$$= \frac{1}{3} \left[-3e^{-x/3} \right]_{0}^{3} = 1 - e^{-1} \approx 0.6321$$ (9-4)

29. $\displaystyle \mu = \int_{0}^{\infty} x f(x)\,dx = \int_{0}^{\infty} \frac{50x}{(x+5)^3}\,dx$

$$= \lim_{R \to \infty} \int_{0}^{R} \frac{50x}{(x+5)^3}\,dx = 50 \lim_{R \to \infty} \left[\frac{1}{(x+5)^2} - \frac{5}{(x+5)^3} \right]\,dx$$

$$= 50 \lim_{R \to \infty} \left(-\frac{1}{(x+5)} + \frac{5}{2} \cdot \frac{1}{(x+5)^2} \right) \Big|_{0}^{R}.$$

$$= 50 \lim_{R \to \infty} \left[\left(-\frac{1}{(R+)} + \frac{5}{2} \cdot \frac{1}{(R+5)^2} \right) - \left(-\frac{1}{5} + \frac{5}{2} \cdot \frac{1}{25} \right) \right]$$

$$= 50 \left(\frac{1}{5} - \frac{1}{10} \right) = 50 \left(\frac{1}{10} \right) = 5$$

Now find the cumulative probability distribution function. When $x < 0$, $F(x) = 0$. When $x \geq 0$, we have:

$$F(x) = \int_{-\infty}^{x} f(t)\, dt = \int_{-\infty}^{0} f(t)\, dt + \int_{0}^{x} f(t)\, dt = 0 + \int_{0}^{x} \frac{50x}{(t+5)^3}\, dt$$

$$= -\frac{50}{2}\left(\frac{1}{(t+5)^2} \right)\Big|_{0}^{x} = -25\left(\frac{1}{(x+5)^2} - \frac{1}{25} \right) = 1 - \frac{25}{(x+5^2}$$

Thus, $F(x) = \begin{cases} 1 - \dfrac{25}{(x+5)^2} & x = 0 \\ 0 & \text{otherwise} \end{cases}$

Next, to find the median, m, we must solve the following for m:

$$F(m) = P(X \le m) = \frac{1}{2}$$

$$1 - \frac{25}{(m+5)^2} = \frac{1}{2}$$

$$\frac{25}{(m+5)^2} = \frac{1}{2}$$

$$(m+5)^2 = 50$$

$$m + 5 = \sqrt{50}$$

$$m + 5 = 5\sqrt{2}$$

Therefore, the median, m, equals $5\sqrt{2} - 5 \approx 2.071$.   (9-3)

30. $f(x) = \begin{cases} \dfrac{0.8}{x^2} + \dfrac{0.8}{x^5} & \text{if } x \ge 0 \\ 0 & \text{otherwise} \end{cases}$

The cumulative distribution function is:

$$F(x) = \begin{cases} 0 & \text{if } x < 1 \\ 1 - \dfrac{0.8}{x} - \dfrac{0.2)x^4}{x} & \text{if } x \ge 1 \end{cases}$$

To find the median m, we solve

$$F(m) = P(X \le m) = \frac{1}{2} \text{ for } m:$$

$$1 - \frac{0.8}{m} - \frac{0.2}{m^4} = \frac{1}{2}$$

Using a graphing utility, we find that $m \approx 1.68$.   (9-3)

31. Consider the integral $\displaystyle\int \frac{e^x}{(1+e^x)^2}\, dx.$

If we let $u = 1 + e^x$, then $du = e^x\, dx$, and $\displaystyle\lim_{x \to \infty} u = \infty$, $\displaystyle\lim_{x \to -\infty} u = 1$. Thus,

$$\int_{-\infty}^{\infty} \frac{e^x}{(1+e^x)^2}\, dx = \int_{1}^{\infty} \frac{1}{u^2}\, du = \lim_{b \to \infty} \int_{1}^{b} \frac{1}{u^2}\, du$$

$$= \lim_{b \to \infty} \left[-\frac{1}{u}\Big|_{1}^{b} \right]$$

$$= \lim_{b \to \infty} \left[1 - \frac{1}{b} \right] = 1$$ (9-1)

32. $\displaystyle\int_{-\infty}^{\infty} (ax^2 + bx + c) f(x) = a \int_{-\infty}^{\infty} x^2 f(x)\, dx + b \int_{-\infty}^{\infty} x f(x)\, dx + c \int_{-\infty}^{\infty} f(x)\, dx$

$$= a(\sigma^2 + \mu^2) + b\mu + c,$$

since $\displaystyle\int_{-\infty}^{\infty} x^2 f(x)\, dx = \sigma^2 + \mu^2$, $\displaystyle\int_{-\infty}^{\infty} x f(x)\, dx = \mu$, and $\displaystyle\int_{-\infty}^{\infty} f(x)\, dx = 1$.   (9-3)

33. $f_1(x) = \begin{cases} 0.25x^2e^{-x/2} & \text{if } x \geq 0 \\ 0 & \text{otherwise} \end{cases}$

$f_2(x) = \begin{cases} 0.0625x^2e^{-x/2} & \text{if } x \geq 0 \\ 0 & \text{otherwise} \end{cases}$

The graphs of f_1 and f_2 are shown at the right.

In comparison with f_1, the graph of f_2 is shifted to the right.
Therefore X_2 should have a greater mean than X_1. (9-3)

34. Refer to the graphs in Problem 33. In comparison with f_1, the area under the graph of f_2 is more spread out. Therefore, the variance of X_2 should be greater than the variance of X_1. (9-3)

35. $\mu_1 = \int_{-\infty}^{\infty} xf_1(x)\,dx = \int_0^{\infty} x(0.25xe^{-x/2})\,dx$

$\approx \int_0^{30} 0.25x^2e^{-x/2}\,dx \approx 4$

$\mu_2 = \int_{-\infty}^{\infty} xf_2(x)\,dx = \int_0^{\infty} x(0.0625x^2e^{-x/2})\,dx$

$\approx \int_0^{30} 0.0625x^3e^{-x/2}\,dx \approx 6$ (9-3)

36. $V_1(x) = \int_{-\infty}^{\infty} x^2f_1(x)\,dx - 4^2 = \int_0^{\infty} x^2(0.25xe^{-x/2})\,dx - 16$

$\approx \int_0^{30} 0.25x^3e^{-x/2}\,dx - 16 \approx 24 - 16 = 8$

$V_2(x) = \int_{-\infty}^{\infty} x^2f_2(x)\,dx - 6^2 = \int_0^{\infty} x^2(0.0625x^2e^{-x/2})\,dx - 36$

$\approx \int_0^{30} 0.0625x^4e^{-x/2}\,dx - 36 \approx 48 - 36 = 12$ (9-3)

37. (A) The total production is given by:

$\int_0^{\infty} R(t)\,dt = \lim_{T \to \infty} \int_0^T R(t)\,dt$

$= \lim_{T \to \infty} \int_0^T (12e^{-0.3t} - 12e^{-0.6t})\,dt$

$= \lim_{T \to \infty} \left[(-40e^{-0.3t} + 20e^{-0.6t}) \Big|_0^T \right]$

$= \lim_{T \to \infty} (-40e^{-0.3T} + 20e^{-0.6T} + 20) = 20$

Thus, the total production is 20 million barrels.

(B) To find when the well will reach 50% of the total production, we must solve

$\int_0^T R(t)\,dt = 10$

for T. Now,

$\int_0^T R(t)\,dt = \int_0^T (12e^{-0.3t} - 12e^{-0.6t})\,dt$

$= [(-40e^{-0.3t} + 20e^{-0.6t})|_0^T]$

$= -40e^{-0.3T} + 20e^{-0.6T} + 20$

Thus, we have

$-40e^{-0.3T} + 20e^{-0.6T} + 20 = 10$

or $\qquad -40e^{-0.3T} + 20e^{-0.6T} = -10$

Using a graphing utility to solve this equation, we find that $T \approx 4.09$ years. (9-1)

38. $f(x) = \begin{cases} 0.02(1-0.01x) & \text{if } 0 \le x \le 100 \\ 0 & \text{otherwise} \end{cases}$

(A) $\displaystyle\int_{40}^{100} f(x)\,dx = \int_{40}^{100} 0.02(1-0.01x)\,dx$

$$= 0.02[(x - 0.005x^2)|_{40}^{100}$$
$$= 0.02[(100 - 50) - (40 - 8)] = 0.36$$

The probability that the weekly demand for popcorn is between 40 and 100 pounds is 0.36.

(B) $\displaystyle P(X \le 50) = \int_0^{50} f(x)\,dx = \frac{1}{50}\int_0^{50}(1-0.01x)\,dx = \frac{1}{50}(x - 0.005x^2)\Big|_0^{50}$

$$= \frac{1}{50}(50 - 0.005 \cdot 50^2) = 1 - 0.25 = 0.75$$

(C) Solve the following for x:

$$\int_0^x f(t)\,dt = 0.96 \qquad (x = \text{number of pounds of popcorn})$$

$$\frac{1}{50}\int_0^x (1 - 0.01t)\,dt = 0.96$$

$$\frac{1}{50}(t - 0.005t^2)\Big|_0^x = 0.96$$

$$\frac{1}{50}(x - 0.005x^2) = 0.96$$

$$x - 0.005x^2 = 48$$
$$5x^2 - 1000x + 48{,}000 = 0$$
$$x^2 - 200x + 9600 = 0$$
$$(x - 80)(x - 120) = 0$$

$$x = 80 \qquad \text{or} \qquad x = 120$$

Thus, 80 pounds of popcorn must be on hand at the beginning of the week. (9-2)

39. Capital Value: $\displaystyle CV = \int_0^\infty 2400e^{-0.06t}\,dt = \lim_{T\to\infty}\int_0^T 2400e^{-0.06t}\,dt$

$$= \lim_{T\to\infty} 2400\frac{e^{-0.06t}}{-0.06}\Big|_0^T$$
$$= \lim_{T\to\infty} -40{,}000(e^{-0.06t} - 1)$$
$$= \$40{,}000 \qquad (9\text{-}1)$$

40. $f(x) = \begin{cases} 6x(1 - x) & \text{if } 0 \le x \le 1 \\ 0 & \text{otherwise} \end{cases}$

(A) $P(X \ge 0.2) = 1 - P(X < 0.2)$

$$= 1 - \int_{-\infty}^{0.2} f(x)\,dx$$
$$= 1 - \int_0^{0.2} f(x)\,dx$$
$$= 1 - \int_0^{0.2} 6x(1 - x)\,dx$$
$$= 1 - \left[(3x^2 - 2x^3)\Big|_0^{0.2}\right]$$
$$= 1 - (0.12 - 0.016) = 0.896$$

(B) The expected value (mean) is given by:

$$\mu = \int_{-\infty}^\infty xf(x)\,dx = \int_0^1 (6x^2 - 6x^3)\,dx$$

$$= \left(2x^3 - \frac{3}{2}x^4\right)\Big|_0^1 = 2 - \frac{3}{2} = 0.5$$

The expected percentage is 50%.

(C) The cumulative distribution function is:

$$F(x) = \begin{cases} 0 & \text{if } x < 0 \\ 3x^2 - 2x^3 & \text{if } 0 \le x \le 1 \\ 1 & \text{if } x > 1 \end{cases}$$

To find the median m, we solve

$$F(m) = P(X \le m) = \frac{1}{2}$$

or $3m^3 - 2m^3 = \dfrac{1}{2}$

for m. Using a graphing utility, we find that $m = 0.5$. The median percentage is 50%.   (9-2, 9-3)

41. Mean failure time $= \mu = 4000$. As an exponential density function, it is expressed by $\lambda = \mu = 4000$. Thus,

$$f(x) = \begin{cases} \dfrac{1}{\lambda} e^{-x/\lambda} = \dfrac{1}{4000} e^{-x/4000} & x \ge 0 \\ 0 & \text{otherwise} \end{cases}$$

The cumulative distribution function is given by:

$$F(x) = \begin{cases} 1 - e^{-x/\lambda} \\ 0 \end{cases} = \begin{cases} 1 - e^{-x/4000} & x \ge 0 \\ 0 & \text{otherwise} \end{cases}$$

(A) $P(X \ge 4000) = \displaystyle\int_{4000}^{\infty} f(x)\, dx = 1 - F(4000)$

$$= 1 - (1 - e^{-4000/4000}) = e^{-1} = 0.3679$$

(B) $P(0 \le X \le 1000) = \displaystyle\int_{0}^{1000} f(x)\, dx = F(1000) - F(0)$

$$= (1 - e^{-1000/4000}) - (1 - e^0) = 1 - e^{-.25} \approx 0.2212$$ (9-4)

42. $\mu = 35{,}000, \sigma = 5{,}000$

z (for $x = 25{,}000$) $= \dfrac{25{,}000 - 35{,}000}{5{,}000} = -2$

Required probability = area A
$ = 0.5 - \text{area } A_1$
$ = 0.5 - 0.4772$
$ = 0.0228$

(9-4)

43. $\mu = 100, \sigma = 10$

(A) z (for $x = 91.5$) $= \dfrac{91.5 - 100}{10} = -0.85$

$\quad\; z$ (for $x = 108$) $= \dfrac{108.5 - 100}{10} = 0.85$

The probability of an applicant scoring between 92 and 108

$\quad = \text{area } A$
$\quad = 2 \cdot \text{area } A_1$

$\quad = 2(\text{area corresponding to } z = 0.85)$
$\quad = 2(0.3023) = 0.6046$

Thus, the percentage of applicants scoring between 92 and 108 is 60.46%.

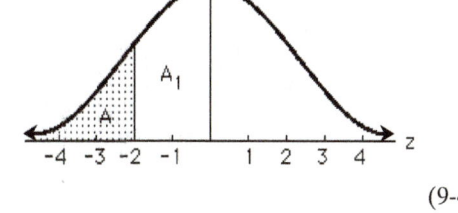

(9-4)

(B) z (for $x = 114.5$) $= \dfrac{114.5 - 100}{10} = 1.45$

The probability of an applicant scoring 115 or higher

$\quad = \text{area } A$

$\quad = 0.5 - (\text{area corresponding to } z = 1.45)$

$\quad = 0.5 - 0.4265$

$\quad = 0.0735$

Thus, the percentage of applicants scoring 115 or higher is 7.35%.

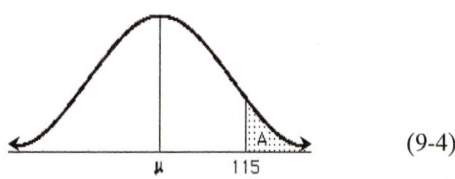

(9-4)

44. $f(x) = \begin{cases} \dfrac{10}{(x+10)^2} & \text{if } x \geq 0 \\ 0 & \text{otherwise} \end{cases}$

(A) $\displaystyle\int_2^8 f(x)\, dx = \int_2^8 \frac{10}{(x+10)^2}\, dx$

Let $u = x + 10$, then $du = dx$; $u = 12$ when $x = 2$, $u = 18$ when $x = 8$. Thus,

$\displaystyle\int_2^8 \frac{10}{(x+10)^2}\, dx = 10 \int_{12}^{18} \frac{1}{u^2}\, du = 10\left[-\frac{1}{u} \Big|_{12}^{18}\right]$

$\qquad = 10\left(\dfrac{1}{12} - \dfrac{1}{18}\right) = \dfrac{5}{18} \approx 0.2778$

The probability that the shelf–life of the drug is between 2 and 8 months is 0.2778.

(B) Probability that the drug is usable after five months is:

$P(X > 5) = \displaystyle\int_5^8 f(x)\, dx = \int_5^8 \frac{10}{(x+0)^2\, dx} = \lim_{R \to \infty} \int_5^R \frac{10}{(x+10)^2}\, dx$

$\qquad = \displaystyle\lim_{R \to \infty} 10\left(-\frac{1}{x+10}\right)\Big|_5^R = -10 \lim_{R \to \infty}\left(\frac{1}{R+10} - \frac{1}{5+10}\right)$

$\qquad = -10\left(-\dfrac{1}{15}\right) = \dfrac{2}{3} \approx 0.6667$

(C) In order to find the median, m, we must solve the following for m:

$P(X \leq m) = \displaystyle\int_0^m f(x)\, dx = \frac{1}{2}$

$\displaystyle\int_0^m \frac{10}{(x+10)^2}\, dx = \frac{1}{2}$

$-\dfrac{10}{(x+10)}\Big|_0^m = \dfrac{1}{2}$

$-\left(\dfrac{10}{m+10} - 1\right) = \dfrac{1}{2}$

$-\dfrac{10}{m+10} + 1 = \dfrac{1}{2}$

$-\dfrac{10}{m+10} = -\dfrac{1}{2}$

$m + 10 = 20$

$m = 10 \text{ months}$

(9-2, 9-3)

45. $f(x) = \begin{cases} \dfrac{1}{\lambda} e^{-x/\lambda} & x \geq 0 \\ 0 & \text{otherwise} \end{cases}$

$P(X > 1) = \dfrac{1}{\lambda} \displaystyle\int_1^8 e^{-x/\lambda}\, dx = e^{-2} \qquad \text{(Given)}$

Thus, $\dfrac{1}{\lambda} \displaystyle\lim_{R \to \infty} \int_1^R e^{-x/\lambda}\, dx = e^{-2}$

$\dfrac{1}{\lambda}(-\lambda) \displaystyle\lim_{R \to \infty} (e^{-x/\lambda})\Big|_1^R = e^{-2}$

$-1 \cdot \displaystyle\lim_{R \to \infty} (e^{-R/\lambda} - e^{-1/\lambda}) = e^{-2}$

$e^{-1/\lambda} = e^{-2}$

Thus, $\qquad -\dfrac{1}{\lambda} = -2$

$\lambda = \dfrac{1}{2}$

Therefore, $f(x)$, with $\lambda = \dfrac{1}{2}$, is given by $f(x) = \begin{cases} 2e^{-2x} & x \geq 0 \\ 0 & \text{otherwise} \end{cases}$

(A) $P(X > 2) = \displaystyle\int_2^8 f(x)\,dx = \int_2^8 2e^{-2x}\,dx = 2\lim_{R\to\infty}\int_2^R e^{-2x}\,dx$

$= 2\lim_{R\to\infty}\left(-\dfrac{1}{2}e^{-2x}\Big|_2^R\right) = -\lim_{R\to\infty}(e^{-2R} - e^{-4}) = -(-e^{-4}) = e^{-4}$

≈ 0.0183

(B) Mean life expectancy:

$\mu = \displaystyle\int_0^8 xf(x)\,dx = \int_0^8 x\cdot 2e^{-2x}\,dx$

$= 2\lim_{R\to\infty}\displaystyle\int_0^R xe^{-2x}\,dx \text{ (integration by parts; } u = x,\, dv = e^{-2x}\,dx)$

$= 2\lim_{R\to\infty}\left[x\left(-\dfrac{1}{2}e^{-2x}\right)\Big|_0^R + \dfrac{1}{2}\int_0^R e^{-2x}\,dx\right] = 2\lim_{R\to\infty}\left(-\dfrac{1}{2}xe^{-2x} - \dfrac{1}{4}e^{-2x}\right)\Big|_0^R$

$= -2\lim_{R\to\infty}\left[\left(\dfrac{1}{2}Re^{-2R} + \dfrac{1}{4}e^{-2R}\right) - \left(0 + \dfrac{1}{4}\right)\right]$

$= -2\left(-\dfrac{1}{4}\right) = \dfrac{1}{2}$ or 0.5 month $or\ \mu = \lambda = \dfrac{1}{2}$ (9-2, 9-3)

46. $R(t) = 15e^{-0.2t} - 15e^{-0.3t}$

(A) The total amount of the drug that is eliminated by the body is given by:

$\displaystyle\int_0^8 R(t)\,dt = \lim_{T\to\infty}\int_0^T R(t)\,dt$

$= \lim_{T\to\infty}\displaystyle\int_0^T (15e^{-0.2t} - 15e^{-0.3t})\,dt$

$= \lim_{T\to\infty}[(-75e^{-0.2t} + 50e^{-0.3t})|_0^T]$

$= \lim_{T\to\infty}[-75e^{-0.2T} + 50e^{-0.3T} + 25]$

$= 25$ milliliters

(B) To find how long it will take for 50% of the drug to be eliminated, we solve

$\displaystyle\int_0^T R(t)\,dt = 0.5(25) = 12.5$

for T:

$\displaystyle\int_0^T (15e^{-0.2t} - 15e^{-0.3t})\,dt = 12.5$

$[(-75e^{-0.2t} + 50e^{-0.3t})|_0^T] = 12.5$

$-75e^{-0.2T} + 50e^{-0.3T} + 25 = 12.5$

$-75e^{-0.2T} + 50e^{-0.3T} + 12.5 = 0$

Using a graphing utility, we find that $T \approx 6.93$ hours.   (9-1)

47. $\mu = 108,\ \sigma = 12$

$z = \dfrac{135 - 108}{12} = \dfrac{27}{12} = 2.25$

$P(X \geq 135) = P(Z \geq 2.25) = 0.5 - P(Z < 2.25)$
$= 0.5 - 0.4878$
$= 0.0122;$

1.22% of the children can be expected to have IQ scores of 135 or more.   (9-4)

48. $N'(t) = \dfrac{100t}{(1 + t^2)^2}$

Therefore, $N(t) = \displaystyle\int \dfrac{100t}{(1 + t^2)^2}\, dt = 50 \int \dfrac{1}{u^2}\, du$

Substitution: $u = 1 + t^2$
$du = 2t\, dt$

$$= \dfrac{50u^{-1}}{-1} + C = \dfrac{-50}{1\ t^2} + C$$

Now, $N(0) = \dfrac{-50}{1 + 0} + C$. Thus, $C = N(0) + 50$ and

$$N(t) = \dfrac{-50}{1 + t^2} + N(0) + 50$$

$$N(3) = \dfrac{-50}{1 + 3^2} + N(0) + 50 = N(0) + 45$$

Therefore, the voting population will increase by 45 thousand in 3 years. If the population grows indefinitely at this rate, then

$$\lim_{T \to \infty} N(t) = \lim_{T \to \infty} \left(\dfrac{-50}{1 + t^2} + N(0) + 50 \right) = N(0) + 50$$

Therefore, the total increase in voting population is 50 thousand.   (9-1)

Appendix A

Exercises A-1

1. vu **3.** $(3 + 7) + y$ **5.** $u + v$ **7.** T   **9.** T   **11.** F   **13.** T   **15.** T   **17.** T   **19.** T   **21.** F   **23.** T   **25.** T   **27.** No
29. (A) F   (B) T   (C) T   **31.** $\sqrt{2}$ and π are two examples of infinitely many.   **33.** (A) N, Z, Q, R (B) R (C) Q, R (D) Q, R
35. (A) F, since, for example, $2(3 - 1) \neq 2 \cdot 3 - 1$ (B) F, since, for example, $(8 - 4) - 2 \neq 8 - (4 - 2)$
   (C) T   (D) F, since, for example, $(8 \div 4) \div 2 \neq 8 \div (4 \div 2)$.
37. $\frac{1}{11}$ **39.** (A) $2.166\,666\,666\ldots$ (B) $4.582\,575\,69\ldots$ (C) $0.437\,500\,000\ldots$ (D) $0.261\,261\,261\ldots$

Exercises A-2

1. 3   **3.** $x^3 + 4x^2 - 2x + 5$ **5.** $x^3 + 1$ **7.** $2x^5 + 3x^4 - 2x^3 + 11x^2 - 5x + 6$ **9.** $-5u + 2$ **11.** $6a^2 + 6a$ **13.** $a^2 - b^2$
15. $6x^2 - 7x - 5$ **17.** $2x^2 + xy - 6y^2$ **19.** $9y^2 - 4$ **21.** $-4x^2 + 12x - 9$ **23.** $16m^2 - 9n^2$ **25.** $9u^2 + 24uv + 16v^2$
27. $a^3 - b^3$ **29.** $x^2 - 2xy + y^2 - 9z^2$ **31.** 1   **33.** $x^4 - 2x^2y^2 + y^4$ **35.** $-40ab$ **37.** $-4m + 8$ **39.** $-6xy$
41. $u^3 + 3u^2v + 3uv^2 + v^3$ **43.** $x^3 - 6x^2y + 12xy^2 - 8y^3$ **45.** $2x^2 - 2xy + 3y^2$ **47.** $x^4 - 10x^3 + 27x^2 - 10x + 1$
49. $4x^3 - 14x^2 + 8x - 6$ **51.** $m + n$ **53.** No change   **55.** $(1 + 1)^2 \neq 1^2 + 1^2$; either a or b must be 0
57. $0.09x + 0.12(10{,}000 - x) = 1{,}200 - 0.03x$ **59.** $20x + 30(3x) + 50(4{,}000 - x - 3x) = 200{,}000 - 90x$
61. $0.02x + 0.06(10 - x) = 0.6 - 0.04x$

Exercises A-3

1. $3m^2(2m^2 - 3m - 1)$ **3.** $2uv(4u^2 - 3uv + 2v^2)$ **5.** $(7m + 5)(2m - 3)$ **7.** $(4ab - 1)(2c + d)$ **9.** $(2x - 1)(x + 2)$
11. $(y - 1)(3y + 2)$ **13.** $(x + 4)(2x - 1)$ **15.** $(w + x)(y - z)$ **17.** $(a - 3b)(m + 2n)$ **19.** $(3y + 2)(y - 1)$
21. $(u - 5v)(u + 3v)$ **23.** Not factorable   **25.** $(wx - y)(wx + y)$ **27.** $(3m - n)^2$ **29.** Not factorable   **31.** $4(z - 3)(z - 4)$
33. $2x^2(x - 2)(x - 10)$ **35.** $x(2y - 3)^2$ **37.** $(2m - 3n)(3m + 4n)$ **39.** $uv(2u - v)(2u + v)$ **41.** $2x(x^2 - x + 4)$
43. $(2x - 3y)(4x^2 + 6xy + 9y^2)$ **45.** $xy(x + 2)(x^2 - 2x + 4)$ **47.** $[(x + 2) - 3y][(x + 2) + 3y]$ **49.** Not factorable
51. $(6x - 6y - 1)(x - y + 4)$ **53.** $(y - 2)(y + 2)(y^2 + 1)$ **55.** $3(x - y)^2(5xy - 5y^2 + 4x)$ **57.** True   **59.** False

Exercises A-4

1. $8d^6$ **3.** $\dfrac{15x^2 + 10x - 6}{180}$ **5.** $\dfrac{15m^2 + 14m - 6}{36m^3}$ **7.** $\dfrac{1}{x(x - 4)}$ **9.** $\dfrac{x - 6}{x(x - 3)}$ **11.** $\dfrac{-3x - 9}{(x - 2)(x + 1)^2}$ **13.** $\dfrac{2}{x - 1}$ **15.** $\dfrac{5}{a - 1}$

17. $\dfrac{x^2 + 8x - 16}{x(x - 4)(x + 4)}$ **19.** $\dfrac{7x^2 - 2x - 3}{6(x + 1)^2}$ **21.** $\dfrac{x(y - x)}{y(2x - y)}$ **23.** $\dfrac{-17c + 16}{15(c - 1)}$ **25.** $\dfrac{1}{x - 3}$ **27.** $\dfrac{-1}{2x(x + h)}$ **29.** $\dfrac{x - y}{x + y}$

31. (A) Incorrect   (B) $x + 1$ **33.** (A) Incorrect   (B) $2x + h$ **35.** (A) Incorrect   (B) $\dfrac{x^2 - x - 3}{x + 1}$ **37.** (A) Correct

39. $\dfrac{-2x - h}{3(x + h)^2 x^2}$ **41.** $\dfrac{x(x - 3)}{x - 1}$

Exercises A-5

1. $2/x^9$ **3.** $3w^7/2$ **5.** $2/x^3$ **7.** $1/w^5$ **9.** $4/a^6$ **11.** $1/a^6$ **13.** $1/8x^{12}$ **15.** 8.23×10^{10} **17.** 7.83×10^{-1} **19.** 3.4×10^{-5}
21. $40,000$ **23.** 0.007 **25.** $61,710,000$ **27.** $0.000\,808$ **29.** 1 **31.** 10^{14} **33.** $y^6/25x^4$ **35.** $4x^6/25$ **37.** $4y^3/3x^5$
39. $\frac{7}{4} - \frac{1}{4}x^{-3}$ **41.** $\frac{5}{2}x^2 - \frac{3}{2} + 4x^{-2}$ **43.** $\dfrac{x^2(x - 3)}{(x - 1)^3}$ **45.** $\dfrac{2(x - 1)}{x^3}$ **47.** 2.4×10^{10}; $24,000,000,000$
49. 3.125×10^4; $31,250$ **51.** 64 **55.** uv **57.** $\dfrac{bc(c + b)}{c^2 + bc + b^2}$ **59.** (A) $\$32,977$ (B) $\$1,484$ (C) 4.50%
61. (A) 9×10^{-6} (B) $0.000\,009$ (C) 0.0009% **63.** $1,417,000$

Exercises A-6

1. $6\sqrt[3]{x^3}$ **3.** $\sqrt[5]{(32x^2y^3)^3}$ **5.** $\sqrt{x^2 + y^2}$ (not $x + y$)   **7.** $5x^{3/4}$ **9.** $(2x^2y)^{3/5}$ **11.** $x^{1/3} + y^{1/3}$ **13.** 5 **15.** 64 **17.** -7 **19.** -16
21. $\frac{8}{125}$ **23.** $\frac{1}{27}$ **25.** $x^{2/5}$ **27.** m **29.** $2x/y^2$ **31.** $xy^2/2$ **33.** $1/(24x^{7/12})$ **35.** $2x + 3$ **37.** $30x^5\sqrt{3x}$ **39.** 2 **41.** $12x - 6x^{35/4}$
43. $3u - 13u^{1/2}v^{1/2} + 4v$ **45.** $36m^{3/2} - \dfrac{6m^{1/2}}{n^{1/2}} + \dfrac{6m}{n^{1/2}} - \dfrac{1}{n}$ **47.** $9x - 6x^{1/2}y^{1/2} + y$ **49.** $\frac{1}{2}x^{1/3} + x^{-1/3}$ **51.** $\frac{2}{3}x^{-1/4} + \frac{1}{3}x^{-2/3}$
53. $\frac{1}{2}x^{-1/6} - \frac{1}{4}$ **55.** $4n\sqrt{3mn}$ **57.** $\dfrac{2(x + 3)\sqrt{x - 2}}{x - 2}$ **59.** $7(x - y)(\sqrt{x} + \sqrt{y})$ **61.** $\dfrac{1}{xy\sqrt{5xy}}$ **63.** $\dfrac{1}{\sqrt{x + h} + \sqrt{x}}$
65. $\dfrac{1}{(t + x)(\sqrt{t} + \sqrt{x})}$ **67.** $x = y = 1$ is one of many choices.   **69.** $x = y = 1$ is one of many choices.
71. False   **73.** False   **75.** False   **77.** True   **79.** True   **81.** False   **83.** $\dfrac{x + 8}{2(x + 3)^{3/2}}$ **85.** $\dfrac{x - 2}{2(x - 1)^{3/2}}$ **87.** $\dfrac{x + 6}{3(x + 2)^{5/3}}$
89. 103.2 **91.** 0.0805 **93.** $4,588$ **95.** (A) and (E);   (B) and (F);   (C) and (D)

Exercises A-7

1. $\pm\sqrt{11}$ **3.** $-\frac{4}{3}, 2$ **5.** $-2, 6$ **7.** $0, 2$ **9.** $3 \pm 2\sqrt{3}$ **11.** $-2 \pm \sqrt{2}$ **13.** $0, \frac{15}{2}$ **15.** $\pm\frac{3}{2}$ **17.** $\frac{1}{2}, -3$ **19.** $\left(-1 \pm \sqrt{5}\right)/2$
21. $\left(3 \pm \sqrt{3}\right)/2$ **23.** No real solution   **25.** $\left(-3 \pm \sqrt{11}\right)/2$ **27.** $\pm\sqrt{3}$ **29.** $-\frac{1}{2}, 2$ **31.** $(x - 2)(x + 42)$
33. Not factorable in the integers   **35.** $(2x - 9)(x + 12)$ **37.** $(4x - 7)(x + 62)$ **39.** $r = \sqrt{A/P} - 1$
41. If $c < 4$, there are two distinct real roots; if $c = 4$, there is one real double root; and if $c > 4$, there are no real roots.
43. $1,575$ bottles at $\$4$ each   **45.** 13.64% **47.** 8 ft/sec; $4\sqrt{2}$ or 5.66 ft/sec

Appendix B

Exercises B-1

1. $5, 7, 9, 11$ **3.** $\frac{3}{2}, \frac{4}{3}, \frac{5}{4}, \frac{6}{5}$ **5.** $9, -27, 81, -243$ **7.** 23 **9.** $\frac{101}{100}$ **11.** $1 + 2 + 3 + 4 + 5 + 6 = 21$ **13.** $5 + 7 + 9 + 11 = 32$
15. $1 + \frac{1}{10} + \frac{1}{100} + \frac{1}{1,000} = \frac{1,111}{1,000}$ **17.** 3.6 **19.** 82.5 **21.** $\frac{1}{2}, -\frac{1}{4}, \frac{1}{8}, -\frac{1}{16}, \frac{1}{32}$ **23.** $0, 4, 0, 8, 0$ **25.** $1, -\frac{3}{2}, \frac{9}{4}, -\frac{27}{8}, \frac{81}{16}$ **27.** $a_n = n - 3$
29. $a_n = 4n$ **31.** $a_n = (2n - 1)/2n$ **33.** $a_n = (-1)^{n+1}n$ **35.** $a_n = (-1)^{n+1}(2n - 1)$ **37.** $a_n = \left(\frac{2}{5}\right)^{n-1}$ **39.** $a_n = x^n$
41. $a_n = (-1)^{n+1}x^{2n-1}$ **43.** $1 - 9 + 25 - 49 + 81$ **45.** $\frac{4}{7} + \frac{8}{9} + \frac{16}{11} + \frac{32}{13}$ **47.** $1 + x + x^2 + x^3 + x^4$ **49.** $x - \dfrac{x^3}{3} + \dfrac{x^5}{5} - \dfrac{x^7}{7} + \dfrac{x^9}{9}$
51. (A) $\sum_{k=1}^{5}(k + 1)$ (B) $\sum_{j=0}^{4}(j + 2)$ **53.** (A) $\sum_{k=1}^{4}\dfrac{(-1)^{k+1}}{k}$ (B) $\sum_{j=0}^{3}\dfrac{(-1)^j}{j + 1}$ **55.** $\sum_{k=1}^{n}\dfrac{k + 1}{k}$ **57.** $\sum_{k=1}^{n}\dfrac{(-1)^{k+1}}{2^k}$ **59.** False   **61.** True
63. $2, 8, 26, 80, 242$ **65.** $1, 2, 4, 8, 16$ **67.** $1, \frac{3}{2}, \frac{17}{12}, \frac{577}{408}$; $a_4 = \frac{577}{408} \approx 1.414\,216$, $\sqrt{2} \approx 1.414\,214$ **69.** $1, 1, 2, 3, 5, 8, 13, 21, 34, 55$

Exercises B-2

1. (A) Arithmetic, with $d = -5$; $-26, -31$ (B) Geometric, with $r = -2$; $-16, 32$ (C) Neither   (D) Geometric, with $r = \frac{1}{3}$; $\frac{1}{54}, \frac{1}{162}$
3. Geometric; 1 **5.** Neither   **7.** Arithmetic; 127.5 **9.** $a_2 = 11, a_3 = 15$ **11.** $a_{21} = 82, S_{31} = 1,922$ **13.** $S_{20} = 930$
15. $a_2 = -6, a_3 = 12, a_4 = -24$ **17.** $S_7 = 547$ **19.** $a_{10} = 199.90$ **21.** $r = 1.09$ or -1.09 **23.** $S_{10} = 1,242, S_\infty = 1,250$
25. $2,706$ **27.** -85 **29.** $1,120$ **31.** (A) Does not exist   (B) $S_\infty = \frac{8}{5} = 1.6$ **33.** $2,400$ **35.** 0.999
37. Use $a_1 = 1$ and $d = 2$ in $S_n = (n/2)[2a_1 + (n - 1)d]$.   **39.** $S_n = na_1$ **41.** No   **43.** Yes
45. $\$48 + \$46 + \cdots + \$4 + \$2 = \$600$ **47.** About $\$11,670,000$ **49.** $\$1,628.89$; $\$2,653.30$

Exercises B-3

1. 720 **3.** 10 **5.** $1,320$ **7.** 10 **9.** 6 **11.** $1,140$ **13.** 10 **15.** 6 **17.** 1 **19.** 816
21. $C_{4,0}a^4 + C_{4,1}a^3b + C_{4,2}a^2b^2 + C_{4,3}ab^3 + C_{4,4}b^4 = a^4 + 4a^3b + 6a^2b^2 + 4ab^3 + b^4$ **23.** $x^6 - 6x^5 + 15x^4 - 20x^3 + 15x^2 - 6x + 1$
25. $32a^5 - 80a^4b + 80a^3b^2 - 40a^2b^3 + 10ab^4 - b^5$ **27.** $3,060x^{14}$ **29.** $5,005p^9q^6$ **31.** $264x^2y^{10}$
33. $C_{n,0} = \dfrac{n!}{0!\,n!} = 1$; $C_{n,n} = \dfrac{n!}{n!\,0!} = 1$ **35.** $1\ 5\ 10\ 10\ 5\ 1$; $1\ 6\ 15\ 20\ 15\ 6\ 1$

INDEX